Exploring the Vertebrate Central Cholinergic Nervous System

Exploring the Vertebrate Central Cholinergic Nervous System

Alexander G. Karczmar, MD

With contributions by

Neelima Chauhan, PhD
Arthur Christopoulos, Bpharm, PhD
Jon Lindstrom, PhD
George J. Siegel, MD

With a Foreword by Victor P. Whittaker, PhD

Alexander G. Karczmar, MD
Professor and Chair Emeritus
Department of Pharmacology and Experimental Therapeutics
Loyola University Medical Center
Maywood, IL 60153
and
Senior Consultant
Edward J. Hines, Jr., VA Hospital
Hines, IL 60141
USA
and
Consultant to the Surgeon General
Department of the Army
Washington, DC
USA

Cover illustration. Several items of cholinergic interes are shown. From left to right, clockwise: Steve Kuffler, John Eccles and Paul Katz, in Canberra, Australia, in the early 1950s (from Karczmar, 2006a, with permission); brain machinery against a shadowy human profile (adopted from Time, March, 25); a common abbreviation for the bonding between a naturally occurring anticholinesterase, huperzine A and acetylcholinesterase of the electric fish, Torpedo; the diagram of circuitry of the Renshaw cell (from Karczmar, 2006a, with permission); a mouse model, used in cholinergic studies of aggression (from Karczmar, 2006b, with permission); leaves of Physostigma venenosum (from Karczmar, 2006b, with permission); a common symbol for a tridimensional structure of a protein.

Library of Congress Control Number: 2005930801

ISBN-10: 0-387-28223-8
ISBN-13: 978-0387-28223-7

Printed on acid-free paper.

9 8 7 6 5 4 3 2 1

springeronline.com

Foreword

Even if the "weapons of mass destruction" (WMD) and, among them, stocks of organophosphorus (OP) agents (also referred to as war gases and nerve gases) were not found in Iraq following the US-Iraq war, the relative ease with which these substances can be made from harmless precursors and the low cost of their manufacture will continue to fascinate power-hungry, ruthless dictators, as well as multinational and international terrorists, particularly as the close relationship between the OP agents and useful insecticides makes it easy to disguise the importation and purchase of small amounts of the precursors. Indeed, the use by Saddam Hussein of a nerve gas against the Kurds and his possible employment of the OP agents during his war with Iran, and the Sarin attack in the Tokyo underground by an extremist religious set magnetized the world with respect to the OP drugs. As these drugs exert their toxicity via their cholinergic action on the nervous, particularly central nervous, system, it is no wonder that the research in the cholinergic field attracts, and merits, our intense attention. These considerations underlie the significance of this book, as Alex Karczmar devotes an entire chapter of *Exploring the Vertebrate Central Cholinergic Nervous System* to anticholinesterases (antiChEs), and as he is an acknowledged expert in the field of cholinergic toxicity as well as a consultant to the Surgeon General of the U.S. Army.

Another and equally cogent reason for our interest in the cholinergic field is the involvement of the cholinergic system in autonomic functions, neuromuscular transmission, control of behavior and higher brain activities that are concerned with cognition, learning, memory and awareness. Furthermore, many neurological and psychiatric disorders such as presenile dementia, Huntington's chorea, schizophrenia, Alzheimer's disease, Parkinsonism, motoneuron disease, as well as a number of peripheral disorders, such as myasthenia gravis, involve cholinergic dysfunction. Accordingly, for several decades now there is no slackening in the production of books and monographs charting the progress of research into cholinergic function and its accessible components such as cholinesterases, acetylcholine and cholinergic receptors. An early, and still very important, volume is the monumental monograph edited by George Koelle, *Cholinesterases and Anticholinesterase Agents*, which was published in 1963 as supplementary volume 15 of the great Heffter-Heubner *Handbuch der Experimentelle Pharmakologie (Handbook of Experimental Pharmacology).* Then, to cite just a few of the early to recent compendia, there is *Le System Cholinergique*, edited by G.-G. Nahas, J.-C. Salamagne Paul Viars, and G. Vourch in 1962; Ann Silver's 1974 *Biology of Cholinesterases*; Alan Goldberg and Israel Hanin's 1976 *Biology of Cholinergic Function*; my own *The Cholinergic Synapse*, published in 1988 as volume 86 of the successor to the *Handbook of Experimental Pharmacology* and my 1992 *The Cholinergic Neuron and Its Target*; and Ezio Giacobini's *Cholinesterases and Cholinesterase Inhibitors* and *Butyrylcholinesterase, Its Function and Inhibitors*, published in 2000 and 2003, respectively. Parenthetically, Alex Karczmar contributed review chapters to all, except the last, of these books and monographs.

Karczmar's *magnum opus* is a very welcome culmination of these efforts, and it covers more territory than most of the monographs concerned with the cholinergic field. It is essentially a one-person effort, as Karczmar wrote eight of the eleven chapters of the book and was the coauthor of another chapter. As a result, the book has a homogeneity of approach—historical and conceptual—as well as a personal bias which adds color and interest to his monograph. Jon Lindstrom, Arthur Christopoulos, and Goerge Siegel with Neelima Chauchan and Alex Karczmar contribute important and up-to-date chapters on nicotinic and muscarinic receptors and on Alzheimer's disease.

The longest of the eleven chapters deals specifically with the physiology and pharmacology of the central cholinergic nervous system. With more than 1500 references, it is a thorough historical and conceptual review as it discusses, *inter alia*, the electrophysiology and pharmacology of cholinergic synapses, cholinoception and loci of release of acetylcholine, and central functions and behaviors exhibiting cholinergic correlates, including the involvement of the cholinergic system in cognition. The chapter even summarizes recent speculations on the metaphysics (in the literal sense of that word!) of the mind-body problem.

The chapter reminds us that the status of acetylcholine as a central neurotransmitter was for many years doubtful. Gradually, however, the electrophysiological evidence provided by John Eccles and the discovery by this author and his associates of nerve terminal localization of acetylcholine established the neurotransmitter—central as well as peripheral—role for acetylcholine. Indeed, me and my associates' work showed that that acetylcholine along with other neurotransmitters is specifically localized in subcellular fraction enriched with detached nerve terminals (synaptosomes) and, within the terminals, in the synaptic vesicles. In addition, our discovery of a cholinergic-specific surface antigen enabled us to isolate synaptosomes derived from cholinergic nerve endings in their pure form by immunoaffinity chromatography as well to enrich the synaptosomal content of acetylcholine; notably, vasoactive intestinal peptide (VIP) was enriched 20-fold in parallel with acetylcholine, serving as acetylcholine's co-transmitter. This co-transmitter role of VIP—and other polypeptides—in central cholinergic function, if thoroughly understood, might well throw a new light on the pharmacology and clinical significance of central cholinergic transmission. Karczmar discusses these matters, as well as recent controversies concerning the role of synaptic vesicles in acetylcholine release, in several chapters (particularly in a section of Chapter 2 entitled "Classical and Unorthodox Hypotheses of ACh Release").

Other chapters survey, with the same historical perspective and comprehensive treatment as the central nerve chapter, cholinergic cells and pathways; metabolism of acetylcholine and choline and cholinesterases; antiChEs, war gases and insecticides; muscarinic and nicotinic receptors (with contributions of Arthur Christopoulos and Jon Lindstrom); cholinergic ontology; and etiology of Alzheimer's disease (prepared by George Siegel, Neelima Chauhan, and Alex Karczmar). Molecular aspects of these subjects are considered as well.

Other chapters with the same historical perspective and comprehensive treatment as Chapter II and IX survey metabolic aspects of cholinergic function, muscarinic and nicotinic receptors, the organophosphorus anticholinesterases, cholinergic ontology and the etiology of Alzheimer's disease. A complete understanding of cholinergic function indeed requires a full integration of its pharmacology, physiology, and molecular biology. As an input to this partnership, this weighty monograph from a distinguished expert with a life-long perspective on the subject, aided by well-known co-authors, is a monumental contribution of enduring value.

Victor P. Whittaker, PhD
Professor Emeritus
Past Director
Arbeits Grüppe Neurochemie
Max Planck-Institut fur Biophysikalische Chemie
Göttingen, Germany

Preface and Acknowledgments

When they heard, some five years ago, of my intent to write, solo, a book on the central cholinergic system and its correlates, my cholinergiker friends told me that I was crazy to even try. Five years later, I agree with their assessment, as without the 3 crucial chapters by Neelima Chauhan, Arthur Christopoulos, Jon Lindstrom, and George Siegel, this book would have never been finished. My friends intimated also, rather sheepishly, that after all, in view of certain aspects of the matter and with some help, perhaps I would be able to complete the book: again, they were right; today the task, for better or worse, is finished.

I dedicate the book to my late, excellent, and old friends George Koelle, Bo Holmstedt, Jack Eccles, and Stanislav Tucek; for many decades we swam together, with joy and enthusiasm, down the mighty cholinergic river.

I wish to thank, for their help and comments, Marsel Mesulam, Palmer Taylor, Peter Waser, Claudio Cuello, Jochen Klein, Giancarlo Pepeu, Chris Gillin, Edson Albuquerque, Larry Butcher, Ezio Giacobini, Pat and Edith McGeer, Israel Hanin, Mona Soreq, Abe Fisher, Victor Whittaker, Mimo Costa, Nigel Birdsall, Ed Hulme, Roger Nitsch, Brian Collier, Maurice Israel, Jean-Pierre Changeux, Agneta Nordberg, Andrzej Szutowicz, Chris Krnjevic, and Ferdinand Hucho; I owe special gratitude to Konrad Loeffelholz, Jean Massoulié, and Yves Dunant for their help with Chapters 2 and 3.

I am particularly thankful to Kyo Koketsu and Syogoro Nishi and their past student Nae Dun for giving me a special insight into the cholinergic neuron and for guiding me for many fruitful years in the laboratory. And I am grateful to Enzo Logo for introducing me to the EEG expressions of the cholinergic system.

Invaluable help was afforded to me by Dr. Logan Ludwig, the director of Media Development and Design, Health Science Library, Loyola University Medical School, and his staff, particularly the reference librarians Janet Mixter and Mary Klatt, M.A.L.S.

Enormous assistance, moral support and warm friendship were afforded to me in the course of the work on this book by Kathleen Lyons, the Scientific Editor, Springer. Painstaking effort was experted in the immense task of preparing the proofs for the book by Barbara Chernord, Chernow Editorial Services.

Jenifer Stelmack, my editorial associate, is a computer whiz; without her computer expertise the difficult job of writing, formatting, and organizing this book could not have been done. Also, she was assiduous in clarifying and editing my text. In addition, Mr. Joseph Tomaszek, computer specialist, Hines VA Hospital, assisted us on many occasions with computer vagaries.

I am also very indebted to Joseph Messer, M.D., who kept me in good health throughout the arduous years of my work on this book.

A special encomium is due to my wife, Marion. She could not have tolerated my preoccupation with the book if she were not very busy on many fronts (and if she were not a swami); even so, she needed much endurance during the years of my writing this opus, and

I am thankful to her for her patience and support. The interest of my sons Chris and Greg in my doings was most supportive, and their urging me on is much appreciated.

I was supported in my research referred to in this book by NIH grants NS6455, NS15858, RR05368, NS16348, and GM 77; VA grant 4380; grants from the Potts, Fidia, and M. Ballweber foundations; Senor Fulbright and Guggenheim fellowships; and CARES of Illinois and AMVETS of Illinois. I wish to thank Ron Flink, CARES executive director, for his unflinching support. I acknowledge also the expert help with the preparation of my grant application to NLM-NIH by Pat Slagle, associate administrator, Research Service, Edward Hines, Jr., VA Hospital, Illinois. The actual writing of the book was supported by NLM-NIH grant G13LMO7249 and by Research Service, Edward Hines, Jr., VA Hospital, Hines, Illinois, and its associate chiefs of staff, Rita Young and Dale Gering.

Alexander G. Karczmar

Contents

8. Cholinergic Aspects of Growth and Development

9. Central Cholinergic Nervous System and Its Correlates

10. Links between Amyloid and Tau Biology in Alzheimer's Disease and Their Cholinergic Aspects

George S. Siegel, Neelima Chauhan, and Alexander G. Karczmar

11. Envoi

Contributors

Neelima Chauhan, MD, PhD
Professor of Neuroanesthesiology
University of Illinois College of Medicine
Chicago, IL 60612
and
Research Scientist
Edward J. Hines, Jr., VA Hospital
Hines, IL 60141
USA

Arthur Christopoulos, PhD
National Health and Medical Council of Australia
Senior Research Fellow
University of Melbourne
Australia

Jon Lindstrom, PhD
Trustee Professor of Neuroscience
and Professor of Pharmacology
Medical College of the University of Pennsylvania
Philadelphia, PA 19104
USA

George J. Siegel, MD
Chief, Neurology Service
Edward J. Hines, Jr., VA Hospital
Hines, IL 60141
USA
and
Professor of Neurology and Cell Biology
Neurobiology and Anatomy
Loyola University Medical Center
Maywood, IL 60153
USA

1

Introduction: History and Scope of This Book

A. Why a Book on the Central Cholinergic System?

Unlike the Roman Empire, the central cholinergic research rises but does not fall. The history of cholinergic research can be traced back to a time when shamans, hunters, medicine men and mystics collected plants, fungi and animal materials endowed with cholinergic ingredients in order to cure mental disease, provide food to the tribe, and identify the culprits. From these exotic beginnings there arose the research on the central cholinergic system that has exploded over the last 150 years. In this book, my associates and I propose to describe these beginnings of the cholinergic lore and to shed light on the recent central nervous system cholinergic research—research that continuously points the way toward distant peaks of new discovery.

No single review of the whole system seems to be available today, which warrants the publication of this book. In addition, this publication is called for, as the discipline of the central cholinergic system occupies today a preeminent and brilliant status among the biological sciences.

Some 85 years ago, the Nobel Prize winner Otto Loewi demonstrated the chemical, cholinergic nature of communication between peripheral nervous system and effector organs at the level of the cardiovagal junction. Twenty years later, Sir John Eccles, another Nobel Prize winner (with Paul Fatt and Kyozo Koketsu), proved that Santiago Ramon y Cajal's neuronal hypothesis translates into chemical, cholinergic communication between central neurons. Eccles also established that this communication, initially excitatory, could change signals (via an interneuron) and become inhibitory in mode (Figure 1-1). These discoveries led to explaining how signals travel along defined central circuitries and culminated with elucidating the generation of functions such as respiration and behaviors such as aggression; beyond that, we are today in the process of defining the cholinergic correlates of learning and cognition and, most excitingly, consciousness or self-awareness. In sum, the cholinergic wisdom has outlined a physical image of animal and human performance, whether corporeal or mental. From neuron to behavior, cognition, and consciousness—what a beautiful road (Table 1-1)!

Several way stations along this path are no less unique and important than the total path itself. One such way station is the system generating and controlling the release of acetylcholine (ACh) from the cholinergic nerve terminals, whether at the periphery or in the central nervous system (CNS). This system, pioneered by Victor Whittaker and Eduardo di Robertis (Figure 1-2; there is a bit of priority battle here), includes organelles; enzymes; transport, uptake, and other active proteins; and ions. These elements participate in ACh synthesis and its component, the choline nerve terminal uptake; generation of synaptic vesicles, their cycling and fusion with the plasmolemma; ACh's cytoplasmic movement; and quantal ACh release (either from the vesicles, the cytoplasm, or both;[1] see Table 1-2) into the synaptic cleft or the junction with the effector organ. To safeguard teleological release of ACh, the elements of this most complex arrangement must work in perfect synchrony and, yet, in a flexible manner; in fact, this interplay is so intricate that, perhaps, a computer program capable of defining this miraculous process and predicting its outcome—ACh release—at any particular moment cannot exist (Karczmar, 1999). Furthermore, this

Figure 1-1. From left to right: Winifred Koelle (wife of cholinergiker George Koelle), Sir John Eccles, and his wife, Helena Tabanikova-Eccles, at the Symposium in Honor of A.G. Karczmar, Loyola Medical Center, Maywood, IL, 1986.

Table 1-1. Brief History of Cholinergic Transmission Studies

Period	Provenance	Individuals	Drug and Materials
Second millennium BC – present	Arab, Egyptian, Roman, South American, African ethnographic medicines	Shamans and healers	Curare, ordeal bean or esere, bella donna, nicotine[1] (see Chapter 6 A-1)
1850–1880	Calabar missionaries, anthropologists, military men, and seamen	William Daniell, Donald Simmons, Hope Wadell	Ordeal bean (esere, eserine, physostigmine)[2] (see Chapter 6 A-1)
1870–1920	Scotch, German, and British investigators (Edinburgh, Dorpat, Halle)	Thomas Fraser, Robert Christison, John Balfour, John Langley, Oswald Schmiedeberg, William Gaskell, Walter Dixon	Nicotine and muscarine; choline derivatives[3] (see Chapter 6 A-1)
1870–1920	British, German, US investigators	Reid Hunt, Walter Dixon, Rudolf Lenz, Roberts Bartholow, Heinrich Winterberg	Choline derivatives, acetylcholine, atropine (atropia), ordeal bean (eserine, physostigmine)[4]
1870–1910	British and Russian investigators	William Gaskell, A.E. Smirnov, John Langley	Peripheral cholinergic pathways[5]
1890–1950	International investigators, including Russian and Spanish	Otto Loewi, Sir John Gaddum, Sir Henry Dale, Sir William Feldberg, Ramon y Cajal, Sir Geoffrey (Lindor) Brown, Alexander Kibjakov, Wilhelm Witanowski, Robert Volle, Kyozo Koketsu, Syogoro Nishi	Autonomic synaptic cholinergic transmission, cholinesterases; nicotinic and muscarinic receptors[6] (see Chapter 6 A-1 and Chapter 8 A-2)
1870–1925	US, French (Polish), and German investigators	Max and Michel Polonowski, Percy Julian, J. Jobst, Otto Hesse, E. and E. Stedman, G. Barger	Isolation, purification, structurization, and synthesis of physostigmine and physostigmine analogs[7] (see Chapter 7 A-1 and BI)

Table 1-1. *Continued*

1850–present	International investigators	Philippe de Clermont, A.W. von Hofmann, P. Nylen, Alexander Arbusov, Willy Lange, Gerhard Schrader	Organophosphorus antiChEs[8] (see Chapter 6 A-1)
1930–1955	International investigators, including Chinese, Italian, Greek, Belgian, and Swedish	Sir Henry Dale, Rene Couteaux, Sir John Gaddum, Sir William Feldberg, Martha Vogt, Sir Geoffrey (Lindor) Brown, P.T. Feng, Paul Fatt Bernard Katz, Rodolfo Miledi, William Paton, Eleanor Zaimis, Daniele Bovet, Stephen Thesleff, Alex Karczmar	Cholinergic neuromyal transmission and its pharmacology; desensitization (receptor inactivation)[9]
1914–1960	International investigators, including Australian, Yugoslav, and Italian	Sir Henry Dale, Sir John Eccles, Sir William Feldberg, Alfred Schweitzer	Central cholinergic transmission[10] (see Chapter 9 A-2)
1960–1975	British and US investigators	Samson Wright, Kasmir (Chris) Krnjevic, David Curtis, Giancarlo Pepeu, H. McLennan, Jose Delgado	Central cholinergic pathways[11] (see Chapter 2 A-1 and DI)
1940–1960	International investigators	George Koelle, M.A. Gerebtzoff, Charles Shute and Peter Lewis, Pat and Edith McGeer	Cholinergic ontogeny, comparative pharmacology, and phylogeny[12] (see Chapter 8 A)
1950–present	British, Australian, US, and Italian investigators	David Nachmansohn, Zenon Bacq, Klas Bertil Augustinsson, K.A. Youngstrom, Theodore Koppanyi, G.A. Buzikov, Alex Karczmar	Retro- and anterograde trophic factors[13] (see Chapter 8 CI-1)
1914–1950	International investigators	Victor Hamburger, Paul Weis, Rita Levi-Montalcini, J.Z. Young, L. Guth, E.D. Bueker, Ian Hendry, Hans Thoenen, E. Cohen, Sir Henry Dale, David Nachmansohn, Klas-Bertil Augustinsson, F. Plattner, H. Hintner, E. and E. Stedman, Bruno Mendel, David Glick, R. Ammon, G. Alles, William Aldridge	Cholinesterases[14] (see Chapter 3 DI)

Sources:
1. Waser, 1983; Levey, 1966.
2. Holmstedt, 1972; Holmstedt and Liljestrand, 1963; Karczmar, 1967a, 1970a, 1988.
3. Holmstedt, 1972; Karczmar, 1986.
4. Holmstedt and Liljestrand, 1963; Karczmar, 1970a.
5. Holmstedt and Liljestrand, 1963; Karczmar, 1986; Dale, 1953.
6. Karczmar, 1970a.
7. Karczmar, 1970a, 1986a, 1986b.
8. Holmstedt, 1959, 1963, 2000.
9. Karczmar, 1967a, 1970a; Zaimis, 1976; Couteaux, 1947; Bovet et al., 1959.
10. Eccles, 1969; Karczmar, 1967a, 1970a, 2001b.
11. Koelle, 1963; Machne and Unna, 1963; Karczmar, 1967a.
12. Karczmar, 1963; Changeux, 1985.
13. Hamburger and Keefe, 1944; Levi-Montalcini, 1975, 1987.
14. Dale, 1953; Karczmar, 1967a; Augustinsson, 1948; Koelle, 1963.

system serves as the best-known model for similar processes involving transmitters other than ACh.

Some of the steps in the evolution of cholinergic research concern the development of surprisingly effective and precise methodologies. To refer to just a few techniques, the measurement of ACh will be commented on first. Progress extended from bioassays capable of measuring micrograms of the transmitter,[2] to chemical measurements culminating in the gas chromatography–mass spectrometry technique that was developed by Israel Hanin in the 1960s and which is sensitive to nanogram levels of ACh, to a more recently developed choline oxidase chemiluminescent method that reacts to picomols of ACh (Israel and Dunant, 1998). Related to these methods was the microdialysis technique carried out by means of intrabrain multiple cannulae; this technique, pioneered and developed by Giancarlo Pepeu (Pepeu et al., 1990), permits measuring the release of ACh from restricted interbrain sites and relating this release to behavioral states of animals. Then there are the electrophysiological techniques. The Canberra and Japanese schools pioneered the use of microelectrodes, micropipettes, and voltage clamp methods capable of measuring electric potentials and currents at restricted brain sites containing just a few neurons; subsequently, a patch-clamp system capable of recording currents originating in single neurons was developed in Germany.

The cholinergic receptology is another province to be made salient. Cholinergic receptology is a classical research area originated by John Langley (see Table 1-1) as he differentiated pharmacologically between the nicotinic and muscarinic receptors. Later, receptors were evaluated via quantitative structure activity relationships and appropriate mathematical formulae by Everhardus Ariens and Robert Furchgott (Table 1-2); this approach yielded an abstract image of receptors and their subtypes. This image became more pragmatic when Peter Waser (Figure 1-3, see color plate) of Zurich employed radiolabeled cholinergic antagonists, principally curarimimetics, to visualize nicotinic receptors (Waser, 1983; Waser et al., 1954). Yet, it was almost a shock to many when, at the 1959 Rio de Janeiro Symposium on "Bioelectrogenesis," Carlos Chagas and Sy Ehrenpreis converted this receptor image to reality by allegedly precipitating and purifying the nicotinic receptor of the Torpedo,[3] enabling molecular methodology and immunological techniques to be applied to the problem. Today, the muscarinic and nicotinic receptor subtypes are identified chemically, and

Figure 1-2. Victor Whittaker and Alex Karczmar as Whittaker delivers a lecture in 1976 at Loyola University Medical Center, Maywood, IL.

Figure 1-3. Peter Waser, Alex Karczmar, and George Koelle during the 1974 Symposium on Cholinergic Mechanisms in Boldern, Switzerland. (See color plate.)

their three-dimensional images are of great esthetic beauty (see Chapters 5 and 6).

Anticholinesterase (antiChE) drugs developed during the cholinergic progress are notorious today. Actually, they exemplify the Pandora's box adage: antiChEs save entire populations from starvation when they are used as pesticides and insecticides, yet they can be also used as war gases and weapons of mass destruction. Indeed, thousands have already been killed by Tabun or VX, and today's political life is filled with references to this matter (see Chapter 7).

Some 35 years ago, the late Edith Heilbronn[4] (Figure 1-4, see color plate) decided that cholinergicity deserves a symposium, and she organized the First International Symposium on Cholinergic Mechanisms (ISCM); it took place in Skokloster, Sweden (Heilbronn and Winter, 1970). This was most timely; during the 1970s, cholinergic knowledge was in an exploding mode, yet only catecholaminergic and serotonergic matters seemed to merit the symposium status. It should be added that earlier, Edith had organized another symposium on cholinergic matters, which was more restricted than the 1970 symposium as it dealt only with "enzymes sensitive to DFP" (Heilbronn, 1967). Edith's initiatives were most fruitful, as they led to continuing ISCMs; altogether, 11 ISCMs have taken place, in Germany, France, England, Switzerland, the United States, and Italy, with the twelfth ISCM booked by Spain. Figure 1-4 shows cholinergikers, including Edith Helbronn, attending the 1983 Oglebay Park, West Virginia, United States, ISCM. These symposia are attended by 100 to 300 cholinergikers from all over the world, from Poland, Finland, Russia, and Australia to the United States and India, China, and Japan. Each ISCM presents the state-of-the-art knowledge of cholinergicity, and serves as a semiofficial report of the current research on the cholinergic system and its pharmacology (see Karczmar, 2004). What this book attempts to do, the ISCMs accomplish every 4 years, and the book would not be possible without the ISCMs and the brotherly and sisterly contacts between these authors and the ISCM cholinergikers.

Figure 1-4. Superimposed note from Edith Heilbronn. From left to right: *row 1* (seated): John Blass, George B. Koelle, Peter G. Waser, Donald J. Jenden, Israel Hanin, Frank C. MacIntosh, Alex Karczmar, Edith Heilbronn, Giancarlo Pepeu, Alan M. Goldberg; *row 2*: Victor J. Nickolson, Nae J. Dun, Stanley M. Parsons, Agneta Nordberg, Ezio Giacobini, B.V. Rama Sastry, Kathleen A. Sherman, Mario Marchi, Michael Stanley, Larry L. Butcher, Fiorella Casamenti, Tsung-Ming Shih, Herbert Ladinsky, Silvana Consolo, Kenneth L. Davis, Darwin L. Cheney, Janusz B. Suszkiw, Michael R. Kozlowski; *row 3*: Dean O. Smith, Steven H. Zeisel, Susan E. Robinson, Barbara Lerer, R. Jane Rylett, Rochelle D. Schwartz, Joan Heller-Brown, Marie-Louise Tjörnhammer, Britta Hedlund, David S. Janowsky, Natraj Sitaram, Linda M. Barilaro, Paul M. Salvaterra, Denise Sorisio, Elias Aizenman, Ileana Pepeu, Aurora V. Revuleta, Felicita Pedata, Clementina Bianchi, Lorenzo Beani, Henry G. Mautner; *row 4*: S. Craig Risch, Guillermo Pilar, E. Sylvester Vizi, Thomas J. Walsh, Sikander L. Katyal, Rob L. Polak, Roni E. Arbogast, Jean Massoulié, Denes Agoston, Brian Collier, Lynn Wecker, Bruce Howard, Richard S. Jope, Bernard Scatton, Matthew Clancy, Paul T. Carroll; *row 5*: William G. VanMeter, Michael Adler, Peter Kasa, Annica B. Dahlström, Gary E. Gibson, Peter C. Molenaar, Ingrid Nordgen, John D. Catravas, Judith Richter, David M. Bowen, Mark Watson, Renato Corradetti, Lorenza Eder-Colli, Marvin Lawson, Ing K. Ho, Jack C. Waymire; *row 6*: Paul L. Wood, Matthew N. Levy, Jean-Claude Maire, Frans Flentge, Richard Dahlbom, Pierre Etienne, George G. Bierkamper, Robert G. Struble, A.J. Vergroessen, Seana O'Reagan, Robert Manaranche, Maurice Israel, Yacov Ashani, Abraham Fisher, Steven Leventer, Alan G. Mallinger; *row 7*: Anders Undén, Edward F. Domino, William D. Blaker, Peteris Alberts, Johann Häggblad, Daniel L. Rickett, Sten-Magnus Aguilonius, Serge Mykita, Hans Selander, Oliver Brown, Henry Brezenoff, Sven-Åke Eckernäs, Frederick J. Ehlert, Björn Ringdahl, Volker Bigl, Duane Hilmas, Clark A. Briggs, Nicolas Morel; *row 8*: Bo Karlén, Michael J. Dowdall, John J. O'Neill, Heinz Kilbinger, Wolf-D. Dettbarn, Konrad J. Martin, Konrad Löeffelholz, Roy D. Schwarz, Jerry J. Buccafusco, Ernst Wulfert, Howard J. Colhoun, Paul Martin, Jack R. Cooper, Crister Larsson, Harry M. Geyer, Michael J. Pontecorvo, William E. Houston, Jurgen von Bredow, Yves Dunant. (From Hanin, 1986. Reprinted by permission of Kluwer Academic/Plenum Press.) (See color plate.)

B. Outline of the History of Central Cholinergic System Research

1. Brief Story of the Studies of Cholinergic Transmission

Central cholinergic research has a brilliant story, which is summarized in Table 1-1.[5] The story begins with the use of plant populations all over the world that contain cholinergic drugs, such as antiChEs, muscarine, atropine, and curares (for details, see Chapter 7 A and Chapter 9 A). We owe this story in part to the studies by the Edinburgh physicians and pharmacognosists of the Calabar bean extract provided by Scotch missionaries to the West African province of Calabar (for references, see footnotes to Tables 1-1, 1-2, and 1-3). But other cholinergic drugs, particularly d-tubocurarine (curari or curare; see Chapter 6-1 and 2, and section B in this Chapter) and atropine (atropia) were known much earlier. Stramonium and hyoscyamine plants and their effects were known by pharaonic Egyptians and then by Arabs (see Levey, 1966, and Karczmar, 1985); nor were the Romans foreign to the consequences of eating atropine-containing plants, and Lucan describes these consequences endured by Roman soldiers during their marches in northern Africa at the time of the civil war between Caesar and Pompey. Actually, the Baroque scholar Antoni Storck stated that "many" or "all authors wrote: Strammonium disrupts the mind, induces madness, erases ideas and memory, brings about convulsions" (Storck, 1757; see also Wassen, 1965).

The heuristic tale that began in the 19th century and which concerns mainly physostigmine deserves special attention. Edinburgh pharmacognosists and an international team of Scottish, German, British, and US investigators worked with the effects of physostigmine (i.e., the active component of Calabar bean extract), muscarine, nicotine, and atropine at mostly the autonomic and neuromyal periphery, and also occasionally on the CNS (in the case of Robert Christison) and provided a basis for the ultimate demonstration of chemical, cholinergic transmission. Additional basic research concerned establishing the autonomic parasympathetic and sympathetic pathways, and is linked with John Newport Langley and Sir Henry Dale. Subsequent significant steps in this peregrination toward the proof of chemical transmission included establishing the evanescence of the effects of systemically administered ACh and the resemblance between the effects of ACh and those of the stimulation of such nerves as the vagus (Dixon, 1907). Sir Henry Dale's name is forever linked with this segment of the cholinergic lore and with the suggestion that there must be an enzyme that accounts for this short-lived action of ACh (Figure 1-5 and Table 1-1).

This research culminated during the 1920s with Otto Loewi's demonstration of cholinergic transmission at the parasympathetic cardiovagal ending (Figure 1-6). The story of Loewi's dreaming about the possible method that could provide this proof and then translating the dream into reality is well known and quoted jointly with heuristic dreams of other researches (e.g., Kekule's dream of accurately portraying the benzene structure); yet, there are those who had a different angle on Loewi's dream (Karczmar, 1996). It must be noted that in the last three decades of the nineteenth century many investigators, in Germany, England, Russia, and United States, including William Gaskell, H.E. Hering, Muskens and, particularly, F.B. Hofmann of Marburg (see his 1883 review) established that, at the mammalian or anuran heart the stimulation of distinct nerves cause inhibition and excitation (dromotropic actions), respectively. Furthermore; in the early years of the twentieth century, Reid Hunt and other investigators demonstrated the presence of choline in tissues, such as the suprarenal gland, and soon thereafter, the presence in various tissues of a "hormone," possibly ACh was reported (see Dale and Dudley, 1924). Then, Hunt with Taveau (1906) and later with J.W. Le Heux (1919: see also chapter 3, Section C3) noted that ACh was many thousand times more potent at the heart and intestine than choline. At the same time, Walter Dixon suggested that muscarine may be the neurohumor of the vagus nerve and Thomas Elliott opined that adrenaline (epinephrine) acts similarly at the sympathetic nerves. Actually, some sixty years earlier Emil Du Bois (1843) listed the release of a chemical agent as one of the options regarding the function of the neuromyal junction. Altogether, the time was very ripe for Loewi's discovery!

A

B

Figure 1-5. (A) Henry Dale in a photograph taken in 1918 at the Lister Institute for Preventive Medicine by Sir Charles Lovatt Evans. (Reprinted by permission of Wellcome Trust.) (B) Sir Henry Dale in a photograph taken at the offices of the Wellcome Trust, 1959. (Reprinted by permission of Wellcome Trust Publishers.)

October 29. 1953

Dear Dr Karczmar,

Thanks for your and Dr Koppanyi's fine contribution to the "Festschrift" I am expecting to get. Would you also kindly forward my thanks to the Medical Faculty and to the Department of Pharmacology.

Very sincerely yours

Figure 1-6. Otto Loewi's photograph and letter of thanks on the occasion of his Festschrift of 1953.

The demonstration of cholinergic transmission existing at parasympathetic sites was later expanded to the sympathetic ganglia and to the skeletal neuromyal junctions. The demonstration was based on neurochemical evidence—the presence of ACh and cholinesterases (ChEs) at the sites in question and the release of ACh at the synapse or junction following presynaptic stimulation—as well as on neurophysiological and neuropharmacological findings. For example, appropriate presynaptic stimulation evoked a postsynaptic (in the ganglia) or endplate (in the neuromyal junction) response that, on pharmacologic analysis, appeared cholinergic in nature. Actually, this research established criteria that must be met before a given junction or synapse is considered cholinergic; Robert Volle (1966) fully defined these criteria.

The neuromyal research[6] led to the explanation of the curarimimetic action and to completing the demonstration of cholinergic transmission at the periphery. The pioneers of this research were Rene Couteaux and George Koelle (Figures 1-3 and 1-7, see color plate), as they established histochemically a "focused" presence of AChE at the neuromyal junction (Couteaux, 1947, 1998). Rene Couteaux also determined the postsynaptic origin of AChE (its presynaptic presence was stressed by Koelle), as it persisted at the endplate following denervation, and demonstrated AChE association with the basal lamina. An important aside: this research also established the phenomena of receptor inactivation (desensitization) and sensitization (Thesleff, 1955; Karczmar, 1967b). Ultimately, these processes result from allosteric modifications of the receptors and their phosphorylations, and these phenomena characterize muscarinic and nicotinic receptors both at the peripheral and central cholinergic sites.

In parallel, the mid-19th century brought about a number of important chemical discoveries and synthetic accomplishments, including the purification of physostigmine as the active ingredient of the Calabar bean's extract and the synthesis of physostigmine and its carbamate analogues. While the first synthesis of organophosphorus (OP) antiChEs dates from the second part of the 19th century (see Table 1-1), the 20th century brought

Figure 1-7. U.J. (Jack) McMahon (left) and Rene Couteaux at the ISCM meeting in Arcachon, 1998. Jack McMahon is a prominent modern investigator of the ultrastructure of the neuromyal junction. (From the Author's private collection.) (See color plate.)

forth a major expansion of OP synthesis and the discovery of their pesticidal as well as toxic actions.

Anticholinesterases, particularly of the OP type, played an important role as tools in the studies of cholinergic transmission. Using these antiChEs as tools was conditioned by the belief common in the 1940s and 1950s that the OP drugs had only one mechanism of action, the inhibition of ChEs (Karczmar, 1970a). Furthermore, in contrast to physostigmine, the OP antiChE action was irreversible in nature, which made it convenient for investigators of cholinergic transmission. Later, it was found that OP and carbamate drugs exhibit actions that are independent of their inhibitory effect on ChEs (see below). Following World War II interest in OP agents, research laboratories that were instituted in England at several sites by the Ministry of Supply and in the United States by the Department of Defense (particularly at Edgewood Arsenal in Maryland) served as teething grounds for a number of prestigious British, American, and Swedish researchers, including George Koelle (see Figure 1-3), Alfred Gilman, Bo Holmstedt, Henry Wills, Irving Wilson, Henry Wilson, Amadeo Marrazzi, Bill Krivoy, Steve Krop, Bernard McNamara, Lord Adrian, Sir William Feldberg, B.A. Kilby, B.C. Saunders, and others. Bo Holmstedt (1963, 2000; see Figure 7-16) presented a detailed account of these events, including those that occurred in Germany.

The first demonstration of the presence of chemical transmission in the central nervous system—which happens to be cholinergic in nature—also took place during the mid-20th century. The omniscient Sir Henry Dale and his students first suggested this presence, and Dale's associate Sir William Feldberg (Figure 1-8) proposed that the selective distribution of ChE sites in the CNS provides evidence for the central presence of cholinergic transmission. And then Sir John Eccles, Paul Fatt, and Kyozo Koketsu proved the point by pharmacologically evaluating responses to Renshaw cell stimulation by spinal motoneuron collateral (see Chapter 9 A-2c, and Figure 9-4). Subsequently, the Canberra team, including David Curtis and Chris (Kasimir) Krnjevic, embarked on long studies of the central cholinoceptivity (Figure 1-9, see color plate). In addition, Krnjevic researched the characteristics of central muscarinic responses and proposed that these responses modulate central activities of non-cholinergic transmitters (see Chapter 9 BIII-1 and BIII-2). Ultimately, scientists of many nations demonstrated the presence of cholinoceptivity throughout the brain and the pharmacological analysis of these postsynaptic responses served to establish the ubiquity of central cholinergic transmission (see Table 1-1).

Figure 1-8. Sir William Feldberg in 1972.

It must be noted that, at the extreme, the criteria established by Robert Volle and others for accepting the presence of cholinergic transmission at any site (see above, this section) requires establishing that a single neuron or a homogeneous (morphologically) group of neurons release ACh upon presynaptic stimulation. While such criteria have been fulfilled for the neuromyal junction, parasympathetic sites, and the ganglia, they cannot be readily met in the CNS. J.F. Mitchell, J.W. Phillis, J.C. Szerb, and others (see Pepeu, 1974, and Figure 1-10, see color plate) collected ACh from cortical slabs of stimulated animals and from the animal cortex surface following reticular stimulation; this procedure would not, of course, meet Volle's criteria and could not identify specific cholinergic neurons involved in the release.

Figure 1-9. Chris (Kazmir) Krnjevic at the Symposium in Honor of A.G. Karczmar, Loyola Medical Center, Maywood, IL, 1986. (See color plate.)

Figure 1-10. From left to right: *row 1*: Doctoressa Ileana Pepeu, Erminio Costa, Marion Allen-Karczmar and Leda Hanin; *row 2*: Israel Hanin, Giancarlo Pepeu, Alex Karczmar. Erminio (Mimo) Costa is a prominent psychoneuropharmacologist and one of the founders of the College of Neuropsychopharmachology. He pioneered the research on the role of serotonin, GABA, and catecholamines. He also contributed to the concept and measurement of the turnover of acetylcholine. (See color plate.)

The research of Henry MacLennan, Jose Delgado, Nicholas Giarman, and particularly Giancarlo Pepeu was closer to the mark. To collect ACh, these investigators employed the intracerebrally implanted push-pull cannula developed by Sir William Feldberg, and Giancarlo Pepeu developed microcannulae and improved methods of measuring ACh (Pepeu, 1974, 1993; Pepeu et al., 1990; see Chapter 9 BIII, and Figures 9-14 and 9-15); the animals used varied from rodents and cats to primates. This sophisticated research served to establish the cholinergicity of brain parts such as the nucleus basalis magnocellularis of Meynert (NBM), as its stimulation releases ACh at the end of its specific radiations. However, to a purist, even this evidence does not fully comply with the Volle criteria. Actually, even Eccles' classic neuropharmacological demonstration of cholinergic transmission existing between the motoneuron and the Renshaw cell falls short of the purist's definitive proof of cholinergicity, as no investigators could demonstrate the release of ACh at a single, identified motoneuron collateral.

The evidence concerning the presence of cholinergic transmission in the CNS closely relates to the identification of central cholinergic pathways. The descriptions of cholinoceptive sites that were obtained during central cholinergic studies were helpful; however, a definitive establishment of central cholinergic pathways depended on the development of methods for visualization within the CNS of the cholinergic system components. The first method, cytochemistry of both butyryl and acetyl cholinesterase (BuChE and AChE), was developed by George Koelle and Jonas Friedenwald. Koelle later adapted the method to electronmicroscopy, which he used to identify major ChE exhibiting nuclei and networks (Koelle, 1963). Michel Gerebtzoff, Peter Lewis, and Charles Shute provided more continuous pictures of the pathways, and they also attempted to obviate a problem inherent in the ChE stain methodology: the problem with the ChE stain as a marker for cholinergic neurons is that ChEs are also present in noncholinergic neurons and their nerve terminals. By judicious use of CNS ablation and lesion techniques, Gerebtzoff, Shute, and Lewis came close to identifying ChE stain sites reflecting radiations from bona fide cholinergic neurons (see Chapter 2 DI and DII).

However, only ACh may serve as a faithful marker of cholinergic cells, via the demonstration, whether by means of a bioassay or chemically, of the presence of ACh in certain single neurons of the CNS. The presence of ACh and other components of the cholinergic system (such as ChEs and cholineacetyl transferase) in homogeneous populations at the periphery and ACh release from such populations was established early in the 20th century, and, in the 1950s, Ezio Giacobini and others showed their presence in single autonomic neurons, but this feat could not be accomplished in the case of single central neurons.

A related demonstration concerned the presence of ACh in several brain parts, and this was carried out via bioassays by British investigators such as Joshua Gaddum and William Feldberg and Russian scientists in support of Henry Dale's and William Feldberg's hypothesis of the central transmitter role of ACh. A particular break was to repeat this demonstration using a chemical method for the identification of ACh; this was carried out by means of a gas chromatography–mass spectrometry (GCMS) method in the 1950s by Don Jenden, Bo Holmstedt, and Israel Hanin in Bo Holmstedt's Toxicology Laboratory of the Karolinska Istitutet of Stockholm (see Chapter 2 A).

Second best was to show the release of ACh from the brain and spinal cord, as described above in this section. An even better approach was to develop ACh histochemistry to characterize as cholinergic certain central neurons. While several histochemical methods were developed to visualize ACh, the most practical and most usable technique depended on the second best, immunohistochemistry of choline acetyltransferase (CAT), the enzyme that synthesizes ACh and is therefore a dependable marker of cholinergic neurons. This method was developed by the McGeers, a husband-and-wife team, and Henry Kimura (McGeer et al., 1986; McGeer and McGeer, 1993; see also Chapter 2 DI, and Figure 1-11, see color plate). Later on, in the hands of Marsel Mesulam, Bruce Wainer, Larry Butcher, Nancy Woolf, and Paul Kasa, this methodology brought about one of the few essentially "finished" tasks of the cholinergic lore: the definitive description of central cholinergic pathways. Without this accomplishment, identification of behaviors endowed with cholinergic correlates would not be possible (see also section 2, below, and Chapter 9 BV.

Figure 1-11. Edith McGeer at the 1986 Symposium in Honor of A.G. Karczmar, Loyola Medical Center, Maywood, IL. (See color plate.)

The question of the cholinergic system's developmental origin arose following the demonstration of central cholinergic transmission and its pathways. Interestingly, attempts at ontogenetic recognition of the cholinergic system and the related comparative identification of its presence were initiated long before Eccles' proof was given (see Table 1-1); in fact, these attempts were intended to achieve the same goal: to prove the existence of central cholinergic transmission. During the 1930s and 1940s, David Nachmansohn and Zenon Bacq proposed that identifying the presence of ChEs at the ontogenetic onset of embryonic or fetal motility, as well as their phylogenetic existence in forms endowed with motility, would demonstrate central cholinergic transmission and its status as a prerequisite for function. Actually, Bacq (1935 and 1947) and Nachmansohn accumulated the ontogenetic and phylogenetic data to support this proposition and, another early cholinergiker, Theodore Koppanyi[7] (Figure 7-22; Chapter 7 DI) was particularly effective in accumulating pertinent evidence in the area of comparative pharmacology (Koppanyi and Sun, 1926; see also Karczmar, 1963, and Whittaker, 1963). Though their evidence cannot be construed as proving the existence of central cholinergic synapses, it is important to relate function such as motility to cholinergic transmission, and this research was a forerunner of intense research involving the relationship of functions and behaviors to the cholinergic system.

Subsequent ontogenetic and phylogenetic cholinergic research provided unexpected evidence for the existence of cholinergic elements such as ACh, ChEs, and CAT in nonnervous tissues of vertebrates (including ephemeral tissues such as the placenta), and their presence in ontogeny long before the appearance of nervous system rudiments (see Chapter 8 CI). These phenomena remain enigmatic today.

The research concerning development yielded still another unexpected discovery: the existence of trophic and growth factors. The conjecture that nerves may play a trophic role was made early (Karczmar, 1946), but it took the technical and conceptual ingenuity of Edward Bueker, Victor Hamburger, and, particularly, the Nobel Prize winner Rita Levi-Montalcini (Levi-Montalcini, 1975, 1987) to demonstrate the existence of trophisms and their capacity to repair neuronal damage and growth deficits (see Table 1-1 and Figure 1-12, see color plate). Actually, today the phenomenon postulated by Karczmar takes the form of anterograde trophism, that is, trophism exerted on a target tissue via the release of a trophic substance by a nerve terminal.

The original form of trophism described by Rita Levi-Montalcini depends on retrograde

Figure 1-12. Nobel Prize winner Rita Levi-Montalcini and Ezio Giacobini at a Vatican Symposium, 1990. (See color plate.)

action on the nerve tissue of a factor present peripherally; Hans Thoenen and Ian Hendry determined the terminology differentiating retrograde and anterograde-acting trophics (see Table 1-1). Today, many trophic factors are recognized and there is an intense search for more, as new trophics are discovered continually. This interest in trophisms results from their potential as cures for human degenerative diseases such as Alzheimer's disease (see Chapter 10), although it may be mildly suggested that the applied expectations raised by research with trophics has not been substantiated so far. At any rate, the trophic action of cholinergic components may underlie their precocious presence during ontogeny (see Chapter 8 CI and CII and Chapter 3 CI).

The story of ChEs and antiChEs flickers in and out during this history of cholinergic research. Cholinesterases are pertinent for the effects of the ordeal bean, its extract, and its active ingredient. It was postulated by Sir Henry Dale that a ChE is a component sine qua non of the cholinergic function, and physostigmine was used as an antiChE in Otto Loewi's studies of the release of ACh from the frog cardiovagal perfusate, although in his first communication on this matter (Loewi, 1921) he did not employ physostigmine (see also Loeffelholz, 1981, 1984; Karczmar, 1996). Yet, physostigmine was employed by Loewi in his subsequent studies as well as by his followers in similar situations involving measurement of ACh efflux from stimulated brain. Sir John Eccles took up physostigmine in his pharmacological analysis of the Renshaw cell responses; Zenon Bacq and K.A. Youngstrom synonymized the presence of ChEs with function; and George Koelle, Peter Lewis, Charles Shute, and Michel Gerebtzoff used the histochemical ChE stain in describing cholinergic pathways.

A concerted effort to elucidate the role and the nature of ChEs must also be recognized. David Nachmansohn described the immense (unique for enzymes) rapidity of ACh-hydrolyzing action of AChE, and Klas-Bertil Augustinsson spent his life characterizing the differential dynamics and substrates of ChEs, their inhibitors, and their animal and human distribution. Then, the Stedmans, Bruno Mendel, and David Glick differentiated the dynamics of AChE and BuChE with regard to their substrates and studied their differential distribution (see Table 1-1; Chapter 3 DI and DII; Chapter 7 BI).

This research was the basis of the contemporary studies of Jean Massoulié, Hermona (Mona) Soreq, Werner Kalow, and Palmer Taylor. It yielded novel notions of variants for both BuChE

Figure 1-13. Hermona Soreq is a prominent scientist from the Hebrew University of Jerusalem. Soreq pioneered molecular studies of cholinesterases. Currently she studies stress-induced variants of acetylcholinesterase. (See color plate.)

and AChE and their polymorphism that include several physical forms and dimensionally differentiated subunits. Hermona Soreq (Figure 1-13, see color plate), a pioneer in the area of genetic and molecular characteristics of ChEs, stresses the genetic flexibility of ChEs that results in production of variants—for example, in response to stress AChE may be mutationally overexpressed, which results in transmittive malfunction (see Chapter 3 DIII).

2. Fanfares for the Central Cholinergic Transmission and Its Correlates

In the light of the significance of the cholinergic, particularly central, cholinergic research and the continuity of its story (Table 1-1 and above), the felicities of its findings are addressed in "fanfares" (Tables 1-2, 1-3, and 1-4), which concern several areas of cholinergic interest. These fanfares are raised with respect to an international field of investigators, though not all cholinergikers who deserve these accolades can be listed in the tables; probably the personal bias of the author was involved in the choices of those that are listed.

a. Metabolism and Synthesis of ACh, and Formation of ACh into a Release Mode

David Nachmansohn initiated this story when he discovered choline acetyl transferase (CAT; originally called choline acetylase by Nachmansohn), which is the enzyme synthesizing ACh;

Table 1-2. Fanfares: Acetylcholine Synthesis, Choline Uptake, and Acetylcholine Release (1930 to 1980)

International investigators	Frank MacIntosh, R. Birks,G.B. Ansell, Richard Wurtman, Stanislaus Tucek, J. Kazimir Blusztajn, Don Jenden, Lynn Wecker, David Nachmansohn, Vincent du Vigneaud, Konrad Loeffelholz	Metabolism of ACh and choline, choline acetyl-transferase (CAT), and acetyl-coenzyme[1]
US and Canadian investigators	Fred Schueler, Frank MacIntosh, T. Haga, Victor Whittaker, S.M. Parsons, Michael Kuchev, Jeffrey Erikson, Louis Hersh, Lee Eiden, Nicolas Morel	Hemicholinium and uptake of choline; vesamicol and neuronal ACh transport; active proteins of the cholinergic neuron[2]
US, Chinese, and Canadian investigators	George Koelle, T.P. Feng, R. Polak, J.C. Szerb, Yves Durant, Laurent Descarries, Jacopo Meldolesi	Presynaptic regulation of acetylcholine release in the CNS and periphery[3] (Chapter 9 BII)

Sources:
1. du Vigneaud, 1941; Jenden et al., 1976; Ansell and Spanner, 1971; MacIntosh et al., 1956.
2. Schrader, 1963; Kuhar, 1978.
3. Koclle, 1963; Szerb, 1977.
4. Karczmar, 1999.

Table 1-3. Fanfares: Cholinergic Behaviors and Cholinergic EEG Phenomena (1940 to 1980)

British, US, and Italian investigators	William Funderburk, Theodore Case, W.H. Gantt, Sir William Feldberg and Stephen Sherwood, Carl Pfeiffer, Daniele Bovet, L.R. Allikmets, Luigi Valzelli, R. Bartus, Alex Karczmar	All behaviors have cholinergic correlates[1] (Chapter 9 BIV and BV)
US and British investigators	Nancy Woolf, Ben Libet, Alex Karczmar, John C. Eccles, Roger Penrose	Self-awareness, organism-environment interaction cholinergic alert non-mobile behavior (CANMB)[2] (Chapter 9 BV and BVI)
International investigators	V. Bonnet and Frederic Bremer, D.B. Tower, K.A.L. Eliott, Giuseppe Moruzzi and Horace W. Magoun, Harold Himwich, Franco Rinaldi, R.G. Hernandez-Peon, Peter Andersen and Sven Andersson, F.T. Brucke, C. Stumpf, Vincenzo Longo, Philip Bradley, Joel Elkes, Michel Jouvet, H. Jasper, R.W. McCarley, J. Allan Hudson, Edward Domino, Chris Gillin, Taddeus Marczynski, Mircea Steriade, Alex Karczmar	EEG, theta rhythms, and rapid eye movement sleep (REM)[3] (Chapter 9 BV-3)

Sources:
1. Karczmar, 1967a, 1970b, 1976; Bovet, 1972; Bartus et al., 1982.
2. Woolf, 1997; Karczmar, 1972; Popper and Eccles, 1977.
3. Longo, 1972; Robinson, 2001; Karczmar, 1976; Jouvet, 1967; Brazier, 1959.

Nachmansohn also identified the coenzyme, acetyl-coenzyme A, that participates in the acetate transfer to choline. During the 1940s and 1950s, Frank MacIntosh discovered the need for choline in this synthesis and for the maintenance of ACh release during synaptic activity (Birks et al., 1956). Subsequently, this role of choline was clarified when Fred Schueler and John Paul Long (Schueler, 1956) developed hemicholiniums, agents that block the nerve terminal transport and uptake of choline. The release mode includes intraneuronal transport and vesicular uptake of ACh, which can be blocked by vesamicol (Table 1-2). Then, there is a vesicular cycling, and, perhaps, cytoplasmic cycling, of ACh. Altogether, the active proteins involved in the choline uptake, ACh transport, and vesicular cycling and fusion of the vesicles with the plasmalemma, jointly with CAT and AChE, define the cholinergic neuron.

The system underlying ACh metabolism and molecular physiology of a cholinergic neuron is most complex, yet it functions smoothly and is flexible and adaptable, and it is entirely teleological that a single "cholinergic gene locus" regulates many of these activities. The description and definition of this system merits a fanfare, and it must be stressed that this definition could not be forthcoming without the development of ultrasensitive and precise biochemical and molecular methods pertinent for the measurement of the components of the cholinergic system and definition of their molecular and genetic properties.

Table 1-4. Fanfares: Molecular and Cellular Aspects of Cholinergic Transmission (from 1940 to the Present)

Argentinean, British, French, and Canadian investigators	Victor Whittaker, Eduardo de Robertis, S.L. Palay, Paul Fatt, J. del Castillo, Maurice Israel, Yves Dunant, Brian Collier, Laurent Descarries, Jacopo Meldolesi	Vesicular and cytoplasmic release of ACh; quantal phenomena and elementary events[1]
US and Japanese investigators	Earle W. Sutherland, Paul Greengard, L.E. and M. Hokin, Yasutoni Nishizuka	Second messengers and phosphorylations[2]
British, US, Swiss, Australian, French, and Brazilian investigators	Sir Henry Dale, Carlos Chagas, Jean-Pierre Changeux, A. Karlin, Jon Lindstrom, Peter Waser, Michael Rafftery, James Patrick, Ed Hulme, Nigel J. Birdsall, Palmer Taylor, Sydney Ehrenpreis, Everhardus Ariens	Cholinergic receptors and their subtypes[3]
Australian, US, Canadian (Yugoslav), and Japanese investigators	John Eccles, Rosamond Eccles, Ben Libet, Kris Krnjevic, Paul Adams, Syogoro Nishi and Kyozo Koketsu	Ionics of cholinergic synaptic potentials[4]
Swedish and US investigators	Thomas Hokfelt, Y.N. and L.Y. Jan, Stephen Kuffler, Nae Dun	Release of noncholinergic transmitters from cholinergic terminals[5]
Australian, French, Israeli, and US investigators	Paul Salvaterra, Michael Raftery, Jean Massoulié, Hermona Soreq, Jon Lindstrom	Molecular biology of components of cholinergic system[6]
British, French, Brazilian, Argentinean, and US investigators	Eduardo de Robertis, Victor Whittaker, David Nachmansohn, Harry Grundfest, D. Albe-Fessard, A. Fessard, Carlos Chagas, R.D. Keynes	Electric organs[7]
US, French, and Israeli investigators	Kyozo Koketsu, Steve Thesleff, Jacques Monod, A. Karlin, Jean-Pierre Changeux, Paul Greengard, Mona Soreq, Alex Karczmar	Pre-and postsynaptic modulations; receptor activation and inactivation[8]

Sources:
1. Whittaker, 1992, 1998; De Robertis, 1961.
2. Greengard, 1978; Sutherland et al, 1968; Robinson, 2001.
3. Waser, 1983; Robinson, 2001; Ehrenpreis, 1961; Ariens and Simonis, 1967.
4. Eccles, 1964; Krnjevic, 1972; Adams, 1981; Robinson, 2001.
5. Hokfelt et al., 1979, 1987; Robinson, 2001.
6. Glick and Soreq, 1999; Salvaterra et al., 1993.
7. Whittaker, 1992, 1998; Albe-Fessard, 1961.
8. Greengard, 1978; Greengard and Browning, 1988; Robinson, 2001; Karczmar et al., 1972.

b. The Processes of ACh Release

First, there is the control of this release, which is exerted by cholinergic and noncholinergic receptors of the cholinergic presynaptic nerve terminals; this control ranges from the nicotinic autoreceptor facilitatory mode leading to "percussive release" demonstrated (or almost demonstrated) by George Koelle (Table 1-2; see Chapter 9 BII) to the muscarinic autoreceptor inhibitory mode described by Rob Polak and John Szerb. And this cholinergic regulation of transmitter release extends to noncholinergic nerve terminals. Furthermore, there is the process of ACh release. Among the many controversies that have raged within the cholinergic lore, the battle over the mode of this release is the most intense: is the release vesicular in nature, as classically maintained by Victor Whittaker and Eduardo de Robertis, or are there modes of cytoplasmic release, as maintained by the unorthodox Franco-Italo-Canadian investigators? (See foreword, Table 1-1 and Chapter 2 C.)

c. Cholinergic Correlates of Behavioral Cognition and Consciousness, and Their Electrophysiological Counterparts That Include the EEG and Evoked Potentials

The cholinergic system was the first neurotransmitter system that was related to behavior and cognition. The 19th century's Robert Christison (see Table 1-1), via self-experimentation, established some of the mental and behavioral effects of ingesting a Calabar bean. William Funderburg, Theodore Case, Sir William Feldberg, Carl Pfeiffer, and Daniele Bovet confirmed Christison's notion in the mid-20th century and extended it to demonstrating the effects of the cholinergic agonists and antagonists on conditioning and learning in animals. Today, a remarkable feature of the cholinergic system is recognized: this system is involved in most measurable functions from respiration to motor movement and endocrine activities, and in behaviors ranging from memory and learning to aggression and addiction; this is true in both humans and animals (see Chapter 9 BIV and BV). The totality of the behavioral roles of the cholinergic system amounts to a syndrome referred to as Cholinergic Alert Behavior (CANMB), which underlies organism-environment interaction and is instrumental in animal adaptability and, speculatively, in the evolution of both humans and animals (Karczmar, 1988; see Chapter 9 BVI).

Finally, perhaps even the mystery of consciousness or self-awareness may have cholinergic correlates. Thus, Nancy Woolf speculates that muscarinic receptors of the small cortical pyramidal cells modules are the sites of quantal (in physical, not cholinergic sense) events that, according to her and Roger Penrose, underlie self-awareness (see Table 1-3, Chapter 9 BVI, and Chapter 11). Of course, since the days of René Descartes, neuroscientists have not come to an agreement on the nature of consciousness and on the special existence of self-awareness; the fact that the cholinergic system enters the fray shows, again, the vitality and the appropriateness of the cholinergic lore (see Chapter 9 BVI and Chapter 11).

The electroencephalogram (EEG) and evoked potential phenomena relate to and sometimes identify cholinergic behaviors. Such phenomena as EEG alerting, described by Giuseppe Moruzzi and Horace Magoun, have strong and unique cholinergic correlates, which were established in the 1950s by Harold Himwich, Franco Rinaldi, Daniele Bovet, and Vincenzo Longo. As already mentioned, the Nobel Prize winner Daniele Bovet (Figure 1-14) was interested in the behavioral effects of nicotinic and nicotinolytics; also, he perceived that EEG may provide an "organic" counterpart of the behavioral effects of cholinergic agents. Accordingly, he induced his associate, Vincenzo Longo (Figure 15, see color plate), to carry out an EEG and evoked potential (such as theta rhythm) evaluation of cholinergic and anticholinergic drugs, a task that Longo performed in great detail. These investigators related the so-called EEG arousal to behavioral arousal and alertness; today, there is an intense discussion of this relationship (see Table 1-3 and Chapter 9 BIV-3) Then, there is the sleep-awakening dipole, and the component of the sleep mode, the rapid eye movement (REM) or dream sleep (see Table 1-3 and Chapter 9 BIV-3). While in the 1960s it was considered that serotonin and norepinephrine are solely involved in REM sleep, the preeminent role of the cholinergic system for REM sleep was stressed subsequently; again, both humans and animals REM-sleep cholinergically.

Figure 1-14. "Murderer's Row." From left to right: Ragnar Grant, Charles Lindsley, Seymour Kety, Sir John Eccles, Daniele Bovet, and Holger Hyden at the Symposium on Brain and Human Behavior, Loyola University Medical Center, Mayood, IL, 1969.

Figure 1-15. Vincenzo Longo and Alex Karczmar in Rome, 1978. Vincenzo Longo, from the Istituto Superiore di Sanità, is a preeminent EEG investigator. He related EEG events and evoked potentials to behavior and REM sleep. (See color plate.)

d. The Discoveries Concerning Molecular and Cellular Aspects of Cholinergic Transmission and the ACh Release System

The discoveries concerning molecular and cellular aspects of cholinergic transmission and the ACh release system define a complex neuronal and nerve terminal cholinergic system, and our comprehension of this system began in the 1950s with the efforts of Victor Whittaker and Eduardo de Robertis. In its classical form, the system deals with the uptake of choline (Table 1-2), formation and axonal transport of vesicles, vesicular uptake of ACh, and vesicular cycling, and vesicle-nerve terminal plasmalemma fusion; these activities relate to and are conditioned by phosphorylation mechanisms and second messengers (see Table 1-4). Recent investigators proposed novel, unorthodox alternatives or additions to the classical system that include the nonvesicular, cytoplasmic release of ACh and/or its release outside of the synaptic specializations (Table 1-4; Chapter 2 C). The ensuing argument adds to the excitement generated within this particular area of cholinergicity.

Altogether, the active proteins and other components of the ACh release system, as well as CAT, acetyl-coenzyme A, and AChE, are the markers of a cholinergic neuron. The synchrony and harmony of the activities of the system's components are consistent with its genomic regulation, which is carried out and expressed by the cholinergic gene locus. Our understanding of the function of the cholinergic system is largely due to our current knowledge of the molecular and genetic characteristics of cholinergic components (Table 1-4; see also Chapter 2 B).

Two additional cellular components contribute to the smooth, point-to-point regulation of synaptic cholinergic transmission, namely, the cholinergic receptors and the ionic conveyers of their action (Table 1-4; see also Chapter 9 BI and BIII, and Chapters 4, 5, and 6). There is a multiplicity of nicotinic and muscarinic receptor subtypes (see Chapters 4, 5, and 6) and specialized couplings exist between these receptor subtypes and the ionic mechanisms that the receptors activate during the initiation of cholinergic transmission. As pointed out above (section BI of this chapter), the unique status of the cholinergic system is well represented by complex molecular, genetic, and cellular processes that underlie the transmittive function of the cholinergic neurons and its synchronization.

There are also applied aspects of the cholinergic system. There are many uses of cholinergic drugs, particularly with respect to peripheral diseases such as myasthenia gravis and ocular disorders, and also in mental diseases such as Alzheimer's (see Chapter 10). Another applied aspect of the cholinergic story is the development of animal models that are useful for developing clinical treatment modes (see Table 1-5 and Chapter 9 BV-4). Several of these models were developed for the study of Alzheimer's disease, including the use of neurotoxins and simultaneous blocking of several transmitter systems. Interesting models that concern a number of disease states and which have a significant future potential for cholinergic studies are models involving immunological techniques and the use of antibodies.

3. Goofs and Boo-Boos, and Recoveries

The success of this brilliant cholinergic pageant resulted from the completion of a long sequence of steps. This success was achieved because many talented individuals were willing to take risks ("to take a flyer," as expressed by Steve [Bernard B.] Brodie [Karczmar, 1989]), thus committing several goofs and boo-boos along the way (Table 1-6). In some cases, the investigators recovered from their errors on their own (Table 1-7), and sometimes they were correct in the long run, even though their experiments were flawed.

David Nachmansohn proposed that axonal conduction—of both cholinergic and noncholinergic axons—is carried out by an AChE-ACh system. His notion was, in part, generated by his artistic sense and by his belief in the unity of all biochemical-energy-related events. This belief was influenced by his teachers, Otto Meyerhoff, Keith Lucas, and A.V. Hill, and Nachmansohn's writings are so well put that one feels like exclaiming: "si non e vero, e ben trovato." Nachmansohn was stubborn in defending his very wrong view, even after it was documented that noncholinergic axons do not contain ACh (Karczmar, 1967a), and he was unduly fond of using excessive doses of drugs to prove his point (see note to Table 1-6). But Nachmansohn compensated for his erroneous

Table 1-5. Fanfares: Models for CNS Diseases with Cholinergic Implications (from 1960 to Present)

US investigators	J.A. Simpson, J. Patrick, and John Lindstrom, Daniel Drachman, A.G. Engel, S. Appel	Immunological models for myasthenia gravis and lateral sclerosis[1]
US investigators	Rita Rudel, J.E. Somers	Chemical models for myotonia cogenita[2]
US investigators	Israel Hanin, Toshio Narahashi	Effects of neurotoxins including AF64A as models for Alzheimer's disease and memory disorders[3] (Chapter 9 BV-d)
US and Czech investigators	Volia (W.T.) Liberson, Charles Scudder, Alex Karczmar, J.O Bures	Nongoal and stereotypic behaviors as models for mental disorders[4] (Chapter 9 BV-d)
Hungarian and US investigators	Vahram Haroutunian, Paul Kasa, D.J. Selkoe, Ben Wolozin, Ken Davis	Neurotransmitter block and chemical (beta-amyloid) models for Alzheimer's disease and cognitive disorders[5] (Chapter 9 BV-d and Chapter 10 A and K)

Sources:
1. Simpson, 1960; Patrick and Lindstrom, 1973; Fisher, 2001.
2. Lehmann-Horn and Rudel, 1995.
3. Narahashi, 1974; Hanin, 1996; Karczmar, 1988.
4. Maier, 1949; Karczmar, 1987a, 1987b, 1988; Bures et al., 1983.
5. Haroutunian et al., 1990; Inestrosa et al., 2003.

Table 1-6. Goofs and Boo-Boos (a sample only)

David Krech, Mark Rosenzweig, and Edward Bennett	"Difference of . . . 3% in the total cortical . . . 2% in the total subcortical AChE . . . correlates . . . with . . . adaptive behavior . . . and . . . learning" (Krech et al., 1966; Rosenzweig et al., 1958).
Walter Riker, Jay Roberts, and Frank Standaert	"The data strongly implicate muscle as the principal tissues of ACh origin, and therefore . . . in . . . neuromuscular structure, formation and release of ACh may not be correlated to synaptic events" (Riker et al., 1957).
Phillipe de Clermont	"TEPP tastes so sweet" (Actually, Clermont lived to be 91; Clermont, 1854).
Andrey Zupancic	"Identity of . . . cholinesterases with cholinoreceptors" (Zupancic, 1967).
David Nachmansohn*	"You can kill me—you can't kill the theory" or "To get incredible results you must use incredible doses" (Nachmansohn, 1961).
John Eccles	"Presynaptic action current (is) responsible for . . . excitatory action at the neuromuscular junction and sympathetic ganglion" (Eccles, 1941–1946; see Eccles, 1946).
Corneille Heymans	"Les deux actions du DFP, l'inactivation des cholinesterases et les actions d'excitation parasympathetique, ne sont pas dependants l'une de l'autre" (Heymans, 1950).
Michel Jouvet	"Les deux états du sommeil . . . sont . . . modulés par des mecanismes monaminergiques" . . . "very close . . . to catecholamines" (Jouvet, 1967, 1972).
Forrest Weight	"Recurrent inhibition from major axon collaterals to motoneurons is a monosynaptic pathway" (there is no Renshaw cell; Weight, 1968).

* David Nachmansohn made these pronouncements at the 1960 Rio de Janeiro Symposium on "Bioelectrogenesis," but these comments did not appear in the 1961 version of the symposium. The symposium's editor, Carlos Chagas, Jr. (the son of Carlos Chagas who, along with Osvaldo Cruz, discovered Chagas-Cruz disease in 1890), was a prominent cholinergiker, a pioneer of studies of the Torpedo electric organ, and one of the first of three investigators (including Sydney Ehrenpreis and David Nachmansohn) to try to isolate the cholinergic receptor (see also Chapter 3).

Table 1-7. Recoveries

Sir John Eccles, Paul Fatt, and Kyozu Koketsu	"Cholinergic . . . Synapses in a Central Nervous System Pathway" (Eccles, Fatt, and Koketsu, 1954).
David Nachmansohn and A.L. Machado	"On addition of ATP to . . . brain . . . and electric organs . . . extracts, the first enzymatic acetylation of choline was achieved" (Nachmansohn and Machado, 1943).
Corneille Heymans and E. Neil	"The Chemical Control and Respiration" (Heymans and Neil, 1958).
Mark Rosenzweig, David Krech, and Edward Bennett	"Relate biochemical activities in the cortex to adaptive behavior pattern . . . plasticity" (Rosenzweig, Krech, and Bennett, 1958).
Michel Jouvet	"Potent inhibitory action of atropine . . . upon PS and PGO spikes" (Jouvet, 1972).

view by discovering, characterizing, and elucidating the role of choline acetyltransferase (CAT) (EC 2.3.1.6) and acetyl coenzyme A (Table 1-7). He was also one of the first to crystallize AChE and made the first attempt to isolate and purify the nicotinic receptor (Nachmansohn's unpublished data). Considering in toto Nachmansohn's goofs and recoveries, he was one of the most brilliant cholinergikers.

Then, there is Sir John Eccles' heresy (Karczmar, 2001a, 2001b; Table 1-6), as for many years he considered and attempted to prove that synaptic transmission is electric in nature (Eccles, 1946; see Chapter 9 A-2c). It is fitting to apply the term "heresy" to this wrong notion, as Sir Henry Dale referred to Eccles' subsequent change of mind as "conversion." This change of mind resulted in experiments on the synapse between the motoneuron collateral and the Renshaw cell carried out by Eccles, Fatt, and Koketsu. These experiments proved the presence of chemical, cholinergic transmission in the CNS and earned Eccles the Nobel Prize (Table 1-7). Eccles commented subsequently (1987) that the manuscript that described these results and which was submitted to the London Journal of Physiology by mail from Australia was the fastest-accepted manuscript in his experience!

Similarly, while Corneille Heymans, a supercilious patrician raconteur and gourmet, may be faulted for believing that all the ganglionic effects of the OP drugs including di-isopropylfluorophosphonate (DFP) and tetraethylpyrophosphate (TEPP) are independent of their AChE inhibition (Heymans, 1950; Table 1-6), he managed to earn the Nobel Prize for his demonstration of the chemical control of respiration via the chemoreceptors of the carotid body and sinus (Heymans and Neil, 1958). In fact, in his studies of these centers, Heymans stressed their exquisite sensitivity to ACh. Actually, Lindgren et al. (1952) proposed a neurotransmitter role for ACh at these sites, which is a role that ultimately could not be justified (Heymans and Neil, 1958). Nevertheless, Heymans was at least partially right in his notions regarding the OP drugs, as these drugs exhibit some actions that are not dependent on inhibition of AChE.

Michel Jouvet initially claimed that catecholamines and serotonin are the sole transmitters regulating the balance between slow and REM sleep (Table 1-6), yet he was probably the first investigator to find that atropine blocks REM sleep (Table 1-7; Jouvet, 1967, 1972). He was also an investigator who used to "play the monamine game" and who pioneered the studies of the multitransmitter nature of the pathways controlling the sleep-wakefulness system (see Chapter 9 BIV-3c).

In experimental terms, Mark Rosenzweig, David Krech, and Edward Bennett never "recovered" from their heresy of imputing functional and learning significance to 2% and 3% differences in brain AChE and to the complex ratios between brain BuChE and AChE, as it seems that they never recanted these experiments (Rosenzweig et al., 1958; Krech et al., 1966; see also Karczmar, 1969). However, this brilliant interdisciplinary team's notions regarding the flexibility and plasticity of the brain and the possibility of demonstrating this flexibility in terms of neurochemical parameters were posited long before these notions became a norm.

Andrey Zupancic always held to his claim that ChEs are identical with cholinergic receptors

(Zupancic, 1967). Of course, his arguments concerning common binding characteristics of the anionic centers of ChEs (which ligated ACh and nicotinic antagonists such as d-tubocurarine after the inhibition of esteratic sites by physostigmine) and of "alleged" receptors were interesting and were state-of-the-art at the time.

The "goof" of Phillipe de Clermont was that he drank in 1854 the potent OP antiChE tetraethylpyrophosphate (TEPP) and erred by not succumbing to this poisonous drug, as this error delayed our understanding of the toxic effects of antiChEs (Table 1-6). Again, this error was intelligible in terms of the state of the art of the time: one of the Edinburgh pioneers of the studies of the Calabar bean, Robert Christison (see Table 1-1) was similarly rash in consuming amounts of Calabar beans equivalent to some 10 mg of physostigmine without first testing the bean in animals, again, without perishing (see Chapter 7 A). However, both Christison and de Clermont redeemed themselves, as they lived to be 91 and 85 years old, respectively.

Forrest Weight (1968) denied the presence of the Renshaw cell some 15 years after the publication of Eccles' papers (Eccles et al., 1953, 1954) and a few years after definitive anatomical identification of the Renshaw cell (Eccles, 1969). What may be said on behalf of Weight is that he managed to survive, apparently in silence. In his deadly critique, Eccles (1969) referred to Weight's paper as "a most audacious attack" on the Renshaw cell and on the existence of synaptic transmission at the site.

Finally, the tale of Walter Riker, Jr., Frank Standaert, and their associates (Table 1-6) is linked, via P.T. Feng as an intermediary, with Eccles' "recovery" from his early heresy when he proposed that neuromyal and ganglionic transmission were generated by "electric action potential." The full story may be reconstituted as follows: at the old Peiping Union Medical College, P.T. Feng and his collaborators studied the neuromyal toad junction, employing physostigmine, guanidine, veratrine, and the like. In these studies, they demonstrated that retrograde discharge could be evoked by means of these drugs from the toad motor nerve endings, and during the 1940s, several investigators confirmed this finding (see Karczmar, 1967a, 1967b). Then, during the 1950s, Riker expanded the notion, adding a novel interpretation to Feng's findings, as he associated the neuromyal transmission with Feng's retrograde discharge and ascribed this transmission to currents originating in the motor nerve terminal (Riker et al., 1957; Standaert and Riker, 1967), which was essentially taking up the position embraced some 20 years earlier by Eccles.

Ultimately, Arthur H. Hayes and Walter Riker (1963) published the data indicating that no more ACh is released at the neuromyal junction during motor nerve stimulation than during rest. To add oil to fire, Standaert and Riker (1967) opined that there was no greater reason to assign the transmitter role to ACh than "to decamethonium, tetraethylammonium, succcinylcholine," as all of these agents exert similar actions at the neuromyal junction. The error worsened when Riker's brother, William Riker, joined the fray and proposed that the same mode of electrical transmission exists at the sympathetic ganglia (Riker and Szreniawski, 1959). These publications sent many pharmacologists scurrying back to the laboratory to retest the transmission mechanism that appeared to have been settled since the days of the knights (Sir Henry Dale, Sir William Feldberg, and Sir John Eccles). They repeated the work of Walter Riker and William Riker, and again obtained the classical results (see, for example, Krnjevic and Straughan, 1964; see also Karczmar, 1967a, 1967b). However, apart from this aberration, William and Walter Riker performed brilliant experimental and educational work and, ultimately, Walter Riker earned the prestigious Sollman Award.

Altogether, the errors made, whether redeemed or not by their perpetrators, were almost always heuristic. Also, they render the cholinergic lore even more interesting than it would have been if its progress had always been smooth. Finally, they make this historical tale more personal.

C. Scope of This Book

The scope of this book is closely linked to its aim to present the cholinergic matters in a historical perspective, and to describe the growth of cholinergic research into its current status—of course, this status is not the end of the cholinergic saga, since its rise continues. Accordingly, the preceding sections of this chapter summarize the history of the cholinergic research and dwell particularly on a number of historical high points (the "fanfares"). Moreover, Table 1-1 also refers to pertinent chapters in this book that further expand the accounts of the cholinergic lore, as each chapter includes its own historical section.

To fully appreciate the central cholinergic system and its functions, one needs to tackle the components of the system. Cholinoceptivity is an important component, as it translates cholinergic transmission into a trans-synaptic effect and, ultimately, into cholinergic function. Aspects of cholinoceptivity, such as receptor subtypes, ensure multiple specializations of the cholinergic function and its adaptability to circumstances, including the environment; Chapters 4, 5, and 6 deal with this preeminent matter. In addition, the central cholinergic function depends on and is demarcated by the topography of the central cholinergic pathways. New methods for identifying cholinergic neurons via the immunochemical marking of CAT made the delineation of these pathways possible, as discussed in Chapter 2. In addition, the earlier sections of this introduction make salient the importance of the functional components and markers of cholinergic neurons, and this area is also described in Chapter 2.

ACh metabolism is of course the condition sine qua non of the existence of the cholinergic system; this metabolism involves synthesis as well as catabolism of ACh, and these cholinergic components are presented in Chapter 3. The novel and important aspects of this area is the question of the limiting step of ACh synthesis—is it choline? Is it CAT? This particular question generated another issue—in view of the agonist actions, under certain circumstances, of choline, could its synthesis and nerve terminal uptake threaten the homeostasis of choline levels? Then there are enigmas with respect to the catabolic role of ChEs; is the role and tissue presence of ChEs restricted to its catabolic activities?

As mentioned in the earlier sections of the introduction, at one time the ontogeny and phylogeny of the cholinergic system were exploited to demonstrate the existence of cholinergic transmission and its importance for functions such as motility. Today, this story may hold only a historical interest, and it may be supposed that the knowledge of cholinergic ontogeny and phylogeny may not relate to our understanding of the central function of the cholinergic system. However, the ontogenetic studies revealed that cholinergic components appear in development prior to forming the rudiments of the nervous system, and the non-nervous appearance of cholinergic components has been demonstrated in phylogenetic research. The role of this precocious appearance of cholinergic components and its presence outside the nervous system may relate to trophisms. And, vice versa, trophic factors are important for cholinergic development, and their discovery constitutes one of the most interesting cholinergic stories. These matters are described in Chapter 8.

Many agents are featured in the cholinergic story, including d-tubocurarine, beta-erythroidine, and curarimimetics; nicotinic and muscarinic agonists and antagonists; choline and ACh uptake inhibitors; and neurotoxins (Narahashi, 1974). They are described, as appropriate, throughout this book. Yet, antiChEs are perhaps the most important—or the most notorious—among these agents. Anticholinesterases are very useful in characterizing cholinergic function, their role as pesticides is extremely important in our agricultural economy, and the use of OP drugs as war gases is of major significance. Altogether, these compounds and their fascinating history merit a special chapter, Chapter 7.

The main focus of Chapter 10 is Alzheimer's disease (AD). Tertiary antiChEs have been used in AD for the last 25 years (see Chapter 10 A), and this use was initiated by the realization of the degeneration of cholinergic neurons in this disease. Even with increased awareness that AD is not, etiologically, a cholinergic disease, the employment of antiChEs in this condition (and in nonspecific geriatric memory loss) is continued, particularly because of the new knowledge with respect to their effects in AD. Chapter 10 dwells on these matters and on the past and present understanding of the cholinergic management of AD.

It must be noted that cholinergic drugs are, or were, employed clinically with respect to neuromyal, autonomic, ocular, and central diseases (Karczmar, 1979, 1981, 1986a). Current cholinergic treatment of peripheral disease is outside of the present scope, although it will be briefly referred to in Chapter 10 A and K; the chapter will also describe briefly several uses of cholinergic drugs for brain diseases besides AD.

Though many aspects of central cholinergicity are introduced and rendered intelligible by Chapters 2 through 8 and Chapter 10, Chapter 9 expands the realm of the central cholinergic discussion by describing characteristic features of the cholinergic central nervous system and its pharmacology. These features range from pre- and postsynaptic physiology and pharmacology, transmittive events and phenomena of central ACh release, to "organic" functions, such as central electrophysiology that include EEG, evoked potentials, REM sleep, seizures, respiration, and endocrine effects. Then there are the behaviors; it must be emphasized that cholinergic correlates were firmly established for a multitude of behaviors, and there may be only very few behaviors which were not evaluated for their cholinergicity (see Chapter 9 BIV and BV). Finally, there are the matters of consciousness and self-awareness (see Chapter 9 BVI). These subjects touch on the millennia-old problems of the body-mind dichotomy. The novel speculations as to the possible contribution of cholinergic synaptic function to these problems constitute an appropriate consummation of the long and brilliant cholinergic saga.

D. Envoi

It was suggested at the beginning of this chapter (section A) that cholinergicity is unique among biological sciences in its importance and in the brilliance of the discoveries made in this field. A few examples were adduced to support this statement, and section B provides additional evidence for this notion; section B also alludes, however incompletely, to the shining researchers who made the cholinergic story so rich and so heuristic. But, besides reminding us of the past, this introduction illustrates the continuity of the progress of the cholinergic story, which extends from molecular and cellular cholinergicity to cognition and self-awareness.

Notes

1. The question of cytoplasmic versus vesicular release of ACh constitutes an ardent controversy (see the foreword and Chapter 2 C-1 and C-2).
2. However, a bioassay utilizing toad lung may measure 10^{-17} molar concentrations of ACh (see Chapter 2 C-2).
3. Apparently, there is something wrong with their results, but nevertheless!
4. The late Edith Heilbronn, German-born Swedish scientist (see Figure 1-4), pioneered the antiChE research, particularly of the OP type. She developed many theorems concerning the mechanisms of their action and many agents that could serve as their antidotes. Among those, she discovered the properties of a fluoride, NaF, that works as an effective reactivator of OP-inhibited phosphorylated AChE. This discovery led to the understanding of many other NaF actions, including its "sensitizing" modulatory effect (Koketsu, 1966).
5. While this book focuses on the central cholinergic system, the research concerning cholinergic periphery is pertinent for the full intelligibility of the progress of cholinergic research and is therefore included in Tables 1-1, 1-2, and 1-3. Table 1-1 shows the approximate dates of the initiation of each particular area of research, but of course there is no such thing as the end of any of the studies in question, as indicated in the table (for example, the period of from 1850 till the present is suggested as the time of studies of organophosphorus antiChEs); sometimes, however, the period of a particular study area is shown as circumscribed, to indicate the time of particular intensity or significance of the studies in question (for example, in the case of the isolation and structurization of physostigmine).
6. One of the pioneers of this research, Rene Couteaux (see Couteaux, 1947), died in 2000; he had surpassed 90 years of age.
7. A leaflet from a Koppanyi Georgetown University Medical Center Lecture states that Theodore Koppanyi would claim priority with respect to just about any discovery in the cholinergic field that has been mentioned; if one doubted it, he could pull out a reprint of his that would prove his point. Altogether, Theodore Koppanyi may be listed among the most eccentric investigators in a field rich in eccentrics. Some of his peculiarities are revealed in Karczmar, 1987a.

References

Adams, P. R.: Acetylcholine receptor kinetics. J. Membr. Biol. *58*: 161–174, 1981.

Albe-Fessard, D.: La synchronization des activités elementaires dans les organes des poissons electriques. In: "Bioelectrogenesis," C. Chagas and A. Paes de Carvalho, Eds., pp. 202–210, Elsevier, Amsterdam, 1961.

Ansell, G. B. and Spanner, S.: Studies on the origin of choline in the brain of the cat. Biochem. J. *122*: 741–750, 1971.

Ariens, A. J. and Simonis, A. M.: Cholinergic and anticholinergic drugs, do they act on common receptors? Ann. N. Y. Acad. Sci. *144*: 842–868, 1967.

Augustinsson, K.-B.: Cholinesterases. A study in comparative anatomy. Acta Physiol. Scand. *15*, Supplt. *52*: 1–182, 1948.

Bacq, Z. M.: Recherches sur la physiologie et la pharmacologie du système nerveux autonome. Arch. Int. Physiol. *42*: 24–60, 1935.

Bacq, Z. M.: L'acetylcholine et l'adrenaline chez les invertebrés. Biol. Rev. *22*: 73–91, 1947.

Bartus, R. T., Dean, R. L. III, Beer, B. and Lippa, A. S.: The cholinergic hypothesis of geriatric memory dysfunction. Sci. *217*: 408–417, 1982.

Birks, R. L., MacIntosh, F. C. and Sastry, P. B.: Pharmacological inhibition of acetylcholine synthesis. Nature *178*: 1181, 1956.

Bovet, D.: Nature, nurture and the psychological approach to learning. In: "Brain and Human Behavior," A.G. Karczmar and J.C. Eccles, Eds., pp. 324–340, Springer Verlag, Berlin, 1972.

Bovet, D., Bovet-Nittti, F. and Martini-Bettolo, G. B.: "Curare and Curare-like Agents," Elsevier, Amsterdam, 1959.

Brazier, M. L.: The historical development of neurophysiology. In: "Neurophysiology," vol. 1, J. Field, Ed., pp. 1–58, Amer. Physiol. Soc., Washington, D.C., 1959.

Bures, J., Buresova, O. and Huston, J.: "Techniques and Basic Experiments for the Study of Brain Behavior." Elsevier, Amsterdam, 1983.

Changeux, J.-P.: David Nachmansohn (1899–1983): A pioneer of neurochemistry. In: "Molecular Basis of Nerve Activity," J.-P. Changeux, F. Hucho, A. Maelicke and E. Neumann, Eds., pp. 1–32, Walter de Gruyter & Co., Berlin, 1985.

Clermont, de P.: Note sur la preparation de quelques ethers. Comptes Rendus *39*: 338–340, 1854.

Couteaux, R.: Contributions a l'étude de la synapse myoneurale. Rev. Canad. Biol. *6*: 563–711, 1947.

Couteaux, R.: Early days in the research to localize skeletal muscle acetylcholinesterases. In: "From *Torpedo* Electric Organ to Human Brain: Fundamental and Applied Aspects," J. Massoulié, Ed., pp. 59–62, 1998.

Dale, Sir Henry H.: "Adventures in Physiology," Pergamon Press, Oxford, 1953.

Dale, H. H. and Dudley, H. W.: The presence of histamine and acetylcholine in the spleen of the ox and the horse. J. Physiol. (Lond.) *68*: 97–123, 1929.

De Robertis, E.: Histophysiology of the synapse and neurosecretion. In: "Bioelectrogenesis," C. Chagas and A. Paes de Carvalho, Eds., pp. 288–294, Elsevier, Amsterdam, 1961.

Dixon, W. E.: On the mode of action of drugs. Med. Mag. (Lond.) *16*: 454–457, 1907.

Du Bois-Reymond, E.: Vorlaeufiger Abriss einer Untersuchung über den elektromotorischen Fische. Ann. Physik. Chem. *58*: 1–27, 1843.

Eccles, J. C.: An electrical hypothesis of synaptic and neuromuscular transmission. Ann. N. Y. Acad. Sci. *47*: 429–455, 1946.

Eccles, J. C.: "The Physiology of Synapses," Springer-Verlag, New York, 1964.

Esscles, J. C.: Historical introduction. Fed. Proc. *28*: 80–94, 1969.

Eccles, J. C.: The story of the Renshaw cell. In: "Neurobiology of Acetylcholine–A Symposium in Honor of Alexander G. Karczmar," N. J. Dun and R. L. Perlman, Eds., pp. 189–194, Plenum Press, New York, 1987.

Eccles, J. C., Fatt, P. and Koketsu, K.: Cholinergic and inhibitory synapses in a central nervous pathway. The Australian J. Sci. *16*: 50–54, 1953.

Eccles, J. C., Fatt, P., Katz, B. and Koketsu, K.: Cholinergic and inhibitory synapses in a pathway from motor-axon collaterals to motoneurones. J. Physiol. (Lond.) *126*: 524–562, 1954.

Ehrenpreis, S.: The isolation and identification of the acetylcholine receptor protein from electric tissue of *Electrophorus electricus*. In: "Bioelectrogenesis," C. Chagas and A. Paes de Carvalho, Eds., pp. 379–393, Elsevier, Amsterdam, 1961.

Fisher, M. A.: Use of cholinesterase inhibitors in the therapy of myasthenia gravis. In: "Cholinesterases and Cholinesterase Inhibitors," E. Giacobini, Ed., pp. 249–261, Martin Dunitz, London, 2001.

Glick, D. and Soreq, H.: Neurobiology. I Drugs: 983–987, 1999.

Greengard, P.: "Cyclic Nucleotides, Phosphorylated Proteins and Neuronal Function." Raven, New York, 1978.

Greengard, P. and Browing, M. D.: Studies of the physiological role of specific neuronal phosphoproteins. Adv. Second Messenger Phosphoprotein Res. *21*: 133–146, 1988.

Hamburger, V. and Keefe, E. L.: The effects of peripheral factors on the proliferation and differentiation in the spinal cord of chick embryos. J. Exper. Zool. *96*: 223–243, 1944.

Hanin, I.: The AF64A model of cholinergic hypofunction: an update. Life Sciences *58*: 1955–1964, 1996.

Haroutunian, V., Santucci, A. C. and Davis,K. L.: Implications of multiple transmitter system lesions for cholinomimetic therapy in Alzheimer's disease. In: "Cholinergic Neurotranmission: Functional and Clinical Studies," S.-M. Aquilonius and P.-G. Gillsberg, Eds., pp. 333–345, Elsevier, Amsterdam, 1990.

Hayes, A. H. and Riker, W. F., Jr.: Acetylcholine release at the neuromuscular junction. J. Pharmacol. Exper. Therap. *142*: 200–205, 1963.

Heilbronn, E., Ed.: "Structure and Reactions of DFP Sensitive Enzymes," Forsvarets Forkskningsanstlt, Stockholm, 1967.

Heilbronn, E. and Winter, A., Eds.: "Drugs and Cholinergic Mechanisms in the CNS," Forsvarets Forskningsanstalt, Stockholm, 1970.

Heymans, C.: Les antidotes du di-isopropylfluorophosphonate (DFP). Arch. Intern. Pharmacodyn. *81*: 230–2237, 1950.

Heymans, C. and Neil, E.: "Reflexogenic Area of the Cardiovascular System," Lee-Brown, Boston, 1958.

Hofmann, F. B.: Über die Einheitlichkeit der Herzhemmungfasern und über die Abhängigkeit ihrer Wirkung vom Zustande des Herzens. Zentralbl. F. Physiol. *4*: 427–452, 1883.

Hokfelt, T., Johansson, O., Ljungdahl, A., Lundberg, J., Schultzberg, M., Terenius, L., Goldstein, M., Elde, R., Steinbusch, H. and Verhofstad, A.: Histochemistry of transmitter interactions–neuronal coupling and coexistence of transmitters. In: "Neurotransmitters," P. Simon, Ed., pp. 31–143, Adv. Pharmacol. Therap., Vol. 2, Pergamon Press, Oxford, 1979.

Hokfelt, T., Millhorn, D., Serrogy, K., Tsuruo, Y., Ceccatelli, S., Lindh, B., Meister, B., Melander, T., Schalling, M. and Bartfai, T.: Coexistence of peptides with classical transmitters. Experientia *43*: 768–780, 1987.

Holmstedt, B.: Pharmacology of organophosphorus cholinesterase inhibitors. Pharmacol. Rev. *11*: 567–688, 1959.

Holmstedt, B.: Structure-activity relationships of the organophosphorus anticholinesterase agents. In: "Cholinesterases and Anticholinesterase Agents," G. B. Koelle, Ed., Handbch. d. exper. Pharmakol., Erganzungswk., vol. 15, pp. 428–485, Springer-Verlag, Berlin, 1963.

Holmstedt, B.: The ordeal bean of old calabar: the pageant of Physostigma venenosum in medicine. In: "Plants in the Development of Modern Medicine," T. Swain, Ed., pp. 303–360, Cambridge University Press, Cambridge, 1972.

Holmstedt, B.: Cholinesterase inhibitors: an introduction. In: "Cholinesterases and Cholinesterase Inhibitors," E. Giacobini, Ed., pp. 1–8, Martin Dunitz, London, 2000.

Holmstedt, B. and Liljestrand, G.: "Readings in Pharmacology," Pergamon Press, Oxford, 1963.

Hunt, R. and Taveau, M.: On the physiological action of certain colin derivatives and new methods for detecting cholin. Brit. Med. J. *11*: 1788–1789, 1906.

Inestrosa, N. C., Ferrari, de G., Opzo, C. and Alvarez, A.: Neurodegenerative processes in Alzheimer's disease: role of A-beta-AChE complexes and Wnt signaling. In: "Cholinergic Mechanisms: Function and Dysfunction," I. Silman, H. Soreq, L. Anglister, D. Michaelson and A. Fisher, Eds., Martin Dunitz, London, 2004.

Israel, M. and Dunant, Y.: Acetylcholine release. Reconstitution of the elementary quantal mechanism. In: "From *Torpedo* Electric Organ to Human Brain: Fundamental and Applied Aspects." J. Massoulié, Ed., pp. 123–128, Elsevier, Amsterdam, 1998.

Jenden, D. J., Jope, R. S. and Weiler, M. H.: Regulation of acetylcholine synthesis: does cytoplasmic acetylcholine control high affinity choline uptake? Science *194*: 635–637, 1976.

Jouvet, M.: Regulation neuro-humorale des états de sommeil. In: "Aspects Biologiques et Cliniques du Système Nerveux Centrale," Symposium de Sandoz, pp. 107–125, Sandoz ltd., Bale, 1967.

Jouvet, M.: Some monoaminergic mechanisms controlling sleep and waking. In: "Brain and Human Behavior," A. G. Karczmar and J. C. Eccles, Eds., pp. 131–161, Springer-Verlag, Berlin, 1972.

Karczmar, A. G.: The role of amputation and nerve resection in the regressing limbs of urodele larvae. J. Exper. Zool. *103*: 401–427, 1946.

Karczmar, A. G.: Ontogenesis of cholinesterases. In: "Cholinesterases and Anticholinesterase Agents," G. B. Koelle, Ed., pp.129–186, Handbch. d. exper. Pharmakol., Erganzungswk., vol. 15, Springer-Verlag, Berlin, 1963.

Karczmar, A. G.: Pharmacologic, toxicologic and therapeutic properties of anti-cholinesterase agents. In: "Physiological Pharmacology," W. S. Root and F. C. Hofman, Eds., vol. 3, pp. 163–322, Academic Press, New York, 1967a.

Karczmar, A. G.: Neuromuscular pharmacology. Annu. Rev. Pharmacol. *7*: 241–276, 1967b.

Karczmar, A. G.: Is the central nervous system overexploited? Fed. Proc. *28*: 147–156, 1969.

Karczmar, A. G.: History of the research with anticholinesterase agents. In: "Anticholinesterase Agents," A. G. Karczmar, Ed., vol. 1, section 13, Intern. Encyclop. Pharmacol. Therap., pp. 1–44, Pergamon Press, Oxford, 1970a.

Karczmar, A. G.: Central cholinergic pathways and their behavioral implications. In: "Principles of Psychopharmacology," W. G. Clark and J. de Giudice, Eds., pp. 57–86, Academic Press, New York, 1970b.

Karczmar, A. G., Introduction: What we know, will know in the future, and possibly cannot ever know in neurosciences. In: "Brain and Human Behavior," A. G. Karczmar and J. C. Eccles, Eds., pp. 1–20, Springer-Verlag, Berlin, 1971.

Karczmar, A.G.: Central actions of acetylcholine, cholinomimetics and related drugs. In: "Biology of Cholinergic Function," A. M. Goldberg and I. Hanin, Eds., pp. 395–449, Raven Press, New York, 1976.

Karczmar, A. G.: New roles for cholinergics in CNS disease. Drug Therapy *4*: 31–42, 1979.

Karczmar, A. G.: Basic phenomena underlying novel use of cholinergic agents, anticholinesterases and precursors in neurological including peripheral and psychiatric disease. In: "Cholinergic Synapses," G. Pepeu and H. Ladinsky, Eds., Adv. in Behav. Biol., vol. 25, pp. 853–569, Plenum Press, New York, 1981.

Karczmar, A. G.: Historical development of concepts of ganglionic and enteric transmission. In: "Autonomic and Enteric Ganglia," A. G. Karczmar, K. Koketsu and S. Nishi, Eds., pp. 3–26, Plenum Press, New York, 1985.

Karczmar, A. G.: Autonomic disease and clinical applications of ganglionic agents. In: "Autonomic and Enteric Ganglia—Transmission and Its Pharmacology," A. G. Karczmar, K. Koketsu and S. Nishi, Eds., pp. 459–475, Plenum Press, New York, 1986a.

Karczmar, A. G.: Historical development of concepts of ganglionic transmission. In: "Autonomic and Enteric Ganglia—Transmission and Its Pharmacology," A. G. Karczmar, K. Koketsu and S. Nishi, Eds., pp. 297–337, Plenum Press, New York, 1986b.

Karczmar, A. G.: Concluding remarks: past, present and future of cholinergic research. In: "Neurobiology of Acetylcholine—A Symposium in Honor of Alexander G. Karczmar," N. J. Dun and R. L. Perlman, Eds., pp. 561–585, Plenum Press, New York, 1987a.

Karczmar, A. G.: Historical aspects of cholinergic transmission. In: "The Basal Forebrain," T. C. Napier, P. W. Kalivas and I. Hanin, Eds., pp. 453–477, 1987b.

Karczmar, A. G.: Schizophrenia and cholinergic system. In: "Receptors and Ligands in Psychiatry," A. K. Sen and T. Lee, Eds., pp. 29–63, Cambridge University Press, Cambridge, 1988.

Karczmar, A. G.: An analysis twenty years later. In: "Neurochemical Pharmacology—A Tribute to B.B. Brodie," E. Costa, Ed., pp. 343–354, Raven Press, New York, 1989.

Karczmar, A. G.: Loewi's discovery and the XXI Century. In: "Cholinergic Mechanisms: From Molecular Biology to Clinical Significance," J. Klein and K. Loeffelholz, Eds., pp. 1–27, Elsevier, Amsterdam, 1996.

Karczmar, A. G.: The Tenth International Symposium on Cholinergic Mechanisms, Arcachon, France, September 1–5, 1998. CNS Drug Revs. *5*: 301–316, 1999.

Karczmar, A. G.: Sir John Eccles, 1903–1997. Part 1. Onto the demonstration of the chemical nature of transmission in the CNS. Persp. Biol. Med. *44*: 76–86, 2001a.

Karczmar, A. G.: Sir John Eccles, 1903–1907. Part 2. The brain as machine or as site of free will. Persp. Biol. Med. *44*: 250–262, 2001b.

Karczmar, A. G.: Conclusions: XI ISCM. In: "Cholinergic mechanism's—Function and Dysfunction," I. Stilman, Ed., pp. 435–451, Martin Dunitz, London, 2004.

Karczmar, A. G., Nishi, S. and Blaber, L. C.: Synaptic modulations. In: "Brain and Human Behavior," A. G. Karczmar and J. C. Eccles, Eds., pp. 63–92, Springer Verlag, Berlin, 1972.

Koelle, G. B.: Cytological distribution and physiological functions of cholinesterases. In: "Cholinesterases and Anticholinesterase Agents," G. B. Koelle, Ed., pp. 187–298, Handbch. d. exper. Pharmakol., Erganzungswk., vol. 15, Springer-Verlag, Berlin, 1963.

Koketsu, K.: Restorative action of fluoride on synaptic transmission blocked by organophosphorous anticholinesterases. Int. J. Neuropharmacol. *3*: 247–254, 1966.

Koppanyi, T. and Sim, K. H.: Comparative studies on pupillary reaction in tetrapoda. II. The effects of pilocarpine and other drugs on the pupil of the rat. Amer. J. Physiol. *78*: 358–363, 1926.

Krech, D., Rosenzweig, M. R. and Bennett, E. L.: Environmental impoverishment, social isolation and changes in brain biochemistry and anatomy. Physiol. Behav. *1*: 99–104, 1966.

Krnjevic, K.: Excitable membranes and anesthetics. In: "Cellular Biology and Toxicity of Anesthetics," R. B. Fink, Ed., pp. 3–9, Williams and Wilkins, Baltimore, 1972.

Krnjevic, K. and Straughan, D. W.: The release of acetylcholine from the denervated rat diaphragm. J. Physiol. (Lond.) *170*: 371–378, 1964.

Kuhar, M. J.: Characteristics and significance of sodium-dependent, high affinity choline uptake. In: "Cholinergic Mechanisms and Psychopharmacology," D. J. Jenden, Ed., pp. 447–456, Plenum Press, New York, 1978.

Le Heux, J. W.: Cholin als Hormon der Darmbewegung. Pflugers Arch. Ges. Physiol. *173*: 8–27, 1919.

Lehmann-Horn, F. and Rudel, R.: Hereditary nondystrophic myotonias and periodic paralyses. Curr. Opin. Neurol. *8*: 402–410, 1995.

Levey, M.: Medieval arabic toxicology. Trans. Amer. Phil. Soc., New Series *56*: 1–130, 1966.

Levi-Montalcini, R.: NGF: an unchartered route. In: "The Neurosciences Paths of Discovery," F. G. Worden, P. Swazey and G. Adelman, Eds., pp. 245–263, MIT Press, Cambridge, Massachusetts, 1975.

Levi-Montalcini, R.: The nerve growth factor 35 years later. Science *237*: 1151–1162, 1987.

Lindgren, S., Liljestrand, G. and Zotterman, Y.: The effect of certain autonomic drugs on the action potentials of the sinus nerve. Acta Physiol. Scand. *26*: 264–290, 1952.

Loeffelholz, K.: Release of acetylcholine in the isolated heart. Am. J. Physiol. *9*: H431–H440, 1981.

Loeffelholz, K.: Hydrolysis, synthesis and release of acetylcholine in the isolated heart. Fed. Proc. *43*: 2603–2606, 1984.

Loewi, O.: Über humorale Übertragbarkeit der Herzenwirkung. I. Mitteilung. Pflugers Arch. Ges. Physiol. *189*: 239–242, 1921.

Longo, V. G.: "Neuropharmacology and Behavior," W. H. Freeman, San Francisco, 1972.

Machne, X. and Unna, K. R. W.: Actions at the central nervous system. In: "Cholinesterases and Anticholinesterase Agents," G. B. Koelle, Ed., pp. 679–700, Handbh. d. exper. Pharmakol., Ergänzungswk., vol. 15, Springer-Verlag, Berlin, 1963.

Maier, N. R. F.: "The Study of Behavior Without a Goal," McGraw-Hill, New York, 1949.

McGeer, P. L. and McGeer, E. G.: Neurotransmitters and their receptors in the basal ganglia. Adv. Neurol. *60*: 93–101, 1993.

McGeer, P. L., Eccles, J. C. and McGeer, E. G.: "Molecular Neurobiology of the Mammalian Brain," Plenum Press, New York, 1986.

Nachmansohn, D.: Chemical and molecular aspects of bioelectrogenesis. In: "Bioelectrogeneis," C. Chagas and A. Paes de Carvalho, Eds., pp. 237–261, Elsevier, Amsterdam, 1961.

Nachmansohn, D. and Machado, A. L.: The formation of acetylcholine. A new enzyme, "choline acetylase." J. Neurophysiol. *6*: 397–403, 1943.

Narahashi, T.: Chemicals as tools in the study of excitable membranes. Physiol. Rev. *54*: 813–889, 1974.

Patrick, J. and Lindstrom, J. M.: Autoimmune response to acetylcholine receptor. Science *180*: 871–872, 1973.

Pepeu, G.: The release of acetylcholine from the brain: an approach to the study of the central cholinergic mechanisms. In: "Progress in Neurobiology," G. A. Kerkut and J. W. Phillis, Eds., vol. 2, part 3, pp. 257–288, Pergamon Press, Oxford, 1974.

Pepeu, G.: Overview and future of CNS cholinergic mechanisms. In: "Cholinergic Neurotransmission: Function and Dysfunction." A. C. Cuello, Ed., pp. 455–458, Elsevier, Amsterdam, 1993.

Pepeu, G., Casamenti, F., Giovannini, M. G., Vannuchi, M. G. and Pedata, F.: Principal aspects of the regulation of acetylcholine release in the brain. In: "Cholinergic Neurotransmission: Functional and Clinical Aspects," S.-M. Aquilonius and P. G. Gillberg, Eds., pp. 273–278, Elsevier, Amsterdam, 1990.

Popper, K. R. and Eccles, J. C.: "The Self and Its Brain." Springer International, Berlin, 1977.

Riker, W. K. and Szreniawski, Z. The pharmacological reactivity of presynaptic nerve terminals in a sympathetic ganglia. J. Pharmacol. Exper. Therap. *126*: 200–238, 233, 1959.

Riker, W. R., Jr., Roberts, J., Standaert, F. G. and Fujimori, H.: The motor nerve terminal as the primary focus for drug-induced facilitation of neuromuscular transmission. J. Pharmacol. Exper. Therap. *121*: 286–312, 1957.

Robinson, J. D.: "Mechanisms of Synaptic Transmission—Bridging the Gaps (1890–1990)," Oxford University Press, Oxford, 2001.

Rosenzweig, M. R., Krech, D. and Bennet, E. L.: Brain chemistry and adaptive behavior. In: "Biological and Biochemical Bases of Behavior," H. F. Harlow and C. N. Woolsey, Eds., pp. 367–400, U. of Wisconsin Press, Madison, Wisconsin, 1958.

Salvaterra, P., Kitamoto, T. and Ikeda, K.: Molecular genetic specification of cholinergic neurons. In: "Cholinergic Neurotransmission: Function and Dysfunction," A. C. Cuello, Ed., pp. 167–174, Elsevier, Amsterdam, 1993.

Schrader, G.: Die Entwicklung neuer Insektizide Phosphorsäure-Ester. Verlag Chemie, Weinheim, 1963.

Schueler, F. W.: Two cyclic analogs of acetylcholine. J. Amer. Pharm. Assoc., Sci. Ed. *45*: 197–199, 1956.

Simpson, J. A.: Myasthenia: a new hypothesis. Scot. Med. J. *5*: 419–436, 1960.

Standaert, F. G. and Riker, W. F., Jr.: The consequences of cholinergic drug action on motor nerve terminals. Ann. N. Y. Acad. Sci. *144*: 517–533, 1967.

Storck, A.: "Libellus, quo Demonstratur Hyosciamum, Aconitum," Typographia Gessari, Napoli, 1757.

Sutherland, E. W., Robinson, G. A. and Butcher, R. W.: Some aspects of the biological role of adenosine 3′,5′-monophosphate (cyclic AMP). Circulation *37*: 279–306, 1968.

Szerb, J. C.: Characterization of presynaptic muscarinic receptors in central cholinergic neurons. In: "Cholinergic Mechanisms and Psychopharmacology," D. J. Jenden, Ed., pp. 49–60, Plenum Press, New York, 1977.

Thesleff, S.: The mode of neuromuscular block caused by acetylcholine, nicotine, decamethonium and succinylcholine. Acta Physiol. Scand. *34*: 218–231, 1955.

Vigneaud, V. du: Interrelationships between choline and other methylated compounds. In: "Comparative Biochemistry. Intermediate Metabolism of Fats. Carbohydrate Metabolism. Biochemistry of Choline," H. B. Lewis, Ed., pp. 234–247, Jacques Cattell Press, Lancaster, Pennsylvania, 1941.

Volle, R. L.: Muscarinic and nicotinic stimulant actions at autonomic ganglia. In: "Ganglionic Blocking and Stimulating Agents," A. G. Karczmar, Ed., pp. 1–106, Internl. Encyclop. Pharnmacol. Therap., section 12, vol. 1, Pergamon Press, Oxford, 1966.

Waser, P. G.: "Cholinerge Pharmakon-Rezeptoren," Orell Fussli Graphishe Betriebe AG, Zurich, 1983.

Waser, P. G.: Schmid, H. and Schmid, K.: Resorption, Verteilung und Ausscheidung von Radio-Calebasssen-Curarin bei Katzen. Arch. Int. Pharmacodyn. *96*: 386–405, 1954.

Wassen, S. H.: The use of some specific kinds of South American Indian snuff and related paraphernalia. Etnologiska Sudier. *28*: 7–116, 1965.

Weight, F. F.: Cholinergic mechanisms in recurrent inhibition of motoneurons. In: "Psychopharmacology," D. H. Efron, Ed., pp. 69–75, Government Printing Office, Public Health Service Publication No. 1836, Washington, D.C., 1968.

Whittaker, V. P.: Identification of acetylcholine and related esters of biological origin. In: "Cholinesterases and Anticholinesterase Agents," G. B. Koelle, Ed., pp. 1–39, Handbch. d. exper. Pharmakol., Ergänzungswk., vol. 15, Springer-Verlag, Berlin, 1963.

Whittaker, V. P.: "The Cholinergic Neuron and Its Target: The Electromotor Innervation of the Electric Ray 'Torpedo' as a Model," Birkhauser, Boston, 1992.

Whittaker, V. P.: Arcachon and cholinergic transmission. In: "From *Torpedo* Electric Organ to Human Brain—Fundamental and Applied Aspects," J. Massoulié, Ed., pp. 53–57, Elsevier, Paris, 1998.

Woolf, N. J.: A possible role for cholinergic neurons in the basal brain and pontomesencephalon in consciousness. Consciousness and Cognition, *6*: 574–596, 1997.

Zaimis, E.: The neuromuscular junction: area of uncertainty. In: "Neuromuscular Junction," E. Zaimis, Ed., pp. 1–21, Springer-Verlag, Berlin, 1976.

Zupancic, A. O.: Evidence for the identity of anionic centers of cholinesterases with cholinoreceptors. Ann. N. Y. Acad. Sci. *144*: 689–693, 1967.

2

Cholinergic Cells and Pathways

A. History of Methodological Development Needed to Define the Cholinergic Neuron, Explain Acetylcholine Release, and Establish Central Cholinergic Pathways

1. What Led to Establishing Cholinergic Pathways?

a. Sir Henry Dale and Sir William Feldberg and the Existence of Cholinergic Pathways

The central presence of cholinergic transmission was first hypothesized by Sir Henry Dale (1937). The Stedmans, Paul Mann, John Quastel, and Maurice Tennenbaum, Dale's associate William (now Sir William) Feldberg, and his coworkers, including Martha Vogt and Catherine Hebb, and Josiah Burn and Edith Bulbring provided additional evidence for Dale's hypothesis; this evidence included the demonstration of the central nervous system (CNS) presence of acetylcholine (ACh) and choline acetyl transferase (CAT), ACh synthesis and ACh release in the CNS, and the central and peripheral effects of muscarinics and anticholinesterases (antiChEs) (Stedman and Stedman, 1937; Feldberg and Vogt, 1948; Feldberg, 1945, 1950; see Eccles, 1964; Karczmar, 1967; and Barker et al., 1972; Mann et al., 1938a, 1938b; see also Chapter 8 A). Also, Henry Dale surmised early the presence of a ChE from his demonstration of the evanescence of the action of ACh (Dale, 1914, 1937; see also Chapters 7A, 8A, and 9A).

Feldberg was struck with the uneven distribution of ACh, sites of ACh release and synthesis, and activities of ChEs in the CNS; these findings led him to postulate that "the central nervous system is built of cholinergic and noncholinergic neurones," distributed in an alternative fashion (Feldberg, 1945). This was the first step toward the notion of a transmitter, including ACh CNS pathways; in fact, Feldberg (1945) was perhaps the first investigator to employ the term "central pathway" to denote "transmission . . . through the mediation of acetylcholine across a number of . . . central . . . synapses." The evidence in question was obtained via the use of several extraction methods and bioassays for extracted ACh, although occasionally chemical identification was attempted (Stedman and Stedman, 1937). Also, collecting released ACh whether from the cerebrospinal fluid or via perfusion of appropriate spinal or brain sites was helpful with formulating Feldberg's notion (Feldberg, 1945; Bulbring and Burn, 1941).

A digression is warranted. Although Zenon Bacq had already employed chemical identification of endogenous ACh in 1935, his method was complex and impracticable. Much earlier ACh bioassays were employed (see, for example, Fuhner, 1918); they were used by Loewi (1921), to identify the "Vagustoff" released by the vagus nerve, and their use continued for decades. The bioassays included Venus heart, frog rectus abdominis, and several other tissues, and generally they were sensitive to ACh concentrations of 10^{-8} to 10^{-10} molar. But one particular bioassay was sensitive to ACh concentrations of 10^{-21} molar (Nishi et al., 1967). It involved the toad lung, but only the Japanese team of Kyozo Koketsu, Syogoro Nishi, and Hiroshi Soeda is capable of employing it successfully (see below, section C). Then an immunocytochemical method was developed for detection of ACh (Geffard et al., 1985); of course, this technique would be most useful in definitive identification of cholinergic neurons and pathways, but there was no follow-up with regard to its employment. Subsequently a number of chemical methods were worked out, including radioisotopic, gas chromatographic–mass spectrometric, fluorometric, and polarographic. The gas chromatography–mass spectrometry (GCMS) method, discovered by Israel Hanin, Don Jenden, and Bo Holmstedt (see Hanin and Goldberg, 1976) is commonly used today; it is sensitive at a nanogram level.[1]

Finally, when Maurice Israel and his associates wished to prove an unorthodox concept of ACh release, they needed an ultrasensitive ACh measurement method to prove their point and developed chemiluminescense to meet this need (Israel and Lesbats, 1981; Israel et al., 1990; see next section). Today, this method is widely used in industry as it allows researchers to deal with a large number of samples.

b. Cholinergic Ascending Reticular Alerting System

The evidence concerning several markers of cholinergic neurons and their CNS locus did not yield a specific description of cholinergic pathways; it suggested only that there may be many such pathways (Eccles, 1964) and that Feldberg's notion of alternative cholinergic-noncholinergic areas or sites may be not quite tenable.

The lucky thought of studying pharmacological effects of cholinergic drugs on the EEG and relating these effects to cholinoceptive sites brought about the first descriptions of specific cholinergic pathways. The alerting EEG effects (fast, low-voltage activity and the appearance of the theta rhythms; see Chapter 9 BIV-3) of ACh and cholinergic muscarinic agonists were noticed early by Frederic Bremer and Jean Chatonnet (1949). Actually, Bremer and Chatonnet ascribed these effects to the direct action of the cholinergic agents on a central cholinergic system, while Darrow and his associates (1944) stated that these phenomena are due the vasodilator actions of the muscarinics on the brain vascularization. But Joel Elkes, Phillip Bradley, and their associates (Bradley and Elkes, 1953), Franco Rinaldi and Harold Himwich (1955a, 1955b), and Vincenzo (Enzo) Longo and Bernardo Silvestrini (1957) obtained similar effects with either ACh or diisopropylfluorophosphonate (DFP) in the rabbit and in the cat, and they blocked these effects by means of atropine (see also Jasper, 1966; Karczmar, 1967). Also, they eliminated the possibilities of the peripheral or vasodilator origin of these effects, or of diffuse actions of the drugs in question on the cortex, as they obtained these effects of ACh or DFP via their carotid injection in the *cerveau isole* preparation but not in the isolated hemisphere preparation. Altogether, these investigators proposed that cholinergic alerting effects are dependent on a cholinergic alerting mesodiencephalic system or ascending reticular activating system (ARAS; Figure 2-1). A similar proposal was made by Kris Krnjevic and J.W. Phillis (1963): they proposed the existence of a cholinergic thalamocortical pathway concerned with projection and augmenting activity as they pointed out that ACh-sensitive cortical cells respond to thalamic or peripheral sensory stimulation with repetitive after-discharges and changes in the EEG. These notions were supported by the finding of Frank (Hank) MacIntosh and Paul Oborin (1953) of the increased release of ACh from the cortex during EEG arousal evoked by brainstem stimulation. In addition, McLennan (1963) proposed, on the basis of ACh release data and on the dependence of this release on functional states of the brain that there is a cholinergic pathway to the basal nuclei that originates in the nucleus ventralis lateralis.

It must be stressed that the very concept of the ARAS is based on the important, early discovery of Giuseppe Moruzzi and Horace Magoun that the stimulation of the reticular formation (within the midbrain tegmentum) induces the general activation, via the thalamus, of the whole forebrain including all cortices; they emphasized that this stimulation causes cortical arousal accompanied by behavioral arousal or awakening (Moruzzi, 1934; Moruzzi and Magoun, 1949). As valid as this discovery is, Moruzzi's and Magoun's identification of behavioral and EEG arousal is not quite correct (see Chapter 9 BIV-3).

Finally, the anatomical description of ascending cholinergic pathways based on cytochemical and immunochemical methods was first provided

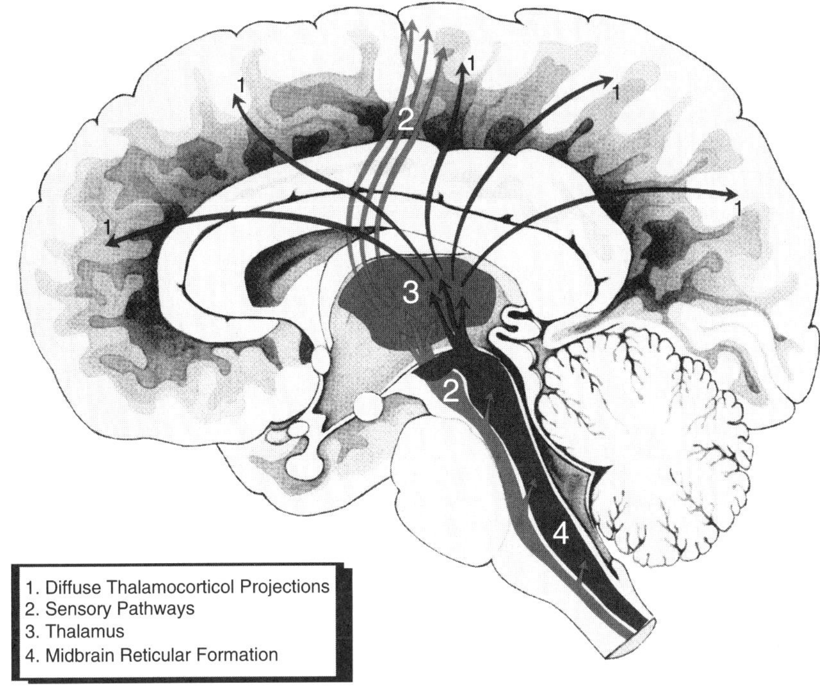

Figure 2-1. Reticular activating system according to Harold Himwich and Franco Rinaldi. (From Himwich, 1963).

by Charles Shute and Peter Lewis and Michel Gerebtzoff, and then by George Koelle; recent studies of the McGeers, Larry Butcher, and Marsel Mesulam are consistent with these findings (see section DI in this chapter). Since the investigations of Bradley, Krnjevic, Phillips, Himwich, Elkes, Bremer, Chatonnet, and Rinaldi, as well as those of Shute and Lewis, the notions of cholinergic alerting actions and their mesodiencephalic origin are generally accepted; it should be pointed out that this concept of a linearly extended cholinergic system is not in accordance with Feldberg's original postulate of alternative cholinergic and noncholinergic pathways.

c. Finally, a Definitive Description of Central Cholinergic Pathways

Histochemical and immunocytochemical means to identify AChE and CAT were the most effective and successful methods to delineate cholinergic pathways. During the 1940s and 1950s, several histochemical methods were developed by Giorgio Gomori, David Glick, and ultimately George Koelle to localize BuChE and AChE (see Koelle, 1963; Karczmar, 1963a, 1963b). All these methods are based on use of tissue slices and application of a substrate (such as a fatty acid ester, for example), hopefully specific for either AChE or BuChE; the substrate, when hydrolyzed by the enzyme, yields a colored or black precipitate, or still another reagent is added to produce the precipitate with the hydrolysate; specific inhibitors of AChE and BuChE are also applied to help, jointly with the use of enzyme specific substrates in identifying the enzyme that is being localized. Koelle and Friedenwald's famous microscopic histochemical method (1949) utilizes acetyl thiocholine or butyryl thiocholine as substrates as well as appropriate inhibitors. This method employs fresh frozen tissues or slices rather than fixed materials (making the method histochemical rather than histological) and produces remarkable resolution of the morphological location of the enzymes. Using his method in rats, rabbits, and cats, Koelle (1954) listed a number of central sites and nuclei that exhibited "intense," "moderate," or "light" staining, including several components of the limbic system, several midbrain and medullary sites, several hypothalamic sites, basal ganglia, and reticular formation, but, surprisingly, he did not refer to ventral horn, although he found some staining in the dorsal horn.

While George Koelle stressed that his findings identified several brain areas that exhibit intense presence of cholinergic synapses, he did not describe, on the basis of these findings, the existence of specific cholinergic pathways. On the other hand, Koelle (1961, 1963) established the important concept of functional (membrane) versus storage or reserve AChE (he employed quaternary antiChEs to differentiate between the two); he also stressed that AChE is present both at the membrane of the soma (the postsynaptic enzyme) and at the nerve terminals (the presynaptic enzyme); also, he modified his original microscopic method so that it could serve for electronmicroscopic investigations.

Koelle's findings were confirmed and expanded by Gerebtzoff (1959) and Shute and Lewis (1967a, 1967b); Gerebtzoff's data may be less dependable than Koelle's, as Gerebtzoff applied Koelle's staining to formalin-fixed rather than fresh-frozen tissues. Shute and Lewis (1967a, 1967b) and Gerebtzoff (1959) also employed lesion techniques to identify the brain site origin of AChE. Similar to Koelle, Gerebtzoff did not use his data to describe specific cholinergic pathways. His sites of intense staining of AChE corresponded to those described by Koelle (i.e., thalamus and hypothalamus, basal ganglia, medullary and pontine sites, including pontine tegmentum, which was later identified as an important source of cholinergic radiation, etc.). Gerebtzoff also stressed a convergence of "cholinergic and non-cholinergic fibres on the Purkinje cell," and, in contradistinction to Koelle, he described the presence of heavy AChE staining in the spinal ventral horn and its motoneurons, as well as in cranial motoneurons.

Charles Shute and Peter Lewis (1963, 1967a, 1967b) also reemployed Koelle's methods to advance significantly the understanding of the cholinergic pathways. They realized that "AChE-containing tracts . . . cannot be unequivocally traced back to their nuclei of origin," and they adopted a novel paradigm to resolve this difficulty: they discovered that "after involvement of AChE-containing tracts in surgical lesion, enzyme accumulated on the cell body side of the cut and disappeared from the opposite side, and that this phenomenon would provide a useful method of determining the polarity of cholinergic pathways"; also, this method allowed tracing a given pathway from the neurons of origin to their terminations. Finally, Shute and Lewis employed special micromethods to be sure that the lesions are applied to appropriate sites.

Their studies led them to define two pathways. The first is "the ascending cholinergic reticular system . . . arising . . . from reticular and tegmental nuclei of the brainstem, and from comparable groups of cells in the fore-brain" and extending to thalamus, subcortical, and cortical areas, hypothalamus, and limbic nuclei; they identified this system with the ARAS (which therefore corresponds to the Rinaldi-Himwich and Krnjevic pathways) and with the alerting EEG phenomena. The second pathway, the cholinergic limbic system, originates from the medial septum and diagonal band and projects to the hippocampal formation and the dentate gyrus, thence to the medial cortex, nuclei of the ascending system (ARAS), and the cerebellum. Again, Lewis and Shute proposed the involvement of this limbic system in such EEG phenomena as the hippocampal theta waves (see Chapter 9 BIV-3). It must be stressed how modern—that is, comparable to the work, 20 years later, of the McGeers, Mesulam, and others—these studies appear (Figures 2-2 and 2-3).

Subsequently, Peter Lewis (with Henderson, 1980) was the first to employ a dual cytochemical technique combining AChE histochemistry with horseradish peroxidase (HRP) procedures. When HRP is injected into the brain it is taken up and transported retrogradely to the neurons, which supply the area of HRP injection (Kristenson et al., 1971); this dual method confidently identifies the sites of origin of cholinergic pathways, and, in the hands of Lewis and Henderson it amply confirmed the Lewis-Shute conclusions. Since the studies of Lewis, Kristenson, and their associates, other agents became available to trace back the origin of axons and neuronal pathways, including certain neurotoxins, fluorescent and radio-autographic tracers such as Fluoro-Gold and Fluoro-Red (see, for example Li and Sakagachi, 1997) and [3H] choline (Jones and Beaudet, 1987).

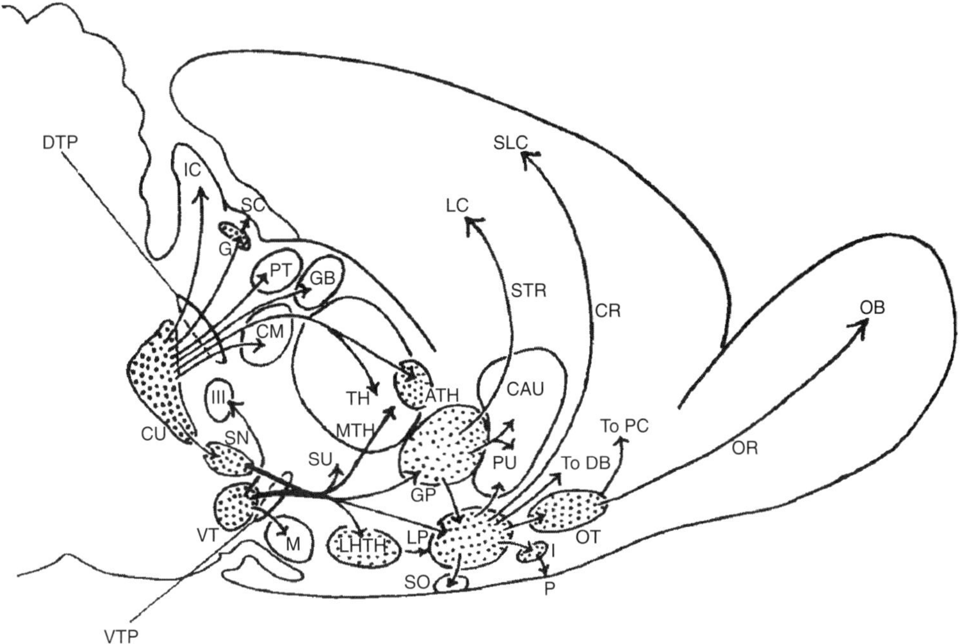

Figure 2-2. Diagram showing the constituent nuclei (stippled) of the ascending cholinergic reticular system in the mid-forebrain, with projections to the cerebellum, tectum, thalamus, hypothalamus, striatum, lateral cortex, and olfactory bulb. ATH, antero-ventral and antero-dorsal thalamic nuclei; CAU, caudate; CM, cetromedian (parafascicular) nucleus; CR, cingulate radiation; CU, nucleus cuneiformis; DB, diagonal band; DTP, dorsal tegmental pathway; G, stratum griseum intermediale of superior colliculus; GB, medial and lateral geniculate bodies; GP, globus pallidus and entopeduncular nucleus; LC, lateral cortex; LHTH, lateral hypothalamic area; LP, lateral preoptic area; M, mammilary body; MTH, mammillo-thalamic tract; OB, olfactory bulb; OR, olfactory radiation; OT, olfactory tubercle; P, plexiform layer of olfactory tubercle; PC, precallosal cells; PT, pretectal nuclei; PU, putamen; SC, superior colliculus; SLC, supero-lateral cortex; SN, substantia nigra pars compacta; SO, supraoptic nucleus; STR, striatal radiation; SU, subthalamus; TH, thalamus; TP, nucleus reticularis tegmenti pontis (of Bechterew); VT, ventral tegmental area and nucleus of basal optic root; VTP, ventral tegmental pathway. (Reprinted from Brain vol. 90, 497–517, 1967, "The Ascending Cholinergic Reticular System: Neocortical, Olfactory and Subcortical Projection" by C.C.D. Shute and P.R. Lewis by permission of Oxford University Press.)

Essentially, the early studies of Shute and Lewis stood the test of time, and the subsequent investigators, while expanding on their data via using different methodology (such as CAT immunocytochemistry; see below), confirmed their conclusions (Kasa, 1971a, 1971b; McGeer et al., 1987a, 1987b). In fact, this confirmation was also obtained by investigators using techniques similar to those employed by Shute and Lewis (see, for example, Krnjevic and Silver, 1965). In addition, in her 1985 study Paula Wilson, who used improved three-dimensional photography and the dual cytochemical technique of Lewis and Henderson (1980) to analyze histochemically the pathways in question, "endorsed . . . the concept . . . of dorsal tegmental projections to most of the nuclei first postulated by Shute and Lewis," although she described additional projections to the thalamus and the cortex, and stressed the significance of the nucleus basalis magnocellularis of Meynert (NBM) and parabrachial nucleus as sources of important cholinergic radiations; these notions were supported by the data obtained via CAT immunohistochemistry method (Fibiger, 1982; Woolf and Butcher, 1986; Bigl et al., 1982; see also next section). Altogether, the Shute-Lewis pathways as modified subsequently are a good approximation of the pathways established some

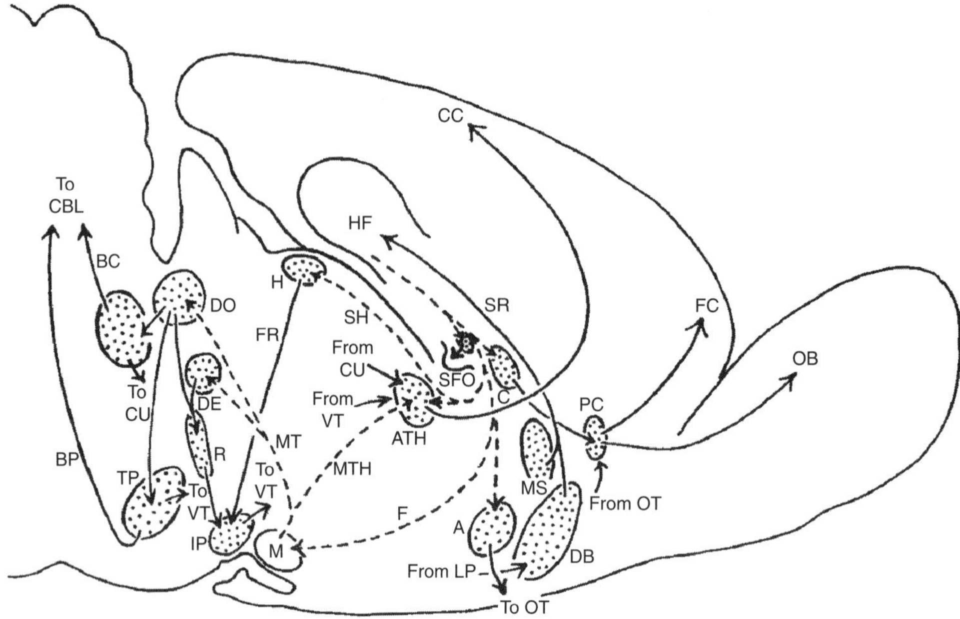

Figure 2-3. Diagram showing cholinesterase-containing nuclei of the midbrain and forebrain (indicated by stipple) connected with the hippocampus, their projections to the medial cortex, and their connections with the ascending cholinergic reticular system. Abbreviations: A, nucleus accumbens; ATH, antero-ventral and antero-dorsal thalamic nuclei; BC, brachium conjunctivum; BP, brachium pontis; C, interstitial nucleus of the ventral hippocampal commissure; CBL, cerebellum; CC, cingulated cortex (cingular and retrosplenial areas); CU, nucleus cuneiformis; DB, diagonal band; DE, deep tegmental nucleus (ventral tegmental nucleus of Gudden); DO, dorsal tegmental nucleus; F, formix; FC, frontal cortex (area infralimbica and anterior limbic area); FR, fasciculus retroflexus (habenula-interpeduncular tract); H, habenular nuclei; HF, hippocampal formation; IP, interpeduncular nucleus; LP, lateral preoptic area; M, mammilary body; MS, medial septal nucleus; MT, mammillo-tegmental tract; MTH, mammillo-thalamic tract; OB, olfactory bulb; OT, olfactory tubercle; PC, precallosal cells; R, dorsal and median nuclei of raphe (nucleus centralis superior); SFO, subfornical organ; SH, stria habenularis; SR, septal radiation; TP, nucleus reticularis tegmenti pontis (of Bechterew); VT, ventral tegmental area. (Reprinted from Brain vol. 90, 521–540, 1967, "The Ascending Cholinergic Reticular System: Neocortical, Olfactory and Subcortical Projection" by C.C.D. Shute and P.R. Lewis by permission of Oxford University Press.)

20 years later by employing CAT immunohistochemistry methodology. It should be stressed that the pathways in question overlap with the central sites, already discussed in this section, of cholinergic receptors, and ACh presence, release, and synthesis.

Nevertheless, is AChE histochemistry a dependable way to identify the cholinergic pathways? The McGeers and Henry Kimura, the pioneers of CAT immunohistochemistry, stressed that AChE "also occurs in non-cholinergic cells" and they adduced as examples "dopaminergic neurons of the substantia nigra, noradrenergic neurons of the locus ceruleus and serotonergic neurons of the raphe, which all stain intensely for AChE" . . . and . . . "are not cholinergic" (Mizukawa et al., 1986); similar arguments were raised by Nancy Woolf and Larry Butcher (1986). In agreement with this notion, several investigators who identified simultaneously AChE and CAT in brain neurons found that a percentage of neurons (usually quite small) exhibited only AChE but not CAT (Eckenstein and Sofroniew, 1981). The matter of the presence of AChE in noncholinergic cells relates to the presence of cholinergic receptors postsynaptically on noncholinergic neurons; this occurrence is characteristic for the synapses between cholinergic radiations and other transmitter systems and is relevant for the McGeers' reference to transmitter interaction. In addition, AChE is present presynaptically on axonal terminals of cholinergic neurons, and this presence may be taken for marker of a cholinergic neuronal soma.

Do these reservations obviate the dependability of AChE histochemistry for the tracing of cholinergic pathways, as implied in the study of Mizukawa et al. (1986)? When AChE histochemistry is carried out without the lesions paradigm, then the point in question is well taken; however, appropriate lesions allow identifying the cholinergic neurons, as explained above, and this safety factor is reinforced by the use of one of the techniques for retrograde identification of the cholinergic neurons. These arguments are supported by the overlap of the pathways based on AChE histochemistry not only with such markers of the cholinergic system as cholinergic receptors and ACh release, but also with the pathways based on the CAT immunohistochemistry; in fact, in the 1980s, several investigators employed AChE histochemistry in conjunction with CAT immunohistochemistry and discovered that the findings obtained by two methods coincided to a great extent (Mizukawa et al., 1986; Woolf and Butcher, 1986;see also below, section IID).

Yet, it is apparent that either ACh or CAT with coenzyme A (CoA), the specific enzyme and coenzyme, which are, respectively, involved in ACh synthesis (see Chapter 3 B) constitute potentially better markers of a cholinergic neuron than AChE. Some attempts at histochemical or immunohistochemical visualization of ACh were made (Geffard et al., 1985; see Kasa, 1986); the histochemical technique employs heteropolyanions that precipitate and visualize ACh and choline, while the immunohistochemical technique uses anti-ACh antibodies. However, these attempts did not lead to any generally accepted methodology. A more successful approach concerned histochemical visualization of the CoA-SH group. This approach was first suggested by Barnett (1968); the actual technique was developed by Catherine Hebb and her associates (1970; see also Hebb and Whittaker, 1958) and Paul Kasa (1971a, 1971b), and Paul Kasa adapted the method for electron microscopy analysis. Using this method, Kasa (1971a, 1971b, 1978; see particularly his detailed and useful review of 1986) identified ascending cholinergic radiations to the several cortical areas, the limbic system, and the spinal cord, as well as intrinsic cholinergic systems in the cortex. Kasa opined that, whenever available, the histochemical CAT and AChE data agree with the more extensive mapping based on CAT immunohistochemistry, and with data concerning other components of the cholinergic system. Generally, this is true; however, his mappings, whether based on histochemistry or immunohistochemistry, differ in several respects from those described by Lewis and Shute (1967) on the basis of AChE histochemistry or by the McGeers, Mesulam, Butcher, Woolf, and others on the basis of CAT immunochemistry. Altogether, as AChE histochemistry is relatively limited in its specificity and discriminatory powers, as recognized by Kasa himself (1986), it is to the immunohistochemical identification of the distribution of CAT that we owe the definitive progress in this area.

The immunohistochemical tracing of CAT was first described by Eng et al. (1974) and Pat and Edith McGeer and Henry Kimura in the 1970s (McGeer et al., 1974); the McGeers and Kimura further developed this method and applied it extensively to the mapping of cholinergic pathways (Kimura et al., 1980, 1981); it is a dependable and most direct procedure for establishing cholinergic pathways, as CAT, per definition, identifies cholinergic neurons. The method is based on producing antibodies to the purified CAT protein, preparation of appropriate antisera, and applying histochemical staining techniques; it became successful only when the purification of the CAT protein became adequate and the specificity of the antibodies achieved. Additionally, the McGeers and the subsequent investigators (see section IIC, below) used lesions, retrograde marking and its visualization (Mesulam, 1978), and antiChEs to achieve precise mapping of the cholinergic pathways. The early, and yet quite advanced mapping was first presented by the McGeers and Henry Kimura (1980, 1981). They stressed the importance of brainstem systems, including cranial motor nuclei, parabrachial complex and tegmental nuclei, as well as forebrain systems including gigantocellular complex and several reticular nuclei; they described thalamic, limbic, and cortical radiations of the brainstem system and the presence of cholinergic cortical interneurons. The subsequent work of particularly Nancy Woolf, Marcel Mesulam, Larry Butcher, Bruce Wainer, and Hans Fibiger described in much detail the sources of origin and the radiations of both descending and ascending branches of the system (see below, section DII) and proposed novel, specific nomenclature for the cholinergic sites of origin (see Chapters 1 through 6; Mesulam et al., 1983a, 1983b).

d. Conclusions

The work of Mesulam, Woolf, and others expanded the vision of Shute, Lewis, Kimura, the McGeers, and their associates and brought about something rare in the cholinergic field: an almost definitive statement concerning at least one area of that field, namely, cholinergic pathways (see below, section IIC). This accomplishment helped link the cholinergic system with central functions and behaviors (see Chapter 9). Furthermore, the early work concerning the identification within the cholinergic neurons of several components of the cholinergic system besides CAT, AChE, and CoA-SH, namely, synaptic vesicles and their dynamics, choline uptake into the cholinergic terminals, and ACh uptake into the vesicles, and so forth, served as a basis for establishing the functional and molecular characteristics of a cholinergic neuron (see section B, below).

2. The Story of Cholinergic Markers and of the Processes Leading to the Release of Acetylcholine

a. What Makes a Cholinergic Neuron Tick?

Besides the early evidence concerning CNS presence of ACh and CAT, additional lines of evidence were established in the 1950s and 1960s; these lines helped identify cholinergic synapses and cholinergic neurons, define their central sites, and explain what makes a cholinergic neuron tick. Of course, the primary line of evidence was initiated by Sir John Eccles as he demonstrated the presence of cholinergic synapses in the CNS; this demonstration included also the findings, important for the definition of the "ticking" of cholinergic neurons, of the cholinoceptivity of a central cholinergic neuron and of its release of ACh (Eccles et al., 1954; Eccles, 1964; see also Chapter 9 A and Karczmar, 2001a, 2001b). Canberra's team of David Curtis, Casmir Krnjevic, John Phillis, and John Crawford continued this line of research, and that of Philip Bradley and others in United Kingdom; it is described in more detail in Chapter 9 A (see also Eccles, 1964; Karczmar, 1967).

Another, related line of evidence concerned cholinoceptivity of brain sites, that is, the presence of cholinergic receptors, whether muscarinic or nicotinic (only monotypes of muscarinic and nicotinic receptors were recognized at the time), and evoked cholinergic potential; this research dealt also with the pharmacology of the receptor responses. This research, carried out initially by the Canberra team, employed the method of electrophoretic application of ACh and drugs such as atropine, beta-erythroidine and d-tubocurarine, muscarine, and nicotine. Their studies in the 1950s and 1960s showed that the central cholinoceptivity is represented mainly and ubiquitously by muscarinic receptors and responses to muscarinic agonists that are potentiated by antiChEs. Indeed, this early research demonstrated that muscarinic receptors are present in the hypothalamus, brainstem and medulla, striate, limbic sites, geniculate, thalamus, and cerebral cortex (see Eccles, 1964; Karczmar, 1967). While today (see Chapter 6 B) many nicotinic sites are distinguished, during the early post-Ecclesian era, besides the Renshaw cells only certain geniculate and cortical sites were recognized as nicotinic (Krnjevic, 1963, 1974; Tebecis, 1970a, 1970b). In addition it became recognized that cholinoceptive responses at cholinergic neurons evoke a "down-the-line" activity that results in the release of ACh from their axon terminals, this released ACh being responsible for the subsequent actions. Subsequently it became apparent that cholinoceptive receptors are present on noncholinergic cells, activation of these receptors resulting in release of other than ACh transmitters, and that cholinergic receptors are located at nerve terminals of cholinergic and noncholinergic neurons (see Chapter 9 BI and BIII). It should be added that, as shown recently (particularly with regard to the neuromyal junction), several proteins (neuregulins such as ARIA) and proteoglycans (such as agrin) mediate the expression, synthesis, and distribution of the cholinergic receptors, and these proteins also serve as markers for cholinergic and cholinoceptine neurons (see Fishbach and Ropsen, 1997).

It was already mentioned that ACh synthesis and its catabolism constituted important items in Dale's and Feldberg's reasoning concerning the role of ACh as a central neurotransmitter. David Nachmansohn provided the crucial evidence in support of this notion, as he and Machado (1943) discovered the synthetic enzyme choline acetyltranferase (CAT; originally termed choline acetylase). Further studies of CAT, its coenzyme A, its central presence and distribution, its role, and its kinetic characteristics were carried out by William Feldberg, Martha Vogt and their associates (see above), Catherine Hebb, David Nachmansohn, Bernard Minz, John Quastel, and others (see Augustinsson, 1948; Quastel et al., 1936; Nachmansohn, 1963; Hebb, 1963).

Then the central (and peripheral) existence and role of a ChE, which was prophesied, as already mentioned, by Dale in 1914, were established by a number of distinguished investigators. The demonstrator of the peripheral cholinergic transmission, Otto Loewi himself, proved Dale's notion. He and Emil Navratil (1924) observed that aqueous extracts of the frog heart destroyed the Vagustoff, that is, ACh; the characteristics of the active extract were those of an enzyme, and Loewi termed it "acetylcholine esterase." Loewi and Navratil (1926) demonstrated subsequently that the potentiation of the vagal effect by physostigmine (they referred to physostigmine as "eserine") is due to physostigmine's antiChE action. Then Stedman et al. (1932) showed that ChEs have as their specific substrates choline esters, and David Glick was probably the first investigator to demonstrate in 1939 the presence of a ChE in the brain (see Glick, 1941). For further information on ChE-focused investigations, in particular on the significance of AChE as contrasted with that of ChEs, see Chapter 3 DI (see also Koelle, 1963; Augustinsson, 1948, 1963) These investigations determined that indeed AChEs are markers of cholinergic neurons and are involved in the question of "what makes the cholinergic neurons tick"; these studies established, however, that this particular marker is not completely reliable, as AChE is present in noncholinergic neurons as well (see Chapter 3 DI).

It should be mentioned that at least some of these findings were made long before Eccles' demonstration of the presence of cholinergic transmission in the central nervous system. As in the case of cholinoceptivity, these sites or markers were ubiquitous; whenever pertinent studies were carried out, the various markers coincided with one another as well as with cholinoceptivity (see, for example, Hebb and Whittaker, 1958; Hebb, 1963).

Additional markers were established subsequently, and their discovery was an important constituent of the evidence for the existence of cholinergic transmission and cholinergic pathways. Palay and Palade (1955) described the presence of synaptic vesicles in the brain, and Victor Whittaker (see Whittaker et al., 1964; Whittaker, 1990) and Eduardo De Robertis (De Robertis and Bennett, 1955) simultaneously expanded on this discovery as well as described definitive and elegant centrifugation methods for obtaining synaptosomes, that is, nerve terminal preparations containing synaptic vesicles.

The neurochemical and cytological analysis of synaptosomes, particularly in Whittaker's laboratory, first in Cambridge and then at Goettingen's Max-Planck-Institut, yielded remarkable results concerning the composition and dynamics of the cholinergic synaptic vesicles, synaptic neurolemmas, and plasma membranes. For example, Whittaker and his associates early purified and isolated cholinergic synaptic vesicles and subsequently developed antisera recognizing the presynaptic plasma membrane (PSPM) to isolate in pure form synaptosomes derived specifically from cholinergic terminals of the Torpedo (see Whittaker and Borroni, 1987; Whittaker, 1990). Note that the PSPM antigens involved in these processes are two gangliosides, Chol-1 alpha and beta, that are present in the mammalian brain (Ferretti and Borroni, 1986). Whittaker and his associates demonstrated also the presence of adenosine triphosphate and other entities subserving the nerve terminal of the cholinergic vesicles (Dowdall et al., 1974; Whttaker, 1992).

This analysis constitutes the basis for the subsequent studies of processes related to the release of ACh from cholinergic nerve terminals. In the modern era, it was shown that these processes require specialized protein systems, which are concerned with generation of ACh, nerve terminal uptake of choline, and vesicular uptake of ACh (see several sections of Chapter 3), as well as with vesicular transport phenomena and recycling of the vesicles; these systems are described in detail in sections B and C of this chapter. Furthermore, besides defining a number of cholinergic markers of the cholinergic neurons, the studies in question led also to description of possibly several modes of ACh release, which I characterize below as classical and unorthodox hypotheses of ACh release (section C, below).

B. Morphology, Cytoanatomy, and Markers of Central Cholinergic Neurons

Is there a specific morphology and cytoanatomy of cholinergic cells that would distinguish them from noncholinergic neurons? Are there any cytoanatomical characteristics of a cholinergic synapse that would provide the basis for such a distinction? Or are the cytoanatomy and/or morphology of either the cholinergic neurons or their synapses not sufficiently specific for such a differentiation? Are additional cholinergic markers, such as the presence of CAT or a choline uptake system needed for the identification of a cholinergic cell? Answering these questions is important not only for the understanding of the characteristics and the function of cholinergic neurons but also for the definition of cholinergic pathways; these matters will be discussed in the two following sections.

1. Morphology and Cytoanatomy of Cholinergic Neurons

Few studies specifically focus on the morphology and cytoanatomy of cholinergic neurons (see Ruggiero et al., 1990; Famiglietti, 1983; Rodieck and Marshak, 1992; Martinez-Rodriquez and Martinez-Murillo 1994); among these investigations, the studies of Famiglietti (1983) and Rodieck with Marshak (1992) concern only 1 cholinergic cell type, namely, the amacrine cells of the retina. In some cases (see, for example, Butcher et al., 1976) the morphology of cholinergic cells is referred to only parenthetically; in other cases (see, for example, Woolf and Butcher, 1986, 1989) this morphology may be deduced from the photomicrographs included with the studies in question. In what follows, the cytoanatomy and morphology are described either on the basis of specific description of the neurons or on the basis of the pertinent photomicrographs. Finally, axons releasing various transmitters (or axons emanating from different nuclei that contain neurons synthesizing the same transmitter) may vary anatomically, but at this time it does not seem possible to differentiate anatomically cholinergic axons from noncholinergic axons. It must be stressed that the neurons are referred to in this section as cholinergic because they were identified as such by means of CAT immunohistochemistry or by means of additional markers (see below).

Altogether, cholinergic cells come in many shapes and sizes. Thus, the alpha motoneurons that supply the striated muscle endplates, whether located in the ventral horns of the spinal cord or in the brainstem, which is the origin of cranial nerves, are among the largest neurons of the nervous system: these polygonal, multipolar neurons may be up to 500 Å in diameter. On the other hand, the gamma motoneurons that supply the spindles are among the smallest neurons, as they range from 18 to 38 μm (Szentagothai and Rajkovits, 1955; see also Brodal, 1981). Other cholinergic neurons are frequently ovoid, round, or oval and elongated; they may be bipolar or multipolar, as in the case of the neurons of the mesencephalic interstitial nucleus of Cajal and nucleus basalis magnocellularis (NBM), respectively. Multipolar ovoid neurons are also present in the nucleus reticularis and in the nucleus ambiguus (Ruggiero et al., 1990). These multipolar neurons are 25 to 40 Å in diameter, although some of the multipolar neurons of NBM, other basal forebrain sites, and the pedunculopontine tegmental nuclei are considerably larger; these larger neurons are usually hyperchromatic (Butcher et al., 1977; Woolf and Butcher, 1989; Mesulam et al., 1983a, 1983b; Martinez-Murillo et al., 1989; Bigl and Arendt, 1992). Smaller (18 to 25 μm) ovoid or fusiform mutipolar neurons were found in the parabrachial complex, basal forebrain substantia nigra, raphe nuclei, periventricular gray, and hypothalamus (Ruggiero et al., 1990; Martinez-Murillo, 1989); also, ovoid or fusiform neurons are the cholinmergic neurons classified by Marsel Mesulam and his associates (Mesulam et al., 1983a, 1983b) as belonging to basal forebrain sectors Ch1 to 3. Martinez-Murillo and his associates (1989) were among the few investigators who described the cholinergic cells of cholinergic complexes of the forebrain, including NBM, in more detail ("cell nucleus showed one or more indentations . . . occupied central position and was surrounded by abundant cytoplasm rich in organelles . . . large lipofuscin granules were also observed . . . the dendrites were thick").

And then there are the amacrine cells of the retina. While the amacrine cells of the innermost nuclear layer are peptidergic, the starburst amacrine cells of the inner plexiform layer are either cholinergic or gabaergic. These large neurons exhibit no obvious polarity and unique morphology (Famiglietti, 1983; Rodieck and Marshak, 1992); they were identified in the retinas of the rabbit, human, and primates (see Giolli et al., 2005).

As can be seen, it is difficult to decide purely on the basis of morphology that a given cell is cholinergic. In fact, neurons subserving noncholinergic transmitters (i.e., catecholaminergic, serotonergic, or peptidergic neurons; perhaps the easily distinguished central histaminergic mast cells may be an exception) may be similar in shape and size to one or another "type" of cholinergic neurons.

May we then look to synaptic morphology for the differentiation in question? In the 1960s, Gray (1969; see also Hutchins, 1987 and Shepherd and Harris, 1998) distinguished morphologically between synapses subserving excitatory and inhibitory transmission; they are referred to as Type 1 and Type 2 synapses, respectively. The excitatory Type 1 synapses have a wider synaptic cleft than Type 2 inhibitory synapses. The postsynaptic membrane of the Type 1 synapses is thick and dense, and occupies a great part of the postsynaptic area, while the Type 2 synapses exhibit dense material both pre- and postsynaptically; thus, Type 2 synapses are symmetrical while Type 1 synapses are asymmetrical. Finally, the synaptic region (synaptic or active zone) of Type 1 synapses is longer than that of Type 2 synapses. Cholinergic synapses are generally Type 1 synapses (see Kimura et al., 1981 and Eccles, 1964; there occasionally may be exceptions to this rule, see Smiley, 1996), but the Type 1 morphology cannot serve for reliable morphological identification of cholinergic transmission, as Type 1 synapses can be activated by noncholinergic excitatory transmitters such as glutamate. Finally, the axons and nerve terminals are characterized by varicosities and boutons (see, for example, Shepherd and Harris, 1998), but it does not appear that these may be used as dependable markers of cholinergic axons and terminals.

Other synaptic markers, the synaptic vesicles, may serve well to identify cholinergic transmission. Following Victor Whittaker's and Eduardo de Robertis' discoveries (see section IIA, above), dynamics of cholinergic vesicles were studied in detail; much of the pertinent research took place first at the Station Biologique of Arcachon, France, and then in Victor Whittaker's laboratory at Max-Planck-Institut in Goettingen, Germany. While this research was conducted with the Torpedo electric organ, the results are consistent with those obtained in mammals, including mammalian brain. Seen via electron microscopy of peripheral or central cholinergic nerve terminals, the ACh-containing cholinergic vesicles were round or oval and had a clear core that exhibited variable degrees of density and variable size (Zimmerman and Whittaker, 1977; see also Prior and Tian, 1955). In mammals, they vary in size from 45 to 50 Å in diameter; they are by far larger in the case of the vesicles of the Torpedo electric organ (Whittaker, 1992; Martinez-Murillo, 1989). Vesicles of similar form and size are seen in Whittaker's synaptosomal preparations (Whittaker, 1992). Occasionally large or fused vesicles are also present; they may be the source of the giant excitatory cholinergic postsynaptic potentials (Eccles, 1964). They usually form clusters throughout the terminals, as well as at the neurolemma of the terminal. Several other modes of cholinergic vesicles are also present: during the process of exocytosis some vesicles fuse with the terminal neurolemma and accordingly change in form from ovoid to flattened; empty vesicles or vesicular ghosts also appear as they are formed in neuronal perikarya as well as at the nerve terminal after the release of their ACh content in the course of vesicular recycling (see next section and Whittaker, 1992). There are also two or more density modes among the vesicles, that is, vesicles may vary in their "molecular acetylcholine content" (MAC; Whittaker, 1990) during the process of cycling (see below); the recycling vesicles that contain freshly synthesized ACh are denser, while the less dense vesicles are present in the axons of the electric fish (Whittaker, 1992).

On the whole, there is a clear distinction among cholinergic vesicles and vesicles containing other transmitters. For example, peptidergic and serotonergic vesicles have a dense core and are larger than cholinergic vesicles, while catecholaminergic vesicles are granular; yet, sometimes clear core vesicles of a size comparable to that of cholinergic vesicles are present in serotonergic or catecholaminergic terminals (Van Bockstaele and Pickel, 1993; Horie et al., 1993; Doyle and Maxwell, 1993). It must be also remembered that peptides, such as the vasoactive intestinal peptide (VIP), are copresent with ACh in cholinergic vesicles (Agoston and Lisiewicz, 1989; see Whittaker, 1990). Altogether, the total picture of cholinergic terminals, cholinergic neurons, and cholinergic vesicles as described in this section is quite diagnostic for cholinergic nerves and synapses.

2. Neurochemical Systems as Markers of Cholinergic Neurons and Intraterminal ACh Motions

Processes of synthesis of ACh include several components such as nerve terminal choline uptake, CAT, acetyl coenzyme A and its synthetase, and synaptic vesicles and their dynamics; these components serve as dependable markers of cholinergic neurons and cholinergic nerve terminals. In addition, ChEs (particularly AChE) help identifying both cholinergic perikarya and nerve terminals of cholinergic neurons. Choline acetyltransferase and AChE characterization of cholinergic cells are discussed in detail in sections DI–DIII, below, and in Chapter 3 B1-3. In this section, the systems linked with the vesicular cycling and storage and release of ACh are specifically considered.

As illustrated by certain aspects of the cytomorphology of cholinergic neurons, described above, synaptic vesicles undergo a cycle that must be subserved by appropriate neurochemical systems. As ACh, CAT, and AChE, these systems define what it is to be a functional cholinergic cell (Weihe et al., 1998); in fact, the systems concerned with vesicular cycling, transport of ACh into the vesicles, and ACh synthesis are regulated by a single cholinergic locus gene (see Eiden, 1998; Mallet et al., 1998).

3. "Cycling" and "Recycling" Processes

A complicated process concerns formation and movement of synaptic vesicles, loading of the vesicles with ACh, the fusion of the vesicles with an endosomal component and the terminal plasmalemma, and the vesicular release of ACh. This process is referred to as cycling—or recycling, if one starts with the empty synaptic vesicles that have released their ACh.

a. Formation and Movement

The process begins with the formation of the empty vesicles and with their movement. Similar to CAT, empty synaptic vesicles are formed within the Golgi organelles and are transported anterograde-fashion at a fast rate (Kiene and Stadler, 1987). This fast transport involves microtubules, actin filaments, and neurofilaments; the vesicles are bound to these organelles by a family of proteins called synapsins (see Whittaker, 1992); mRNAs coded for these and other proteins are contained in the Golgi bodies of the perikaryon, and Whittaker (1992) suggests that these proteins subserve generally vesicular transport and related processes with regard to both cholinergic and noncholinergic vesicles. Phosphorylations and dephosphorylations serve to link with and liberate the vesicles from the elements of cytoskeleton and to mobilize the vesicles for exocytosis.

b. Acetylcholine Loading into the Vesicles

At the nerve terminal, a specific protein facilitates loading ACh into empty vesicles (this process was called "concentrative uptake" by Whittaker, 1992). The protein is referred to as vesicular ACh transporter (VAChT; Bahr and Parsons, 1986). Phenyl piperidines inhibit this process, (–) 2-(4-phenylpiperidino) cyclohexanol (vesamicol) being the most powerful and specific inhibitor of VAChT action; it is interesting and teleological that vesamicol shows a much higher affinity for empty vesicles (vesicle ghosts) than for loaded vesicles (Noremberg and Parsons, 1989; see also Whittaker, 1992). Sophisticated molecular studies of VAChT by Varoqui and Erickson (1998) indicated that the vesicular transport of ACh requires cholinergic-specific amino acids within the N-terminal portion of VAChT, and that this is the site of action of vesamicol. The important aspect of the transporter mechanism is that the genes for VAChT and CAT are colocalized: "the gene encoding the vesicular acetylcholine transporter has been localized within the first intron of the gene encoding acetycholinetransferase and is in the same transcriptional orientation" (Mallet et al., 1998). In fact, certain polypeptide factors involved in cholinergic ontogeny such as cholinergic differentiation factor/leukemia inhibitory factor concomitantly increase VAChT and CAT mRNA levels. Furthermore, the regulation of expression of CAT and VAChT is coregulated by the cholinergic gene locus that contains genes both for CAT and for VAChT (Mallet et al., 1998; Wu and Hersh, 2004; Lim et al., 2000; for further details of this coregulation, see Chapter 3 B-1); thus, CAT and VAChT gene transcriptions share common promoters (Mallet et al., 1998) . It should be stressed that the transporter in question is not necessarily specific for cholinergic neurons and vesicles (Cervini et al., 1995; Mallet et al., 1998) and that different transporters subserve other transmitter systems such as monoaminergic and serotonergic systems (see, for example, Zucker et al., 2001).

Whittaker and his associates discovered early that ATP is copackaged with ACh into the cholinergic vesicles (as it is with catecholamines in granular catecholaminergic vesicles; Whittaker et al., 1964). The specific carrier for this uptake is saturable and of the high-affinity type (Whittaker, 1992); it was identified as the vesicle component 11 or vesicular ATP translocase, which is a protein-binding active factor (Lee and Witzemann, 1983). The ATPase, which is present in the vesicular wall, assists the ATP translocase, which maintains the proton gradient stimulating the translocation; this gradient also facilitates the VAChT-activated vesicular uptake of ACh (Whittaker, 1992, 1998). According to Whittaker (1992), similar to ACh, ATP is taken up preferentially into the pool of recycling vesicles, that is, into the pool of vesicles released empty from the nerve terminal plasma membrane after their fusion with the membrane and after the release of ACh; this is consistent with the notion of a "readily releasable pool" of vesicles (vesicle pool immediately available for release; see below). Acetylcholine and ATP do not course freely inside the vesicles but are adsorbed to an intravesicular proteoglycan matrix (Reigada et al., 2003); in fact, when the vesicular membrane is treated with distilled water, ACh and ATP remain attached to the matrix as long as the cations are not added; this explains why vesicular ACh is so stable upon stimulation of the terminal (at least until enzymically hydrolyzed) or upon purification (Yves Dunant, personal communication).

c. Docking and Fusion

Docking of the vesicles and the fusion of synaptic vesicle plasma membranes directly lead to ACh release (see Whittaker, 1992). This is a most important step, as the demonstration that the fusion of vesicles and the release of vesicular contents of ACh—or quanta—are linked is a part of the proof of the quantal nature of ACh release; subsequently, this demonstration was further helped by development of novel, refined techniques (see below, this section). The fusion factors include presynaptic membrane proteins, syntaxin, synaptotagmins and attachment proteins, SNAPs (particularly SNAP 25 and synaptobrevin), and synaptic vesicle proteins (VAMP1 and 2; Hou and Dahlstrom, 2000; Morel at al., 1998; Chapman et al., 1955; Robinson et al., 2004); these proteins interact with Ca^{2+} during processes of fusion and ACh release. Some or all of these proteins form fusion attachment protein receptor complexes called SNAREs; additional proteins of the vesicles such as synaptophysin (P38), Spring (a finger protein), and VMG also play a role in docking and fusion processes (Whittaker, 1992; Li et al., 2005; Fasshauer et al., 2003). SNAREs are cleaved by clostridial neurotoxins, "most potent inhibitors of neurotransmitter release known" (El Far et al., 1998; see Figure 2-4). Importantly, cholinergic stimulation—such as nicotinic, via alpha7 recptors—may activate attachment proteins (Liu et al., 2005).

It was proposed that SNAREs act as synapse-specific membrane recognition molecules, acceptors for docking and fusion catalysts (Morel et al., 1998; El Far et al., 1998; Rothman, 1994). In addition, other vesicular proteins, the RABs, as well as internal vesicular matrix (Reigada et al., 2003), subserve the vesicular plasmalemma fusion and ACh release; however, some RABs are concerned with an entirely different function, the endoplasmic fusion (Sudhof, 2000, 2005). The fusion and release processes produce empty synaptic vesicles, ready for recycling; this recycling—that is, reformation of ACh-loaded vesicles—may occur "directly" or via an endosomal intermediary, activated by a RAB protein (the role of several synaptic vesicle RAB proteins, such as rabphilin, is

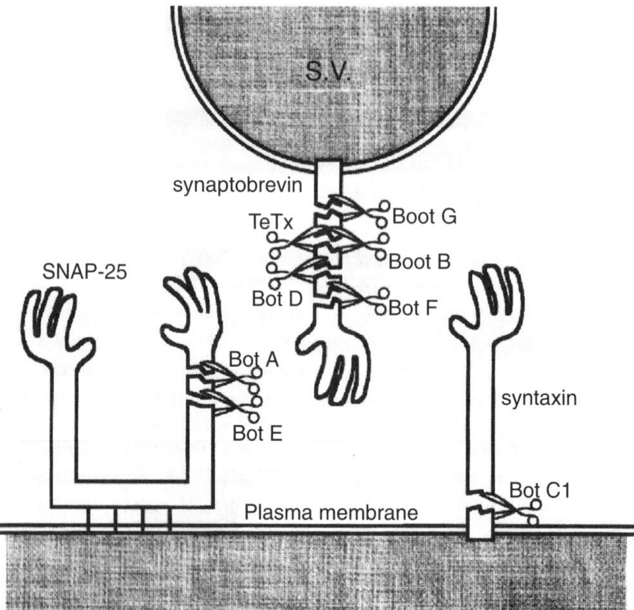

Figure 2-4. Clostridial toxin cleavage sites. The proteins cleaved by the clostridial neurotoxins are shown to highlight their juxtaposed membrane orientations in the synaptic vesicles (synaptobrevin) and plasma (syntaxin and SNAP-25) membranes. This oversimplifies the *in vivo* situation, as fractions of cellular syntaxin and SNAP-25 are found on synaptic vesicles. Transmembrane domains insert syntaxin and synaptobrevin into the membrane, whereas palmitoylation of cysteine residues is responsible for the membrane localization of SNAP-25. Note that Bot C1 may cleave both syntaxin and SNAP-25. (From El Far et al., 1988; reprinted by permission from Elsevier Press.)

at this time not clear; Sudhof, 2000; Lonart and Sudhof, 2001).

It must be added that fusion, whether between vesicles, between vesicles and presynaptic membranes, or between vesicles and the endosomal intermediate, was demonstrated *in vitro* (Whittaker, 1992; Sudhof, 2000). Several technical innovations contributed to this demonstration. Thus, the quick-freezing technique employed in the Bruno Ceccarelli Center of the University of Milan (see, for example, Torri-Tarelli et al., 1985, 1990) stabilizes the pertinent events within 1 ms. Then, precise monitoring of exo- and endocytosis of the vesicles was made possible by the use of lipophilic fluorescent probes such as FM dyes and the optic recording of their location within the cell and vesicular membranes (Cochilla et al., 1999; Sudhof, 2000, 2005). The employment of these techniques led to the description of exocytosis as consisting of the fusion of vesicles with the plasma membrane (plasmalemma) which creates small indentations in the latter and coated vesicles; after the emptying of the vesicles and the release of Ach, the empty, uncoated vesicle is ready for recycling and ACh uptake. The use of this technique jointly with appropriate immunocytochemistry methods linked time-wise the protein-activated fusion and the emptying of vesicles and added support for the theorem of the quantal hypothesis of the release of ACh. It should be stressed that, unlike the transporter proteins, fusion and exocytosis proteins (and other release proteins such as the mediatophore, which is discussed below, in section B2) might not be specific for the cholinergic vesicles (see, for example, Bacci et al., 2001; Israel and Dunant, 1998).

d. Postsynaptic Membranes and Cholinergic Receptors

A number of active proteins related to neurotrophins are engaged in the regulation and formation of cholinergic receptors and membrane ion channels. Neuroregulins, including ARIA, and laminin chains are among these proteins. While

ARIA is predominantly displayed at the neuromyal junctions, it may be also present centrally, as are the laminins (Fischbach and Rosen, 1997; Yin et al., 2003). Also, electrotactins and related proteins are involved in synaptogenesis; they were identified by Israel Silman, Joel Sussman, and their associates as adhesion proteins, which are members of the AChE family and which form a pattern of electrostatic potential that participates in the formation of active synapses (Botti et al., 1998; Rydberg et al., 2004).

While not listed generally as markers of the cholinergic neuron, it appears that these proteins could serve in this capacity (Sanes and Truccolo, 2003).

e. Cycling and Recycling of Synaptic Vesicles

The course of exocytosis involves several kinds of synaptic vesicles; Victor Whittaker remarked early on the heterogeneity of the vesicles as well as proposed the notion of vesicle cycling, if the total life cycle of the vesicles is considered, or recycling, if exocytosis is taken as the starting point of the phenomenon (Whittaker, 1992, 1998). This heterogeneity and cycling of vesicles were originally established in terms of the differences, before and after cholinergic nerve terminal stimulation, in the density, size, and specific radioactivity of ACh of the vesicles, following their loading with radioactive choline and their release of ACh; other means, such as fluorescent dyes, laser scanning, and so on, are also used for the analysis of vesicle heterogeneity and their cycling (Ryan, 2001). Cultured hippocampal neurons yielded much evidence with respect to this matter (see, for example, Murthy and Stevens, 1999).

As already referred to, essentially, there exist two kinds of cholinergic vesicles: empty and full (or ACh loaded). The empty vesicles are formed either in the neuronal perikarya or at the nerve terminal; they are generated at the terminal following ACh release and the emptying of the ACh-loaded vesicles (exocytosis). The empty vesicles formed in the perikarya of the cholinergic neurons migrate down the axon, carried by the axoplasmic flow. They become loaded in or near the nerve terminal as they incorporate ACh via the action of VAChT proteins (see above, this section) and the processes of ACh synthesis (see Chapter 3 B); the vesicles that become empty via ACh release are also subject to ACh loading. Following Whittaker's original description of the processes of recycling or cycling, he and others (see, for example, Prior and Tian, 1995; Pyle et al., 2000; Sudhof, 2000) included several pools of synaptic vesicles in the cycle.

Originally, Whittaker (1992) recognized four pools and defined their different roles during recycling and exocytosis. The empty vesicles arriving from the cell body represent the first pool. Then there is the reserve pool of ACh-loaded vesicles. The next pool is the pool of docked vesicles; Sudhof (2000) and others refer to this pool as "the readily releasable pool" that contains only relatively few vesicles docked in the active presynaptic zone. Finally, there is the stimulus-induced recycling pool of empty vesicles that preferentially takes up ACh that is newly synthesized in the cytoplasm (Barker et al., 1972). Pyle et al. (2000) determined that the latter pool of vesicles might be rapidly converted into ACh-loaded vesicles ("immediately after exocytosis"). Pyle used FM dyes in this instance (see also Sudhof, 2000; Ryan, 2001).

Some investigators recognize additional pools and processes. Thus, Thomas Sudhof (2000) writes of a "recycling pool," which he defines as labeled, by the FM-143 dye, under conditions of intensive synaptic stimulation; according to Sudhof, this pool consists of the "readily releasable pool" and a "reserve pool"; both include ACh-loaded vesicles. Then several proteins, in addition to those that regulate vesicular cycling via their role in docking in fusion (see above, this section), such as dynamine, are directly important in the organization of cycling (see Weible et al., 2004).

4. False Transmitters

The issue of false transmitters is of interest in the present context. The false transmitters are ACh analogues that are taken up by the vesicles and released in lieu of ACh. However, in general, false transmitters cannot be taken up by the nerve terminal choline uptake system; therefore, appropriate precursors, that is, choline analogues capable of being taken up by the terminal and then acetylated in the terminal by CAT, are used to form the false transmitters. Such precursors are, for example, pyrrolidinecholine, homocholine, and triethylcholine (Whittaker, 1992); the generation of false transmitters by these precursors was established for the Torpedo electric organ and for the mammalian brain (von Schwarzenfeld, 1979; von Schwarzenfeld et al., 1979). Depending on the false transmitter, they can render the cholinergic transmission either ineffective (Jenden, 1990; Jenden et al., 1989) or partially effective, forming small endplate potentials (see next section).

5. Release Processes

The process of release is activated by the endogenous or electric presynaptic stimulation and ensuing depolarization of the membrane; this process is Ca^{2+} dependent. As is well known, Bernard Katz, Rodolfo Miledi, and Juan Del Castillo (Del Castillo and Katz, 1954; Katz, 1966; Katz and Miledi, 1965) demonstrated this dependence; they could even relate in situ the concentration of Ca^{2+} to the size of the endplate potentials. The phenomena of depolarization and Ca^{2+}-dependent ACh release may be duplicated with brain synaptosomes (see Whittaker, 1992). The saturable Ca^{2+} uptake is catalyzed by ATP and calmodulin. Calmodulin is a cytosolic protein regulating the formation of the cytoskeleton (Kretsinger, 1987; see also Whittaker, 1992), but it also facilitates, in the presence of ATP, Ca^{2+} uptake, presumably via activation of many protein phosphokinases. As calcium uptake and calmodulin dynamics can be visualized in situ or in the synaptosomes (see Whittaker, 1992), these components of ACh release may also serve as markers of a cholinergic nerve terminal (for additional, modern insights into the processes of ACh release, see below, section C).

It is important to remember that, since the days of Thomas Hokfelt and Victor Whittaker, it has been well known that ACh is coreleased with other transmitters, including catechol and indoleamines, and particularly peptides (see Chapter 9 BIII-2).

6. Cholinergic Markers in Noncholinergic Cells and Tissues

The markers of cholinergic cells appear in noncholinergic cells and tissues of the adult vertebrates, in ephemeral organs such as placenta (see Sastry, 1997), during preneurogenetic stages of ontogenesis (Karczmar, 1963a, 1963b; Chapter 8 BI and BII), as well as in monocellular and metazoan species devoid of nervous system (or of contractile systems, present in flagellates, which exhibit cholinergic correlates (see below, this section, and Chapter 8 BV). These phenomena pose a problem with regard to the identification of cholinergic neurons. The related question is that of the place of BuChE in this identification process.

The question of the presence of cholinergic markers such as SNAREs or RABs in noncholinergic or nonneuronal cells is moot, as this matter was not investigated to any extent in such cells. Yet, cholinoceptive receptors appear at the nerve terminals of noncholinergic neurons such as glutaminergic cells and/or the postsynaptic sites of noncholinergic neurons, including inhibitory interneurons; are other cholinergic markers present in these instances? (See section C, this chapter, and Chapter 9 BI; see also Atkinson et al., 2004).

Then, may other cholinergic components such as ChEs, CAT, VAChT, and ACh serve as cholinergic markers? Some of these components, particularly ChEs, appear in noncholinergic neurons and in nonnervous cells and tissues (for example, in blood), in preneurogenesis stages of development, in nonnervous tissues of invertebrates including bacteria, and in plants. Their presence and role in these cases is, to say the least, mysterious and piquant: What is the role of ChEs in bacteria and plants? Why do certain mammalian species exhibit AChE in the red blood cells and BuChE in the plasma while a reverse situation exists in other species? Why do platelets of some species contain AChE, while those of other species do not? Why are both AChE and BuChE as well as ACh present in the cerebrospinal fluid, as established early by Sir William Feldberg (Feldberg and Schriever, 1936; see also Koelle, 1963; Karczmar, 1967; Augustinsson, 1948; Fischer, 1971; Goedde et al., 1967)? These matters are discussed at length in section DIII of Chapter 3.

Of particular interest is the presence of BuChE in cholinergic neurons (Koelle, 1963; Giacobini, 2000, 2002; it is also present in the glia and in the plaques of Alzheimer's disease; see Chapter 10). Should BuChE be considered a cholinergic marker? Ezio Giacobini seems to think so, as he opines that BuChEs of the central neurons protect against excess ACh that may occur in some "physiological" conditions. He bases this notion on his finding (Giacobini, 2002; see Chapter 3 DIII) that a specific antiBuChE augmented markedly rat cortical levels of ACh; does it follow from this evidence that BuChE may serve as a protector? What would be the physiological conditions under which BuChE could play this role? This is not to deny the possibility that both BuChE and AChE may play this scavenger role in the blood, where it would rapidly hydrolyze ACh that may be released into the blood from various tissues (Karczmar and Koppanyi, 1956; Chapter 3 DIII).

It appears altogether that a neuron may be deemed dependably as cholinergic when it exhibits the preponderance of cholinergic markers, while the presence of cholinergic components in nonnervous cells such as the erythrocytes, preneurogenetic embryos, and uninnervated organisms such as bacteria is both a mystery and a problem.

Certain speculations were already raised with regard to the role of BuChE (see above and Chapter 3 DIII). Furthermore, the ontogenetic presence of cholinergic components prior to neurogenesis and in species devoid of nerve cells may relate to the demonstrated function of cholinergic components as morphogens (see Chapter 8 BV and CIII and Chapter 11 A) or, perhaps, as metabotropes.

7. Conclusions

There are several problems with respect to the processes described in this section. For instance, what determines the ratio between empty vesicles generated in the cholinergic neuron perikaryon and the empty vesicles recovered at the nerve terminal plasma membrane after the release of ACh? Then, which mechanism moves the empty vesicles away from the terminal (see section C5)? It must also be stressed that while very many markers of the cholinergic neurons were identified, many more will be discovered in the future.

It seems clear that we cannot describe at this time a parsimonious mechanism that would link sequentially the proteins in question during the processes that involve the vesicles, ACh synthesis and vesicular uptake, and the preparation to release and the release of ACh. Are all the present and future SNAREs, SNAPs, RABs, and other paladins of the phenomena described necessary (Ybe et al., 2001)? Do myelin protein expressions of genes such as Nogo, which are inhibitory during neuronal growth and regeneration, assume a regulatory role in adulthood, as proposed by Roger Nitsch and his associates with respect to the hippocampus (Meier et al., 2003; see also Chapter 8 BII)? Is there some redundancy to this system? Are all these proteins and phospholipids specific for ACh release and for the cholinergic system, or are they needed for the release of other transmitters? These unsolved problems and unanswered questions will multiply when we consider the process of ACh release; indeed, a number of speculations do not fit within the picture presented in this section (see next section).

C. Classical and Unorthodox Hypotheses of Acetylcholine Release

There are several models of the release of ACh from a presynaptic cholinergic terminal: the classical model established in the 1950s by Victor Whittaker, Eduardo De Robertis, Bernard Katz, Ricardo Miledi, Paul Fatt, John Hubbard, Steve Thesleff, and Jose del Castillo (see Karczmar, 1967; Katz, 1966; Eccles, 1964; McLennan, 1963; Whittaker, 1992; see also section B5, above), and three current, neoclassical, or unorthodox pictures of the release of ACh. ACh release is regulated by cholinergic presynaptic receptors, and it is linked with postsynaptic responses. Electro- and neurophysiological details of the postsynaptic and presynaptic cholinergic receptor responses and detailed knowledge of synaptic potentials and currents and their ionic mechanisms are out of the scope of this section (see, however, Chapters 3, 6, and 9 BI). Nevertheless, certain aspects of postsynaptic responses are relevant to the quantal phenomena and hence are pertinent for this section.

1. The Classical ACh Release Model and Postsynaptic Responses

While much of the pertinent information concerning the classical model of ACh release deals with the skeletal neuromyal junction and its endplate, vertebrate autonomic neurons, the Torpedo electric organ, and invertebrate neurons such as the giant synapses of the squid (Eccles, 1964; Karczmar et al., 1986; Whittaker, 1992), ample evidence confirms the validity of the model for the central cholinergic synapses.

The release occurs when the nerve terminal is depolarized by endogenous or electric presynaptic stimulation and ensuing depolarization of the terminal in the presence of Ca^{2+} and, presumably, calmodulin, as well as a number of fusion and related proteins (see above, sections B3 and B5). Then the presynaptic cholinoceptive receptors modulate ACh release when they abut on cholinergic terminals, and regulate the release of other transmitters when they are present at noncholinergic terminals; in fact, either the same or different proteins are involved in these two types of release regulation (see, for example, Atkinson et al., 2004).

As is well known, Bernard Katz, Ricardo Miledi, and Jose Del Castillo (Del Castillo and Katz, 1954; Katz, 1966; Katz and Miledi, 1965) demonstrated this dependence for the neuromyal junction; they could relate in situ the concentration of Ca^{2+} to the size of the endplate potentials. Subsequent work demonstrated this relationship with respect to excitatory postsynaptic potentials of the central or peripheral neurons. Further details of this phenomenon are presented in Chapter 9 BI.

The release is generally multiquantal; that is, it involves a number of synaptic vesicles, particularly when it elicits a productive response, such as a spike. A productive response may be a postsynaptic response originating at either the soma or dendrites of a cholinoceptive neuron, or it may originate at a presynaptic site of a cholinergic or, for that matter, a noncholinergic neuron. There is a delay of 1 to 2 ms before the postsynaptic response appears; in fact, this delay is a characteristic phenomenon of a chemically, transmitter-operated synapse, and its presence is a part of the proof of the existence of chemical transmission, since it does not occur at the sites of electric transmission (see Eccles, 1963; McGeer et al., 1987b).

The postsynaptic cholinergic response may be excitatory or inhibitory in nature; the excitatory and inhibitory responses are depolarizing and hyperpolarizing, respectively. The productive postsynaptic excitatory response is initiated, whether at the dendritic or somatic membranes, by an excitatory postsynaptic slow or fast potential (sEPSP and fEPSP; an endplate potential [EPP] is evoked at the striated muscle endplate). The sEPSP and fEPSP are generated at muscarinic and nicotinic neuronal sites, respectively. Ultimately, as it reaches the critical threshold (at about—40 mv), the fEPSP generates the postsynaptic spike, which is the definitive signal underlying interneuronal communication. The sEPSPs do not generate spikes, unless in combination with a facilitatory transmitter or compound, or with an antiChE; therefore, the sEPSP is a modulatory rather than a transmittive potential (see Krnjevic, 1969, 1974). The inhibitory and excitatory potentials represent the movement of appropriate ions and, therefore, specific currents (Eccles, 1964).

The nicotinic fEPSP may appear in the absence of a spike, if the receptive postsynaptic membrane is partially obtunded by postsynaptic blockers such as d-tubocurarine, curarimimetics, and certain toxins; if the axonal conduction is partially blocked by inhibitors of presynaptic, axonal conduction such as blockers of Na channels (e.g., tetradotoxin); if ACh synthesis is attenuated by blockers of choline uptake and/or ACh synthesis; and if the release of ACh is partially inhibited by blockers of presynaptic Ca^{2+} channels, including high concentrations of Mg^{2+}, and of calmodulin-Ca^{2+} interaction (McGeer et al., 1987b). It may be also generated by electrophoretic application of ACh at a concentration insufficient to generate a spike. Also, it is possible to visualize the fEPSP by applying a cellular microelectrode or a patch clamp electrode at an angstrom-small distance from the receptor site; in this case, following effective presynaptic stimulation the fEPSP appears as a shoulder of the rising spike (Kuffler, 1949; Katz, 1966; Kuffler et al., 1984; Thesleff, 1960).

During the 1950s and 1970s, Paul Fatt, Bernard Katz, and Ricardo Miledi proceeded to "miniaturize" the postsynaptic responses. First, Fatt and Katz (1952) demonstrated the presence at the quiescent neuromyal junction of spontaneously and randomly occurring "subthreshold" excitatory responses, which they referred to as miniature endplate potentials (mEPPs) or miniature endplate currents (mEPCs). After appropriate analysis it was posited that these responses are due to spontaneous release of single "packets" of ACh that "might" correspond to the contents of single synaptic vesicles (Katz, 1966; notice the conditional mode of Sir Bernard's statement; Sir Bernard was careful to distinguish proven fact from hypothesis; see below, next section). Ultimately, this phenomenon, originally described for the neuromyal junction, was shown to extend to all cholinergic, peripheral, and central synapses.

Is there a relation between the "miniatures" and the fEPSPs? Such a relation, if demonstrated, would cement the notion of the "quantal" nature of postsynaptic responses. Bernard Katz and Ricardo Miledi provided the necessary proof. While of similar size, the EPSPs fluctuate randomly in size, and this fluctuation can be augmented *in vitro* by diminishing Ca^{2+} concentration or increasing Mg^{2+} concentration. These fluctuations may be analyzed statistically by the use of Poisson's theorem; using this analysis in his striated-muscle experiments Katz (1958) established that fEPSPs are composed of a certain number of quanta, that is, synaptic vesicles. Also, other methods may be used to evaluate quantal fEPSP content, including comparing the miniature fEPSPs or miniature EPPs size to fEPSPs, establishing the relationships among the elementary events, the "miniatures," and the fEPSPs (see below, this section), considering the size of the vesicles and the concentration of the vesicular ACh, and so on (McGeer et al., 1987b; Eccles, 1964; Thesleff et al., 1984). Investigators employing these various methods of analysis reported relatively similar values for the quantal contents of fEPSPs recorded in central or peripheral neurons; these values range from 100 to 200.

Syogoro Nishi, Hideho Soeda, and Kyozo Koketsu (1967) carried out a pertinent study on the toad sympathetic ganglion. These investigators employed Corsten's (1940) frog lung bioassay for ACh—which they rendered more sensitive by adapting it for the toad lung—for the measurement of ACh; ultimately, they related ACh release via a single volley (from a single synaptic knob), the fEPSP, and the "miniatures"; they estimated the quantal content of the fEPSPs at 100 to 200 and the ACh contents of the fEPSPs at 6,000 to 8,000 molecules. It should be pointed out that the bioassay employed by the Japanese investigators is most sensitive, as it responds to 10^{-21} M concentrations of ACh and as it is logarithmically linear over the range of 10^{-6} to 10^{-21} M concentrations. Yet, this bioassay is used very infrequently as it may be employed by only the most precise workers endowed with the most patient and delicate hands; Syogoro Nishi and his associates spent entire, long, and exhausting days to combine, for the purpose of a single experiment, electrophysiology, ACh collection, measuring the transmitter, and fixing the ganglion for the necessary microanatomy evaluation. These values are in agreement with those obtained at the central synapses, and even with those obtained recently by means of the ultrasensitive chemiluminescent method for the measurement of ACh release (see below, next section).

There is a great difference between synapses with regard to quantal size. In the case of neuromuscular and electroplaque junctions, the quantum size is 2 to 3 nÅ; this size is due to the release of 6,000 to 10,000 ACh molecules. This constitutes a large quantal size, and the quanta in question are composed of about 10 subunits of about 1,000 molecules each. The ganglionic and probably central synapses exhibit smaller quanta; in the case of the former, the quantum size comprises only about 1,000 ACh molecules (Yves Dunant, personal communication; see also Nishi et al., 1967; Bennet, 1995).

Then Katz and Miledi performed the second phase of "miniaturization"; it concerned the synaptic "noise." Many investigators noticed the microvolt "noise" in their neuromyal preparations following ACh electrophoresis or in quiescent preparations and attributed it to instrumental imperfections. Only Katz and Miledi (1970, 1973) had the serendipity to perceive that the "noise" is a physiological phenomenon; they proposed that there is a spontaneous leakage of ACh into the synaptic cleft (amounting to a 10^{-8} M concentration and inducing a low-level—a few microvolts—depolarization) and that a few molecules of this ACh generate single-channel responses. Katz and Miledi (1977) opined that this phenomenon obtains not only pharmacologically via ACh electrophoresis but also naturally; they referred to it as an "elementary" event, that is, the current generated by opening a single nicotinic receptor (a direct measurement of individual receptor openings was achieved by Neher and Sakmann [1992] via the patch clamp technique). Again, the phenomenon in question obtains not only at the neuromyal junction but also at the central synapses (see also Masukawa and Albuquerque, 1978).

Still another miniaturization may be possible. Rene Couteaux identified at the frog myoneural junctions "active zones," which he described as thickenings of the junctional membranes and zones of concentrations of the synaptic vesicles; there may be from 100 to 300 active zones at the frog and mammalian myoneural junctions, and one quantum may react with each "active zone" (Couteaux and Pecot-Dechavassine, 1970; see also Kuno et al., 1971).

2. Nonclassical Notions on Release of Acetycholine

The notion of ACh leakage, proposed by Fatt and his associates, should be related to the unorthodox hypotheses concerning nonquantal release of ACh. There are three such hypotheses. The earliest one was posited in the 1970s as the "kiss and run" model of exocytosis by the late Bruno Ceccarelli; it was presented in a definitive form by Jacopo Meldolesi (1998). Meldolesi actually contributed earlier (Torri-Tarelli et al., 1985, 1990; Meldolesi and Ceccarelli, 1981) to the classical lore of quantal release of ACh by analyzing the role of fusion and transport proteins in the quantal phenomena (see above, section BI). More recently, Meldolesi (Valtorta et al., 1990, 2001) assumed a prudent position: he described Ceccarelli's concept as "fascinating" and proposed that the "kiss and run" process "operates in parallel with the classical . . . vesicle recycling." Meldolesi and Ceccarelli used the quick-frozen technique to demonstrate that the processes of plasmalemmal invagination, fusion, and vesicle recycling are not the rule for exocytosis: "many of the vesicles more intimately continuous with the plasmalemma . . . seem . . . to appear not as invaginations, open to extracellular space, but still as vesicles, sealed . . . by thin diaphragms . . . in direct continuity with the cell surface . . . [therefore,] there is incomplete fusion" (Meldolesi, 1998); this leads to effective and indeed rapid exocytosis and recycling, or to "kiss and run" (Stevens and Williams, 2000). Additional, supportive evidence indicates that exocytosis—perhaps a portion of the total process at any time—may be dissociated from recycling of membrane invaginations and vesicles (Henkel and Betz, 1995).

It should be noted that Meldolesi refers to "many," not all, vesicles as participating in the "kiss and run" process. Also, Henkel and Betz (1995) have only indirect evidence as to the amount of ACh that may be released independently of classical recycling (they consider the cycling of the membrane-ligated fluorescent dye FM1-43 as a monitor of exocytosis). Altogether, Meldolesi and his associates (Meldolesi, 1998; Valtorta et al., 2001) suggest that the "regulated" exocytosis with its complicated assembly of regulatory proteins and organelles is a phenotypic phenomenon, and that not all its components—"secretion competence factors"—may be expressed molecularly; therefore, he speculates that the two processes may run in parallel.

The second of these drastic hypotheses was posited in the 1970s and 1980s by Maurice Israel, Nicolas Morel, Bernard Lesbats, and Yves Dunant (see, for example, Israel et al., 1983; Israel, 2004; Israel and Dunant, 1998, 2004; Dunant, 2000).[2]

They isolated from cholinergic nerve terminals—but not from postsynaptic membranes—a protein, which they associated with Ca^{2+}-dependent release of ACh and which they called the mediatophore. Then they showed that among a number of cell lines, including glioma, fibroblast, and several neuroblastoma lines that were loaded with Ach, only those rich in the mediatophore could release the transmitter; furthermore, the release capacity could be given to cell lines incapable of release by transfecting them with plasmids encoded with mediatophore. Furthermore, antisense probes hybridizing the mediatophore messenger blocked the release of ACh from these preparations as well as from synaptosomes, "naturally" capable to release ACh.

All this could simply signify that the mediatophore belongs to the proteins, such as soluble N-ethylmaleimide attachment proteins (SNAPs), that are involved in docking of synaptic vesicles and in the exocytosis (see above, sections B-2 and B-3), and that it does not mediate any vesicle-independent release processes. In fact, mediatophore is a homo-oligomer of a 16-kDa subunit which is associated in a sector of the nerve terminal membrane which includes other proteins linked with release mechanisms such as vesicular ATPase (V-ATPase; Israel and Dunant, 1998); then the difference between mediatophore and other cholinergic release proteins such as SNAPs would be only that the mediatophore plays a more general role than the latter, as it subserves translocations and release mechanisms in noncholinergic cells such as glia.

But there is another angle to the story that supports the notion of the mediatophore being essential for ACh release which is independent of the vesicles, and, in fact, the mediatophore subserves the main process of ACh release. Indeed, Israel's team presents data that suggest that this is indeed so in the case of the cholinergic cells. In addition, other investigators (see Prior and Tian, 1995) showed that, following labeling of cholinergic nerve terminals with radioactive choline, isotopic composition of released ACh matches closely the cytoplasmic (free) rather than vesicular (bound) transmitter. Also, this release relates to the decrease in the concentration of cytosolic ACh (Dunant and Israel, 1985). The French-Montreal-Geneva team suggested also that during the release process a constant number of mediatophore molecules may be activated "close to a calcium channel" (Israel and Morel, 1990), this phenomenon resulting in numerically quantal release, although the release is not originating from synaptic vesicles; thus, Maurice Israel and his colleagues can have their cake and eat it too. Ultimately, Israel and his associates (see Israel and Dunant, 1998) suggested that mediatophore molecules represent "elementary pores that translocate ACh from . . . either . . . the cytosol, or synaptic vesicles." It may be added that for the measurement of ACh, Israel, Dunant, Morel, and their associates used the elegant, ultrarapid, and ultrasensitive choline oxidase chemiluminescent method; the technique is sensitive at picomolar concentrations of ACh (see section A).

As pointed out by Maurice Israel and Yves Dunant (Israel and Dunant, 2004; see also Peters et al., 2001), mediatophore has a more general role than that of transmitter release. It appears to be needed (with Ca^{2+} and SNAREs) for membrane fusion (such as the fusion between vesicular and postsynaptic membranes) and proton translocation in V-ATPase processes.

Laurent Descarries and Daniel Umbriaco presented the third and final "radical" hypothesis (or speculation) of nonsynaptic, diffuse release of ACh in 1995. Descarries, Mircea Steriade, and associates expanded the hypothesis subsequently (Descarries, 1998; Descarries and Mechawar, 2000). These investigators employed CAT immunostaining and advanced electronmicroscopy to study cholinergic, monoaminergic, and serotonergic nerve terminals and their axonal varicosities in several rat brain areas. They claimed that in rat cerebral cortex, nestriatum and hippocampus 70% to 80% [*sic*] of varicosities showed "no hint" of synaptic differentiation" or "junctional complex." Altogether, Descarries and associates (Descarries et al., 1997, 2004; Descarries, 1998) proposed that these nonsynaptic varicosities subserve an evoked or spontaneous (diffuse) release of ACh (and other monoamines) into the synaptic cleft. This ACh release is nonsynaptic and results in ambient, low ACh concentrations in the synaptic cleft (or "extracellular space"; Descarries, 1998).

Descarries and associates felt that several findings supported their notion. For example, muscarinic and nicotinic receptors are present at non- or extrasynaptic sites, whether at somatic locations, dendritic spines, or dendritic branches (Mrzlyak et al., 1993). Also, many investigators (e.g. Newton and Justice, 1993, and Israel and Morel, 1990) apparently demonstrated that low ACh levels in the nanomolar range continually exist in the synaptic cleft. Finally, Victor Gisiger, Laurent Descarries, and others provided evidence suggesting that some AChE forms (i.e., G4) may preserve (not eliminate) ambient ACh levels in the synaptic cleft (see Descarries et al., 1997 and Chapter 3 D, III). A related hypothesis—or shall we call it a speculation?—posited by Coggan and his associates (2005), is that, at least at the parasympathetic ganglia transmission occurs not only synaptically but also ectopically, i.e., outside opf synaptic specializations; this mode would involve alpha7 nicotinic postsynaptic receptors.

3. Conclusions

Intense discussions are held at various meetings among adherents to the various hypotheses of nonvesicular release of ACh, such as Maurice Israel and his associates on the one hand, and Victor Whittaker on the other. Sometimes, positions taken by the adversaries seem to soften. Thus, Maurice Israel proposed a mechanism via which their mediatophore process yields a quantal (although not vesicular) release of ACh. Victor Whittaker is more intransigent and retains his position. Actually, in the 1970s he suggested that the since the recycling vesicles (Vesicles Peak

2, or VP2 vesicles) obtained in his experiments with radioactive choline were newly created following the intense stimulation, they must necessarily exhibit the isotopic ACh composition that matches that of the cytosol; this would explain Israel's process as effective only under special conditions.

Altogether, it is not easy to relate the three unorthodox ideas on the mode of ACh release to the classical, quantal hypothesis of release. There is no denying that much evidence supports one or more of the novel ideas, such as the presence of ambient levels of ACh in the synaptic cleft, the absence of synaptic specializations at certain nerve terminal sites where ACh release does occur, and the presence of mediatophore at cholinergic nerve terminals and its capacity for ACh release. Yet the evidence for the vesicular release of ACh is very strong (see Victor Whittaker's comment on this matter in his Foreword to this book). Thus, the relation between synaptic vesicles, mEPPs, and EPPs is well established, for both the central and the peripheral nervous system. Also well demonstrated are the relationships among recycling of the vesicles, the proteins involved in this process, and the release of ACh. Even certain minutiae of the release and vesicular processes are demonstrably connected; thus, the fast vesicle recycling supports and accompanies intense stimulation (Sara et al., 2002). Perhaps then we should try to combine the various release models. Do they all operate under physiological conditions? Do they represent alternative release modes? Do they operate under special circumstances only? Do the pertinent mechanisms and entities—and particularly the mediatophore—belong to the markers of the cholinergic neuron, such as VAChT, SNAPs, and SNAREs?

The classical model alone is immensely complex. It contains multiple components and processes of synthesis of ACh: concentrative—to use Whittaker's term—vesicular uptake of acetylcholine and the activating proteins involved in this uptake and regulated by the cholinergic gene locus; vesicular docking and plasmalemma fusion processes, again including activating proteins; and actual, Ca^{2+}-mediated release of acetylcholine. Processes of vesicular cycling and recycling accompany these phenomena, as the vesicles move from the empty to the loaded form, constituting several pools of vesicles. In part at least, the so-called motor proteins that move along the cytoskeleton filaments propagate this movement; they contribute to muscle contraction and to vesicular movement (see also Whittaker, 1992, 1998; Sudhof, 2000; Howard, 2001). The conceptualization involved in formulating these processes and their components is flabbergasting; it still leaves unanswered questions. How do the various proteins link (we have at this time just barely a notion of the formation of protein complexes such as SNAREs that activate vesicular fusion)? What promotes and organizes temporally the cycling motion of the vesicles? What is the precise molecular and genetic control of the expression of these multiple processes? What determines the ratio between empty vesicles and the vesicles generated in the cholinergic neuron perikaryon? Which is the mechanism that moves the empty vesicles away from the terminal?

ACh release would become even more complex if, aside from the classical, the unorthodox processes also participated in the release. Only a most advanced computer program could describe this phenomenon (or could it?). I once suggested that only a very advanced computer program might be capable of describing another phenomenon, the causative transit from synaptic transmission via the multiple cholinergic and noncholinergic pathways to specific functional or behavioral events (Karczmar, 1993, 2004; see Chapter 9 BVI-BIV and V). Can a superprogram be devised to describe the combined phenomena?

It must be added that the cholinergic markers of the cholinergic neuron are not pertinent only for the transmittive function of this neuron, and this chapter's discussion does not necessarily imply that transmission constitutes the only role of the cholinergic neuron. Several nontransmittive roles were proposed, beginning in the late twentieth century. For example, it is well established that the cholinergic system and its components have a trophic role and a regenerative role, with regard to both the adult nervous system and development; when enacting this role, they are referred to as "morphogens." Then John Eccles and the McGeers proposed that the cholinergic system and its components exert a metabotropic role, as ACh (as well as certain other transmitters) triggers the neuronal membrane to precipitate second messenger–generated effects leading to metabolic changes in the neuron (McGeer et al., 1987b). These and related matters are discussed in detail in Chapters 8 BIV, 3 DIII, and 11.

DI. Central Cholinergic Pathways—An Introduction

The work of Kimura, the McGeers, Mesulam, Woolf, Kasa, Butcher,Wainer, Woolf, Wenk, and others expanded the vision of Shute and Lewis, Koelle and Gerebtzoff, and their associates and brought about something rarely heard of in the cholinergic field: an almost definitive statement concerning at least one area of that field, namely cholinergic pathways, even though there is a need to clean up discrepancies between the various investigators' pathway maps and to clarify the inconsistencies in nomenclature they used (see section A, above, and sections DI, DII, and DIII, below). The main consequence of this accomplishment is that, combined with lesion studies and pharmacological investigations of central functions and behaviors, this work established the cholinergic correlates of functions and behaviors, as well of certain disease states (see Chapters 9 BIV, BV, and BVI and 10 A; details concerning distribution of ChEs that are discussed in Chapter 3 DIII are also pertinent in the present context).

The cholinergic pathways were studied within the last 30 years by modern methods—which include CAT immunocytochemistry and histochemistry, various methods of measurements of AChE and of nicotinic and muscarinic receptors, and lesion and retrograde staining techniques (see section A, above, and Chapter 9 BIV–BVI)—in monkeys, apes, marmosets, raccoons, chickens, humans, rabbits, rodents, and cats (the rat brain was most particularly investigated). Generally, there is a remarkable similitude among the cholinergic neuronal groups and pathways of these species; thus, cholinergic neuronal clusters of the raccoon's forebrain are remarkably similar to those described for the rat and monkey by Mesulam and his team (Mesulam et al., 1983a, 1983b; Mesulam, 1976, 1990, 2003, 2004). Yet, species differences exist and will be pointed out.

It was already mentioned (see section A) that maps obtained in the past via Koelle's histochemical method for staining AChE are not as dependable for the identification of cholinergic neurons as the CAT staining method (for example, the Koelle method identifies both cholinergic and cholinoceptive cells; see section IIA). It was also mentioned that, nevertheless, maps provided decades ago by Shute and Lewis (1966, 1967a, 1967b) and Krnjevic and Silver (1966; see section IIA) essentially agree with the mapping obtained by means of the CAT stain provided more recently (Wilson, 1985; Kasa, 1986; Mesulam, 2000). For example, Pamela Wilson's (1985) maps of an AChE-containing tegmental pathway correspond closely to the McGeers' maps obtained by means of CAT immunocytochemistry visualizing the parabrachial system. Furthermore, some investigators who employed jointly in single studies methods for detection of both CAT and AChE obtained parallel results (Kasa and Silver, 1969; Kasa, 1971a, 1971b; Eckenstein and Sofroniew, 1983; Satoh et al., 1983).

The McGeers, Mesulam, Fibiger, Wenk, Butcher, Woolf, Wainer, and their associates described CAT immunocytochemistry–based maps of cholinergic pathways for many species. There is considerable overlap among these maps, but there are also many differences—in substance, in terminology, and in the mode of subdividing the cholinergic system, as will appear clearly in this section.[3] The CAT immunohistochemistry maps of Kimura and the McGeers were prepared first (see section A). These maps, together with amplifications and improvements provided by Wainer, Woolf, Butcher, and Fibiger, and comparisons with the maps prepared later by Mesulam and his associates, are presented first; the Mesulam maps are described subsequently.

Chronology is not the only reason for this two-punch mode of presentation. In view of the manifold differences among the various maps, it is well-nigh impossible to present a single pathway description that would smoothly amalgamate the work of all the pertinent teams; separate presentation of two main systems and their juxtaposition should present a clear picture of the cholinergic system as a whole.

DII. The Cholinergic Pathways Presented by the McGeers, Kimura, and Kasa and Expanded by Butcher, Woolf, and Wainer

Upon developing the immunohistochemical staining method for CAT, the McGeers and Kimura stressed that the cholinergic marking obtained by this method is reliable and leads to a clear-cut identification of cholinergic neurons, their pathways, and radiations (McGeer and

McGeer, 1989, 1993; McGeer et al., 1983, 1984a, 1984b, 1987a, 1987b; Kimura et al., 1980, 1981); this is particularly true when the CAT marking is combined with retrograde staining techniques and other means of cholinergic identification. On the basis of these methods, the McGeers identified a number of cholinergic pathways. They distinguished, for several species including humans, 5 major and several minor cholinergic pathways or systems (McGeer et al., 1984b, 1987a). Their major systems include the medial forebrain system, parabrachial complex, reticular formation and its giganto- and magnocellular fields, motor nuclei subserving the peripheral nerves, and striatal interneurons (Figures 2-5 and 2-6; see below).

The inconvenience here is that the nomenclatures used by the McGeers and the subsequent investigators in the description of the pathways differ.[4] It also complicates the matter that nomenclature for the pertinent nuclei—such as those of the brainstem—is frequently revised (see, for example, Reiner et al., 2004). The major systems recognized by Nancy Woolf, Larry Butcher, Hans Fibiger, and Bruce Wainer include the motor nuclei and the striatal interneurons, but they refer to McGeers' medial forebrain and its nucleus basalis of Meynert (NBM) as either magnocellular forebrain or basal forebrain, and to the McGeers' parabrachial complex and its pediculopontine and lateral tegmental nuclei as components of the brainstem and the spinal cord, namely pontomesencephalic tegmentum (Woolf, 1991; Butcher, 1995; Woolf and Butcher, 1986 and 1989; Butcher et al., 1993; Butcher and Woolf, 2003; Wainer et al., 1993; to include the tegmentum in the brainstem jointly with the spinal cord may not be felicitous). Also, Woolf and others distinguish, in addition to the McGeers' systems, the diencephalic complex, while the McGeers distinguish, outside of the Woolf-Butcher systems, the

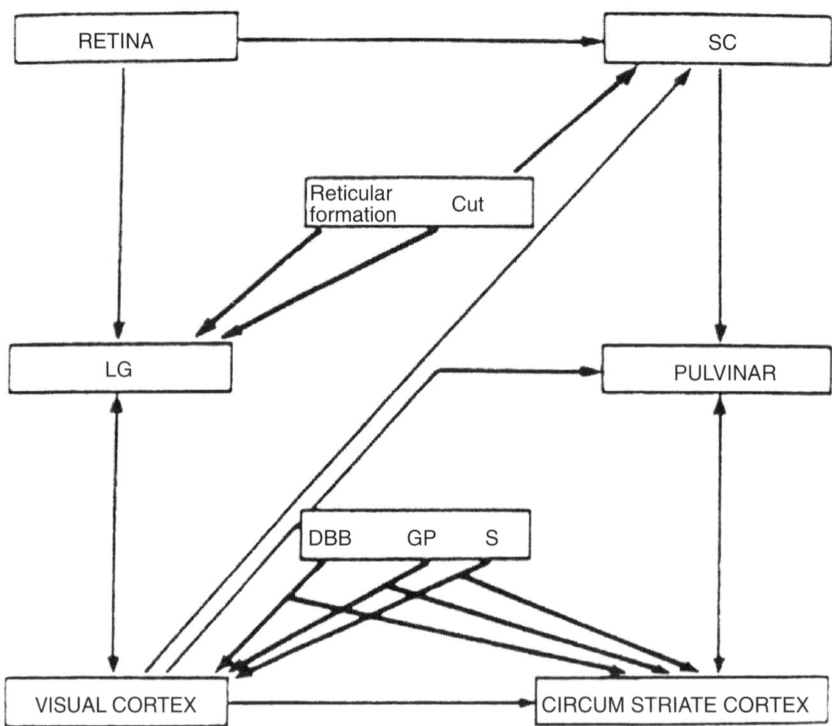

Figure 2-5. Diagram summarizing some of the connections between various parts of the visual pathways and visual cortical areas. The cholinergic routes are indicated by thick arrows. DBB, nucleus of the diagonal band (Broca); GP, golbus pallidus; LG, lateral geniculate nucles; S, stria terminalis; SC, superior colliculus. (From Kasa, 1986.)

reticular formation in their classification. Then Paul Kasa embraces still another approach (see Kasa, 1986 and Figure 2-5): he lists the major brain divisions and subdivisions (i.e., telencephalon, diencephalons, mesencephalon, etc., and their main subdivisions) and then describes the cholinergic projections to these locations. However, as described below, there are many similarities among the radiations described by the McGeers, Nancy Woolf, and Larry Butcher for the forebrain (or basal forebrain) and NBM.

1. The McGeers' Medial Forebrain System

The McGeers (see, for example, McGeer et al., 1987a) described the medial forebrain complex or system and its magnocellular component as consisting of a sheet of "giant" cholinergic cells; it "starts just anterior to the anterior commisure and extends in a caudolateral direction . . . to terminate . . . at . . . the caudal aspect of the lentiform nucleus." Divac described these projections in 1975, but not as cholinergic in nature. The system includes several nuclei; in a rostral to caudal order, they are: medial septum nucleus, nuclei of the vertical and horizontal limbs of the band of Broca, and the NBM. Woolf (1991) also includes substantia innominata and nucleus ansae lenticularis in this series. The system is essentially afferent in nature, and its descending radiation to the brainstem is mostly not cholinergic (Semba et al., 1989).

Nucleus basalis of Meynert is of primary importance as it projects to the frontal, occipital, parietal, and temporal cortical lobes, or the entire cortical mantle, as well as to the amygdala, the thalamus, and its reticular nuclei; the presence of the NBM radiations was stressed not only by the McGeers but also by Mesulam, Kasa, and the Woolf-Butcher team (Mesulam and Geula, 1988; Mesulam et al., 1983a, 1983b; Kasa, 1986; Wainer et al., 1993; see also Wenk, 1997; Woolf, 1991). The NBM is one of the earliest recognized brain structures, as it was identified (as *ganglion basale*) in 1872 by Meynert; it was described in detail by the Swiss biologist and neurologist Rudolf A. von Kolliker and named by him after Meynert. While it is well formed and extensive in several mammals including the dolphin and the human (Mesulam and Geula, 1988), it is frequently not listed specifically in neuroanatomy texts, except as a part of the ventral basic ganglia or substantia innominata (e.g., Brodal, 1981). It is absent in the rabbit, muskrat, and several other mammalian species; Gorry (1964) did not describe this site as cholinergic; he defined it morphologically as poorly organized and sparsely distributed within substantia innominata in the cat, dog, and rat. This notion was confirmed for the rat by investigators employing the CAT stain (see Wenk, 1997; Mesulam et al., 1983a; Mesulam and Geula, 1988); accordingly, Wenk (1997) referred to this structure in the rat as "homologous" to that of the nucleus basalis of, say, human; actually it is referred to frequently in the case of the rat as nucleus basalis magnocellularis (NBMC); other complications of the matter are discussed in section D2, below. Finally, in the zebra finch, the nucleus basalis has as its analog the ventral paleostriatum, which radiates in the bird to the forebrain and its song control nuclei; presumably, other bird species exhibit similar cholinergic systems (Li and Sakaguchi, 1997).

The cholinergic cells of the medial septum and vertical and horizontal limb of the forebrain's diagonal band are intensely projected to the limbic system and several cortical areas, including cingulate, pyriform, and entorhinal cortices. The McGeers, the Butcher-Woolf-Fibiger team, and other investigators (e.g., Wainer et al., 1993) emphasized the morphologic preponderance and behavioral significance of the diagonal band and projections from other regions of the forebrain to the limbic system, including the septum, amygdale, and hippocampus. The McGeers and others also described cholinergic diagonal band neuron projections to the olfactory bulb, the nucleus interpeduncularis, and the entorhinal and perirhinal cortex (Kasa, 1986; Dohanich and McEwen, 1986; Woolf and Butcher, 1989; Woolf, 1991; McGeer et al., 1987a, 1987b). The horizontal band extends, according to the McGeers (1987a, 1987b; see also Dohanich and McEwen, 1986), to the parietal, occipital, temporal, and frontal lobes, habenula, and amygdala; according to Woolf (1991), the horizontal band and magnocellular preoptic area extend efferently to, besides the nuclei and areas referred to by the McGeers (1987a), the olfactory bulb and pyriform nuclei, hypothalamus, tegmentum (peduculopontine and pediculopontine tegmental areas), raphe, and locus ceruleus. Some of this radiation is frequently referred to as septohippocampal pathway (see Woolf, 1991, 1997, 1998; Butcher et al., 1993). Also, Woolf and Butcher used Mesulam's nomenclature when they referred to the neurons of the medial septum and the vertical limb of the diagonal band (which also contribute to the septohippocampal radiation) as Ch1 and Ch2 neurons (see the next section). While the preponderance—or at least a good proportion—of the cholinergic cells of the septum and the diagonal band, as well as of the cholinergic cells of NMBC, send ascending projections to the cortex and the limbic system, only a small proportion of the cells of these systems that send descending projection to the brainstem are cholinergic (Semba et al., 1989).

As referred to above, the McGeers and the Woolf-Butcher team have described cholinergic forebrain radiations to the olfactory bulb and to the entorhinal cortex (see also Halasz and Shepherd, 1983). While such radiations imply the existence of a cholinergic olfactory system, the two teams do not explicitly propose the presence of such a system; it is, however, proposed by Kasa (1986). Kasa stresses that CAT and AChE activity is present in several layers of the olfactory bulb; CAT levels are particularly high in the glomerular and internal plexiform layers. Besides the forebrain, this afferent activity originates in the olfactory tubercle and, perhaps, in the islands of Calleja (Shute and Lewis, 1967a, 1967b). On the other hand, it is not clear from the evidence presented by the McGeers, the Woolf-Butcher team, and Kasa whether or not cholinergic neurons contribute to the efferent outflow from the olfactory tubercle to the olfactory cortex (including enthorhinal and pyriform cortices); a nucleus basalis–olfactory pathway is, however, defined by Mesulam and his team (see Selden et al., 1998, and below, section D2).

It was already mentioned that the McGeers and their associates and the Woolf-Butcher team use the terms "medial forebrain complex" and "basal forebrain" (the "magnocellular forebrain"), respectively, for systems that are relatively similar but differ somewhat with respect to the nuclei or groups of origin and their radiations' targets. The nuclei and the radiations of the Woolf-Butcher system are somewhat richer than those proposed by the McGeers, and in her 1991 review Nancy Woolf describes in detail both efferents and afferents of the 7 nuclei that she recognizes in the case of her "basal forebrain" (see also Butcher and Woolf, 2003). Similarly, the Woolf-Butcher description of the functional significance of their system ranges more widely than the description offered by the McGeers for their complex. However, the McGeers, just as other investigators, assign major importance for loss of cognition and Alzheimer's disease to the loss of neurons and of the cholinergic systems in the "medial basal forebrain" and the nucleus basalis (see, e.g., McGeer et al., 1987a, 1987b). Indeed, Nancy Woolf (1997, 1998) assigns a major role for cognition and consciousness and self-awareness (see Chapter 9 BVI) to the basal forebrain and the reticular formation and its radiation to the cerebral cortex modules.

2. The McGeers' Parabrachial Complex

The McGeers and associates recognized the parabrachial complex as the second major system and the "most intense and concentrated cholinergic cell group in the brain stem." It surrounds conjunctivum commencing in the most rostral aspects of the pons and radiates along the brachium conjunctivum and superior cerebellar peduncles in the caudodorsal direction. The pathway contains several nuclei such as the pedunculopontine tegmental nucleus, medial and lateral parabrachial nuclei, and Kolliker-Fuse nuclei (occasionally there are disagreements between the McGeers and Mesulam's team [Mesulam et al., 1984] regarding the presence of CAT in some of the nuclei of the McGeers' parabrachial complex; see below, next section). According to the McGeers and other 1980s investigators, including Bruce Wainer and his associates (Saper and Loewy, 1982; Lee et al., 1988), the pedunculotegmental-parabrachial system ascends to all thalamic nuclei (including anterior nuclear, reticular nuclear, and posterior nuclear areas), the substantia nigra, and the cortex; Saper and Loewy (1982) also describe projections of this system to the hypothalamus and amygdala (see also Kasa, 1986). In 1987 (McGeer et al., 1987a), the McGeers opined, "this region is a major supplier of afferents to all parts of the diencephalon . . . including hypothalamus . . . and the limbic system and possibly to the cortex." It should be noted that the McGeers' term "parabrachial system" corresponds to an extent—but not entirely—to Nancy Woolf's term "pontomesencephalic formation"; this matter is discussed further below. Subsequently, Edith McGeer (Semba et al., 1989) stressed the importance of brainstem (pedunculopontine tegmental and dorsal raphe nuclei) in projecting to the cortex and receiving afferents from the magnocellular regions of the basal forebrain. Nancy Woolf and her associates (see, e.g., Woolf, 1991) add tectum and basal forebrain as destinations of the tegmental nuclei described by the McGeers, and this is an important addition to the McGeers' complex.

Nancy Woolf and Larry Butcher (see, e.g., Woolf, 1991; Woolf and Butcher, 1986; Butcher and Woolf, 2003; and Woolf et al., 1986) depart from the McGeers in not employing the term "parabrachial complex"; they instead localize a somewhat similar complex in the brainstem and refer to it as "pontomesencephalic tegmentum"; they emphasize the presence within this site of laterodorsal tegmental and pedunculopontine nuclei. Palkovits and Jacobowitz (1974) and Hoover and Jacobowitz (1979) include these neurons within the cuneiform nucleus. Butcher, Woolf, and others (Woolf and Butcher, 1986, 1989; Woolf, 1991) describe its efferent (ascending) radiations, which include, in addition to those enumerated by the McGeers, subthalamus, habenula, striatal areas and basal ganglia, reticular formation, medium septum, lateral geniculate, and stria terminalis. Woolf and other investigators opine that the descending projections of the pontomesencephalic tegmentum also reach the cranial nuclei, including the trigeminal, facial, and hypoglossal motor nuclei, vestibular nuclei, raphe, locus ceruleus, and pontine and medullary and pontine reticular nuclei (Jones and Yang, 1985; for further references, see Kasa, 1986; Woolf, 1991; Butcher and Woolf, 1986; 1989). Nancy Woolf (1991, 1996, 1997, 1998) emphasizes the role of the radiation from the pontomesencephalic tegmental system (the McGeers' parabrachial complex) to the basal forebrain, relaying thence to the cerebral cortex for processes of memory and consciousness; it is interesting that she states that this notion "replaces the older notion of a nonstop pathway originating in the cholinergic reticular formation" or system (Shute and Lewis, 1967a, 1967b). This matter is addressed again in the next section.

3. The McGeers' Reticular System

The McGeers' third major system, the reticular system, is "a scattered collection of very large cells," the giganto- and magnocellular neurons (McGeer et al., 1987). These cells aggregate medially with respect to the raphe and ventrally with respect to the inferior olive; longitudinally, they extend from the rostral pons to the caudal medulla. They contain nuclei reticularis pontis oralis and caudalis, reticularis tegementis pontis, reticularis gigantocelluralis and reticularis lateralis, as well as the formatio reticularis centralis (or medularis) and cuneiform nucleus. Particularly the hypoglossal and gigantocelluralis nuclei contain high levels of CAT and of ACh located in the soma and nerve terminals (Kimura et al., 1981), perhaps indicating the presence of cholinergic-cholinergic relays (see Woolf and Butcher, 1986). The system's neurons radiate to the thalamus and other rostral nuclei, cerebellum, superior colliculus, and spinal cord (see Kasa, 1986). Again, there is a substantial difference between this system as described by the McGeers and the same system described by others. While Woolf (1991) refers to the medicular reticular nuclei and the medullary tegmentum as the source of the medullary reticular radiation, she ascribes only afferent projections to this source and defines them as coursing to the cerebellar cortex; she adds as the medullary origin of this reticular outflow the prepositus hypoglossal nucleus. Mesulam and Jacobowitz include nucleus cuneiformis with the pedunculopontine nuclei, and they refer to these nuclei as the pontomesencephalic reticular formation, or the Sector Ch5 cluster (see next section), which the McGeers list under their parabrachial system (Mesulam et al., 1983a, 1983b; Mesulam, 1990; Palkovits and Jacobowitz, 1974; Jacobowitz and Palkovits, 1974).

Apparently, the McGeers' account of the reticular system and pontine tegmentum does not jibe with the original description of the reticular system and its identification with the ARAS by Shute and Lewis, Himwich and Rinaldi, and Krnjevic (see section A, above). But Woolf, Butcher, and Wainer emphasize the cortical radiations of the components of their parabrachial complex (tegmental nuclei) and assign to them a role in arousal, memory, and the sleep-wakefulness cycle; that is, they synonymize this system with the classical ARAS (Woolf and Butcher, 1986, 1989; Woolf, 1991).

4. The McGeers' Striatal Interneurons

Striatal cholinergic interneurons constitute the fourth system of the McGeers; they identified it very early (McGeer et al., 1971). These neurons are present in the caudate, putamen, and accumbens; following their cholinergic identification by McGeer et al. (1971), they were recognized as such ("intrinsically-organized local circuits") by Woolf (1991), Kasa (1986), and Mesulam (1984). But, efferents also radiate from the striate and the basal ganglia, including substantia innominata and globus pallidum, to superior colliculus, cortex, and, possibly, substantia nigra (for references, see Kasa, 1986). The basal ganglia constitute a problem. As the rest of the striate, some components (e.g., substantia nigra) may or may not contain cholinergic interneurons (see Kasa, 1986). However, the literature agrees on the presence of CAT and AChE in the substantia nigra, but it is not clear whether the CAT stain represents afferents to this site, such as the postulated striatonigral path, or innervation radiating from peduculopontine tegmentum and nucleus accumbens (Kasa, 1986; McGeer et al., 1987a).

5. The McGeers' Motor Nuclei

The motor nuclei of the peripheral nerves are the McGeers' fifth system. These include cranial nerve nuclei, which are the source of the efferent nerves to the autonomic system, motor nuclei innervating the facial muscle, and motor counterparts in the anterior and lateral horns of the spinal cord. This is in agreement with the work of Kasa (1986), Woolf (1991), and others; similar to the McGeers, Woolf, Butcher, and their associates distinguish the cholinergic motor neurons as a separate component, which is a part of their brainstem and spinal cord system. The spinal cord should be mentioned in this context; it was not reviewed by the McGeers or the Butcher-Woolf team. AChE and CAT stain is obtained in the motor neurons, as referred to above, as well as in other parts of the ventral horn (see also Kasa, 1986; Dun et al., 2001); according to Kasa (1986) and Barber et al. (1984), cholinergic neurons may be also present in the dorsal and intermediate spinal cord, and the muscarinic nature of the dorsal horn receptors (of the M2, M3, and M4 type) was demonstrated by Zhang et al. (2005; see, however, Dun et al., 2001). The cholinergic innervation of these sites may originate in the nucleus ruber and from the reticular system of the McGeers, particularly the hypoglossal and magnocellular nuclei. Other descending cholinergic pathways originating in the hypothalamus and the brainstem (parabrachial system of the McGeers) may contribute to the spinal innervation (Loewy, 1990).

6. The Other (Sometimes Minor) Systems

The McGeers and others also recognize several additional cholinergic systems. For example, cholinergic interneurons are present in the cortex, striate, hypothalamus, interpeduncular nuclei, and hippocampus (Tago et al., 1987); however, the McGeers did not describe any cholinergic radiations emanating from the cholinergic diencephalic (hypothalamic) neurons. The intrinsic nature of some of the cortical and striatal cholinergic neurons was also evidenced in the studies of Butcher and Woolf (e.g. Butcher et al., 1975; Butcher and Woolf, 2004), Bolam and Wainer (e.g. Bolam et al., 1984), and Kasa (1971 and 1986); a variety of techniques including CAT and AChE staining were employed. Interpeduncular nucleus may contain the highest content of CAT, even higher that that of putamen and caudate (Kasa, 1986); this may be mostly derived from habenular and other cholinergic afferents abutting on the nucleus interpeduncularis. There also may be intrinsic cholinergic cells in the nucleus ruber, as well as cholinergic efferents descending spinally from this nucleus (rubrospinal cholinergic tract (see Kasa, 1986; Woolf, 1991). The axons and the dendrites of the intrinsic cortical cholinergic interneurons "elaborately intertwine and interlap" (Kasa, 1986), which is in agreement with the notions of Woolf and her functional interpretations of the role of the cholinergic system in cognition (1991, 1997; see Chapter 9 BV and BVI).

The cerebellum is another, probably minor locus of the cholinergic system that is rarely mentioned in the Woolf-Butcher or McGeers studies (see, however, Kasa, 1986). But CAT and AChE are present in the Golgi cells, intracerebral nuclei, and granular and molecular layers; they are also present in the mossy fibers and in the terminals of mossy and parallel fibers, as well as other cerebellar terminals (for literature, see Kasa, 1986). The source of this CAT and AChE is not known. Shute and Lewis (1967a, 1967b) opined on the basis of their AChE stain that cholinergic afferents reach the cerebellum via the 3 cerebellar peduncles. Furthermore, cholinergic receptors, notably of several nicotinic types, are present in all lobules of the cerebellar cortex, although their density may be small (De Filippi et al., 2005). Cholinergic origins of the afferents are not clear, although the red nucleus and the spinal afferent pathways may serve as such. Furthermore, cerebellar cortex neurons seem to exhibit cholinoceptivity (for references, see Kasa, 1986). Altogether, the cerebellum seems to have a cholinergic system that is ready to be defined.

Cholinergic cells are also found in the retina (amacrine cells) and other components of the visual system (Domino, 1973; Domino, et al., 1973). CAT activity is high in amacrine cells and the geniculate, lateral reticular nucleus, superior olivary complex, magnocellular nucleus, vestibular nuclei, inner ear spiral ganglion, and several auditory system components (see McGeer et al., 1987; Kasa, 1986a, 1986b; Woolf, 1991.)

Cholinergic contributions to visual and auditory systems require additional comments. Several components of the visual system (i.e., retina, optic nerve, lateral geniculate, superior colliculus, and visual cortex) contain CAT and AChE (Rasmusson, 1993; see also Chapter 9 BV and Kasa, 1986). Though there is ample evidence that retinal neurons send cholinergic projections to the geniculate, there seems to be no direct radiation from the geniculate or colliculus to the visual cortex (see Figure 19 in Woolf, 1991). Furthermore, forebrain areas, including the substantia innominata, corpus pallidus, and diagonal band, innervate several laminae of the visual cortex and also exhibit high AChE concentrations (Bear et al., 1985; see Kasa, 1986, and Woolf, 1991, for further references). The geniculate and superior colliculus are also innervated by the dorsal tegmentum and the reticular formation or the parabigeminal nucleus (see above, and Woolf, 1991; Kasa, 1986; Mesulam, 1990, McGeer et al., 1987a, 1987b). However, constructing a full diagram of the cholinergic visual system, though partial and speculative constructions have been offered, is impossible (Figure 2-5).

The auditory cholinergic system is infrequently described, yet cholinergic contribution to this system is undeniable (see Kasa, 1986). The denervation method and using CAT and AChE markers demonstrate a cholinergic presence in the superior olive, organ of Corti hair cells (or their base), spiral ganglion, and cochlea. Altogether, the generally accepted auditory pathway is the olivocochlear bundle (see also Chapter 9 BV). However, CAT and/or AChE are present in several auditory system components (i.e., the geniculate, inferior colliculus, and nucleus cuneiformis), and Kasa (1986) diagrammatized a cholinergic auditory pathway that unites the auditory apparatus with the auditory cortex via the nuclei in question.

Similar to the McGeers' "major" systems or complexes, the "minor" systems exhibit afferent terminals originating in many brain parts. For example, the interpeduncular nucleus is probably innervated by the basal forebrain, parabrachial (pontomesencephalic), and septum. All these findings agree with the demonstration of ACh release and the presence of cholinergic (muscarinic or nicotinic) receptors in these loci (for references, see Kasa, 1986; McGeer et al., 1987a, 1987b; Woolf, 1991.)

7. Comparing and Commenting on the McGeer and Woolf-Butcher Pathway Maps

The McGeer and the Woolf-Butcher pathway maps may be compared in several ways (see Figure 2-6). Butcher, Woolf, and their associates and the McGeers use the same terms and replicate to a great extent the McGeers' descriptions of certain pathways described by the McGeers. This is the case with the motor nuclei and the striatum, except that Butcher and Woolf (see Woolf, 1991; Butcher, 1995) add to the McGeers' striatum the islands of Calleja (of the olfactory tubercle) and the olfactory tubercle (anterior perforated substance). The cholinergic neurons of the islands may supply the olfactory band (Kasa, 1986). Also, Butcher and Woolf classify their striatum, somewhat in the developmental sense, under telencephalon. Then they employ different nomenclature but define similar—but not identical—morphology as they describe the pontomesencephalic tegmentum of the brainstem and the spinal cord in lieu of the McGeers' parabrachial complex, and they refer to the basal forebrain rather than to the medial forebrain systems of the McGeers.

Finally, Butcher and Woolf adduce additional systems to those of the McGeers, such as the diencephalon complex with its hypothalamic nuclei and medial habenula (similar to Kasa, 1986). According to Woolf, the hypothalamic nuclei radiate to the cortex, while according to Kasa (1986), the habenula, a thalamic component, may radiate to the interpeduncular nucleus. Actually, several investigators assert that cholinergic cells are present in all parts of the hypothalamic nuclei (for references, see Woolf, 1991). These cells are frequently interneurons and form intrinsically organized local circuits. They form also the tuberoinfundibular pathway involved in the anterior pituitary hormone release; this classical pathway was discovered on pharmacological grounds by Mary Pickford and defined as cholinergic by George Koelle on the basis of his histochemical AChE stain (see Chapter 9 BIV-2); this notion was confirmed on the basis of CAT immunocytochemistry by Pat McGeer and his team (Tago et al., 1987).

Several general comments are appropriate. Comparing the pertinent figures and tables presented by the McGeers (McGeer et al., 1987a, 1987b; Semba et al., 1989), on the one hand, and Woolf with Butcher and Wainer, on the other (Woolf and Butcher, 1986, 1989; Wainer et al., 1993; see Figures 2-6 and 2-7), it is apparent that the cholinergic radiations depicted by Woolf and their associates are more ubiquitous and farreaching than those described by the McGeers; on the other hand, they correspond rather closely to those included in Mesulam's system (see next section).

Then there is the matter of targets and their afferents and efferents. Described by the McGeers or Woolf and Butcher, the afferent pathways supply cholinergic radiation to, and release ACh at, cholinergic or noncholinergic but cholinoceptive targets. Such pathways constitute cholinergic-noncholinergic or cholinergic-cholinergic relays similar to those existing in the autonomic sympathetic and parasympathetic system (see also Chapter 9 BI; see Figure 9-8). Moreover, the neuronal groups described by either the McGeers or the Woolf-Butcher team as sending afferents to cholinergic or noncholinergic targets generally also receive afferents from these targets. Altogether, the McGeers (Semba et al., 1989) and the Woolf-Butcher team (see Woolf, 1991; see also Kasa, 1986) describe mutual, multiple, and ubiquitous connections among all the systems or pathways that they describe, including the cholinergic hypothalamic and habenular (thalamic) neurons, pontomesencephalic complex (McGeers' parabrachial system), forebrain (or basal forebrain, including diagonal band nucleus), reticular system, and cortex. It is somewhat misleading that these and other investigators also mention efferent pathways. This is the case of the "efferents" of the motoneurons and autonomic preganglionic neurons, which radiate to the skeletal muscle and the ganglia, respectively, while the McGeers state that the brainstem parabrachial system (pedunculopontine tegmental nuclei) receives afferents from the magnocellular regions of their basal forebrain complex, which projects efferents to the parabrachial system (Semba et al., 1989); in either case, these "efferents" are, at the same time, afferents of the neurons in question.

Nancy Woolf (1991, 1997, 1998) adduces important generalizations to her and Larry Butcher's work. She emphasizes that the cholinergic system exhibits rich "interdigitation, interconnection" via the cholinergic dendrites and axon collaterals "and re-entrant circuits." Indeed, in her 1991 review she describes in detail not only radiations of her cholinergic systems of origins (including the basal brain and the brainstem) to ubiquitous target nuclei and cell groups, but also the afferents from noncholinergic and cholinergic complexes to these groups of origin. These hallmarks—interdigitations, reentrant circuits, and interconnections—of the cholinergic system contribute to its global character, and Nancy Woolf contrasts this global arrangement of the cholinergic system with the modular arrangement of the sensory and cortical circuits. She also attaches special importance to the innervation by individual cholinergic forebrain cells of the functional cortical units termed "macrocolumns" as she relates this feature to the phenomena of memory, learning, and consciousness (see Chapter 9 BV

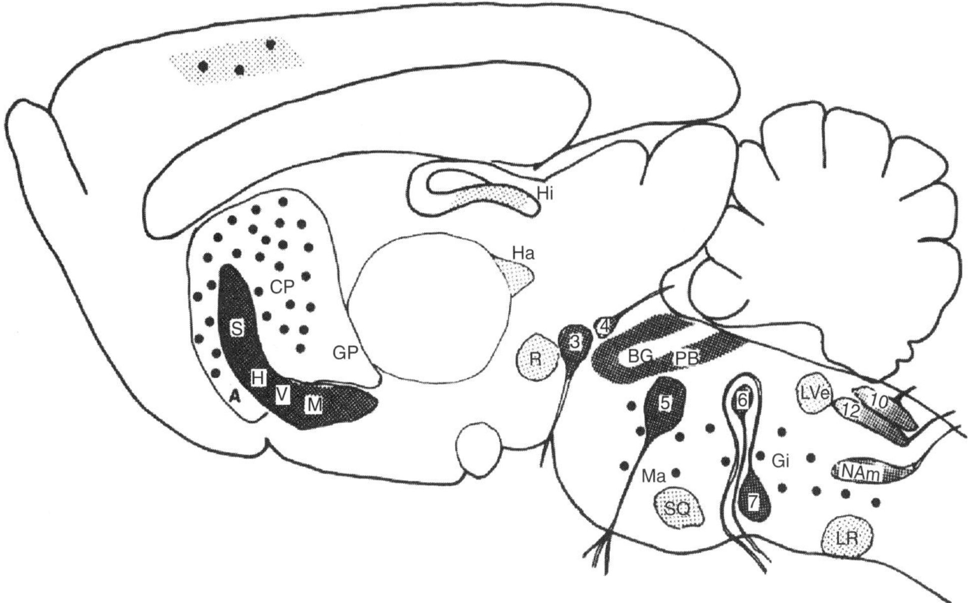

Figure 2-6. Sagittal view of rat brain illustrating ChAT-containing neurons. Major systems are indicated by black dots or heavy stippling, minor ones by light stippling. A, nucleus accumbens; Am, amygdala; BC, brachium conjectivum; CP, caudate-putamen; Gi, gigantocellular division of the reticular formation; GP, globus pallidus; H, Horizontal limb of diagonal band; Ha, habenula; Hi, hippocampus; IC, inferior colliculus; IP, interpenduncular nucleus; LR, lateral reticular nucleus; LVe, lateral vestibular nucleus; M, nucleus basalis of Maynert; Ma, magnocellular division of the reticular formation; NAm, nucleus ambiguous; PB, parabrachial complex; R, red nucleus; S, medial septum; SN, substantia nigra; SO, superior olive; V, vertical limb of diagonal band. (From McGeer et al., 1985. Reprinted by permission from Kluwer Academic/Plenum Publishers.)

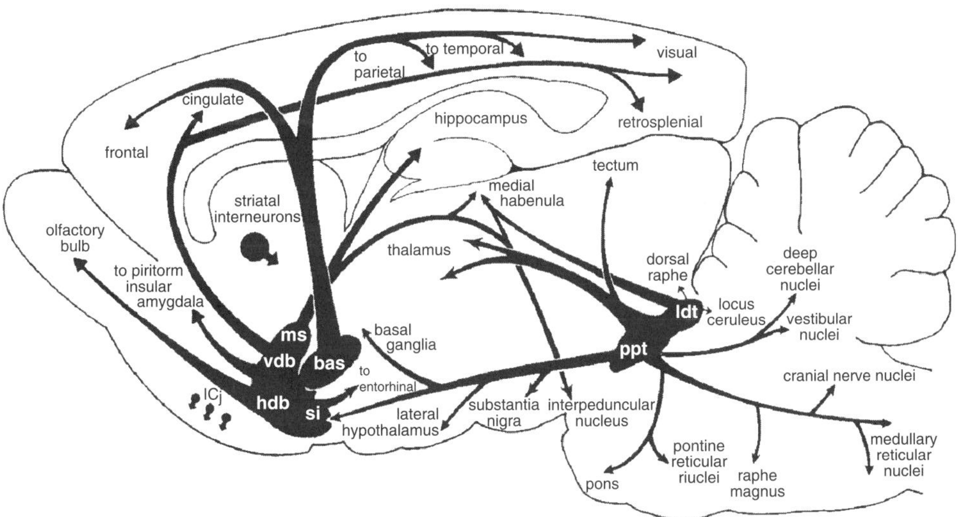

Figure 2-7. Cholinergic neurons in the basal forebrain and pontomesencephalon have widespread projections. Cholinergic neurons in the basal forebrain, including those in the medial septal nucleus (ms), vertical diagonal band nucleus (vdb), horizontal diagonal band nucleus (hdb), substantia innominata (si), and nucleus basalis (bas), project to the entire cerebral cortex, hippocampus, and amygdala. Cholinergic neurons in the pontomesencephalon include those in the pedunculopontine nucleus (ppt) and laterodorsal tegmental nucleus (ldt) and have ascending projections to the basal forebrain and thalamus. (From Woolf, 1997.)

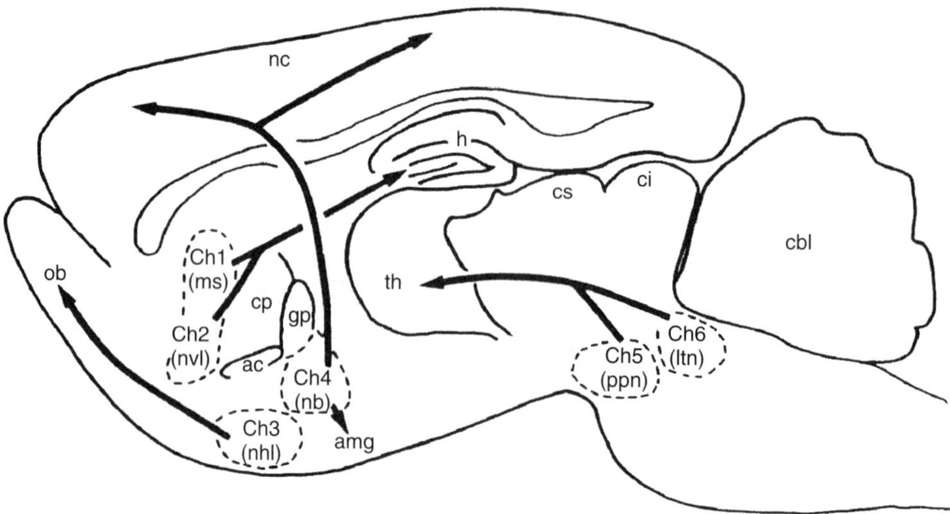

Figure 2-8. Schematic representation of some ascending cholinergic pathways. The traditional nuclear groups, which most closely correspond to the Ch subdivisions, are indicated in parentheses. However, the correspondence is not absolute. ac, anterior commissure; amg, amygdala; cbl, cerebellum; ci, inferior colliculus; cp, caudate-putamen complex; cs, superior colliculus; gp, globus pallidus; h, hippocampus; ltn, lateral dorsal tegmental nucleus; ms, medial septum; nb, nucleus basalis; nc, neocortex; nhl, horizontal limb nucleus; nvl, vertical limb nucleus; ob, olfactory bulb; ppn, pedunculopontine nucleus; th, thalamus. (From Mesulam et al., 1983a.)

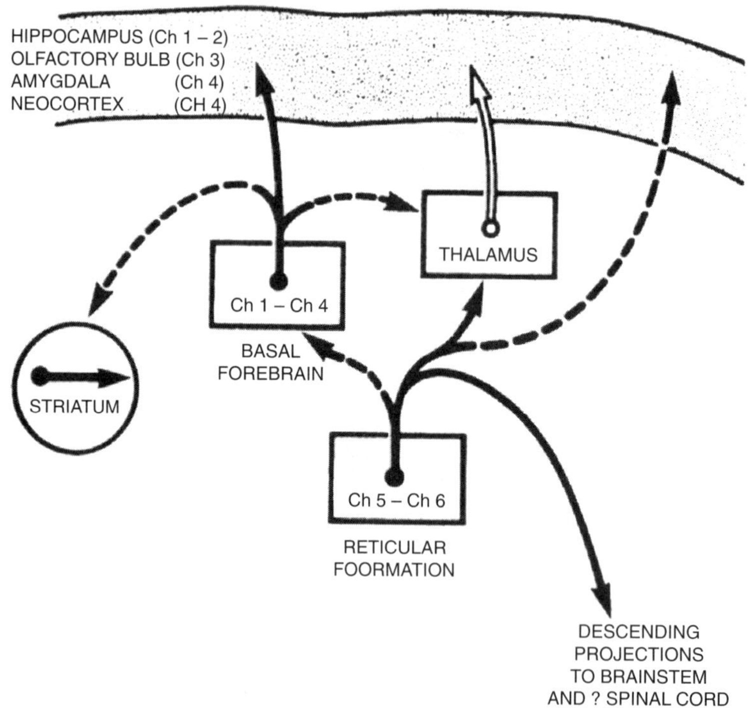

Figure 2-9. Diagrammatic representation of some cholinergic pathways. The solid arrows indicate major pathways and the broken arrows minor pathways. The open circle and arrow indicate that the thalamocortical pathway is noncholinergic. (From Mesulam, 1990. Reprinted by courtesy of Marsel Mesulam.)

and BVI and Figure 9-35 and 9-36). Also, the cholinergic neurons receive complete sets of somatosensory and proprioceptive information, and this juxtaposition prevents the neurons "ordinarily from being intensely driven by any individual input." Interestingly and perhaps inconsistently, Woolf also opines that "the bombardment of multiple weak inputs and the re-entrant circulation of activation . . . makes the cholinergic systems unstable . . . and . . . chaotic," but she feels also that this characteristic makes the system "exquisitely sensitive . . . to variations in initial input."

DIII. Cholinergic Pathways as Defined by Marsel Mesulam, His Associates, and Other Recent Investigators

Marsel Mesulam became involved in cholinergic pathways early, as in 1976 he had already embarked on the marking of central cholinergic afferents by the retrograde horseradish peroxidase method that he developed, combined with AChE cytochemistry. Subsequently, he begun to employ CAT immunocytochemistry as well, and also the classical lesion method, to accentuate somatic CAT visualization; moreover, he used DFP pretreatment to cause a hyperexpression of AChE in the perikarya (Butcher and Bilezikjian, 1975). Altogether, for more than 25 years he has been engaged in defining maps of the cholinergic system in rodents and primates, including humans.

Mesulam contributed new findings and much detail to the definition of cholinergic pathways; he also provided the cholinergikers with a useful subdivision of the cholinergic system into 8 "major sectors" (or "constellations," as sometimes referred to by Mesulam), which he labeled Ch1 to Ch8 (Mesulam et al., 1983a, 1983b; Mesulam, 1990). Mesulam claims that "the nomenclature for the nuclei that contain . . . cholinergic cells has engendered considerable inconsistency and confusion" (a sentiment expressed here by this author as well), and that his terminology eliminates the inconsistency (Mesulam et al., 1983a). The additional advantage of Mesulam's categorizing is that, as the sectors refer to circumscribed sites rather than brain parts (e.g., medial septum versus the McGeers' forebrain) their projections are also relatively limited and easy to define. Actually, only a few investigators appear to use Mesulam's terminology, although many refer to the nuclei listed within Mesulam's sectors without using his nomenclature (e.g., Jones and Cuello, 1989).

It is important that the Ch1 to Ch8 sectors are present as relatively homologous entities in rodents, cats, primates, and humans (Mesulam et al., 1983a, 1983b, 1984; Mesulam, 1990; Selden et al., 1998; Mesulam, 2003, 2004). There are, of course, species differences with respect to Mesulam's maps; some investigators opine that Mesulam's system applies more to primates than to rodents (for example, raccoon; Bruckner et al., 1992; see also Butcher and Semba, 1989), and Mesulam himself states that there are "potential difficulties for the Ch nomenclature," particularly for the Ch1 to Ch4 sectors, and he refers to species differences between the sectors. However, according to many investigators, the discrepancies appear small (see Vincent and Reiner, 1987, for references).

It is notable that the neurons of the various sectors differ in size, shape, or chromicity (coloring). It is also important that the sectors are, without exception, heterogeneous; that is, both cholinergic and noncholinergic neurons are present in the sectors; the ratio of cholinergic versus noncholinergic neurons differs among the sectors. Finally, the perikarya of the Ch1 to Ch8 neurons as well as the nerve terminals of cholinergic or noncholinergic radiations to these neurons contain AChE.

Four important sectors, Ch1 to Ch4, are present in the basal forebrain (Mesulam et al., 1983a, 1983b). Ch1 neurons are contained in the medial septal nucleus, particularly along the midline raphe and the outer edge of the septum. Mesulam et al. (1983a, 1983b) state that about 80% of Ch1 neurons are noncholinergic; the opinion of Mesulam and his team that the cholinergic neurons constitute a small minority of the Ch1's neurons is not shared by other teams (see, for example, Senut et al., 1989). Ch1's main outflow is the hippocampus, and this description agrees with that of the McGeers, Kasa, the Butcher-Woolf team, and Senut et al. (1989).

The boundaries between the Ch1 and Ch2 sectors are not well defined and, in Mesulam's diagrams, the Ch1 and Ch2 sectors form a bean-shaped continuum (see Table 2-1 and Figures 2-8 and 2-9). The Ch2 sector consists of neurons located within the vertical limb of the diagonal band; 70% or more of Ch2's neurons are cholinergic. The cholinergic and noncholinergic neurons of the diagonal band form distinct clusters. According to Mesulam and his associates (1983a, 1983b), Ch2 neurons project to the hippocampus—similar to the Ch1 neurons—as well as to the hypothalamus and occipital cortex. The McGeer and Woolf-Butcher teams add to these projections of the vertical diagonal band projections to several cortices and to the interpeduncular nucleus, and their description agrees with that of Senut et al. (1989).

The Ch3 sector comprises neurons of the horizontal limb of the diagonal band. Mesulam states that he does not differentiate the horizontal limb from the preoptic magnocellular nucleus or area; he also opines, jointly with the Bigl-Butcher-Woolf team, that there may be an overlap between the Ch3 and Ch4 sectors (see Bigl et al., 1982). Depending on the site within this nucleus, the frequency of cholinergic cells varies between 25% and 75%. The main outflow of the Ch3 neurons is, according to Mesulam, the olfactory bulb; there is a general consensus between Mesulam and the Woolf-Butcher team as to this generalization. As the Ch3 and Ch4 sectors appear to overlap, the radiations from the overlap area extend to several neocortical areas. It may be added that the Woolf-Butcher team identifies the preoptic magnocellular nucleus as a separate area, and they describe the radiations from this nucleus as extending, similar to the Ch3 to Ch4 overlap area, to several neocortices.

Marsel Mesulam described the Ch4 sector of the monkey as "providing the major source of cholinergic projections to the cortical mantle" (Mesulam et al., 1983a, 1983b). Subsequently, Mesulam defined it as nucleus basalis magnocellularis and considered it homologous with the nucleus basalis of the rat (e.g., Mesulam and Geula, 1988). However, Shute and Lewis (1967a, 1967b), Jacobowitz and Palkovits (1974), and Emson et al. (1979) considered the nucleus basalis to be a part of the entopeduncular nucleus and globus pallidus, and Mesulam et al. (1983a, 1983b) suggest that neurons listed within substantia innominata and the preoptic magnocellular nucleus "should probably" also be considered parts of the nucleus basalis and of the Ch4 sector.

In monkeys and rats, about 90% of the neurons of nucleus basalis are cholinergic. However, Shute and Lewis (1967a, 1967b), Jacobowitz and

Table 2-1. Nomenclature for Cholinergic Projections of the Basal Forebrain and Upper Brainstem in the Rat

Cholinergic Cell Groups	Traditional Nomenclature for the Nuclei That Contain the Cholinergic Neurons	Major Source of Cholinergic Innervation for
Ch1	Medial septal nucleus	Hippocampus
Ch2	Vertical limb nucleus of the diagonal band	Hippocampus
Ch3	Lateral part of the horizontal limb nucleus of the diagonal band	
Ch4	Nucleus basalis of Meynert, globus pallidus	
Ch5	Substantia innominata	
	Nucleus of the ansa lenticulars, neurons lateral to the vertical limb nucleus and those on the medial parts of the horizontal limb nucleus of the diagonal band	
Ch6	(including parts of the preoptic magnocellular nucleus)	
	Nucleus pedunculopontinus, neurons within the parabrachial area	
	Laterodorsal tegmental nucleus	

See Woolf, 1997, for further details.

Palkovits (1974) and Emson et al. (1979) considered the nucleus basalis as being a part of the entopeduncular nucleus and globus pallidus (as defined by Mesulam and his team). Mesulam and his team opine also that Ch4 sector is the principal source of cholinergic projections to the neocortex and that it extends to the amygdala (Mesulam et al., 1983a, 1983b; Mesulam and Geula, 1988; Mesulam, 1990; Selden et al., 1998).

Mesulam and his team (e.g., Selden et al., 1998) distinguish two "discrete, organized" pathways, medial and lateral, that originate in the Ch4 sector (the lateral pathway is further subdivided by Mesulam into two "divisions"). Mesulam emphasizes that these pathways contain AChE, CAT, and nerve growth factor, but little NAPDH activity, while NAPDH is present in the Ch5 and Ch6 neurons (Selden et al., 1998).

The medial Ch4 pathway supplies the cingulate, parolfactory, percingulate, and retrospinal cortices, and it merges with the lateral pathway within the occipital lobe. The lateral pathway projects to frontal, parietal, temporal, and occipital cortices. The Ch4 sector (or nucleus basalis) increases in size and differentiation in the course of vertebrate evolution, culminating with its status in the human (Gorry, 1964); in fact, Mesulam (1990, 1998) distinguishes 4 "divisions" of the human Ch4, rather than the 2 "divisions" he differentiates in the rat and monkey. The notion that Ch4 (i.e., the nucleus basalis, including substantia innominata) is a major neocortex supplier is shared by the McGeers, the Woolf-Butcher team, Kasa, and Mesulam. These and other investigators also agree that Ch4 is important for memory, learning, and, more generally, cognitive function, as well as for cognitive disorders, aging, and Alzheimer's disease.

Sectors Ch5 and Ch6 are present in the human and animal reticular formation (Mesulam et al., 1989). Contrary to the forebrain sectors, these constellations exhibit high NAPDD activity levels. On the other hand, Ch4 neurons and other forebrain sectors are rich in nerve growth factor (NGF) protein, while the Ch5 and Ch6 sectors are not (Mesulam et al., 1989). How does this finding jibe with the information that Ch4 neurons are much more affected in Alzheimer's disease than the Ch5 and Ch6 neurons (Zweig et al., 1987)?

The Ch5 sector is localized in the pontomesencephalic reticular formation that comprises neurons within the pedunculopontine nucleus; some of the Ch5 neurons also extend into the cuneiform and parabrachial nuclei, as well central tegmental tract and other adjacent sites. As are other sectors, Ch5 is heterogeneous, as it contains noncholinergic neurons as well. The Ch5 sector radiates to the thalamus and its several nuclei, including anterior, lateral, and reticular nuclei, and it sends a minor projection to the neocortex. Mesulam emphasizes that the Ch5 (and Ch6) radiations correspond to the reticulothalamic pathway of Shute and Lewis (1967a, 1967b) and to the ascending reticular activating system of Moruzzi and Magoun (1949; see above, section A1–A3, and Chapter 9 BIV-3). Mesulam and his team (see Mesulam et al., 1983a, 1983b; Mesulam, 1990) do not exclude the possibility that the Ch5 sector also sends weak projections to the habenula and hypothalamus, to the olfactory bulb, and to the spinal cord and the brainstem.

Radiations from Ch5 and particularly the pedunculopontine nucleus are similar to the radiations of this nucleus described by the McGeers and the Woolf-Butcher team, although additional projections are described for the Mesualm system, or complex. However, there are divergences between the systems proposed by the McGeers and Mesulam. Thus, Mesulam's sector Ch5 corresponds to the parabrachial radiation of the McGeers rather than to their reticular system, and his adjudication to the Ch5 sector of the parabrachial and cuneiform nuclei of the pontomesencephalic reticular formation (with its pedunculopontine nuclei) relates the Ch 5 sector the Woolf-Butcher team's pontomesencephalic tegmentum (see above, section DII).

The Ch6 sector neurons are localized in the laterodorsal tegmental nucleus, which is confined within the periventricular gray. According to Mesulam, the Ch6 projections are similar to the Ch5 projections, as both Ch5 and Ch6 sectors innervate various thalamic neurons; the concern here is still with a reticulothalamic system (Mesulam et al., 1989). Moreover, Mesulam opined that, similar to the Ch5 neurons, the Ch6 neurons might project to the habenula, hypothalamus, neocortices, hippocampus, and olfactory bulb. The Ch6 sector seems to correspond more closely to the Woolf-Butcher pontomesencephalic tegmentum system with its laterodorsal tegmental nuclei than to the McGeers' parabrachial complex. The Woolf-Butcher pontomesencephalic tegmental complex radiates, similar to Mesulam's sector Ch6, to the thalamic nuclei, parts of the limbic system, the cortices, and the habenula; however, Woolf and Butcher propose that their pontomesencephalic tegmental complex radiates also to additional brain parts, including basal ganglia and striate, lateral geniculate, and pontine reticular nuclei—a major difference between their complex and Mesulam's Ch6 sector (see section DII, above).

Marsel Mesulam is quite brief when describing sectors Ch7 and Ch8. He opined that the cholinergic neurons of sector Ch7 were located in the medial habenula and projected to the interpeduncular nucleus. This is in accordance with the Woolf-Butcher team's and Kasa's classifications. In fact, Kasa (1986) suggested that the lateral habenular nuclei also project to the interpeduncular nucleus and may serve as a relay for the diagonal band cholinergic cell communication with the nucleus. On the other hand, the McGeers did include the habenula in their projection system.

Finally, the Ch8 sector is localized in the parabigeminal nucleus of the pontomesencephalic region, and its neurons project to the superior colliculus and the lateral geniculate (Mesulam et al., 1989). The innervation of the colliculus constitutes a bewildering situation; Kasa (1986) refers to a number of origins of cholinergic innervation of the superior colliculus, including reticular nuclei, striate, and reticular, peduculopontine tegmental, or cuneiform nuclei. This cholinergic innervation of the superior colliculus and geniculate is important because these two constellations are a part of what may be a cholinergic visual system (see above, section DII).

DIV. Conclusions

When localized, CAT and AChE act as cholinergic markers, adding to the identification of cholinergic neurons and cholinergic pathways (see sections 2 B, above). Moreover, descriptions of these pathways establish the cholinergic system's ubiquity and its presence in brain parts crucial for sensory processing, functions, and behaviors (Karczmar, 2004). This role of the cholinergic system in sensory function needs stressing: while the importance of the cholinergic system in function and behavior has been recognized since the work of William Feldberg, it was denied with regard to the sensorium by Michail Michelson (1974; Michelson and Zeymal, 1970) and others.

Certain special features characterize the cholinergic networks and pathways. Nancy Woolf (1991) emphasized that they form complex interconnections, as they consist of afferents to both cholinergic and noncholinergic neurons and as they receive noncholinergic and cholinergic relays. Altogether, cholinergic pathways are not modular or linear in pattern, but global, and this ensures the subtle, point-to-point control of transmission across the networks in question.

Following the early work of Shute and Lewis, Gerebtzoff, and Koelle, assiduous investigations by the McGeers, Larry Butcher, Nancy Woolf, Marsel Mesulam, Paul Kasa, and Bruce Wainer almost definitely established the cholinergic pathways and networks. That is not to say that there are no differences in their views, and these differences were outlines in this section. On the whole, these divergences are small and do not interfere with our understanding of the major cholinergic pathways such as the forebrain (Mesulam's Ch1 to Ch4 sectors) and its nucleus basalis of Meynert, and the pontomesencephalic formation of the Woolf-Butcher team (Mesulam's Ch5 and Ch6 sectors). Also, this general understanding of the cholinergic pathways, nuclei, and networks helps in associating these entities with functions and behaviors that are endowed with cholinergic correlates (see Chapter 9). Yet, certain areas require more analysis, as in the case of descending pathways connecting supraspinal and spinal sites such as spinal motor and autonomic nuclei.

An important point must be raised. This section focuses on cholinergic sites of origin of cholinergic pathways and networks, yet the cholinergic pathways connect everywhere with pathways manned by other transmitters—in fact, peptides, monoamines, indoleamines, and amino acids interact throughout the brain. These multitransmitter interconnections are not clear, as studies that would simultaneously involve connections among several transmitters are difficult and, therefore, rare (see, however, Senut et al., 1989; Lee et al., 1988; Luppi et al., 1988; and Dun et al.,

1993). Yet, culling from studies concerning different single transmitters as well as those concerning several transmitters simultaneously, it became obvious that interconnections among all the existing transmitter systems are present throughout the brain. This notion is supported by the demonstration that all central functions and behaviors are regulated by multiple transmitters, even though some of them may exhibit preponderant cholinergic correlates (see Chapter 9 BIV–BVI).

Notes

1. This digression leads to a story. As a part of his PhD dissertation as a University of California, Los Angeles, graduate student in Don Jenden's laboratory, Israel Hanin developed the gas chromatography method in the 1960s and went with Jenden as a postdoc to develop, in Bo Holmstedt's laboratory at the Karolinska Institutet, the complete gas chromatography–mass spectrometry method using the equipment available in Holmstedt's laboratory. Thereupon, Jenden, Holmstedt, and Hanin proceeded to use the method for the first chemical identification of ACh in the brain. As they succeeded, Jenden and Holmstedt sent their results, by wireless, to Henry Dale: they felt that he would be happy to learn that his belief in the presence of ACh could be vindicated via a chemical brain method for identifying and measuring ACh. This was a few months before Dale's death in 1968 but, with his usual courtesy, Dale managed to tell Holmstedt and Jenden, via a letter, how much he enjoyed getting this news. Apparently, Hanin's name does not appear in that exchange (see Holmstedt, 1975).
2. This part of the story is quite piquant. The distinguished French neuroscientist, the late Rene Couteaux, who with David Nachmansohn was an early student of neuromyal AChE (see Couteaux, 1953, 1998), sent a promising young French scientist, Maurice Israel, to Whittaker's Cambridge laboratory to help Whittaker in the purification of synaptic vesicles of the Torpedo. When back in France, Maurice Israel continued the work on the purification of the vesicles and, subsequently, he became one of the exponents of the hypothesis of the nonvesicular release of ACh, and a vigorous opponent of Whittaker's classical image of the vesicular release of ACh (see below, this section, and section IIB5).
3. Frequently, the various teams do not refer to one another in their reviews; for example, the reviews of Butcher et al. (1993), Wainer et al. (1993), Woolf (1991), and Mesulam et al. (1983) do not refer to the work of the McGeers, and the 1988 paper of Lee and Wainer (Lee et al., 1988) manages not to quote either the Mesulam team or the Woolf-Butcher team. The review of Semba et al. (1989; with Edith McGeer as the coauthor) and of Butcher (1995) may be exceptional as they quote Hans Fibiger's and Mesulam's work, and Fibiger's, Mesulam's and the McGeers and Kimura's work, respectively.
4. This statement reminded Yves Dunant of Sir William Feldberg's dictum that "there is a type of scientist who, if given the choice, would rather use his colleague's toothbrush than his [or her] terminology" (cited by Katz, 1969).

References

Agoston, D. D. and Lisiewicz, J.: Calcium uptake and protein phosphorylation in myenteric neurons, like the release of vasoactive intestinal olypeptide and acetylcholine, are frequency-dependent. J. Neurochem. *52*: 1637–1640, 1989.

Atkinson, L., Batten, T. F., Moores, T. S., Varoqui, H., Erickson, J. D. and Dechars, J.: Differential co-localization of the P2X7 receptor subunit with glutamate transporters VGLUT1 and VGGLUT2 in rat CNS. Neurosci. *123*: 761–768, 2004.

Augustinsson, K. B.: Cholinesterases. Acta Physiol. Scand. *15*, Supplt. 52: 1–182, 1948.

Augustinsson, K. B.: Classification and comparative enzymology of the cholinesterases and methods for their determination. In: "Cholinesterases and Anticholinesterase Agents," G. B. Koelle, Ed., pp. 89–128, Handbch. d. exper. Pharmacol., Ergänzungswk., vol. 15, Springer-Verlag, Berlin, 1963.

Bacci, A., Coco, S., Pravettoni, E., Schenk, U., Armano, S., Frassoni, C., Verderio, C., De Camilli, P. and Matteoli, M.: Chronic blockade of glutamate receptors enhances presynaptic release and downregulates the interaction between synaptophysin-synaptobrevin-vesicle-associated membrane protein 2. J. Neurosci. *21*: 6588–6596, 2001.

Bahr, B. A. and Parsons, S. M.: Demonstration of a receptor in Torpedo synaptic vesicles for the acetylcholine storage blocker L-trans-2-(4-phenyl[3,4-3H]-piperidino) cyclohexanol. Proc. Natl. Acad. Sci. USA. *83*: 2267–2270, 1986.

Barber, R. P., Phelps, P. E., Houser, C. R., Crawford, G. D., Salvaterra, P. M. and Vaughn, J. E.: The morphology and distribution of neurons containing choline acetyltransferase in the adult rat spinal cord: an immunocytochemical study. J. Comp. Neurol. *229*: 329–346, 1984.

Barker, L. A., Dowdall, M. J. and Whittaker, V. P.: Choline metabolism in the cerebral cortex of guinea pig. Biochem. J. *130*: 1063–1075, 1972.

Barnett, R. J. and Higgins, J. A.: Proc. Electron Microscopy Soc., 26th Ann. Meeting, p. 1, 1968.

Bear, M. F., Carnes, K. M. and Ebner, F. F.: Postnatal changes in distribution of acetylcholinesterase in kitten striate cortex. J. Comp. Neurol. *237*: 519–532, 1985.

Bennet, M. R.: The origin of Gaussian distributions of synaptic potentials. Prog. Neurobiol. *45*: 331–350, 1995.

Bigl, V., Woolf, N. J. and Butcher, L.L.: Cholinergic projections form basal forebrain to frontal, parietal, temporal, occipital, and cingulate cortices: a combined fluorescent tracer and acetylcholinesterase analysis. Brain Res. Bull. *8*: 727–749, 1982.

Bolam, J. P., Wainer, B. H. and Smith, A. D.: Characterization of cholinergic neurons in the rat neostriatum. A combination of choline acetyltransferases immunocytochemistry, Golgi-impregnation and electron microscopy. Neuroscience. *12*: 711–718, 1984.

Botti, S. A., Felder, C. E., Sussman, J. L and Silman, I.: Electrotactins: a class of adhesion proteins with conserved electrostatic and structural motifs. Protein Eng. *11*: 415–420, 1998.

Bradley, P. B. and Elkes, J.: The effect of atropine, hyoscyamine, physostigmine, and neostigmine on the electrical activity of the brain of conscious cat. J. Physiol. *120*: 14P, 1953.

Brodal, A.: "Neurological Anatomy," Third Edition, Oxford University Press, Oxford, 1981.

Bruckner, G., Schober, W., Hartig, W., Ostermann-Latif, C., Webster, H. H., Dykes, R. W., Rasmusson, D. D. and Biesold, D.: The basal forebrain cholinergic system in the raccoon. J. Chem. Neuroanat. *5*: 441–451, 1992.

Bulbring, E. and Burn, J. H.: Observations bearing on synaptic transmission by acetylcholine in the spinal cord. J. Physiol. *100*: 337–368, 1941.

Butcher, L. L.: Cholinergic neurons and networks. In: "The Rat Nervous System," George Patinos, Ed., Second Edition, Academic Press, New York, 1995.

Butcher, L. L. and Bilezikjian, L.: Acetylcholinesterase-containing neurons in the neostriatum and substantia nigra after punctate intracerebral injection of d-isopropyl fluorophosphate. Eur. J. Pharmacol. *37*: 115–125, 1975.

Butcher, L. L. and Semba, K.: Reassessing the cholinergic basal forebrain nomenclature schemata and concepts. Trends in Neurosc. *12*: 483–485, 1989.

Butcher, L. L. and Woolf, N. J.: Cholinergic neurons and networks revisited. In: "The Rat Nervous System," Third Edition, Elsevier, Amsterdam, 2004.

Butcher, S. H., Butcher, L. L. and Cho, A. K.: Modulation of neostriatal acetylcholine in the rat by dopamine and 5-hydroxytryptamine afferents. Life Sci. *18*: 733–743, 1976a.

Butcher, S. H., Butcher, L. L. and Harms. Fast fixation of brain in situ by high intensity microwave irradiation: application to neurochemical studies. J. Microw. Power. *11*: 61–65, 1976b.

Butcher, L. L., Marchand, R., Parent, A. and Poirier, L. J.: Morphological characteristics of acetylcholinesterase-containing neurons in the CNS of DFP-treated monkeys. Part 3. Brain stem and spinal cord. J. Neurol. Sci. *32*: 169–185, 1977.

Butcher, L. L., Oh, J. D. and Woolf, N.: Cholinergic neurons identified by in situ hybridization histochemistry. In: "Cholinergic Function and Dysfunction," A. C. Cuello, Ed., pp. 1–8, Elsevier, Amsterdam, 1993.

Cervini, R., Houhou, L., Pradat, P. F., Bejanin, S., Mallet, J. and Berrard, S.: Specific vesicular acetylcholine transporter promoters lie within the first intron of the rat choline acetyltransferase gene. J. Biol. Chem. *270*: 24654–24657, 1995.

Chapman, E., Hanson, P. I., An, S. and Jahn, R.: Calcium regulates the interaction between synaptotagmin and syntaxin. J. Biol. Chem. *270*: 23667–23671, 1955.

Cochilla, A. J., Angleson, J. K. and Betz, W. J.: Monitoring secretory membrane with FM1–43 fluorescence. Annu. Rev. Neurosci. *22*: 1–10, 1999.

Coggan, J. S., Bartol, T. M., Esquenazi, E., Stiles, J. R., Lamont, S., Martone, M. E., Berg, D. K., Ellisman, M. H. and Sejnowski, T. J.: Evidence for ectopic neurotransmission at a neuronal synapse. Science *309*: 387–388, 2005.

Corsten, M.: Bestimmung kleinster Acetylcholinmengen am Lungenpräparat des Frosches. Arch. Ges. Physiol. *244*: 281–291, 1940.

Couteaux, R.: Particularites histochimiques des zones d'insertion du muscle strié. C. R. Soc. Biol. (Paris) *147*: 1974–1976, 1953.

Couteaux, R.: Early days in the research to localize skeletal muscle acetylcholinesterase. In: "From *Torpedo* Electric Organ to Human Brain: Fundamental and Applied Aspects," J. Massoulié, Ed., pp. 59–62, Elsevier, Amsterdam, 1998.

Couteaux, R. and Pecot-Dechavassine, M.: Vesicules synaptiques et poches au niveau des "zones acives" de la jonction neuromusculaire. C. R. Sci. (Paris) *271*: 2346–2349, 1970.

Dale, H. H.: The action of certain esters and ethers of choline and their relation to muscarine. J. Pharmacol. Exper. Therap. *6*: 147–190, 1914.

Dale, H. H.: Transmission of nervous effects by acetylcholine. Harvey Lect. *32*: 229–245, 1937.

Darrow, C. W., McCulloch, W. S., Green, J. R., Davis, E. W. and Garol, H. W.: Parasympathetic regulation of high potential in the electroencephalogram. Arch. Neurol. Psychiat. *52*: 337–338, 1944.

De Robertis, E. and Bennett, H. S.: Some features of the submicroscopic morphology of the synapses in frog and earthworm. J. Biophys. Biochem. Cytol. *1*: 47–58, 1955.

Del Castillo, J. and Katz, B.: Action, and spontaneous release, of acetylcholine at an inexcitable nerve-muscle junction. J. Physiol. *126*: 27, 1954.

Descarries, L.: The hypothesis of an ambient level of acetylcholine in the central nervous system. In: "From *Torpedo* Electric Organ to Human Brain: Fundamental and Applied Aspects," J. Massoulié, Ed., pp. 215–220, Elsevier, Amsterdam, 1998.

Descarries, L. and Mechawar, N.: Ultrastructure evidence for diffusion transmission by monoamine and acetylcholine neurons of the central nervous system. Prog. Brain Res. *125*: 27–47, 2000.

Descarries, L., Gisiger, V. and Steriade, M.: Diffuse transmission by acetylcholine in the CNS. Prog. Neurobiol. *53*: 603–625, 1997.

Descarries, L., Mechawar, N., Aznavour, N. and Watkins, K. C.: Structural determinants of the roles of acetylcholine in cerebral cortex. Prog. Brain Res. *145*: 45–58, 2004.

Divac, I.: Magnocellular nuclei of the basal forebrain project to neocortex, brain stem, and olfactory bulb as determined by a retrograde tracing technique. Review of some behavioral correlates. Brain Res. *93*: 38–98, 1975.

Dohanich, G. P. and McEwen, B. S.: Cholinergic limbic projections and behavioral role of basal forebrain nuclei in the rat. Brain Res. Bull. *16*: 472–482, 1986.

Domino, E. F.: Evidence for a functional role of acetylcholine in the lateral geniculate nucleus. In: "The James E. P. Toman Memorial Volume," H. Sabelli, Ed., pp. 95–109, Raven Press, New York, 1973.

Domino, E. F., Krause, R. R. and Bowers, J.: Regional distribution of some enzymes involved with putative neurotransmitters in the human visual system. Brain Res. *58*: 167–171, 1973.

Doyle, C. A. and Maxwell, D. J.: Direct catecholaminergic innervation spinal dorsal horn neurons with axons ascending the dorsal columns in cat. J. Comp. Neurol. *331*: 434–444, 1993.

Dowdall, M. J., Boyne, A. F. and Whittaker, V. P.: Adenosine triphosphate. A constituent of cholinergic synaptic vesicles. Biochemo J. *140*: 1–12, 1974.

Dun, N. J., Karczmar, A. G., Wu, S.-Y. and Shen, E.: Putative transmitter systems of mammalian sympathetic preganglionic neurons. Acta Neurobio. Exp. *53*: 53–63, 1993.

Dun, S. L., Brailoiu, G. C., Yang, J., Chang, J. K. and Dun, N. J.: Urotensin II immunoreactivity in the brain stem and spinal cord of the rat. Neurosci. Lettr. *305*: 9–12, 2001.

Dunant, Y.: Quantal acetylcholine release: vesicle fusion or intramembrane particles? Microsc. Res. Tech. *49*: 38, 2000.

Dunant, Y. and Israel, M.: The release of acetylcholine. Sci. Amer. *252*: 58–66, 1985.

Eccles, J. C., Fatt, P. and Koketsu, K.: Cholinergic and inhibitory synapses in a pathway from motor-axon collaterals to motoneurones. J. Physiol. (Lond.) *126*: 524–562, 1954.

Eccles, J. C.: "The Physiology of Synapses." Springer-Verlag, New York, 1964.

Eckenstein, F. and Sofroniew, M. V.: Identification of central cholinergic neurons containing both choline acetyltransferase and acetycholinesterase and of central neurons containing only acetylcholinesterase. J. Neurosci. *3*: 2286–2291, 1983.

Eiden, L. E.: The cholinergic gene locus. J. Neurochem. *70*: 2227–2240, 1998.

El Far, O., O'Connor, V., Dresbach, T., Pellegrini, L., DeBello, W., Schweizer, F., Augustine, G., Heuss, C., Schafer, T., Charlton, M. and Betz, H.: Protein interactions implicated in neurotransmitter release. In: "From *Torpedo* Electric Organ to Human Brain: Fundamental and Applied Aspects," J. Massoulié, Ed., pp. 129–133, 1998.

Emson, P. C., Paxinos, G., Le Gal Lasalle, G., Ben-Ari, Y. and Silver, A.: Cholineacetyl transferase- and acetylcholinesterase-containing projections from basal forebrain to the amyloid complex of the rat. Brain Res. *165*: 271–281, 1979.

Eng, L. F., Uyeda, C. T., Chao, L.-F. and Wolfgram, F.: Antibody to bovine choline acetyltransferase and immunofluorescent localization of the enzyme in neurones. Nature *250*: 243–245, 1974.

Famiglietti, E. V., Jr.: On and off pathways through amacrine cells in mammalian retina: the synaptic connections of "starburst" amacrine cells. Vision Res. *23*: 1265–1279, 1983.

Fasshauer, D., Antonin, W., Margittai, M., Pabst, S. and Jahn, R. T.: Structural and functional conservation of SNARE complexes. In: "Cholinergic Mechanisms—Function and Dysfunction," H. Soreq, L. Anglister, D. Michaelson and A. Fisher, Eds., Martin Dunitz, London, 2004.

Fatt, P. and Katz, B.: The electric activity of the motor end-plate. Proc. R. Soc. Lond. B. Biol. Sci. *140*: 183–186, 1952.

Feldberg, W.: Present views on the mode of action of acetylcholine in the central nervous system. Physiol. Rev. *25*: 596–642, 1945.

Feldberg, W.: The role of acetylcholine in the central nervous system. Brit. Med. Bull. *6*: 312–321, 1950.

Feldberg, W. and Schriever, H.: The acetylcholine content of the cerebro-spinal fluid of dogs. J. Physiol. *86*: 277–284, 1936.

Feldberg, W. and Vogt, M.: Acetylcholine synthesis in different regions of the central nervous system. J. Physiol. *107*: 372–381, 1948.

Ferreti, P. and Borroni, E.: Putative cholinergic-specific gangliosides in guinea pig forebrain. J. Neurochem. *46*: 1888–1894, 1986.

Fibiger, H. C.: The organization and some projections of cholinergic neurons in the mammalian forebrain. Brain. Res. Rev. *4*: 327–388, 1982.

Fischbach, G. D. and Rosen, K. M.: ARIA: A neuromuscular junction neuregulin. Annu. Rev. Neurosci. *20*: 429–458, 1997.

Fischer, H.: Vergleichende Pharmakologie von Übertragensubstanzen in tiersystematischer Darstellung. Handbch. d. exper. Pharmakol., vol. 26, Springer-Verlag, Berlin, 1971.

Fühner, H.: Untersuchunger über die periphere Wirkuny des physostigmins. Arch. exp. Pathol. Pharmakol. *82*: 205–220, 1918.

Giacobini, E.: Cholinesterase inhibitors: from Calabar bean to Alzheimer therapy. In: "Cholinesterases and Cholinesterase Inhibitors," E. Giacobini, Ed., pp. 181–226, Martin Dunitz, London, 2000.

Giacobini, E.: Cholinesterases: new roles in brain function and Alzheimer's disease. Neurochem. Res., *28*: 515–522, 2003.

Giolli, R. A., Clarke, R. J., Torigoe, Y., Blanks, R. H. and Fallon, J. H.: The projection of GABA-ergic neurons of the medial terminal accessory optic nucleus to the pretectum in the rat. Braz. J. Med. Biol. Res. *23*: 1349–1352, 1990.

Glick, D.: The nature and significance of cholinesterase. In: "Comparative Biochemistry. Intermediate Metabolism of Fats. Carbohydrate Metabolism. Biochemistry of Choline," H. B. Lewis, Ed., pp. 213–233, Jaques Cattell Press, Lancaster, Pennsylvania, 1941.

Geffard, M., McRae-Deguerce, A. and Souan, M. L.: Immunocytochemical detection of acetylcholine in the rat central nervous system. Science *229*: 77–79, 1985.

Gerebtzoff, M. A: "Cholinesterases, a Histochemical Contribution to the Solution of Some Functional Problems," MacMillan (Pergamon), New York, 1959.

Goedde, H. W., Doenike, A. and Altland, K.: "Pseudocholinesterasen." Springer-Verlag, Berlin, 1967.

Gorry, J. D.: Studies of the comparative anatomy of the ganglion basalis of Meynert. Acta Anat. *55*: 51–104, 1964.

Gray, E. G.: Electron microscopy of excitatory and inhibitory synapses: a brief review. Prog. Brain Res. *31*: 141–155, 1969.

Halasz, N. and Shepherd, G. M.: Neurochemistry of the vertebrate olfactory bulb. Neurosci. *10*: 579–617, 1983.

Hanin, I. and Goldberg, A. M.: Appendix I: Quantitative assay methodology for choline, acetylcholine, choline acetyltransferase, and acetylcholinesterase. In: "Biology of Cholinergic Function," A. M. Goldberg and I. Hanin, Eds., pp. 647–659, Raven Press, New York, 1976.

Hebb, C. O.: Formation, storage and liberation of acetylcholine. In: "Cholinesterases and Anticholinesterase Agents," G. B. Koelle, Ed., Handbch. d. exper. Pharmakol., Ergänzungswk., vol. 15, pp. 55–88, Springer-Verlag, Berlin, 1963.

Hebb, C. O. and Whittaker, V. P.: Intracellular distribution of acetylcholine and choline acetyltransferase. J. Physiol. (Lond.) *142*: 187–196, 1958.

Hebb, C. O., Kasa, P. and Mann, S. P.: Nature *226*: 814–816, 1970.

Henkel, A. W. and Betz, W. J.: Staurosporine blocks evoked release of FM1–43 but not acetylcholine from frog motor nerve terminals. J. Neurosci. *15*: 8255–8257, 1995.

Himwich, H. E.: Some specific effects of psychoactive drugs. In: "Specific and Non-specific Factors in Psycho-pharamcology," M. Rinkel, Ed., pp. 3–71, Philosophical Library, New York, 1963.

Holmstedt, B.: Progress from the history of research on cholinergic mechanisms. In: "Cholinergic Mechanisms," P. G. Waser, Ed., pp. 1–21, Raven Press, New York, 1975.

Hoover, D. B. and Jacobowitz, D. M.: Neurochemical and histochemical studies of the effect of a lesion of the nucleus cuneiformis on the cholinergic innervation of discrete areas of the rat brain. Brain Res. *170*: 113–122, 1979.

Horie, S., Shioda, S. and Nakai, Y.: Catecholaminergic innervation of oxytocin neurons in the paraventricular nucleus of the rat hypothalamus as revealed by double-labeling immunoelectron microscopy. Acta. Anat. (Basel). *147*: 184–192, 1993.

Hou, X. E. and Dahlstrom, A.: Synaptic vesicle proteins and neuronal plasticity in adrenergic neurons. Neurochem. Res. *25*: 1275–300, 2000.

Howard, J.: "Mechanics of Motor Proteins and the Cytoskeleton." Sinauer Associates, Sunderland, Massachusetts, 2001.

Hutchins, J. B.: Acetylcholine as a neurotransmitter in the vertebrate retina. Exp. Eye Res. *45*: 1–38, 1987.

Israel, M. and Dunant, Y.: Acetylcholine release. Reconstitution of the elementary quantal mechanism. In: "From *Torpedo* Electric Organ to Human Brain: Fundamental and Applied Aspects," J. Massoulié, Ed., pp. 123–128, Elsevier, Amsterdam, 1998.

Israel, M. and Dunant, Y.: Mediatophore no longer an artefact. In: "Cholinergic Mechanisms—Function and Dysfunction," I. Silman, H. Soreq, L. Anglister, D. Michaelson and A. Fisher, Eds., pp. 91–97, Martin Dunitz, London, 2004.

Israel, M. and Lesbats, B.: Continuous determination by a chemiluminscent method of acetylcholine release and compartmentation in *Torpedo* electric organ synaptosomes. J. Neurochem. *37*: 1475–1483, 1981.

Israel, M. and Morel, N.: Mediatophore: a nerve terminal membrane protein supporting the final step of the acetylcholine release process. Prog. Brain Res. *84*: 101–110, 1990.

Israel, M., Lesbats, B., Malaranche, R. and Morel, N.: Acetylcholine release from proteoliposomes equipped with synaptosomal membrane constituents. Biochem. Biophys. Acta *728*: 438–448, 1983.

Israel, M., Lesbats, B., Sbia, M. and Morel, N.: Acetylcholine translocating protein: mediatophore at rat neuromuscular synapses. J. Neurochem. *55*: 1758–1762, 1990.

Jacobowitz, D. M. and Palkovits, M. Topographic atlas of catecholamine and acetylcholinesterase-containing neurons in the rat brain. 1. Forebrain (telencephalon, diencephalon). J. Comp. Neurol. *157*: 13–28, 1974.

Jasper, H. H.: Pathophysiological studies of brain mechanisms in different states of consciousness. In: "Brain and Conscious Experience," J. C. Eccles, Ed., pp. 256–282, Springer-Verlag, Berlin, 1966.

Jenden, D. J., Russell, R. W., Booth, R. A., Knusel, B. J., Lauretz, S. D., Rice, K. M. and Roch, M.: Effects of chronic *in vivo* replacement of choline with a false cholingergic precursor Exs. *57*: 229–235, 1989.

Jones, B. E. and Beaudet, A.: Retrograde labeling of neurones in the brain stem following injections of [3H] choline into the brain stem of rats. Exper. Brain Res. *65*: 437–448, 1987.

Jones, B. E. and Cuello, A. C.: Afferents to the basal forebrain cholinergic cell area from pontomesencephalic—catecholamine, serotonin, and acetylcholine—neurons. Neurosci. *31*: 37–61, 1989.

Jones, B. E. and Yang, T. Z.: The efferent projections from the reticular formation and the locus coeruleus studied by anterograde and retrograde axonal transport in the rat. J. Comp. Neruol. *242*: 56–92, 1985.

Karczmar, A. G.: Ontogenesis of cholinesterases. In: "Cholinesterases and Anticholinesterase Agents," G. B. Koelle, Ed., Handbch. d. exper. Pharmakol., Ergänzungswk., vol. 15: 799–832, Springer-Verlag, Berlin, 1963a.

Karczmar, A. G.: Ontogenetic effects of anti-cholinesterase agents. In: "Cholinesterases and Anticholinesterase Agents," G. B. Koelle, Ed., Handbch. d. exper. Pharmakol., Ergänzungswk., vol. 15: 179–186, Springer-Verlag, Berlin, 1963b.

Karczmar, A. G.: Pharmacologic, toxicologic and therapeutic properties of anticholinesterase agents. In: "Physiological Pharmacology," W. S. Root and F. G. Hofman, Eds., vol. 3, 163–322, Academic Press, New York, 1967.

Karczmar, A. G.: Brief presentation of the story and present status of studies of the vertebrate cholinergic system. Neuropsychopharmacology. *9*: 181–199, 1993.

Karczmar, A. G.: Sir John Eccles, 1903–1997. Part 1. Onto the demonstration of the chemical nature of transmission in the CNS. Perspectives in Biol. & Med. *44*: 76–86, 2001a.

Karczmar, A. G.: Sir John Eccles, 1903–1997. The brain as machine or as a site of free will? Perspectives in Biol. & Med. *44*: 250–262, 2001b.

Karczmar, A. G.: Conclusions. In: "Cholinergic Mechanisms—Function and Dysfunction," I. Silman, Ed., pp. 435–451, Martin Dunitz, London, 2004.

Karczmar, A. G. and Koppanyi. T.: Changes in "transport" cholinesterase levels and responses to intravenouslyadministeredacetylcholineandbenzoylcholine. J. Pharm. Expt. Therap. *116*: 245–253, 1956.

Kasa, P.: Ultrastructural localization of choline acetyltransferase and acetylcholinesterase in central and peripheral tissue. In: "Histochemistry of Nervous Transmission," O. Eranko, Ed., Progr. Brain Res. *34*: 337–344, Elsevier, Amsterdam, 1971.

Kasa, P.: Distribution of cholinergic neurones within the nervous system revealed by histochemical methods. In: "Synaptic Transmission," D. Beisold, Ed., pp. 45–55, Karl Marx U. Press, Leipzig, 1978.

Kasa, P.: The cholinergic systems in brain and spinal cord. Progr. in Neurobiol. *28*: 211–271, 1986.

Kasa, P. and Silver, A.: The correlation between choline acetyltransferase and acetylcholinesterase activity in different areas of the cerebellum of rat and guinea pig. J. Neurochem. *16*: 389–396, 1969.

Katz, B.: Microphysiology of the neuromuscular junction; a physiological quantum of action at the myoeneural junction. Bull Johns Hopkins Hosp. *102*: 275–295, 1958.

Katz, B.: "Nerve, Muscle and Synapse," McGraw-Hill, New York, 1966.

Katz, B.: "The Release of Neural Transmitter Substances," University Press, Liverpool, 1969.

Katz, B. and Miledi, R.: The effect of the calcium on acetylcholine release from motor nerve terminals. Proc. R. Soc. Lond. B. Biol. Sci. *161*: 496–503, 1965.

Katz, B. and Miledi, R.: Membrance noise produced by acetylcholine. Nature. *226*: 962–963, 1970.

Katz, B. and Miledi, R.: The characteristics of "end-plate noise" produced by different depolarizing drugs. J. Physiol. *230*: 707–717, 1973.

Katz, B. and Miledi, R.: Suppression of transmitter release at the neuromuscular junction. Proc. R. Soc. Lond. B. Biol. Sci. *196*: 465–469, 1977.

Kiene, M. L. and Stadler, H.: Synaptic vesicles in electromotoneurones. I. Axonal transport, site of transmitter uptake and processing of a core proteoglycan during maturation. Embo. J. *6*: 2209–2215, 1987.

Kimura, H., McGeer, P. L., Peng, F. and McGeer, E. G.: Choline acetyltransferase-containing neurons in rodent brain demonstrated by immunohistochemistry. Science *208*: 1057–1059, 1980.

Kimura, H., McGeer, P. L., Peng, J. H. and McGeer, E. G.: The central cholinergic system studied by choline acetylase immunohistochemistry. J. Comp. Neurol. *200*: 151–201, 1981.

Koelle, G. B.: The histochemical localization of cholinesterases in the central nervous system of the rat. J. Comp. Neurol. *100*: 211–228, 1954.

Koelle, G. B.: A proposed dual neurohumoral role of acetylcholine: its function at the pre- and post-synaptic sites. Nature (Lond.) *190*: 208–211, 1961.

Koelle, G. B.: Cytological distributions and physiological functions of cholinesterases. In: "Cholinesterases and Anticholinesterase Agents," G. B. Koelle, Ed., pp. 187–298, Handbch. d. exper. Pharmakol., Ergänzungswk., vol. 15, Springer-Verlag, Berlin, 1963.

Koelle, G. B. and Friedenwald, J. S.: A histochemical method for localizing cholinesterase activity. Proc. Soc. Exp. Biol. *70*: 617–622, 1949.

Kretsinger, R. H.: Calcium coordination and the calmodulin fold: divergent versus convergent evolution. Cold Spring Harb. Symp. Quant. Biol. *52*: 499–510, 1987.

Kristenson, K., Olsson, Y. and Sjostrand, J.: Axonal uptake and retrograde transport of exogenous proteins in the hypoglossal nerve. Brain Res. *12*: 339–406, 1971.

Krnjevic, K.: Chemical nature of synaptic transmission in vertebrates. Physiol. Rev. *54*: 418–540, 1974.

Krnjevic, K. and Phillis, J. W.: Pharmacological properties of acetylcholine-sensitive cells in the cerebral cortex. J. Physiol. (Lond.) *166*: 328–350, 1963.

Krnjevic, K. and Silver, A.: A histochemical study of chlinergic fibres in the cerebral cortex. J. Anat. (Lond.), *99*: 711–759, 1965.

Krnjevic, K. and Silver, A.: Acetylcholinesterase in the developing forebrain. J. Anat. *100*: 63–89, 1966.

Kuffler, S. W.: Transmitter mechanisms at nerve-muscle junction. Arch. Sci. Physiol. *3*: 585–561, 1949.

Kuffler, S. W., Nicholls, J. G. and Martin, A. R.: "From Neuron to Brain," Second Edition, Sinauer Associates, Sunderland, Massachusetts, 1984.

Kuno, M., Turkanis, S. A. and Weakly, J. N.: Correlations between nerve terminal size and transmitter release at the neuromuscular junction of the frog. J Physiol. (Lond.) *213*: 545–556, 1971.

Lee, D. A. and Witzemann, V.: Photoaffinity labeling of a synaptic vesicle specific nucleotide transport system from Torpedo marmorata. Biochemistry. *22*: 6123–6130, 1983.

Lee, H. J., Rye, D. B., Hallanger, A. E., Levey, A. I. and Wainer, B. H.: Cholinergic vs. Noncholinergic efferents from the mesopontine tegmentum to the extrapyramidal motor system nuclei. J. Comp. Neurol. *275*: 469–492, 1988.

Lewis, P. R. and Henderson, Z.: Tracing putative cholinergic pathways by a dual cytochemical technique. Brain Res. *196*: 489–493, 1980.

Lewis, P. R. and Shute, C. C. D.: The cholinergic limbic system: Projections to hippocompal formation, medial cortex, nuclei of the ascending cholinergic reticular system and the subformial organ and supraoptic crest. Brain *90*: 521–540, 1967.

Loewi, O.: Über humorale Übertragbarkeit der Herzenwirkung. I. Mitteilung. Pflügers Arch. Ges. Physiol. *189*: 239–242, 1921.

Li, R. and Sakaguchi, H.: Cholinergic innervation of the song control nuclei by the ventral paleostriatum in the zebra finch: a double labeling study with retrograde fluorescent tracers and choline acetyltransferase immunochemistry. Brain Res. *763*: 239–246, 1997.

Li, Z., Burrone, J., Tyler, W. J., Hartman, K. N., Albeanu, D. F. and Murthy, V. N.: Synaptic vesicle recycling studied in transgenic mice expressing synaptophluorin. Proc. Natl. Acad. Sci. USA. *102*: 6131–6136, 2005.

Lian, W., Chen, G., Wu, D., Brown, C. K., Madauss, K., Hersh, L. B. and Rodgers, D. W.: Crystallization and preliminary analysis of neurolysin. Acta Crystallogr. D. Biol. Crystallogr. *56*: 1644–1646, 2000.

Liu, Z., Tearle, A. W., Nai, Q. and Berg, D. K.: Rapid activity-driven SNARE dependent trafficking of nicotinic receptors on somatic spines. J. Neurosci. *25*: 1159–1168, 2005.

Loewi, O. and Narratil, E.: Über humorale Übertragbarkeit des Herznerven-Wirkung, VII. Mitteilung. Pflügers Arch. ges. Physiol. *206*: 135-140, 1924.

Loewi, O. and Narratil, E.: Über humorale Übertragbarkeit des Herznerven-Wirkung, X. Mitteilung. Über das Schicksal des Vagusstoffes, Pflügers Arch. ges. Physiol. *214*: 678-688, 1926.

Loewy, A. D.: Central autonomic pathways. In: "Central Regulation of Autonomic Function," A. D. Loewy and K. M. Spyer, Eds., pp. 88–103, Oxford University Press, Oxford, 1990.

Lonart, G. and Sudhof, T. C.: Characterization of rabphilin phosphorylation using phosphospecific antibodies. Neuropharmacol. *41*: 643–649, 2001.

Longo, V. G. and Silvestrini, B.: Action of eserine and amphetamine on the electric activity of the rabbit brain. J. Pharmacol Exper. Therap. *120*: 160–170, 1957.

Luppi, P. H., Sakai, K., Fort, P., Salvert, D. and Jouvet, M.: The nuclei of origin of monoaminergic, peptidergic, and cholinergic afferents to the cat nucleus reticularis magnocellularis: a double-labeling study with cholera toxin as a retrograde tracer. J. Comp. Neurol. *277*: 1–20, 1988.

Mallet, J., Houhou, L., Pajak, F., Oda, O., Cervini, R., Bejanin, S. and Berrard, S.: The cholinergic locus: ChAT and VAChT genes. In: "From *Torpedo* Electric Organ to Human Brain: Fundamental and Applied Aspects," J. Massoulié, Ed., pp. 145–147, Elsevier, Amsterdam, 1998.

Mann, P. J. G., Tennenbaum, M. and Quastel, J. H.: On the mechanism of acetylcholine formation in brain *in vitro*. Biochem. J. *32*: 243–261, 1938.

Martinez-Rodriguez, R. and Martinez-Murillo, R.: Molecular and cellular aspects of neurotransmission and neuromodulation. Int. Rev. Cytol. *149*: 217–292, 1994.

Masukawa, L. M. and Albuquerque, E. X.: Voltage- and time-dependent action of histrionicotoxin on the endplate current of the frog muscle. J. Gen. Physiol. *72*: 351–367, 1978.

McGeer, E. G. and McGeer, P. L.: Effect of fixatives on rat brain acetylcholinesterase activity. J. Neurosci. Methods. *28*: 235–237, 1989.

McGeer, P. L. and McGeer, E. G.: Neurotransmitters and their receptors in the basal ganglia. Adv. Neurol. *60*: 93–101, 1993.

McGeer, P. L., McGeer, E. G., Fibiger, H. C. and Wickson, V.: Neostriatal choline acetylase and cholinesterase following selective brain lesions. Brain Res. *35*: 308–314, 1971.

McGeer, P. L., McGeer, E.G., Singh, V.K. and Chase, W. H.: Choline acetyltransferase localization in the central nervous system by immunchistochemistry. Bran. Res. *80*: 211–217, 1974.

McGeer, P. L., McGeer, E. G. and Peng, J. H.: Choline acetyltranferase purification and immunohistochemical localization. Life Sci. *34*: 2319–2338, 1984a.

McGeer, P. L., McGeer, E. G., Suzuki, J., Dolman, C. E. and Nagai, T.: Aging, Alzheimer's disease and the cholinergic system of the basal forebrain. Neurol. *34*: 741–745, 1984b.

McGeer, P. L., McGeer, E. G., Mizukawa, K., Tago, H. and Peng, J. H.: Distribution of cholinergic neurons in human brain. In "Neurobiology of Acetylcholine, a Symposium in Honor of A. G. Karczmar," N. J. Dun and R. L. Perlman, Eds., pp. 3–16, Plenum Press, New York, 1987a.

McGeer, P. L., Eccles, J. C. and McGeer, E. G.: "Molecular Neurobiology of the Mammalian Brain," Second Edition, Plenum Press, New York, 1987b.

MacIntosh, F. C. and Oborin, P. E.: Release of acetylcholine from intact cerebral cortex. Proc. XIX Intern. Physiol. Congr., Abstracts, Montreal, pp. 580–581, 1953.

McLennan, H.: "Synaptic Transmission," Saunders, New York, 1963.

Meier, S., Brauer, A. U., Heimrich, B., Schwab, M. E., Nitsch, R. and Savaskan, N. E.: Molecular analysis of Nogo expression in the hippocampus during development and following lesion and seizure. FASEB J. *17*: 1153–1155, 2003.

Meldolesi, J. and Ceccarelli, B.: Exocytosis and membrane recycling. Philos. Trans. R. Soc. Lond. B. Biol. Sci. *296*: 55–65, 1981.

Meldolesi, J.: Calcium signalling. Oscillation, activation, expression. Nature. *392*: 863, 865–866, 1998.

Mesulam, M. M.: A horseradish peroxidase method for the identification of the efferents of acetylcholinesterase-containing neurons. J. Histochem. Cytochem. *24*: 1281–1284, 1976.

Mesulam, M. M.: Tetramethylbenzidine for horseradish peroxidase neurochemistry: a non-carcinogenic blue reaction product with superior sensitivity for visualizing neural afferents, and efferents. J. Histochem. Cytochem. *26*: 106–117, 1978.

Mesulam, M. M.: Human brain cholinergic pathways. Prog. Brain Res. *84*: 231–241, 1990.

Mesulam, M. M.: Some cholinergic themes related to Alzheimer's disease: Synaptology of the nucleus basalis, location of m2 receptors, interactions with amyloid metabolism, and perturbations of cortical plasticity. In: "From *Torpedo* Electric Organ to Human Brain: Fundamental and Applied Aspects," J. Massoulié, Ed., pp. 293–298, Elsevier, Amsterdam, 1998.

Mesulam, M. M.: Neuroanatomy of cholinesterases in the normal human brain and in Alzheimer's disease. In: "Cholinesterases and Cholinesterase Inhibitors," E. Giacobini, Ed., pp. 121–137, Martin Dunitz, London, 2000.

Mesulam, M. M.: The cholinergic innervation of the human cerebral cortex. Prog. Brain Res. *145*: 67–68, 2003.

Mesulam, M. M.: The cholinergic innrevation of the human cerebral cortex. Prog. Brain Res. *145*: 67–78, 2004.

Mesulam, M. M. and Geula, C.: Nucleus basalis (Ch4) and cortical innervation in the human brain: observations based on distribution of acetylcholinesterase and choline acetyltransferase. J. Comp. Neurol. *275*: 216–240, 1988.

Mesulam, M. M., Mufson, E. J., Wainer, B. H. and Levey, A. I.: Central cholinergic pathways in the rat: an overview based on an alternative nomenclature (Ch1–Ch6). Neurosci. *10*: 1185–201, 1983a.

Mesulam, M. M., Mufson, E. J., Levey, A. J and Wainer, B. H.: Cholinergic innervation of cortex by the basal forebrain: cytochemistry and cortical connections of the septal area, diagonal band nuclei, nucleus basalis (substantia innominata) and hypothalamus in the rhesus monkey. J. Comp. Neurol. *214*: 170–197, 1983b.

Mesulam, M. M., Mufson, E. J., Levey, A. I. and Wainer, B. H.: Atlas of cholinergic neurons in the forebrain and upper brainstem of the macaque based on monoclonal choline acetyltransferase immunohistochemistry and acetylcholinesterase histochemistry. Neurosci. 669–686, 1984.

Mesulam, M. M., Geula, C., Bothwell, M. A. and Hersh, L. B.: Human reticular formation: cholinergic neurons of the pedunculopontine and laterodorsal tegmental nuclei and some cytochemical comparisons to forebrain cholinergic neurons. J. Comp. Neurol. *283*: 611–33, 1989.

Michelson, M. J.: Some aspects of evolutionary pharmacology. Biochem. Pharmacol. *23*: 2211–2224, 1974.

Michelson, M. J. and Zeymal, E. B.: "Acetycholina," Hayka, Leningrad, 1970 (in Russian).

Mizukawa, K., McGeer, P. L., Tago, H., Peng, J. H., McGeer, E. G. and Kimura, H.: The cholinergic system of the human hindbrain studied by choline acetyltransferase immunohistochemistry and acetylcholinesterase histochemistry. Brain Res. *379*: 39–55, 1986.

Morel, N., Tabenblatt, P., Synguelakis, M. and Shiff, G.: A syntaxin-SNAP 25 blocking complex is formed without docking of synaptic vesicles. In: "From *Torpedo* Electric Organ to Human Brain: Fundamental and Applied Aspects," J. Massoulié, Ed., pp. 289–392, Elsevier, Amsterdam, 1998.

Moruzzi, G.: Contribution à l'électrophysiologie de cortex moteur: facilitation àpres decharge et epilepsie corticales. Arch. Intern. Physiol. *49*: 33–100, 1934.

Moruzzi, G. and Magoun, H. W.: Brain stem reticular formation and activation of the EEG. Electoenceph. Clin. Neurophysiol. *1*: 455–473, 1949.

Murthy, V. N. and Stevens, C. F.: Reversal of synaptic vesicle docking at central synapses. Nat. Neurosci. *2*: 503–507, 1999.

Mrzlyak, L., Levey, A. I. and Goldman-Rakic, P. S.: Association of m1 and m2 muscarinic receptor proteins with asymmetric synapses in the primate cerebral cortex: morphological evidence for cholinergic modulation of excitatory neurotransmission. Proc. Natl. Acad. Sci. USA *90*: 5194–5198, 1993.

Nachmansohn, D.: Choline acetylase. In: "Cholinesterases and Anticholinesterase Agents," G. B. Koelle, Ed., pp. 40–54, Handbch. d. exper. Pharmakol., Ergänzungswk., vol. 15, Springer-Verlag, Berlin, 1963.

Nachmansohn, D. and Machado, A. L.: The formation of acetylcholine. A new enzyme, "choline acetylase." J. Neurophysiol. *6*: 397–403, 1943.

Neher, E. and Sakmann, B.: The patch clamp technique. Sci. Amer. *266*: 44–51, 1992.

Newton, P. and Justice, J. B.: Quantitative microdialysis of acetylcholine under basal conditions and during acetylcholine inhibition in the rat. Soc. Neurosci. Abstracts *21*: 67, 1993.

Nishi, S., Soeda, H. and Koketsu, K.: Release of acetylcholine from sympathetic pregangliomic nerve terminals. J. Neurophysiol. *30*: 114–134, 1967.

Noremberg, K. and Parsons, S. M.: Regulation of the vesamicol receptor in cholinergic synaptic vesicles by acetylcholine and an endogenous factor. J. Neurochem. *52*: 913–920, 1989.

Palay, S. L. and Palade, G. E.: Fine structure of the neurons. J. Biophys. Biochem. Cytol. *1*: 69–88, 1955.

Palkovits, M. and Jacobowitz, D. M.: Topographic atlas of catecholamine- and acetylcholinesterase-containing neurons in the rat brain—II. Hindbrain (mesencephalon, rhombencephalon). J. Comp. Neurol. *157*: 29–42, 1974.

Peters, K. W., Qi, J., Johnson, J. P., Watkins, S. C. and Frizzell, R. A.: Role of snare proteins in CFTR and ENaC trafficking. Pflügers Arcr. *443*, Supplt. 1: S65–S69, 2001.

Prior, C. and Tian, L.: The heterogeneity of vesicular acetylcholine storage in cholinergic nerve terminals. Pharmacol. Res. *32*: 345–353, 1995.

Pyle, J. L., Kavalali, E., Piedras, T., Renteria, E. S. and Tsien, R. W.: Rapid reuse of readily releasable pool vesicles at hippocampal synapses. Neuron *28*: 221–231, 2000.

Quastel, J. H., Tennenbaum, M. and Wheatley, A. H. M.: Choline ester formation in, and choline esterase activities of tissues *in vitro*. Biochem. J. *30*: 668–681, 1936.

Rasmusson, D. D.: Cholinergic modulation of sensory information. In: "Cholinergic Function and Dysfunction," A. C. Cuello, Ed., pp. 357–364, Elsevier, Amsterdam, 1993.

Reigada, D., Diez-Perez, I., Gorostiza, P., Verdaguer, A., Gomez de Arnada, I., Pineda, O., Vilarrasa, J., Marsal, J., Blasi, J., Aleu, J. and Solsona, C.: Control of neurotransmitter release by an internal gel matrix in synaptic vesicles. Proc. Natl. Acad. Sci. USA *100*: 3485–3490, 2003.

Reiner, A., Perkel, D. J., Bruce, L. L., Butler, A. B., Csillag, A., Kuenzel, W., Medina, L., Paxinos, G., Shimizu, T., Striedter, G., Wild, M., Ball, G. F., Durand, S., Guturkun, O., Lee, D. W., Mello, C. V., Powers, A., White, S. A., Hough, G., Kubikova, L., Smulders, T. A., Wada, K., Dugas-Ford, J., Husband, S., Yamamoto, K., Yu, J., Siang, C. and Jarvis, E.

D.: Revised nomenclature for avian telencephalon and some related brainstem nuclei. J. Comp. Neurol. *473*: 377–414, 2004.

Rinaldi, F. and Himwich, H. E.: Alerting responses and actions of atropine and cholinergic drugs. A.M.A. Arch. Neurol. Psychiat. *73*: 387–395, 1955a.

Rinaldi, F. and Himwich, H.: Cholinergic mechanisms involved in function of mesodiencephalic activating system. Arch. Neurol. Psychiatry *73*: 396–402, 1955b.

Robinson, I. M., Ranjam, R. and Schwartz, T. L.: Synaptotagmins I and IV promote transmitter release independent of Ca^{2+}-binding in the C2A domain. Wature *418*: 336–340, 2002.

Rodieck, R. W. and Marshak, D. W.: Spatial density and distribution of choline acetyltransferase immunoreactive cells in human, macaque, and baboon retina. J. Comp. Neurol. *321*: 46–64, 1992.

Rothman, J. E.: Mechanisms of intracellular protein transport. Nature *372*: 55–63, 1994.

Ruggiero, D. A., Giuliano, R., Anwar, M. Stornetta, R. and Reis, D. J.: Anatomical substrates of cholinergic-autonomic regulation in the rat. J. Comp. Neurol. *292*: 1–53, 1990.

Ryan, T. A.: Presynaptic imaging techniques. Curr. Opin. Neurobiol. *11*: 544–549, 2001.

Rydberg, E. H., Macion, R., Ben-Mordehai, T. Z., Solomon, A., Rees, D. M., Toker, L., Botti, S., Auld, V. J., Silman, I. and Pussman, J. L.: Overexpression of the extracellular and cytoplasmic domains of the Drosophila adhesion protein, gliotactin. In: "Cholinergic Mechanisms, Function and Dysfunction," I. Silman, H. Soreq, L. Anglister, D. Michaelson and A. Fisher, Eds., pp. 687–689. Martin Dunitz, London, 2004.

Sanes, J. N. and Truccolo, W.: Motor "binding": do functional assemblies in primary cortex have a role? Neuron *38*: 3–5, 2003.

Saper, C. B. and Loewy, A. D.: Projections of the pedunculopontine tegmental nucleus in the rat: evidence of additional extrapyramidal circuitry. Brain Res. *252*: 367–372, 1982.

Sara, Y., Mozhayeva, M. G., Liu, X. and Kavalali, E. T.: Fast vesicle recycling supports neurotransmission during sustained stimulation at hippocampal synapses. J. Neurosci. *22*: 1608–1617, 2002.

Sastry, B. V.: Human placental cholinergic system. Biochem. Pharmacol. *53*: 1577–1586, 1997.

Satoh, K., Armstrong, D. M. and Fibiger, H. C.: A comparison of the distribution of central cholingergic neurons as demonstrated by acetylcholinesterase pharmacohistochemistry and choline acetyltransferase immunohistochemistry. Brain Res. Bull. *11*: 693–720, 1983.

Selden, N. R., Gitelman, D. R., Parrish, T. B. and Mesulam, M. M.: Trajectories of cholinergic pathways within the cerebral hemispheres of the human brain. Brain *121*: 2249–2257, 1998.

Semba, K., Reiner, P. B., McGeer, E. G. and Fibiger, H. C.: Brainstem projecting neurons in the rat basal forebrain: neurochemical, topographical, and physiological distinctions from cortically projecting cholinergic neurons. Brain Res. Bull. *22*: 501–509, 1989.

Senut, M. C., Menetrey, D. and Lamour, Y.: Cholinergic and peptidergic projections from the medial septum and the nucleus of the diagonal band of broca to dorsal hippocampus, cingulate cortex and olfactory bulb: a combined wheatgerm agglutinin-apohorseradish peroxidase-gold immunohistochemical study. Neurosci. *30*: 385–403, 1989.

Shepherd, G. M. and Harris, K. M.: Three-dimensional structure and composition of CA3 _ CA1 axons in rat hippocampal slices: implications for presynaptic connectivity and compartmentalization. J. Neurosci. *18*: 8300–8310, 1998.

Shute, C. C. D. and Lewis, P. R.: AChE distribution following lesions in the brain fibre tracts: several tract systems in the upper brain stem of the rat are cholinergic. Nature *199*: 1160, 1963.

Shute, C. C. D. and Lewis, P. R.: The ascending cholinergic reticular system: neocortical, olfactory and subcortical projections. Brain *90*: 197–518, 1967a.

Shute, C. C. D. and Lewis, P. R.: The cholinergic limbic system: projections to hippocampal cortex, nuclei of the ascending cholinergic reticular system, and the subfornical organ and supra-optic crest. Brain *90*: 521–540, 1967b.

Smiley, J. F.: Monoamines and acetylcholine in primate cerebral cortex: what anatomy tells us about function. Rev. Bras. Biol. *56*: 153–164, 1996.

Stedman, E. and Stedman, E.: The mechanism of biological synthesis of acetylcholine. The isolation of acetylcholine produced by brain tissue *in vitro*. Biochem. J. *31*: 817–827, 1937.

Stedman, E., Stedman, E. and Easson, L. H.: Cholineesterase. An enzyme present in the serum of the horse. Biochem. J. *26*: 2056–2066, 1932.

Stevens, C. F. and Williams, J. H.: "Kiss and Run" Exocytosis at Hippocampal Synapses. Proc. Natl. Acad. Sci. USA, *97*: 12828–33, 2000.

Sudhof, T. C.: The synaptic vesicle revisited. Neuron *28*: 317–320, 2000.

Sudhof, T. C.: CAPS in search of lost function. Neuron. *46*: 2–4, 2005.

Szentagothai, J. and Rajkovits, K.: Effect of specific neural function on structure of neural elements. Acta. Morphol. Acad. Sci. Hung. *5*: 253–274, 1955.

Tago, H., McGeer, P. I., Bruce, G. and Hersh, L. B.: Distribution of choline acetyltransferase-containing neurons of the hypothalamus. Brain Res. *415*: 49–62, 1987.

Tebecis, A. K.: Properties of cholinoceptive neurones in the medial geniculate nucleus. Brit. J. Pharmacol. *38*: 117–138, 1970a.

Tebecis, A. K.: Studies on cholinergic transmission in the medial geniculate nucleus. Brit. J. Pharmacol. *38*: 138–147, 1970b.

Thesleff, S.: Effects of motor innervation on the chemical sensitivity of skeletal muscle. Physiol. Rev. *40*: 734–752, 1960.

Thesleff, S.: Different kinds of acetylcholine release from the motor nerve. Int. Rev. Neurobiol. *28*: 59–88, 1986.

Torri-Tarelli, F., Grohovaz, F., Fesce, R. and Ceccarelli, B.: Temporal coincidence between synaptic vesicle fusion and quantal secretion of acetylcholine. J. Cell Biol. *101*: 1386–1399, 1985.

Torri-Tarelli, F., Valtorta, F., Villa, A. and Meldolesi, J.: Functional morphology of the nerve terminal at the frog neuromuscular junction: recent insights using immunocytochemistry. In: "Cholinergic Neurotransmission: Functional and Clinical Aspects," S.-M. Aquilonius and P.-G. Gillberg, Eds., pp. 83–92, Elsevier, Amsterdam, 1990.

Valtorta, F., Fesce, R., Grohovaz, F., Haimann, C., Hurblut, W. P., Iezzi, N., Torri-Tarelli, F., Villa, A. and Ceccarelli, B.: Neurotransmitter release and synaptic vesicle recycling. Neurosci. *35*: 477–89, 1990.

Valtorta, F., Meldolesi, J. and Fesce, R.: Synaptic vesicles: is kissing a matter of competence? Trends Cell. Biol. *11*: 3324–328, 2001.

Van Bockstaele, E. J. and Pickel, V. M.: Ultrastructure of serotonin-immunoreactive terminals in the core and shell of the rat nucleus accumbens: cellular substrates for interactions with catecholamine afferents. J. Comp. Neurol. *334*: 603–617, 1993.

Varoqui, H. and Erickson, J. D.: Dissociation of the vesicular acetylcholine transporter domains important for high-affinity transport recognition, binding of vesamicol and targeting to synaptic vesicles. J. Physiol. Paris. *92*: 141–144, 1998.

Vincent, S. R. and Reiner, P. B.: The immunohistochemical localization of choline acetyltransferase in the cat brain. Brain Res. Bull. *18*: 371–415, 1987.

von Schwarzenfeld, I.: Origin of transmitters released by electrical stimulation from a small metabolically very active vesicular pool of cholinergic synapses in guinea-pig cerebral cortex. Neuroscience. *4*: 477–493, 1979.

von Schwarzenfeld, I., Sudlow, G. and Whittaker, V. P.: Vesicular storage and release of cholinergic false transmitters. Prog. Brain Res. *49*: 163–174, 1979.

Wainer, B. H., Steininger, T. L., Roback, J. D., Burke-Watson, M. A., Mufson, E. J. and Kordower, J.: Ascending cholinergic pathways: functional organization and implications for disease models. Prog. Brain Res. *98*: 9–30, 1993.

Weible, M. W., 2nd, Ozsarac, N., Grimes, M. L. and Hendry, I. A.: Comparison of nerve terminal events effecting retrograde transport of vesicles containing neurotrophins or synaptic vesicle components. J. Neurosci. Rev. *75*: 771–181, 2004.

Weihe, E., Schafer, M. K., Schutz, B., Anlauf, M., Depboylu, C., Brett, C., Chen, L. and Eiden, L. E.: From the cholinergic gene locus to the cholinergic neuron. J. Physiol. (Paris) *92*: 385–388, 1998.

Wenk, G. L.: The nucleus basalis magnocellularis cholinergic system: one hundred years of progress. Neurobiol. Learn Mem. *67*: 85–95, 1997.

Whittaker, V.P.: The cell and molecular biology of the cholinergic synapse: twenty years of progress. In: "Cholinergic Neurotransmission: Functional and Clinical Aspects," S.-M. Aquilonius and P.-G. Gillberg, Eds., pp. 419–436, Elsevier, Amsterdam, 1990.

Whittaker, V. P.: "The Cholinergic Neuron and its Target: the Electromotor Innervation of the Electric Ray 'Torpedo' as a Model," Birkhauser, Boston, 1992.

Whittaker, V. P.: Arcachon and cholinergic transmission. In: "From *Torpedo* Electric Organ to Human Brain: Fundamental and Applied Aspects," J. Massoulié, Ed., pp. 53–57, 1998.

Whittaker, V. P. and Borroni, E.: The application of immunochemical techniques to the study of cholinergic function. In: "Neurobiology of Acetylcholine—A Symposium in Honor of Alexander G. Karczmar," N. J. Dun and R. L. Perlman, Eds., pp. 53–71, Plenum Press, New York, 1987.

Whittaker, V. P., Michaelson, I. A. and Kirkland, R. J.: The separation of synaptic vesicles from nerve-endings particles ("synaptosomes"). Biochem. J. *90*: 293–303, 1964.

Wilson, P. M.: A photographic perspective on the origin, form, course and relations of acetylcholinesterase-containing fibres of the dorsal tegmental pathway in the rat brain. Brain Res. Rev. *10*: 85–118, 1985.

Woolf, N. J.: Cholinergic systems in mammalian brain and spinal cord. Prog. Neurobiol. *37*: 475–524, 1991.

Woolf, N. J.: Global and serial neurons form a hierarchically arranged interface proposed to underlie memory and cognition. Neuroscience. *74*: 625–651, 1996.

Woolf, N. J.: A possible role for cholinergic neurons of the basal forebrain and pontomesencephalon in consciousness. Conscious Cogn. *6*: 574–596, 1997.

Woolf, N. J.: A structural basis for memory storage in mammals. Prog. Neurobiol. *55*: 59–77, 1998.

Woolf, N. J. and Butcher, L. L.: Cholinergic systems in the rat brain: III. Projections from the pontomesencephalic tegmentum to the thalamus, tectum, basal ganglia, and basal forebrain. Brain Res. Bull. *16*: 603–637, 1986.

Woolf, N. J. and Butcher, L. L.: Cholinergic systems in the rat brain: IV. Descending projections of the pontomesencephalic tegmentum. Brain Res. Bull. *23*: 519–540, 1989.

Woolf, N. J., Hernit, M. C. and Butcher, L. L.: Cholinergic and non-cholinergic projections from the rat basal forebrain revealed by combined choline acetyltransferase and Phaseolus vulgaris leucoagglutinin immunohistochemistry. Neurosci. Lett. *66*: 281–286, 1986.

Wu, D. and Hersh, J. B.: Choline acetyltransferase: celebrating its fiftieth year. J. Neurochem. *62*: 1653–1663, 1994.

Ybe, J. A., Wakeham, D. E., Brodsky, F. M. and Hwang, P. K.: Molecular structures of proteins involved in vesicle fusion. Traffic *1*: 474–479, 2000.

Yin, Y., Kikkava, Y., Mudd, J. L., Skarnes, W. C., Sanes, J. R. and Miner, J. H.: Expression of laminin chains by central neurons: analysis with gene and protein trapping techniques. Genesis *36*: 114–127, 2003.

Zhang, H. M., Li, D. P., Chen, S. R. and Pan, H. L.: M2, M3 and M4 receptor subtypes contribute to muscarinic potentiation of GABAergic inputs to spinald orsal horn neurons. J. Pharmacol. Exp. Therap., *3*: 697–704, 2005.

Zimmerman, H. and Whittaker, V. P.: Morphological and biochemical heterogeneity of cholinergic synaptic vesicles. Nature *267*: 633–635, 1977.

Zucker, M., Weizman, A. and Rehavi, M.: Characterization of high-affinity [3H]TBZOH binding to the human platelet vesicular monoamine transporter. Life Sci. *69*: 2311–2317, 2001.

Zweig, R. M., Whitehouse, P. J., Casanova, M. F., Walker, L. C., Jankel, W. R. and Price, D. L.: Loss of pedunculopontine neurons in progressive supranuclear palsy. Ann. Neurol. *22*: 18–25, 1987.

3

Metabolism of Acetylcholine: Synthesis and Turnover

A. Historical Introduction

According to John Eccles (1936), the precondition for accepting acetylcholine (ACh) as the cholinergic transmitter is that there should be a way to destroy it or otherwise terminate its action rapidly. The corollary of this concept is that a mechanism must be available for maintaining ACh synthesis during the course of synaptic activity. Lindor Brown and William Feldberg formulated this corollary at the same time. They stated, "If it is true that acetylcholine is liberated by nerve impulse and not synthesized . . . by this impulse . . . resynthesis must occur at some time if the tissue is to retain its acetylcholine" (Brown and Feldberg, 1936).

The first attempts at confirming these hypotheses pinpointed a mechanism for ACh synthesis: rapid hydrolysis of ACh may provide needed and sufficient choline for rapid ACh synthesis. Brown and Feldberg (1936) provided indirect evidence for this mechanism: they bioassayed ACh and choline in the Locke perfusate of the cat superior cervical ganglion and demonstrated that "stimulation of the preganglionic fibers causes . . . a 2 to 5 total . . . increase in total choline present in the venous outflow" over the choline value in quiescent ganglion outflow. Also, they found that in the course of ganglionic stimulation, "the choline in the venous outflow can all be accounted for by acetylcholine liberated by stimulation." However, Feldberg and Brown were not explicit in formulating the hypothesis (expressed later by Perry, 1953) that ganglionic "output of acetylcholine during stimulation remains consistently high . . . and ACh . . . is immediately hydrolyzed and the choline used in the re-synthesis of a 'bound' form of acetylcholine." A further basis for this hypothesis was provided in 1961 by Birks and MacIntosh: they found in the *in situ* perfusate of the cat superior cervical ganglion small (700 μg/ml) amounts of choline (for similar earlier data see Bligh, 1952); thereupon, they calculated that to maintain the constant release of ACh during a 20 second stimulation of the ganglion, the ganglion must be "remarkably efficient in extracting choline from the extracellular fluid." Ultimately, Hank MacIntosh (with P. Oborin) employed 3-hemicholinium, a compound developed by Fred Schueler that blocks choline nerve terminal uptake, to prove their point. Thus, the research of Brown, Feldberg, MacIntosh, Birks, and Perry established a strict relation between released ACh and choline, as well as the concept of a highly efficient choline uptake system in the terminal.

In addition, this early research led to another ACh dynamic. Feldberg, Brown, Birks, and MacIntosh stressed that ACh release in ganglionic tissues is highly variable and related to synaptic activity (as expressed experimentally by the rate of preganglionic stimulation). Using extrapolation, Brown and Feldberg (1936a, 1936b) and Birks and MacIntosh (1961) linked ACh release and synaptic activity to the rate of ACh synthesis. Accordingly, Birks and MacIntosh (1961) stated that "during prolonged stimulation at high frequency the rate of release goes up by a factor of perhaps 70 and the rate of synthesis by a factor of 7" (for similar data obtained earlier, see Bligh, 1952). At the time, Birks and MacIntosh were prescient; they antedated Victor Whittaker (see Chapter 2 B), as they proposed that the release also depended on traffic between several ACh "pools," such as "stored" ACh and "available" ACh; they suggested that the "available" ACh pool was a pool of "newly synthesized depot

ACh." These concepts relate to even earlier demonstrations by Mann et al. (1938) and MacIntosh (1938) that ACh synthesis varies in brain slices depending on the availability of glucose. A similar phenomenon was demonstrated for the neuromyal junctions, when Alkon et al. (1970) and Potter (1970) later demonstrated the relation between synaptic activity and ACh synthesis. Altogether, this concept of flexible, activity-dependent ACh synthesis and the related idea of ACh pools (which function differently during synaptic activity) constitute the basis for the current research on ACh pools, cycling of synaptic vesicles, and so forth (see Chapter 2 B).

Brown and Feldberg forged another link or two in the story of choline's relation to ACh. When fatigue attenuated the contractions of stimulated nictitating membrane, the contractions were restored after minute quantities of whole blood were added to the perfusate of the superior cervical ganglion (Brown and Feldberg, 1936a, 1936b). The investigators concluded that an "essential" blood factor was needed for ACh synthesis; when they demonstrated that applying small concentrations of choline to the perfusate restored decreased ACh output and nictitating membrane responses, they suggested that the blood factor in question was choline. Then they showed that choline is present in ganglionic tissue,[1] that its levels increase in the perfusate of the stimulated ganglion, and that it maintains high ACh output from a stimulated ganglion when added to the perfusate. They concluded that "the remarkable activity of choline . . . suggested that there might exist in the ganglion a metabolism of choline, related to preganglionic stimulation, and comparable to that of acetylcholine." This notion was supported by Gerard and Tupikova (1939) and Alexander von Muralt,[2] as they demonstrated that the nerve activity induced phospholipid breakdown and the generation of an acid-soluble phosphatide fraction; this fraction yielded free choline that was available for acetylation into ACh.

Almost simultaneously, Quastel, Mann, and their associates observed ACh synthesis under aerobic conditions and felicitously suggested that this process may require an enzyme system (Quastel et al., 1936), which appeared very effective, as it yielded up to 50 μg/g/tissue/hour of ACh (Mann et al., 1938, 1939). This enzyme proved to be choline acetyltransferase (the abbreviation for this enzyme used in this book is CAT; elsewhere, the abbreviation ChAT is frequently employed), and it was discovered by David Nachmansohn and his associates in the course of their studies of the Torpedo electric organ. They called it "choline acetylase" (Nachmansohn and Machado, 1943; the enzyme was renamed "choline acetyltransferase" in 1961). Before proving the enzymic nature and identity of the pertinent factor present in the electric organ extract, Nachmansohn (Nachmansohn and Machado, 1943) first established the energy requirement of the CAT-activated synthesis. He showed that Torpedo's electric organ's phosphate bonds yield energy for the action of CAT, similar to the phosphate bonds providing energy for the action potential of the muscle and for the muscle contraction (Meyerhof, 1941), and he used ATP in the ACh-forming extracts. The extract was sensitive to temperature, iodoacetate, HCN, pH, K^+ and certain quaternary ions (see Nachmansohn's 1963a review of this subject), which is characteristic for enzymes. In their paper "A New Enzyme: Choline Acetylase," Nachmansohn and Machado (1943) concluded, "Energy rich phosphate bonds are used . . . in conjunction with CAT . . . for resynthesis . . . following its hydrolysis upon release . . . of ACh."

Jointly with the discovery of CAT, Nachmansohn and Machado described for the first time an acetylating system that is required to complete the system needed for ACh synthesis from choline: their choline acetyltransferase–containing extract lacked the capacity to synthesize ACh after dialysis (Nachmansohn and Machado, 1943; Nachmansohn 1943, 1963a). Then, Lipmann and Lipton (Lipmann and Kaplan, 1946) identified the missing factor as a nonprotein organic cofactor or coenzyme called "the acetyl CoA" and demonstrated that acetyl grouping is transferred to choline when ATP and choline acetyltransferase are present.

Altogether, the 1930s and 1940s studies of William Feldberg, Lindor Brown, Frank MacIntosh, David Nachmansohn, P. Mann, J. Quastel, F. Lipmann, and their associates established that a system consisting of choline derived by hydrolysis from released ACh, choline generated by neuronal phospholipid metabolism, choline nerve terminal uptake process, choline acetyltransferase, coenzyme A, and glucose may account for synaptic (and also metabotropic) ACh. This was a solid

Figure 1-3. Peter Waser, Alex Karczmar, and George Koelle during the 1974 Symposium on Cholinergic Mechanisms in Boldern, Switzerland.

Figure 1-4. Superimposed note from Edith Heilbronn. From left to right: *row 1* (seated): John Blass, George B. Koelle, Peter G. Waser, Donald J. Jenden, Israel Hanin, Frank C. MacIntosh, Alex Karczmar, Edith Heilbronn, Giancarlo Pepeu, Alan M. Goldberg; *row 2*: Victor J. Nickolson, Nae J. Dun, Stanley M. Parsons, Agneta Nordberg, Ezio Giacobini, B.V. Rama Sastry, Kathleen A. Sherman, Mario Marchi, Michael Stanley, Larry L. Butcher, Fiorella Casamenti, Tsung-Ming Shih, Herbert Ladinsky, Silvana Consolo, Kenneth L. Davis, Darwin L. Cheney, Janusz B. Suszkiw, Michael R. Kozlowski; *row 3*: Dean O. Smith, Steven H. Zeisel, Susan E. Robinson, Barbara Lerer, R. Jane Rylett, Rochelle D. Schwartz, Joan Heller-Brown, Marie-Louise Tjörnhammer, Britta Hedlund, David S. Janowsky, Natraj Sitaram, Linda M. Barilaro, Paul M. Salvaterra, Denise Sorisio, Elias Aizenman, Ileana Pepeu, Aurora V. Revuleta, Felicita Pedata, Clementina Bianchi, Lorenzo Beani, Henry G. Mautner; *row 4*: S. Craig Risch, Guillermo Pilar, E. Sylvester Vizi, Thomas J. Walsh, Sikander L. Katyal, Rob L. Polak, Roni E. Arbogast, Jean Massoulié, Denes Agoston, Brian Collier, Lynn Wecker, Bruce Howard, Richard S. Jope, Bernard Scatton, Matthew Clancy, Paul T. Carroll; *row 5*: William G. VanMeter, Michael Adler, Peter Kasa, Annica B. Dahlström, Gary E. Gibson, Peter C. Molenaar, Ingrid Nordgen, John D. Catravas, Judith Richter, David M. Bowen, Mark Watson, Renato Corradetti, Lorenza Eder-Colli, Marvin Lawson, Ing K. Ho, Jack C. Waymire; *row 6*: Paul L. Wood, Matthew N. Levy, Jean-Claude Maire, Frans Flentge, Richard Dahlbom, Pierre Etienne, George G. Bierkamper, Robert G. Struble, A.J. Vergroessen, Seana O'Reagan, Robert Manaranche, Maurice Israel, Yacov Ashani, Abraham Fisher, Steven Leventer, Alan G. Mallinger; *row 7*: Anders Undén, Edward F. Domino, William D. Blaker, Peteris Alberts, Johann Häggblad, Daniel L. Rickett, Sten-Magnus Aguilonius, Serge Mykita, Hans Selander, Oliver Brown, Henry Brezenoff, Sven-Åke Eckernäs, Frederick J. Ehlert, Björn Ringdahl, Volker Bigl, Duane Hilmas, Clark A. Briggs, Nicolas Morel; *row 8*: Bo Karlén, Michael J. Dowdall, John J. O'Neill, Heniz Kilbinger, Wolf-D. Dettbarn, Konrad J. Martin, Konrad Löeffelholz, Roy D. Schwarz, Jerry J. Buccafusco, Ernst Wulfert, Howard J. Colhoun, Paul Martin, Jack R. Cooper, Crister Larsson, Harry M. Geyer, Michael J. Pontecorvo, William E. Houston, Jurgen von Bredow, Yves Dunant. (From Hanin, 1986. Reprinted by permission of Kluwer Academic/Plenum Press.)

Figure 1-7. U.J. (Jack) McMahon (left) and Rene Couteaux at the ISCM meeting in Arcachon, 1998. Jack McMahon is a prominent modern investigator of the ultrastructure of the neuromyal junction. (From the Author's private collection.)

Figure 1-9. Chris (Kazmir) Krnjevic at the Symposium in Honor of A.G. Karczmar, Loyola Medical Center, Maywood, IL, 1986.

Figure 1-10. From left to right: *row 1*: Doctoressa Ileana Pepeu, Erminio Costa, Marion Allen-Karczmar (Swami Sharadananda) and Leda Hanin; *row 2*: Israel Hanin, Giancarlo Pepeu, Alex Karczmar. Erminio (Mimo) Costa is a prominent psychoneuropharmacologist and one of the founders of the College of Neuropsychopharmachology. He pioneered the research on the role of serotonin, GABA, and catecholamines. He also contributed to the concept and measurement of the turnover of acetylcholine.

Figure 1-11. Edith McGeer at the 1986 Symposium in Honor of A.G. Karczmar, Loyola Medical Center, Maywood, IL.

Figure 1-12. Nobel Prize winner Rita Levi-Montalcini and Ezio Giacobini at a Vatican Symposium, 1990.

Figure 1-13. Hermona Soreq is a prominent scientist from the Hebrew University of Jerusalem. Soreq pioneered molecular studies of cholinesterases. Currently she studies stress-induced variants of acetylcholinesterase.

Figure 1-15. Vincenzo Longo and Alex Karczmar in Rome, 1978. Vincenzo Longo, from the Istituto Superiore di Sanità, is a preeminent EEG investigator. He related EEG events and evoked potentials to behavior and REM sleep.

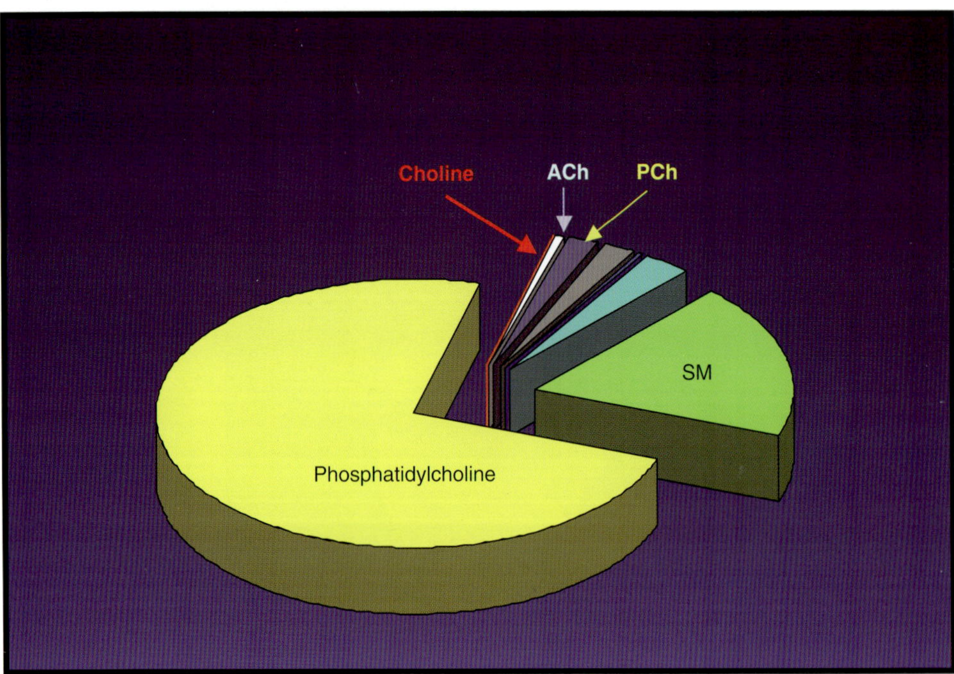

Figure 3-3. Distribution of choline in brain tissue. (From Loeffelholz and Klein, 2004, with permission.) ACh = acetylcholine; PCh = phosphatidylcholine; SM = sphingomyelin.

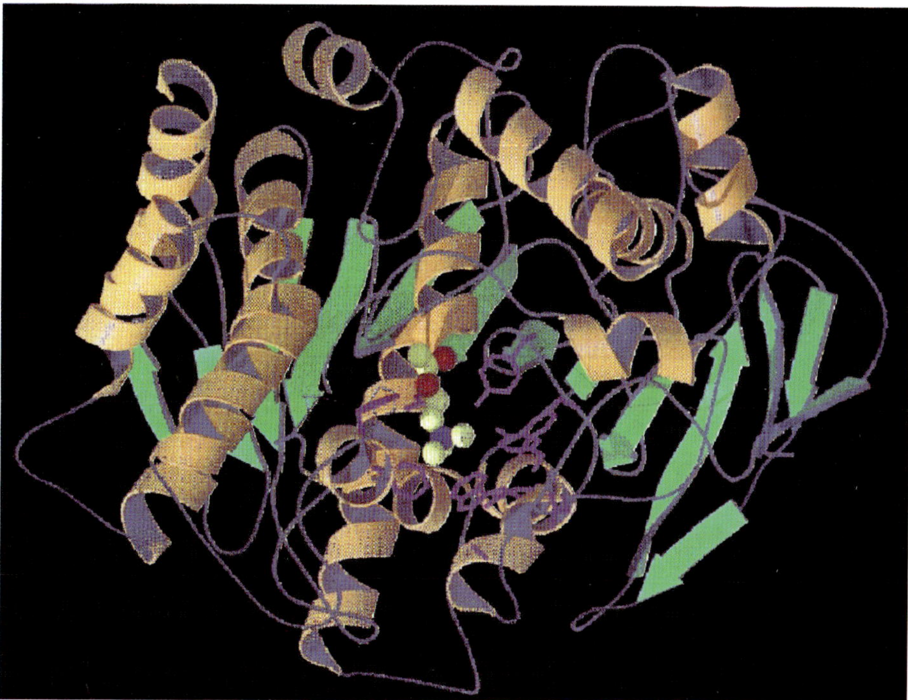

Figure 3-8. Ribbon diagram of the three-dimensional structure of TcAChE. The side chains of the catalytic triad and of key aromatic residues in the active-site forge are indicated as purple stick figures. Acetylcholine, manually docked in the active site, is represented as a space-filling model, with carbon atoms shown in yellow, oxygen atoms in red, and nitrogen in blue. The quaternary group of the ACh faces the indole of Trp84. (From Silman and Sussman, 2000, with permission.)

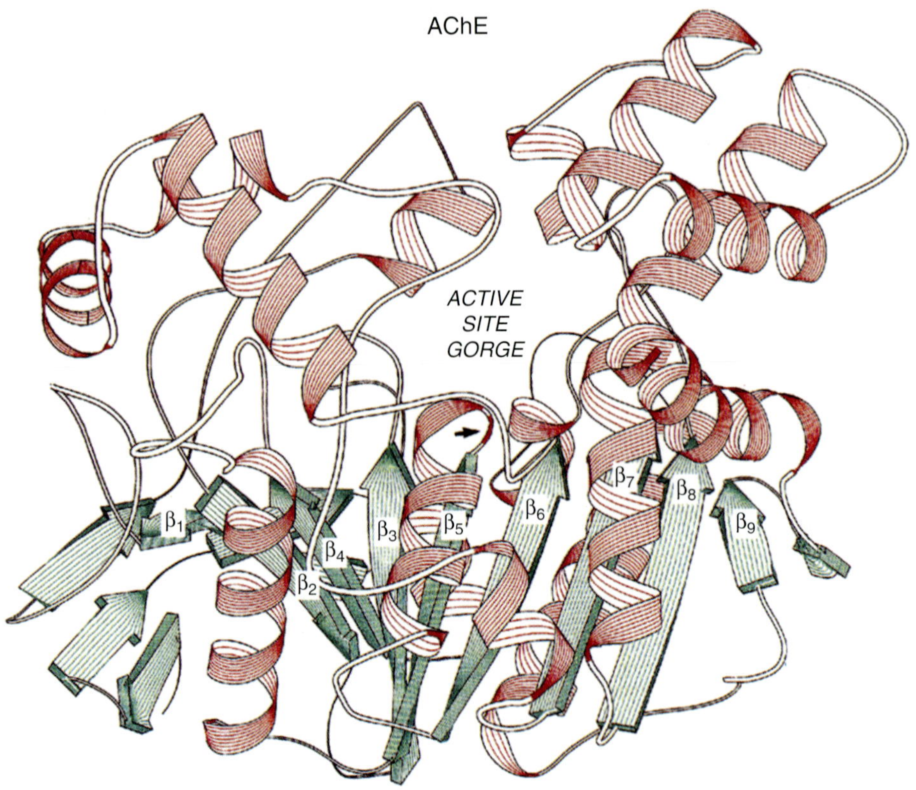

Figure 3-9. Three-dimensional structure of G_2^a AChE from Torpedo. The structure is represented in a ribbon diagram showing β strands (green) and α as helices (red). The active gorge is located above the central β sheet and the arrow marks the location of the active-site serine, Ser 200. (From Massoulie et al., 1993b. with permission.)

Figure 7.2. The plant *Physostigma venenosum Balforii*. (From Karczmar et al., 1970.)

Figure 11-1. Distant peaks. (From Goodwin Ahlberg, 1998, with permission.)

Figure 7.2. The plant *Physostigma venenosum Balforii*. (From Karczmar et al., 1970.)

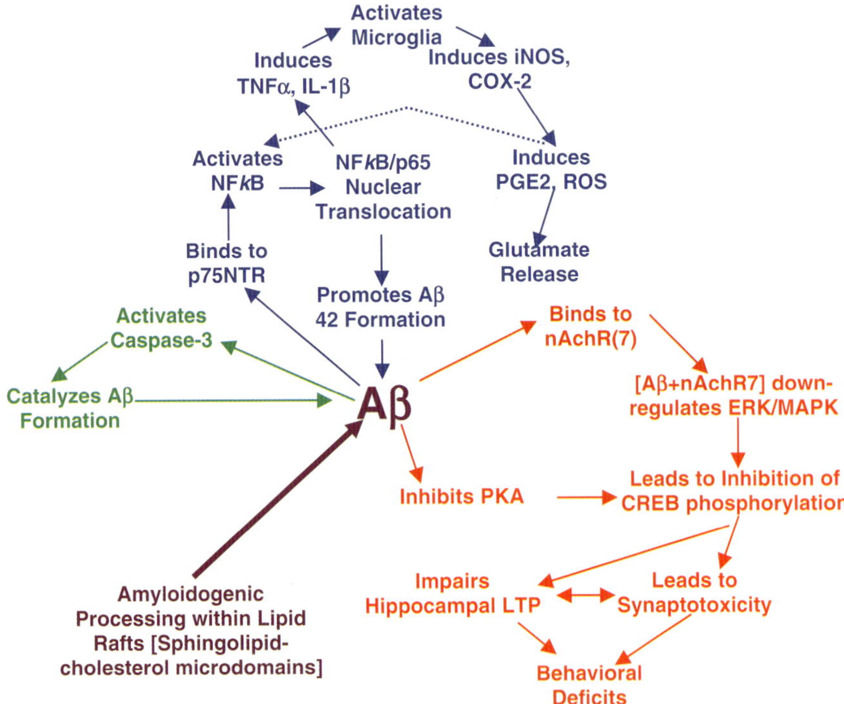

Figure 10-2. Major events of amyloidogenic vicious cycle include: (top) Aβ-induced activation of nuclear transcription factor-NF*k*B, which in turn activates the pro-inflammatory cytokines (TNFα, IL-1β, IFNγ), leading to subsequent induction of inflammatory molecules such as inducible nitric oxide (iNOS), cyclooxygenase 2 (COX-2), and prostaglandin E2 (PGE2). These inflammatory molecules produce reactive oxygen species (ROS), leading to oxidative damage. In addition, COX-2 and iNOS trigger further activation of NF*k*B. NF*k*B binds to the p65 subunit, translocates to the nuclear compartment, and promotes the formation of $A\beta_{42}$. Besides being a part of this vicious cycle, all these inflammatory molecules have other detrimental effects on neurons and neuronal connections, thus participating in neurodegeneration and synaptic deficits in AD; (left) activation of caspase-3, which leads to apoptosis and simultaneously potentiates further Aβ generation; (bottom right) impairment of PKA- and ERK/CREB-signaling, leading to synaptic and LTP deficits in the hippocampus that result in impairment of memory and learning; (bottom left) cholesterol, lipid, and lipoprotein receptor interactions with Aβ.

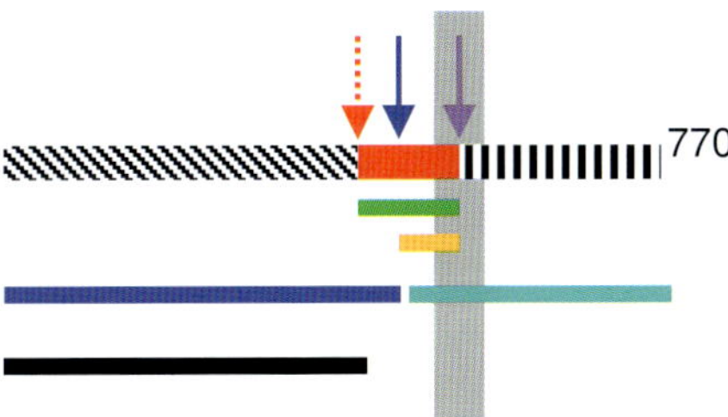

Figure 10-3. Schema showing proteolytic cleavage sites on the APP_{770} molecule and the different peptides resulting from proteolytic cleavages: bar with diagonal lines, extracellular N-terminal APP fragment; solid bar between diagonal- and vertical-line bars, Aβ-segment; bar with vertical lines, intracellular C-terminal fragment; dashed arrow, β-secretase site (after residue 671); center arrow, α-secretase site (after residue 687); right arrow, γ-secretase site (after residues 710 or 712). Second horizontal line represents full-length Aβ 40/42/43. Third horizontal line represents p3 fragment. Fourth horizontal line (left) represents sAPPα fragment (N-terminal-687). Fourth horizontal line (right) represents carboxyterminal fragment (CTF). Bottom horizontal line represents sAPPβ fragment (N-terminal-671). Vertical bar represents the membrane-spanning segment, which comprises residues 700 to 723.

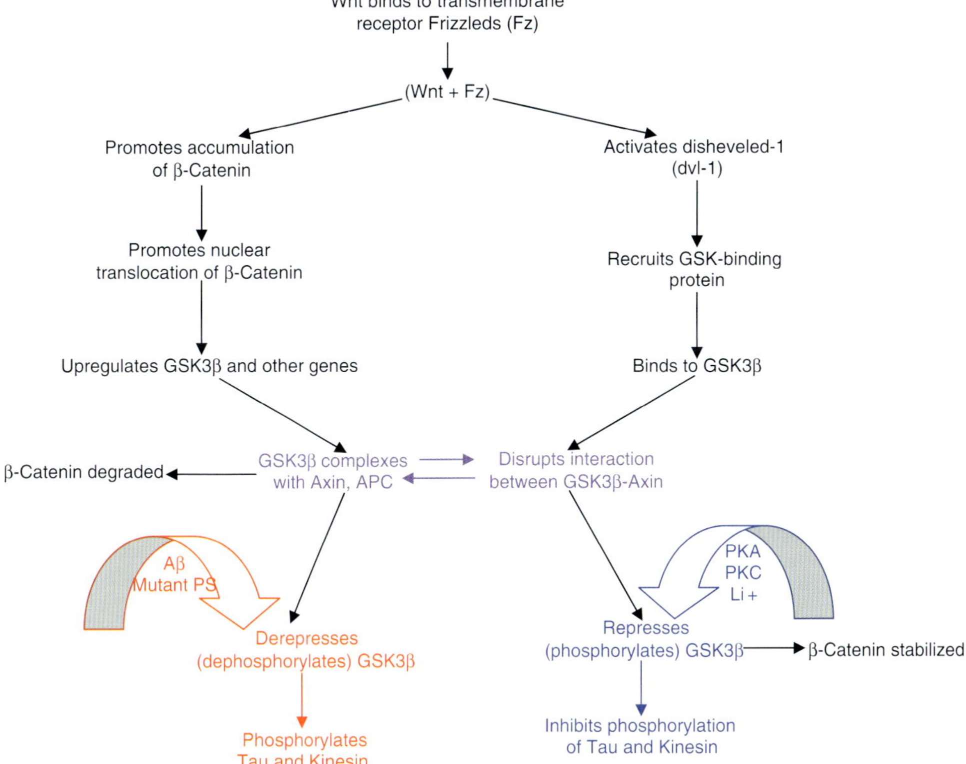

Figure 10-4. Schema showing GSK3β signaling involved in linking amyloid, tau, and kinesin biology. Signaling pathway in black indicates Wnt-GSK3β cascade under normal circumstances. Signaling events in lavender indicate dynamic balance between phosphorylation and dephosphorylation of tau under normal circumstances. Signaling events at bottom left indicate overactivation of GSK3β leading to overphosphorylation of tau and kinesin under abnormal conditions. Signaling events at bottom right indicate factors that attenuate overphosphorylation of tau and kinesin.

Figure 11-1. Distant peaks. (From Goodwin Ahlberg, 1998, with permission.)

base for further studies that established details and mechanisms of this system and generalizations concerning the source of choline needed for ACh synthesis and release and homeostasis of choline pools in the brain and in the plasma.

B. ACh Synthesis

As stated in the preceding section, the main components of ACh synthesis are acetyl coenzyme A, metabolisms generating acetate and choline, choline's uptake system, and "choline acetyltransferase (EC-2.3.1.6). The International Union of Biochemistry's Enzyme Commission recommended in 1961 the name "choline acetyltransferase" as the official substitution for the original term "choline acetylase" proposed by Nachmansohn and Machado (1943; see previous section). I will discuss the metabolic processes that concern these components and consider which of these may be considered as a limiting step in the biosynthesis of ACh. Also, the choline system and its uptake cannot be considered alone but must be discussed in conjunction with choline-containing lipids and their metabolism. It must be added that the choline uptake and metabolism must be linked with choline-containing lipids and their metabolism. This matter was clearly expressed in 1979, when Ansell and Spanner commented, "For the first 50 years of this century the main interest in choline was as a constituent of acetylcholine . . . this interest and . . . the study of the biochemistry of acetylcholine . . . were . . . totally divorced from that of choline-containing lipids (notably glycerophospholipids) . . . ." The link between choline lipids and the cholinergic system was not forged until the 1970s.

1. Choline Acetyltransferase

Similar to other neuronal proteins, choline acetyltransferase (CAT; EC2.3.1.6) appears to be synthesized in the neuronal perikarya; this has been established via CAT extraction and subsequent measurement of its activity or via histochemical immunoassay techniques (see Tucek, 1970, 1990; Ichikawa and Hirata, 1986). This neuronal CAT generation is characteristic for cholinergic neurons; nevertheless, CAT may exist in nonneuronal vertebrate or invertebrate tissues (see Chapter 2 B).

The gene for CAT, initially described for Drosophila (Greenspan, 1980), was identified subsequently for other species (see Tucek, 1990); it was isolated in the human by Louis Hersh and M. Hahn (see Hersh et al., 1996; see also Mallet et al., 1990). The initiation codon was established for rodent and human genes (Hersh et al., 1996). The human gene contains an enhancer and a repressor element; the genes of other species may also include a silencer element (Hersh et al., 1996). The CAT genes are a part of the "cholinergic gene locus" that also contains other cholinergic genes, particular the vesicular ACh transporter (VAChT; Kim and Hersh, 2004; see also Chapter 2 B2). The CAT genes expression can be modulated by a number of growth and differentiation factors, which also increase the multiple species of CAT mRNA (Mallet et al., 1998). According to Jacques Mallet, this modulatory capacity of CAT gene expression constitutes the basis for the "phenotypic plasticity" of the CAT gene expression. This plasticity is consistent with the notion of multiple forms of CAT that are described below in this section.

Choline acetyltransferase is involved in the following reaction:

$$\text{Acetyl Coenzyme A} + \text{Choline} \xrightarrow{\text{CAT}} \text{ACh} + \text{Coenzyme A}$$

This reaction follows a random Theorell-Chance mechanism in which a low number of ternary complexes participate. It is reversible, though under equilibrium conditions the concentrations of ACh and coenzyme A are higher than concentrations of choline and acetyl coenzyme A. Indeed, K_{eq} values reported by various investigators are high, and it is consistent with these high values that CAT has a much higher affinity for acetyl coenzyme A than for choline (Glover and Potter, 1971). However, Stanislaus Tucek (1978, 1984, 1990) stated that "the published values are surprisingly variable," ranging from 12.3 (Pieklik and Buynn, 1975) to 514 (Potter et al., 1968). This is because many factors influence K_{eq}: ionic composition of the media, the pH, the enzyme's source, its physical form or isoenzymic character, the assay system conditions, and so on, and these factors vary from one laboratory to another. K_{eq} and free energy change are related; accordingly, the reaction is exergenic.

An interesting question is, How crucial is CAT for the synthesis of ACh? Is CAT the rate-limiting factor for this synthesis? The other candidates for this exalted position are choline uptake, factors involved in choline metabolism, and acetyl CoA. As far as CAT is concerned, this question may be resolved as follows. The actual intracellular reaction rate of CAT, *v*, relates to CAT's maximal reaction rate, *V*, via the dissociation constants (K_m's) for the 2 CAT substrates, choline and acetyl CoA (400 μM and 10 μM, respectively) and the intracellular concentrations of these substrates; this relationship is given by an equation proposed by Gutfreund (1965; see also Tucek, 1984, 1985, 1990). The K_m values of the substrates are significantly higher than their intracellular concentrations (i.e., 50 μM for choline and 5 μM for acetyl CoA; White and Wu, 1973); this means that CAT's intracellular activity is far less than its potential maximum rate (*V*), which obtains when the substrate concentration is saturated. This suggests, although it does not exclude the possibility, that CAT is not a rate-limiting factor.

Stanislaus Tucek felt that this notion is reinforced by Potter's (1970) results. Potter worked with the rat nerve and used [^{3}H] ACh and the leech bioassay of ACh. As he measured the release of ACh, the replacement of endogenous ACh with the exogenous, tritiated compound, and CAT activity, he found that "ACh synthesis rates . . . were . . . sufficient . . . so that . . . fully replaced stores were maintained . . . (in appropriate segments of the diaphragm) . . . during nerve stimulation." On the basis of Potter's and other investigators' data, Tucek calculated that "the highest rate of release . . . of ACh . . . (at 20H$_z$ rate of stimulation) was 17 times lower than the maximum activity of CAT." However, Tucek (1984) had a reservation to make. At the physiological intracellular concentrations of acetyl CoA and choline, the activity of CAT is many times lower than its maximal rate of activity (*V*) at saturation concentrations of the substrates. Thus, it is possible that under certain conditions CAT activity may not be sufficient to maintain the rate of ACh synthesis that would be needed to compensate for ACh release at high levels of synaptic activity; accordingly, Tucek (1984) states that "the possibility of CAT becoming a rate limiting factor in synthesis of ACh . . . should not be . . . excluded *a priori*," particularly when there is a stress on ACh synthesis caused by depletion of ACh under the conditions of the experiments of Kasa et al. (1982) and Wecker et al. (1978; see below). Similarly, Potter (1970) warned that "it is not possible to say whether the higher release rates observed with . . . the nerve-diaphragm preparation and with the sympathetic ganglia (Birks and MacIntosh, 1961) . . . are accompanied by correspondingly high rates of transmitter synthesis." Nevertheless, Potter's and Tucek's calculations suggest that CAT's maximal capacity exceeds requirements for ACh synthesis and that CAT is not the rate-limiting factor during fast neuromyal function; does this hold for central cholinergic transmission?

The purification and molecular and biophysical analysis of CAT helped establish its structure, its active centers, and their catalytic groupings. The imidazole group of histidine plays an important role, as it is involved in the binding of the acetyl group of acetyl coenzyme A: immobilized on a gel, CAT binds the label from [14C] acetyl-CoA, and this binding is prevented when histidine residues are masked with diethylpyrocarbonate (Malthe-Sorensen, 1976; for further evidence supporting this concept, see Tucek, 1978 and Baker and Dowdall, 1976). Furthermore, CAT contains a number of thiol groups, and its activity and ACh synthesis are blocked by SH reagents such as iodoacetate or methane thiosulphonate (cf. Tucek, 1978). However, some of this inhibition may result from an effect of SH reagents on coenzyme A (Currier and Mautner, 1976). Moreover, this inhibition is not specific, because it is common for many enzymes. Finally, CAT is subject to phosphorylations at its serine and threonine residues, and these phosphorylations may regulate the activity of the enzyme as well as its membrane-binding properties (see Dobransky and Rylett, 2003, 2005).

Choline acetyltransferase exhibits an anionic site subserving its binding to choline's quaternary nitrogen. This site may be similar to the anionic site of most ChEs (see sections DI–DII), and it contains a dicarboxylic amino acid. Several schemes for the reaction between CAT and acetyl CoA were postulated. According to Malthe-Sorenssen (1976), acetyl CoA attaches to an imidazole grouping of CAT; subsequently it undergoes a nucleophilic attack by CAT and forms the acetylated imidazole. Then choline binds the anionic site and receives the acetyl group. The first product released is ACh, followed by CoA. Other mechanisms were also proposed (Tucek, 1978). Similar to ChEs that hydrolyze many choline esters (see section DII), CAT transfers the acyl groups of propionyl CoA and butyryl CoA to choline. However, this transfer is slower than that of the acetyl group (Banns et al., 1977; cf. Tucek, 1978). Similarly, aliphatic and other substitutions on the nitrogen of choline lower the rate of acetylation of these compounds by CAT (Baker and Dowdall, 1976; see Tucek, 1978).

The analytical procedures revealed some other characteristics of CAT. Choline acetyltransferase is a basic globular protein with a radius of 3.39 nm (Rossier, 1976a, 1976b; Rossier and Benda, 1978). McGeer et al. (1987) stressed, "Regardless of its source, there is remarkable consistency regarding the molecular weight of native CAT, possibly signifying the homogeneity of the gene through many species." This weight is approximately 68,000 daltons for most species. Choline acetyltransferase is also present as monomeric variants with a molecular weight of 58,000 to 70,000 daltons, their aggregates, and isoenzymes (cf. Tucek, 1978; Blusztajn and Wurtman, 1983). However, the isoenzymes may be artifacts or may represent aggregates (Hersh et al., 1984; Hirata and Axelrod, 1980).

Cellular localization of CAT and its various forms was primarily studied in nerve terminal preparations. The pioneering research of Catherine Hebb and Victor Whittaker (see Whittaker, 1965) established that CAT, which is present in their synaptosome preparations, while bound in the particulate matter, can be readily released into the solution. This finding is consistent with subsequent studies that established that CATs are located in the cytoplasm, are water soluble, bind reversibly with membranes, and are easily extracted and purified by centrifugation; on fractionation of hypo-osmotically treated nerve endings, they are found in fractions containing cytoplasm and its soluble components (Smith and Carroll, 1980). However, following some of these treatments, certain forms of CATs remain associated with particulate neuron matter, including various membranes (McCaman et al., 1965; see Tucek, 1984, 1985, 1990; Blusztajn and Wurtman, 1983). The particulate and soluble CAT fractions may exhibit different physical characteristics or they may represent different isoenzymes; they may exhibit different properties, such as kinetics and affinities for their substrates (choline and acetyl CoA; cf. Blusztajn and Wurtman, 1983). Some of these differences may be readily demonstrated by changing the ionic strength of the milieu. For example, at low ionic strength, membrane-bound CAT and soluble CAT show high and low affinity for choline, respectively; the opposite is true at high ionic strength (Smith and Carroll, 1980; cf. Blusztajn and Wurtman, 1983; see also below).

As there is only a single gene for CAT, these characteristics and properties of CAT may simply be variations derived from posttranslational or postexpression modifications (Jean Massoulié, personal communication), and it is more important to dwell on the basic regulatory mechanisms underlying CAT (and VAChT, since these two genes are both members of the cholinergic gene locus; see Chapter 2 B-2 and B-3) expression than on the variations. The regulatory mechanisms became clear with the advent of molecular biology methods comprising, among others, transient cell transfection analysis, splicing, cloning, deoxyribonuclease-sensitive site mapping, and transgenic mice models, as used particularly by Louis Hersh and Jacques Mallet (Mallet et al., 1990; Cervini et al., 1995; Wu and Hersh, 1994). Expression of CAT is regulated by molecular mechanisms operated by several exons, promoters, silencing factors, repressors and alternative splicings; these regulations offer a number of alternatives and lead to the generation of several mRNA types (Mallet et al., 1998; Shimojo and Hersh, 2004). This concatenation of regulatory mechanisms is not the end of the matter, as the CAT (and VAChT; see Chapter 2 B-2 and B-3) expression is regulated by additional mechanisms that include, among others, protein kinase A and differentiation and growth factors (Berse and Blusztajn, 1995; Tanaka et al., 1998).

Speculations were raised with regard to relations between the various forms of CAT and their functions. Original forms of CAT may be synthesized and released into the cytoplasm. They may include forms that exhibit positive charges; these forms could readily attach to the inner nerve terminal membrane and membranes of synaptic vesicles (cf. Tucek, 1985, 1990). Another species of CAT may be nonionically associated with membranes. The various CAT forms may be differentially sensitive to inhibition by excess of their substrates or metabolites. Their activities and activation characteristics also differ. Choline acetyltransferases ionically bound to membranes have low choline and coenzyme A affinity at high ionic strengths that may prevail at or near the membranes. This binding may mitigate inhibition of these CATs by substrates or metabolites, and this phenomenon may be teleological for the situations that may develop at synaptic vesicles and nerve terminal membranes. Nonionically membrane-bound CATs may be activated and/or released from the membrane to take on the brunt of synthesis, such as when an ion (i.e., Ca^{2+} or Cl^{-}) enters the terminal (Rossier and Benda, 1978). These activated CAT forms may also exhibit high activity or high affinity to the substrates and may then assume characteristics of highly active cytoplasmic forms of CAT (Malthe-Sorenssen, 1979). Finally, the role and location of cytoplasmic and membrane-bound CATs may be dependent on the location of ACh (and vice versa). Acetylcholine may be vesicular and cytoplasmic (a controversial area; see Chapter 2 B and C and section CIV, below). It would be tempting to associate vesicular and cytoplasmic ACh with membrane-bound and cytoplasmic species of CAT, respectively, although membrane-bound CAT seems to be associated with plasma rather than vesicular membrane.

It may also be speculated that the various forms of CAT may relate to the "phenotypic plasticity" of the CAT gene expression; the actual genetic lineage of the various forms of CAT and its isoenzymes was not established yet. As other components of the cholinergic system and of cholinergic neurons, CAT is a flexible enzyme that can adapt, via posttranslational modifications or regulatory changes, to varying states and functions of the cholinergic synapse.

2. Acetyl Coenzyme A

Like CAT, acetyl coenzyme A is necessary for ACh synthesis. The structure of this enzyme was elucidated early by Fritz Lipmann; the coenzyme contains pantothenic acid linked by a phosphate to a nucleotide (for details, see Nachmansohn, 1963a). Subsequent investigations of several aspects of acetyl coenzyme A were carried out particularly by the Prague team of Stanislaus Tucek, Vladimir Dolezal, and their associates, and by the Gdansk team of Andrzej Szutowicz, Aniela Jankowska, and their colleagues.

Acetyl coenzyme A is generated by acetyl-CoA synthetase from acetate in the presence of ATP; glucose, pyruvate, citrate, and butyrate are also involved in the generation of acetate, glucose being converted to pyruvate (see, for example, Szutowicz et al., 1982; Tomaszewicz et al., 1997: Starai et al., 2002; see also Browning, 1976). Separate genes are involved in the generation of acetyl coenzyme A from these precursors, yielding several pools of coenzymes (Nikolau et al., 2000). The coenzyme needed for ACh synthesis is formed at the inner mitochondrial membrane. While it is not clear how it accesses the cytoplasm and the cytoplasmic CAT, several explications were posited to elucidate this problem; good evidence indicates that carrier systems and/or Ca^{2+}-dependent transport across hydrophilic channels in the mitochondrial membrane are involved in the transfer in question (Tucek, 1990; Bielarczyk and Szutowicz, 1989; Szutowicz et al., 1998).

While needed for ACh synthesis, acetyl coenzyme A is usually not considered a cholinergic neuron marker (Jope, 1979; Tucek, 1990). This is because it is present in many nonneuronal tissues, including liver and heart, and, in fact, in invertebrate organisms (including bacteria and plants) as well. Also, in vertebrates acetyl coenzyme A is involved in noncholinergic phenomena, including brain energy metabolism and cholesterol and fatty acids synthesis (Agranoff et al., 1999; Tomaszewicz et al., 1998). Yet, acetyl coenzyme A is colocalized with CAT and AChE in the cholinergic brain sites, including cholinergic nerve terminals, as well as in the peripheral cholinergic nervous system (Shea and Aprison, 1977; Tucek et al., 1982; Yip et al., 1991). It should be added that at such sites as the neuromuscular junction and the electric organ, external acetate can also be used in ACh synthesis, as it is transported into the nerve terminals and converted into acetyl coenzyme A (Tucek, 1978).

The activity of acetyl coenzyme A is generally not believed to be the limiting step for ACh synthesis (Budai et al., 1986; Tucek, 1990), although it was speculated that the release of the coenzyme from the mitochondria into the cytoplasm may be a rate-limiting factor for ACh synthesis (Taylor and Brown, 1999). At any rate, the rate of synthesis of the coenzyme and some of its characteristics seem to change under different conditions, as demonstrated by Stanislav Tucek, Andrzej Szutowicz,

Aniela Jankowska, and their associates. Thus, the coenzyme is able to adapt to conditions of high demand for ACh: in the presence of excessive ACh release, the utilization of acetyl coenzyme A is increased to augment ACh synthesis (Ricny and Tucek, 1980; Jankowska et al., 2000). On the other hand, under conditions of its relative shortage, the coenzyme may be particularly vulnerable to neurotoxins such as Al^{3+}. As already mentioned, coenzyme A is involved in energy production as well as in ACh synthesis, and Andrzej Szutowicz and his associates proposed that there is a competition for the coenzyme between the energy requirements and the demands for ACh synthesis. This competition, combined with vulnerability of the coenzyme to neurotoxins, including naturally occurring toxins, may affect not only synthesis of ACh but also the energy metabolism of the neuron, contributing to its apoptosis (Tomaszewicz et al., 1998; Szutowicz et al., 2000); in fact, similar mechanisms involving the coenzyme may contribute to the generation of encephalopathies, including Alzheimer's disease, as suggested by Szutowicz (Figure 3-1; see Tomaszewicz et al., 1998).

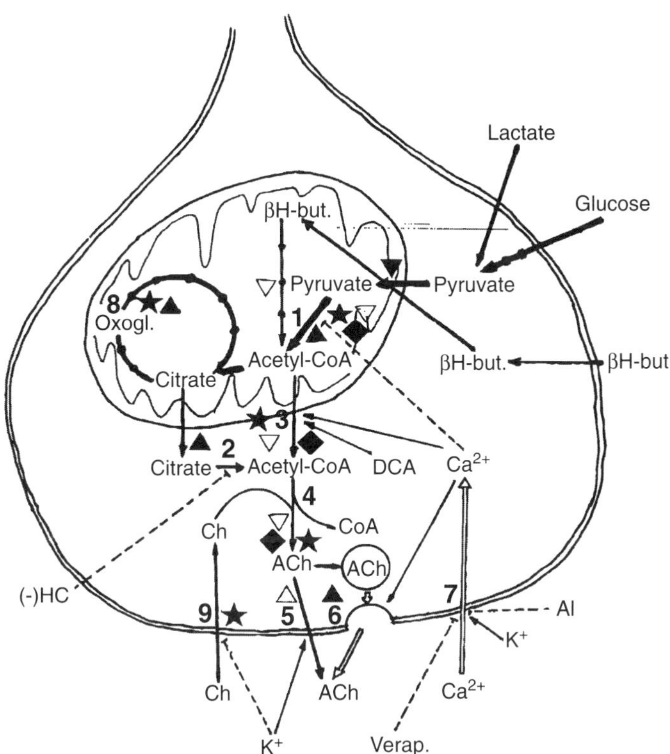

Figure 3-1. Putative sites of disturbances in acetyl CoA metabolism in cholinergic terminals in different encephalopathies. 1, pyruvate dehydrogenase (EC 1.2.4.1.); 2, ATP-citrate hydrolase (EC 4.1.3.8.); 3, direct transport of acetyl CoA through the mitochondrial membrane; 4, choline acetyltransferase (EC 2.3.1.6.); 5, Ca-independent ACh release; 6, Ca-dependent (quantal) ACh release; 7, calcium influx; 8, 2-oxoglutarate dehydrogenase (EC 1.2.4.2.); 9, high-affinity choline uptake; βH-but., β-hydroxybutyrate; DCA, dichloroacetate; (-)HC, (-)hydroxycitrate; Oxogl., 2-oxoglutarate; Verap., verapamil. Thick and medium-thick arrows indicate different enzymatic and transport steps. Thin solid arrows and thin dashed indicators point to activatory and inhibitory influences, respectively, for cations and drugs. Possible sites of metabolic disturbances in different encephalopathies: star, Alzheimer's disease: triangle, thiamine deficiency: diamond, inherited pyruvate dehydrogenase deficiency; inverted triangle, diabetes. Open and filled symbols indicate possible activatory and inhibitory influences, respectively. (From Szutowicz et al., 1998, with permission.)

3. Axoplasmic Transport of CAT

If CAT generation is similar to that of other enzymes and proteins that are used intracellularly rather than for "export," then CAT should be synthesized in the free, cytosol-located ribosomes rather than in the membrane-bound polysomal ribosomes, and then discharged into the internal space of the endoplasmic reticulum (see also above, section BI). However, it is not known as yet whether this picture applies as well to the proteins "exported" toward the synapse by axoplasmic anterograde transport (see Morfini et al., 2006). If, nevertheless, the picture of CAT synthesis in the free ribosomes is true, then the CAT of the perikaryon must be either present in the cytoplasmic, soluble form or it should be readily releasable from its bound status. It is not easy to differentiate between granular, membrane-bound, and soluble CAT; as will be seen, an indirect approach may help in resolving the problem.

Actually, CAT belongs to the category of enzymes that are transported axonally toward the synapse. Paul Weiss first formulated the concept of axonal transport. In 1947, he demonstrated a swelling of a ligated nerve on the proximal side of the ligation and a thinning on its distal side; he hypothesized brilliantly that the perikaryon biosynthesized new axoplasm—or protein—which is siphoned into the axon by an "axomotile" mechanism. Then William Feldberg and Martha Vogt (1948) found CAT in the axons; consistently with Paul Weiss' hypothesis they suggested that "the presence of CAT in the axon might be most simply accounted for . . . by assuming that it is in transit between the cell body, where it is manufactured, and the nerve endings where it has its function"; thus, this transport is anterograde in character.

Then, Catherine Hebb showed that when the goat sciatic nerve was cut *in vivo*, CAT activity and protein content increased proximal to the transection site (above the cut) and were gradually eliminated from distal portions of the nerve (Hebb and Waites, 1956; Hebb, 1959); similar findings were presented by Whittaker (1958). In toto, these findings indicated to Catherine Hebb that "the proximo-distal movement of the enzyme is related to the growth of the axon . . . ; as . . . CAT is attached to some part of the axon structure such as the endoplasmic reticulum . . . it . . . is . . . a fixed constituent of the continually growing axon . . . and is kept in transit . . . jointly with the growth-related proteins . . . by the growth of this structure"; this signifies that CAT is transported from a cholinergic perikaryon to the synapse via the axons (Hebb, 1963; see also Whittaker, 1965, 1990) and constitutes a modern rendition of Paul Weiss' "axomotile" mechanism. However, later it was shown that CAT transport is bidirectional, as it can be anterograde or retrograde, and it is interesting that synaptic vesicles may also be transported via the latter (Weible et al., 2004; von Bartheld, 2004).

Additional evidence seems to support Hebb's notion. For example, cellular membrane or membrane-bound materials such as phospholipids show a downflow (Abe et al., 1973; Rostas et al., 1975); accordingly, "fixed constituents"—growth-related proteins—do not have to be dissolved in the cytosol to be transported. Then, while CAT initially moves in isolated axon preparations, this transport is considerably slower than the CAT transport in the axons that are cut *in vivo*; after three days CAT seems to disappear altogether (Tucek, 1970). This suggests that as the isolated axons no longer benefit from perikaryon-based protein manufacture (including CAT), the proximodistal push diminishes to zero and CAT disappears.

In spite of the data in question, Hebb's notion of transport of CAT as being linked with axonal flow and endoplasmic reticulum may be questioned today. Choline acetyltransferase histochemistry shows that it is localized in neuronal cytoplasm, readily extractable, and never found inside cell organelles such as the endoplasmic reticulum (Kasa, 1971). Furthermore, while CAT is easily bound to membranes at low pH or in solutions with low ionic concentrations, under the physiological conditions of pH values and ionic levels, it should not be trapped inside organelles (including the synaptic vesicles) or strongly bound to membranes; it is most probably freely dissolved in and/or readily released into the cytosol. Finally, CAT accumulates, in the divided or ligated nerves, well above the accumulated particulate organelle-rich material (Zelena, 1969).

The axoplasmic flow originally observed by Weiss (1947) was slow (approximately 1 mm/day). He may have related it, as did Hebb (1963) to the axonal growth since he wrote of the axoplasmic flow as a "translocation" of the "solid axonal matrix . . . moving as a single, cohesive, viscous, somewhat plastic body within the Schwann (or glia) cell" (Weiss, 1947). However, faster rates were reported soon thereafter (Miani, 1960; Lubinska, 1964; cf. Grafstein and Forman, 1980). At this time, the two types of transport, fast and slow, are recognized (cf. Grafstein and Forman, 1980; Lubinska, 1975; Tucek, 1978; Dahlstrom et al., 1985; Brady, 1995; Zimmerman et al., 1998). Which of them is responsible for CAT anterograde transport?

The fast transport does not depend on protein synthesis or growth, since it is not blocked by protein synthesis inhibitors. On the other hand, it is markedly sensitive to metabolic inhibitors and temperature—high temperature may disrupt fibrous axonal structures, such as microtubules—and it is readily blocked by tubulin blockers, colchicine, and vincristine (see Grafstein and Forman, 1980). The block of tubulin would interfere with the fast axonal transport, which is linked with microtubules, and Schmitt (1968) proposed that the fast transport depends on a "mechanochemical . . . interaction" between the microtubules and the membrane of the organelles transported by the flow. The slow flow may concern axonal growth, as it is blocked by protein synthesis inhibitors. The sensitivity of the slow transport to tubulin blockers is controversial; at any rate, this system does not appear to be as potently inhibited by these agents, as is the fast system.

It is generally believed that most constituents of small organelles move fast, "at velocities ranging from 40 to over 400 mm/day" (Lubinska, 1975); synaptic vesicles are moved by the fast transport (Whittaker, 1990). Fast transport of these constituents may continue after ligation or destruction of the cell body (Ochs, 1974); this again suggests that this flow does not concern the axonal growth. On the other hand, slow transport involves cytoplasmic materials including soluble proteins and enzymes such as CAT and catechol-o-methyltransferase (Grafstein and Forman, 1980; Whittaker, 1990). The rates of the slow transport vary from fractions of 1 mm to a few mm/day.

Some evidence is inconsistent with the notion of the slow axonal flow of the cytosol-located CAT. For example, Dahlstrom et al. (1985) reported that vinblastine or colchicine, the tubulin blockers, block the transport of CAT. Then some investigators found that CAT moves in the axons at a fast pace (cf. Tucek, 1978). This variability of results concerning CAT transport rate and its axonal localization may be due to several factors that may have differed from one exprimenter to another. Thus, the rate of CAT transport depends on the age of the animal (cf. for instance, Jablecki and Brimijoin, 1975), the axons' physiological activity and activity of other than cholinergic nerves (Lubinska, 1975; Booj et al., 1981), temperature, pathology, species, and nerve type (von Bartheld, 2004).

4. Conclusions and Questions

Perikaryon-generated CAT is present in the cell or in the axon in several molecular forms. These forms may be solubilized in the cytosol or enter into reversible bonding with the cell membranes or the membranes of the organelles. In the latter case they are readily released into the cytosol. The various forms of CAT may have different functions, substrate affinity, and kinetics, and separate forms of CAT may subserve vesicular versus cytoplasmic ACh synthesis (see Chapter 2 C).

It is generally accepted that CAT is transported via the slow axonal transport, independently of the proteins involved in axonal growth. The axonal transport of CAT, similar to the transport of phospholipids and proteins biosynthesized in the perikaryon, depends on, and is linked with, an axonal substructure. What is the need for two types of axonal transport? Why do two cholinergic components, synaptic vesicles and CAT, move down the axon at two different—in fact highly different—rates?

There may be several forms and pools of acetyl coenzyme A. The mitochondrial acetyl coenzyme A should not participate in the CAT activity, as CAT is a cytosolic enzyme; which of the nonmitochondrial pools of the coenzyme participates in ACh synthesis? What is the mode and molecular site or sites of bonding of CAT with acetyl CoA? What is the contribution to this matter of the genetic and molecular characteristics of CAT, its primary peptidic structure, as well as its three-dimensional characteristics? What about any posttranslational modifications of CAT and its attachment to microtubules and vesicles that carry CAT down the axons?

Also, ACh-generating coenzyme seems to be involved in both the synthesis and energy processes—does this occur only in extreme ACh demand conditions?

CI. Choline Metabolism

1. Introduction

When Feldberg and Brown (1936a, 1936b), Perry (1953) and Birks and MacIntosh (1961) realized that choline was needed to maintain ACh release during synaptic activity, they suggested that choline uptake and choline anabolism must occur in the synapses. However, until the early 1980s, cholinergikers working in this area looked at choline's relation to ACh release strictly from the viewpoint of synaptic ACh synthesis, and disregarded choline's involvement in the synthesis of membrane phospholipids; this viewpoint was emphasized in the reviews, such as those of Tucek (1978, 1984, 1985). The contrasting view emerged with the work of the late Brian Ansell, Roger Wurtman, Kazimir Blusztajn, Steven Zeisel, Jochen Klein, and Konrad Loeffelholz; these studies concerned a broad picture of choline metabolism and homeostasis of choline levels in the nervous tissues and the blood. Sections CII and CIII describe choline metabolism and its synaptotropic, metabotropic, and phospholipid-generating significance; the uptake of choline is dealt with separately in section CIV.

CII. Choline Metabolism and Choline Transport into the Brain

Choline is present in cellular and nuclear membranes and membranes of the endoplasmic reticulum and myelin sheaths of all cells; it is a component of phosphatidylcholine (lecithin), sphingomyelin, the plasmalogens, and other phospholipids. Thus, choline must participate in the metabolism leading to the formation and the maintenance of choline-containing membrane phospholipids (Agranoff et al., 1999); therefore, it is also present in metabolites involved in this formation, such as phosphorylcholine and lysolecithin (lysophosphatidylcholine). Altogether, choline metabolism is important for all cells and their organelles.

While Kazimir Blusztajn stated that "cholinergic neurons are unique, since they alone utilize choline for an additional purpose, synthesis of their neurotransmitter, ACh" (Blusztajn et al., 1987), it must be remembered that ACh is also present in nonneuronal vertebrate tissues and nonneurogenic organisms and plants, and that Blusztajn's statement should be combined with Konrad Loeffelholz's reminder that "the cholinergikers world" (generally) "does not include the biochemistry of phospholipids" (Loeffelholz, personal communication; see Chapter 8 BI and section A in this chapter). In addition, it happens that phospholipids can generate choline and, thus, ACh.

Altogether, there are several sources of choline generation, and this matter, the choline uptake into the nervous tissues, and the relative importance of these sources for the generation and release of synaptic ACh are discussed in the following section.

1. Choline Originating from Released ACh

Nerve terminals take up, via a carrier system, choline originating from hydrolysis of released ACh (see section CIV of this chapter and Chapter 2 B). There is a general agreement that this uptake provides a considerable portion of the choline needed for ACh synthesis (Collier and Katz, 1974), although the extent of this particular choline pool's contribution to ACh synthesis is unclear. There are few references to this matter in classical literature; for instance, Perry (1953) showed that when ACh hydrolysis is inhibited by physostigmine—thus diminishing the levels of intersynaptic choline—the release of ACh from the ganglion is abated (this effect, however, could be due to accumulated ACh's action at the nerve terminal). Later, this problem was briefly alluded to in the reviews of Blusztajn and Wurtman (1983) and Tucek (1978), but several reviews do not refer to the matter at all (Blusztajn et al., 1988; Zeisel, 1986; Tucek, 1984, 1985, 1990; Loeffelholz, 1996; Loeffelholz and Klein, 2004).

2. Dietary Choline

Until the1980s, it was thought that nervous tissue could not supply choline in amounts sufficient for effective release of ACh, and that dietary choline is needed for nerve function. For example, Best and Huntsman (1932) demonstrated that choline and/or phosphatidylcholine (lecithin), which serve as choline precursors, were needed, vitaminlike, in the diet for function; in fact, their deficiency causes human malfunctions (for review, see Zeisel, 1981 and 1988; see also Barbeau, 1979 and Barbeau et al., 1979). Later, it was established that high levels of dietary choline increase plasma choline levels, and that this augmented plasma choline moves across the blood-brain barrier (BBB; by means of transport mechanisms) into the brain (Wurtman, 1987; Loeffelholz et al., 1993; Tucek, 1990; Loeffelholz, 1996; Loeffelholz and Klein, 2004). The transport mechanisms in question are not related to the choline uptake systems of the nerve terminal (see section CIV of this chapter and Chapter 2 B2; see also Rylett, 1987; Rylett and Schmidt, 1993). This dietary augmentation of plasma and brain levels of choline can be mimicked by the administration of choline systemically, intracerebrally or via the cerebrospinal fluid (CSF; see Loeffelholz and Klein, 2004). The brain penetration of either dietary or administered choline is consistent with the findings that increased intake of choline leads to behavioral and functional effects (see section CIII in this chapter).

3. Brain Phospholipids and Choline

Phospholipids present in the brain include phosphatidylethanolamine, phosphatidylcholine, and other related compounds (Blusztajn et al., 1979; Tucek, 1990). They are generated in the brain from choline and other sources via complex pathways that involve acetyl coenzyme A and formation of fatty acids (Agranoff et al., 1999), as well as transported into the brain from the blood; indeed, like choline, phosphatidylcholine may be supplied by the blood to the brain (see Tucek, 1990; see also section CIV-2). Also, phosphatidylcholine can be generated, via transmethylation (Hirata and Axelrod, 1980), from phosphatidylethanolamine and phosphatidylserine (Blusztajn and Wurtman, 1983; Blusztajn et al., 1979, 1988). In fact, somewhat similar to phosphatidylcholine, certain cell lines (not necessarily neuronal cultures) require these phospholipid components not only for generation of phosphatidylcholine but also for the maintenance of their life cycle. Altogether, there is an exchange among phosphatidylethanolamine, phosphatidylserine, and phosphatidylcholine. In this exchange, an enzyme facilitates substituting one phospholipid "base" (serine, ethanolamine, or choline) for another; this process is referred to as "base exchange" (Salerno and Beeler, 1973; Zeisel, 1988; Blusztajn and Wurtman, 1983).

While the generation of phosphatidylcholine from phosphatidylethanolamine and/or the "base exchange" occurs significantly in liver (Zeisel, 1988), these processes also take place in other tissues, including the brain and other nervous tissues (Hirata and Axelrod, 1980; Blusztajn et al., 1988). In fact, it occurs at very high rates in the brain microsomal fraction (Gaiti et al., 1974), and choline may be liberated in this exchange (see Figure 3-2); thus, choline may be generated not only from phosphatidylcholine. Also, choline induces *in vivo* phosphatidylcholine formation in tissues such as the heart, ganglion, neuromuscular junction, and brain (Hendelman and Bunge, 1969; Gomez et al., 1970; Chang and Lee, 1970; McCarty et al., 1973; Tucek, 1990; Loeffelholz, 1989; Loeffelholz and Klein, 2004); as stressed by Blusztajn et al. (1987), this may occur in the absence of the formation of ACh. However, the application of choline to brain slices frequently yielded ACh as well as phosphorylated choline derivatives and phosphatidylcholine; phosphorylated cholines and phospholipids predominated, and the exact ratio of ACh over choline metabolites depended on the brain location (Saito et al., 1986). Of course, it is teleological that choline should be used predominantly for the formation of phospholipids, compounds needed for all cells including the noncholinergic cells, rather than solely for the synthesis of ACh. However, this is a controversial topic; for example, Zeisel (1988) does not think that "base exchange" may form significant amounts of free choline.

Whatever its sources, that is, endogenous phospholipid metabolism (Agranoff et al., 1999; see also section CIV, below) and transport processes involving choline and liver-generated phospholipids, phosphatidylcholine constitutes 30% to 60% of membrane phospholipids and some 70% of all brain lipids (Figure 3-3, see color plate; Blusztajn et al., 1988; Loeffelholz and Klein, 2004; see also section CIV-2); this abundance of brain phospholipids and phosphatidylcholine underlies

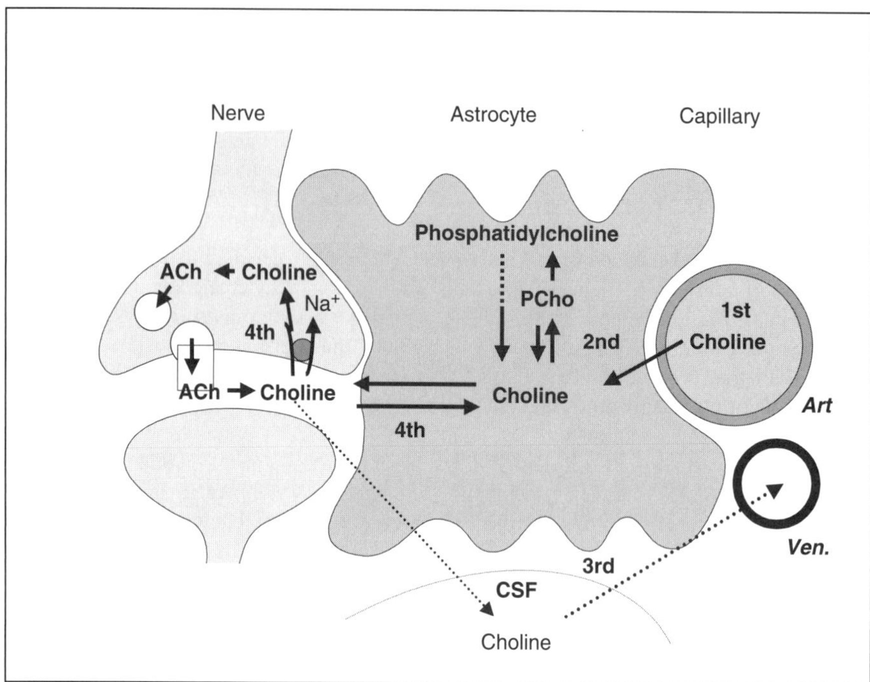

Figure 3-2. The traffic of choline, acetylcholine, phosphatidylcholine, and phosphoryl choline across the nerve terminal, the third ventricle, the astrocyte, the capillary, and the ventricle. PCho, phosphoryl choline. (From Loffelholz and Klein, 2004, with permission.)

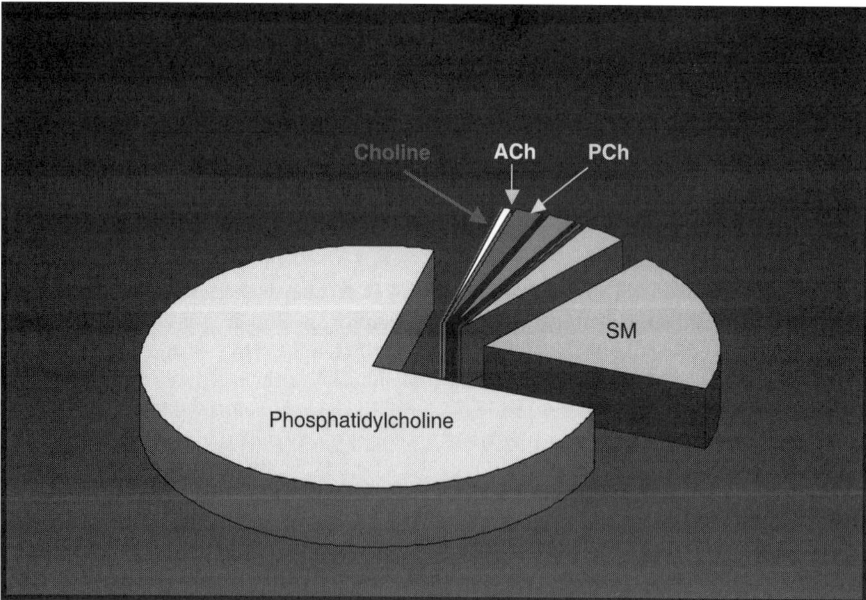

Figure 3-3. Distribution of choline in brain tissue. (From Loeffelholz and Klein, 2004, with permission.) ACh = acetylcholine; PCh = phosphatidylcholine; SM = sphingomyelin. (See color plate.)

hypotheses that these materials are one of the sources of choline subserving synaptotropic ACh generation (for full discussion of this matter, see next section; Tucek, 1978; Blusztajn et al., 1988; Loeffelholz and Klein, 2004; Blusztajn et al., 1987; Maire and Wurtman, 1985; Zeisel, 1986).

The liver phospholipid metabolism is also important in the context of the derivation of brain phospholipids. That choline generates phosphatidylcholine in the liver has been known for a long time (Kennedy and Weiss, 1956). The process in question is ATP dependent and involves the formation of phosphorylated products of choline, phosphorylcholine and cytidine diphosphocholine. Actually, the liver is the major source of phosphatidylcholine of the brain (McMurray, 1964), and this became known soon after the 1956 demonstration of the liver pathway for phosphatidylcholine (cf. Ansell, 1973, 1979).

It must be stressed that the case of a number of peripheral junctions (see, e.g., McCarty et al., 1973), in brain slices (Saito et al., 1986) and neuronal cultures, the addition of choline also causes the formation of phosphorylated derivatives of choline and of phospholipids (see Blusztajn and Wurtman, 1983; Reinhardt and Wecker, 1987).

4. The Role of Phospholipids and Choline Uptake in the Generation of Brain Choline

The important matter of the role of phospholipids and choline uptake in the generation of brain choline concerns the primary origin of choline as the generator of ACh. In other words, are the brain phospholipids (and, indirectly, phospholipids of the liver) or the uptake of free choline across the BBB the major source of choline needed for the synthesis and release of ACh? This dilemma was presented early by Stanislaus Tucek (1978) and studied extensively by Konrad Loeffelholz with Jochen Klein, and Krzystof Blusztajn and his associates.

Free choline is generated from phosphatidylcholine in several tissues, including the heart, intestine, kidney, lungs, and—important in the present context—brain (Bhatnagar and MacIntosh, 1967). In fact, there is a rapid postmortem release of choline from these tissues (Dross and Kewitz, 1972; see also Ansell and Spanner, 1979).

Related phospholipids such as sphingomyelin also liberate choline; while the liver is a major source of this catabolism it is present also in the brain (Loeffelholz, 1998).

In the liver, this formation of choline is carried out in at least 2 steps; phospholipases A1 and A2 and lypophospholipase form glycerophosphorylcholine, which is then hydrolyzed by glycerophosphocholine diesterase to choline (see Ansell and Spanner, 1979). Phospholipases D and C may be also involved. A related liver pathway generates choline from sphingomyelin (Ansell and Spanner, 1979). However, the enzymes in question are also present in the brain, as known since 1957 (Webster et al., 1957; cf. Ansell and Spanner, 1979); as shown subsequently, brain catabolism of phosphatidylcholine via the action of phospholipases A1 and A2 and a diesterase is similar to the choline-generating processes carried out by the liver. In addition, in the brain, phospholipase C may form choline from phosphatidylcholine with phosphorylcholine and diethylglycerol as intermediaries acted on by phosphatase (Zeisel, 1988). However, Zeisel (1988) believes that neither this route nor the formation of choline via glycerophosphorylcholine should be considered as a major source of choline. He quotes as relevant findings that the fall of glycerophosphate that occurs is too small to account for the simultaneous formation of choline by the brain, and that zinc does not inhibit the glycerophosphorylcholine diesterase, yet it potently inhibits the formation of choline. Zeisel (1988) stresses that "all membrane-containing subcellular fractions of brain produced free choline at approximately the same rate," while the soluble fractions did not contain obligatory or facilitatory factors for choline formation; he does not identify the ultimate, major intermediate for choline formation from phosphatidylcholine.

Phosphatidylcholine is hydrolyzed when the neuronal activity is high, and synaptic ACh and available free choline levels low (Blusztajn and Wurtman, 1980); under these circumstances, the degradation of phosphatidylcholine to choline may be particularly rapid. Related findings are that phosphatidylcholine is capable of increasing ACh synthesis under conditions of stimulated—or excessive—cholinergic activity (Bierkamper and Golderg, 1980; Ulus et al., 1977; see Growdon, 1987) and that *in vitro* vigorous electrical stimulation of rat striate slices depletes membrane phospholipids, while choline "fully protects . . . phospholipid depletion . . . and . . . enhances ACh release" (Wurtman et al., 1988). As Krzystof Blusztajn stated picturesquely, "cholinergic neurons cannibalize their own membranes to ensure adequate supplies of choline for acetylcholine synthesis" (Blusztajn et al., 1985). However, the concentrations of choline, its phosphorylated derivatives, and phospholipids increase during incubation of the brain with choline; this may suggest that choline is not directly or readily utilized as a source of ACh and/or that phospholipids are not the main source of synaptic choline (Tucek, 1978; Zeisel, 1988; Saito et al., 1986; Loeffelholz and Klein, 2004). These conclusions are consistent with the demonstration that choline-enriched media always generate phosphatidylcholine but not always ACh (Blusztajn and Wurtman, 1983; Loeffelholz and Klein, 2004).

Thus, under certain conditions, phosphatidylcholine, whether as the constituent of membrane lipids (cf. Zeisel, 1988) or present interstitially, may constitute an important source of choline (Blusztajn and Wurtman, 1983; Parducz et al., 1976). As commented by Loeffelholz (Loeffelholz and Klein, 2004), the synaptic choline compartment or pool gets its choline from all choline-containing molecules, including ACh and phospholipids; the relative contributions vary depending on circumstances, and choline phospholipids, normally, may not necessarily be used for the synthesis of ACh (Blusztajn and Wurtman, 1983; Parducz et al., 1976).

The question of whether the free synaptotropic brain choline is derived from brain phospholipids or the brain uptake of choline relates to Helmuth Kewitz's finding of the negative arterial-venous difference in brain choline, which signifies that choline leaves the brain in greater amounts than it enters the brain (Dross and Kewitz, 1972; Kewitz and Pleul, 1981). This ominous finding seemed to suggest that there may be a leak of choline that would lead to its deficiency, unless there is, besides choline brain uptake, its *de novo* formation of choline in the brain. Increased blood phosphatidylcholine uptake into the brain and *de novo* synthesis of choline in the brain would explain Kewitz's findings. Yet, the de novo synthesis of choline in the brain is negligible (Tucek, 1990), phospholipids pass the BBB poorly, and Stanislaus Tucek (1990), Konrad Loeffelholz with Jochen Klein (2004), and Zeisel (1988) all opine that the choline liberation from brain phospholipids cannot account for synthesis and release of acetylcholine.

Loeffelholz and Klein stressed, first, that Kewitz's findings were obtained under conditions of fasting; under those of normal diet he would have found a positive arteriovenous difference; that is, under these conditions, the uptake of choline into the brain is higher than its efflux. Then they presented data indicating that choline originating from released ACh and dietary choline present in the blood and taken up across the BBB constitute the major source of synaptic choline; indeed, this source of choline is, under most conditions, sufficient to maintain cholinergic transmission. They also emphasized that a cycle is formed between components of the lipid metabolism, that is, choline, ACh, phosphatidylcholine, and phospholipids, and their fluxes, and this cycle is affected by conditions such as fasting and neuronal activity. Parenthetically, Krzysztof Blusztajn et al. (1987) found that a choline supplement afforded to rats during gestation improves memory function and ACh turnover of young rats, and both Blusztajn and Loeffelholz refer to choline as exhibiting vitaminlike, essential nutrientlike activities.

Importantly, Konrad Loeffelholz with Jochen Klein formulated, on the basis of their studies of phospholipid, choline, and ACh dynamics, the notion of choline homeostasis (Loeffelholz, 1989). They were concerned for many years with the possibility that excessive levels of choline may be detrimental in mammals, as choline is a relatively potent cholinergic agonist. This was described severally for the periphery between 1890 and 1920 (Le Heux, 1919) and confirmed more recently by Hans Kosterlitz[3] (see also Karczmar, 1967; Loeffelholz and Klein, 2004). This is true as well for the CNS (see Chapter 9 BI-2 and Albuquerque et al., 2004). In fact, Helmuth Kewitz's findings could indicate that such a situation may obtain relatively readily. Accordingly, Klein and Loeffelholz wished to describe processes maintaining this homeostasis and protecting the receptors from excess of choline and its possible agonist action (Loeffelholz and Klein, 2004).

Konrad Loeffelholz and Jochen Klein's scheme includes rapid metabolic clearance of brain choline that follows its uptake as the first line of defense against accumulation of choline. The saturability of the BBB with respect to uptake and transport of choline constitutes the second line of defense (the same restriction exists for glucose). Loeffelholz and Klein refer to the arteriovenous difference (AVD) between the arterial and venous levels of choline as the third line of defense, since the AVD reflects the balance between the net uptake of choline into the brain and the net efflux of choline from the brain; indeed, Loeffelholz and Klein (2004) point out that under conditions of high intake of dietary (nutritional) choline, excess choline is released into the venous blood (see above, and Figures 3-2 and 3-4; Loeffelholz and Klein, 2004).

The phospholipids-choline cycle and choline influx-efflux relationship involve homeostatic interactions. In turn, these phenomena interact with the cholinergic system, creating altogether a complex balancing act. Several instances of this act were already adduced. For example, stimulation of the cholinergic system and a high activity level may interact with the phospholipid metabolism (see Wurtman et al., 1988). Then phospholipase A2 modulates the choline uptake (see Rylett and Schmidt, 1993). Jochen Klein and Konrad Loeffelholz (see Loeffelholz, 1998; Loeffelholz and Klein, 2004) demonstrated that muscarinic agonists and antiChEs activate phospolipase D, thus increasing phospholipid hydrolysis, and provide choline for increased synthesis and release of ACh.

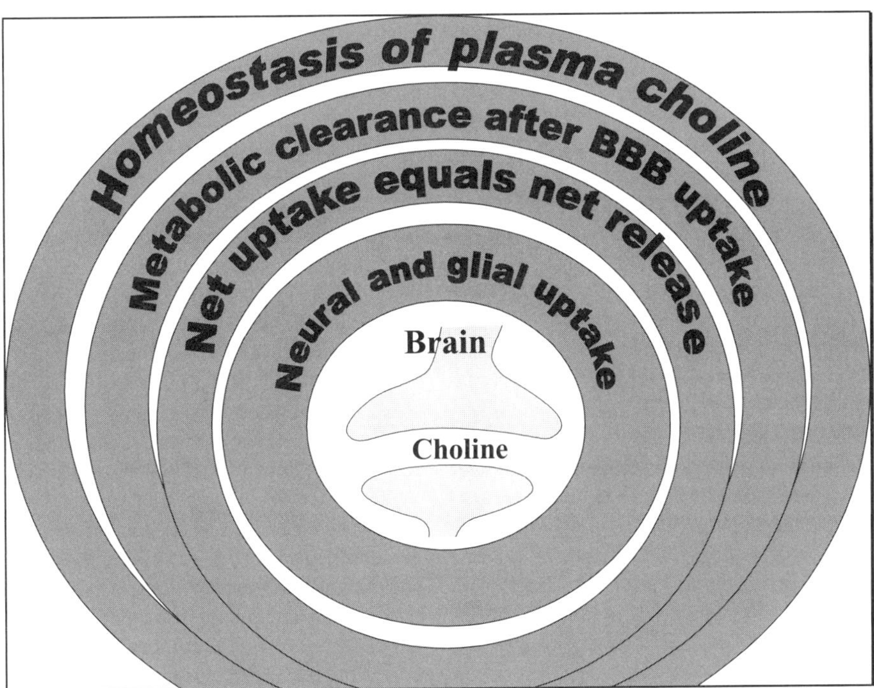

Figure 3-4. Four "lines of defense" against excess choline. (From Loeffelholz and Klein, 2004, with permission.)

CIII. Choline and Phosphatidylcholine as Precursors of Acetylcholine and Phospholipids

The concepts described above should be related to the effects of the administration of nondietary choline and choline precursors on ACh levels, turnover, and release; thus, whenever precursors are administered to affect ACh synthesis and release and cholinergic transmission, the findings described in the previous section concerning the metabolic origin of choline and ACh must be considered. This matter is particularly important, as choline and related compounds have been used empirically for many years in humans to improve cognitive function and alertness, as well as combat depression and schizophrenia. Among the first investigators to utilize this approach were Carl Pfeiffer and his associates (Pfeiffer and Jenney, 1957; Pfeiffer, 1959), as they proposed that cholinergic excitation is useful in at least certain forms (hebephrenic, particularly) of schizophrenia and for inducing alertness, and as they urged the use of Deanol, which they perceived to be an ACh precursor, for these purposes.

The effects of the administration of choline and its precursors are quite clear with respect to choline levels. Thus, either phosphatidylcholine or choline increased levels of choline in the blood and/or cerebrospinal fluid of humans as well as increased brain levels of choline in several animal species (Kindel and Karczmar, 1981, 1982; Davis et al., 1977a, 1977b, 1978; Aquilonius and Eckernas, 1975; see also Wurtman, 1979; Bartus et al., 1982; Jenden, 1979). In fact, the dietary choline content may regulate blood levels of choline in the rat as well as human—and choline levels in the brain as well (Growdon et al., 1977; Cohen and Wurtman, 1976; Tucek, 1990; Loeffelholz et al., 1993; see also section CII, above).

But the notion of the effect of the choline and choline precursor administration on ACh levels and turnover is controversial and, sometimes, unduly enthusiastic. For example, Davis and his associates stated that "the evidence that choline administration can increase brain levels of acetylcholine is overwhelming" (Davis et al., 1979; see also Davis et al., 1977a, 1977b). And many investigators reported or took for granted that choline augmented ACh synthesis in the brain and in neuronal cultures (see, for example, Cohen and Wurtman, 1976, 1977; Growdon, 1988; Ulus et al., 1978; for further references, see Bartus et al., 1982). However, there is good evidence showing that in starved or choline-depleted animals, administration of choline did raise ACh levels (Kuntscherova, 1971, 1972; Kuntscherova and Vlk, 1968; cf. Tucek, 1978; Loeffelholz and Klein, 2004). This last information relates to the important work of Lynn Wecker (cf., for example, Wecker et al., 1978), who found that choline increases brain ACh of animals in which ACh was depleted by atropine (see below). Finally, reports that choline or phosphatidylcholine increases cholinergic transmission (see below) are consistent with the proposition that choline and/or phosphatidylcholine increase ACh synthesis and release.

However, warning signals were raised, as when Bartus et al. (1982) stated that the "findings . . . as to whether . . . the increases in brain choline . . . can induce a concomitant increase in the synthesis (and presumably release of ACh) . . . generate controversy." In fact, some of the evidence quoted by the reviewers in support of the statement of Davis et al. (1979) concerns cholinergic agonists and their effect on ACh dynamics rather than the matter of the effectiveness of ACh precursors (cf. for example, Haubrich et al., 1972). Actually, a number of investigators found that under normal conditions choline or phosphatidylcholine administration had no effect on brain levels of ACh (Pedata et al., 1977; Kindel and Karczmar, 1981; see Tucek, 1985, for additional references). For example, Jope (1986) did not find any increase of ACh with phosphatidylcholine administration whether in the cortex, hippocampus, or striate, similar to the earlier data of Kindel and Karczmar (1981).

The controversy in question is important, since some investigators believe that phosphatidylcholine or choline may be used in certain neurological and mental conditions to increase the turnover and release of ACh and augment the activity of cholinergic synapses; this would be analogous to the use of dopa as precursor of dopamine in parkinsonism (see for example, Wurtman, 1979, as he speaks eloquently of the "precursor control of transmitter synthesis"). Accordingly, it is important to resolve the problem of inconsistency among the findings. Besides the methodological differences among the investigations in question and/or possible bias of the investigators involved[4] there may be additional reasons for the diversity of the findings. Special conditions surrounding the experiments in question may be particularly important. As well phrased by Wecker (1986), "Choline may be used to support the synthesis of ACh when there is increased demand for the precursor, i.e. when the activity of central cholinergic neurons is increased . . . while . . . choline does not enhance synthesis or metabolism of ACh under 'normal' physiological and biochemical conditions," and this sentiment is in agreement with the findings of Konrad Loeffelholz and Stanislaus Tucek; it must be added that negative arteriovenous difference with respect to brain choline (Dross and Kewitz,1968) was obtained in fasting or avitaminotic animals and that the effects of precursors of ACh such as choline and phosphatidylcholine depend on the nutritional state of the animal and its synaptic activity (see section CII, above). To induce this demand, Wecker and her associates (Wecker et al., 1977; Wecker and Dettbarn, 1979; Wecker and Goldberg, 1981) caused depletion of brain ACh via various means. These means include facilitating ACh release either by blocking with atropine or with phenothiazines the muscarinic nerve terminal receptors that inhibit ACh release (see Chapter 9 BI-1), by increasing neuronal (not necessarily uniquely cholinergic) activity by convulsant drugs and opiate withdrawal from addicted animals, or by neurotoxin-induced degeneration of glutaminergic neurons (London and Coyle, 1978). Under these conditions Lynn Wecker found that choline supplementation is prophylactic vis-à-vis the depletion of ACh. Wecker's particular point is that this effect is "not . . . mediated through an enhanced steady-state concentration of free choline in the brain, but rather may involve alterations in the metabolism of phospholipids," including formation of phosphatidylcholine and choline (Wecker et al., 1977). This is interesting, as it suggests that augmentation of ACh synthesis in the brain is not achievable via increase in the choline levels in the brain—even though up to a 3-fold increase in the brain levels of choline may be achieved with multiple choline dosing (Kindel and Karczmar, 1981, 1982)—but via changes in phospholipid and choline metabolism, whether caused by increased neuronal activity and/or by choline-evoked neuronal stimulation.

Of course, change in ACh levels does not necessarily reflect an increase in ACh synthesis or turnover; in fact, just the reverse may be conjectured, as increased ACh levels may indicate its decreased utilization and therefore a decrease in turnover (or "re-synthesis"; Hanin and Costa, 1976). It is regrettable that only a few studies of the effect of precursors on ACh turnover are available; most of these studies indicate that choline and/or phosphatidylcholine do not increase ACh turnover. This was true for the rodent brain (Kindel and Karczmar, 1981, 1982; Brunello et al., 1982; Eckernas, 1977), and in the rat nerve-diaphragm study of Potter (1970). Actually, Kindel and Karczmar (1981, 1982) found a significant decrease in ACh turnover in some of their experiments; they pointed out that in their experiments the precursors elevated brain choline concentrations to levels that may cause a stimulation of the cholinergic and/or cholinoceptive neurons, and that this stimulation may decrease via a negative feedback ACh synthesis and/or turnover. In fact, such decrease in turnover was caused by cholinergic agonists and antiChEs (Choi et al., 1973; for further references, see Hanin and Costa, 1976). A similar effect would be expected from choline stimulation of cholinoceptive nerve terminal receptors (see Chapter 9 BI-1). Of course, poor or null effects of precursors on ACh levels and/or turnover may also result from utilization of choline for purposes other than ACh synthesis, such as generation or maintenance of the membranes and energy requirements. On the other hand, some data suggest that ACh synthesis and turnover may be augmented by neuronal stimulation, whether behavioral, electrical, or chemical (Kindel and Karczmar, 1981, 1982), and increased turnover of ACh was found to be accelerated by high concentrations of choline in slices of Torpedo electric tissue (Morel, 1976).

Another possible consequence of precursor loading of the cholinergic system is an increase in ACh release. As early as 1936, Brown and Feldberg demonstrated that the presence of choline in a ganglion preparation increased ACh output during stimulation, while Mann, Quastel, and their associates (Mann et al., 1938) made a similar finding with regard to an *in vitro* brain preparation. It must be added that the choline effect was small compared with that of pyruvate or glucose in the case of the experiments of Mann et al. (1938), while Brown and Feldberg (1936) could not duplicate the choline effect on ACh efflux in the ganglion in every instance. Then, Frank MacIntosh (1979; Birks and MacIntosh, 1961) commented that the role of choline on ACh release was easy to demonstrate in the presence of hemicholinium or excessive ganglionic stimulation, and when choline-free media were used prior to the addition of choline. Also, Bierkamper and Goldberg (1979) found that when diisopropyl fluorophosphate was used to completely inhibit AChE, high concentrations of choline markedly increased the release of ACh in a rat nerve-diaphragm preparation; interestingly enough, phosphatidylcholine had no such effect. It should be noted that in these experiments augmentation of the release of ACh was achieved either with concentrations of choline much higher than physiological, or under special, "nonphysiological" circumstances, while Potter, Birks, MacIntosh, and others (see above) were concerned with demonstrating the need of physiological concentrations of choline for "normal" synaptic function rather than with the effect of choline on ACh release.

Altogether, it appears that, as in the case of ACh turnover, choline affects release of ACh only under conditions of a depleted or stressed cholinergic system; furthermore, this effect may involve the choline uptake by the terminal while metabotropic choline (see Chapter 2 B) may not participate in the phenomenon of choline-driven ACh release. Finally, it is difficult to interpret the action of choline on ACh release—particularly under conditions of stress—as it may result from the action of choline on ACh turnover. It must be added that the facilitatory effects of choline on ACh release—if present—do not result from stimulation of nerve terminal cholinoceptive receptors, as such stimulation generally leads to a decrease of ACh release.

Another consequence of increased levels of choline or phosphatidylcholine, consistent with increased ACh turnover, is the increased activity of cholinergic synapses. Much of the pertinent evidence concerns choline stimulation at the sympathetic sites of the synthesis of tyrosine hydroxylase, the limiting factor for the generation of catecholamines. Cholinergic stimulation of this synthesis was studied extensively by Julius Axelrod (see, for example, Mueller et al., 1969) and Erminio Costa (Chuang and Costa, 1974); they termed this phenomenon a "neuronally mediated" or "trans-synaptic" effect. It is consistent with this notion that the phenomenon in question requires the presence of presynaptic cholinergic nerves (Thoenen, 1974). Choline causes this trans-synaptic effect at the dopaminergic neurons of the rat brain, in the adrenal medulla, and in the sympathetic ganglia (Ulus and Wurtman, 1976; Ulus et al., 1977); for example, oral administration of choline to rats increased their catecholamine secretion at the appropriate sites (Scally et al., 1978).

Furthermore, Wurtman and his associates used either an agent that increased splanchnic firing, or choline, or both, and found that combined treatment resulted in an "increase . . . in tyrosine hydroxylase activity . . . that was . . . greater than the sum of increases caused by the two individual treatments." The results of this difficult experiment need further analysis. Altogether, Wurtman (1979) and Ulus (Ulus et al., 1979) argued that this event is due to acceleration of the presynaptic firing induced by stress of choline administration or by choline itself (a more likely occurrence) and to an "increase in the amount of transmitter released per firing" (Wurtman, 1979).

In a similar vein, choline potentiates the increase in tyrosine hydroxylase activity caused by drugs that augment sympathetic impulses or by cold stress (cf. Wurtman, 1979; Ulus et al., 1977, 1979). This potentiation depended on cholinergic presynaptic input, since it was blocked by cholinergic blockers (Mueller et al., 1970). These results are analogous to those obtained by Lynn Wecker and others with choline under conditions of ACh depletion, and they also relate to the data indicating that very active transmission rather than "normal" transmission is dependent on choline (see above, this section). On the other hand, this evidence does not signify that choline must increase ACh turnover under "normal" or physiological circumstances; furthermore, the investigators in question do not consider as significant the action of choline as a cholinergic agonist on, for example, tyrosine hydroxylase activity.

Ulus et al. (1988) argued that choline is too weak an agonist to be effective at the concentration that obtains in the experiments in question. However, to measure the agonistic action of choline, Wurtman, Ulus, and their associates used peripheral (cardiac and smooth muscle) preparations that may not be relevant for the problem at hand. Indeed, these preparations are muscarinic in nature, yet, in a much earlier study, Kosterlitz et al. (1968) showed that choline is a relatively potent nicotinic agonist at the sympathetic ganglia, a site very pertinent for the work of Ulus and Wurtman, and that the tyrosine hydroxylase effect of choline may be nicotinic (Mueller et al., 1970). Similar views as to the nicotinic potency of choline were embraced by Konrad Loeffelholz and Jochen Klein (see above, section 4) and Edson Albuquerque (Albuquerque et al., 1998; see Chapter 9 BI); but these views as well as those of Roger Wurtman and others (Wurtman, 1979) must be reconciled with Konrad Loeffelholz and Jochen Klein's postulate of choline homeostasis (see above).

There is also a hidden factor; it involves the phosphatidylcholine or choline contribution to the synthesis and maintenance of membrane phospholipids versus their contribution to ACh synthesis (Growdon, 1988). As already described, dietary choline and phosphatidylcholine may be converted respectively into phosphatidylcholine and choline; this information generally concerns nonbrain tissues (plasma, liver, intestinal mucosa, etc.; see Houtsmuller, 1979) and does not include evidence as to the generation of ACh (see above, this section). In addition, Jope (1986) found that either chronic or acute administration of phosphatidylcholine to rats led to high levels of plasma fatty acids, cholesterol, triglycerides, and phosphocholine, while this treatment did not generate brain ACh, whether in the cortex, hippocampus, or striate. On the other hand, Rossiter and Strickland (1970) found that intravenous choline is rapidly converted to phosphorylated choline and phosphatidylcholine; this pathway may lead ultimately to ACh. Similarly, Wecker (1986) showed that phosphatidylcholine and total lipids increased more than 2-fold in the microsomal fraction of the striatum of rats maintained on a high choline diet and that choline did not effect ACh synthesis unless under conditions of depleted ACh.

The effects of dietary supplement or high levels of choline on phospholipids and ACh synthesis should be distinguished from their use as markers of the various metabolisms. In this case, radioactive choline was used. As already reviewed, these studies led mainly to the concept of choline-induced formation of phosphatidylcholine; they concerned such diverse tissues as the heart, ganglion, neuromuscular junction, and brain (Hendelman and Bunge, 1969; Gomez et al., 1970; Chang and Lee, 1970; and McCarty et al., 1973; for the relation of these findings to the question of the origin of synaptic choline and ACh, see section CII, above). It is of interest that some investigators traced the resulting phospholipid into the myelin sheath of nerve endings (McCarty et al., 1973). However, in some of these studies ACh formation was measured as well. Choline applied to brain slices of several mouse brain parts became incorporated into the pools of endogenous choline, phospholipids, and ACh. The extent of choline incorporation into the ACh pool depended on the brain part. Generally, it was limited compared to its entry into choline and phospholipids; however, the percentage of choline incorporation into ACh was significantly higher (30%) in the striate than in the cortex (Saito et al., 1986; Kindel and Karczmar, 1981, 1982). The slices in question were, of course, heterogeneous with respect to the transmitter nature of their synapses; accordingly, the ratio of choline incorporation into ACh could be higher if only cholinergic synapses were taken into account. It is of interest to note in this context that Lynn Wecker (cf. Reinhardt and Wecker, 1987) showed that the increase in total lipids occurred both with microsomal and with synaptosomal membrane fraction, but phosphatidylcholine was augmented by choline only in the microsomes. All this evidence taken together suggests that administered choline is used only to a limited extent for synthesis of ACh and that the fatty substances and choline-containing lipids serve as a "sink" for phosphatidylcholine or choline, although subsequently this process may lead to generation of free choline and ACh (Karczmar, 1987); these points are relevant with regard to the hypothesis of choline homeostasis (see above, this section).

An important point must be made. We should look at the matter of synthesis of ACh versus that of phospholipids from the viewpoint of the total dynamics of the system, as did Wecker, Ulus, and Wurtman. Choline may be used for maintenance of synaptic activity and "trans-synaptic mediation" under conditions of ACh depletion and stress, because under these circumstances distribution of choline to phospholipids versus ACh is regulated by action of high-affinity choline carriers and a homeostatic mechanism. Also, the change in the composition of phospholipids caused by choline or other precursors may affect the formation of ACh (Atweh et al., 1975; Jope, 1986; Rylett, 1987; Loeffelholz, 1996; Loeffelholz and Klein, 2003).

CIV. Choline Transport and Uptake

1. Historical Introduction

There are two important aspects of choline transport, distribution and uptake. First, choline and choline-generating compounds, such as lecithin (phosphatidylcholine), are present in many foods (see Wurtman, 1979) and, as has been known for many years (see, for example, Vigneaud, 1941), choline is a constituent of most tissues, including the nervous tissue; accordingly, it is of interest to trace choline transport into the brain.

Second, the realization that choline participates in cholinergic transmission and metabolism of ACh led to the studies of a unique aspect of choline transport, namely, its high-affinity, nerve terminal uptake; this interest dates from the investigations of the 1930s to the 1950s of Brown, Feldberg, Birk, MacIntosh, and their associates. It appears that the transports of choline into the nerve terminal and across the BBB and the red blood cell (RBC) membrane offer some similarities, and the studies of their mechanisms helped clarify the mechanism involved in these choline transfers.

In the early 1940s David Nachmansohn (with Machado, 1943) was motivated toward the discovery by CAT by his finding that choline accelerates ACh synthesis; even earlier, Lindor (Jeffrey) Brown and William Feldberg (1936a, 1936b) demonstrated that choline rather than ACh can be detected in the effluent of presynaptically stimulated ganglia *not* pretreated with antiChEs. These findings should have led readily to the concept that, besides CAT, choline uptake by the nerve terminal is a factor in ACh synthesis. Yet, as Brown and Feldberg (1936a, 1936b) could not, at the time, obtain a balance sheet between the kinetics of ACh and choline and, particularly, between the kinetics of their storage and depletion, they did not conceptualize that the uptake of choline derived from released ACh is involved in the resynthesis of the latter. While they commented that they "have no reason to suppose that the choline obtained by extraction" from the ganglion "*is not* available for synthesis" of ACh, they did not expressly state that choline of the ganglionic perfusate constitutes a portion of ganglionic "choline obtained by extraction."

Birks, Perry, MacIntosh, and their colleagues resolved the problem. Birks and MacIntosh (1961; MacIntosh, 1959) developed improved methods for the measurement of "ACh . . . and choline output and content of . . . sympathetic . . . ganglia perfused with eserinized Locke's solution." Using these methods they demonstrated the presence of 2 specific pools of ACh in the nerve terminals, the "smaller more readily available for release by nerve impulses," and the larger pool, which is depleted when the ganglion is not "stationary . . . and which is . . . for the most part located in the extra-synaptic portions of the preganglionic axons"; furthermore, they linked these ACh pools with the choline uptake. Altogether, they improved the balance sheet that Brown and Feldberg left unfinished (1936a, 1936b); moreover, the data led Frank MacIntosh to predict that the cholinergic transmitter system must include a fourth specific component in addition to the three (i.e., ACh, CAT, and cholinesterases): "a choline carrier located in some membrane at the synapse" (MacIntosh, 1959).

The next step undertaken by Birks and MacIntosh (1961) was their demonstration of the "remarkably efficient" choline uptake system. They based this conclusion on an interesting calculation: they measured ACh release during stimulation and at the steady state of ACh pools in the nerve terminal, as well as the response of the nictitating membrane and concluded that, in order to maintain the response, the nerve endings "must be . . . able to take up and acetylate some 20% of the choline supplied to the ganglion during the few seconds required for the plasma to pass through the ganglionic vessels." They further commented that "since choline as a quaternary base diffuses slowly into most cells and since the nerve endings can form only a small part of the bulk of the ganglion, this fact is rather remarkable . . . and . . . the endings . . . must be provided with a special mechanism for the entry of choline ions"; a most heuristic statement. Further basis for the mechanism in question was provided soon after by Birks (1963), as he demonstrated the need for Na in ACh synthesis.

The story of hemicholiniums should be mentioned at this time, as the use of these compounds was most helpful for clarification of choline uptake, and as these compounds are still frequently employed in cholinergic research. Birks and MacIntosh showed that the depletion of the ACh pool and the attenuation of the response of the nictitating membrane induced by ganglionic activity were identical whether in the absence of choline in the perfusate or with the use of hemicholinium in the presence of choline; furthermore, the effects of hemicholinium could be readily antagonized by choline. Hemicholinium (HC-3) was the compound developed earlier by Fred Schueler. Schueler coined the name, as HC-3 contains two cholinelike moieties, which undergo spontaneous change leading to the formation of hemiacetal. As described by Fred Schueler and Paul Long, his graduate student at the time, hemicholinium causes a characteristic toxicity that could be expected to arise from slowly developing interference with cholinergic transmission (Schueler, 1956; Long and Schueler, 1954; Long, 1963): hemicholinium, its analogs, and related compounds (Bowman et al., 1962) act as "paralyzants" and produce "slowly-developing, frequency-dependent transmission failure in nerve muscle preparations . . . and muscle paralysis of many species . . . ; this paralysis . . . is specifically and readily reversed by choline." Schueler's data and their own results suggested to Birks and MacIntosh (1961) that choline is taken up by the nerve terminal, and that HC-3 competes with this transport. Further characterization of this uptake is provided below (this section 3, below). It is of interest that when some reviewers (e.g., Tucek, 1978, 1984, 1985, 1990; Speth and Yamamura, 1979) describe the choline uptake system and the results obtained with hemicholinium, they do not quote Schueler's (1956) contributions.

As for the early studies of the distribution of choline in the brain and other tissues: choline was discovered as early as 1862 by Strecker as a constituent of lecithin (phosphatidylcholine) and sphingomyelin, and the presence of free choline in the brain was known almost as long (see Ansell and Spanner, 1979; section CIII, above). As observed by Blusztajn and Wurtman (1983), it was believed in the 1970s (see, for example, Ansell and Spanner, 1971; Diamond, 1971) that the "free choline molecules in the brain and the bloodstream . . . constitute separate metabolic pools, the positive charge of choline's quaternary nitrogen precluding its passage across the blood-brain barrier" (Blusztajn and Wurtman, 1983). Thus, it was suspected that cholinergic neurons of the brain could either utilize, via an extremely efficient uptake, choline generated in the course of cholinergic transmission, or obtain choline from lecithin or isolecithin, which serve as vehicles transporting choline into the brain (Hendelman and Bunge, 1969). However, subsequently it was noticed that plasma contains choline and that choline readily crosses the BBB. In this context, Tucek (1984) pointed out a paradox. While "choline passes the hematoencephalic barrier . . . so rapidly that after intravenous injection of labelled choline to rats the specific radioactivity of choline in the brain . . . could be traced along . . . the descending but not the ascending part of the . . . choline brain levels . . . curve (Dross and Kewitz, 1972)," yet, after intraarterial infusion of choline in the rat, the brain concentration of choline is about 100 times lower than that of the blood, suggesting that "the blood-brain barrier *does* represent a very serious obstacle to . . . choline . . . movement." This paradox was resolved with the discovery of the role of choline carriers (e.g., Cornford et al., 1978) and active, high- and low-affinity and BBB choline uptake systems as these phenomena relate to the distribution of choline to brain cells and to the cerebrospinal fluid (CSF) following the intravenous route of choline administration (see Tucek, 1978, 1990; Blusztajn and Wurtman, 1983; see also below, this chapter, sections CIV-2 and CIV-4).

Subsequent research (see sections 2 to 4, below) established the characteristics of the choline carriers and uptake systems; these findings help reconcile brain and CSF distribution of plasma choline—derived from liver metabolism and diet—with the brain generation of choline and with the mechanisms of choline influx/efflux ratio and homeostasis of choline (Blusztajn and Wurtman, 1983; Loeffelholz, 1996; Loeffelholz and Klein, 2004; see also this Chapter, sections CIII and CIV-3).

2. Choline Distribution

a. Blood Plasma

Under normal circumstances, concentrations of choline in blood plasma range from 5 to 20 nM (Tucek, 1978, 1984); values of about 10 μM were found in man, although much lower (0.81 μM; Spanner et al., 1976) and much higher (22–63 μM) values were reported as well. Altogether, widely inconsistent values were obtained in different laboratories (cf. Tucek, 1978), although on the whole these values are much higher in newborns (65 μM at birth) and they decrease (to about 10 μM) in the 20th postnatal day (see, for example, Zeisel and Wurtman, 1981, for this interesting information). As already mentioned, there is a negative arteriovenous difference in plasma concentration in many, but not all, species; it seems to be barely present in man (Aquilonius and Eckernas, 1975). It should be added that choline concentrations in plasma are relatively stable on repeated measurement in the same animal or subject (cf. Tucek, 1978; see also the question of choline homeostasis, section CII).

Do the levels change on providing choline by infusion or in the diet? Initially, it was thought that even after infusions of choline the latter is taken up by visceral organs including liver or kidney (in preference to blood and other tissues), so that the blood levels, after initial increase, return rapidly to the initial state (cf. Tucek, 1978). More recently however, it was found in man and animals that plasma concentrations of choline may increase "several fold . . . postprandially" and/or following ingestion of phosphatidylcholine or choline-supplemented meals (Zeisel, 1988; Blusztajn and Wurtman, 1983; Growdon, 1988), although large amounts of the precursors are generally required. A particularly extensive effect of subcutaneously administered choline on rat plasma choline levels was reported by Kuntscherova (1972); this increase may last for hours (Hirsch and Wurtman, 1978; Zeisel et al., 1980; Zeisel, 1988; Fischer et al., 2005).

Conversely, malnourished patients or patients maintained on a low choline, phosphatidylcholine, or methionine diet showed low levels of plasma choline (Sheard et al., 1986), although this finding is not universal (cf. Blusztajn and Wurtman, 1983); the same seems true for animals, and levels as low as one half of control values were found in rats kept on a choline-deficient diet (Haubrich, et al., 1976). Besides choline itself and lecithin, other choline-containing phospholipids (including lysophosphatidylcholine and sphingomyelin) and related substances that are present in the diet increase plasma levels of choline (Wurtman, 1979). This matter involves the absorption; intestinal, plasma, and liver metabolism; and transport of these compounds, as well as the transport of choline that they may generate (Houtsmuller, 1979). It is another matter that choline availability during fetal development is important for neuronal differentiation and development of certain kinases (see Chapter 8 CIII).

This subject is of a special, although peripheral, interest in the current context: choline formation occurs during lipoprotein, cholesterol, and fat metabolism in the intestinal epithelium (enterocytes). This metabolism leads to the formation in the enterocytes of chylomicrons and very low density lipoproteins (VLDLs); thus, fat metabolism and degradation of lipoprotein is involved in the release of choline into blood (Houtsmuller, 1979). This is a special example of the relationship between phospholipid and choline metabolism.

b. Erythrocytes

One would imagine that the discovery of either facilitated or low-affinity uptake of choline into the RBCs (Askari, 1966) would include its identification in that tissue, particularly in view of the earlier demonstration of the presence of both choline and ACh in the plasma or whole blood of several species (see above). Yet, the early students of this uptake (Askari, 1966; Martin, 1968) did not mention the presence of endogenous choline in the RBCs. Only in 1972 did Israel Hanin and his associates (Hanin et al., 1979) demonstrate this presence in human RBCs by means of a gas chromatography–mass spectrometry (GCMS) method. Hanin (see Hanin et al., 1978) stressed that while in many subjects the range of concentrations of RBC choline was relatively narrow (8–10 nM/ml), in some subjects "choline levels in the RBC's exhibited a tremendous variation." This finding was noteworthy, as the method itself was reliable, and as the reproducibility of the choline levels was well documented (o.c.). While Israel Hanin and his associates found also that the levels could be influenced by the diet and/or intake of precursors (e.g., Hanin et al., 1978), this particular factor apparently was not involved in the variation in question, and Hanin (cf. Green et al., 1972; Hanin et al., 1978) suggested heuristically that "there appeared to be . . . clinical significance to the observed values of RBC Ch." The original finding of Hanin and his associates of the presence of choline in the RBCs was confirmed for humans (see, e.g., Jope, 1986; Domino et al., 1982) and animals. As pointed out, RBCs may serve as models of choline transport mechanism, and choline of human RBCs has a diagnostic value for several clinical conditions, including abnormal lipid profile; yet, many recent reviews of choline metabolism do not refer to RBC choline and its significance (see, e.g., Tucek, 1978, 1984, 1985, 1990).

c. Cerebrospinal Fluid

It has been consistently demonstrated that choline concentration in the CSF, while significant, is lower, in man and animals, than in the blood plasma (cf. Tucek, 1978, 1985). The question of choline (and ACh) concentration in the human CSF is of particular importance, as the CSF is the only tissue—other than blood—available in the human man for studies of choline and ACh metabolism in the course of the precursor or antiChE treatment of Alzheimer's disease. Furthermore, CSF values of these compounds may reflect relatively precisely their brain levels (Giacobini, 2000).

d. Brain

Generally, choline brain concentrations quoted in the older literature refer to concentrations in the "whole" brain tissue, which includes interstitial (extracellular) fluid and plasma. When this "whole brain" choline concentration is measured continually, following decapitation or other methods of sacrifice in animals, or postmortem in humans, this concentration rises rapidly. This phenomenon is due to the continued presence, under these conditions, of ChEs and consequent hydrolysis of ACh. Thus, the brain choline values obtained by Stavinoha and his associates (e.g., Stavinoha and Weintraub, 1974), who used the microwave oven to sacrifice the animals, and by Mann and Hebb (1977 see also Mann et al., 1938), who quickly froze the brains immediately after sacrifice, are lower (20–30 nM/g) than those obtained by others. Choline values for isolated brain tissues or slices are relatively high because under these circumstances there is an anoxia that stimulates hydrolysis of choline esters, and choline cannot be removed from the tissue by blood (although it is released into the medium; cf. Tucek, 1985).

Direct measurement of intracellular brain choline is not possible at this time; however, when choline values for "whole" brain are corrected for the expected concentrations of choline in the brain plasma and brain interstitial water (which may be similar to choline concentration in the CSF; for discussion, see Tucek, 1985, 1990), the "whole" brain value of choline appears to be mostly due to intracellular choline. This notion agrees with the data and concepts presented by Konrad Loeffelholz (1996; Loeffelholz and Klein, 2004).

An important issue concerns the dependence of brain choline levels on the availability—via appropriate diet or choline and choline precursor supplements—of choline. This matter was already discussed (see above, sections CII and CIII); while the dependence of brain ACh concentration on the "normal" supply of choline, choline-containing phospholipids, and related substances is somewhat controversial (see above, sections CII and CIII), the increase of brain choline levels following the increase of its supply is indubitable. Independently of the route employed for supplying choline, brain choline levels are significantly augmented. Perhaps the lowest (although significant) increase was reported in the earliest of the pertinent studies (Kuntscherova et al., 1972; more recent results indicate that this increase may be severalfold (Kindel and Karczmar, 1981, 1982; Blusztajn and Wurtman, 1983). This increase is temporary because of the choline homeostasis processes (Loeffelholz, 1996). Indeed, in mice, the increase was relatively short-lived, and even after repeated dosing each day for several days with barely tolerated amounts of choline (120–200 mg/kg per dose; intestinal and autonomic effects were evoked by the high dosing, and large amounts were not readily consumed), the increase lasted for only a few hours (Kindel and Karczmar, 1981, 1982). Similar results were obtained with administering high doses of phosphatidylcholine or related compounds (cf. Tucek, 1985; Growdon, 1987).

The choline contents of separate brain parts were compared in only a few studies. Most data were obtained in the rat and mouse (see Saelens and Simke, 1976); the rodent striate and the hypothalamus exhibit more choline than the other brain parts (cf. Tucek, 1978; Sethy et al., 1973). However, these differences were relatively moderate (amounting to no more than 50%); sporadically, relatively high values were also reported for the rat telencephalon (Smith et al., 1975). This is not surprising: the basal ganglia, the striate, the hypothalamus, and the medial forebrain are rich in cholinergic pathways (see Chapter 2 D I-III), which is consistent with their levels of ACh being many times higher than those of, for example, the cortex (Saelens and Simke, 1976) and with the presence of high-affinity choline uptake systems in at least some of these areas (see below); thus, choline should be preferentially localized in these brain regions.

Moreover, the administration of choline or choline-containing phospholipids should lead to preferential distribution of choline to these brain parts, the assumption being that a high degree of the utilization of choline for ACh synthesis versus its utilization for the phospholipid and related metabolism occurs in the areas of high density of cholinergic function. The few pertinent studies that are available concern only a few brain parts and differ with respect to techniques and the species employed (Racagni et al., 1976; Eckernas, et al. 1977; Pedata et al., 1977; Hirsch et al., 1977). The results suggest that on its administration, choline distributes somewhat preferentially to the caudate versus either the cortex or the hippocampus; yet, as in the case of the distribution of endogenous choline in the various brain parts, the differences are moderate and/or inconsistent (see Saelens and Simke, 1976).

e. Other Tissues

As can be expected, uptake and metabolism in the liver, the major site of choline utilization, exhibit the highest concentrations of choline; for example, in the cat, liver content of choline amounts to 812.5 nM/g (Shaw, 1938) versus 1–30 nM/g levels in the blood and in the brain. While choline levels were rarely evaluated in the liver, blood, and brain by the same laboratory and by the same methods, high liver levels as compared to the values obtained for other tissues were generally reported (see Saelens and Simke, 1976). Other parenchymatous tissues exhibit choline levels lower than those of the liver but considerably higher than those of the brain and blood (Saelens and Simke, 1976). It appears that data are still lacking with regard to the contribution of these choline-rich sites to the blood and, particularly, to the choline homeostasis processes (Loeffelholz, 1996); could the liver and the parenchyma serve as a sink in these processes?

2. Choline High-Affinity Uptake and Transport Systems

a. Choline Uptake of Neuronal Tissues

The tissue distribution of endogenous choline as well as of exogenous, supplied choline is linked with choline uptake, whether into cholinergic and noncholinergic neurons, or nonnervous tissues. This subject was already alluded to in the framework of the history of the choline-CAT relationship and of the need of choline for the maintenance of the function of cholinergic neurons (see above, section A); it was pointed out that Birks and MacIntosh (1961) employed hemicholiniums to prove the hypothesis of the need of "intense" choline uptake for the maintenance of this function. It must be stressed: "the ability of a cell to transport choline is not unique to the nervous tissue. Choline transport seems to be a general finding for virtually all cell types . . . the apparent ubiquity of choline transport systems probably reflects an evolutionary adaptation for the transport of choline, a positively charged ion, across cell membranes as a means to supply choline for cellular metabolic needs." This fruitful statement of Barker (1976) relates to the need of choline both for ACh synthesis ("synaptotropic choline") and for the synthesis of phospholipids and cell membranes ("metabotropic choline").

Hank Yamamura, a pioneer in the area of choline uptake investigations, indicated that the early choline uptake studies were hampered by a lack of appropriate methodology (see Yamamura and Snyder, 1973). Ultimately, availability of ^{3}H-choline of high specific radioactivity "permitted the study of the uptake at low concentrations of choline at which a high affinity" Na-dependent system must operate as predicted by Haga (1971; see also Okuda and Haga, 2003). Using these low concentrations of choline, Victor Whittaker and his associates (Whittaker, 1972; see also Potter, 1968) finally demonstrated the existence of a high-affinity neuronal choline uptake system. Since that time, many investigators, notably Michael Kuhar, Hank Yamamura, and Sol Snyder (see Tucek, 1978; Speth and Yamamura, 1979), established the characteristics of the high-affinity neuronal choline uptake (sodium-dependent high-affinity choline uptake, SDHACU; Speth and Yamamura, 1979). Sodium-dependent high-affinity choline uptake is HC-3 sensitive; it is also temperature and energy dependent, thus being sensitive to metabolic inhibitors (Speth and Yamamura, 1979; Tucek, 1978, 1984, 1985). As expected for a high-affinity system, which is efficacious at low concentrations of choline, its K_m is quite low (less than 5 nM; for references, see Tucek, 1978, 1984, 1985). The Na^+ activates the system possibly by increasing the affinity of the carrier to choline and, thereby, decreasing the K_m and increasing the *V*max. The $Ca2^+$ and chloride also seem to be necessary (Martin, 1972; Ksiezak-Reding and Goldberg, 1982), although the role of $Ca2^+$ is controversial (see Tucek, 1984, 1985, for further references).

Membrane potential is important for SDHACU, and Tucek (1985) stated straightforwardly that "the main and undoubted driving force for the transport of choline is the electric potential across the neuronal membranes." He quoted the data of Vaca and Pilar (1979) and Beach et al. (1980) concerning K^+-induced depolarization, which indicated "that linear relationship . . . obtains . . . between the rate of choline uptake and the logarithm of K^+ concentration . . . (in the range of 3–55 mM) . . . in the incubation medium . . . i. e., between membrane potential and the rate of transport." Other depolarizing agents exerted similar inhibitory effects on choline transport (Murrin and Kuhar, 1976). Accordingly, Tucek (1985) used the Nernst equation, as he calculated the ratio between the intracellular and extracellular concentration of choline to be 16.4.

There are a few complications with this story. First, while SDHACU is decreased in the course of depolarization, it is activated by depolarization when measured after the depolarization subsided. This was noticed with respect to synaptosomes and cortical slices (for references, see Tucek, 1978, 1985), and in semiartificial systems such as liposomes (King and Marchbanks, 1982). Is this increase of uptake due to a hyperpolarization that could have followed the depolarization, in analogy to poststimulation hyperpolarization (Tucek, 1985)? This explanation is consistent with Tucek's stress on the relationship between the "electric potential across . . . neuronal membranes and choline transport." The other explanation is that "depolarization-induced . . . increase in the high affinity uptake of choline is not initiated by the depolarization as such but by the release of ACh and the consequent change of ACh concentration in the nerve terminals" (Tucek, 1985). The other complication arises from the finding that rather than being linear, the relationship between K^+ and SDHACU has an optimum at the K^+ concentration of approximately 0.35 mM, higher K^+ concentrations lowering the choline uptake (Beach et al., 1980); this finding is not quite consistent with the decrease of choline uptake that occurs during depolarization.

Is SDHACU specific for cholinergic neurons? Barker (1979) stated succinctly that the "primary function of . . . the SDHACU . . . is to supply choline for the synthesis of ACh . . . and the uptake appears to be coupled to the synthesis of ACh"; this particular ACh "is also preferentially released during synaptic transmission." Several lines of evidence support this concept. First, SDHACU is localized in nerve terminals (in their cytosol but not in the mitochondrial fraction; Yamamura and Snyder, 1973); this was shown, for example, for the presynaptic terminals of ganglion cells (Suszkiw and Pilar, 1976). It is uncertain whether or not it is also present in the perikarya (see Suszkiv and Pilar, 1976, versus Richelson and Thompson, 1973). Second, this system is eliminated by destruction of cholinergic nerve terminals in many central and peripheral cholinergic pathways; in brain sites rich in SDHACU (such as interpeduncular neurons; Kuhar et al., 1973; see also Amara and Kuhar, 1993), presynaptic denervation lowers SDHACU activity, the activity of CAT, and ACh content simultaneously. In this context, SDHACU appears to be closely correlated, location-wise and during development, with CAT activity, ACh content, and/or ACh turnover. Finally, SDHACU seems to correlate with the presence and distribution of the muscarinic receptors of the brain (Yamamura et al., 1974); it is not clear at this time whether or not SDHACU correlates equally well with the nicotinic receptors.

A minor problem arises at this point. Sodium-dependent high-affinity choline uptake correlates with brain sites that are rich in cholinergic pathways and synapses and in CAT, as it is high in interpeduncular nucleus and in the caudate, whereas it is low in frontal cortex. However, choline uptake and CAT and AChE activity do not change in parallel between the sites that are cholinergically rich and poor, as would be expected; there is a 10- to 200-fold difference between the alterations of these three entities (see Figure 5-4 in Tucek, 1978).

The demonstration that choline transported by SDHACU into neurons is entirely acetylated into ACh would constitute striking evidence for the link between SDHACU and the cholinergic function. The pertinent evidence was first provided for brain synaptosomes by Haga (1971), and then by Simon, Kuhar, Yamamura and Tucek for synaptosomes, isolated ganglia, minced brain and brain slices (Simon et al., 1976; Tucek, 1978, 1984, 1988; Cuello, 1996). It was shown also that in this synaptosomal system, hemicholiniums or lack of Na^+ inhibits ACh synthesis (Yamamura and Snyder, 1973; Saito et al., 1986). Finally, SDHACU seems to correlate with the presence and distribution of muscarinic receptors in the brain (Yamamura et al., 1974), though whether SDHACU correlates with the nicotinic receptors is unclear.

Related evidence concerns the relation between the SDHACU system and cholinergic activity. Choline uptake and/or SDHACU was increased in synaptosomes obtained from animals treated with convulsant and atropinics, that is, following augmented CNS activity or ACh depletion (Atweh et al., 1975; see Tucek, 1978, 1985). Similarly, SDHACU was augmented following depolarization or electric stimulation of brain slices and synaptosomes (Antonelli et al., 1981). Certain experiments are particularly convincing: in the early 1960s, i.e., even before the identification of SDHACU, Birks, MacIntosh, Brown and Feldberg (see above, this chapter, section C IV-1) demonstrated the relationship between choline and its uptake, ganglionic activity and transmission, on the one hand, and ACh release, on the other. Then Friesen and Khatter (1971) demonstrated that choline content increases temporarily above control values following prolonged ganglionic stimulation—presumably because of increased choline uptake. Similar evidence implicating SDHACU more specifically in cholinergic activity was provided later by Collier and his associates (Collier et al., 1983; for further references, see Tucek, 1985). Yet, certain findings are not consistent with a rigid linking of SDHACU and the cholinergic neurons and their activity. Thus, SDHACU is present in cholinergic neurons that are denervated, lack nerve terminals, and do not produce ACh, or in noncholinergic cells such as glia (Richelson and Thompson, 1973), fibroblasts (Barald and Berg, 1978) and the RBCs (Martin, 1968; see Tucek, 1978, 1984 for further references). The matter of the presence of SDHACU in the RBCs is expanded on below.

An interesting aspect of the relation between choline uptake and cholinergic neuronal activity concerns ACh effect on the uptake. It was first noticed by Potter (1968) that ACh inhibits, even at low concentrations, the high-affinity uptake (for further references, see Tucek, 1978, 1985). These results were obtained *in vitro* only, but it is teleological to assume that this phenomenon occurs also *in vivo*, as it would serve as a useful negative feedback. It must be pointed out in this context that ACh is present in the synaptic cleft (see Chapter 2 C); this ACh may affect the potential of both the postsynaptic and presynaptic membrane—could this presynaptic effect relate to the negative action of ACh on choline uptake?

How is the neuronal activity linked with SDHACU? Is this link related to release and consequent depletion of ACh, as indicated by the effect on SDHACU of slice depolarizers and, *in situ*, of the augmenters of ACh release? Yet, some of the studies quoted indicate that SDHACU activation may occur independently of augmented ACh release (cf. Tucek, 1985) and be due to a hyperpolarization. Still another possibility is that neuronal activity increases nerve terminal uptake and accumulation of choline via its effect on Ca^{2+}, that is, via stimulation-dependent augmentation of Ca^{2+} influx, described classically by Bernard Katz; indeed, uptake of choline (and its analogues) is Ca^{2+} dependent and, in fact, antagonized by Mg^{2+} (Barker, 1976). Indeed, it may be generalized that SDHACU is necessary to maintain cholinergic function and that its expression and activity may be modulated by conditions of cholinergic hypofunction (Brandon et al., 2004).

As can be seen, the activity-uptake relationship takes the form of a complex feedback system; for example, neuronal activity may stimulate uptake via a number of mechanisms such as change in membrane polarity, ACh depletion, Ca^{2+} influx, SDHACU expression, and so on, while it may depress the uptake via the inhibitory action of cleft ACh on the choline uptake. Finally, this feedback system may be important for the choline homeostasis processes (see this Chapter, section CIV-3).

b. Active (High-Affinity) Choline Uptake System of the RBCs

The presence of endogenous choline in the RBCs and choline uptake into the RBCs were already alluded to. Besides choline, AChE, phosphatidylcholine (Galli et al., 1992), and CAT (and ACh in very small concentrations; Chang and Gaddum, 1933) are present in the RBCs of at least some species, such as the dog and the human. The presence of AChE in the RBCs has been known since the 1940s (Mendel et al., 1943), and the presence of BuChE, CAT, and/or ACh synthesis since the 1930s (Quastel et al., 1936; cf. Augustinsson, 1948). Thus, the RBCs contain all the components of the cholinergic system, including choline.

The RBC uptake system is a part of "an exchange flux phenomenon" (Martin, 1968). The presence of the high-affinity component was indicated as early as 1966 by Askari; that its characteristics are, at least in part, homologous with those of the SDHACU system of the neurons was established by Martin in 1967 and 1968. Indeed, Martin (1967, 1968) demonstrated that the human RBC uptake system is saturated at relatively low external concentrations of choline. In fact, the K_m of the process in question (0.5 mM) is similar to that of the neuronal SDHACU and is Na^+ dependent. Furthermore, certain ions such as Li^+ inhibit the RBC choline uptake as they do in the cases of other SDHACU systems. Martin (1967, 1968) demonstrated also that the system involves actual influx of choline rather than "a sodium-dependent adsorption of choline to the erythrocyte membrane." However, the question of the sensitivity of the uptake to hemicholinium was not resolved, as Martin confined his evaluation of hemicholinium action to the study of the effect of high concentrations of hemicholinium on combined choline influx and efflux—which were markedly inhibited by these concentrations. Finally, it is likely that a slow, low-affinity uptake system may exist in the RBCs as well (Askari, 1966; see also below).

What is the role of cholinergic components in the RBCs? Greig and Holland (1949) proposed that RBC "fragility" and ionic permeability are controlled by the interaction between AChE and the ACh (see also Cullumbine, 1963). This concept was based on their observation that ACh and physostigmine decrease Na^+/K^+ permeability of the RBC membrane. Subsequently, working with the OP antiChE, di-isopropyl fluorophosphonate (DFP), Thompson and Whittaker (1952) found that DFP antiChE reduce the "passive" (low-affinity?) but not "active" (high-affinity?) uptake of Na^+. There are a number of problems with this interesting concept (cf. Koelle, 1963). First, the concentrations of antiChEs needed to cause these effects were "several magnitudes higher than those which produce complete inhibition of AChE" (Koelle, 1963); is this effect analogous to the noncholinergic actions of antiChEs, which are discussed in Chapter 7 DI? Second, Na^+ and K^+ transport does not appear to be handicapped in erythrocytes that are devoid of or poor in AChE, such as those of the cat (see Koelle, 1963).

To what extent is the high-affinity RBC uptake of choline related to the physiological ionic fluxes of the RBCs, which involve Na^+, K^+, and Cl^-? The transport in question includes several processes, such as the ($Na^+ + K^+$)-ATPase system, which is ouabain sensitive (cf., for example, Martin, 1968) and the Cl^-- dependent, loop-diuretics-sensitive transport system, referred to as a "symporter," as it couples anionic and cationic transports. External choline behaves in the RBCs as a cation and is used as a Na^+ or K^+ substituent in the evaluation of the role of the inorganic ions in the transport in question. It appears that, so far, no studies have been carried out as to the significance of the relationship between choline and $Na^+/K^+/Cl^-$ transport systems of the RBCs; in fact, the basic question of the differential sensitivity of the 2 systems to agents such as hemicholiniums, loop diuretics (for example, furosemide), and ouabain were not studied in depth. Altogether, the question of the similarities and dissimilarities between the neuronal and RBC choline high-affinity uptake systems should be further clarified.

It is of particular interest that the RBC levels and dynamics of choline may serve as clinical markers, particularly in diagnosis of certain mental and nutritional malfunctions (Compher et al., 2002). For example, Israel Hanin suggested that the RBCs constitute a model of lithium effect on the brain and depressive mental disease (for references, see Stoll et al., 1991). Finally, it should be emphasized that the cholinergic regulation of the ionic fluxes across the RBC membrane is an illustration of the metabotropic role of the cholinergic system (see Chapter 2 C-3).

3. Low-Affinity Choline Uptake

a. Sharing of Two Uptake Systems

Haga (1971) was the first to propose the existence of another choline uptake system. This system is effective at higher concentrations of choline than SDHACU, as expressed by its K_m value, which is much higher than that of the SDHACU (Haga, 1971). This is a unanimous finding, although actual values vary widely depending on the investigator, species, site, and so on. For example, in the homogenate of brain striatum, the K_m values for SDHACU and the low-affinity system amounted to 1.4 and 93 nM, respectively, while the corresponding values for the rat cortex were 3.1 and 33 nM, respectively (Yamamura and Snyder, 1973). Furthermore, the low-affinity carrier (or uptake) system (LAUS) has a lower dependence on Na^+, and lower sensitivity to metabolic inhibitors and hemicholiniums (it can be blocked by high concentrations of HC-3) than the SDHACU (Haga and Noda, 1973; see also Barker, 1976; Tucek, 1978, 1985).

An important point is that the K_m value for the two systems could be obtained at the same sites; that is, many sites are endowed with both systems. This is true for several brain parts; interestingly, the two systems are present in areas relatively rich—such as the striate—and relatively poor—such as the cortex—in cholinergic synapses. In addition, the two systems are present also at the periphery, including the sympathetic ganglia (Bowery and Neal, 1975) and intestinal muscle (assumedly, in the plexi of the latter, Pert and Snyder, 1974). They also exist in cultures of nerve cells, the invertebrate nervous system, minced preparations, slices, and synaptosomes (for references, see Tucek, 1978).

In view of the similarity of the distribution of these two systems, the question is whether or not the systems differ functionally. As already pointed out, the SDHACU system is correlated with the cholinergic function, and Haga and Noda (1973) demonstrated that in the rat cortex, which is endowed with these two systems, only SDHACU was associated with ACh synthesis. However, this functional distinction may not always be warranted. After all, the low-affinity system is present in many neuronal tissues; moreover, some data suggest that in the presence of synaptic activity—for example, when nerve terminals are depolarized—the low-affinity system supplies choline for the synthesis of ACh (Carroll and Goldberg, 1975).

The important aspect of this question is that, of course, choline uptake is needed for neuronal and nonneuronal cells, and for the metabotropic as well as synaptotropic needs; does this signify that the uptake of choline for a great many cells and tissues should be of the low-affinity type? Or does SDHACU sometimes subserve the transport of metabotropic choline as, for instance, in the case of RBCs?

b. Blood-Brain Barrier Transport of Choline

Choline is transported across the blood-brain barrier (BBB) into the brain and in spite of rapid choline movement from the blood to the brain, the BBB is a “serious” (Tucek, 1984) obstacle for this transport. At least some of its characteristics suggest that a LAUS is involved here. For example, the brain choline uptake system has a quite high K_m (Barker, 1976; Conford et al., 1978), in fact, its K_m value is considerably higher than the K_m values for the LAUSs of the various neuronal systems described above (Pardridge, 1983; Tucek, 1990; Loeffelholz and Klein, 2004). Altogether, Konrad Loeffelholz and Jochen Klein (2004) opine that the LAUS is involved in the choline transport across the BBB. This is teleological, since the LAUS is more effective than SDHACU at the relatively high concentrations of choline that are present in the blood (although the LAUS of the BBB is saturable).

It appears that used at high concentrations, hemicholiniums block brain synthesis of ACh and brain choline uptake (Gardiner and Gwee, 1977; Loeffelholz and Klein, 2004). Hemicholiniums are particularly effective in blocking SDHACU; however, their inhibition of the BBB transport of choline is not inconsistent with the notion that a LAUS is involved in this transport, since the K_i of hemicholinium is high, which is characteristic for hemicholinium action on a LAUS (Loeffelholz and Klein, 2004; Konrad Loeffelholz, personal communication).

DI. Cholinesterases

1. Cholinesterases, Anticholinesterases, and Cholinergic Transmission: from Dale to Loewi

The first intimations that there is such an entity as cholinesterase (ChE) are linked with the story of ACh. First, there is the matter of the kinetics of ACh action. In 1914 Sir Henry Dale, noticing the evanescence of ACh's action when administered systemically, opined that "it seems not improbable . . . [note the double-negative phraseology: careful Sir Henry!] . . . that an esterase contributes to the removal of the active ester from circulation, and the restoration of the original condition of sensitiveness" (this sentence implies that the omniscient Dale felt that there is such a thing as desensitization). This matter of evanescence was followed up by Sir John Eccles; as mentioned in the introduction to this chapter, Eccles opined (in 1936, that is, during the electric transmission phase of his career) that ACh may be considered as a transmitter only if its action could be terminated rapidly, to prevent its clogging the synapse. Actually, ChE and in fact AChE were known at the time, but Eccles did not perceive what a fast enzyme AChE is. Second, there was the matter of ACh synthesis. Sir William Feldberg and Sir Lindor (Geoffrey) Brown (the knights of the cholinergic Table Ronde, jointly with Sir John Eccles and Sir Henry Dale) realized that choline, liberated by the ChE action, is important for synthesis of ACh (see introduction to this chapter).

Studies of the effects of physostigmine, which actually preceded Dale's proposition, contributed to the affirmation of Dale's postulate. As early as 1872, Heidenhain demonstrated that the extract of Calabar bean increases salivary secretion. The subsequent work of Dixon (1907), Hunt (1915), Führer (1918), and others showed that physostigmine or the Calabar bean extract potentiates autonomic functions as well as the effects of ACh (see Karczmar, 1967, 1970, 1986; Chapter 9 A). These studies preceded the demonstration of the antiChE action of physostigmine; however, in 1918 Fuhner made a suggestion closely paralleling Dale's postulate, namely, that physostigmine prolongs vagal bradycardia by inactivating an enzyme capable of splitting a neurohumoral substance. The final links among endogenous release of ACh, its hydrolysis by ChEs, and the antiChE effect of physostigmine were forged by Otto Loewi (1921, 1960; Loewi and Navratil, 1926). Loewi showed that "eserine" (physostigmine) potentiated both the vagal and the ACh-induced bradycardia; he demonstrated also by using bioassay and pharmacological analysis that the "Vagusstoff" is, indeed, ACh (Karczmar, 1996). Since the days of Loewi, the role of ChEs in nervous function has been studied incessantly, and these studies, jointly with those of antiChEs, established, once and forever, the presence of cholinergic transmission whether in the CNS or in the peripheral nervous system (see Chapters 7 and 9).

2. Cholinesterases: Classifications, Types, Forms, and Variants

Beginning in the 1920s, it was shown that a great variety of species and tissues exhibit ChEs; this is also the case with the other components of the cholinergic system, such as ACh and CAT (see Chapter 8 BV). Several methods were used for this purpose, including manometric or gasometric, titrimetric, colorimetric, and fluorometric (see Augustinsson, 1948, 1963; Usdin, 1970; see below, this section). For example, in 1925 Abderhalden et al. demonstrated the presence of a ChE in the intestine, Stedman and hisassociates (1932) in the blood, and Plattner and Hintner (1930) found it in several glands, blood, and brain, but not in plants (this latter finding was invalidated subsequently; see Chapter 8 BV). David Glick, who in the 1930s and 1940s studied intensely the characteristics of "serum" ChEs, published in 1941 a very complete survey of the presence of ChE in the tissues of "swine"; blood, skeletal, and smooth muscle, several glands (including testes), organs such as spleen and kidney, ganglia, and several parts of the brain were listed. In his 1941 paper Glick did not differentiate among various ChEs, but he did so a bit later (Glick, 1945). A few years later, Klas-Bertil Augustinsson (1948), another important pioneer in the ChE area, published a similarly long list, which included placenta, sperm, and eye; this presence of a ChE in nontransmittive, nonsynaptic tissues such as blood, but also in the ephemeral organ the placenta, constitutes till today one of the mysteries of the field. The novelty of Augustinsson's list is that it referred to two enzymes: Type I and Type II ChEs, or "specific" and "non-specific" ChEs. Finally, in 1970 Earle Usdin prepared the most complete listing in existence of the presence of ChEs in species ranging from mammals, insects, and echinoderms to amphibia and so on (from bat to elephant to leafhopper), including tissues and types of ChE involved.

Ammon and Chytrek (1939), Glick (1941), and Augustinsson (1948) were the first investigators to refer to the low levels of ChEs in disease states. In comparison with the current listings of the position of ChEs in Alzheimer's disease and myasthenia gravis, the references in question are exotic: the diseases listed as showing abnormal levels of ChEs (generally, the serum enzyme is referred to) include hypo-and hyperthyroidism, epilepsy, schizophrenia, parkinsonism, allergy, tuberculosis, and so forth. More appropriately, Glick (1941, 1945) also refers to myasthenia gravis (he states that the results concerning ChEs of myasthenia are inconsistent), and all three investigators list liver disease as linked with low levels of ChE.

As already mentioned, Glick referred mainly to serum ChE in his 1941 list of the presence of the enzyme in various tissues and did not recognize that ChEs are a family of enzymes, although he quotes in his paper the 1930s studies of David Nachmansohn concerning the CNS ChE; he does not quote the earlier work of Galehr and Plattner (1927) and Alles and Hawes (1940), both dealing with the erythrocytic enzyme. Actually, while Galehr and Plattner (1927) recognized the difference in the ChE activities of the serum versus the whole blood, as well as the rapid hydrolysis of ACh by the erythrocytes, they explained this effect by the nonenzymatic surface catalysis of ACh by the RBCs (see also Karczmar, 1970). It was Alles and Hawes who in 1940 differentiated the erythrocytic enzyme from the serum enzyme by the ability of the former to hydrolyze acetyl β-methylcholine (it is notable that they carried out this investigation on the human blood). On the other hand, the Stedmans with Easson (1932) purified from the horse serum an enzyme that they named choline esterase.

Ultimately, Augustinsson (1944) separated electrophoretically the serum and erythrocytic enzyme. Then Mendel and Rudney (1943, 1944) proposed a general classification of enzymes studied by Glick, the Stedmans, and Alles with Hawes; this classification was based on their substrate specificity and sensitivity to inhibitors, as they referred to pseudocholinesterase of the serum and true cholinesterase of the erythrocytes; the former hydrolyzed benzoylcholine and butyrylcholine preferentially, while the latter splits propionylcholine and acetyl β-methylcholine preferentially. Alles and Hawes (1940) and Glick (1945) referred to the term "pseudocholinesterase," used by Mendel and Rudney, "as unfortunate" and they felt that this term should be blocked from scientific literature. They considered, quite correctly, that there is nothing "pseudo" about the enzyme of the serum and certain other tissues and that, just as other ChEs, the blood enzyme splits ACh and other choline esters. They preferred the term "nonspecific ChE" for the enzyme in question. And, lastly, Nachmansohn and Rothenberg (1944) spoke of "specific cholinesterase" (or acetylcholinesterase) and "cholinesterase," the former characterizing, besides erythrocytes, the nervous tissues. Through the 1950s these and other names were employed by the various investigators (see Table 4 in Usdin, 1970); the terms "acetylcholinesterase" (AChE) and "butyrylcholinesterase" (BuChE) seem to be generally used today. And, in 1961 (see Usdin, 1970) EC numbers were established for the two enzymes by the Enzyme Commission of the International Union of Biochemistry; they were EC 3.1.1.7 for AChE and EC 3.1.1.8 for BuChE (the "official" names established by the commission are acetylcholine acetyl-hydrolase for AChE and acylcholine acyl-hydrolase for BuChE; these names are just about never used in the literature).

Nachmansohn and Wilson (1955) stressed the velocity of ChEs' action; this finding, subsequently confirmed by others (Rosenberry, 1975; Quinn, 1987), underlay Eccles' acceptance of the chemical transmission hypothesis (Chapter 9 A). Cholinesterases are among the fastest enzymes known; as stated by Massoulié et al., (1993a), "the hydrolysis of acetylcholine by acetylcholinesterase approaches the maximal theoretical limit set by diffusion of substrate," and Quinn (1987) commented that in AChE, "the evolution created a 'perfect enzyme.' "

The presence of both AChE and BuChE is characteristic for vertebrates; on the other hand, the presence of ali- and aryl esterases is characteristic for insects. Augustinsson, Glick, Sir Arnold Burgen (1949), and subsequently William Aldridge (1953a, 1953b) defined cholinesterases as "a group of esterases which hydrolyze choline esters at a higher rate than other esters"(Augustinsson, 1963). It was recognized early (Augustinsson, 1948) that ChEs are also characterized by being inhibited by low (i.e., 10–5 M) concentrations of physostigmine or prostigmine, and by their sensitivity to quaternary ammonium ions (an indication of the presence of an anionic site within the ChE moiety); a similar definition was offered by Engelhard et al. (1967). Cholinesterases belong to a wider group of esterases, which are readily inhibited by organophosphorus (OP) compounds. These esterases are called B-esterases by Aldridge (1951, 1953), and they include also the so-called ali-esterases. Contrary to ChEs, aliesterases are unable to hydrolyze choline esters, but they are able to hydrolyze certain other esters; they are resistant to physostigmine or prostigmine. Finally, there are also A-esterases (Aldridge, 1953b), or aryl-esterases (Augustinsson, 1959), which are resistant to OP agents and which hydrolyze aryl esters at a high rate. This classification is similar to that adapted by Mounter and Whittaker (1953) and by subsequent investigators; it differs from that presented by Bergmann and Rimon (1957; see also Usdin, 1970). It may be noted that ali-esterases and aryl-esterases are present in the nervous system of some but not all insects (Chadwick, 1963; see also Zupancic, 1967).

These studies of the hydrolytic actions of various ChEs needed the development of methods for the measurement of these hydrolytic effects and of the different substrate specificities of the enzymes in question. It should be stressed that the same methodologies were and are used for the measurement of ChE inhibition by antiChEs. Ammon's (1933; Ammon and Chytrek, 1939) manometric method was the first to be widely employed, although ACh bioassays were also employed to quantify ChEs in terms of the disappearance of ACh exposed to ChE-containing tissues (see Augustinsson, 1948). This latter methodology was subsequently improved by utilizing the gas chromatography–mass spectrometry method for measuring hydrolysis of ACh and other substrates (Hanin and Godberg, 1976). The Ammon method was felicitously modified by Ezio Giacobini, who utilized Cartesian microdiver to measure the CO_2 evolution via the ChE action in single cells (Giacobini, 1959). Fluorometric and radiometric techniques were employed as well. Also, George Koelle collaborated with Jonas Friedenwald (see Koelle, 1963) to develop his famous histochemical method for the qualitative and semiquantitative measurement of ChEs and their localization in tissues, including nervous tissue (Friedenwald, 1955; Koelle, 1963). Koelle subsequently modified this method to make it amenable for electron microscopy. Among the newest additions to the ChE measurement are the use of positron-emission tomography (PET) and appropriate tracers and of "micro-immobilized-enzyme reactors" (Zhang et al., 2003; Bartolini et al., 2004). Quantification of ACh can also be used for the measurement of the efficacy of antiChEs (see Chapter 2 B).

Kalow employed the term "atypical" ChE to describe a genetically determined BuChE. "Atypical" BuChE is present in certain human cohorts at levels that are lower than those of "normal" BuChE in the general population; moreover, the "atypical" BuChE is specifically sensitive to inhibition by a number of inhibitors, including dibucaine, and it reveals decreased substrate affinity as compared to the "wild" variant (Kalow and Gunn, 1959; see also Usdin, 1970; Kalow, 2005). This BuChE variant became well known when Werner Kalow demonstrated that it characterizes a human cohort (about 2% of the white population) that is hypersensitive to the muscle relaxant succinylcholine, as their blood cannot efficiently hydrolyze succinylcholine. This cohort is at risk for anesthetic procedures that include the use of muscle relaxants. Subsequently, it was proposed that 4 alleles are involved in the regulation of the appearance of atypical versus typical BuChE (Kalow and Gunn, 1957; 1959; Foldes, 1957; see also Usdin, 1970). As demonstrated particularly by Mary Whittaker, Bert La Du, and their associates, many more allelic variants of BuChE became known subsequently (Hangaard et al., 1991; La Du et al.,

1991; La Du, 1995; Primo-Parmo et al., 1996, 1997; Satoh et al., 2002).

Butyryl cholinesterase "isozymes" can be defined as differing in their activities and/or substrate specificities from the wild or "normal" forms of BuChE; it must be noted that, so far, no allelic variants of AChE that would differ in their activity have been described. Butyryl cholinesterase isozymes also reveal special physical properties: they exhibit different mobilities in an electrophoretic field as compared to the "normal" forms. According to the International Union of Biochemistry, they should be numbered in accordance with the sequence of their mobility (Markert, 1968; see also Usdin, 1970). This approach is impracticable with respect to AChE molecular forms, because their mobility depends on glycosylation and on oligomeric association, not only on peptidic composition. At any rate, classification of any enzyme should be based primarily on genetic differences between the various forms, whether of BuChE or AChE (Jean Massoulié, personal communication).

Another type of variability results from posttranslational modifications and quaternary associations (oligomerization) of ChEs. The term "polymorphism" was first employed by Jean Massoulié (see Massoulié et al., 1970), who differentiated several molecular, physical forms of ChEs, sometimes referred to as size-isomers. Prior to Massoulié's demonstration, the terms "isozymes," "variants," "subtypes," "atypical forms," and "size isomers" were used (see Usdin, 1970). The recognition of polymorphism of ChEs brought forth a new aspect of ChE complexity and led to new ways of classifying these enzymes.

A series of studies by the teams of J. Massoulié, B. Wermuth, T.L. Rosenberry, I. Stilman, and P. Taylor initiated in the 1960s and 1970s the identification of "molecular forms" of ChEs (Massoulié and Rieger, 1969; Wermuth et al., 1975; Rosenberry et al., 1999). These studies led dramatically to the description of not only the "forms" of ChEs and their active, catalytic sites but also their three-dimensional structure and even amino acid sequence. Massoulié's "molecular forms" or "size isomers" differ primarily in the numbers of their subunits, posttranslational modifications, and presence or absence of structural noncatalytic proteins, such as the collagen tails; they differ also in their functional localization (in the neuronal membrane versus the basal lamina of the neuromyal junction). They do not differ in their enzymatic activity or substrate specificity, since they share an identical catalytic domain.

The multimethodology that was involved included differential solubilization and precipitation, chromatography, purification and isolation, crystallization and crystallography, and sequencing analysis (see below, section D II). Studies of the quantitative structure-activity relationship between ChEs and inhibitors or ChEs and substrates also contributed to this knowledge (see Chapter 7 BI and BII). It should be emphasized that until these three-dimensional images were obtained, the researchers represented ChEs and their active sites as being linear and relatively flat, although occasionally active sites were presented as simple "pockets" (see Usdin, 1970; Long, 1963; Wilson, 1967; Karczmar, 1967).

Crystallization of ChEs was an important step in imaging the three-dimensionality of ChEs. It appears that David Nachmansohn as well as some Russian investigators were first to appropriately solubilize, purify, and crystallize via precipitation procedures the Torpedo enzyme; while Nachmansohn did show at some meetings pertinent photographs (for example, at the 1959 Rio de Janeiro Comparative Bioelectrogenesis Symposium; Chagas and Paes de Carvalho, 1961), it seems that there is no publication by him or by the Russians to this effect. As far as the literature goes, the first crystallization of the Torpedo electric organ AChE was credited to Walo Leuzinger and Peter Waser (Leuzinger and Baker, 1967; Hopff et al., 1973); the first three-dimensional organization of a ChE was obtained by Joel Sussman and his collaborators (Sussman et al., 1991).

3. Active Sites of Cholinesterases and Cholinesterase Reactions with Inhibitors and Substrates

Koshland (1960) defined the "active site" or "active center" of the enzyme as "a collection of contact and auxiliary aminoacids . . . responsible for the enzymatic action," and this "collection" exhibits an "induced fit" with respect to the substrate (or inhibitor). This notion of "induced fit" is similar to Irv Wilson's concept of "complementarity" (see Chapter 7 A). In this sense, the active center of a ChE is composed of the anionic and catalytic (or esteratic) sites; Zeller and Bissegger introduced these terms in 1943. It should be pointed out that the early structure (or "molecular form") of ChE was essentially linear (Wilson, 1967), and the cationic and catalytic sites are, in the pertinent diagrams, clumsily superimposed on the linear or flat enzymic molecule (see, for example, Wilson, 1967); what else could the investigators of the 1960s do? While Krupka and Main depict a "pocket" configuration in their diagrams that explain the ChE-inhibitor or ChE-substrate interactions, they do not show any direct evidence for this configuration (Krupka, 1966; Krupka and Laidler, 1961; Main, 1976). The presence of an anionic site and the size of the distance between anionic and catalytic sites were proposed early on the basis of the inhibitory effects of trimethylammonium on choline ester hydrolysis by AChE and by the analysis of the structure-activity relationship of several series of choline esters (see Usdin, 1970; Long, 1963).

The amino acid sequence, which is immediately adjacent to the catalytic site, was established by the use of radioactive di-isopropyl fluorophosphate (DF32P) and the isolation of 32P-labeled peptides after hydrolysis of the DF32P-enzyme molecule (Oosterbaan, 1967). This procedure was facilitated by using not only AChE and BuChE but also simpler, closely chemically (but not evolutionarily) related enzymes such as chymotrypsin (see Usdin, 1967; Cohen and Oosterbaan, 1963). The amino acid sequence in question contains serine, its hydroxyl being particularly involved in the catalytic process. Furthermore, after analyzing the effects of pH on the effectiveness of AChE inhibitors, Irv Wilson and Fred Bergmann concluded that the imidazole radical of histidine is a facilitatory subsite of the esteratic center acting via the increase of the nucleophilicity of the site (see, for example, Wilson and Bergmann, 1950). According to these early investigations, BuChE shares with AChE the catalytic site and its serine and imidazole moieties. However, BuChE may differ from AChE with respect to the ionic site, or BuChE may contain additional anionic sites (Usdin, 1970; Cohen and Oosterbaan, 1963).

Of course, all this evidence and all these notions deal with ChEs as enzymes concerned with the hydrolysis of ACh and the transmittive function. Yet, in the 1960s Andrej Zupancic, the son of the famous Yugoslav poet (see Chapter 1, Tables 1-6 and 1-7) proposed that AChE acts also as cholinergic receptor, thus fulfilling the functions of muscarinic or nicotinic receptors. Subsequently, he modified this viewpoint somewhat by proposing that the receptor and ACh hydrolyzing functions are located on one macromolecule (see, for example, Zupancic, 1967). Then there are other nonenzymic actions of ChEs; these are discussed below, in section DIII).

The Columbia team of Irving Wilson originally proposed mechanisms and equations depicting the reaction of AChE with substrates such as ACh (Wilson and Bergman, 1950; Wilson, 1967). The equations included factor G, a component of the esteratic site that exhibits electron transfer properties and the serine-oxygen and imidazole facilitator component. The site exhibits high nucleophilicity. Altogether, Wilson postulated that the electrophilic C of the ester group of the substrate forms a coulombic bond with the nucleophilic, basic group of the esteratic site; similar schemes were proposed by Krupka (1966) and others (see also Cohen and Oosterbaan, 1963; Usdin, 1970). It was also realized early that the reaction is stereospecific, although the molecule of ACh is not asymmetric.

The reactions of ChEs with inhibitors are discussed in detail in Chapter 7 BII and BIII. Historically, the pioneers already mentioned, such as Klas-Bertil Augustinsson, Sir Arnold Burgen, William Aldridge, Francis Hobbiger, Paul Long, and Bo Holmstedt, defined the antiChE effects of carbamates and related antiChEs as dependent on carbamylation, and of OP agents as due to phosphorylation; both reactions involve the serine hydroxyl component of the esteratic site (Burgen, 1949; Burgen and Hobbiger, 1951; Aldridge, 1953a, 1953b; Holmstedt, 1959). This common effect was referred to by Wilson (1967) as "acid transference." These investigators also proposed that ChEs may be inhibited via ligand action at the anionic site. Finally, they were concerned with the differentiations among irreversible, reversible, pseudoreversible, and noncompetitive ChE inhibitions. For example, Wilson et al. (1961) demonstrated the formation of a carbamylated enzyme following the antiChE action of a carbamate; they pointed out that the generation of carbamylated ChE is analogous to the formation of phosphorylated enzyme and, while there is slow hydrolysis of the complex and recovery of the free enzyme, carbamic acid rather than the original carbamate moiety is recovered. Accordingly, carbamates cannot be considered truly reversible inhibitors since reversibility is defined as recovery of both the intact enzyme and the unchanged inhibitor. On the other hand, Winteringham (1966) considered the recovery of the free enzyme and the carbamate decomposition products as the proof of reversibility of the carbamate-ChE interaction, an interpretation that Usdin (1970) considered "unfortunate." I agree with Usdin, and I regret that since the time of Wilson's dictum to the present era some 40 years later, most investigators refer, without caveat, to carbamates as reversible inhibitors.

4. Conclusions

Almost all modern trends and topics of ChE research are represented in this 1914–1970 period of studies: from classification of ChEs and their significance in transmission to their structures, active sites, and reactions with inhibitors and substrates. Ann Silver's 1974 book is an excellent review of the state of knowledge during this period. Only one major subject was barely attacked during that time: the molecular and genetic regulation of ChEs and their subunits. Nevertheless, we went very far during this 1914–1970 period, from Sir Henry Dale's surmise about the existence of a ChE to establishing the active moiety of the esteratic site. Yet did this progress prepare us sufficiently to face the current three-dimensional, multicurved amino acid sequence representation of an AChE?

DII. Structures, Active Sites, and Mechanism of Action

1. Structures and Active Sites

The modern era of the studies of the three-dimensional structure of ChEs and their active sites began with the adaptation to these investigations of the spectrophotometric and radiometric assays, analysis of solubility and sedimentation parameters of hydrodynamic and physical characteristics, molecular biology approaches including site-directed mutagenesis, and the X-ray and related analysis of crystallized forms of ChEs (see Silver, 1974; Massoulié et al., 1993a, 1993b, 2000; Richardson, 1985; Soreq et al., 1990; Shafferman and Velan, 1992; Shafferman et al., 1994). Another approach is based on quantitative structure activity analysis of the ligand-ChE interactions, including interactions with antiChEs and neurotoxins (Endo and Tamiya, 1987); this approach is particularly fruitful in clarifying the three-dimensional image of active, enzymic sites of ChEs ("pockets," "gorges," or "cavities," depending on the investigator); the pertinent findings are discussed in detail in Chapter 7 BI and BII.

Altogether, the efforts over the last 30 years of Jean Massoulié, Avigdor Shafferman, Israel Stilman, Joel Sussman, Terry Rosenberg, Mona Soreq, Richard Rotundo, Nibaldo Inestrosa, Zoran Radic, and Palmer Taylor among others yielded an elegant three-dimensional picture of ChEs and brought about a good understanding of their molecular biology and genetics. These studies started with the description of the analysis of hydrodynamic properties via sedimentation techniques, leading to the distinction between "globular" and "asymmetric"; the studies continued with a more detailed analysis of composition of these forms, leading to the characterization of "collagen-tailed" and "GPI (glycophosphatidylinositol)-anchored" forms; then there was the advent of cloning techniques and analysis of splice variants (R, H, and T), followed by site-directed mutagenesis and genetic modification of mice. These efforts were reported at a number of symposia and conferences, including the 11th International Symposium on Cholinergic Mechanisms (Silman, 2004; see also Massoulié et al., 1993a, 1993b). A nomenclature concerning the subunits and molecular forms of ChEs was established at the Israeli Oholo Conference of 1992 (see Massoulié et al., 1993a, 1993b and Massoulié, 2004), and this nomenclature is essentially employed in this chapter.

The richness of the possible combinations of subunits, molecular forms, or size-isomers is such that any simple description and classification of this jungle is most difficult; Jean Massoulié made a valiant attempt at such a description and organization of the forms, particularly of AChE (see below, this section). It does not help the matter that the investigators use multiple terms—such as types, molecular forms and molecular species, size-isomers and variants; subunits and domains—

to indicate one and the same entity and may employ several criteria when referring to any of these terms. Parsimony is pushed even further away as several ways of classifying the molecular forms (or molecular species, or size-isomers or types) are utilized by various investigators and as they employ complex classifications that include several levels of categories; also, there are chronological shifts in the classifications or terminology used by the investigators over periods of time. For example, Palmer Taylor and his colleagues (see Taylor and Radic, 1994; Taylor et al., 2000) refer to heteromeric and homomeric molecular species of AChE, and the terms "asymmetric," "hydrophilic," and "glycophospholipid-linked" constitute for them subclassifications of AChE (see also Tsim et al., 1988). On the other hand, Jean Massoulié and his associates did not use Taylor's classifications, as they categorized their molecular forms as either asymmetric or globular (Toutant and Massoulié, 1987; Massoulié et al., 1993a, 1993b). Subsequently, Massoulié classified these forms by their types of subunits (R, H, or T, produced by alternative splicing of their C-terminal peptides) and, in the case of type T, by their "homomeric" or "heteromeric" composition. "Homomeric" forms comprise monomers, dimers, and tetramers of ChE subunits, while heteromeric forms are associated with anchoring proteins, the collagen ColQ, or the transmembrane protein PRiMA (Perrier et al., 2003). Jean Massoulié opines that the earlier classification as "globular" and "asymmetric" forms is less informative than the new classification, because it only distinguished the collagen-tailed molecules containing ColQ ("asymmetric") from all others (Massoulié, 2000, 2004).

ChEs are glycoproteins that carry a net negative charge; an electrostatic dipole is oriented along the axis of the active, catalytic "gorge" of the ChE molecule (see below, and Taylor and Brown, 1999). Cholinesterases show "the basic structural motif" of hydrolases, that is, the alpha, beta fold (Richardson, 1985), which consists of large central beta sheets or strands surrounded by 15 alpha helices; the fold comprises the catalytic domain, which is accessible via a gorge, or cavity (Ollis et al., 1992; the alpha/beta hydrolase fold). The same or similar fold, the domain, and the gorge appear in the three-dimensional molecular forms of all ChEs.

Jean Massoulié and his CNRS, Paris, team described the mRNA encoding of R, H, and T variants that generates the molecular forms of AChE, their catalytic and terminal domains, and their secretion peptides; this description is generally accepted today. There are only T-type forms of BuChE. Acetylcholinesterase variants may not differ in size; for example, the dimers of subunits H are GPI-anchored, while dimers of subunits T are not, but their sizes cannot be readily distinguished. The signal peptides are removed during biosynthesis of ChEs (see next section, and Massoulié, 2000). Massoulié and his team defined the molecular forms of ChEs as monomers, oligomers, or polymers of catalytic subunits with molecular weights ranging from 70 to 110 kDa; the weight of the subunits differs from one species to another (70 kDa for mammals and 110 kDa for chickens and quails (Rotundo, 1984; Massoulié et al., 1993a, 1993b). Recently, several investigators, including Jean Massoulié with Susanne Bon, Palmer Taylor, and Mona Soreq with David Glick (see, for example, Soreq and Zakut, 1990; Soreq et al., 1990; Soreq and Glick, 2000; Soreq et al., 2004; Cousin et al., 1998). Soreq et al., (2004) related the subunits and domains with their molecular biology and genetics, and it is convenient and, indeed, necessary to define and describe these entities with respect to their genetic derivation.

The coding region of the vertebrate gene for ChEs comprises a variable number of exons (depending on species) transcribing the catalytic domain; either one exon or several alternative regions[5] generate the short carboxy (C)-terminal domains. The noncatalytic associated proteins, ColQ and PRiMA, are produced by distinct genes; they are attached to the catalytic AChE and BuChE subunits by disulfide bridges. The gene regions encoding various C-terminal domains are referred to functionally as H (hydrophobic), S (soluble or present in snake venoms), T (tailed), and R (readthrough). The physiological significance of AChE of type R is doubtful, and the S variant is present as a monomeric soluble AChE species in the Elapid snake venom. Accordingly, in the context of this book we are concerned with transcripts H (containing "exon" or coding region 5) and T (containing "exon" or coding region 6) that engender in the vertebrates including humans the subunits AChE-H and AChE-T (Massoulié, 2000; Massoulié et al., 1993a, 1993b). The combinations of the variant subunits, with or without associated proteins ColQ and PRiMA, generate the different size-isomers (types, molecular forms).

The size-isomers are similar in their catalytic properties but differ hydrodynamically, in their ionic interaction, and particularly in the mode of their attachment to the neuronal membrane or to the neuromuscular basal lamina; collagen-tailed molecules are not anchored in any membrane. While today, both Palmer Taylor and Jean Massoulié refer to two major categories of size-isomers (or molecular forms), homomeric and heteromeric, Massoulié employs a subclassification of size-isomers that includes the asymmetric (A) and globular (G) forms; these forms can be distinguished on the basis of their quaternary structures (Figures 3-5 and 3-6; Taylor et al., 2000; Massoulié, 2000; Massoulié et al., 1993a, 1993b, 1998; Zhang et al., 2005). Furthermore, Massoulié

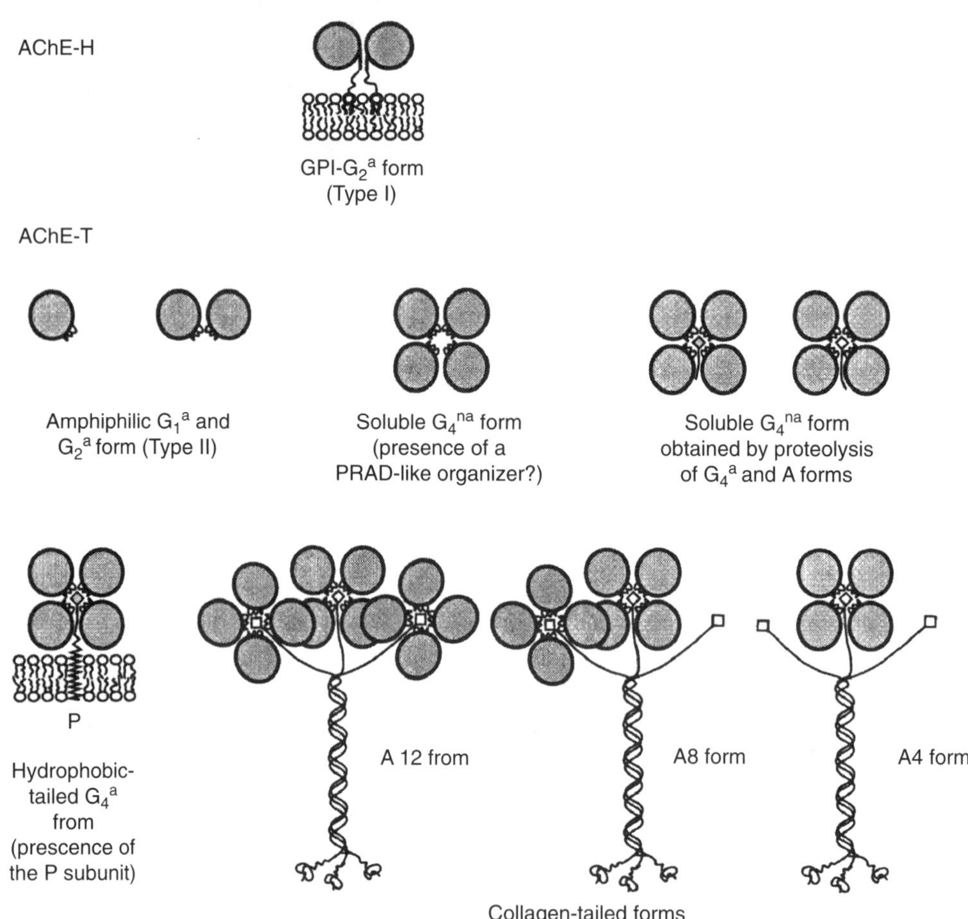

Figure 3-5. Quaternary structure of AChE molecular forms. AChE-H subunits generate only GPI-anchored dimers, while AChE-T subunits generate a wide diversity of molecular forms. In A forms, AChE-T tetramers are organized around the proline-rich attachment domain (PRAD), represented as a white square; the quaternary organization of membrane-anchored G_4^a that contains a hydrophobic P subunit is probably similar. Proteolysis of hydrophobic-tailed and collagen-tailed forms produces soluble G_4^{na} forms that contain a PRAD or PRAD-like "organizer"; the existence of Himeric G_4^{na} tetramers that do not possess an organizer is debated. (From Massoulié, 2000, with permission.)

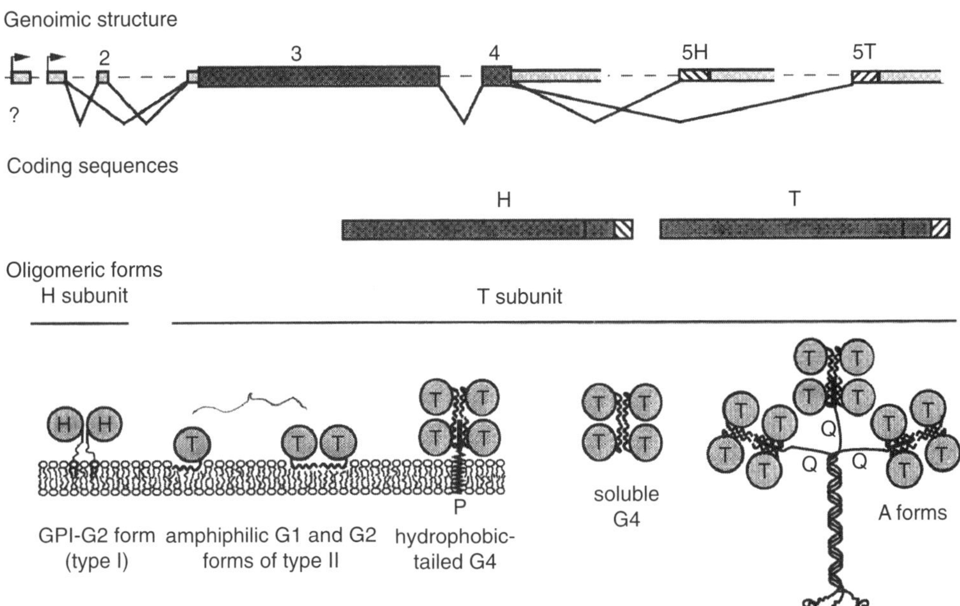

Figure 3-6. Genomic structure of the AChE gene in Torpedo differential splicing of mRNA, and quaternary structure of the major oligomeric forms of cholinesterases in vertebrates. (From Massoulié et al., 1993, with permission.)

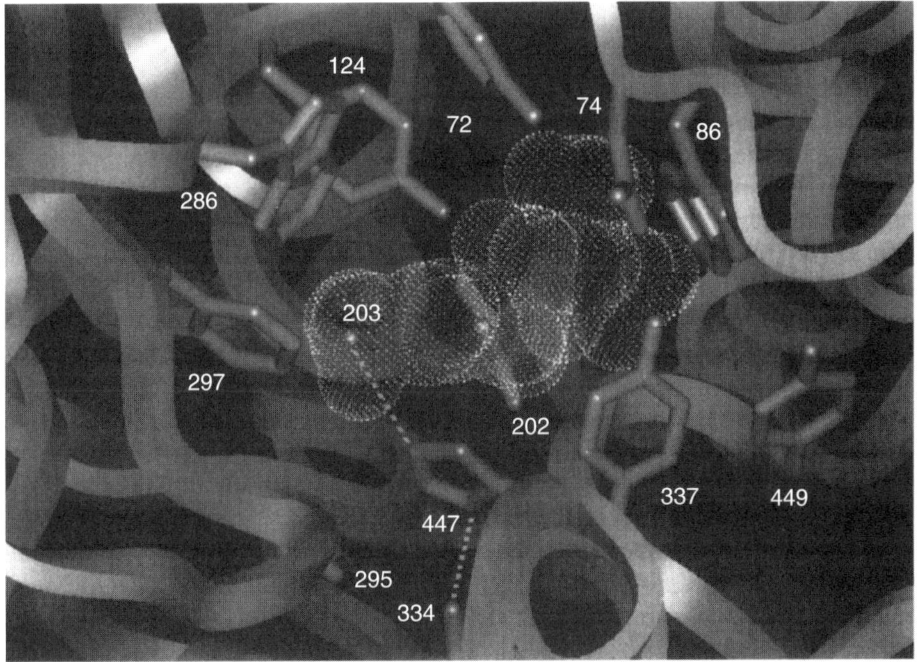

Figure 3-7. View of the active gorge of mammalian acetylcholinesterases looking into the gorge cavity. The gorge is 18 to 20 Å in depth in a molecule of 40 Å diameter and is heavily lined with aromatic amino acid side chains. Side chains from several sets of critical residues are shown emanating from the α carbonamide backbone: (1) A catalytic triad of Glu334, His447, and Ser203 is shown by dotted lines to denote the hydrogen bonding pattern. This renders Ser203 more nucleophilic to attack the carbon of acetylcholine (shown in white with the van der Waals surface). This leads to formation of an acetyl enzyme, which is deacetylated rapidly. (2) The acyl pocket outline by Phe295 and Phe297 is of restricted size in acetylcholinesterase. In butyryl cholinesterase, these side chains are aliphatic, increasing the size and flexibility in the acyl pocket. (3) The choline subsite lined by the aromatic residues Trp86, Tyr337, and Tyr449 and the ionic residue Glu202. (4) A peripheral site, which resides at the rim of the gorge, encompasses Trp286, Tyr72, Tyr124, and Asp74. This site modulates catalysis by binding inhibitors or, at high concentrations, a second substrate molecule. (From Palmer Taylor, 1999, with permission.)

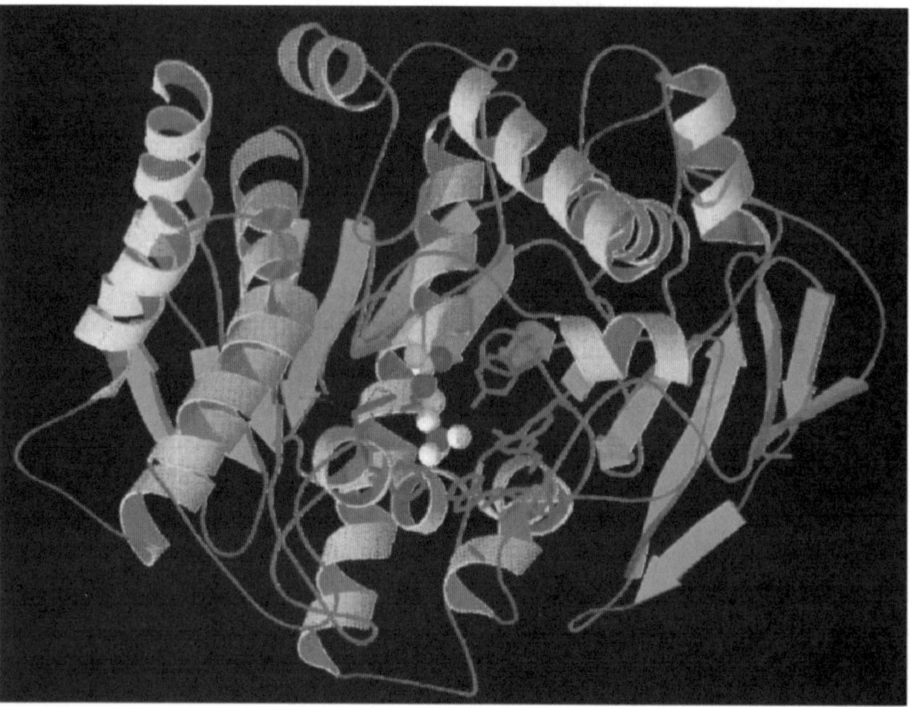

Figure 3-8. Ribbon diagram of the three-dimensional structure of TcAChE. The side chains of the catalytic triad and of key aromatic residues in the active-site forge are indicated as purple stick figures. Acetylcholine, manually docked in the active site, is represented as a space-filling model with carbon atoms shown in yellow, oxygen atoms in red, and nitrogen in blue. The quaternary group of the ACh faces the indole of Trp84. (From Silman and Sussman, 2000, with permission.) (See color plate.)

portrays the subunit composition (subunits AChE-T or AChE-H) of the molecular forms, whether globular or asymmetric-tailed (Figures 3-7 and 3-8); Taylor also depicts the globular and asymmetric "molecular species" (Figure 5-3 in Taylor et al., 2000). Heteromers are tetramers of AChE-T subunits that are di-sulfide-bonded to either the membrane-bound PRiMA protein or a triple helix of collagen ColQ. Homomers are monomers, dimers, or tetramers with or without membrane linkages; the linkages differ from those exhibited by heteromers. Acetylcholinesterase-H subunits produce dimers that are attached to cell membranes by a glycophosphatidylinositol (GPI) anchor, posttranslationally added at their C terminus (Silman and Futerman, 1987). Monomers and dimers of AChE-T subunits are amphiphilic, in the sense that they interact with detergent micells *in vitro*. It is not clear whether these species are physiologically attached to the membranes at the cell surface, or represent precursors of the functional heteromeric oligomers inserted into cell membranes or the extracellular matrix by their anchoring proteins.

According to Jean Massoulié, there are 5 G and 3 A forms, which contain 1, 2, or 3 tetramers of catalytic subunits attached to the ColQ strands of the triple collagen. While several classification modes may be—and were—applied to these isomers or molecular forms, the classification generally accepted today is based on (1) the type of the variant (AChE-T or AChE-H) and (2) the type of the quaternary structure; this classification includes homomeric forms without an anchoring protein and heteromeric forms containing ColQ (asymmetric or collagen tailed) or PRiMA

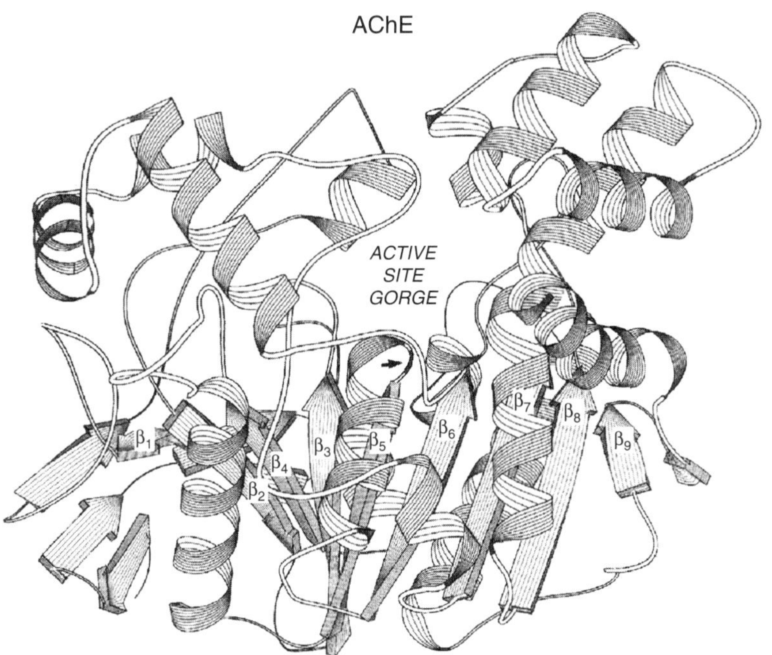

Figure 3-9. Three-dimensional structure of G_2^a AChE from Torpedo. The structure is represented in a ribbon diagram showing β strands (green) and α as helices (red). The active gorge is located above the central β sheet and the arrow marks the location of the active-site serine, Ser 200. (From Massoulié et al., 1993b, with permission.) (See color plate.)

(membrane bound or hydrophobic tailed forms). The following description (Figure 3-9, see color plate) further clari-fies this issue. The 5 G size-isomers (GPI-G [glycophosphatidylinositol]-anchored-G2, AChE-H dimer; amphiphilic-G1, AChE-T monomer; amphiphilic G2, AChE-T dimer; hydrophobic tailed G4, AChE-T tetramer; and soluble G4, AChE-T tetramer) are globular and symmetric. The sixth size-isomer is collagen tailed and asymmetrical. On the other hand, the first 4 G size-isomers (GPI-G2 [or type I amphiphilic], AChE-H dimer; G1 AChE-T monomer; G2 AChE-T dimer; and hydrophobic G4 AChE-T tetramer) are homomeric, while the last G size-isomer and the A size-isomer are heteromeric. The 5 G molecular forms (GPI, AChE-H dimer; amphiphilic-G1 [type II amphiphilic], AChE-T monomer; amphiphilic [type II amphiphilic] G2, AChE-T dimer; hydrophobic tailed G4, AChE-T tetramer; and soluble G4, AChE-T tetramer) are globular and symmetric; the sixth size-isomer is collagen tailed and asymmetrical (as the structures of these forms is by now known, the distinction based on the globular and asymmetric nature of the molecular forms may be not necessary). On the other hand, the first 4 G size-isomers (GPI-G2, AChE-H dimer; G1 AChE-T monomer; G2 AChE-T dimer; and soluble G4 AChE-T tetramer) are homomeric, while the last G size-isomer and the A size-isomer are heteromeric.

As stated, the G1 and G2 forms of type II are the monomers and dimers of ACHE-T subunits; they are probably not anchored in membranes, but represent unassembled precursors of the functional heteromeric oligomers, collagen tailed and hydrophobic tailed. Of the two types of amphiphilics, the type I G1 isomer possesses a GPI anchor; it is a dimer containing two H subunits extracellularly; this molecular form is referred to as GPI-G2 amphiphilic (GPI-G2a) type I size-isomer. There are two type II size-isomers, a monomer and a dimer (amphiphilic G1a and G2a forms). The last membrane-bound G size-isomer is a tetramer containing T extracellular subunits and a P (now known as PRiMA) protein hydrophobic anchor (Perrier et al., 2003); it is referred to by Massoulié and his team as the G4a isomer; this latter isomer corresponds to Palmer Taylor's heteromeric, lipid-linked heteromer. This species is the predominant form of vertebrate CNS AChE. The fifth G size-isomer lacks the anchor and is soluble; it is a T subunit tetramer. The last two forms are not readily distinguishable by their size.

Asymmetric or collagen-tailed forms of ChEs are characterized by the presence of a collagen tail formed by the triple-helix association of three collagen Q (ColQ) subunits; this association is based on interaction of four C-terminal t peptides of AChE-T subunits with proline-rich attachment domain (PRAD) located in the N-terminal, non-collagenous region of each ColQ subunit (more acronymics than U.S. army manuals). Either one, two, or three catalytic tetramers of T subunits (AChE-t) are attached to the three subunits of ColQ, generating the three A4, A8, and A12 forms, catalytic tetramers containing T subunits; depending on the number of ColQ attachments to the noncatalytic subunits, Massoulié and his collaborators (1993; Massoulié, 2000) as well as Palmer Taylor (Taylor, 2004a, 2004b; Radic et al., 2005), Terry Rosenberry (see Dvir et al., 2004), and Israel Silman (see Zeev-Ben-Mordechai et al., 2004) distinguish among A4, A8, and A12 forms. Jean Massoulié (Massoulié et al., 1993a, 1993b) stresses that the A forms are present in all vertebrates, including the prevertebrate amphioxus, but not in invertebrates; he makes the interesting comment that it is likely that these forms "arose at some early stage in evolution of vertebrates, when a single cholinesterase existed" (the question is, How did the other size-isoforms of ChEs arise?). The A forms are particularly characterized by their capacity to readily interact with polyanionic compounds such as heparin and glycosaminoglycans at low ionic strength. This phenomenon is responsible for the ready attachment of the A forms to cellular matrices such as basal laminae of the neuromuscular endplate and of the synapses (see Low, 1989; Massoulié et al., 1993a, 1993b).

The H regions of AChE-H subunits encode C-terminal domains, which contain a signal for addition of a GPI; GPI acts as a membrane anchor for the AChE-H. The signal induces a transamidation reaction at the omega cleavage/addition site. The peptide sequence of the C-terminal domain was elucidated by Palmer Taylor, Jean Massoulié, and their associates and is available on the ESTHER Web site database (see Massoulié et al., 1993a; Massoulié, 2000; Taylor et al., 2000; Cousin et al., 1998). While the catalytic domains contain three pairs of cysteines forming intracatenary disulfide loops for their active conformation, the C-terminal h and t peptides include cystein residues, which are important, as they establish intercatenary disulfide bridges that allow the formation of dimers in the case of AChE-H and AChE-T subunits, and/or association with ColQ or PRiMA in the case of AChE-T subunits (Massoulié et al., 1993a, 1993b).

In the case of the asymmetric, collagen-tail size-isomers, the two AChE-T subunits are linked together, while the two other AChE-T subunits connect with one ColQ strand by disulfide bonds. Besides cysteine residues, the C-terminal t peptides contain aromatic residues that play an essential role in their interaction with the PRAD motives of the anchoring proteins, ColQ and PRiMA.

In 2004, Massoulié and his associates demonstrated spectroscopically that the t peptide of the C terminal assumes an alpha-helical configuration, its aromatic residues forming a hydrophobic domain (Bon et al., 2004; Belbeoc'h et al., 2004). This peptide contains a WAT (still another acronym), "tryptophan amphiphilic tetramerization" domain. The domain is quite pluripotent; WAT helps form tetramers that can attach to a structural protein; it associates with the PRAD of collagen ColQ and of the transmembrane 20kDa protein PRiMA, generating the collagen-tailed and hydrophobic-tailed forms. As shown by I. Silman, J. Sussman, J. Massoulié, and their associates a three-dimensional structure of this region shows a complex of four t peptides (or WATs) surrounding the PRAD peptide (Dvir et al., 2004). Jean Massoulié regards PRAD and WAT as "organizers," as the t peptide is particularly active in tetramerization of the size-isomers and in their bonding processes (Massoulié, 2002).

As to the active, enzymic, catalytic site and its gorge (or pocket), the modern, three-dimensional picture of this domain differs dramatically from the relatively flat figure that was common some 25 years ago (see section DI). Today's image of the gorge originates from the crystallographic studies of Palmer Taylor and Joel Sussman and subsequent crystallography related to the quantitative structure-activity analysis of the bonding between ChEs and ligands, including antiChEs (Sussman et al., 1991; Bourne et al., 1995; see also Chapter 7 BI and BII). The catalytic site is accessed through a narrow "gorge" ("pocket," or "cavity") located within the large central beta sheet and the appended alpha helices (Richardson, 1985; see also above, this section). Joel Sussman and his associates (1991) found it surprising that these very fast enzymes afford such a limited entry of the substrate ACh into the catalytic site. In the case of AChE, the gorge is 18 to 20 Å in depth and some 40 Å in diameter; thus, the gorge reaches as far as halfway through the ChE protein. Some portions of the gorge are narrower than 20 Å, and some are wider. Butyryl cholinesterase gorge is wider than AChE gorge (Massoulié, 2000; Radic et al., 2005; see also Chapter 7 BI and BII). In the case of AChE, the gorge is lined with 14 aromatic amino acid residues that constitute about 40% of the gorge's area; they are partially replaced by aliphatic residues in the case of BuChE (Sussman et al., 1991; Massoulié et al., 1993b). Thus, the two phenylalanines of the AChE gorge are replaced by aliphatic hydrophobic residues in the case of the BuChE gorge. These aromatics particularly limit the size of the acyl pocket of AChE, so, due to this change, the BuChE gorge has a wider diameter than that of AChE, and it can hydrolyze choline esters larger than ACh (Harel et al., 1992; Ordentlich et al., 1993; Vellom et al., 1993).

At the bottom of the gorge is located the catalytic triad, which consists of glutamate, serine, and histidine, arranged in a spatial configuration resembling that of the aspartate-histidine-serine triad of serine hydrolases such as trypsin and chymotrypsin; this triad is all that remains of the historical notion of the active site. Additional residues of phenylalanine, alanine, and glycine as well as several glycans (4 for Torpedo electric organ AChE and 7 for human BuChE) link serially with the catalytic site; they are not directly involved in the catalytic activity but contribute to the binding of the substrate or the ligands. The second important residue is present in the lining of the gorge. In the case of AChE it contains aromatic residues that restrict the diameter of the gorge; in the case of BuChE this site contains aliphatic amino acid residues, yielding a gorge of a wider diameter (see also Chapter 7 BI and BII).

As proposed already in the 1960s and 1970s, in addition to the catalytic site or domain, there is an anionic peripheral site that is localized on the surface of the enzyme; this dipole character of the enzyme may facilitate the appropriate orientation of the quaternary substrate ACh—and other quaternary ligands—along the ChE surface (see Chapter 7 BI and BII; see, however, Shafferman et al., 1994). Also, the peripheral site may contribute to the allosteric inhibition of the enzyme (Changeux, 1966). As shown by Ferdinand Hucho and his associates and Joel Sussman and his team, this domain contains an indole ring and several other hydrogen-bonding aromatic residues (Weise et al., 1990; Sussman et al., 1991). Still other bonding sites are present in the acyl pocket of ChEs (Taylor and Radic, 1994); some of these underlie bonding and inhibitory effects of certain non-OP, noncarbamate inhibitors such as para-dimethylaminobenzene diazonium fluronborate (DDF; Langenbuch-Cachat et al., 1988; see Massoulié et al., 1993a, 1993b).

There are relatively minor differences between size-isomers and active sites of BuChE and AChE, although two distinct genes control BuChE and AChE synthesis. In fact, the basic structures of these enzymes are very similar; indeed, during development there may appear hybrid asymmetric forms with catalytic subunits of both BuChE and AChE attached to the same collagen tail (Tsim et al., 1988; see also Massoulié et al., 1993a, 1993b). On the other hand, the "gorge," or "pocket," of BuChE is larger than that of AChE (see above; see also Chapter 7 BI and BII). Also, a tryptophan residue is located at the entrance of the gorge in the case of AChE but not BuChE. Altogether, it was hypothesized that the duality of AChE and BuChE is specific for the vertebrates and that it originated via the duplication of the ancestral single ChE gene (Massoulié et al., 1993a, 1993b; see also Chatonnet and Lockridge, 1989). Altogether, both BuChEs and AChEs present a rich array of homo-and hetero-oligomeric molecular forms (or size-isomers) ranging in complexity from monomers to large collagen-tailed structures that may include as many as 12 catalytic subunits.

2. The Genetics and Sequence of ChE Subunits

AChE and BuChE were first cloned from Torpedo electric organ and human blood (Schumacher et al., 1986; Sikorav et al., 1987, 1988; Arpagaus et al., 1990; see also Massoulié, 2000; Taylor et al., 2000). This molecular cloning as well as amino acid sequencing established the amino acid composition and amino acid sequence of AChE and BuChE. There are a number of hydrolases and nonenzymic proteins that are characterized, just like ChEs are, by the alpha/beta hydrolase fold and, generally, by the presence of serine and a pentapeptide sequence around the active serine center (not in the case, for example, of *C. elegans* ChEs). Among the noncatalytic proteins containing ChE-like domains are thyroglobulin, tactins that are present in Drosophila membranes, and neuroligin, a mammalian cell adhesion molecule; interestingly, these ChE homologues lack the catalytic components and do not hydrolyze choline esters (Krejci et al., 1991; Ollis et al, 1992; Taylor and Radic, 1994; Taylor et al., 2000). The database of the sequences of these proteins is available from the ESTHER Web site (Cousin et al., 1998; see also Taylor et al., 2000).

Single genes encode AChE of the electric fish, mammals, aves, and humans. In the human, the gene is located on the long arm of chromosome 5 (Getman et al., 1992). The matter is more complicated in the case of BuChE. The gene encoding the catalytic subunit of BuChE is located in the human on chromosome 3, but additional genes located on other chromosomes may encode the combination of this catalytic unit with other subunits (Soreq et al., 1987a, 1987b; Arpagaus et al., 1990; Masson et al., 1990; see also Massoulié et al., 1993a, 1993b). As already mentioned, the protochordate *Amphioxus* has two genes encoding a BuChE-like and AChE-like protein. Curiously, many invertebrates exhibit single genes whether for AChE or BuChE; actually, Drosophila' s single gene expresses a ChE that exhibits substrate specificity between that of AChE and BuChE, while the nematode *C. elegans* has four genes encoding AChE-like proteins closely related in their peptide sequence (Combes et al., 2001).

The Torpedo and the human AChE genes have 4 exons encoding the catalytic domain of the enzyme and a 3' region containing alternative splice acceptor sites, generating variants R, H, and T, as indicated in the preceding section. Their localization is listed on the genetic maps with respect to the Cap sites (start sites for transcription). During transcription, alternative splicing at the gene's reading frame gives rise to the various molecular forms of AChE of the Torpedo and of the human, and to conversion of AChE to BuChE (Harel et al., 1992). This alternative maturation of the transcripts appears to be absent in the case of BuChE (Massoulié et al., 1993a, 1993b; Taylor and Radic, 1994; Taylor et al., 2000).

In fact, whereas the sequences of the Torpedo and mammalian t peptides (encoded in region T) are very similar (75%, which is more than the similarity between the sequences of the catalytic domains), the sequences of the h peptide (encoded by region H) are different. Their common features are constituted by the presence of one or two cysteines near the catalytic domain, allowing dimerization, and a C-terminal stretch of hydrophobic residues, which induce cleavage and the addition of a GPI. The lack of sequence conservation and the fact that teleost fishes, amphibians, reptiles, and birds do not exhibit the H form of AChE suggest that this form appeared independently in the evolution of Torpedo and mammals.

Many alleles of the human BuChE gene were described (see La Du et al., 1991, and the previous section). The mutations affect protein sequence and diminish or destroy catalytic activity. One or two mutations may be present; the isozyme discovered by Werner Kalow (see section DI, above), which exhibits diminished capacity for succinylcholine hydrolysis, results from a bimutational phenomenon (Neville et al., 1990a, 1990b). There are also mutational variants of AChE. In insects, mutations that confer resistance to pesticides generally depend on the change in a single amino acid (Pravalorio and Fournier, 1992). Other variants that do not depend on genic mutations may occur: in humans they appear following multiple transfusions in patients suffering from hemoglobinuria or, according to Mona Soreq, in subjects following stress (see below; and Massoulié et al., 1993a, 1993b; Soreq and Glick, 2000).

3. Mechanism of Enzymic ChE Action

Several steps are necessary for ACh hydrolysis. Step 1 involves penetration of the substrate into the gorge and its bonding in an appropriate configuration. While the mechanism of the substrate motion down the gorge is not clearly understood, the bonding of the quaternary ACh to the active and to the anionic site was clarified; it is due to the character of the active and anionic sites (and perhaps additional sites; see the previous section). As already mentioned, the dipole character of the gorge may facilitate bonding and allosteric phenomena. Both AChE and BuChE exhibit glycosylation sites at the surface of their molecules; they characteristically contain the asparagine residue. According to Velan et al. (1993), glycosylation affects folding and secretion, but it probably does not directly affect enzymic activity. The bonding of a substrate molecule at the peripheral site may be followed by its transfer to the active site in an active fit phenomenon; this indicates that the active site may display conformational or allosteric adaptability (Rosenberry, 1975). Inhibitors such as fasciculin, which bind to the peripheral site, may block enzymic activity in this manner (Eastman et al., 1995; Mallender et al., 2000; see also Chapter 7 BI).

The second step in ACh hydrolysis engages acylation of the serine of the active site—serine acylation as a part of ACh hydrolysis was proposed early in this story (see section A, above). Acetylation (a "nucleophilic" attack) is due to the nucleophilic character of serine, which is caused by the proton transfer from glutamate to histidine; in fact, Quinn (1987) finds that AChE acts as an acid-base proton catalyst. The need of nucleophilicity for the ChE catalysis constitutes, again, the classical notion of Irwin Wilson (see section A). The acylation step proceeds through the formation of a tetrahedral transition state (see also Chapter 7 BI and BII); choline is released at this step. The acylated state is very short (approximately 10 microseconds), and this contributes to the fast character of AChE as an enzyme. Finally, the last step of ACh hydrolysis is the deacetylation process and release of choline and acetate. Again, the nucleophilic character of the catalytic triad that facilitates acetylation also speeds up deacetylation (see Quinn, 1987; Rosenberry, 1975; Main, 1976).

It must be added that since the catalytic domains of ChE are identical for all the AChE and BuChE size-isomers of a given enzyme (AChE or BuChE), and since the difference in the dimension of the "gorge" is not crucial for ACh hydrolysis, whether by AChE or by BuChE, kinetics and mechanism of ACh hydrolysis by all these enzyme variants or size-isomers are similar. Finally, while choline esters other than ACh are hydrolyzed by AChE and its size-isomers, and by BuChE via mechanisms identical to those described here, slower rates of hydrolysis obtain generally for these esters. The exceptions are butyrylcholine and propionylcholine, which are split faster than ACh by BuChE, and methacholine, which is hydrolyzed as fast as ACh by AChE. As already affirmed, butyryl- and propionyl-choline are preferred and specific substrates for BuChE, and methacholine for AChE; these esters are not endogenous (see section DI, above). Finally, the carbamylation and phosphorylation of the catalytic site's serine by carbamate and OP antiChEs, respectively, are processes analogous to the phenomenon of acylation; they are described in detail in Chapter 7 BI and BII.

4. Synthesis, Stabilization, Assembly, and Secretion of ChEs

Like CAT and other secreted proteins (see section B-1 in this chapter), ChEs are synthesized in the endoplasmic reticulum, and lose their signal peptide cotranslationally. Their multiple mRNA translation generates the various subunits and various molecular forms. In birds, inactive precursors have been observed; they represent as much as 80% of the newly synthesized subunits (Rotundo, 1984, 1990); this probably does not occur in mammals (Brimijoin and Hammond, 1996; Bertrand et al., 2001). The role of inactive molecules is unknown at this time (Valette et al., 1991; Valette and Massoulié, 1991). The precursors acquire activity in the reticulum, perhaps via the folding effect due to the so-called molecular chaperon proteins. The inactive molecules and some active molecules are degraded through the endoplasmic reticulum associated degradation (ERAD), which involves retrotranslocation from reticulum into cytoplasm and hydrolysis by proteasomes (Belbeoc'h et al., 2003). Acetylcholinesterase-H subunits acquire their GPI anchor in the reticulum; oligomers also assemble in this compartment, and this process includes the association of AChE-t subunits with collagen ColQ into collagen-tailed forms (Bon et al., 2003, 2004).

N-glycans are added cotranslationally to the nascent polypeptide chains; they are modified in the reticulum and perhaps also in the Golgi apparatus, after the transfer of the enzyme to this latter compartment (Rotundo, 1984; Rotundo and Carbonetto, 1987).

Central and ganglionic neurons almost exclusively express the AChE-T subunit, and the predominant molecular forms are G1, G2, and G4, with A forms being very weakly represented. In the case of the motor unit, the motor axons carry the various molecular forms of AChE by the fast axonal and slow flow induced by the motor proteins, similar to CAT (see section B-3 in this chapter). While the A forms and tetramers are transported by the fast flow, the G1 and G2 forms move with the slow flow (Couraud et al., 1980; Goemere-Vannests et al., 1988).The postganglionic axons carry mostly the G4 and A forms of AChE.

The final expression and localization of AChE, whether at synapses or junctions or in erythrocytes and other tissues, is a complex process. It is facilitated by AChE mRNA "stabilization," Ca^{2+} mobilization at the so-called L-type channels, and several trophic factors (Fernandes et al., 1998; see also Taylor et al., 2000); it is also regulated by certain protein phosphatases. In particular, these processes are involved in the formation of the neuromyal junction (see, for example, Jasmin et al., 2002); it can be only surmised that parallel events are present at the cholinergic synapses, whether peripheral or central (Deshenes-Furry et al., 2005; Rotundo et al., 2005). It is known, however, that GPI-G2a as well as the G4 AChE form are present presynaptically in the Torpedo electric organ (Eichler et al., 1992). The postsynaptic distribution of the A forms is not clear and the pertinent description is inconsistent. Massoulié (2000) states that they are abundant in the central nervous system of Torpedo and other fishes and in amphibians, but not in reptiles, birds, and mammals. These forms are attached to the basal lamina of the synaptic cleft of the mammalian neuromuscular junction (McMahan et al., 1978; Wallace et al., 1985). The collagen tail contains heparin-binding sites (Deprez et al., 2000, 2003). The tail may be associated with glycosaminoglycans of perlecan (Peng et al., 1999). Abramson, Taylor, Doctor, and their associates assign it to the synaptic cleft, while some unpublished work localizes it in the basal lamina; basal lamina may also incorporate the G2 forms (Abramson et al., 1989).

Besides being transported down the axon, AChE—and probably also BuChE—is secreted from nerve cells, both at the periphery (for example, in autonomic ganglia) and in the CNS (Greenfield, 1985; for further references, see Massoulié et al., 1993a, 1993b). This process occurs prior to the bonding of the enzyme at the basal lamina, and it requires glycosylation (Bon et al., 1991). Actually, nonnervous tissues that contain AChE or BuChE, such as skeletal muscle, hematopoietic cells, and cell lines with transfected AChE, also secrete the enzyme. The biosynthetic chain and positioning of AChE or BuChE in the erythrocytes, serum, and other nonnervous tissues remains an enigma; it is also not known whether biosynthesis and localization of AChE differ at the pre- and postsynaptic sites—AChE being present at both.

DIII. Distribution and Roles of Cholinesterases

1. Distribution of ChEs

The image of the central nervous system distribution of AChE within the central cholinergic pathways was obtained by the use of George Koelle's histochemical method (see also Friedenwald, 1955); this image was described in detail in Chapter 2 DI; it should be noted that the modern view of these pathways is based on employment of the CAT immunohistochemistry (see Chapter 2 DII and DIII).[6] Even in the central nervous system, AChE, BuChE, or both are present in nonneuronal cells such as glia, meninges, and capillaries (for references, see Koelle, 1963; Soreq and Glick, 2000). Acetylcholinesterase or BuChE also appears at several sensory sites, such as sensory corpuscles including Pacinian corpuscles of the human; cutaneous and mucous sensory nerve terminals; gustatory papillae of animals and humans; and the olfactory epithelium (see Koelle, 1963). Mikhail Michelson (see Michelson and Zeimal, 1970) felt that the cholinergic system is not involved in sensory transmission and that nonsensory distribution of the cholinergic system is characteristic for vertebrates. The presence of ChEs in the peripheral sensory sites contradicts Michelson's notion; the CAT presence in pathways concerned with sensorium (see Chapter 2 DII and DIII) and the cholinergic correlates of sensory perception (see Chapter 9 BV) are also inconsistent with Michelson's view.

While major significance of AChE concerns transmission, the appearance of AChE (and BuChE) outside of the nervous system must be emphasized. Glial and capillary presence of ChEs is just one example of the nonneuronal presence of these enzymes; in fact, they are quite ubiquitous. Of course, some of the nonnervous sites that exhibit AChE are cholinergically innervated by parasympathetic nerve endings or sympathetic autonomic ganglia; therefore, the presence of AChE in these tissues or organs, including heart, blood vessels, intestinal and bronchiolar smooth muscle, liver, salivary, lacrimal, and parotid glands, pancreas, and hypophysis relates to its role in transmission at the effector organs and in gland secretion (although other humoral elements contribute to generation of pancreatic secretion; see Karczmar et al., 1986). For example, in the pancreas, AChE and other cholinergic components are involved in the cholinergic secretion of insulin. Mary Pickford's classical studies in the 1940s demonstrated that AChE in the hypothalamic-hypophyseal tract is involved in cholinergic secretion of oxytocin and vasopressin (see Chapter 9 BV-2; see also Koelle, 1963; Lucinei Balbo et al., 2000). The problem is that these sites also contain BuChE. For example, BuChE appears in the adenohypophysis (see Koelle, 1963). The liver in particular is a rich source of BuChE that is, depending on species, localized in the parenchyma or the sinusoidal cell lining (Augustinsson, 1948). The presence and function of BuChE in the pancreas is also enigmatic. Structural and molecular characteristics of rat and human pancreatic cholesterol esterase are akin to those of ChEs. In humans, the cholesterol esterase may be structurally related to bile salt activators (Feaster et al., 1997), and BuChE may be involved in this activation in the pancreas and related enzymic activities of the liver. However, BuChE appears in the pancreas of only some species (i.e., dog) and not in others (i.e., rat, rabbit, and sheep; see Koelle, 1963).

Also, the presence of AChE in the cardiac auricles and ventricles (Gerebtzoff, 1959; Koelle, 1963) may relate, at least in part, to its role in cardiovagal transmission; however, BuChE is also present at these cardiac sites, where its role is enigmatic (see below, section 2). Finally, the presence of AChE in the circulatory system (i.e., the endothelium of blood vessels) and in capillaries of most organs, including the brain, may relate to the autonomic innervation of these sites (see Gerebtzoff, 1959; Koelle, 1963; Pakasi and Kasa, 1992; Kasa, 1971). Again, the appearance of BuChE at these sites is more difficult to explain.

Similarly, smooth muscle of the bronchioles, gastrointestinal tract and its mucosal cells, and urogenital system contains AChE, which is also present, as expected, in enteric neurons (see Karczmar et al., 1986). Again, the presence of AChE may be linked with the parasympathetic innervation of these sites. However, these sites also contain BuChE, and this presence may be difficult to explain, although certain hypotheses were posited in this respect (see section 2, below).

But ChEs are present in nonnervous tissues that are not targets of autonomic or neuromyal innervation. First, there is the perennial, enigmatic matter of ChEs of the blood elements and lymphatic tissues. Platelets and erythrocytes of some species contain AChE (the GPI-anchored AChE dimers are present in the erythrocytes); megakaryocytes may be the source of the platelet enzyme (Zajicek, 1955; Usdin, 1970; Paoletti et al., 1992); and BuChE is generally present in the serum. However, in some ruminants, the chicken, and the rabbit, AChE is present in the serum; it was claimed that BuChE may be present in erythrocytes of some species (see Koelle, 1963; Goedde et al., 1967); then, the human platelets lack AChE (Koelle, 1963; Zajicek, 1956; Augustinsson, 1948; Fischer, 1971).

The presence of AChE and BuChE in blood elements may relate to their role in splitting blood lipids and "spilled" ACh, as suggested, speculatively, by Karczmar and Koppanyi (1956), or certain toxins (Neville et al., 1990a, 1990b). Yet, these speculations do not resolve the matter. Why is AChE sometimes present in the erythrocytes and sometimes in the serum? Why is it sometimes present in, and sometimes absent from, the platelets? Then, why should the platelet AChE be in inverse relationship to the erythrocytic AChE, as reported by Zajicek (1957)?

AChE exists in the spleen, lymph nodes, tonsils, and other lymphatic tissues (see Koelle, 1963). AChE, BuChE, or both compounds exist in the thyroid of several species and during the early development of the thymus, though these enzymes disappear almost completely after maturity

(Koelle, 1963). While the thyroid activity is not linked with cholinergic or any other transmission, the thyroid's AChE was linked with the presence of the perivascular innervation of that gland. Again, what is BuChE doing at these sites?

Surprisingly, aside from the motor neuromyal junction, many muscle sites exhibit AChE, BuChE, or both. For example, Rene Couteaux (1953), the French pioneer of skeletal muscle research, showed that there are high AChE concentrations in the musculotendinous junctions of the striatal muscle fibers. Yet, this junction is not cholinergic and does not respond to ACh. In some (but not all) species, AChE exists in noninnervated portions of the sarcolemma and in the sarcoplasm, including the M bands (myosincholinesterase of the sarcoplasm could have been mistakenly taken for AChE; see Koelle, 1963; Beckett and Bourne, 1957). Finally, while the presence of AChE in the skeletal muscle spindles and intrafusal fibers can be expected (they are innervated by the small motor nerve system), BuChE's presence in the cytoplasm of these structures is perplexing (see Koelle, 1963; Gerebtzoff, 1959). Moreover, AChE and BuChE are present in the epithelium of aquatic animals (for example, frogs) and the mucous membranes of the rabbit trachea (see Augustinsson, 1948, 1963; Koelle, 1963; Karczmar et al., 1986).

This matter of the presence of ChEs in nonnervous and noninnervated tissues should be related to the precocious occurrence of AChE and/or BuChE in development of many vertebrates and invertebrates[6]: they are present in the fertilized oocyte and in the two- or four-cell ontogenetic stages of several species—long before the neurula stage and neurogenesis (Karczmar, 1963a). In fact, AChE and BuChE both appear in the gametes, including human sperm and oocytes (Soreq et al., 1987a, 1987b). In a related matter, antiChEs and cholinergic agonists and antagonists exert morphogenetic effects prior to neurogenesis (see Chapter 8 D and Karczmar, 1963b). The possible role of the cholinergic components prior to neurogenesis is discussed below, in section 2). Section 2 (see also Chapter 8 BI; see also Karczmar, 1963a). Also, ChEs appear during early development of avian cartilage (see Chapter 8).

Then, there is the similarly enigmatic presence of ChEs in ephemeral organs. That a ChE exists in placenta and in the amnion of the chick has been known since the 1940s (Comline, 1947; see Koelle, 1963; Gerebtzoff, 1959); later, the presence of a ChE in the chorionic villi was also established (Zakut et al., 1991). The placental enzyme appears to be, in animals and humans, AChE; amnion may exhibit both BuChE and AChE, and BuChE is present in chorionic villi; amnion also contains ACh and CAT (Sakuragawa et al., 1997; Sastry, 1997). It must be stressed that these ephemeral tissues are not innervated.

It appears, then, that ChEs are omnipresent in vertebrate organs and tissues other than the central and peripheral nervous system, as well as in preneurogenesis stages of development. A related, important issue is that ChEs are not just vertebrate enzymes (in fact, they appear in protochordates such as *Ciona intestinalis*), since they are present throughout the animal, plant, and bacterial kingdoms. Thus, they are present in protozoa including paramecium, bryozoa, bacteria, and Cnidaria; in echinoderms, sponges, and snails; and in mollusks, insects, and worms (Karczmar, 1963a; Chadwick, 1963; Koelle, 1963; Fischer, 1971; see also below, section 2). For some of these species, the presence of ChEs relates to the presence of the cholinergic innervation and to the motor function of the animals in question; in other species, this presence may concern ciliary activity (see below, section 2), but what is the need for AChE or ChE in sponges and in the blood of snails? Actually, Klas-Bertil Augustinsson (1948) emphasized that AChE activity in the blood of the snail is some ten times higher than in the blood of the rat or dog. It was already stressed (section DI, above) that this phylogenetic omnipresence of the components of the cholinergic system was of great interest to the cholinergic pioneers, including David Nachmansohn, Theodore Bullock, and Zenon Bacq. This phylogenetic omnipresence of the components of the cholinergic system strongly suggests that they exert a beneficiary effect on the organism and that they possess a major evolutionary significance; this hypothesis is returned to in Chapter 9 BVI.

A modern aspect of the distribution of ChEs concerns the localization of transcripts and size-isomers (the various physical forms of AChEs; see Massoulié, 2000, 2004; and section DII, above). The H subunits originating from the H transcripts are present in the form of GPI-G2 amphiphilic size-isomers in many embryonic tissues, but in adult mammals they are present only in haematopoetic tissues and erythrocytes (and in the Torpedo*)*. The T subunits are present in the adult mammalian muscle and its neuromyal junction, and they include globular amphiphilic and nonamphiphilic forms as well as collagen-tailed forms. The T subunits also exist in the mammalian central nervous system, and they include amphiphilic and nonamphiphilic globular (G) tetramers. While in the rat and in the quail the G1 amphiphilic form predominates during early CNS development, the hydrophobic-tailed G4 amphiphilic size-isomer predominates in the adult brain (Massoulié, 2000). The differential organ localization of these size-isomers is not completely understood, but it is clear that their C-terminal peptides (see above, section DII) regulate their localization at the various sites, such as basal lamina; the differences between these domains of the various size-isomers may underlie their specific organ and tissue distribution.

The understanding of the evolutionary and ontogenetic role of the cholinergic components awaits the explanation of the functional significance of distribution and role of the molecular forms of AChE, as well as of the nicotinic and muscarinic receptor subtypes. Besides the enigma of the presence of ChEs throughout phylogenesis, there is also the question of dimorphism of ChEs with respect to gender (see, for example, Maul and Farris, 2004).

2. Roles of ChEs

In the 1940s and1950s, after the early studies of the OP drug DFP and after the demonstration of cholinergic transmission at the peripheral and central synapses, it was thought that DFP exhibits a single effect and that AChE has a single role, both concerned with synaptic transmission. Even then, the role of BuChE was a mystery (see, for example, Augustinsson, 1963), and soon it became apparent that there might be more than one role for AChE. There are many reasons for considering BuChE to be an enigmatic enzyme; actually, Ezio Giacobini (2000) referred to it as "an enzyme in search of the substrate." Butyryl cholinesterase is present jointly with AChE in the neurons, as well as in the white brain matter, and it can hydrolyze ACh (Karczmar and Koppanyi, 1956), but it is much less efficient and much slower than AChE as an ACh hydrolase; accordingly, BuChE is not involved physiologically in synaptic hydrolysis of ACh (although it may be involved in regulation of ACh release at the neuromyal junction (Minic et al., 2003). Moreover, both BuChE and AChE appear very early in nonnervous stages of ontogenesis (Karczmar, 1963a, 1963b; Layer, 1990; see also above, section 1, and Chapter 8 BI and BII), and, again, there is no explanation for this phenomenon. Then, the ratio between the two enzymes varies without apparent reason from one tissue to another, as does the ratio and the distribution from one species to another, particularly during development (Edwards and Brimijoin, 1982). Finally, ChEs are present in many nonvertebrates, including monocellular species, and in plants; while the presence of both ChEs in paramecium and *Trypanosoma*, within or at the basis of their cilia, may subserve interciliary conduction and ciliary movement, there is no definitive explanation for the presence of ChEs, BuChE, and/or AChE in other invertebrates, bacteria, and plants (see section 1, above; Beyer and Wense, 1936; Koelle, 1963).

Besides the problems related to BuChE distribution, there is no explanation with respect to BuChE rhythms; similarly to AChE, BuChE shows diurnal and seasonal patterns of activity, at least in some species (Goedde et al., 1967). Still another problem is that differential inhibition of BuChE in animals does not affect cholinergic transmission or general behavior (Tunek and Svensson, 1988), and individuals who lack or are partially deficient in BuChE seem to function normally (except in special conditions, i.e., succinylcholine treatment; see above)

It is not that attempts at elucidating the role or roles of BuChE are lacking (see section 1, above) An old hypothesis proposed that BuChE protects the organism from excessive amounts of ACh or choline esters. This notion relates to the finding that nonnervous tissues (e.g., intestine) produce ACh constantly; therefore, there may occasionally be the danger of ACh overproduction and its undue action (Collumbine, 1963). In a similar vein, Dirnhuber and Lovatt Evans (1954) posited that in the submaxillary gland, BuChE has a "general mopping up function" that is needed following the catalytic action of AChE, and Karczmar and Koppanyi (1956) demonstrated that

BuChE and AChE of the erythrocytes and the serum cooperate in destroying choline esters (including ACh) whether administered pharmacologically or present endogenously in the blood (for example, a butyrylcholine derivative; Kewitz, 1959); they proposed that AChE and BuChE of the blood serve as "transport" ChEs and that the "transport" enzyme protects the organism from the excess ACh that may escape from other tissues into blood and from other choline esters present in the blood. Then, the high levels of BuChE in the lung and liver, which are major detoxification sites of vertebrates including humans, led several investigators to suggest that this BuChE serves as a detoxificant (Neville et al., 1990a, 1990b; Jbilo et al., 1994) and that other BuChE may serve a similar purpose with respect to certain drugs and toxins. Also, Neville et al. (1990a, 1990b) suggested that the resistance of some BuChE variants to plant alkaloids may protect these enzymic variants and preserve them as detoxifiers, thus affording selective, evolutionary advantage to the organisms exhibiting these variants. In a similar vein, Hermona Soreq and David Glick (2000) emphasized that BuChE may act as a scavenger that binds antiChEs as well as, possibly, some naturally occurring toxins.

In addition, Ezio Giacobini (2000, 2003a, 2003b) showed that a specific inhibitor of BuChE, given intracortically in the rat, markedly increased cortical levels of ACh, and he felt that this finding supports the validity of the "mopping up function" that BuChE may exercise under certain circumstances. It is of course expected that, since BuChE hydrolyzes ACh, antiBuChEs increase the levels of ACh, yet Giacobini's finding does not necessarily signify that AChE requires BuChE's help in splitting synaptically released ACh. Yet, he may have a point: it may be that certain brain parts, such as some nuclei of the thalamus, preferentially contain BuChE, while others especially exhibit AChE; thus, at these sites BuChE may be an important regulator of the duration of action of ACh and/or participate in cholinergic transmission (Darvesh and Hopkins, 2003); this point is, however, controversial. It should be added that Giacobini relates this capacity of antiBuChEs to increase ACh levels to the treatment of Alzheimer's disease, and this point is discussed in Chapter 10 K.

Certain hypotheses propose a functional role of BuChE in tissues or organs that are nonnervous in nature. Thus, George Koelle and his associates (1950) suggested that the BuChE of the small intestine smooth muscle controls the tone and the activity of the small intestine, as antiChEs affected intestinal function after selective inhibition of BuChE (see Koelle et al., 1950; Koelle, 1963; Collumbine, 1963). On a similar basis, Joshua Burn (with Kottegoda, 1953) suggested that BuChE controls the rate and beat amplitude of the rabbit auricles, and Dirnhuber and Lovatt Evans (1954) opined that BuChE regulates the secretion of the submaxillary gland. Shelley (1955) commented that these findings regarding the auricles and the small intestine relate to the relatively high level of ACh in these tissues (Shelley's old data warrant rechecking). An interesting speculation concerns the reactivation of the hypodynamic frog heart by BuChE (and/or AChE); this effect could be observed even in the presence of physostigmine and was attributed to the activation by the enzymes of the sodium pump (Beznak, 1958; see also Goedde et al., 1967); it must be added that similarly it was proposed that the ionic transport of erythrocytes is regulated by ACh and AChE (see next paragraph). Finally, as BuChE and AChE are present both precociously prior to neurogenesis and during early neurogenesis, it was speculated that these enzymes might play a trophic and differentiative role in development (Layer, 1990; see below and Chapter 8 D).

To turn now to the role or roles of AChE: AChE was conceived at an early date as having the sole action of a synaptic and junctional enzyme, involved in the transmission of central and peripheral cholinergic synapses and in the transmission at the neuromyal junctions. George Koelle (1963) discussed at length the transmittive (humoral) role of AChE and posited that, depending on its site, the enzyme has two roles. The presynaptic enzyme protects the terminal from retrograde firing action of the released ACh, whereas the postsynaptic AChE terminates the action of released ACh and prevents it from clogging the synapse (Eccles, 1936, 1964). Also, George Koelle (1963) proposed that not all of the presynaptic enzyme is immediately functional. He used quaternary and tertiary antiChEs and his histochemical method for specific localization of AChE, particularly at the ganglia; the quaternaries inhibited the membrane-located enzyme, and, when the membrane enzyme was protected by a reversible antiChE, the tertiaries inhibited specifically the cytoplasmic enzyme. Employing this strategy Koelle and his associates determined that the presynaptic control exerted by AChE is due to the "external," or "functional," enzyme located in the membrane, while the cytoplasmic enzyme acted as the "reserve enzyme" (see also Karczmar, 1967).

An exception to the notion of the synaptic, transmittive role of AChE was posited by David Nachmansohn (1993b). Mostly on the basis of his studies of the effect of antiChE drugs on axonal conduction, he proposed that ACh is locally released in both cholinergic and noncholinergic axons, changing the axonal membrane permeability and inducing the nerve potential. He postulated further that AChE regulates this conductive process and that it is involved in axonal conduction no less than in synaptic transmission. Finally, he sought to demonstrate that antiChEs affect both processes. Much evidence was collected indicating that this hypothesis, pursued stubbornly by Nachmansohn,[7] is not tenable, as ACh and AChE are nearly or completely absent from noncholinergic—particularly sensory—neurons and their axons and as concentrations of antiChEs that Nachmansohn was forced to employ to block axonal conduction were some 100 times higher than those needed to completely inhibit AChE of other tissues. Yet, David Nachmansohn claimed, "One needs unorthodox concentrations to induce unorthodox effects" (see Chapter 1 and Table 1-6; for further arguments on the matter, see Koelle, 1963; and Karczmar, 1967).

Nevertheless, there are problems with the "single role" concept of AChE. A particular dilemma is constituted by its presence in nonneuronal, nontransmittive tissues. One of these tissues is blood. Not only is AChE presence in the blood mysterious, but, additionally, why is AChE found in erythrocytes of some species and the serum of some other species (see section DIII-1, above)? It must be added that erythrocytes contain other components of the cholinergic system, including ACh and the choline uptake system (Chapter 2 B and section CIV, above). And, like BuChE, AChE is present precociously during development as well in ephemeral tissues such as placenta (see above, this section, and Goutier-Pirotte and Gerebtzoff, 1955). The enzyme appears to be synthesized locally in the placental syncytium, and at the end of gestation it moves to the labyrinthine lacunae that contain the fetal blood.

It must also be remembered that AChE and BuChE are secreted from neurons and certain other tissues (see preceding section). These findings compound the problem; does this release relate to the nontransmittive role of these enzymes in the nonnervous tissues? Does the release of AChE and BuChE from neurons also reflect a nontransmittive role of these enzymes in the nervous tissue?

George Koelle (1963) stated that "it . . . is . . . not possible to assign even a tentative function to the enzyme or its substrate" with regard to the presence of AChE in nontransmittive tissues such as erythrocytes or musculotendinous junctions of the skeletal muscle, or in the ephemeral tissues, or precociously in ontogenesis. Actually, tentative notions and speculations were raised with reference to these cases (for reviews, see Greenfield, 1991; Appleyard, 1992; Massoulié et al., 1993a, 1993b). Thus, Michail Gerebtzoff (Goutier-Pirotte and Gerebtzoff, 1955) pointed to the shift, toward the end of gestation, of AChE from the placental syncytium to the prefetal blood in the placental lacunae and speculated that this reflects the placental hemopoietic activity that is taken over by other tissues of the embryo following gestation. Indeed, Rossi et al. (1991) and Hermona Soreq and her team (Patinkin et al., 1990) presented evidence indicating that cholinergic mechanisms involving certain splice variants of AChE regulate and evoke proliferation and differentiation of hematopoietic lineages (Soreq particularly emphasizes the importance in this matter of the C-terminal peptide of the "readthrough" variant, which she refers to as ARP; see also Massoulié et al., 1993a, 1993b). Is placental AChE precociously formed for this particular aim, then?

AChE and ACh (and sometimes BuChE; see above) of the erythrocytes, the epithelium of the frog, or the crab gills were implicated more than half a century ago in ion transport (see, for example, Greig and Holland, 1949; Kirschner, 1953; see also Karczmar, 1967; Koelle, 1963). These hypotheses were based mostly on the permeability effects in these tissues of antiAChEs, and Mary and Lowell Hokin (1960; see also Hokin and Dixon, 1993) proposed that at the membranes in question, phosphatidic acid functions as a sodium pump, ACh inducing a structural modification in the membrane, thus activating the pump. This putative function might be pertinent with respect to the presence of AChE in musculotendinous junctions or in the placental syncytium. Also, these ionic effects may underlie the role of AChE in protecting the BBB and explain the damage inflicted on this barrier by antiChEs (see Chapter 7 DI).

Moreover, Botti et al. (1998; see also Massoulié et al., 1993a, 1993b) proposed an "electrotactin" role for AChE (see also Chapter 2 B-3). Their thought was based on their demonstration of the homology of AChE with neurotactin (of Drosophila) and mammalian neuroligin; neurotactin and neuroligin are adhesion proteins (electrotactins) that are capable of interaction or bonding with other proteins and of conferring adhesive properties on cells that express the electrotactins. The homology in question relates to the electric dipole character and certain peptide sequences exhibited by ChEs as well as other electrotactins (see above, section DII-1, and Massoulié et al., 1993a, 1993b). In this sense, ChEs, like other electrotactins, may subserve a "structural" capacity for the cells that express them, independently of transmittive or hydrolytic functions.

Also, whether in the nervous system or in other tissues, AChE—and BuChE—may play a trophic role, and recent data strengthen this notion. The trophic and differentiation role of the components of the cholinergic system was suggested some 40 years ago (Karczmar, 1963a, 1963b; Karczmar, 1987; see Chapter 8 C). It was speculated subsequently that the precocious, preneurogenesis presence of ChEs in the fetus may reflect their trophic effects, and the ontogenetic and teratogenic effects of antiChEs have a bearing on this speculation (see Chapter 8 CIII and Massoulié et al., 1993a, 1993b). In addition, ChEs seem to play a part in the neuritic movement, in the regulation of axonal growth, and in the differentiation of synapses and junctions (Layer, 1990; see also Massoulié et al., 1993a, 1993b). This trophic role of ChEs may persist postnatally and into adulthood. Thus, a robust augmentation of AChE activity and AChE mRNA levels and their subsequent decline were noticed during the early postnatal development of the rat thalamus and related to axonal outgrowth (Brimijoin and Hammond, 1996; see Chapter 8 BIII). An example of the adult role was adduced by Csillik and Savay (1958), who found that residual AChE of the neuromyal denervated subneuronal apparatus acts trophically on ingrowth of regenerating axon terminals. George Koelle also demonstrated this adult trophic function of ChEs. Koelle showed that the resynthesis of BuChE following its inhibition by sarin was accelerated by the AChE when the latter was protected from inhibition, and vice versa (Koelle et al., 1979; Koelle, personal communication). Is it possible that this trophic role of ChEs is related to their presence not only during development but also in adult, nonnervous tissues? May it relate to their original, morphogenetic role in ancestral organisms? (See section 3, below.)

Current hypotheses do not directly relate to the matters raised so far, such as the role of ChEs in nonnervous tissues, and so forth, and, in fact, open an entirely new vista on the roles of ChEs. Hermona Soreq and David Glick (2000, 2002) suggested that AChE plays a role in certain behavioral states, particularly stress. First, they found that AChE-R ("readthrough" AChE; see above) accumulates in the long term after stress, administration of cortisol or the stress hormone, or exposure to OP antiChEs. The underlying mechanism may be a c-fos-mediated increase in AChE gene expression. This increase may have short-range benefit, for example, enhancing hydrolysis of ACh accumulated by antiChEs, but it may also have long-range deleterious behavioral (including decrease of survival) and pathological effects (on neurites; Layer's notion on this matter was already mentioned). Recently, the Israeli team introduced an antisense oligonucleotide (EN101) that eliminates the stress-induced AChE-R transcript; accordingly, it may have a therapeutic effect on stress (Brenner et al., 2003).

These concepts and other notions presented in this section illustrate the possibility that, just as ChEs are a multiform family of enzymes and enzymic variants or isoforms, they also are multifunctional. How far we are from the early concepts of a single role for AChE!

3. Conclusions

Altogether, much remains to be done to unravel the role of BuChE, an "enzyme without natural substrate," and to clear up the mysteries of AChE as a multifunction enzyme. The matter was much simpler when AChE was thought of as an enzyme with a single, transmittive function at the synapses and junctions and when it was not known that it can be secreted from the nerves as if it were a transmitter! Today, after many years of efforts of distinguished investigators and in spite of many ingenious speculations, it is still not clear what BuChE and AChE do in the serum of some species and what AChE does in the erythrocytes of other species, what the role is of either enzyme during early stages of development, and what the significance is of the ubiquity of these enzymes throughout the plant and animal kingdoms, including noninnervated forms.

Speculatively, we can say that ChEs descend from an ancestral form that was not originally involved in transmission, but rather in trophic and morphogenic phenomena, the transmission role of these enzymes being an adaptation (for criticism of this notion, see Chapter 11); demonstrated flexibility and adaptability of ChEs is consistent with this speculation.

Notes

1. Choline's and phosphatidylcholine's presence in the brain was known much earlier (Shimizu, 1921).
2. The Swiss Alexander von Muralt is an early cholinergiker who delivered one of the first comprehensive texts on the metabolism of choline and ACh, and on AChE. He was also a distinguished mountain climber and at one time the president of the Alpine Mountaineering Society.
3. Hans Kosterlitz was a prominent pharmacologist, an escapee from Nazi Germany who served for many years as head of pharmacology in Aberdeen, Scotland. While he is cited here for one of his many contributions to the cholinergic lore, he is particularly known as a pioneer in the area of naturally occurring opiates, the endorphins.
4. There is wide-ranging use—whether as prescription medicines or over-the-counter drugs and food supplements—of ACh precursors (such as choline and several phospholipids) in human depression and memory disorders; there is also the patent potential in the pertinent research.
5. Many investigators refer to the regions encoding the variable C-terminal peptides of AChE as exons. Jean Massoulié (personal communication) feels that the term "exons" as applied to these regions is incorrect, because these regions represent coding sequences that are not defined by splicing limits. Massoulié points out also that the mRNA encoding the R subunit also contains the H and T regions, which are noncoding because of a preceding stop codon; similarly, the mRNA encoding the H subunit contains the T region (Taylor et al., 2000).
6. Jean Massoulié commented (personal communication) that, as these enzymes can be measured with great sensitivity, and as they were studied very extensively, their presence was reported even in cases where their levels were very low.
7. David Nachmansohn's studies of the role of ACh in conduction are stressed and criticized here. It should not be forgotten, however, that Nachmansohn was a great and prolific scientist, a pioneer in the studies of ChEs and their characteristics, and the discoverer (with Machado) of CAT. He merited a Nobel Prize and probably was not awarded one because of his longtime focus on the lost cause of ACh as the axonal conductor substance (Nashmansohn, 1963b).

References

Abderhalden, E., Pafrath, H. and Sickel, H.: Beitrag zur Frage der Inkre-(Hormon-) Wirkung des Cholins auf die motorischen Funktionen des Verdauungstkanales. II. Mitteilung. Arch. Ges. Physiol. *207*: 241–253, 1925.

Abe, T., Haga, T. and Kurokawa, M.: Rapid transport of phosphatidylcholine occurring simultaneously with protein transport in frog sciatic nerve. Biochem. J. *136*: 731–740, 1973.

Abramson, S. N., Ellisman, M. H., Deerink, T. J., Maulet, Y., Gentry, M. K., Doktor, B. P. and Taylor, P.: Differences in structure and distribution of the molecular forms of acetylcholinesterase. J. Cell Biol. *108*: 2301–2311, 1989.

Adibhatla, R. M., Hatcher, J. F. and Dempsey, R. J.: Citicoline: neuroprotective mechanisms in cerebral ischemia. J. Neurochem. *80*: 12–23, 2002.

Agranoff, B. W., Benjamins, J. A. and Hajra, A. K.: Lipids. In: "Basic Neurochemistry," G. J. Siegel, Ed., pp. 47–67, Lippincott-Raven, Philadelphia, 1999.

Albuquerque, E. X., Pereira, E. F. R., Braga, M. F. M. and Alkondon, M.: Contribution of nicotinic receptors to the function of synsapses in the central nervous system: The action of choline as a selective agonist of alpha7 receptors. In: "From *Torpedo* Electric Organ to Human Brain: Fudamental and Applied Aspects," J. Massoulié, Ed., pp. 309–316, Elsevier, Amsterdam, 1998.

Aldridge, W. N.: Some observations on the characteristics of serum esterases with special reference to hydrolysis of diethyl p-nitrophenyl phosphate (E 600). Ph. D. Thesis, London University, 1951.

Aldridge, W. N.: Serum esterases. I. Two types of esterases (A and B) hydrolyzing p-nitrophenyl acetate, propionate, butyrate, and a method for their determination. Biochem. J. *53*: 110–117, 1953a.

Aldridge, W. N.: The differentiation of true and pseudocholinesterase by organophosphorus compounds. Biochem J. *53*: 62–67, 1953b.

Alkon, D. L., Schmidt, D. E., Green, J. P. and Szilagyi, P. I. A.: The release of acetylcholine from the innervated and denervated diaphragm. J. Pharmacol. Exper. Therap. *174*: 346–349, 1970.

Alles, G. A. and Hawes, R. C.: Cholinesterase in the blood of man. J. Biol. Chem. *133*: 375–390, 1940.

Amara, S. G. and Kuhar, M. J.: Neurotransmitter transporters: recent progress. Annu. Rev. Neurosci. *16*: 73–93, 1993.

Ammon, R.: Die fermentative Spaltung des Acetylcholins. Arch. Ges. Physiol. *233*: 486–491, 1933.

Ammon, R. and Chytrek, E.: Die Bedeutung der Enzyme in der klinischen Diagnostik. Ergebn. Enzymforsch. *8*: 91–187, 1939.

Ansell, G. B. and Spanner, S.: Studies on the origin of choline in the brain of the rat. Biochem. J. *122*: 741–750, 1971.

Ansell, G. B. and Spanner, S.: Sources of choline for acetylcholine synthesis in the brain. In: "Choline and Lecithin in Brain Disorders," A. Barbeau, J. H. Growdon and R. J. Wrutman, Eds., pp. 35–56, Nutrition and the Brain, Raven Press, vol. 5 New York, 1979.

Antonelli, T., Beani, L., Bianchi, C., Pedata, F. and Pepeu, G.: Changes in synaptosomal high affinity choline uptake following electrical stimulation of guinea-pig cortical slices: effect of atropine and physostigmine. Br. J. Pharmacol. *74*: 525–531, 1981.

Appleyard, M.: Secreted acetylcholinesterase: non-classical aspects of a classical enzyme. Trends in Neurosci. *15*: 485–490, 1992.

Aquilonius, S. and Eckernas, S. A.: Plasma concentrations of free choline in patients with Huntington's chorea on high doses of choline chloride. N.E.J. Med. *293*:1105–1106, 1975.

Arpagaus, M., Kott, M., Vatsis, K. P., Bartels, C. F., La Du, B. N. and Lockridge, O.: Structure of the gene for human butyrylcholinesterase. Evidence for a single copy. Biochem. *29*: 1124–131, 1990.

Askari, A.: Uptake of some quaternary ammonium ions by human erythrocytes. J. Gen. Physiol. *49*: 1147–1160, 1966.

Atweh, S., Simon, J. R. and Kuhar, M. J.: Utilization of sodium dependent high affinity choline uptake *in vitro* as a measure of the activity of cholinergic neurons *in vivo*. Life Sci. *17*:1535–1544, 1975.

Augustinsson, K.-B.: Studies on blood cholinesterases. Ark. Kemi Miner. Geol. *18A*: 1–16, 1944.

Augustinsson, K.-B.: Cholinesterases. A study in comparative anatomy. Acta Physiol. Scand. *15*, Supplt. 52: 1–182, 1948.

Augustinsson, K.-B.: Electrophoresis studies on blood plasma esterases. II. Avian, reptilian, amphibian and piscine plasmata. Acta Chem. Scand. *13*: 1081–1096, 1959.

Augustinsson, K.-B.: Classification and comparative enzymology of the cholinesterases and methods for their determination. In: "Cholinesterases and Anticholinesterase Agents," G. B. Koelle, Ed., pp. 89–128, Handbch. d. exper. Pharmakol., Ergänzwk., vol. 15, Springer-Verlag, Berlin, 1963.

Bacq, Z. M.: Physiologie comparee de la transmission chimique des excitations nerveuses. Ann. Soc. Roy. Zool. Belg. *72*: 181–203, 1941.

Baker, R. R. and Dowdall, M. J.: Some properties of choline acetyltransferase isolated from the electric organ of Torpedo marmorata: specificity for choline analogues and inhibition by certain amines and amino acids. Neurochem. Res. *1*: 153–169, 1976.

Banns, H., Hebb, C. and Mann, S. P.: The synthesis of ACh from acetyl CoA, acetyl dephospho CoA and acetylpantetheine phosphate by choline acetyltransferase. J. Neurochem. *29*: 433–437, 1977.

Barald, K. F. and Berg, D. K.: High affinity choline uptake by spinal cord neurons in dissociated cell culture. Dev. Biol. *65*: 90–99, 1978.

Barbeau, A.: Lecithin in movement disorders. In: "Choline and Lecithin in Brain Disorders," A. Barbeau, J. H. Growdon and R. J. Wurtman, Eds., pp. 263–271, Nutrition and the Brain, vol. 5, Raven Press, New York, 1979.

Barbeau, A., Growdon, J. H. and Wurtman, R. J., Eds.: "Choline and Lecithin in Brain Disorders," Nutrition and the Brain, vol. 5, Raven Press, New York,1979.

Barker, L. A.: Modulation of synaptosomal high affinity choline transport. Life Sci. *18*: 725–731, 1976.

Bartheld, von, C. S.: Axonal transport and neuronal transcytosis of trophic factors, tracers, and pathogens. J. Neurobiol. *58*: 295–314, 2004.

Bartolini, M., Cavrini, V. and Andrisano, V.: Monolithic micro-immobilized-enzyme reactor with human recombinant acetylcholinesterase for on-line inhibition studies. J. Chromatography *1031*: 27–34, 2004.

Bartus, R. T., Dean, R. L., Beer, B. and Lippa, A. S.: The cholinergic hypothesis of geriatric memory dysfunction. Sci. *217*: 408–417, 1982.

Beach, R. I., Vaca, K. and Pilar, G.: Ionic and metabolic requirements for high-affinity choline uptake and acetylcholine synthesis in nerve terminals at a neuromuscular junction. J. Neurochem. *34*: 1387–1398, 1980.

Beckett, E. B. and Bourne, G. H.: The histochemistry of normal and abnormal human muscle. Proc. Roy. Soc. Med. *50*: 308–312, 1957.

Belbeoc'h, S., Massoulié, J. and Bon, S.: The C-terminal T peptide of acetylcholinesterase enhances degradation of unassembled active subunits through the ERAD pathway. EMBO J. *22*: 3536–3545, 2003.

Belbeoc'h, S., Falasca, C., Leroy, J., Ayon, A., Massoulié, J. and Bon, S.: Elements of the C-terminal t peptide of acetylcholinesterase that determine amphiphilicity, homomeric and heteromeric associations, secretion and degradation. Eur. J. Biochem. *271*: 1476–1487, 2004.

Bergmann, F. and Rimon, S.: A new type of esterase in hog-kidney extract. Biochem. J. *67*: 481–486, 1957.

Berse, B. and Blusztajn, J. K.: Co-ordinated up-regulation of choline acetyltransferase and vesicular acetylcholine transporter gene expression by retinoid acid receptor alpha, cAMP, and CNTF/LIF signaling pathways in a murine septal cell line. J. Biol. Chem. *270*: 22101–22104, 1995.

Bertrand, C., Chatonnet, A., Takke, C., Yan, Y. L., Postlethwait, J., Toutant, J. P. and Cousin, X.: Zebrafish acetylcholinesterase is encoded by a single gene localized on linkage group 7. Gene structure and polymorphism; molecular forms and expression pattern during development. J. Biol. Chem. *276*: 464–474, 2001.

Best, C. H. and Huntsman, M. E.: The effect of components of lecithin upon disposition of fat in the liver. J. Physiol. (Lond.) *75*: 405–412, 1932.

Beznak, A. B. L.: Effect of acetylcholinesterqses on the surviving frog heart. Nature *181*: 1190–1192, 1958.

Bhatnagar, S. P. and MacIntosh, F. C.: Effects of quaternary bases and inorganic cations on acetylcholine synthesis in nervous tissue. Can. J. Physiol. Pharmacol. *45*: 249–268, 1967.

Bielarczyk, H. and Szutowicz, A.: Evidence for the regulatory function of synaptoplasmic acetyl CoA in acetylcholine synthesis in nerve endings. Biochem. J. *262*: 377–380, 1989.

Bierkamper, G. C. and Goldberg, A. M.: Effect of choline on the release of acetylcholine from the neuromuscular junction. In: "Choline and Lecithin in Brain Disorders," A. Barbeau, J. H. Growdon and R. J. Wurtman, Eds., 243–251, Nutrition and the Brain, vol. 5, Raven Press, New York, 1979.

Bierkamper, G. G. and Goldberg, A. M.: Release of acetylcholine from the vascular perfused rat phrenic nerve hemidiaphragm. Brain Res. *202*: 234–237, 1980.

Birks, R. I.: The role of sodium ions in the metabolism of acetylcholine. Can. J. Biochem. Physiol. *41*: 2573–2597, 1963.

Birks, R. and MacIntosh, F. C.: Acetylcholine metabolism of a sympathetic ganglion. Can. Biochem. Physiol. *39*: 787–827, 1961.

Bligh, J.: The level of free choline in plasma. J. Physiol. (Lond.) *117*: 234–240, 1952.

Blusztajn, J. K. and Wurtman, R. J.: Choline and cholinergic neurons. Science *221*: 614–620, 1983.

Blusztajn, J. K. and Wurtman, R. J.: Choline and cholinergic neurons. Science. *221*: 614–620, 1983.

Blusztajn, J. K. and Wurtman, R. J.: Choline and cholinergic neurons. Science. *221*: 614–620, 1983.

Blusztajn, J. K., Zeisel, S. H. and Wurtman, R. J.: Synthesis of lecithin (phosphatidylcholine) from phosphatidylethanolamine in bovine brain. Brain Res. *179*: 319–327, 1979.

Blusztajn, J. K., Richardson, U. I., Liscovitch, M., Mauron, C. and Wurtman, R. J.: Phospholipids in cellular survival and growth. In: "Lecithin: Technological, Biological and Therapeutic Aspects," I. Hanin, and G. B. Ansell, Eds., pp. 85–94, Plenium Press, New York, 1987.

Blusztajn, J. K., Cermak, J. M., Holler, T. and Jackson, D. A.: Imprinting of hippocampal metabolism of choline by its viability during gestation: implications for cholinergic neurotransmission. In: "From *Torpedo* Electric Organ to Human Brain: Fundamental and Applied Aspects," J. Massoulié, Ed., pp. 199–203, Elsevier, Amsterdam, 1998.

Bon, S., Rosenberry, T. I. and Massoulié, J.: Amphiphilic, glycophosphatidylinositol-specific phospholipase C (PI-PLC)-insensitive monomers and dimers of acetylcholinesterase. Cell Mol. Neurobiol. *11*: 157–172, 1991.

Bon, S., Ayon, A., Leroy, J. and Massoulié, J.: Trimerization domain of the collagen tail of acetylcholinesterase. Neurochem. Res. *28*: 523–535, 2003.

Bon, S., Dufourcq, J., Leroy, J., Cornut, I. and Massoulié, J.: The C-terminal t peptide of acetylcholinesterase forms an alpha helix that supports homomeric and heteromeric interactions. Eur. J. Biochem. *271*: 33–47, 2004.

Booj, S., Dahkoff, A. G., Larsson, P. A. and Dahlstrom, A.: Influence of descending bulbospinal monoamine neurons on atonal transport of acetylcholine and cholinergic enzymes. J. Neural. Transm. *52*: 213–215, 1981.

Botti, S. A., Felder, C. E., Sussman, J. L. and Silman, I.: Electrotactins: a class of adhesion proteins with conserved electrostatic and structural motifs. Protein Eng. *11*: 415–420, 1998.

Bourne, Y., Taylor, P. and Marchot, P.: Acetylcholinesterase inhibition by fasciculin. Crystal structure of the complex. Cell *83*: 503–512, 1995.

Bowery, N. G. and Neal, M. J.: Proceedings: Failure of denervation to influence the high affinity uptake of choline by sympathetic ganglia. Br. J. Pharmacol. *55*: 278, 1975.

Bowman, W. C., Hemsorth, B. A. and Rand, M. J.: Myasthenic-like features of the neuromuscular transmission failure produced by triethylcholine. J. Pharm. Pharmacol. *14*: 37T–41T, 1962.

Boyd, W. D., Graham-White, J., Blackwood, G., Glen, I. and McQueen, J.: Clinical effects of choline in Alzheimer senile dementia. Lancet *2*: 711, 1977.

Brandon, E. P., Mellott, T., Pizzo, D. P., Coufal, N., D'Amour, K. A., Gobeske, K., Lortie, M., Lopez-Coviella, I., Berse, B., Thal, L. J., Gage, F. H. and Blusztajn, J. K.: Choline transporter 1 maintains cholinergic function in choline acetyltransferase haploinsufficiency. J. Neurosci. *24*: 5459–5466, 2004.

Brenner, T., Hamra-Amitay, Y., Evron, T., Boneva, N., Seidman, S. and Soreq, H.: The role of readthrough acetylcholinesterase in the pathophysiology of myasthenia gravis. FASEB J. *17*: 214–222, 2003.

Brimijoin, S. and Hammond, P.: Transient expression of acetylcholinesterase messenger RNA and enzyme activity in developing rat thalamus studied by quantitative histochemistry and *in situ* hybridization. Neurosci. *71*: 555–565, 1996.

Brown, G. L. and Feldberg, W.: The acetylcholine metabolism of a sympathetic ganglion. J. Physiol. (Lond.) *88*: 265–283, 1936a.

Brown, G. L. and Feldberg, W.: The action of potassium on the superior cervical ganglion of the cat. J. Physiol. (Lond.) *86*: 290–305, 1936b.

Browning, E. T.: Acetylcholine synthesis: Substrate availability and the synthetic reaction. In: "Biology of Cholinergic Function," A. M. Goldberg and I. Hanin, Eds., pp. 187–201, Raven Press, New York, 1976.

Brunello, N., Cheney, D. L. and Costa, E.: Increase in exogenous choline fails to elevate the content or turnover rate of cortical, striatal or hippocampal acetylcholine. J. Neurochem. *38*: 1160–1163, 1982.

Budai, D., Ricny, J., Kasa, P. and Tucek, S.: 4-(1-naphthylvinyl) pyridine decreases brain acetylcholine *in vivo* but does not alter the level of acetyl-CoA. J. Neurochem. *46*: 990–992, 1986.

Bullock, Th. H. and Nachmansohn, D.: Cholinesterase in primitive nervous systems. J. Cell. Comp. Physiol. *20*: 1–4, 1942.

Burgen, A. S. V.: The mechanism of action of anticholinesterase drugs. Brit. J. Pharmacol. *4*: 219–228, 1949.

Burgen, A. S. V. and Hobbiger, F.: The inhibition of cholinesterases by alkylphosphates and alkyl phenol phosphates. Brit. J. Pharmacol. *4*: 592–605, 1951.

Burn, J. H. and Kottegoda, S. R.: Action of eserine on the auricles of the rabbit heart. J. Physiol. (Lond.) *121*: 360–373, 1953.

Carroll, P. T. and Goldberg, A. M.: Relative importance of choline transport to spontaneous and potassium depolarized release of ACh.

Cervini, R., Houhou, L., Pradat, P. F., Bejanin, S., Mallet, J. and Berrard, S.: Specific vesicular acetylcholine transporter promoters lie within the first intron of the rat choline acetyltransferase gene. J. Biol. Chem. *270*: 24654–24657, 1995.

Chadwick, L. E.: Actions on insects and other invertebrates. In: "Cholinesterases and Anticholinesterase Agents," G. B. Koelle, Ed., pp. 741–798, Handbch. d. exper. Pharmakol., Ergänzwk., vol. 15, Springer-Verlag, Berlin, 1963.

Chagas, C. and Paes de Carvalho, A. Eds.: "Bioelectrogenesis," Elsevier, Amsterdam, 1961.

Chang, C. C. and Lee, C.: Studies on the [3H] chloine uptake in rat phrenic nerve-diaphragm preparations. Int. J. Neuropharmacol. *9*: 223–233, 1970.

Chang, C. H. and Gaddum, J. H.: Choline esters in tissue extracts. J. Physiol, (Lond.) *79*: 255–285, 1933.

Changeux, J.-P.: Responses of acetylcholinesterase from *Torpedo marmorata* to salts and curarizing drugs. Molec. Pharmacol. *2*: 339–392, 1966.

Chatonnet, A. and Lockridge, O.: Comparison of butyrylcholinesterase and acetylcholinesterase. Biochem. J. *260*: 625–634, 1989.

Choi, R. L., Roch, M. and Jenden, D. J.: A regional study of acetylcholine turnover in rat brain and the effect of oxotremorine. Proc. West. Pharmacol. Soc. *16*: 188–190, 1973.

Chuang, D. M. and Costa, E.: Biosynthesis of tyrosine hydroxylase in rat adrenal medulla after exposure to cold. Proc. Nat. Acad. Sci. USA *71*: 4570–4574, 1974.

Cohen, E. L. and Wurtman, R. J.: Brain acetylcholine: control by dietary choline. Sci. *191*: 561–562, 1976.

Cohen, E. L. and Wurtman, R. J.: Brain acetylcholine: increase after systemic choline administration. Life Sci. *16*: 755–762, 1977.

Cohen, J. A. and Oosterbaan, R. A.: The active site of acetylcholinesterase and related esterases and its reactivity toward substrates and inhibitors. In: "Cholinesterases and Anticholinesterase Agents," G. B. Koelle, Ed., pp. 299–373, Handbch. d. exper. Pharmakol., Ergänzwk., vol. 15, Springer-Verlag, Berlin, 1963.

Collier, B. and Katz, B.: Acetylcholine synthesis from recaptured choline in a sympathetic ganglion. J. Physiol. (Lond.) *238*: 639–655, 1974.

Collier, B., Kwok, Y. N. and Welner, S. A.: Increased acetylcholine synthesis and release following presynaptic activity in a sympathetic ganglion. J. Neurochem. *40*: 91–98, 1983.

Collumbine, H.: Actions at autonomic effector sites. In: "Cholinesterases and Anticholinesterase Agents," G. B. Koelle, Ed., pp. 505–529, Handbch. d. exper. Pharmakol., Ergänzwk., vol. 15, Springer-Verlag, Berlin.

Combes, D., Fedon, Y., Toutant, J. P. and Arpagaus, M.: Acetylcholinesterase genes in the nematode Caenorhabditis elegans. Int. Rev. Cytol. *209*: 207–239, 2001.

Combes, D., Fedon, Y., Toutant, J.-P. and Arpagaus, M.: Four acetylcholinesterase genes in the nematode *Ceaenorhabditis elegans.* In: "Symposium on Cholinergic Mechanisms—Function and Dysfunction," I. Silman, H. Soreq, L. Anglister, D. Michaelson and A. Fisher, Eds., Martin Dunitz, London, 2004.

Comline, R. S.: Synthesis of acetylcholine by non-nervous tissue. J. Physiol. (Lond.) *105*: 650–710, 1947.

Compher, C. W., Kinosian, B. P., Stoner, N. F., Lentine, D. C. and Buzby, G. P.: Choline and vitamin B12 deficiency are involved in folate-deplete long-term parenteral nutrition patients. JPEN J. Parenter. Enteral Nutr. *26*: 57–52, 2002.

Cornford, E. M., Braun, L. D. and Oldendorf, W. H.: Carrier mediated blood-brain barrier transport of choline and certain choline analogs. J. Neurochem. *30*: 299–308, 1978.

Couraud, J. Y., Koenig, H. L. and Di Giamberardino, L.: Acetylcholinesterase molecular forms in chick ciliary ganglion: pre- and postsynaptic distribution derived from denervation, axotomy, and double section. J Neurochem. *34*: 1209–1218, 1980

Cousin, X., Bon, S., Massoulié, J. and Bon, C.: Identification of a novel type of alternatively spliced exon from the acetylcholinesterase gene of Bungarus fasciatus. Molecular forms of acetylcholinesterase in the snake liver and muscle. J. Biol. Chem. *2173*: 8812–9820, 1998.

Couteaux, R.: Particularités histochimiques des zones d'insertion du muscle strié. C. R. Soc. Biol. (Paris) *147*: 1974–1976, 1953.

Crews, F. T., Hirata, F. and Axelrod, J.: Identification and properties of methyltransferases that synthesize phosphatidycholine in rat brain synaptosomes. J. Neurochem. *34*: 1491–1498, 1980.

Cuello, A. C.: Effects of trophic factors on the CNS cholinergic phenotype. In: "Cholinergic Mechanisms: From Molecular Biology to Clinical Significance," J. Klein and K. Loeffelholz, Eds., pp. 347–358, Elsevier, Amsterdam, 1996.

Currier, S. F. and Mautner, H. G.: Evidence for a thiol reagent inhibiting choline acetyltransferase by reacting with the thiol group of coenzyme A forming a potent inhibitor. Biochem. Biophysic. Res. Commun. *69*: 431–436, 1976.

Dahlstrom, A. A., Larsson, P.-A., Carlsson, S. S. and Booj, S.: Localization and axonal transport of immunoreactive cholinergic organelles in rat motor neurons—an immunofluorescent study. Neurosci. *14*: 607–625, 1985.

Darvesh, S. and Hopkins, D. A.: Differential distribution of butyrylcholinesterase and acetylcholinesterase in the human thalamus. J. Comp. Neurol. *463*: 25–43, 2003.

Davis, K. L., Berger, P. L., Hollister, L. E., Deamoral, J. R. and Barchas, J. D.: Cholinergic dysfunction in mania and movement diseases. In: "Cholinergic Mechanisms and Psychopharmacology," D. J. Jenden, Ed., pp. 755–780, Plenum Press, New York, 1977a.

Davis, K. L., Berger, P. A. and Hollister, L. E.: Deanol for tardive dyskinesia. J. Psychiat. *134*: 807, 1977b.

Davis, K. L., Berger, P. L., Hollister, L. E. and Barchas, J. D.: Cholinergic involvement in mental disorders. Life Sci. *22*: 1865–1872, 1978.

Davis, K. L., Berger, P. A. and Hollister, L. E.: Clinical and preclinical experience with choline chloride in Huntington's disease and tardive dyskinesia: unanswered questions. In: "Choline and Lecithin in Brain Disorders," A. Barbeau, J. H. Growdon and R. J. Wurtman, Eds., pp. 305–315, Nutrition and the Brain, vol. 5, Raven Press, New York, 1979.

Deprez, P., Doss-Pepe, E., Brodsky, B. and Inestrosa, N. C.: Interaction of the collagen-like tail of asymmetric acetylcholinesterase with heparin depends on the triple-helical conformation, sequence and stability. Biochem J. *350*, pt. *1*: 283–290, 2000.

Deprez, P., Inestrosa, N. C. and Krejci, E.: Two different heparin-binding domains in the triple-helical domains of ColQ the collagen tail subunit of synaptic acetylcholinesterase. J. Biol. Chem. *278*: 23233–23242, 2003.

Deschenes-Furry, J., Belanger, G., Mwanjewe, J., Lunde, J. A., Parks, R. J., Perrone-Bizzozero, N. and Jasmin, B. J.: The RNA-binding protein HuR binds to acetylcholinesterase transcripts and regulates their expression in differentiating skeletal muscle cells. J. Biol. Chem. *280*: 25361–25368, 2005.

Diamond, I.: Choline metabolism in brain: The role of choline transport and the effects of phenobarbital. Arch. Neurol. *24*: 333–339, 1971.

Dirnhuber, P. and Lovatt Evans, C.: The effects of anticholinesterases on humoral transmission in the submaxillary gland. Brit. J. Pharmacol. *9*: 441–458, 1954.

Dixon, W. E.: On the mode of action of drugs. Med. Mag. (Lond.) *16*: 454–457, 1907.

Dobransky, T. and Rylett, R. J.: Functional regulation of choline acetyltransferase by phosphorylation. Neurochem. Res. *28*: 537–542, 2003.

Dobransky, T. and Rylett, R. J.: A model for dynamic regulation of choline acetyltransferase by phosphorylation. J. Neurochem. *95*: 305–313, 2005.

Domino, L. E., Domino, S. E. and Domino, E. F.: Toxicokinetics of 2,2′, 4,4′, 5,5′-hexabromobiphenyl in the rat. J. Toxicol. Environ. Health. *9*: 815–833, 1982.

Dross, K. and Kewitz, H.: Concentration of acetylcholine, choline and glycerylphosphorylcholine in the brain during anesthesia and in DFP poisoning. Naunyn. Schmiedebergs Arch. Exp. Pathol. Pharmakol. *260*: 107–109, 1968.

Dross, K. and Kewitz, H.: Concentration and origin of choline in the rat brain. Naunyn Schmiedeberg's Arch. Pharmacol. *274*: 91–106, 1972.

Dvir, H., Kryger, G., Johnson, J. L., Rosenberry, T. L., Silman, I. and Sussman, J. L.: Crystallization and determination of the X-Ray structure of human AChE. In: "Cholinergic Mechanisms—Function and Dysfunction," I. Silman, H. Soreq, L. Anglister, D. Michaelson and A. Fisher, Eds., Martin Dunitz, London, 2004a.

Dvir, H., Harel, M., Bon, S., Liu, W. O., Vidal, M., Garbay, C., Sussman, J. L., Massoulié, J. and Silman, I.: The synaptic acetylcholinesterase tetramer assembles around a polyproline II helix. Embo. J. *23*: 4394–4405, 2004.

Eastman, J., Wilson, E. J., Cervenansky, C. and Rosenberry, T. L.: Fasciculin 2 binds to the peripheral site on acetylcholinesterase and inhibits substrate hydrolysis by slowing a step involving proton transfer during enzyme acylation. J. Biol. Chem. *270*: 19644–19701, 1995.

Eccles, J. C.: Synaptic and neuronmuscular transmission. Ergebn. Physiol. *38*: 399–444, 1936.

Eccles, J. C.: "The Physiology of Synapses," Springer-Verlag, New York, 1964.

Eckernas, S. A.: Plasma choline and cholinergic mechanisms in the brain. Methods, function, and role in Huntington's chorea. Acta Physiol. Scand. Suppl. *449*: 1–62, 1977.

Eckernas, S. A. and Aquilonius, S. M.: Plasma choline in healthy subjects and in Huntington's chorea patients on high oral doses of choline chloride. In: "Cholinergic Mechanisms and Psychopharmacology," D. J. Jenden, Ed., pp. 747–754, Plenum Press, New York, 1977.

Eckernas, S. A. and Aquilonius, S. M.: Use of choline in five patients with Huntington's Disease. In: "Choline and Lecithin in Brain Disorders," A. Barbeau, J. H. Growdon and R. J. Wurtman, Eds., pp. 325–330, Nutrition and the Brain, vol. 5, Raven Press, New York, 1979.

Edwards, J. A. and Brimijoin, S.: Divergent regulation of acetylcholinesterase and butyrylcholinesterase in tissues of the rat. J. Neurochem. *38*: 1393–1408, 1982.

Eichler, J., Silman, I. and Anglister, L.: G2-Acetylcholinesterase is presynaptically localized in *Torpedo* electric organ. J. Neurocytol. *21*: 707–716, 1992.

Endo, T. and Tamiya, N.: Current view on the structure-function relationship of postsynaptic neurotoxins from snake venom. Pharmac. Ther. *34*: 403–451, 1987.

Engelhard, N., Prchal, K. and Nenner, M.: Acetylcholinesterase. Angew. Chem. Int. Ed. *6*: 615–626, 1967.

Feaster, S. R., Quinn, D. M. and Barnett, B. L.: Molecular modeling of the structure of human and rat cholesterol esterases. Protein Sci. *6*: 73–79, 1997.

Feldberg, W. and Vogt, M.: Acetylcholine synthesis in different regions of the central nervous system. J. Physiol. *107*: 372–381, 1948.

Fernandes, K. J. L., Kobayashi, N. R., Jasmin, B. J. and Tetzlaff, W.: Acetylcholinesterase gene expression in axotomized rat facial motoneuron is differentially regulated by neurotropins: correlation with TrkB and TrkCmRNA levels and isoforms. J. Neurosci. *18*: 9936–9947, 1998.

Fischer, H.: "Vergleichende Pharmakologie von Überträgersubstanzen in tiersystematischen Darstellung," Springer-Verlag, Berlin, 1971.

Fischer, L. M., Scearce, J. A., Mar, M. H., Patel, J. R., Blanchard, R. T., Macintosh, B. A., Busby, M.G. and Zeisel, S. H.: Ad libitum choline intake in healthy individuals meets or exceeds the proposed adequate intake level. J. Nutr. *135*: 826–829, 2005.

Foldes, F. F.: Fate of muscle relaxants in man. Acta Anaesth. Scand. *1*: 63–69, 1957.

Fournier, D., Karch, F., Bride, J.-M., Hall, L. M. C., Berge, J.-B. and Spierer, P.: *Drosophila melanogaster* acetylcholinesterase gene: structure, evolution and mutations. J. Mol. Biol. *210*: 15–22, 1990.

Friedenwald, J. S.: Histochemistry: a review. Pharmacol. Rev. *7*: 83–96, 1955.

Friesen, A. J. and Khatter, J. C.: The effect of preganglionic stimulation on the acetylcholine and choline content of a sympathetic ganglion. Can. J. Physiol. Pharmacol. *49*: 375–381, 1971.

Fühner, H.: Untersuchungen über den Synergismus von Giften. IV. Die chemische Ervegbarkeitssteigerung glatter muskulatur. Arch. Exp. Pathol. Pharmakol. *82*: 51–85, 1918.

Gaiti, A., De Medio, G. E., Brunetti, M., Amaducci, L. and Porcellati, G.: Properties and function of calcium-dependent incorporation of choline, othanolamine and serine into the phospholipids of isolated rat brain microsomes. J. Neurochem. *33*: 1153–1163, 1974.

Galli, C., Sirtori, C. R., Mosconi, C., Medini, L., Gianfranceschi, G., Vaccarino, V. and Scolastico, C.: Prolonged retention of doubly labeled phosphatidylcholine in human plasma and erythrocytes after oral administration. Lipids. *27*: 1005–1012, 1992.

Galehr, O. and Plattner, F.: Über das Schicksal des Azetylcholins im Blute. Arch. Ges. Physiol. *218*: 488–505, 1927.

Gardiner, J. E. and Gwee, M. C.: Effect of two hemicholiniums on the disposition and distribution of endogenous free choline in anaesthetized rabbits. Br. J. Pharmacol. *60*: 351–356, 1977.

Gerard, B. W. and Tupikova, N.: J. Cell. Comp. Physiol. *13*: 1–13, 1939.

Gerebtzoff, M. A: "Cholinesterases: A Histochemical Contribution to the Solution of Some Functional Problems," Pergamon Press, New York, 1959.

Getman, D. K., Eubanks, J. H., Camp, S., Evans, G. A. and Taylor, P.: The human gene encoding acetylcholinesterase is located on the long arm of chromosome 7. Am. J. Hum. Genet. *51*: 170–177, 1992.

Giacobini, E.: Determination of cholinesterase in the cellular components of neurons. Acta Physiol. Scand. *45*: 311–327, 1959.

Giacobini, E.: Cholinesterase inhibitors: from Calabar bean to Alzheimer therapy. In: "Cholinesterases and Cholinesterase Inhibitors," E. Giacobini, Ed., pp. 181–226, Martin Dunitz, London, 2000.

Giacobini, E.: Butyrylcholinesterase: its role in brain function. In: "Butyrylcholinesterase, Its Function and Inhibitors," E. Giacobini, Ed., pp. 1–20, Martin Dunitz, London, 2003a.

Giacobini, E.: Drugs that target cholinesterase. In: "Cognitive Enhancing Drugs," J. J. Buccafusco, Ed., pp. 11–22, Bikhauser Publ., Berlin, 2003b.

Glick, D.: The nature and significance of cholinesterase. In: "Comparative Biochemistry. Intermediate Metabolism of Fats. Carbohydrate Metabolism, Biochemistry of Choline," H. B. Lewis, Ed., pp. 213–233, Jaques Cattell Press, Lancaster, Pennsylvania, 1941.

Glick, D.: The controversy on cholinesterase. Science *102*: 100–101, 1945.

Glover, V. A. S. and Potter, L. T.: Purification and properties of choline acetyltransferase from ox brain striate nuclei. J. Neurochem. *18*: 571–580, 1971.

Goedde, H. W., Doenicke, A. and Altland, K.: "Psudocholinesterasen," Springer-Verlag, Berlin, 1967.

Goemaere-Vanneste, J., Couraud, J. Y., Hassig, R., Di Giamberardino, L. and van den Bosch de Aguilar, P.: Reduced axonal transport of the G4 molecular form of acetylcholinesterase in the rat sciatic nerve during aging. J. Neurochem. *51*: 1746–1754, 1988.

Gomez, M. V., Domino, E. F. and Sellinger, O. Z.: The effect of hemicholiniuar-3 on the *in vivo* formation of cerebral phosphaticlylcholine. Biochem. Biophys. Acta. *202*: 153–162, 1970.

Goutier-Pirotte, M. and Gerebtzoff, M. A.: Acetylcholinesterase in the guinea pig placenta; initial results of histochemical and biochemical research. Arch. Int. Physiol. *63*: 445–447, 1955.

Grafstein, B. and Forman, D. S.: Intracellualr transport in neurons. Physiol. Rev. *60*: 1167–1283, 1980.

Greenfield, S. A.: The significance of dendritic release of transmitter and protein in the substantia negra. Neurochem. Int. *7*: 877–901, 1985.

Greenfield, S. A.: A noncholinergic action of acetylcholinesterase (AChE) in the brain: from neuronal secretion to generation of movement. Cell. Molec. Neurobiol. *11*: 55–78, 1991.

Greenspan, R. J.: Mutations of choline acetyltransferase and associated neural defects in Drosophila melanogaster. Comp. Physiol. *137*: 83–92, 1980.

Greig, M. E. and Holland, W. C.: Studies on the permeability of erythrocytes. I. The relation between cholinesterase activity and permeability of dog erythrocytes. Arch. Biochem. *23*: 370–384, 1949.

Growdon, J. H.: Use of phosphatidylcholine in brain diseases: an overview. In: "Lecithin: Technological, Biological and Therapeutic Aspects," I. Hanin, Ed., and G. B. Ansell, Eds., pp. 121–136, Plenum Press, New York, 1987.

Growdon, J. H., Cohen, E. L. and Wurtman, R. J.: Treatment of brain diseases with dietary precursors of neurotransmitters. Ann. Int. Med. *86*: 337–339, 1977.

Gutfreund, H.: "An Introduction to the Study of Enzymes," Blackwell Scientific, Oxford, 1965.

Haga, T.: Synthesis and release of (14 C)acetylcholine in synaptosomes. J. Neurochem. *18*: 781–798, 1971.

Haga, T. and Noda, H.: Choline uptake systems of rat brain synaptosomes. Biochim. Biophys. Acta. *291*: 564–575, 1973.

Hangaard, J., Whittaker, M., Loft, A. G. and Norgaard-Pedersen, B.: Quantification and phenotyping of serum cholinesterase by enzyme antigen immunoassay: methodological aspects and clinical applicability. Scand. J. Clin. Lab. Invest. *51*: 349–358, 1991.

Hanin, I. and Costa, E.: Approaches used to estimate brain acetylcholine turnover rate *in vivo*: effects of drugs on brain acetylcholine turnover rate. In: "Biology of Cholinergic Function," A. M. Goldberg and I. Hanin, Eds., pp. 355–378, Raven Press, New York, 1976.

Hanin, I. and Goldberg, A. M.: Appendix I: quantitative assay methodology for choline, acetylcholine, choline acetyltransferase, and acetylcholinesterase. In: "Biology of Cholinergic Function," A. M. Goldberg and I. Hanin, Eds., pp. 647–654, Raven Press, New York, 1976.

Hanin, I., Massarelli, R. and Costa, E.: An approach to the study of the biochemical pharmacology of cholinergic function. Adv. Biochem. Psychopharmacol. *6*: 181–202, 1972.

Hanin, I., Merikangas, J. R., Merikangas, K. R. and Kopp, U.: Red-cell choline and Gilles de la Tourette syndrome. N. Engl. J. Med. *301*: 661–662, 1979.

Harel, M., Sussman, J. L., Krejci, E., Bon, S., Chanal, P. and Massoulié, J.: Conversion of acetylcholinesterase to butyrylcholinesterase: modeling and mutagenesis. Proc. Natl. Acad. Sci. U.S.A. *89*: 10827–10831, 1992.

Haubrich, D. R., Reid, W. D. and Gilette, J. R.: Acetylcholine formation in mouse brain: an effect of cholinergic drugs. Nature New Biol. *238*: 88–89, 1972.

Haubrich, D. R., Wang, P. F., Chippendale, T. and Proctor, E.: Choline and acetylcholine in rats: effect of dietary choline. J. Neurochem. *27*: 1305–1313, 1976.

Hebb, C.: Chemical agents of the nervous system. Int. Rev. Neurobiol. *1*: 165–193, 1959.

Hebb, C.: Formation, storage and liberation of acetylcholine. In: "Cholinesterases and Anticholinesterase Agents," G. B. Koelle, Ed., 55–88, Handbch. d. exper. Pharmakol., Ergänzungswk, vol. 15, Springer-Verlag, Berlin, 1963.

Hebb, C. O. and Waites, G. M.: Choline acetylase in antero and retrograde degeneration of a cholinergic neuron. J. Physiol. (Lond.) *132*: 667–671, 1956.

Hebb, C. O. and Whittaker, V. P.: Intracellular distribution of acetylcholine and choline acetyltransferase. J. Physiol. (Lond.) *142*: 187–196, 1958.

Heidenhain, R.: Über die Wirkung eimiger Gitte auf die nerven der Glaudula Supermaxillaris. Pflügers Arch. ges. Physiol. J. 309–318, 1872.

Hell, C. O., Krause, M. and Silver, A.: Intracellular binding of choline acetylase in the ventral spinal roots. J. Physiol. (Lond.) *148*: 69–70, 1959.

Hendelman, W. J. and Bunge, R. P.: Radioantographic studies of choline incorporation into peripheral nerve myelin. J. Cell Biol. *40*: 190, 1969.

Hersh, L. B., Wainer, B. H. and Andrews, C. P.: Multiple isoelectric and molecular weight variants of choline acetyltransferase. Artifact or real? J. Biol. Chem. *259*: 1253–1258, 1984.

Hersh, L. B., Inoue, H. and Li, Y.-P.: Transcriptional regulation of human choline acetyltransferase gene. In: "Cholinergic Mechanisms: From Molecular Biology to Clinical Significance," J. Klein and K. Loeffelholz, Eds., pp. 47–54, Elsevier, Amsterdam, 1996.

Hersh, L. B. and Shimojo, M.: Regulation of cholinergic gene expression by NRSF/REST. In: "Cholinergic Mechanisms: Function and Dysfunction," I. Silman, H. Soreq, L. Anglister, D. Michaelson and A. Fisher, Eds., Martin Dunitz, London, 2004.

Hirata, F. and Axelrod, J.: Phospholipid methyltransferase asymmetry in synaptosomal membranes. Sci. *209*: 1082–1090, 1980.

Hirsch, M. J. and Wurtman, R. J.: Lecithin consumption increases acetylcholine concentrations in rat brain and adrenal gland. Science. *202*: 223–225, 1978.

Hirsch, M. J., Growdon, J. H. and Wurtman, R. J.: Increase in hippocampal acetylcholine after choline administration. Brain Res. *125*: 383–385, 1977.

Hokin, L. E. and Dixon, J. F.: The phosphoinositide signalling system. I. Historical background. II. Effects of lithium on the accumulation of second messenger inositol 1,4,5-triphosphate in brain cortex slices. In: "Cholinergic Function and Dysfunction," A. C. Cuello, Ed., pp. 309–315, Elsevier, Amsterdam, 1993.

Hokin, M. R. and Hokin, L. E.: The role of phosphatidic acid and phosphoinositide in transmembrane transport elicited by acetylcholine and other humoral agents. Intern. Rev. Neurobiol. *2*: 99–136, 1960.

Holmstedt, B.: Pharmacology of organophosphorus cholinesterase inhibitors. Pharmacol. Rev. *11*: 567–688, 1959.

Hopff, W. H., Riggio, G. and Waser, P. G.: Affinity chromatography of acetylcholine acetyl-hydrolase. FEBS Letters *35*: 220–222, 1973.

Houtsmuller, U. M. T.: Metabolic fate of dietary lecithin. In: "Choline and Lecithin in Brain Disorders, A. Barbeau, J. H. Growdon and R. J. Wurtman, Eds., pp. 83–108, Nutrition and the Brain, vol. 5, Raven Press, New York, 1979.

Hunt, R.: Some physiological actions of the homocholins and some of their derivatives. J. Pharmacol. Exper. Therap. *6*: 477–525, 1915.

Ichikawa, T. and Hirata, Y.: Organization of choline acetyltransferase containing structures in the forebrain of the rat. J. Neurosci. *6*: 281–292, 1986.

Jablecki, C. and Brimijoin, S.: Axoplasmic transport of choline acetyltransferase activity in mice: effect of age and neurotomy. J. Neurochem. *25*: 583–593, 1975.

Jankowska, A., Madziar, B., Tomaszewicz, M. and Szutowicz, A.: Acute and chronic effects of aluminum on acetyl-CoA and acetylcholine metabolism in differentiated and nondifferentiated SN56 cholinergic cells. J. Neurosci. Res. *62*: 615–622, 2000.

Jbilo, O., Bartels, C. F., Chatonnet, Q., Toutant, J. P. and Lockridge, O.: Tissue distribution of human acetylcholinesterase and butyrylcholinesterase messenger RNA. Toxicon *32*: 445–457, 1994.

Jenden, D. J.: Precursor control of transmitter synthesis. In: "Choline and Lecithin in Brain Disorders," A. Barbeau, J. H. Growdon and R. J. Wurtman, Eds., pp. 1–12, Nutrition and the Brain, vol. 5, Raven Press, New York, 1979.

Jope, R. S.: High affinity choline transport and acetyl-CoA production in brain and their roles in the regulation of acetylcholine synthesis. Brain Res. *180*: 313–344, 1979.

Jope, R. S.: Biochemical effects of lecithin administration. In: "Dynamics of Cholinergic Function," I.

Hanin, Ed., pp. 827–835, Plenum Press, New York, 1986.

Kahlson, G. and MacIntosh, F. C.: Acetylcholine synthesis in a sympathetic ganglion. J. Physiol. (Lond.) *96*: 277–292, 1936.

Kalow, W.: Pharmacogenomics: historical perspective and current status. Methods Mol. Biol. *311*: 3–15, 2005.

Kalow, W. and Gunn, D. R.: The relation between dose of succinylcholine and duration of apnea in man. J. Pharamcol. Exper, Therap. *120*: 2203–2214, 1957.

Kalow, W. and Gunn, D. R.: Some statistical data on atypical cholinesterase of human serum. Ann Hum Genet. *23*: 239–250, 1959.

Karczmar, A. G.: Ontogenesis of cholinesterases. In: "Cholinesterases and Anticholinesterase Agents," G. B. Koelle, Ed., pp. 799–832, Handbch. d. Exper. Pharmakol., Ergänzungswk., vol. 15, Springer-Verlag, Berlin, 1963a.

Karczmar, A. G.: Ontogenetic effects of anti-cholinesterase agents. In: "Cholinesterases and Anticholinesterase Agents," G. B. Koelle, Ed., pp. 179–186, Handbch. d. Exper. Pharmakol., Ergänzungswk., vol. 15, Springer-Verlag, Berlin, 1963b.

Karczmar, A. G.: Pharmacologic, toxicologic and therapeutic properties of anticholinesterase agents. In: "Physiological Pharmacology," W. S. Root and F. G. Hofman, Eds., vol. 3, pp. 163–322, Academic Press, New York, 1967.

Karczmar, A. G.: History of the research with anticholinesterase agents. In: "Anticholinesterase Agents," A. G. Karczmar, Ed., pp. 1–44, Int. Encyclop. Pharmacol. Therap., section 13, vol. 1, Pergamon Press, Oxford, 1970.

Karczmar, A. G.: Brain acetylcholine and seizures. In: "Psychobiology of Convulsive Therapy," M. Fink, S. Kety, J. McGaugh and T. A. Williams, Eds., pp. 251–270, W. H. Winston and Sons, Washington, D.C., 1974.

Karczmar, A. G.: Concluding remarks: past, present and future of cholinergic research. In: "Neurobiology of Acetylcholine," N. J. Dun and R. L. Perlman, Eds., pp. 561–585, Plenum Press, New York, 1987.

Karczmar, A. G.: Loewi's discovery and the XXI Century. In "Cholinergic Mechanisms: From Molecular Biology to Clinical Significance," J. Klein and K. Loeffelholz, Eds., pp. 1–27, Elsevier, Amsterdam, 1996.

Karczmar, A. G. and Koppanyi, K.: Changes in "transport" cholinesterase levels and responses to intravenously administered acetylcholine and benzoylcholine. J. Pharm. Expt. Therap. *116*: 245–253, 1956.

Karczmar, A. G., Koketsu, K. and Nishi, S., Eds.: "Autonomic and Enteric Ganglia," Plenum Press, New York, 1986.

Kasa, P., Szepesy, G., Gulya, K., Bansaghy, K. and Rakonczay, A.: The effect of 4-(1-naphthylvinyl) pyridine on the acetylcholine system and the number of synaptic vesicles in the central nervous system of the rat. Neurochem. Int. *4*: 185–193, 1982.

Kasa, P.: Ultrastructural localization of choline acetyltransferase and acetylcholinesterase in central and peripheral tissue. In: "Histochemistry of Nervous Transmission," O. Eranko, Ed., pp. 337–344, Progr. Brain Res., vol. 34, Elsevier, Amsterdam, 1971.

Kennedy, E. P. and Weiss, S. B.: The function of fytidine coenzymes in the biosynthesis of phospholipids. J. Biol. Chem. *222*: 193–214, 1956.

Kewitz, H.: Nachweis von 4-Amino-n-butyrylcholin in Warmblutergehirn. Naunyn-Schmiedeberg's Arch. Exp. Path. Pharmakol. *237*: 308–315, 1959.

Kewitz, H. and Pleul, O.: Synthesis of choline in the brain. In: "Cholinergic Mechanisms," G. Pepeu and H. Ladinsky, Eds., pp. 405–413, Plenum Press, New York, 1981.

Khandelwal, J. K., Szilagyi, P. I., Barker, L. A. and Green, J. P.: Simultaneous measurement of acetylcholine and choline in brain by pyrolysis-gas chromatography-mass spectrometry. Eur. J. Pharmacol. 1981 Dec 3;76(2–3): 145–156.

Kim, M.-H. and Hersh, L. B.: Vesicular acetylcholine transporter trafficking and adaptor proteins. In: "Cholinergic Mechanisms—Function and Dysfunction," I. Silman, H. Soreq, L. Anglister, D. Michaelson and A. Fisher, Eds., Martin Dunitz, London, 2004.

Kindel, G. H. and Karczmar, A. G.: Effect of single and repeated choline administration on brain choline, acetylcholine and acetylcholine turnover of CF 1 mice. Fed. Proc. *40*: 269, 1981.

Kindel, G. H. and Karczmar, A. G.: Effect of choline administration on brain choline acetylcholine and acetylcholine turnover of young and old mice. Fed. Proc. *41*: 1323, 1982.

King, R. G. and Marchbanks, R. M.: The incorporatioan of solubilized choline-transport activity into liposomes. Biochem. J. *204*: 565–576, 1982.

Kirschner, L. B.: Effect of cholinesterase inhibitors and atropine on active ion transport across frog skin. Nature *172*: 348–349, 1953.

Klein, J., Koppen, A. and Loeffelholz, K.: Small rises in plasma choline reverse the negative arteriovenous difference of brain choline. J. Neurochem. *55*: 1231–1236, 1990.

Klein, J., Koppen, A. and Loeffelholz, K.: Regulation of free choline in rat brain: dietary and pharmacological manipulations. Neurochem. Int. *32*: 479–485, 1998.

Klein, J., Buchholzer, M.-L., Kopf, S. and Loeffelholz, K.: Control of acetylcholine release under stimulatory conditions by its biosynthetic precursors, glucose and choline. In: "XIth Cholinergic

Symposium on Cholinergic Mechanisms—Function and Dysfunction," I. Silman, A. Soreq, L. Anglister, D. Michaelson, and A. Fisher, Eds., Martin Dunitz, London, 2004.

Koc, H., Mar, M. H., Ranasinghe, A., Swenberg, J. A. and Zeisel, S. H.: Quantitation of choline and its metabolites in tissues and foods by liquid chromatography/electrospray ionization-isotope dilution mass spectrometry. Anal. Chem. *74*: 4734–4740, 2002.

Koelle, G. B.: Cytological distribution and physiological functions of cholinesterases. In: "Cholinesterases and Anticholinesterase Agents," G. B. Koelle, Ed., pp. 187–298, Handbh. d. exper. Pharmakol., Ergänzungswk., vol. 15, Springer-Verlag, Berlin, 1963.

Koelle, G. B., Koelle, E. S. and Friedenwald, J. S.: The effect of inhibition of specific and non-specific cholinesterase on the motility of the isolated ileum. J. Pharmacol. Exper. Therap. *100*: 180–191, 1950.

Koelle, G. B., Rickard, K. K. and Ruch, G. A.: Interrelationships between ganglionic acetylcholinesterase and non-specific cholinesterase of the cat and rat. Proc. Natl. Acad. Sci. U.S.A. *76*: 60612–60616, 1979.

Koshland, D. E.: Correlation of function and structure in enzyme action. Science *142*: 1533–1541, 1963.

Kosterlitz, H. W., Lees, G. M. and Wallis, D. I.: Resting and action potentials recorded by the sucrose-gap method in the superior cervical ganglion of the rabbit. J. Physiol. *195*: 39–53, 1968.

Kovarik, Z., Radic, Z., Berman, H. A., Simeon-Rudolf, V., Reiner, E. and Taylor P.: Mutant cholinesterases possessing enhanced capacity for reactivation of their phosphonylated conjugates. Biochem. *43*: 3222–3229, 2004.

Krejci, E., Duval, N., Chatonnet, A., Vincent, P. and Massoulié, J.: Cholinesterase-like domains in enzymes and structural proteins: functional and evolutionary relationships and identification of catalytically essential aspartic acid. Proc. Natl. Acad. Sci. U.S.A. *88*: 6647–6651, 1991.

Krupka, R. M.: Chemical structure and function of the active center of acetylcholinesterase. Biochem. *5*: 1988–1998, 1966.

Krupka, R. M. and Laidler, K. J.: Molecular mechanisms for hydrolytic enzyme action. I. Apparent noncompetitive inhibition, with special reference to acetylcholinesterase. II. Inhibition of acetylcholinesterase by excess substrate. III. A general mechanism for inhibition of acetylcholinesterase. IV. The structure of the active center and the reaction mechanism. J. Amer. Chem. Soc. *83*: 1445–1460, 1961.

Ksiezak-Reding, H. J. and Goldberg, A. M.: The role of chloride in acetylcholine metabolism. J. Neurochem. *38*: 121–129, 1982.

Kuhar, M. J., Sethy, V. H., Roth, R. H. and Aghajanian, G. K.: Choline: selective accumulation by central cholinergic neurons. J. Neurochem. *20*: 581–593, 1973.

Kuntscherova, J.: Ein Beitrag zur Klarstellung der Acetylcholinumsatzstorung in den Geweben weisser Ratten mit partiellen Pankreatektomie. Acta Physiol. Hung. *39*: 57–61, 1971.

Kuntscherova, J.: Effect of short-term starvation and choline on the acetylcholine content of organs of albino rats. Physiol. Bohemoslov. *21*: 655–660, 1972.

Kuntscherova, J. and Vlk, J.: Über die Bedeutung der Cholin und Glukosezufuhr für die Normalisierung der durch Hungern herabgesetzten Acetylcholinvorrate in den Geweben. Plzen. lek. Sborn. *31*: 5–12, 1968.

Kushland, D. E.: The active sites and enzyme action. Adv. Enzymol. *22*: 45–97, 1960.

La Du, B. N.: Butyrylcholinesterase variants and the new methods of molecular biology. Acta Anaesthesiol. Scand. *39*: 11139–11141, 1995.

La Du, B. N., Bartels, C. F., Nogueira, C. P., Arpagaus, M. and Lockridge, O.: Proposed nomenclature for human butyrylcholinesterase genetic variants identified by DNA sequencing. Cell Mol. Neurobiol. *11*: 79–89, 1991.

Langenbuch-Cachat, J., Bon, C., Mulle, C., Goeldner, M., Hirth, C. and Changeux, J.-P.: Photoaffinity labeling of the acetylcholine binding sites on the nicotinic receptor by an aryldiazonium derivative. Biochem. *27*: 2337–2345, 1988.

Lauder, J. M. and Schambra, U. B.: Morphogenetic roles of acetylcholine. Environ. Health Perspect. *107,* Supplt.: 65–69, 1999.

Le Heüx, J. W.: Cholin als Hormon der Darmbewegung Pflüger. Arch. Ges. Physiol. *173*: 8–27, 1919.

Leuzinger, W. and Baker, A. L.: Acetylcholinesterase. Proc. Natl. Acad. Sci. U.S.A. *57*: 446–451, 1967.

Li, J. Y., Volknandt, W., Dahlstrom, A., Herrmann, C., Blasi, J., Das, B. and Zimmermann, H.: Axonal transport of ribonucleoprotein particles (vaults). Neuroscience. *91*: 1055–1065, 1999.

Lipmann, F. and Kaplan, N.O.: A common factor in the enzymatic acetylation of sulfanilamide and of choline. J. Biol. Chem. *162*: 743–744, 1946.

Loeffelholz, K.: Receptor regulation of choline phospholipid hydrolysis. A novel source of diacylglycerol and phosphatidic acid. Biochem. Pharmacol. *38*: 1543–1549, 1989.

Loeffelholz, K.: Brain choline has a typical precursor profile. In: "From *Torpedo* Electric Organ to Human Brain: Fundamental and Applied Aspects," J. Massoulié, Ed., pp. 235–239, Elsevier, Amsterdam, 1998.

Loeffelholz, K. and Klein, J.: How is the brain supplied with choline? . . . But how is it protected against excess choline concentrations? In: "XIth Cholinergic Symposium on Cholinergic Mechanisms—Function and Dysfunction," I. Silman, H. Soreq, L. Anglister, D. Michaelson, and A. Fisher, Eds., pp. 395–398, Martin Dunitz, London, 2004.

Loeffelholz, K., Klein, J. and Koeppen, A.: Choline, a precursor of acetylcholine and phospholipids in the brain. Progr. Brain Res. *98*: 199–201, 1993.

Loewi, O.: Uber die humorale Ubertragbarkeit der Herzenwirkung. I. Pfluegers Arch. Geasamte Physiol. *189*: 239–242, 1921.

Loewi, O. and Navratil, E.: Über humorale Übertragbarkeit der Herzenswirkung. X. Über das Schicksal des Vagusstoffes. Pflügers Arch. Ges. Physiol. *214:* 678–688, 1926.

Loewi, O.: An autobiographic sketch. Perspect. Biol. Med. *4*: 3–25, 1960.

London, E. D. and Coyle, J. T.: Pharmacological augmentation of acetylcholine levels in rainats lesioned rat striatumo Biochem. Pharmacol. *27*: 2962–2965, 1978.

Long, J. P. and Schueler, F. W.: A new series of cholinesterase inhibitors. J. Am. Pharm. Assoc. Am. Pharm. Assoc. (Baltim). *43*: 79, 1954.

Long, J. P.: Structure-activity relationships of the reversible anticholinesterase agents. In: "Cholinesterases and Anticholinesterase Agents," G. B. Koelle, Ed., pp. 374–427, Handbch. d. exper. Pharmakol., Ergänzungswk., vol. 15, Springer-Verlag, Berlin, 1963.

Low, M. G.: Glycosylphosphatidylinositol: a versatile anchor of cell surface proteins. FASEB J. *3*: 1600–1608, 1989.

Lubinska, L.: Axoplasmic streaming in regenerating and in normal nerve fibres. Prog. Brain Res. *13*: 1–71, 1964.

Lubinska, L.: On axoplasmic flow. Int. Rev. Neurobiol. *17*: 241–296, 1975.

Lucinei Balbo, S., Gravena, C., Bonfleur, M. L. and de Freitas Mathias, P. C.: Insulin secretion and acetylcholinesterase activity in monosodium-l-glutamate-induced obese mice. Horm. Res. *54*: 1186–1191, 2000.

MacIntosh, F. C.: L'effet de la section des fibres préganglionnaires sur la teneur en acétylcholine du ganglion sympathique. Arch. Int. Physiol. *47*: 321–324, 1938.

MacIntosh, F. C: Formation, storage, and release of acetylcholine at nerve endings. Can. J. Biochem. Physiol. *37*: 3343–3356, 1959.

MacIntosh, F. C.: Are acetylcholine levels related to acetylcholine release? In: "Choline and Lecithin in Brain Disorders," A. Barbeau, J. H. Growdon and R. J. Wurtman, Eds., pp. 201–217, Nutrition and the Brain, vol. 5, Raven Press, New York, 1979.

Main, A. R.: Structure and inhibitors of cholinesterase. In: "Biology of Cholinergic Function," A. M. Goldberg and I. Hanin, Eds., pp. 269–353, Raven Press, New York, 1976.

Maire, J.-C. and Wurtman, R. J.: Effects of electrical stimulation and choline availability on the release and contents of acetylcholine and choline in superfused slices from rat striatum. J. Physiol. (Paris) *80*: 189, 1985.

Maler, L., Lollins, M. and Mathieson, W. B.: The distribution of acetylcholinesterase and choline acetyl transferase in the cerebellum and posterior lateral line lobe of weably electric fish (Gymnotidae). Brain Res. *226*: 320–325, 1981.

Mallender, W. D., Szegletes, T. and Rosenberry, T. L.: Acetylthiocholine binds to asp74 at the peripheral site of human acetylcholinesterase as the first step in the catalytic pathway. Biochemistry. *39*: 7753–7763, 2000.

Mallet, J., Berrard, S., Brice, A., Habert, E., Raynaud, B., Vernier, P. and Weber, M.: Choline acetyltransferase: a molecular genetic approach. In: "Cholinergic Neurotransmission: Functional and Clinical Aspects," S.-M. Aquilonius and P.-G. Gillberg, Eds., pp. 3–10, Elsevier, Amsterdam, 1990.

Mallet, J., Houhou, L., Pajak, F., Oda, O., Cervini, R., Bejanin, S. and Berrard, S.: The cholinergic locus: ChAT and VAChT genes. In: "From *Torpedo* Electric Organ to Human Brain: Fundamental and Applied Aspects," J. Massoulié, Ed., pp. 145–147, Elsevier, Amsterdam, 1998.

Malthe Sorensen, D.: Choline acetyltransferase evidence for acetyl transfer by a residence residue. J. Neurochem. *27*: 873–881, 1976.

Malthe Sorenssen, D.: Recent progress in the biochemistry of choline acetyltransferase. In: "The Cholinergic Synapse," S. Tucek, Ed., pp. 45–58, Progr. Brain Res., vol. 49, Elsevier, Amsterdam, 1979.

Mann, P. J. G., Tennenbaum, M. and Quastel, J. H.: On the mechanism of acetylcholine formation in brain *in vitro*. Biochem. J. *32*: 243–261, 1938.

Mann, P. J. G., Tennenbaum, M. and Quastel, J. H.: Acetylcholine metabolism in the central nervous system. The effects of potassium and other cations on acetylcholine liberation. Biochem. J. *33*: 822–835, 1939.

Mann, S. P. and Hebb, C.: Free choline in the brain of the rat. J. Neurochem. *28*: 241, 1977.

Markert, C. L.: The molecular basis for isozymes. Ann. N.Y. Acad. Sci. *115*: 14–40, 1968.

Martin, K.: Active transport of choline into human erythrocytes. J. Physiol. *191*: 105P–106P, 1967.

Martin, K.: Concentrative accumulation of choline by human erythrocytes. J. Gen. Physiol. *51*: 497–516, 1968.

Martin, K.: Extracellular cations and the movement of choline cross the erythrocyte membrane. J. Physiol. *224*: 207–230, 1972.

Masson, P., Chatonnet, A. and Lockridge, O.: Evidence for single butyrylcholinesterase gene in individuals carrying the plasma cholinesterase variant (CHE2). FEBS Lett. *262*: 115–118, 1990.

Massoulié, J.: Molecular forms and anchoring of acetylcholinesterase. In: "Cholinesterases and Cholinesterase Inhibitors," E. Giacobini, Ed., pp. 81–101, Martin Dunitz, London, 2000.

Massoulié, J. and Rieger, F.: L'acétylcholinèsterase des organs electriques de poisons (torpille et gymnote), complex membranaires. Eur. J. Biochem. *11*: 441–445, 1969.

Massoulié, J. and Bon, S.: Processing and anchoring of cholinesterases in muscle and brain. In: "Cholinergic Mechanisms. Function and Dysfunction," I. Silman, H. Soreq, L. Anglister, D. Michaolson, and A. Fisher, Eds., pp. 155–169, Martin Dunitz, London, 2004.

Massoulié. J., Rieger, F. and Bon, S.: Relations between molecular complexes of acetylcholinesterase. C. R. Acad. Sci. Hebd. Seances Acad. Sci. *270*: 1837–1840, 1970.

Massoulié, J., Pezzementi, L., Bon, S., Krejci, E. and Vallette, F.-M.: Molecular and cellular biology of cholinesterases. Progr. Neurobiol. *41*: 31–91, 1993a.

Massoulié, J., Sussman, J., Bon, S. and Silman, I.: Structure and functions of acetylcholinesterase and butyrylcholinesterase. In: "Cholinergic Function and Dysfunction," A. C. Cuello, Ed., pp. 139–146, Elsevier, Amsterdam, 1993b.

Massoulié, J., Anselmet, A., Bon, S., Krejci, E., Legay, C., Morel, N. and Simon, N.: Acetylcholinesterase: C-terminal domains, molecular forms and functional localization. In: "From *Torpedo* Electric Organ to Human Brain: fundamental and Applied Aspects," J. Massoulié, Ed., pp. 183–190. Elsevier, Amsterdam, 1998.

Massoulié, J., Anselmet, A., Bon, S., Krejci, E., Legay, C., Morel, N. and Simon, S.: The polymorphism of acetylcholinesterase: post-translational processing, quaternary associations and localization. Chem. Biol. Interact. *119–120*: 29–42, 1999.

Maul, J. D. and Farris, J. L.: The effect of sex on avian plasma cholinesterase activity: a potential source of variation in avian biomarker endpoint. Arch. Environmt. Contamin. & Toxicol. *47*: 253–258, 2004.

McCaman, R. E., Rodriguez de Lores Arnaiz, G. and De Robertis, E.: Species differences in subcellular distribution of choline acetylase in the CNS. J. Neurochem. *13*: 927–935, 1965.

McCarty, L. P., Knight, A. S. and Chenoweth, M. B.: Incorporation of [14C] choline into phospholipids in the isolated phrenic nerve diaphragm preparation of the rat. J. Neurochem. *20*: 487–494, 1973.

McGeer, P. L., McGeer, E. G. and Feng, J. H.: Choline acetyltransferase: purification and immunohistochemical localization. Life Sci. *34*: 2319–2338, 1987.

McMahan, U. J., Sanes, J. R. and Marshall, L. M.: Cholinesterase is associated with the basal lamina at the neuromuscular junction. Nature *271*: 172–174, 1978.

McMurray, W. C.: Metabolism of phosphatides in developing rat brain. Incorporation of radio active precursors. J. Neurochem. *11*: 287–299, 1964.

Mendel, B. and Rudney, H.: Studies on cholinesterase. I. Cholinesterase and pseudo-cholinesterase. Biochem. J. *37*: 59–63, 1943.

Mendel, B. and Rudney, H.: The cholinesterases in the light of recent findings. Science *100*: 499–500, 1944.

Meyerhof, O.: Zur Energetik der Zellvorgänge. Vandenhoeck und Ruprecht, Berlin, 1913.

Meyerhof, O.: "Die chemischen Vorgänge im Muskel und ihr Zusammenhang mit Arbeitleistung und Warmebildung," Berlin, 1930.

Meyerhof. O.: Oxidoreduction in carbohydrate breakdown. Biol. Symp. *5*: 141–156, 1941.

Miani, N.: Evidence of a proximo-distal movement along the axon of phospholipid synthesized in the nerve-cell body. Nature. *193*: 887–888, 1962.

Michelson, M. J. and Zeimal, E. V.: Acetylcholin. Nauka, Leningrad, (in Russian), 1970.

Minic, J., Chatonnet, A., Krejci, E. and Molgo, J.: Butyrylcholinesterase and acetylcholinesterase activity and quantal transmitter release at normal and acetylcholinesterase knockout mouse neuromuscular junction. Brit. J. Pharmacol. *11138*: 177–187, 2003.

Morel, N.: Effect of choline on the rates of synthesis and of release of acetylcholine in the electric organ of Torpedo. J. Neurochem. *27*: 779–784, 1976.

Morfini, G., Stenoien, D. L. and Brady, S. T.: Axonal transport. In: "Basic Neurochemistry", G. J. Siegel, Ed., in Chief, pp. 485–501, Elsevier, Amsterdam, 7th Edition, 2006.

Mounter, L. A. and Whittaker, V. P.: The hydrolysis of esters of phenol by cholinesterases and other esterases. Biochem J. *54*: 551–559, 1953.

Mueller, R. A., Thoenen, H. and Axelrod, J.: Inhibition of trans-synaptically increased tyrosine hydroxylase activity by cyclohexinoide and actinomycin D. Molec. Pharmacol. *5*: 463–469, 1969.

Mueller, R. A., Thoenen, H. and Axelrod, J.: Inhibition of neurally induced tyrosine hydroxylase by nicotinic receptor blockade. Eur. J. Pharmacol. *10*: 51–56, 1970.

Muralt, V. A.: "Die Signalübermittlung im Nerven," Verlag Birkhauser, Basel, 1946.

Murrin, L. C. and Kuhar, M. J.: Activation of high-affinity choline uptake *in vitro* by depolarizing agents. Mol. Pharmacol. *12*: 1082–1090, 1976.

Nachmansohn, D.: Choline acetylase. In: "Cholinesterases and Anticholinesterase Agents," G. B. Koelle, Ed., pp. 40–54, Handbch. d. exper. Pharmakol., Ergänzungswk, vol. 15, Springer-Verlag, Berlin, 1963a.

Nachmansohn, D.: Actions on axons, and evidence for the role of acetylcholine in axonal conduction. In: "Cholinesterases and Anticholinesterase Agents," G. B. Koelle, Ed., pp. 799–832, Handbch. d. exper. Pharmakol., Ergänzungswk., vol. 15, Springer-Verlag, Berlin, 1963b.

Nachmansohn, D. and Machado, A. L.: The formation of acetylcholine. A new enzyme, "choline acetylase." J. Neurophysiol. *6*: 397–403, 1943.

Nachmansohn, D. and Rothenberg, M. A.: On the specificity of cholinesterase in nervous tissue, Science. *100*: 454–455, 1944.

Nachmansohn, D. and Wilson, I. B.: Acetylcholinesterase. In: "Methods in Enzymology," S. P. Colowick and N. O. Kaplan, Eds., vol. 1, pp. 259–333, Academic Press, New York, 1955.

Neville, L. F., Gnatt, A., Loewenstein, Y. and Soreq, H.: Aspartate-to-glycine substitution confers resistance to naturally occurring and synthetic anionic-site ligands on *in ovo* produced butyrylcholinesterase. J. Neurosci. Res. *27*: 452–460, 1990a.

Neville, L. F., Gnatt, A., Paddam, R., Seidman, S. and Soreq, H.: Anionic site interactions in human butyrylcholinesterase disrupted by single point mutation. J. Biol. Chem. *265*: 20735–20738, 1990b.

Nikolau, B. J., Oliver, D. J., Schnable, P. S. and Wurtele, E. S.: Molecular biology of acetyl-CoA metabolism. Biochem. Soc. Trans. *28*: 591–593, 2000.

Ochs, S.: Systems of material transport in nerve fibers (axoplasmic transport) related to nerve function and trophic control. Ann. N.Y. Acad. Sci. *228*: 202–223, 1974.

Okuda, T. and Haga, T.: High-affinity choline transporter. Neurochem Res. *28*: 483–488, 2003.

Ollis, D. L., Cheah, E., Cygler, M., Dijkstra, B., Frolow, F., Franken, S. M., Harel., M., Remington, S. J., Silman, I. and Schrag, J.: The alpha/beta hydrolase fold. Protein Eng. *5*: 197–211, 1992.

Oosterbaan, R. A.: Constitution of DFP enzymes. In: "Proceedings of the Conference on the Structure and Reactions of DFP Sensitive Enzymes," E. Heilbronn, Ed., pp. 25–36, Swedish Research Institute of National Defense, Stockholm, 1967.

Ordentlich, A., Kronman, C., Barak, D., Stein, D., Ariel, N., Marcus, D., Velan, B. and Shafferman, A.: Engineering resistance to 'aging' of phosphorylated human acetylcholinesterase. Role of hydrogen bond network in the active center. FEBS Lett. *334*: 215–220, 1993.

Pakasi, M. and Kasa, P.: Glial cells in coculture can increase the acetyl cholinesterase activity in human brain endothelial cells. Neurochem. Int. *21*: 129–133, 1992.

Paoletti, F., Mocali, A. and Vannucchi, A. M.: Acetylcholinesterase in murine erythroleukemia (Friend) cells: evidence for megakaryocyte-like expression and potential growth-regulatory role of enzyme activity. Blood. *79*: 2873–2879, 1992.

Pardridge, W. M.: Transport of nutrients and hormones through the blood-brain barrier. Fed. Proc. *43*: 201–204, 1984.

Parducz, A., Kiss, Z. and Joo, F.: Changes of phosphatidylcholine content and the number of synaptic vesicles in relation to neurohumoral transmission in sympathetic ganglia. Experimentia *32*: 1520, 1976.

Patinkin, D., Seidman, S., Eckstein, F., Benseler, F., Zakut, H. and Soreq, H.: Manipulations of cholinesterase gene expression modulate murine megakaryocytopoiesis *in vitro*. Mol. Cell Biol. *10*: 6046–6050, 1990.

Pedata, F., Wieraszko, A. and Pepeu, G.: Effect of choline, phosphorylcholine and diethylaminoethanol on brain acetylcholine level in the rat. Pharmacol. Res. Commun. *9*: 755–762, 1977.

Peng, H. B., Xie, H., Rossi, S. G. and Rotundo, R. L.: Acetylcholinesterase clustering at the neuromuscular junction involves perlecan and dystroglycan. J. Cell. Bio. *145*: 911–921, 1999.

Perrier, N. A., Kherif, S., Perrier, A. L., Dumas, S., Mallet, J. and Massoulié, J.: Expression of PRiMA in the mouse brain: membrane anchoring and accumulation of "tailed" acetylcholinesterase. Eur. J. Neurosci. *18*: 1837–1847, 2003.

Perry, W. L. M.: Acetylcholine release in the cat's superior cervical ganglion. J. Physiol. (Lond.) *119*: 439–454, 1953.

Pert, C. B. and Snyder, S. H.: High affinity transport of choline into the myenteric plexus of guinea-pig intestine. J. Pharmacol. Exp. Ther. *191*: 102–108, 1974.

Pfeiffer, C. C.: Parasympathetic neurohormones: possible precursors and effect on behavior. Int. Rev. Neurobiol. *1*: 195–244, 1959.

Pfeiffer, C. C. and Jenney, E. H.: The inhibition of conditioned response and the counter action of schizophrenia by muscarinic stimulation of the brain. Ann. N.Y. Acad. Sci. *66*: 753–754, 1957.

Pieklik, J. R. and Buynn, R. W.: Equilibrium constants of the reactions of choline acetyltransferase, carnitine acetyltransferase, and acetylcholinesterase under physiological conditions. J. Biol. Chem. *259*: 4445–4450, 1975.

Plattner, F. and Hintner, H.: Die Spaltung von Acetylcholin durch Organextracte and Korperflüssigkeiten. Pflugers Arch. Ges. Physiol. *255*: 19–25, 1930.

Potter, L. T.: Synthesis, storage and release of [14C] acetylcholine in isolated rat diaphragm muscles. J. Physiol. (Lond.) *206*: 145–166, 1970.

Potter, L. T., Glover, V. A. S. and Saelens, J. K.: Choline acetyltransferase from rat brain. J. Biol. Chem. *243*: 3864–3870, 1968.

Pravalorio, M. and Fournier, D.: *Drosophila* acetylcholinesterase: characterization of different mutants resistant to insecticides. Biochem. Genet. *30*: 77–83, 1992.

Primo-Parmo, S. L., Bartels, C. F., Wiersema, B., van der Spek, A. F., Innis, J. W. and La Du, B. N.: Characterization of 12 silent alleles of the human butyrylcholinesterase (BCHE) gene. Am. J. Hum. Genet. *58*: 52–64, 1996.

Primo-Parmo, S. L., Lighstone, H. and La Du, B.: Characterization of an unstable variant (BChE115D) of human butyrylcholinesterase. Pharmacogen. *7*: 27–34, 1997.

Quastel, J. H., Tennenbaum, M. and Wheatley, A. H. M.: Choline ester formation in, and choline esterase activities of tissues *in vitro*. Biochem. J. *30*: 1668–1681, 1936.

Quinn, D. M.: Acetylcholine: enzyme structure, reaction dynamics and virtual transition states. Chem. Rev. *87*: 955–979, 1987.

Racagni, G., Cheney, D. L., Zsilla, G. and Costa. E.: The measurement of acetylcholine turnover rate in brain structures. Neuropharmacology. *15*: 723–736, 1976.

Radic, Z., Manetsch, R., Krasinski, A., Raushel, J., Yamauchi, J., Garcia, C., Kolb, H., Sharpless, K. B. and Taylor, P.: Molecular basis of interactions of cholinesterases with tight binding inhibitors. Chem. Biol. Interact. *157–158*: 133–141, 2005.

Reinhardt, R. R. and Wecker, L.: Regulation of choline phosphorylation in rat striatum. In: "Neurobiology of Acetylcholine," N. J. Dun and R. L. Perlman, Eds., pp. 145–157, Plenum Press, New York, 1987.

Richardson, J. S.: Describing patterns of protein tertiary structure. Methods Enzymol. *115*: 341–358, 1985.

Richelson, E. and Thompson, E. J.: Transport of neurotransmitter precursors into cultured cells. Nat New Biol. 1973 Feb 14; *241*(111): 201–204.

Ricny, J. and Tucek, S.: Relation between the content of acetyl-coenzyme A and acetylcholine in brain slices. Biochem. J. *188*: 683–688, 1980.

Rosenberry, T. L.: Acetylcholinesterase. Adv. Enzym. *43*: 103–218, 1975.

Rosenberry, T. L., Mallender, W. D., Thomas, P. J. and Szegeltes, T.: A steric model for inhibition of acetylcholinesterase by peripheral site ligands and substrate. Chem. Biol. Interact. *119–120*: 85–97, 1999.

Rossi, A., Vicini, F., Scarsella, G. and Biagioni, S.: Acetylcholinesterase distribution in subpopulations of murine thymocyte. J. Neurosci. Res. *29*: 201–206, 1991.

Rossier, J.: Purification of rat brain choline acetyltransferase. J. Neurochem. *26*: 543–548, 1976a.

Rossier, J.: Biophysical properties of rat brain choline acetyltransferase. J. Neurochem. *26*: 555–559, 1976b.

Rossier, J. and Benda, P.: Activation of choline acetyltransferase by chloride: a possible regulatory mechanism. In: "Cholinergic Mechanisms and Psychopharmacology," D. J. Jenden, Ed., pp. 207–221, Plenum Press, New York, 1978.

Rossiter, R. J. and Strickland, K. P.: Metabolism of phosphoglycorides. Handbk. Neurochem. A. Lajtha, Ed., pp. 467–489, vol. 3, Plenum Press, 1970.

Rostas, J. A. P., McGregor, A., Jeffrey, P. L. and Austin, L.: Transport of cholesterol in the chick optic system. J. Neurochem. *24*: 295–302, 1975.

Rotundo, R. L.: Purification and properties of the membrane-bound form of acetylcholinesterase from chicken brain. Evidence for two distinct polypeptide chains. J. Biol. Chem. *259*: 13186–13194, 1984.

Rotundo, R. L.: Nucleus-specific translation and assembly of acetylcholinesterase in multinucleated muscle cells. J. Cell. Biol. *110*: 715–719, 1990.

Rotundo, R. L. and Carbonetto, S. T.: Neurons segregate clusters of membrane-bound acetylcholinesterase along their neurites. Proc. Natl. Acad. Sci. U.S.A. *84*: 2063–2067, 1987.

Rotundo, R. L., Rossi, S. G., Kimbell, L. M., Ruiz, C. and Marrero, E.: Targeting acetylcholinesterase to the neuromuscular synapse. Chem. Biol. Interact. *157–158*: 15–21, 2005.

Rylett, R. J.: Further studies on the actions of choline mustard aziridinium ion at the cholinergic nerve terminal. In: "Cellular and Molecular Basis of Cholinergic Function," M. J. Dowdall and J. N. Hawthorne, Eds., pp. 646–650, Ellis Horwood, Chichester, 1987.

Rylett, R. J. and Schmidt, B. M.: Regulation of the synthesis of acetylcholine. In: "Cholinergic Function and Dysfunction," A. C. Cuello, Ed., pp. 161–166, Elsevier, Amsterdam, 1993.

Saelens, J. K., Simke, J. P., Schuman, J. and Allen, M. P.: Studies with agents which influence acetylcholine metabolism in mouse brain. Arch. Int. Pharmacodyn. Ther. *209*: 250–258, 1974.

Saito, M., Kindel, G., Karczmar, A. G. and Rosenberg, A.: Metabolism of choline in brain of the aged CBF1 mouse. J. Neurosci. Res. *15*: 197–204, 1986.

Sakuragawa, N., Misawa, H., Ohsugi, K., Kakishita, K., Ishii, T., Thangavel, R., Tohyama, J., Elwan, M., Yokoyama, Y., Okuda, O., Arai, H., Oguno, L. and Sato, K.: Evidence for active acetylcholine metabolism in human amniotic cells: applicable to intracerebral allografting for neurological disease. Neurosci. *232*: 53–56, 1997.

Salerno, D. M. and Beeler, D. A.: The biosynthesis of phospholipids and their precursors in rat liver involving de novo methylation, and base-exchange pathways, *in vivo*. Biochim. Biophys. Acta. *326*: 325–338, 1973.

Sasaki, T., Kawamura, K., Tanaka, Y., Aando, S. and Senda, M.: Assessment of choline uptake for the synthesis and release of acetylcholine in brain slices by a dynamic autoradiographic technique using [11C] choline. Brain Res. Brain Res. Protoc. *10*: 1–11, 2002.

Sastry, B. V.: Human placental cholinergic system. Biochem. Pharmacol. *53*: 1577–1586, 1997.

Satoh, T., Taylor, P., Bosron, W. F., Shanghani, S. P., Hosokawa, M. and La Du, B. N.: Current progress on esterases: from molecular structure to function. Drug Metab. Dispos. *30*: 488–493, 2002.

Scally, M. C., Ulus, I. H. and Wurtman, R. J.: Enhancement by choline of the induction of adrenal tyrosine hydroxylase by phenoxybenzamine, 6-hydroxydopamine, insulin or exposure to cold. J. Neural Transm. *204*: 676–682, 1978.

Schmitt, F. O.: Fibrous proteins–neuronal organelles. Proc. Natl. Acad. Sci. USA. *60*: 1092–1092, 1968.

Schober, A., Minichiello, L., Keller, M., Huber, K., Layer, P. G., Raig-Lopoz, J. L.; Garcia-Arraws, J. E., Wein, R. and Unsisher, K.: Reduced acetylcholinesterase activity in adrenal medulla and loss of sympathetic preganglionic neurons in TrKA-deficient but not TrKB-deficient mice. J. Neurosci. *17*: 891–903, 1997.

Schueler, F. W.: Two cyclic analogs of acetylcholine. J. Amer. Pharm. Ass. Sci E. *45*: 197–199, 1956.

Schumacher, M., M., Camp, S., Maulet, Y., Newton, M. MacPhee-Quigley, K., Taylor, S. S., Friedman, T. and Taylor, P.: Primary structure of *Torpedo californica* acetylcholinesterase deduced from cDNA sequence. Nature *319*: 407–409, 1986.

Sethy, V. H., Kuhar, M. J., Roth, R. H., Van Woert, M. H., Aghajanian, G. K.: Cholinergic neurons: effect of acute septal lesion on acetylcholine and choline content of rat hippocampus. Brain Res. *55*: 481–484, 1973.

Shafferman, A. and Velan, B., Eds.: "Multidisciplinary Approaches to Cholinesterase Functions," Plenum Press, New York, 1992.

Shafferman, A., Velan, B., Ordentlich, A., Kronman, C., Grosfeld, H., Leitner, M., Flashner, Y., Cohen, S., Barak, D. and Ariel, N.: Substrate inhibition of acetylcholinesterase: residues affecting signal transduction from the surface to the catalytic center. EMBO J. *11*: 3561–3568, 1992.

Shafferman, A., Ordentlich, A., Barak, D., Kronman, C., Ber, R. and Bino, B.: Electrostatic attraction by surface charge does not contribute to the catalytic efficiency of acetylcholinesterase. EMBO J. *13*: 3338–3455, 1994.

Shea, P. A. and Aprison, M. H.: The distribution of acetyl-CoA in specific areas of the CNS of the rat as measured by a modification of a radiometric assay for acetylcholine and choline. J. Neurochem. *28*: 51–58, 1977.

Sheard, N. F., Tayek, J. A., Bistrian, B. R., Blackburn, G. L. and Zeisel, S. H.: Plasma choline concentration in humans fed parenterally. Am. J. Clin. Nutr. *43*: 219–224, 1986.

Shelley, H.: A correlation between cholinesterase inhibition and increase in the muscle tone in rabbit duodenum. Brit. J. Pharamcol. *10*: 26–35, 1955.

Siegel, G. J., Ed.-in-Chief: "Basic Neurochemistry," Sixth Edition, Lippincott-Raven Publisher, Philadelphia 1999.

Sikorav, J. L., Krejci, E. and Massoulié, J.: cDNA sequences of Torpedo marmorata acetylcholinesterase: primary structure of the precursor of a catalytic subunit; existence of multiple 5 untranslated regions. EMBO J. *6*: 1865–1873, 1987.

Sikorav, J. L., Duval, N., Anselmet, A., Bon, S., Krejci, E., Legay, C., Osterlund, M., Reimund, B. and Massoulié, J.: Complex alternative splicing of acetylcholinesterase transcripts in Torpedo electric organ: primary structure of the precursor of the glycolipid-anchored dimeric form. EMBO J. *7*: 2983–2993, 1988.

Silman, I., H. Soreq, L. Anglister, D. Michaelson, and A. Fisher, Eds., "Cholinergic Mechanisms—Function and Dysfunction," Martin Dunitz, London, 2004.

Silman, I. and Futerman, A. H.: Modes of attachment of acetylcholinesterase to the surface membrane. Eur. J. Biochem. *1170*: 11–22, 1987.

Silman, I. and Sussman, J. L.: Structural studies on cholinesterases. In: "Cholinesterases and Cholinesterase Inhibitors," E. Giacobini, Ed., pp. 9–25, Martin Dunitz, London, 2000.

Silver, A.: The biology of cholinesterases. In: "Frontiers of Biology," A. Neuberger and E. L. Tatum, Eds., pp. 1–596, North-Holland, Amsterdam, 1974.

Simon, J. R., Atweh, S. and Kuhar, M. J.: Sodium-dependent high affinity choline uptake: a regulatory step in the synthesis of acetylcholine. J. Neurochem. *26*: 909–922, 1976.

Sitaram, N., Weingartner, H. and Gillin, J. C.: Choline chloride and arecoline: effects on memory and sleep in man. In: "Choline and Lecithin in Brain Disorders," A. Barbeau, J. H. Growdon and R. J. Wurtman, Eds., pp. 367–375, Nutrition and the Brain, vol. 5, Raven Press, New York, 1979.

Smith, C. P. and Carroll, P. T.: A comparison of solubilized and membrane bound forms of choline-o-acetyltransferase (EC2.3.1.6) in mouse brain nerve endings. Brain Res. *185*: 363–371, 1980.

Smith, J. E., Lane, J. D., Shea, P. A., McBride, W. J. and Aprison, M. H.: A method for concurrent measurement of picomole quantities of acetylcholine, choline, dopamine, norepinephrine, serotonin, 5-hydroxytryptophan, 5-hydroxyindoleacetic acid, tryptophan, tyrosine, glycine, aspartate, glutamate, alanine, and gamma-aminobutyric acid in single tissue samples from different areas of rat central nervous system. Anal. Biochem. *64*: 149–169, 1975.

Soreq, H. and Glick, D.: Novel studies for cholinesterases in stress and inhibitor responses. In: "Cholinesterases and Cholinesterase Inhibitors," E. Giacobini, Ed., pp. 47–61, Martin Dunitz, London, 2000.

Soreq, H., Meshorer, E., Cohen, O., Yirmiya, R., Ginzburg, D. and Glick, D.: The molecular neurobiology of acetylcholine variants: from stressful insults to antisense intervention. In: "Cholinergic Mechanisms—Function and Dysfunction," I. Silman, H. Soreq, L. Anglister, D. Michaelson, and A. Fisher, Eds., pp. 119–124, Martin Dunitz, London, 2004.

Soreq, H. and Zakut, H.: Expression and *in vivo* amplification of the human acetylcholinesterase and butyrylcholinesterase genes. In: "Cholinergic Neurotransmission: Functional and Clinical Aspects," S.-M. Aquilonius and P.-G. Gillberg, Eds., pp. 51–61, Elsevier, Amsterdam, 1990.

Soreq, H., Malinger, G. and Zakut, H.: Expression of cholinesterase genes in human oocytes revealed by in-situ hybridization. Hum. Reprod. *2*: 689–693, 1987a.

Soreq, H., Zamir, R., Zevin-Sotkin, D. and Zakut, H.: Human cholinesterase genes localized by hybridization to chromosomes 3 and 16. Human Genet. *77*: 325–328, 1987b.

Soreq, H., Ben-Aziz, R., Prody, C. A., Seidman, S., Gnatt, A., Neville, L., Lieman-Hurwitz, J., Lev-Lehman, M. E., Ginzberg, D., Lapidot-Lipson, Y. and Zakut, H.: Molecular cloning and construction of coding region for human acetylcholinesterase reveals a G + C-rich attenuating structure. Proc. Natl. Acad. Sci. U.S.A. *87*: 9688–9692, 1990.

Spanner, S., Hall, R. C. and Ansell, G. B.: Arteriovenous differences of choline and choline lipids across the brain of rat and rabbit. Biochem. J. *154*: 133–140, 1976.

Speth, R. C. and Yamamura, H. I.: On the ability of choline and its analogues to interact with muscarinic cholinergic receptors in the rat brain. Eur. J. Pharmacol. *58*: 197–201, 1979.

Starai, V. J., Celic, I., Cole, R. N., Boeke, J. D. and Escalante-Semerena, J. C.: Sir2-dependent activation of acetyl-CoA synthetase by deacetylation of active lysine. Science. *298*: 2390–2392, 2002.

Stavinoha, W. B. and Weintraub, S. T.: Choline content of rat brain. Science. *183*: 964–965, 1974.

Stedman, E., Stedman, E. and Easson, L. H.: Cholinesterase. An enzyme present in the blood seruin of the horse. Biochem. J. *26*: 2056–2066, 1932.

Stoll, A. L., Cohen, B. M. and Hanin, I.: Erythrocyte choline concentrations in psychiatric disorders. Biol. Psychiatry. *29*: 309–321, 1991.

Sussman, J. L., Harel, M., Frolow, F., Oefner, C., Goldman, A., Toker, L. and Silman, I.: Atomic structure of acetylcholinesterase from *Torpedo californica*: a prototypic acetylcholine-binding protein. Science *253*: 872–879, 1991.

Suszkiw, J. B. and Pilar, G.: Selective localization of a high affinity choline uptake system and its role in ACh formation in cholinergic nerve terminals. J. Neurochem. *26*: 1133–1138, 1976.

Szutowicz, A., Kabata, J. and Bielarczyk, H.: The contribution of citrate to the synthesis of acetyl units in synaptosomes of developing rat brain. J. Neurochem. *38*: 1196–1204, 1982.

Szutowicz, A., Tomaszewicz, M., Bielarczyk, H. and Jankowska, A. L: Putative significance of shifts in acetyl-Co A compartmentalization in nerve terminals for disturbances of cholinergic transmission in the brain. Dev. Neurosci. *20*: 485–492, 1998.

Szutowicz, A., Tomaszewicz, M., Jankowska, A., Madziar, B. and Bielarczyk, H.: Acetyl-CoA metabolism in cholinergic neurons and their susceptibility to neurotoxic inputs. Metab. Brain Dis. *15*:29–44, 2000.

Szutowicz, A., Bielarczyk, H., Gul, S., Zielinski, P., Pawelczyk, T. and Tomaszewicz, M.: Nerve growth factor and acetyl-L-carnitine evoked shifts in acetyl-CoA and cholinergic SN56 cell vulnerability to neurotoxic inputs. J. Neurosci. Res. *79*: 185–192, 2005.

Tanaka, H., Shimojo, M., Wu, D. and Hirsh, L. B.: Regulation of the cholinergic gene locus. In: "From *Torpedo* Electric Organ to Human Brain: Fundamental and Applied Aspects," J. Massoulié, Ed., pp. 149–152, Elsevier, Amsterdam, 1998.

Taylor, P.: René Couteaux Lecture: Ligand recognition in the cholinergic nervous system analysed by structural templates and protein dynamics. In: "Cholinergic Mechanisms: Function and Dysfunction," I. Silman, H. Soreg, L. Anglister, D. Michaelson, and A. Fisher Eds., pp. 81–86, Martin Dunitz, London, 2004.

Taylor, P. and Radic, Z.: The cholinesterases: from genes to proteins. Annu. Rev. Pharmacol. Toxicol. *34*: 281–320, 1994.

Taylor, P. and Brown, J. H.: Acetylcholine. In: "Basic Neurochemistry. Molecular, Cellular and Medical Aspects," G. J. Siegel, Ed., pp. 214–242, Lippincott-Raven, Philadelphia, 1999.

Taylor, P., Luo, Z. D. and Cap, S.: The genes encoding the cholinesterases: structure, evolutionary relationships, and regulation of their expression. In: "Cholinesterases and Cholinesterase Inhibitors," E. Giacobini, Ed., pp. 63–79, Martin Dunitz, London, 2000.

Thoenen, H.: Trans-synaptic enzyme induction. Life. Sci. *14*: 223–235, 1974.

Thompson, E. H. and Whittaker, V. P.: Cholinesterase activity and sodium transport in the human red cell. Biochim. Biophys. Acta. *9*(6): 700–701. 1952.

Tomaszewicz, M., Bielarczyk, H., Jankowska, A. and Szutowicz, A.: Pathways of beta-hydroxybutyrate contribution to metabolism of acetyl-CoA and acetylcholine in rat brain terminals. Folia Neuropath. *35*: 244–266, 1997.

Tomaszewicz, M., Jankowska, A., Madziur, B., Blusztajn, J. K. and Szutowicz, A.: Effect of nerve growth factor and other differentiating agents on expression of enzymes of acetyl-CoA and acetylcholine metabolism in SN56 cholinergic cells. In: "From *Torpedo* Electric Organ to Human Brain: fundamental and Applied Aspects," J. Massoulié, Ed., pp. 504–505, Elsevier, Amsterdam, 1998.

Toutant, J.-P. and Massoulié, J.: Acetylcholinesterase. In: "Mammalian Ectoenzymes," A. J. Kenny and A. J. Turner, Eds., pp. 290–328, Elsevier, Amsterdam, 1987.

Toutant, J.-P. and Massoulié, J.: Cholinesterases: tissue and cellular distribution of molecular forms and their physiological regulation. Handbk. Exp. Pharmac. *86*: 225–266, 1988.

Tsim, T. W. K., Randall, W. R. and Barnard, E. A.: An asymmetric form of muscle acetylcholinesterase contains three subunit subtypes and two enzymic activities in one molecule. Proc. Natl. Acad. Sci. U.S.A. *85*: 1262–1966, 1988.

Tucek, S.: Transport and changes of activity of choline acetyltransferase in the peripheral stump of an interrupted nerve. Brain Res. *82*: 249–270, 1970.

Tucek, S.: Acetylcholine synthesis in neurons. Chapman and Hall, London, 1978.

Tucek, S.: Problems in the organization and control of acetylcholine synthesis in brain neurons. Progr. Biophys. Molec. Biol. *44*: 1–46, 1984.

Tucek, S.: Regulation of acetylcholine synthesis in the brain. J. Neurochem. *44*: 11–24, 1985.

Tucek, S.: The synthesis of acetylcholine: twenty years of progress. In: "Cholinergic Neurotransmission: Functional and Clinical Aspects," S.-M. Aquilonius and P.-G. Gilberg, Eds., pp. 467–477, Elsevier, Amsterdam, 1990.

Tunek, A. and Svensson, L. A.: Bambuterol, a carbamate ester prodrug of terbutaline, as inhibitor of cholinesterases in human blood. Drug Metab. Dispos. *16*: 759–764, 1988.

Tucek, S., Ricny, J. and Dolezal, V.: Acetylcoenzyme A and the control of the synthesis of acetylcholine in the brain. Acta Neurobiol. Exper. *42*: 59–68, 1982.

Ulus, I. H. and Wurtman, R. J.: Choline administration: activation of tyrosine hydroxylase in dopaminergic neurons of rat brain. Science *194*: 1060–1061, 1976.

Ulus, I. H., Hirsch, M. J. and Wurtman, R. J.: Transsynaptic induction of adrenal tyrosine hydroxylase activity by choline: evidence that choline administration increases cholinergic transmission. Proc. Natl. Acad. Sci. U.S.A. *74*: 798–800, 1977.

Ulus, I. H., Wurtman, R. J., Scally, M. C. and Hirsch, M. J.: Effect of choline on cholinergic function. In: "Cholinergic Mechanisms and Psychopharmacology," D. J. Jenden, Ed., pp. 525–538, Plenum Press, New York, 1978.

Ulus, I. H., Arslan, Y., Tanrisever, R. and Kiran, B. K.: Postsynaptic effects of choline administration. In: "Choline and Lecithin in Brain Disorders," A. Barbeau, H. J. Growdon and R. J. Wurtman, Eds., pp. 219–226, Nutrition and the Brain, vol. 5, Raven Press, New York, 1979.

Ulus, I. H., Millington, W. R., Buyukuysal, R. L. and Kiran, B. K.: Choline as an agonist: determination of its agonistic potency on cholinergic receptors. Biochem. Pharmacol. *37*: 2747–2755, 1988.

Usdin, E.: Reactions of cholinesterases with substrates, inhibitors and reactivators. In: "Anticholinesterase Agents," A. G. Karczmar, Ed., 47–354, Intern. Encyclop. Pharmacol. Therap., vol. 1, Pergamon Press, Oxford, 1970.

Vaca, K. and Pilar G.: Mechanisms controlling choline transport and acetylcholine synthesis in motor nerve terminals during electrical stimulation. J. Gen. Physiol. *73*: 605–628, 1979.

Valette, F. M. and Massoulié, J.: Regulation of the expression of acetylcholinesterase by muscular activity in avian primary cultures. J. Neurochem. *56*: 1518–1525, 1991.

Valette, F. M., De La Porte, S. and Massoulié, J.: Regulation and distribution of acetylcholinesterase molecular forms *in vivo* and *in vitro*. In: "Cholinesterases: Structure, Function, Mechanism, Genetics and Cell Biology," J. Massoulié, F. Bacou, E. A. Barnard, A. Chatonnet, B. P. Doctor and D. M. Quinn, Eds., pp. 98–102, American Chemical Society, Washington, D.C., 1991.

Velan, B., Kronman, C., Ordentlich, A., Flashner, Y., Leitner, M., Cohen, S. and Shafferman, A.: N-glycosylation of human acetylcholinesterase: effects on activity, stability aand biosynthesis. Biochem. J. *296*: 649–656, 1993.

Vellom, D. C., Radic, Z., Li, Y., Pikering, N. A., Camp, S. and Taylor, P.: Amino acid residues controlling acetylcholinesterases and butyrylcholine specificity. Biochem. *32*: 12–17, 1993.

Vigneaud, du V.: Interrelationship between choline and other methylated compounds. In: "Comparative Biochemistry. Intermediate Metabolism of Fat. Carbohydrate Metabolism Biochemistry of Choline," H. B. Lewis, Ed., pp. 234–247; The Jaques Cattell Press, Lancester, Pennsylvannia, 1941.

Wallace, B. G., Nitkin, R. M., Reist, N. E., Fallon, J. R., Moayeri, N. N. and McMahan, U. J.: Aggregates of acetylcholinesterase induced by acetylcholine receptor-aggregating factor. Nature *315*: 574–577, 1985.

Webster, G. R., Marples, E. A. and Thompson, R. H. S.: Blycerylphosphorylcholine diesterase activity of nervous tissue. Biochem. J. *65*: 374–377, 1957.

Wecker, L.: The utilization of supplemental choline by brain. In: "Dynamics of Cholinergic Function," I. Hanin, Ed., pp. 851–858, Plenum Press, New York, 1986.

Wecker, L. and Dettbarn, W. D.: Relation between choline availability and acetylcholine synthesis in discrete brain parts. J. Neurochem. *32*: 961–967, 1979.

Wecker, L. and Goldberg, A. M.: Regulation of acetylcholine release during increased neuronal activity. In: "Cholinergic Mechanisms: Phylogenetic Aspects, Central and Peripheral Synapses, and Clinical Significance," G. Pepeu and H. Ladinsky, Eds., pp. 451–461, Plenum Press, New York, 1981.

Wecker, L., Dettbarn, W. D. and Schmidt, E. D.: Choline administration: Modification of central actions of atropine. Sci. *199*: 86–87, 1978.

Weise, C., Kreienkamp, H. J., Raba, R., Pedak, A., Aaviksaar, A. and Hucho, F.: Anionic subsites of the acetylcholinesterase from *Torpedo californica.* EMBO J. *9*: 3885–3888, 1990.

Weiss, P.: Protoplasm synthesis and substance transfer in neurons. Abstracts, XVIIIth Int. Physiol. Congress, Oxford U. Press, p. 101, 1947.

Wermuth, B., Ott, R., Gentinetta, R. and Brodbeck, U.: Oligomeric forms of acetylcholinesterase and allosteric properties. In: "Cholinergic Mechanisms," P. G. Waser, Ed., pp. 299–308, Raven Press, New York, 1975.

Whittaker, V. P.: The storage and release of acetylcholine. Biochem. J. *128*: 73P–74P, 1972.

White, H. L. and Wu, J. C.: Kinetics of choline acetyltransferase (E.C.2.3.1.6) from human and other mammalian central and peripheral nervous tissues. J. Neurochem. *20*: 297–307, 1973.

Whittaker, V. P.: The application of subcellular fractionation techniques to the study of brain function. Progr. Biophys. Molec. Biol. *15*: 39–96, 1965.

Whittaker, V. P.: The cell and molecular biology of the cholinergic synapse: twenty years of progress. In: "Cholinergic Neurotransmission: Fundamental and Applied Aspects," S.-M. Aquilonius and P.-G. Gillberg, Eds., pp. 419–436, 1990.

Wilson, I. B.: Conformation changes in acetylcholinesterase. Ann. N.Y. Acad. Sci. *144*: 664–674, 1967.

Wilson, I. B. and Bergmann, F.: Studies on cholinesterase. VII. The active surface of acetylcholine esterase derived from effects of pH on inhibitors. J. Biol. Chem. *185*: 683–692, 1950.

Wilson, I. B., Harrison, M. A. and Ginsburg, S.: Carbamyl derivatives of acetylcholinesterase. J. Biol. Chem. *236:* 11498–11500, 1961.

Winteringham, F. P. W.: Dilution effects on the measurement of blood cholinesterase inhibition by carbamates. Nature *212*: 1368–1369, 1966.

Wu, D. and Hersh, J. B.: Choline acetyltransferase: celebrating its fiftieth year. J. Neurochem. *62*: 1653–1663, 1994.

Wurtman, R. J.: Precursor control of transmitter synthesis. In: "Choline and Lecithin in Brain Disorders," A. Barbeau, J. H. Growdon and R. J. Wurtman, Eds., pp. 1–12, Nutrition and the Brain, vol. 5, Raven Press, New York, 1979.

Wurtman, R. J.: Nutrients affecting brain composition and behavior. Integr. Psychiat. *55*: 226–238, 1987.

Wurtman, R. J., Blusztajn, J. K., Growdon, J. H. and Ulus, I. H.: Cholinesterase inhibitors increase the brain's need for free choline. In "Current Research in Alzheimer Therapy," E. Giacobini and R. Becker, Eds., pp. 95–100, Taylor and Francis, New York, 1988.

Yamamura, H. J. and Snyder, S. H.: High affinity transport of choline into synaptosomes of rat Brain J. Neurochem. *21*: 1355–1374, 1973.

Yamamura, H. I., Kuhar, M. J. and Snyder, S. H.: *In vivo* identification of muscarinic cholinergic receptor binding in rat brain. Brain Res. *80*: 170–176, 1974.

Zeisel, S. H. and Wurtman, R. J.: Developmental changes in rat blood choline concentration. Biochem. J. *198*: 565–570, 1981.

Zeisel, S. H., Growdon, J. H., Wurtman, R. J., Magil, S. G. and Logue, M.: Normal plasma choline responses to ingested lecithin. Neurology. *30*: 1226–1229, 1980.

Yip, V., Carter, J. G., Pusateri, M. E., McDougal, D. B., Jr. and Lowry, O. H.: Distribution in the brain and retina of four enzymes of acetyl CoA synthesis in relation to choline acetyltransferase and acetylcholine esterase. Neurochem Res. *16*: 629–635, 1991.

Zajicek, J.: Studies on histogenesis of blood platelets and megakaryocytes. Acta Physiol. Scand. *138,* Supplt. *40*: 1–32, 1957.

Zakut, H., Lieman-Hurwitz, J., Zamir, R., Sidell, L., Ginzburg, D. and Soreq, H.: Chorionic villus cDNA library displays expression of butyrylcholinesterase: putative genetic disposition for ecological disaster. Prenat. Diagnos. *11*: 597–607, 1991.

Zeev-Ben-Mordehai, T., Silman, I. and Sussman, J. L.: Use of the morphing graphics to visualize conformational differences between ACHEs from different species and inhibitor-induced conformational changes. In: "Cholinergic Mechanisms—Function and Dysfunction," I. Silman, H. Soreq, L. Anglister, D. Michaelson and A. Fisher, Eds., pp. 747–749, Martin Dunitz, London, 2004.

Zeisel, S. H.: Dietary choline: biochemistry, physiology and pharmocology. Ann. Rev. Nutr. *1*: 95–121, 1981.

Zeisel, S. H.: Formation of unesterified choline by rat brain. Biochem. Biophys. Acta *835*: 331–343, 1985.

Zeisel, S. H.: Factors which influence the availability of choline to brain. In: "Dynamics of Cholinergic Function," I. Hanin, Ed., pp. 837–849, Plenum Press, New York, 1986.

Zeisel, S. H.: Phosphatidylcholine: endogenous precursor of choline. In: "Lecithin: Technological, Biological and Therapeutic Aspects," I. Hanin, and G. B. Ansell, Eds., Plenum Press, New York, 1987.

Zelena, J.: Bidirectional shift of mitochondria in axons after injury. Symp. Int. Soc. Cell Biol. *8*: 73–94, 1969.

Zeller, E. and Bissegger, A.: Über die Cholinesterase des Gehirn und der Erythrocyten. III. Mitteilung über die Beeinflussung von Fermentreaktionen durch Chemotherapeutica und der Farmaka. Helv. Chim. Acta *26*: 1619–1630, 1943.

Zhang, D., Suen, J., Zhang, Y., Song, Y., Radic, Z., Taylor, P., Holst, M. J., Bajaj, C., Baker, N. A. and McCammon, J. A.: Tetrameric mouse acetylcholinesterase: continuum diffusion rate calculations by solving the steady-state Smoluchowski equation using finite element methods. Biophys. J. *88*: 1659–1665, 2005.

Zhang, M. R., Furutsuka, K., Maeda, J., Kikuchi, T., Kida, T., Okauchi, T., Irie, T. and Suzuki, K.: N-[18F]fluoroetoethyl-4-piperidyl acetate ([18F]FEtP4A): A PET tracer for imaging brain acetylcholinesterase *in vivo*. Bioorganic & Medic. Chem. *11*: 519–527, 2003.

Zimmermann, H., Volknand, W., Hausinger, A. and Herrmann, C.: Molecular properties and cellular distribution of cholinergic synaptic proteins. In: "Cholinergic Neurotransmission Functional and Clinical Aspects," S.-M. Aquilonius and P.-G. Gillberg, Eds., Elsevier, Amsterdam, pp. 32–40, 1990.

Zupancic, A. O.: Evidence for the identity of anionic centers of cholinesterases with cholinoreceptors. Ann. NY Acad. Sci. *144*: 689–693, 1967.

4

History of Research on Nicotinic and Muscarinic Cholinergic Receptors

Alexander G. Karczmar, Jon Lindstrom, and Arthur Christopoulos

The story of receptors is bound with the history of the recognition of chemical transmission at the periphery and in the central nervous system (CNS). In turn, this recognition is based on the progress in the description of the anatomy of the peripheral and central nervous systems. This progress was described in detail several times (see, for example, Brazier, 1959; Karczmar, 1967, 1986; Pick, 1970); at this time, it suffices to list the main steps of the early history of this progress and the principal authors who were involved (Table 4-1). As shown in the table, by the 18th century the main characteristics of the central, autonomic, and peripheral motor systems were laid down; in fact, even 250 years earlier the great da Vinci had rendered in his figures with much precision and few errors the details of the nervous systems (and other tissues).

A. Curare, Muscarine, Enzymes, and Bacteria

The story of the receptors was initiated long before the demonstration that the peripheral and central nervous systems both function via chemical transmission (rarely, via electric transmission) and, in the case of the autonomic and neuromyal sites as well as at important central pathways, via cholinergic signaling (see Chapter 9A and Karczmar, 1986, 1996). Actually, this initiation occurred on two early fronts: the first concerned curare and the pharmacology of the neuromyal junction, while the second concerned the enzymes and microorganisms.

The curare story includes the identification of its peripheral action, ultimately defined as localized at the receptor of the endplate. Curare, a derivative of the South American plant *Strychnos toxifera* and related South American plants (such as the genus *Chododendron*), was historically and prehistorically employed in hunting and in war by South American tribes (Holmstedt et al., 1983, 1984; Wassen and Holmstedt, 1963). Its first mention in Western literature is by Pietro Martire D'Anghera, an Italian companion of the conquistadores who describes it and its use in his 1516 book *De Orbe Novo Decades* (see Bovet and Bovet-Nitti, 1949). Sir Walter Raleigh and other explorers and seamen became interested in curare, and samples of curare-containing materials were brought to Europe (see also Koelle, 1970; Taylor, 1995).

These materials got into the hands of Arnold von Koelliker in Germany and Claude Bernard in France. In 1856, von Koelliker described at a session of the French Academy of Sciences the "extinguishing" effect of curare at the motor nerves; he was apparently unaware of the earlier, more precise results obtained by Claude Bernard, so Claude rushed to present, again, his results at the next session of the academy (Bernard, 1856; see also Holmstedt and Liljenstrand, 1963).[1] Bernard's sense that curare acts at the endings of the motor nerves, the muscles remaining excitable, was based on the use, in frogs, of appropriate ligations and the subcutaneous injections of a piece of material containing curare *in vivo*, as well as on the *in vitro* experiments with muscle-nerve preparations. Actually, Bernard came near to suggesting that curare acts at the neuromyal junction, as he stated that "curare must act on terminal plates of motor nerves" and that "curare does no more than interrupt something motor which puts the nerve and the muscle into electrical

Table 4-1. Early History of Anatomical Studies of the Peripheral and Central Nervous System

Herophilus (300 BCE), Erasistratus (290 BCE)	Describe peripheral nerves, including autonomic, and brain convolutions
Galen (129–200)	Describes 7 pairs of "swellings" (autonomic ganglia)
da Vinci (1452–1519)	Draws complete pictures of human anatomy, including central and autonomic nervous systems
Andreas Vesalius (1514–1565?)	Illustrates in his tables peripheral and central nervous systems, including, possibly, the vagosympathetic nerve (under "septi transversi nervus")
C. Estienne, B. Eustachio, and C. Reid (1550–1630)	Describe sympathetic postganglionic nerves as "intercostals"
J.B. Winslow (1669–1760)	Renames "intercostals" "nervi sympathici maximi"; they generate "animal spirits" or "sympathins"

relationship . . . required . . . for the movement" (from Bennett, 2000; see also Fessard, 1967); similar notions were developed by another Frenchman, E.F. Vulpian (1966). Emile Du Bois-Reymond should be mention in the context of Bernard's reference to the electrical origin of muscle movement; Du Bois-Reymond concluded in his studies of the neuromyal junction that there are several options to explain the nerve-muscle relationship; while he preferred the electric potential and the "electromotive particles" current as the source of movement, he conceded that a chemical—such as lactic acid or ammonia—may be involved (Du Bois-Reymond, 1877). It must be added that the German microanatomist W. Kuhne described the endplate of the muscle in 1888.

An important step in the present context was made by A.C. Crum Brown and T.R. Fraser of Edinburgh (Crum Brown and Fraser, 1868).[2] They established the first evidence indicating that a structure-activity relationship (SAR) clarifies the mode of action of curare, as they showed that the quaternary ammonium bases exhibit the common property of exerting curarelike action at the neuromyal junctions. This finding readily lent itself to the concept that a pharmacological agent must act by its conformity with the site of its action, namely, a receptor. The work of Crum Brown and Fraser also constitutes the beginning of the use of SAR in the studies of receptors and in the development of site-effective pharmacological agents (see Chapter 7 BI and BII).

An important step was made by John Langley[3] and William Dickinson as they found that curare, similar to large doses of nicotine, paralyzes the autonomic ganglia; this finding helped in the notion of similar effects of curare—or d-tubocurarine; see below—at a number of peripheral and central sites (Langley and Dickinson, 1989).[3] It is even more important in the present context that in his work Langley referred to a "a receptive substance" which consists of "radicles of the protoplasmic molecule" of the muscle to explain the actions of nicotine and curare on the muscle (Langley, 1905, 1906; see also the elegant book of Robinson, 2001). Also pertinent is the discovery by Langley of the antagonism at the neuromyal junction between nicotine, when used in small, excitatory doses, and curare—another finding consistent with the notion of the receptors.

It must be noted that several native South American tribes possessed in the middle of the last century (see Holmstedt et al., 1983/1984) methods for concocting curare-rich preparations and techniques for further purifications of these preparations were developed in Europe and South America in the nineteen forties, culminating with the efforts of O. Wintersteiner, Venancio Deulofeu, Heinrich and Theodor Wieland and others (see Bovet and Bovet-Nitti, 1949; Bovet et al., 1959). Even earlier, H. King (1935) sought to establish the structure of the pure component of the curare materials; however, his formula contained two quaternary ammonium groupings, and a long time passed before it was established (see Bovet, 1972) that the d-tubocurarine molecule contains one quaternary and one tertiary nitrogen grouping. This progress culminated in the work of

Daniele Bovet and his wife, Filomena Bovet-Nitti, on synthetic derivatives of d-tubocurarine and their SAR (Bovet and Bovet-Nitti, 1949). Parenthetically, the Bovets used the term "placca motrice" and "sostanza recettrice" to explain the neuromyal effects of curare and d-tubocurarine.

In parallel with the studies of Langley, the concept of receptors received a boost on a micromolecular level. Working with several enzymes and such compounds as glucosides, the German biochemist Emil Fischer (1894) established that enzymes possess "active sites" and the compounds, whether inhibitors, accelerators, or substrates, act on these sites. To be effective at these sites, they must fit like "a lock and key" via a "template principle"—a phenomenon subsequently referred to by Irwin Wilson as obeying "the law of complementarity" (see Chapter 7BII). Then the famous German microbiologist Paul Ehrlich (1909) posited "the side-chain theory." It was based on his studies of trypanosomes and arsenicals, which he developed for "chemotherapy" treatment of trypanosomiasis. Ehrlich stated that a "combining group of the protoplasmic molecule to which the introduced group is anchored will hereafter be termed receptor"; this group possessed "chemical side chains." Ehrlich also felt that the receptors contain an sulfhydryl (SH) group as an active anchor, and he extended his theory into the side-chain theory of immunity.

It is exciting that these two scientific giants, Langley and Ehrlich made a contact! In his earlier work Langley did not refer to Ehrlich's work, but, already in 1908 he stated that his notion of drugs or "secretins" action on the "receptive" substance "is in general on the lines of Ehrlich's theory of immunity", that is, the part of the theory concerning the drug action on the receptors. Ehrlich (1914) graciously acknowledged "the brilliant investigations by Langley". Ehrlich stated that Langley's work "on the effects of alkaloids . . . caused my doubts to disappear and made the existence of chemoreceptors seem probable . . ." (see also Bennett, 2000).

It should be noted that, in the cholinergic story, muscarine and atropine occupy almost as an important position as curare. Their sources, Amanita muscaria and solanacea plants, were used prehistorically and today by African and South American tribes for ther magic properties as hallucinogens (see Wasson, 1973; Lewin, 1924; Henry, 1949). In the present context of the research concerning cholinergic (muscarinic) receptors, atropine is less relevant than muscarine (for the story of atropine see several chapters in Holmstedt's and Liljenstrand's book, 1969, and Karczmar, 1986). Oswald Schmiedeberg, the father of pharmacology, was the first scientist to study the Amanita materials, and he and Koppe (1869) pointed out that muscarine and vagal stimulation produced identical cardiac effects, these effects being both antagonized by atropine. In due time, Schmiedeberg's findings and additional XIXth Century investigations of muscarine made Walter Dixon suggest that muscarine may be the substance which is involved in the vagal effect on the heart (see Holmstedt and Liljesrand, 1963; Waser, 1962; Karczmar, 1986). Of course the work of Reid Hunt, Sir Henry Dale and, ultimately Otto Loewi established ACh as the endogenous agent that evokes the vagodepressant effect at the heart (see Holmstedt and Liljestrand, 1963 and Karczmar, 1986). Subsequently, the presence of muscarinic receptors was established for the ganglia and the enteric nervous system, and studies of SARs of muscarinic and atropinic drugs were initiated with respect to the muscarinic smooth muscle sites (see Karczmar, 1986).

As to the central muscarinic receptors, Chris Krnjevic and other members of the Canberra team[4] demonstrated in the nineteen sixties that muscarinic receptors are present in the brain and that, actually, they predominate in the brain compared to the nicotinic receptors (see also Karczmar, 1963 and 1993). In fact, even the classical nicotinoceptice neuron, the Renshaw cell, contains muscarinic receptors (many central neurons, including the Renshaw cell, exhibit both muscarinic and nicotinic receptors; see Chapter 9A). It appeared early that central and peripheral muscarinic receptors resemble each other: the SARs of atropinics used as antagonists of peripheral and central muscarinic effects were found to be closely correlated (Karaczmar and Long, 1958).

Ultimately, as in the case of the nicotinic receptors, the use of ligands and toxins, their radioactive derivatives and of molecular biology methods helped in establishing the structures of the subtypes of muscarinic receptors (Burgen, 1989; see Chapter 5D).

B. Receptors and the Demonstration of the Cholinergic Transmission

As already pointed out, the notion of cholinergic receptors is bound with the demonstration of cholinergic transmission; indeed, synaptically released ACh requires receptors to be effective, unless it generates electric current to evoke postsynaptic and junctional transmission. In fact, John Langley's notice of the "receptive" substances was logical outcome of the reasoning such as this; furthermore, his demonstration of the curare-nicotine antagonism at the ganglia and the neuromyal junctions and subsequent demonstrations of ACh-atropine antagonism at muscarinic peripheral and central sites and ACh-curarimimetic or nicotine-curaremimetic antagonism at the cholinergic central sites militated against the notion of the generation of electric current by ACh and favored the receptor concept.

Then, Sir Henry Dale distinguished two different actions of ACh, the depressor actions at the cardiovagus terminals and other parasympathetic sites, and the pressor actions at the ganglia that ACh could exert after its depressor action was blocked by atropine. Previously John Langley showed that in small, non-paralysant doses nicotine exhibits excitatory effects. Dale referred to these two phenomena as "muscarine actions" and "nicotine actions." And Dale postulated that these actions occur at a "excitable structure" which is endowed with a "specific constitution"—a neat definition of a receptor (Dale, 1914, 1934, 1965; the term "excitable structure" was also used in the nineteen thirties by Lord E.D. Adrian). Indeed, Dale felt that in the nineteen thirties the stage was set for the direct demonstration of chemical—in this case, cholinergic—transmission at these sites.

In fact, the proofs of peripheral and central cholinergic transmission at the parasympathetic nerve endings and at the spinal Renshaw cell were delivered by Otto Loewi and John Eccles, respectively, relatively soon after Dale's prediction in the case of the cardiovagus and many decades after his dictum in that of the CNS (see Chapter 8A and Karczmar, 1967a, 1986, 1996, 2001). In the nineteen thirties and forties A.W. Kibjakov, Geoffrey Brown, Sir William Brown and Sir William Feldberg provided the demonstration of cholinergic transmission at the ganglia and several parasympathetic sites (Karczmar, 1967a). Subsequently, the presence of muscarinic and nicotinic transmission at several CNS sites was established by Chris Krnjevic, Davis Curtis, J.W. Phillis, R.W. Ryall, and other (see above, Section 1, and Karczmar, 1967).

An interesting aspect of the receptor story concerns the developmental appearance of nicotinic and muscarinic receptors; the novelty of this aspect is the apparent desynchrony between the ontogenetic appearance of the receptors and of other cholinergic components (see several Subchapters of Chapter 8, and Salpeter, 1987).

C. The "Abstract" Phase of the Receptor Research

The demonstration of the presence of cholinergic transmission at the peripheral and central nervous systems and the notion that receptors are necessary for cholinergic action did not lead immediately to the direct demonstration of receptors as chemical entities. In fact, Robert Furchgott stated that "It must be admitted . . . that, with rare exceptions, we can neither identify the receptor as an individual chemical entity nor study the primary chemical or physical change which occurs when drug and receptor interact. Nevertheless, there is impressive *indirect* evidence, particularly from studies on quantitative aspects of drug antagonism, for the validity of the concept that receptors are specific and real (Schild, 1962)" (from Furchgott, 1964; italics mine). This statement constitutes a very good motto for the "abstract phase" of receptor research, i.e., a phase when receptors were described in terms of equations and physicochemical concepts. It should be added that this "abstract" work was carried out in terms of dose-effect relationships and SARs established with respect to smooth and skeletal muscle of frogs and other species (see Schild, 1979). An interesting preparation was the rat intestinal preparation of John Vane (1964), a Nobelist: the preparation is arranged vertically and, as the agonist or antagonists descends the preparation, it reacts, depending on the agent, with specific parts of the intestine that are sensitive to various peptides, catecholamines, histamine, etc. Another useful preparation is the oesophagus of the opossum as it discriminates among several muscarinics (Goyal and Rattan, 1978). While much of the SAR and related work quoted in this section was carried out *in vitro* on the smooth and skeletal muscle preparations, some early investigations concerned the central muscarinic SAR (Karczmar and Long, 1958; see also Bebbington and Brimblecombe, 1965; Taylor, 1995; and several chapters of book edited by Joan Heller Brown, 1989).

Actually, this era started already before the demonstration of the presence of cholinergic transmission whether at the periphery or central. In 1926, Alfred Clark posited the concept of competitive actions at the receptors of agonists and antagonists. Clark came to the notion of receptors as he calculated, on the basis of the heart mass and the effective dose of digitalis preparations that only very small fraction of the heart's muscle cell surface is occupied by the drug when the latter is effective. Subsequently, he became cognizant of the notions of his Cambridge predecessor, A.V. Hill that the of nicotine-induced frog muscle contraction is reflected by the mass equation relating the height of the contraction to the combination between the agonist and "some constituent of the muscle" i.e. the receptor (Hill, 1909; see also Robinson, 2001). Clark, studying the effects of ACh on the frog heart and muscle, proposed that, essentially, the response to a "competitive" agonist is a positive function of the number of receptors occupied by the agonist, as represented in his equations by the RA element ("occupation theory"). Clark and John Gaddum defined competitive antagonism and agonism as receptor-agonist and receptor-antagonist complexes such that their components can fully dissociate. And Gaddum developed an equation that describes the reversible competitive antagonism that occurs when the antagonist competes with the agonist by reacting reversibly with the same receptor with which the agonists reacts.

Robert Furchgott (1955, 1964[6]), and E.J. Ariens (1954, 1964) corrected Clark's equation by introducing the concepts of intrinsic activity and dissociation constants, while Stephenson (1956) amended Ariens' equation by introducing the efficacy factor. Then, H. Schild (see Schild, 1957) developed plots that determined the agonist association constants. Next, Paton (1961) introduced "rate theory" of agonist and antagonist action as he produced evidence indicating that the receptor response is a function of the rate of association between drug molecules and receptors. And corrections concerning muscarinic responses were advanced by Sir Arnold Burgen and his associates (Burgen and Hiley, 1975; see also Burgen, 1989). Then, E.J. Ariens stressed the selective role of stereoselectivity as he demonstrated that only one of stereoisomers of active agonists and antagonists may be effective at a receptor—a notion that is, of course, very consistent with the receptor concept (see, for example, Beld and Ariens, 1974). In addition, Ariens and others stressed that only only one of the stereoisomers of an agent is effective at a receptor. In this context Ariens (1993) posited the terms "homochiral" and "chiral" to denote drugs that contain single isomers and "mixtures" of isomers, respectively.

It was soon realized that agonists differ in their maximal activity and that certain other agonist and antagonist characteristics did not fit the earlier notions. Accordingly, additional definitions were introduced by Furchgott and the Nijmegen team of E.J. Ariens and J.M. Van Rossum. Thus, partial agonists exert partial effect compared to "full agonists," even after saturating the available receptors (Ariens, 1964). Also, they developed the concept of "spare receptor." This term is to be understood as follows. All receptors theoretically contribute to a given effect. The term "spare receptors" signifies that the receptor-signal transduction is attainable with submaximal receptor occupancies, which, in turn, indicates that the receptor-signal transduction system of a tissue is very efficiently coupled; and, since the maximal effect is attained at submaximal occupancies, the tissue is said to possess "spare receptors." Actually, the importance of this concept is that there is a significant amplification built into the stimulus-response machinery of the particular tissue. Finally, Ever hardus Ariens used the term "autoinhibition" which was defined as a secondary action of the agonists at the site different from the specific site of the agonist effect.

An important concept that was referred to by Furchgott and the Nijmegen investigators was that of noncompetitive antagonism. In this case, the antagonists react with a site on the receptor different from the site of action of the agonists; while attached there, the antagonist prevents the action of the agonist. Furchgott (1955, 1964) claimed that the noncompetitive antagonism does not exist and the noncompetition can only arise when the antagonist acts on another receptor. He felt also that evidence exists only for irreversible competitive antagonism from which there is a "extremely slow recovery" and he posited equations illustrating this effect (see also Robinson, 2001). This notion was in agreement with the findings of Mark Nickerson (1956) concerning the relation at a smooth muscle between histamine and dibenamine, an irreversible antagonist and with the notion that protection is possible with regard to irreversible antagonists (Furchgott, 1955, 1964). Yet, the concept of noncompetitive antagonism which occurs when the antagonist and the agonist act at two different sites of the same receptor received recent confirmation with respect to AChE (see Chapter 3D1 and 3D2) and receptors; in fact, cholinergic (nicotinic and muscarinic) and non-cholinergic receptors possess allosteric binding sites that can mediate non-competitive antagonism.

Ultimately, Robert Furchgott presented a number of SAR curves, each being characteristic for a specific type of AR interaction. Importantly, he also referred to the phenomena of desensitization as originating at the receptor (see also Chapter 8B3-2)

Peter Pauling, the son of the double Nobelist, Linus Pauling employed a different approach in this era of "abstract" receptor. Pauling and his associates used atomic models, SARs and diffraction analysis of the crystals of pertinent compounds to establish their structures and their conformation and conformational parameters with regard to the activity of these compounds at the cholinergic sites (he and his associates rarely use the term receptor). It is important to note that "by 'structure . . . he . . . refers to the three dimensional structure of molecules" . . . "rather than two-dimensional chemical line diagrams" (Pauling, 1975)! Pauling stresses also that, outside of the nervous system, ACh structure may be "labile" and therefore analysis of ACh crystals may not be as useful as that of more rigid cholinergic drugs. He applied this methodology to muscarinic and nicotinic agonists and antagonists and curarimemetics. He also studied the conformational characteristics of ACh as relevant to its role as the substrate for AChE. Ultimately, he defined, with respect to these various types of cholinergic drugs and to ACh the distances between key atoms or groups (such as the distance between the nitrogen and the methyl radical in the case of ACh) as well angles and rotation coefficients of the linkages between various radicals.

It is surprising how the parameters that Pauling established for the active cholinergic drugs agree with the modern results and modern models posited for the molecules in question. And his notion of the importance for the effectiveness of cholinergic agents of three-dimensional conformity between the receptors and the agents in question is analogous with Irwin Wilson's law of "complimentarity" with Wilson used for the rational design of oxime reactivators (see Chapter 7B2)

D. The Reality of the Receptors

The first inkling that "the impressive indirect evidence" for the existence of receptors may be turned into direct evidence was provided by Peter Waser. In the 1950s and 1960s, he employed radiaoactive nicotinic and curarimimetic ligands (particularly decamethonium and curarine and radioautograpy methods; Waser, 1960). Using Koelle's AChE stain and comparing the locale of the stain and of the ligand he demonstrated that the ligands bind almost exclusively to the endplates of skeletal muscles (such as intercostals of the mouse). Also, curarine bound weakly to "non-specific mucopolysaccharide receptors". In subsequent work concerning the "precipitation" of the receptors (see infra, this Section) ACh and curarine bound to polysaccharides causing sometimes erroneous interpretation of the results. Similar work followed, utilizing radioactive toxins derived from the cobra and kait venoms such as alpha-bungarotoxin that proved very specific for the nicotinic receptors (Lee et al., 1967; Raftery et al., 1971; see also Lindstrom, 1998). And radioactive toxins derived from the snake mamba as well as radioactive agonists such as oxotremorine and antagonists such dexetimide (Bel and Ariens, 1974) were used to demonstrate the presence and distribution of muscarinic receptors (Birdsall et al., 1978; see Wolfe, 1989 and Birdsall and Hulme, 1989 P; see also Chapter 5B and 5D1).

To one of us (AGK); the real drama occurred at a symposium held in 1959. At this symposium three investigators showed the audience a "real" material which they claimed were the nicotinic or curareceptive receptors. Two of these investigators were Carlos Chagas and Miss A. Hasson of the Rio de Janeiro Instituto de Biofisica. These investigators, who are the pioneers of the research on the electric fish, employed extraction methods described originally by Chagas. The purified extract, obtained as a powder, was a polysaccharide that could bind ammonium bases, ACh and succinylcholine (Chagas and Hasson, 1961; see also Hasson-Voloch, 1968). Then, Sy Ehrenpreis (1961) used the ammonium sulfate fraction of the extract from the electroplaque material to precipitate by means of d-tubocurarine a protein; this protein could bind, relatively specifically, several ammonium bases as well as an oxime.

At the symposium, these two papers were vigorously discussed. While "seeing" the receptor was exciting, then and now, the materials obtained by Hasson and Ehrenpreis were not accepted as receptors (see also Robinson, 2001). It is interesting that while Ehrenpreis worked at the time in David Nachmansohn's laboratory, Nachmansohn did not accept Ehrenpreis' protein as the receptor. There was subsequently an attempt by Jean-Pierre Changeux to obtain the nicotinic receptor via solubilizing the electroplaque membrane with a detergent; he demonstrated the binding of the resulting protein with radioactive decamethonium (Changeux et al., 1970). Additional progress concerning nicotinic receptors was made in the seventies of the past Century by A. Karlin and Ricardo Miledi, two pioneers in research on nicotinic receptors. Miledi and his associates (1971) used electric tissues of *Torpedo* and employed purification, toxin binding and membrane dissolving methods to isolate a 80,000 MW protein which they considered to be "the acetylcholine receptor". Karlin used an irreversible sulhydryl-labelling reagent maleimido benzyl trimethyl ammonium to identify a 40,000 MW peptide in receptor containing preparations of the electroplaque (see Karlin and Cowburn, 1973).

In due time, the approachs, employed by the pioneers in cholinergic receptor isolation yielded to different methodology that proved more fruitful both with respect to nicotinic and muscarinic receptors. For either receptor, these methods involved ligand binding sequential purification steps, immunization, affinity chromatography, aminoacid sequencing, crystallography of purified receptors and of their prototypes, and, ultimately molecular methodology; these methods are described in Chapters 5D and 6B.

It should be noted that the pertinent research that concerns the nicotinic receptors has had an advantage over the studies of muscarinic receptors. There are several sources that are very rich in niconitic receptors, including the electroplaques of several species of *Torpedo* (see several articles in Chagas and Carvalho, 1961). On the contrary, "the solubilization and purification of mAChR (muscarinic ACh receptor) has proven to be rather difficult because of a low density of receptor sites," . . . of available the tissues, organs and species . . . "poor yields of solubilized protein upon dergent extraction, and ". . . mscarinic . . . "receptor instability in many commonly used . . . detergents" (Schimerlik, 1989; see also Schimerlik, 1990). Porcine atria and porcine cerebullum are used currently with good success. The modern information concerning ligand binding, molecular biology methodology, and identification of post-translational modifications are discussed in detail in Chapter 5D.

E. Receptor Subtypes

Langley and Dale did not envisage that the nicotinic and muscarinic receptors are heterogeneous, and this state of the matter persisted till the 1950s of the previous Century. Thus, William Paton and Eleanor Zaimis (1951) described striking differences between nicotinic receptors of the ganglia and of the neuromyal junction, this difference being reflected, for example, in the differences in the potency of alpha-bungarotoxin at the brain, ganglionic and neuromyal sites. Actually, a single organ may contain a variety of nicotinic receptors, as stressed by Agneta Nordberg (see Adem and Nordberg, 1988). These nicotinic receptor differences relate to the diversity of the nicotinic receptor genes (see Patrick et al., 1993).

Modern methods of labeling the nicotinic receptor, multiple purification steps (see Raftery et al., 1971), chromatographic and ultracentrifuge techniques, aminoacid sequencing, electron-microscopy (Kistler et al., 1982) and, finally, molecular biology techniques brought about the identification of the structure and the distribution of nicotinic receptor subtypes (see Chapter 6, and Robinson, 2001). An important step in this direction was made by A. Karlin. He defined the four protein subunits of the purified nicotinic receptor of the Electrophorus plaque (Karlin and Cowburn, 1973). It turned out that nicotinic receptor subtypes represent different combinations of subunits. This diversification of nicotinic (and muscarinic) subtypes contributes to the plasticity and versatility of the brain function (see Chapters 5 and 6; see also Robinson, 2001 and Changeux et al., 1996).

As in the case of nicotinic receptors, it was thought originally that there is only one type of muscarinic receptor; and, as already mentioned (see above, Section 1) pharmacological evidence indicated that peripheral and central muscarinic receptors are similar if not identical (Karczmar and Long, 1958). Indeed, similar indication was derived from the nineteen seventies studies of Sir Arnold Burgen, Henry Yamamura, Sol Snyder, Ed Hulme and Nigel Birdsall of the relationship between the binding properties and the analysis of smooth muscle responses (see Burgen, 1989; Birdsall and Hulme, 1989, and Wolfe, 1989). Yet, soon evidence begun emerging indicating that, as expressed by Nigel Birdsall and Ed Hulme, "... the picture ... is very complex" (Birdsall and Hulme, 1989). Already in 1951, Riker and Wescoe found that the curarimimetic gallamine blocks the muscarinic cardiovagal muscarinic receptor, but not the muscarinic receptors elsewhere. And, an atropinic, 4-diphenyl-acetoxy-N-Methyl piperidine methiodide (4-DAMB) showed similar discrimination with regard to smooth muscle and the cardiac muscarinic receptor (Barlow et al., 1976). And, with respect to agonists, in 1961 Peter Roszkowski described an unusual muscarinic synthetic, McN-A-343, which stimulated the ganglia, inducing a vasopressor effect, but which has no effects on the heart or the smooth muscle! This riddle was solved when it was demonstrated that this compound stimulates selectively the M1 receptors (other explanations were offered by Bjorn Ringdahl, 1989).

These early intimations served as an introduction to the notion of the multiplicity of muscarinic receptors; the introduction of pirenzepine (Leitold et al., 1977) was an important step in this direction, as it appeared that pirenzepine discriminated among at least three receptor subtypes (see Burgen, 1989 and Birdsall and Hulme, 1989). Chapter 5 describes 5 muscarinic receptor subtypes, and the list may not be closed as yet!

F. Brief Conclusions

As in the case of other areas of cholinergic research, the development of our understanding of the cholinoceptive receptors was rapid and dramatic. It progressed from the demonstration of peripheral site of action of curare by Claude Bernard to the linking of this discovery with the notions of John Langley and Sir Henry Dale and with the definitions of the receptor posited by Paul Ehrlich and Emil Fischer. Later, this understanding advanced rapidly from the days of the "abstract receptors" to "real receptors", their aminoacid sequence, their structure and subtypes. Methodologically, we moved from the SAR of the frog rectus or frog atrial response to the QSARs, molecular biology methods and to the concept of allostericity.

Our current understanding of the receptors and their subtypes completes our knowledge of the mechanisms of transmission, of the messenger-connected signalling by the receptors, and of the specificity of transmission combined with the versatility and plasticity of peripheral and central responses. And, from the applied viewpoint, the definition of the receptors and their subtypes leads to advances in specific treatment of disease and affords hope for success in this area.

Notes

1. Claude Bernard is a good example of the glory of the French science in the XIXth Century, the French competing with the Germans and the English for scientific hegemony. Claude was a pupil of Francois Magendie, the identifier of dorsal roots as site of sensation and ventral roots as a site of motor action (there is a controversy as to whether Magendie or Charles Bell has the priority with resect to this discovery). Besides his epochal studies of curare, Claude Bernard is credited with the discovery that the toxicity of carbon monoxide is due to its binding with hemoglobin and with the development of the concept of the "milieu interieur" (see Mauriac, 1954).

2. For the contribution of Fraser to the studies of Calabar bean and physostigmine, see Chapter 9A.
3. For the great John Newport Langley's contributions to the understanding of the autonomic nervous system (he actually coined the terms "autonomic nervous system" and "parasympathetic nervous system"), see Fletcher (1926), Holmstedt and Liljestrand (1963), and Karczmar (1986).
4. These investigators established central cholinoceptivity via direct, iontophoretic application of ACh and pertinent drugs to various CNS sites. For the full story of the demonstration of central, autonomic, and neuromyal cholinergic transmissions, see Chapter 9 A; Karczmar, 1986; and Zaimis, 1962, respectively.
5. Raymond Alquist (1948) and a number of distinguished pharmacologists, such as Ulf Svante von Euler and R.J. Lefkowitz, worked mostly with catecholamines as they contributed to the concept of receptors. But, Lee Limbird (2004) could have been more balanced if he referred even once to cholinoceptivity in his 2004 general review of the receptor concept.
6. While Furchgott reviewed (1964) the types of cholinergic, muscarinic, curarimimetic, and nicotinic receptors, their agonists and antagonists (see also Trendelenburg, 1961), and while he is a distinguished theoretician of receptor dynamics and kinetics, unfortunately, he is not a cholinergiker. His research, on which he based his generalizations, concerns the smooth intestinal and vascular muscle, the actions of catecholamines and their antagonists, and the energy requirements for their relaxation or dilation and contraction (Furchgott, 1955). His 1955 review, elegant and conceptually rich, is more than 80 pages long and contains 404 references (one of us may own a record for both categories; see Karczmar, 1967). He earned his Nobel Prize (1998) for his experimental a as well as theoretical work and for his demonstration that blood vessel endothelia release nitric oxide—a retrograde messenger—which dilates the blood vessels. Once, when he was a candidate for the presidency of the American Society for Pharmacology and Therapeutics, Furchgott told the ASPET members, "Vote for me if you want a bad president."
7. For early differentiation between muscarinic receptor subtypes, see Roszkowski, 1961.

References

Adem, A. and Nordberg, A.: Nicotinic receptor heterogeneity in mammalian brain. In: "The Pharmacology of Nicotine," M. J. Rand and Thurau, Eds., pp. 227–247, IRL Press, Washington DC, 1988.

Alquist, R. P.: A study of adrenotropic receptors. Am. J. Physiol. *155*: 586–600, 1948.

Ariens, E. J.: Affinity and intrinsic activity in the theory of competitive inhibition. Intern. Pharmacodyn. *99*: 32–39, 1954.

Ariens, E. J.: "Molecular Pharmacology," Academic Press, New York, 1964.

Ariens, E. J.: Nonchiral, homochiral and composite chiral drugs. Trends Pharmacol. Sci. *14*: 68–75, 1993.

Babbington, A. and Brimblecombe, R. W.: Muscarinic receptors in the peripheral and central nervous system. Adv. Drug Res. *2*: 143–172, 1955.

Barlow, R. B., Berry K. J., Glenton, P. A. M., Nickolaou, N. M. and Soh, K. S.: A comparison of affinity constants for muscarinic-sensitive acetylcholine receptors in guinea-pig atrial pacemaker cells at 29 degrees C and in ileum at 29 degrees C and 37 degrees C. Br. J. Pharmacol. *58*: 613–620, 1976.

Bennett, M. R.: The concept of transmitter receptors: 100 years on. Neuropharmacol. *39*: 523–546,

Bernard, C.: Analyse physiologique des propriétés des systèmes musculaire et nerveux au moyen du curare, C. R. Acad. Sci. *43*: 825–829, 1856.

Birdsall, N. J. M. and Hulme, E. C.: The binding properties of muscarinic receptors. In: "The Muscarinic Receptors," J. Heller Brown, Ed., pp. 31–92, The Humana Press, Clifton, NJ, 1989.

Birdsall, N. J. M., Burgen, A. S. V. and Hulme, E. C.: The binding of agonists to brain muscarinic receptors. Mol. Pharmacol. *14*: 723–726, 1978.

Birdsall, N. J., Chan, S. C., Eveleigh, P., Hulme, E. C. and Miller, K. W.: The models of binding of ligands to cardiac muscarinic receptors. Trends Pharmacol. Sci. 31–34, 1989.

Bovet, D.: Synthetic inhibitors of neuromuscular transmission, chemical structures and structure activity relationships. In: "Neuromuscular blocking and stimulating agents," Intern. Ecyclop. Pharmacol. Theryp. J. Cheymol, Ed., pp. 243–294, vol. 1, Sect. 14, Pergamon Press, Oxford, 1972.

Bovet, D. and Bovet-Nitti, F.: Il curaro. Reconditi Dell'Istituto Sup. Di Santita *12*: 7–269, 1949.

Bovet, D., Bovet-Nitti, F. and Marini-Bettola, G. B.: Curare and Curare-Like Agents." Elsevier Amsterdam, 1959.

Brazier, M. A. B.: The historical development of neurophysiology. In: "Handbook of Physiology. Section I: Neurophysiology," J. Field, Ed., and H. W. Magoun, Section Ed., pp. 1–58, American Physiological Society, Washington, D.C., 1959.

Brown, D. A.: The history of the neuronal nicotinic receptors. In: "Neuronal Nicotinic Receptors," F. Clementi, D. Fornasari and C. Giotti, Eds., pp. 3–11, Springer-Verlag, Berlin, 2000.

Burgen, A. S. V. The role of ionic interactions at the muscarinic receptor. Br. J. Pharmacol. *25*: 4–7, 1965.

Burgen, A. S. V.: History and basic properties of the muscarinic cholinergic receptor. In: "The Muscarinic Receptors," J. Heller Brown, Ed., pp. 3–27, The Humana Press, Clifton, NJ, 1989.

Burgen, A. S. V.: Evolution and acetylcholine receptors. In: "Cholinergic Function and Dysfunction," A. C. Cuello, Ed., pp. 129–131, Elsevier, Amsterdam, 1993.

Burgen, A. S. V. and Hiley, C. R.: The use of an alkylating antagonist in investigating the properties of muscarinic receptors. In: "Cholinergic Mechanisms," P. G. Waser, Ed., pp. 381–385, Raven Press, New York, 1975.

Chagas, C. and Carvalho, Paes de, Eds. "Bioelectrogenesis," pp. 379-396, Elsevier, Amsterdam, 1961.

Changeux, J.-P., Kasai, M., Huchet, M. and Meunier, J. C.: Extraction à partir du tissue electrique de gymnote d'une proteine presentant plusiers proprietés characteristiques du recepteur physiologique d'acetylcholine. C. R. Acad. Sci. *270D*: 2864–2867, 1970.

Changeux, J. P., Bessis, A., Bourgeois, J. P., Corringer, P. J., Devillers-Thiery, A., Eisele, I. L., Kerszberg, M., Lean, C., Le Novere, N., Picciotto, M. and Zoli, M.: Nicotinic receptors and brain plasticity. Cold Spring Harb. Symp. Quant. Biol. *61*: 343–362, 1996.

Clark, A. J.: The reaction between acetylcholine and muscle cells. J. Physiol. *61*: 530–561, 1926.

Clark, A. J.: "General Pharmacology." Handbch. d. exper. Pharmacol., Ergänzungswk. vol. 4, Springer-Verlag, Berlin, 1937.

Crum Brown, A. and Fraser, T. R.: "Connection Between Chemical Constitution and Physiological Action." Neill and Company, Edinburgh, 1868.

Dale H. H.: The action of certain esters and ethers of choline, and their relation to muscarine. J. Pharmacol. Exper. Therap. *6*: 147–190, 1914.

Dale, H. H.: Pharmacology and nerve endings. Proc. Roy. Soc. Med. *28*: 319–322, 1934.

Du Bois-Reymond, E.: "Gesammelte Abhandlungen zur allgemeinen Muskel- und Nervenphysik," vol. 2, Veit, Leipzig, 1877.

Ehrenpreis, S.: The isolation and identification of the acetylcholine receptor protein from electric tissue of Electrophorus electricus (L). In: "Bioelectrogenesis," C. Chagas and A. Paes de Carvalho, Eds., pp. 379–396, Elsevier, Amsterdam, 1961.

Ehrlich, P.: Üeber den jedtzigen Stand der Chemotherapie. Ber. Dtsch. Chem. Ges. *42*: 17–47, 1909.

Ehrlich, P.: Chemotherapy. In: "The Collected Papers of Paul Ehrlich," F. Himmelweit, Ed., vol. 1, pp. 505–518, Pergamon Press, London, 1915.

Fastier, F. N.: Modern concepts in relationship between structure and biological activity. Ann. Rev. Pharmacol. Toxicol. *4:* 51–68, 1964.

Fessard, A.: Claude Bernard and the physiology of junctional transmission. In: "Claude Bernard and Experimental Medicine," F. Grande and M. B. Visscher, Eds., pp. 105–124, Schenkmn Publ., Cambridge, 1967.

Fischer, E.: Einfluss der Configuration auf die Wirkung der Enzyme. Ber. Dtsch. Chem. Ges. *27*: 2985–2993, 1894.

Fletcher, W. M.: John Newport Langley. In memoriam. J. Physiol. *61*: 1–27, 1926.

Furchgott, R. F.: The pharmacology of vascular smooth muscle. Pharmacol. Rev. *7*: 183–265, 1955.

Furchgott, R. F.: Receptor mechanism. Ann. Rev. Pharmacol. *4*: 21–50, 1964.

Gaddum, J. H.: The quantitative effects of antagonistic drugs. J. Physiol. *89*: 7P–8P, 1937.

Goyal, R. K. and Rattan, S.: Neurohumaoral, hormonal and drug receptors of the lower oesophageal sphincter. Gastroenterol. *74*: 598–617, 1978.

Hasson, A. and Chagas, C.: Purification of macromolecular components of the aqueous extracts of the electric organ [E. electricus (L)] with binding capacity *in vitro*, for quaternary ammonium bases. In: "Bioelectrogenesis," C. Chagas and A. Paes de Carvalho, Eds., pp. 362–378, Elsevier, Amsterdam, 1961.

Hasson-Voloch, A.: Curare and acetylcholine receptor substances. Nature *218*: 330–333, 1968.

Heller Brown, J., Ed.: "The Muscarinic Receptors," Humana Press, Clifton, NJ, 1989.

Henry, T. A.: "The Plant Alkaloids," Blakiston Co., Philadelphia, 1949.

Hill, A. V.: The mode of action of nicotine and curari, determined by the form of contraction curve and the method of temperature coefficients. J. Physiol. *39*: 361–372, 1909.

Holmstedt, B. and Liljenstrand, G.: "Readings in Pharmacology." Pergamon Press, Oxford, 1963.

Holmstedt, B. R., Wassen, S. H., Chanh, P. H., Clavel, P. and Lasserre, B.: Alleged native antidote to curare. Goteborgs Etnografiska Museum Ann., 19–25, 1983/1984.

Hulme, E. C., Birdsall, N. J. M., Burgen, A. S. V. and Mehta, P.: The binding of antagonists to brain muscarinic receptors. Mol. Pharmacol. *14*: 737–750, 1978a.

Karczmar, A. G.: Pharmacologic, toxicologic, and therapeutic properties of anticholinesterase agents. In: "Physiological Pharmacology," W. S. Root and F. G. Hofman, Eds., vol. 3, pp. 163–322, Academic Press, New York, 1967a.

Karczmar, A. G.: Neuromuscular Pharmacology. Annual Review of Pharmacol., H. W. Elliott, Ed., *7*: 241–276, 1967b.

Karczmar, A. G.: Historical development of concepts of ganglionic transmission. In: "Autonomic and Enteric Ganglia. Transmission and Pharmacology," A. G. Karczmar, K. Koketsu and S. Nishi, Eds., pp. 3–26, Plenum Press, New York, 1986.

Karczmar, A. G.: Brief presentation of the story and present status of studies of the vertebrate cholinergic system. Neuropsychopharmacol. *9*: 181–199, 1993.

Karczmar, A. G.: Loewi's discovery and the XXI Century. In: "Ninth International Symposium on Cholinergic Mechanisms," J. Klein and K. Loeffelholz, Eds., pp. 1–27, Elsevier, Amsterdam, 1996.

Karczmar, A. G.: Synaptic, behavioral and toxicological effects of cholinesterase inhibitors in animals and humans. In: "Cholinesterases and Cholinesterase Inhibitors", E. Giacobini, Ed., pp. 157–180, Martin Dunitz, London, 2000.

Karczmar, A. G.: Sir John Eccles, 1903–1997. Part 1. Onto the demonstration of the chemical nature of transmission in the CNS. Perspectives in Biol. & Med. *144*: 76–86, 2001a.

Karczmar, A. G.: Sir John Eccles, 1903–1997. Part II. The brain as machine or as a site of free will? Perspectives in Biol. & Med. *144*: 250–262, 2001b.

Karczmar, A. G. and Long, J. P.: Relationship between peripheral cholinolytic potency and tetraethylpyrophosphate antagonism of a series of atropine substitutes. J. Pharm. Expt. Therap. *123*: 230–237, 1958.

Karlin, A. and Cowburn, D. W.: The affinity-labeling of partially purified acetylcholine receptor from *Electrophorus*. Proc. Natl. Acad. Sci. U.S.A. *70*: 3636–3640, 1973.

King, H.: Curare alkaloids. I. Tubocurarine. J. Chem. Soc. *2*: 1381–1383, 1935.

Kistler, J., Stroud, R. M., Klymkowsky, M. W., Lalancette, R. A. and Fairclough, R. H.: Structure and function of an acetylcholine receptor. Biophys. J. *37*(1): 371–83, 1982.

Koelle, G. B.: Neuromuscular blocking agents. In: "Pharmacological Basis of Therapeutics," L. S. Goodman and A. Gilman, Eds., Fourth Edition, pp. 601–619, Macmillan, London, 1970.

Kuhne, W.: On the origin and the causation of vital movement. Proc. Roy. Soc. *44*: 427–448, 1888.

Langley, J. N.: On the reaction of cells and nerve endings to certain poisons, chiefly as regards the reaction of striated muscle to nicotine and to curare. J. Physiol. *33*: 374–413, 1905.

Langley, J. N.: On nerve endings and on special excitable substances in cells. Proc. Roy. Soc. Section B *78*: 170–194, 1906.

Langley, J. N.: On the contraction of muscle, chiefly in relation to the presence of "receptive" substances. J. Physiol. *36*: 347–348, 1907.

Langley, J. N.: On contraction of muscle, chiefly in relation to the presence of 'receptive' substance. Part III. The reaction of frog muscle to nicotine after denervation. J. Physiol. *37*: 285–300, 1908.

Langley, J. N.: On contraction of muscle chiefly in relation to the presence of 'receptive substances.' Part IV. The effect of curari and some other substances on the nicotine response of the sartorius and gastrocnemius muscles of the frog. J. Physiol. *39*: 235–239, 1909.

Langley, J. N. and Dickinson, W. L.: On the local paralysis of peripheral ganglia and on the connection of different classes of nerve fibers with them. Proc. Roy. Soc. *46*: 423–421, 1989.

Lee, C., Tseng, L. and Chiu, T.: Influence of denervation on localization of neurotoxins from clapid venoms in rat diaphragm. Nature *215*: 1177–1178, 1967.

Letold, M., Engelhon, R., Balhause, H., Kuhn, F. J., Ziegler, H., Schierok, H. J., Seidel, H., Guderlei, B. and Milotzke, W.: General pharmacodynamic properties of secretion and ulcer inhibition by pirenzepine (LS 519). Therapiewoche *27*: 1551–1654, 1977.

Lewin, L.: "Phantastica," Goerg Stilke Verlag, Berlin, 1924.

Limbird, L. E.: The receptor concept. A continuing evolution. Molec. Intervention *4*: 326–336, 2004.

Lindstrom, J.: Purification and cloning of nicotinic acetylcholine receptors. In: "Neuronal Nicotinic Receptors: Pharmacology and Therapeutic Opportunities," S. P. Arneric and J. D. Brioni, Eds., pp. 3–23, Wiley-Liss, New York, 1998.

Lukas, R. J., Changeux, J. P., Le Novere, N., Albuquerque, E. X., Balfour, D. J., Berg, D. K., Bertrand, D., Chiappinelli, V. A., Clarke, P. B., Collins, A. C., Dani, J. A., Grady, S. R., Kellar, K. J., Lindstrom, J. M., Marks, M. J., Quik, M., Taylor, P. W., and Wonnacot, S.: International Union of Pharmacology. XX. Current status of the nomenclature for nicotinic acetylcholine receptors and their subunits. Pharmacol. Rev. *51*: 397–401, 1999.

Mauriac, P.: "Claude Bernard," Paillard, Paris, 1954.

Miledi, R., Molinoff, P. and Potter, L. T.: Isolation of the cholinergic receptor protein of Torpedo electric tissue. Nature *229*: 554–557, 1971.

Nickerson, M.: Receptor occupancy and tissue response. Nature *178*: 697–698, 1957.

Paton, W. D. M.: A theory of drug action based on the rate of drug-receptor combination. Proc. Roy. Soc. (London), Section B *154*: 21–69, 1961.

Paton, W. D. M. and Rang, H. P.: The uptake of atropine and related drugs by intestinal smooth muscle of the guinea pig in relation to acetylcholine receptors. Proc. Roy. Soc. (London), Section B *163*: 1–44, 1965.

Paton, W. D. M. and Zaimis, E.: Paralysis of autonomic ganglia by methonium salts. Brit. J. Pharmacol. *6*: 155–168, 1951.

Patrick, J. and Lindstrom, J.: Autoimmune response to acetylcholine receptor. Science *180*: 871–872, 1973.

Patrick, J., Sequela, P., Vermino, S., Amador, V., Luetje, C. and Dani, J. A.: Functional diversity of neuronal acetylcholine receptors. In: "Cholinergic Function and Dysfunction," A. C. Cuello, Ed., pp. 113–120, Elsevier, Amsterdam, 1993.

Pauling, P.: The shapes of cholinergic molecules. In: "Cholinergic Mechanism," P. G. Waser, Ed., pp. 241–249, 1975.

Pick, J.: "The Autonomic Nervous System," Lippincott, Philadelphia, 1970.

Raftery, M. A., Schmidt, J., Clark, D. G. and Wolcott, R. G.: Demonstration of a specific alpha-bungarotoxin binding component in *Electrophorus electricus* electroplax membranes. Biochem. Biophys. Res. Commun. *45*: 1622–1629, 1971.

Ringdahl, B.: Structural determinants of muscarinic agonist activity. In: "The Muscarinic Receptors," J. Heller Brown, Ed., pp. 151–218, The Humana Press, Clifton, NJ, 1989.

Ringdahl, B., Roch, M., Katz, E. D. and Frankland M. C.: Alkylating partial muscarinic agonists related to oxotremorine. N-[4-[(2-haloethyl)methylamino]-2-butynyl]-5-methyl-2-pyrrolidones. J. Med. Chem. Mar; *32*(3): 659–663, 1989.

Robinson, J. D.: "A Century of Synaptic Transmission," Oxford University Press, Oxford, 2001.

Roszkowski, A. P.: An unusual type of ganglionic stimulant. J. Pharmacol. Exper. Therap. *132*: 156–170, 1961.

Salpeter, M. M.: Development and neural control of the neuromuscular junction and of the junctional acetylcholine receptor. In: *The Vertebrate Neuromuscular Junction*, Alan R. Liss, Inc. pp. 55–115, 1987.

Schild, H. O.: Drug antagonism and pAx. Pharmacol. Rev. *9*: 242–246, 1957.

Schild, H. O.: Recollections and reflections of an autopharmacologist. In: "Trends in Autonomic Pharmacology," S. Kalsner, Ed., vol. 1, pp. 1–18, Urban and Schwarzenberg, Baltimore, 1979.

Schimerlik, M. I.: Structure and regulation of muscarinic receptors. Annu. Rev. Physiol. *51*: 217–227, 1989.

Schimerlik, M. I.: Structure and function of muscarinic receptors. Prog. Brain Res. *84*: 11–9, 1990.

Stephenson, R. P.: A modification of receptor theory. Brit. J. Pharmacol. *11*: 379–393, 1956.

Taylor, P.: Neuromuscular blocking agents. In "Goodman and Gilman's The Pharmacological Basis of Therapeutics," A. G. Gilman, L. S. Goodman, T. W. Rall and Ferid Murad, Eds., Seventh Edition, pp. 222–235, Macmillan, New York, 1995.

Trendelenburg, U.: Pharmacology of antonomic ganglin. Ann. Rev. Pharmacol. *1*: 219–238, 1961.

Vane, J. R.: The use of isolated organs for detecting active substances in the circulating blood. Brit. J. Pharmacol. Chemother. *23*: 360–373, 1964.

Vulpian, E. F.: "Lecons sur la Physiologie Génèrale et Comparée du Systeme Nerveux," Bailliere Editeurs, Paris, 1966.

Waser, P. G.: The cholinergic receptor. J. Pharm. Pharmacol. *12*: 577–594, 1960.

Waser, P. G.: "Cholinerge Pharmakon-Rezeptoren." Naturforschended Gesellschaft, Zurich, 1983.

Wassen, S. H. and Holmstedt, B.: The use of Parica, an ethnological and pharmacological review. Ethnos, 1963, 5–45, 1963.

Wilson, I. B.: Molecular complementarity and antidotes for alkylphosphate poisoning. Fed. Proc. *18*: 752–758, 1959.

Wolfe, B.: Subtypes of cholinergic muscarinic receptors. In: "The Muscarinic Receptors", J. Heller Brown, Ed., pp. 125–150, The Humana Press, Clifton, New Jersey, 1989.

Yeh, J. J., Yasuda, R. P., Davila-Garcia, M. I., Xiao, Y., Ebert, S., Gupta, T., Kellar, K. J. and Wolfe, B. B.: Neuronal nicotinic acetylcholine receptor alpha3 subunit protein in rat brain and sympathetic ganglion measured using a subunit-specific antibody: regional and ontogenic expression. J. Neurochem. *77*: 336–346, 2001.

Zaimis, E.: Experimental hazards and artifacts in the study of neuromuscular blocking drugs. In: "Curare and Curare-like Agents", A. V. S. De Reuck, Ed., pp. 75–82, Churchill, London, 1962.

5

Muscarinic Acetylcholine Receptors in the Central Nervous System: Structure, Function, and Pharmacology

Arthur Christopoulos

A. Introduction

Muscarinic acetylcholine (ACh) receptors (mAChRs) are members of the 7 transmembrane-spanning, guanine nucleotide-binding protein (G protein)-coupled receptor (GPCR) superfamily. There are 5 subtypes of mAChRs, and these mediate the majority of the actions of ACh in both the central nervous system (CNS) and the periphery. For example, peripheral mAChR activation results in a reduction in heart rate and vasodilatation of vascular beds, increases in exocrine secretions from sweat and lacrimal glands, and contraction of smooth muscle in the gastrointestinal tract, airways, ciliary body, and iris sphincter. All 5 subtypes of mAChRs are expressed in the CNS and are postulated to play a role in learning and memory, arousal, REM sleep, psychotic states, control of movement, thermoregulation, reward behaviors, and the generation of epileptic foci. This chapter provides an overview of current knowledge regarding the structure, function, and pharmacology of the mAChRs, with a particular emphasis on their roles in the CNS. For a historical introduction, see Chapter 4 A–E.

B. Pharmacological Classification

Historically, the studies of Riker and Wescoe (1951) on the cardioselective inhibition of mAChR function by gallamine gave the first indication of the existence of more than one subtype of mAChR. A decade later, Roszkowski (1961) demonstrated that the agonist 4-(3-chlorophenylcarbamoyloxy)-2-butynyltrimethylammonium chloride (McN-A-343) had a preferential effect on mAChRs located in sympathetic ganglia while having practically no activity on mAChRs in the heart and smooth muscle, and Barlow et al. (1976) also provided early functional evidence for differences between the pharmacology of mAChRs located in the ileum and those in the atrium. However, it was not until the introduction of the antagonist pirenzepine and the demonstration that this compound could clearly differentiate between types of mAChRs in both radioligand binding (Hammer et al., 1980) and functional (Brown et al., 1980; Hammer and Giachetti, 1982) assays, that the existence of at least 2 distinct mAChR subtypes was formally proposed (Hammer and Giachetti, 1982). The subsequent advent of molecular cloning techniques (see section D, below) has of course confirmed the existence of 5 distinct subtypes of mAChR protein, but the pharmacological classification of mAChRs remains of paramount importance in confirming the functional relevance of any expressed gene product, as well as validating novel therapeutic agents that target mAChRs.

In recent years, the advent of newer, high-throughput functional screening assays promises to yield novel mAChR agonists with the potential for improved selectivity over existing agents. In addition, the use of a variety of snake toxins, predominantly from the black and green mamba, that appear to show quite impressive degrees of selectivity for 1 or 2 mAChR subtypes relative to all others (see below) are also beginning to allow for a more definitive pharmacological classification of mAChR subtypes. However, mAChR pharmacology is still generally characterized by a lack of truly selective agonists, as well as a lack of readily available antagonists that can differentiate any one mAChR subtype to the exclusion of all others.

1. Agonists

The use of agonist ligands to classify mAChR subtypes is unreliable for two reasons. First, the orthosteric site, that is, the domain on the receptor recognized by its endogenous agonist (ACh), is highly conserved across all 5 mAChR subtypes (see Section D), making it especially difficult to design small molecule agonists that discriminate one mAChR subtype from another. Second, the binding of an agonist ligand to any GPCR promotes coupling of the receptor to its cognate G protein and changes the affinity state of the receptor for the agonist relative to that determined in the absence of G protein (Birdsall and Lazareno, 1997; Christopoulos and El-Fakahany, 1999b; Ehlert, 1985); measures of agonist affinity can therefore vary considerably between different systems expressing the same receptor but different complements of G proteins. Additionally, the binding behavior of GPCR agonists can also be influenced by other post-receptor-binding phenomena such as desensitization, interaction with accessory cellular proteins, and events linked to the signal transduction process (Kenakin, 1997a). Such phenomena are well recognized for the mAChRs. For example, Figure 5-1 illustrates the effects of the guanine nucleotide GTP on the competition between the classic mAChR agonist carbachol and the antagonist [^{3}H]N-methylscopolamine ([^{3}H]NMS) at native M_2 mAChRs in rat myocardial membranes. It can be seen that increasing concentrations of GTP lead to a significant reduction in the apparent affinity of carbachol for the M_2 mAChR by over 3 orders of magnitude due to the uncoupling of the receptor from its G protein by the GTP (Ehlert and Rathbun, 1990). Although the common use of recombinant expression systems has led to an explosion in our ability to study individual mAChRs in common cellular backgrounds, the overexpression of receptors, as often occurs in such systems, can lead to pleiotropic receptor coupling to multiple effector pathways, which again can impact profoundly on agonist pharmacology (Kenakin, 1997a).

In addition to the difficulties associated with the use of agonists to characterize mAChR subtypes, the design of selective agonists for the

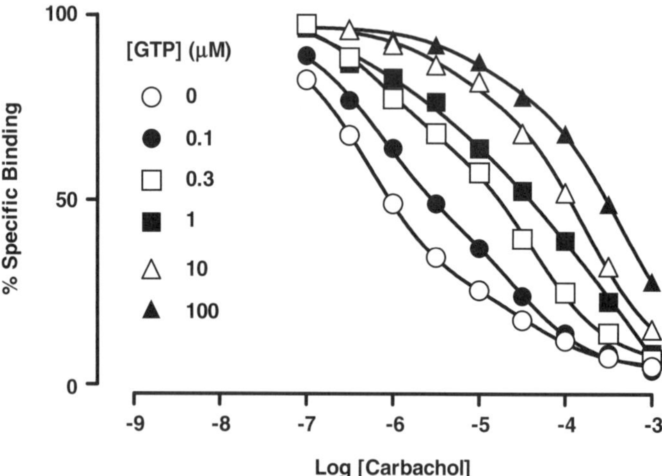

Figure 5-1. Dependence of muscarinic acetylcholine receptor agonist affinity on G protein coupling. Effects of GTP on the competition between [^{3}H]N-methylscopolamine and the agonist carbachol at M_2 mAChRs in rat myocardial membranes. (Data replotted from Ehlert and Rathbun, 1990.)

treatment of various disorders remains a significant challenge. A very common observation with many mAChR agonists is the display of "functional" selectivity, which exploits differences in agonist efficacy between mAChR subtypes, rather than affinity differences. The rationale behind this approach is that the agonist will activate only the receptor at which it has highest efficacy, while mediating minimal or no response for those subtypes at which it has negligible efficacy. In general, most functionally selective mAChR agonists are partial agonists, and a major difficulty with the use of partial agonists as selective therapeutic agents is that the expression of agonist efficacy, or lack thereof, is highly dependent on the functional status of the target receptor in its native environment. This cannot be readily gauged from functional studies on recombinant expression systems. Furthermore, changes in receptor expression and/or coupling in disease states are often unpredictable, and can readily alter or even nullify any functional selectivity.

Classic mAChR agonists include small molecules such as carbachol, pilocarpine, arecoline, oxotremorine, bethanecol, and McN-A-343 (Mitchelson, 1988). Of the classical agonists, McN-A-343 (Figure 5-2) has long been thought to display a preference for the M_1 mAChR, but more recent studies have suggested a slightly higher preference for the M_4 mAChR (Lazareno and Birdsall, 1993). Since the 1990s, a number of novel mAChR agonists have been introduced that possess some degree of mAChR subtype selectivity. An obvious approach to the development of such agents has been to manipulate the structure of ACh itself, particularly through the development of rigid analogues of the neurotransmitter. Talsaclidine (Figure 5-2) is one example of this approach, its structure based on the rigid ACh analogue aceclidine. Although talsaclidine shows minimal selectivity in terms of binding assays, it has been claimed to possess significant functional selectivity for the M_1 mAChR (Ensinger et al., 1993).

Another popular structure that has resulted in much structure-activity work is that of the agonist arecoline (Figure 5-2); some of the earliest work in designing M_1 mAChR-selective agonists was based on this agonist as a template (Gloge et al., 1966). Replacement of the ester group of

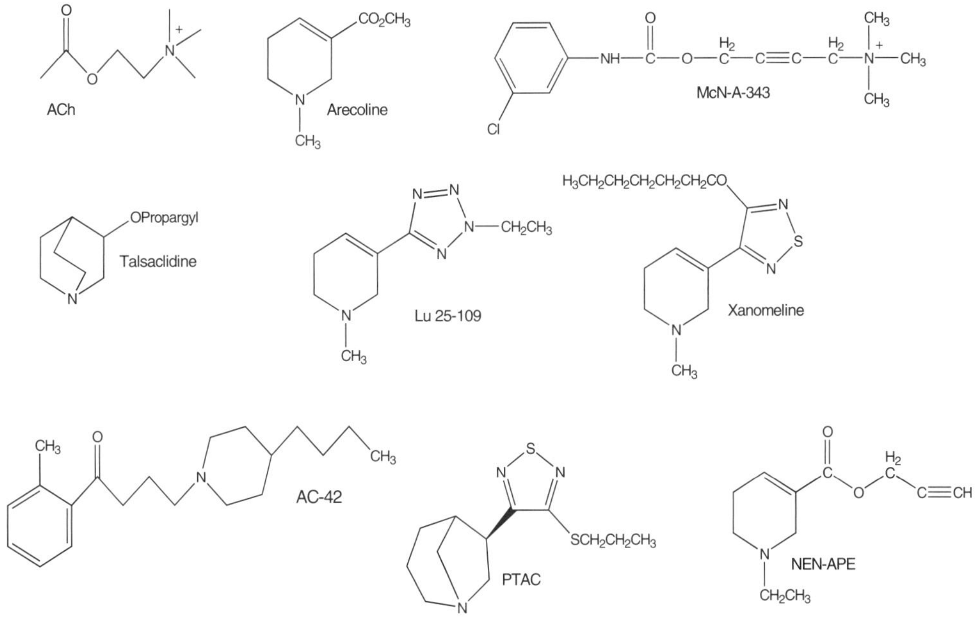

Figure 5-2. Agonists of mAChRs.

arecoline, or of conformationally more rigid azabicyclic analogues, with bioisosteric heterocycles has more recently led to compounds such as 2-ethyl-5-(1-methyl-1,2,5,6-tetrahydropyridyl)-*2H*-tetrazole (Lu 25-109; Meier et al., 1997) and xanomeline (Sauerberg et al., 1992; Shannon et al., 1994; Figure 5-2), both claimed to possess functional selectivity for M_1 mAChRs, although xanomeline also has significant activity at M_4 mAChRs (Bymaster et al., 1997; Shannon et al., 2000). Subsequent studies with the latter agonist have revealed additional pharmacological properties. Specifically, xanomeline is able to bind to the M_1 mAChR in a persistent manner that is resistant to extensive washout but still capable of interacting with the orthosteric site and classic antagonists in a reversible, competitive manner (Christopoulos and El-Fakahany, 1997; Christopoulos et al., 1998b, 1999a). By analogy with the β_2 adrenoceptor "captive agonist" salmeterol (Coleman et al., 1996), it has been proposed that xanomeline may form a persistent attachment with the M_1 mAChR by utilizing additional "anchor" sites on the receptor that are outside the classic orthosteric domain (Christopoulos et al., 1998b). More recent work has garnered further support for the interaction of this compound with multiple epitopes on the M_1 mAChR, including a potential allosteric mode of interaction (Jakubík et al., 2002).

Because the orthosteric domain in the mAChRs shows high conservation across receptor subtypes, the development of larger mAChR agonists, which may interact with less conserved receptor regions, has been pursued as another means of obtaining selectivity. A number of highly potent mAChR agonists have been discovered from this approach. Studies by Sauerberg and colleagues have demonstrated the utility of the ester isostere 1,2,5-thiadiazole group (Sauerberg et al., 1998a, 1998b) as a useful template for developing potent mAChR agonists, such as the compound phenylpropylargyloxy-1,2,5-thiadiazole-quinuclidine (NNC 11-1314; Figure 5-3). A particularly important property of this series of compounds is the ability to attach extended alkyl groups to the molecules while still retaining agonist activity. These findings have led to further studies investigating the effects of dimerization of such compounds (Christopoulos et al., 2001). The development of bivalent receptor ligands is a very powerful approach to attaining enhanced ligand potency and selectivity for a variety of receptor systems, including mAChRs (Christopoulos et al., 1999b; Melchiorre et al., 1989; Perez et al., 1998a, 1998b; Portoghese, 1989; Rajeswaran et al., 2001). Accordingly, the dimerization of NNC 11-1314 led to a marked enhancement of binding affinity for the dimeric agonist NNC 11-1585 or a profound alteration in the functional selectivity profile for the dimeric NNC 11-1607 (Figure 5-3; Christopoulos et al., 2001).

Additional approaches that have led to functionally selective mAChR agonists include the replacement of ester groups with oximes (e.g., sabcomeline; Bromidge et al., 1997; Loudon et al., 1997) and the use of oxotremorine analogues (Conti et al., 1997).

Despite significant advances in the development of highly potent mAChR agonists, absolute mAChR subtype selectivity remains elusive. However, a recent study by Spalding et al. (2002) has identified the first mAChR agonist that shows unprecedented functional selectivity for the M_1 mAChR over all other subtypes. The compound, 4-n-butyl-1-[4-(2-methylphenyl)-4-oxo-1-butyl]-piperadine hydrogen chloride (AC-42; Figure 5-2), was discovered using a high-throughput functional screen of 145,000 structurally diverse small molecule ligands. Although radioligand binding assays revealed a similar affinity for AC-42 at all 5 mAChR subtypes, agonist activity was observed only at the M_1 mAChR. Additional experiments with chimeric mAChRs identified the likely site of interaction of AC-42 to include regions of the receptor close to transmembrane (TM) domains TM1 and TM7 that are not conserved across mAChR subtypes. This finding suggests that, at least for the M_1 mAChR, an "ectopic" activation domain may exist outside the classic orthosteric site. However, it remains to be determined whether this ectopic site constitutes part of an allosteric binding domain that is also known to exist on mAChRs (see below).

From most studies conducted to date, it is clear that the development of mAChR agonists targeting the CNS has focused on M_1 mAChRs, driven predominantly by the "cholinergic hypothesis of dementia" that postulates a potential role for postsynaptic M_1 mAChRs in providing some relief from the cognitive deficits associated with Alzheimer's disease (Giacobini, 1992; Whitehouse et al., 1982). Unfortunately, none of the functionally selective M_1 mAChR agonists that have made it to clinical trials for this purpose

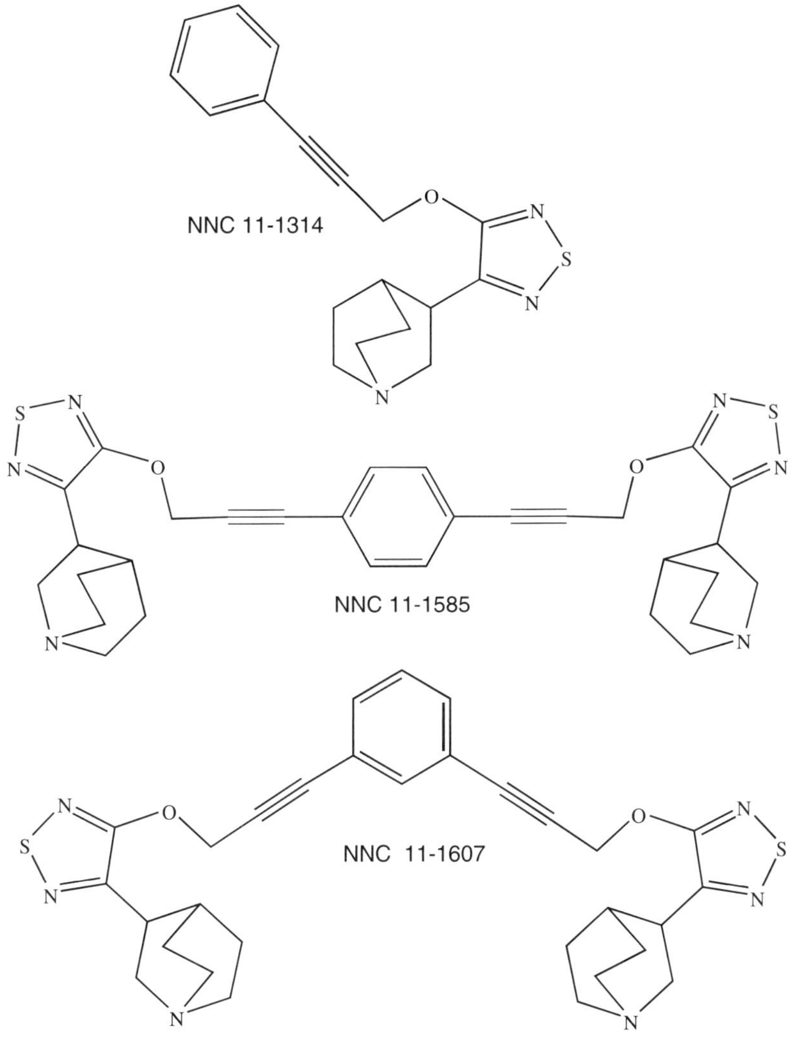

NNC 11-1314

NNC 11-1585

NNC 11-1607

Figure 5-3. Agonists of mAChRs (continued).

have been approved for clinical use (Felder et al., 2000). In addition to drug failure due to unacceptable side effects, this disappointing result highlights two important considerations. First, the validity of the cholinergic hypothesis of dementia remains to be more strenuously investigated. Second, the functional selectivity ascribed to mAChR agonists ex vivo may not translate to the same type of selectivity *in vivo* due to modifications in the neuronal environment within the brain, perhaps due to the disease progression itself.

More recently, studies have begun to evaluate the role of mAChR agonists in the treatment of other CNS disorders, such as schizophrenia and pain. Interestingly, investigations of the role of mAChRs in these disorders suggest that the M_1 mAChR is unlikely to prove the main therapeutic target (Felder et al., 2000). For example, agonist activity at M_4 mAChRs has been noted for a number of atypical antipsychotic agents, such as clozapine (Zeng et al., 1997; Zorn et al., 1994). In agreement with this observation, the novel ligand (5R, 6R) 6-(3-propylthio-1, 2,5-thiadiazol-4-yl)-1-azabicyclo [3.2.1] octane (PTAC; Figure 5-2), which is a partial agonist at M_2 and M_4 mAChRs and an antagonist at M_1, M_3, and M_5 mAChRs, displays antipsychotic properties that are similar in nature to those of established atypical antipsychotic agents such as olanzapine and clozapine (Bymaster et al., 1998). N-ethyl-guvacine propargyl ester (NEN-APE; Figure 5-2) is another agonist that shows highest functional selectivity for M_2 mAChRs, while acting as an antagonist at the M_1 and M_3 mAChRs (Mutschler et al., 1995).

2. Antagonists

It is now well recognized that the pharmacological classification of receptors based on antagonist affinity profiles is a generally more preferable pharmacological approach than the use of agonist affinity profiles (Christopoulos and El-Fakahany, 1999b; Kenakin, 1997b). However, there are a number of methodological considerations that must always be borne in mind when using antagonists to pharmacologically define the involvement of a particular receptor subtype. First, if radioligand binding assays are used, it is paramount that measurements be made at equilibrium; this is particularly pertinent when studying allosteric modulators of the mAChRs, as these compounds can significantly delay the time taken for the system to reach an equilibrium state (Christopoulos and Kenakin, 2002; Lazareno and Birdsall, 1995; Proska and Tucek, 1994). Second, concordance between radioligand binding–based estimates of antagonist affinity profiles and functional (null method)–based estimates of antagonist affinity profiles is desirable but not always attainable. This is usually due to variations in the composition and ionic strength of the buffer used for *in vitro* experiments, but large discrepancies should always be investigated further. These issues have all arisen in past studies of mAChRs, but overall there has been an extraordinary degree of concordance between many studies of ligand binding and studies of ligand effects on mAChR agonist function for a wide range of antagonists (Caulfield, 1993). Thus, the most widespread and reliable approach to the pharmacological characterization of mAChR subtypes remains the use of antagonist ligands. One caveat to this approach is that, due to the high degree of sequence homology between the orthosteric binding sites across the 5 mAChRs, a battery of antagonists is required to characterize each mAChR subtype; no antagonist is yet readily available that can conclusively demonstrate the involvement of one mAChR to the exclusion of all others.

Pharmacologically, M_1 mAChRs are generally defined by a high affinity for pirenzepine and 4-diphenylacetoxy-N-methyl-piperidine methiodide (4-DAMP), an intermediate affinity for *para*-fluoro-hexahydrosiladifenidol (pFHHSiD), and a low affinity for the polymethylene tetraamine methoctramine or the alkaloid himbacine (Table 5-1; Figures 5-4 and 5-5). More recently, MT7, a toxin derived from the venom of the black mamba (*Dendroaspis polylepsis*), has been shown to possess a remarkable degree of selectivity for the M_1 mAChR over all other subtypes (Caulfield and Birdsall, 1998). M_2 mAChRs are defined by a high affinity for tripitramine (Table 5-1; Figure 5-5), 5,11-dihydro-11-[2-[2-[(N, N-dipropylaminom ethyl)piperidin-1-yl]ethylamino]-carbonyl] 6H-pyrido[2,3-b][1,4]benzodiazepin-6-one (AF-DX 384; Table 5-1; Figure 5-4), methoctramine, and himbacine, and a low affinity for pirenzepine, 4-DAMP, and pFHHSiD. In addition, the snake venom m2-toxin can also be used to pharmacologically define the M_2 mAChR with a high degree of selectivity over other mAChR subtypes (Carsi et al., 1999). M_3 mAChRs are defined by a high affinity for 4-DAMP, pFHHSiD, and darifencacin, and a low affinity for pirenzepine, methoctramine, tripitramine, and himbacine (Table 5-1). Useful tools for studying the M_4 mAChR include (S)-(+)-(4aR, 10bR)-3,4,4a, 10b-tetrahydro-4-propyl-2H, 5H-[1]benzopyrano-[4,3-b]-1,4-oxazin-9-ol (PD102807; Table 5-1; Figure 5-4), which has highest affinity for the M_4 mAChR relative to all other subtypes, and the toxin MT3, isolated from the venom of the green mamba, *Dendroaspis angusticeps*, which shows high-affinity binding to M_4 mAChRs, reasonable binding to M_1 mAChRs, and minimal binding to the other subtypes; this toxin was originally called "m4-toxin" (Liang et al., 1996). The use of himbacine and pFHHSiD also helps to distinguish the M_4 from the M_2 mAChR, as both compounds have a similar (high) affinity for the M_4 mAChR, but only himbacine retains a high affinity for the M_2 mAChR (Table 5-1).

The pharmacological characterization of M_5 mAChRs has traditionally been difficult, especially in terms of differentiating this subtype from the M_3 mAChR. However, the use of antagonists such as 11-((4-[4-(diethylamino)butyl]-1-piperidinyl)acetyl)-5,11-dihydro-6H-pyrido(2,3-b)(1,4)benzodiazepine-6-one (AQ-RA 741; Figure 5-5), which has *lowest* affinity for the M_5 mAChR,

Table 5-1. Selected Antagonist Dissociation Constants (pK_B Values) for Mammalian Muscarinic Acetylcholine Receptors

Antagonist	Receptor Subtype				
	M_1	M_2	M_3	M_4	M_5
Atropine	9.0–9.7	8.8–9.3	8.9–9.8	8.9–9.6	8.9–9.7
Pirenzepine	7.8–8.5	6.3–6.7	6.7–7.1	7.1–8.1	6.2–7.2
Methoctramine	7.1–7.8	7.8–8.7	6.3–7.0	7.4–8.1	6.9–7.2
4-DAMP	8.6–9.4	7.8–8.4	8.9–9.3	8.4–9.4	8.3–9.0
Himbacine	6.8–7.2	7.7–8.3	6.9–7.4	7.5–8.8	6.1–6.3
AF-DX 384	7.3–7.5	8.2–9.0	7.2–7.8	8.0–8.7	6.3
Tripitramine	8.4–8.9	9.4–9.9	7.1–7.8	7.8–8.5	7.3–7.5
Darifenacin	7.5–7.8	7.0–7.4	8.4–8.9	7.7–8.0	8.0–8.1
PD 102807	5.3	5.7–5.8	6.2–6.3	7.3	5.2
AQRA741	7.6	8.9	7.5	8.0	6.0
*p*FHHSiD	7.4–7.7	6.7–6.9	7.7–7.8	7.2–7.5	6.9
MT3	7.1	<6	<6	8.7–9.0	<6
MT7	6.7–7.1	<6	<6	<6	<6

Sources: Adapted from Caulfield and Birdsall (1998); Eglen (1998); Eglen et al. (2001); Eglen and Watson (1996); Liang et al. (1996); Potter (2001).

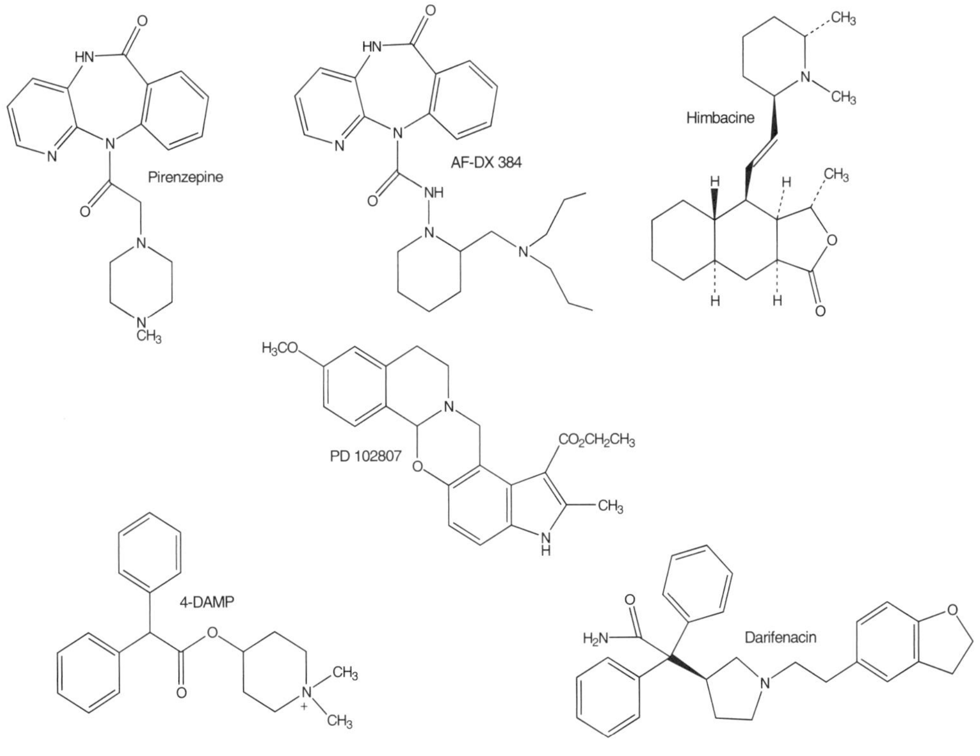

Figure 5-4. Selective mAChR antagonists.

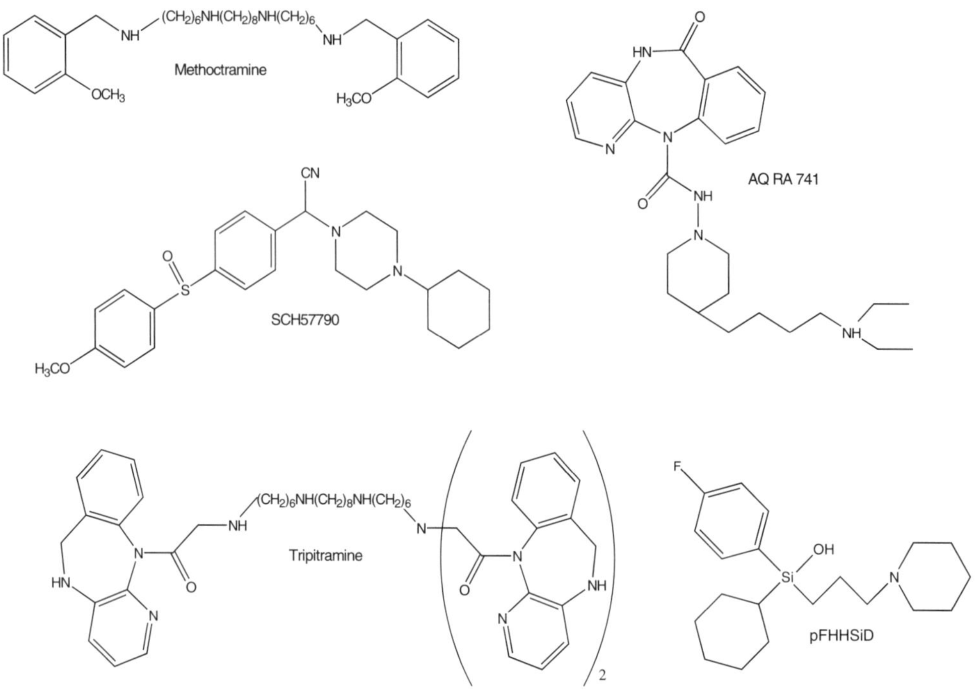

Figure 5-5. Selective mAChR antagonists (continued).

while retaining a reasonable affinity for the M_3 mAChR (Table 5-1), now allows for some *in vitro* discrimination. The toxin Mtα, isolated from *Dendroaspis angusticeps*, has also been shown to possess a high affinity for M_3 and M_5 mAChRs (Potter, 2001).

In contrast to the large focus on M_1 mAChRs with respect to agonist therapy, the major CNS targets for mAChR antagonists are likely to be the M_2 and M_4 mAChRs, particularly due to their association with presynaptic terminals and the control of neurotransmitter release. This rationale formed the basis for the development of 4-cyclohexyl-a-[4-[[4-methoxyphenyl] sulfinyl] - phenyl]-1-piperazineacetonitrile (SCH57790; Figure 5-5), a centrally active M_2-selective antagonist that has a 40–50-fold selectivity over M_1 mAChRs and has shown some promise in animal studies of cognition (Carey et al., 2001; Lachowicz et al., 1999). Although tripitramine also possesses a high degree of M_2 selectivity, particularly over M_3 mAChRs, its separation from M_1 and M_4 affinity is modest to weak (Felder et al., 2000), and its physicochemical properties are likely to ensure poor brain penetration *in vivo*.

3. Allosteric Modulators

In addition to the orthosteric site recognized by agonists and competitive antagonists, mAChRs possess at least one allosteric binding site located at a more extracellular level relative to the orthosteric domain (Christopoulos et al., 1998a). This finding has led to the exciting prospect of developing better mAChR subtype-selective agents, because the receptor epitopes thought to be involved in the binding of allosteric ligands show greater sequence divergence across receptor subtypes than does the orthosteric site (Christopoulos et al., 1998a; Ellis, 1997). Indeed, the mAChRs are generally considered a model system with which to illustrate the properties of allosteric modulation of many different types of GPCRs (Christopoulos, 2002; Christopoulos and Kenakin, 2002).

An important feature of allosteric modulation of receptors is that the binding of ligands to the orthosteric site may be elevated (allosteric enhancement) as well as reduced (allosteric inhibition). For example, the neuromuscular blocker alcuronium (Figure 5-6) is able to allosterically

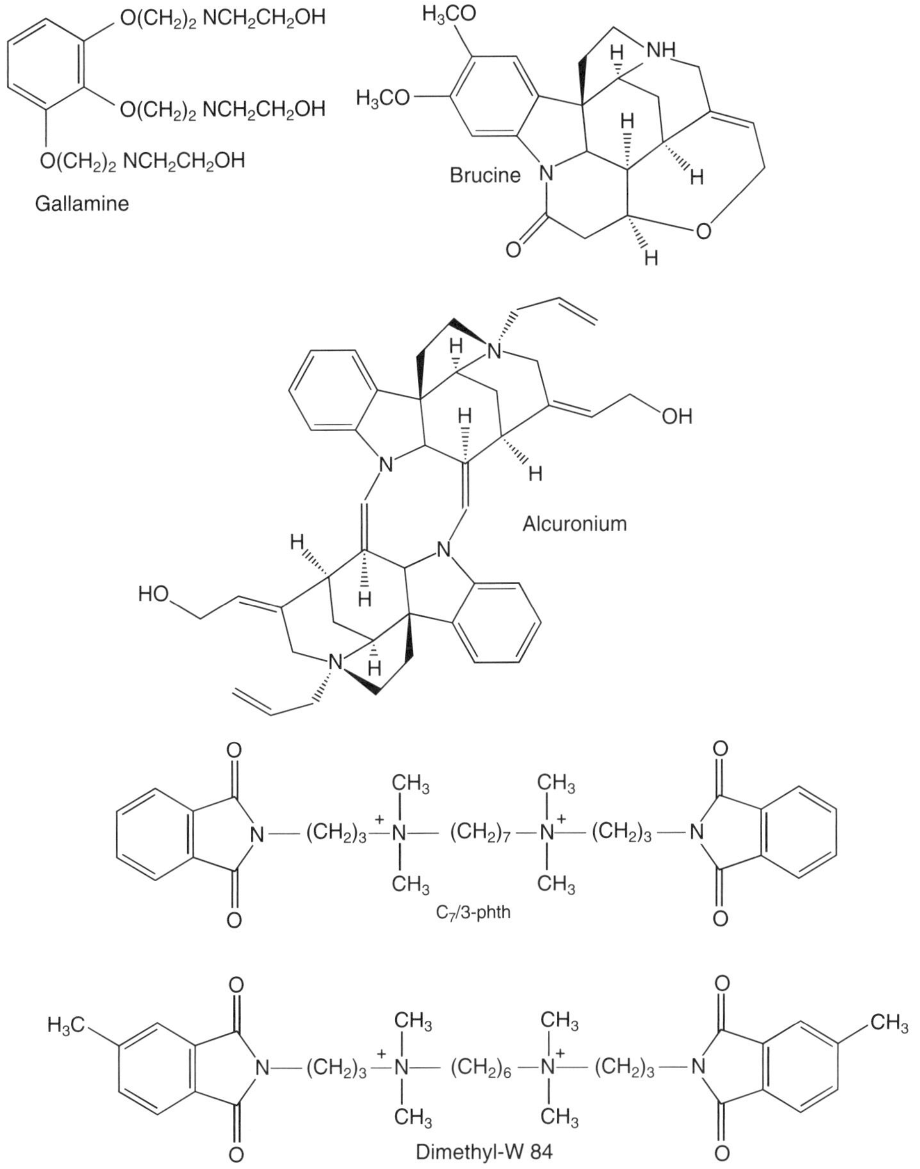

Gallamine

Brucine

Alcuronium

C_7/3-phth

Dimethyl-W 84

Figure 5-6. Allosteric modulators of mAChRs.

enhance the binding of the antagonist [^{3}H] NMS at the M_2 and M_4 mAChRs while inhibiting the same ligand at the M_1, M_3, and M_5 mAChRs (Jakubík et al., 1995b). This finding also illustrates another important concept in the design and exploitation of allosteric ligands, namely, that the same modulator tested in combination with the same orthosteric ligand can yield markedly different effects depending on the receptor subtype.

The magnitude and direction of the allosteric interaction between orthosteric and allosteric binding sites at equilibrium is often referred to as the "cooperativity" between the sites, and can be quantified using a simple ternary complex model of ligand-receptor interactions (Christopoulos and Kenakin, 2002; Ehlert, 1988; Lazareno and Birdsall, 1995; Stockton et al., 1983). There is no correlation between a modulator's affinity for the allosteric site on the free receptor and its cooperativity with an orthosteric ligand (Christopoulos et al., 1999b), so each of these parameters can, in theory, be manipulated independently of the others in terms of structure-activity studies.

Mechanistically, most allosteric modulators of mAChRs mediate their cooperative effects on orthosteric affinity by altering the rates of orthosteric ligand association with, and/or dissociation from, the receptor (Christopoulos, 2000). Indeed, with a limited number of exceptions, the ability to retard orthosteric radioligand dissociation is a hallmark of allosteric modulators of the mAChRs (Ellis, 1997); the dissociation kinetic binding assay thus serves as a sensitive means for detecting allosteric drugs (Christopoulos, 2002; Lazareno and Birdsall, 1995).

A general structural feature of many allosteric modulators of the mAChRs is the presence of a positively charged nitrogen (Holzgrabe and Mohr, 1998). This is certainly the case with the three best-studied examples of mAChR allosteric modulators, gallamine (Clark and Mitchelson, 1976; Stockton et al., 1983), alcuronium (Jakubík et al., 1995b; Proska and Tucek, 1994; Tucek et al., 1990), and bis(ammonium)alkane derivatives such as heptane-1,7-bis-(dimethly-3′-phthalimidopropyl)-ammonium bromide (C_7/3-phth; Figure 5-6) and its hexamethylene congener, W84 (Christopoulos and Mitchelson, 1994; Christopoulos et al., 1999b; Holzgrabe and Mohr, 1998; Kostenis et al., 1994; Lanzafame et al., 1996; Mohr et al., 1992; Tränkle and Mohr, 1997). Although gallamine is considered a poor lead compound for further structure-activity studies, the remaining compounds have provided impetus for additional refinements in structure-activity studies. For example, alkaloids structurally related to alcuronium, such as strychnine, brucine, and additional derivatives, have been shown to exert differential cooperative effects not only against [^{3}H] NMS but, importantly, with ACh itself at different mAChR subtypes (Gharagozloo et al., 1999; Jakubík et al., 1997; Lazareno and Birdsall, 1995; Lazareno et al., 1998; Proska and Tucek, 1995). These findings suggest that allosteric enhancers of ACh binding are an attainable therapeutic strategy in the treatment of Alzheimer's disease (Birdsall et al., 1997).

Studies on bis (ammonium) alkane derivatives, originally derived from the structure of hexamethonium, have also yielded positive initial results. Historically, these compounds are among the first identified examples of allosteric modulators of mAChRs, and have the additional advantage of proven tolerability in animals (Kords et al., 1968; Lüllman et al., 1969). In addition, these substances show minimal cross-interaction at other GPCR types (Franken et al., 2000; Pfaffendorf et al., 2000) and have resulted in the discovery of dimethyl-W84 (Figure 5-6), to date the only ligand of sufficient affinity to radiolabel an allosteric site on the M_2 mAChR (Tränkle et al., 1998, 1999). An elegant series of studies by Mohr and colleagues has eludicated many structure-activity requirements for the interaction of these compounds with the M_2 mAChR allosteric site (Holzgrabe and Mohr, 1998). Other compounds that have had reasonable systematic investigation as allosteric ligands of mAChRs include esters of α-truxillic acid (Lysikova et al., 1999; Urbansky et al., 1999).

Most recently, Lazareno, Birdsall, and colleagues have provided evidence for a second allosteric site on M_1–M_4 mAChRs, distinct from that recognized by the compounds described in the preceding paragraphs, that interacts with indolocarbazoles and a series of androstane derivatives (Lazareno et al., 2000, 2002). The nature of potential interactions between this second allosteric site and the better-characterized gallamine-binding site remain to be determined.

Interestingly, there are some studies that have raised the possibility of allosteric agonists of the mAChRs, that is, compounds that can activate the receptor in their own right in the absence of orthosteric ligand. An early study by Birdsall et al. (1983) revealed that the binding properties of the agonist McN-A-343 at the M_2 mAChR receptor in the heart are consistent with an allosteric mechanism. More recently, Angeli et al. (2002) have described the effects of a series of deoxamuscaroneoxime derivatives as both agonists of the M_2 mAChR and allosteric modulators of the same receptor. Intriguingly, Jakubík et al. (1996, 1998) have demonstrated agonistlike properties of "classical" modulators, such as gallamine and alcuronium, in recombinant expression and receptor reconstitution systems. These latter findings have not been confirmed by others, however, and may reflect specific properties of the stoichiometry of receptors to G proteins in artificial expression systems.

Despite a long-standing interest in allosteric modulators of mAChRs by some researchers, the overall concept of GPCR allosteric modulation remains relatively underdeveloped compared to the far larger body of drug discovery efforts that have targeted the traditional, orthosteric site (Christopoulos, 2002). In large part, this reflects the bias of high-throughput drug discovery toward assays that rely on orthosteric ligands as probes, but this is now changing with the advent of more sensitive, functional screening methods (Christopoulos and Kenakin, 2002; May and Christopoulos, 2003). The pursuit of allosteric ligands of GPCRs is thus timely and worthwhile because allosteric modulators offer a number of advantages that are not readily attainable with orthosteric drugs, including the potential for greater receptor subtype selectivity and the ability to fine-tune, either positively or negatively, physiological processes while maintaining the spatial and temporal profile associated with endogenous neurohumoral signaling (Birdsall et al., 1996; Christopoulos and Kenakin, 2002).

C. Localization and Distribution

The main hindrance to the identification and localization of mAChR subtypes has traditionally been the lack of subtype-specific markers. This is particularly so for studies in the brain, which expresses all 5 mAChRs in different regions and to differing extents. Nevertheless, significant progress has been made over the last decade or so due to the application of three approaches to mapping the distribution and localization of mAChRs: radioligand binding studies, immunoprecipitation using subtype-selective antibodies, and in situ hybridization histochemistry. Overall, the CNS distribution of mAChRs determined by these three approaches has shown general concordance, although each technique possesses specific advantages and disadvantages with respect to the fidelity of the information it provides.

The earliest studies on the distribution of central mAChRs relied on radioligand binding using the nonselective, high-affinity radioligand [^{3}H] quinuclidinyl benzilate ([^{3}H] QNB; Snyder et al., 1975). These studies found that the density of mAChRs was highest in the forebrain, and tended to parallel the distribution of other cholinergic markers such as acetylcholinesterase and choline acetyltransferase. The distribution of mAChRs in more caudal regions of the brain was significantly less than that in the forebrain. Subsequent refinements in the radioligand-based approach to mapping mAChRs in the CNS involved the use of the more hydrophilic antagonist radioligand [^{3}H] NMS as well as [^{3}H] QNB, in conjunction with relatively subtype-selective antagonists (see Ehlert et al., 1995). Quantitative autoradiography has also been used to study mAChR distribution in the brain (Smith et al., 1991). An advantage of this approach, compared to the standard homogenate-based binding assays, is that it allows for a better visualization of anatomical receptor localization.

In situ hybridization studies have also yielded useful information regarding central mAChR distribution. M_1 mAChR mRNA is readily detected in the cerebral cortex, hippocampus, thalamus, caudate-putamen, amygdala, olfactory bulb, olfactory tubercle, and dentate gyrus (Buckley et al., 1988; Caulfield, 1993; Vilaro et al., 1994). M_2 mRNA is detected in the basal forebrain, caudate-putamen, hippocampus, hypothalamus, amygdala, and pontine nuclei. M_3 mRNA is found in the olfactory tubercle, cerebral cortex, hippocampus, thalamus, caudate-putamen, and amygdala (Buckley et al., 1988), whereas M_4 mRNA is highest in the olfactory bulb, olfactory tubercle, hippocampus, and striatum (Caulfield, 1993). M_5 mRNA was detected by Weiner et al. (1990) in the substantia nigra pars compacta.

In a recent study, Krejci and Tucek (2002) adopted a quantitative approach to determining the absolute levels of mAChR mRNA in rat cortex using competitive RT-PCR. They were able to accurately resolve the message into the following percentages: M_1 (36%), M_2 (21%), M_3 (25%), M_4 (11%) and M_5 (7%). The relatively high percentage of M_3 mAChR mRNA and low percentage of M_4 mAChR mRNA is at odds with previous studies using other approaches and requires further investigation. It should be noted, however, that although mRNA-based techniques are very sensitive, the study of mRNA levels alone cannot be used to quantify the absolute levels and distribution of expressed receptor protein; the presence of message does not mean that the protein is expressed at an equivalent level.

The generation of mAChR subtype-selective antibodies was spearheaded by the work of Levey and colleagues (Dorje et al., 1991; Levey et al., 1991, 1994) and Wolfe and coworkers (Li et al., 1991; Wall et al., 1991a, 1991b; Yasuda et al., 1993) and has provided useful information on relative mAChR subtype abundance in the brain and on cellular and subcellular distribution of some of the mAChR subtypes. The subtype-specific antisera in these studies were raised against the third intracellular loop of each mAChR, because it is within this region that the greatest sequence divergence occurs between subtypes (see section D). However, two drawbacks of the immunoprecipitation approach remain the variability in the sensitivity of some of the polyclonal antisera, and in the accessibility, or lack thereof, of antibody epitopes to intracellular receptor domains (Flynn et al., 1997). For example, an early study by Levey et al. (1991) found no immunoprecipitation with an M_5 mAChR antibody, whereas a later study by Yasuda et al. (1993) could identify low levels of M_5 immunoreactivity in the striatum, hippocampus, pons medulla, and cerebellum.

Overall, standard radioligand-based approaches remain the most extensively applied methods to study the distribution of mAChRs in the CNS. Although these approaches are able to resolve M_1 and M_2 mAChRs from the other 3 subtypes in the brain, they show less success in differentiating M_3, M_4, and M_5 mAChRs due to the overlapping affinities of various ligands for the 5 mAChRs. However, elegant adaptations that exploit differences in the dissociation kinetic profile of antagonists such as [^{3}H] NMS, in the absence or presence of selective receptor antagonists, have been used to provide better discrimination among all 5 subtypes (Flynn et al., 1997; Waelbroeck et al., 1990).

Recently, Flynn and colleagues (Flynn et al., 1997; Reever et al., 1997) used quantitative autoradiography with optimized labeling conditions to provide an exhaustive study of the distribution of the 5 mAChRs in the rat brain. The key to this particular series of studies was the use of specific cocktails of subtype-selective antagonists to "mask" the presence of all mAChR subtypes except the desired receptor under study. These experiments revealed regions exhibiting the highest levels of M_1 mAChRs to be the hippocampus, nucleus accumbens, and caudate putamen. Labeling was also noted in the molecular layer of the dentate gyryus. Some M_1 mAChRs were found in the olfactory tubercle in the basal forebrain. These regions of labeling are generally consistent with previous observations using autoradiography (Mash and Potter, 1986; Spencer et al., 1986), immunocytochemistry (Levey et al., 1991), and in situ hybridization (Buckley et al., 1988; Vilaro et al., 1994), but some discrepancies were also noted, particularly in the diagonal band of Broca, which failed to reveal M_1 labeling in the binding studies (Flynn et al., 1997). For the M_2 mAChRs, labeling was observed in the occipital region of the cerebral cortex, the dorsal region of the caudate, the olfactory tubercle and the nucleus accumbens. Intense labeling was observed in the superficial layers of the superior and inferior colliculi. Additional labeling was found in the pontine and parabrachial nuclei, the motor trigeminal, and the facial nuclei in the brainstem, and some labeling was also present in the cerebellum. These findings were in general agreement with previous studies (Mash and Potter, 1986; Spencer et al, 1986; Levey et al, 1991), but disagreed with in situ experiments that had failed to find M_2 mRNA in the cortex and striatum (Buckley et al, 1988; Vilaro et al., 1994). In general, binding for the M_3 mAChR revealed lower intensities of labeling than the M_1, M_2, and M_4 mAChRs (Flynn et al., 1997; Reever et al., 1997). Diffuse M_3 mAChR was found across cortical laminae, as well as the CA1, CA2, and CA3 regions of the hippocampus (Flynn et al, 1997). This is generally consistent with previous reports using different methodologies (Buckley et al., 1988; Levey et al., 1994), with the exception of the striatum, which does not reveal M_3 mRNA but does display immunoreactivity and radiolabeling to the M_3 mAChR. Labeling of M_4 mAChRs was found in the caudate putamen, the nucleus accumbens, posterior regions of the cortex, CA1 the hippocampal region, and the outer layer of the anterior olfactory nucleus. Within the basal forebrain, dense labeling was noted in the olfactory tubercle. In the pontine, facial, and trigeminal motor nuclei of the brainstem, there was significant overlap of M_4 labeling with M_2 labeling. For the M_5 mAChRs, the most intense labeling was observed in the outermost layer of the cortex, but also seen in the caudate putamen, nucleus accumbens, CA1 and CA2 hippocampal regions, and polymorphic layer of the dentate gyrus. The olfactory tubercle was the most intensely stained region in the basal forebrain, and some labeling was also noted in the superior colliculus (Flynn et al., 1997; Reever et al., 1997). Figure 5-7 summarizes some of the general findings from this series of studies with respect to the relative abundance of mAChR subtypes in various brain regions.

Given that the M_1 and M_4 mAChRs are most abundant in the cerebral cortex, basal ganglia and hippocampus, it is possible that these receptors play a role in cognitive disorders, schizophrenic symptoms, and movement disorders. Based on their putative localization on cholinergic neurons in the basal forebrain, striatum, and cranial nerves, M_2 mAChRs have been generally classified as autoreceptors (Mash et al., 1985; Mash and Potter, 1986; Spencer et al., 1986), but also act as heteroreceptors that modulate the release of other transmitters in addition to ACh. M_3 mAChRs have been postulated to play a role in mediating GABA release in the striatum (Raiteri et al., 1990), and to inhibit glutamate release in the accumbens, amygdala and striatum (Sugita et al., 1991), suggesting some role in dyskinesias and perhaps epilepsy. The presence of M_5 mAChRs in the striatum, nucleus accumbens, substantia nigra pars compacta, and ventral tegmental area is consistent with their postulated role of controlling dopamine release in these regions of the brain. Importantly, the potential physiological roles of the mAChRs gleaned from studies of receptor localization and distribution are now being supplemented by findings made using mAChR knockout mice (see Section F).

D. Molecular Biology

1. Receptor Subtypes

The initial cloning of the M_1 and M_2 mAChRs by Numa and colleagues (Kubo et al., 1986a, 1986b), followed by the cloning of the M_3, M_4, and M_5 genes (Bonner et al., 1987, 1988; Peralta et al., 1987), confirmed that the mAChRs are glycoproteins that share a number of structural features associated with the superfamily of GPCRs (Hulme et al., 1990; Wess, 1993a, 1993b). All such receptors are believed to exist as 7 transmembrane-spanning, α-helical domains connected by 3 extracellular (e1, e2, e3) and 3 intracellular (i1, i2, i3) loops. The N-terminal of the receptors is also located extracellularly while the C-terminal is intracellular (Hulme et al., 1990). All vertebrate mAChR genes cloned to date are intronless within the coding regions (Caulfield and Birdsall, 1998). The chromosomal localization of the human mAChR genes are as follows (Bonner et al.,

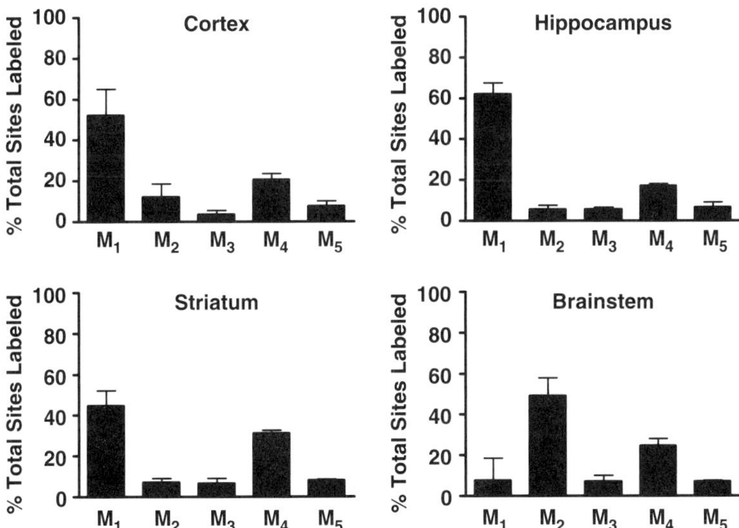

Figure 5-7. Percentages of total [3H]N-methylscopolamine binding sites in selected regions of rat brain. (Data replotted from Flynn et al., 1997.)

1991): M_1, 11q12–13; M_2, 7q35–36; M_3, 1q43–44; M_4, 11p12–11.2; M_5, 15q26. Phylogenetic analysis reveals that the mAChRs have evolved from a common ancestor (Horn et al., 2000). The first major split in the phylogenetic tree corresponds with the preferential coupling of the odd-numbered mAChRs to $G_{q/11}$ proteins, and the even-numbered mAChRs to the $G_{i/o}$ proteins (Figure 5-8).

The amino acid composition of the human mAChR subtypes varies from 460 amino acids for the M_1 mAChR to 466 for the M_2 mAChR, 479 for the M_4 mAChR, 532 for the M_5 mAChR, and 589 for the M_3 mAChR. Primary sequence alignment reveals that the individual subtypes share approximately 145 invariant amino acid residues (Wess, 1993a, 1993b) and show between 89% and 98% common amino acid identity across various mammalian species (Bonner, 1989a, 1989b; also, see Figure 5-8). Most differences between mAChR subtypes reside in the extracellular amino termini, the i3 loop, and the cytoplasmic carboxyl terminus. The greatest divergence occurs within the i3 loop, which varies from 156 amino acids for the M_1 mAChR to 239 amino acids for the M_3 mAChR, and accounts for 34% to 45% of the total amino acids in the mAChRs (Ehlert et al., 1995). Within the TM segments, there is 63% identity across the 5 mAChR subtypes, but it is even greater when the M_1, M_3, and M_5 mAChRs are grouped together as one homologous group, and the M_2 and M_4 mAChRs are grouped together as another.

The high degree of homology within the TM domains may certainly explain some of the difficulty in designing highly selective mAChR agonists and antagonists. It should be noted, however, that similar findings for other GPCR families have often been followed by the demonstration that even subtle differences in amino acid composition, as occur between species, for instance, can cause profound effects on ligand pharmacology (see Kenakin, 1996). Although this has generally not been observed for the mAChRs, one should remain cognizant of the possibility that novel ligands targeting less conserved receptor sequences can show markedly different pharmacologies across subtype and species. One such approach already being pursued is the targeting of extracellular mAChR allosteric sites (see section B, above).

A recent milestone in the field of structural biology was the determination of the crystal structure of the GPCR bovine rhodopsin in its ground state at 2.8 Å resolution (Palczewski et al., 2000). Rhodopsin is the prototypical member of the Family A GPCR superfamily, which is the largest subgrouping of all GPCRs and includes the mAChRs. The crystal structure of rhodopsin con-

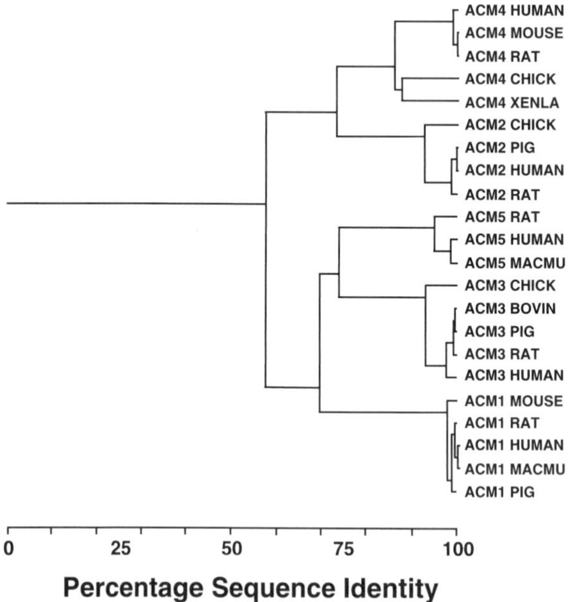

Figure 5-8. Phylogenetic tree for the mAChRs. MACMU, rhesus macaque; BOVIN, bovine; XELA, xenopus laevis. (Adapted from Horn et al., 2000.)

firmed the 7 transmembrane-spanning topology of this receptor and many other predicted structural properties, but also revealed additional interesting features, such as a fourth intracellular loop that forms an eighth cytoplasmic helix at right angles to the C terminus, a "kink" in TM2, and distortions from ideal α-helicity in TMs 5 and 7. Sequence conservation across many GPCR families suggests that these features are likely to contribute to the structure and function of the mAChRs. Thus, the determination of the crystal structure of rhodopsin allows further testing and refinement, for the first time, of the various structure-function models of the mAChRs that have been developed over the last decade and a half.

2. Structural Determinants of Ligand Binding

The 7 transmembrane helices (TM1–TM7) of the mAChRs are arranged in a ringlike, "helical-wheel," structure (Hulme et al., 1990). This model shows many of the conserved amino acids to face the inner, hydrophilic "pore" of this arrangement, and experimental evidence of such an arrangement has been provided (Pittel and Wess, 1994). ACh is thought to bind within the transmembrane-spanning regions of the mAChRs about 10 to 15 Å from the extracellular surface (Hulme et al., 1990). Early studies on the structural requirements for mAChR ligand-receptor recognition highlighted the role of a positive charge on the ligand for activity at the receptor (e.g., Burgen, 1965). This requirement has been borne out for many classic mAChR agonists and antagonists, and suggests the presence of an anionic site on the receptor that would serve a complementary recognition role for the positive ligand charge. The mAChRs all posses a conserved aspartate residue within TM3 (Asp147 in the rat M_3 mAChR sequence; Figure 5-9) that is believed to act as the counterion for the positively charged amino headgroup of ACh and other amine ligands. This feature is also common for other GPCRs that bind biogenic amine agonists (Strader et al., 1994). Covalent affinity-labeling experiments using the irreversibly binding agonist acetylcholine mustard or the antagonist propylylbenzilylcholine mustard have found this aspartate residue to be specifically alkylated (Curtis et al., 1989; Spalding et al., 1994), and mutagenesis of this amino acid leads to a marked reduction or loss of agonist binding (Fraser et al., 1989; Hulme et al., 1995; Schwarz et al., 1995).

Despite its obvious importance to agonist binding, the conservation of the TM3 aspartate

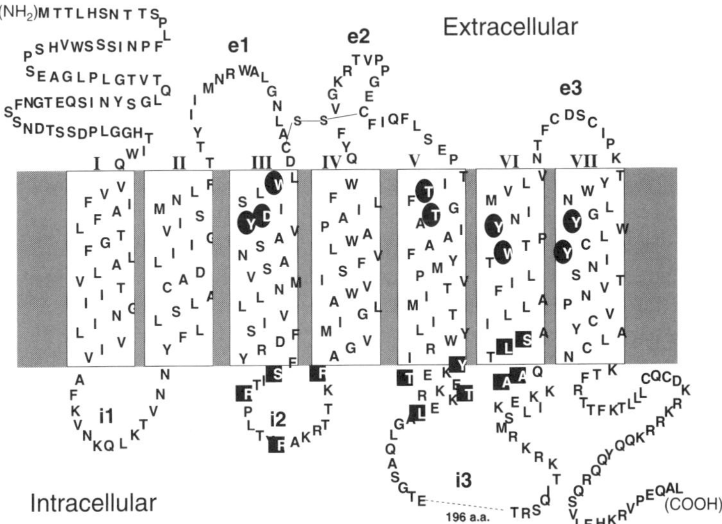

Figure 5-9. Snake diagram of the rat M_3 mAChR. Amino acids in circles have been implicated in defining the binding domain for ACh. Amino acids in squares have been shown to determine G protein coupling specificity.

residue across most GPCRs that recognize biogenic amines implies that this residue alone cannot account for the specificity of binding of ACh to the mAChR family. Accordingly, there are a number of other critical residues specific to the mAChRs that help determine the binding of the endogenous agonist. Using the rat M_3 mAChR as a reference (Figure 5-9), Tyr148, Thr231, Thr234, Tyr506, Tyr529, and Tyr533 have all been implicated in ACh binding specificity (Wess et al., 1995). These 6 amino acids are absolutely conserved across the mAChRs but not most other GPCRs (Wess et al., 1995). In addition, there are 2 conserved tryptophan residues (Trp143 and Trp503, using the rat M_3 mAChR as a reference) that can also affect ACh binding through an interaction that utilizes the energy provided by the aromatic character of these residues (Wess et al., 1995). Some of the amino acids important for ACh binding (e.g., Asp147, Trp143, Trp503, Tyr529) are also important for the binding of tropanelike antagonists, such as atropine and NMS, but it is acknowledged that antagonists utilize additional contact points that do not appear to influence agonist binding, and that other amino acids important for agonist binding do not contribute significantly to the binding of antagonists (Wess et al., 1995).

Extensive homology modeling of the M_1 mAChR, supplemented by the use of the crystal structure of bovine rhodopsin as a template, has recently been performed by Hulme and colleagues (Lu et al., 1997, 2001, 2002; Lu and Hulme, 1999, 2000). The ground state of the receptor is characterized by a number of intramolecular hydrogen-bonding networks and van der Waals interactions between highly conserved amino acid residues in TMs 1-4 and 6-7 (Lu et al., 2001, 2002). Molecular docking studies, using ACh and NMS, confirmed much of the mapping data of the orthosteric binding site obtained previously from studies using point and/or scanning mutagenesis (Allman et al., 2000; Lu et al., 1997; Lu and Hulme, 1999; Ward et al., 1999; Wess, 1996). This includes a positioning of the quaternary ammonium headgroup of both agonist and antagonist within a charge-stabilized aromatic cage formed by contacts in TM3, TM6, and TM7. However, differences between the mode of agonist and antagonist binding were also predicted from the docking studies, again in agreement with experimental observations. For instance, the smaller acetoxy side chain of ACh cannot form the same type of stabilizing inter-helical interactions that are available to the bulkier tropic acid and phenyl ring side chains of NMS; ACh thus relies on additional stabilizing interactions provided by amino acids near the top of TM5, whereas NMS does not (Allman et al., 2000; Lu et al., 2002; Ward et al., 1999).

Although there is abundant pharmacological evidence for the presence of distinct allosteric binding sites on all 5 mAChRs (see section B, above), far less is known about the molecular composition of these sites relative to the orthosteric domain. Nevertheless, there have been a number of investigations, primarily based on site-directed mutagenesis and/or the construction of receptor chimeras that have begun to reveal a likely arrangement of the allosteric domain recognized by prototypical modulators such as gallamine and alcuronium. For example, since allosteric interactions are evident at all 5 subtypes, it is likely that some conserved residues are involved, such as Trp101, Trp400, and possibly Asp71 (using the M_1 mAChR sequence) (Christopoulos et al., 1998a; Matsui et al., 1995). A quartet of acidic amino acids (Glu172, Asp173, Gly174, Glu175), unique to the second extracellular loop of the M_2 mAChR, may provide a further extracellular point of attraction and stabilization for charged modulators, and this may explain why the M_2 mAChR appears to be the most readily modulated of all the mAChRs (Leppik et al., 1994). It is also likely that other epitopes in the second and third outer loops of the receptors that contain acidic amino acids play a fundamental role in providing subtype selectivity for different muscarinic allosteric modulators (Ellis and Seidenberg, 2000). Recent studies by Mohr, Ellis and colleagues (Buller et al., 2002; Voigtlander et al., 2003) have identified 2 key amino acids (Tyr177, Thr423) responsible for the selectivity of alkane bis-ammonium-type allosteric modulators of the M_2 mAChR. Preliminary molecular modeling of the M_1 (Birdsall et al., 2001) and M_2 mAChRs (A. Christopoulos unpublished data), based on 2.8 Å crystal structure of rhodopsin (Palczewski et al., 2000), has found general agreement with the preceding speculations. In particular, a region of conserved extracellular residues above TMs 5–7 form a cleft that appears to act as an "entrance" to the deeper orthosteric binding site; this spatial arrangement can explain the dramatic slowing effects many muscarinic allosteric modulators have on orthosteric ligand kinetics (Christopoulos and Kenakin, 2002).

3. Structural Determinants of Receptor Activation and G Protein Coupling

There is a large body of experimental evidence in favor of the i3 loop of the mAChRs as a major determinant of intracellular coupling to G proteins. Much of this work was initially based on chimeric receptor studies (England et al., 1991; Kubo et al., 1988; Lechleiter et al., 1990; Wess et al., 1989, 1990; Wong et al., 1990). For example, the construction of a receptor chimera where the i3 loop of the $G_{i/o}$-coupled dopamine D_2 receptor was replaced with the i3 loop of the $G_{q/11}$-coupled M_1 mAChR led to the generation of classic M_1 mAChR responses, such as intracellular calcium mobilization, by dopamine receptor ligands; this response is not normally observed at the wild type D_2 receptor (England et al., 1991). Similar exchange of the i3 loops of the M_1 mAChR with the β_2 adrenoceptor (G_s-coupled) leads to an M_1 chimera that responds to mAChR agonists with a typical β_2 adrenoceptor response, the stimulation of adenylyl cyclase activity (Wong et al. 1990). Extension of these studies to include chimeras constructed using regions from different subtypes of mAChR have further confirmed the role of the i3 loop as important in G protein coupling.

Mutational analysis of mAChRs has also been used extensively to identify a series of amino acids, located in the i2 and i3 loops, that are major contributors to receptor-G protein coupling specificity (Kostenis et al., 1998b; Figure 5-9). For instance, a major hydrophobic surface that is critical for receptor-G protein recognition is formed by amino acids at the N-terminus of the i3 loop; this is common to more than just the mAChR family of GPCRs (Bluml et al., 1994a, 1994b; Strader et al., 1994). A second hydrophobic surface that helps determine coupling specificity is comprised of a set of 4 amino acids (Ala488, Ala489, Leu492, and Ser493 in the rat M_3 mAChR) at the C terminal of the i3 loop (Blin et al., 1995; Burstein et al., 1995b; Liu et al., 1995). Within the i2 loop, there is also a set of 4 amino acids (Ser168, Arg171, Arg176, and Arg183 of the rat M_3 mAChR) that help determine $G_{q/11}$-protein coupling specificity. Substitution of either set (i2 or i3) of the aforementioned amino acid quadruplets from the M_3 mAChR into the M_2 mAChR yielded a mutant receptor that was functionally similar to the wild type M_3 mAChR (Blin et al., 1995). These M_3 mAChR G protein coupling residues are highly conserved in the M_1 and M_5 mAChRs, consistent with the known coupling preference of these receptors to the $G_{q/11}$-protein signaling pathway (Kostenis et al., 1997b, 1997c; Kunkel and Peralta, 1993; Lee et al., 1996).

In a series of studies on the $G_{i/o}$-coupled M_2 mAChR, Liu et al. (1995) identified 4 amino acids (Val385, Thr386, Ile389, and Leu390; "VTIL" motif) at the junction of the i3 loop with TM6 that appeared to contain a specific sequence element recognized by the C terminus of the Gα subunit. This finding was consistent with the work described above that mapped receptor-G protein contact points for the $G_{q/11}$-coupled mAChRs. Furthermore, the insertion of a series of alanine residues immediately C terminal to the VTIL motif led to a mutant receptor that was constitutively active in the absence of ligand. The insertion of these extra amino acids thus appeared to have led to a downward translational movement of TM6 out of the transmembrane space and, hence, an exposure of some of the critical residues in the i3/TM6 junction to the cytoplasm. This mechanism was thus postulated to mimic a final step of the conformational change caused by agonist activation of the wild-type receptor (Liu et al., 1996).

Despite the impressive body of work elucidating important structural motifs on the receptors that help govern G protein coupling selectivity, the major determinant of the specificity of this interaction is actually encoded in the C terminus of the Gα subunit itself (Bourne, 1997). Recent studies of mAChR-G protein coupling have highlighted how single point mutations in this region of the G protein are sufficient to govern receptor selectivity. For example, Kostenis et al. (1997a) identified 2 mutant Gα_q subunits where a single G$\alpha_q \rightarrow$ Gα_i point mutation in each led to a productive interaction with the wild-type M_2 mAChR; this receptor preferentially couples to the $G_{i/o}$ family of G proteins and normally shows only a weak interaction with the $G_{q/11}$ family. Similarly, although the M_3 mAChR does not normally couple to Gα_s protein subunits, an extension of the studies of Kostenis et al. (1997a) found G$\alpha_s \rightarrow$ Gα_q point mutations that were sufficient to couple the M_3 mAChR to the mutant Gα_s (Kostenis et al., 1997c). In addition to these studies of the C terminal region of Gα proteins, mutagenesis studies of the N-terminal region of Gα protein subunits also suggest a role for this domain in governing mAChR-G protein interactions (Kostenis et al., 1997b, 1998a, 1998b). Additional interactions have been reported between the i3 loop of M_2 or M_3 mAChRs and $\beta\gamma$ subunits of the G protein heterotrimers (Taylor et al., 1996; Wu et al., 2000), suggesting further points of contact that help define coupling selectivity beyond the $G\alpha$ subunit.

Recent molecular modeling studies, again using the 2.8 Å resolution crystal structure of ground-state bovine rhodopsin, have provided further support for previously postulated mechanisms associated with activation of the mAChRs. Using the M_1 mAChR as a model protein, Lu et al. (2002) have proposed that activation "begins" with a closure of a charge-stabilized aromatic cage (see above) around the quaternary ammonium headgroup of ACh, leading to a disruption of a hydrophobic "latch" structure formed by intramolecular contacts of amino acids in TMs 3, 6, and 7; an important contribution to this series of contacts is provided by the conserved NSxxNPxxY motif. A subsequent step in the conformational change that stabilizes an active receptor state is the disruption of a charge-reinforced H-bond to the arginine in the highly conserved DRY motif found at the cytoplasmic end of TM3 (Lu et al., 1997; Lu and Hulme, 2000). The final global effect of these changes in the conformational state of the receptor is likely to be an increased mobility of TM7 and an outward movement of TM6, which expose important G protein contact residues that would otherwise remain inaccessible to the cytoplasmic milieu. This postulated mechanism is consistent with the chimeric receptor and mutagenesis studies of Wess and colleagues (described above) on receptor-G protein contacts.

In addition to the conformational changes mediated by agonist binding, contemporary models of GPCR activity postulate that these receptors are able to adopt a number of distinct conformational states even in the absence of ligand (Christopoulos and Kenakin, 2002). Some of these conformations can also mediate agonist-independent cellular signaling, a phenomenon referred to as constitutive receptor activity. Within these "multistate" models of GPCR action, ligand binding is considered to play a stabilizing role, leading to an enrichment in the abundance of specific receptor states and a diminution in others. Ligands that stabilize active (i.e., signaling) receptor states are classed as agonists, whereas ligands that stabilize inactive states are classed as inverse agonists. Neutral antagonists constitute a relatively rare class of ligand that shows equal preference for binding to both active and inactive states (Kenakin et al., 1995). Important support for this model of GPCR activity comes from the large number of experimental studies over the past decade or so that have directly revealed constitutive (ligand-independent) activation of various GPCRs, including the mAChRs (Seifert and Wenzel-Seifert, 2002).

A common method for investigating constitutive GPCR activation is by introducing activating mutations in the receptor structure. For example, substitution of Glu360 with Ala in the i3 loop of the M_1 mAChR leads to a significant enhancement in the ability of the mutant receptor to signal through the phosphoinositide (PI) pathway in the absence of agonist (Hogger et al., 1995). Furthermore, agonist potency is significantly enhanced when tested at this mutant receptor, whereas atropine behaves as an inverse agonist, causing a reduction in basal PI hydrolysis. A similar approach by Spalding et al. (1995), who mutated residues in the TM6 domain, led to the creation of a constitutively active M_5 mAChR, again revealing that classic mAChR antagonists such as atropine, NMS, QNB, pirenzepine, and 4-DAMP are all capable of acting as inverse agonists.

Importantly, inverse agonism of classic mAChR antagonists has also been demonstrated in studies using wild-type receptors in both native (Daeffler et al., 1999; Hilf and Jakobs, 1992) and recombinant expression systems (Jakubík et al., 1995a). An interesting approach by Burstein et al. (1997, 1995a) utilized the overexpression of $G\alpha_q$ subunits in a recombinant system to cause constitutive activation of wild-type M_1, M_3, and M_5 mAChRs, and revealed that *all* antagonists tested in this system behaved as inverse agonists.

4. Posttranslational Modifications and Receptor Regulation

Various residues have been identified on the mAChRs as possible sites of posttranslational modifications. For example, a series of Asn residues in the N-terminal domain may be targets for

N-glycosylation but do not appear to play an important role in cell-surface expression, ligand binding, or G protein coupling (Ohara et al., 1990; van Koppen and Nathanson, 1990). A pair of conserved Cys residues in e1 and e2 are believed to participate in the formation of a disulfide bond that may play a functional role in stable protein folding (Kurtenbach et al., 1990; Savarese et al., 1992), while a conserved Cys in the C-terminal tail may represent a possible palmitoylation site (Wess, 1993b). Indeed, Hayashi and Haga (1997) have shown that palmitoylation of this cysteine in the M_2 mAChR affects receptor-G protein coupling.

Like all GPCRs, the mAChRs are dynamic proteins that are able to cycle through cells continuously, and their relative distributions between intracellular and extracellular compartments govern the magnitude and extent of signaling. This process is commonly referred to as "GPCR trafficking," and it includes the synthesis of new receptors from the Golgi complex, the sequestration of receptors away from the cell surface, and the endocytosis of receptors to internal compartments, where they are either recycled back to the surface or are degraded (Bunemann et al., 1999; Koenig and Edwardson, 1997) (Figure 5-10).

The dynamic GPCR trafficking cycle is influenced by a number of regulatory mechanisms and posttranslational modifications that modulate the response of the receptor following exposure to agonists and, often, antagonists. For example, the second messenger-activated protein kinases, protein kinase A (PKA) and protein kinase C (PKC), have long been known to mediate heterologous desensitization of various GPCRs, including members of the mAChR family (Bunemann et al., 1999; Haddad and Rousell, 1998). This form of desensitization, however, shows a minimal dependence on agonist occupancy (Bunemann et al., 1999); activation of PKA or PKC by one type of GPCR is often sufficient to lead to phosphorylation of consensus sites (Ser and Thr residues) within the intracellular loops and carboxyl-terminal tail of another type of GPCR, irrespective of the activation status of the latter receptor. Heterologous desensitization, therefore, represents a general mechanism associated with receptor cross talk and, currently, remains relatively poorly understood when compared to the wealth of knowledge generated over the last decade on homologous receptor desensitization.

The ability of specific GPCRs to be phosphorylated as a direct consequence of occupancy by their cognate agonists initiates the phenomenon of homologous receptor desensitization. This type of phosphorylation is mediated by the family of G protein-coupled receptor kinases (GRKs), of which there are currently 7 known members. GRKs 1, 4, and 7 display very discrete, tissue-specific localization, whereas GRKs 2, 3, 5, and 6 are more widely distributed (Ferguson, 2001). In common with the second messenger-activated kinases, the GRKs also phosphorylate GPCRs on serine and threonine residues, but these are likely to be residues that become exposed only after the receptor has been occupied by an agonist. The M_1–M_4 mAChRs are all phosphorylated by different GRKs, *in vitro* and/or *in vivo* (Bunemann and Hosey, 1999). Extensive studies on the M_2 mAChR, in particular, have revealed a number of specific serine/threonine-rich motifs in the i3 loop that are sites for GRK-mediated phosphorylation (Pals-Rylaarsdam and Hosey, 1997). Comparatively less is known about the interaction of the M_5 mAChR with GRKs, although overexpression

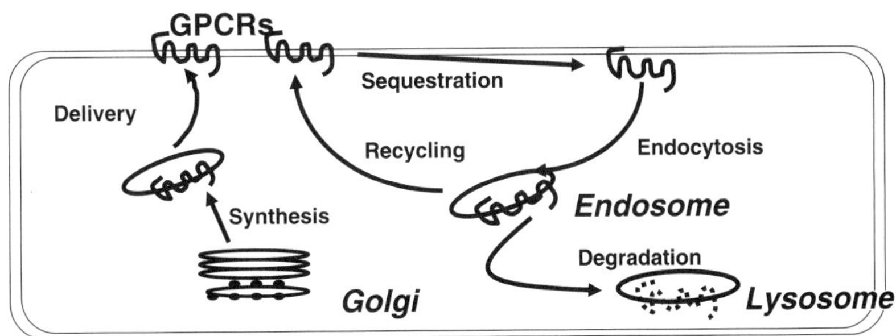

Figure 5-10. Schematic diagram of the cellular trafficking cycle of mAChRs.

of GRK2 has been found to increase the internalization of this receptor (Tsuga et al., 1998).

Receptor phosphorylation by GRKs is necessary, but generally not sufficient, to lead to complete homologous receptor desensitization. Rather, the consequence of agonist-dependent phosphorylation is to increase the receptor affinity for, and the recruitment of, soluble cytosolic proteins referred to as "arrestins," which then lead to uncoupling of the receptor from its G protein by steric inhibition (Krupnick and Benovic, 1998; Pals-Rylaarsdam et al., 1997). There are 4 known isoforms of arrestin proteins. Two of these (arrestin 1 and cone arrestin) are localized to the visual system, whereas arrestin 2 (β-arrestin1) and arrestin 3 (β-arrestin 2) are widely expressed and involved in the regulation of most GPCRs, including the mAChRs (Krupnick and Benovic, 1998). Acute desensitization is thus achieved as the intracellular interaction between the activated receptor and G protein is "arrested" (Bunemann et al., 1999).

Another important level of regulation following agonist exposure is endocytosis (internalization), in which receptors translocate from the cell surface to intracellular compartments. In addition to inhibiting G protein-mediated signaling, the nonvisual arrestins act as adaptor proteins and can target phosphorylated GPCRs to clathrin-coated pits (Luttrell and Lefkowitz, 2002). Clathrin is a trimeric scaffold protein that facilitates a major physiological endocytic pathway (Kirchhausen et al., 1997). The study of GPCR endocytosis represents an area of exhaustive current research, and most discoveries made to date highlight the fact that receptor endocytosis, and trafficking in general, is highly dependent on the cell type in which it is investigated as well as the specific receptor involved. For example, in HEK-293 cells, the internalization of M_1, M_3, and M_4 mAChRs is severely retarded in the presence of a dominant-negative β-arrestin mutant, whereas the internalization of the M_2 mAChR under the same conditions is hardly affected (Vögler et al., 1999), even though the M_2 mAChR can bind arrestin *in vitro* (Pals-Rylaarsdam et al., 1997). In contrast, the internalization of the M_1, M_3, and M_4 mAChRs in HEK-tsA201 cells appears to be arrestin independent, whereas internalization of M_2 mAChRs is arrestin dependent (Lee et al., 1998).

The internalization via clathrin-coated vesicles of the mAChRs is also dependent on a monomeric GTPase protein, dynamin, which regulates the budding of clathrin-coated pits from the plasma membrane (Vögler et al., 1999; Zhang et al., 1996). As with the case of the arrestins, however, the role of dynamin in the internalization of different mAChRs is not clear-cut, most likely because of the various host cell environments that have been used in the different studies (Pals-Rylaarsdam and Hosey, 1997; Pals-Rylaarsdam et al., 1995; Roseberry and Hosey, 1999; Schlador and Nathanson, 1997; Vögler et al., 1998; Werbonat et al., 2000). In general, there are at least three different modes for mAChR internalization identified to date: (1) arrestin-, dynamin-, and clathrin-dependent, (2) arrestin-independent, dynamin-dependent, and (3) arrestin- and dynamin-independent (Bunemann et al., 1999).

Following internalization, phosphorylated receptors are delivered to endosomes, which provide an acidic environment to allow specific GPCR phosphatases to associate with and dephosphorylate the receptors (Bunemann et al., 1999; Lefkowitz, 1998). These receptors may be recycled back to the cell surface to respond to agonist stimulation, or they may be degraded by lysozomes. In comparison to M_1 mAChRs, M_2 mAChRs recycle back to the cell surface very slowly (Vögler et al., 1998).

Apart from the role of desensitization, internalization of receptors also contributes to down-regulation of receptors, a process that leads to a more prolonged attenuation of signal transduction. Down-regulation is a phenomenon characterized by ligand-induced decreases in the total cellular receptor number following long-term ligand exposure and is often quantified by use of a lipophilic radioligand such as the antagonist [^{3}H] QNB for the mAChRs. A change in total cellular receptor number thus represents the third level of regulation that GPCRs employ to adapt to prolonged agonist exposure. The exact mechanisms underlying this process still remain unclear, but the reduction in receptor number is generally associated with lysozomal receptor degradation and/or a decrease in new protein expression and posttranscriptional mechanisms (Bunemann et al., 1999; Pitcher et al., 1998).

E. Signaling Pathways

1. G Proteins

The classical view of mAChR signal transduction posits that signaling is initiated by the interaction of these receptors with a variety of G protein heterotrimers. This view is well borne out by two decades of research, and has led to the generalization that the M_1, M_3, and M_5 mAChRs preferentially couple to G proteins of the $G_{q/11}$ family, whereas M_2 and M_4 mAChRs preferentially couple to G proteins of the $G_{i/o}$ family (Caulfield and Birdsall, 1998). It is now recognized, however, that this generalization belies the more sophisticated aspects of G protein coupling specificity (Wess, 1998). Furthermore, estimates based on the sequenced human genome predict that there are at least 27 $G\alpha$, 5 $G\beta$, and 14 $G\gamma$ isoforms of G protein subunits (Albert and Robillard, 2002), resulting in a theoretically bewildering possibility of heterotrimer combinations that can regulate different GPCR pathways.

Within a particular G protein heterotrimer, the $G\alpha$ subunit has always been considered the important determinant of G protein-effector specificity. For example, after coupling of $G_{q/11}$ proteins to M_1, M_3, or M_5 mAChRs, subsequent dissociation of the $G\alpha_{q/11}$ subunit universally activates all phospholipase C-β (PLC-β) subtypes (Exton, 1996). Similarly, dissociated $G\alpha_i$ inhibits adenylyl cyclase (AC) after activation via M_2 or M_4 mAChRs (Sunahara et al., 1996). However, there is ample evidence that mAChRs also utilize $\beta\gamma$ subunits in signal transduction (Caulfield, 1993; Katz et al., 1992). Furthermore, the actual isoform composition of the $\beta\gamma$ dimer has been shown to directly influence the effectiveness of mAChR activation. For instance, in Sf9 insect cells expressing the M_2 mAChR, it was found that $\beta_4\gamma_2$ and $\beta_1\gamma_2$ were able to directly activate a variety of downstream effectors with equal efficacy, although M_2 mAChR stimulation of nucleotide exchange within the heterotrimer $G\alpha_o\beta_4\gamma_2$ was significantly greater than within $G\alpha_o\beta_1\gamma_2$ (Hou et al., 2001). Similarly, reconstitution studies of the M_2 mAChR with G_o heterotrimers that varied only in their γ subunit also revealed striking differences in M_2 mAChR-stimulated GTP hydrolysis (Hou et al., 2000).

One of the most commonly used methods for directly measuring GPCR-mediated guanine nucleotide exchange is the $[^{35}S]$-GTPγS binding assay (Hilf et al., 1989; Lazareno and Birdsall, 1993), and recent refinements of this technique have revealed that mAChRs possess differing G protein activation profiles that are dependent not only on the receptor subtype involved but also on the nature of the agonist used for activation (Akam et al., 2001). This finding implies that it is possible for agonists to selectively "traffic" the stimulus imparted to the same mAChR via different G protein subtypes (Kenakin, 1995a, 1995b), introducing an additional layer of complexity to the study of receptor signaling specificity.

The process of mAChR-mediated G protein coupling can be further influenced by the regulator of G protein signaling (RGS) family of proteins (Neubig and Siderovski, 2002). RGS proteins can accelerate the intrinsic GTPase activity of $G\alpha$ subunits, thus helping terminate GPCR signaling (De Vries et al., 2000). Depending on the cellular complement of receptors, $G\alpha$ proteins, and RGS proteins, selective tuning of GPCR-mediated signals can occur. For example, Rumenapp et al. (2001) used M_3 mAChRs expressed in HEK-293 cells to determine which G proteins were involved in M_3 mAChR-mediated phospholipase D (PLD) activation in comparison to M_3 mAChR-mediated PLC activity. By expressing various $G\alpha$ subunits and RGS proteins, the latter with differing selectivities for G_q or G_{12} proteins, it was found that the M_3 mAChR signaled to PLD via G_{12} proteins but not by G_q proteins. This study also showed the usefulness of G protein subtype-selective RGS proteins in determining specific roles of pertussis (PTX)-insensitive G proteins in receptor-effector coupling.

The coupling of mAChRs to G proteins is not only limited to the well-studied example of the heterotrimeric variety traditionally associated with these receptors. A number of GPCRs, including mAChRs, activate Rho, a member of the subfamily of small monomeric G proteins (RhoA, Rac, Cdc42) involved in growth regulation, actin cytoskeletal organization, and other cellular functions (Sah et al., 2000; Seasholtz et al., 1999). Rho activation by several subtypes of mAChR appears to be mediated by $G\alpha_{12}$ and $G\alpha_{13}$ (Sagi et al., 2001). M_1 mAChRs activate Rho to induce gene transcription (Fromm et al., 1997; Hirabayashi and Saffen, 2000) and to inhibit the delayed rectifier K^+ channel, $K_{v1.2}$ (Cachero et al., 1998). M_2 mAChR-mediated Rho activation couples these receptors to the PLD pathway (Schmidt et al., 1995b).

Another recently described interaction is that between the M_4 mAChR and the elongation factor 1A2 (eEFIA2), a GTP-binding protein involved in protein synthesis (McClatchy et al., 2002). In rat striatal neurons, eEFIA2 and the M_4 mAChR are colocalized in the soma and neutropil. At the moment, the functional significance of this interaction remains unknown.

2. Enzymes

a. Phospholipases

An important class of effector targets for the mAChRs is the phospholipase enzymes C, A, and D (PLC, PLA, PLD). Activation of PLC catalyzes the formation of inositol 1,4,5-trisphosphate (IP_3) and diacylglycerol (DAG), which are involved in the regulation of intracellular calcium flux, secretion, cell transformation, cellular growth, and differentiation (Berridge and Irvine, 1989; Fain et al., 1988; Rhee et al., 1989). In the brain, the principal PLC isozyme is PLC-β and is primarily associated with the G protein-inositol phosphate signaling mechanism. Although the M_1 mAChR is the most abundant $G_{q/11}$-coupled receptor subtype in the brain, M_3 and M_5 mAChRs are also likely to contribute to muscarinic PI hydrolysis in cerebral cortex and hippocampus (Ayyagari et al., 1998). Using recombinant systems, it has also been shown that the M_2 mAChR can also mediate a weak stimulation of PLC in a PTX-insensitive manner (Offermanns et al., 1994; Schmidt et al., 1995a, 1995b, 1998). M_2 and M_4 mAChRs can also couple to PLC-β pathways via a PTX-sensitive G protein using PLC-β2 or PLC-β3 and Gβγ, rather than PLC-β1 and $G\alpha_{q/11}$ (Katz et al., 1992).

PLA_2 catalyses the hydrolysis of membrane phospholipids leading to the generation of free arachidonic acid and the corresponding lysophospholipid (Felder, 1995). The M_1, M_3, and M_5 mAChRs have been linked to PLA_2 activation through agonist-stimulated inositol phosphate formation (Conklin et al., 1988). Even though the M_2 or M_4 mAChRs have not been linked to activation of PLA_2, they have been shown to facilitate PLA_2 activity (Felder, 1995; Felder et al., 1991). A reduction in PLA-catalyzed reactions in the nervous system may be associated with some pathophysiological conditions, including functional and degenerative changes in Alzheimer's disease (Gattaz et al., 1995).

Many GPCRs cause activation of PLD with hydrolysis of phosphatidylcholine to yield phosphatidic acid and choline (Exton, 1997, 1999; Jones et al., 1999; Liscovitch et al., 2000). Activation of receptors in HEK293 cells expressing the M_3 mAChR subtype led to stimulation of both PLC and PLD (Rumenapp et al., 2001; Schmidt et al., 1995b), and a similar effect occurred with M_1, M_2, and M_4 subtypes (Pepitoni et al., 1991; Schmidt et al., 1995b). Coupling of receptor activation to PLD is via PTX-insensitive G proteins, of the G_{12} type (Rumenapp et al., 2001), and may involve Rho kinase (Schmidt et al., 1999). There is also evidence that ADP-ribosylation factor (ARF) proteins and their nucleotide-exchange factor are involved in the signaling pathway leading from mAChR activation to PLD stimulation in HEK cells (Rumenapp et al., 1995).

b. Adenylyl Cyclases

The adenylyl cyclase (AC) family of enzymes can respond to the actions of extracellular effectors either by direct interaction with subunits of membrane-anchored G proteins such as $G\alpha_s$, and/or indirectly through alterations of intracellular ionic composition and kinase activity (Simonds, 1999). Molecular cloning has identified at least 10 isoforms of AC (AC1–10) with different expression patterns that allow AC to integrate stimulatory or inhibitory signals through GPCRs and regulate levels of cAMP within cells (Tucek et al., 2001). Signaling of all mAChR subtypes may involve modification of the activity of various isoforms of AC for regulation of cAMP production.

Inhibition of cAMP synthesis results from the direct coupling of M_2 or M_4 mAChRs to AC through $G_{i/o}$ proteins (Dittman et al., 1994; Haga et al., 1986; Nathanson, 1987; Onali and Olianas, 1995). A reduction in cAMP levels via M_2 mAChR activation may also occur through activation of phosphodiesterase (PDE) (Han et al., 1998). However, in HEK-293 cells M_4 mAChRs exhibited both inhibition and stimulation of AC1 and

AC3 depending on receptor density and agonist concentration (Dittman et al., 1994). In CHO cells transfected with human M_2 mAChR, the inhibition of cAMP synthesis caused by low concentrations of mAChR agonists becomes weaker if the agonist concentration is raised (Michal et al., 2001). In native neuronal tissue, a bimodal control on cAMP formation was observed with activation of rat olfactory bulb M_4 mAChRs (Olianas et al., 1998; Olianas and Onali, 1996; Onali and Olianas, 1995).

It has also been shown that activation of mAChRs can elevate intracellular levels of cAMP in both neuroblastoma cells expressing endogenous mAChRs, such as the M_3 subtype (Baron and Siegel, 1989; Baumgold and Fishman, 1988; Nakagawa-Yagi et al., 1991), and cells transfected with the M_1, M_3, and M_5 subtypes (Dittman et al., 1994; Felder et al., 1989; Gurwitz et al., 1994). For the M_1 mAChRs, stimulation of AC activity was via G_s protein coupling (Dittman et al., 1994; Gurwitz et al., 1994). In other cases, cAMP stimulation has paralleled increased PI hydrolysis and Ca^{2+}/calmodulin activation of AC activity (Buck and Fraser, 1990; Shapiro et al., 1988).

c. Nitric Oxide Synthase

Nitric oxide (NO) is an important neuromodulator. Concentrations of NO under normal physiological conditions within the body are maintained at low levels under the influence of the enzyme, NO synthase (NOS) (Bredt and Snyder, 1992, 1994). In the CNS, the constitutively expressed neuronal and endothelial types of NOS isozymes, nNOS (NOS-I) and eNOS (NOS-III), respectively, have been implicated in memory processing (Son et al., 1996). The generation of NO through the actions of either nNOS or eNOS requires the release of intracellular Ca^{2+} (Christopoulos and El-Fakahany, 1999a). Receptors that are linked to the mobilization of intracellular Ca^{2+} stores, therefore, are most commonly associated with NO formation.

M_1, M_3, and M_5 mAChRs each promote a robust mobilization of intracellular calcium via the $G_{q/11}$/PLC pathway, and, not surprisingly, each receptor has been shown to activate nNOS in stably cotransfected CHO-K1 cells (Wang et al., 1996). However, it is possible that additional mechanisms exist for nNOS activation in the presence of minimal cytosolic Ca^{2+} mobilization. For example, stimulation of M_2 mAChRs in transfected CHO-K1 cells with carbachol produced little intracellular Ca^{2+}, yet nNOS activity in these same cells was of comparable magnitude to that observed in M_1-expressing CHO-K1 cells (Wang et al., 1997).

In the brain, NO can modulate mAChR-mediated rhythmic slow activity in rat hippocampal slices (Bawin et al., 1994) and inhibition of NOS can cause impairment of learning and memory that can be reversed by mAChR activation (Kopf and Baratti, 1996), although a direct link between mAChR-mediated NO production and memory processing has yet to be demonstrated.

d. Mitogen-Activated Protein Kinase Pathways

The mitogen-activated protein kinase (MAPK) family of enzymes is divided into a number of subgroups, including extracellular signal regulated kinases (ERK1 and 2), the stress-activated protein kinases, or c-Jun NH_2 terminal kinases (SAPKs/ JNKs), the p38 kinases, and ERK5 (Garrington and Johnson, 1999). Stimulation of mAChRs can lead to activation of each of these pathways.

ERK1 (p44 MAPK) and ERK2 (p42 MAPK) may be activated by all 5 mAChR subtypes, but the time course varies depending on the cellular background. Activation of ERK1 or 2 by M_1, M_3, and M_5 mAChRs typically involves PKC following $G_{q/11}$ activation of the PLC pathway with production of DAG (Kim et al., 1999). Utilization of the Raf family of serine-threonine kinases is also common, although involvement of Ras is not (Crespo et al., 1994; Guo et al., 2001; Hawes et al., 1995; van Biesen et al., 1996). Alternative pathways for activation of ERK1 or 2 by M_1 or M_3 mAChRs also exist in some cells. These appear to involve either (1) G proteins other than $G_{q/11}$ or (2) PKC-independent processes (Berkeley et al., 2001; Haring et al., 1998; Kumahara et al., 1999). Although best-characterized for the M_1, M_3, and M_5 mAChRs, the M_2 and M_4 mAChRs can also activate MAPK by utilizing the $\beta\gamma$ subunits of G proteins (Koch et al., 1994; Lopez-Ilasaca et al., 1997). A similar pathway operates in CNS neurones for M_1–M_4 mAChRs (Rosenblum et al., 2000).

JNK may be involved in neuronal cell differentiation (Heasley et al., 1996) and in the regulation of gene expression (Marinissen et al., 1999; Yamauchi et al., 1999). Activation of JNK in COS-7 cells transfected with M_1 or M_2 mAChRs showed that both receptor subtypes could activate JNK via a Ras and Rac 1-dependent route (Coso et al., 1996).

The enzyme p38 MAPK is involved in regulation of cell movement and development as well as the production of tumor necrosis factor α (TNFα) and interleukin 1 (IL1). M_1 mAChR activation of p38 MAPK in HEK-293 cells appears to involve both $G\alpha_q$ and $G\beta\gamma$ regulating MKK3 and MKK6 through parallel signaling pathways (Nagao et al., 1998; Yamauchi et al., 2001).

ERK5, or big mitogen-activated protein kinase 1 (BMK1), appears to be activated by oxidative stress and to be involved in the regulation of a number of nuclear transcription factors (Abe et al., 1997; Fukuhara et al., 2000). In transiently transfected COS-7 cells, activity of ERK5 may be modulated by mAChRs that utilize α subunits of $G_{q/11}$ or $G_{12/13}$, such as M_1 mAChRs, but not by those that couple to G_i, such as M_2 mAChRs (Fukuhara et al., 2000). Carbachol, acting via M_1 mAChRs in NIH 3T3 cells, has also been shown to activate ERK5 (Marinissen et al., 1999).

Epidermal growth factor (EGF) and mAChR agonists, together with their receptors, interact in complex ways to modulate cellular function. Transactivation of the EGF receptor can occur following mAChR activation, promoting EGF receptor dimerization and tyrosine kinase activity along with mutual potentiation or inhibition of their actions (Daub et al., 1996, 1997). Transactivation of the EGF receptor has been shown to be an essential part of ERK activation by GPCRs, both $G_{q/11}$ and $G_{i/o}$ coupled, in Rat 1 fibroblasts and COS-7 cells (Daub et al., 1996, 1997). Activation of either M_1 or M_2 mAChRs by carbachol in COS-7 cells caused increases in phosphorylation of the EGF receptor although the effect of M_1 mAChR activation was more pronounced.

Given the plethora of signaling modes associated with the MAPK and EGF pathways, it is perhaps unsurprising that there are a variety of cholinergic functions that have been associated with mAChR-mediated kinase signaling. These include gene induction, long-term potentiation, electrolyte transport, the secretion of amyloid precursor protein, cellular growth and apoptosis, ion channel modulation, and modification of GRK activity (Lanzafame et al., 2003).

3. Ion Channels

The coupling of mAChRs to ion channels has been extensively covered previously (see reviews by Beech, 1997; Bolton et al., 1999; Caulfield, 1993; Janssen and Sims, 1997; Jones, 1993; Wickman and Clapham, 1995). The following discussion summarizes observations that have been made with particular respect to modulation of neuronal ion channel activity by mAChRs.

a. Potassium Channels

The "M current" is a slowly activating and deactivating potassium conductance that was first described in sympathetic neurons (Brown and Adams, 1980) and subsequently shown to be expressed in a variety of neuronal cell types (Marrion, 1997). The activation of mAChRs, and other GPCRs, linked to $G_{q/11}$ proteins has long been known to modulate M current activity, although it was not until recently that molecular biological approaches identified the channel mediating the M current as belonging to the KCNQ subtype of potassium channel (Wang et al., 1998). The subject of some contention, however, is the identity of the intracellular messenger(s) linking receptor activation to M current modulation. Recently, Suh and Hille (2002) provided elegant experimental evidence that the activation of PLC by the M_1 mAChR initiates M current modulation by depleting cellular stores of PIP_2, which then leads to channel inhibition. Recovery of the channel was dependent on intracellular ATP and phosphoinositide 4-kinase activity.

G protein inward-rectifying potassium (GIRK) channels are another class of potassium channels modulated by mAChR activation. GIRK channels are characterized by an initial low basal level of channel activity that increases on activation by $G_{i/o}$-coupled receptors. The activation of these channels causes a decrease in cellular excitability by hyperpolarization and, accordingly, extends the duration between action potentials (Hill and Peralta, 2001). In general, neuronal GIRK channels play a role in the resting membrane potential.

Recent experiments on mAChRs in rat superior cervical ganglion (SCG) neurons, heterologously coexpressed with GIRK1/2 channels, found that stimulation of endogenous M_2 mAChRs, but not endogenous M_4 mAChRs, selectively activated GIRK currents (Fernandez-Fernandez et al., 1999, 2001). GIRK channels were preferentially activated by $\beta\gamma$ subunits freed from M_2 mAChR-stimulated $G\alpha_i$, whereas Ca^{2+} channels were preferentially inhibited by $\beta\gamma$ subunits released from M_4 mAChR-activated $G\alpha_o$ (Fernandez-Fernandez et al., 2001). It was suggested that these effects occurred because of the topographical arrangement of receptor subtypes and G proteins in the SCG rather than because of any fundamental selectivity of the receptor subtypes for particular G proteins.

Stimulation of a number of $G_{q/11}$-coupled receptors and the resultant cellular signaling pathways can also inhibit GIRK channels (Rogalski et al., 1999; Sharon et al., 1997; Stevens et al., 1999). In *Xenopus* oocytes, Hill and Peralta (2001) found that GIRK1/4 currents were suppressed by M_1 mAChR stimulation. Further studies showed that the M_1 mAChR potently suppressed a novel delayed rectifier K^+ channel, termed RAK, through a pathway involving PLC activation and direct tyrosine phosphorylation of the RAK protein.

b. Calcium Channels

In many cells, elevation of intracellular calcium ($[Ca^{2+}]_i$) via mAChRs is required for stimulation of PLA_2, PLD, $PLC\gamma$, tyrosine kinases, Ca^{2+}/calmodulin-dependent ACs, and NO formation. Additionally, Ca^{2+} is involved in the regulation of structural plasticity in neurons by influencing the cytoskeleton and associated proteins (see Felder, 1995). While mAChR activation can lead to elevation of $[Ca^{2+}]_i$ via the IP_3 pathway and release of Ca^{2+} from intracellular stores, mAChRs also affect Ca^{2+} entry through other types of ion channels, including L-type voltage-operated Ca^{2+} channels (Lanzafame et al., 2003). In addition, mAChR agonists have been shown to depress N-type Ca^{2+} channel currents in rat SCG neurons by 2 G protein-coupled signaling pathways (Hille et al., 1995). One of these pathways is rapid and membrane delimited and shows sensitivity to PTX, suggesting that M_2 or M_4 mAChRs are involved via liberation of $G\beta\gamma$ subunits. The second pathway of SCG neurons also affects L-type channels, is slower in nature, and is PTX insensitive, suggesting the involvement of diffusible second messenger linked to M_1 mAChR activation. A recent study by Shapiro et al. (1999), using mAChR knockout mice, has generally confirmed these observations, but has also revealed that the fast, PTX-sensitive component of channel modulation in the latter animal model is mediated by M_2, and not M_4, mAChRs.

In the rat striatum, activation of muscarinic autoreceptors present on cholinergic nerves inhibits both N- and P/Q-type Ca^{2+} channels but appears more efficiently coupled to the former channel type (Dolezal et al., 2001).

T-type Ca^{2+} channel activity was not affected by M_1 mAChRs expressed in NIH 3T3 cells, whereas M_3 and M_5 mAChRs enhanced channel activity (Pemberton et al., 2000). The 3 subtypes all increased cAMP levels in the cells, and the effect on the T-type channel appeared to be associated with a subsequent activation of PKA arising from the effect on cAMP. The failure of M_1 mAChR activation to enhance channel activity was attributed to concomitant activation of PKC, opposing the effect of PKA. Activation of M_2 and M_4 mAChRs had no effect alone but after pretreatment of the cells with forskolin to elevate cAMP levels, M_2 mAChR activation induced an inhibition in T channel activity.

R-type Ca^{2+} channels appear to be formed from α_{1E} Ca^{2+} channel subunits and are found in neurons in the CNS and periphery (Jeong and Wurster, 1997; Tottene et al., 2000; Wu et al., 1998). They contribute to influx of Ca^{2+} in dendrites (Kavalali et al., 1997), providing transient surges of Ca^{2+} rather than a steady influx (Randall and Tsien, 1997), and secretion of ACh from central cholinergic neurones (Allen, 1999). The M_1 (Melliti et al., 2000) and M_2 (Meza et al., 1999) mAChRs have each been shown to both facilitate and inhibit R-type channels. In HEK-293 cells, M_1 mAChRs utilized $G\alpha_q$ to produce facilitation, and this effect was blocked by RGS2 and by the C-terminal region of $PLC\beta1$, which acts as a GTPase activator of $G\alpha_q$. Channel inhibition by M_1 mAChRs was a weaker effect, involved $G\beta_\gamma$, and was unaffected by either inhibitor. M_2 mAChR-induced channel activation was weaker than that observed with the M_1 mAChR, but channel inhibition was more pronounced, suggesting the involvement of different $G_{\beta\gamma}$ subunits to those utilized by the M_1 mAChR (Melliti et al., 2000). The M_2 mAChR-mediated inhibitory effect was inhibited by PTX but the facilitatory response was not (Meza et al., 1999). There may be differences in the susceptibility of various R-type channels to mAChR modulation. In rat basal forebrain cholinergic neurons, muscarine or ACh failed to inhibit R-type Ca^{2+} channels present on the soma, but muscarine inhibited ACh release induced by Ca^{2+} influx through R- as well as N- and Q-type channels on neurites (Allen, 1999).

Finally, a Ca^{2+}-dependent nonspecific cation current (I_{cat}) activated by mAChR agonists in mouse hippocampus, as well as a mixed Na^+/K^+ current (I_H), appear to be mediated by M_1 mAChRs, as the responses are not present in M_1 mAChR knockout mice (Fisahn et al., 2002). A non-selective cation current was also found to be responsible for a mAChR agonist-induced slow after-depolarization in rat prefrontal cortex (Haj-Dahmane and Andrade, 1998) and for carbachol-induced plateau potentials in the rat limbic system (Kawasaki et al., 1999).

c. Chloride Channels

Activation of mAChRs, although not directly opening Cl^- channels, may initiate a Ca^{2+}-dependent opening of such channels leading to efflux of Cl^- and a resulting depolarization contributing to further influx of Ca^{2+} via VOCCs (Janssen and Sims, 1997). In rat SCG, Marsh et al. (1995) identified a delayed Ca^{2+}-dependent Cl^- current that was induced by the synergistic action of Ca^{2+} and DAG and was blocked by inhibitors of PKC. The resulting Cl^- current was induced by ACh through the simultaneous activation of nicotinic receptors to produce a rise in $[Ca^{2+}]_i$ and mAChRs to generate DAG.

4. Receptor Dimerization/Oligomerization

For many decades, the classic view of signaling via GPCRs has always assumed that each receptor exists and interacts with other membrane proteins as a monomer. This is in contrast to the activation of all other known receptor superfamilies, which characteristically require some form of oligomerization to mediate their effects. For example, all ligand-gated ion channels exist as oligomeric complexes, while both the nuclear receptor and enzyme-linked receptor superfamilies require dimerization for activation (Changeux and Edelstein, 1998). The view of GPCRs as monomeric proteins is now giving way to the realization that these receptors can also come together to form dimers or higher-order oligomers, and that this type of receptor-receptor interaction can lead to markedly different physiological and/or pharmacological properties of the resulting receptor complex.

Although the concept of GPCR dimerization has really come into its own only within the last decade, provocative evidence from radioligand-binding studies of the mAChRs has actually existed for some time to suggest that classic orthosteric ligands could bind in a cooperative fashion, as would be expected for interaction with an oligomeric receptor complex (Boyer et al., 1986; Chidiac et al., 1997; Hirschberg and Schimerlick, 1994; Mattera et al., 1985; Potter et al., 1991; Potter and Ferrendelli, 1989; Wregget and Wells, 1995). For example, Figure 5-11 shows the binding of the agonist oxotremorine-M in competition with the antagonist [^{3}H]AF-DX 384 at native, purified, M_2 muscarinic receptors. In the presence of G protein coupling, the competition curve is inhibitory and biphasic, but when the nonhydrolyzable GTP analog Gpp(NH)p is included in the assay to uncouple receptor-G protein complexes, a bell-shaped binding curve is obtained for the agonist-antagonist interaction. This behavior cannot be reconciled within the standard ternary complex mechanism of ligand-receptor-G protein, but it can be rationalized if it is assumed that M_2 mAChRs can exist as dimers or higher-order oligomers within the cell membrane (Wregget and Wells, 1995).

Molecular biological evidence for mAChR oligomerization has been obtained from co-expression studies using chimeric receptor constructs or receptor fragments. For example, Maggio et al. (1993a, 1993b), constructed a series of α_{2C}-adrenoceptor/M_3 mAChR chimeras that contained the first 5 transmembrane domains of 1 receptor type linked to the last 2 of the other type of receptor and then studied their properties in a recombinant expression system. When transfected alone into COS-7 cells, neither chimera showed significant ligand-binding activity. However,

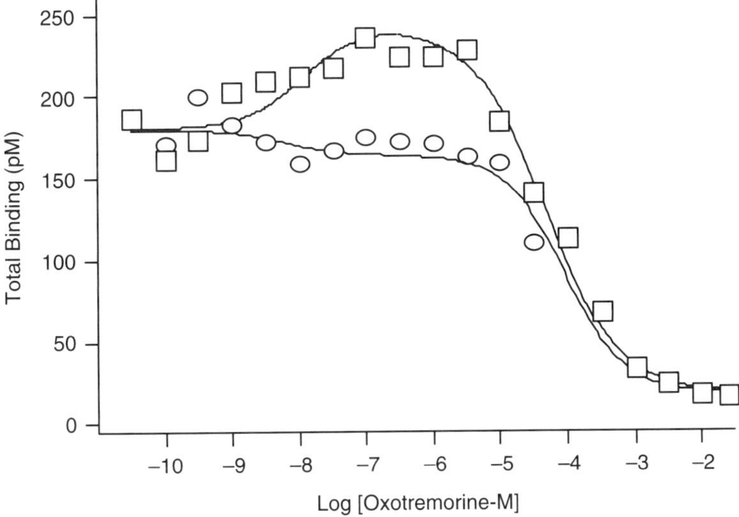

Figure 5-11. Agonist competition binding at an oligomeric receptor complex. Interaction between oxotremorine-M and [^{3}H]AF-DX 384 at the M_2 mAChR copurified with G proteins from porcine sarcolemmal membranes. Data were obtained in the absence (circles) or presence (squares) of the nonhydrolyzable GTP analog, Gpp(NH)p. Curves through the data represent the best fit based on a model of receptor oligomerization. (Data replotted from Wregget and Wells, 1995.)

when they were coexpressed, significant numbers of both α_{2C} and M_3 binding sites were detected. Furthermore, this phenomenon was functionally relevant, as the cotransfected cells were able to respond to stimulation with an mAChR agonist. This "functional rescue" of receptor activity on coexpression of the 2 different chimeric constructs could be explained if an intermolecular rearrangement of transmembrane domains between the 2 receptor chimeras occurs, thus highlighting the possibility of GPCR-GPCR interactions. Similar results have also been observed in experiments using N- and C-terminally fragmented M_3 mAChRs; coexpression of the receptor fragments led to significant restoration of [^{3}H]NMS binding (Schöneberg et al., 1995), again suggesting that mAChRs are composed of multiple autonomous folding units that may "swap" between receptor pairs to form dimers (Gouldson et al., 1998).

Another very common technique used to provide biochemical evidence for GPCR dimerization involves the co-immunoprecipitation of differentially epitope-tagged receptors (Hebert and Bouvier, 1998). This approach was recently used to investigate the homodimerization of the M_3 mAChR (Zeng and Wess, 1999). In that study, the receptor was C-terminally epitope-tagged with the last 9 amino acids of the rhodopsin receptor or the last 29 amino acids of the V2 vasopressin receptor. Receptor homodimers were subsequently revealed by immunoprecipitation using an antibody to one of the epitope tags followed by Western blotting using an antibody to the other tag. The study by Zeng and Wess (1999) revealed two interesting properties of the M_3 mAChR homodimers. First, the sensitivity of the dimers to the effects of the reducing agent dithiothreitol suggested that part of the dimeric linkage involves the formation of disulfide bonds across the interacting receptors. Second, the formation of M_3 mAChR homodimers appeared to be specific, as co-immunoprecipitation studies of M_3 mAChRs with either M_1 or M_2 mAChRs failed to find any evidence for mAChR heterodimerization. This latter finding is in contrast to a recent study by Maggio et al. (1999), who proposed the existence of M_2/M_3 mAChR heterodimers based on the appearance of novel pharmacological properties of ligands binding to these receptors when the latter are coexpressed. The reasons for these conflicting findings remain undetermined.

Co-immunoprecipitation was also used recently by Park et al. (2001) to investigate the homodimerization of M_2 mAChRs, which were

N-terminally tagged with either c-myc or FLAG epitopes and coinfected into Sf9 insect cells. This study revealed that M_2 mAChRs might exist not only as dimers, but potentially as higher-order oligomers, such as trimers and even tetramers. Another important observation made in that study was that the degree of co-immunoprecipitation of M_2 mAChR homodimers was unaffected by ligand treatment, suggesting that mAChR dimers are formed constitutively and are insensitive to regulation by ligands. A similar finding was made by Zeng and Wess (1999) in their study of M_3 mAChR homodimers.

An important issue in current studies of GPCR oligomerization remains the physiological and/or pathophysiological relevance of the phenomenon. Some studies have demonstrated that GPCR ligands may actively regulate the monomer/oligomer ratio, but the majority of studies to date either have reported the opposite finding, as described above for M_2 and M_3 mAChR homodimers, or have not investigated the potential for oligomer regulation by ligands (Christopoulos and Kenakin, 2002). The advent of newer biophysical techniques, such as fluorescence and/or bioluminescence resonance energy transfer assays, are beginning to provide more sensitive measurements of GPCR dimerization in living cells (Bouvier, 2001) but, to date, there have been no reports of these approaches having been applied to the study of mAChR dimers. Nevertheless, there is at least 1 GPCR (the $GABA_B$ receptor) for which a heterodimeric state represents the native, functional state (White et al., 1998), and at least one example of where a disease (preeclampsia) may be mediated, in part, by the formation of AT_1 angiotensin—B_2 bradykinin GPCR heterodimers (AbdAlla et al., 2001). Thus, the elucidation of the functional relevance of dimers/oligomers of other GPCRs, including the mAChRs, remains an important issue.

F. Functional Insights Gained from the Use of Muscarinic Acetylcholine Receptor Knockout Mice

For many decades, a recurring theme in the study of the physiology of mAChR function has been the difficulty in assigning functions to specific subtypes; this is especially so for studies conducted in the CNS, which expresses all 5 subtypes of mAChR. In the last few years, a significant advance in addressing this problem has been made with the use of gene-targeting technology to generate transgenic mice that lack, either individually or in certain combinations, the M_1–M_5 mAChR genes.

The standard approach for the generation of mAChR knockout mice has generally been the same in most studies reported to date, and involved the initial isolation of the mAChR genes from a mouse genomic library followed by the generation of targeting vectors with replacements of functionally essential segments of the receptor coding sequences. These vectors were then introduced into embryonic stem cells and microinjected into blastocysts to generate male offspring, which were then mated with female offspring. A subsequent series of intermating of offspring heterozygous for the desired mAChR mutation produced homozygous mAChR knockout ($^{-/-}$) mice (see, e.g., Gomeza et al., 1999a, 1999b; Matsui et al., 2000; Miyakawa et al., 2001; Yamada et al., 2001a, 2001b). In all instances reported to date, deficiency in any of the mAChR subtypes does not appear to affect fertility and longevity of these mice compared to their wild-type littermates.

$M_1^{-/-}$ mAChR mice do not appear to display any significant alterations in centrally mediated processes such as nociception, sensory-motor gating, motor coordination, or anxiety-related behavior (Miyakawa et al., 2001). Not surprisingly, however, these mice do reveal severe reductions in mAChR-mediated PI hydrolysis, assessed both *in vivo* and in primary neuronal cultures, as well as a complete abolition of MAPK activation in CA1 pyramidal hippocampal neurons (Berkeley et al., 2001; Bymaster et al., 2003; Felder et al., 2001; Hamilton et al., 1997; Porter et al., 2002). These deficits in signaling are consistent with the known role of M_1 mAChRs in mediating these pathways (see section E).

Given the large body of previous studies implicating the M_1 mAChR as playing an important role in learning and memory (see Chapter 9 BV), one particularly surprising finding noted with the $M_1^{-/-}$ mAChR mice was an overall minimal effect on working memory performance in the 8-arm radial maze test, in the Morris water maze test, and in fear-conditioning studies, although an increase in the number of revisiting errors was noted during trials without delay (Miyakawa et al., 2001). Another study that investigated effects on working memory in the radial arm maze and fear conditioning, however, was able to demonstrate a significant deficit of $M_1^{-/-}$ mAChR mice in these tasks relative to wild-type mice (Anagnostaras et al., 2003). These findings suggest that M_1 mAChRs probably do play a role in some aspects of learning and memory, but perhaps not to the extent that had been suggested previously by studies using more traditional pharmacological approaches (see Chapter 9 BV). This raises the possibility that the more severe cognitive deficits associated with mAChR antagonism in the CNS may involve other mAChR subtypes in addition to the M_1 mAChR and/or are offset by compensatory developmental changes in the $M_1^{-/-}$ mAChR mice.

A more notable abnormality observed in studies of $M_1^{-/-}$ mAChR mice is a significant elevation in locomotor activity (Gerber et al., 2001; Miyakawa et al., 2001), which is associated with an increase in extracellular dopamine release in the striatum (Gerber et al., 2001). These findings are consistent with previous work suggesting a role for M_1 mAChR-mediated striatal control of locomotion (Di Chiara et al., 1994; see also Chapter 9 BIV). Another aspect of central mAChR function likely to be mediated by the M_1 mAChR is the occurrence of pilocarpine-induced epileptic seizures, as this phenomenon is absent in $M_1^{-/-}$ mAChR mice (Hamilton et al., 1997). Taken together, these findings suggest that it is possible for M_1 mAChRs to become novel targets for the treatment of Parkinson's disease, certain forms of schizophrenia, or epilepsy (see also Chapter 10).

It has been known for some time that mAChR agonists and antagonists can regulate the release of acetylcholine from cholinergic nerve terminals, with agonists decreasing neurotransmitter release and antagonists increasing release (Kilbinger et al., 1984; Starke et al., 1989; see also Chapter 9 BII and BIII). Given that M_2 and M_4 mAChRs are able to inhibit voltage-sensitive presynaptic calcium channels that are involved in neurotransmitter exocytosis, the M_2 and M_4 mAChR subtypes have for some time been considered the primary candidates for modulating ACh release. However, pharmacological studies to date have resulted in conflicting evidence with respect to the actual identity of presynaptic mAChR subtypes. Important new insights into this function of mAChRs in the CNS have most recently been obtained in $M_2^{-/-}/M_4^{-/-}$ mAChR double knockout mice, which show a complete abolition of oxotremorine inhibition of potassium-evoked ACh release in tissue slices from cortex, hippocampus, and striatum (Zhang et al., 2002a). Interestingly, the use of single mAChR knockout mice in similar experiments indicated that the M_2 mAChR is the major autoinhibitory receptor in the cortex and hippocampus, whereas the M_4 mAChR is the major autoinhibitory receptor in the striatum (Zhang et al., 2002a).

Another long-standing conundrum in the field has been the identity of the mAChR(s) mediating mAChR agonist-promoted analgesia. Although the ability of various cholinergic agonists to effect both spinal and supraspinal analgesia has been reported for some time, classic pharmacological approaches have implicated multiple mAChR subtypes in mediating this effect, including the M_1, M_2, and M_4 subtypes (Eisenach, 1999). Recent knockout studies have confirmed a role for both M_2 and M_4 mAChRs, but no other mAChR subtype, as being the receptors responsible for muscarinic agonist-mediated analgesic effects (Duttaroy et al., 2002). Studies on $M_2^{-/-}$ or $M_4^{-/-}$ mAChR single knockout mice have further refined this observation to suggest that the predominant analgesic role is played by the M_2 mAChR, assessed using classic tail-flick and hot plate tests (Gomeza et al., 1999a) as well as electrophysiological and neurochemical studies on peripheral cutaneous nerve endings (Bernardini et al., 2002).

Related, but nonoverlapping, roles for M_2 and M_4 mAChRs in the CNS have also been identified from knockout mouse studies. Specifically, $M_2^{-/-}$ mAChR mice showed a complete loss of oxotremorine-induced tremor (see also Chapter 9 BIV), a finding that suggests that the M_2 mAChR is the receptor mediating the cholinergic tremor associated with Parkinson's disease. In $M_4^{-/-}$ mAChR mice, a small but significant increase in basal locomotor activity was noted, relative to wild-type littermates, and these knockout mice were also hyper-responsive to stimulation with a D_1 dopamine receptor agonist (Gomeza et al., 1999b; see also Chapter 9 BIV). Since most spiny projection neurons originating in the striatum and projecting to the substantia nigra coexpress D_1 and M_4 receptors (Ince et al., 1997), it appears that M_4 mAChRs play a predominant role in dampening D_1-mediated effects on locomotor activity. These M_4 mAChRs are most likely located on neuronal cell bodies, rather than terminals, and probably mediate their effects on dopamine release through an indirect mechanism involving striatal GABAergic pathways (Zhang et al., 2002a). Specific antagonism of this M_4-mediated pathway may prove useful in the treatment of Parkinson's disease.

Although the M_3 mAChR is generally expressed in low abundance in the CNS, it is nevertheless expressed widely. To date, however, most studies of this receptor have generally focused on its roles in the periphery, with little known about its CNS functions. A recent study of the $M_3^{-/-}$ mAChR mouse has revealed that the M_3 mAChR, located on nerve terminals in the striatum, is involved in inhibiting dopamine release through an action on the striatal GABA system (Zhang et al., 2002a). Perhaps the most important finding from studies on the $M_3^{-/-}$ mAChR mice, however, is with respect to the potential role of the M_3 mAChR in centrally mediated regulation of feeding behavior and body weight (see Chapter 9 BIV-2). $M_3^{-/-}$ mAChR mice are characterized by significantly reduced body weight and peripheral fat deposits that persist through the life of the animals, a reduction in food intake, and low serum levels of total triglycerides and the hormones leptin and insulin (Yamada et al., 2001b). Radioligand-binding studies in the $M_3^{-/-}$ mAChR mice revealed that one of the sites of profound loss of this receptor was the hypothalamus, consistent with the known role of this region in the regulation of feeding behavior. The same study also revealed alterations in a number of peptides that modulate feeding, including increases in the expression of the orexigenic agouti-related peptide, reductions in the anorectic proopiomelanocortin, and a decrease in the expression of melanin-concentrating hormone (MCH). Although the former two findings are consistent with effects noted previously in studies of fasting animals, the decreased expression of MCH is paradoxical because levels of this peptide usually increase in fasting animals (Flier and Maratos-Flier, 1998). Nevertheless, the $M_3^{-/-}$ mAChR mice did retain responsiveness to the orexigenic effects of MCH (Yamada et al., 2001b). These findings provide a wealth of novel information regarding the link between central cholinergic pathways and the leptin/melanocortin system, and may lead to novel therapeutic approaches to the treatment of eating disorders.

As discussed previously, the most enigmatic mAChR in the CNS has been the M_5 mAChR, given its low and discrete levels of expression, and the lack of selective pharmacological tools with which to delineate its physiological actions. The generation of $M_5^{-/-}$ mAChR mice has thus led to a renewed interest in this receptor as a potential therapeutic target. $M_5^{-/-}$ mAChR mice showed a blunted response to the effects of oxotremorine-mediated increases in dopamine release in the striatum, although the maximal effect of high-dose oxotremorine on dopamine release was maintained (Yamada et al., 2001a; Zhang et al., 2002b). This finding suggests that the M_5 mAChR plays a minor role in facilitating dopamine release in the striatum relative to the more prominent role of the M_4 mAChR. More striking effects, however, were noted in the ventral tegmental area (VTA), a region known to be involved in mediating the effects of opiate withdrawal and addiction behavior. A previous study using transgenic mice that expressed functionally impaired, truncated, M_5 mAChRs found that the M_5 mAChR is important in mediating a prolonged facilitation of dopamine release in the VTA (Forster et al., 2002). The authors suggested that the M_5 mAChR thus plays an important role in motivational behavior associated with drugs of abuse via its effects on dopamine release. In support of this hypothesis, genetic ablation of the M_5 mAChR in the VTA led to a significant reduction in the rewarding effects of morphine, as well as an attenuation of naloxone-induced morphine withdrawal symptoms (Basile et al., 2002). Importantly, the analgesic properties of morphine remained unaltered, thus identifying the M_5 mAChR as a novel target for the treatment of opiate addiction.

Finally, an important physiological role that appears to be mediated by the M_5 mAChR is that of cerebrovascular vasodilatation; the powerful vasodilator effects of ACh are almost completely abolished in cerebral arterioles (pial vessels) and the basilar artery of $M_5^{-/-}$ mAChR mice (Yamada et al., 2001a). Interestingly, this did not appear to be the case for extracerebral vessels, such as the $M_5^{-/-}$ mAChR mouse coronary or carotid arteries, indicating that the role for M_5 mAChRs in the vasculature is restricted to cerebral blood vessels.

From the preceding discussion, it is evident that some of the findings made using mAChR knockout mice have confirmed previous studies based on pharmacological characterization of mAChRs (see Chapter9 BIV and BV). However, it is also clear that the mAChR gene ablation approach has led to a significant number of novel findings that have identified new physiological and potential therapeutic roles for mAChRs.

G. Conclusions

Recent advances in molecular and structural biology have led to a better understanding of the structure, function, and regulation of mAChRs and their associated signaling pathways. The generation of mAChR knockout mice, in particular, is beginning to unravel discrete functions for each mAChR subtype in the CNS as well as pointing to novel therapeutic areas. However, the translation of such fundamental mAChR research into improved therapeutic outcomes in man still relies on the identification of novel pharmacological agents that display superior selectivity to most currently available mAChR ligands. This challenge is particularly relevant to the CNS, which expresses all 5 mAChRs, but can result in significant advances in the treatment of debilitating disorders such as schizophrenia, pain, cognitive dysfunction, Parkinson's disease, and epilepsy.

Acknowledgments

The author is a Senior Research Fellow of the National Health and Medical Research Council (NHMRC) of Australia. Work in the author's laboratory is funded by NHMRC project grant numbers 209083 and 251538.

References

AbdAlla, S., Lother, H., el Massiery, A. and Quitterer, U.: Increased AT_1 receptor heterodimers in preeclampsia mediate enhanced angiotensin II responsiveness. Nat. Med. *7*: 1003–1009, 2001.

Abe, J., Takahashi, M., Ishida, M., Lee, J. D. and Berk, B. C.: c-Src is required for oxidative stress-mediated activation of big mitogen-activated protein kinase 1. J. Biol. Chem. *272*: 20389–20394, 1997.

Akam, E. C., Challiss, R. A. and Nahorski, S. R.: $G_{q/11}$ and $G_{i/o}$ activation profiles in CHO cells expressing human muscarinic acetylcholine receptors: dependence on agonist as well as receptor-subtype. Br. J. Pharmacol. *132*: 950–958, 2001.

Albert, P. R. and Robillard, L.: G protein specificity. Traffic direction required. Cell. Signal. *14*: 407–418, 2002.

Allen, T. G.: The role of N-, Q- and R-type Ca2+ channels in feedback inhibition of ACh release from rat basal forebrain neurones. J. Physiol. 515: 93–107, 1999.

Allman, K., Page, K. M., Curtis, C. A. and Hulme, E. C.: Scanning mutagenesis identifies amino acid side chains in transmembrane domain 5 of the M_1 muscarinic receptor that participate in binding the acetyl methyl group of acetylcholine. Mol. Pharmacol. *58*: 175–184, 2000.

Anagnostaras, S. G., Murphy, G. G., Hamilton, S. E., Mitchell, S. L., Rahnama, N. P., Nathanson, N. M. and Silva, A. J.: Selective cognitive dysfunction in acetylcholine M1 muscarinic receptor mutant mice. Nat. Neurosci. *6*: 51–58, 2003.

Angeli, P., Marucci, G., Buccioni, M., Piergentili, A., Giannella, M., Quaglia, W., Pigini, M., Cantalamessa, F., Nasuti, C., Novi, F., Maggio, R., Qasem, A. R. and Spampinato, S.: Deoxamuscaroneoxime derivatives as useful muscarinic agonists to explore the muscarinic subsite: demox, a modulator of orthosteric and allosteric sites at cardiac muscarinic M_2 receptors. Life Sci. *70*: 1427–1446, 2002.

Ayyagari, P. V., Gerber, M., Joseph, J. A. and Crews, F. T.: Uncoupling of muscarinic cholinergic phosphoinositide signals in senescent cerebral cortical and hippocampal membranes. Neurochem. Int. *32*: 107–115, 1998.

Barlow, R. B., Berry, K. J., Glenton, P. A. M., Nikolaou, N. M. and Soh, K. S.: A comparison of affinity constants for muscarine-sensitive acetylcholine receptors in guinea-pig atrial pacemaker cells at 29°C and in ileum at 29°C and 37°C. Br. J. Pharmacol. *58*: 613–620, 1976.

Baron, B. M. and Siegel, B. W.: Alpha 2-adrenergic and muscarinic cholinergic receptors have opposing actions on cyclic AMP levels in SK-N-SH human neuroblastoma cells. J. Neurochem. *53*: 602–609, 1989.

Basile, A. S., Fedorova, I., Zapata, A., Liu, X., Shippenberg, T., Duttaroy, A., Yamada, M. and Wess, J.: Deletion of the M_5 muscarinic acetylcholine receptor attenuates morphine reinforcement and withdrawal but not morphine analgesia. Proc. Natl. Acad. Sci. U.S.A. *99*: 11452–11457, 2002.

Baumgold, J. and Fishman, P. H.: Muscarinic receptor-mediated increase in cAMP levels in SK-N-SH human neuroblastoma cells. Biochem. Biophys. Res. Commun. *154*: 1137–1143, 1988.

Bawin, S. M., Satmary, W. M. and Adey, W. R.: Nitric oxide modulates rhythmic slow activity in rat hippocampal slices. Neuroreport *5*: 1869–1872, 1994.

Beech, D. J.: Actions of neurotransmitters and other messengers on Ca^{2+} channels and K^+ channels in smooth muscle cells. Pharmacol. Ther. *73*: 91–119, 1997.

Berkeley, J. L., Gomeza, J., Wess, J., Hamilton, S. E., Nathanson, N. M. and Levey, A.: I. M_1 muscarinic acetylcholine receptors activate extracellular signal-regulated kinase in CA1 pyramidal neurons in mouse hippocampal slices. Mol. Cell. Neurosci. *18*: 512–524, 2001.

Bernardini N., Roza C., Sauer S. K., Gomeza J., Wess J. and Reeh P. W.: Muscarinic M_2 receptors on peripheral nerve endings: a molecular target of antinociception. J. Neurosci. *22*: RC229, 2002.

Berridge, M. J. and Irvine, R. F.: Inositol phosphates and cell signalling. Nature *341*: 197–205, 1989.

Birdsall, N. J. M. and Lazareno, S.: To what extent can binding studies allow the quantification of affinity and efficacy? Ann. N.Y. Acad. Sci. *812*: 41–47, 1997.

Birdsall, N. J. M., Burgen, A. S. V., Hulme, E. C., Stockton, J. M. and Zigmond, M. J.: The effect of McN-A-343 on muscarinic receptors in the cerebral cortex and heart. Br. J. Pharmacol. *78*: 257–259, 1983.

Birdsall, N. J., Lazareno, S. and Matsui, H.: Allosteric regulation of muscarinic receptors. Prog. Brain Res. *109*: 147–151, 1996.

Birdsall, N. J., Farries, T., Gharagozloo, P., Kobayashi, S., Kuonen, D., Lazareno, S., Popham, A. and Sugimoto, M.: Selective allosteric enhancement of the binding and actions of acetylcholine at muscarinic receptor subtypes. Life Sci. *60*: 1047–1052, 1997.

Birdsall, N. J., Lazareno, S., Popham, A. and Saldanha, J.: Multiple allosteric sites on muscarinic receptors. Life Sci. *68*: 2517–2524, 2001.

Blin, N., Yun, J. and Wess, J.: Mapping of single amino acid residues required for selective activation of Gq/11 by the m3 muscarinic acetylcholine receptor. J. Biol. Chem. *270*: 17741–17748, 1995.

Bluml, K., Mutschler, E. and Wess, J.: Functional role of a cytoplasmic aromatic amino acid in muscarinic receptor-mediated activation of phospholipase C. J. Biol. Chem. *269*: 11537–11541, 1994a.

Bluml, K., Mutschler, E. and Wess, J.: Identification of an intracellular tyrosine residue critical for muscarinic receptor-mediated stimulation of phosphatidylinositol hydrolysis. J. Biol. Chem. *269*: 402–405, 1994b.

Bolton, T. B., Prestwich, S. A., Zholos, A. V. and Gordienko, D. V.: Excitation-contraction coupling in gastrointestinal and other smooth muscles. Annu. Rev. Physiol. *61*: 85–115, 1999.

Bonner, T. I.: The molecular basis of muscarinic receptor diversity. Trends Neurosci. *12*: 148–151, 1989a.

Bonner, T. I.: New subtypes of muscarinic acetylcholine receptors. Trends Pharmacol. Sci., Supplt.: 11–15, 1989b.

Bonner, T. I., Buckley, N. J., Young, A. C. and Brann, M. R.: Identification of a family of muscarinic acetylcholine receptor genes. Science *237*: 527–532, 1987.

Bonner, T. I., Young, A. C., Brann, M. R. and Buckley, N. J.: Cloning and expression of the human and rat m5 acetylcholine receptor genes. Neuron *1*: 403–410, 1988.

Bonner, T. I., Modi, W. S., Seuanez, H. N. and O'Brien, S. J.: Chromosomal mapping of the five human genes encoding the muscarinic acetylcholine receptors. Cytogenet. Cell. Genet. *58*: 1850–1851, 1991.

Bourne, H. R.: How receptors talk to trimeric G proteins. Curr. Opin. Cell Biol. *9*: 134–142, 1997.

Bouvier, M.: Oligomerization of G-protein-coupled transmitter receptors. Nat. Rev. Neurosci. *2*: 274–286, 2001.

Boyer, J. L., Martinez-Carcamo, M., Monroy-Sanchez, J. A., Posadas, C. and Garcia-Sainz, J. A.: Guanine nucleotide-induced positive cooperativity in muscarinic-cholinergic antagonist binding. Biochem. Biophys. Res. Comm. *134*: 172–177, 1986.

Bredt, D. S. and Snyder, S. H.: Nitric oxide, a novel neuronal messenger. Neuron *8*: 3–11, 1992.

Bredt, D. S. and Snyder, S. H.: Nitric oxide: a physiologic messenger molecule. Annu. Rev. Biochem. *63*: 175–195, 1994.

Bromidge, S. M., Brown, F., Cassidy, F., Clark, M. S., Dabbs, S., Hadley, M. S., Hawkins, J., Loudon, J. M., Naylor, C. B., Orlek, B. S. and Riley, G. J.: Design of [R-(Z)]-(+)-alpha-(methoxyimino)-1-azabicyclo[2.2.2]octane-3-acetonitrile (SB 202026), a functionally selective azabicyclic muscarinic M_1 agonist incorporating the N-methoxy imidoyl nitrile group as a novel ester bioisostere. J. Med. Chem. *40*: 4265–4280, 1997.

Brown, D. A. and Adams, P. R.: Muscarinic suppression of a novel voltage-sensitive K+ current in a vertebrate neurone. Nature *283*: 673–676, 1980.

Brown, D. A., Forward, A. and Marsh, S.: Antagonist discrimination between ganglionic and ileal muscarinic receptors. Br. J. Pharmacol. *71*: 362–364, 1980.

Buck, M. A. and Fraser, C. M.: Muscarinic acetylcholine receptor subtypes which selectively couple to phospholipase C: pharmacological and biochemical properties. Biochem. Biophys. Res. Commun. *173*: 666–672, 1990.

Buckley, N. J., Bonner, T. I. and Brann, M. R.: Localization of a family of muscarinic receptor mRNAs in rat brain. J. Neurosci. *8*: 4646–4652, 1988.

Buller, S., Zlotos, D. P., Mohr, K. and Ellis, J.: Allosteric site on muscarinic acetylcholine receptors: A single amino acid in transmembrane region 7 is critical to the subtype selectivities of caracurine V derivatives and alkane-bisammonium ligands. Mol. Pharmacol. *61*: 160–168, 2002.

Bunemann, M. and Hosey, M. M.: G-protein coupled receptor kinases as modulators of G-protein signalling. J. Physiol. (Lond.) *517*: 5–23, 1999.

Bunemann, M., Lee, K. B., Pals-Rylaarsdam, R., Roseberry, A. G. and Hosey, M. M.: Desensitization of G-protein-coupled receptors in the cardiovascular system. Annu. Rev. Physiol. *61*: 169–192, 1999.

Burgen, A. S. V.: The role of ionic interactions at the muscarinic receptor. Br. J. Pharmacol. *25*: 4–7, 1965.

Burstein, E. S., Spalding, T. A., Brauner-Osborne, H. and Brann, M. R.: Constitutive activation of muscarinic receptors by the G-protein Gq. FEBS Lett. *363*: 261–263, 1995a.

Burstein, E. S., Spalding, T. A., Hill-Eubanks, D. and Brann, M. R.: Structure-function of muscarinic receptor coupling to G proteins. Random saturation mutagenesis identifies a critical determinant of receptor affinity for G proteins. J. Biol. Chem. *270*: 3141–3146, 1995b.

Burstein, E. S., Spalding, T. A. and Brann, M. R.: Pharmacology of muscarinic receptor subtypes constitutively activated by G proteins. Mol. Pharmacol. *51*: 312–319, 1997.

Bymaster, F. P., Whitesitt, C. A., Shannon, H. E., DeLapp, N., Ward, J. S., Calligaro, D. O., Shipley, L. A., Buelke-Sam, J. L., Bodick, N. C., Farde, L., Sheardown, M. J., Olesen, P. H., Hansen, K. T., Suzdak, P. D., Swedberg, M. D. B., Sauerberg, P. and Mitch, C. H.: Xanomeline: a selective muscarinic agonist for the treatment of Alzheimer's disease. Drug. Dev. Res. *40*: 158–170, 1997.

Bymaster, F. P., Shannon, H. E., Rasmussen, K., Delapp, N. W., Mitch, C. H., Ward, J. S., Calligaro, D. O., Ludvigsen, T. S., Sheardown, M. J., Olesen, P. H., Swedberg, M. D. B., Sauerberg, P. and Finkjensen, A.: Unexpected antipsychotic-like activity with the muscarinic receptor ligand (5R,6R)6-(3-propylthio-1,2,5-thiadiazol-4-Yl)-1-azabicyclo[3.2.1]octane. Eur. J. Pharmacol. *356*: 109–119, 1998.

Bymaster, F. P., McKinzie, D. L., Felder, C. C. and Wess, J.: Use of M1-M5 muscarinic receptor knockout mice as novel tools to delineate the physiological roles of the muscarinic cholinergic system. Neurochem. Res. *28*: 437–442, 2003.

Cachero, T. G., Morielli, A. D. and Peralta, E.G.: The small GTP-binding protein RhoA regulates a delayed rectifier potassium channel. Cell *93*: 1077–1085, 1998.

Carey, G. J., Billard, W., Binch, H., 3rd, Cohen-Williams, M., Crosby, G., Grzelak, M., Guzik, H., Kozlowski, J. A., Lowe, D. B., Pond, A. J., Tedesco, R. P., Watkins, R. W. and Coffin, V. L.: SCH 57790, a selective muscarinic M_2 receptor antagonist, releases acetylcholine and produces cognitive enhancement in laboratory animals. Eur. J. Pharmacol. *431*: 189–200, 2001.

Carsi, J. M., Valentine, H. H. and Potter, L. T.: m2-toxin: A selective ligand for M_2 muscarinic receptors. Mol. Pharmacol. *56*: 933–937, 1999.

Caulfield, M. P.: Muscarinic receptors—characterization, coupling and function. Pharmacol. Ther. *58*: 319–379, 1993.

Caulfield, M. P. and Birdsall, N. J.: International Union of Pharmacology. XVII. Classification of muscarinic acetylcholine receptors. Pharmacol. Rev. *50*: 279–290, 1998.

Changeux, J. P. and Edelstein, S. J.: Allosteric receptors after 30 years. Neuron *21*: 959–980, 1998.

Chidiac, P., Green, M. A., Pawagi, A. B. and Wells, J. W.: Cardiac muscarinic receptors: cooperativity as the basis for multiple states of affinity. Biochemistry *36*: 7361–7379, 1997.

Christopoulos, A.: Quantification of allosteric interactions at G protein-coupled receptors using radioligand binding assays. In: "Current Protocols in Pharmacology," S. J. Enna, Ed., pp. 1.22.1–1.22.40, Wiley and Sons, New York, 2000.

Christopoulos, A.: Allosteric binding sites on cell-surface receptors: novel targets for drug discovery. Nat. Rev. Drug. Discov. *1*: 198–210, 2002.

Christopoulos, A. and El-Fakahany, E. E.: Novel persistent activation of muscarinic M_1 receptors by xanomeline. Eur. J. Pharmacol. *334*: R3–R4, 1997.

Christopoulos, A. and El-Fakahany, E. E.: The generation of nitric oxide by G protein-coupled receptors. Life Sci. *64*: 1–15, 1999a.

Christopoulos, A. and El-Fakahany, E. E.: Qualitative and quantitative assessment of relative agonist efficacy. Biochem. Pharmacol. *58*: 735–748, 1999b.

Christopoulos, A. and Kenakin, T.: G Protein-coupled receptor allosterism and complexing. Pharmacol. Rev. *54*: 323–374, 2002.

Christopoulos, A. and Mitchelson, F.: Assessment of the allosteric interactions of the bisquaternary heptane-1,7-bis(dimethyl-3′-pthalimidopropyl)ammonium bromide at M_1 and M_2 muscarine receptors. Mol. Pharmacol. *46*: 105–114, 1994.

Christopoulos, A., Lanzafame, A. and Mitchelson, F.: Allosteric interactions at muscarinic cholinoceptors.

Clin. Exp. Pharmacol. Physiol. *25*: 184–194, 1998a.

Christopoulos, A., Pierce, T. L., Sorman, J. L. and El-Fakahany, E. E.: On the unique binding and activating properties of xanomeline at the M_1 muscarinic acetylcholine receptor. Mol. Pharmacol. *53*: 1120–1130, 1998b.

Christopoulos, A., Parsons, A. M. and El-Fakahany, E. E.: Pharmacological analysis of the novel mode of interaction between xanomeline and the M_1 muscarinic acetylcholine receptor. J. Pharmacol. Exp. Ther. *289*: 1220–1228, 1999a.

Christopoulos, A., Sorman, J. L., Mitchelson, F. and El-Fakahany, E. E.: Characterization of the subtype selectivity of the allosteric modulator heptane-1,7-bis-(dimethyl-3′-pthalimidopropyl) ammonium bromide (C_7/3-phth) at cloned muscarinic acetylcholine receptors. Biochem. Pharmacol. *57*: 171–179, 1999b.

Christopoulos, A., Grant, M. K., Ayoubzadeh, N., Kim, O. N., Sauerberg, P., Jeppesen, L. and El-Fakahany, E. E.: Synthesis and pharmacological evaluation of dimeric muscarinic acetylcholine receptor agonists. J. Pharmacol. Exp. Ther. *298*: 1260–1268, 2001.

Clark, A. L. and Mitchelson, F.: The inhibitory effects of gallamine on muscarinic receptors. Br. J. Pharmacol. *58*: 323–331, 1976.

Coleman, R. A., Johnson, M., Nials, A. T. and Vardey, C. J.: Exosites: their current status, relevance to the duration of action of long-acting b_2-adrenoceptor agonists. Trends Pharmacol. Sci. *17*: 324–330, 1996.

Conklin, B. R., Brann, M. R., Buckley, N. J., Ma, A. L., Bonner, T. I. and Axelrod, J.: Stimulation of arachidonic acid release and inhibition of mitogenesis by cloned genes for muscarinic receptor subtypes stably expressed in A9 L cells. Proc. Natl. Acad. Sci. U. S.A. *85*: 8698–8702, 1988.

Conti, P., Dallanoce, C., De Amici, M., De Micheli, C. and Ebert, B.: Synthesis and binding affinity of new muscarinic ligands structurally related to oxotremorine. Bioorg. Med. Chem. Lett. *7*: 1033–1036, 1997.

Coso, O. A., Teramoto, H., Simonds, W. F. and Gutkind, J. S.: Signaling from G protein-coupled receptors to c-Jun kinase involves beta gamma subunits of heterotrimeric G proteins acting on a Ras and Rac1-dependent pathway. J. Biol. Chem. *271*: 3963–3966, 1996.

Crespo, P., Xu, N., Daniotti, J. L., Troppmair, J., Rapp, U. R. and Gutkind, J. S.: Signaling through transforming G protein-coupled receptors in NIH 3T3 cells involves c-Raf activation. Evidence for a protein kinase C-independent pathway. J. Biol. Chem. *269*: 21103–21109, 1994.

Curtis, C. M., Wheatley, M., Bansal, S., Birdsall, N. J. M., Eveleigh, P., Pedder, E. K., Poyner, D. and Hulme, E. C.: Propylbenilylcholine mustard labels an acidic residue in transmembrane helix 3 of the muscarinic receptor. J. Biol. Chem. *264*: 489–495, 1989.

Daeffler, L., Schmidlin, F., Gies, J. P. and Landry, Y.: Inverse agonist activity of pirenzepine at M_2 muscarinic acetylcholine receptors. Br. J. Pharmacol. *126*: 1246–1252, 1999.

Daub, H., Weiss, F. U., Wallasch, C. and Ullrich, A.: Role of transactivation of the EGF receptor in signalling by G-protein-coupled receptors. Nature *379*: 557–560, 1996.

Daub, H., Wallasch, C., Lankenau, A., Herrlich, A. and Ullrich, A.: Signal characteristics of G protein-transactivated EGF receptor. Embo. J. *16*: 7032–7044, 1997.

De Vries, L., Zheng, B., Fischer, T., Elenko, E. and Farquhar, M. G.: The regulator of G protein signaling family. Annu. Rev. Pharmacol. Toxicol. *40*: 235–271, 2000.

Di Chiara, G., Morelli, M. and Consolo, S.: Modulatory functions of neurotransmitters in the striatum: ACh/dopamine/NMDA interactions. Trends Neurosci.*17*: 228–233, 1994.

Dittman, A. H., Weber, J. P., Hinds, T. R., Choi, E. J., Migeon, J. C., Nathanson, N. M. and Storm, D. R.: A novel mechanism for coupling of m4 muscarinic acetylcholine receptors to calmodulin-sensitive adenylyl cyclases: crossover from G protein- coupled inhibition to stimulation. Biochemistry *33*: 943–951, 1994.

Dolezal, V., Lisa, V., Diebler, M. F., Kasparova, J. and Tucek, S.: Differentiation of NG108–15 cells induced by the combined presence of dbcAMP and dexamethasone brings about the expression of N and P/Q types of calcium channels and the inhibitory influence of muscarinic receptors on calcium influx. Brain Res. *910*: 134–141, 2001.

Dorje, F., Levey, A. I. and Brann, M. R.: Immunological detection of muscarinic receptor subtype proteins (m1–m5) in rabbit peripheral tissues. Mol. Pharmacol. *40*: 459–462, 1991.

Duttaroy, A., Gomeza, J., Gan, J. W., Siddiqui, N., Basile, A. S., Harman, W. D., Smith, P. L., Felder, C. C., Levey, A. I. and Wess, J.: Evaluation of muscarinic agonist-induced analgesia in muscarinic acetylcholine receptor knockout mice. Mol. Pharmacol. *62*: 1084–1093, 2002.

Eglen, R. M.: Muscarinic receptor antagonists. Pharmacological and therapeutic utility. In: "Receptor-Based Drug Design," P. Leff, Ed., pp. 273–296, Marcel Dekker, New York, 1998.

Eglen, R. M. and Watson, N.: Selective muscarinic receptor agonists and antagonists. Pharmacol. Toxicol. *78*: 59–68, 1996.

Eglen, R. M., Choppin, A. and Watson, N.: Therapeutic opportunities from muscarinic receptor research. Trends Pharmacol. Sci. *22*: 409–414, 2001.

Ehlert, F. J.: The relationship between muscarinic receptor occupancy and adenylate cyclase inhibition in the rabbit myocardium. Mol. Pharmacol. *28*: 410–421, 1985.

Ehlert, F. J.: Estimation of the affinities of allosteric ligands using radioligand binding and pharmacological null methods. Mol. Pharmacol. *33*: 187–194, 1988.

Ehlert, F. J. and Rathbun, B. E.: Signaling through the muscarinic receptor-adenylate cyclase system of the heart is buffered against GTP over a range of concentrations. Mol. Pharmacol. *38*: 148–158, 1990.

Ehlert, F. J., Roeske, W. R. and Yamamura, H. I.: Molecular biology, pharmacology, and brain distribution of subtypes of the muscarinic receptor. In: "Psychopharmacology: The Fourth Generation of Progress," F. E. Bloom and D. J. Kupfer, Eds., pp. 111–124, Raven Press, New York, 1995.

Eisenach, J. C.: Muscarinic-mediated analgesia. Life Sci. *64*: 549–554, 1999.

Ellis, J.: Allosteric binding sites on muscarinic receptors. Drug. Dev. Res. *40*: 193–204, 1997.

Ellis, J. and Seidenberg, M.: Interactions of alcuronium, TMB-8, and other allosteric ligands with muscarinic acetylcholine receptors: studies with chimeric receptors. Mol. Pharmacol. *58*: 1451–1460, 2000.

England, B. P., Ackerman, M. S. and Barrett, R. W.: A chimeric D2 dopamine/m1 muscarinic receptor with D2 binding specificity mobilizes intracellular calcium in response to dopamine. FEBS Lett. *279*: 87–90, 1991.

Ensinger, H. A., Doods, H. N., Immel-Sehr, A. R., Kuhn, F. J., Lambrecht, G., Mendla, K. D., Muller, R. E., Mutschler, E., Sagrada, A., Walther, G. et al.: WAL 2014—a muscarinic agonist with preferential neuron-stimulating properties. Life Sci. *52*: 473–480, 1993.

Exton, J. H.: Regulation of phosphoinositide phospholipases by hormones, neurotransmitters, and other agonists linked to G proteins. Annu. Rev. Pharmacol. Toxicol. *36*: 481–509, 1996.

Exton, J. H.: Phospholipase D: enzymology, mechanisms of regulation, and function. Physiol. Rev. *77*: 303–320, 1997.

Exton, J. H.: Regulation of phospholipase D. Biochim. Biophys. Acta *1439*: 121–133, 1999.

Fain, J. N., Wallace, M. A. and Wojcikiewicz, R. J.: Evidence for involvement of guanine nucleotide-binding regulatory proteins in the activation of phospholipases by hormones. FASEB J. *2*: 2569–2574, 1988.

Felder, C. C.: Muscarinic acetylcholine receptors: signal transduction through multiple effectors. FASEB J. *9*: 619–625, 1995.

Felder, C. C., Kanterman, R. Y., Ma, A. L. and Axelrod, J.: A transfected m1 muscarinic acetylcholine receptor stimulates adenylate cyclase via phosphatidylinositol hydrolysis. J. Biol. Chem. *264*: 20356–20362, 1989.

Felder, C. C., Williams, H. L. and Axelrod, J.: A transduction pathway associated with receptors coupled to the inhibitory guanine nucleotide binding protein Gi that amplifies ATP-mediated arachidonic acid release. Proc. Natl. Acad. Sci. U.S.A. *88*: 6477–6480, 1991.

Felder, C. C., Bymaster, F. P., Ward, J. and DeLapp, N.: Therapeutic opportunities for muscarinic receptors in the central nervous system. J. Med. Chem. *43*: 4333–4353, 2000.

Felder, C. C., Porter, A. C., Skillman, T. L., Zhang, L., Bymaster, F. P., Nathanson, N. M., Hamilton, S. E., Gomeza, J., Wess, J. and McKinzie, D. L.: Elucidating the role of muscarinic receptors in psychosis. Life Sci. *68*: 2605–2613, 2001.

Ferguson, S. S.: Evolving concepts in G protein-coupled receptor endocytosis: the role in receptor desensitization and signaling. Pharmacol. Rev. *53*: 1–24, 2001.

Fernandez-Fernandez, J. M., Wanaverbecq, N., Halley, P., Caulfield, M. P. and Brown, D. A.: Selective activation of heterologously expressed G protein-gated K+ channels by M2 muscarinic receptors in rat sympathetic neurones. J. Physiol. (Lond.) *515*: 631–637, 1999.

Fernandez-Fernandez, J. M., Abogadie, F. C., Milligan, G., Delmas, P. and Brown, D. A.: Multiple pertussis toxin-sensitive G-proteins can couple receptors to GIRK channels in rat sympathetic neurons when expressed heterologously, but only native G_i-proteins do so in situ. Eur. J. Neurosci. *14*: 283–292, 2001.

Fisahn, A., Yamada, M., Duttaroy, A., Gan, J. W., Deng, C. X., McBain, C. J. and Wess, J.: Muscarinic induction of hippocampal gamma oscillations requires coupling of the M_1 receptor to two mixed cation currents. Neuron *33*: 615–624, 2002.

Flier, J. S. and Maratos-Flier, E.: Obesity and the hypothalamus: novel peptides for new pathways. Cell *92*: 437–440, 1998.

Flynn, D. D., Reever, C. M. and Ferrari-Dileo, G.: Pharmacological strategies to selectively label and localize muscarinic receptor subtypes. Drug. Dev. Res. *40*: 104–116, 1997.

Forster, G. L., Yeomans, J. S., Takeuchi, J. and Blaha, C. D.: M_5 muscarinic receptors are required for prolonged accumbal dopamine release after electrical stimulation of the pons in mice. J. Neurosci. *22*: RC190, 2002.

Franken, C., Trankle, C. and Mohr, K.: Testing the specificity of allosteric modulators of muscarinic receptors in phylogenetically closely related histamine H1-receptors. Naunyn Schmiedebergs Arch. Pharmacol. *361*: 107–112, 2000.

Fraser, C. M., Wang, C. D., Robinson, D. A., Gocayne, J. D. and Venter, J. C.: Site-directed mutagenesis of m1 muscarinic acetylcholine receptors: conserved aspartic acids play important roles in receptor function. Mol. Pharmacol. *36*: 840–847, 1989.

Fromm, C., Coso, O. A., Montaner, S., Xu, N. and Gutkind, J. S.: The small GTP-binding protein Rho links G protein-coupled receptors and Galpha12 to the serum response element and to cellular transformation. Proc. Natl. Acad. Sci. U.S.A. *94*: 10098–10103, 1997.

Fukuhara, S., Marinissen, M. J., Chiariello, M. and Gutkind, J. S.: Signaling from G protein-coupled receptors to ERK5/Big MAPK 1 involves Gaq and Ga12/13 families of heterotrimeric G proteins. Evidence for the existence of a novel Ras and Rho-independent pathway. J. Biol. Chem. *275*: 21730–21736, 2000.

Garrington TP and Johnson GL: Organization and regulation of mitogen-activated protein kinase signaling pathways. Curr. Opin. Cell. Biol. *11*: 211–218, 1999.

Gattaz, W. F., Maras, A., Cairns, N. J., Levy, R. and Forstl, H.: Decreased phospholipase A2 activity in Alzheimer brains. Biol. Psychiatry *37*: 13–17, 1995.

Gerber, D. J., Sotnikova, T. D., Gainetdinov, R. R., Huang, S. Y., Caron, M. G. and Tonegawa, S.: Hyperactivity, elevated dopaminergic transmission, and response to amphetamine in M1 muscarinic acetylcholine receptor-deficient mice. Proc. Natl. Acad. Sci. U.S.A. *98*: 15312–15317, 2001.

Gharagozloo, P., Lazareno, S., Popham, A. and Birdsall, N. J.: Allosteric interactions of quaternary strychnine and brucine derivatives with muscarinic acetylcholine receptors. J. Med. Chem. *42*: 438–445, 1999.

Giacobini, E.: Cholinomimetic replacement of cholinergic function in Alzheimer's disease. In: "Treatment of Dementias," E. M. Meyer, J. W. Simpkins, J. Yamamoto and F. T. Crews, Eds., pp. 19–34, Plenum Press, New York, 1992.

Gloge, H., Lullmann, H. and Mutschler, E.: The action of tertiary and quaternary arecaidine and dihydroarecaidine esters on the guinea pig isolated ileum. Br. J. Pharmacol. *27*: 185–195, 1966.

Gomeza, J., Shannon, H., Kostenis, E., Felder, C., Zhang, L., Brodkin, J., Grinberg, A., Sheng, H. and Wess, J.: Pronounced pharmacologic deficits in M2 muscarinic acetylcholine receptor knockout mice. Proc. Natl. Acad. Sci. U.S.A. *96*: 1692–1697, 1999a.

Gomeza, J., Zhang, L., Kostenis, E., Felder, C., Bymaster, F., Brodkin, J., Shannon, H., Xia, B., Deng, C. and Wess, J.: Enhancement of D1 dopamine receptor-mediated locomotor stimulation in M_4 muscarinic acetylcholine receptor knockout mice. Proc. Natl. Acad. Sci. U.S.A. *96*: 10483–10488, 1999b.

Gouldson, P. R., Snell, C. R., Bywater, R. P., Higgs, C. and Reynolds, C. A.: Domain swapping in G-protein coupled receptor dimers. Protein Eng. *11*: 1181–1193, 1998.

Guo, F. F., Kumahara, E. and Saffen, D.: A CalDAG-GEFI/Rap1/B-Raf cassette couples M_1 muscarinic acetylcholine receptors to the activation of ERK1/2. J. Biol. Chem. *276*: 25568–25581, 2001.

Gurwitz, D., Haring, R., Heldman, E., Fraser, C. M., Manor, D. and Fisher, A.: Discrete activation of transduction pathways associated with acetylcholine m1 receptor by several muscarinic ligands. Eur. J. Pharmacol. *267*: 21–31, 1994.

Haddad, E. and Rousell, J.: Regulation of the expression and function of the M_2 muscarinic receptor. Trends Pharmacol. Sci. *19*: 322–327, 1998.

Haga, K., Haga, T. and Ichiyama, A.: Reconstitution of the muscarinic acetylcholine receptor. Guanine nucleotide-sensitive high affinity binding of agonists to purified muscarinic receptors reconstituted with GTP-binding proteins (G_i and G_o). J. Biol. Chem. *261*: 10133–10140, 1986.

Haj-Dahmane, S. and Andrade, R.: Ionic mechanism of the slow after-depolarization induced by muscarinic receptor activation in rat prefrontal cortex. J. Neurophysiol. *80*: 1197–1210, 1998.

Hamilton, S. E., Loose, M. D., Qi, M., Levey, A. I., Hille, B., McKnight, G. S., Idzerda, R. L. and Nathanson, N. M.: Disruption of the m1 receptor gene ablates muscarinic receptor-dependent M current regulation and seizure activity in mice. Proc. Natl. Acad. Sci. U.S.A. *94*: 13311–13316, 1997.

Hammer, R. and Giachetti, A.: Muscarinic receptor subtypes: M_1 and M_2 biochemical and functional characterization. Life Sci. *31*: 2991–2998, 1982.

Hammer, R., Berrie, C. P., Birdsall, N. J., Burgen, A. S. and Hulme, E. C.: Pirenzepine distinguishes between different subclasses of muscarinic receptors. Nature *283*: 90–92, 1980.

Han, X., Kobzik, L., Severson, D. and Shimoni, Y.: Characteristics of nitric oxide-mediated cholinergic modulation of calcium current in rabbit sino-atrial node. J. Physiol. *509*: 741–754, 1998.

Haring, R., Fisher, A., Marciano, D., Pittel, Z., Kloog, Y., Zuckerman, A., Eshhar, N. and Heldman, E.: Mitogen-activated protein kinase-dependent and protein kinase C-dependent pathways link the m1 muscarinic receptor to beta-amyloid precursor protein secretion. J. Neurochem. *71*: 2094–2103, 1998.

Hawes, B. E., van Biesen, T., Koch, W. J., Luttrell, L. M. and Lefkowitz, R. J.: Distinct pathways of G_i- and G_q-mediated mitogen-activated protein kinase activation. J. Biol. Chem. *270*: 17148–17153, 1995.

Hayashi, M. K. and Haga, T.: Palmitoylation of muscarinic acetylcholine receptor m2 subtypes: reduction in their ability to activate G proteins by mutation of a putative palmitoylation site, cysteine 457, in the carboxyl-terminal tail. Arch. Biochem. Biophys. *340*: 376–382, 1997.

Heasley, L. E., Storey, B., Fanger, G. R., Butterfield, L., Zamarripa, J., Blumberg, D. and Maue, R. A.: GTPase-deficient G alpha 16 and G alpha q induce PC12 cell differentiation and persistent activation of cJun NH2-terminal kinases. Mol. Cell. Biol. *16*: 648–656, 1996.

Hebert, T. E. and Bouvier, M.: Structural and functional aspects of G protein-coupled receptor oligomerization. Biochem. Cell Biol. *76*: 1–10, 1998.

Hilf, G. and Jakobs, K. H.: Agonist-independent inhibition of G protein activation by muscarinic acetylcholine receptor antagonists in cardiac membranes. Eur. J. Pharmacol. (Molec. Pharmacol. Sec.) *225*: 245–252, 1992.

Hilf, G., Gierschik, P. and Jakobs, K. H.: Muscarinic acetylcholine receptor-stimulated binding of guanosine 5′-*O*-(3-thiotriphosphate) to guanine-nucleotide-binding proteins in cardiac membranes. Eur. J. Biochem. *186*: 725–731, 1989.

Hill, J. J. and Peralta, E. G.: Inhibition of a G_i-activated potassium channel (GIRK1/4) by the Gq-coupled m1 muscarinic acetylcholine receptor. J. Biol. Chem. *276*: 5505–5510, 2001.

Hille, B., Beech, D. J., Bernheim, L., Mathie, A., Shapiro, M. S. and Wollmuth, L. P.: Multiple G-protein-coupled pathways inhibit N-type Ca channels of neurons. Life Sci. *56*: 989–992, 1995.

Hirabayashi, T. and Saffen, D.: M_1 muscarinic acetylcholine receptors activate zif268 gene expression via small G-protein Rho-dependent and lambda-independent pathways in PC12D cells. Eur. J. Biochem. *267*: 2525–2532, 2000.

Hirschberg, B. T. and Schimerlick, M. I.: A kinetic model for oxotremorine M binding to recombinant porcine m2 muscarinic receptors expressed in Chinese hamster ovary cells. J. Biol. Chem. *269*: 26127–26135, 1994.

Hogger, P., Shockley, M. S., Lameh, J. and Sadee, W.: Activating and inactivating mutations in N- and C-terminal i3 loop junctions of muscarinic acetylcholine Hm1 receptors. J. Biol. Chem. *270*: 7405–7410, 1995.

Holzgrabe, U. and Mohr, K.: Allosteric modulators of ligand binding to muscarinic acetylcholine receptors. Drug Discov. Today *3*: 214–222, 1998.

Horn, F., van der Wenden, E. M., Oliveira, L., Ijzerman, A. P. and Vriend, G.: Receptors coupling to G proteins: is there a signal behind the sequence? Proteins *41*: 448–459, 2000.

Hou, Y., Azpiazu, I., Smrcka, A. and Gautam, N.: Selective role of G protein gamma subunits in receptor interaction. J. Biol. Chem. *275*: 38961–38964, 2000.

Hou, Y., Chang, V., Capper, A. B., Taussig, R. and Gautam, N.: G Protein beta subunit types differentially interact with a muscarinic receptor but not adenylyl cyclase type II or phospholipase C-beta 2/3. J. Biol. Chem. *276*: 19982–19988, 2001.

Hulme, E. C., Birdsall, N. J. M. and Buckley, N. J.: Muscarinic receptor subtypes. Ann. Rev. Pharmacol. Toxicol. *30*: 633–673, 1990.

Hulme, E. C., Curtis, C. A., Page, K. M. and Jones, P. G.: The role of charge interactions in muscarinic agonist binding, and receptor-response coupling. Life Sci. *56*: 891–898, 1995.

Ince, E., Ciliax, B. J. and Levey, A. I.: Differential expression of D1 and D2 dopamine and m4 muscarinic acetylcholine receptor proteins in identified striatonigral neurons. Synapse *27*: 357–366, 1997.

Jakubík, J., Bacaková, L., El-Fakahany, E. E. and Tucek, S.: Constitutive activity of the M_1-M_4 subtypes of muscarinic receptors in transfected CHO cells and of muscarinic receptors in the heart cells revealed by negative antagonists. FEBS Lett. *377*: 275–279, 1995a.

Jakubík, J., Bacáková, L., El-Fakahany, E. E. and Tucek, S.: Subtype selectivity of the positive allosteric action of alcuronium at cloned M_1-M_5 muscarinic acetylcholine receptors. J. Pharmacol. Exp. Ther. *274*: 1077–1083, 1995b.

Jakubík, J., Bacakova, L., Lisa, V., El-Fakahany, E. E. and Tucek, S.: Activation of muscarinic acetylcholine receptors via their allosteric binding sites. Proc. Natl. Acad. Sci. U.S.A. *93*: 8705–8709, 1996.

Jakubík, J., Bacakova, L., El-Fakahany, E. E. and Tucek, S.: Positive cooperativity of acetylcholine and other agonists with allosteric ligands on muscarinic acetylcholine receptors. Mol. Pharmacol. *52*: 172–179, 1997.

Jakubík, J., Haga, T. and Tucek, S.: Effects of an agonist, allosteric modulator, and antagonist on guanosine-g-[^{35}S]thiotriphosphate binding to liposomes with varying muscarinic receptor/G_o protein stoichiometry. Mol. Pharmacol. *54*: 899–906, 1998.

Jakubík, J., Tucek, S. and El-Fakahany, E. E.: Allosteric modulation by persistent binding of xanomeline of the interaction of competitive ligands with the M_1 muscarinic acetylcholine receptor. J. Pharmacol. Exp. Ther. *301*: 1033–1041, 2002.

Janssen, L. J. and Sims, S. M.: Muscarinic regulation of ion channels in smooth muscle. In: "Cellular Aspects of Smooth Muscle Function," C. Y. Kao and M. E. Carsten, Eds., pp. 132–168, Cambridge University Press, Cambridge, 1997.

Jeong, S. W. and Wurster, R. D.: Muscarinic receptor activation modulates Ca2+ channels in rat intracardiac neurons via a PTX- and voltage-sensitive pathway. J. Neurophysiol. *78*: 1476–1490, 1997.

Jones, D., Morgan, C. and Cockcroft, S.: Phospholipase D and membrane traffic. Potential roles in regulated exocytosis, membrane delivery and vesicle budding. Biochim. Biophys. Acta *1439*: 229–244, 1999.

Jones, S. V.: Muscarinic receptor subtypes: modulation of ion channels. Life Sci. *52*: 457–464, 1993.

Katz, A., Wu, D. and Simon, M. I.: Subunits beta gamma of heterotrimeric G protein activate beta 2 isoform of phospholipase C. Nature *360*: 686–689, 1992.

Kavalali, E. T., Zhuo, M., Bito, H. and Tsien, R. W.: Dendritic Ca2+ channels characterized by recordings from isolated hippocampal dendritic segments. Neuron *18*: 651–663, 1997.

Kawasaki, H., Palmieri, C. and Avoli, M.: Muscarinic receptor activation induces depolarizing plateau potentials in bursting neurons of the rat subiculum. J. Neurophysiol. *82*: 2590–2601, 1999.

Kenakin, T.: Agonist-receptor efficacy I: mechanisms of efficacy and receptor promiscuity. Trends Pharmacol. Sci. *16*: 188–192, 1995a.

Kenakin, T.: Agonist-receptor efficacy II: agonist trafficking of receptor signals. Trends Pharmacol. Sci. *16*: 232–238, 1995b.

Kenakin, T.: The classification of seven transmembrane receptors in recombinant expression systems. Pharmacol. Rev. *48*: 413–463, 1996.

Kenakin, T.: Differences between natural and recombinant G protein-coupled receptor systems with varying receptor/G protein stoichiometry. Trends Pharmacol. Sci. *18*: 456–464, 1997a.

Kenakin T. P.: "Pharmacologic Analysis of Drug-Receptor Interaction," Lippincott-Raven, Philadelphia, 1997b.

Kenakin, T., Morgan, P. and Lutz, M.: On the importance of the "antagonist assumption" to how receptors express themselves. Biochem. Pharmacol. *50*: 17–26, 1995.

Kilbinger, H., Halim, S., Lambrecht, G., Weiler, W. and Wessler, I.: Comparison of affinities of muscarinic antagonists to pre- and postjunctional receptors in the guinea-pig ileum. Eur. J. Pharmacol. *103*: 313–320, 1984.

Kim, J. Y., Yang, M. S., Oh, C. D., Kim, K. T., Ha, M. J., Kang, S. S. and Chun, J. S.: Signalling pathway leading to an activation of mitogen-activated protein kinase by stimulating M_3 muscarinic receptor. Biochem. J. *337*: 275–280, 1999.

Kirchhausen, T., Bonifacino, J. S. and Riezman, H.: Linking cargo to vesicle formation: receptor tail interactions with coat proteins. Curr. Opin. Cell. Biol. *9*: 488–495, 1997.

Koch, W. J., Hawes, B. E., Allen, L. F. and Lefkowitz, R. J.: Direct evidence that Gi-coupled receptor stimulation of mitogen-activated protein kinase is mediated by G beta gamma activation of p21ras. Proc. Natl. Acad. Sci. U.S.A. *91*: 12706–12710, 1994.

Koenig, J. and Edwardson, J. M.: Endocytosis and recycling of G protein-coupled receptors. Trends Pharmacol. Sci. *18*: 276–287, 1997.

Kopf, S. R. and Baratti, C. M.: Enhancement of the post-training cholinergic tone antagonizes the impairment of retention induced by a nitric oxide synthase inhibitor in mice. Neurobiol. Learn. Mem. *65*: 207–212, 1996.

Kords, H., Lullmann, H., Ohnesorge, F. K. and Wassermann, O.: Action of atropine and some hexane-1,6-bis-ammonium derivatives upon the toxicity of DFP in mice. Eur. J. Pharmacol. *3*: 341–346, 1968.

Kostenis, E., Holzgrabe, U. and Mohr, K.: Allosteric effect on muscarinic M_2-receptors of derivatives of the alkane-bis-ammonium compound W84. Comparison with bispyridinium-type allosteric modulators. Eur. J. Med. Chem. *29*: 947–953, 1994.

Kostenis, E., Conklin, B. R. and Wess, J.: Molecular basis of receptor/G protein coupling selectivity studied by coexpression of wild type and mutant m2 muscarinic receptors with mutant Ga_q subunits. Biochemistry *36*: 1487–1495, 1997a.

Kostenis, E., Degtyarev, M. Y., Conklin, B. R. and Wess, J.: The N-terminal extension of Gaq is critical for constraining the selectivity of receptor coupling. J. Biol. Chem. *272*: 19107–19110, 1997b.

Kostenis, E., Gomeza, J., Lerche, C. and Wess, J.: Genetic analysis of receptor-Galphaq coupling selectivity. J. Biol. Chem. *272*: 23675–23681, 1997c.

Kostenis, E., Zeng, F. Y. and Wess, J.: Functional characterization of a series of mutant G protein aq subunits displaying promiscuous receptor coupling properties. J. Biol. Chem. *273*: 17886–17892, 1998a.

Kostenis, E., Zeng, F. Y. and Wess, J.: Structure-function analysis of muscarinic acetylcholine receptors. J. Physiol. (Paris) *92*: 265–268, 1998b.

Krejci, A. and Tucek, S.: Quantitation of mRNAs for M_1 to M_5 subtypes of muscarinic receptors in rat heart and brain cortex. Mol. Pharmacol. *61*: 1267–1272, 2002.

Krupnick, J. G. and Benovic, J. L.: The role of receptor kinases and arrestins in G protein-coupled receptor regulation. Annu. Rev. Pharmacol. Toxicol. *38*: 289–319, 1998.

Kubo, T., Fukuda, K., Mikami, A., Maeda, A., Takahashi, H., Mishina, M., Haga, T., Haga, K., Ichiyama, A., Kangawa, K. et al.: Cloning, sequencing and expression of complementary DNA encoding the muscarinic acetylcholine receptor. Nature *323*: 411–416, 1986a.

Kubo, T., Maeda, A., Sugimoto, K., Akiba, I., Mikami, A., Takahashi, H., Haga, T., Haga, K., Ichiyama, A.,

Kangawa, K. et al.: Primary structure of porcine cardiac muscarinic acetylcholine receptor deduced from the cDNA sequence. FEBS Lett. *209*: 367–372, 1986b.

Kubo, T., Bujo, H., Akiba, J., Mishina, M. and Numa, S.: Location of a region of the muscarinic receptor involved in selective effector coupling. FEBS Lett. *241*: 119–125, 1988.

Kumahara, E., Ebihara, T. and Saffen, D.: Protein kinase inhibitor H7 blocks the induction of immediate-early genes zif268 and c-fos by a mechanism unrelated to inhibition of protein kinase C but possibly related to inhibition of phosphorylation of RNA polymerase II. J. Biol. Chem. *274*: 10430–10438, 1999.

Kunkel, M. T. and Peralta, E. G.: Charged amino acids required for signal transduction by the m3 muscarinic acetylcholine receptor. Embo. J. *12*: 3809–3815, 1993.

Kurtenbach, E., Pedder, E. K., Curtis, C. A. and Hulme, E. C.: The putative disulphide bond in muscarinic receptors. Biochem. Soc. Trans. *18*: 442–443, 1990.

Lachowicz, J. E., Lowe, D., Duffy, R. A., Ruperto, V., Taylor, L. A., Guzik, H., Brown, J., Berger, J. G., Tice, M., McQuade, R., Kozlowski, J., Clader, J., Strader, C. D. and Murgolo, N.: SCH 57790: a novel M2 receptor selective antagonist. Life Sci. *64*: 535–539, 1999.

Lanzafame, A., Christopoulos, A. and Mitchelson, F.: Interactions of agonists with an allosteric antagonist at muscarinic acetylcholine M_2 receptors. Eur. J. Pharmacol. *316*: 27–32, 1996.

Lanzafame, A. A., Christopoulos, A. and Mitchelson, F.: Cellular signaling mechanisms for muscarinic acetylcholine receptors. Receptors and Channels *9*: 241–260, 2003.

Lazareno, S. and Birdsall, N. J. M.: Pharmacological characterization of acetylcholine-stimulated [^{35}S]-GTPgS binding mediated by human muscarinic m1-m4 receptors: antagonist studies. Br. J. Pharmacol. *109*: 1120–1127, 1993.

Lazareno, S. and Birdsall, N. J. M.: Detection, quantitation, and verification of allosteric interactions of agents with labeled and unlabeled ligands at G protein-coupled receptors: interactions of strychnine and acetylcholine at muscarinic receptors. Mol. Pharmacol. *48*: 362–378, 1995.

Lazareno, S., Gharagozloo, P., Kuonen, D., Popham, A. and Birdsall, N. J. M.: Subtype-selective positive cooperative interactions between brucine analogues and acetylcholine at muscarinic receptors: Radioligand binding studies. Mol. Pharmacol. *53*: 573–589, 1998.

Lazareno, S., Popham, A. and Birdsall, N. J.: Allosteric interactions of staurosporine and other indolocarbazoles with N-[methyl-^{3}H]scopolamine and acetylcholine at muscarinic receptor subtypes: identification of a second allosteric site. Mol. Pharmacol. *58*: 194–207, 2000.

Lazareno, S., Popham, A. and Birdsall, N. J.: Analogs of WIN 62,577 define a second allosteric site on muscarinic receptors. Mol. Pharmacol. *62*: 1492–1505, 2002.

Lechleiter, J., Hellmiss, R., Duerson, K., Ennulat, D., David, N., Clapham, D. and Peralta, E.: Distinct sequence elements control the specificity of G protein activation by muscarinic acetylcholine receptor subtypes. Embo. J. *9*: 4381–4390, 1990.

Lee, K. B., Pals-Rylaarsdam, R., Benovic, J. L. and Hosey, M. M.: Arrestin-independent internalization of the m1, m3, and m4 subtypes of muscarinic cholinergic receptors. J. Biol. Chem. *273*: 12967–12972, 1998.

Lee, N. H., Geoghagen, N. S., Cheng, E., Cline, R. T. and Fraser, C. M.: Alanine scanning mutagenesis of conserved arginine/lysine-arginine/lysine-X-X-arginine/lysine G protein-activating motifs on m1 muscarinic acetylcholine receptors. Mol. Pharmacol. *50*: 140–148, 1996.

Lefkowitz, R. J.: G protein-coupled receptors. III. New roles for receptor kinases and beta-arrestins in receptor signaling and desensitization. J. Biol. Chem. *273*: 18677–18680, 1998.

Leppik, R. A., Miller, R. C., Eck, M. and Paquet, J. L.: Role of acidic amino acids in the allosteric modulation by gallamine of antagonist binding at the m2 muscarinic acetylcholine receptor. Mol. Pharmacol. *45*: 983–990, 1994.

Levey, A. I., Kitt, C. A., Simonds, W. F., Price, D. L. and Brann, M. R.: Identification and localization of muscarinic acetylcholine receptor proteins in brain with subtype-specific antibodies. J. Neurosci. *11*: 3218–3226, 1991.

Levey, A. I., Edmunds, S. M., Heilman, C. J., Desmond, T. J. and Frey, K. A.: Localization of muscarinic m3 receptor protein and M3 receptor binding in rat brain. Neuroscience *63*: 207–221, 1994.

Li, M., Yasuda, R. P., Wall, S. J., Wellstein, A. and Wolfe, B. B.: Distribution of m2 muscarinic receptors in rat brain using antisera selective for m2 receptors. Mol. Pharmacol. *40*: 28–35, 1991.

Liang, J. S., Carsi-Gabrenas, J., Krajewski, J. L., McCafferty, J. M., Purkerson, S. L., Santiago, M. P., Strauss, W. L., Valentine, H. H. and Potter, L. T.: Anti-muscarinic toxins from Dendroaspis angusticeps. Toxicon *34*: 1257–1267, 1996.

Liscovitch, M., Czarny, M., Fiucci, G. and Tang, X.: Phospholipase D: molecular and cell biology of a novel gene family. Biochem. J. *345*: 401–415, 2000.

Liu, J., Conklin, B. R., Blin, N., Yun, J. and Wess, J.: Identification of a receptor/G-protein contact site critical for signaling specificity and G-protein activation. Proc. Natl. Acad. Sci. U.S.A. *92*: 11642–11646, 1995.

Liu, J., Blin, N., Conklin, B. R. and Wess, J.: Molecular mechanisms involved in muscarinic acetylcholine receptor-mediated G protein activation studied by insertion mutagenesis. J. Biol. Chem. *271*: 6172–6178, 1996.

Lopez-Ilasaca, M., Crespo, P., Pellici, P. G., Gutkind, J. S. and Wetzker, R.: Linkage of G protein-coupled receptors to the MAPK signaling pathway through PI 3-kinase gamma. Science *275*: 394–397, 1997.

Loudon, J. M., Bromidge, S. M., Brown, F., Clark, M. S., Hatcher, J. P., Hawkins, J., Riley, G. J., Noy, G. and Orlek, B. S.: SB 202026: a novel muscarinic partial agonist with functional selectivity for M1 receptors. J. Pharmacol. Exp. Ther. *283*: 1059–1068, 1997.

Lu, Z. L. and Hulme, E. C.: The functional topography of transmembrane domain 3 of the M_1 muscarinic acetylcholine receptor, revealed by scanning mutagenesis. J. Biol. Chem. *274*: 7309–7315, 1999.

Lu, Z. L. and Hulme, E. C.: A network of conserved intramolecular contacts defines the off-state of the transmembrane switch mechanism in a seven-transmembrane receptor. J. Biol. Chem. *275*: 5682–5686, 2000.

Lu, Z. L., Curtis, C. A., Jones, P. G., Pavia, J. and Hulme, E. C.: The role of the aspartate-arginine-tyrosine triad in the m1 muscarinic receptor: Mutations of aspartate 122 and tyrosine 124 decrease receptor expression but do not abolish signaling. Mol. Pharmacol. *51*: 234–241, 1997.

Lu, Z. L., Saldanha, J. W. and Hulme, E. C.: Transmembrane domains 4 and 7 of the m1 muscarinic acetylcholine receptor are critical for ligand binding and the receptor activation switch. J. Biol. Chem. *276*: 34098–34104, 2001.

Lu, Z. L., Saldanha, J. W. and Hulme, E. C.: Seven-transmembrane receptors: crystals clarify. Trends Pharmacol. Sci. *23*: 140–146, 2002.

Lüllman, H., Ohnesorge, F. K., Schauwecker, G.-C. and Wasserman, O.: Inhibition of the actions of carbachol and DFP on guinea pig isolated atria by alkane-bis-ammonium compounds. Eur. J. Pharmacol. *6*: 241–247, 1969.

Luttrell, L. M. and Lefkowitz, R. J.: The role of beta-arrestins in the termination and transduction of G-protein-coupled receptor signals. J. Cell Sci. *115*: 455–465, 2002.

Lysikova, M., Fuksova, K., Elbert, T., Jakubik, J. and Tucek, S.: Subtype-selective inhibition of [*methyl-*^{3}*H*]-N-methylscopolamine binding to muscarinic receptors by a-truxillic acid esters. Br. J. Pharmacol. *127*: 1240–1246, 1999.

Maggio, R., Vogel, Z. and Wess, J.: Coexpression studies with mutant muscarinic/adrenergic receptors provide evidence for intermolecular "cross-talk" between G-protein-linked receptors. Proc. Natl. Acad. Sci. U.S.A. *90*: 3103–3107, 1993a.

Maggio, R., Vogel, Z. and Wess, J.: Reconstitution of functional muscarinic receptors by co-expression of amino- and carboxyl-terminal receptor fragments. FEBS Lett. *319*: 195–200, 1993b.

Maggio, R., Barbier, P., Colelli, A., Salvadori, F., Demontis, G. and Corsini, G. U.: G protein-linked receptors: Pharmacological evidence for the formation of heterodimers. J. Pharmacol. Exp. Ther. *291*: 251–257, 1999.

Marinissen, M. J., Chiariello, M., Pallante, M. and Gutkind, J. S.: A network of mitogen-activated protein kinases links G protein-coupled receptors to the c-jun promoter: a role for c-Jun NH2-terminal kinase, p38s, and extracellular signal-regulated kinase 5. Mol. Cell. Biol. *19*: 4289–4301, 1999.

Marrion, N. V.: Control of M-current. Annu. Rev. Physiol. *59*: 483–504, 1997.

Marsh, S. J., Trouslard, J., Leaney, J. L. and Brown, D. A.: Synergistic regulation of a neuronal chloride current by intracellular calcium and muscarinic receptor activation: a role for protein kinase C. Neuron *15*: 729–737, 1995.

Mash, D. C. and Potter, L. T.: Autoradiographic localization of M_1 and M_2 muscarine receptors in the rat brain. Neuroscience *19*: 551–564, 1986.

Mash, D. C., Flynn, D. D. and Potter, L. T.: Loss of M2 muscarine receptors in the cerebral cortex in Alzheimer's disease and experimental cholinergic denervation. Science *228*: 1115–1117, 1985.

Matsui, H., Lazareno, S. and Birdsall, N. J.: Probing of the location of the allosteric site on m1 muscarinic receptors by site-directed mutagenesis. Mol. Pharmacol. *47*: 88–98, 1995.

Matsui, M., Motomura, D., Karasawa, H., Fujikawa, T., Jiang, J., Komiya, Y., Takahashi, S. and Taketo, M. M.: Multiple functional defects in peripheral autonomic organs in mice lacking muscarinic acetylcholine receptor gene for the M_3 subtype. Proc. Natl. Acad. Sci. U.S.A. *97*: 9579–9584, 2000.

Mattera, R., Pitts, B. J., Entman, M. L. and Birnbaumer, L.: Guanine nucleotide regulation of a mammalian myocardial muscarinic receptor system. Evidence for homo- and heterotropic cooperativity in ligand binding analyzed by computer-assisted curve fitting. J. Biol. Chem. *260*: 7410–7421, 1985.

May, L. T. and Christopoulos, A.: Allosteric modulators of G protein-coupled receptors. Curr. Opin. Pharmacol. *3*: 551–556, 2003.

McClatchy, D. B., Knudsen, C. R., Clark, B. F., Kahn, R. A., Hall, R. A. and Levey, A. I.: Novel interaction between the M4 muscarinic acetylcholine receptor and elongation factor 1A2. J. Biol. Chem. *277*: 29268–29274, 2002.

Meier, E., Frederiksen, D., Nielsen, M., Lemboel, H. L., Pedersen, H. and Hyttel, J.: Pharmacological *in vitro* characterization of the arecoline bioisostere, Lu 25–109-T, a muscarinic compound with M1-agonist and M2/M3-antagonist properties. Drug. Dev. Res. *40*: 1–16, 1997.

Melchiorre, C., Minarini, A., Angeli, P., Giardina, D., Gulini, U. and Quaglia, W.: Polymethylene tetraamines as muscarinic receptor probes. Trends Pharmacol. Sci., Supplt.: 55–59, 1989.

Melliti, K., Meza, U. and Adams, B.: Muscarinic stimulation of alpha1E Ca channels is selectively blocked by the effector antagonist function of RGS2 and phospholipase C-beta1. J. Neurosci. *20*: 7167–7173, 2000.

Meza, U., Bannister, R., Melliti, K. and Adams, B.: Biphasic, opposing modulation of cloned neuronal alpha1E Ca channels by distinct signaling pathways coupled to M_2 muscarinic acetylcholine receptors. J. Neurosci. *19*: 6806–6817, 1999.

Michal, P., Lysikova, M. and Tucek, S.: Dual effects of muscarinic M_2 acetylcholine receptors on the synthesis of cyclic AMP in CHO cells: dependence on time, receptor density and receptor agonists. Br. J. Pharmacol. *132*: 1217–1228, 2001.

Mitchelson, F.: Muscarinic receptor differentiation. Pharmacol. Ther. *37*: 357–423, 1988.

Miyakawa, T., Yamada, M., Duttaroy, A. and Wess, J.: Hyperactivity and intact hippocampus-dependent learning in mice lacking the M1 muscarinic acetylcholine receptor. J. Neurosci. *21*: 5239–5250, 2001.

Mohr, K., Staschen, C. M. and Ziegenhagen, M.: Equipotent allosteric effect of W84 on [^{3}H]NMS-binding to cardiac muscarinic receptors from guinea-pig, rat, and pig. Pharmacol. Toxicol. *70*: 198–200, 1992.

Mutschler, E., Moster, U., Wess, J. and Lambrecht, G.: Muscarinic receptor subtypes—pharmacological, molecular biological and therapeutical aspects. Pharma. Acta Helv. *69*: 243–258, 1995.

Nagao, M., Yamauchi, J., Kaziro, Y. and Itoh, H.: Involvement of protein kinase C and Src family tyrosine kinase in Galphaq/11-induced activation of c-Jun N-terminal kinase and p38 mitogen-activated protein kinase. J. Biol. Chem. *273*: 22892–22898, 1998.

Nakagawa-Yagi, Y., Saito, Y., Takada, Y. and Takayama, M.: Carbachol enhances forskolin-stimulated cyclic AMP accumulation via activation of calmodulin system in human neuroblastoma SH-SY5Y cells. Biochem. Biophys. Res. Commun. *178*: 116–123, 1991.

Nathanson, N. M.: Molecular properties of the muscarinic acetylcholine receptor. Annu. Rev. Neurosci. *10*: 195–236, 1987.

Neubig, R. R. and Siderovski, D. P.: Regulators of G-protein signalling as new central nervous system drug targets. Nat. Rev. Drug. Discov. *1*: 187–197, 2002.

Offermanns, S., Wieland, T., Homann, D., Sandmann, J., Bombien, E., Spicher, K., Schultz, G. and Jakobs, K. H.: Transfected muscarinic acetylcholine receptors selectively couple to Gi-type G proteins and Gq/11. Mol. Pharmacol. *45*: 890–898, 1994.

Ohara, K., Uchiyama, H., Haga, T. and Ichiyama, A.: Interaction of deglycosylated muscarinic receptors with ligands and G proteins. Eur. J. Pharmacol. *189*: 341–346, 1990.

Olianas, M. C. and Onali, P.: Characterization of the G protein involved in the muscarinic stimulation of adenylyl cyclase of rat olfactory bulb. Mol. Pharmacol. *49*: 22–29, 1996.

Olianas, M. C., Ingianni, A. and Onali, P.: Role of G protein betagamma subunits in muscarinic receptor-induced stimulation and inhibition of adenylyl cyclase activity in rat olfactory bulb. J. Neurochem. *70*: 2620–2627, 1998.

Onali, P. and Olianas, M. C.: Bimodal regulation of cyclic AMP by muscarinic receptors. Involvement of multiple G proteins and different forms of adenylyl cyclase. Life Sci. *56*: 973–980, 1995.

Palczewski, K., Kumasaka, T., Hori, T., Behnke, C. A., Motoshima, H., Fox, B. A., Le Trong, I., Teller, D. C., Okada, T., Stenkamp, R. E., Yamamoto, M. and Miyano, M.: Crystal structure of rhodopsin: a G protein-coupled receptor. Science *289*: 739–745, 2000.

Pals-Rylaarsdam, R. and Hosey, M. M.: Two homologous phosphorylation domains differentially contribute to desensitization and internalization of the m2 muscarinic acetylcholine receptor. J. Biol. Chem. *272*: 14152–14158, 1997.

Pals-Rylaarsdam, R., Xu, Y., Witt-Enderby, P., Benovic, J. L. and Hosey, M. M.: Desensitization and internalization of the m2 muscarinic acetylcholine receptor are directed by independent mechanisms. J. Biol. Chem. *270*: 29004–29011, 1995.

Pals-Rylaarsdam, R., Gurevich, V. V., Lee, K. B., Ptasienski, J. A., Benovic, J. L. and Hosey, M. M.: Internalization of the m2 muscarinic acetylcholine receptor. Arrestin-independent and -dependent pathways. J. Biol. Chem. *272*: 23682–23689, 1997.

Park, P., Sum, C. S., Hampson, D. R., Van Tol, H. H. and Wells, J. W.: Nature of the oligomers formed by muscarinic m2 acetylcholine receptors in Sf9 cells. Eur. J. Pharmacol. *421*: 11–22, 2001.

Pemberton, K. E., Hill-Eubanks, L. J. and Jones, S. V.: Modulation of low-threshold T-type calcium channels by the five muscarinic receptor subtypes in NIH 3T3 cells. Pflugers Arch. *440*: 452–461, 2000.

Pepitoni, S., Mallon, R. G., Pai, J. K., Borkowski, J. A., Buck, M. A. and McQuade, R. D.: Phospholipase D activity and phosphatidylethanol formation in stimulated HeLa cells expressing the human m1 muscarinic acetylcholine receptor gene. Biochem. Biophys. Res. Commun. *176*: 453–458, 1991.

Peralta, E. G., Ashkenazi, A., Winslow, J. W., Smith, D. H., Ramachandran, J. and Capon, D. J.: Distinct primary structures, ligand-binding properties and tissue-specific expression of four human muscarinic acetylcholine receptors. Embo. J. *6*: 3923–3929, 1987.

Perez, M., Jorand-Lebrun, C., Pauwels, P. J., Pallard, I. and Halazy, S.: Dimers of $5HT_1$ ligands preferentially bind to $5HT_{1B/1D}$ receptor subtypes. Bioorg. Med. Chem. Lett. *8*: 1407–1412, 1998a.

Perez, M., Pauwels, P. J., Fourrier, C., Chopin, P., Valentin, J. P., John, G. W., Marien, M. and Halazy, S.: Dimerization of sumatriptan as an efficient way to design a potent, centrally and orally active $5\text{-}HT_{1B}$ agonist. Bioorg. Med. Chem. Lett. *8*: 675–680, 1998b.

Pfaffendorf, M., Batink, H. D., Trankle, C., Mohr, K. and van Zwieten, P. A.: Probing the selectivity of allosteric modulators of muscarinic receptors at other G-protein-coupled receptors. J. Auton. Pharmacol. *20*: 55–62, 2000.

Pitcher, J. A., Freedman, N. J. and Lefkowitz, R. J.: G protein-coupled receptor kinases. Annu. Rev. Biochem. *67*: 653–692, 1998.

Pittel, Z. and Wess, J.: Intramolecular interactions in muscarinic acetylcholine receptor studied with chimeric m2/m5 receptors. Mol. Pharmacol. *45*: 61–64, 1994.

Porter, A. C., Bymaster, F. P., DeLapp, N. W., Yamada, M., Wess, J., Hamilton, S. E., Nathanson, N. M. and Felder, C. C.: M1 muscarinic receptor signaling in mouse hippocampus and cortex. Brain Res. *944*: 82–89, 2002.

Portoghese, P. S.: Bivalent ligands and the message-address concept in the design of selective opioid receptor antagonists. Trends Pharmacol. Sci. *10*: 230–235, 1989.

Potter, L. T.: Snake toxins that bind specifically to individual subtypes of muscarinic receptors. Life Sci. *68*: 2541–2547, 2001.

Potter, L. T. and Ferrendelli, C. A.: Affinities of different cholinergic agonists for the high and low affinity states of hippocampal M1 muscarine receptors. J. Pharmacol. Exp. Ther. *248*: 974–978, 1989.

Potter, L. T., Ballesteros, L. A., Bichajian, L. H., Ferrendelli, C. A., Fisher, A., Hanchett, H. E. and Zhang, R.: Evidence for paired M_2 muscarinic receptors. Mol. Pharmacol. *39*: 211–221, 1991.

Proska, J. and Tucek, S.: Mechanisms of steric and cooperative actions of alcuronium on cardiac muscarinic acetylcholine receptors. Mol. Pharmacol. *45*: 709–717, 1994.

Proska, J. and Tucek, S.: Competition between positive and negative allosteric effectors on muscarinic receptors. Mol. Pharmacol. *48*: 696–702, 1995.

Raiteri, M., Marchi, M., Paudice, P. and Pittaluga, A.: Muscarinic receptors mediating inhibition of gamma-aminobutyric acid release in rat corpus striatum and their pharmacological characterization. J. Pharmacol. Exp. Ther. *254*: 496–501, 1990.

Rajeswaran, W. G., Cao, Y., Huang, X. P., Wroblewski, M. E., Colclough, T., Lee, S., Liu, F., Nagy, P. I., Ellis, J., Levine, B. A., Nocka, K. H. and Messer, W. S., Jr.: Design, synthesis, and biological characterization of bivalent 1-Methyl- 1,2,5,6-tetrahydropyridyl-1,2,5-thiadiazole derivatives as selective muscarinic agonists. J. Med. Chem. *44*: 4563–4576, 2001.

Randall, A. D. and Tsien, R. W.: Contrasting biophysical and pharmacological properties of T-type and R-type calcium channels. Neuropharmacology *36*: 879–893, 1997.

Reever, C. M., Ferrari-DiLeo, G. and Flynn, D. D.: The M5 (m5) receptor subtype: fact or fiction? Life Sci. *60*: 1105–1112, 1997.

Rhee, S. G., Suh, P. G., Ryu, S. H. and Lee, S. Y.: Studies of inositol phospholipid-specific phospholipase C. Science *244*: 546–550, 1989.

Riker, W. F. and Wescoe, W. C.: The pharmacology of flaxedil with observations on certain analogs. Ann. N.Y. Acad. Sci. *54*: 373–394, 1951.

Rogalski, S. L., Cyr, C. and Chavkin, C.: Activation of the endothelin receptor inhibits the G protein-coupled inwardly rectifying potassium channel by a phospholipase A2-mediated mechanism. J. Neurochem. *72*: 1409–1416, 1999.

Roseberry, A. G. and Hosey, M. M.: Trafficking of M_2 muscarinic acetylcholine receptors . J. Biol. Chem. *274*: 33671–33676, 1999.

Rosenblum, K., Futter, M., Jones, M., Hulme, E. C. and Bliss, T. V.: ERKI/II regulation by the muscarinic acetylcholine receptors in neurons. J. Neurosci. *20*: 977–985, 2000.

Roszkowski, A. P.: An unusual type of sympathetic ganglionic stimulant. J. Pharmacol. Exp. Ther. *132*: 156–170, 1961.

Rumenapp, U., Geiszt, M., Wahn, F., Schmidt, M. and Jakobs, K. H.: Evidence for ADP-ribosylation-factor-mediated activation of phospholipase D by m3 muscarinic acetylcholine receptor. Eur. J. Biochem. *234*: 240–244, 1995.

Rumenapp, U., Asmus, M., Schablowski, H., Woznicki, M., Han, L., Jakobs, K. H., Fahimi-Vahid, M., Michalek, C., Wieland, T. and Schmidt, M.: The M3 muscarinic acetylcholine receptor expressed in HEK-293 cells signals to phospholipase D via G12 but not Gq-type G proteins: regulators of G proteins as tools to dissect pertussis toxin-resistant G proteins in receptor-effector coupling. J. Biol. Chem. *276*: 2474–2479, 2001.

Sagi, S. A., Seasholtz, T. M., Kobiashvili, M., Wilson, B. A., Toksoz, D. and Brown, J. H.: Physical and functional interactions of Gaq with Rho and its exchange factors. J. Biol. Chem. *276*: 15445–15452, 2001.

Sah, V. P., Seasholtz, T. M., Sagi, S. A. and Brown, J. H.: The role of Rho in G protein-coupled receptor signal transduction. Annu. Rev. Pharmacol. Toxicol. *40*: 459–489, 2000.

Sauerberg, P., Olesen, P. H., Nielsen, S., Treppendahl, S., Sheardown, M. J., Honore, T., Mitch, C. H., Ward, J. S., Pike, A. J., Bymaster, F. P., Sawyer, B. D. and Shannon, H. E.: Novel functional M1 selective agonists. Synthesis and structure-activity relationships of 3-(1,2,5-thiadiazolyl)-1,2,5,6-tetarhydro-1-methylpyridines. J. Med. Chem. *35*: 2274–2283, 1992.

Sauerberg, P., Jeppesen, L., Olesen, P. H., Rasmussen, T., Swedberg, M. D. B., Sheardown, M. J., Fink-Jensen, A., Thomsen, C., Thogersen, H., Rimvall, K., Ward, J. S., Calligaro, D. O., DeLapp, N. W., Bymaster, F. P. and Shannon, H. E.: Muscarinic agonists with antipsychotic-like activity: structure- activity relationships of 1,2,5-thiadiazole analogues with functional dopamine antagonist activity. J. Med. Chem. *41*: 4378–4384, 1998a.

Sauerberg, P., Jeppesen, L., Olesen, P. H., Sheardown, M. J., Fink-Jensen, A., Rasmussen, T., Rimvall, K., Shannon, H. E., Bymaster, F. P., DeLapp, N. W., Calligaro, D. O., Ward, J. S., Whitesitt, C. A. and Thomsen, C.: Identification of side chains on 1,2,5-thiadiazole-azacycles optimal for muscarinic M1 receptor activation. Bioorg. Med. Chem. Lett. *8*: 2897–2902, 1998b.

Savarese, T. M., Wang, C. D. and Fraser, C. M.: Site-directed mutagenesis of the rat m1 muscarinic acetylcholine receptor. Role of conserved cysteines in receptor function. J. Biol. Chem. *267*: 11439–11448, 1992.

Schlador, M. L. and Nathanson, N. M.: Synergistic regulation of m2 muscarinic acetylcholine receptor desensitization and sequestration by G protein-coupled receptor kinase-2 and b-arrestin-1. J. Biol. Chem. *272*: 18882–18890, 1997.

Schmidt, M., Bienek, C., van Koppen, C. J., Michel, M. C. and Jakobs, K. H.: Differential calcium signalling by m2 and m3 muscarinic acetylcholine receptors in a single cell type. Naunyn Schmiedebergs Arch. Pharmacol. *352*: 469–476, 1995a.

Schmidt, M., Fasselt, B., Rumenapp, U., Bienek, C., Wieland, T., van Koppen, C. J. and Jakobs, K. H.: Rapid and persistent desensitization of m3 muscarinic acetylcholine receptor-stimulated phospholipase D. Concomitant sensitization of phospholipase C. J. Biol. Chem. *270*: 19949–19956, 1995b.

Schmidt, M., Lohmann, B., Hammer, K., Haupenthal, S., Nehls, M. V. and Jakobs, K. H.: Gi- and protein kinase C-mediated heterologous potentiation of phospholipase C signaling by G protein-coupled receptors. Mol. Pharmacol. *53*: 1139–1148, 1998.

Schmidt, M., Voss, M., Weernink, P. A., Wetzel, J., Amano, M., Kaibuchi, K. and Jakobs, K. H.: A role for rho-kinase in rho-controlled phospholipase D stimulation by the m3 muscarinic acetylcholine receptor. J. Biol. Chem. *274*: 14648–14654, 1999.

Schöneberg, T., Liu, J. and Wess, J.: Plasma membrane localization and functional rescue of truncated forms of a G protein-coupled receptor. J. Biol. Chem. *270*: 18000–18006, 1995.

Schwarz, R. D., Spencer, C. J., Jaen, J. C., Mirzadegan, T., Moreland, D., Tecle, H. and Thomas, A. J.: Mutations of aspartate 103 in the Hm2 receptor and alterations in receptor binding properties of muscarinic agonists. Life Sci. *56*: 923–929, 1995.

Seasholtz, T. M., Majumdar, M. and Brown, J. H.: Rho as a mediator of G protein-coupled receptor signaling. Mol. Pharmacol. *55*: 949–956, 1999.

Seifert, R. and Wenzel-Seifert, K.: Constitutive activity of G-protein-coupled receptors: cause of disease and common property of wild-type receptors. Naunyn Schmiedebergs Arch. Pharmacol. *366*: 381–416, 2002.

Shannon, H. E., Bymaster, F. P., Calligaro, D. O., Greenwood, B., Mitch, C. H., Sawyer, B. D., Ward, J. S., Wong, D. T., Olesen, P. H., Sheardown, M. J., Swedberg, M. D. B., Suzdak, P. D. and Sauerberg, P.: Xanomeline: a novel muscarinic receptor agonist with functional selectivity for M_1 receptors. J. Pharmacol. Exp. Ther. *269*: 271–281, 1994.

Shannon, H. E., Rasmussen, K., Bymaster, F. P., Hart, J. C., Peters, S. C., Swedberg, M. D., Jeppesen, L., Sheardown, M. J., Sauerberg, P. and Fink-Jensen, A.: Xanomeline, an M_1/M_4 preferring muscarinic cholinergic receptor agonist, produces antipsychotic-like activity in rats and mice. Schizophr. Res. *42*: 249–259, 2000.

Shapiro, M. S., Loose, M. D., Hamilton, S. E., Nathanson, N. M., Gomeza, J., Wess, J. and Hille, B.: Assignment of muscarinic receptor subtypes mediating G-protein modulation of Ca(2+) channels by using knockout mice. Proc. Natl. Acad. Sci. U. S.A. *96*: 10899–108904, 1999.

Shapiro, R. A., Scherer, N. M., Habecker, B. A., Subers, E. M. and Nathanson, N. M.: Isolation, sequence, and functional expression of the mouse M1 muscarinic acetylcholine receptor gene. J. Biol. Chem. *263*: 18397–18403, 1988.

Sharon, D., Vorobiov, D. and Dascal, N.: Positive and negative coupling of the metabotropic glutamate receptors to a G protein-activated K+ channel, GIRK, in Xenopus oocytes. J. Gen. Physiol. *109*: 477–490, 1997.

Simonds, W. F.: G protein regulation of adenylate cyclase. Trends Pharmacol. Sci. *20*: 66–73, 1999.

Smith, T. D., Annis, S. J., Ehlert, F. J. and Leslie, F. M.: N-[3H]methylscopolamine labeling of non-M1, non-M2 muscarinic receptor binding sites in rat brain. J. Pharmacol. Exp. Ther. *256*: 1173–1181, 1991.

Snyder, S. H., Chang, K. J., Kuhar, M. J. and Yamamura, H. I.: Biochemical identification of the mammalian muscarinic cholinergic receptor. Fed. Proc. *34*: 1915–1921, 1975.

Son, H., Hawkins, R. D., Martin, K., Kiebler, M., Huang, P. L., Fishman, M. C. and Kandel, E. R.: Long-term potentiation is reduced in mice that are doubly mutant in endothelial and neuronal nitric oxide synthase. Cell *87*: 1015–1023, 1996.

Spalding, T. A., Birdsall, N. J., Curtis, C. A. and Hulme, E. C.: Acetylcholine mustard labels the binding site aspartate in muscarinic acetylcholine receptors. J. Biol. Chem. *269*: 4092–4097, 1994.

Spalding, T. A., Burstein, E. S., Brauner-Osborne, H., Hill-Eubanks, D. and Brann, M. R.: Pharmacology of a constitutively active muscarinic receptor generated by random mutagenesis. J. Pharmacol. Exp. Ther. *275*: 1274–1279, 1995.

Spalding, T. A., Trotter, C., Skjaerbaek, N., Messier, T. L., Currier, E. A., Burstein, E. S., Li, D., Hacksell, U. and Brann, M. R.: Discovery of an ectopic activation site on the M_1 muscarinic receptor. Mol. Pharmacol. *61*: 1297–1302, 2002.

Spencer, D. G., Jr., Horvath, E. and Traber, J.: Direct autoradiographic determination of M_1 and M_2 muscarinic acetylcholine receptor distribution in the rat brain: relation to cholinergic nuclei and projections. Brain Res. *380*: 59–68, 1986.

Starke, K., Gothert, M. and Kilbinger, H.: Modulation of neurotransmitter release by presynaptic autoreceptors. Physiol. Rev. *69*: 864–989, 1989.

Stevens, E. B., Shah, B. S., Pinnock, R. D. and Lee, K.: Bombesin receptors inhibit G protein-coupled inwardly rectifying K+ channels expressed in Xenopus oocytes through a protein kinase C- dependent pathway. Mol. Pharmacol. *55*: 1020–1027, 1999.

Stockton, J. M., Birdsall, N. J. M., Burgen, A. S. V. and Hulme, E. C.: Modification of the binding properties of muscarinic receptors by gallamine. Mol. Pharmacol. *23*: 551–557, 1983.

Strader, C. D., Fong, T. M., Tota, M. R. and Underwood, D.: Structure and function of G protein-coupled receptors. Annu. Rev. Biochem. *63*: 101–132, 1994.

Sugita S., Uchimura N., Jiang Z. G. and North R. A.: Distinct muscarinic receptors inhibit release of gamma-aminobutyric acid and excitatory amino acids in mammalian brain. Proc. Natl. Acad. Sci. U.S.A. *88*: 2608–2611, 1991.

Suh, B. C. and Hille, B.: Recovery from muscarinic modulation of M current channels requires phosphatidylinositol 4,5-bisphosphate synthesis. Neuron *35*: 507–520, 2002.

Sunahara, R. K., Dessauer, C. W. and Gilman, A. G.: Complexity and diversity of mammalian adenylyl cyclases. Annu. Rev. Pharmacol. Toxicol. *36*: 461–480, 1996.

Taylor, J. M., Jacob-Mosier, G. G., Lawton, R. G., VanDort, M. and Neubig, R. R.: Receptor and membrane interaction sites on Gb. A receptor-derived peptide binds to the carboxyl terminus. J. Biol. Chem. *271*: 3336–3339, 1996.

Tottene, A., Volsen, S. and Pietrobon, D.: a_{1E} subunits form the pore of three cerebellar R-type calcium channels with different pharmacological and permeation properties. J. Neurosci. *20*: 171–178, 2000.

Tränkle, C. and Mohr, K.: Divergent modes of action among cationic allosteric modulators of muscarinic M_2 receptors. Mol. Pharmacol. *51*: 674–682, 1997.

Tränkle, C., Mies-Klomfass, E., Cid, M. H. B., Holzgrabe, U. and Mohr, K.: Identification of a [^{3}H]Ligand for the common allosteric site of muscarinic acetylcholine M_2 receptors. Mol. Pharmacol. *54*: 139–145, 1998.

Tränkle, C., Weyand, O., Schroter, A. and Mohr, K.: Using a radioalloster to test predictions of the cooperativity model for gallamine binding to the allosteric site of muscarinic acetylcholine M_2 receptors. Mol. Pharmacol. *56*: 962–965, 1999.

Tsuga, H., Okuno, E., Kameyama, K. and Haga, T.: Sequestration of human muscarinic acetylcholine receptor hm1-hm5 subtypes: Effect of G protein-coupled receptor kinases GRK2, GRK4, GRK5 and GRK6. J. Pharmacol. Exp. Ther. *284*: 1218–1226, 1998.

Tucek, S., Musilkova, J., Nedoma, J., Proska, J., Shelkvnikov, S. and Vorlicek, J.: Positive cooperativity in the binding of alcuronium and N-methylscopolamine to muscarinic acetylcholine receptors. Mol. Pharmacol. *38*: 674–680, 1990.

Tucek, S., Michal, P. and Vlachova, V.: Dual effects of muscarinic M2 receptors on the synthesis of cyclic AMP in CHO cells: background and model. Life Sci. *68*: 2501–2510, 2001.

Urbansky, M., Proska, J., Ricny, J. and Drasar, P.: Truxillic acid derivatives, neuromuscular blocking agents with very high affinity for the allosteric site of muscarinic acetylcholine receptors. Collect. Czech. Chem. Commun. *64*: 1980–1992, 1999.

van Biesen, T., Hawes, B. E., Raymond, J. R., Luttrell, L. M., Koch, W. J. and Lefkowitz, R. J.: G(o)-protein alpha-subunits activate mitogen-activated protein kinase via a novel protein kinase C-dependent mechanism. J. Biol. Chem. *271*: 1266–1269, 1996.

van Koppen, C. J. and Nathanson, N. M.: Site-directed mutagenesis of the m2 muscarinic acetylcholine receptor. Analysis of the role of N-glycosylation in receptor expression and function. J. Biol. Chem. *265*: 20887–20892, 1990.

Vilaro, M. T., Palacios, J. M. and Mengod, G.: Multiplicity of muscarinic autoreceptor subtypes? Comparison of the distribution of cholinergic cells and cells containing mRNA for five subtypes of muscarinic receptors in the rat brain. Brain Res. Mol. Brain Res. *21*: 30–46, 1994.

Vögler, O., Bogatkewitsch, G. S., Wriske, C., Krummenerl, P., Jakobs, K. H. and van Koppen, C. J.: Receptor subtype-specific regulation of muscarinic acetylcholine receptor sequestration by dynamin. Distinct sequestration of m2 receptors. J. Biol. Chem. *273*: 12155–12160, 1998.

Vögler O., Nolte, B., Voss, M., Schmidt, M., Jakobs, K. H. and van Koppen, C. J.: Regulation of muscarinic acetylcholine receptor sequestration and function by b-Arrestin. J. Biol. Chem. *274*: 12333–12338, 1999.

Voigtlander, U., Johren, K., Mohr, M., Raasch, A., Tränkle, C., Buller, S., Ellis, J., Holtje, H.-D. and Mohr, K.: Allosteric site on muscarinic acetylcholine receptors: Identification of two amino acids in the M_2 muscarinic receptor that account entirely for the M_2/M_5 subtype selectivities of some structurally diverse allosteric ligands in N-methylscopolamine occupied receptors. Mol. Pharmacol. *64*: 21–31, 2003.

Waelbroeck, M., Tastenoy, M., Camus, J. and Christophe, J.: Binding of selective antagonists to four muscarinic receptors (M_1 to M_4) in rat forebrain. Mol. Pharmacol. *38*: 267–273, 1990.

Wall, S. J., Yasuda, R. P., Hory, F., Flagg, S., Martin, B. M., Ginns, E. I. and Wolfe, B. B.: Production of antisera selective for m1 muscarinic receptors using fusion proteins: distribution of m1 receptors in rat brain. Mol. Pharmacol. *39*: 643–649, 1991a.

Wall, S. J., Yasuda, R. P., Li, M. and Wolfe, B. B.: Development of an antiserum against m3 muscarinic receptors: distribution of m3 receptors in rat tissues and clonal cell lines. Mol. Pharmacol. *40*: 783–789, 1991b.

Wang, H. S., Pan, Z., Shi, W., Brown, B. S., Wymore, R. S., Cohen, I. S., Dixon, J. E. and McKinnon, D.: KCNQ2 and KCNQ3 potassium channel subunits: molecular correlates of the M-channel. Science *282*: 1890–1893, 1998.

Wang, S. Z., Lee, S. Y., Zhu, S. Z. and el-Fakahany, E. E.: Differential coupling of m1, m3, and m5 muscarinic receptors to activation of neuronal nitric oxide synthase. Pharmacology *53*: 271–280, 1996.

Wang, S. Z., Lee, S. Y., Zhu, S. Z., Wotta, D. R., Parsons, A. M. and el-Fakahany, E. E.: Activation of neuronal nitric oxide synthase by M2 muscarinic receptors associated with a small increase in intracellular calcium. Pharmacology *55*: 10–17, 1997.

Ward, S. D., Curtis, C. A. and Hulme, E. C.: Alanine-scanning mutagenesis of transmembrane domain 6 of the M_1 muscarinic acetylcholine receptor suggests that Tyr381 plays key roles in receptor function. Mol. Pharmacol. *56*: 1031–1041, 1999.

Weiner, D. M., Levey, A. I. and Brann, M. R.: Expression of muscarinic acetylcholine and dopamine receptor mRNAs in rat basal ganglia. Proc. Natl. Acad. Sci. U.S.A. *87*: 7050–7054, 1990.

Werbonat, Y., Kleutges, N., Jakobs, K. H. and van Koppen, C. J.: Essential role of dynamin in internalization of M_2 muscarinic acetylcholine and angiotensin AT_{1A} receptors. J. Biol. Chem. *275*: 21969–21974, 2000.

Wess, J.: Molecular basis of muscarinic acetylcholine function. Trends Pharmacol. Sci. *14*: 308–313, 1993a.

Wess, J.: Mutational analysis of muscarinic acetylcholine receptors: Structural basis of ligand/receptor/G protein interactions. Life Sci. *53*: 1447–1463, 1993b.

Wess, J.: Molecular biology of muscarinic acetylcholine receptors. Crit. Rev. Neurobiol. *10*: 69–99, 1996.

Wess, J.: Molecular basis of receptor/G-protein-coupling selectivity. Pharmacol. Ther. *80*: 231–264, 1998.

Wess, J., Brann, M. R. and Bonner, T. I.: Identification of a small intracellular region of the muscarinic m3 receptor as a determinant of selective coupling to PI turnover. FEBS Lett. *258*: 133–136, 1989.

Wess, J., Bonner, T. I. and Brann, M. R.: Chimeric m2/m3 muscarinic receptor: role of carboxyl terminal domains in selectivity of ligand binding and coupling to phosphoinositide hydrolysis. Mol. Pharmacol. *38*: 82–87, 1990.

Wess, J., Liu, J., Bluml, K., Yun, N., Schoneberg, T. and Blin, N.: Mutational analysis of muscarinic receptors: structural basis of ligand binding and G protein recognition. In: "Molecular Mechanisms of Muscarinic Acetylcholine Receptor Function," J. Wess, Ed., pp. 1–18, R.G. Landes Company, Austin, 1995.

White, J. H., Wise, A., Main, M. J., Green, A., Fraser, N. J., Disney, G. H., Barnes, A. A., Emson, P., Foord, S. M. and Marshall, F. H.: Heterodimerization is required for the formation of a functional $GABA_B$ receptor. Nature *396*: 679–682, 1998.

Whitehouse, P. J., Price, D. L., Struble, R. G., Clark, A. W., Coyle, J. T. and Delon, M. R.: Alzheimer's disease and senile dementia: loss of neurons in the basal forebrain. Science *215*: 1237–1239, 1982.

Wickman, K. and Clapham, D. E.: Ion channel regulation by G proteins. Physiol. Rev. *75*: 865–885, 1995.

Wong, S. K., Parker, E. M. and Ross, E. M.: Chimeric muscarinic cholinergic: beta-adrenergic receptors that activate Gs in response to muscarinic agonists. J. Biol. Chem. *265*: 6219–6224, 1990.

Wregget, K. A. and Wells, J. W.: Cooperativity manifest in the binding properties of purified cardiac muscarinic receptors. J. Biol. Chem. *270*: 22488–22499, 1995.

Wu, G., Bogatkevich, G. S., Mukhin, Y. V., Benovic, J. L., Hildebrandt, J. D. and Lanier, S. M.: Identification of Gbetagamma binding sites in the third intracellular loop of the M(3)-muscarinic receptor and their role in receptor regulation. J. Biol. Chem. *275*: 9026–9034, 2000.

Wu, L. G., Borst, J. G. and Sakmann, B.: R-type Ca2+ currents evoke transmitter release at a rat central synapse. Proc. Natl. Acad. Sci. U.S.A. *95*: 4720–4725, 1998.

Yamada, M., Lamping, K. G., Duttaroy, A., Zhang, W., Cui, Y., Bymaster, F. P., McKinzie, D. L., Felder, C. C., Deng, C. X., Faraci, F. M. and Wess, J.: Cholinergic dilation of cerebral blood vessels is abolished in M(5) muscarinic acetylcholine receptor knockout mice. Proc. Natl. Acad. Sci. U.S.A. *98*: 14096–14101, 2001a.

Yamada, M., Miyakawa, T., Duttaroy, A., Yamanaka, A., Moriguchi, T., Makita, R., Ogawa, M., Chou, C. J., Xia, B., Crawley, J. N., Felder, C. C., Deng, C. X. and Wess, J.: Mice lacking the M3 muscarinic acetylcholine receptor are hypophagic and lean. Nature *410*: 207–212, 2001b.

Yamauchi, J., Kaziro, Y. and Itoh, H.: Differential regulation of mitogen-activated protein kinase kinase 4 (MKK4) and 7 (MKK7) by signaling from G protein beta gamma subunit in human embryonal kidney 293 cells. J. Biol. Chem. *274*: 1957–1965, 1999.

Yamauchi, J., Tsujimoto, G., Kaziro, Y. and Itoh, H.: Parallel regulation of mitogen-activated protein kinase kinase 3 (MKK3) and MKK6 in Gq-signaling cascade. J. Biol. Chem. *276*: 23362–23372, 2001.

Yasuda, R. P., Ciesla, W., Flores, L. R., Wall, S. J., Li, M., Satkus, S. A., Weisstein, J. S., Spagnola, B. V. and Wolfe, B. B.: Development of antisera selective for m4 and m5 muscarinic cholinergic receptors: distribution of m4 and m5 receptors in rat brain. Mol. Pharmacol. *43*: 149–157, 1993.

Zeng, F. Y. and Wess, J.: Identification and molecular characterization of m3 muscarinic receptor dimers. J. Biol. Chem. *274*: 19487–19497, 1999.

Zeng, X. P., Le, F. and Richelson, E.: Muscarinic m4 receptor activation by some atypical antipsychotic drugs. Eur. J. Pharmacol. *321*: 349–354, 1997.

Zhang, J., Ferguson, S. S., Barak, L. S., Menard, L. and Caron, M. G.: Dynamin and beta-arrestin reveal distinct mechanisms for G protein- coupled receptor internalization. J. Biol. Chem. *271*: 18302–18305, 1996.

Zhang, W., Basile, A. S., Gomeza, J., Volpicelli, L. A., Levey, A. I. and Wess, J.: Characterization of central inhibitory muscarinic autoreceptors by the use of muscarinic acetylcholine receptor knock-out mice. J. Neurosci. *22*: 1709–1717, 2002a.

Zhang, W., Yamada, M., Gomeza, J., Basile, A. S. and Wess, J.: Multiple muscarinic acetylcholine receptor subtypes modulate striatal dopamine release, as studied with M1-M5 muscarinic receptor knock-out mice. J. Neurosci. *22*: 6347–6352, 2002b.

Zorn, S. H., Jones, S. B., Ward, K. M. and Liston, D. R.: Clozapine is a potent and selective muscarinic M_4 receptor agonist. Eur. J. Pharmacol. *269*: R1–R2, 1994.

6

Neuronal Nicotinic Receptors: History, Structure, and Functional Roles

Jon Lindstrom

A. Scope of the Chapter

1. Introduction

Nicotinic pharmacology goes back to the 1844 discovery by Claude Bernard that curare could paralyze rabbit skeletal muscles without affecting the heart (Bennett, 2000; see Chapter 4 A). John Langley developed the concept of transmitter receptors through studies of neuromuscular transmission in 1905 to 1907. Otto Loewi demonstrated chemical transmission on heart muscle in 1921, and Dale and coworkers identified this transmitter as acetylcholine in 1936.

In 1914, Sir Henry Dale initially reported that two alkaloids, nicotine and muscarine, acted as agonists to reproduce different aspects of the effects of acetylcholine and proposed that these effects were mediated by two different types of acetylcholine receptors, nicotinic and muscarinic (Dale, 1914, 1954). This initial distinction between what have come to be recognized as nicotinic acetylcholine receptor (nAChR) ligand-gated cation channels and muscarinic acetylcholine receptor (mAChR) GTP protein-linked receptors has stood the test of time. Concepts about AChRs started developing at the very beginning of the concept of drug receptors, and the sophistication of ideas about AChRs and receptors in general coevolved.

Over the following half century, nAChRs became the archetype for a ligand-gated ion channel neurotransmitter receptor (Changeux, 1990; Karlin, 1991). Skeletal muscle provided a homogeneous type of synapse on accessible tissue with large, easily studied cells and high local concentrations of nAChRs whose role in neurotransmission was clearly recognized. It was appreciated that binding of ACh released from nerve endings to nAChRs in the postsynaptic membrane triggered transient opening of a nonselective cation channel. Passive current flow through the nAChR channel resulted in depolarization of the membrane that triggered action potentials, which were propagated along the sarcolemma. However, the nAChR molecule remained a black box. It was clear that neuronal nAChRs at autonomic ganglia were pharmacologically distinct from those in skeletal muscle, but that they played a similar role in neurotransmission. By contrast, the pharmacologically distinct, slow-onset, long-duration responses mediated by mAChRs in smooth muscle and other tissues were fundamentally different. It was not fully evident prior to molecular studies that the two types of AChRs had no common evolutionary history. Then it became obvious that there was no structural homology between nAChRs and mAChRs. Muscarinic acetylcholine receptors required a GTP binding protein to link them to their effector, while nAChRs comprised both ACh binding sites and the cation channel whose opening they regulate as intrinsic parts of the same transmembrane protein. However, even now, there can occasionally be ambiguities in distinguishing among nAChR, mAChR, and other receptor subtypes pharmacologically (Fuchs, 1996; Rothlin et al., 1999, 2003; Parker et al., 2003).

Molecular studies of nAChRs in the late 1960s and early 1970s preceded those of other neurotransmitter receptors as a result of two gifts of nature: fish electric organs and snake venom toxins (Changeux, 1990; Karlin, 1991; Lindstrom, 1999). Electric organs from electric eels (*Electrophorus electricus*) and, later, electric rays (*Torpedo californica*) provided a rich and homogeneous source of nAChR protein. Electric organ cells evolved from skeletal muscle through loss of contractile proteins and gain of synapses from one per cell to thousands per cell. Using the same mechanisms involved in neuromuscular transmission to produce voltages per cell around 0.1 V, but large currents in cells up to the size of a dime, and batteries of these cells stacked in series and parallel in kilograms of electric organs, discharges could be produced sufficient to stun a horse. Peptide toxins from the venoms of kraits and cobras block skeletal neuromuscular transmission, thereby paralyzing their victims, by binding nearly irreversibly to the ACh binding sites of muscle nAChRs. These toxins are easily purified and labeled with ^{125}I to quantitate and localize nAChRs, or labeled with fluorescent tags or other markers for histological localization of nAChRs. Affinity columns made with cobra toxin coupled to agarose permitted specific binding of AChRs solubilized from electric organs using the detergent Triton X-100 and one-step purification by competitive elution with cholinergic ligands.

Muscle-type nAChRs affinity purified from electric organs revealed that muscle nAChRs were composed of four kinds of subunits (termed α, β, γ, and δ molecular weight) (Karlin, 1991, 2002; Taylor et al., 2000). Only the lowest molecular weight subunit (α) reacted with an affinity label directed at the ACh binding site (which reacted only after a unique disulfide bond was reduced). N-terminal amino acid sequence determinations on these subunits revealed that all the AChR subunits were homologous proteins (Raftery et al., 1980). This sequence data also permitted the initial cloning of cDNAs for these subunits using molecular biological techniques, which were just developing at that time (Noda et al., 1982). Repeated cycles of low-stringency hybridization initially using these electric organ cDNAs revealed that not only were there muscle $\alpha 1$ subunits with a characteristic pair of adjacent cysteines near the ACh binding site but also neuronal $\alpha 2$ to $\alpha 10$ subunits with homologous pairs (Heinemann et al., 1991; LeNovere and Changeux, 1995; LeNovere et al., 2002; Lindstrom, 2000a; Karlin, 2002). Neuronal homologues of muscle $\beta 1$ subunits termed $\beta 2$-4 were also found, as was an ε subunit characteristic of nAChRs at mature neuromuscular junctions.

Initial purification of electric organ AChRs also resulted in development of antibodies to AChR subunits and the realization that autoantibodies to nAChRs cause myasthenia gravis (MG) (Patrick and Lindstrom, 1973; Lindstrom 2000b, 2003). When electric organ nAChRs were first affinity purified, it was necessary to prove that the protein in fact was the real nAChR. Later this was definitively accomplished by expressing cloned subunit cDNAs in Xenopus oocytes. Initially, purified nAChR was injected into rabbits in order to raise antibodies, with the intention that demonstrating that antibodies to the purified protein would block the function of nAChRs in electric organ cells, thereby proving that the protein was the AChR. It turned out that the rabbits got weak and died, exhibiting the signs of MG (Patrick and Lindstrom, 1973). Detailed studies of this experimental autoimmune MG (EAMG) in rats and MG in humans revealed the pathological mechanisms by which autoantibodies impair neuromuscular transmission in MG, provided an immunodiagnostic assay for MG, and focused therapy for MG on specific immune suppression (Lindstrom, 2000b, 2003; Vincent et al., 2001). Monoclonal antibodies (mAbs) to subunits of muscle-type nAChRs were initially developed as model autoantibodies for use in defining the antigenic structure and pathological roles of these antibodies (Tzartos and Lindstrom, 1980; Tzartos et al., 1982, 1998). These mAbs also became valuable for studying nAChR synthesis and location. Monoclonal antibodies to the main immunogenic region (MIR) of muscle nAChR were first found to cross-react with neuronal nAChRs in chickens (Swanson et al., 1983). Immunoaffinity purified chicken nAChRs were used to produce mAbs to other subunits and species (Whiting and Lindstrom, 1986a; Whiting et al., 1987). Repeated cycles of this process followed by production of mAbs to synthetic and bacterially-expressed nAChR subunit cDNAs has led to large libraries of nAChR subunit mAbs (Whiting and Lindstrom, 1986b, 1987, 1988). These mAbs have proven useful in immunohistological localization of neuronal nAChRs as well as in studies of their synthesis and regulation by nicotine (Lindstrom, 2000a).

Recently an antibody-mediated autoimmune response to autonomic α3 AChRs has been discovered (Vernino et al., 2000), and immunization with bacterially expressed human AChR peptides was shown to produce an animal model of this autoimmune autonomic neuropathy (AAN). It may be that in the future other autoimmune diseases involving other nAChR subunits will be found (Lindstrom, 2000b).

Several decades ago, the roles of brain nAChRs were little known or appreciated. Central effects of mAChR drugs in vertebrate brains were appreciated, and it was evident that, whereas nAChRs were the predominant excitatory neurotransmitter receptors in the periphery, glutamate receptors were the predominant receptors in the central nervous system (CNS; Madden, 2002). The roles of these two kinds of excitatory ligand-gated ion channel receptors could as easily have been reversed, as they are in insects. In insects, glutamate receptors are used for neuromuscular transmission (DiAntonio et al., 1999), and nAChRs are the predominant excitatory receptor in the CNS (Matsuda et al., 2001; Tomizawa et al., 2003).

Now it is evident that CNS nAChRs play important roles in humans (LeNovere et al., 2002; Sargent, 2000; Clementi et al., 2000; Romanelli and Gaultieri, 2003). The most medically important of these roles is in the addiction to tobacco mediated by nicotine (Mansvelder and McGehee, 2002), which contributes to the premature deaths of more than 400,000 Americans per year and more than 3.6 million per year worldwide (Peto et al., 1999). The beneficial effects of nicotine have made nAChRs potential drug targets for treating Alzheimer's disease, Parkinson's disease, Tourette's syndrome, depression, chronic pain, schizophrenia, and other problems (Lloyd and Williams, 2000; Romanelli and Gaultieri, 2003).

In addition, neuronal nAChRs are also being discovered to play previously unexpected roles in nonneuronal cholinergic signaling between nonneuronal cells such as keratinocytes.

2. Scope

This chapter briefly summarizes contemporary knowledge of the structures of nAChRs and their functional roles.

B. Structures of Nicotinic Acetylcholine Receptors

1. Nicotinic Acetylcholine Receptor Subtypes Are Defined by Their Subunit Compositions

Nicotinic acetylcholine receptors are part of a superfamily of homologous ligand-gated ion channels that includes vertebrate glycine receptors, $GABA_A$ and $GABA_C$ γ amino butyric acid receptors, and $5HT_3$ serotonin receptors, as well as some invertebrate glutamate and histidine receptors (Karlin, 2002). All of these receptors have families of subunits that can be assembled in various combinations to produce receptor subtypes. Complex families of subunits are characteristic of receptors from early in evolution. Sequencing of the *C. elegans* genome revealed 42 nAChR-like subunits potentially capable of assembling into a huge number of nAChR subtypes to serve a nervous system containing only 302 neurons (Bargman, 1998). Sequencing of the genome of the pufferufish *Fugu rubripes* has revealed 28 AChR subunits, making it the largest vertebrate nAChR family known to date, and suggesting that additional subunits may remain to be discovered in the human genome (Jones et al., 2003).

Nicotinic acetylcholine receptor subtypes are defined by their subunit composition (Figure 6.1). Subtypes differ in sensitivity to activation, desensitization, nicotine-induced upregulation, channel properties, localization, turnover rate, and probably other properties such as transcriptional and posttranslational regulatory mechanisms. Many of these differences may be critical for their physiological roles. Some differences in agonist or antagonist affinities may be incidental to their normal physiological role, but critical for distinguishing them as drug targets.

Nicotinic acetylcholine receptors are formed by 5 homologous subunits organized like barrel staves around a central cation channel (Karlin, 2002). The primordial nAChR presumably was homomeric, and gene duplication led to families of homologous subunits, which assembled in various combinations to produce heteromeric nAChRs (LeNovere et al., 2002). The simplest

Prominent AChR Subtypes

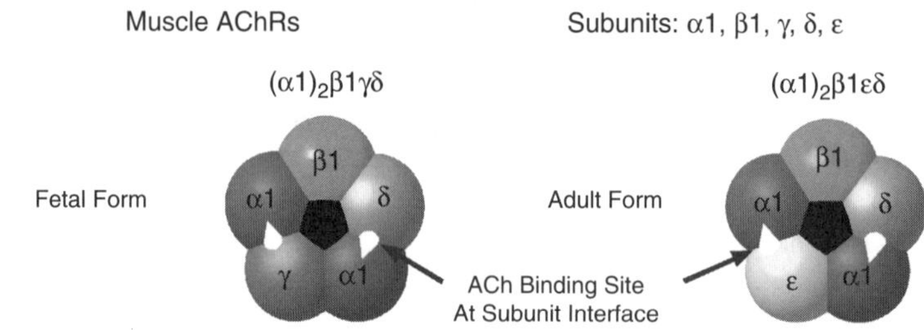

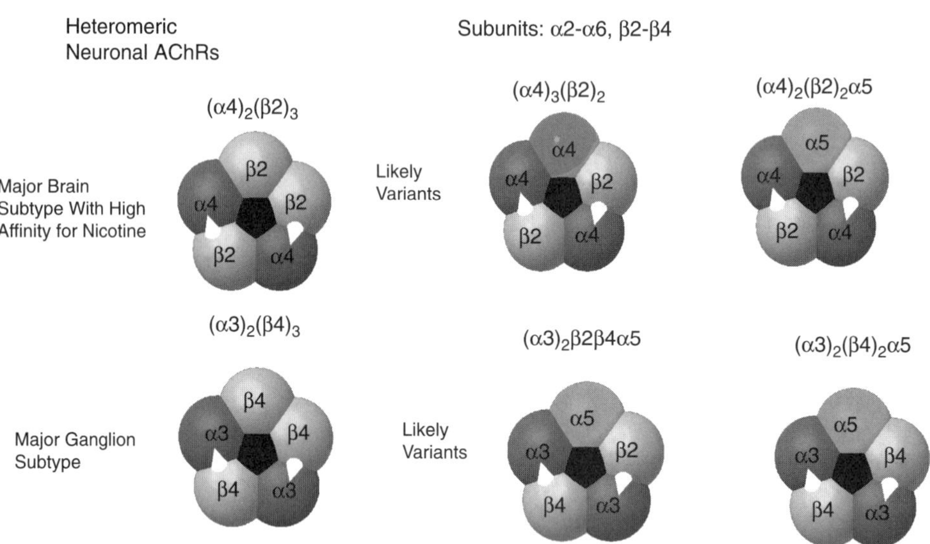

α5 and β3 may occupy only positions camparable to β1 in which they do not participate in forming ACh binding sites.

α6 can participate in forming ACh binding sites, but can also assemble in AChRs with, for example, α3 or α4 subunits.

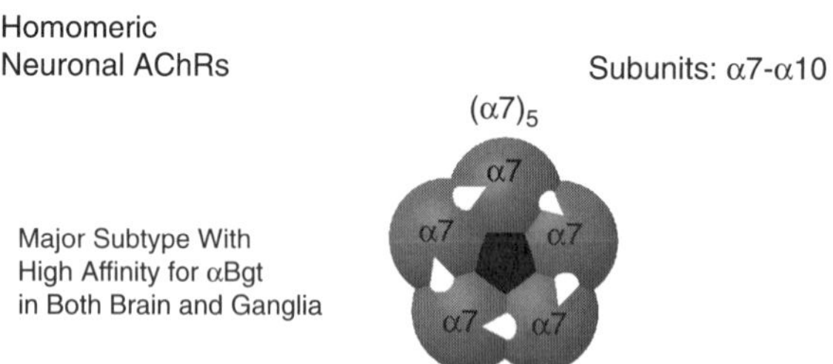

heteromeric nAChRs contain only 2 kinds of subunits. The best characterized heteromeric nAChRs from electric organs and muscle contain 4 kinds of subunits. These have provided a model for the organization of subunits within an nAChR. The most complex nAChR subtypes might contain 5 different subunits.

Subunits α7, α8, and α9 can form homomeric AChRs (Schoepfer et al., 1990; Elgoyhen et al., 1994, 2001; Lindstrom, 2000a). In mammals, α7 AChRs are the predominant homomeric nAChR. Subunit α8 has been found in chickens but not in mammals (Keyser et al., 1993). It is found both as homomers and as heteromers with α7. Subunit α9 was initially found in cochlear hair cells (Elgoyhen et al., 1994), but it has also been found in lymphocytes (Lustig et al., 2001) and keratinocytes (Grando, 2000). It may usually exist as a heteromer with α10. Like α1 muscle-type AChRs, α7, α8, and α9 are competitively inhibited by α bungarotoxin and cobra toxin. Unlike α1 AChRs, which exhibit subnanomolar K_D values for binding these snake venom toxins, α7 exhibits a K_D of 2 nM, α8 a K_D of 17 nM; α9 has even substantially lower affinity (Keyser et al., 1993). Unlike α1 nAChRs, whose cation channels are nonselective among cations and are not voltage sensitive, α7, α8, and α9 channels are especially permeable to Ca^{2+} and, like all other neuronal nAChRs, exhibit inward rectification (closing when the membrane is depolarized) (Gerzanich et al., 1994; Elgoyhen et al., 1994).

All nAChRs in the continued presence of agonists first open their channels in bursts of millisecond openings, then go into a desensitized conformation characterized by a closed channel and higher binding affinity (Quick and Lester, 2002). Subunits α7, α8, and α9 in AChRs desensitize more rapidly than others, perhaps because homomers have 5 ACh binding sites rather than the 2 ACh binding sites found in heteromeric AChRs (Utesher et al., 2002). Desensitization plays little role in the normal function of muscle nAChRs, which are exposed to 1 mM ACh for only about 1 millisecond, due to the high local concentration of acetylcholinesterase (AChE). The normal role of desensitization with other nAChR subtypes is less clear (Utesher et al., 2002). Desensitization can be important in pathological conditions. For an extreme example, a nerve gas such as sarin inhibits AChE, resulting first in convulsions and fasciculation of muscle, followed by flaccid paralysis due to accumulation of desensitized nAChRs and depolarizing block of Na^+ channel activation (Gunderson et al., 1992). A more common example is exposure to nicotine resulting from tobacco use. Whereas nAChRs in muscle are exposed to ACh for a millisecond at a time, nicotine is present in a high concentration (approximately μM) bolus for many seconds then in a lower concentration (approximately 0.1 μM in serum) for many hours (Benowitz, 1996). Muscle α1 nAChRs have very low affinity for nicotine (EC_{50} for activation ≥ 100 μM), but other AChR

Figure 6.1. Major nAChR subtypes. nAChR subtypes are defined by their subunit compositions. Five homologous subunits are arranged around a central cation channel. The primordial nAChR is thought to have been homomeric (LeNovere et al., 2000a). α7, α8, and α9 subunits can form homomeric nAChRs. Most α7 nAChRs are probably homomeric, but in chickens α7 together with α8 forms heteromeric nAChRs (Keyser et al., 1993). α8 *in vivo* forms both heteromeric and homomeric nAChRs. α8 has not been found in mammals. α9 is probably usually in heteromeric AChRs with α10 (Elgoyhen et al., 2001). The muscle type nAChR of Torpedo electric organ is the heteromeric nAChR whose subunit composition and organization is best known, and the organization of subunits in other heteromeric nAChRs is patterned after it (Karlin, 2002), α2, 3, 4, or 6 subunits can form functional nAChRs when paired with β2 or β4 subunits (Lindstrom, 2000a). More complex stoichiometries are common *in vivo*, and may be necessary for proper assembly or function. For examples, the α6β2 combination does not function when expressed in Xenopus oocytes, the α6β4β3 combination functions much better than α6β4, (Kuryatov et al., 2000), and α6α4β2β3 combinations are found *in vivo* (Champtiaux et al., 2003; Vailati et al., 2003, Zoli et al., 2002). α5 and β3 are closely related subunits which, like the β1 subunits of muscle nAChRs, appear not to be able to form ACh binding sites in combination with other subunits, but which can assemble in a position equivalent to β1 subunits (Gerzanich et al., 1998). In the figure, ACh binding sites are depicted as cones at specific subunit interfaces. Notes that α4β2 nAChRs, at least in heterologous expression systems, can exist in two stoichiometries, $(\alpha4)_2(\beta2)_3$ on which nicotine has high potency, and $(\alpha4)_3(\beta2)_2$ on which nicotine has low potency (Nelson et al., 2003, Zhou et al., 2003).

subtypes are much more sensitive, for example, α4β2 nAChRs (K_D approximately 2 nM, EC_{50} for activation approximately 2 μM) (Lentje and Patrick, 1991; Olale et al., 1997). Subunit α7 AChRs have low sensitivity to activation by ACh by (EC_{50} = 79 μM) and nicotine (EC_{50} = 40 μM), but are quite sensitive to desensitization by nicotine (IC_{50} = 0.003 μM).

Nicotinic acetylcholine receptors of fetal or denervated muscle have the subunit stoichiometry $(\alpha 1)_2\beta 1\gamma\delta$ (Karlin, 2002). In mature muscle, γ subunits are replaced by ε subunits. Subunits of Torpedo electric organ nAChRs are organized around the cation channel in the order α1, γ, α1, δ, β1. All the subunits contribute equally to the lining of the cation channel. There are 2 ACh binding sites located at the interfaces between α1 and γ subunits and between α1 and δ (Sine, 2002; LeNovere et al., 2002b). By convention, the plus side of the α subunit is defined as contributing to the formation of the ACh binding site in combination with the minus side of the adjacent subunit. The β1 subunit does not participate in forming an ACh binding site, but because it takes part in the conformation changes associated with activation and desensitization, it can influence both of these processes.

Heteromeric neuronal nAChRs can be formed by combinations of α2, α3, or α4 subunits in combination with β2 or β4 subunits (Karlin, 2002; LeNovere et al., 2002a; Lindstrom, 2000a). Subunits β3 or α5 can assemble in the β1 position, but these closely related subunits cannot form functional homomeric nAChRs by themselves or functional nAChRs in combination with just one other α or β subunit. Unlike homomeric or muscle nAChRs, these heteromeric nAChRs do not bind α bungarotoxin. Subunits α2 and α4 are closely related subunits. In rodent brains, α4β2 nAChRs form the predominant subtype with high affinity for nicotine (Flores et al., 1992; Whiting and Lindstrom, 1988). In rodents, α2 is a minor component, but it may play a larger role in primate brains (Han et al., 2003). About 36% of human brain α4β2 AChRs contain α5 subunits (Gerzanich et al., 1998). Transfected cell lines expressing human α4β2 AChRs contain a mixture of 2 stoichiometries. The $(\alpha 4)_2(\beta 2)_3$ nAChRs exhibit high sensitivity to ACh (EC_{50} = 0.7 μM), whereas $(\alpha 4)_3(\beta 2)_2$ nAChRs, in which an α4 subunit appears to have taken the β1 position, exhibit much lower sensitivity to ACh (EC_{50} = 70 μM) and desensitize more rapidly (Nelson et al., 2003). This mixture of stoichiometries may not be an artifact of transfection, since mouse thalamus nAChRs thought to be composed of only α4 and β2 subunits exhibit a mixture of 2 similar sensitivities to ACh (Kim et al., 2003). Autonomic ganglia neurons typically express α3, β2, β4, α5, and α7 subunits (Conroy and Berg, 1995; DeBiasi, 2002). The postsynaptic nAChRs are a mixture of α3β4, α3β2, and α3β4β2 nAChRs, some of which also contain α5, presumably in the β1 position (Nai et al., 2003). Mice lacking α3 subunits lack autonomic transmission (Xu et al., 1999a). Autonomic ganglia nAChRs typically reflect the relatively low ACh (EC_{50} = 59 μM) and nicotine (EC_{50} = 48 μM) sensitivities and other properties characteristic of α3β4 nAChRs (Nelson et al., 2001), but the situation is complex, and due to differences in channel kinetics, α3β2 nAChRs may contribute to transmission out of proportion to their fraction of the total α3 nAChRs (Nai et al., 2003). Knockout mice lacking either β2 or β4 subunits retain autonomic function, so there are sufficient amounts of either α3β2 or α3β4 subtypes to sustain transmission (Xu et al., 1999b). Subunit β2 is widely distributed in the CNS, but β4 is restricted to a few areas, especially the medial habenula, fasciculus retroflexus, interpeduncular nuclei, and pineal (Sargent, 2000). Subunit α3 is found in relatively small amounts in several brain regions; α6 is found in only dopaminergic or aminergic neurons in the brain, usually in combination with a variety of other nAChR subunits (Goldner et al., 1997; Champtiaux et al., 2002; Lena et al., 1999). Subunit α6 nAChRs have been particularly difficult to express (Gerzanich et al., 1997). In Xenopus oocytes, α6β4 combinations form functional nAChRs and α6β4β3 combinations are more efficiently expressed, while α6β2 subunit combinations do not form functional AChRs (Kuryatov et al., 2000). Subunit α6 will coassemble with α3 or α4 and β2. Immunoisolation studies of brain and retina have revealed the proportions of a complex mixture of nAChR subtypes expressed in these tissues, and some of the changes in nAChR subtypes that occur with development (Zoli et al., 2002: Vailati et al., 2003). Single neurons can express and differentially localize several nAChR subtypes, but the significance of this is not yet really clear (Berg and Conroy,

2002). No single neuronal nAChR subtype appears to be used, so far as is known, exclusively for a postsynaptic, presynaptic, or extrasynaptic functional role. There is much to be learned about the overlapping functional roles of the many nAChR subtypes.

2. Structures of AChR Subunits

All nAChR subunits share several features (Karlin, 2002; Lindstrom, 2000a). To produce the mature polypeptide sequence, a signal sequence of about 20 amino acids is removed from the N-terminus during translation (Keller and Taylor, 1999). The large N-terminal domain of each subunit consists of about 210 amino acids. This contains a disulfide-linked loop corresponding to Cys128 to Cys142 of α1 subunits, which is a signature feature of subunits of all receptors in this superfamily. The extracellular domain forms the ACh binding sites and the vestibule of the cation channel. Following the extracellular domain are 3 closely spaced transmembrane domains, termed M1 to M3, which are formed by amino acids about 220 to 310. The M2 domain forms most of the lining of the channel. There is a large cytoplasmic domain between the C-terminus of M3 and the N-terminus of the final transmembrane domain, M4. This domain varies from 110 to 270 amino acids, with α4 having by far the largest cytoplasmic domain. The M4 domain is about 20 amino acids leading to a short extracellular C-terminal domain of 10 to 20 amino acids.

The high-resolution (2.7 Å) x-ray crystal structure of the ACh binding protein (AChBP) provides a detailed model for the structure of the nAChR extracellular domain (Brejc et al., 2001) (Figure 6.2). The AChBP is a soluble protein secreted by certain molusc glia to modulate transmission by providing a sink for binding ACh where it neither activates nAChRs nor is destroyed by AChE (Smit et al., 2001). The sequence of the AChBP shows 24% identity to that of α7 nAChRs. Viewed from the top, the AChBP is an 80 Å-diameter pentameric ring of subunits around a central 18 Å-diameter hole corresponding to the vestibule of the nAChR channel. The subunits forming the walls of this ring are 31 Å thick. Viewed from the side, AChBP is a cylinder 62 Å high. The N-terminus is at the extracellular tip on the outer surface. The C-terminus, corresponding to the beginning of M1, is at what would be the outer edge of the membrane surface. The signature loop is near what would be the surface of the membrane. The AChBP signature loop is much more hydrophilic than the conserved sequence characteristic of nAChR subunits, probably because this adaptation helps to provide water solubility. The 5 ACh binding sites are formed halfway up the subunits at the interfaces between the plus side of the subunits, which is marked by the protuberance of a loop, and the adjacent cysteine pair characteristic of all α subunits (corresponding to α1 192, 193), located on the right-hand side of subunits in the cylinder viewed from the side, and the minus side of adjacent subunits.

3. Structures of nAChRs

a. Shape of the nAChR Protein

The overall shape of nAChR proteins is illustrated by low-resolution (9 Å and 4.6 Å) electron diffraction structures of Torpedo α1 nAChRs frozen in two-dimensional crystalline arrays in tubular membrane fragments (Unwin, 1993, 2000; Miyazawa et al., 1999). The overall shape of the extracellular domain reflects that seen in the AChBP. The nAChR viewed from the side is basically cylindrical, narrowing somewhat as it extends across the membrane, then widening slightly on the cytoplasmic surface. The nAChR extends approximately 65 Å above the membrane, approximately 35 Å across it, and approximately 20 Å below it, for an overall length of approximately 120 Å. The approximately 20 Å diameter of the vestibule narrows to closed at this resolution shortly below the membrane surface as the channel per se narrows to less than 10 Å.

On the cytoplasmic surface of Torpedo nAChR, a rapsyn protein is aligned in contact with the large cytoplasmic domain (Unwin, 1993, 2000; Miyazawa et al., 1999). This 43 kD extrinsic membrane protein is associated on a 1:1 basis with muscle-type nAChRs, serving as a link to the cytoskeleton and involved in aggregating nAChRs in the postsynaptic membrane. Rapsyn is not associated with neuronal nAChRs.

Recent evidence suggests that members of the PSD-95 family of postsynaptic density proteins

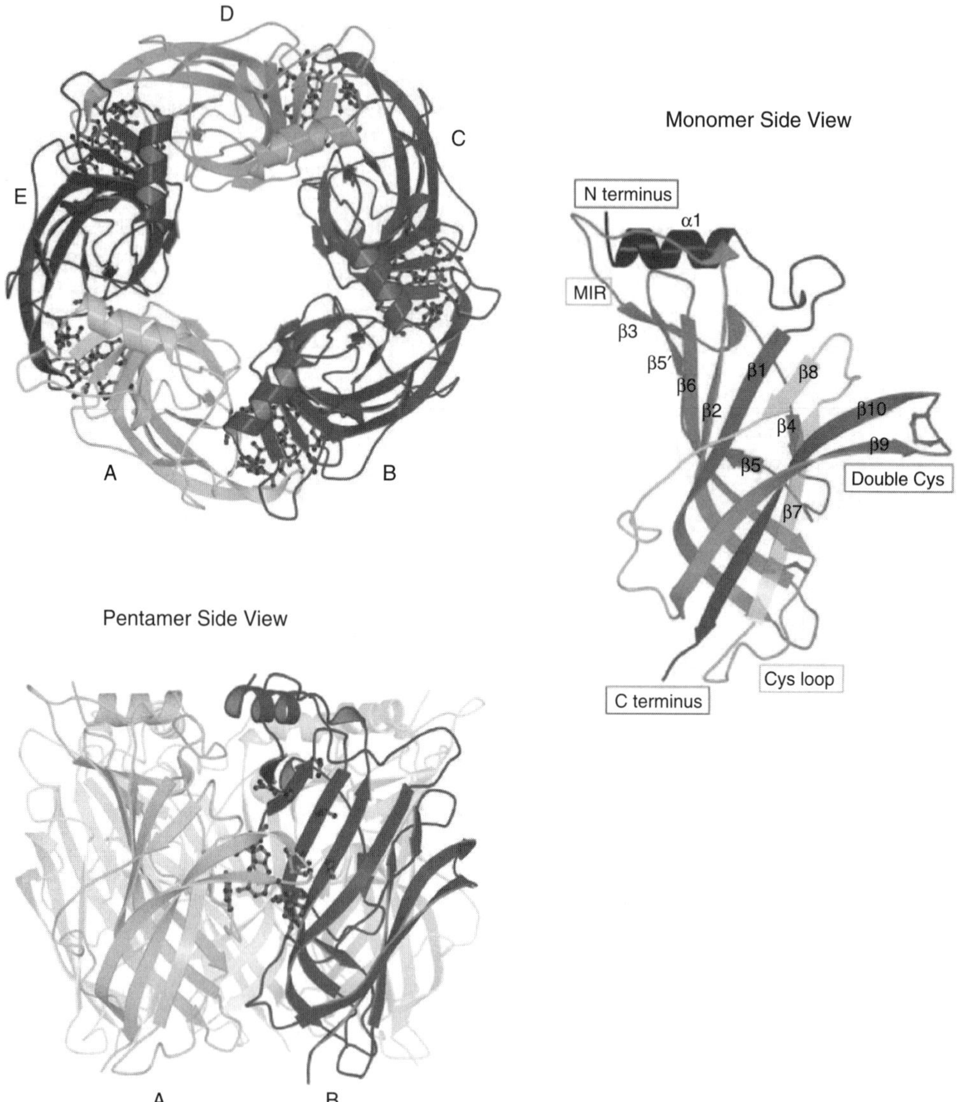

Figure 6.2. ACh binding protein structure. This water soluble protein secreted by mollusc glia corresponds in structure to the extracellular domain of homomeric α7 nAChRs. It provides a model for the organization of the polypeptide chain within the extracellular domain of all members of the cysteine loop superfamily of receptors. The structure of the cloned binding protein expressed in yeast was determined at 2.7Å resolution by x-ray crystallography (Brejc et al., 2001). It is a 62Å high cylinder 80Å in diameter with an 18Å diameter central hole corresponding to the vestibule of the nAChR channel. The five ACh binding sites contain the buffer component HEPES (N-2-hydroxy-ethylpiperazine-N-2 ethansulfonic acid). The adjacent disulfide-linked cysteine pair corresponding to α1 192–193, which is characteristic of all α subunits, is on a projection of what can be defined as the "+" side of the subunit. This interacts with the "–" side of the adjacent subunit to form the ACh binding site. The loop corresponding to part of the main immunogenic region (MIR) of α1 subunits is located at the extracellular tip (but the AChBP does not bind to antibodies directed at the MIR on α1 subunits). The signature loop characteristic of all subunits of receptors in this superfamily is located at the base near where the extracellular loops of the transmembrane domains and the outer leaflet of the lipid bilayer would be located. Modified from Brejc et al., (2001) with permission.

known to be associated with glutamate receptors may be involved in anchoring α3 and α7 nAChRs and tethering Ca^{2+} sensitive effector proteins where they could be most influenced by ion flow through the nAChR channel (Conroy et al., 2003). Different PDZ-domain proteins in this family are associated with α3 nAChRs (which are found postsynaptically in chick ciliary ganglion neurons) and α7 nAChRs (which are found perisynaptically in these cells). The large cytoplasmic domain of α7 and α3 AChRs is known to be important in targeting their localization to particular parts of ciliary ganglion neurons (Williams et al., 1998).

b. Arrangement of AChR Subunits

In Torpedo nAChRs, the arrangement of subunits around the molecule is α1γα1δβ1 (Karlin, 2002). There are 2 distinct ACh binding sites at the α1γ and α1δ interfaces (Sine, 2002; LeNovere et al., 2002b). Other heteromeric AChRs are thought to reflect a similar pattern, for example, α3β4α3β4β4 with 2 identical ACh binding sites at α3β4 interfaces, or α3β2α3β4β4 with 2 different ACh binding sites. Subunits α5 and β3 are thought to be able to assemble only in the β1-like position, for example, α3β4α3β4α5. Nicotinic acetylcholine receptors containing, for example, both α3 and α6 subunits would have 2 distinct ACh binding sites, for example, α3β2α6β2β2, with ACh binding sites of both α3β2 and α6β2 interfaces. Such an nAChR would be expected to be blocked by antagonists selective for either α3 or α6 nAChRs. For α4β2 nAChRs, because of the fundamental homologies in nAChR subunit structure, if sufficient concentrations of particular subunits are present, either β2 or α4 can compete for assembly in the β1 position to produce an α4β2α4β2β2 stoichiometry when excess β2 is present or an α4β2α4β2α4 stoichiometry when excess α4 is present (Nelson et al., 2003; Zhou et al., 2003).

c. ACh Binding Site

The ACh binding sites on nAChRs are composed of amino acids from 3 regions (loops A, B, C) of the extracellular domain on the plus side of α1-4 and α6-9 subunits and 3 regions (loops D, E, F) on the minus side of β2, β4, γ,δ,ε or α7, α8, α9, or α10 subunits (LeNovere et al., 2002b; Sine, 2002; Karlin, 2002). Contact amino acids identified by affinity labeling and mutagenesis experiments are readily recognized in the crystal structure of the AChBP (Brejc et al., 2001). The simplest agonist for AChRs is tetramethylammonium. Movements of amino acids within the ACh binding sites initiated by binding of the ammonium on cholinergic ligands must be critical to initiating the conformation changes of the AChR through which ligands cause opening of the channel gate and, on long exposure, desensitization of nAChRs. In vertebrate AChRs, as in AChE, the quaternary amine is bound not through interactions with negatively charged amino acids but rather through interactions with the π electrons of aromatic amino acids. Insects, again, may do things in reverse. Nicotine presumably evolved as a plant insecticide. At one time it was extensively used for this purpose in agriculture. As insects have become resistant to ACh inhibitors, and because of fears of ACh inhibitors' vertebrate toxicities, insect-selective neonicotinoid agonist insecticides have become important (Matsuda et al., 2001). Their selectivity for insects versus vertebrates appears to result from the presence in the cation-binding subsite of insect nAChRs of positively charged amino acids, which can bind to unprotonatable electron-withdrawing substituents of neonicotinoids located near where tertiary or quaternary amines are found in other cholinergic ligands (Tomizawa et al., 2003).

High-resolution crystal structures of nicotinic ligands bound to the AChBP or to nAChRs are not yet available, but the structure of the AChBP has been used to computer model docking of cholinergic ligands, and spectroscopic and mutational methods are being used to study in detail the binding of ligands to AChBP (LeNovere et al., 2002b; Fruchart-Gaillard et al., 2002; Grutter et al., 2003; Hansen et al., 2002; Gao et al., 2003). Location of the ACh binding site to subunit interfaces is an ideal place for controlling the conformational changes associated with activation and desensitization. Large curariform antagonists that bridge the interface are well positioned to stabilize the resting conformation. Smaller agonists that initiate conformation changes within the site may trigger concerted sliding or twisting along subunit interfaces to accomplish the concerted conformation changes associated with activation and desensitization. ACh is a small, flexible molecule. The nAChR subtype-selective ligands have particularly rigid conformations that have high affinity for only one or a few types of ACh binding sites.

Most of the highest-affinity nAChR ligands have been natural products, but there has been increasing development of subtype-selective ligands that provide useful research tools and potential drugs (Romanelli and Gaultieri, 2003). Snake venom toxins have provided excellent tools for studying α1 and α7-9 nAChRs. Epibatidine has provided an excellent high-affinity ligand for quantifying α2 to α6 nAChRs. Conotoxins have been found that are selective antagonists of many of these nAChR subtypes, and more of these small, rigid peptide toxins continue to be discovered (MacIntosh, 2000; Dowell et al., 2003).

d. Cation Channel

Amino acids that line the nAChR cation channel have been identified by the substituted cysteine accessibility method (SCAM), in which amino acids on putative transmembrane domains are systematically replaced by cysteines through *in vitro* mutagenesis, then mutated nAChRs are tested for susceptibility to blockage by a positively charged reagent highly reactive with the cysteine thiol (e.g., methanethiosulphonate ethylammonium) (Karlin, 2002; Akabas et al., 1994, 1995; Zhang and Karlin, 1997; 1998). Often substitution of the cysteine does not greatly alter function. Then the ability of the reagent to block the nAChR channel before or after opening or desensitization can be quite informative. Such experiments suggest that much of the M2 transmembrane domain (approximately 75%) is α helical, that the N-terminal third of M1 (which is not α helical) and most of M2 line the cation channel, and that in the resting state the open lumen of the channel extends nearly to the cytoplasmic surface where the gate is formed by the highly conserved linking sequence between M1 and M2 (α1 240 to 251). Other mutagenesis and affinity-labeling experiments are consistent with this interpretation. The M3 and M4 domains are in contact with lipid.

The fundamental homologies among receptors in this superfamily were elegantly illustrated by the demonstration that the selectivity of the α7 nAChR could be changed from cations to anions by changing 2 M2 channel-lining amino acids to those characteristic of the chloride-selective glycine and GABA receptors in these positions and adding a proline to lengthen the M1–M2 loop (Galzi et al., 1992). The reverse changes in the glycine receptor changed its selectivity from anionic to cationic (Keramidas et al., 2000). Although only 3 amino acids were changed in these subunits, in a homopentamer each change produced a ring of 5 amino acids lining the receptor channel. Another elegant demonstration of the structural homologies in this superfamily was construction of a chimera between the extracellular domain of α7 and the remainder of a $5HT_3$ receptor subunit to produce ACh-gated receptors with the channel properties of $5HT_3$ receptors (Elsele et al., 1993)

Higher-resolution (4 Å) cryoelectronmicroscopy studies of Torpedo nAChRs have recently been reported in which the results are interpreted to suggest that the gate may be a hydrophobic girdle in the middle of the channel (Miyazawa et al., 2003) (Figure 6.3). The sequence near the cytoplasmic surface linking M1 and M2 that was identified as the gate by the SCAM was not resolved in these studies (Wilson and Karlin, 1998, 2001). Resolution of the conflicting conclusions of the two methods will require higher-resolution structural studies of open and closed channels. The current cryoelectronmicroscopy images show all the transmembrane domains as being α helical, with a central ring of M2 domains surrounded by a 5-pointed starlike wall of M1, M3, and M4 domains with which the ring of M2 domains makes limited contact (Miyazawa et al., 2003). It is proposed that agonist-induced rotations of a socket formed by the loop between the β1 and β2 components of the α1 extracellular domain interact through a pin formed by the end of the α1 M2 α helix extending nearly 10 Å above the lipid bilayer to induce a twisting motion in the M2 α helix that opens the hydrophobic girdle. Whatever turns out to be the absolute truth, it is clear that movements initiated by the quaternary amine of ACh in ACh binding sites at subunit interfaces 30 Å above the lipid bilayer propagate through the nAChR structure to regulate opening of a channel gate 45 Å to 60 Å away (depending on whether the gate is in the middle or cytoplasmic end of the channel).

Electrophysiologists have proposed that there are 2 gates, the activation gate discussed thus far

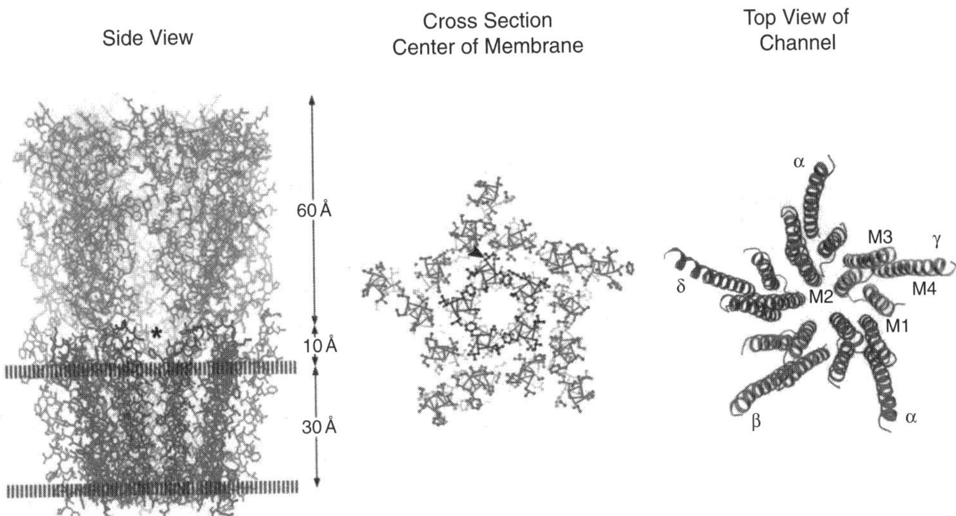

Figure 6.3. Structure of the Torpedo electric organ nAChR. Electron microscopic analysis of frozen two dimensional tubular crystalline arrays of nAChRs was refined to 4Å resolution (Miyazawa et al., 2003). The side view depicts the outer 10Å of the transmembrane domains as located above the lipid bilayers. Virtually nothing is depicted as extending below the lipid bilayer, but the size of the large cytoplasmic domain formed by the loop between the M3 and M4 transmembrane domains is significant, and in other cryoelectronmicroscopic images is shown to extend down as five 30Å fingers to grasp the rapsyn molecule located beneath each electric organ nAChR linking it to the cytoskeletom (Miyazawa et al., 1999). The images shown here emphasize the transmembrane pore structure of the AChR. The cross section through the membrane exphasizes a central hydrophilic channel lining region formed by M2 transmembrane domains surrounded by other transmembrane domains that interface with lipids. The top view of the transmembrane domains shows the symmetric organization of the four transmembrane domains within each subunit. Modified from Miyazawa et al. (2003) with permission.

and a desensitization gate (Auerbach and Akk, 1998). In the resting state, the activation gate would be closed and the desensitization gate open. In the desensitized state the desensitization gate would be closed and the activation gate would be open. The cryoelectronmicroscopy, SCAM, and electrophysiological models might be resolved if the hydrophobic girdle in the middle of the channel were the desensitization gate and the gate formed at the cytoplasmic surface by the M1-M2 linker were the activation gate.

Cryoelectronmicroscopy at 4.6 Å resolution led to the suggestion that gaps less than 10 Å wide in the wall of the cytoplasmic vestibule to the channel might provide ion-selective filters through which cations could approach the channel (Miyazawa et al., 1999). Five fingerlike projections appear to extend, one from the cytoplasmic domain of each subunit, to grasp the rapsyn centered below. In $5HT_3$ receptors, replacement of 3 arginine residues in sequences thought to be homologous to those forming these fingers increased single-channel conductance 28-fold (Kelly et al., 2003). Lateral access portals to ion channels ("hanging gondolas") have been described for sodium and potassium channels (Kobertz et al., 2000; Sokolova et al., 2001; Sato et al., 2001). Such an extended structure for AChR large cytoplasmic domains might explain why mAbs to the large cytoplasmic domain and its constituent synthetic peptides often react with both native AChRs and denatured subunits on western blots, by contrast to more conformation-dependent mAbs directed at the compact, rigidly constrained structure of the extracellular domain (Das and Lindstrom, 1991).

e. Large Cytoplasmic Domain

The large cytoplasmic domain between M3 and M4 is the most variable domain in sequence among nAChR subunits and among species (Lindstrom, 2000a). Consequently, many subunit-specific antibodies bind to this domain. This domain interacts with rapsyn, in the case of muscle nAChRs (Maimone and Merlie, 1993; Miyazawa et al., 1999), or PSD-95 and perhaps other proteins, in the case of neuronal nAChRs (Conroy et al., 2003), to anchor to the cytoskeleton and organize secondary effectors nearby. The chaperone protein 14-3-3η binds to serine 441 in the large cytoplasmic domain of α4, especially when this serine is phosphorylated (Jeanclos et al., 2001). The calcium sensor protein VILIP binds to α4 302 to 339, resulting in increased surface expression and ACh sensitivity (Lin et al., 2002). The chaperone COPI interacts at lysine 341 of α1 subunits. The large cytoplasmic domain contains phosphorylation sites which regulate the rate of desensitization (Fenster et al., 1999). It is involved in targeting intracellular transport (Williams et al., 1998). Despite its distance from both the ACh binding sites and the cation channel, it contains sequences that influence channel-gating kinetics (Wang et al, 2000).

f. The Main Immunogenic Region

Immunization with native α1 nAChRs produces antibodies directed primarily at the extracellular surface, whereas immunization with denatured nAChRs or subunits produces antibodies directed primarily at the cytoplasmic surface (Das and Lindstrom, 1991; Froehner, 1981).

More than half the antibodies to native α1 nAChRs in an immunized animal with EAMG or of autoantibodies in a human with MG are directed at the MIR (Tzartos and Lindstrom, 1980; Tzartos et al., 1982, 1998) (Figure 6.4). A single mAb to the MIR can prevent binding of more than half the serum antibodies in MG or EAMG. The MIR is a highly conformation-dependent epitope; thus, immunization with denatured subunits by default provokes antibodies to the cytoplasmic surface. Some mAbs to the MIR retain some affinity for denatured subunits, which lead to mapping of α1 66 to 76 as containing amino acids contributing to the MIR. Mutagenesis experiments in which amino acids in this sequence of α1 from Xenopus (which does not bind mAbs to the MIR) were substituted into α1 of Torpedo nAChRs revealed that amino acids 68 and 71 were especially important (Saedi et al., 1990). Replacement of either Torpedo amino acid by the Xenopus equivalent prevented binding of mAbs to the MIR, independent of whether they were raised to nAChRs from electric organ or human muscle or whether they were absolutely conformation dependent or able to bind on western blots.

The AChBP does not bind mAbs to the MIR, but amino acids corresponding to α1 66 to 76 are found in a loop at the extracellular tip of the AChBP on the extracellular surface angled away from the central axis (Brejc et al., 2001), just as expected from cryoelectronmicroscopic studies of MIR mAbs bound to Torpedo α1 nAChR (Beroukhim and Unwin, 1995) and other studies (Conti-Tronconi et al., 1981). This location of the MIR makes sense (Lindstrom, 2003, 2000b). It explains why mAbs to the MIR do not affect nAChR function. The MIR is far from the ACh binding sites, the cation channel, and subunit interfaces, where movements might be involved in activation. The MIR angled away from the central axis explains why single mAbs cannot cross-link the 2 α1 subunits in α1 nAChR but are extremely efficient at binding to the extracellular surface and cross-linking adjacent α1 nAChRs. Autoantibodies impair neuromuscular transmission in MG and EAMG by cross-linking nAChRs (thereby increasing their rate of internalization and destruction) and by fixing complement (thereby targeting the postsynaptic membrane for focal lysis).

Some mAbs to the MIR on α1 subunits also bind to similar sequences on α3, α5, and β3 subunits, but do not bind to other AChR subunits (Conroy et al., 1992; Wang et al., 1996). In MG patients, there is no cross-reaction of autoantibodies to α1 AChRs with α3 nAChRs of autonomic ganglia (Vernino et al., 2000). In patients with autoimmune autonomic neuropathy, there is no cross-reaction of autoantibodies to ganglionic α3 nAChRs with muscle α1 AChRs (Vernino et al., 2000). The basis of the difference in recognition of the MIR by rats producing mAbs and human autoantibodies remains unknown.

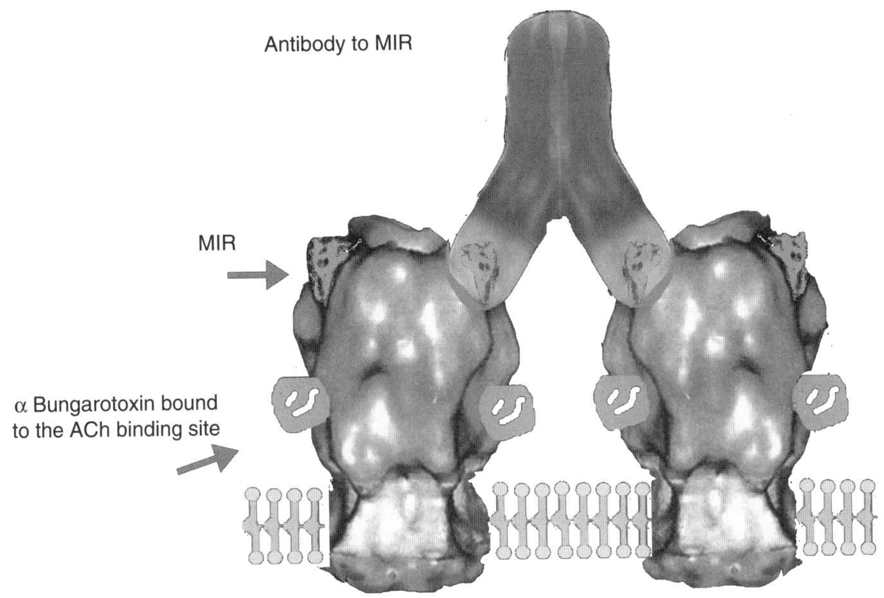

Figure 6.4. The main immunogenic region. What is known about the structure of the MIR helps to explain the pathological significance in myasthenia gravis. The MIR is easily accessible on the extracellular surface. This permits autoantibodies to bind and fix complement leading to focal lysis of the postsynaptic membrane. The MIR at the extracellular end of each α1 subunit is oriented outward so that an antibody can easily link adjacent nAChR but not crosslink the two α1 subunits with a single nAChR. Crosslinking of patches of nAChRs increases their rate of endocytosis and destruction (this is called antigenic modulation). The MIR is distant from the ACh binding sites, subunit interfaces that might move during activation, and the cation channel, explaining why antibodies bound to the MIR do not inhibit nAChR function or interfere with the binding of iodinated α bungarotoxin that is used to quantitate nAChRs in immunodiagnostic assays. The precise structure of the MIR which accounts for its high immunogenicity, the conformation dependence of most antibody binding, or the differences in binding between those MIR antibodies which recognize only α1 or both α1 and α3, remain to be determined.

C. Functional Roles of Nicotinic Receptors

1. Skeletal Muscle α1 nAChRs

At neuromuscular junctions on skeletal muscle, α1 nAChRs in the postsynaptic membrane provide a critical link in transmission (Ruff, 2003). Invasion of the nerve ending by an action potential evokes release of ACh at active zones in the presynaptic membrane, which are located immediately across from AChRs concentrated in semicrystalline arrays at the tips of folds in the postsynaptic membrane. The current passing through the AChRs, in the millisecond before destruction of ACh by AChE, terminates the response, greatly amplifies the current involved in the action potential sufficiently to ensure a substantial safety factor for depolarization of the postsynaptic membrane, and triggers an action potential in the large muscle fiber. The concentration of ACh transiently reaches 1 mM.

By contrast, neuronal nAChRs in postsynaptic roles are often not present in such large synapses with such large safety factors. Central synapses in general are likely to exhibit a 0.1 probability of transmission, compared to the 1.0 probability at neuromuscular junctions. Many neuronal AChRs act presynaptically to modulate transmitter release (Kaiser et al., 2000). Some act extrasynaptically (Berg and Conroy, 2002). Neuronal nAChRs in nonneuronal tissue appear to be involved in cholinergic signaling, which does not involve highly specialized synapses

(Grando and Horton, 1997; Wang et al., 2003).

2. Ganglionic AChRs

At autonomic ganglia, α3 nAChRs play a postsynaptic functional role similar to that of α1 AChRs (Berg et al., 2000; Debiasi, 2002). However, rather than a single $(\alpha1)_2\beta1\varepsilon\delta$ subtype, as at mature neuromuscular junctions, there are a variety of α3 nAChRs containing β2, β4, and α5 subunits.

Ciliary ganglia of chickens have been especially well studied, but may be atypical in several aspects (Berg and Conroy, 2002; Conroy and Berg, 1995). These form specialized calyx synapses. At early periods during development, transmission can be mediated either by postsynaptic α3 nAChRs or by extremely abundant perisynaptic α7 nAChRs. In adult chickens, transmission can be largely electrotonic and bypass nAChRs. It has been suggested that the perisynaptic α7 nAChRs' predominant role may be trophic. This might involve responding to choline diffusing from the synapse subsequent to AChE activity to provide feedback on the level of synaptic activity. Calcium ions entering through α7 nAChRs could regulate signaling cascades that could regulate gene expression of synaptic components.

The enteric nervous system contains as many neurons as the spinal cord, and most of these enteric ganglionic neurons express α3. This can be imaged nicely using mAbs to the MIR (Kirchgessner and Liu, 1998; Obaid et al., 1999).

3. Central nAChRs

The α4β2 nAChRs are the major subtype with high affinity for nicotine in rodent brains (Picciotto et al., 2001; Flores et al., 1992). It has been suggested that in primates α2 AChRs may play a more important role than in rodents (Han et al., 2003). Immune precipitation studies using mAbs to the MIR indicate that 20% of rat brain α4 AChRs and 36% of human neocortex α4 AChRs contain α5 or β3 subunits (Gerzanich et al., 1998).

The unusually high ACh affinity of α4β2 AChRs suggests that they may be especially well suited to participate in volume transmission, where ACh may act on AChRs distant from sites of release (Zoli et al., 1999). It has been suggested that most brain nAChRs are located presynaptically and function to modulate the release of various transmitters (Kaiser et al., 2000).

The α7 AChRs are as abundant in brain as are α4β2 AChRs. The α7 AChRs have been found in both presynaptic and postsynaptic functional roles (Berg and Conroy, 2002).

In hippocampal CA1 interneurons, both α4β2 and α7 AChRs have been found to modulate the release of GABA (Alkondon and Alquerque, 2001). The α4β2 AChRs were involved in tonic inhibition, while α7 AChRs were involved in shorter, more phasic inhibition.

Minor subtypes, for example, α6 nAChRs, can play significant roles, often in combination with other AChR subtypes. Nicotine-induced release of dopamine from striatal synaptosomes is nearly equally divided between α6 . . . and α4β2 AChRs; thus, α6 AChRs may play a significant role in addiction to nicotine (Kulak et al., 1997). The α6 nAChRs on dopaminergic neurons of the substantia nigra may be especially sensitive to loss in Parkinson's disease and its animal models, more so than α4β2 AChRs on these neurons, suggesting that α6 nAChRs may be especially important in this disease and might be potential drug targets (Kulak et al., 2002; Quick et al., 2003). A variety of nAChR subunits are expressed in the locus ceruleus and other aminergic neurons (Lena et al., 1999; Klink et al., 2001), and electrophysiological, immunoprecipitation, and knockout mouse studies (Zoli et al., 2002; Champtiaux et al., 2003) suggest that quite a complex variety of AChR subtypes are expressed. How these are distributed within the neurons is being discovered. On dopaminergic terminals in rat striatum, 4 subtypes of nAChRs were found: $(\alpha4)_2(\beta2)_3$ (30%), $(\alpha4)_2(\beta2)_2\alpha5$ (30%), $(\alpha6)_2(\beta2)_2\beta3$ (25%), and $\alpha4\alpha6(\beta2)_2\beta3$ (15%) (Zoli et al., 2002). On non-dopaminergic terminals in the same preparation, both α4β2 and a small amount of α2α4β2 nAChRs were immunoisolated. Knockout mouse experiments indicated that on the cell bodies of dopaminergic neurons in the ventral tegmental area are α7, α4β2, and α4α6β2 nAChRs (Champtiaux et al., 2003). The endings of these neurons contain both α4β2 and α4α6β2 nAChRs. To add further complexity, activation of these neurons by glutamate is modulated by presynaptic α7 nAChRs, while inhibition of these neurons by GABA is modulated by α4β2 nAChRs.

Retina contains a complex mixture of nAChR subtypes, which changes during development (Vailati et al., 2003). This is best known in the case of chickens, which have α8 AChRs, not seen in mammals. In chicken brain, α8 homomers are a minor component compared to α7, and much of this α8 is present as heteromers with α7 (Keyser et al., 1993). In chick retina, α8 predominates over α7, and α8 homomers are common. Among conventional heteromeric nAChRs, 84% contain β2 on embryonic day 7, but only 32% on postnatal day 1 (Vailati et al., 2003). During this interval, the incidence of β4 goes from 22% to 78%. Long before synapses develop, α3, α4, and α7 are expressed, whereas α2, α6, α8, β3, and β4 are expressed only late in development. The major heteromeric subtype early in development is α4β2 nAChRs, but by postnatal day 1 the array of subtypes is quite complex. On postnatal day 1, of β2-containing nAChRs, 28% contained α2, 11% α3, 23% α4, and 19% α6. At that time, of β4-containing AChRs, 15% contained α2, 35% α3, 40% α4, and 19% α6.

Nicotinic acetylcholine receptors can be involved in unusual forms of synaptic transmission. For example, in chicken cochlear hair cells, postsynaptic α9α10 nAChRs mediate entry of Ca^{2+} in a small, rapid excitatory current, which triggers Ca^{2+} activated K^{+} channels to produce a large, delayed, sustained inhibitory current (Fuchs, 1996; Elgoyhen et al., 1994, 2001; Vetter et al., 1999). Because these nAChRs have low affinity for α bungarotoxin and significant affinity for atropine, transmission at this synapse initially appeared to be mediated by a muscarinic rather than a nicotinic AChR.

4. Effects of Nicotine

In tobacco users, nicotine is present for prolonged periods, in boluses of μM concentration for seconds or minutes after inhaling and sustained in serum at concentrations up to 0.1 or 0.2 μM for many hours (Benowitz, 1996). Thus, its effects on nAChRs can be quite complex (Mansvelder and McGehee, 2002; Dani, 2003; Dani and Heinemann 1996; Dani et al., 2001). The many nAChR subtypes located at several places on interacting pathways make the effects of nicotine even more complex (Champtiaux et al., 2003). The α1 nAChRs are negligibly affected by concentrations of nicotine associated with tobacco use. Subtypes with higher affinity might be transiently activated by high-concentration boluses, then either desensitized or left at a smoldering level of activation, depending on the affinity and susceptibility to reversible desensitization or longer-lasting inactivation of the subtype. Nicotine causes increases in the amount of nAChRs (Flores et al., 1992, 1997; Perry et al., 1999; Davila-Garcia et al., 2003; Peng et al., 1994, 1997; Wang et al., 1998; Nelson et al., 2001, 2003). The extent of upregulation in brain areas is typically 50 to 100% (Flores et al., 1997), but in some brain areas and in tissues such as adrenal gland, pineal, superior cervical ganglia, or retina (which are rich α3 AChRs), upregulation is not seen at the nicotine concentrations found in smokers (Davila-Garcia et al., 2003). Acetylcholine receptor subtypes with high affinity for nicotine (α4β2) are more sensitive to upregulation than are lower-affinity subtypes (α3β2 and α7) (Peng et al., 1994, 1997). Human α3 AChRs that contain β2 are sensitive to upregulation, whereas those with β4 subunits are not (Wang et al., 1998; Nelson et al., 2001). The stoichiometry $(\alpha4)_2(\beta2)_3$ is more sensitive to nicotine-induced upregulation than is the less ACh-sensitive $(\alpha4)_3(\beta2)_2$ stoichiometry (Nelson et al., 2003). In transfected cell lines expressing nAChRs, the extent of upregulation can be 10- to 20-fold (Wang et al., 1998). Mechanisms of upregulation include both increased assembly of preexisting subunits and increased lifetime of surface AChRs, but not usually increased transcription or translation of AChR subunits (Peng et al., 1994; Wang et al., 1998). The multiple nAChR subtypes produced by single neurons complicate the analysis of upregulation. Mechanisms of nicotine-induced upregulation of α3 and α7 AChRs in the same cell may differ (Ridley et al., 2001). The different nAChR-containing neurons involved in circuits regulating the release of dopamine, which is thought to mediate addiction to nicotine and other drugs of abuse, complicates the analysis of nicotine's effects. Further, nicotine has many effects because nAChRs are found throughout the nervous system and in nonneuronal cells. These effects include not only acute effects enhancing or impairing transmission but also long-term effects on neuroprotection and gene regulation.

Despite these complexities, it is possible to relate effects of nicotine on cloned nAChR subtypes to behaviors characteristic of tobacco users. The 0.2 μM concentration of nicotine in a smoker's serum would have some acute activating effect on human α4β2 AChRs (EC_{50} = 0.3 μM), little effect on α3β2 AChRs (EC_{50} = 3 μM), and negligible acute activating effect on α7 AChRs (EC_{50} = 40 μM) (Olale et al., 1997). However, after 3 hours in 0.2 μM nicotine, desensitization and inactivation eliminate virtually all response to 100 μM ACh of α4β2 AChRs (IC_{50} = 0.017 μM), 90% of that of α7 AChRs (IC_{50} = 0.003 μM), but only 20% of that of α3 AChRs (IC_{50} = 0.87 μM). Thus, in a smoker during the day normal signaling through brain α4β2 AChRs should be essentially completely inhibited while signaling through autonomic ganglion α3 AChRs should be little affected. After overnight without smoking, the first cigarette of the day is the most rewarding. In the morning, an upregulated amount of α4β2 AChRs (EC_{50} for upregulation = 0.2 μM) recovered from desensitization should produce an acute response to nicotine before settling in to a desensitized state (Peng et al., 1994). The amounts of α3 nAChRs (EC_{50} for upregulation approximately100 μM) and α7 AChRs (EC_{50} for upregulation approximately 65 μM) would not be upregulated significantly in smokers. In the morning the α7 nAChRs, like the α4β2 AChRs, should be susceptible to transient activation by nicotine before becoming desensitized.

5. Neuronal AChRs in Nonneuronal Tissues

Nicotinic acetylcholine receptors have been found to be involved in signaling between a variety of nonexcitable cells (Sharma and Vijayaraghavan, 2002). Such cell types range from astrocytes to vascular endothelia to bronchial epithelia.

Nicotinic acetylcholine receptors and ACh have been found in lung tumor cells and in other lung tissues (Song et al., 2003). These are especially relevant to the effects of nicotine, since virtually all lung tumors are the result of tobacco smoking (Clementi et al., 2000). In some cases nicotine promotes proliferation of these tumor cells. It has also been reported that the tobacco-associated carcinogen NNK (4-methylnitrosamino)-1-(3 pyridyl-1-butanone) acts as an α7-selective agonist to trigger the Akt signaling pathway epithelial cells (West el al., 2003).

Nicotinic acetylcholine receptors and a complete cholinergic signaling system have also been found in keratinocytes (Grando and Horton, 1997). Several AChR subtypes have been found in skin. In regulating apoptosis and extracellular matrix regulations, α7 AChRs are involved (Nguyen et al., 2001; Arredondo et al., 2002). Decreased collagen, elastin, and metalloproteinase are found in α7 knockout mouse skin (Arredondo et al., 2003). NAChRs have also been found in bronchial and vascular epithelia (Maus et al., 1998).

Several nAChRs have been reported in the immune system. The most dramatic effect was the demonstration that stimulation of the vagus nerve inhibits inflammation induced by bacterial lipopolysaccharide by acting on α7 AChRs in macrophages to prevent release of tumor necrosis factor F (Wang et al., 2003). In α7 knockout mice this effect is lost. Thus, excitatory "neuronal" AChRs, in this case on nonneuronal cells under the control of the nervous system, can act in an inhibitory fashion.

The presence of nAChRs and a cholinergic signaling system in epidermal tissue and other tissues suggests the possibility that primordial nAChRs may have been involved in primordial intercellular signaling before evolution of the nervous system refined and specialized this system for more rapid signaling.

6. Genetic Approaches to Understanding Physiological Roles of nAChRs

a. Congenital Myasthenic Syndromes

Mutations in nAChR subunits, rapsyn, AChE, CAT, and other proteins involved in neuromuscular transmission cause muscular weakness resembling that of autoimmune diseases such as MG and Lambert Eaton's myasthenic syndrome (Engel et al., 2003). Many congenital myasthenic syndromes (CMS) have been studied in detail. These provide an elegant model for potential types of diseases involving neuronal nicotinic signaling. The myasthenic syndromes are much more easily recognized and characterized than neuronal syndromes would be because, unlike neuronal nicotinic signaling, the physiology of neuromuscular transmission is well known, impairment is easily identifiable, and both functional assays and biopsies are accessible. The effects on human neuromuscular transmission and synaptic morphology can be explained by the functional effects of subunit mutations on expressed cloned human AChRs and the emerging knowledge of nAChR structure.

Examples of CMS caused by mutations in nAChR subunits can be quite instructive.

Common forms of CMS involve "slow-channel" syndromes in which mutations in the M2 channel lining sequence and elsewhere produce hyperexcitable nAChRs with prolonged channel openings (Engel et al., 2003). The nAChRs can be spontaneously activated by ambient choline (Zhou et al., 1999). Excitotoxicity from such mutations causes degeneration of the junctional folds due probably to excess entry of Ca^{2+}. The unusually long channel openings in these syndromes result from mutations that increase affinity for agonists or mutations in M2 or elsewhere that destabilize the resting state of the nAChR.

Most "fast-channel" syndromes exhibit unusually rapid decay of the synaptic response as a result of mutations in the extracellular domain, which reduce affinity for binding of ACh to the open channel state or impair transition to the active conformation (Engel et al., 2003). However, some mutations in M3, M4, or the large cytoplasmic domain near M4 were also found to produce fast-channel syndromes. All of these patients had both the mutation that caused the channel defect and a null mutation in the other allele. It is striking that a large number (more than 60) of AChR mutations have been found and that patients often have more than one mutation. This suggests that there may similarly be many disease-associated mutations in neuronal AChRs waiting to be discovered.

The effects of many CMS mutations on α nAChR assembly, affinity, conductance, and channel conductance have been elegantly characterized. The functional effects of these mutations have been interpreted in detail in terms of the structure of the AChBP and other models of nAChR structure, and the effects on the physiology and ultrastructure of human neuromuscular junctions accounted for in detail. The many papers involved are reviewed in detail in Engel et al. (2003) and elsewhere. These have provided new insights into the structure and function of nAChRs, but a detailed review is beyond the scope of this chapter.

Many CMS mutations are in ε subunits (Engel et al., 2003). These often cause premature termination of protein synthesis or otherwise prevent formation of functional ε subunits. However, neuromuscular transmission is sustained despite the loss of ε subunits as a result of induction of synthesis of γ subunits characteristic of the fetal form of muscle nAChRs.

Congenital myasthenic syndromes due to deficiencies in the amount of nAChRs are known to be caused by mutations in nAChR subunits, which hinder assembly, and by mutations in rapsyn, which hinder aggregation of nAChRs at junctions. Mutations in other proteins that regulate the expression and localization of these nAChRs probably will be found.

b. Diseases Caused by Mutations in Neuronal AChRs

The wide variety of CMS caused by mutations in muscle AChRs and associated proteins suggests that there are potentially many similar disease-causing mutations in neuronal nAChRs and their associated proteins, but most have not yet been discovered. However, neuronal nAChR mutations have been implicated in an unusual mild epilepsy and a rare lethal bowel disease.

Autosomal dominant nocturnal frontal lobe epilepsy (ADNFLE) patients suffer from brief partial epileptic seizures during sleep (Steinlein, 2000; Raggenbass and Bertrand, 2002; Sutor and Zolles, 2001). Three mutations in $\alpha 4$ and 2 in $\beta 2$ subunits have been found to cause ADNFLE. All of these mutations are in or near M2. Some produce use-dependent activation, reduce Ca^{2+} permeability, or increase ACh sensitivity. It has been suggested that what all of these mutations have in common is dominant inhibition of Ca^{2+}-dependent potentiation of nAChR activation (Rodrigues-Pinguet et al., 2003). It is known that presynaptic nAChRs facilitate release of both excitatory and inhibitory transmitters. It has been proposed that inhibition of the $\alpha 4\beta 2$ nAChR response to high ACh concentrations by reducing Ca^{2+} from 2 mM to 0 is a negative feedback mechanism to inhibit ACh stimulation of glutamate release by presynaptic nAChRs when Ca^{2+} flow through postsynaptic glutamate receptors has depleted the local Ca^{2+} concentration. Then mutations preventing this feedback might lead to excessive glutamate release if the synchronous repetitive firing associated with sleep spindles depleted Ca^{2+} from the extracellular space around these synapses.

Megacystis-microcolon-intestinal hypoperistalsis syndrome is a rare, lethal fetal disease whose symptoms resemble those of mice in which α3 subunits have been knocked out (Xu et al., 1999a). There is some evidence from a limited number of patients that α3 nAChRs are missing in such patients (Richardson et al., 2001). Given the extensive expression of α3 nAChRs in the enteric nervous system (Kirchgessner and Liu, 1998; Obaid et al., 1999), their loss would be expected to be catastrophic.

c. Effects in Mice of nAChR Subunit Knockouts and Knockins

Nicotinic acetylcholine receptor subunit knockout and knockin mice can be very useful for suggesting potential physiological roles for nAChR subtypes (Champtiaux and Changeux, 2002; Picciotto et al., 2001), but there are several limitations to such studies. One is that human brains might express more α2 subunits than do those of mice or otherwise differ in AChR subtype expression. Knockout of a subunit might underestimate its normal functional significance, if compensation from other subunits minimized the effects of loss of that subunit (e.g., consider compensation by γ for loss of ε in CMS). Conversely, knockin of an excitotoxic mutant subunit might overestimate the functional significance of a subunit if many neurons that even transiently express it during development were killed by the excitotoxicity. Below are some examples of interesting insights gained from studies of nAChR subunit knockout and knockin studies in mice.

Knockout of α3 nAChR subunits is perinatal lethal (Xu et al., 1999a). As might be expected from loss of all signaling through autonomic ganglia, the mice have impaired growth, bladder enlargement and infection, and dilated pupils. Knockout of neither β2 nor β4 subunits produced this effect, due to compensation of one for the other, but knockout of both was similarly lethal (Xu et al., 1999b). This might be expected if ganglionic postsynaptic membranes usually contained a mixture of α3β4, α3β2, and α3β2β4 nAChR subtypes, which all contributed to transmission with a reasonable safety factor (Nai et al., 2003).

Knockout of α4 eliminated most high-affinity nicotine binding in brain and greatly reduced nicotine-induced antinociception (Marubio et al., 1999; see also Chapter 9 BV). It is known that both α4 and α6 presynaptic nAChRs contribute to nicotine-induced release of dopamine from striatal synaptosomes (Kulak et al., 1997; Zoli et al., 2002; Mansvelder and McGehee, 2002). Thus, it was surprising to discover that α4 knockout mice have striatal dopamine levels twice as high as controls (Marubio et al., 2003). These results suggest that α4 nAChRs are also involved in tonic inhibition of dopamine release and that the net effect of α4 knockout, or of the sustained desensitization expected of α4β2 AChRs in the presence of nicotine concentrations known to be sustained in smokers, is to inhibit this inhibition, resulting in increased dopamine release.

Knockin of an α4 subunit with an M2 mutation that makes it hyperactive showed increased sensitivity to nicotine-induced seizures (Labarca et al., 2001; Fonck et al., 2003). These mice showed progressive loss of dopaminergic neurons in the substantia nigra, presumably due to excitotoxic activation induced by activation through endogenous choline. They also exhibited increased anxiety and decreased learning.

Mouse strains that had been initially selected as long-sleep and short-sleep forms were found to have an A529T polymorphism in the large cytoplasmic domain of their α4 subunits (Kim et al., 2003). Mouse thalamus appears to contain a mixture of high-nicotine-sensitivity $(\alpha4)_2(\beta2)_3$ and low-nicotine-sensitivity $(\alpha4)_3(\beta2)_2$ stoichiometry AChRs. Strains with the T529 variant associated with an increased proportion of the high-sensitivity stoichiometry exhibited increased sensitivity to nicotine-induced seizures, respiration rate, body temperature, and Y maze performance.

Knockout of the α5 subunit dramatically reduced sensitivity to nicotine-induced behaviors and seizures (Salas et al., 2003a: see also Chapter

9 BIV-3). This suggests that nAChR subtypes containing α5, perhaps $(\alpha4)_2(\beta2)\alpha5$ and $\alpha4\alpha6(\beta2)_2\alpha5$, are especially important for these behaviors. In autonomic ganglia, $(\alpha3)_2(\beta4)_2\alpha5$ and $(\alpha3)_2(\beta2)_2\alpha5$ postsynaptic nAChRs would be expected to be present. Knockout of α5 also had modest autonomic effects (Wang et al., 2002a, 2002b). Due to impaired ganglionic transmission, α5 knockout mice were resistant to cardiac arrest at high-frequency vagal stimulation. Ganglionic transmission was more sensitive to channel block by hexamethonium in the knockout mice.

Knockout of α6 subunits produced no gross behavioral or anatomical effects but eliminated binding of α conotoxin M2 in the brain (Champtiaux et al., 2002). It was suggested that substitution of $(\alpha4)_2(\beta2)_2\alpha5$ nAChRs for $\alpha6\alpha4(\beta2)_2\alpha5$ nAChRs in the knockout compensated for expected losses in nicotine binding sites. More extensive behavior characterization of these mice should prove interesting.

Knockout of α7 AChRs produced remarkably few obvious effects given the wide distribution of this subunit in the central and peripheral nervous systems (Orr-Urtreger et al., 1997). Brain α bungarotoxin binding sites and rapidly desensitizing nicotinic currents in the hippocampus were lacking in α7 knockout mice. Small autonomic effects were seen (Franceschine et al., 2000). Decreased baroreflex-mediated tachycardia in α7 knockouts indicated participation of α7 in the autonomic reflex that maintains blood pressure homeostasis. The α7 knockout mice do not breed well. Perhaps this is related to the observation that α7 AChRs on sperm are involved in the acrosome reaction required for egg fertilization (Bray et al., 2002).

Knockin mice with α7 subunit M2 mutations that produce hyperactive AChRs died at birth (Orr-Urtreger et al., 2000). They exhibited extensive apoptotic cell death throughout the somatosensory cortex.

Knockout of α9 subunits prevents cholinergic modulation of cochlear hair cell function (Vetter et al., 1999).

The most widely expressed nAChR subunits in brain are β2 subunits, and β2 knockouts have exhibited some of the most interesting behavioral effects (Picciotto et al., 1995, 1998; Zoli et al., 1999b; Rossi et al., 2001; Cohen et al., 2002; Granon et al., 2003). These mice appear basically healthy but show increased neuronal cell death and loss of cognitive function with age (Zoli et al., 1999b; see also Chapter 9 BV). This suggests the possibility of normal neuroprotective effects of these nAChRs. Neuroprotective effects of nAChRs can be demonstrated against excitotoxicity and loss of trophic factors. The β2 knockouts lack high-affinity nicotine binding sites (i.e., α4β2 AChRs and others), and they are resistant to developing nicotine self-administration (Picciotto et al., 1998). Nicotine-induced release of striatal dopamine does not occur in the knockouts. They show reduced nicotine-induced antinociception (Marubio et al., 1999). Arousal from sleep and breathing drives were accentuated in the knockout (Cohen et al., 2002). It was suggested that this reflected β2-containing AChRs in the carotid body, which are usually involved in breathing responses in response to hypoxia, and that nicotine stimulation of these nAChRs suppresses breathing and may contribute to sudden infant death syndrome. Abnormally high passive avoidance (fear-associated learning) is present in β2 knockout mice, and β2 knockout mice lack a nicotine-induced increase in this behavior shown by normals (Picciotto et al., 1995). It has been reported that β2 knockout mice exhibit deficits in executive and social behavior that resemble autism and attention deficit hyperactivity disorder (Granon et al., 2003). Inducible expression of β2 in corticothalamic efferents of β2 knockout mice reveal presynaptic α4β2 AChRs that mediate normal passive avoidance behavior (King et al., 2003).

Autonomic effects are observed with the knockout of β2 and β4 subunits. Knockout of only β2 subunits has no obvious effects on autonomic transmission. Knockout of only β4 subunits prevents nicotine-induced contraction of bladder smooth muscle but does not cause obvious autonomic impairment *in vivo* (Wang et al., 2003). Effects were similar to those in the α5 knockout (Wang et al., 2002a). Knockout of both β2 and β4 prevents all transmission through autonomic postsynaptic α3 AChRs, causing perinatal death by mechanisms similar to those seen with the α3 knockout (Xu et al., 1999b).

Central effects are also found in β4 knockout mice (Salas et al., 2003b). They appear less anxious than controls on elevated-plus and staircase mazes, but more anxious during social isolation.

Increasingly sophisticated knockout, knockin, and conditional expression studies in combination with ligand binding, histological, and electrophysiological studies and a sophisticated battery of behavioral tests should provide increasingly detailed clues to the physiological roles of nAChR subtypes in mice. These studies will help to guide histological, pharmacological, and clinical studies needed to reveal the physiological roles of nAChR subtypes in humans.

References

Akabas, M. and Karlin, A.: Identification of acetylcholine receptor channel-lining residues in the M1 segment of the α subunit. Biochemistry *34*: 12496–12500, 1995.

Akabas, M., Kaufmann, C., Archdeacon, P. and Karlin, A.: Identification of acetylcholine receptor channel-lining residues in the entire M2 segment of the α subunit. Neuron *13*: 919–927, 1994.

Alkondon, M. and Alquerque, E.: Nicotinic acetylcholine receptor α7 and α4β2 subtypes differentially control GABA ergic input to CA1 neurons in rat hippocampus. J. Neurophysiology *86*: 3043–3055, 2001.

Arredondo, J., Nguyen, V., Chernyavsky, A., Bercovich, D., Orr-Urtreger, A., Kummer, W., Lips, K., Veter, D. and Grando, S.: Central role of α7 nicotinic receptor differentiation of the stratified squamous epithelium. J. Cell Biol. *159*: 325–336, 2002.

Arredondo, J., Nguyen, V., Chernyavsky, A., Bercovich, D., Orr-Urtreger, A., Vetter, D. and Grando, S.: Functional role of α7 nicotinic receptor in physiological control of cutaneous homeostasis. Life Sci. *72*: 2063–2067, 2003.

Auerbach, A. and Akk, G.: Desensitization of mouse nicotinic acetylcholine receptor channels: a two gate mechanism. J. Gen. Physiol. *112*: 181–197, 1998.

Bargman, C.: Neurobiology of the C. elegans genome. Science *282*: 2028–2033, 1998.

Bennett, M.: The concept of transmitter receptors: 100 years on. Neuropharmacology *39*: 523–546, 2000.

Benowitz, N.: Pharmacology of nicotine: addiction and therapeutics. Annu. Rev. Pharmacol. Toxicol. *36*: 597–613, 1996.

Berg, D. and Conroy, W.: Nicotinic α7 receptors: synaptic options and downstream signaling in neurons. J. Neurobiol. *53*: 512–523, 2002.

Berg, D., Shoop, R., Chang, K. and Cuevas, J.: Nicotinic acetylcholine receptors in ganglionic transmission. Handbook Exp. Pharmacol. *144*: 247–267, 2000.

Beroukhim, R. and Unwin, N.: Three dimensional location of the main immunogenic region of the acetylcholine receptor. Neuron *15*: 323–331, 1995.

Bray, C., Son, J.-H. and Meizel, S.: A nicotinic acetylcholine receptor is involved in the acrosome reaction of human sperm initiated by human recombinant ZP3. Biol. Reprod. *67*: 782–788, 2002.

Brejc, K., van Dyk, W., Klaassen, R., Schurmmans, M., van der Oost, J., Smit, A. and Sixma, T.: Crystal structure of an ACh-binding protein reveals the ligand-binding domain of nicotinic receptors. Nature *411*: 269–276, 2001.

Champtiaux, N. and Changeux, J.-P.: Knock-out and knock-in mice to investigate the role of nicotinic receptors in the central nervous system. Current Drug Targets—CNS and Neurological Disorders *1*: 319–330, 2002.

Champtiaux, N., Han, Z.-Y., Rossi, F., Zoli, M., Marubio, L., MacIntosh, J. and Changeux, J.-P.: Distribution and pharmacology of α6-containing nicotinic acetylcholine receptors analyzed with mutant mice. J. Neurosci. *22*: 1208–1217, 2002.

Champtiaux, N., Gotti, C., Cordero-Erausquien, M., David, D., Przbylski, C., Lena, C., Clementi, F., Moretti, M., Rossi, F., LeNovere, N., MacIntosh, J., Gardier, A. and Changeux, J.-P.: Subunit composition of functional nicotinic receptors in dopaminergic neurons investigated with knock-out mice. J. Neurosci. *23*: 7820–7829, 2003.

Changeux, J.-P.: Functional architecture and dynamics of the nicotinic acetylcholine receptor: an allosteric ligand-gated ion channel. In: "Fidia Research Neuroscience Award Lectures," vol. 4, pp. 21–168, Raven Press, New York, 1990.

Clementi, F., Court, J. and Perry E.: Involvement of neuronal nicotinic receptors in disease. Handbook Exp. Pharmacol. *144*: 751–778, 2000.

Cohen, G., Han, Z.-Y., Grailhe, R., Gallego, J., Gaultier, C., Changeux, J.-P. and Lagercrantz, H.: β2 nicotinic acetylcholine receptor subunit modulates protective responses to stress: a receptor basis for sleep-disordered breathing after nicotine exposure. Proc. Natl. Acad. Sci. U.S.A. *9*: 13272–13277, 2002.

Conroy, W. and Berg, D.: Neurons can maintain multiple classes of nicotinic acetylcholine receptors distinguished by different subunit compositions. J. Biol. Chem. *270*: 4424–4431, 1995.

Conroy, W., Vernallis, A. and Berg, D.: The α5 gene product assembles with multiple acetylcholine receptor subunits to form distinctive receptor subtypes in brain. Neuron *9*: 1–20, 1992.

Conroy, W., Liu, Z., Nai, Q., Coggan, J. and Berg, D.: PDZ-containing proteins provide a functional postsynaptic scaffold for nicotinic receptors in neurons. Neuron *38*: 759–771, 2003.

Conti-Tronconi, B., Tzartos, S. and Lindstrom, J.: Monoclonal antibodies as probes of acetylcholine receptor structure. II. Binding to native receptor. Biochemistry *20*: 2181–2191, 1981.

Dale, H.: The action of certain esters and ethers of choline, and their relation to muscarine. J. Pharmacol. Exp. Ther. *6*: 147–190, 1914.

Dale, H.: The beginning and the prospects of neurohumoral transmission. Pharmacol. Rev. *4*: 7–13, 1954.

Dani, J.: Roles of dopamine signaling in nicotine addiction. Mol. Psychiatry *8*: 255–256, 2003.

Dani, J. and Heinemann, S.: Molecular and cellular aspects of nicotine abuse. Neuron *16*: 905–908, 1996.

Dani, J., Ji, D. and Zhou, F.-M.: Synaptic plasticity and nicotine addiction. Neuron *31*: 349–352, 2001.

Das, M. K. and Lindstrom, J.: Epitope mapping of antibodies to acetylcholine receptor alpha subunits using peptides synthesized on polypropylene pegs. Biochemistry *30*: 2470–2477, 1991.

Davila-Garcia, M., Musachio, J. and Kellar, K.: Chronic nicotine administration does not increase nicotinic receptors labeled by [^{125}I] epibatidine in adrenal gland, superior cervical ganglia, pineal or retina. J. Neurochem. *85*: 1237–1246, 2003.

Debiasi, M.: Nicotinic mechanisms in the autonomic control of organ systems. J. Neurobiol. *53*: 568–579, 2002.

DiAntonio, A., Peterson, S., Heckmann, M. and Goodman, C.: Glutamate receptor expression regulates quantal size and quantal content at the Drosophila neuromuscular junction. J. Neurosci. *19*: 3023–3032, 1999.

Dowell, C., Olivera, B., Garrett, J., Staheli, S., Walkins, M., Kuryatov, A., Yoshikami, D., Lindstrom, J. and MacIntosh, M.: A conotoxin PIA is selective for α6 subunit-containing nicotinic acetylcholine receptors. J. Neurosci., 2003.

Elgoyhen, A., Johnson, D., Boulter, J., Vetter, D. and Heinemann, S.: α9: an acetylcholine receptor with novel pharmacological properties expressed in rat cochlear hair cells. Cell *79*: 705–715, 1994.

Elgoyhen, A., Vetter, D., Katz, E., Rothlin, C., Heinemann, S., and Boulter, J.: α10: A determinant of nicotinic cholinergic receptor function in mammalian vestibular and cochlear mechanosensory hair cells. Proc. Natl. Acad. Sci. U.S.A. *98*: 3501–3506, 2001.

Elsele, J.-L., Bertrand, S., Galzi, J.-L., Devilliers-Thiery, A., Changeux, J.-P. and Bertrand, D.: Chimaeric nicotinic-serotinergic receptor combines distinct ligand binding and channel specificities. Nature *366*: 479–483, 1993.

Engel, A., Ohno, K. and Sine, S.: Sleuthing molecular targets for neurological diseases at the neuromuscular junction. Nature Rev. Neurosci. *4*: 339–352, 2003.

Ezzati, M. and Lopez, A.: Estimates of global mortality attributable to smoking in 2002. Lancet *362*: 847–852, 2003.

Fenster, C., Beckman, M., Parker, J., Sheffield, E., Whitworth, M., Quick, M. and Lester, R.: Regulation of α4β2 nicotinic receptor desensitization by calcium and protein kinase C. Mol. Pharm. *55*: 432–443. 1999.

Flores, C., Rogers, S., Pabreza, L., Wolfe, B. and Kellar, K.: A subtype of nicotinic cholinergic receptor in rat brain is composed of α4 and β2 subunits and is upregulated by chronic nicotine treatment. Mol. Pharm. *41*: 31–37, 1992.

Flores, C., Davila-Garcia, M., Ulrich, Y. and Kellar, K.: Differential regulation of neuronal nicotinic receptor binding sites following chronic nicotine administration. J. Neurochem. *69*: 2216–2219, 1997.

Fonck, C., Nashmi, R., Deshpande, P., Damaj, M., Marks, M., Riedel, A., Schwarz, J., Collins, A., Labarca, C. and Lester, H.: Increased sensitivity to agonist-induced seizures, strab tail, and hippocampal theta rhythm in knock-in mice carrying hypersensitive α4 nicotinic receptors. J. Neurosci. *23*: 2582–2590, 2003.

Franceschine, D., Orr-Urtreger, A., Yu, W., Mackey, L., Bond, R., Armstrong, D., Patrick, J., Beaudet, A. and DeBiasi, M.: Altered baroreflex responses in α7 deficient mice. Behav. Brain Res. *1113*: 3–10, 2000.

Froehner, S.: Identification of exposed and buried determinants of the membrane bound acetylcholine receptor from *Torpedo californica*. Biochemistry *20*: 4905–4915, 1981.

Fruchart-Gaillard, C., Gilquin, B., Antil-Delbeke, S., LeNovere, N., Tamiya, T., Corringer, P.-J., Changeux, J.-P., Menez, A. and Servent, D.: Experimentally based model of a complex between a snake toxin and the α7 nicotinic receptor. Proc. Natl. Acad. Sci. U.S.A. *99*: 3216–3221, 2002.

Fuchs, P.: Synaptic transmission at vertebrate hair cells. Current Opinion in Neurobiology *6*: 514–519, 1996.

Galzi, J.-L., Devillers-Thiery, A., Hussy, N., Bertrand, S., Changeux, J.-P. and Bertrand D.: Mutations in the channel domain of a neuronal nicotinic receptor convert ion selectivity from cationic to anionic. Nature *359*: 500–505, 1992.

Gao, F., Bren, N., Little, A., Wang, H.-L., Hansen, S., Talley, T., Taylor, P. and Sine, S.: Curariform antagonists bind in different orientations to acetylcholine-binding protein. J. Biol. Chem. *278*: 23020–23026, 2003.

Gerzanich, V., Anand, R. and Lindstrom, J.: Homomers of α8 and α7 subunits of nicotinic receptors exhibit similar channel but contrasting binding site properties compared to α7 homomers. Mol. Pharm. *45*: 212–220, 1994.

Gerzanich, V., Kuryatov, A., Anand, R. and Lindstrom, J.: "Orphan" α6 nicotinic AChR subunit can form a functional heteromeric acetylcholine receptor. Mol. Pharm. *51*: 320–327, 1997.

Gerzanich, V., Wang, F., Kuryatov, A. and Lindstrom, J.: α5 subunit alters desensitization, pharmacology, Ca^{++} permeability and Ca^{++} modulation of human neuronal α3 nicotinic receptors. J. Pharmacol. Exp. Ther. *286*: 311–320, 1998.

Goldner, F., Dineley, K. and Patrick, J.: Immunohistochemical localization of the nicotinic acetylcholine receptor subunit α6 to dopaminergic neurons in the substantia nigra and ventral tegmental area. Neuroreport *8*: 2739–2742, 1997.

Grando, S.: Autoimmunity to keratinocyte acetylcholine receptors in pemphigus. Dermatology *201*: 290–295, 2000.

Grando, S. and Horton, R.: The keratinocyte cholinergic system with acetylcholine as an epidermal cytotransmitter. Curr. Opin. Dermatol. *4*: 262–268, 1997.

Granon, S., Fauro, P. and Changeux, J.-P.: Executive and social behaviors under nicotinic regulation. Proc. Natl. Acad. Sci. U.S.A. *100*: 9596–9601, 2003.

Grutter, T., Carvalho, L., LeNovere, N., Corringer, P.-J., Edelstein, S. and Changeux, J.-P.: An H-bond between two residues from different loops of the acetylcholine binding site contributes to the activation mechanism of nicotinic receptors. EMBO J. *22*: 1990–2003, 2003.

Gunderson, C., Lehmann, C., Sidell, F. and Jabbari, B.: Nerve agents: a review. Neurology *42*: 946–950, 1992.

Han, Z.-Y., Zoli, M., Cardona, A., Bourgeois, J.-P., Changeux, J.-P. and LeNovere, N.: Localization of $[^3H]$ nicotine, $[^3H]$ cytisine, $[^3H]$ epibatidine, and $[^{125}I]$ α bungarotoxin binding sites in the brain of *Macaca mulatta*. J. Comp. Neurol. *461*: 49–60, 2003.

Hansen, S., Radic, Z., Talley, T., Molles, B., Deerinck, T., Tsigelny, I. and Taylor, P.: Tryptophan fluorescence reveals conformational changes in acetylcholine binding protein. J. Biol. Chem. *277*: 41299–41302, 2002.

Heinemann, S., Boulter, J., Connolly, J., Deneris, E., Duvoisin, R., Hartley, M., Hermans-Borgmeyer, I., Hollmann, M., O'Shea-Greenfield, A., Papke, R., Rogers, S. and Patrick, J.: The nicotinic receptor genes. Clin. Neuropharmacol. *14*: S45–S61, 1991.

Jeanclos, E., Lin, L., Treuiel, M., Rao, J., DeCoster, M. and Anand, R.: The chaperone protein 14-3-3h interacts with the nicotinic acetylcholine receptor α4 subunit. J. Biol. Chem. *276*: 28281–28290, 2001.

Jones, A., Elgar, G. and Satelle, D.: The nicotine acetylcholine receptor gene family of the pufferfish, *Fugu rubripes*. Genomics *82*: 441–451, 2003.

Kaiser, S., Soliokov, L. and Wonnacott, S.: Presynaptic neuronal nicotinic receptors: pharmacology, heterogeneity, and cellular mechanisms. Handbook Exp. Pharmacol. *144*: 193–211, 2000.

Karlin, A.: Explorations of the nicotinic acetylcholine receptor. Harvey Lectures Series *85*: 71–107, 1991.

Karlin, A.: Emerging structure of the nicotinic acetylcholine receptors. Nature Rev. Neurosci. *3*: 102–114, 2002.

Keller, S. and Taylor, P.: Determinants responsible for assembly of the nicotinic acetylcholine receptor. J. Gen. Physiol. *113*: 171–176, 1999.

Kelly, S., Dunlop, J., Kirkness, E., Lambert, J. and Peters, J.: A cytoplasmic region determines single channel conductance in $5\text{-}HT_3$ receptors. Nature *424*: 321–324, 2003.

Keramidas, A., Moorhouse, A., French, C., Schofield, P. and Barry, P.: M2 pore mutations convert the glycine receptor channel from being anion- to cation-selective. Biophys. J. *79*: 247–259, 2000.

Keyser, K., Britto, L., Schoepfer, R., Whiting, P., Cooper, J., Conroy, W., Brozozowska-Prechtl, A., Karten, H. and Lindstrom, J.: Three subtypes of α bungarotoxin-sensitive nicotinic acetylcholine receptors are expressed in chick retina. J. Neurosci. *13*: 442–454, 1993.

Kim, H., Flanagin, B., Qin, C., McDonald, R. and Stitzel, J.: The mouse Chrna4 A529T polymorphism alters the ratio of high to low affinity α4β2 nAChRs. Neuropharmacology *45*: 345–354, 2003.

King, S., Marka, M., Grady, S., Caldarone, B., Koren, A., Mukhin, A., Collins, A. and Picciotto, M.: Conditional expression in corticothalamic efferents reveals a developmental role for nicotinic acetylcholine receptors in modulation of passive avoidance behavior. J. Neurosci. *23*: 3837–3843, 2003.

Kirchgessner, A. and Liu, M.-T.: Immunohistochemical localization of nicotinic acetylcholine receptors in the guinea pig bowel and pancreas. J. Comp. Neurol. *390*: 497–514, 1998.

Klink, R., deKerchove d'Exaerde, A., Zoli, M. and Changeux, J.-P.: Molecular and physiological diversity of nicotinic acetylcholine receptors in the mid brain dopaminergic nuclei. J. Neurosci. *21*: 1452–1463, 2001.

Kobertz, W., Williams, C. and Miller, C.: Hanging gondola structure of the T1 domain in a voltage-gated K^+ channel. Biochemistry *39*: 10347–10352, 2000.

Kulak, J., Nguyen, T., Olivera, B. and MacIntosh, J.: a Conotoxin MII blocks nicotine-stimulated dopamine release in rat striatal synaptosomes. J. Neurosci. *17*: 5263–5270, 1997.

Kulak, J., MacIntosh, J. and Quick, M.: Loss of nicotinic receptors in monkey striatum after 1-methyl-4-phenyl-1,2,3,6-tetrahydropyridine treatment is due to a decline in α-conotoxin MII sites. Mol. Pharm. *61*: 230–238, 2002.

Kuryatov, A., Olale, F., Cooper, J., Choi, C. and Lindstrom, J.: Human α6 AChR subtypes: subunit composition, assembly, and pharmacological responses. Neuropharm. *39*: 2570–2590, 2000.

Labarca, C., Schwarz, J., Deshpande, P., Schwarz, S., Nowak, M., Fonck, C., Nashmi, R., Kofuji, P., Dang, H., Shi, W., Fidan, M., Khakh, B., Chen, Z., Bowers, B., Boulter, J., Wehner, J. and Lester, H.: Point mutant mice with hypersensitive α4 nicotinic receptors show dopaminergic deficits and increased anxiety. Proc. Natl. Acad. Sci. U.S.A. *98*: 2786–2791, 2001.

Lena, C., deKerchove d'Exaerde, A., Cordero-Erausquin, M., LeNovere, N., del Mar Arroyo-Jiminez, M. and Changeux, J.-P.: Diversity and distribution of nicotinic acetylcholine receptors in the locus ceruleus neurons. Proc. Natl. Acad. Sci. U.S.A. *96*: 12126–12131, 1999.

Lennon, V., Ermilov, L., Szursewski, J. and Vernino, S.: Immunization with neuronal nicotinic acetylcholine receptor induces neurological autoimmune disease. J. Clin. Invest. *111*: 907–913, 2003.

LeNovere, N. and Changeux, J.-P.: Molecular evolution of the nicotinic acetylcholine receptor: an example of multigene family in excitable cells. J. Mol. Evol. *40*: 155–172, 1995.

LeNovere, N., Corringer, P.-J. and Changeux, J.-P.: The diversity of subunit composition in nAChRs: evolutionary origins, physiologic and pharmacologic consequences. J. Neurobiol. *53*: 447–456, 2002a.

LeNovere, N., Grutter, T. and Changeux, J.-P.: Models of the extracellular domain of the nicotinic receptors and of agonist and Ca^{++} binding sites. Proc. Natl. Acad. Sci. U.S.A. *99*: 3210–3215, 2002b.

Lentje, C. and Patrick, J.: Both α and β subunits contribute to the agonist sensitivity of neuronal nicotinic acetylcholine receptors. J. Neurosci. *11*: 837–845, 1991.

Lin, L., Jeanclos, E., Treuil, M., Braunwell, K.-H., Gundelfinger, E. and Anand, R.: The calcium sensor protein visinin-like protein-1 modulates the surface expression and agonist sensitivity of the α4β2 nicotinic acetylcholine receptor. J. Biol. Chem. *277*: 41872–41878, 2002.

Lindstrom, J.: Purification and cloning of nicotinic acetylcholine receptors. In: "Neuronal Nicotinic Receptors: Pharmacology and Therapeutic Opportunities," S. Arneric and D. Brioni, Eds., pp. 3–23, John Wiley and Sons, New York, 1999.

Lindstrom, J.: The structures of neuronal nicotinic receptors. In: "Neuronal Nicotinic Receptors. Handbook of Experimental Pharmacology," F. Clementi, C. Gotti, D. Fornasari, Eds., vol. 144, pp. 101–162, New York, Springer, 2000a.

Lindstrom, J.: Acetylcholine receptors and myasthenia. Muscle Nerve *23*: 453–477, 2000b.

Lindstrom, J.: Autoimmune diseases involving nicotinic receptors. J. Neurobiol. *53*: 656–665, 2002.

Lindstrom, J.: Acetylcholine receptor structure. In: "Myasthenia Gravis and Related Disorders," H. Kaminski, Ed., pp. 15–52, Humana Press, Totowa, NJ, 2003.

Lloyd, G. and Williams, M.: Neuronal nicotinic acetylcholine receptors as novel drug targets. J. Pharmacol. Exp. Ther. *292*: 461–467, 2000.

Lustig, L., Peng, H., Hiel, H., Yamamoto, T. and Fuchs, P.: Molecular cloning and mapping of the human nicotinic acetylcholine receptor a10 (CHRNA 10). Genomics *73:* 272–283, 2001.

Madden, D.: The structure and function of glutamate receptor ion channels. Nature Rev. Neurosci. *3*: 91–100, 2002.

Maimone, M. and Merlie, J.: Interaction of the 43 kd postsynaptic protein with all subunits of the muscle nicotinic acetylcholine receptor. Neuron *11*: 53–66,1993.

Mansvelder, H. and McGehee, D.: Cellular and synaptic mechanisms of nicotine addiction. J. Neurobiol. 53: 606–617, 2002.

Marubio, L., del Mar Arroyo-Jiminez, M., Cordero-Erausquin, M., Lena, C., LeNovere, N., de Kerchove d-Exaerde, A., Huchet, M., Damaj, M. and Changeux, J.-P.: Reduced antinociception in mice lacking neuronal nicotinic receptor subunits. Nature *398*: 805–810, 1999.

Marubio, L., Gardier, A., Durier, S., David, D., Klink, R., Arroyo-Jiminez, M., MacIntosh, J., Rossi, F., Champtiaux, N., Zoli, M. and Changeux, J.-P.: Effects of nicotine in the dopaminergic system of mice lacking the α4 subunit of neuronal nicotinic acetylcholine receptors. Eur. J. Neurosci. *17*: 1329–1337, 2003.

Matsuda, K., Buckingham, S., Kleier, D., Rauh, J., Grauso, M. and Sattelle, D.: Neonicotinoids: insecticides acting on nicotinic acetylcholine receptors. Trends Pharmacol. Sci. *22*: 573–580, 2001.

Maus, A., Pereira, E., Karachunski, P., Horton, R., Navaneetham, D., Macklin, K., Cortes, W., Albuquerque, E. and Conti-Fine, B.: Human and rodent bronchial epithelial cells express functional nicotinic acetylcholine receptors. Mol. Pharm. *54*: 779–788, 1998.

McIntosh, J.: Toxin antagonists of the neuronal nicotinic acetylcholine receptor. Handbook Exp. Pharm. *144*: 455–476, 2000.

Miyazawa, A., Fujiyoshi, Y., Stowell, M. and Unwin, N.: Nicotinic acetylcholine receptor at a 4.6 Å resolution: transverse tunnels in the channel wall. J. Mol. Biol. *288*: 756–786, 1999.

Miyazawa, A., Fijiyoshi, Y. and Unwin, N.: Structure and gating mechanism of the acetylcholine receptor pore. Nature *423*: 949–955, 2003.

Nai, Q., MacIntosh, M. and Margiotta, J.: Relating neuronal nicotinic acetylcholine receptor subtypes defined by subunit composition and channel function. Mol. Pharm. *63*: 311–324, 2003.

Nelson, M., Wang, F., Kuryatov, Choi, C., Gerzanich, V. and Lindstrom, J.: Functional properties of human nicotinic AChRs expressed by IMR-32 neuroblastoma cells resemble those of α3β4 AChRs expressed in permanently transfected HEK cells. J. Gen. Physiol. *118*: 563–582, 2001.

Nelson, M., Kuryatov, A., Choi, C., Zhou, Y. and Lindstrom, J.: Alternate stoichiometries of α4β2 nicotinic acetylcholine receptors. Mol. Pharm. *63*: 332–341, 2003.

Nguyen, V., Ndoye, A., Hall, L., Zia, S., Arrendo, J., Chernyavsky, A., Kist, D., Zelickson, B., Lowry, M. and Grando, S.: Programmed cell death of keratinocytes culminates in apoptotic secretion of a humectant upon secretagogue action of acetylcholine. J. Cell. Sci. *114*: 1189–1204, 2001.

Noda, M., Takahashi, H., Tanabe, T., Toyosato, M., Furutani, Y., Hirose, T., Asai, M., Inayama, S., Miyata, T. and Numa S.: Primary structure of a-subunit precursor of *Torpedo californica* acetylcholine receptor deduced from cDNA sequence. Nature *299*: 793–797, 1982.

Obaid, A., Koyano, T., Lindstrom, J., Sakai, T. and Salzberg, B.: Spatio-temporal patterns of activity in an intact mammalian network with single cell resolution: optical studies of nicotinic activity in an enteric plexus. J. Neurosci. *19*: 3073–3093, 1999.

Olale, F., Gerzanich, V., Kuryatov, A., Wang, F. and Lindstrom, J.: Chronic nicotine exposure differentially affects the function of human α3, α4, and α7 neuronal nicotinic receptor subtypes. J. Pharmacol. Exp. Ther. *283*: 675–683, 1997.

Orr-Urtreger, A., Goldner, F., Saeki, M., Lorenzo, J., Goldberg, L., DeBiasi, M., Dani, J., Patrick, J. and Beaudet, A.: Mice deficient in the α7 neuronal nicotinic acetylcholine receptor lack α bungarotoxin binding sites and hippocampal fast nicotinic currents. J. Neurosci. *17*: 9165–9171, 1997.

Orr-Urtreger, A., Broide, R., Kasten, M., Dang, H., Dani, J., Beaudet, A. and Patrick, J.: Mice homozygous for the L250T mutation in the α7 nicotinic acetylcholine receptor show increased neuronal apoptosis and die within 1 day of birth. J. Neurochem. *74*: 2154–2166, 2000.

Parker, J., Sarkar, D., Quick, M. and Lester, R.: Interactions of atropine with heterologously expressed and native α3 subunit-containing nicotinic acetylcholine receptors. Br. J. Pharmacol. *138*: 801–810, 2003.

Patrick, J. and Lindstrom, J.: Autoimmune response to acetylcholine receptor. Science *180*: 871–872, 1973.

Peng, X., Anand, R., Whiting, P. and Lindstrom, J.: Nicotine-induced increase in neuronal nicotinic receptors results from a decrease in the rate of receptor turnover. Mol. Pharm. *46*: 523–30, 1994.

Peng, X., Gerzanich, V., Anand, R., Wang, F. and Lindstrom, J.: Chronic nicotine treatment up-regulates α3 and α7 acetylcholine receptor subtypes expressed by the human neuroblastoma cell line SH-SY5Y. Mol. Pharm. *51*: 776–784, 1997.

Perry, D., Davila-Garcia, M., Stockmeier, C. and Kellar, K.: Increased nicotinic receptors in brains from smokers: membrane binding and autoradiography studies. J. Pharm. Exp. Ther. *289*: 1545–1552, 1999.

Peto, R., Chen, Z. and Boreham, J.: Tobacco—the growing epidemic. Nature Med. *5*: 15–17, 1999.

Picciotto, M., Zoli, M., Lena, C., Bessis, A., Lallemand, Y., LeNovere, N., Vincent, P., Pich, M., Brulet, P. and Changeux, J.-P.: Abnormal avoidance learning in mice lacking functional high affinity nicotine receptor in the brain. Nature *374*: 65–67, 1995.

Picciotto, M., Zoli, M., Rimondini, R., Lena, C., Marubio, L., Pich, E., Fuxe, K., Changeux, J.-P.: Acetylcholine receptors containing the β2 subunit are involved in the reinforcing properties of nicotine. Nature *391*: 173–177, 1998.

Picciotto, M., Caldarone, B., Bruzell, D., Zacharion, V., Stevens, T. and King, S.: Neuronal nicotinic acetylcholine receptor knockout mice: physiological and behavioral phenotypes and clinical implications. Pharmacol. Ther. *92*: 89–108, 2001.

Quick, M. and Lester, R.: Desensitization of neuronal nicotinic receptors. J. Neurobiol. *53*: 457–478, 2002.

Quick, M., Sum, J., Whiteaker, P., McCallum, S., Marks, M., Musachio, J., MacIntosh, M., Collins, A. and Grady, S.: Differential declines in striatal nicotinic receptor subtype function after nigrostriatal damage in mice. Mol. Pharm. *63*: 1169–1179, 2003.

Raftery, M., Hunkapillar, M., Strader, C. and Hood, L.: Acetylcholine receptor: complex of homologous subunits. Science *208*: 1454–1457, 1980.

Raggenbass, M. and Bertrand, D.: Nicotinic receptors in circuit excitability and epilepsy. J. Neurobiol. *53*: 580–589, 2002.

Richardson, C., Morgan, J., Jasani, B., Green, J., Rhodes, J., Williams, G., Lindstrom, J., Wonnacott, S., Thomas, G. and Smith, V.: Megacystis-microcolon-intestinal hypoperistalsis syndrome (MMIHS) is associated with absence of the α3 nicotinic acetylcholine receptor subunit. Gastroenterology *121*: 350–357, 2001.

Ridley, D., Rogers, A. and Wonnacott, S.: Differential effects of chronic drug treatment on α3 and α7 nicotinic receptor binding sites, in hippocampal neurons and SH-SY5Y cells. Br. J. Pharmacol. *133*: 1286–1295, 2001.

Rodrigues-Pinguet, N., Jia, L., Li, M., Fige, A., Klassen, A., Troung, A., Lester, H. and Cohen, B.: Five ADNFLE mutations reduce the Ca^{++} dependence of the mammalian α4β2 acetylcholine response. J. Physiol. *550*: 11–26, 2003.

Romanelli, M. and Gualtieri, F.: Cholinergic nicotinic receptors: competitive ligands, allosteric modulators, and their potential applications. Med. Res. Rev. *23*: 393–426, 2003.

Rossi, F., Pizzorusso, T., Porciatti, V., Marubio, L., Maffei, L. and Changeux, J.-P.: Requirement of the nicotinic acetylcholine receptor β2 subunit for the anatomical and functional development of the visual system. Proc. Natl. Acad. Sci. U.S.A. *98*: 6453–6458, 2001.

Rothlin, C., Katz, E., Verbitsky, M. and Elgoyhen, B.: The α9 nicotinic acetylcholine receptor shares pharmacological properties with type A γ-aminobutyric acid, glycine, and type 3 serotonin receptors. Mol. Pharm. *55*: 248–254, 1999.

Rothlin, C., Lioredyno, M., Silbering, A., Plazas, P., Casati, M., Katz, E., Guth, P. and Elgoyhen, B.: Direct interaction of serotonin type 3 receptor ligands with recombinant and native α9α10-containing nicotinic cholinergic receptors. Mol. Pharm. *63*: 1067–1074, 2003.

Ruff, R.: Neuromuscular junction physiology and pathophysiology. In: "Current Clinical Neurology: Myasthenia Gravis and Related Disorders," H. Kaminski, Ed., pp. 1–13, Humana Press, Totowa, NJ, 2003.

Saedi, M., Anand, R., Conroy, W. and Lindstrom, J.: Determination of amino acids critical to the main immunogenic region of intact acetylcholine receptors by *in vitro* mutagenesis. FEBS Lett. *267*: 55–59, 1990.

Salas, R., Orr-Urtreger, A., Broide, R., Beaudet, A., Paylor, R. and DeBiasi, M.: The nicotinic acetylcholine receptor subunit α5 mediates short-term effects of nicotine *in vivo*. Mol. Pharm. *63*: 1059–1066, 2003a.

Salas, R., Pieri, F., Fung, B., Dani, J. and DeBiasi, M.: Altered anxiety-related responses in mutant mice lacking the β4 subunit of the nicotinic receptor. J. Neurosci. *23*: 6255–6263, 2003b.

Sargent, P.: The distribution of neuronal nicotinic acetylcholine receptors. Handbook of Exp. Pharm. *144*: 163–192, 2000.

Sato, C., Veno, Y., Asai, K., Takahashi, K., Satoo, M., Engel, A. and Fujiyoshi, Y.: The voltage sensitive sodium channel is a bell-shaped molecule with several cavities. Nature *409*: 1047–1051, 2001.

Schoepfer, R., Conroy, W., Whiting, P., Gore, M. and Lindstrom, J.: Brain α bungarotoxin-binding protein cDNAs and MAbs reveal subtypes of this branch of the ligand-gated ion channel gene superfamily. Neuron *5*: 35–48, 1990.

Sharma, G. and Vijayaraghavan, S.: Nicotinic receptor signaling in nonexcitable cells. J. Neurobiol. *53*: 512–534, 2002.

Sine, S.: The nicotinic receptor ligand binding domain. J. Neurobiol. *53*: 431–446, 2002.

Smit, A., Syed, N., Schaap, D., vanMinnen, J., Klumperma, J., Kits, K., Lodder, H., van der Schors, R., van Elk, R., Sorgendrager, B., Brejc, K., Sixma, T. and Geraerts, W.: A glia-derived acetylcholine-binding protein that modulates synaptic transmission. Nature *411*: 261–268, 2001.

Sokolova, O., Kolmakova-Partensky, L. and Grigorieff, N.: Three dimensional structure of a voltage-gated potassium channel at 2.5 nm resolution. Structure *9*: 215–220, 2001.

Song, P., Sekhon, H., Proskocil, B., Bluztajn, J., Mark, G. and Spindel, E.: Synthesis of acetylcholine by lung cancer. Life Sci. *72*: 2159–2168, 2003.

Steinlein, O.: Neuronal nicotinic receptors in human epilepsy. Eur. J. Pharmacol. *393*: 243–247, 2000.

Sutor, B. and Zolles, G.: Neuronal nicotinic acetylcholine receptors and autosomal dominant nocturnal frontal lobe epilepsy: a critical review. Eur. J. Physiol. *442*: 642–651, 2001.

Swanson, L., Lindstrom, J., Tzartos, S., Schmued, L., O'Leary, D. and Cowan, W.: Immunohistochemical localization of monoclonal antibodies to the nicotinic acetylcholine receptor in chick midbrain. Proc. Natl. Acad. Sci. U.S.A. *80*: 4532–4536, 1983.

Taylor, P., Osaka, H., Molles, B., Keller, S. and Malany, S.: Contributions of studies of the nicotinic receptor from muscle to defining structural and functional properties of ligand-gated ion channels. Handbook Exp. Pharmacol. *144*: 79–100, 2000.

Tomizawa, M., Zhang, N., Durkin, K., Olmstead, M. and Casida, J.: The neonicotinoid electronegative pharmacophore plays the crucial role in the high affinity and selectivity for the Drosophila nicotinic receptor: an anomaly for nicotinoid cation-p interaction model. Biochemistry *42*: 7819–7827, 2003.

Tzartos, S. and Lindstrom, J.: Monoclonal antibodies used to probe acetylcholine receptor structure: localization of the main immunogenic region and detection of similarities between subunits. Proc. Natl. Acad. Sci. U.S.A. *77*: 755–759, 1980.

Tzartos, S., Seybold, M. and Lindstrom, J.: Specificities of antibodies to acetylcholine receptors in sera from myasthenia gravis patients measured by monoclonal antibodies. Proc. Natl. Acad. Sci. U.S.A. *79*: 188–192, 1982.

Tzartos, S., Barkas, T., Cung, M., Mamalaki, A., Marraud, M., Orlewski, P., Papanastasious, D., Sakarellos, C., Sakarellos-Daitsiotis, M., Tsantili, P. and Tsikaris, V.: Anatomy of the antigenic structure of a large membrane autoantigen, the muscle-type nicotinic acetylcholine receptor. Immunol. Rev. *163*: 89–120, 1998.

Unwin, N.: Nicotinic acetylcholine receptor at 9 Å resolution. J. Mol. Biol. *229*: 1101–1124, 1993.

Unwin, N.: Nicotinic acetylcholine receptor and the structural basis of fast synaptic transmission. Philos. Trans. R. Soc. Lond., B, Biol. Sci. *1404*: 1813–1829, 2000.

Utesher, V., Meyer, E. and Papke, R.: Activation and inhibition of native neuronal α-bungarotoxin sensitive nicotinic ACh receptors. Brain Res. *948*: 33–46, 2002.

Vailati, S., Moretti, M., Langhi, R., Rovati, G., Clementi, F. and Gotti, C.: Developmental expression of heteromeric nicotinic receptor subtypes in chick retina. Mol. Pharm. *63*: 1329–1337, 2003.

Vernino, S., Low, P., Fealey, R., Stewart, J., Farreigia, G. and Lennon, V.: Autoantibodies to ganglionic acetylcholine receptors in autoimmune autonomic neuropathies. N. Engl. J. Med. *343*: 847–855, 2000.

Vetter, D., Liberman, M., Mann, J., Barhanin, J., Boulter, J., Brown, M., Saffiote-Kolman, J., Heinemann, S. and Elgoyhen, A.: Role of α9 nicotinic ACh receptor subunits in the development and function of cochlear efferent innervation. Neuron *23*: 93–103, 1999.

Vincent, A., Palace, J. and Hilton-Jones, D.: Myasthenia gravis. Lancet *357*: 2122–2128, 2001.

Wang, F., Gerzanich, V., Wells, G., Anand, R., Peng, X., Keyser, K. and Lindstrom, J.: Assembly of human neuronal nicotinic receptor α5 subunits with α3β2, and β4 subunits. J. Biol. Chem. *271*: 17656–17665, 1996.

Wang, F., Nelson, M., Kuryatov, A., Keyser, K. and Lindstrom, J.: Chronic nicotine treatment up-regulates human α3β2 but not α3β4 acetylcholine receptors stably transfected in human embryonic kidney cells. J. Biol. Chem. *273*: 28721–28732, 1998.

Wang, H.-L., Ohno, K., Milone, M., Brengman, J., Evoli, A., Batocchi, A.-P, Middleton, L., Christodoulou, K., Engel, A. and Sine, S.: Fundamental gating mechanism of nicotinic receptor channel revealed by mutation causing a congenital myasthenic syndrome. J. Gen. Physiol. *116*: 449–462, 2000.

Wang, N., Orr-Urtreger, A., Chapman, J., Rabinowitz, R., Nachman, R. and Korczyn, A.: Autonomic function in mice lacking α5 neuronal nicotinic acetylcholine receptor subunit. J. Physiol. *542*: 347–354, 2002a.

Wang, N., Orr-Urtreger, A. and Korczyn, A.: The role of neuronal nicotinic acetylcholine receptor subunits in autonomic ganglia: lessons from knockout mice. Prog. Neurobiol. *68*: 341–360, 2002b.

Wang, N., Orr-Urtreger, A., Chapman, J., Rabinowitz, R. and Korczn, A.: Deficiency of nicotinic acetylcholine receptor β4 subunit causes autonomic cardiac and intestinal dysfunction. Mol. Pharm. *63*: 574–580, 2003a.

Wang, H., Yu, M., Ochani, M., Amella, C., Tanovic, M., Susaria, S., Li, J., Wang, H., Yang, H., Ulloa, L., Al-Abed, Y., Czura, C. and Tracey, K.: Nicotinic acetylcholine receptor α7 subunit is an essential regulator of inflammation. Nature *421*: 384–388, 2003b.

West, K., Brognard, J., Clark, A., Linnoida, I., Yang, X., Swain, S., Harris, C., Belinsky, S. and Dennis, P.: Rapid Akt activation by nicotine and a tobacco carcinogen modulates the phenotype of normal human airway epithelial cells. J. Clin. Invest. *111*: 81–90, 2003.

Whiting, P. and Lindstrom, J.: Purification and characterization of nicotinic acetylcholine receptor from chick brain. Biochemistry *25*: 2082–2093, 1986a.

Whiting, P. and Lindstrom, J.: Pharmacological properties of immuno-isolated neuronal nicotinic receptors. J. Neurosci. *6*: 3061–3069, 1986b.

Whiting, P. and Lindstrom, J.: Purification and characterization of a nicotinic acetylcholine receptor from rat brain. Proc. Natl. Acad. Sci. U.S.A. *84*: 595–599, 1987.

Whiting, P. J. and Lindstrom, J. M.: Characterization of bovine and human neuronal nicotinic acetylcholine receptors using monoclonal antibodies. J. Neurosci. *8*: 3395–3404, 1988.

Whiting, P., Liu, R., Morley, B. and Lindstrom, J.: Structurally different neuronal nicotinic acetylcholine receptor subtypes purified and characterized using monoclonal antibodies. J. Neurosci. *7*: 4005–4016, 1987.

Williams, B., Temburni, M., Levy, M., Bertrand, S., Bertrand, D. and Jacob, M.: The long internal loop of the α3 subunit targets nAChRs to subdomains within individual synapses on neurons *in vivo*. Nat. Neurosci. *1*: 557–562, 1998.

Wilson, G. and Karlin, A.: The location of the gate of the acetylcholine receptor channel. Neuron *20*: 1269–1281, 1998.

Wilson, G. and Karlin, A.: Acetylcholine receptor channel structure in the resting, open, and desensitized states probed with the substituted-cysteine-accessibility method. Proc. Natl. Acad. Sci. U.S.A. *98*: 1241–1248, 2001.

Xu, W., Gelber, S., Orr-Urtreger, A., Armstrong, D., Lewis, R., Ou, C.-N., Patrick, J., Role, L., DeBiasi, M. and Beaudet, A.: Megacystis, mydriasis, and ion channel defect in mice lacking the α3 neuronal nicotinic acetylcholine receptor. Proc. Natl. Acad. Sci. U.S.A. *96*: 5746–5751, 1999a.

Xu, W., Orr-Urtreger, A., Nigro, F., Gelber, S., Sutcliffe, C., Armstrong, D., Patrick, J., Role, L., Beaudet, A. and DeBiasi, M.: Multiorgan autonomic dysfunction in mice lacking the β2 and the β4 subunits of neuronal nicotinic acetylcholine receptors. J. Neurosci. *19*: 9298–9305, 1999b.

Zhang, H. and Karlin, A.: Identification of acetylcholine receptor channel-lining residues in the M1 segment of the β-subunit. Biochemistry *36*: 15856–15864, 1997.

Zhang, H. and Karlin, A.: Contribution of the β-subunit M2 segment to the ion-conducting pathway of the acetylcholine receptor. Biochemistry *37*: 7952–7964, 1998.

Zhou, M., Engel, A. and Auerbach, A.: Serum choline activates mutant acetylcholine receptors that cause slow channel congenital myasthenic syndromes. Proc. Natl. Acad. Sci. U.S.A. *96*: 10466–10471, 1999.

Zhou, Y., Nelson, M., Kuryatov, A., Choi, C., Cooper, J. and Lindstrom, J.: Human α4β2 AChRs formed from linked subunits. J. Neurosci., 2003.

Zoli, M., Jansson, Sykova, E., Agnati, L. and Fuxe, K.: Volume transmission in the CNS and its relevance for neuropsychopharmacology. Trends Pharmacol. Sci. *20*: 142–150, 1999a.

Zoli, M., Picciotto, M., Ferrari, R., Cocchi, D. and Changeux, J.-P.: Increased neurodegeneration during ageing in mice lacking high-affinity nicotine receptors. EMBO J. *18*: 1235–1244, 1999b.

Zoli, M., Moretti, M., Zanardi, A., MacIntosh, A., Clementi, F. and Gotti, C.: Identification of the nicotinic receptor subtypes expressed on dopaminergic terminals in the rat striatum. J. Neurosci. *22*: 8785–8789, 2002.

7

Anticholinesterases and War Gases

A. Historical Introduction

Cholinesterase inhibitors, or antiChE agents, are widely used in research. In addition, their medicinal use in humans is prehistoric. AntiChEs were employed as active ingredients in botanical and animal materials for primitive and folk medicine, witchcraft, and ethnotoxicology and used as hunting devices. They were derived from certain Chinese mosses such as *Huperzine serrata*, from the Calabar bean (the runner bean), which contains *Physostigma venenosum*, and from the snowdrop and related Amaryllidaceae and *Galanthus* plants. From the 19th century on, *Physostigma* has been used in research and medicine (see Chapter 11) as galenicals, purified preparations, and synthetic physostigmine. Eventually, antiChEs became the preferred tool in studies aimed at proving cholinergic transmission and exploring its range.

David Nachmansohn, a life-long student of the cholinergic system (Figure 7-1; see Chapters 3, 4 and 9), opined that the ACh transmitter role resulted from pharmacological observations obtained with antiChE agents (Nachmansohn, 1963). Another pioneer of cholinergic studies, John Henry Gaddum (1954), expounded similar views as he stated that antiChE research was the basis for the understanding cholinergic transmission. During the late 19th century, the discovery of organophosphorus (OP) antiChEs and their intense development forwarded these studies.

As inhibitors of cholinesterases, the OP, carbamate, and other antiChE compounds exert pharmacological actions, which can be expected from the potentiation of endogenous or applied ACh and ChE-hydrolyzable cholinomimetics. Initially, contrary to carbamates, investigators thought that OP drugs including di-isopropyl fluorophosphonate (DFP) exert their actions solely via their inhibitory effects on cholinesterases (ChEs; see Koelle and Gilman, 1949; see also section DI of this chapter.) However, current research shows that OP and antiChEs exert direct effects independent of inhibiting ChEs. These effects include morphopathological and morphogenetic actions, and synaptic modulations. The modulatory actions are exerted on transmitters and on cholinomimetics that are not hydrolyzable by cholinesterases.

The present chapter focuses on the classification and structural activity of OP, carbamates, and other antiChEs and describes the mechanisms involved in their inhibition of ChEs. With particular focus on OP war gases, this chapter discusses the toxicity of these drugs and their antidoting as part of the central pharmacology of these compounds. Another topic concerns early and modern views on antiChE action, including the findings concerning effects that are independent of their antiChE action. Finally, the chapter describes the use of OP antiChEs as insecticides and vermicides (Holmstedt, 1972). The toxicological and environmental aspects of this matter are also considered.

1. History of Anticholinesterase Agents Research

a. The Pageant of Physostigma Venenosum, Other Naturally Occurring Anticholinesterases, and Related Compounds

Prehistorically, physostigmine or other antiChEs were used medicinally in Egypt, China,

Figure 7-1. Sir John Eccles and David Nachmansohn, Rio de Janeiro, 1959, at the Symposium on "Bioelectrogenesis." (From Chagas and De Carvalho, 1961.)

and elsewhere (Levey, 1966; Hanin et al., 1991). Their toxic properties were exploited for hunting, and they were employed as antidotes for the ethnographic agent curare. Actually, Holmstedt et al. (1985) tested ethnographic material exhibited in Chicago's Museum of Natural History as "containing a cure for arrow poison [or curare]" and found that it possessed d-tubocurarine rather than anti-curare activities. However, the best-known ethnographic employment of antiChEs was during tribal trials; Calabar bean (*Physostigma venenosum*) seeds (Figure 7-2, see color plate; this taxonomic identification was made by John Hutton Balfour [1861]; see Holmstedt, 1972 and Wassen and Holmstedt, 1963) were used by the Efik people of Old Calabar, a province of Nigeria. A British army medical officer reported that 6 of the ordeal beans, or essere, ground and "macerated in water," were given to those accused of criminal conduct. The criminals who died were presumed guilty and those who survived, innocent (Daniell, 1846; Holmstedt, 1972). Most likely, a person considered "innocent" swallowed the liquid material quickly and without hesitation; a bolus would be formed and could elicit life-saving vomiting.

Figure 7-2. The plant *Physostigma venenosum Balforii*. (From Karczmar et al., 1970.) (See color plate.)

Scottish missionaries such as H. M. Wadell (see Holmstedt, 1972) constituted a link between ethnographic uses of an ordeal bean and its subsequent research. They published their findings in Scotland's Missionary Record and provided the Edinburgh Medical School with a crude ordeal bean. These events constituted the origin of the "scientific phase" concerning the studies of the bean that were initiated by the Edinburgh faculty (see Gaddum, 1962; Holmstedt, 1972). They also were the origin of the audacious self-experimentation with the bean carried out by Robert Christison (1855), who at the time was the Professor of Materia Medica and Clinical Medicine at the University of Edinburgh.

In his first attempt, Christison swallowed 6 beans in the afternoon. The following morning, after noticing a "morphia"-like soporific action, the nutty professor swallowed 12 grains or "the fourth of a seed . . . of . . . ordeal bean . . . which originally weighed forty-eight grains." The effects were nearly disastrous! Within minutes of ingesting the bean, Christison felt a "very decided giddiness," "torpidity," and sleepiness, which strongly resembled effects from using "opium" or "Indian hemp," though "all the while [his mind was so active] that . . . he . . . was not conscious of sleep." Christison proceeded to evacuate "the very energetic poison" by means of "swallowing the shaving water which . . . he [had] just used." Fortunately, this home emetic worked, even though the absorption of the active element of the seed was sufficient to cause alarming symptoms. As ascertained by Christison and two of his medical friends, "The pulse and the action of the heart . . . were . . . very feeble, frequent, and most irregular . . . the prostration great." He also suffered muscular and cardiovascular effects.

Christison was saved by the short-lived nature of the bean's active ingredient; indeed, he was unaware of the seriousness of consuming "essere." While he did carry out some animal experiments using the bean, Christison "did not consider it advisable to study . . . the details of the action . . . of Calabar bean . . . by means of experiments on animals" prior to his self-experimentation. Altogether, the 78-year-old "energetic old man" barely recovered. (Holmstedt, 1972; for de Clermont's 1854 experience with an OP drug, see the following section).

Thomas Fraser, Christison's successor to the Edinburgh chair of Materia Medica, initiated thorough animal studies using the Calabar bean (see Fraser, 1964, 1972). He described respiratory depression, miosis, and muscle paralysis evoked by the bean's extract and was the first investigator to establish the atropine-physostigmine (or extract of the Calabar bean) antagonism. These studies also led to the bean's potential use as a treatment for mydriasis induced by atropine (Robertson, 1863) and possibly treating cholera via its gastrointestinal effects. These early studies (see Hudson, 1873) led subsequently to multiple clinical uses of antiChEs (see Chapter 11). (For a description of Christison's and Fraser's studies concerning the central action of the bean, see Chapter 9).

These medical applications led to an early epidemiological incident concerning the bean. The Calabar bean was brought to Liverpool in 1864 to be used in some kind of medical treatment. Accidentally, 42 children consumed the bean, present in the "heap of rubbish—the sweepings of a ship which recently brought a considerable quantity of . . . the beans . . . from the West Coast of Africa" (Cameron, 1864). Cameron treated the resulting toxicity (which resembled Christison's experience) with "emetics and a plentiful supply of warm water and brandy." Ultimately, one child died. Commenting on this treatment, Holmstedt (1972) remarked, "Dr. Cameron seems to have been entirely unfamiliar with the work in Edinburgh"; otherwise, he would have used atropine in the treatment of the poisoning (see also Kleinwaechter, 1864). Additional cases of human poisoning from the bean were documented in the late 19th century (see Holmstedt, 1972). The central toxicity of the bean or physostigmine and its antagonism by atropine were described not only by Christison and Fraser but also by Bourneville, Gubler and Labbee, and Hudson and Bartholow (see Karczmar, 1970, 1986; Holmstedt, 1959, 1963, 1972).

The two other natural nonphysostigmine antiChEs, huperzine and galanthamine, can also be considered "prehistoric" agents. Huperzines were used for centuries in China as folk medicines in the form of the club moss *Huperzia serrata*, for alleviating aging and memory problems relating to aging (Hanin et al., 1991; Tang et al., 1988). While there is little evidence that *Galanthus* or other Amaryllidaceae were used as folk medicines, a literary search involving a study of the *Odyssey* convinced Plaitakis and Duvoisin (1983) that *Galanthus* or its extract were used by Odysseus and his crew as an antagonist as they become intoxicated by Circe with *Strammonium.*

The late-19th-century experimental work in Germany was facilitated when Jobst and Hesse (1864) isolated the bean's active agent; they called it "physostigmine." A year later, the agent was crystallized by Vee and Leven (1865) in France and called "eserine." There seems to be no legal or chemical pronouncement regarding the employment of the two terms; both "physostigmine" and "eserine" are used, although the term "physostigmine" is more widely recognized today.

It took 50 years following the purification and crystallization of physostigmine to elucidate its structure. While the structural delineation of physostigmine's ring system is credited to the work of Stedman and Barger (1925), the actual presentation of the structure was made10 years earlier by Polonowski and Nitzburg (1915). However, Stedman and Barger (1925) defined the structure and provided a detailed structure-activity relationship for physostigmine and its derivatives. Ten years later, their work led to the first synthesis of physostigmine[1] (Julian and Pike, 1935). Physostigmine structure was definitively settled when Peter Pauling (Petcher and Pauling, 1973)[2] elucidated its crystalline structure. Knowledge of physostigmine's structure and the capability for its synthesis led to the synthesis of many carbamates and related antiChEs. Today, this effort vigorously continues. Actually, Klaus-Bertil Augustinsson (1948), a pioneer of ChE and antiChE studies, demonstrated that ChEs are very sensitive enzymes and many substances (for example, methylene blue, caffeine, choline, gum arabic, and cysteine) inhibit ChEs.

The studies that followed the Edinburgh investigations became international. While German scientists were primarily involved, French, English, Russian, Austrian, and US investigators also conducted important investigations during the 19th century. This work clearly illustrates physostigmine's effectiveness at a number of peripheral sites such as (following Henry Dale's nomenclature) parasympathetic terminals, sympathetic ganglia, and the neuromyal junctions.

Several historical aspects of cholinergic transmission are pertinent for antiChE research. Following the work of Harnack and Witkowski in early 1870's, Heidenhain (1872) Loewi and Mansfeld (1910), Hunt (1915), Führer (1917–1918), and others (see also Karczmar, 1967, 1970) demonstrated that physostigmine and/or the bean extract sensitized parasympathetic and motor sites to subthreshold nerve stimulation; and great John Langley (with T. Kato, 1915; see also chapters 1 B and 4 B) compared the effects of physotigmine on innervated and denervated muscle. Some of these investigators also described the potentiating effect of physostigmine on the response to exogenously administered ACh or to the stimulation of parasympathetic nerves. The central effects of physostigmine were reported by several investigators in the nineteen thirties and forties (see for example Schwietzer and Wright, 1937), following their original description some eighty years earlier by Christison (see above, and Schweitzer and Wright, 1937). For the history of these phenomena and the demonstration of cholinergic transmission at the periphery by Otto Loewi (1921) see chapter 9 and Karczmar (1970). For the proof of central cholinergic transmission and for the 19th- and 20th-century studies of the central actions of physostigmine and other antiChE, see Eccles et al. (1954) and Chapter 9.)

The other aspect of this story concerns ChEs. In 1914, Dale suggested that the evanescence of the effects for intravenously administered ACh resulted from fast-acting, ACh-hydrolyzing enzymes that are present in the blood. Otto Loewi (Engelhart and Loewi, 1930), Stedman (1926), and Mathes (1930) then supported Dale's notion when they demonstrated how physostigmine's effects on the peripheral synapses result from an inhibition of the ChE in the tissues. Loewi (Engelhart and Loewi, 1930) also showed physostigmine's potent inhibition of this enzyme, while Plattner and Hintner (1930) demonstrated that ACh is split by tissue extracts.

Subsequently, ChEs were classified as "a family of enzymes" including synaptic AChE or propionyl ChE and pseudo ChE or butyryl ChE (BuChE; see Chapters 3, 5, and 6). As synthetic analogs of physostigmine and other synthetic antiChEs became available, Zeller and Bisseger (1943), Nachmansohn and Schneeman (1945), Hawkins and Gunther (1946), and Hawkins and Mendel (1949) made an important discovery: they demonstrated that various inhibitors have selective effects on these and other ChEs. For example, Hawkins and Gunther showed that m-hydroxyphenylbenzyltriethyl ammonium bromide (Nu-683) was a selective BuChE inhibitor, while methylhydroxy purines and certain carbamates preferentially inhibited AChE. This early work established a few hundred- or even a few thousand-fold ratios between antiBuChE and antiAChE potency of these compounds. Furthermore, this ratio was greatly extended with subsequently developed compounds (Myers, 1954; Long, 1963; Taylor et al., 2000).

A related aspect involves elucidating the ChE inhibition mechanism by the carbamate and other antiChEs including OP antiChEs. Stedman (1926) demonstrated that the antienzymic action of physostigmine resulted from its carbamate moiety rather than 2 linked pyrrolidine rings (see also Long, 1963). Irwin Wilson (Wilson et al., 1960), David Adams and Victor Whittaker (1950) initially considered physostigmine inhibition of ChE to be a reversible process with a relatively short half-life, as the inhibited ChE recovered its activity upon dilution, which is typical of reversible inhibition. This may have led Koster (1946) and Koelle (1946) to employ physostigmine as a protector of the enzyme from OP antiChEs. Koster's and Koelle's early discovery initiated studies regarding the use of physostigmine in antidoting toxic actions of irreversible OP inhibitors. Today physostigmine and related compounds are components of a preventive "cocktail" used to protect ChEs enzymes against OP and war gas toxicity (see below).

Nevertheless, Wilson et al. (1961) subsequently demonstrated that ChE carbamylation occurs when physostigmine inhibits the enzyme, and following inhibition neither the carbamate moiety nor physostigmine was reconstituted in water; instead, carbamic acid was formed. The absence of recovery in unchanged modes of the carbamate-physostigmine complex was consistent with the definition of irreversible inhibition (see below; see also Usdin, 1970) and current findings support this notion (Holmstedt, 1959, 1963; Taylor et al., 2000).

However, reversible antiChEs also exist. Oximes and NaF were the first reversible compounds discovered; they were described in the 1960s (Krupka, 1966; Usdin, 1970; see below for these compounds' reactivating effects on phosphorylated or acylated enzymes). In addition, many of the miscellaneous compounds (i.e., gum arabic or cysteine) listed by Augustinsson (1948; see above) as antiChEs are reversible inhibitors. Finally, many tissues, including the brain, contain chemically unidentified "naturally occurring" reversible antiChEs (Karczmar and Koppanyi, 1954).

The demonstration of physostigmine's effects and the mode of action for physostigmine and related carbamates were followed by their clinical application in ophthalmology, gastroenterology, myasthenia gravis, anesthesia, and Alzheimer's disease (Peters and Levin, 1977; Remen, 1932; see also Chapters 10 and 11). The use of these compounds as pesticides was initiated in the 1950s (Holmstedt, 1972; Kuhr and Dorough, 1976). Note that physostigmine is a poor insecticide, as repeatedly shown since Thomas Fraser (1864), after noticing that the beans shipped to him from Calabar showed traces of excrement and cocoon material from the *Deiopeia pulchella* moth, demonstrated that the Calabar bean extract did not affect caterpillars. In view of the ordeal bean's "extreme activity," Fraser could not believe that it was possible that "any animal form . . . could be . . . subjected to its influence and still retain hold on life." After injecting the caterpillars with the bean's active principle, he found that indeed the caterpillars were immune to the poison.

This paradox was based on the principle of selective toxicity. In other words, there are carbamates nontoxic to insects but toxic to humans and other animals, and vice versa. The differences in the toxicity result from differences among ChEs in various species and the differences in their response to physostigmine and other inhibitors (see Holmstedt, 1972; Chadwick, 1963). In fact, successfully using carbamate antiChEs as pesticides resulted from developing compounds with low or absent mammalian toxicity (sometimes caused by their rapid metabolism in mammals; see section C, below), high pesticidal action, and rapid environmental degradability. Gysin at Geigy Co. in Switzerland first developed these compounds as insect repellents including both weak repellents and active insecticides (Buxtorf and Spindler, 1954; Gysin, 1955). While carbamates were initially used as repellents, subsequently hundreds of carbamates and related pesticidal compounds were developed as vermicides and fungicides, and thousands of tons of these compounds are used throughout the world (Hayes, 1982; Eto, 1974).

It must be emphasized that certain synthetics that became a source for potent antiChEs were developed long before the synthesis of physostigmine. These compounds are diaminoacridines; they were synthesized in Germany at the beginning of the twentieth century as anthelminthics and antiseptics (see Summers et al., 1988; Thornton and Gershon, 1988). During the Second World War, Rubbo, Albert, and their Australian team (Albert and Glendhill, 1945; Albert et al., 1945) discovered that the antibacterial capacity of the compounds might play a significant role in their country's war efforts. When synthesizing compounds related to diaminocoacridines, they developed monoaminoacridines, which included tetrahydro-5-aminoacridine (THA, tacrine). This compound proved to have no antibacterial activity yet exhibited a central excitatory capacity. Its basicity and capacity of passing from the oil to water phase, and their central nervous system activity, were related to its antiChE action. The Australian team (Shaw and Bentley, 1949; Christie et al., 1958) attempted to exploit the antiChE property of aminoacridines (particularly THA) by using them as antagonists for coma induced by atropinics and barbiturates, ketamine, and morphine-induced depression. (For further information on the relationship between antiChEs and morphine, see Chapter 9 BV.) The Australian team included Sam Gershon, who, following his move from Australia to the Missouri Institute of Psychiatry in St. Louis, extended his THA studies to investigate its action on atropinic psychomimetics (Gershon, 1966). In addition, while in St. Louis, Gershon suggested that Summers use THA for treating Alzheimer's disease, which proved novel and fruitful (Summers et al., 1988).

Demonstrating physostigmine's action, its potential as a tool for cholinergic studies and the rapid development of carbamate and other non-OP antiChEs (motivated in part by their usefulness as insecticides and pesticides, as well as in the treatment of Alzheimer disease and myasthenia gravis (Fisher, 2000; see also Chapter 11 A) led to intense research of these compounds' peripheral or central actions.

b. OP Drugs

OP drugs were first synthesized at the beginning of the 19th century, though there was a lapse of 100 years between this synthesis and the demonstration of the toxic insecticidal and pharmacological potential of the OP agents.

Lassaigne synthesized the first OP drugs as phosphate esters in 1820 (see Eto, 1974; Fest and Schmidt, 1970; and Chambers, 1992), and the Arbusovs, a father-and-son duo living in Sweden, developed a number of methods for OP synthesis (Arbusov, 1906; Arbusov and Arbusov, 1931; Nylen, 1930). Moreover, a certain M. Moschnine, in the laboratories of M. Wurtz (a great French chemist), was the first to synthesize a prototype of OP drugs and war gases, tetraethylpyrophosphate (TEPP; there is no further identification of M. Moschnine, but judging by the name which, before it was franchised, must have been Moschnin, he must have been a Russian). Though the Arbusovs also synthesized TEPP, the credit for Moschnine's original work was given to a student of Wurtz, de Clermont, after he published his own interpretation of the studies (de Clermont, 1954).

De Clermont tasted the clear TEPP liquid. According to Holmstedt (1963), "it is difficult to explain why de Clermont did not succumb to the TEPP he tasted." Bo Holmstedt seemed to regret de Clermont's escape, as he stated, "If de Clermont's death . . . had happened, the toxicity of anticholinesterase agents of the organophosphorus type would have been discovered much earlier than it actually was." In fact, de Clermont lived to the ripe age of 91 years! Holmstedt (1963) also remarked that among those investigating OP drugs, only a few scientists succumbed to their effects, despite the volatility and high toxicity when inhaled. For example, Nylen (1930) worked with TEPP for many years without any knowledge of its toxicity (Holmstedt, 2000). The secret of the matter probably lies in OP instability and easy hydrolysis in water.

German investigators were the main force behind the synthesis of OP drugs (Holmstedt, 1963, 2000). This began with von Hoffman (1873), who actually worked on OP chemistry in England but contributed the synthesis of methylphosphoryl dichloride (the first C-P linked compound) upon his return to Germany. Another German investigator, Michaelis (1903), synthesized OP drugs with important P-CN bonds. Later, Willy Lange of the Berlin University and his student Gerda von Krueger (Lange and Krueger, 1932; see also Holmstedt, 1963, 2000) developed P-F linked OP compounds. Lange also noticed an insecticidal action among the compounds; he also described how he and his coworkers exhibited their toxic effects, which included headaches, a sensitization to light, respiratory depression, hallucinations, and distortions of consciousness ("Bewusztseins-Trübungen"; Lange, personal letter to Bo Holmstedt, April 7, 1952; see Holmstedt, 1963, 2000). Gerhard Schrader of I.G. Farbenindustrie then recorded this toxic action in the 1930s when he was developing OP compounds for their insecticidal potential (see 1952 and 1963 reviews by Schrader). Among these compounds were OP drugs containing fluorines such as DFP (Wirth, 1949).

When the toxic effects of OP agents on animals (and, presumably, on humans) became known to the German scientists (in about 1935; see Holmstedt, 1963), the German government classified the pertinent results as secret, and instigated further work, both synthetic and experimental, at the Militärärzliche Akademie; the department in question was headed by Wolfgang Wirth.

Wolfgang Wirth's story merits special attention. Wirth's doctoral thesis (Habilitation) in Wurzburg dealt with toxicity of OP drugs. At that time, Wirth was a student of Ferdinand Flury, a devoted Hitler admirer and follower, who was interested in and published on, euthanasia for the unworthy (*Lebensunwerten*). Subsequent to his appointment to the academy Wirth became an important leader in the Army Armaments Department (Heereswaffenamt), where his duties included the supervision of Gas Protection Laboratories (Gasschützlaboratorien) in Spandau and in other war-related offices. He was made by Hitler a member of the Scientific Senate for the Army Health Activities (Senat des Heeressanitätwesen), and he was awarded a number of commendations, including the War Merit Service Cross (Kriegsverdienstkreuz; see Klee, 2003).

There seems to be no doubt that in his capacities Wirth was aware of and/or witnessed human experiments with war gases; the question is, did he participate in these experiments (Woelk, 2003; Klee, 2003; Konrad Loeffelholz, letter to Wolfgang Woelk, January 12, 2004)? It was argued by Woelk that Wirth declined active participation in the experiments because he was critical of their scientific merit (Woelk, 2003). But was his reticence due only to the lack of scientific merit of the experiments (Loeffelholz, 1999a, 2000)? In addition, there are Wirth's direct statements concerning experiments in Spandau with skin applications of OP agents (Wirth does not say whether he directed or carried out these experiments; Loeffelholz, 2004). Altogether, there is considerable controversy as to Wirth's moral values and his activities during the Third Reich; indeed, according to some German scientists an attempt has been made to whitewash Wirth's behavior (see Habermann, 1999; Loeffelholz, 1999, 2000; Loeffelholz, letter to E. Habermann, Steptember 16, 1999).

Wirth lived to be 98; after the Second World War and after a one's year "US Internierung," he resumed, very successfully, his academic career, and earned a number of medals and honorary memberships in several German scientific associations (Klee, 2003).

As mentioned, during the late 1930s, the information with regard to OP agents became classified and the German industry was no longer free to develop the OP drugs for commercial use. However, just prior to this clandestine phase of OP drug research, pharmacological investigations of OP compounds were initiated in animals and "perhaps" their antiChE action was recognized (see Holmstedt, 1959, 1963).

After the Germans invaded Poland in September 1939, initiating the Second World War, the German government built a large factory complex in Duhernfurt, in east Germany. This factory concentrated on manufacturing OP agents as war gases, known as G or GB agents. At the peak of its war gas production, some 3,000 workers were employed at the factory (Holmstedt, 1963). After World War II ended, Duhernfurt installations became a part of Russian-occupied Poland, and the Russian army supposedly transported thousands of tons of the nerve gas tabun, a P-CN compound, and smaller amounts of sarin to Russia (Holmstedt, 1963). It must be added that there were long negotiations between Churchill and Roosevelt on one side and Stalin on the other regarding the exact position of the border between Poland and East Germany, and George Koelle (1981) speculated that the bone of contention was actually whether the factory at Durhenfurst should belong to Russia or to the Western alliance (Figure 7-3).

Organophosphorus agents were developed and studied in England and the United States during the 1940s and1950, as these agents became known

Figure 7-3. Alleged argument between the allies at the Yalta peace conference regarding the location of the boundary between Poland and Germany; the question was, Should the Duhernfurt factory belong to Poland or to Germany? (From Koelle, 1981.)

as nerve or war gases. However, whether or not the development of nerve and war gases was an independent effort or was generated by the OP research in Germany still remains in question. Bernard Kilby (1949) stated that the research was initiated by the Ministry of Supply at several sites in England on the basis of reports (published in open literature) by Lange and Krueger (1932); this paper referred to the toxic effects of the compounds. However, Holmstedt (1963) felt that the impetus resulted from a leak of information from Germany, as the British teams were set up by the ministry to pay particular attention to the P-F compounds.

This research and development led to the synthesis of DFP by Saunders (see Saunders, 1957) and to a number of basic findings including demonstrating the antiChE action of OP agents. This finding was obtained by a British team including Adrian (now Lord Adrian), William Feldberg, and Bernard Kilby, along with the American teams working in Edgewood Arsenal, Maryland, at Reed Hospital, and at Georgetown University in Washington, DC. Their work and earlier work by German investigators L. Landle, Eberhard Gross, and others (Holmstedt, 2000; Schrader, 1952) at Farbenfabriken Bayer in Elberfeld and Militärärztliche Akademie in Berlin proved that similar to physostigmine, the pharmacological and toxic effects of OP agents are caused by their inhibitory action on ChEs; potentiation of the endogenous and exogenous ACh action; and the potentiation of response to the stimulation of the appropriate nerves (Koelle and Gilman, 1949; Koppanyi and Karczmar, 1951; Burgen, 1949; see Karczmar, 1967; Bodansky, 1945). During World War II investigations, the irreversible nature of this inhibition was recognized by German and Allied scientists. Mackworth and Webb (1948) were the first to publish a report on the phenomenon. The second step of this inhibition, the "aging" process, contributes to the irreversibility, which was recognized much later (Hobbiger, 1955; Michel, 1958).

The early German and English work established that the initial inhibition was due to phosphorylation (today referred to as acylation; see below, section BI, and Reiner and Radic, 2000) of the esteratic site of either AChE or BuChE molecules (see Chapter 3 DI and DII). Arnold Burgen (1949; now Sir Arnold) was the first to publish research describing the phenomenon.

Organophosphorus antiChEs are easy to develop. During and after the Second World War, thousands of OP drugs were synthesized for their potential as insecticides, G agents, or war gases, and (rarely) as therapeutic drugs such as those used in the treatment of Alzheimer's disease (see Chapter 10). This synthetic explosion led to nomenclative chaos which may be remedied by adherence to terminology developed by Larsson et al. (1954) and Holmstedt (1959; see below). During this development it appeared that, similar to carbamates and other non-OP antiChEs, OP compounds had a selective effect on the various ChEs. For example, Michel (1955) and Jandorf et al. (1955) showed that DFP is more effective against BuChE than against AChE. The reverse is true for G agents such as sarin and soman; the war gas tabun is equally effective against these enzymes (Holmstedt, 1963; Usdin, 1970).

Whether prophylactically or as a treatment for poisoning, atropine's use as an antidote for physostigmine toxicity became significant for German, English, and US investigators when they learned of OP's potential use in war. They studied atropine's antagonism on OP poisoning and several atropine analogs were developed and tested (Wills, 1963). The analogs proved effective, but only to a limited degree, since the best atropinics could protect against only 2 LD50s of potent OP agents such as sarin. Just about any drugs, from barbiturates to ganglionic blockers to tranquilizers and muscle relaxants (Wills, 1963, 1970), were tested and proved ineffective when used with or without atropinics. However, the first break occurred when Koelle (1946) and Koster (1946) employed physostigmine prophylactically as a protector of ChEs from phosphorylation.

An even more important occurrence was the discovery of reactivation, which stemmed from Hestrin's demonstration (1949) that intact AChE and hydroxymates were released from acetylated enzymes upon incubation with hydroxylamine. Irwin Wilson (1951) then used hydroxylamine (which also hydrolyzes OP agents; Hackley et al., 1955) as a "nucleophilic reactivator," a new concept. The reactivation resulted in freeing the enzyme in its intact state and the formation of a phosphoryl product that differed from the original OP drug. However, hydroxylamines were not very effective as prophylactic or antidotal agents.

Wilson, Childs, and Davies (Childs et al., 1955; Wilson et al., 1955) made a significant step when they developed oximes. These investigators, including Edith Heilbronn-Wikstroem, the German-Swedish pioneer in this area (1963, 1964, 1965), worked out schemes for these agents' reactivating action. When working in David Nachmansohn's Columbia University laboratory, Irwin Wilson tailor-designed very active oximes in accordance with considerations of structure-activity relationship; he referred to this relationship as "the law of complementarity" (see Wilson, 1959).[3]

Subsequently, a number of mono- and bisquaternary oximes were developed (see Wills, 1970; Usdin, 1970) and it was found tht unrelated compounds acted as reactivators. Among these them was NaF. Sodium fluoride shows a number of activities. As discovered by Heilbronn (1964), the compound is a reactivator, an antiChE that acts on both AChE and BuChE, and an ACh sensitizer (that is, it facilitates ACh action on cholinoreceptors independently of its antiChE effect). However, only oximes are practical when used as reactivators. Today, oximes are components of a therapeutic "cocktail" (with atropine or atropinics and reversible antiChEs) and are employed by the armies of several nations as prophylactic and therapeutic antidotes of war gas poisoning (see below, sections DI, DII and E).

As to the uses of OP agents, similar to carbamates these agents are employed as insecticides, vermicides, and pesticides. Lange and Krueger (1932) were the first to describe the insecticidal action of these drugs. In 1937, when Gerard Schrader also described the action, I.G. Farbenindustrie patented general structures for OP compounds endowed with potent insecticidal action (Schrader, 1952; see also Holmstedt, 1963). Bladan (tetraethyl pyrophosphate) was the first insecticide used agriculturally, and the popular insecticide parathion was among the early compounds synthesized by Schrader in the 1930s and 1940s. Bladan's use, however, was curtailed because of its potential toxicity to "all forms of life," including humans (Ecobichon, 1996). Among the first OP pesticides synthesized outside of Germany was the insecticide Malathion, which was developed by the American Cyanamide Co. in 1950 (see Eto, 1974); similar to other newer OP synthetics, Malathion showed selective parasite toxicity with, assumedly, no or little human toxicity.

Nevertheless, the post–World War II question concerning dangers of OP drug use as nerve gases, insecticides, or pesticides still remains with us. Though OP agents were not employed as war gases during World War II, they were used in terrorist actions and in local wars, which presently leaves us with potential exposure to this danger (see, for example, Balali-Mood and Shariat, 1998; Yokoyama et al., 1998). Furthermore, the possible use of OP drugs and their antidotes was connected with Gulf War syndrome related to the Persian Gulf War of 1990–1991, a very controversial subject (see section DI).

Although chemists attempted to develop insecticides that are only toxic to pests, cases of human and animal toxicity from OP and carbamate agents were and are frequently reported. During the 1960s, such cases were described in Iran, Egypt, Nigeria, El Salvador, Russia, Canada, Japan, and Australia. Sometimes, mass toxicities occurred, as with antimalarial use of carbamates in Nigeria and other African countries. A similar large-scale poisoning was reported in Egypt: ascaridial epidemics occurred due to flooding resulting from building the Aswan High Dam; the disease was treated with an OP drug, and the treatment caused toxicities (Bo Holmstedt, personal communication). This led the World Health Organization (WHO) and several national and US organizations to establish vigorous evaluation stages for old and new pesticides, and appropriate treatment of their toxicity was proposed. (Oximes may be not effective in carbamate poisoning; see Mileson et al., 1998, and below, section C.)

The United States reacted to the episodes of mass toxicity by reorganizing the Federal Insecticide Fungicide and Rodenticide Act (FIFRA) established by Congress in 1947. The act's administrative authority was then turned over to the newly established Environmental Protection Agency (EPA) in 1972. The EPA and its Office of Pesticide Control have established a number of requirements regarding mandatory assessments concerning the potential toxicity and environmental impact of pesticides prior to commercially producing and marketing any new pesticides (Mileson et al., 1998; Ecobichon, 1996). Recently, the EPA established a need for evaluating the potential toxicity and environmental impact of pesticide combinations.

Ecobichon (1996) stressed that even though insecticides of the OP type are chemically related to OP war gases, the insecticides used today are at least 4 generations of development away from those highly toxic chemicals. He also deemed certain reports as "controversial," including Gershon and Shaw's (1961) report of human toxicity resulting from exposure to OP insecticides. However, Ecobichon (2000, 2001) stated that there was a lack of regulations for OP insecticides in several countries and a lack of execution of these regulations when existent in developing countries where the uses of older, more toxic, and environmentally dangerous pesticides abound. Altogether, Ecobichon and others admitted that toxic and environmental accidents continue to occur frequently (see, for example, Marrs, 1993).

OP agents used as pesticides are extremely valuable for agriculture throughout the world, and this use may be the difference between life and death in underdeveloped countries (Ecobichon, 2000, 2001; Casida and Quistad, 1998). Altogether, the development of antiChEs of OP and other types not only constitute a cornucopia of plenty, but also open up a Pandora's box of calamities: Did Gerhard Schrader foresee this outcome when he was developing many of the OP compounds? Was he aware that the toxic potential of OP drugs was exploited by the German government for war purposes? Perhaps his statement ("Mein innigster Wunsch its es, dass diese Stoffklasse 'Tabun' . . . sich nur segensreich in Form neuer Heilmittel auswirkt. Möge ein guetiges Geschick uns alle davor zu bewahren, dass diese and ähnliche Stoffe zur Zerstörung von Menschenleben eingesetzt werden")? (Schrader, 1950; see Holmstedt, 1959) was only a convenient afterthought.

BI. Classification, Chemical Structures, Structure-Activity Relationships, and Bonding Relationships of Anticholinesterases

1. Naturally Occurring Anticholinesterases, Carbamates, and Related Synthetic Anticholinesterases

Initially, physostigmine, synthetic carbamates, and related synthetic compounds were classified as reversible inhibitors, while OP drugs were considered irreversible antiChEs (Koelle and Gilman, 1949; Holmstedt, 1959; Winteringham, 1966). In 1990, the notion of classifying carbamates as reversible inhibitors was accepted in a textbook chapter (Taylor 1990) and again in a review by Kaur and Zhang (2000). Other naturally occurring antiChEs, huperzines and galanthamine (also spelled "galantamine"; see Brufani and Filocamo, 2000), were also classified as reversible inhibitors (Brufani and Filocamo, 2000). Though some of the compounds are relatively short acting, when compared to OP antiChEs, it is more accurate to consider them irreversible antiChEs (see section A).

Physostigmine is the mother compound of carbamate and related synthetic antiChEs (the old, by now abandoned name for the alkaloid is "eserine"). Polonowski and Nitzburg (1915) and Stedman and Barger (1925) determined physostigmine's structure, and Julian and Pike (1935) accomplished its synthesis. The carbamate moiety of physostigmine is attached to a benzene ring coupled with 2 pyrrolidine rings. The seeds of the Calabar bean contain several physostigmine-related compounds, including geneserine, which has 1 oxidized pyrrolidine ring (Figure 7-4; Polonowski and Nitzburg, 1915).

Figure 7-4. Structures of several carbamate antiChE's.

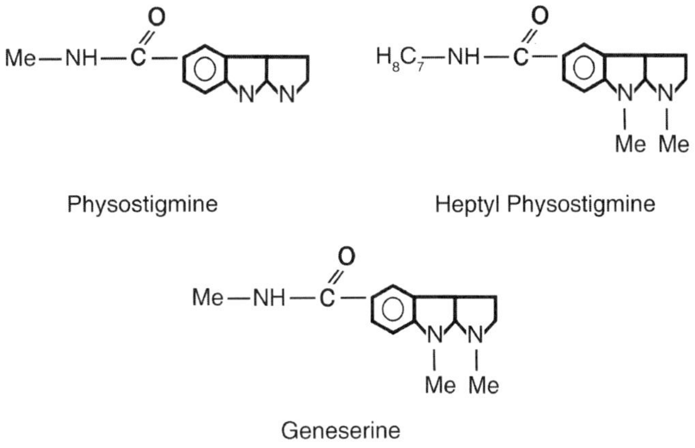

Physostigmine

Heptyl Physostigmine

Geneserine

Stedman demonstrated that pyrrolidine moiety is not essential for physostigmine's activity, and that methyl carbamates with substituted phenols are just as active as, if not more active than, physostigmine (Stedman and Barger, 1925; Stedman, 1926, 1929; see also Long, 1963). These derivatives were characterized by substitutions on the carbamate moiety and included compounds such as nonyl and dimethyl-physostigmine, phenserin, toserin, and heptylphysostigmine (heptastigmine; see Figure 7-4). Subsequently, many years after Stedman's initial studies, several synthetic physostigmine derivatives were developed via substitutions on the carbamate moiety and nucleophilic substitutions at the tricyclic physostigmine framework. These compounds showed a long-lasting antiAChE effect and weaker antiBuChE effects (Brufani et al., 1987, 1988; Kamal et al., 2000; Rege and Johnson, 2003). Other carbamate synthetics have the Calabar alkaloid geneserine as their basis. The Italian team Claudio and Pietra (Trabace et al., 2000) recently synthesized several geneserine derivatives that exhibit potent antiAChE action. One of the derivatives, CHF2819, has been tested for its possible use in Alzheimer's disease (see Chapter 10).

At first, geneserine, pyrrolidine, and nonpyrrolidine drugs were tested in animals as miotics or intestinal stimulants. Then, using methylbutyrate or tributyrin as substrates, the agents were tested as ChE inhibitors and evaluated against BuChEs of the guinea pig liver or serum (see Long, 1963; Usdin, 1970; Main, 1976). Avram Goldstein (1951) evaluated physostigmine's antiChE mechanism of action for plasma ChE (a BuChE), and in the 1960s, Wilson and his associates carried out the first biochemical studies of physostigmine's action on AChEs.

When Stedman and his coworkers (see Stedman, 1926; Long, 1963) realized that pyrrolidine moiety is not essential for antiChE action and that the phenyl carbamate structure of physostigmine can be readily modified, they proceeded to synthesize a series of carbamate antiChEs lacking pyrrolidine rings. There were differences among these compounds. The reciprocal position of the carbamate moiety and the amino N could be varied and there could be different substitutions on the 2 moieties. Finally, a charge on the amino N was present in some compounds and absent in others. These investigations led to Blaschko's development (1949) of the first useful miotic, miotine, or Mestinon, and pyridostigmine (Figure 7-5). Twenty years later, pyridostigmine became a significant AChE protector against OP drugs and an important prophylactic against war gas toxicity (Figure 7-6; see section E). Further synthesis of miotinelike compounds conducted in the 1960s and 1970s led to the development of potent antiChEs (including carbaryl, or Sevin, carbofuran, and bambuterol) that are used as insecticides (Reiner and Radic, 2000). Today, the development of antiChE insecticides continues unabated (Casida and Quistad, 1998).

Some of physostigmine's synthetic analogs (i.e., carbaryl, carbofuran, and rivastigmine) and related compounds (i.e., phenserin or tolserin) are selective inhibitors of AChE (Kamal et al., 2000). However, bambuterol, another related compound, is a selective inhibitor of BuChE. Furthermore,

Phenyl Carbamates and Substituted Benzylamines

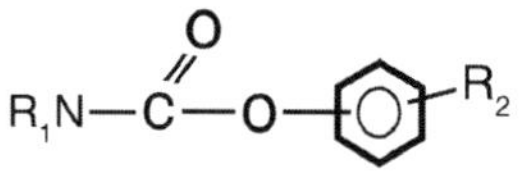

General Structure
R_1 = Aliphatic Substituents
R_2 = Substituted Amines

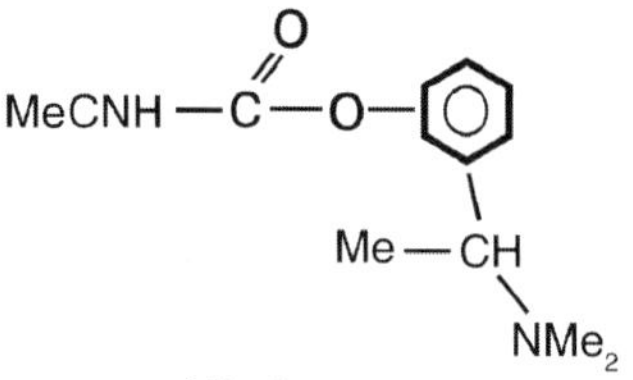

Miotine

Figure 7-5. General structure of substituted benzylamines and structure of miotine.

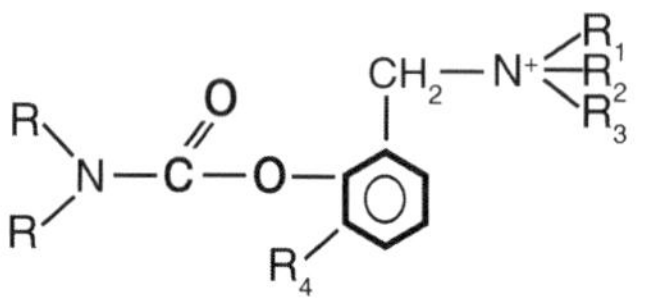

General Structure

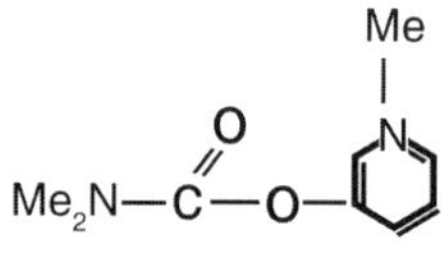

Pyridostigmine

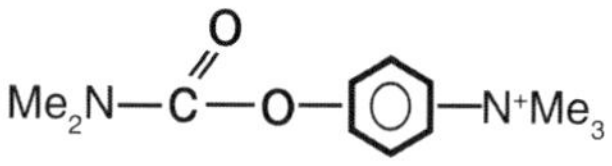

Neostigmine

Figure 7-6. General structure of additional benzylamines and structure of pyridostigmine and neostigmine (prostigmine.).

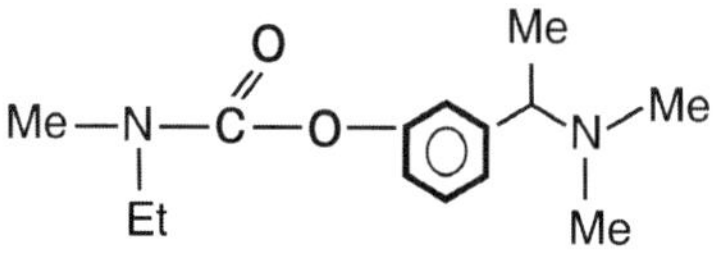

Rivastigmine

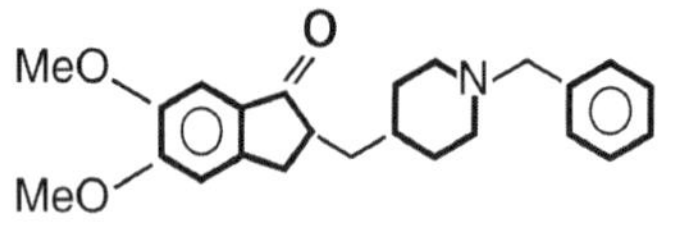

Donepezil (Aricept)

Figure 7-7. Structures of rivastigmine and donepezil (Aricept), drugs used in the therapy of Alzheimer's disease (see Chapter 11).

while carbofuran and carbaryl are currently employed as pesticides, nonpyrrolidine carbamates such as rivastigmine donepezil (Figure 7-7), and ENA 713, (Exelon), are being used in Alzheimer's disease therapy (Anand et al., 1996).

The studies conducted by Stedman, Blaschko, and their associates led to another important finding: the quaternization of m-posited nitrogen intensifies miotic and antiAChE activity (Stedman, 1926; Blaschko et al., 1949; Aeschlimann and Reinert, 1931). The carbamate antiChE neostigmine (prostigmine) is a result of this quaternization (see Figure 7-6). Quaternary compounds such as neostigmine do not penetrate the CNS. However, when overused, these compounds may damage the blood-brain barrier and facilitate their own CNS penetration (see below, section DI). While N-methyl carbamates hydrolyze readily in solution, neostigmine, an N-dimethyl carbamate, is much more stable.

Further modifications of carbamate structures represented by physostigmine, miotine, and neostigmine incorporated replacing the carbamate moiety with a hydroxyl. Then, the hydroxyl was esterified with various organic fatty and nonfatty (i.e., phenyl) radicals (Randall and Lehman, 1950; Hobbiger, 1952; Burgen and Hobbiger, 1951). Other changes involved substituting carbamate groupings with phosphates, replacing methyl groups on the N-cationic head with aliphatic and/or nonaliphatic radicals, and distancing the quaternary head from the phenyl ring.

In a radical departure from earlier investigations, Aeschlimann and Reinert (1931), Wuest and Sakal (1950), and others synthesized and tested compounds lacking the carbamate moiety. Some of these compounds (i.e., pyridines) exhibit nitrogen incorporated in a heterocyclic ring. Much later, Peter Waser and his associates synthesized additional heterocyclic antiChEs (see Waser et al., 1988). Generally, these noncarbamate compounds are weak and short-lasting inhibitors of either AChE or BuChE. One of these compounds, a phenyl ammonium edrophonium, proved to be useful clinically as a diagnostic agent in myasthenia gravis because of its short duration of action. Additionally, several aminopyridines exhibit direct non-antiChE effects on the cholinoceptive receptors of the central nerve terminals (presynaptic receptors; Figure 7-8).

Among the heterocyclics are quinoline and pyridyl (or piperidine) derivatives. They may have either carbamate or noncarbamate substitutions on their heterocyclic ring and may be either tertiary or quaternary compounds (Wuest and Sakal, 1950; Stempel and Aeschlimann, 1954). A heterocyclic antiChE compound with a long and diversified history is an aminoacridine called tetrahydro-5-amino acridine (THA, tacrine; also see section A-1). Ultimately, it was shown that THA is an antiAChE; this finding led to this compound's use in Alzheimer's disease (Summers et al., 1988; see Chapters 9 and 10).

Current studies indicate that a number of compounds related to THA exhibit considerable antiAChE activity (Recanatini et al., 2000). The development of other heterocyclics (i.e., the piperidine derivatives) led to N-benzylpiperidine agents (Sugimoto et al., 1990). Some of these agents were obtained as benzylpiperidine amino acids by a novel procedure of lipase-mediated amidation (Martinez et al., 2000). E2020 or donepezil (Aricept) is a piperidine currently recommended for treating early or moderately advanced Alzheimer's disease (see Figure 7-9 and see Chapters 9 BV and 10 K).

Some of the newer piperidine heterocylic agents (i.e., benzylamino piperidinyl or phthalimide derivatives) include TAK-147. TAK-147 is a compound that allegedly exhibits both a potent antiChE action and neurotrophic effects; therefore, Ishihara et al. (2000) suggested that TAK-147 might be useful for treating Alzheimer's disease. Other piperidine heterocyclics include benzylindoles and benzisoxazoles (Ishihara et al., 1991, 2000; Villalobos et al., 1994; see also Kaur and Zhang, 2000). Among other interesting synthetic heterocyclics are morpholino derivatives of carbamates characterized by a long alkyl side chain that imparts potency and long, quasi-irreversible (or pseudo-reversible) duration of action to the

Noncarbamate Alkylamino and Alkylated Analogues of Edrophonium

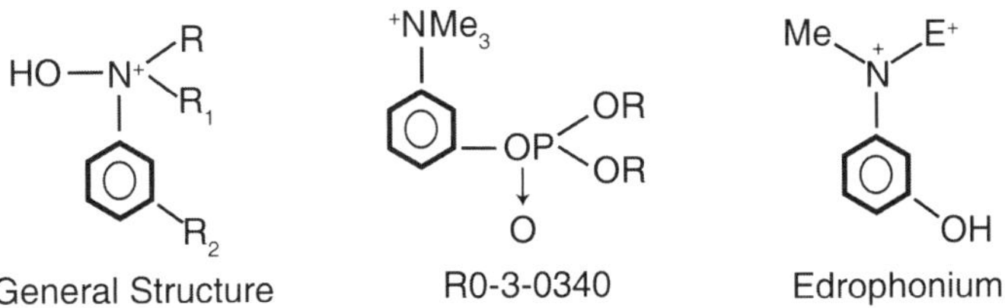

Figure 7-8. General structure of edrophonium, RO-3-0340 and edrophonium.

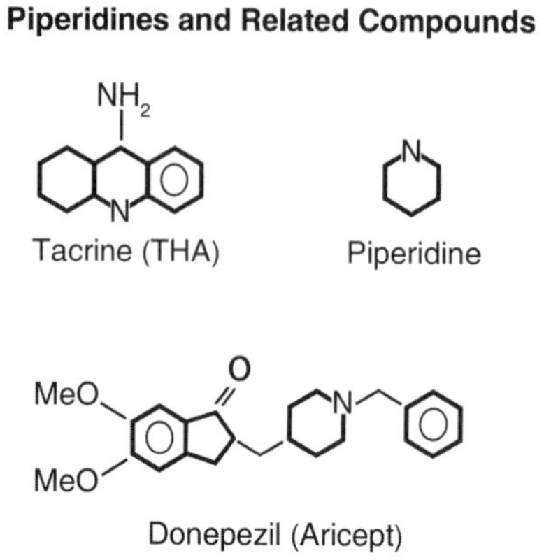

Piperidines and Related Compounds

Tacrine (THA) Piperidine

Donepezil (Aricept)

Figure 7-9. Piperidine structure and several piperidine antiChE's.

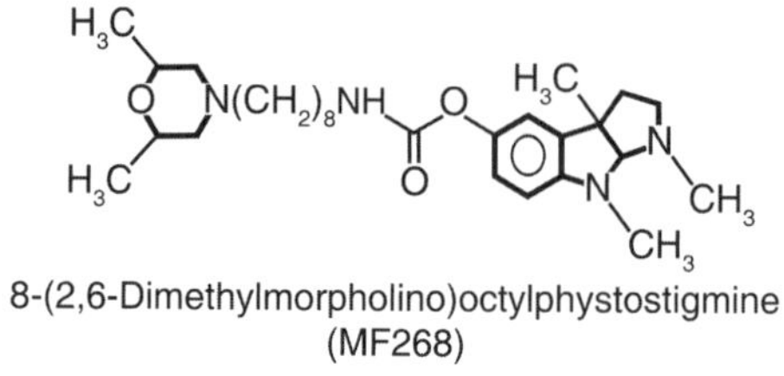

8-(2,6-Dimethylmorpholino)octylphystostigmine (MF268)

Figure 7-10. Amorpholine physostigmine MF268.

molecule. Among the morpholine-physostigmines are potent, brain-penetrating compounds, MF-217 and MF-268, which are being tested for Alzheimer's disease treatment (Perola et al., 1997; Brufani and Filocamo, 2000; see Figure 7-10).

Among the heterocyclics derived from natural plants are huperzines and galanthamines (see also section A-1; Figure 7-11). Huperzines are present in several herbs and plants, including *Huperzia serrata* and *Lycopodium selago* (Felgenhemer et al., 2000). *Huperzia* is a Chinese herb that was used since presumably 2000 BC in China for medicinal purposes (Hanin et al., 1991). Though several huperzines were synthesized, huperzines A and B were most thoroughly investigated (Xia and Kozikowski, 1989; Hanin et al., 1991). Huperzine A is a more potent antiChE than huperzine B; both compounds are more potent against AChE than against BuChE. Moreover, huperzine A penetrates into the brain. Huperzine A was clinically assessed in China for the treatment of human senile memory impairment (not necessarily in Alzheimer's disease). However, huperzine A has recently been used in the United States and Europe in the treatment of Alzheimer's disease (see Chapter 10 K).

The isolation of galanthamine and the demonstration of its antiAChE action occurred in the 1940s. Galanthamine was first isolated in the Soviet Union (Proskurnina and Yakolena, 1947) from the Caucasian snowdrop *Galanthus woronowil* and then in Bulgaria and Japan from related plants of the amaryllidaceae family (see Thomsen et al., 1991; Domino, 1988). Galanthamine is a heterocyclic phenantridine derivative. It can be readily synthesized, and its analogs, including bisquaternaries and compounds with one of its rings being open, can be easily developed (Baron and Kirby, 1962; Guillou et al., 2000; Herlem et al., 2003; see also Domino, 1988). Galanthamine's structure resembles the structure of opioids. In fact, morphine, codeine, and other

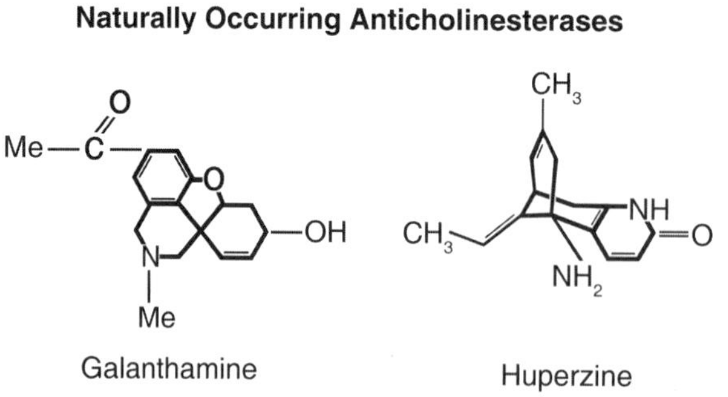

Naturally Occurring Anticholinesterases

Galanthamine Huperzine

Figure 7-11. Structures of naturally occurring antiChE's, galanthamine (galantamine) and huperzine.

opioids are both relatively potent and specific AChE inhibitors (see below). However, galanthamine is several times more potent against antiAChE than morphine (Irwin and Smith, 1960; Boissier et al., 1960). Galanthamine is currently being studied for its potential use in Alzheimer's disease (Thomsen et al., 1990, 1991; Tariot et al., 2000; see also Chapter 10 K).

Ultimately, carbamate and pyrrolidine moieties are not needed for antiChE action. Also, the presence of nitrogen in a heterocyclic ring or a side chain is not required. Among the nonnitrogen compounds are synthetics and naturally occurring antiChEs. The synthetics include phenyl sulfonyl fluorides and trifluoroacetophenones (Figure 7-12). Early on, Fahrney and Gold synthesized phenyl sulfonyl fluorides (1963; see also Barnett and Rosenberry, 1978 and Kraut et al., 2000). Benzenesulfonyl fluoride (BSF) inhibits both mouse brain and *Torpedo electrica* electric organ AChE. Phenylmethylsulfonyl fluorides have just one additional methylene in the bridge between the benzene and the sulfonyl fluoride moiety; they inactivate not only the mouse brain AChE but also mouse brain BuChE (for the significance of this finding, see below). Other synthetic fluorides are the trifluoroacetophenones, which are among the most potent antiAChEs (Reiner and Radic, 2000).

Among naturally occurring nonnitrogen antiChEs are fasciculins. They are the purified toxins of the green mamba snake that are highly potent and highly specific for AChE. Chemically, fasciculins are unusual among the antiChEs; they are "three-finger" peptides (Radic et al., 1994).

Synthesis of bisquaternaries was an important development. French investigators Funke and Depierre (Funke et al., 1952) initiated this development as they synthesized biscarbamate and bisquaternary compounds. Actually, they were not interested in developing bisquaternary agents; they were focused on antiChE properties of the urethane moiety and wished to see whether "deux fonctions urethanes" would not be better than "une fonction urethane." When they linked 2 neostigmines or 2 related carbamates with a hydrocarbon bridge (i.e., 1,3-dihydroxypropane), they immediately noticed that *in vitro*, the compounds exhibited an antiChE activity higher than their monoquaternary analogues. In fact, it was originally reported that these bisquaternaries were active at nanomolar concentrations; this value was downward corrected later. The brain penetrability of bisquaternaries is even more limited than that of monoquaternaries such as neostigmine. Their *in vivo* inhibition of AChE may be demonstrated only at the peripheral cholinergic sites, particularly at the myoneural junction (Karczmar, 1955 unpublished data). Some of these compounds include bisneostigmine and demecarium (Humorsol, BC-48).

Subsequent developments (see Long, 1963) led to the discovery of very active antiChEs (i.e., bisquaternary biscarbamates, bisquinolines, benzoquinoniums and bisquinones, oxamides, bispyridines, biscoumaranyl ketones, and related

Phenyl Sulfonyl Fluorides

Benzene Sulfonyl Fluoride (BSF)

Trifluoroacetophenone

Phenyl Methyl Sulfonyl Fluoride (PMSF)

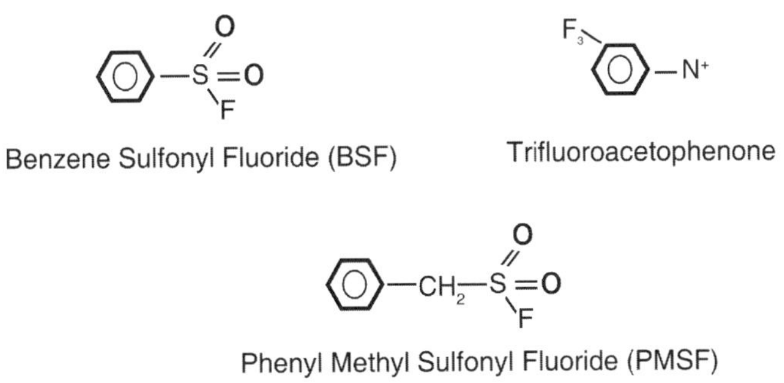

Figure 7-12. Three phenyl sulfonyl antiChEs.

derivatives of disubstituted benzyl derivatives of methylene, ketone, and carbinol). The biscoumarynyl ketones and oxamides are uniquely potent and selective antiAChEs. Oxamides constitute one of the most complete series of bisquaternaries ever studied. This series was developed at Sterling-Winthrop Research Institute (SWRI) in Rensselaer, NY. The development originated with James Hoppe's discovery at SWRI (Hoppe and Arnold, 1952) that a neuromuscular blocking agent called benzoquinonium (Mytolon) exhibited considerable antiAChE activity. A logical transformation of the benzoquinonium structure was its "expansion" into oxamide moiety. Of the oxamides, the benzyl-substituted (especially 2-chlorobenzyl) quaternaries were particularly active as AChE inhibitors. Note that quaternary nitrogens of the benzoquinonium and oxamide derivatives are at a 14 Å distance. This characteristic endows the compounds with the capacity to activate or block (depending on dose) the neuromyal junction, and to exhibit high antiAChE activity. One of these compounds, ambenonium, was and still is employed as a drug useful in myasthenia gravis (see Chapter 10; Lands et al., 1958).

2. Structure-Activity Relationships and Bonding of Carbamate and Related Anticholinesterases

The structure-activity relationship (SAR) investigations dating from the 1920s through 1950s failed to establish a valid structure-activity relationship for carbamate and related compounds, that is, an SAR that would precisely identify the exigencies of bonding between the AChE and BuChE molecules. Part of the problem was that the compounds were not tested in a standard mode. As already indicated, many of the original compounds were tested as miotics or smooth-muscle stimulants rather than ChE inhibitors (see, for example, Stedman, 1926, 1929). When the antiChE activity of the compounds was measured, the ChEs used in the test were either not identified (see Lehmann, 1946) or different ChEs were employed in the assays by different investigators. Some agents were assayed on plasma or liver BuChE, while others were tested with erythrocyte, eel, or brain AChE (see Long, 1963; Main, 1976; Usdin, 1970). Therefore, it was impossible to compare their potencies as antiChE agents and to reach any conclusions regarding their SAR. Altogether, the qualitative SAR methodology employed in the 1950s and 1960s does not approach the exigencies of bona fide quantitative structure-activity relationship (QSRA; see below).

However, some of the insights gained during this early work were meritorious. For example, the work of Stedman (1926), Wuest and Sakal (1950), and others showed that pyrrolidine moiety and carbamate groupings are not necessary for antiChE action. (See later in this section for the modern view of the SAR for physostigmine and its congeners.) Then Wilson and Quan (1958) opined that strong hydrogen bonding to the anionic or active (esteratic) site of AChE was needed for potent antiAChE activity. (Today, the anionic and esteratic sites are considered 2 subsites of the catalytic center of AChE; see Chapter 3 DI and DII). For Irwin Wilson, this SAR illustrated the law of "complementarity" (see above, section A).

Another early insight concerned the potent antiAChE action of bisquaternary compounds. A crucial factor for the bisquaternary compounds (i.e., bisneostigmines, bisisoquinoliniums, and bissubstituted oxamides) was the 14 Å distance between their quaternary heads (see Lands et al., 1957, 1958). To explain this finding, John-Paul Long (1963) suggested that each AChE molecule exhibits 2 bonding sites at a 14 Å distance. The modern view is that bisquaternaries, like ambenonium, combine with the 2 anionic sites: the central site is a component of the catalytic center or gorge, and the peripheral site is located near the gorge (see Chapter 3 DI and DII and below, section C).

John-Paul Long had another early notion: antiChEs (whether bisquarternaries, tertiary, or quaternary monocarbamates) bear some structural resemblance to the ACh molecule. Additionally, Stedman, Aeschlimann and Reinert, Blaschko, Burgen and Hobbiger, Randall and Lehman, and others (see Long, 1963; Holmstedt, 1959; Usdin, 1970) developed SAR generalizations concerning specific classes (e.g., substituted neostigmines and bisneostigmines, pyridine carbamates, and variants of bisquinoliniums and oxamides) of carbamate and related compounds.

In modern studies, both QSAR analysis and stereochemical concepts of anticholinesterase bonding give investigators a better understanding of the attachment of antiChEs to AChE and BuChE.

They also clarify the structures of ChEs. In these studies, the QSARs are established via "Hansch analysis" (Recanatini et al., 2000). Hansch analysis employs a correlation equation for linking ChE inhibitory activity with the parameters that define hydrophobic, electronic, and steric properties of the antiChE molecule. The steric properties relate the three-dimensional shape of the inhibitor's molecule to its binding (or inhibitory) properties. The QSAR also employs a correlation equation for linking the activity with specific substituents. Comparative molecular field analysis (CoMFA) is also relevant for the QSAR (Cramer et al., 1988; see also Recanatini et al., 2000).

Aside from the QSAR equations, software programs such as SYSDOC software (Pang et al., 1994; Kaur and Zhang, 2000) are used to identify the binding sites of antiChEs that are instrumental in "docking." Docking explains how antiChE compounds penetrate and bind within the AChE or BuChE catalytic ("aromatic"; see Chapter 3 DI and DII) gorge. Once AChE structure was visualized by means of these quantitative analyses, the methods of molecular biology were used to modify this structure to represent human AChE, as in the case of the Torpedo electric organ AChE (TcAChE). This in turn made it possible to measure the inhibitory activity of newly developed antiChE compounds with the human enzyme. As the number, role, and dimension of parameters regulating bonding differed from one basic chemical structure to another as well as among series of compounds, a special QSAR equation was developed for each series of antiChEs (Kaur and Zhang, 2000). However, in some cases the application of procedures listed and the QSAR equations failed to explain or predict the antiChE activity for certain series of compounds (Kaur and Zhang, 2000).

In addition to defining the bonding of ligands to AChE and BuChE via QSARs and their equations, various methods concerning the docking and enzymic gorge penetration of antiChEs are employed for visualizing bonds between ChEs and inhibitors. Some of these methods include crystallo-optics and spectroscopic analyses of the enzymes and inhibitor molecules (Sussman et al., 1991), crystallographic x-ray determination, and topographic resolution using appropriate computer programs (DOCK program; Yamamoto et al., 1994). The methods led to the creation of dramatic and esthetic computerized superimpositions illustrating the three-dimensional views of AChE or BuChE molecules binding with their inhibitors (see, for example, Brufani and Filogamo, 2000; Recanatini et al., 2000; Silman and Sussman, 2000; Sugimoto et al., 2000). Altgether, these findings constitute the ultimate vindication of Wilson's law of complementarity.

QSAR studies and the three-dimensional reconstructions of bonding between ChEs and their inhibitors defined the role of substituents within the basic structures of various antiChEs with respect to their antienzymic potency. They also identified the sites involved in binding of antiChEs with AChE and BuChE. Additionally, QSAR studies explored relations between the inhibitors' structure and catalytic moieties of enzymes located at the bottom of the aromatic gorge (see Chapter 3 DI and DII).

At the onset, difficulties in interpreting data arose from our notion of the enzymic gorge. The gorge is 20 Å deep and Sussman et al. (1991) established that the gorge contains an active site about 4.4 Å in diameter (see Chapter 3 DI and DII). So, theoretically only a few compounds may enter the gorge. Actually, Axelsen et al. (1994) suggested that even a rather potent antiAChE, ACh or tetramethylammonium, could not penetrate the gorge. However, penetration could be obtained by trimming the side chains from molecules such as ethopropazine and bambuterol (Reiner and Radic, 2000). It is possible that specific amino acid residues disable certain compounds from penetrating into the gorge, as these residues may block their entry (Kraut et al., 2000).

Sussman, Silman, and associates (Kraut et al., 2000) also proposed that phenylmethylsulfonyl fluoride (PMSF), a potent inhibitor of mouse AChE, cannot inhibit Torpedo AChE (see above, this section), as certain residues in the Torpedo AChE gorge block the entry of PMSF. These residues are absent from the mouse AChE gorge. Since benzenesulfonyl fluoride (BSF) differs for PMSF by lacking 1 methylene, its capability of entering the Torpedo AChE gorge is not hindered; accordingly, BSF inhibits both the Torpedo and the mouse enzyme. This also explains why certain hydrophobic inhibitors of AChE larger than PMSF are capable of entering the gorge and inhibiting the Torpedo AChE.

Since the active BuChE gorge is wider than the AChE gorge (see Chapter 3 DI and DII), ethopropazine and bambuterol are able to penetrate the active BuChE gorge (see Reiner and Radic, 2000). In addition, there are binding sites located outside the active gorge. The presence of highly positive charges and aromatic or hydrophobic substituents on large compounds (i.e., fasciculines and trifluoroacetophenones) facilitate the rapprochement among these compounds and binding sites outside the active gorge. This binding is the first step in the inhibition process, and these compounds are among the most potent antiChEs (Reiner and Radic, 2000). The need for a concentrated positive charge within the inhibitor's molecule relates to the electrostatic dipole character of the catalytic AChE site. The site forms an electric field that is colinear with, or parallel to, the axis of the active gorge (Porschke et al., 1996; Reiner and Radic, 2000; see also Chapter 3 DII and DIII). Accordingly, these compounds are bound (or "trapped"; Reiner and Radic, 2000) at the peripheral anionic site and/or at hydrophobic sites of the extra gorge (Chapter 3 DII and DIII). The binding may also result in inhibition by either restricting access to the gorge or through allosteric interference with the catalytic activity (see also Chapter 3 DI and DII).

Altogether, a number of teams proposed that aside from small carbamate derivatives, larger compounds may affect the catalytic site within the gorge. Their conclusions seem to be at odds with the views of Reiner and Radic (2000). For example, Ishihara and associates (1993, 2000; see also Kaur and Zhang, 2000) analyzed the QSARs of several large benzylaminoindoles and phtalidimide compounds that are potent inhibitors of AChE. The optimal activity, with respect to TcAChE, was associated with C6 chain length and hydro-philic and electron-withdrawing substituents on phtalidimide moiety. Additionally, incorporating a 3-MeNHCO2 group on the phenyl ring of the benzylamino moiety facilitated the carbamylation of the esteratic AChE site.

Also, other modifications were "supposed" to contribute the compounds binding with the hydrophobic sites outside of the catalytic center or gorge and with sites in the gorge (see below). In fact, studies at the Takeda laboratories of 50 related compounds allowed Ishihara and associates (2000) to identify (within the TcAChE) the sites involved in the binding of their compounds. The binding sites (the Trp rings) in the catalytic gorge included the hydrophobic sites and the second peripheral anionic site, which also exhibited hydrophobic properties (Figure 7-13). Takeda Industry's team aimed at developing compounds that would exhibit selective activity within the CNS and would be

A Hypothetical Diagram of the Active Sites of AChE

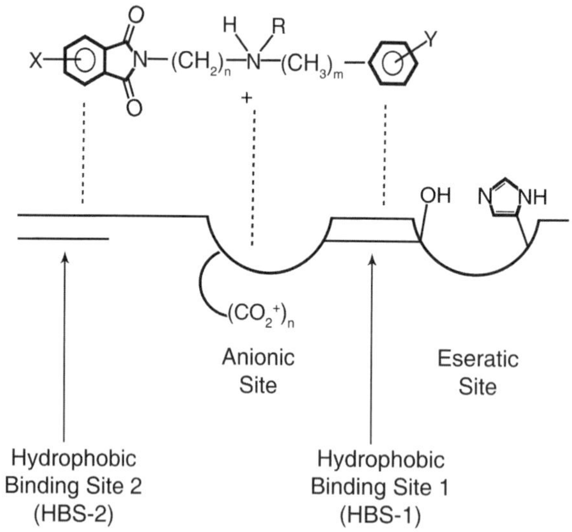

Figure 7-13. A hypothetical diagram of the active site of AChE. (From Ishihara et al., 2000 with permission.)

potentially used in treating Alzheimer's disease. Apparently, the heterocyclic derivatives of the Takeda Industry phtalidimides, such as TAK-147, exhibited a favorable central-peripheral activities ratio without losing antiAChE potency.

The phthalimide moiety of this series was replaced by other complex rings (i.e., indanone) in a series of antiAChEs including the piperidine donepezil (Aricept), a drug used in the treatment of Alzheimer's disease (see Chapter 10 K). With donepezil, the presence of carbonyl grouping on the indanone moiety favors antiAChE action, as the grouping probably promotes hydrogen bonding in the enzyme's gorge. Docking and QSAR analyses (Brufani and Filocamo, 2000; Kaur and Zhang, 2000) indicated that the piperidine ring and its protonated nitrogen, the benzyl substituent, and the inandanone carbonyl in donepezil's structure formed bonds at several levels of the gorge. These moieties interact with the gorge's aromatic residue rings and with the peripheral anionic site, "spanning the entire length of the active-site gorge of the enzyme" (Sugimoto et al., 2000; Saxena et al., 2003; see also Chapter 3 DI and DII).

In a related series of benzisoxazoles (Villalobos et al., 1994; Kaur and Zhang, 2000), the indanone ring of donepezil and other piperidines was replaced by the benzisoxazole moiety. Again, residues, which strengthen hydrogen bonding of the benzisoxazole oxygen within the enzymic gorge, improved the antiChE action. Also, bonding in the gorge and electron-withdrawing moieties were proved important for the antiChE activity of various benzisoxazoles.

Sugimoto et al. (1990, 1995) and Ishihara et al. (1993; see also Kaur and Zhang, 2000) also carried out studies demonstrating the importance of electron-withdrawing and hydrophobic substituents. These studies involved a series of benzylpiperidines, including donepezil, which were tested against purified mouse or rat brain AChE (see Figure 7-7) However, the penetration and bonding of several rings within the gorge did not lead to a direct action of piperidines on the catalytic site, and AChE specificity of benzylpiperidines resulted from the differences between AChE's and BuChE's peripheral anionic sites (Silman and Sussman, 2000). With benzylpiperidines and benzisoxazoles, benzylaminoindoles, phthalimides, or indanones, investigators used the QSAR to illustrate how the bulkiness of the substituents on the benzyl or piperidyl rings; the chain length that separates benzyl ring from the heterocyclic constructs; and the basicity of the nitrogen affect the compounds' bonding to AChE (or BuChE). The exact enzymic sites responsible for these links have not been identified.

Similarly, QSARs were established for several series of aminoacridines or analogues of tacrine (Kaul, 1988; Proctor and Harvey, 2000; and Kaur and Zhang, 2000; P.N. Kaul established also antiChE activities of several marine organisms). Seemingly, the planar ring system is important for the antiChE potency of these compounds. Substitutions that distort tacrine's conformation weaken antiChE activity, while substitutions at the amino nitrogen have no such effect. A more exact QSAR analysis emphasized the importance of the R1 substituents' length. Therefore, the potent aminoacridine derivatives may enter into a hydrophobic interaction with the nongorge, noncatalytic peripheral anionic site of AChE. As posited by the Paris-Rehovot team (see Massoulié et al., 1993), this site is characterized by a tryptophan moiety. Another finding supports this conclusion, as tacrine affected the spectrum of AChE after its catalytic site was phosphorylated by sarin (Dawson, 1989; see also Proctor and Harvey, 2000).

Altogether, Silman, Massoulié, and associates suggested that tacrine and its active congeners bind to the main anionic and catalytic (esteratic) sites (Silman et al., 1994; Massoulié et al., 1993; Massoulié, 2000; see also Brufani and Filocamo, 2000; Ishihara et al., 2000). Two additional compounds, bistacrines and tacrine-huperzine A hybrid, were synthesized with "docking" analysis (Kaur and Zhang, 2000) showing that these compounds may bind simultaneously at both the catalytic gorge and the peripheral bonding sites. Indeed, these new synthetics exhibited antiAChE potency that was many times higher than the potency of the parent compound. A distance of 16 Å between the ring nitrogen was most favorable for the antiAChE activity.

Additionally, there are several analogs of huperzine A, an alkaloid, which structurally resemble morphinoids. (More precisely, aminoethlidine-tetrahydro-7-methyl-5, 9-methanocycoocta, 9[b] pyridin-2-one.) The rigidity of huperzine's chain structure is conformed differently than the flexible chain structure of ACh. However, computer-generated superimpositions of huperzine A and ACh revealed a basic similarity between the two agents (see Silman and Sussman, 2000). It is chemically easy to prepare analogs of the huperzines, and many of these compounds were synthesized (Bai et al., 2000; Kaur and Zhang, 2000). The fused, nonflexible ring system of huperzine A, its steric/hydrophobic nature, the 3-carbon bridge, and the capacity of pyridone and the corresponding heterocyclic ring of huperzine to form hydrogen bonds with the catalytic gorge of AChE are all important for antiAChE activity. (Apparently, only 1 hydrogen bond is actually formed; Silman and Sussman, 2000.)

Also, on the basis of the QSAR and the "docking" assessment (Raves et al., 1997; Silman and Sussman, 2000), the carbonyl grouping of huperzine may interact with the aromatic rings in the catalytic gorge. One of the bondings is close to the bottom of the gorge near the catalytic domain, and another bonding is near the opening of the gorge close to the peripheral anionic site. (See Figure 3-7; Chapter 3 DII) However, huperzine A may not bond directly with the peripheral site (Brufani and Filocamo, 2000). Altogether, Silman and Sussman (2000) felt that the orientation of huperzine A to AChE (at least with TcAChE) is quite unique.

Returning to the classic compound, physostigmine and its congeners, as gathered from modern QSAR analyses, hydrophobicity, the bulk of the substituents, and their polarity play a role in the antiAChE potency. In a series of synthetics, the carbamate moiety was replaced by a series of compounds exhibiting second inonizable nitrogen at the carbamate terminal where the distance between the nitrogens varied in length (Chen et al., 1992; see also Kaur and Zhang, 2000). After QSAR analysis, the compounds supported the past conclusion (see above, this section) regarding increased potency of bisquaternary compounds when compared with their monoquaternary analogs. The results also agreed with the notion that bisquaternary inhibitors bind with 2 anionic AChE sites where one site represents the aromatic ring within the gorge (see above, this section, and Chapter 3 DII). Moreover, investigators used crystallographic studies of long-chain physostigmine derivatives, such as MF268, to provide evidence that (aside from the entrance to the gorge) alternate routes of entrance may exist ("back door" hypothesis; Bartolucci et al., 1999).

3. Mechanism of Action of Carbamate and Related "Reversible" Anticholinesterases

During the 1940s and1950s, the notion of carbamates and related "reversible" antiChEs mechanism of action was vague. Klas-Bertil Augustinsson (1948) referred to a competitive action in which "urethanes [carbamates] were at an active center of . . . cholinesterase." Much later John-Paul Long (1963) suggested that it was possible for these drugs to act by "removing or altering charges on the enzyme center." He also suggested that quaternary biscarbamates act via binding 2 cationic sites that normally bond ACh receptors, thus shielding the esteratic site of the enzyme from ACh. The modern explanation of the increased antiAChE potency of bisquaternary carbamates relates to Long's notion (1963). Currently, investigators have proposed that these compounds do not enter the active AChE gorge (Reiner and Randic, 2000). Instead, they inhibit the enzyme through links between their quaternary nitrogens and the electric field adjacent to the active gorge. Accordingly, the biscarbamates and related compounds inhibit the catalytic activity of the enzyme by restricting the access of ACh to its active center (Eastman et al., 1995; Reiner and Radic, 2000).

Carbamylation with the esteratic site of ChEs active center was established in the early 1960s by Irwin Wilson, Richard Winterigham, and others (see Usdin, 1970). Considering carbamates and related compounds "reversible" inhibitors is perplexing. Reversibility of enzymic inhibition implies recovery of intact enzymes and unchanged inhibitors upon dilution. Wilson and associates (1961) demonstrated that diluting carbamylated AChE yields an intact enzyme *and* carbamic acid instead of the original carbamate (Usdin, 1970; Brufani and Filogamo, 2000). However, the kinetics of carbamic moiety dissociation differs among the carbamates, particularly the newer carbamates developed for use in Alzheimer's disease therapy (see Chapter 10 K). With these agents, the dissociation may be quite slow. This is illustrated when comparing the kinetics of the action of tacrine and donepezil (Aricept, E2020) with physostigmine (Reiner and Radic, 2000; Giacobini, 2000). Donepezil-AChE moiety dissociation has a slow rate, which results from a close bonding of donepezil at several sites on its gorge (see above, section BI-1), as well as at the peripheral anionic site (Saxena et al., 2003). Accordingly, some investigators refer to these carbamate and biscarbamate agents as "pseudo-reversible." Additionally, the slow metabolism of carbamates and piperidine antiChEs contribute to the long half-life of their inhibitory effect (Giacobini, 2000).

The non-OP quaternary antiChEs that do not carbamylate the enzyme include edrophonium and a potent bisquaternary compound, ambenonium (see above, and section BI). These compounds may be considered reversible antiChEs (Reiner and Radic, 2000), even though ambenonium dissociates from the enzyme at a very slow rate because of its tight bonding in the AChE gorge (see above). However, with edrophonium, this process is rapid. Other reversible inhibitors include oximes and NaF, and the mechanism for their inhibitory effect is unclear. Oximes may bind to both the catalytic and peripheral binding sites (Reiner et al., 1996; Reiner and Radic, 2000), while NaF may be ligated near the catalytic site (Usdin, 1970). Sodium fluoride's sensitizing synaptic action may also be pertinent for its antiChE action (see Chapter 9 BIV; see also Wang et al., 2001, Karczmar et al., 1972 and Koketsu, 1966, 1984). Finally, some inhibitors, such as fasciculin, may regulate the activity of AChE via conformational or allosteric mechanisms, which may be exerted outside of the active gorge and the esteratic site (see Chapter 3 DII and Radic et al., 1995).

Similar to the OP drugs (see section BI-1), carbamates generally inhibit both BuChE and AChE, as the two enzymes are structurally similar (Chapter 3 DI and DII; section BII; see below). Many carbamates (including the "original" carbamate, physostigmine) inhibit AChE more potently than BuChE, while carbamates such as Nu0683 (a dimethyl carbamate of 2-OH-5-phenylbenzyl trimethylammonium bromide) and MF-8622 (a morphinoid carbamate) preferentially inhibit BuChE (Long, 1963; Giacobini, 2000). Furthermore, bisquaternary oxamides and bisneostigmine compounds (see above, section BI and BI-1) are almost specific AChE inhibitors. Recently, compounds developed for the treatment of Alzheimer's disease, such as donepezil (Aricept; see Chapter 10 K), are also near-specific AChE inhibitors (Giacobini, 2000). Sometimes, a small structural change in a compound induces a major change in the inhibitory preference. For example, pheneserine (a phenyl derivative of physostigmine) is AChE selective (see Brufani and Filogamo, 2000).

The relation between structural differences among carbamates, their inhibitory capacity with respect to AChE and BuChE, and the structural differences between these two enzymes was defined for some antiChEs. For example, donepezil and the bisquaternary agent BW284C51 occupy the entire active gorge of AChE as they interact with the aromatic groups at the rim and inside the gorge. This finding explains donepezil's "tight" bonding with AChE. It also clarifies donepezil's specificity for AChE, as BuChE does not possess these residues (Radic et al., 1993; see also Massoulié, 2000; Reiner and Radic, 2000). Moreover, comparing the larger BuChE gorge (or acyl pocket) to AChE's gorge may explain the specific antiBuChE action of compounds endowed with large substituents such as Nu683 and MF-8622.

BII. Organophosphorus Compounds, Their Structure-Activity Relationships, and Bonding

1. A General Description of Organophosphorus Agents

Organophosphorus (OP) compounds contain a P-C bond. Though the trivalent phosphorus may form P-C compounds with a pyramidal configuration, these compounds are unstable and undergo a tautomeric change into pentavalent, tetrahydric C-P compounds (see Abou-Donia and Lapadula, 1990). In 1903, Michaelis provided the general structure of OP agents. These compounds are stable and can be readily synthesized. Later, Gerhard Schrader (1937, 1950, 1952; Figure 7-14) developed hundreds of these compounds and expanded upon Michaelis' formulation (see Figure 7-15). The basic structure refers to phosphorus acid, which has either two similar or two dissimilar substituents (R1 and R2), with either one or both of these substituents containing a carbon atom bound to the phosphorus atom. An organic or inorganic residue is also present (X).

Moreover, the oxygen of the phosphorus acid may be replaced by sulfur. In 1959, Bo Holmstedt

Figure 7-14. Dr. Gerhard Schrader. (From Koelle, 1963; reprinted by permission from Springer-Verlag GmbH & Co. KG).

(Figure 7-16) stated that R1and R2 are "capable of almost infinite variation" and there are many possible substitutions for X. Bo Holmstedt (1959, 1963; Figure 7-17) must be credited with a valid classification of OP drugs that many investigators still employ. [Some of these investigators do not give appropriate acknowledgement to Holmstedt (see, for example, Brufani and Filocamo, 2000; Ecobichon, 1996; and Hayes, 1982.]

However, some compounds did not fit Holmstedt's classification, such as the toxogonine derivative ethoxylmethylphosphonyl toxogonine, which has an R substituent with a complex double piperidine structure and halogenated iminophosphate insecticides (Van Wazer, 1961; Luo et al., 1999; see also Reiner and Radic, 2000; Makhayeva et al., 1998). Also, Holmstedt did not distinguish a separate category for compounds exhibiting aromatic radicals such as nitrophenyl. (He listed them among type B and C OP agents; Figure 7-17.) These radicals may impart special potency to certain OP drugs (see Eto, 1974; section BII-2).

Finally, several investigators employed nomenclatures that differed from those employed by Holmstedt. (See, for example, Abou-Donia, 1992; Hayes, 1982; Chambers, 1992; Figure 7-15; Table 7-1). The table includes categories of OP agents, such as war gases and insecticides. The war gases sarin and soman are represented in the A-1 class of OP drugs. The war gas tabun is an A-3 compound, and VX belongs to class C-1 of OP drugs. Di-isopropyl fluorophosphonate (DFP) is a potential war gas and was studied more than any other OP agent. Di-isopropyl fluorophosphonate is an A-2 class agent. Several insecticides are also included in these classes: Dimefox and mipafox are class A-4 drugs; Dipterex is a class B-1 agent; phosdrin and paraoxon are class B-2 antiChEs; EPN, Systox, and parathion are class D-2 compounds; and Thimet and OMPA are class D-3 and class E insecticides.

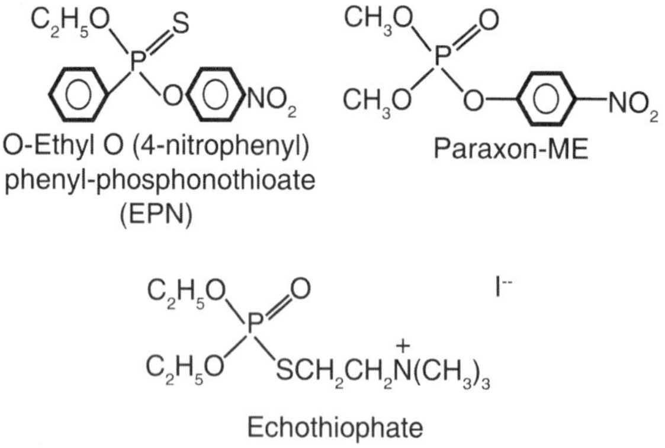

Figure 7-15. Schrader's expansion of Michaelis' formulation.

Figure 7-16. Bo Holmstedt with natives of Amazonas, 1963.

Table 7-1. General Classification of OP Compounds

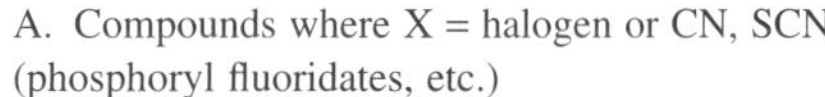

A. Compounds where X = halogen or CN, SCN (phosphoryl fluoridates, etc.)
   (1) R1 and R2 = alkyl
   (2) R1 = alkoxy, R2 = aryl
   (3) R1 and R2 = alkoxy
   (4) R1 = alkamido, R2 = alkoxy (and similar compounds)
B. Compounds where X = alkyl, alkoxy, or aryloxy (phosphlinates)
   (1) Alkoxydialkyl or dialkoxyalkyl compounds
   (2) Trialkoxy compounds and dialkoxy, aryloxy compounds
C. Compounds where X = imino or toxogonin radical
D. Thiol- and thiono-phosphorus compounds
   (1) Thiol-phosphorus compounds
   (2) thionophosphoryl and thionophosphoryl fluoridate compounds
   (3) thiolthionophosphoryl compounds
E. Derivatives of pyrophosphorus acid
F. Compounds containing quaternary nitrogen such as phosphorylcholines

Figure 7-17. General Structure of OP agents (with an attempt to represent their tridimentionality).

In 1952, on behalf of Bayer Leverkusen Co. and the I.G. Farbenindustrie, Schrader patented a class B-1 compound, Metrifonate (Dipterex, trichlorfon), as an insecticide and an anthelminthic (see Holmstedt, 1963; Hayes, 1982). Metrifonate's molecule is nonenzymically rearranged in tissues or solutions, and the resultant dichlorvos is the active antiChE. Therefore, Metrifonate acts as a slow-release agent, making it selectively toxic to pests and harmless to animals or humans. Metrifonate is also a specific AChE inhibitor. Because of these properties, Metrifonate was considered as potentially contributing to the treatment of Alzheimer's disease, though it was eventually withdrawn from clinical use (Brufani and Filocamo, 2000; see Chapter 10 K).

Metabolic routes, including oxidation, generally detoxify the OP drugs. However, several OP drugs must be metabolized to become active. These compounds are frequently referred to in the literature as acting "indirectly," but this term does not constitute a felicitous description of their action. For example, parathion and Malathion must be oxidized into paraoxon and malaoxon to become active. EPN, another popular insecticide, is also oxidized. However, instead of being activated, EPN is detoxified by this process (under certain circumstances EPN oxidation results in increased toxicity).

Moreover, class C compounds include a number of halogenated iminophosphate insecticides, such as OP toxogonin derivatives or phosphoryl oximes (Luo et al., 1999) and O-o-dialkyl-o-chlormethylchlorformimino phosphates, which were recently developed in Russia (Makhayeva et al., 1998). These potent antiAChEs are formed when phosphorylated AChEs are reactivated by oximes.

Some of the class D-1 compounds include phosphorothioate 217 AO and its quaternary phospholine derivative, 217 MI (echothiophate). Echothiophate was used in the therapy for myasthenia gravis and is currently used in ophthalmology to attenuate intraocular pressure. In a basic experiment, George Koelle discovered that 217 AO is capable of inhibiting both the external and cytoplasmic AChE, while only the external enzyme is inhibited with 217 MI. He used this difference to distinguish the functional role of cytoplasmic versus external AChE (Koelle and Steiner, 1956; Koelle, 1963; see Chapter 3 DIII).

2. Structure-Activity and Bonding of Organophosphorus Anticholinesterases

In 1950, Gerhard Schrader (see Schrader, 1963; see also section A-Ib and Holmstedt, 1959, 1963) made the first contribution to the OP drugs' SAR. He proposed that active OP inhibitors should include an "acid anhydride linkage" (Schrader's "acyl rule"). The acyl radical did not have an R-C = O- grouping. Instead, it had a moiety that included protonic acids with HF or phosphoric acid (see also Eto, 1974). Schrader's "rule" suggested that the stronger the acidic character of the OP agent, the quicker the "acyl group" is displaced and phosphorylation of the enzyme occurs.

Then Clark et al. (1963) expanded this rule when suggesting that one of the radicals linked to the phosphorus atom was capable of accommodating electrons from the P-X bond. Additionally, the phosphorus atom must exhibit a strong positive charge (Eto, 1974). The radicals and groups may be aromatic rings (i.e., strong inhibitors parathion and paraoxon) or chlorine atoms (for example, dichlorfos; see Eto, 1974). Note that the "Clark-Schrader" rules do not always apply to war gases (i.e., tabun and soman) or to compounds that "should" be equipotent (i.e., EPN and parathion).

In view of the Clark-Schrader rule's inconsistencies, other SAR notions were posited. In 1957, Tammelin designed OP compounds that represented analogies to ACh and neostigmine or physostigmine including several quaternary thiocholines. Then Main (1976) stressed the importance of the leaving group's structure. However, he did not define the characteristics optimal for inhibition.

A related concept refers to shape and electronic characteristics of the P atom when bonding to the adjacent groups. This bonding characterizes trivalent OP compounds with a pyramidal symmetry, while the pentavalents show a tetrahedral symmetry (Van Wazer, 1961; Hudson, 1965; see also Eto, 1974). These considerations may involve the OP agents' ability to penetrate the AChE gorge (Chapter 3 DI and DII) and the properties of the agents' ligands at the active esteratic site, as the alkyl and other groupings affect the electron transfer within the tetrahedral or pyramidal configurations of the OP drugs. However, these notions regarding the characteristics of the P atom when bonding to adjacent groups have not been translated into specific SAR considerations.

The approaches of Tammelin, Eto, Van Waser, and Hudson reflect the obvious: antiChEs should fit the enzyme similar to its substrate, ACh. This notion was fully embraced by Elsa Reiner, Zoran Radic, Palmer Taylor, and associates when they employed a three-dimensional modeling of OP ligand-AChE moiety based on the amino acid sequence and other analyses of crystalline AChE (Radic et al., 1984; Reiner and Radic, 2000; see section BI-1 and Chapter 3 DI and DII). Ultimately, they proposed that successful bonding depended on characteristics of the P-atom involved in the covalent phosphorylation, the capacity of the phosphonyl or carbonyl oxygen atom to fit in the oxyanion "hole" of AChE, and the potentiality of the leaving group to point toward the opening of the enzyme's gorge (see section BI-1). Also, the binding induced a conformational change in certain amino acid residue systems of AChE ("Omega Loop"; see Chapter 3 DI and DII) that neighbor the enzyme's active site (Shih et al., 2001). Unfortunately, aside from proposing that the excessive weight and dimension of alkyls within the OP moiety prevent the entry of an OP drug into the gorge, Radic, Reiner, Palmer, and associates did not manage to study a sufficient number of OP agents to describe an OP structure optimal for inhibiting AChE (or BuChE).

3. Mechanism of Action of Organophosphorus Anticholinesterases and Reactivation of Phosphorylated Enzyme

a. Inhibition of AChE by OP Agents

Similar to carbamates and related "reversible" or "pseudo-reversible" antiChEs (see Usdin, 1970; see also above, section A-I), the OP compounds act upon the active ACh-hydrolyzing esteratic site located in the gorge formed within the AChE moiety. Other AChE bonding sites participate in the localization of the OP drugs within the AChE molecule (see Chapter 3 DI and DII). In the course of inhibition, the OP agent (or other antiChEs) first penetrates into the AChE gorge, and then it inhibits the enzyme (see Chapter 3 and section BII-2, below). Irwin Wilson, David Nachmansohn, and Felix Bergmann (Wilson et al., 1950; Wilson, 1955) showed that cationic substrates were hydrolyzed faster than their uncharged analogs. Hence, they proposed a model of the hydrolytic or "anionic" site containing a carboxylic acid or related moiety. Zeller and Bissegger coined the term "anionic" site in 1943. (The site is also referred to as "aromatic" or "catalytic." See above, section BI-2, and Chapter 3 DII and DIII.)

In 1961, Krupka and Laidler included an acid site in their model that also contained a basic position represented by an imidazole nitrogen. The investigators identified this site by binding the area containing active AChE with radioactive compounds (i.e., DF32P), cleaving the bound AChE (and other, related esterases) at amide and other links and chemically analyzing the cleft fragments. This process enabled them to establish that the active site contained serine as the major ligand for OP drugs, carboxylic amino acids, and other amino acids such as histidine, which contains an imidazole grouping (Usdin, 1970). Serine's ligand role results in stechiometric phosphorylation. One serine moiety is phosphorylated by one molecule of the OP agent, as the "leaving group" from the OP compound is released (see above, section B-2). The presence of carboxylic amino acids and histidine's imidazole radical vindicated Wilson's, Nachmansohn and Bergmann's, and Krupka and Laidler's hypotheses.

Since BuChE and its variants and AChE and its subtypes are structurally similar (see Chapter 3 DI and DII), many OP agents (and carbamates; see section BI-1, above) frequently inhibit both enzymes. However, there are differences among the enzymes: comparing the larger acyl or esteratic pocket of BuChE with AChE's esteratic pocket (see Chapter 3 DII and Taylor et al., 2000) explains why some antiChEs are specific or semispecific to certain enzymes. This notion has been accepted since the 1940s and 1950s (see Hawkins and Gunther, 1946; Davison, 1955; see also Augustinsson, 1948, Karczmar, 1967; Goede et al., 1967).

DFP is from 10 to 300 times (depending on the investigator) more active against BuChE than against AChE (see Holmstedt, 1963). Several OP insecticides (including isoOMPA and mipafox) share this quality with DFP. In fact, some investigators found that isoOMPA inhibits BuChE 1,000 times more than AChE. However, quaternary OP agents (i.e., methylfluorophosphorylcholine and homocholine) are significantly more active against AChE than against BuChE.

Note that various investigators reported different I50 values for many of the antiChEs that inhibit either BuChE or AChE in the same or different species. In fact, this is the secret for insecticidal uses of mipafox or isoOMPA. While they are effective against insect AChE (or other ChEs involved in vital insect functions; see Chadwick, 1963), they are ineffective against human AChE, which underlies their safety in humans. (Of course, the mammalian capacity to hydrolyze and detoxify these OPs also contributes to their usefulness as pesticides or insecticides.)

There are several additional notions regarding SAR differences for the OP drugs' inhibitory potency. The quaternary phosphoryl-choline drugs are more potent AChE inhibitors than BuChE, while bulky alkyl substitutions on the OP molecule favor the BuChE inhibition (see section BII-1 and Eto, 1974). Then certain considerations were specifically posited for insect AChE. Aryl substitutions and the small size of the substituents favored the inhibition of the insect enzymes (Eto, 1974). Finally, the SAR considerations can sometimes be explained by considering the enzymes' structural characteristics. For example, isoOMPA is a more potent inhibitor of BuChE than AChE because the large acyl or esteratic pocket of BuChE is larger than that of AChE (Taylor et al., 2000; see also Chapter 3 DI and DII). Also, the presence of several anonic sites on or near AChE's active gorge may explain the high antiAChE activity of the quaternary OP drugs. This is an attraction that may also result from the electric field generated at the surface of the AChE (or BuChE) gorge by the active site's dipolar character (Reiner and Radic, 2000).

A great deal of the quoted work was carried out with chymotrypsin and other hydrolases, as these enzymes were purified, crystallized, and readily available. In 1991, teams from the CNRS of Paris, Rehovot, and Jerusalem (Massoulié et al., 1993; Sussman et al., 1991) managed to obtain crystallized forms of AChE amenable to appropriate analysis. They solubilized TcAChE with phosphatidylinositol-specific phosphates C (PIPLC), crystallized the solubilized enzyme, and analyzed it by means of x-ray crystallography. Their use of molecular biology methods for analyzing the peptidic fractions of the enzyme was assisted by current molecular knowledge regarding vertebrate AChE regulated by a single gene (in contradistinction to certain nonvertebrate species; see Chapter 3 DIII).

This multidisciplinary methodology brought about more precise descriptions of both AChE's phosphorylation and active site. The investigators established that the active site, located at the bottom of the AChE gorge (see above, section BI-2, and Chapter 3), was actually a triad containing residues of serine, glutamate, and histidine. The efficacy and rapidity of the OP-AChE bonding results from high nucleophilicity of catalytic serine, which is correlated with the triad's ability to form a short, strong hydrogen bond (SSHB). This notion was posited by Cassidy et al. (1997) with chymotrypsin and also appeared true for AChE where the hydrogen bonding distance is shortened. The formation of SSHB takes place on the catalytic triad between glutamate and histidine residues (Viragh et al., 2000). The anionic site, particularly its aspartate moiety, also may play a part in the OP-induced inhibition of the enzyme, even if the agent in question is not a quartenary (Ordentlich et al., 2004).

b. Phosphorylated AChE Aging

Hobbiger (1955) was the first investigator to describe a process called "aging" for the phosphorylated "plasma" ChE (BuChE). Then Hobbiger found that aging also occurs with the phosphorylated "true" ChE or AChE (see Usdin, 1970): "Aging . . . a reaction with first-order kinetics . . . is a process in which inhibited enzymes . . . initially reactivateable by . . . reactivators . . . become refractory to reactivation" (Usdin, 1970; see also Hobbiger, 1955; Davies and Green, 1956; Heilbronn, 1963; Berends, 1964; Benshop and Keijer, 1966; reactivation and reactivators will be described in the next section.) In addition, the aged phosphorylated enzyme cannot reactivate spontaneously in solutes such as water or biological media.

Altogether, demonstrating this process completed our knowledge of the inhibitory phenomena. During the 1950s and 1960s, the Dutch team of Oosterbaan, Jansz, and Cohen proposed that the inhibition included bonding, phosphorylation, and "aging," which consisted of dealkylation of the phosphorylated enzyme (see Oosterbaan et al., 1961). Several mechanisms for the "internal" dealkylation were also proposed (see, for example, Usdin, 1970; Ecobichon, 1996; Keijer and Wolring, 1969; Masson et al., 1999). One of these mechanisms included a "nucleophile attack" or acid catalysis at the P-0-alkyl radical (Usdin, 1970).

Later, the Walter Reed Hospital, in Washington, DC, and the U.S. Army Chemical Defense Institute team of Saxena, Viragh, Doctor, and others (Viragh et al, 1997; Saxena et al., 1998) suggested that two amino carboxylic acids and several other amino acids of the active AChE (and BuChE) site provide "an electrostatic push and electrostatic pull" for the dealkylation process. Finally, the Weizman Institute team of Joel Sussman and Israel Silman (see Silman and Sussman, 2000) analyzed the crystal structures of aged OP-AChE conjugates. They concluded that 4 potential donors in the catalytic AChE site stabilized the aged OP residue and that the site itself underwent a conformational change: both phenomena underlay the residue's refractoriness. This transformation was also reflected when the OP inhibitors changed topography in the AChE molecule (Yingge et al., 2001), and was demonstrated by means of atomic force microscopy and "force" spectroscopy (Vinckier and Semenza, 1998, see also Millard et al., 1999).

c. Rate of Aging

The discoveries of aging phenomena were described in the 1950s by Wilson, Davies, Hobbiger, and others (see previous section), and the vast difference in the rate of aging among phosphorylated ChEs has been recognized. For example, AChE phosphorylated by soman, sarin, or DFP ages more rapidly than AChE phosphorylated by tabun or TEPP (see Hobbiger, 1963).

Also, rate of aging may be affected, and aging prevented by carbamate antiChE's, NaF and choline (Dawson, 1989; Dehlawi et al., 1994).

d. Reactivation of Phosphorylated AChE

The irreversibility of OP drug action is not absolute. Reactivation may occur with the originally inhibited enzyme, as it restores the enzyme's activity and functional structure. Reactivation may occur spontaneously in the solvent, or involve the action of specific compounds or "reactivators." Though it is difficult after the enzymes have aged, spontaneous reactivation may take place with aged phosphorylated enzymes. The process is slow, especially for phosphorylated AChE as opposed to phosphorylated BuChE, which reactivates relatively fast. However, OP drugs with heavy alkyl substitutions in positions R1 and R2 (see above, section BII) generally exhibit slower spontaneous reactivation than those with lighter substitutions (Usdin, 1970).

Prominent investigators in this field include Irving Wilson, Edith Heilbronn, Frank Hobbiger, Edward Poziomek, Bernard Jandorf, William Aldridge, Heinrich Kewitz, and Henry Wills (see Usdin, 1970; Wills, 1970; Wilson, 1959; Aldridge, 1957, 1976; Aldridge and Reiner, 1972; Davies and Green, 1956; Holmstedt, 1959). The first "break" in the search for reactivators was in 1951, when Bernard Jandorf discovered that, *in vitro*, hydroxylamine hydrolyzed OP compounds. In addition, Irving Wilson demonstrated that hydroxylamine reactivates phosphorylated AChE (Wilson, 1951; Jandorf et al., 1955), which related to Hestrin's (1949) earlier studies of hydroxylamine actions on acetylated AChE (see section A).

Wilson's, Jandorf's, and Hobbiger's investigations (see Cohen and Oosterbaan, 1963) were followed by the rapid development of hydroxamic derivatives, especially several types of oximes. When Wilson realized that quaternary oximes are more potent reactivators than their tertiary analogs, he developed very active oximes such as pralidoxime, which is a mono-oxime pyridinium compound (2-PAM; Wilson and Ginsburg, 1955). Additionally, Edith Heilbronn (1965) showed that 2-PAM reactivates phosphorylated enzymes 50,000 times faster than hydroxylamines. Other mono-oximes included imidazolinium oximes (Reiner et al., 1996). Until today, pralidoxime remained the most potent reactivator and is only rivaled by the bisquaternary di-oximes, such as bispyridinium compounds including obidoxime and toxogonin, TMB-4, and HI6 (and hydroxylamines) that reactivate both *in vivo* and *in vitro* (Table 7-2; Engelhard and Erdmann, 1963).

Even active oximes (i.e., pralidoxime) may not be equally potent against all OP agents. Accordingly, pralidoxime might be less effective

Table 7-2. Structures of representative reactivators. (From Karczmar et al., 1970.)

Common Designation	Trivial Name	Structure
DAM	Diacctylmonoxime	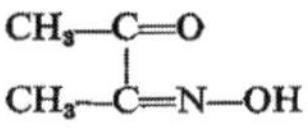 CH_3—C=O CH_3—C=N—OH
DINA	Diisonitrosoncatone	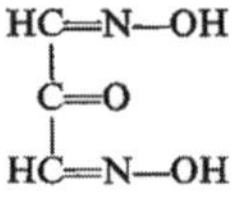 HC=N—OH C=O HC=N—OH
LUH6	Toxogenin	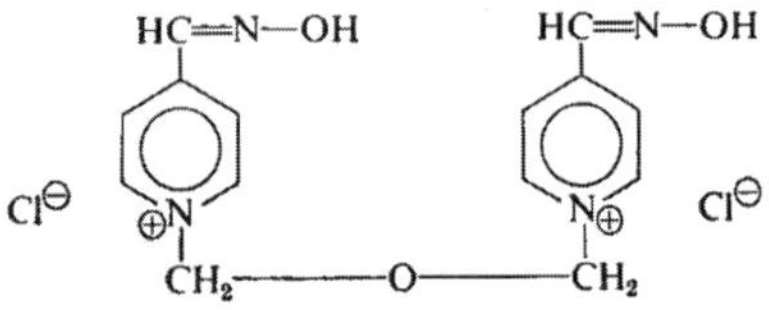 HC=N—OH HC=N—OH $Cl^{\ominus}$ $\oplus$N N$\oplus$ $Cl^{\ominus}$ CH_2—O—CH_2
MINA	Monoisonltrosoncatone	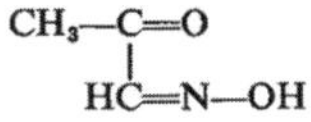 CH_3—C=O HC=N—OH
2-PAM	Pyridine-2-aldoxime metholodide	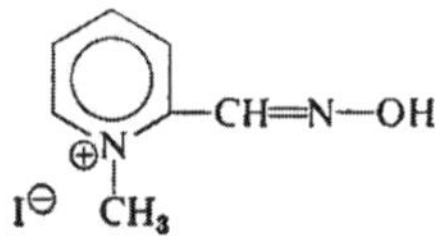 —CH=N—OH $\oplus$N $I^{\ominus}$ CH_3
3-PAM	Pyridine-3-aldoxime methiodide	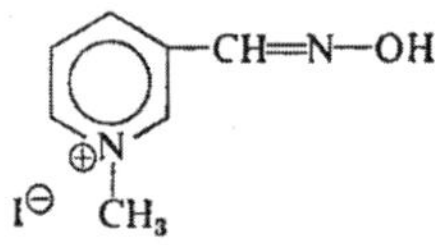 —CH=N—OH $\oplus$N $I^{\ominus}$ CH_3
4-PAM	Pyridine-4-aldoxime methiodide	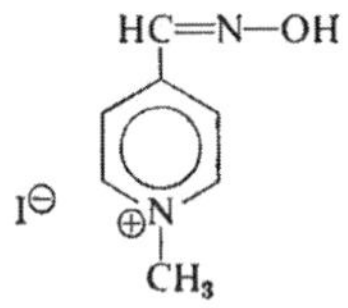 HC=N—OH $I^{\ominus}$ $\oplus$N CH_3
P-2-S	Pyridine-2-aldoxime mathyl mathanesulfonate	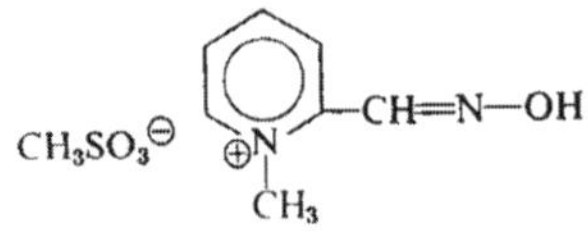 —CH=N—OH $CH_3SO_3^{\ominus}$ $\oplus$N CH_3
TMB-4	*N,N'*-Trimethylene *bls* (pyridine-4-aldoxime bromide)	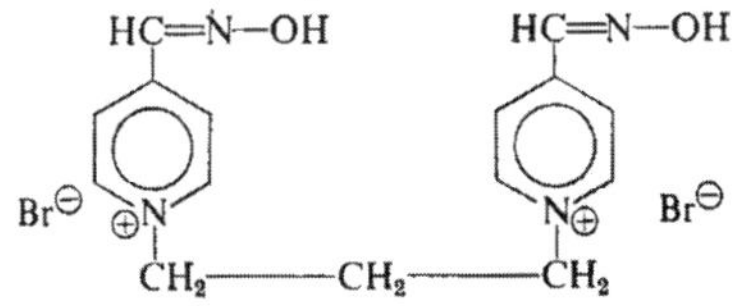 HC=N—OH HC=N—OH $Br^{\ominus}$ $\oplus$N N$\oplus$ $Br^{\ominus}$ CH_2—CH_2—CH_2
HI-6		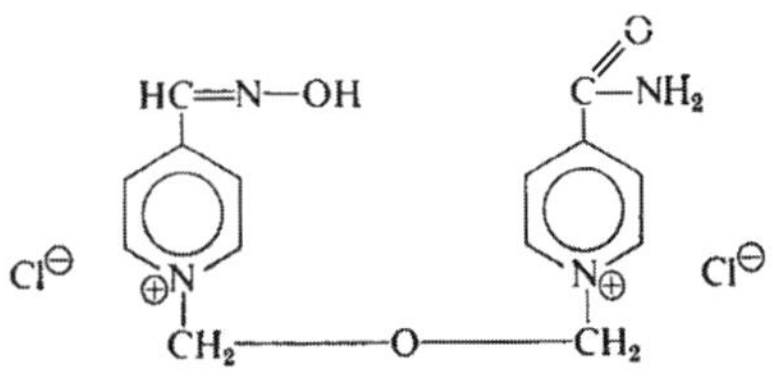 HC=N—OH O=C—NH_2 $Cl^{\ominus}$ $\oplus$N N$\oplus$ $Cl^{\ominus}$ CH_2—O—CH_2

against sarin than against tabun. However, the bispyridinium compounds (obidoxime and HI6) seem effective against many OP agents, though they may have problems reactivating to soman-inhibited AChE. (See Wills, 1970; Eto, 1974; Ecobichon, 1996; see also section E, below.) These problems with reactivation may be related to rapid aging of the phosphorylated moiety, which results from enzyme inhibition by certain OP drugs (i.e., soman and/or OP agents endowed with large alkyl radicals).

Also, reactivation may not readily occur with quaternary OP drugs, such as the phosphorylcholines (type F OP compounds; see section BII-1), as the quaternary nitrogen of these compounds prevents the oxime from bonding to the enzyme's anionic site (Eto, 1974). Investigators showed other compounds, including certain oxamides, choline, and NaF, had reactivating capacities (Karczmar, 1965; Wilson and Froede, 1971; Heilbronn, 1963, 1964). Oxamides act as reactivators, "sensitizers," and antiChEs. Sodium fluoride is a sensitizer (see Chapter 9 BIII-2f; Heilbronn, 1964).

In 1965, Edith Heilbronn presented a diagram based on her own investigations and those of others (Aldridge, 1976; Wilson, 1959; see also Karczmar, 1967; Eto, 1974; Ecobichon, 1996; see also Manetsch et al., 2004). Heilbronn used the diagram to explain that reactivity is dependent on the nonaged status of the phosphorylated enzyme and the nucleophilic nature of the reactivator, and that the nucleophilic oxime group reacts with the electrophilic P atom of the phosphorylated AChE. The oxime-phosphate residue then splits off, freeing the regenerated enzyme. However, the quaternary oximes have an additional advantage: the oximes' quaternized N combines with the anionic site of the enzyme, orienting them so nucleophilic attack can occur. This orientation occurs more readily with bisquaternary oximes, as the second quaternary N attaches to additional anionic AChE sites (see Chapter 3 DII and DIII).

Then Wilson (1959) described other structural requirements for the reactivators when he posited the law of "complimentarity" (see above) to explain a rational design of reactivators that would optimally fit the phosphorylated enzyme. This notion is substantiated by the stereoisometry nature of the reactivator's action. Oximes are planarly rigid because a double bond links their molecule's carbon and nitrogen. Depending on the oxime, its syn or anti configurations "complement" the topography of the phosphorylated AChE when the oxime attaches itself to the enzyme's anionic site (Wilson, 1959; see also Karczmar, 1967). Similar mechanisms explain the reactivator actions of oxamides. However, with NaF, the reactivation depends on the strong affinity of fluoride to the P of the phosphorylated enzyme (Heilbronn, 1964; see also Usdin, 1970).

Note that phosphorylated oximes caused by reactivating the phosphorylated enzyme are potent antiChEs. In fact, some of these compounds are more potent inhibitors than the OP agents causing the original phosphorylations (i.e., sarin; see Hackley et al., 1959; Schoene, 1971; Luo et al., 1999). Initially, investigators thought the half-life of these compounds was short, thus concluding that using the oximes as antidotes would not be dangerous. However, though toxogonin and H6 are safe, obdoxime and pralidoxime are not (Figure 7-18; Luo et al., 1999; see also Reiner and Radic, 2000).

BIII. Anticholinesterases Other Than Organophosphorus, Carbamate, and Related Anticholinesterases

Cholinesterases are sensitive enzymes that are inhibited by many substances besides OPs and carbamates. Long (1963) and Augustinsson (1948) presented a long list of substances that inhibit BuChE, AChE, or both (see also Karczmar and Koppanyi, 1954; Koppanyi and Karczmar, 1949), such as amino acids and several amines, protamines, and histones; opioids, antipyretic analgesics, and anti-inflammatory agents; several antibiotics; cholinergic agonists and antagonists; anticoagulants; convulsants; barbiturates, chloral hydrate and ethanol; local and general anesthetics; ketones (Pereira et al., 2004); certain dyes; some vitamins; cystine and other sulphides; quinine; and even gum arabic. Oxime and NaF reactivators also inhibit ChEs (Usdin, 1970), and they are truly reversible inhibitors (see sections BI-1 and BI-2). Finally, blood, brain and other tissues contain natural antiChEs, though their nature has not been established (Karczmar and Koppanyi, 1954).

Reactivation of Alkylphosphorylated Acetylcholinesterase

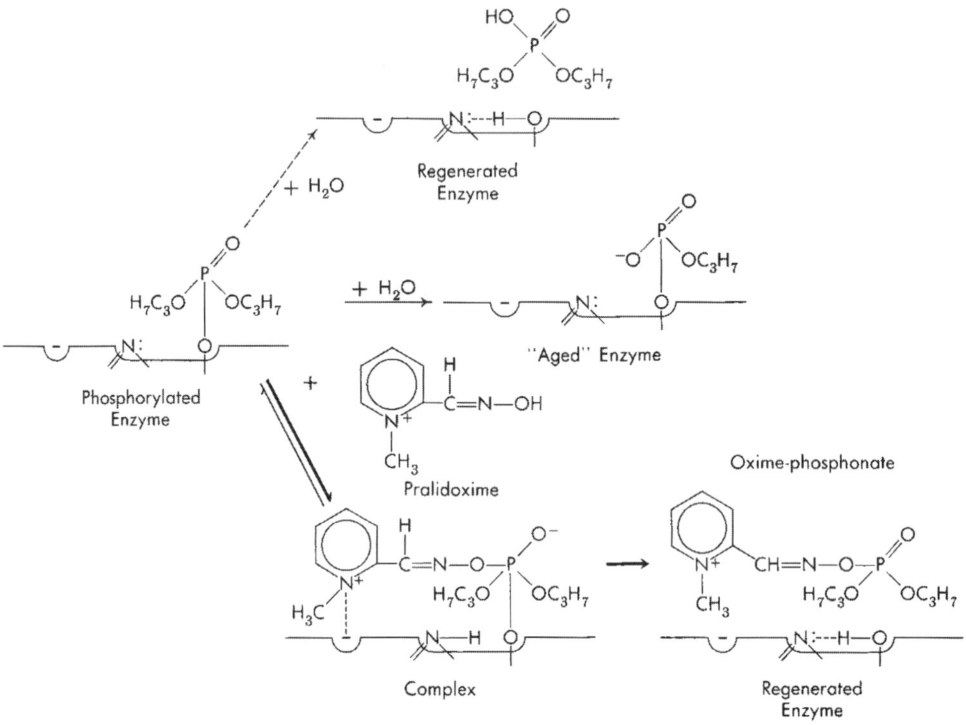

Figure 7-18. Reactivation of alkylphosphorylated acetylcholinesterase (AChE). (From Taylor, 1990, with permission.)

Generally, these compounds are weak BuChE and AChE inhibitors. Though Augustinsson (1948) considered methylene blue "a strong inhibitor of serum ChE," methylene blue was active against the "serum" enzyme at concentrations of only 4×10^{-4} M, whereas choline's I50 value against AChE was 2×10 M (Stovner, 1956). Note that there may be exceptions. Although early data concerning atropine and related tropanes or glycoalkaloids indicated that these agents were weak AChE or BuChE inhibitors, the latest results suggest that some of these compounds may inhibit either enzyme at nanomolar concentrations (McGehee et al., 2000).

Strychnine, thiamine, and the curarimimetics are unique. Nachmansohn (1938) suggested that strychnine's convulsant action might be caused by its antiAChE effect. However, strychnine's I50 value against eel AChE amounts from only 10^{-3} to 10^{-4} M. This amount of inhibition is not sufficient to explain any effect of strychnine in the CNS, including convulsions (see also Chapter 9 BIV). In 1941 Glick attempted to exploit the antiChE action of thiamine to explain its metabolic "preserving effect" on choline esters and its alleged sensitizing action on ACh. However, this speculation was not borne out by subsequent studies. Finally, William Bowman (1982) intensely studied curarimimetics and depolarizing neuromyal junction blockers, emphasizing that their antiChE action is dependent on a structural resemblance to ACh moiety. His point was well taken, and related to Zupancic's (1967) speculation that ChEs serve as cholinergic receptors mediating the cholinergic actions. However, Zupancic's supposition is no longer tenable, as the antiChE action of curarimimetics is not related to their main pharmacological action. Today, several potent antiChEs structurally resemble ACh and/or cholinergic receptors. The prime example is fasciculin, a congener of the nicotinic blockers, bungarotoxins (see above).

C. Metabolism and Detoxification of Anticholinesterases

1. Metabolism, Detoxification, and Activation of Organophosphorus Anticholinesterases

Latent toxicity of OP pesticides and the toxicological potential of OP war gases make OP chemical detoxification routes extremely practical and important. During the last 50 years, intense research in this area has been carried out. (See reviews of O'Brien, 1960; Menzie, 1972; Usdin, 1970; Eto, 1974; Hayes, 1982; Kulkarni and Hodgson, 1984; Matsumura, 1985; Chambers and Levi, 1992; Ecobichon, 1996; Satoh and Hosokawa, 1998.) However, perhaps as a result of security problems limiting open literature publications concerning war gases and the intense use of insecticides, studies of insecticidal metabolism overwhelm studies regarding war gas metabolism. There is also a cliquishness regarding the two literatures, as publications concerning war gases warily refer to the insecticides and vice versa. In addition to detoxification, there are two other important metabolic routes. The first route leads to activating inert OP antiChEs and the second route causes OP-induced delayed neuropathy (OPIDN).

a. Detoxification

One of the detoxification processes follows the inhibitory phosphorylation of AChE or BuChE by the OP drug. At the time of phosphorylation, the "leaving" or "labile" group is detached from the OP molecule while either the phosphorylated enzyme is degraded or the phosphorylated moiety is liberated via spontaneous or chemical reactivation of the inhibited enzyme (Holmstedt, 1959, 1963; Usdin, 1970). The phosphorylation process is a degradation phenomenon, since it removes the free OP agent. This process may not be very important for *in vivo* detoxification of OP drugs. However, it is a basis for a prophylactic mode: esterases related to ChEs are phosphorylated by OP agents and remove ChEs from the blood. Therefore, they experimentally serve as antidotes to OP agents for humans or animals (Karczmar, 1984). Actually, a number of other proteins may detoxify by bonding the OP drugs (Cohen, 1957). These types of antidotes are referred to as "scavengers" or "sponges" (see section E).

In 1974, Eto described that the detoxification of OP agents, phosphoric acid derivatives (see above, section B3), was a result of cleavage on the phosphorus ester and could be accomplished in a variety of ways. During the 1940s and 1950s an important cleavage process was shown: a number of hydrolases or esterases present in animal tissues (blood, kidneys, liver, and lungs) hydrolyze OP drugs such as DFP and paraoxon (Mazur, 1946; Aldridge, 1951; see also Mounter, 1963; Ecobichon, 1996). The metabolites formed are de-esterified, alkylated phosphoric acids (di-*iso*-propyl phosphoric acid in the case of DFP). The alkylated phosphates and other OP metabolites are excreted into the urine and feces. Also, the detoxifications described by Mazur (1946) and Aldridge (1951) may have been carried out with hydrolases classified as aryl- or aliesterases.

Arylesterases represented by hydrolase EC 3.1.3 (see Johnson and Talbot, 1983; also referred to as "esterase" and "phosphotriesterase"; see also Chapter 3 DI) hydrolyze carboxylic esters (with the exception of choline esters) and split aromatic esters at a high rate. They are resistant to many OP agents. In fact, they hydrolyze many OP compounds, and when present in animal or human tissue, both aryl- and aliesterases may detoxify OP drugs. Therefore, they may be used as their antidotes. In addition, there is an entirely different detoxification involving the cytochrome P450 (monooxygenase) system that also results in dearylation (see below, this section). However, aliesterases split aliphatic esters (with the exception of choline esters) and are very sensitive to OP chemicals. Both types of esterases are readily inhibited by physostigmine.

Paraoxon and other aryl OP agents are hydrolyzed by arylesterases (see Mounter, 1963; Aldridge, 1951, 1953) and Mazur's DFPase could be an aliesterase. Aryl- and aliesterases main significance is their presence in insect heads and other tissues. They are vital for the insects (Chadwick, 1963), yet many OP drugs do not attack these enzymes. Another major metabolic route involves an oxidative process dependent on microsomal systems and cytochrome P450. Originally, this route was related to the activation of OP compounds and subsequently associated with detoxification (O'Brien, 1959, 1960; see also below).

Finally, our knowledge of OP detoxification involves glutathione (Hollingworth, 1969; Motoyama and Dauterman, 1974). The early investigations concerning hydrolases, glutathione, and the oxidative processes clarified that OP molecules offered many points and modes for detoxifying attacks. Currently, 15 mechanisms of detoxification have been described (see below, and Figure 7-19). Investigators have had difficulties making generalizations and classifications for OP detoxification because of these mechanisms and problems with nomenclature. For example, arylesterase is frequently referred to as either A-esterase, aromatic esterase, carboxylesterase, phosphotriesterase, or hydrolase (see Eto, 1974; Mileson et al., 1998; Lassiter et al., 1999; Pond et al., 1998; Satoh and Hosokawa, 1998; and Karanth and Pope, 2000). In addition, the term arylesterase may refer to P450-dependent detoxification or to

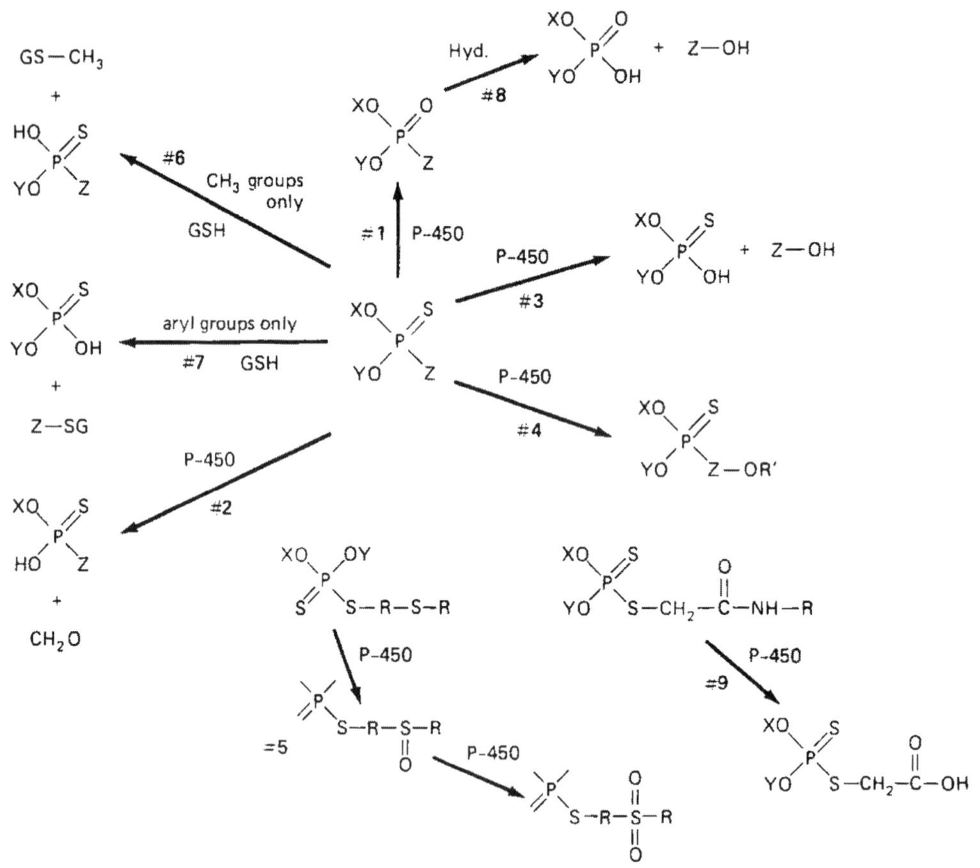

Figure 7-19. A schematic diagram depicting the various phase 1 and 2 biotransformation pathways of an organophosphorus ester and the nature of the products formed as a consequence of oxidative, hydrolytic, GSH-mediated transfer and conjugation of intermediate metabolites in mammals. (From Ecobichon, 1990, with permission.)

a nonmicrosomal hydrolase (compare Pond et al., 1998 and Ma and Chambers, 1994; see also below). Mileson et al. (1998) list carboxylases and arylesterases as hydrolases, whereas Karanth and Pope (2000) consider carboxylesterases and A-esterases to be two separate categories for detoxifying enzymes. Lassiter et al. (1999) and others classified the carboxylase capable of detoxifying chlorpyrifos and the chlorpyrifos-oxon (its active metabolite) as EC 3.1.1.1 and refer to chlorpyrifos-oxon-detoxifying enzymes as arylesterase, listing the latter as EC 3.1.1.2. These investigators do not commit themselves to further categorizations (Pond et al., 1998). Anhydrolases, Aldridge's hydrolases, and Mazur's DFPases are all capable of hydrolyzing DFP and related OP compounds.

Similar inconsistencies existed with phosphotriesterases (PTEs), paraoxonase, and parathion-hydrolase (see Johnson and Talbot, 1983; Ecobichon, 1996; Rauschel, 1998; see also section E, below). In 1992, Maxwell cited these and other similar nomenclature problems in his review. Recently, all of the detoxifying enzymes described were shown to exhibit several isozymes and to be polymorphic, which adds to the complications (see, for example, Guengerich and Liebler, 1985; Nakatsugawa, 1992).

Ecobichon (1996) undertook a meritorious attempt to classify metabolic routes of OP insecticides, though earlier Eto (1976) had already described some of these routes (Figure 7-19). Ecobichon stressed that OP metabolism used many modes and mechanisms for detoxifying or activating that were both species specific and "dependent on substituent groups attached to the basic 'backbone' of the OP structure." He further pointed out that his categorization was primarily concerned with phase 1 of detoxification, which included oxidative, reductive, and hydrolytic reactions leading to the toxicology of either inert or activated metabolites. Phase 2 mostly concerned conjugations converting the metabolite into a polar compound that can be readily excreted by mammals, including humans.

Ecobichon delegated 9 mechanisms for phase 1. Mechanism 1 relates to cytochrome's P450 oxidative (monooxygenase) desulfurations, which mainly lead to activating relatively nontoxic compounds. Other than cytochrome P450, the mechanism involves enzymes that utilize coenzyme-reduced nicotinamide adenine, dinucleotide phosphate (NADPH), and an NADPH-regenerating system to provide oxygen and electrons needed to produce polar-excretable metabolites. This monooxygenase system, also referred to as the mixed- functions oxidases system, was located in the hepatic microsomal endoplasmic reticulum of mammals and is present in other tissues and in insects.

Mechanisms 2 and 3 also depend on a process related to cytochrome P450, which leads to detoxification rather than activation. Mechanism 2 was a dealkylating reaction, and demethylation and deethylation occurred readily with the formation of an aldehyde. However, when the mechanism dealt with larger alkyl radicals, dealkylation was impossible. The insecticide chlorfenvinphos is a good example of this detoxification (Levi and Hodgson, 1992). Mechanism 3 was a dearylation reaction, resulting in the formation of the phenol and dialkylphosphoric acid. Diazinon, parathion and chlorpyrifos are the insecticides de-arylated by the P450-mediated processes.

Note that this system is capable of both activating and detoxificating these and other OP chemicals (Ma and Chambers, 1994; Larkin and Tjeerdema, 2000). Additional catalyses dependent upon the cytochrome P450 monooxygenase system involved an attack on side chains, which included the aromatic ring hydroxylation (mechanism 4), thioether oxidation (mechanism 5), deamination (mechanism 6), alkyl and N-hydroxylation (mechanism 7), and n-oxide formation (mechanism 8) (see Figure 7-19).

A process related to the cytochrome P450 monooxygenase-dependent metabolism is carried out by flavin-containing monooxygenases (FMO; Levi and Hodgson, 1992). Both systems are mainly localized in the hepatic microsomal endoplasmic reticulum, though they are also present in other tissues. Mammalian FMO has not been identified in insects (Levi and Hodgson, 1992). The FMO system was capable of carrying out a number of oxygenations and desulfurations (mechanism 9), which was exemplified by the oxidative desulfuration of the insecticide fonofos. Frequently, the FMO and the cytochrome P450 systems overlapped within the detoxification process of a particular OP agent. So, the effectiveness and the contribution of the two systems appeared moot (Levi and Hodgson, 1992).

A number of hepatic transferases (glutathione S-transferases) use glutathione as a cofactor and an acceptor of o-alkyl and o-aryl radicals to yield monodesmethyl, dialkylphosphoric or dialkylphosphorothioic metabolites (mechanisms 10 and 11). Parathion, methyl parathion, paraoxon, and methy paraoxon are detoxified by mechanisms 10 and 11 (see Sultatos, 1992). Strangely, while diazinon and mevinphos are metabolized in several species by glutathione-dependent processes, methyl parathion and additional OP compounds in the mouse are metabolized *in vitro* only by glutathione-dependent processes (Sultatos, 1992).

When expanding his classification, Ecobichon (1996) distinguished tissue hydrolases including carboxylesterases, arylesterases (versus the cytochromec P450-dependent arylase; see above), phosphorylphosphatases, phosphotriesterases, and carboxyamidases. Carboxylesterases were represented by EC 3.1.1.1, which was capable of detoxifying the important insecticide Malathion and chlorpyrifos (Lassiter et al., 1999; Dauterman and Main, 1996; mechanism 12). Arylesterase was described by Aldridge in 1951 and then designed like EC 3.1.1.2, the chlorpyrifos-oxon detoxifying enzyme (mechanism 13).

Phosphotriesterases (PTEs), also described as parathion or parathion hydrolases or paraoxonases, were hydrolases characterized by a unique binuclear metallic active center. The bivalent metal was zinc, though cadmium could be used to replace the zinc in a series of apo-enzymes (Rauschel, 1998; mechanism 14). When derived from certain *Pseudomonas* or *Flavobacterium* species, the PTE was defined as EC 3.1.1.3. Actually, while this mechanism involved the inactivation of paraoxon and the formation of a diethoxy phosphoric acid, other detoxification routes are also possible (Rauschel, 1998; Kasai et al., 1992). PTE, a versatile enzyme that uses OP drugs substrates and plays a minor role in hydrolyzing the drugs, exhibited "extraordinary" (Rauschel, 1998) turnover values. Altogether, PTE was a versatile enzyme capable of hydrolyzing a number of OP agents, ranging from insecticides such as paraoxon and EPN to war gases such as DFP, soman, sarin, VX, and tabun.

PTE mutates readily, and some of these mutations exhibit increased detoxifying potency toward war gases such as VX (Gopal et al., 2000). PTE is polymorphic, and several genotypes express its forms (Brophy et al., 2001). Currently, the enzyme is found in many insects, microorganisms, and mammalian tissues, including human serum, where it is referred to as serum paraoxonase, PON1 (see, for example, Furlong et al., 2000).

In 1998, Rauschel stressed that the ubiquitous presence of PTE is enigmatic and though OP chemicals cannot be considered natural substances, PTE (especially in view of its high turnover value with respect to many OP agents) is "an enzyme in search of its natural substrate." Of course, the same may be said for many of the enzymes capable of hydrolyzing OP agents. Bernard Brodie (1956) speculated that "detoxifying enzymes have been developed by a process of evolution to protect the organism from a multitude of foreign compounds that the organism may encounter during its lifetime." Therefore, they should not be expected to search for their natural substrates. Note that according to studies of paroxonases, DFPases, OP hydrolases, esterases, phosphatases, and arylesterases, the enzymes are not clearly identified following the categories outlined above and/or defined by Ecobichon (1996). (See, for example, Johnson and Talbot, 1983; Pond et al., 1998; Gill and Ballesteros, 2000; Mulbry, 2000; Mackness et al., 2000; Furlong et al., 2000; Qiao et al., 1999.) This "category" includes an important family of enzymes represented by OP anhydrolases (i.e., prolidases) that are capable of splitting the peptide links; prolidases include EC 3.1.8.2, an enzyme capable of splitting a wide range of OP war gases (Cheng et al., 1999).

Altogether, the routes and enzymology of OP insecticide and war gas detoxification are complex and manifold, as a number of attacks may be directed at a single OP chemical, as stressed by Mounter (1963) and then by Eto (1974). For example, fonofos may be detoxified by both FMA and cytochrome P450 systems (Levi and Hodgson, 1992). Parathion (besides being activated) is detoxified by a number of systems including the cytochrome P450 and FMO systems, glutathione-dependent transferase, several hydrolases, and so forth (Nakatsugawa, 1992). The major location for these systems is the liver. The detoxifying importance of the hepatic system is enhanced by the rapid and possibly active uptake of some OP drugs into the liver. Also, the circulatory characteristics of the liver and liver lobule prolong the presence of the poison in the liver and prevent its elution before significant detoxification can take place (hepatic release threshold; Nakatsugawa, 1992).

We should not be anthropomorphic with the detoxification system and its routes in mammals, including humans. Insects and pests, including worms, exhibit almost all of the systems described above. (See, for example, Qiao et al., 1999; Scharf et al., 2001; Siegfried and Scott, 1992, 1997) Chadwick (1963) and others proposed that these enzymes played a role in pests' resistance to OP compounds. Early on, investigators doubted this notion (see O'Brien, 1959, 1960). Today, the enzyme's role in pest resistance relates to the adaptive increase in pesticide resistance caused by selective genetic development of effective detoxifying enzymic systems in pests exposed to OP and carbamate pesticides for long periods of time (Eto, 1974). The ideal situation: pests are sensitive to OP drugs when they do not contain the detoxifying enzymes, whereas humans and animals have the detoxifying enzymes and are insensitive when accidentally exposed to the drugs.

The systems are also present in microorganisms, including carboxylesterases (EC 3.1.1.1), phosphoric triester hydrolases (PTE; EC 3.1.8) that include paraoxonases (EC 3.8.2.1 and EC 3.1.8.1), DFPase (EC 3.1.8.2), and a phosphoesterase (EC 3.1.3). (See Johnson and Talbot, 1983; Raushel, 1998.) Note: PTEs are present in soil bacteria (i.e., *Pseudomonas diminuta* and *Flavobacterium sp.*) and other microorganisms; like other hydrolases, the PTEs de-esterify the OP agents, yielding phosphoric acids; they are capable of inactivating a number of OP agents such as pesticides and war gases, including EPN, paraoxon, sarin, soman, and VX (Raushel, 1998).

Frank Raushel (1998) established the tridimensional structure of PTEs, their amino acid sequence, and the amino acid composition of their active sites, and he stressed that PTEs contain a binuclear metal center (similar to a number of phosphatases) that facilitates the action of water molecules (Wilcox, 1996). Reiner and Radic (2000) consider these divalent cations as cofactors of PTEs. Therefore, chelation inhibits the PTE activity. DFPases are present in a great variety of soil and infectious bacteria, and the DFPase of *Proteus vulgaris* and *Salmonella* are particularly active (Mounter, 1963).

b. Metabolic Activation

Several OP agents must undergo metabolic transformation before becoming active inhibitors. The oxidative desulfuration of phosphorothioate esters (type D OP drugs; see section BII-1, above) was the only metabolic system reviewed that produces this activation (Diggle and Gage, 1951; see also Ecobichon, 1996). For example, oxidative desulfuration activates phosphorothioates such as parathion, fenthion, and Malathion. A spontaneous rearrangement of some OP insecticides also produces activation (see Eto, 1974).

c. Metabolic Potentiation of Toxicity

Though the mechanism for potentiation of toxicity (rather than an additive effect) is not clear, it often results from combined administration of 2 antiChEs, whether they are 2 carbamates, 2 OP drugs, or an OP and a carbamate compound (see Hayes, 1975; Karczmar, 1967). However, when potentiation of toxicity results from administering a combination of a quaternary and tertiary antiChE, combining the 2 drugs' central and peripheral effects causes the potentiation. A special case of toxicity potentiation arises from a joint administration of 2 insecticides, Malathion and EPN (Frawley et al., 1957). Cook et al. (1957; see also Dubois, 1963) presented data suggesting that the phenomenon would result from EPN inhibition of Malathion hydrolysis by carboxylase. Ken Dubois (who was very active in this area) identified 4 pairs of OP insecticides that inhibited one another's metabolism and exhibited mutual potentiation of toxicity when simultaneously administered to animals. Others identified additional pairs (Dubois et al., 1949; Dubois, 1963; see also Chambers, 1992; Chambers and Levi, 1992).

There is an alternative explanation that agrees with Ken Dubois' notion: strong potentiation of toxicity occurs in dogs when first administering EPN and then Malathion. However, potentiation of toxicity also arises when the sequence is reversed. Furthermore, robust toxicity arises when Malathion is administered in subeffective doses 45 minutes prior to subliminal doses of EPN. Since Malathion is rapidly metabolized, it is inconceivable that toxicity could arise from EPN's ability to protect small amounts of Malathion, which remain present during EPN administration (Karczmar et al., 1962). Potentiation was also present when the effects of the 2 drugs were measured at specific neuroeffectors (ganglia and neuromyal junction; Karczmar et al., 1962; see also Wescoe et al., 1950).

DI. Toxic Effects of Anticholinesterases

Two important issues concern the matter of the toxicity of antiChE: (1) Used as pesticides, OP and carbamate agents are frequently toxic to humans, via skin contact or via inhalation. (2) OP drugs may be used as war gases, which are inhalable as well as absorbed on contact. This danger is prominent during wars and terrorist attacks and is with us today. Though during the 1940s the United States, Germany, Russia, and England were capable of using OP compounds, the compounds were not employed during World War II. However, following the war, these countries continued to develop new OP war gases and their antidotes. Additionally, many Middle Eastern and Asian countries, such as Israel, China, Pakistan, and Iraq, and some terrorist groups were engaged in both researching and producing OP agents. These drugs are easy to multiply because their synthesis is both simple and inexpensive; they can even be synthesized in your garage!

The studies involving OP and carbamate toxicity intensified as the countries competed to gain knowledge and develop the agents. Unfortunately, the majority of data are restricted. In both its own laboratories (at the Walter Reed Hospital in Washington, DC, or at the Army Chemical Center, Edgewood, Maryland) and through subcontracting, the US Department of Defense (DOD) researched the OP drugs and their antidotes (see Karczmar et al., 1962; Moore, 1998). These studies led to the development of a treatment "cocktail" (see below, section E). Prior to an expected attack with OP agents, pyridostigmine is administered as part of the "cocktail"; this use of pyridostigmine has a bearing on controversy regarding Gulf War syndrome. (See below, this section and section DII.) When this kind of research was conducted in the United States and other countries, there were accidental poisonings with OP and carbamate drugs (see below, section E).

The combat use of OP drugs is considered a gray area, and for many episodes solid evidence is not available. However, there is good documentation regarding Iraq's use of the agents during the 1983–1984 war with Iran and Iraq's attempt to eliminate its Kurdish minority (Balali-Mood and Shariat, 1998). During the 1990 conflict in Yugoslavia, Bosnians, Serbs, and Croatians were accused of using OP agents, though the information is mostly anecdotal. (See Jane's Defense Weekly, August 21, 1993; E.J. Hogendoorn, personal communication.) Other examples are Operation Desert Shield (ODS; August 1990 to late February 1991) and Gulf War syndrome (PGWS) or "mystery illness." There is a possibility that Gulf War syndrome is related to either Iran's use of OP agents or the use of pyridostigmine as an OP antidote. However, this issue is controversial. (For a review, see section D3.)

OP war gases are used by rogue nations and by terrorist organizations. There were two incidents in which a religious fanatic group in Japan used sarin. A more serious incident, the Tokyo subway tragedy of 1995, involved around 5,500 people and resulted in serious poisonings and fatalities (Yokoyama et al., 1998). After the September 11, 2001, bombings of the Twin Towers and the Pentagon, the possible use of OP agents as terrorist weapons became alarming, although these agents were not used in the attacks in question. In fact, it appears that the terrorist organization Al Qaeda may possess and/or tried to develop both biological and chemical weapons of mass destruction (WMD; Clarke, 2004). It should be added that, beginning with President Clinton's administration, the US government (and indubitably other governments as well) has pursued via several agencies and also under a national coordinator studies concerning the detection of terrorist attempts at the use of WMD and the reaction to these attempts if they occur (Clarke, 2004).

OP and carbamate toxicity actions mainly result from their inhibition of synaptic AChEs along with protection and accumulation of naturally released ACh. ACh accumulation at the cholinergic synapses translates into cholinergic activities and behavior. Also, BuChE inhibition may contribute to the toxicity. There are different degrees for antiChE inhibition of BuChEs and

AChEs (see sections BI-2 and BII-2). Even war gases like G-compounds may exert a significant BuChE action. For example, DFP is almost equipotent against the 2 ChE families (Holmstedt, 1959, 1963; Radic et al., 1993). Furthermore, selective BuChE inhibition increases the brain's ACh level (see Giacobini, 2000). However, the physiological and pharmacological role of BuChEs is unclear, or "mysterious" (Mesulam, 2000), and pharmacological and toxic antiChE effects are mainly the result of AChE inhibition. (See Chapter 3 DIII. Findings concerning AChE inhibition and ACh accumulation are presented in this chapter.)

Since the central as well as peripheral (neuromyal and ganglionic) toxicological effects of antiChEs depend mainly on their accumulation of ACh, they mimic the toxic effects of ACh and/or cholinergic agonists at these sites; in fact, the toxic effects are preceded by actions similar to pharmacological (or physiological), central and peripheral, actions of ACh (see Chapter 9). The toxicology of antiChEs goes beyond their central actions, as, under certain circumstances, antiChE toxicity includes peripheral, autonomic, and motor phenomena.

Additional phenomena augment antiChE toxicity. Many OP antiChEs and related OP compounds that are weak ChE inhibitors induce several types of neuropathies including neuronal hyperactivity-induced neuropathy and OPIDN. Another OP-induced pathology is delayed pulmonary toxicity (DPT; Chambers, 1992). AntiChEs also exert direct neuronal actions at muscarinic and nicotinic receptors that contribute to their toxicity or delayed actions but do not depend on ChE inhibition. These direct actions may result in delayed cognitive toxicity (DCT). Then, either directly or by inhibiting ChEs, antiChEs exhibit morphogenetic and teratogenic actions.

Recently, the possibility was raised that some of the chronic toxicity of OP agents is due to the stimulation of free radical production, induction of oxidative stress, lipid peroxidation, and disturbance of the antioxidant capacity of the organism; these effects may or may not depend on AChE inhibition and on resulting neurological disturbances, and they were demonstrated both in humans and in animals (Ranjbar et al., 2002; see also Abdollahi et al., 2004).

1. Cholinergic Toxicity of Anticholinesterases

a. Acute Toxicity of OP Anticholinesterases

The toxicity and fatal actions of OP antiChEs discussed in this section mainly result from the inhibition of AChE (with a contribution from the inhibition of BuChE) and accumulation of ACh. In view of their lipid solubility and excellent penetration across the blood-brain barrier, tertiary OP antiChEs (i.e., G-agents, TEPP, and DFP) and all of the OP insecticides (i.e., Malathion, EPN, parathion, diazinon, chlorpyrifos, etc.) exert toxic and pharmacological actions through combined autonomic parasympathetic, sympathetic, neuromyal, and central effects. Even quaternary OP agents, such as echothiophate, may penetrate the CNS (particularly when in the presence of stress) by damaging the blood-brain barrier and following the advent of peripheral actions, thus exhibiting central effects (Karczmar, 1967; Soreq and Glick, 2000).

AChE is present in high concentrations in cholinergic neurons (and is released from the neurons, (Chapter 3 D) and its inhibition by the OP drugs induces ACh accumulation at the synapse. Evoking the effects of OP (and carbamate) antiChEs requires considerable inhibition of central and peripheral neuronal enzymes. During evolution the "excessive" amounts of the enzyme may have developed to protect the organism from the consequences of accidental inhibition by certain foods and by naturally occurring antiChEs (see above, section BI). An inhibition of 50% or more (Monnet-Tschudi et al., 2000) of the neuronal AChE may be needed before any pharmacological effects occur; an even higher level of inhibition is required for toxic signs (Karczmar, 1967).

Inhibition of blood AChE is frequently measured when examining antiChE poisoning in humans (Winteringham, 1966). The relationship between that inhibition and a neuronal inhibition (particularly in the CNS) depends on the agent's tissue distribution and varies from one agent to another. Blood AChE inhibition of 50% may be pathognomic for OP toxicity. This degree of inhibition may correspond to either higher or lower levels of brain AChE inhibition (Mileson et al., 1998). Altogether, it is difficult to predict the relationship between AChE inhibitions in the blood or CNS and ACh accumulation on the one hand and the toxicity or fatality resulting from OP drug exposure on the other (Sheets et al., 1997; see also Mileson et al., 1998). The problem is compounded when, in the cases of poisoning, plasma BuChE or the combined whole blood BuChE and AChE are assayed, rather than the red blood cell AChE, the latter being a better marker of antiChE toxicity. Contrary to Ecobichon (1996), monitoring potential OP intoxication in this manner is not reliable.

The OP agents' route of penetration is important. Since war gases (the G-agents) are dispensed as gases, they become toxic to humans when they are inhaled and absorbed through the skin and the lung epithelium. Local war-related toxicity from G agents occurred repeatedly, and there were cases of terrorists using the G agents (see section A). Finally, both human voluntary and forced (in Nazi Germany) studies of OP and carbamate drugs (the latter being studied as potential antidotes) occasionally led to behavioral toxicity and persisted for months (see below).

Dermal and inhalation routes of intoxication are characteristic of the OP insecticides (Gershon and Shaw, 1961; Ecobichon, 1996). Sometimes, an accidental drenching of agricultural workers with insecticides may occur (Ecobichon, 1996). Currently, some 100 cases of accidental toxicity caused by OP insecticides occur each year in just the United States, and these accidents are extremely common in underdeveloped countries (K. Hamerlink, W. Chen, personal communication). In Japan and Australia, outbreaks of epidemiological insecticidal toxicity took place when the agents were used agriculturally. Moreover, workers involved in producing OP insecticides or G agents in Germany, other European countries, and the United States have suffered from OP intoxication (Spiegelberg, 1963; Duffy and Burchfiel, 1980; see below, next section). On record, there are suicidal attempts with OP insecticides. There were also cases of fatal sheep intoxication with OP insecticides in Australia, the western United States, and elsewhere.

Although OP toxicity can occur readily upon contact (see section A), OP investigators did not suffer from severe toxicity (even those who have worked in the field for many decades). However, there were cases of occasional miosis and bronchiolar constriction. A perfect example was De Clermont's survival from ingesting TEPP (Figure 7-20; see section A; Holmstedt, 1963). In animals such as cats, dogs, rats, rabbits, guinea pigs, and primates, acute toxicity of lipid-soluble OP agents (effects of doses near or above an LD50 dose of the drug) induced by inhalation of OP gases, ingestion, or systemic administration present an alarming picture that is unyielding in seriousness to the toxicity induced by strychnine or cyanide: peripheral autonomic and neuromuscular (somatic) effects and central actions are present (see Dubois, 1963; Wills, 1963, 1970; Karczmar, 1967, 2000; Ecobichon, 1996).

The parasympathetic symptoms involve the exocrine glands, eyes, gastrointestinal and respiratory smooth muscle, bladder, gall duct, and peripheral cardiovascular system. They include miosis, salivation, lacrimation including chromodochryroea (bloody tears, in rats), sweating, bronchorrhea, laryngospasm, micturition and defecation, bronchoconstriction, bradycardia, and hypotension (frequently observed following initial increase of heart rate and blood pressure). The salivation,

Figure 7-20. Philippe De Clermont, 1830–1921. (From Koelle, 1963, with permission.)

lacrimation, urination, and defecation constitute, in rodents, the so-called SLUG syndrome; the effects upon salivation and the respiratory smooth muscle contribute to the accumulation of fluids in the upper airways.

The sympathetic actions include those on the sympathetic ganglia and adrenals. While early effects may include ganglionic stimulation and release of norepinephrine or epinephrine, ganglionic paralysis ensues, which contributes to hypotension. The neuromuscular phenomena include early muscle fasciculations, which are particularly visible in the eyelids, tongue, facial muscles, and muscle cramps. Following these phenomena, the tendon reflexes are diminished and muscles including the intercostals and the diaphragm become flaccid. The central effects begin with restlessness, increased motor activity, respiratory and sensory hyper-responsiveness, which is followed by ataxia and tremors and then by attenuation of reflexes; extrapyramidal syndrome is present as well (Hsieh et al., 2001). Cheyne-Stokes respiration and later dyspnea and respiratory depression and convulsions follow; then there is coma and respiratory arrest as miosis yields to mydriais.

Toxicity and lethality of lipid-soluble OP have been well documented in humans. (For the early description of this toxicity, see, for example, Bidstrup et al., 1953; Comroe et al., 1946.) The documentation involves human volunteers, the use of OP drugs in war, exposure to OP pesticides, suicides with OP drugs, toxicity arising from industrially manufacturing OP agents, and victims of terrorist attacks (the Tokyo subway tragedy). Moreover, sarin was used in the Tokyo subway. The peripheral and central OP actions noticed in survivors (9 subjects died) were classical, though they did not include seizures, which are more likely to characterize fatal incidents (Yokoyama et al., 1998; Balali-Mood and Shariat, 1998; Okumura et al., 1996; Moore, 1998; Wills, 1970; Karczmar, 1981, 1998, 2000; Ecobichon, 1996; M. Gaon and J. H. Holmes, unpublished data culled from 525 cases of industrial exposure to sarin). In 1966, sarin escaped from a building manufacturing the drug, causing mass toxicity. Fortunately, there were no deaths among the 41 victims (Wills, 1970).

During the 1960s OP toxicity was tested in the United States in groups of volunteers. One group of volunteers was administered physostigmine, carbamate protectors (i.e., pyridostigmine) of AChE from phosphorylation, large doses of atropine, and additional agents (Karczmar, 1981). Another volunteer group was sprayed with VX and sarin. After the event was disclosed in 2001, the Veterans Administration sought to compensate those that may have suffered from ill effects (Anonymous, New York Times, May 26, 2002). Apparently, aside from the immediate acute effects (i.e., parasympathetic hyperactivity and muscle twitches), mood, cognition, and sleep changes occurred and persisted at length. Additionally, testing or manufacturing OP drugs led to accidents in humans and animals. These incidents were reported in the US press, on the radio, and on television (for example, Tom Brokaw, NBC Nightline, November 11, 2002; Michel Martin, NBC Nightline, February 18, 2003; see also Wheelwright, 2002).[4]

In cases of human carbamate or OP intoxication, general acute symptoms of peripheral nicotinic and muscarinic intoxication are undoubtedly apparent. These symptoms include pathognomonic miosis; impaired accommodation and scleral injection; sweating, rhinorrhea, lacrimation, and salivation; abdominal cramps and other gastrointestinal symptoms; respiratory difficulties and cough; dyspnea, constriction sensation in the chest, wheezing, and rales; twitching of facial muscles and tongue, tremors, and fasciculations; bradycardia and EKG changes, pallor, and cyanosis; anorexia, nausea, vomiting, diarrhea, and urination. These signs and symptoms are accompanied by central effects such as dizziness, tremulousness, and confusion; ataxia; headache, fatigability, and paresthesia. Finally, seizures, convulsions (with appropriate EEG changes), twitching, coma, and respiratory failure may occur (Wills, 1970; Okumura et al., 1996; Balali-Mood and Shariat, 1998). If the subject survives past the day of poisoning, there are personality changes, mood swings, aggressive events and psychotic episodes including schizoid reactions, paranoid and religious delusions, and exacerbations of preexisting psychiatric problems. Sleep is poor from nightmares and hallucinations; disturbances or deficits in memory and attention, and additional delayed effects also occur (see below, section DI-3; Ward et al., 1952; Karczmar, 1984).

The cause of death upon acute intoxication with the lipid-soluble OP drugs depends on species. Lethality is mainly muscarinic and results from bronchoconstriction in the rabbit and rat ("asthmatic" death). For both cats and humans, central and neuromyal actions involving respiratory functions are significant. Irwin Wilson (personal communication) stressed that an impaired relationship between inspiratory and expiratory medullary centers is crucial for toxicity.

There is also the question of convulsions, their relation to respiratory collapse, and their role in the lethality of OP drugs. In cats and possibly in humans, convulsions are not induced by respiratory depression and anoxia. Instead, OP drugs directly induce convulsions (Glenn et al., 1987; Lebeda and Rutecki, 1987; see also Chapter 9 BIV). However, the CNS depression that follows prolonged convulsions induced by OP drugs is not the cause of OP death in cats or humans, as the immediate cause of death in these species is the direct depressant action of the drugs on the respiratory centers (secondary to initial respiratory stimulation; see Chapter 9 BIV). This block occurs before the circulatory collapse and does not always follow a seizure (Oberst et al., 1956; Wills, 1970; Karczmar, 2000). Many of the toxic effects (including ganglionic and neuromyal depression and the block of respiratory centers) are a result of the OP drug-induced synaptic depolarization or desensitization (see Chapter 9 BIII).

b. Chronic Toxicity of OP Drugs and Development of Tolerance

Under certain circumstances, chronic, sublethal exposure to OP drugs induces cumulative toxicity in animals (sheep and cows). The ChE inhibition following the first exposure to sublethal air concentrations of OP agents can be augmented by subsequent exposures that result in chronic toxicity (see Callaway and Davies, 1957 and Hayes, 1982). Chronic toxicity might have occurred among agricultural workers using OP pesticides and workers employed in OP-related industries (Grob, 1963). Delayed symptoms (i.e., lung damage or DCT) may accompany chronic toxicity (see below).

This cumulative toxicity occurs with repeated doses or concentrations of OP drugs (i.e., 1/3 of the LD50 or IC50). Each dose induces a significant AChE inhibition, distinct effects (e.g., intestinal discomfort and muscle twitches), or minor toxic phenomena (e.g., bronchiospasm, respiratory difficulties). It is not clear whether smaller doses or concentrations of OP agents (i.e., those that induce miosis) applied once a day are conducive to accumulation (Moore, 1998).

Developing tolerance is an unexpected aspect of chronic OP drug exposure, and during World War II German investigators discovered this phenomenon in mammals and, perhaps, in humans (Bo Holmstedt, personal communication). Contrary to the amounts of OP drugs that cause cumulative actions, tolerance occurs in mammals with distinct subtoxic doses of OP drugs (approximately 1/15 and 1/8 of the LD50 dose of OMPA) and DFP or parathion (Dubois et al., 1949; Glow and Richardson, 1967; Rider et al., 1952).

Barnes and Denz (1951, 1954) demonstrated that tolerances to the insecticides Schradan and Systox occurred when signs of toxicity were overt, such as tremors and convulsions (see also Dubois, 1963; Hoskins and Ho, 1992). Bombinski and Dubois (1958; Dubois, 1963) showed that "typical cholinergic effects" occurred with a daily 1/4 or 1/8 LD50 dose of Di-syston (Disulfoton) in rats during the first week of treatment and disappeared during the following 3 weeks of study. Furthermore, when Di-syston was given in a daily 1.0-mg/kg dose (1/4 LD50), brain and blood levels of AChE gradually decreased by 80% and were maintained at this level, though serious toxic effects were absent. This level, when achieved in acute experiments, is more than ample for associating the effects with toxicity. Maintaining this low level of AChE following OP exposure was confirmed for other tissues (see, for example, Gupta et al., 1985). Accordingly, the animals became tolerant to an abnormally low level of AChE, which might persist for the significant portion of a rat's life (one year). Interestingly, aquatic larval urodeles convulse with doses of DFP sufficient for complete ChE inhibition; however, as they breathe through their gills and skin, the urodeles can survive, completely recover from convulsions, and swim normally while their ChEs are still completely inhibited (Karczmar and Koppanyi, 1953).

In the nineteen seventies, Russell, Jenden, and Overstreet (see, for example, Russell et al., 1958, 1971) and Lim, Ho, and Hoskins (see Hoskins and Ho, 1992) conducted studies utilizing consummatory, motor activity, and operant techniques to demonstrate OP tolerance. Some of these studies revealed that insecticides and war gases (i.e., soman, sarin and DFP) induce tolerance. However, Hoskins and Ho (1992) felt that tolerance to G agents is "of low level." These investigators also concluded that only certain behaviors or effects might be subject to tolerance. Brodeur and Dubois (1964) and Hoskins and Ho (1992) also showed that there was a cross-tolerance among various OP agents, OP drugs, and cholinomimetics.

Several mechanisms enabling organisms to develop tolerance to OP drugs were explored, such as the enhancement of OP drug disposition. OP drugs are metabolized by cytochrome P450 systems, and inducing these systems by phenobarbital and other agents helps develop tolerance to OP drug toxicity (Dubois, 1963; see also Hoskins and Ho, 1992). Investigators also proposed that inducing other detoxifying systems or enzymes, such as OP phosphatases and the neurotoxin esterase, an enzyme involved in OP-induced delayed neurotoxicity (CDPIDN: see below, next section), may also facilitate the development of OP tolerance (see below, section 1-c and DII; see Hoskins and Ho, 1992).

However, these propositions are not consistent with the finding that tolerance can develop when AChE levels in blood and other tissues are low. Woolf Dettbarn and associates attempted to resolve this inconsistency (Gupta et al., 1985; Dettbarn et al., 1999; Milatovic and Dettbarn, 1996). They found that, similar to AChE, carboxylesterases and aliesterases are inhibited by OP drugs (i.e., paraoxon) and recover rapidly. Furthermore, chronic treatment increases brain and diaphragm AChE affinity to ACh. Dettbarn felt that these two phenomena contribute to the development of tolerance on chronic treatment. He also pointed out that AChE of the muscle (but not brain) recovers when tolerance is developed. Ultimately, the mechanism proposed by Dettbarn and others does not apply to *all* OP drugs and is not consistent with prolonged low levels of AChE activities following OP drug exposure.

Tolerance may result from appropriate changes in storage, synthesis, and/or release of ACh and/or choline uptake (Brodeur and Dubois, 1964; Wecker et al., 1977; Hoskins and Ho, 1992). The data are not consistent with the values of stored, free, and total ACh, choline uptake, and brain parts involved in these changes. Overstreet and associates (1974) proposed that changes in affinity, sensitivity to cholinergic agonists or antagonists, binding, and number (downgrading) of nicotinic or muscarinic receptors in the CNS underlie tolerance (Gupta et al., 1985; Schwartz and Kellar, 1983, 1986; see Hoskins and Ho, 1992). However, the data do not explain the development of tolerance, and it is unclear how increased sensitivity to atropinics or nicotinics relates to tolerance (Overstreet et al., 1974; Fernano et al., 1986).

Another aspect of chronic antiChE toxicity concerns interaction among transmitter systems. Hoskins and Hoh (1992) proposed that during the development of tolerance, increased cholinergic activity is associated with increases in dopaminergic *and* GABAergic activity, thus the balance between the transmitter systems remains unchanged (see also Chapter 9 BIII-2). Hoskins and his associates (see Hoskins and Ho, 1992) based their tenet of increased cholinergic activity on finding increased levels of ACh. However, increased levels of ACh, GABA, or dopamine are not tantamount to increased turnover and increased activity of these transmitter systems. Also, it is unclear why muscarinic receptors are downgraded during tolerance and why the balance between transmitters achieved at abnormally high activity levels results in tolerance (Hoskins and Hoh, 1992). Finally, repeated (or perhaps even single) administration of antiChEs, particularly to neonate or prenatal rodents and rabbits, induces changes in catecholamines and corresponding behavioral effects (see, for example, Karczmar, 2000; Glisson et al., 1974; Slotkin et al., 2002).

Finally, genetic mechanisms may be involved in resistance or tolerance. Therefore, in several species ranging from flies and larval flies to man, c-fos and gene expression mechanisms may induce tolerance upon chronic administration of OP drugs. This development of tolerance involves the augmentation of detoxification systems (Osei-Atweneboana et al., 2001; Yamada et al., 2001). Roger Nitsch, Heinrich von der Kammer, and associates (Kammer et al., 1998, 2001) demonstrated a related mechanism where antiChEs or ACh activation of M1 receptors increased the transcription of promoters, including the AChE gene promoter. In fact, AChE activity is upregulated by chronic administration of carbamate or OP antiChEs (see Giacobini, 2000).

In addition to acquired tolerance, resistance to OP agents is native to certain species. For example, some mammals (i.e., guinea pigs) and some pests (i.e., several species of flies) resist OP agents (see O'Brien, 1960; Dubois, 1963; Eto, 1974). Many factors cause the resistance, including skin or cuticle absorption, block of distribution, and the presence of detoxifying enzymes (i.e., carboxylases and OP hydrolases). Since enzyme levels are genetically controlled, mechanisms involved in the native resistance to OP drugs are related to mechanisms regulating tolerance and acquired resistance (see, for example, Osei-Atweneboana et al., 2001; Wallace, 1992).

c. Delayed Toxicities, Including Gulf War Syndrome, Delayed Cognitive Toxicity, OP-Induced Delayed Neurotoxicity and Delayed Pulmonary Toxicity

Early on, the general opinion among investigators was that sublethal effects of antiChEs were relatively short lived and related to AChE inhibition in humans or animals (see, for example, Grob et al., 1947; Grob, 1963; Ward et al., 1952). However, it was soon observed that post-OP drug administration changes, EEG abnormalities, insomnia, and other behavioral effects lasted from 2 to 9 weeks. This long duration was inconsistent with AChE's short resynthesis time and the rapid behavioral adaptation of the organism (see this Chapter, section DI1-b and Chapter 3 DII). Furthermore, even longer-lasting effects in humans and animals extending long past ChE resynthesis were reported during the 1940s, 1950s, and 1960s (Gershon and Shaw, 1961; Spiegelberg, 1963; Marrs, 1993, Karczmar, 1981, 2000; Ecobichon, 1996). These early cases involving industrial workers engaged in OP manufacturing took place during World War II in Germany (Spiegelberg, 1963) and probably in the United States. Also, Mohamed Abou-Donia and his associates (Abou-Donia and Preissig, 1976; Abou-Donia, 1981, 1982; Abou-Donia and Lapadula, 1990) reported that several insecticides, including Malathion and permethrin, administered chronically to rats alone or in combinations, produced deficits of sensorimotor function, while AChE activity was decreased in the midbrain but not in the cortex; also, they found that this chronic treatment induced neurodegeneration in several brain parts. This latter finding is somewhat unusual, particularly as it was obtained by dermal application of moderate doses ("real-life doses," as stated by the authors.)

Alarming effects occurred among agricultural workers in Australia long after their exposure to OP pesticides (Gershon and Shaw, 1961). The symptoms included the following array of effects: tinnitus and nystagmus, pyrexia, ataxia, paresthesia, polyneuritis, tremor and paralysis, slurred speech, loss of memory, confusion, dissociative cognition, and loss of concentration. Drowsiness combined with insomnia, somnambulism, excessive dreaming and nightmares, generalized weakness and lassitude, emotional liability, restlessness, anxiety, and schizoid episodes also occurred. It is of interest that long-term emotional effects of OP drugs may be induced via their effect on synthesis and release of cytokines (Carvajal et al., 2005). Also, it was occasionally suggested that immunopathology (such as changes in lymphocytes) may be caused by OP agents, including OP insecticides (see, for example, Navarro et al., 2001; Tarkowski et al., 2004). It should be noted that, in animals at least, pre- or neonatal exposure to small, nontoxic doses of OP drugs (particularly OP insecticides) may sensitize their CNS to antiChE exposure in adulthood (see, for example, Eriksson and Talts, 2000).

Ecobichon (1996) felt that other investigators did not duplicate this "array of symptoms." However, the results obtained after Gershon and Shaw's study supported the notion that toxicity (including DCT) was delayed past AChE recovery. Important cases involved delayed long-lasting DCT that affected agricultural workers dealing with OP pesticides and workers involved in manufacturing OP drugs. In 1963, Spiegelberg reported on these occurrences, which were further described by Whorton and Obrinsky (1983), Rosenstock et al. (1991), and Duffy and Burchfiel (Duffy et al., 1976; Duffy and Burchfiel, 1980; Burchfiel and Duffy, 1982). In fact, Duffy and Burchfiel reported that quantitative EEG changes and changes in sleep, memory, and personality were present even 2 years after exposure. These investigators replicated their findings in the monkey (see also Steenland et al., 1994; Karczmar, 2000).

Following suicidal attempts with OP insecticides, a similar delayed syndrome may occur. Senanayake and Karalliedde (1987) described it as the "intermediate syndrome," which follows early cholinergic symptomology. Also, several months or years after victims were exposed to sarin during the Tokyo subway terrorist attack, EEG and EKG changes, deficits of psychomotor and motor performance, memory decline, fatigue and evidence of post-traumatic stress, imbalance, and general health deterioration were still present (Yokoyama et al., 1998; Nishiwaki et al., 2001). According to Yokoyama's and other's studies, classic cholinergic symptoms and/or diminution of blood AChE activity took place before the delayed toxic syndrome. Note that delayed toxicity as well as tolerance appear to be independent of AChE activity.

During the 1940s, Corneille Heymans (a Nobel Prize man, Figure 7-21), Heinrich Schaeffer, Theodore Koppanyi (Figure 7-22) and this author proposed that direct receptor action of antiChEs (independent of their inhibition of ChEs) underlies the delayed and prolonged effects of OP agents (Heymans, 1951; Koppanyi et al., 1947; Koppanyi and Karczmar, 1951; Schaeffer, 1947). There are several reasons for this conclusion: (1) These investigators found a poor relationship among ChE inhibition and overt and pharmacological actions of OP drugs. (2) The OP drugs' pharmacological actions were augmented when using doses of OP agents from 5 to 10 times higher than those needed for complete BuChE or AChE inhibition. (3) Finally, OP agents potentiated pharmacological actions of nonhydrolyzable cholinergic drugs such as nicotine. Later, VanMeter et al. (1978) gave cats tetraethyl pyrophosphate (TEPP) to induce complete AChE inhibition, EEG arousal, and occasionally EEG seizures. After a few hours, the EEG returned to normal while AChE inhibition remained complete. Subsequently, TEPP dosing induced repeated EEG arousal patterns. These findings indicated that antiChEs of the OP (and also carbamate type) elicit direct actions, unrelated to their inhibition of ChEs.

Figure 7-22. Theodore Koppanyi in 1952.

Figure 7-21. Corneille Heymans in the nineteen fifties.

Tissue lesions are direct actions of OP drugs, though it is not clear whether this effect is cholinergic receptor mediated or dependent on other target sites. Wills (1970) reported that OP drugs might cause tissue damage, including neurotoxic action, specifically OPIDN (see below). However, aside from OPIDN, other phenomena concerning both nervous and nonnervous tissues also occur with OP drugs, especially the drugs that do not appear capable of causing OPIDN. Thus, in animals and in humans, postmortem lung lesions were found in lethal cases of poisoning from OP drugs (Grob et al., 1950; Wills, 1970). There were also cases of sublethal poisonings and DPT, which most likely occurred with phosphorothionates (Aldridge and Nemery, 1984; Aldridge et al., 1985; Verschoyle and Aldridge, 1987). The pathology included changes in pulmonary gas exchange, interstitial thickening, and necrosis of type 1 alveolar epithelial cells in the rat. Aldridge and associates proposed that these changes resulted from the inhibition of certain lung enzymes and toxic metabolites of phosphorothionates.

OP agents (GB chemicals) may induce a number of muscle and heart pathologies, neuropathies, and neurotoxicities. Several explanations were offered for the direct actions of OP agents, other antiChEs, and the tissue damage that they inflict. These effects may involve the second messenger systems, nicotinic and muscarinic receptors and/or channels (see Chapter 9 BIII). For example, the Eldefrawis and their associates proposed that direct actions of OP drugs on pre and pos-synaptic muscarinic receptors might contribute to their toxicity (see for example Katz et al., 1977). However, they acknowledged that these compounds' receptor affinity is several magnitudes lower than their AChE affinities.

Many investigators attributed the direct actions of OP drugs and/or tissue damage to hypoxia that results from seizures. For example, this was Britt et al.'s (2000) explanation for muscle and heart lesions induced by OP drugs in the macaque monkey. However, Woolf Dettbarn and associates induced myopathies and brain neurotoxity in rats with sublethal doses of DFPP that did not cause convulsions (Dettbarn, 1984; Yang and Dettbarn, 1998). Therefore, Dettbarn attributed myopathies inflicted by OP antiChEs to cholinergically induced muscle hyperactivity and/or fasciculations, Ca^{2+} mobilization, and formation of free radicals from changes in cytochrome c-oxids/ase and xanthine oxidase.

Organophosphorus-induced delayed neurotoxicity is a specific delayed neurotoxic syndrome (unrelated to the syndromes discussed above). The first known occurrence of OPIDN, paralysis, or polyneuritis ("polynevrites") was discovered in 1896 and resulted from treating tuberculosis with phosphocreosote (Roger and Recordier, 1934; Lorot, 1996; see also Davies, 1963; Richardson, 1983; Abou-Donia, 1992; Abou-Donia and Lapadula, 1990). Subsequent outbreaks of OPIDN resulted from a pentavalent OP agent called triorthocresylphosphate (TOCP) that, when in the form of an oily substance, could be employed as an adulterant or additive. In the 1930s, TOCP was added to the alcoholic extract of Jamaican ginger to augment the extract's intoxicating factor. The resulting neurotoxicity was called "ginger jake paralysis" or "jake leg" (Smith et al., 1930; see also Ecobichon, 1996). At the same time, TOCP was found to be an adulterant of a parsley extract, which was employed in France and Holland as an abortifacient (Ter Braak, 1931). Ingesting cooking oils containing TOCP as an additive also caused OPIDN. This intoxication was common in Germany during World War II and occurred in other countries. Organophosphorus-induced delayed neurotoxicity also affected industrial workers producing paints containing TOCP (see Davies, 1963).

Many organophosphorus agents can cause OPIDN or similar syndromes. For example, Koelle and Gilman (1946) described that prolonged paralysis occurred in dogs after administering sublethal doses of DFP. Bidstrup et al. (1953) reported that OPIDN-like symptoms in humans resulted from mipafox exposure. Later, many investigators related OPIDN to several (but not all) OP drugs (Johnson, 1982; Olajos et al., 1978). Altogether, as contaminants, insecticides, tools for suicides, or oil ingredients, certain OP drugs induced OPIDN all over the world (Abou-Donia and Lapadula, 1990). Since the late 19th century, around 100,000 cases of OPIDN have been identified.

The neurodegenerative phenomena of OPIDN in humans or susceptible animals include sensory loss and ataxia related to distal degeneration of sensory and motor axons in ascending and descending spinal tracts and peripheral sensory and motor nerves (US Environmental Protection Agency [EPA], 1985; see also Richardson, 1992; Abou-Donia, 1992). The first signs of OPIDN in humans are burning, cramping, and stinging pains followed by decreased sensitivity in the hands and legs (Abou-Donia and Lapadula, 1990). Besides ataxia, the motor disturbances may include initial flaccidity followed by spasticity and hyperreflexia or clonus. Originally, OPIDN was considered a demyelinating disease (see, for example, Davies, 1963). Today the nerve damage is identified as Wallerian, "dying back" degeneration of axons and their myelin sheath with demyelination as a secondary event. However, currently there is controversy involving the nature of the nerve damage.

At any rate, the pathognomic features of clinical OPIDN signs are aggregation and partial condensation of the axons, neurofilaments, and neurotubules accompanied by proliferation of the smooth endoplasmic reticulum (SER). Then, axonal degeneration causes SER and partially condensed neurofilaments and tubules to form granular, electron-dense masses (Bishoff, 1967, 1970, 1980; see also Abou-Donia and Lapadula, 1990 and Damodoran et al., 2001).

Davies (1963) and Abou-Donia (1992) distinguished between two subtypes of OPIDN. The damage caused by pentavalent OP insecticides is restricted to brainstem and peripheral nerves (type I neurotoxicity), while trivalent OP agents (referred to by Abou-Donia as triaryl phosphates) may cause Wallerian degeneration and chromatolysis, and necrosis of brain and ganglionic neurons. Also, both the Wallerian and neuronal lesions extend to the cortex and the thalamus (type II neurotoxicity).

Pentavalent OP drugs and phosphates have different degenerative action mechanisms and different OPIDN-inducing potencies in a given species of pests. Many pentavalents may cause OPIDN, including the popular insecticides mipafox, parathion, Malathion EPN, and Leptophos, along with some GB agents. Currently, few trivalents induce type II neurotoxicity. They are not used as insecticides, though the trivalent merphos is employed as a defoliant (Abou-Donia and Lapdula, 1990; Hayes, 1982).

The onset of OPIDN in humans may be delayed for weeks or months following exposure. Some OP compounds must be metabolized before they can induce OPIDN, part of the delay being due to this process; however, most of the OP compounds capable of causing OPIDN induce OPIDN directly (e.g., TOCP; see section C1). In animal models, the delay varies from several days to weeks. Complete recovery from full-blown cases of OPIDN in humans does not happen, though partial recovery begins with the hands and legs 16 to 18 months after the syndrome's onset. Spasticity and hyperreflexia seem irreversible, which sometimes leads to the misdiagnosing of persistent OPIDN as multiple sclerosis or encephalitis (Abou-Donia and Lapadula, 1990).

OPIDN is species specific; classical OPIDN (type I neurotoxicity) may be induced in hens (a favorite animal model for OPIDN), chickens, ferrets, monkeys, and farm animals. This induction excludes rats, mice, gerbils, rabbits, and guinea pigs, though type II neurotoxicity may be induced in rats. Symptoms such as length of delay and (see below) mechanism of action of syndromes caused by trivalents differ among the species. Another characteristic of OPIDN is age susceptibility. Young chicks and rats are more resistant to OPIDN induced by OP drugs. Repeated dosing by the appropriate compounds may induce OPIDN in chicks (Abou-Donia and Lapadula, 1990; Richardson, 1992; Abou-Donia, 1992).

Prospective OP insecticides are synthesized daily, and it is important to identify their OPIDN potential; OPIDN-SAR of these compounds may be used for this purpose. In 1963, Davies pointed out that certain thionophosphoryl fluoridates, bi-alkoxy phosphoryl fluoridates and alkoxy aryl phosphoryl fluoridates are active OPIDN-inducing compounds, and di-alkyl phosphofluoridates, alkoxy di-alkyl phosphinates, and derivatives of pyrophosphorus acid are not. (See above, section BII-1, and Table 7-1 for the structures in question and their classification status.) Abou-Donia (1992) stressed that phosphoryl fluoridates and di-alkoxy alkyl phosphinates are potent OPIDN inducers. (See section BII-1, Table 7-1, compounds A and B; Abou-Donia refers to these compounds as phosphonates and phosphofluoridates.) He found trialkoxy and thiol compounds to be weak OPIDN inducers. (See section BII-1, Table 7-1, compounds B and D; according to Abou-Donia's classification, these compounds are phosphorothioates and phosphates, respectively.) Indeed, dimethon, a thiol or phosphorothioate insecticide and dichlorvos, and trialkoxy (phosphate insecticide) are either incapable of causing OPIDN or are weak OPIDN inducers. Abou-Donia emphasized also that the presence of an ortho-methyl group in alkyl derivatives of phosphoric acid endows these compounds with considerable OPIDN induction power (Abou-Donia and Lapdula, 1990).

However, pentavalents other than methyl substituents at the R1 or R2 positions (see above, section BII-1) induce OPIDN. According to Richardson (1992), compounds containing at least one O or NH group linking an R group to phosphorus, or agents belonging to several of the A, B and other classes (various alkoxy compounds, see section BII-1), may cause OPIDN. Furthermore, when compared to imino phosphates with bulkier or longer substitutions, (Makhaeva et al., 1998) the imino phosphates (see section BII-1) endowed with short alkyl substituents are weak OPIDN inducers. Makhaeva and her associates used QSAR methods (see section BI-1) to analyze the OPIDN potentials for their imino compounds and stressed that OPIDN inductability was related to the high hydrophobicity of these agents.

Currently, the SARs of OP agents are not very clear. This may result from various investigators employing different classes of compounds for analyses, and generalizations valid for one class of agents may not have been valid for another. Also, comparing the various studies is difficult, as investigators employed different chemical nomenclatures with the compounds. While our knowledge of various pesticide capacities to induce OPIDN is quite extensive (probably more than 300 hundred pesticides were assessed for this capacity; Abou-Donia, 1992; Hayes, 1982), there are lacunae of pertinent information concerning insecticides, pesticides, and especially war gases. (DFP induces OPIDN, but the data are only anecdotal with other GB agents).

Predicting an OP agent's OPIDN-inducing capacity relates to the mechanisms involved in OPIDN. These mechanisms cannot include AChE inhibition because OPIDN occurs long after OP drugs cause cholinergic crisis and after spontaneous recovery or reactivation of the inhibited enzyme (Davies, 1963). In fact, the original culprit for jake paralysis TOCP does not inhibit AChE at all; rather, it inhibits BuChE. Moreover, early studies established that there is a dissociation between OPIDN induction capacity on the one hand and AChE inhibition of OP compounds on the other (see Wills, 1970; Davies, 1963). For example, Davies et al. (1960) found that similar to DFP, many potent AChE inhibitors (i.e., phosphofluoridates) caused OPIDN, while other potent inhibitors (i.e., pyrophosphates) did not. Then, OPIDN occurred with small doses of OP compounds, which caused limited AChE inhibition and did not evoke cholinergic symptoms (Leptophos is an example; Abou-Donia and Preissig, 1976). Similar discrepancies occurred in other studies but were not referred to by investigators (Tables 7-3 and 7-4; Casida et al., 1963 and Lankaster, 1960).

Kamata's findings (Kamata et al., 2001a, 2001b) are important: they clarify OPIDN independence from AChE inhibition. Kamata et al. (2001a, 2001b) reported that DFP induced OPIDN in hens after an 8-day delay, and following the setback, AChE activity in the membrane and the cytosol of the brain and spinal cord recovered. Since George Koelle's histochemical studies, AChE is mainly considered a membrane enzyme, thus the reference by Kamata and associates to the cytosol enzyme is rare (Koelle, 1963; see, however, Carrington and Abou-Donia, 1985). However,

Table 7-3. Insecticides that Produce Type I OPIDN in Humans

Compound	Chemical Designation
TOCP	Tri-o-cresyl phosphate
Omethoate	O-O-Dimethyl S-methylcarboxymethyl phosphorothiate
Leptophos	O-4-Bromo-2, 5-dichlorolhenyl O-methyl phenylphosphorothioate
Mipafox	N_1N'-Diisopropylphosphotodiamidic fluoride
Trichloronate	O-Ethyl O-2, 4,5-trichlorophenylethyl phosphorothiate
Parathion	O, O-Diethyl O-4-nitrophenyl phosphate
Methamidophos	O, S-Dimethyl phosphoramidothioate
Fenthion	O, O-Dimethyl O-4-methylthio-m-tolyl phosphorothiate
Chlorpyrifos	O, O-Diethyl O- (3,5,6-trichloro-2-pyridinyl) phosphorothioate

Table 7-4. Characteristics of Type I and Type II OPIDN. (From Abou-Donia, 1992, with permission.)

Characteristic	Type I	Type II
1. Chemical structure	Pentavalent phosphorus atom, e.g., TOCP, DFP	Trivalent phosphorus atom, e.g., TPP_1, $TOCP_1$
2. Species selectivity	Rodents are less sensitive	Rodents are sensitive
3. Clinical signs	Hen, flaccid paralysis Cat, flaccid paralysis Rat, no clinical signs	Hen, flaccid paralysis Cat, extensor rigidity Rat, partial flaccid paralysis, tail kinking, bidirectional circular motion
4. Length of latent period before onset of clinical signs	Hen, 6–14 days Cat, 14–21 days Ferret, 4 days Rat, no clinical signs	Hen, 4–6 days Cat, 4–7 days Ferret, 10–14 days Rat, 7 days
5. Age Sensitivity	Young chicks are insensitive	Young chicks are more sensitive
6. Neuropathological lesions	Wallerian-type degeneration of specific ascending and descending tracts of sensorimotor pathways of the brainstem and spinal cord and in peripheral nerves. No changes in nerve cell body or dorsal ganglia.	In the addition to the Wallerian-type degeneration, there is chromatolysis and necrosis of nerve cell body and ganglia; in addition to brainstem lesions, lesions also occur in corticol and thalamic regions.
7. Protection with phenylmethyl sulfonyl fluoride PMSF	Full protection against small doses, i.e., 125 mg TOCP per kg; partial protection against higher doses.	Protects against small doses, i.e., 250 mg TPP_1 per kg; synergizes neurotoxicity of large doses, i.e., 1000 mg TPP_1 per kg.
8. Inhibition of neurotoxic esterase NTE	Hen, 65–70% inhibition Rat, 65–70% inhibition	Hen, 70% inhibition Rat, 39% inhibition
9. Chromaffin cells	a. No effect on catecholamine secretion b. No effect on ^{45}Ca uptake evoked by 10 μM nicotine or 56 mM K^+ c. No effect on ATP synthesis via ^3H-adenosine Incorporation d. No morphological changes	a. Inhibition of catecholamine secretion b. Inhibition of ^{45}Ca uptake c. Inhibits ATP synthesis d. Swollen and disrupted mitochondria

their data indicated that, similar to the membrane enzyme, cytosolic AChE recovers from OP-induced inhibition before signs of OPIDN appear.

Though TOPC does not inhibit AChE, it is a potent butyryl enzyme blocker. Thompson and colleagues linked this finding with BuChE's white-matter localization because myelin is the site of OPIDN damage (see, for example, Earl and Thompson, 1952; see also Davies, 1963). Therefore, they suggested that OP-induced BuChE inhibition causes OP neurotoxicity and stressed that DFP and mipafox (i.e., TOCP) are more potent

BuChE than AChE inhibitors. However, the duration of BuChE inhibition by OP compounds does not overlap with OPIDN. Compounds such as pyrophosphates inhibit BuChE without causing OPIDN (see Davies, 1963; Abou-Donia, 1981; Class E of OP agents; see Table 7-1 in section BII, above). Today, BuChE induction of OPIDN is no longer pursued.

During the 1960s, Martin Johnson and William Aldridge came up with a novel mechanism for OPIDN. These investigators proposed that OPIDN resulted from inhibiting a protein they called neurotoxic esterase (NTE). During the next 15 years, Johnson provided a plethora of data supporting this notion (Aldridge et al, 1969; Johnson, 1969a, 1969b, 1970). However, the idea is controversial. For example, after his notion concerning OPIDN was attacked at the 1981 Symposium on Organophosphate Agents, Johnson (Johnson, 1981) was allowed only a brief comment.

In their early work, Johnson and Aldridge (Johnson, 1969b) showed that neurotoxic OP drugs were bound to protein constituents in hen brains that were not ligands of nonneurotoxic OP compounds. In addition, Johnson (1969b) carried out a series of experiments concerning interaction at various esteratic protein sites with nonneurotoxic or neurotoxic OP agents. Johnson concluded that the neurotoxic OP drugs bind with a large molecule that is different from AChE, BuChE, and other esterases, NTE. Other investigators concluded that OPIDN-inducing OP agents bind to a hydrolase or esterase that differs from ChEs (Kamata et al., 2001a, 2001b). Also, Kamata et al. (2001a, 2001b) showed that, in the cytosol and the membrane, binding NTE and AChE sites are different for physostigmine and several OP compounds. Finally, Johnson (1982) defined NTE as a phenyl valerate hydrolase (or esterase) resistant to paraoxon inhibition and sensitive to mipafox inhibition, which he related to the appearance of OPIDN.

Several methods for concentrating, purifying, and lyophilizing NTE were developed (see, for example, Makhaeva et al., 1998; Veronesi and Padilla, 1992). In the neurons, NTE is a membrane-bound enzyme, and NTE is present in the brain, spinal cord, peripheral nerves, and other tissues (Richardson, 1992; Abou-Donia and Lapdula, 1990). Note that similarly to OP drugs that inhibit ChEs without causing OPIDN, OPIDN-inducing antiChE agents promote classic ontiChE toxicity before the appearance of the symptoms of OPIDN.

Rudy Richardson (1992) stressed that NTE can be inhibited irreversibly or reversibly. He proposed that only OP agents capable of inhibiting NTE irreversibly might form aging links with NTE and induce OPIDN. For example, the reversible NTE inhibitors, phosphinates, carbamates, and phenylmethylsulfonyl fluoride (PMSF), do not induce NTE aging or OPIDN. However, Richardson's notion of the characteristics of OP drugs may not be tenable, as several compounds containing alkoxy groupings that irreversibly inhibit NTE (i.e., demeton, diazinon, and others) do not cause OPIDN (see Table 7-3 and Hayes, 1975, 1982). Then, it is doubtful whether or not there exists a dependable, inverse relationship between the capacity to induce aging AChE and aging NTE.

Reversible antiAChEs protect AChE from OP agents and protect NTE from OP drugs that induce OPIDN. As a result, these antiAChEs are prophylactic for OPIDN. Carbamates and PMSF also exhibit this effect (Johnson and Lauwerys, 1969; see also Richardson, 1992). However, when the reversible NTE inhibitors are given after subtoxic doses of NTE-inducing OP agents, OPIDN "promotion" or potentiation occurs (Pope et al., 1991; Richardson, 1992).

The notion that NTE phosphorylation (or the formation of aged NTE) and marked NTE inhibition (such as 70%) lead to OPIDN is supported by the following evidence: Makhaeva et al. (1998) related the ratio between imino OP drugs (class C compounds; see section BII) inhibiting NTE and AChE to their OPIDN-inducing capacity, concluding that the OPIDN capacity of a class C compound can be determined by this ratio.

Rudy Richardson (Jianmongkol et al., 1996; Richardson, 1995) followed a similar reasoning when he related the OPIDN potential of an OP compound to the ratio between its neuropathic and LD50 dose. Similar evidence was provided for several OP insecticide series (see, for example, Jianmongkol et al., 1996; Wu and Casida, 1996; for review, see also Ecobichon, 1996. "Promotion" and protection by reversible NTE inhibitors (see above) also seemed to support the hypothesis of OPIDN's NTE-based nature.

Altogether, the evidence for Martin Johnson and William Aldridge's hypothesis is strong and Rudy Richardson supported it vigorously. However, the hypothesis is controversial. The following is a paraphrase of an objection by Mohamed Abou-Donia (Abou-Donia and Lapadula, 1990) and others: NTE is a somewhat mystical enzyme because it has no known physiological function. Rudy Richardson (1992) stated that the lacuna might relate to several unknowns concerning NTE. While NTE's cytosolic and membrane location, axonal transport, and molecular mass were described (see Abou-Donia and Lapadula, 1990), its chemical structures, the constitution of its active site, and its molecular characteristics have not been identified. Abou-Donia also stated that evidence for the hypothesis is correlative, and no specific mechanism was proposed to describe how the inhibition and aging of phosphorylated NTE leads to OPIDN (Abou-Donia and Lapadula, 1990).

Additional problems with the hypothesis may be raised. As mentioned, young chicks and rats are not susceptible to OPIDN, yet NTE is present in their nervous system (Richardson, 1992). Then, NTE, ChEs, and several hydrolases or phosphatases are stereochemically similar, which mitigates the NTE hypothesis of OPIDN generation (see Doorn et al., 2001; Chapter 3 DI). Also, while maximum inhibition and aging of phosphorylated NTE occur within hours after administrating pertinent OP or carbamate agents, OPIDN does not occur until NTE activity almost returns to normal, days or weeks later (Kamata et al., 2001a, 2001b). Finally, while several war gases significantly inhibit NTE, they do not cause OPIDN in experimental animals.

An alternative explanation for OPIDN was proposed by Abou-Donia (Abou-Donia and Lapadula, 1990; Abou-Donia, 1992) and it relates to Bischoff's (1967, 1970, 1980) pathognomic studies. The hypothesis describes a series of events for OPIDN induction. First, OP compounds phosphorylate and activate protein kinases (in particular Ca^{2+}/calmodulin kinase II; Ca^{2+}/CAM-KII) by competing with ATP as a phosphoryl group donor and by phosphorylating serine or threonine hydroxyl residues of the kinases. Activated calmodulin enhances the phosphorylation of cytoskeletal protein (i.e., alpha and beta tubulins) and hyper-releases Ca^{2+}. Then, cytoskeletal phosphorylation and the abundance of free Ca^{2+} induce cytoskeleton pathology and axonal degeneration (see Abou-Donia and Lapadula, 1990; Abou-Donia, 1992).

Abou-Donia and associates accumulated evidence for their hypothesis. They demonstrated the OP-phosphorylation's specificity for Ca^{2+}/CAM-KII, as OPIDN-inducing OP drugs do not inhibit ATPase. (However, this may not constitute sufficient specific evidence.) These investigators also showed altered expression and levels of both Ca^{2+}/CAM-KII and cytoskeletal proteins during OPIDN, and expanded their hypothesis by suggesting that stress related to OPIDN induces oxidative and cytochrome c changes that contribute to cytoskeletal protein pathology and apoptosis (Gupta and Abou-Donia, 2001; Xie et al., 2001; Abu-Quare and Abou-Donia, 2001a, 2001b).

Mohamed Abou-Donia' s hypothesis has weaknesses; while he called Johnson and Aldridge's hypothesis correlative, so is his. Abou-Donia (see Damodoran et al., 2001) describes the complex kinetics of mRNA expression for cytoskeletal proteins induced by DFP. However, these data are not consistent with his hypothesis. Moreover, Abou-Donia's studies concern DFP and TOCP, and it is unclear whether his hypothesis applies to other OP inducers of OPIDN. Also, it would be worthwhile to show that OP agents that do not cause OPIDN have no effect on the phenomena. Furthermore, the kinetics of DFP effects on Ca^{2+}/CAM-KII are not parallel with the progress and signs of OPIDN (see, for example, Gupta and Abou-Donia, 2001).

Additional unproven hypotheses exist. As already suggested, OP agents may cause OPIDN by releasing fluoride or interfering with vitamin E metabolism (see Wills, 1970), and they may affect glia and its development, thus interfering with myelination (Garcia et al., 2002). Altogether, more research is needed to provide a definitive description of OPIDN mechanisms.

DII. Toxic Effects Involving the US 1991 War Effort in the Persian Gulf

Operation Desert Storm (ODS), or the Persian Gulf War (1990 to 1991) and the related Gulf War syndrome (GWS) or "mystery illness" bring forth another potential form of delayed OP toxicity.

The total number of US military personnel involved in this operation was nearly 700,000 (Department of Veterans Affairs [DVA], 1996, 1997). Shortly after the war, a number of the returning veterans complained of fatigue, muscular and neuromyal abnormalities, headache, neuropsychological complaints (i.e., cognition defects and forgetfulness), sleep disturbances, and personality changes.

Initially, the US Department of Defense spokesperson denied chemical warfare or exposure to toxic chemicals during ODS. However, academic and government sources in France and the Czech Republic presented opposing evidence. Their data included new information from the US press, and criticism from the US General Accounting Office (GAO) and pressure from the public, the media, and Congress caused the DOD to change its posture.

In 1996 and 1997 the DOD suggested that allied bombings of Iraqi plants and ammunition depots might have released sarin and/or other chemical and biological warfare tools. Furthermore, a spokesman for the Presidential Blue Ribbon Committee stated that approximately 100,000 veterans might have been exposed to war gases, fuel fumes, and other toxic chemicals. Also, either OP agents that could have been used by Iraq, or the prophylactic administration of pyridostigmine by the United States may have caused GWS (Jamal, 1995; Haley et al., 1997a, 1997b).

As the controversy grew, it was augmented by copious media reports and coverage. (See the interesting bibliography of newspaper and magazine articles published by the DVA, 1996, 1997 and the New York Times article by Philip Shenon, April 18, 1997.) The number of complaints by ODS veterans increased and the US government, Congress, and the DVA became involved. A number of successive registries were set up by the US Registry Agency, and between 1993 and 2001, the National Health Surveys conducted voluntary referral programs for ODS veterans at Veterans Administration hospitals throughout the United States. The registry and/or referral programs used screening tests that included physical and cognitive examinations and questionnaires to ascertain the veterans' location and type of service during ODS. The data collected were collated and analyzed by the Office of Public Health and Environmental Hazards and related DVA agencies. Currently, there are approximately 60,000 ODS veterans registered at VA hospitals throughout the United States; the number is growing (compared to approximately 700,000 military personnel deployed in the theater of operations).

Furthermore, Congress established communications and follow-up procedures among the DVA, DOD, US Institute of Medicine, National Institutes of Health (NIH), and several blue ribbon committees and workshops. Additionally, President Bill Clinton issued an executive order in 1993 concerning mandatory registration for ODS veterans and support for their treatment (see Karczmar, 1998). These directives will probably not handle problems generated by the veritable bureaucratic plethora surrounding ODS (see DVA, 1996, 1997; Gerrity and Feussner, 1997, 1999).

Finally, on August 31, 1993, Congress authorized President Bill Clinton to request the DVA to coordinate GWS research. Funds were allocated to establish multiproject centers at several VA hospitals and to conduct research at both federal and academic levels. The research involved clinical evaluation of ODS veterans and animal studies.

Currently, there is no general consensus involving data from the clinical studies. Some studies, including evaluations carried out in England, supported "the hypothesis that clusters of symptoms experienced by many Gulf War veterans represent discrete, factor analysis–derived syndromes that apparently reflect a spectrum of neurological injuries involving the central, peripheral, and autonomic nervous system" (Haley et al., 1997a, 1997b; Haley and Kurt, 1997; Jamal, 1995; Haley et al., 2001). All of the evaluations (including those conducted by the registry agency) reported similar data: muscular and neuromyal abnormalities (Jamal et al., 1995), headache and neuropsychological complaints including cognition defects, forgetfulness (see also Haley et al., 1997a, 1997b), sleep disturbances, anxiety and personality changes. Approximately 30% of the registrants experienced the triad of headache, neuropsychological complaints, and sleep disturbances. (see Black et al., 2004).

Sometimes, GWS symptoms appear 2 or 3 years after exposure. Since stress (i.e., post-traumatic stress syndrome) is a delayed condition, the GWS characteristics imply that Gulf War syndrome is caused by ODS-related stress. This phenomenon is under investigation.

Other studies made salient certain features that characterized the ODS environment and may cause GWS. These studies concern OP drugs and AChE protectors (i.e., pyridostigmine), pesticides, DEET-containing insect repellants, biological warfare agents and viruses, fumes resulting from jet fuel and burning oil-well fires, and sand particles (silica).

The multiplicity of causative factors adds complexity to the issue. In fact, many variables must be recognized: demographics, type and length of service, length of exposure, health, age, sex, and other individual characteristics. Also, overall evaluation is difficult because of the differences among tests and questionnaires (not standardized between evaluations). Altogether, the statistics are forbidding. It is extremely difficult for evaluators to reach valid conclusions, especially for the 2001 registry, initiated 10 years after the Gulf War.

Different conclusions were gained from each study, and every conclusion was challenged. Though many investigators concluded that OP-related GWS existed (e.g., Haley et al., 1997a, 1997b), the NIH Technology Assessment Workshop Panel (1994) denied that GWS occurred. Mahoney (2001) described GWS as a "social story" rather than an etiologically defined disease or health problem. The problem may not be solved in the near future, as several cohorts of ODS veterans will be identified, each cohort exhibiting a distinct syndrome (see, for example, Karczmar, 1998).This author predicts that one of these cohorts will show signs of delayed OPIDN or DCT (see above, sections DI-1C and DII). In vindication of this opinion, results from a recent registration (2001) suggest that the occurrence ratio of Lou Gehrig's disease (muscle dystrophy, amyotrophic lateral sclerosis [ALS]) among Gulf War veterans and controls is 2:1, and ALS resembles, in many respects, OPIDN (Stolberg, 2001; see also Time, "Old War, New Victims," October 6, 2003). However, some doubts were raised with regard to the reliability of this result, and a 2004 study of Davis et al., which included approximately 2,000 veterans and 3,000 "non-deployed" personnel, found that the prevalence of distal symmetric polyneuropathy (DSP; this syndrome is related to but not necessarily identical to OPIDN) was not higher among deployed than nondeployed personnel; the neurological tests used in the study were carried out some 10 years following the Gulf War (see also Rose, 2003). And then there are attempts to side step the issue by referring to the Persian Gulf Syndrome as the chronic multisyndrome illness (CMI; Dominic Reda, personal information, 2005).

The question of genetic and teratologic delayed effects that may arise in veterans exposed to war gases was frequently raised. Generally, it appears that such effects did not take place, yet some data indicate that chromosomal effects may be present following such exposure (see, for example, Hirata et al., 2004); these findings relate to those obtained in animals that suggest that early postnatal exposure to OP may induce developmental changes (see Chapter 8D; Richardson and Chambers, 2005).

Aside from programs dealing with clinical evaluations of ODS veterans, animal research has been initiated and carried out at a number of environmental hazard VA centers, and academic and federal laboratories. These studies test pyridostigmine or OP agents with direct (channel) effects and neurotoxic phenomena. Furthermore, researchers investigate interactions among OP drugs, excitatory transmitters, and their exocytotoxicity, and other noncholinergic transmitters. Additional programs concern stress, pulmonary pathologies, genetic and morphogenetic phenomena (see, for example, Behra et al., 2004), viruses, infections and parasites, OP agents alone or combined with other compounds, silica, herbicides and other toxicants, and immunological and endocrine factors (DVA, 1996, 1997). Currently, several dozens of pertinent animal research projects are being carried out; some of these studies deal with animal models in which combined effects of several agents, including prostigmine, and stress are evaluated (see, for example, Abdel-Rahman et al., 2004a, 2004b). And, in many cases, it was found that many of the agents tested produce in animals effects corresponding to those characteristic for GWS.

Whether as war gases and terrorist weapons or as insecticides, antiChEs of the OP type constitute a source of significant danger; the treatment of antiChE-induced poisoning is of paramount importance, especially with regard to the worldwide use of antiChE insecticides, which entails frequent toxicity (see Eto, 1974; Hayes, 1982; Karczmar, 1998; Singh and Sharma, 2000), and in view of the possibility that they can be used in terrorist attacks and future wars as a fatal weapon.

E. Treatment of Anticholinesterase Poisoning

Christison's (1855) formidable self-experiment treating toxicity from the Calabar bean (by using shaving water as an emetic) is the earliest Western record for using antidotes in antiChE poisoning. Following Christison's study, Fraser (1872) studied antidotal properties of atropine less impetuously by conducting experiments on animals. Additional animal experiments illustrated atropine's antidoting properties. These experiments were carried out long before the bean's cholinergic action was demonstrated. Ironically, Kleinwaechter (1864) used the Calabar bean extract to counteract atropine (see section A).

It is unclear when atropine was first used in humans as an antiChE antidote, but German, British, and US scientists working with OP drugs probably conducted experiments with atropine and other antimuscarinics during World War II (Holmstedt, 1959, 1963, 2000; Karczmar, 1970). In fact, aside from atropine and antimuscarinics, blocking and depressant drugs (i.e., curarimimetics and ganglionic blockers, local anesthetics, benactazines, anticonvulsants, barbiturates, and phenothiazines) were studied as antagonists of OP and carbamate toxicity (see Karczmar, 1967; Wills, 1970). The nonatropinic agents are ineffective antidotes (when combined with other drugs they may show some protection). However, anticonvulsant agents (i.e., NMDA antagonists) are being explored (see, for example, Lallement et al., 1998 and 1999).

Atropine is an effective antidote and prophylactic for OP and carbamate toxicity. When used for treating OP poisoning, atropine must be administered within 1 hour before exposure to the antiChE. Note that effective atropine doses for antiChE treatment are much larger (e.g., 5 to 10 mg in repeated doses) than doses used in routine therapy.

Whether as war gases, terrorist weapons, or insecticides, antiChEs including the OP agents are a source of significant danger. As mentioned previously, discovering a treatment for anticholinesterase-induced poisoning is of paramount importance, especially since the worldwide use of antiChE insecticides entails frequent toxicity (see Eto, 1974; Hayes, 1982; Karczmar, 1998; Singh and Sharma, 2000). It should be added that, particularly in the case of insecticides and herbicides, it is important to be able to detect their presence—this may be an ultimate prophylactic means. Several sensors were developed, and hybrid "biosensors" are available that appear to be sensitive at nanomolar levels (see, for example, Lei et al., 2004).

1. Treatments and Prophylaxis of OP Drugs

It should be stated first that when a case of human toxicity is suspected to be due to exposure to OP drugs (and carbamates), antiChEs as the cause of the toxicity have to be first established. Conditions (such as agricultural or a pertinent industrial milieu, and war or terror circumstances) and certain specific symptoms may lead to such a verdict, yet testing for inhibited ChEs constitutes final identification of antiChE toxicity (probably appropriate blood analysis—which, nowadays, can be very fast—is the only test of any practical use). Besides ChEs, other markers, such as phsophorylated albumin, may serve the same purpose (Peeples et al., 2005).

Many investigators attempted to prevent or antidote antiChE toxicity by using reversible antiChEs for AChE protection, atropinics, and oximes (see section A1), and workshops exploring new antidotal and prophylactic methods were organized by the DOD and other organizations (see, for example, Karczmar, 1984). "Sponging" or "sink" compounds are among the antidotes proposed. These compounds eliminate or sidetrack OP drugs before they arrive at their targets (i.e., AChEs, ChEs, and polyanions). Albumins and related compounds act as sponges, and blood albumin is a natural sponge (Silva et al., 2004).

Other prophylactics are compounds that hydrolyze OP agents ("scavengers"). Among those are naturally occurring arylesterases such as praoxonases, studied particularly by La Du, Sorenson, and Primo-Parmo. Their antiatherogenic action may constitute their normal function (Sorenson et al,. 1995); however, La Du and his associates stressed their anhydrolase action on OP agents and proposed that these agents may serve as endogenous guardians against toxic agents, including, besides OP drugs, oxidized lipids and bacterial ebdotoxins (La Du et al., 1999). Several meetings were devoted to the development of paraoxonase and related agents for treatment and prophylaxis of OP poisoning; an interesting aspect of this possibility is that sera of certain animals such as rabbits are particularly rich in serum paraoxonase and thus may serve as good providers of paraoxonase (Kuo and La Du, 1995). A refined "scavenger" process combines OP hydrolyzing enzymes (anhydrolases) with delivery systems, encapsulating the "scavengers" into circulating liposomes (see, Petrikovics et al., 2000a, 2000b and Li et al., 2000).

Early, it was proposed that infusions of ChEs may antagonize OP or carbamate poisoning (Karczmar et al., 1953). Some 50 years later, this notion is not quite dead (Guven et al., 2004; Ashani and Pistinner, 2004).

Other experimental antidotes include immunizing agents, specific cholinergic receptor antagonists, and drugs preventing delayed behavioral responses to OP agents (see Karczmar, 1984; Gunderson et al., 1992; and Smythies and Golomb, 2004). These novel antidotes have not been tested in field conditions.

War gases easily penetrate animal and human skin. Currently, agents (i.e., hypochlorites and alkali) decontaminating the environment or the skin (Cabal et al., 2004) are used to prevent war-gas toxicity. Masks and protective clothing are also used (Wills, 1963).

a. Animal Experimentation

Specific OP antidotes were intensely explored in animals. First, reversible antiChEs were tested. Prophylaxis using reversible tertiary antiChEs (i. e., physostigmine) and quaternary antiChEs (i.e., neostigmine and pyridostigmine) protect animals. When antidotal efficacy of the compounds was tested, they protected rodents, guinea pigs, and primates against 2 to 4 median lethal doses (LD50) of OP agents such as isofluorophate (DFP) and sarin (Karczmar, 1967; Wills, 1970; see also US Army Chemical Defense Bioscience Reviews, 1985, 1990). The antiChEs must be administered in doses inhibiting at least 50% of AChE in the animals' blood, CNS, and peripheral nervous system.

Oximes are OP antidotes that may be used prophylactically or antidotally. Wilson showed that the oxime antidotal potency is similar or superior to carbamate potency in mice (Wilson et al., 1955; Wilson, 1959; see also section A). Pralidoxime was the most effective oxime synthesized and tested by Wilson. Following his discovery, more oximes were synthesized, yet pralidoxime is still one of the most effective reactivators (see Hayes, 1975). Note that when prophylactically employing oximes, they are not useful after phosphorylated AChE is aged (see section BII-2). In addition, oximes are not equally efficient against all OP drugs. For example, pralidoxime is not effective against sarin (see Wills, 1970). Among the pesticides, ethoxy pyrophosphorus derivatives (i.e., tetraethyl pyrophosphate, TEPP) and ethoxy compounds (i.e., parathion) are more sensitive to oxime reactivation than methoxy derivatives (i.e., isodemeton). However, the validity of these reports is controversial (see Hayes, 1975, 1982; Wills, 1970).

Furthermore, the effectiveness of oximes in treatment may be limited because, as quaternaries, they cannot penetrate the CNS (see, for example, Waser et al., 1987; Bismuth et al., 1992; Balali-Mood and Shariat, 1996). However, the quaternary nature of the oximes does not prevent them from reactivating peripheral AChE in the neuromyal junction, autonomic nervous system, and red blood cells; these effects should exert an antidotal action.

Actually, the notion that oximes cannot penetrate the CNS may not be valid. In 1960, Vincenzo Longo, David Nachmansohn, and Daniele Bovet used electroencephalographic data to prove that oximes are centrally effective against sarin.

Drug combinations used prophylactically ("prophylactic cocktail") or as a treatment are the most effective way to deal with OP drug toxicity (Balali-Mood and Shariat, 1998). Many combinations were tested and usually included a reversible antiChE, an oxime, and atropine. The most effective cocktail components are pyridostigmine, atropine, and either pralidoxime or obidoxime. This cocktail may protect animals against 10 to 20 LD50 doses of most OP agents (see Wills, 1970; Karczmar, 1984; Eto, 1974).

b. Human Experience

After studying cases of OP poisonings in Iran, Balali-Mood and Shariat (1998) found that pralidoxime and atropine were very effective for treating the poisonings. They also found some reactivation of red blood cell AChE when administering the oximes within 6 hours of exposure.

Altogether, the Iranian investigators idiosyncratically concluded, "The inhibition of central AChE . . . of the CNS . . . is less important . . . with regard to OP toxicity." Their report is questionable. The insignificance of CNS AChE inhibition is arguable, as 6 hours after OP exposure the phosphorylated enzyme should have aged and become resistant to reactivation.

Ecobichon (1996) espoused a view similar to that of Balali-Mood and Shariat (1998): he suggested that serious poisoning from exposure to organophosphorus insecticides required treatment with atropine-oxime combinations. Nevertheless, OP toxicity and its symptoms will occur when prophylactic measures are not employed or fail (e.g., when oximes are given after phosphorylated AChE has aged). Once OP toxicity occurs, the only recourse is to antagonize the symptoms, making atropine, anticonvulsants, and other symptomatic treatment modes effective (i.e., artificial respiration). Prophylactic cocktails may also be employed. Note that oximes are toxic (i.e., liver toxicity from obidoxime), and atropine should be used alone in poisonings that are less severe.

It is of interest that La Du and his associates (Haley et al., 1999) found an association between ODS symptomatology and low serum activity of the paraoxonase of the patients exhibiting the syndrome; this finding suggests that the development of paraoxonase treatment may be indicated for the therapy of OP toxicity (see above, section E-1).

c. Treatment of OPIDN Toxicity

Cholinergic antagonists used to treat OP drug toxicity are not effective against OPIDN; thus, other avenues have been explored. Nicotinic acid derivatives were tested. These derivatives delayed triorthocresyl phosphate (TOCP)-induced OPIDN, though the mechanism for the effect was unclear (Rogers et al., 1964). Then, Richardson (1992) found that NTE inhibitors link either reversibly or irreversibly with NTE (see section BII-3, above). On this basis, he and others tested reversible NTE inhibitors (i.e., PMSF, its analogs, and carbamates) as OPIDN preventers (Richardson, 1992; Johnson and Read, 1993; Piao et al., 1995; Massicotte et al., 1999). However, using reversible NTE inhibitors is problematic, as they may promote OPIDN development (see section BII-3). Currently, their OPIDN promotion seems to nullify their use (Piao et al., 1999).

2. Treatment of Toxicity Induced by Non-OP AntiChEs

When considering treatments for carbamate antiChE poisoning, it is important to note how the poisoning occurs. The poisoning may develop when agricultural workers are exposed to carbamate insecticides. These drugs are also used in suicide attempts (Ecobichon, 1996). Another unexpected danger relates to plant-based beverages containing naturally occurring antiChEs. For example, there was a case involving two individuals who drank tea mistakenly prepared with the dried herb fir club moss, *Lycopodium selago*, instead of the herb *Lycopodium calvum*. Since *Lycopodium selago* contains huperzine A (see section BI-1), the two individuals exhibited typical cholinergic symptoms, and huperzine A was identified in their blood. Fortunately, they recovered.

Note that AChE carbamylation is a short-lived phenomenon, as carbamates are reversible inhibitors (see section BII-2). Dimethyl compounds are a special case: they produce a carbamylated anticholinesterase, which may be reactivated with oximes. However, oximes are harmful when employed in animals poisoned with monomethyl carbamate, and a man who ingested carbaryl died 6 hours after a 2-PAM treatment (Farago, 1969; Hayes, 1982).

Usually, toxicity from carbamate and related "reversible" antiChEs does not require treatment. Rest and supportive measures are recommended. Atropine is used in severe cases of poisoning. Artificial respiration and/or oral or intravenous doses of atropine (around 10 mg) are needed for more serious cases. Anticonvulsants or sedatives are used to treat the rare occurrence of convulsions.

F. Cholinergic Teratology

The morphogenetic and teratologic effects of OP compounds came to intense consideration within the context of Gulf War, and several registries were and still are concerned with this subject.

On the whole, the evidence for the occurrence of teratology as a result of possible exposure of humans to war gases in the course of the Gulf War is inconclusive or absent.

This is not to state that antiChEs of either carbamate or OP type cannot produce morphogenetic effects; in fact, the presence of such effects was established as early as at the beginning of the 19th century, that is, long before the discovery of cholinergic transmission (see section A, above; Chapter 8 D; Karczmar, 1963; Karczmar et al., 1973). Subsequently, it was demonstrated that these effects are caused by carbamates and both insecticide and war gas OP agents in many species, both vertebrate and invertebrate, ranging from echinoderms and arthropods to fishes and mammals. The effects in question vary from antigrowth, neurotoxic, spermicidal and antiovarian action to extrogastrulation, arrest of neurogenesis, and malformations, whether of body symmetry, the skeleton, or the endplate (Karczmar, 1963; Karczmar et al., 1973; Luca and Balan, 1987; Vernagy, 1992; Schuytema et al., 1994; Crumpton et al., 2000). For example, the skeletal effects included, in the chick, the occurrence of "wry neck," "rumplessness," micromelia, and syndactilism, while the effects on sperm or ova included gametocidal actions as well as morphological effects (Luca and Balan, 1987); furthermore, DNA damage was also noticed (Richards and Kendall, 2002), while the neurotoxic action involved the inhibition of neurite growth (Crumpton et al., 2000).

It should be noted that in many of the pertinent experiments, high doses of antiChEs were needed to obtain the effects in question, sometimes above those that produce complete inhibition of AChE; however, this was not a universal finding. Accordingly, direct rather than antiChE actions may have been involved in some of the teratologic, morphogenetic, and toxic actions of these compounds.

Interesting consequences were the postnatal behavioral effects of prenatal exposure to antiChEs. Thus, in mice, prenatal (7 to 14 day of development) exposure to OP drugs and OP insecticides in concentrations that inhibited significantly but not completely AChE activity increased in the adults aggression and agonistic behavior in the absence of visible teratogenic and neurogenetic effects (Karczmar and Srinivasan, reported at the 1973 Symposium on Fetal Pharmacology, unpublished). Somewhat similarly, early postnatal exposure of mice to an insecticide during 11 to 14 postnatal days caused an increase in agonistic and novelty-seeking behavior (Ricceri et al., 2003); the problem here is that this treatment did not affect AChE, while an earlier postnatal treatment that reduced AChE activity had no behavioral effects.

It should be added that antiChEs, particularly the OP insecticides, may induce developmental teratology via acting on entities other than the cholinergic system, for example, on serotonergic transmission (Aldridge et al., 2003).

An interesting aspect of the matter of teratology concerned with cholinergic pharmacology is that while antiChEs and cholinomimetics induce malformations, so do, paradoxically, cholinolytics, such as atropine and drugs that block ACh synthesis, such as hemicholinium (although the pertinent studies are relatively few; see Karczmar et al., 1973; Buznikov, 1990). The speculative explanations for this paradox may be as follows. Neurotransmitters, including ACh, act as synaptogens, trophics, and developmental facilitators (see Chapter 8 CIII), so antiChEs may produce teratologic effects via unbalancing neurotransmitters involved in development, or via blocking, at high doses, cholinergic transmission and thus inhibiting cholinergic synaptogenic and developmental effect, while cholinergic blockers and hemicholiniums may produce this latter action (see Karczmar et al., 1973).

It should be added that cholinergic teratology and related conditions can be caused as well by dietary deficiencies in such cholinergic substances as choline (see Albright et al., 2001 and Chapter 8 BII).

Notes

1. The late Percy Julian, my friend and Oak Park neighbor, was a great synthetic chemist; he is particularly known for his synthesis of corticosteroids. This synthesis was patented and Percy became a rich man.
2. Peter Pauling was the son of Linus Pauling, a double Nobel Prize winner. Linus Pauling's race with Crick and Watson to establish the structure for DNA is famous; Watson and Crick won when they demonstrated that DNA's structure is the double helix. As a son of a genius, Peter was not satisfied being "only" a distinguished biochemist, and the short life of Peter was not a happy one.

3. Wilson tested the oximes in mice dying from DFP poisoning. After he observed their miraculous recovery, he ran into Nachmansohn's office with the happy news. Nachmansohn only grumbled, "You were lucky."
4. Tom Brokaw quoted a reliable source stating that Saddam Hussein purchased large amounts of atropine in 2002. This report indicated that Hussein might have been planning to use war gases in the war with the United States; he would then have needed atropine to protect his own soldiers in the event. However, subsequent developments threw doubt on Hussein's possession of war gases at the outset of the second war with Iraq, or his intent to use them.

References

Abdel-Rahman, A., Abou-Donia, S., El-Masry, E., Shetty, A. and Abou-Donia, M.: Stress and combined exposure to low doses of pyridostigmine bromide, DEET, and permethrin produce neurochemical and neuropathological alterations in cerebral cortex, hippocampus and cerebellum. J. Toxicol. Environ. Health, Part A, *67*: 163–192, 2004a.

Abdel-Rahman, A., Dechkovskaia, A. M., Goldstein, L. B., Bullman, S. H., Khan, W., El-Mastry, E. M. and Abou-Donia, M. B.: Neurological deficits induced by malathion, DEET, and permethrin, alone or in combination in rats. J. Toxicol. Environ. Health *67*: 331–356, 2004b.

Abdollahi, M., Ranjbar, A., Shadn, S., Nikfar, S. and Rezaiee, A.: Pesticides and oxidative stress: a review. Med. Sci. Monit. *10*: 144–147, 2004.

Abou-Donia, M. B.: Organophosphorus ester-induced delayed neurotoxicity. Annu. Rev. Pharmacol. Toxicol. *21*: 511–548, 1931.

Abou-Donia, M. B.: Triphenyl phosphite: A type II organophosphorus compound-induced delayed neurotoxic agent. In: "Organophosphates: Chemistry, Fate and Effects," J. E. Chambers and P. E. Levi, Eds., pp. 327–351, Academic Press, San Diego, 1992.

Abou-Donia, M. B. and Lapadula, D. M.: Mechanisms of organophosphorus ester-induced delayed neurotoxicity: type I and type II. Annu. Rev. Pharmacol. Toxicol. *30*: 405–440, 1990.

Abou-Donia, M. and Preissig, S. H.: Delayed neurotoxicity of Leptophos: Toxic effects on the nervous system of hens. Toxicol. Appl. Pharmacol. *35*: 269–282, 1976.

Abu-Quare, A. W. and Abou-Donia, M. B.: Combined exposure to DEET (N, N-diethyl-m-toluamide) and permethrin-induced release of rat brain mitochondrial cytochrome c. J. Toxicol. Environ. Health *63*: 243–252, 2001a.

Abu-Quare, A. W. and Abou-Donia, M. B.: Combined exposure to sarin and pyridostigmine bromide increased levels of rat urinary 3-nitrotyrosine and 8-hydroxy-2'deoxyguanosine, biomarkers of oxidative stress. Toxicol. Lett. *123*: 51–58, 2001b.

Adams, D. H. and Whittaker, V. P.: The cholinesterases of human blood. II. The forces acting between enzymes and substrate. Biochimi Biophys. Acta. *4*: 543–558, 1950.

Aeschlimann, J. A. and Reinert, M.: Pharmacological action of some analogues of physostigmine. J. Pharmacol. *43*: 413–444, 1931.

Albert, A. and Glendhill, W.: Improved synthesis of amino acridines. Part IV. Substituted 5-aminoacridines. J. Soc. Chem. Ind. *44*: 169–172, 1945.

Albert, A., Rubio, S. D., Goldacre, R. J., Davey, M. E. and Stone, J. D.: The influence of chemical constitution on antibacterial activity. Part II: A general survey of the acridine series. Br. J. Exp. Pathol. *26*: 160–162, 1945.

Albright, C. D., Mar, M.-H., Friedrich, C. B. and Brown, E. C.: Maternal choline availability alters the localization of p15Ink4B and p27Kip 1 cyclin-dependent kinase inhibitors in the developing fetal rat brain hippocampus. Dev. Neurosci. *23*: 100–106, 2001.

Aldridge, J. E., Seidler, F. J., Meyer, A., Thillai, I. and Slotkin, T. A.: Serotonergic systems targeted by developmental exposure to chlorpyrifos during different critical periods. Environ. Health Perspect. *111*: 1736–1743, 2003.

Aldridge, W. N.: Some observations on the characteristics of serum esterases with special reference to hydrolysis of diethyl p-nitrophenyl phosphate (E 600). Ph.D. thesis, London University, 1951.

Aldridge, W. N.: Two types of esterases (A and B) hydrolyzing p-nitrophenyl acetate, propionate, butyrate, and a method for their determination. Biochem. J. *53*: 110–117, 1953.

Aldridge, W. N.: Organophosphorus compounds and esterases. Annu. Rep. Chem. Soc. 1956, *53*: 294–305, 1957.

Aldridge, W. N.: Survey of major points of interest about reactions of cholinesterases. Croat. Chem. Acta *47*: 225–233, 1976.

Aldridge, W. N. and Nemery, B.: Toxicology of triethylphosphorothioates with particular reference to lung toxicity. Fundam. Appl. Toxicol. *4*: S215–S223, 1984.

Aldridge, W. N. and Reiner, E.: "Enzyme Inhibitors as Substrates," North-Holland, Amsterdam, 1972.

Aldridge, W. N., Barnes, J. M. and Johnson, M. K.: Studies on delayed neurotoxicity produced by some organophosphorus compounds. Ann. N. Y. Acad. Sci. *160*: 314–322, 1969.

Aldridge, W. N., Dinsdale, D., Nemery, B. and Verschoyle, R. D.: Some aspects of the toxicology of trimethyl and triethyl phosphorothioates. Fundam. Appl. Toxicol. *5*: S47–S60, 1985.

Anand, R., Hartman, R. D. and Hayes, P. E.: An overview of the development of SDZ ENA 713, a brain selective cholinesterase inhibitor. In: "Alzheimer Disease: From Molecular Biology to Therapy," R. Becker and E. Giacobini, Eds., pp. 239–243, Birkhaeser, Boston, 1996.

Arbusov, A. E.: Uber die Struktur der phosphorigen Säure und ihre Derivate. IV. Isomerisation und Übergang der Verbindungen der dreiwertigen Phosphors in solche der fünfwertigen. Chem. Zbl. *2*: 1640, 1906.

Arbusov, A. E. and Arbusov, B. A.: Über die Ester der pyrophosphorigen, der Unterphosphor- und der Pyrophosphorsaure. J. Prakt. Chem. *238*: 103–132, 1931.

Ashani, Y., and Pistinner, S.: Estimation of the upper limit of human butyrylcholinesterase dose required for protection against organophosphates toxicity: a mathematically based toxicokinetic model. Toxicol. Sci. *77*: 185–187, 2004.

Augustinsson, K.-B.: Cholinesterases. Acta Physiol. Scand. *15*, Supplt. *52*: 1–182, 1948.

Axelsen, P., Harel, M., Silman, I. and Sussman, J.: Structure and dynamics of the active site gorge of acetylcholinesterase. Synergictic use of dynamics simulation and X-ray crystallography. Protein Sci. *3*: 188–197, 1994.

Bai, D. L., Tang, X. C. and He, X. C.: Huperzine A, a potential therapeutic agent for treatment of Alzheimer's disease. Curr. Med. Chem. *7*: 355–374, 2000.

Balali-Mood, M. and Shariat, M.: Effects of oximes in acute organophosphate pesticide poisoning—a retrospective study. Proceedings of the Chemical and Biological Medical Treatment Symposium, Spiez, Switzerland, pp. 119–124, 1996.

Balali-Mood, M. and Shariat, M.: Treatment of organophosphorus poisoning. Experience of nerve agents and acute pesticide poisoning on the effects of oximes. In: "From *Torpedo* Electric Organ to Human Brain: Fundamental and Applied Aspects," J. Massoulié, Ed., pp. 375–378, Elsevier, Amsterdam, 1998.

Balfour, J. H.: Description of the plant, which produces the ordeal bean of Calabar. Trans. R. Soc. Edinb. *22*: 305–312, 1861.

Barnes, J. H., and Denz, F. A.: The chronic toxicity of p-nitrophenyl diethylthiophosphate (E605), a long-term feeding experiment with rats. J. Hyg. *49*: 430–441, 1951.

Barnes, J. M. and Denz, F. A.: The reaction of rats to diets containing octamethyl pyrophosphoramide (Schradan) and o, o-diethyl-S-ethylmercaptoethanol thiophosphate ("Systox"). Br. J. Ind. Med. *11*: 11–19, 1954.

Baron, D. H. R. and Kirby, G. W.: Phenol oxidation and biosynthesis. V. The synthesis of galanthamine. Proc. Chem. Soc. *1962*: 806–817, 1962.

Bartolucci, C., Perola, E., Cellai, L., Brufani, M. and Lamba, D.: 'Back door' opening implied by the crystal structure of carbamoylated acetylcholinesterase. Biochemistry *38*: 5714–5719, 1999.

Behra, M., Etard, C., Cousin, X. and Strahle, U.: The use of zebra fish mutants to identify secondary target effects of acetylcholine esterase inhibitors. Toxicol. Sci. *77*: 325–333, 2004.

Benshop, H. P. and Keijer, J. H.: On the mechanism of aging of phosphorylated cholinesterases. Biochem. Biophys. Acta. *128*: 586, 1966.

Berends, F.: Reactivering en "veroudering" van Esterasen Geremd met Organische Fosforverbindengen. Ph.D. Thesis, University of Leiden, 1964.

Bidstrup, P. L., Bonnell, J. A. and Beckett, A. G.: Paralysis following poisoning by a new organic phosphorus insecticide (Mipafox). Br. Med. J. *1*: 1068–1072, 1953.

Bischoff, A.: The ultrastructure of tri-ortho-cresyl phosphate in poisoning. I. Studies of myelin and axonal alterations in sciatic nerve. Acta Neuropathol. *9*: 158–174, 1967.

Bischoff, A.: Ultrastructure of the ortho-cresyl phosphate poisoning in the chicken. Acta Neuropathol. *15*: 142–155, 1970.

Bischoff, A.: Clinical and experimental work in neurotoxicity. Dev. Toxicol. Environ. Sci. *8*: 39–51, 1980.

Bismuth, C., Inns, R. H. and Marrs, T. C.: Efficacy, toxicity and clinical use of oxime in anticholinesterase poisoning. In: "Clinical and Experimental Toxicology of Organophosphates and Carbamates," B. Ballentyne, T. C. Marrs and W. N. Aldridge, Eds., pp. 555–577, Butterworth-Heinmann, Oxford, 1992.

Black, D. W., Carney, C. P., Peloso, P. M. Woolson, R. F., Schwartz, D. A., Voelker, M. D., Barrett, D. H. and Doebbeling, B. N.: Gulf War veterans with anxiety: prevalence, comorbidity, and risk factors. Epidemiology. *15*: 135–142, 2004.

Blaschko, H., Bulbring, E. and Chou, T. C.: Tubocurarine antagonism and inhibition of cholinesterases. Br. J. Pharmacol. *2*: 1116–1120, 1949.

Bodansky, O.: Contribution of medical research in chemical warfare to medicine. Science *102*: 517–521, 1945.

Boissier, J. R., Combes, G. and Pagny, J.: La galanthamine, puissant cholinergique naturel. I. Sources. Structure chimique. Characterisation. Extraction. Toxicite. Action sur les fibres lisses. Extrait Ann. Pharmac. Fran. *18*: 888–900, 1960.

Bombinski, T. J. and Dubois, K. P.: Toxicity and mechanism of action of Di-Syston. A. M. A. Arch. Ind. Health *17*: 192–199, 1958.

Bowman, W. C.: Non-relaxant properties of neuromuscular blocking drugs. Br. J. Anaesth. *53*: 147–160, 1982.

Britt, J. O., Jr., Martin, J. L., Okerberg, C. V. and Dick, E. J., Jr.: Histopathologic changes in the brain, heart, and skeletal muscle of rhesus macaques, ten days after exposure to soman (an organophosphorus nerve agent). Comp. Med. *50*: 133–139, 2000.

Brodeur, J. and Dubois, K. P.: Studies on the mechanism of acquired tolerance by o, o-diethyl-s-2-(etylthio) ethyl phosphorodithioate (Di-syston). Arch. Int. Pharmacodyn. *149*: 560–570, 1964.

Brodie, B. B.: Pathways of drug metabolism. J. Pharm. Pharmacol. *8*: 1–17, 1956.

Brophy, V. H., Jampsa, R. L., Clendenning, J. B., McKinstry, L. A., Jarvik, G. P. and Furlong, C. E.: Effects of 5' regulatory-region polymorphisms on paraoxonase-gene (PON1) expression. Am. J. Hum. Genet. *86*: 1428–1436, 2001.

Brufani, M. and Filocamo, L.: Rational design of cholinesterase inhibitors. In: "Cholinesterases and Cholinesterase Inhibitors," E. Giacobini, Ed., pp. 27–46, Martin Dunitz, London, 2000.

Brufani, M., Castellano, C., Marta, M., Oliverio, A., Pagella, P. G., Pavone, M. and Rugarli, P. L.: A long-lasting cholinesterase inhibitor affecting neural and behavioral processes. Pharmacol. Biochem. Behav. *26*: 625–629, 1987.

Brufani, M., Castellano, C., Marta, M., Murroni, F., Oliverio, A., Pagella, P. G., Pavone, F., Pomponi, M. and Rugarli, P. L.: From physostigmine to physostigmine derivatives as new inhibitors of cholinesterases. In: "Current Research in Alzheimer Therapy," E. Giacobini and R. Becker, Eds., pp. 343–352, Taylor and Francis, New York, 1988.

Burchfiel, J. L. and Duffy, F. H.: Organophosphate neurotoxicity: chronic effects of sarin on the electroencephalogram of monkey and man. Neurobehav. Toxicol. Teratol. *4*: 767–778, 1982.

Burgen, A. S. V.: The mechanism of action of anticholinesterase drugs. Br. J. Pharmacol. *4*: 219–228, 1949.

Burgen, A. S. V. and Hobbiger, F.: The inhibition of cholinesterases by alkylphosphates and alkyphenolphosphates. Br. J. Pharmacol. *6*: 593–601, 1951.

Buxtorf, A. and Spindler, M., Eds.: "Fifteen Years of Geigy Pest Control," Buchdruckerei Karl Werner AG, Basel, 1954.

Buznikov, G. A.: "Neurotransmitters in Embryogenesis," Academic Press, New York, 1990.

Cabal, J., Kuca, K., Sevelova-Bartosova, L. and Dohnal, V.: Cyclodextrines as functional agents for decontamination of the skin. Acta Medica (Hradec Kralove) *47*: 115–118, 2004.

Callaway, S. and Davies, D. R.: The association of blood cholinesterase levels with the susceptibility of animals to sarin and ethyl pyrophosphate poisoning. Br. J. Pharmacol. *12*: 382–387, 1957.

Cameron, J.: The recent cases of poisoning by Calabar bean. Med. Tms. Gaz. (Lond.): 406–410, 1864.

Carrington, C. D. and Abou-Donia, M. B.: Characterization of [3H]diisopropyl phosphofluoridate-binding proteins in hen brain. Rates of phosphorylation and sensitivity to neurotoxic and non-neurotoxic organophosphorus compounds. Biochem. J. *228*: 537–544, 1985.

Carvajal, F., Sanchez-Amate, M., Sanchez-Santed, F. and Cubero, I.: Neuroanatomical targets of the organophosphate chlorpyrifos by c-fos immunolabeling. Toxicol. Soc.: *84*: 360–367, 2005.

Casida, J. F. and Quistad, G. B.: Golden age of insecticide research: past, present or future? Annu. Rev. Entomol. *43*: 1–16, 1998.

Casida, J. E., Baron, R. L., Eto, M. and Engel, J. L.: Potentiation and neurotoxicity induced by ceratin organophosphates. Biochem. Pharmacol. *12*: 73–83, 1963.

Cassidy, C. S., Lin, J. and Frey, P. A.: A new concept for the mechanism of action of chymotrypsin: the role of the low-barrier hydrogen bond. Biochem. *36*: 4576–4584, 1997.

Chadwick, L. E.: Actions on insects and other invertebrates. In: "Cholinesterases and Anticholinesterase Agents," G. B. Koelle, Ed., Handbch d. exper. Pharmakol., Erganezungswk., vol. 15, pp.740–798, Springer-Verlag, Berlin, 1963.

Chambers, H. W.: Organophosphorus compounds: an overview. In: "Organophosphates: Chemistry, Fate and Effects," J. E. Chambers and P. E. Levi, Eds., pp. 3–17, Academic Press, San Diego, 1992.

Chambers, J. E. and Levi, P. E., Eds.: "Organophosphates: Chemistry, Fate and Effects," Academic Press, San Diego, 1992.

Chen, Y. L., Nielsen, J., Hedberg, K., Dunaiskis, A., Jones, S., Russo, L., Johnson, J., Ives, J. and Liston, D.: Syntheses, resolution, and structure-activity relationships of potent acetylcholinesterase inhibitors: 8-carbaphysostigmine analogues. J. Med. Chem. *35*: 1429–1433, 1992.

Cheng, T. C., DeFrank, J. J. and Rastogi, V. K.: Alteromonas prolidase for organophosphorus G-agent decontamination. Chemico-Biol. Interactions *119–120*: 455–462, 1999.

Childs, A. F., Davies, D. R., Green, A. and Rutland, I. P.: The reactivation by oximes and hydroxamic acids of cholinesterases inhibited by organophosphorus compounds. Br. J. Pharmacol. *10*: 462–465, 1955.

Christie, G., Gershon, S., Gray, R., Shaw, F. H., McCance, I. and Bruce, D. W.: Treatment of certain side effects of morphine. Br. Med. J. *1*: 675–680, 1958.

Christison, R.: On the properties of the ordeal bean of Old Calabar. Monthly Mer. J. (Lond.) *20*: 193–204, 1855.

Clark, V. M., Todd, L. and Warren, S. G.: Studies on phosphorylatious XXV: Metal ion catalyzed reactions of phosphagens. Biochem. J. *338*: 591–598, 1963.

Clarke, R. A.: "Against All Enemies," Free Press, New York, 2004.

Clermont, P. de: Note sur la preparation de quelques ethers. C. R. *39*: 338–340, 1854.

Cohen, J. A.: Purification and properties of dialkylfluorophosphatases. Biochim. Biophys. Acta. *26*: 29–39, 1957.

Cohen, J. A. and Oosterbaan, R. A.: The active site of acetylcholinesterase and related cholinesterases and its reactivity towards substrates and inhibitors. In: "Cholinesterases and Anticholinesterase Agents," G. B. Koelle, Ed., Handbch. d. exper. Pharmakol., Erganzungswk., vol. 15, pp. 298–373, Springer-Verlag, Berlin, 1963.

Comroe, J. H., Jr., Todd, J. and Koelle, G. B.: The pharmacology of di-isopropyl fluorophosphate (DFP) in man. J. Pharmacol. Exp. Ther. *87*: 414–420, 1946.

Cook, J. W., Blake, J. R. and Williams, M. W.: The enzymic hydrolysis of Malathion and its inhibition by EPN and other organic phosphorus compounds. J. Assoc. Agric. Chem. *40*: 664–665, 1957.

Cramer, R. D., III, Patterson, D. E. and Bunce, J. D.: Comparative Molecular Field Analysis (CoMFA) . I. Effect of shape on binding of steroids to carrier proteins. J. Am. Chem. Soc. *110*: 5959–5967, 1988.

Crumpton, T. L., Seidler, F. J. and Slotkin, T.A.: Is oxidative stress involved in developmental neurotoxicity of chlorpyrifos? Brain *121*: 191–195, 2000.

Dale, H. H.: The action of certain esters and ethers of choline and their relation to muscarine. J. Pharmacol. Exp. Ther. *6*: 147–190, 1914.

Damodoran, T. V., Abdel-Rahman, A. and Abou-Donia, M. B.: Altered time course of MRNA expression of alpha tubulin in the central nervous system of hens treated with diisopropyl phosphofluoridate (DFP). J. Neurochem. Res. *26:* 43–50, 2001.

Daniell, W. F.: On the natives of Old Callebar, West Coast of Africa. Proc. Ethnol. Soc. *40*: 313–327, 1846.

Dauterman, W. C. and Main, W. R.: Relationship between acute toxicity and *in vitro* inhibition and hydrolysis of a series of homologs of malathion. Toxicol. Appl. Pharmacol. *9*: 40–48, 1996.

Davies, D. R.: Neurotoxicity of organophosphorus compounds. In: "Cholinesterases and Anticholinesterase Agents," G. B. Koelle, Ed., Handbch. d. exper. Pharmakol., Erganzungswk., vol. 15, pp. 861–882, Springer-Verlag, Berlin, 1963.

Davies, D. R. and Green, A. L.: The kinetics of reactivation by oximes of cholinesterase inhibited by organophosphorus compounds. Biochem. J. *63*: 520–535, 1954.

Davies, D. R., Holland, P. and Rumens, M. J.: The relationship between the chemical structure and neurotoxicity of alkyl organophosphorus compounds. Br. J. Pharmacol. *15*: 271–278, 1960.

Davis, L. E., Eisen, S. A., Murphy, F. M., Alpern, R., Parks, B. J., Blanchard, M., Reda, D. J., King, M. K., Mithen, F. A. and Kang, H. K.: Clinical and laboratory assessment of distal peripheral nerves in Gulf War veterans and spouses. Neurology *63*: 1070–1077, 2004.

Davison, A. N.: Return of cholinesterase activity in the rat after inhibition by organophosphorus compounds. Biochem. J. *60*: 339–346, 1955.

Dawson, R. M.: Tacrine slows the rate of ageing of sarin-inhibited acetylcholinesterase. Neurosci. Lett. *227*: 227–230, 1989.

Dehlawi, M. S., Eldefrawi, A. T., Eldefrawi, M. E., Anis, N. A. and Valdes, J. J.: Choline derivatives and sodium fluoride protect acetylcholinesterase against irreversible inhibition and aging by DFP and paraoxon. J. Biochem. Toxicol. *9*: 261–268, 1994.

Dettbarn, W. D.: Pesticide-induced muscle necrosis: mechanisms and prevention. Fundam. Appl. Toxicol. *4*: 18–26, 1984.

Dettbarn, W. D., Yang, Z. P. and Milatovic, D.: Different role of carboxylases in toxicity and tolerance to paraoxon and DFP. Chem. Biol. Interact. *119–120*: 445–454, 1999.

Diggle, W. M. and Gage, J. C.: Cholinesterase inhibition *in vitro* by 0,0-diethyl-0-p-nitrophenyl thiophosphate (parathion, E605). Biochem. J. *49*: 491–494, 1951.

Domino, E. F.: Galanthamine: another look at an old cholinesterase inhibitor. In: "Current Research in Alzheimer Therapy," E. Giacobini and R. Becker, Eds., pp. 295–303, Taylor and Francis, New York, 1988.

Doorn, J. A., Talley, T. T., Thompson, C. M. and Richardson, R. J.: Probing the active sites of butyryl cholinesterase and cholesterol esterase with isomalathion-stereoselective inactivation of serine hydrolases structurally related to acetylcholinesterase. Chem. Res. Toxicol. *14*: 807–813, 2001.

Dubois, K. B.: Toxicological evaluation of anticholinesterase agents. In: "Cholinesterases and Anticholinesterase Agents," G. B. Koelle, Ed., Handbch. d. exper. Pharmakol., Erganzungswk., vol. 15, pp. 832–859, Springer-Verlag, Berlin, 1963.

Dubois, K. B., Doul, P. R., Salerno, P. R. and Coon, J. M.: Studies on the toxicity and mechanism of action of p-nitrophenyl thionophosphate (Parathion). J. Pharmacol. Exp. Ther. *95*: 79–91, 1949.

Duffy, F. H. and Burchfiel, J. L.: Long term effects of the organophosphate Sarin on EEG in monkeys and man. Neurotoxicology *1*: 667–689, 1980.

Duffy, F. H., Buchfiel, J. L. and Sim, V. M.: Persistent effects of organophosphate exposure as evidenced by electroencephalographic measurements. In: "Pesticide Induced Delayed Neurotoxicity," R. L. Baron, Ed., p. 102, US Environmental Protection Agency, EPA-600-/1–76–025, July 1976.

Earl, C. J. and Thompson, R. H. S.: The inhibitory action of tri-*ortho*-cresyl phosphate on cholinesterases. Br. J. Pharmacol. *7*: 261–269, 1952.

Eastman, E., Wilson, J., Cervenansky, C. and Rosenberry, T. L.: Fasciculin 2 binds to the peripheral site on acetylcholinesterase and inhibits substrate hydrolysis by slowing a step involving proton transfer during enzyme acylation. J. Biol. Chem. *270*: 19694–19701, 1995.

Eccles, J. C., Fatt, P. and Koketsu, K.: Cholinergic and inhibitory synapses in a pathway from motor-axon collaterals to motoneurones. J. Physiol. (Lond.) *126*: 524–562, 1954.

Ecobichon, D. J.: Toxic effects of pesticides. In: "Casarett and Doull's Toxicology," C. D. Klaasen, Ed., pp. 643–689, Fifth Edition, McGraw-Hill, New York, 1996.

Ecobichon, D. J.: Our changing perspectives on benefits and risks of pesticides—an overview. Neurotoxicology *21*: 211–218, 2000.

Ecobichon, D. J.: Pesticide use in developing countries. Toxicology *160*: 27–33, 2001.

Engelhard, N. and Erdmann, W. D.: Ein neuer Reaktivator für durch alkyl-phosphatgehemmte Azetylcholinesterase. Klin. Wochenschr. *41*: 870–875, 1963.

Engelhart, E. and Loewi, O.: Fermentative Azetylcholinespaltung im Blut und ihre Hemmung durch Physostigmine. Arch. Exp. Pathol. Pharmakol. *150*: 1–13, 1930.

Eriksson, P. and Talts, U.: Neonatal exposure to neurotoxic pesticides increases adult susceptibility: a review of current findings. Neurotoxicology *21*: 37–48, 2000.

Eto, M.: "Organophosphorus Pesticides: Organic and Biological Chemistry," CRC Press, Cleveland, 1974.

Fahrney, D. E. and Gold, A. M.: Sulfonyl fluorides as inhibitors of esterases. I. Rates of reaction with anticholinesterase, α-chymotrypsin, and trypsin. J. Amer. Chem. Soc. *85*: 997–1000, 1963.

Farago, A.: Suicidal, fatal Sevin (1-naphtyl-N-methylcarbamate) poisoning. Arch. Toxicol. *24*: 309–315, 1969.

Felgenhauer, N., Zilker, T., Worek, F. and Eyer, P.: Intoxication with huperzine A, a potent anticholinesterase found in the fir club moss. J. Toxicol. Clin. Toxicol. *38*: 803–808, 2000.

Fernando, J. C. R., Hoskis, B. and Ho, I. K.: The role of dopamine in behavioral supersensitivity to muscarinic antagonist following cholinesterase inhibition. Life Sci. *29*: 2169–2176, 1986.

Fest, C. and Schmidt, K. J.: Insektizide Phosphorsäureester. In: "Chemie der Pflanzenschutz- und Schädlingsbekämpfungsmittel," R. Weigler, Ed., vol. 1, pp. 246–291, Springer-Verlag, Berlin, 1970.

Fisher, M.: Use of cholinesterase inhibitor in the therapy of myasthenia gravis. In: "Cholinesterases and Cholinesterase Inhibitors," E. Giacobini, Ed., pp. 249–262, Martin Dunitz, London, 2000.

Fraser, T. R.: On the characters, actions and therapeutic uses of the bean of Calabar. Edinburgh Med. J.: *9*: 36–56, 1863.

Fraser, T. R.: On the moth of esere, or ordeal bean of Old Calabar. Ann. Nat. Hist., 3rd Series, *13*: 389–393, 1864.

Fraser, T. R.: An experimental research on the antagonism between the actions of physostigma and atropia. Trans. R. Soc. Edinb. *26*: 529–713, 1877.

Frawley, J. P., Fuyat, H. N., Hagan, E. C., Blake, J. R. and Fitzhugh, O. G.: Marked potentiation in mammalian toxicity from simultaneous administration of two anticholinesterase compounds. J. Pharmacol. Exp. Ther. *121*: 96–106, 1957.

Fühner, H.: Untersuchungen uber die periphere Wirkung des Physostigmins. Arch. Exp. Pathol. Pharmakol. *85*: 205–220, 1917–1918.

Funke, A., Depierre, F. and Krucker, M. W.: Exaltation de l'activité anticholinesterasique des sels d'ammonium quaternaire des phenoxyalcanes par l'introduction de groupements urethanes. C. R. Acad. Sci. *7*: 762–764, 1952.

Furlong, C. E., Li, W. F., Brophy, V. H., Jarvik, G. P., Richter, R. J., Shih, D. M., Lusis, A. J. and Costa, L. G.: The PON1 gene and detoxification. Neurotoxicology *21*: 581–587, 2000.

Gaddum, J. H.: Anticholinesterases—the history of work on anticholinesterases. Chem. Ind. *1954*: 266–268, 1954.

Gaddum, J. H.: The pharmacologists of Edinburgh. Ann. Rev. Pharmacol. *2*: 1–9, 1962.

Garcia, S. J., Seidelr, F. J., Qiaao, D. and Slotkin, T. A.: Chlorpyrifos targets developing glia: effects on glial fibrillary acidic protein. Brain *133*: 1151–1161, 2002.

Gerrity, T. R. and Feussner, J. R.: Emerging research on the treatment of Gulf War veterans' illnesses. J. Occup. Environ. Med. *41*: 440–442, 1999.

Gershon, S.: Behavioral effects of anticholinergic psychotomimetics and their antagonism in man and animals. Recent Adv. Biol. Psychiatry *8*: 1441–1149, 1966.

Gershon, S. and Shaw, F. H.: Psychiatric sequelae of chronic exposure to organophosphorus insecticides. Lancet *1*: 1371–1374, 1961.

Giacobini, E.: Cholinesterase inhibitors: from Calabar bean to Alzheimer therapy. In: "Cholinesterases and Cholinesterase Inhibitors," E. Giacobini, Ed., pp. 181–226, Martin Dunitz, London, 2000.

Gill, I. and Ballesteros, A.: Degradation of organophosphorous nerve agents by enzyme-efficient biocatalytic materials for personal protection and large-scale detoxification. Biotechnol. Bioeng. *70*: 400–410, 2000.

Glenn, J. F., Hinman, D. J. and McMaster, S. B.: Electroencephalographic correlates of nerve agent poisoning. In: "Neurobiology of Acetylcholine. A Symposium in Honor of A. G. Karczmar," N. J. Dun and R. L. Perlman, Eds., pp. 503–534, Plenum Press, New York, 1987.

Glick, D.: The nature and significance of cholinesterase. In: "Comparative Biochemistry. Intermediate Metabolism of Fats. Carbohydrate Metaboli. Biochemistry of Cholne," H. B. Lewis, Ed., pp. 213–233, Jacques Cattell Press, Lancaster, PA, 1941.

Glisson, S. N., Karczmar, A. G. and Barnes, L.: Effects of diisopropyl phosphofluoridate on acetylcholine, cholinesterase and catecholamines of several parts of rabbit brain. Neuropharmacology *13*: 623–632, 1974.

Goldstein, A.: Properties and behaviors of purified human plasma cholinesterase. III. Competitive inhibition by prostigmine and other alkaloids with special reference to differences in kinetic behavior. Arch. Biochem. *34*: 169–188, 1951.

Giacobini, E.: Cholinesterase inhibitors: from the Calabar bean to Alzheimer therapy. In: "Cholinesterases and Cholinesterase Inhibitors," E. Giacobini, Ed., pp. 181–226, Martin Dunitz, London, 2000.

Glow, P. H. and Richardson, A. J.: Control of a response after chronic reduction of cholinesterase. Nature *214*: 629–630, 1967.

Goede, H. W., Doenike, A. and Altland, K.: "Pseudocholinsterasen." Springer-Verlag, Berlin, 1967.

Gopal, S., Rastogi, V., Ashman, W. and Mulbry, W.: Mutagenesis of organophosphorus hydrolase to enhance hydrolysis of the nerve agent VX. Biochem. Biophys. Res. Commun. *279*: 516–519, 2000.

Grob, D.: Anticholinesterase intoxication in man and its treatment. In: "Cholinesterases and Anticholinesterase Agents," G. B. Koelle, Ed., Handbch. d. exper. Pharmakol., Erganzungswk., vol. 15, pp. 989–1027, Springer-Verlag, Berlin, 1963.

Grob, D., Harvey, A. M., Langworthy, O. R. and Lilienthal, J. L., Jr.: The administration of diisopropyl fluorophosphate (DFP) to man. III. Effect on the central nervous system with special reference to the electrical activity of the brain. Bull. Johns Hopkins Hosp. *81*: 257–266, 1947.

Grob, D., Garlick, W. L. and Harvey, A. McG.: The toxic effects in man of the anticholinesterase insecticide parathion (p-nitrophenyl diethyl thiophusphate). Bull Johns Aopkins Hosp. *87*: 106–129, 1950.

Guengerich, F. P. and Liebler, D. C.: Enzymatic activation of chemicals to toxic metabolites. Crit. Rev. Toxicol. *14*: 259–307, 1985.

Guillou, C., Mary, A., Renko, D. Z., Gras, E. and Thal, C.: Potent acetylcholinesterase inhibitors: design, synthesis and structure-activity relationships of alkylene linked bis galanthamine and galanthamine-galanthaminium salts. Bioorg. Med. Chem. Lett. *10*: 637–639, 2000.

Gunderson, C. H., Lehmann, C. R., Sidell, F. R. and Jabbari, B.: Nerve agents: a review. Neurology. *42*: 946–950, 1992.

Gupta, R. C., Patterson, G. T. and Dettbarn, W. D.: Mechanisms involved in the development of tolerance to DFP toxicity. J. Fundam. Appl. Toxicol. *6* (Pt. 2): S17–S28, 1985.

Gupta, R. P. and Abou-Donia, M. B.: Enhanced activity and level of protein kinase A in the spinal cord supernatant of diisopropylphosphofluoridate (DFP)-treated hens. Distribution of protein kinases and phosphatases in spinal cord subcellular fractions. Mol. Cell Biochem. *220*: 15–23, 2001.

Guven, M., Sungur, M., Eser, B., Sari, I. and Altuntas, F.: The effects of fresh frozen plasma on cholinesterase levels and outcomes in patients with organophosphate poisoning. J. Toxicol. Clin. Toxicol. *42*: 617–623, 2004.

Gysin, H.: Über einige neue Insektizide. Chimia *8*: 205–210, 1955.

Habermann, E.: Gechichtliches. Wolfgang Wirth (1898–1996) als Figur der Zeitgeschichte. DGPT Forum *25*: 60–61, 1999.

Hackley, B. E., Jr., Plapinger, R., Stolberg, M. and Wagner-Jauregg, T.: Acceleration of hydrolysis of organic fluorophosphates and fluorophosphonates with hydroxamic acids. J. Am. Chem. Soc. *77*: 5653, 1955.

Hackley, B. E., Jr., Steinberg, G. M. and Lamb, J. C.: Formation of potent inhibitors of AChE by reaction of pyridinaloximes with isopropyl methylphosphonofluoridate (GB). Arch. Biochem. Biophys. *80*: 21–114, 1959.

Haley, R. W. and Kurt, T. L.: Self-reported exposure to neurotoxic chemical combinations in the Gulf War. A cross-sectional epidemiologic study. JAMA. *277*: 231–237, 1997.

Haley, R. W., Hom, J., Roland, P. S., Bryan, W. W., Van Ness, P. C., Bonte, F. J., Devous, M. D. Sr, Mathews,

D., Fleckenstein, J. L., Wians, F. H. Jr., Wolfe, G. I. and Kurt, T. L.: Evaluation of neurologic function in Gulf War veterans. A blinded case-control study. JAMA. *277*: 223–230, 1997.

Haley, R. W., Kurt, T. L. and Hom, J.: Is there a Gulf War Syndrome? Searching for syndromes by factor analysis of symptoms. JAMA. *277*: 215–222, 1997b. Erratum in: JAMA *278*: 388, 1997b.

Haley, R. W., Billecke, S. and La Du, B. N.: Association of low pON1 type Q (type A) arylesterase activity with neurological symptom complexes in Gulf War veterans. Toxicol. Appl. Pharmacol. *157*: 227–233, 1999.

Haley, R. W., Luk, G. D. and Petty, F.: Use of structural equation modeling to test the construct validity of a case definition of Gulf War syndrome: invariance over developmental and validation samples, service branches and publicity. Psychiatry. Res. *102*: 175–200, 2001.

Hanin, I., Tang, X. C. and Kozikowski, A. P.: Clinical and preclinical studies with huperzine. In: "Cholinergic Basis for Alzheimer Therapy," R. Becker and E. Giacobini, Eds., pp. 305–313, Birkhauser, Boston, 1991.

Harnack, E. and Witkowski, L.: Pharmakologische Untersuchungen über das Physostigmin und Calabarin. Arch. Exp. Pathol. Pharmakol. *5*: 401–454, 1876.

Hawkins, R. D. and Gunther, J. M.: Studies on cholinesterase. 5. The selective inhibition of pseudocholinesterase *in vivo*. Biochem. J. *40*: 192–197, 1946.

Hawkins, R. D. and Mendel, B.: Studies on cholinesterase. 6. The selective inhibition of true cholinesterase *in vivo*. Biochem. J. *44*: 260–264, 1949.

Hayes, W. J., Jr.: "Toxicology of Pesticides," Williams and Wilkins, Über Baltimore, 1975.

Hayes, W. J., Jr.: "Pesticides Studied in Man," Williams and Wilkins, Baltimore, 1982.

Heidenhain, R.: Uber die Wirkung einiger Gifte auf die Nerven der Glandula submaxillaris. Pflügers Arch. Ges. Physiol. *5*: 309–318, 1872.

Heilbronn, E.: *In vitro* reactivation and "ageing" of tabun-inhibited blood cholinesterases. Biochem. Pharmacol. *12*: 25–36, 1963.

Heilbronn, E.: The effect of sodium fluoride on sarin-inhibited cholinesterases. Acta Chem. Scand. *18*: 2410, 1964.

Heilbronn, E.: Action of fluoride on cholinesterases. II. *In vitro* reactivation of cholinesterases inhibited by organophosphorus compounds. Biochem. Pharmacol. *14*: 1363–1373, 1965.

Herlem, D., Martin, M. T., Tahl, C. and Guillou, C.: Synthesis and structure-activity relationships of open D-Ring galanthamine analogues. Bioorg. Med. Chem. Lett. *13*: 2389–2391, 2003.

Hestrin, S.: The reaction of acetylcholine and carboxylic acid derivatives with hydroxylamine, and its analytical applications. J. Biol. Chem. *180*: 249–261, 1949.

Heymans, C.: Les substances anticholinesterasiques. Expos. Annu. Biochim. Med. *12 Serie*: 21–53, 1951.

Hirata, Y., Kawada, T. and Minami, M.: Elevated frequency of sister chromatid exchanges of lymphocytes in sarin-exposed victims of the Tokyo sarin disaster 3 years after the event. Toxicology *201*: 209–217, 2004.

Hobbiger F.: The mechanism of anticurare action of certain neostigmine analogues. Br. J. Pharmacol. *7*: 223–236, 1952.

Hobbiger, F.: Effect of nicotinhydroxamic acid methiodide on human plasma cholinesterase inhibited by organophosphate containing a dialkylphosphate group. Br. J. Pharmacol. *11*: 295–303, 1955.

Hobbiger, F. W.: Reactivation of phosphorylated acetocholinesterase. In: "Cholinesterases and Anticholinesterase Agents," G. B. Koelle, Ed., pp. 921–988, Hndbch, d. exptl. Pharmakol, Evgänzungsiok, vol. 15, Springer, Berlin, 1963.

Hoffman, V. A. W.: Weitere Beobachtungen über die Phosphinsäuren. Berichte d. deutoschen chem. Gesellschaft, Berlin *6*: 303–308, 1873.

Hollinworth, R. M.: Dealkylation of organophsophorus esters by mouse liver enzymes *in vitro* and *in vivo*. J. Agric. Food Chem. *17*: 987–996, 1969.

Holmstedt, B.: Pharmacology of organophosphorus cholinesterase inhibitors. Pharmacol. Rev. *11*: 567–688, 1959.

Holmstedt, B.: Structure-activity relationships of the organophosphorus anticholinesterase agents. In: "Cholinesterases and Anticholinesterase Agents," G. B. Koelle, Ed., pp. 428–485, Handbch. d. exper. Pharmakol., Springer-Verlag, Berlin, 1963.

Holmstedt, B.: The ordeal bean of old calabar: the pageant of Physostigma venenosum in medicine. In: "Plants in the Development of Modern Medicine," T. Swain, Ed., pp. 303–360, Cambridge University Press, Cambridge, 1972.

Holmstedt, B.: Cholinesterase inhibitors: an introduction. In: "Cholinesterases and Anticholinesterase Agents," E. Giacobini, Ed., pp. 1–7, Martin Dunitz, London, 2000.

Holmstedt, B. R., Wassen, S. H., Chanh, P. H., Clavel, P. and Lasserre, B.: Alleged native antidote to curare. Goteborgs Etnografiska Museum Ann. *1983/1984*: 19–25, 1985.

Hoppe, J. O. and Arnold, A.: Influence of alkyl substitution in the onium nitrogen centers of Mytolon. Fed. Proc. *11*: 358–359, 1952.

Hoskins, B. and Ho, I. J.: Tolerance to organophosphorus cholinesterase inhibitors. In: "Organophosphates:

Chemistry, Fate and Effects," J. E. Chambers and P. E. Levi, Eds., pp. 285–297, Academic Press, San Diego, 1992.

Hsieh, B. H., Deng, J. F., Ger, J. and Tsai, W. J.: Acetylcholinesterase inhibition and the extrapyramidal syndrome: A review of neurotoxicity of organophosphate. Neurotoxicology *22*: 423–427, 2001.

Hudson, J. Q. A.: Remarks on the physiological action and therapeutic uses of Physostigma venenosum or the ordealbean of Calabar. Sothern. Med. Rec. *3*: 705–721, 1873.

Hudson, R. F.: "Structure and Mechanism in Organo-Phosphorus Chemistry," Academic Press, New York, 1965.

Hunt, R.: Some physiological actions of the homocholins and of some of their derivatives. J. Pharmacol. Exp. Ther. *6*: 477–525, 1915.

Hunt, R. and Taveau, R. de M. On the physiological action of certain cholin derivatives and new methods for detecting cholin. Br. Med. J. *2*: 1788–1791, 1906.

Irwin, R. L. and Smith, H. J. III: Cholinesterase inhibition by galanthamine and lycoramine. Biochem. Pharmacol. *3*: 147–148, 1960.

Ishihara, Y., Kato, K. and Goto, G.: Central cholinergic agents. II. Synthesis and acetylcholinesterase inhibitory activities of N-[omega-[N-alkyl-N-(phenylmethyl)amino]alkyl]-3-arylpropenamides. Chem. Pharm. Bull. (Tokyo). *39*: 3236–3243, 1991.

Ishihara, Y., Miyamoto, M., Nakayama, T. and Goto, G.: Central cholinergic agents. IV. Synthesis and acetylcholinesterase inhibitory activities of omega-[N-ethyl-N-(phenylmethyl)amino]-1-phenyl-1-alkanones and their analogues with partial conformational restriction. Chem. Pharm. Bull. *41*: 529–538, 1993.

Ishihara, Y., Goto, G. and Miyamoto, M.: Central selective acetylcholinesterase inhibitor with neurotrophic activity: structure-activity relationships of TAK-147 and related compounds. Curr. Med. Chem. *7*: 341–354, 2000.

Jamal, G. A.: Long term neurotoxic effects of chemical warfare organophosphate compounds (Sarin) Adverse Drug React. Toxicol. Rev. *14*: 83–84, 1995.

Jandorf, B. J., 1951. As quoted by W. H. Sumerson in: "Progress in the Biochemical Treatment of Nerve Gas Poisoning." U. S. Armed Forces Chem. J. *9*: 24–26, 1955.

Jandorf, B. J., Michel, H. O., Schaffer, N. K., Egan, R. and Summerson, W. H.: The mechanism of reaction between esterases and phosphorus-containing antiesterases. Discuss. Faraday Soc. *20*: 134–142, 1955.

Jianmongkol, S., Berkan, C. E. Thompson, C. M. and Richardson, R. J.: Relative potencies of the four stereoisomers of isomalathion for inhibition of brain acetylcholinesterase and neurotoxic esterase in vitro. Toxicol. Appl. Pharmacol. *139*: 342–348, 1996.

Jobst, J. and Hesse, O.: Über die Bohne von Calabar. Ann. der Chem. und Pharm. *129*: 115–121, 1864.

Johnson, L. M. and Talbot, H. W., Jr.: Detoxification of pesticides by microbial enzymes. Experientia *39*: 1236–1246, 1983.

Johnson, M. K.: The delayed neurotoxic effect of some organophosphorus compounds. Identification of the phosphorylation site as an esterase. Biochem. J.: *19*: 711–719, 1969a.

Johnson, M. K.: A phosphorylation site in brain and the delayed neurotoxic effect of some organophosphorus compounds. Biochem. J. *114*: 711–717, 1969b.

Johnson, M. K.: Organophosphorus and other inhibitors of brain "neurotoxic esterase" and the development of delayed neurotoxicity in hens. Biochem. J. *114*: 523–531, 1970.

Johnson, M. K.: Comment. Fund. Appl. Toxicol. *1*: 244, 1981.

Johnson, M. K.: The target for initiation of delayed neurotoxicity by organophosphorus esters: Biochemical studies and toxicological applications. In: "Reviews in Biochemical Toxicology," E. Hodgson, J. R. Bend, and R. M. Philipot, Eds., vol. 4, pp. 141–212, Elsevier, New York, 1982.

Johnson, M. K. and Lauwerys, R.: Protection by some carbamates against the delayed neurotoxic effects of diisopropylphosphofluoridate. Nature *222*: 1066–1067, 1969.

Johnson, M. K. and Read, D. J.: Prophylaxis against and promotion of organophosphate-induced delayed neuropathy by phenyl di-n-pentyphosphinate. Chem. Biol. Interact. *87*: 449–455, 1993.

Julian, P. L. and Pike, J.: Studies in indole series. V. Complete synthesis of physostigmine (eserine). J. Amer. Chem. Soc. *57*: 755–757, 1935.

Kamal, M. A., Greig, N. H., Alhomida, A. S. and Al-Jafari, A. A.: Kinetics of human acetylcholinesterase inhibition by the novel experimental therapeutic agent, tolserine. Biochem. Pharmacol. *60*: 561–570, 2000.

Kamata, R., Saito, S. Y., Suzuki, T., Takewaki, T. and Kobayashi, H.: Correlation of binding sites for diisopropyl phosphofluoridate with cholinesterase and neuropathy target esterase in membrane and cytosol preparations from hen. Neurotoxicology *22*: 203–214, 2001a.

Kamata, R., Saito, S. Y., Suzuki, T., Takewaki, T., Kofujita, H., Ota, M. and Kobayashi, H.: A comparative study of binding sites for diisopropyl phosphofluoridate in membrane and cytosol preparations form spinal cord and brain of hens. Neurotoxicology *22*: 191–202, 2001b.

Kammer, von der H., Mayhaus, M., Albrecht, C., Enderich, J., Wegner, M. and Nitsch, R. M.: Muscarinic acetylcholine receptors activate expression of the EGR gene family of transcription factors. J. Biol. Chem. *23*: 14538–14544, 1998.

Kammer, von der H., Demiralay, C., Andresen, B., Albrecht, C., Mayhaus. M. and Nitsch, R. M.: Regulation of gene expression by muscarinic acetylcholine receptor Biochem. Soc. Symp. *67*: 131–140, 2001.

Karanth, S. and Pope, C.: Carboxylesterase and A-esterase activities during maturation and aging: relationship to the toxicity of chlorpyrifos and parathion in rats. Toxicol. Sci. *58*: 282–289, 2000.

Karczmar, A. G.: Ontogenetic effects. In "Cholinesterases and Anticholinesterase Agents," G. B. Koelle, Ed., Handbch. d. exper. Pharmakol., Ergänzungswk., vol. 15, pp. 799–832, Springer-Verlag, Berlin, 1963.

Karczmar, A. G.: Pharmacology and antagonism of four organophosphorus agents. DOD Subcontract SU-630505, Melpar, Inc., Falls Church, Virginia, 1965.

Karczmar, A. G.: Pharmacologic, toxicologic, and therapeutic properties of anticholinesterase agents. In: "Physiological Pharmacology," W. S. Root and F. G. Hofman, Eds., vol. 3, pp. 163–322, Academic Press, New York, 1967.

Karczmar, A. G.: History of the research with anticholinesterase agents. In: "Anticholinesterase Agents," vol. 1, Section 13, A. G. Karczmar, Ed., Intern. Encyclop. Pharmacol. Ther., pp. 1–44, Pergamon Press, Oxford, 1970.

Karczmar, A. G.: Central acute and delayed effects of anticholinesterases with emphasis on behavior sleep phenomena and psyche. Report to National Academy of Sciences Committee on Toxicology, Anticholinesterase Panel, Washington, DC, 1981.

Karczmar, A. G., Convener: Workshop on New Conceptual Approaches to Prophylaxis or Therapy of Organophosphorus Poisoning, Port St. Lucie, Florida. US Department of Defense, Washington, DC, 1984.

Karczmar, A. G.: Historical development of concepts of ganglionic transmission. In: "Autonomic and Enteric Ganglia—Transmission and Its Pharmacology," A. G. Karczmar, K. Koketsu, and S. Nishi, Eds., pp. 297–337, Plenum Press, New York, 1986.

Karczmar, A. G.: Brief presentation of the story and present status of studies of the vertebrate cholinergic system. Neuropsychopharmacology *9*: 181–199, 1993.

Karczmar, A. G.: Anticholinesterases: dramatic aspects of their use and misuse. Biochem. Intern. *32*: 401–411, 1998.

Karczmar, A. G.: Synaptic, behavioral and toxicological effects of cholinesterase inhibitors in animals and humans. In: "Cholinesterases and Cholinesterase Inhibitors," E. Giacobini, Ed., pp. 157–180, Martin Dunitz, London, 2000.

Karczmar, A. G. and Koppanyi, T.: Central effects of diisopropylfluorophosphonate (DFP) in urodele larvae. Arch. Exp. Pathol. Pharmacol. *219*: 263–272, 1953.

Karczmar, A. G. and Koppanyi, T.: Possible occurrence of natural cholinesterase inhibitors in animal tissues. Fed. Proc. *13*: 372–373, 1954.

Karczmar, A. G., Koppanyi, T. and Sheatz, G. C.: Further studies on intravenously injected cholinesterase preparations. J. Pharmacol. Exp. Ther. *107*: 501–518, 1953.

Karczmar, A. G., Awad, O. and Blachut, K.: Toxicity arising from joint intravenous administration of EPN and Malathion to dogs. Toxicol. Appl. Pharmacol. *4*: 133–147, 1962.

Karczmar, A. G., Blachut, K., Ridlon, S., Gothelf, B. and Awad, O.: Pharmacological actions at various neuroeffectors of single and combined administration of EPN and Malathion. Int. J. Neuropharmacol. *2*: 163–181, 1963.

Karczmar, A. G., Nishi, S. and Blaber, L. C.: Synaptic modulations. In: "Brain and Human Behavior," A. G. Karczmar and J. C. Eccles, Eds., Springer-Verlag, Berlin, 1972.

Karczmar, A. G., Srinivasan, R. and Bernsohn, J.: Cholinergic function in the developing fetus. In: "Fetal Pharmacology," L. O. Boreus, Ed., pp. 127–177, Raven Press, New York, 1973.

Kasai, Y., Konno, T. and Dauterman, W. C.: The role of phosphorotriester hydrolases in the detoxification of organophsophorus insecticides. In: "Organophosphates: Chemistry, Fate and Effects," J. E. Chambers and P. E. Levi, Eds., pp. 169–182, Academic Press, San Diego, 1992.

Katz, E. J., Cortes, V. I., Eldefrawi, M. E. and Eldefrawi, A. T.: Chlorpyrifos, parathion, and their oxons bind to and desensitize a nicotinic acetylcholine receptor: relevance to their toxicities. Toxicol. Appl. Pharmacol. *146*: 227–236, 1997.

Kaul, P. N.: Drug discovery: past, present and future. Prog. Drug Res. *50*: 9–105, 1998.

Kaur, J. and Zhang, M. Q.: Molecular modeling and QSAR of reversible acetylcholinesterase inhibitors. Curr. Med. Chem. *7*: 273–294, 2000.

Keijer, J. H. and Wolring, G. Z.: Stereospecific aging of phosphonylated cholinesterases. Biochim. Biophys. Acta *185*: 465–468, 1969.

Kilby, B. A.: Alkyl fluorophosphonates and related compounds. Research *2*: 417–422, 1949.

Klee, E.: "Das Personen Lexikon zum Dritten Reich," p. 681, S. Fischer Verlag, Leipzig, 2003.

Kleinwaechter, L.: Beobachtung ueber die Wirkung des Calabar-Extracts gegen Atropin Vergiftung. Berl. Klin. Wschr. *1*: 369–371, 1864.

Koelle, G. B.: Protection of cholinesterase against irreversible inactivation by DFP *in vitro*. J. Pharmacol. Exp. Ther. *88*: 153–171, 1946.

Koelle, G. B.: Cytological distributions and physiological functions of cholinesterases. In: "Cholinesterases and Anticholinesterase Agents," G. B. Koelle, Ed., Handbch. d. exper. Pharmakol., Ergänzungswk., vol. 15, pp. 189–298, Springer-Verlag, Berlin, 1963.

Koelle, G. B.: Organophosphate poisoning—an overview. Fund. Appl. Toxicol. *1*: 129–134, 1981.

Koelle, G. B. and Gilman, A.: The chronic toxicity of DFP in dogs, monkeys and rats. J. Pharmacol. Exp. Ther. *87*: 435–448, 1946.

Koelle, G. B. and Gilman, A.: Anticholinesterase drugs. J. Pharmacol. Exp. Ther. (Pt. II) *95*: 166–216, 1949.

Koelle, G. and Steiner, E. C.: The cerebral distribution of tertiary and quaternary anticholinesterase agent following intravenous and intraventricular injection. J. Pharmacol. Exp. Ther. *118*: 420–434, 1956.

Koketsu, K.: Restorative action of fluoride on synaptic transmission blocked by organophosphorous anticholinesterases. Int. J. Neuropharmacol. *5*: 247–254, 1966.

Koketsu, K.: Modulation of receptor sensitivity and action potentials by transmitters in vertebrate neurons. Jap. J. Physiol. *34*: 945–960, 1984.

Koppanyi, T. and Karczmar, A. G.: Eserine-like activity of adenylic acid and of water-soluble adenosinetriphosphate (ATP). Am. J. Physiol. *159*: 576–577, 1949.

Koppanyi, T. and Karczmar, A. G.: Contribution to the study of action of the mechanism of action of cholinesterase inhibitors. J. Pharmacol. Exp. Ther. *101*: 327–344, 1951.

Koppanyi, T., Karczmar, A. G. and King, T. O.: The effect of tetraethyl pyrophosphate on sympathetic ganglionic activity. Science *106*: 492–493, 1947.

Koster, R.: Synergisms and antagonisms between physostigmine and diisopropyl fluorophosphate in cats. J. Pharmacol. Exp. Ther. *88*: 39–46, 1946.

Kraut, D., Goff, H., Pai, R. K., Hosea, N. A., Silman, I., Sussman, J. L., Taylor, P. and Voet, J.: Inactivation studies of acetylcholinesterase with phenylmethylsulfonyl fluoride. Mol. Pharmacol. *57*: 1243–1248, 2000.

Krupka, R. M.: Fluoride inhibition of acetylcholinesterase. Mol. Pharmacol. *2*: 1183–1190, 1966.

Krupka, R. M. and Laidler, K. J.: Molecular mechanisms for hydrolytic enzyme action. I. Apparent noncompetitive inhibition, with special reference to acetylcholinesterase. II. Inhibition of acetylcholinesterase by excess substrate. III. A general mechanism for inhibition of acetylcholinesterase. IV. The structure of the active center and the reaction mechanism. J. Am. Chem. Soc. *83*: 1445–1460, 1961.

Kryger et al., 1998, in Massolié, 2000.

Kuhr, R. J. and Dorough, H. W.: "Carbamate Insecticides: Chemistry, Biochemistry and Toxicology." CRC Press, Boca Raton, Florida, 1976.

Kulkarni, A. P. and Hodgson, E.: The metabolism of insecticides: the role of monooxygenase enzymes. Annu. Rev. Pharmacol. *24*: 19–42, 1984.

Kuo, C. L. and La Du, B. N.: Comparison of purified human and rabbit serum paraoxonases. Drug Metab. Dispos. *23*: 935–944, 1995.

La Du., B. N., Aviram, M., Billecke, S., Navab, M., Primo-Parmo, S., Sorenson, R. C. and Standiford, T. J.: On the physiological role(s) of the paraoxonases. Chem. Bio. Interact. *119–120*: 379–388, 1999.

Langley, J. N. and Kato, T.: The physiological actions of physostigmine and its action on denervated skeletal muscle. J. Physiol. *49*: 410–415, 1915.

Lallement, G., Dorandeu, F., Filliat, P., Carpentier, P., Baille, V. and Blanchet, G.: Medical management of organophosphate-induced seizures. In: "From *Torpedo* Electric Organ to Human Brain: Fundamental and Applied Aspects," J. Massoulié, Ed., pp. 369–373, Elsevier, Amsterdam, 1998.

Lallement, G., Baubichon, D., Clarencon, D., Galonnier, M., Peoc'h, M. and Carpentier, P.: Review of the value of gacyclidine (GK-11) as adjuvant medication to conventional treatments of organophosphate poisoning: primate experiments mimicking various scenarios of military terrorist attack by soman. Neurotoxicology *20*: 675–684, 1999.

Lands, A. M., Hoppe, J. O., Arnold, A. and Kirchner, F. K: An investigation of the structure-activity correlations within a series of ambenonium analogs. J. Pharmacol. Exp. Ther. *123*: 121–127, 1958.

Lands, A. M., Karczmar, A. G. and Arnold, A.: The action of ambenonium (Win 8077) and related compounds in the mouse. J. Pharmacol. Exp. Ther. *119*: 541–549, 1957.

Lange, W. and Krueger, von G.: Über Ester der Monofluorphsophorsäure. Ber. dtsch. chem Ges. *65*: 1958–1961, 1932.

Lankaster, M. C.: A note on the demyelination produced in hens by dialkylfluoridates. Br. J. Pharmacol. *15*: 279–281, 1960.

Larkin, D. J. and Tjeerdema, R. S.: Fate and effects of diazinon. Rev. Environ. Contam. Toxicol. *166*: 49–82, 2000.

Larsson, L., Holmstedt, B. and Tjus, E.: Some considerations regarding the nomenclature of organic phosphorus compounds. Acta Chem. Scand. *8*: 1563–1569, 1954.

Lassiter, T. L., Barone, S., Jr., Moser, V. C. and Padilla, S.: Gestational exposure to chlorpyrifos: dose response profiles for cholinesterase and carboxylesterase activity. Toxicol. Sci. *52*: 92, 1999.

Lebeda, F. J. and Rutecki, P. A.: Organophosphorus anticholinesterase-induced epileptiform activity in the hippocampus. In: "Neurobiology of Acetylcholine. A Symposium in Honor of A. G. Karczmar," N. J. Dun and R. L. Perlamn, Eds., pp. 437–450, Plenum Press, New York, 1987.

Lehmann, G.: The "curare-like" action of inhibitors of cholinesterase. In: "Jubilee Volume Dedicated to Emil Christoph Barell," scientific workers of Roche Companies, Eds., pp. 314–326, Frederick Reinhardt, Basel, 1946.

Lei, Y., Mulchandani, P., Chen, W., Wang, J. and Mulchandani, A.: Whole cell-enzyme hybrid amperometric biosensor for direct determination of organophosphorous nerve agents with p-nitrophenyl substituent. Biotech. Bioeng. *85*: 796–713, 2004.

Levey, M.: Medieval arabic toxicology. Trans. Am. Phil. Soc., New Series *56*: 1–130, 1966.

Levi, P. E. and Hodgson, E.: Metabolism of organophosphorus compounds by the flavin-containing monooxygenase. In: "Organophosphates: Chemistry, Fate and Effects," J. E. Chambers and P. E. Levi, Eds., pp. 141–154, Academic Press, San Diego, 1992.

Li, W. F., Costa, L. G., Richter, R. J., Hagen, T., Shih, D. M., Tward, A., Lusis, A. J. and Furlong, C. E.: Catalytic efficiency determines the in-vivo efficacy of PON1 for detoxifying organophosphorus compounds. Pharmacogenetics *10*: 767–779, 2000.

Loeffelholz, K.: Leserbrief. DGPT-Forum *26*: 53, 2000.

Loewi, O.: Über humorale Übertraqbarkeit des Hernervenverwirkung. I. Mitteilung. Pflügers Arch. Ges. Physiol. *189*: 239–242, 1921.

Loeffelholz, K.: Letter to W. Woelk, June 12, 2004. Loewi, O.: Über humorale Übertraqbarkeit des Hernervenverwirkung. I: Mitteilung. Pflügers Arch. Ges. Physiol. *189*: 239–242, 1921.

Loewi, O. and Mansfeld, G.: Über den Wirkungsmodus des Physostigmins. Arch. Exp. Path. Pharmakol. *62*: 180–185, 1910.

Long, J. P.: Structure-activity relationships of the reversible anticholinesterase agents. In: "Cholinesterases and Anticholinesterase Agents," G. B. Koelle, Ed., Handbch. d. exper. Pharmakol., Ergänzungswk., vol. 15, pp. 374–427, Springer-Verlag, Berlin, 1963.

Longo, V. G., Nachmansohn, D. and Bovet, D.: Aspects electroencephalographiques de phosphate d'isopropyl (sarin). Arch. Intern. Pharmacodyn. *123*: 282–293, 1960.

Lorot, C.: Les combinaisons de la creosote dans le traitement de la tuberculose pulmonaire. These, Paris, 1996.

Luca, D. and Balan, M.: Sperm abnormality assay in the evaluation of genotoxic potential of carbaryl in rats. Morphol. Embryol. *33*: 19–22, 1987.

Luo, C., Saxena, A., Smith, M., Garcia, G., Radic, Z., Taylor, P. and Doctor, B. P.: Phosphoryl oxime inhibition of acetylcholinesterase during oxime reactivation is prevented by edrophonium. Biochemistry *38*: 9937–9947, 1999.

Ma, T. and Chambers, J. E.: Kinetic analysis of desulfuration and dearylation of parathion and chlorpyrifos by rat liver microsomes. Food Chem. Toxicol. *32*: 763–767, 1994.

Mackness, B., Durrington, P. N. and Mackness, M. I.: Low paraoxonase in Persian Gulf War veterans self-reporting. Biochem. Biophys. Res. Commun. *276*: 729–733, 2000.

Mackworth, J. F. and Webb, E. F.: Inhibition of serum cholinesterase by alkyl fluophosphates. Biochem. J. *42*: 91–95, 1948.

Mahoney, D. B.: A normative construction of Gulf War syndrome. Persp. Biol. Med. *44*: 575–583, 2001.

Main, A. R.: Structure and inhibitors of cholinesterase. In: "Biology of Cholinergic Function," A. M. Goldberg and I. Hanin, Eds., pp. 269–353, Raven Press, New York, 1976.

Makhaeva, G. F., Filolenko, I. V., Yankovskaya, V. L., Fomicheva, S. and Malygin, V. V.: Comparative studies of o,o dialkyl-o-chloromethylchloroformino phosphates: interaction with neuropathy target esterase and acetylcholinesterase. Neurotoxicology *19*: 623–628, 1998.

Manetsch, R., Krasinski, A., Radic, Z., Raushel, J., Taylor, P., Sharpless, K. B. and Kolb H. C.: In situ click chemistry: enzyme inhibitors made to their own specifications. J. Am. Chem. Soc. *126*: 12809–12818, 2004.

Marrs, T. C.: Organophosphate poisoning. Pharmacol. Ther. *58*: 51–66, 1993.

Martinez, A., Lanot, C., Perez, C., Castro, A., Lopez-Serrano, P. and Conde, S.: Lipase-catalyzed synthesis of new acetylcholinesterase inhibitors: N-benzylpiperidine amino acid derivatives. Bioorg. Med. Chem. *8*: 731–738, 2000.

Massicotte, C., Inzana, K. D., Ehrich, M. and Jortner, B. S.: Neuropathologic effects of phenylmethylsulfonyl fluoride (PMSF)-induced promotion and protection in organophosphorus ester-induced neuropathy (OPIDN) in hens. Neurotoxicology *20*: 749–759, 1999.

Masson, P., Clery, C., Guerra, P., Redslob, A., Albaret, C., Fortier, P. L.: Hydration change during the aging of phosphorylated human butyrylcholinesterase: importance of residues aspartate-70 and glutamate-197 in the water network as probed by hydrostatic and osmotic pressures. Biochem. J. *343*: 361–369, 1999.

Massoulié, J.: Molecular forms and anchoring of acetylcholinesterase. In: "Cholinesterases and Cholinesterase Inhibitors," E. Giacobini, Ed., pp. 81–101, Martin Dunitz, London, 2000.

Massoulié, J., Sussman, J., Bon, S. and Silman, I: Structure and function of acetylcholinesterase and butyrylcholinesterase. In: "Cholinergic Function and Dysfunction," A. C. Cuello, Ed., pp. 139–146, Elsevier, Amsterdam, 1993.

Matsumura, F.: "Toxicology of Insecticides," Plenum, New York, 1985.

Matthes, K.: The action of blood on acetylcholine. J. Physiol. (Lond.) *70*: 338–348, 1930.

Maxwell, D. M.: Detoxification of organophosphorus compounds by carboxylesterase. In: "Organophosphates: Chemistry, Fate and Metabolism," J. E. Chambers and P. E. Levi, Eds., pp. 183–199, Academic Press, San Diego, 1992.

Mazur, A.: An enzyme in animal tissues capable of hydrolyzing the phosphorusfluorine bond of alkyl fluorophosphates. J. Biol. Chem. *164*: 271–289, 1946.

McGehee, D. S., Krasowski, M. D., Fung, D. L., Wilson, B., Gronert, G. A. and Moss, J.: Cholinesterase inhibition by potato glycoalkaloids slows mivacurium metabolism. Anesthesiology *93*: 510–519, 2000.

Menzie, C. M.: "Metabolism of Pesticides," Bureau of Sport Fisheries and Wildlife. Special Scientific Report. Wildlife no. 127, Washington, DC, 1972.

Mesulam, M.: Neuroanatomy of cholinesterases in the normal human brain and in Alzheimer's disease. In: "Cholinesterases and Cholinesterase Inhibitors," E. Giacobini, Ed., pp. 121–137, Martin Dunitz, London, 2000.

Michaelis, C. A. A.: Über die organischne Verbindungen des Phosphors mit Stickstoff. Liebigs Ann. Chem. 129–258, 1903.

Michel, H. O.: Kinetics of the reactions of cholinesterase, chymotrypsin and trypsin with organophosphorus inactivators. Fed. Proc. *14*: 255, 1955.

Michel, H. O.: Development of resistance of alkylphosphorylated cholinesterase to reactivation by oximes. Fed. Proc. *17*: 275, 1958.

Mileson, B. E., Chambers, J. E., Chen, W. L., Dettbarn, W., Ehrich, M., Eldefawi, A. T., Craylor, D. W., Hamernik, K., Hodgson, E., Karczmar, A. G., Padilla, S., Pope, C. N., Richardson, R. J., Saunders, D. R., Sheets, L. P., Sultates, L. J. and Wallace, K. B.: Common mechanism of toxicity: a case study of organophosphorus pesticides. Toxicol. Sci. *41*: 8–20, 1988.

Milatovic, D. and Dettbarn, W. D.: Modification of acetylcholinesterase during adaptation to chronic, subacute paraoxon application in rat. Toxicol. Appl. Pharmacol. *136*: 20–28, 1996.

Millard, C. B., Kryger, G. and Ordentlich, A.: Crystal structures of aged phosphonylated acetylcholinesterase: nerve agent reaction products at the atomic level. Biochemistry *38*: 7032–7039, 1999.

Monnet-Tschudi, F., Zurich, M. G., Schilter, B., Costa, L. G. and Honneger, P.: Maturation-dependent effects of chlorpyrifos and parathion on acetylcholinesterase and neuronal and glial markers in aggression. Toxicol. Appl. Pharmacol. *165*: 175–183, 2000.

Moore, D. H.: Long term health effects of low dose exposure to nerve agent. In: "From *Torpedo* Electric Organ to Human Brain: Fundamental and Applied Aspects," J. Massoulié, Ed., pp. 325–328, Elsevier, Amsterdam, 1998.

Motoyama, N. and Dauterman, W. C.: The role of nonoxidative metabolism in organophosphorus resistance. J. Agric. Food Chem. *22*: 350–356, 1974.

Mounter, L. A.: Metabolism of organophosphorus anticholinesterase agents. In: "Cholinesterases and Anticholinesterase Agents," G. B. Koelle, Ed., Handbch. d. exper. Pharmakol., Ergänzungswk., vol. 15, pp. 486–504, Springer-Verlag, Berlin, 1963.

Mulbry, W.: Characterization of a novel organophosphorus hydrolase from NRRL B-24074. Microbiol. Res. *154*: 285–288, 2000.

Myers, D. K.: "Studies on Selective Esterase Inhibitors," Uitgeverij Excelsior, s'Gravenhage, 1954.

Nachmansohn, D.: Sur l'action de la strychnine. C. R. Soc. Biol. *129*: 941–943, 1938.

Nachmansohn, D.: Actions on axons, and evidence for the role of acetylcholine in axonal conduction. In: "Cholinesterases and Anticholinesterase Agents," Handbch. d. exper. Pharmakol., Ergänzungswk., G. B. Koelle, Ed., vol. 15, pp. 799–832, Springer-Verlag, Berlin, 1963.

Nachmansohn, D. and Schneemann, H.: On the effect of drugs on cholinesterase. J. Biol. Chem. *159*: 239–240, 1945.

Nakatsugawa, T.: Hepatic disposition of organophosphorus insecticides: a synthesis of *in vitro*, *in situ* and *in vivo* data. In: "Organophosphates: Chemistry, Fate and Metabolism," J. E. Chambers and P. E. Levi, Eds., pp. 201–227, Academic Press, San Diego, 1992.

Navarro, H. A., Basta, P. V., Seidler, F. J. and Slotkin, T. A.: Neonatal chlorpyrifos administration elicits deficits in immune function in adulthood: a neural effect? Res. Dev. Brain Res. *130*: 249–252, 2001.

Nishiwaki, Y., Mackawa, K., Ogawa, Y., Ansukai, N., Minami, M. and Omae, K.: Effects of sarin on the nervous system in rescue team staff members and police officers 3 years after the Tokyo subway sarin attack. Environ. Health Persp. *109*: 1169–1173, 2001.

Nylen, P.: Studien über organische Phosphorverbindungen. Doctoral Diss., Uppsala, 1930.

Oberst, F. W., Ross, R. S., Christensen, M. K., Crook, J. W., Cresthull, P. and Unland, C. W.: The use of atropine and artificial respiration in the resuscitation of dogs exposed by inhalation to lethal concentration of GB vapro. J. Pharmacol. Exp. Ther. *116*: 44, 1956.

O'Brien, R. D.: Activation of thionophosphates by liver microsomes. Nature *183*: 121–121, 1959.

O'Brien, R. D.: "Toxic Phosphorus Esters," Academic Press, New York, 1960.

Okumura, T., Nobukatsu, T., Ishimatsu, S., Miyanoki, S., Mitsuhashi, A., Kumda, K., Tanaka, T. and Hinohara, S.: Report on 640 victims of the Tokyo subway sarin attack. Ann. Emerg. Med. *28*: 129–135, 1996.

Olajos, E. J., DeCaprio, A. P. and Rosenblum, I.: Central and peripheral esterase activity and dose-response relationship in adult hens after acute and chronic oral administration of diisopropyl fluorophosphate. Ecotoxicol. Environ. Safety *2*: 383–399, 1978.

Oosterbaan, R. A., Jansz, H. S. and Cohen, J. A.: Studies with 18O on the mechanism of hydrolytic enzyme reaction. Abstracts, 5th Intern. Congr. Biochem., p. 119, Moscow, August 10–16, 1961.

Ordentlich, A., Barak, D., Sod-Moriah, G., Kaplan, D., Mizrahi, D., Segall, Y., Kronman, C., Karton, Y., Lazar, A., Marcus, D., Velan, B. and Shaferman, A.: Stereoselectivity toward VX is determined by interactions with residues of the acyl pocket as well as the peripheral anionic site of AChE. Biochemistry *43*: 11255–11265, 2004.

Osei-Atweneboana, M. Y., Wilson, M. D., Post, R. J. and Boakye, D. A.: Temephos-resistant larvae of Simulium sanctipauli associated with a distinctive new chromosome inversion in untreated rivers of southwestern Ghana. Med. Vet. Entymol. *15*: 113–116, 2001.

Overstreet, D. H., Russell, R. W., Vasquez, B. J. and Dalglish, F. W.: Involvement of muscarinic and nicotinic receptors in behavioral tolerance to DFP. Pharmacol. Biochem. Behav. *2*: 45–54, 1974.

Pang, Y. P. and Kozikowski, A. P.: Prediction of the binding site of 1-benyl-4-[(5,6-dimethoxy-1-indanon-2-41)methyl] piperidine in acetylcholinesterase by docking studies with the SYSDOC program. J. Comp. Aided Mol. Design *5*: 683–693, 1995.

Peeples, E. S., Schopfer, L. M., Duysen, E. G., Spauldng, R., Voelker, T., Thompson, C. M. and Lockridge, O.: Albumin, a new biomarker of organophosphorus toxicant exposure, identified by mass spectrometry. Toxicol. Sci. *83*: 303–312, 2005.

Pereira, M. E., Adams, A. I. and Silva, N. S.: 2,5-Hexanedione inhibits rat brain acetylcholinesterase activity *in vitro*. Toxicol. Lett. *146*: 2629–274, 2004.

Perola, E., Cellai, L., Lamba, D., Filocamo, L. and Brufani, M.: Long chain analogs of physostigmine as potential drugs for Alzheimer's disease: new insights into the mechanism of action in inhibition of acetylcholinesterase. Biochem. Biophys. Acta *1343*: 41–50, 1997.

Petcher, T. J. and Pauling, P.: Cholinesterase inhibitors: structure of eserine. Nature. *241*: 277, 1973.

Peters, B. and Levin, H. S.: Effects of physostigmine and lecithin on memory in Alzheimer's disease. Ann. Neurol. *6*: 219–221, 1977.

Petrikovics, I., Cheng, T. C., Papahadjopoulos, D., Hong, K., Yin, R., DeFrank, J. J., Jaing, J., Song, Z. H., McGuinn, W. D., Sylvester, D., Pei, L., Madec, J., Tamulinas, C., Jaszberenyi, J. C., Barcza, T. and Way, J. L: Long circulating liposomes encapsulating organophosphorus acid anhydrolase in diisopropylfluorophosphate antagonism. Toxicol. Sci. *57*: 16–21, 2000a.

Petrikovics, I., McGuinn, W. D., Sylvester, D., Yuzapavik, P., Jiang, J., Way, J. L., Papahadjopoulos, D., Hong, K., Yin, R., Cheng, T. C. and DeFrank, J. J.: *In vitro* studies on sterically stabilized liposomes (SL) as enzyme carriers in organophosphorus (OP) antagonism. Drug Deliv. *7*: 83–89, 2000b.

Piao, F. Y., Kitabatake, M., Xie, X. K. and Yamauchi, T.: Effects of various post-treatment by phenylmethysulfonyl fluoride on delayed neurotoxicity induced by leptophos. J. Toxicol. Sci. *20*: 609–617, 1995.

Plaitakis, A. and Duvoisin, R. C.: Homer's moly identified as *Galanthus nivalis L.* Physiologic antagonist to stramonium poisoning. Clin. Neuropharmacol. *6*: 1–5, 1983.

Plattner, F. and Hintner, H.: Die Spaltung von Acetylcholin durch Organextracte and Korperflüssigkeiten. Pflügers Arch. Ges. Physiol. *255*: 19–25, 1930.

Polonowski, M. and Nitzburg, C.: Etudes sur les alcaloides de la feve de Calabar (II). La geneserine nouvelle alcaloide de la fève. Bul. Soc. Chimies *17*: 244–256, 1915.

Pond, A. L., Chambers, H. W., Coyne, C. P. and Chambers, J. E.: Purification of two rat hepatic proteins with A-esterase activity toward chlorpyrifos-oxon and paraoxon. J. Pharmacol. Exp. Ther. *286*: 1404–1411, 1998.

Pope, C., Chapman, M., Arthun, D., Tanaka, D. and Padilla, S.: Age and sensitivity to organophosphorus-induced delayed neurotoxicity (OPIDN): effects of phenylmethysulfonyl fluoride (PMSF). Toxicologist *11*: 304, 1991.

Porschke, D., Creminon, C., Cousin, X., Bon, C., Sussman, J. and Silman, I.: Electrooptical measurements demonstrate a large permanent dipole moment associated with acetylcholinesterase. Biophys. J. *70*: 1603–1608, 1996.

Proctor, G. R. and Harvey, A. L.: Synthesis of tacrine analogues and their structure-activity relationships. Curr. Med. Chem. *7*: 295–302, 2000.

Proskurnina, N. F. and Yakovlena, A. P.: On alkaloid *Galanthus woronowi.* Zhur. Obshchei. Khim. *17*: 1216, 1947.

Qiao, C. L., Sun, Z. Q. and Liu, J. E.: New esterase enzymes involved in organophosphate resistance in Culex pipiens (Diptera: Cullicidae) from Guang Zhou. China. J. Med. Entomol. *36*: 666–670, 1999.

Radic, Z., Reiner, E. and Simeon, V.: Binding sites on acetylcholinesterase for reversible ligands and phosphorylating agents: a theoretical model tested on haloxon and phosphostigmine. Biochem. Pharmacol. *33*: 671–677, 1984.

Radic, Z., Pickering, N., Vellom, D. C., Camp, S. and Taylor, P.: Three domains in the cholinesterase molecule confer selectivity for acetyl- and butyrylcholinesterase inhibitors. Biochem. *32*: 12074–12084, 1993.

Radic, Z., Duran, R., Vellom, D. C., Li, Y., Cervenasky, C. and Taylor, P.: Site of fasciculin interaction with acetylcholinesterase. J. Biol. Chem. *269*: 11233–11239, 1994.

Radic, Z., Quinn, D. M., Vellom, D. C., Camp, S. and Taylor, P.: Allosteric control of acetylcholinesterase catalysis by fasciculin. J. Biol. Chem. *270*: 20391–20399, 1995.

Randall, L. D. and Lehmann, G.: Pharmacological properties of some neostigmine analogs. J. Pharmacol. *99*: 16–32, 1950.

Ranjbar, A., Pasalar, P., and Abdollahi, M.: Induction of oxidative stress and acetylcholinesterase inhibition in organophosphorous pesticide manufacturing workers. Hum. Exp. Toxicol. *21*: 179–182, 2002.

Rauschel, F. M.: Phosphotriesterase: an enzyme in search of its natural substrate. Adv. Enzymol. Related Areas Mol. Biol. *74*: 51–93, 1998.

Raves, M. L., Harel, M., Pang, Y.-P., Silman, I., Kozikowski, A. P. and Sussman, J. L.: 3D structure of acetylcholinesterase complexed with the nootropic alkaloid (-) huperzine A. Nature Struct. Biol. *4*: 57–63, 1997.

Recanatini, M., Cavalli, A., Belluti, F., Piazzi, L., Rampa, A., Biosi, A., Gobbi, S., Valenti, P., Andrisano, V., Bartolini, M. and Cavrini, V.: SAR of 9-amino-1, 2,3,4-tetrahydroacridine-based acetylcholinesterase inhibitors: synthesis, enzyme inhibitory activity, QSAR, and structure-based CoMFA of tacrine analogues. J. Med. Chem. *43*: 2007–2018, 2000.

Rege, P. D. and Johnson, F.: Application of vicarious nucleophilic substitutions to the total synthesis of dl-physostigmine. J. Org. Chem. *68*: 6133–6139, 2003.

Reiner, E. and Radic, Z.: Mechanism of action of cholinesterase inhibitors. In: "Cholinesterases and Anticholinesterase Agents," E. Giacobini, Ed., pp. 103–119, Martin Dunitz, London, 2000.

Reiner, E., Skrinjaric-Spoljar, M. and Simeon-Rudolf, V.: Binding sites on acetylcholinesterase and butyrylcholinesterase for pyridinium and imidazolinium oximes, and other reversible ligands. Perod. Biol. *98*: 325–329, 1996.

Remen, L.: Zur Pathogenese und Therapie der myasthenia gravis pseudoparalytica. Dtsch. Z. Nervenheilk. *128*: 66–78, 1932.

Ricceri, L., Markina, N., Valanzano, A., Fortuna, S., Cometa, M. F., Meneguz, A. and Calamandrei, G.: Developmental exposure to chlorpyrifos alters reactivity to environmental and social cues in adolescent mice. Toxicol. Appl. Pharmacol. *191*: 189–201, 2003.

Richards, S. M. and Kendall, R. J.: Biochemical effects of chlorpyrifos on two developmental stages of Xenopus laevis. Environ. Toxicol. Chem. *21*: 1826–1835, 2002.

Richardson, R. J.: Neurotoxic esterase: research trends and prospects. Neurotoxicology *4*: 157–162, 1983.

Richardson, R. J.: Interactions of organophosphorus compounds with neurotoxic esterase. In: "Organophosphates: Chemistry, Fate and Effects," J. E. Chambers and P. E. Levi, Eds., pp. 299–323, Academic Press, San Diego, 1992.

Richardson, R. J.: Assessment of the neurotoxic potential of chlorpyrifos relative to other organophosphorus compounds: a critical review of literature. J. Toxicol. Environ. Health *44*: 135–165, 1995.

Richardson, R. J. and Chambers, J. E.: Effects of repeated oral postnatal exposure to chlorpyrifos on cholinergic neurochemistry in developing rats. Toxicol. Sci.: *84*: 352–359, 2005.

Rider, J. A., Ellinwood, L. E. and Coon, J. M.: Production of tolerance in the rat to octamethyl pyrophosphoramide (OMPA). Proc. Soc. Exper. Biol. Med. *81*: 455–459, 1952.

Robertson, A. D.: The Calabar bean as a new agent in ophthalmic medicine. Edinb. Med. J.: *8*: 193–204, 1863.

Roger, H. and Recordier, M.: Les polynevrites phosphocreositiques (phosphate) de creosote. Ann. Med. *35*: 44, 1934.

Rogers, J. C., Chambers, H. and Casida, J. E.: Nicotinic acid analogs: effect on response of chick embryos and hens to organophosphate toxicants. Science *144*: 539, 1964.

Rose, M. R.: Gulf War service is an uncertain trigger for ALS. Neurology *6*: 730–731, 2003.

Rosenstock, L., Keifer, M. and Daniell, W. E.: Chronic central nervous system effects of acute pesticide intoxication. Lancet *338*: 223–227, 1991.

Russell, R. W., Vasquez, B. J., Overstreet, D. H. and Dalglish, F. W.: Effects of cholinolytic agents on behavior following development of tolerance to low cholinesterase activity. Psychopharmacologia *20*: 32–41, 1958.

Russell, R. W., Vasquez, B. J., Overstreet, D. H. and Dalglish, F. W.: Consummatory behavior during tolerance to and withdrawal from chronic depression of cholinesterase activity. Physiol. Behav. *7*: 523–528, 1971.

Satoh, T. and Hosokawa, M.: The mammalian carboxylesterases: from molecules to functions. Annu. Rev. Pharmacol. Toxicol. *38*: 257–288, 1998.

Saunders, B. C.: "Phosphorus and Fluorine, the Chemistry and Toxic Actions of Their Organic Compounds," First Edition, Cambridge University Press, Cambridge, 1957.

Saxena, A., Viragh, C., Frazier, D. S., Kovach, I. M., Maxwell, D. M., Lockridge, O. and Doctor, B. P.: The pH dependence of dealkylation in soman-inhibited cholinesterases and their mutants: further evidence for a push-pull mechanism. Biochemistry *37*: 15086–15096, 1998.

Saxena, A., Fredorko, J. M., Vinayaka, C. R., Medhekar, R., Radic, Z., Taylor, P., Lockridge, O. and Doctor, B. P.: Aromatic amino-acid residues at the active and peripheral anionic sites control the binding of E2020 (Aricept) to cholinesterases. Eur. J. Biochem. *270*: 4447–4458, 2003.

Schaeffer, H.: Über das normale Verhalten der Cholinesterase im Blut. Pflugers Arch. *249*: 405–430, 1947.

Scharf, M. E., Parimi, S., Meinke, L. J., Chandler, L. D. and Siegfried, B. D.: Expression and three family 4 cytrochrome P450 (CYP4)* genes identified from insecticide-resistant and susceptible western corn rootworms, Diabrotica virigifera virigifera. Insect Mol. Biol. *10*: 139–146, 2001.

Schmiedeberg, O. and Koppe, R.: "Das giftige Alkaloid des Fliegenpilzes," F. C. W. Vogel, Leipzig, 1869.

Schoene, K.: Reaktivierung von 0,0 diaethylphosphoryl Azetylcholinesterase. Reaktivierungs-Rephosphorylierungs Gleichgewicht. Biochem. Pharmacol. *21*: 163–170, 1972.

Schrader, G. Belgian patent 556,009, 1937.

Schrader, G.: Organische Phosphor-Verbindungen als neuartige Insektizide (Auszung), Angea. Chem. *62*: 471–473, 1950.

Schrader, G.: Die Entwicklung neuer Insekitizide auf Grundlage von organischen Fluor- und Phosphorverbindungen. Monographie No. 62, *2*, Aufl. Winheim, Verlag Chemie, 1952.

Schrader, G.: "Die Entwicklung neuer Insektizide Phosphorsäure-Ester," Verlag Chemie, Weinheim, 1963.

Schwartz, R. D. and Kellar, K. J.: Nicotinic cholinergic binding sites in the brain: Regulation *in vivo*. Science *220*: 214–216, 1986.

23. Schwartz, R. D. and Kellar K. J.: Nicotinic cholinergic receptor binding sites in the brain: regulation *in vivo*. Science. *220*: 214–216, 1983.

Schuytema, G. S., Nebeker, A. V. and Griffis, W. L.: Toxicity of Gthion and Guthion 2S to *Xenopus laevis* embryos. Arch. Environ. Contamination Toxicol. *27*: 250–255, 1994.

Schweitzer, A. and Wright, S.: The action of eserine and related compounds and of acetylcholine on the central nervous system. J. Physiol. (Lond.) *89*: 165–197, 1937.

Senanayake, N. and Karalliedde, L.: Neurotoxic effects of organophosphorus insecticides. N. Engl. J. Med. *316*: 761–763, 1987.

Shaw, F. H. and Bentley, G.: Some aspects of the pharmacology of morphine with special reference to its antagonism by 5-aminoacridine THA and other chemically related compounds. Med. J. Austral. *2*: 868, 1949.

Sheets, L. P., Hamilton, B. F., Sangha, G. K. and Thyssen, J. H.: Subchronic neurotoxicity screening studies with six organophosphate insecticides: an assessment of behavior and morphology relative to cholinesterase inhibition. Fund. Appl. Toxicol. *35*: 101–119, 1997.

Shih, J., Bond, A. E., Radic, Z. and Taylor, P.: Reversibly bound and covalently attached ligands induce conformational changes in the omega loop, cys69-cys96, of mouse acetylcholinesterase. J. Biol. Chem. *276*: 42196–42204, 2001.

Siegfried, B. D. and Scott, J. G.: Biochemical characterization of hydrolytic and oxidative enzymes in insecticide resistant and susceptible strains of the German cockroach (Dictyoptera: Blattellidae). J. Econ. Entomol. *85*: 1092–1098, 1992.

Siegfried, B. D., Ono, M. and Swanson, J. J.: Purification and characterization of carboxylesterase associated with resistance of the greenbug, Schizaphis graminum (Homoptera: Aphididae). Arch. Insect Biochem. Physiol. *36*: 229–240, 1997.

Silman, I., Harel, M., Axelsen, P., Raves, M. and Sussman, J. L.: Three-dimensional structures of acetylcholinesterase and of its complexes with anticholinesterase agents. Biochem. Soc. Trans. *22*: 745–749, 1994.

Silman, I. and Sussman, J. L.: Structural studies on acetylcholinesterase. In: "Cholinesterases and Cholinesterase Inhibitors," E. Giacobini, Ed., pp. 9–25, Martin Dunitz, London, 2000.

Silva, D., Cortez, C. M., Cunha-Bastos, J. and Louro, S. R.: Methyl parathion interaction with human and bovine serum albumin. Toxicol. Lett. *147*: 53–61, 2004.

Singh, S. and Sharma, N.: Neurological syndromes following organophosphate poisoning. Neurology India *48*: 308–313, 2000.

Slotkin, T. A., Tate C. A., Cousins, M. M. and Seidler, F. J.: Functional alterations in CNS catecholamine systems in adolescence and adulthood after neonatal chlorpyrifos exposure. Brain *133*: 163–173, 2002.

Smith, M. I., Elvove, R., and Frazier, W. H.: The pharmacological actions of certain phenol esters, with special reference to etiology of the so-called ginger paralysis. Publ. Health Reports *45*: 2509, 1930.

Smythies, J. and Golomb, B.: Nerve gas antidotes. J. Roy. Soc. Med. *97*: 32, 2004.

Sorenson, R. C., Primo-Parmo, S. L., Kjuo, C. L., Adkins, S., Lockridge, O. and La Du, B. N.: Reconsideration of catalytic center and mechanism of mammalian paraoxonase/arylesterase. Proc. Natl. Acad. Sci. U.S.A. *92*: 7187–7191, 1995.

Soreq, H. and Glick, D.: Novel studies for cholinesterases in stress and inhibitor responses. In: "Cholinesterases and Cholinesterase Inhibitors," E. Giacobini, Ed., pp. 47–61, Martin Dunitz, London, 2000.

Spiegelberg, U.: Psychopathologisch-neurologische Dauerschaden nach gewerblicher Intoxikation durch Phosphorsäre-ester (Alkylphosphate). Proc. Int. Congr. Occup. Health. Excerpta Med. Found. Intern. Congr. Ser. 62, pp. 1778–1780, 1963.

Stedman, E.: Studies on the relationship between chemical constitution and physiological action. I. Position of isomerism in relation to miotic activity of some synthetic urethanes. Biochem. J. *20*: 719–734, 1926.

Stedman, E.: Studies on the relationship between chemical constitution and physiological action. II. The miotic action of urethanes derived from the isomeric hydroxymethylbenzyldimethylamines (HBDM). Biochem. J. *23*: 17–24, 1929.

Stedman, E. and Barger, G.: Physostigmine (eserine). Part III. J. Chem. Soc. *127*: 247–258, 1925.

Steenland, K., Jenknis, B., Ames, R. G., O'Malley, M., Chrislip, D. and Russo, J.: Chronic neurological sequelae to organophosphate pesticide poisoning. Am. J. Public Health *84*: 731–736, 1994.

Stempel, A. and Aeschlimann, J. A.: Esters of basically substituted 3-pyrridols with physostigmine-like activity. J. Am. Chem. Soc. *74*: 3323–3326, 1954.

Stolberg, S. G.: Decade after the Gulf War, GI illness tied to Lou Gehrig's disease. Chicago Tribune, December 11, 2001.

Stovner, J.: The effect of choline on the action of anticholinesterases. Acta Pharmacol. (Kbh.) *12*: 175–186, 1956.

Sugimoto, H., Tsuchiya, Y., Sugumi, H., Higurashi, K., Karibe, N., Imura, Y., Sasaki, A., Kawakami, Y., Nakamura, T., Araki, S., Yamanishi, Y. and Yamatsu, K.: Novel piperidine derivatives. Synthesis and anti-acetylcholinesterase activity of 1-benzyl-4-[2-(benzoylamino) ethyl] piperidine derivatives. J. Med. Chem. *37*: 1880–1887, 1990.

Sugimoto, H., Iimura, Y., Yamanishi, Y. and Yamatsu, K.: Synthesis and structure-activity relationships of acetylcholinesterase inhibitors: 1-benzyl-4-[(5,6-dimethoxy-1-oxoindan-2-yl)methyl] piperidine hydrochloride and related compounds. J. Med. Chem. *38*: 4821–4829, 1995.

Sugimoto, H., Yamanishi, Y., Iimura, Y. and Kawakami, Y.: Donepezil hydrochloride (E2020) and other acetylcholinesterase inhibitors. Curr. Med. Chem. *7*: 303–339, 2000.

Sultatos, L. G.: Role of glutathione in the mammalian detoxification of organophosphorus insecticides. In: "Organophosphates: Chemistry, Fate and Effects," J. E. Chambers and P. E. Levi, Eds., pp. 155–168, Academic Press, San Diego, 1992.

Summers, W. K., Majovski, L. V., Marsh, G. M., Tachiki, K. and King, A.: Tacrine, back to the future? In: "Current Research in Alzheimer Therapy," E. Giacobini and R. Becker, Eds., pp. 259–265, Taylor and Francis, New York, 1988.

Sussman, J. L., Harel, M., Frolow, F., Oefner, C., Goldman, A., Toker, L. and Silman, I.: Atomic structure of acetylcholinesterase from *Torpedo californica*: a prototypic acetylcholine-binding protein. Science *253*: 872–879, 1991.

Tammelin, L. E.: Dialkoxy-phosphorylthiocholines, alkoxy-methyl-phosphorylthiocholines and analogous choline esters. Acta Chem. Scand. *11*: 1340–1349, 1957.

Tang, X. C., Zhu, X. D., and Lu, W. H.: Studies on nootropic effects of huperzine A and B: two selective AchE inhibitors. In: "Current Research in Alzheimer Therapy," E. Giacobini and R. Becker, Eds., pp. 289–293, 1988.

Tariot, P. N., Solomon, P. R., Morris, J. C., Kershaw, P., Lilienfeld, S. and Ding, C: A 5-month, randomized, placebo-controlled trial of galantamine in AD. The Glantamine USA-10 Study Group. Neurology *54*: 2269–2276, 2000.

Tarkowski, M., Luz, W. and Birindelli, S: The lymphocytic cholinergic system and its modulation by organophosphorus insecticides. Int. J. Occup. Med. Environ. Health *17*: 325–337, 2004.

Taylor, P.: Anticholinesterase agents. In: "The Pharmacological Basis of Therapeutics," A. G. Gilman, T. N. Rall, A. S. Niess, and P. Taylor, Eds., pp. 131–149, Eighth Edition, Pergamon Press, New York, 1990.

Taylor, P., Luo, Z. D. and Camp, S.: The genes encoding the cholinesterases: structure, evolutionary relationships and regulation of their expression. In: "Cholinesterases and Cholinesterase Inhibitors," E. Giacobini, Ed., pp. 63–79, 2000.

Ter Braak, J. W. C.: L Een epidemie van polyneuritis van bijzondere oorsprong: A polyneuritis epidemic of peculiar origin. Ned. T. Geneesk. *75*: 2323, 1931.

Thomsen, T., Bickel, U., Fischer, J. P. and Kewith, H.: Galanthamine hydrobromide in a long-term treatment of Alzheimer's disease. Dementia *1*: 46–51, 1990.

Thomsen, T., Kewitz, H., Bickel, U., Straschill, M. and Holl, G.: Preclinical and clinical studies with galanthamine. In: "Cholinergic Basis for Alzheimer Therapy," R. Becker and E. Giacobini, Eds., pp. 329–336, 1991.

Thornton, J. E. and Gershon, S.: The history of THA. In: "Current Research in Alzheimer Therapy," E. Giacobini and R. Becker, Eds., pp. 267–278, Taylor and Francis, New York, 1988.

Trabace, L., Cassano, T., Steardo, L., Pietra, C., Villetti, G., Kendrick, K. M. and Cuomo, V.: Biochemical and neurobehavioral profile of CHF2819, a novel, orally active acetylcholinesterase inhibitor for Alzheimer's disease. J. Pharmacol. Exp. Ther. *294*: 187–194, 2000.

US Army Chemical Defense Bioscience Reviews, US Army Medical Research Institute of Chemical Defense, Edgewood Arsenal, Maryland, 1985, 1990.

Usdin, E.: Reactions of cholinesterases with substrates, inhibitors and reactivators. In: "Anticholinesterase Agents," A. G. Karczmar, Ed., pp. 49–354, Intern. Encyclop. of Pharmacol. and Therapeutics, Section 13, Pergamon Press, Oxford, 1970.

US Environmental Protection Agency. Registration of pesticides in United States: proposed guidelines. Subpart G. Neurotoxicity Fed. Regist. *50*: 39, 439–458, 470, 1985.

VanMeter, W. G., Karczmar, A. G. and Fiscus, R. R.: CNS effects of anticholinesterases in the presence of inhibited cholinesterases. Arch. Intern. Pharmacodyn. Ther. *23*: 249–260, 1978.

Van Wazer, J. R., Ed.: "Phosphorus and Its Compounds," Vols. 1 and 2, Interscience, New York, 1958, 1961.

Vee, M. and Leven, M.: De L'alcaloïde de la fève de Calabar et éxperiences physiologigues avec a même alcaloïde. J. Pharm. et Chimie. *1*: 70–72, 1865.

Vernagy, L.: Teratological examination of the insecticide methylparathion (Wofatox 50 EC) on pheasant embryos.2. Biochemical study. Acta. Vet. Hung. *40*: 203–206, 1992.

Veronesi, B. and Padilla, S.: Rodent models of organophosphorus-induced delayed neuropathy. In: "Organophosphates: Chemistry, Fate and Effects," J. E. Chambers and P. E. Levi, Eds., pp. 353–366, Academic Press, San Diego, 1992.

Verschoyle, R. D. and Aldridge, W. N.: The interaction between phosphorotionate insecticides, pneumotoxic phosphorothiolates and effects on ling 7-ethoxycoumarin. Arch. Toxicol. *60*: 311–318, 1987.

Villalobos, A., Blake, J. F., Biggers, C. K., Butler, T. W., Chapin, D. S., Chen, Y. L., Ives, J. L., Jones, S. S. and Liston, D. R.: Novel benzi soxazole derivative as poterat and selective inhibitors of acetylcholinesterase J. Med. Chem. *37*: 2731–2734, 1994.

Vinckier, A. and Semenza, G.: Measuring elasticity of biological materials by atomic force microscopy. FEBS Lett. *430*: 12–16, 1998.

Viragh, C., Akhmetshin, R., Kovach, I. M. and Broomfield, C.: Unique push-pull mechanism of dealkylation in soman-inhibited cholinesterases. Biochem. *36*: 8243–8242, 1997.

Viragh, C., Harris, T. K., Reddy, P. M., Massiah, M. A., Mildvan, A. S. and Kovach, I. M.: NMR evidence for a short, strong hydrogen bond at the active site of a cholinesterase. Biochemistry *39*: 16200–16205, 2000.

Wallace, K. B.: Species-selective toxicity of organophosphorus insecticides: a pharmacodynamic phenomenon. In: "Organophosphates: Chemistry, Fate and Effects," J. E. Chambers and P. E. Levi, Eds., pp. 79–105, Academic Press, San Diego, 1992.

Wang, P., Verin, A. D., Birukova, A., Gilbert-McClain, L. I., Jacobs, K. A. and Garcia, J. G.: Mechanisms of sodium fluoride–induced endothelial cell barrier dysfunction: role of MLC phosphorylation. Am. J. Physiol. Lung Cell Mol. Physiol. *281*: L1472–L1483, 2001.

Ward, J. R., Gosselin, R., Comstock, J., Stagg, J. and Blanton, B. R.: Case report of a severe human poisoning by GB. MLRR 151, Edgewood, Maryland, 1952.

Waser, P. G., Caratsch, C. G., Chang, S. R., Hopff, W., Gotheil, A., Kaiser, E., Sammet, R. and Streichenberg, C.: Organophosphate poisoning and its treatment with oximes. In: "Neurobiology of Acetylcholine. A Symposium in Honor of A. G. Karczmar," N. J. Daw and R. Perlman, Eds., pp. 459–471, Plenum Press, New York, 1987.

Waser, P. G., Berger, S., Haas, H. L. and Hofman, A.: 4-Aminopyridine (4-AP)-derivatives as central cholinergic agents. In: "Current Research in Alzheimer Therapy," E. Giacobini and R. Becker, Eds., pp. 337–342, Taylor and Francis, New York, 1988.

Wassen, S. H. and Holmstedt, B.: The use of Parica, an ethnological and pharmacological review. Ethnos *1963*: 5–45, 1963.

Wecker, L., Mobley, P. L. and Dettbarn, W. D.: Central cholinergic mechanisms underlying adaptation to reduced cholinesterase activity. Biochem. Pharmacol. *26*: 633–637, 1977.

Wescoe, W. C., Riker, W. F. and Beach, V. L.: Studies on the inter-relationship of certain cholinergic compounds. III. The reaction between 3-acetoxy phenyltrimethylammonium methylsulfate, 3-hydroxy phenyltrimethylammonium bromide and cholinesterase. J. Pharmacol. *99*: 265–276, 1950.

Wheelwright, J.: "The Irritable Heart," W. W. Norton, New York, 2002.

Whorton, M. C. and Obrinsky, D. L.: Persistence of symptoms after mild to moderate acute organophosphate poisoning among 19 farm field workers. J. Toxicol. Environ. Health *11*: 347–354, 1983.

Wilcox, D. E.: Binuclear metallohydrolases. Chem. Rev. *96*: 2435–2458, 1996.

Wills, J. H.: Pharmacological antagonists of the anticholinesterase agents. In: "Cholinesterases and Anticholinesterase Agents," G. B. Koelle, Ed., Handbch. d. exper. Pharmakol., vol. 15, pp. 883–920, Springer-Verlag, Berlin, 1963.

Wills, J. H.: Toxicity of anticholinesterases and treatment of poisoning. In: "Anticholinesterase Agents," A. G. Karczmar, Ed., pp 353–469, Intern. Encyclop. of Pharmacol. and Therapeutics, Section 13, Pergamon Press, Oxford, 1970.

Wilson, I. B.: Acetylcholinesterase. XI. Reversibility of tetraethyl pyrophosphate inhibition. J. Biol. Chem. *190*: 4286–4291, 1951.

Wilson, I. B.: The interaction of tensilon and neostigmine with acetylcholinesterase. Arch. Int. Pharmacodyn. Ther. *104*:204–213, 1955.

Wilson, I. B.: Molecular complementarity and antidotes for alkylphosphate poisoning. Fed. Proc. *18*: 752–758, 1959.

Wilson, I. B. and Froede, H. C.: The design of reactivators for irreversibly blocked acetylcholinesterase. In: "Drug Design," E. J. Ariens, Ed., pp. 213–221, Academic Press, New York, 1971.

Wilson, I. B. and Ginsburg, S.: A powerful reactivator of alkylphosphate-inhibited acetylcholinesterase. Biochem. Biophys. Acta. *18*: 168–170, 1955.

Wilson, I. B. and Quan, C.: Acetylcholinesterase studies of molecular complimentariness. Arch. Biochem. *73*: 131–143, 1958.

Wilson, I. B., Bergmann, F. and Nachmansohn, D.: Acetylcholinesterase. X. Mechanism of catalysis of acylation reactions. J. Biol. Chem. *186*: 781–790, 1950.

Wilson, I. B., Hatch, M. A. and Ginsburg, S.: Carbamylation of acetylcholinesterase. J. Biol. Chem.: *235*: 2312–2315, 1960.

Wilson, I. B., Ginsburg, S. and Meislish, E. K.: The reactivation of acetylcholinesterase inhibited by tetraethyl pyrophosphate and di-isopropyl fluorophosphate. J. Am. Chem. Soc. *77*: 4286–4291, 1955.

Wilson, I. B., Harrison, M. A. and Ginsburg, S.: Carbamyl derivatives of acetylcholinesterase. J. Biol. Chem. *236*: 1498–1300, 1961.

Winteringham, F. P. W.: Blood cholinesterase inhibition as an index of exposure to insecticidal carbamates. Bull. Wld. Hlth. Org. *35*: 452–453, 1966.

Wirth, W.: Zur Pharmakologie der Phosphorsäure-ester. Diathyl-p-nitrophenylphosphat ("Mintakol"). Naunyn Schmiedeberg's Arch. Exp. Pathol. Pharmakol. *207*: 547–568, 1949.

Wolfgang, W.: Der Pharmakologe und Toxikologe Wolfgang Wirth (1898–1996) und die Giftgasforschung im Nationalsozialismus. In: *Nach der Diktatur. Die Medizinische Akademie Düsseldorf vom Ende des Zweiten Weltkriegs bis in die 1960er Jahre*, W. Woelk, F. Sparing, K. Bayer and M. G. Esch, Eds., pp. 269–287. Essen: Klartext Verlag, 2003.

Wu, S. Y. and Casida, J. E.: Subacute toxicity induced in mice by potent neuropathy target esterase inhibitors. Toxicol. Appl. Pharmacol. *139*: 195–202, 1996.

Wuest, H. M. and Sakal, E. H.: Some derivatives of 3-pyridol with para-sympathomimetic properties. J. Am. Chem. Soc. *73*: 1210–1216, 1950.

Xia, Y. and Kozikowski, A. P.: A practical synthesis of the Chinese "nootropic" agent Huperzine A: a possible lead in the treatment of Alzheimer's disease. J. Am. Chem. Soc. *111*: 4116–4117, 1989.

Xie, K., Gupta, R. P. and Abou-Donia, M. B.: Alteration in cytoskeletal protein levels in sciatic nerve on post-treatment of diisopropyl phosphofluoridate (DFP)-treated hen with phenylmethylsulfonyl fluoride. J. Neurochem. Res. *26*: 235–243, 2001.

Yamada, A., Shoji, T., Tahara, H., Emoto, M. and Nishizawa, Y.: Effect of insulin resistance on serum paraoxonase activity in a nondiabetic population. Metabolism *50*: 805–811, 2001.

Yamamoto, Y., Ishihara, Y. and Kuntz I. D.: Docking analysis of a series of benzylamino acetylcholinesterase inhibitors with a phthalimide, benzoyl, or indanone moiety. J. Med. Chem. *37*: 3141–3153, 1994.

Yang, Z. P. and Dettbarn, W. D.: Lipid peroxidation and changes in cytochrome C oxidase and Xanthine oxidase activity in organophosphorus anticholinesterase induced myopathy. In: "From *Torpedo* Electric Organ to Human Brain: Fundamental and Applied Aspects," J. Massoulié, Ed., pp. 157–161, Elsevier, Amsterdam, 1998.

Yingge, Z., Chunli, B., Chen, W. and Delu, Z.: Force spectroscopy between acetylcholine and single acetylcholinesterase molecules and the effects of inhibitors and reactivators studied by atomic force microscopy. J. Pharmacol. Exp. Ther. *297*: 798–803, 2001.

Yokoyama, K., Araki, S., Murata, K., Nishikitani, M., Okumura, T. and Ishimatsu, S.: Chronic neurobehavioral and central and autonomic system effects of Tokyo subway sarin poisoning. In: "From *Torpedo* Electric Organ to Human Brain: Fundamental and Applied Aspects," J.

Massoulié, Ed., pp. 317–324, Elsevier, Amsterdam, 1998.

Zeller, E. A. and Bissegger, A.: Üeber die Cholinesterase des Gehirn und der Erythrocyten. 3. Mitteilung ueber die Beeinflussung von Fermentreaktionen durch Chemotherapeutica und Farmaka. Helv. Chimische Acta *26*: 1619–1630, 1943.

Zupancic, A. O.: Evidence for the identity of anionic centers of cholinesterases with cholinoreceptors. Ann. N.Y. Acad. Sci. *144*: 689–693, 1967.

8

Cholinergic Aspects of Growth and Development

A. Historical Introduction

This chapter concerns three related cholinergic subjects (1) development, or ontogenesis, and the related subject of phylogenesis; (2) trophic activities, and growth and nerve growth factors; and (3) teratology. In addition, ontogenetic aspects are akin to phylogenetic and comparative considerations, and will also be touched on.

Development involves ontogenesis of the cholinergic components (i.e., choline acetyltransferase [CAT], various cholinesterases [ChEs], acetylcholine [ACh], and muscarinic and nicotinic receptors) and the relationship between their development and transmission and function. In fact, this relationship was the subject of the earliest investigations pertinent to this chapter. While the cholinergic system's significance for peripheral autonomic nervous function was proven by Loewi and associates in the early 1920s (Loewi, 1921; see also Loewi, 1960), its central role was not clear at the time and, during the 1930s, Nachmansohn, Youngstrom, and Bacq (see Nachmansohn, 1938, 1939; Youngstrom, 1938a, 1941; Bacq, 1975; see also Bacq, 1975) hypothesized that establishing a link between the ontogenetic and phylogenetic appearance of AChE and the onset of motility will support the notion of the central transmittive function of AChE and of the cholinergic system (see also Chapter 1 B-1 and section BII-6, below). An interesting feature of these early studies is that some of Youngstrom's studies involved a 56-day human embryo. (For a photograph of Nachmansohn, see Figure 7-1.)

Accordingly, Zenon Bacq (1935) and Bullock and Nachmansohn (1942) related the phylogenesis of motility to the presence of a ChE: no enzyme was found in a protozoan, *Paramecium multimicronucleolatum*, or in the sponge *Scypha*, while significant amounts were found in the *Hydrozoa* "coinciding with the first appearance of a differentiated nervous system" (Bullock and Nachmansohn, 1942). The problem with this notion is that, as is well known today, ChEs and other components of the cholinergic system appear in nonnervous tissues as well as in monocellular organisms devoid of neurons and incapable of motility (see also below, section BIV); also, they relate to conductile systems such as cilia and, accordingly, are present in *Teterahymena gelii* and *Trypanosoma rhodesiense* (see Karczmar, 1963a, 1963b). David Nachmansohn (1938, 1939) related the motility of chick embryo to the ontogenetic arrival of ChE and to the rate of its developmental increase. With more precision, Charles Sawyer (1943, 1944) related the appearance of specific motor activities (i.e., "the early coil," "flutter," and various types of swimming) and autonomic function (i.e., cardiac beat of the urodeles and fish) to the appearance and rapid increase of ChE levels (see Figures 8-1 and 8-2). During Sawyer's studies, considerable motor and autonomic activity was recorded prior to the appearance of significant enzymic activity (although some activity was noticed in the premotile stages of the urodele *Amblystoma*). Most likely, the activity resulted from the lack of sensitivity in the methods available in the 1930s and 1940s for evaluating ChEs. In fact, Sawyer and Nachmansohn did not differentiate between AChE and BuChE in their studies. However, on the whole, Nachmansohn, Youngstrom, Bacq, and others (for additional references, see Karczmar, 1963a, 1963b) did demonstrate a link between ChEs and function, and this finding elegantly complements the proof of the synaptic, transmittive role of the central cholinergic system pro-

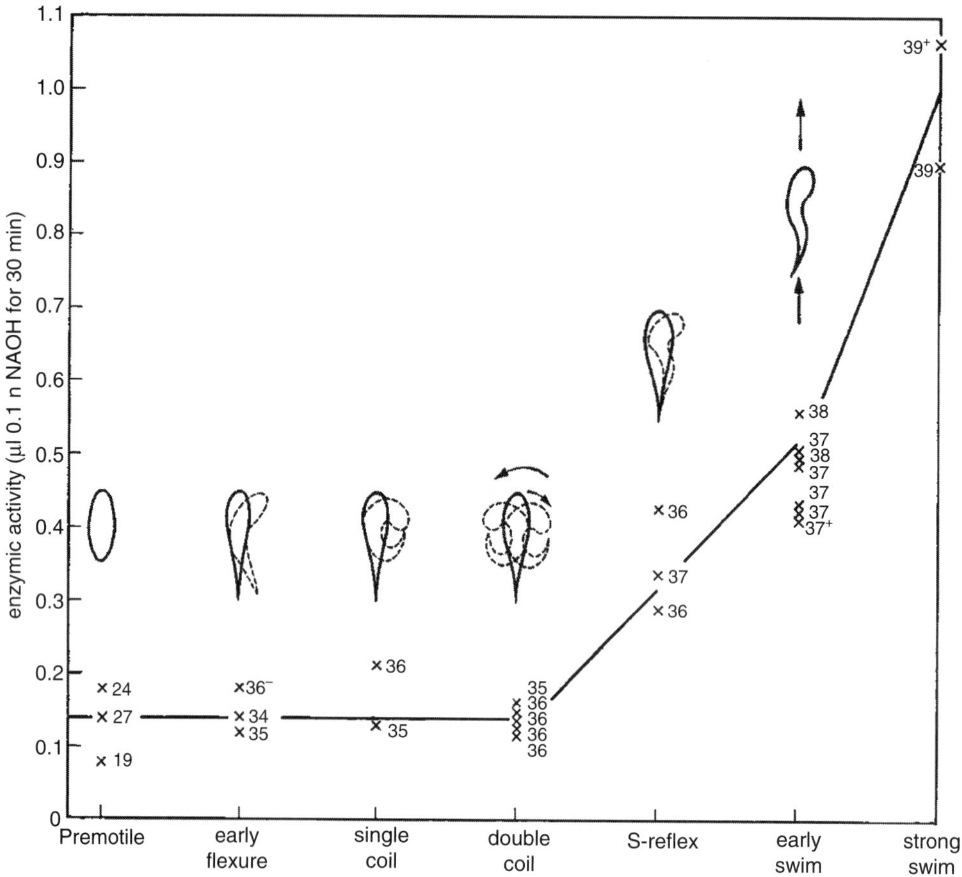

Figure 8-1. Relationship between function and cholinesterase activity during development of *Ambylstoma punctatum* (from Sawyer 1943). Numerals opposite points indicate morphological stages. At room temperature, approximately 10 days of development separated stages 19 and 39+. (From Karczmar, 1963a, with permission.)

vided by Sir John Eccles some 20 years later (Chapter 9 A).

Further investigations of cholinergic development included studies of the ontogenesis for all cholinergic components (i.e., cholinergic receptors, uptake systems, CAT, etc.) and of the relationship between their developmental manifestation and the appearance of specific synaptic events (i.e., synaptic potentials and synaptic transmission) and specific functions. These later studies, conducted by Victor Hamburger, Metzler and Humm, Bonichon, and others more clearly depicted the ontogenetic appearance of AChE and ACh or CAT, the rapid rate of their development, and the onset and subsequent fast rate of neurogenesis and certain types of function (see, for example, Bonichon, 1957; Hamburger, 1948; see also Karczmar, 1963a). Particularly detailed were the studies of Shen et al. (1956) of the appearance of AChE during the development of the chick embryo retina. This study led to the early demonstration of the existence of cholinergic synapses between ganglion and amacrine cells, and, combined with the earlier work of Lindeman (1947) related the presence of the papillary constrictor reflex with the cholinergic maturation of the retina.

An important finding of these early studies was the presence in many species, both vertebrate and invertebrate, of AChE, BuChE, and, in some

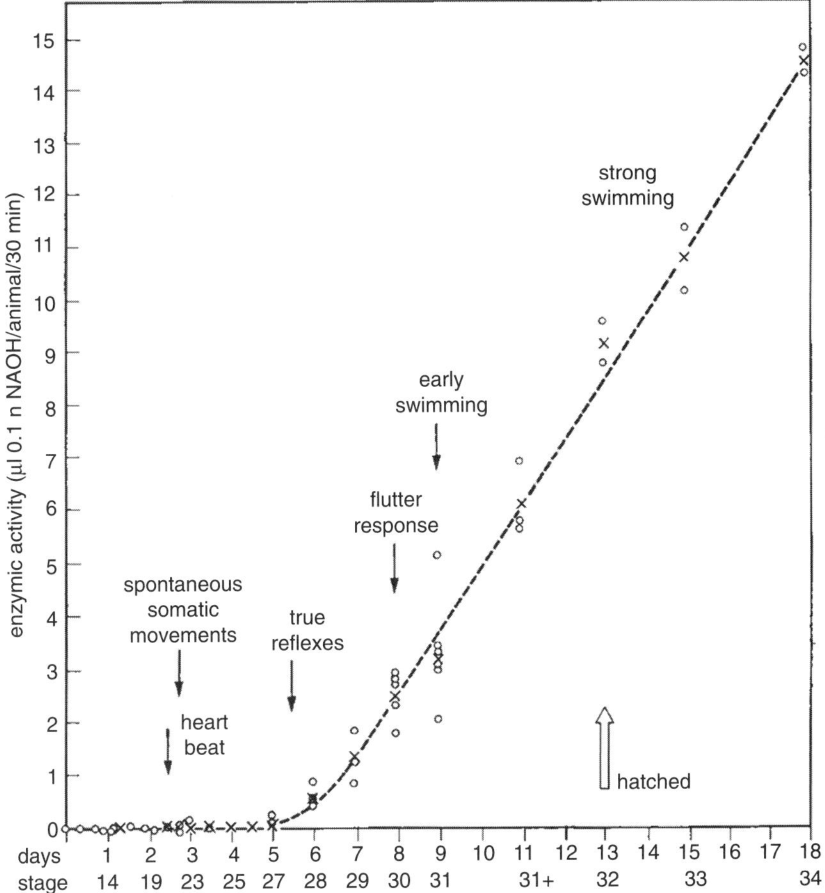

Figure 8-2. Relationship between function and ChE activity during development of *Fundulus hetercolitus* (from Sawyer, 1944). The animals are classified according to stages of Oppenheimer, and the approximate time of onset of various movements is listed. The mean ChE level for each stage is represented by an "x." Spontaneous movements occur while there is as yet very little enzyme present. (From Karczmar, 1963a, with permission.)

studies, of ACh prior to neurogenesis. In the chick, ACh and AChE appear very early in neurogenesis, before the onset of transmission, function, or synaptogenesis. Actually, in many invertebrate or vertebrate forms, ChEs are present in the gastrula or blastula. In fact, these components of the cholinergic system were found in unfertilized egg of *Drosophila melanogaster,* sea urchins, and amphibia (see, for example, Augustinsson and Gustafson, 1949; Numanoi, 1957; Poulson and Boell, 1946; for further references, see Karczmar, 1963a; Karczmar et al., 1973; Buznikov, 1984; Buznikov et al., 1996; see Chapter 9 BI).

There are several bases for the relationship between cholinergic ontogenesis and the second topic of this chapter, cholinergic trophic activities and the growth factors and nerve growth factor (NGF; Levi-Montalcini and Amprino, 1947). The first hint of a relationship was offered by the early evidence for the precocious developmental appearance of the cholinergic components. These findings led Karczmar (1963a; see also Karczmar, 1946) to speculate, "The cholinergic system or its components not only condition transmission and function, but also induce anatomic substrates of these processes and act as inductors or evocators."

This speculation relates to Eccles' concept (cf. McGeer et al., 1978) that, besides its "synaptotropic" role, the cholinergic system also carries out a "metabotropic" role. In fact, the nervous system generally and the cholinergic system specifically may exert a trophic or morphogenetic influence not only during ontogenesis but also under entirely different circumstances. For example, these nerve-induced trophisms occur during tissue or limb regeneration (Karczmar, 1946). Also, trophisms emanate from the motor nerves at the neuromyal junction, where they are involved in receptor regulation and denervation sensitization (Guth, 1968). Gennady Buznikov (1984) referred to this precocious appearance of ChEs or ACh as "prenervous" and linked the appearance with a regulatory and trophic role of the cholinergic system. Buznikov and his associates (2003, 2005) (Lauder, 1993, Lauder and Schambra, 1999) pointed out that in addition to ACh, other transmitters (particularly serotonin) also play a role. They referred to these transmitters as "morphogens."[1]

Ontogenesis of the cholinergic system relates to trophisms via the trophic function of the cholinergic components during their precocious, prenervous appearance and via the converse phenomena of the biochemical and morphological dependence of cholinergic ontogenesis on trophic factors. The earliest known is the NGF. This factor was discovered by Rita Levi-Montalcini (Levi-Montalcini and Amprino, 1947; Levi-Montalcini, 1965; and Bueker (1948), Victor Hamburger playing a significant role in this discovery (Levi-Montalcini, 1982). In this context, Victor Hamburger made an interesting point as he analogized the role of trophics with the role of chemical embryonic induction of structures such as neural tube; the discovery of this induction and of its chemical nature is a classical event in the field of embryogenesis (Spemann, 1938; Hamburger, 1997). Today, we distinguish a number of trophic factors, originating from sources including venoms and the CNS (Gage and Bjorklund, 1987). Accordingly, the mutual phenomena of trophisms affecting and induced by the cholinergic system are also discussed in this chapter.

The third topic of this chapter, cholinergic teratology, links with both cholinergic ontogenesis and its trophic aspects. The early fetal and embryonic onset of cholinergic ontogenesis implies that cholinergic and anticholinergic drugs should affect development. Furthermore, as the cholinergic system exerts trophic and morphogenetic influences, the interference with this system should affect development as well. In fact, teratology caused by cholinergic and anticholinergic drugs was described long before cholinergic transmission and cholinergic ontogenesis were established (Mathews, 1902; Sollmann, 1902; for further references, see Karczmar, 1963b and Buznikov, 1984).

BI. Cholinergic Ontogeny: Precocious Developmental Appearance of the Components of the Cholinergic System

1. Gametal and Preneuronal Appearance of the Components of the Cholinergic System

The first studies of the ontogenetic appearance of the components of the cholinergic system were carried out in the 1930s and 1940s; they involved the measurement of ChEs as the pertinent methodology became available at that time, antedating the techniques for the quantitative measurement of CAT and ACh (see Karczmar, 1963a). These early findings involve both vertebrates and invertebrates, and they deal with neurogenetic stages of development. However, some of this work concerned earlier ontogenetic stages and led to the important discovery of the precocious, or "prenervous," appearance of ChEs (Karczmar, 1963a; Buznikov, 1988; Buznikov et al., 1972).

These early concepts of the precocious appearance of cholinergic components were expanded to even include gametes. ChEs were found in sperm and unfertilized eggs (oocytes) of both the invertebrates and the few vertebrates that were studied (Poulson and Boell, 1946; Tibbs, 1960; for further references, see Karczmar, 1963a; Koelle, 1963; Fischer, 1971; Kusano et al., 1982; Buznikov, 1990; Buznikov et al., 1996). In addition, Hermona Soreq and associates (1987) established the presence of AChE in human oocytes using state-of-the-art molecular biology techniques. It is interesting that this gametal appearance of ChEs and other cholinergic components (see below, this section) may be reduced or perhaps even eliminated after fertilization (Buznikov et al., 1996; Eusebi et al., 1979; Kusano et al., 1982; see also below, this section).

The postfertilization, preneural presence of ChEs was demonstrated during the 1930s and 1940s for many invertebrate species, such as a number of insects (Poulson and Boell, 1946) and the sea urchin; in fact, AChE appears in the urchin embryo soon after fertilization (Augustinsson and Gustafson, 1949; see Figure 8-3). In addition, AChE activity was found to be high in the primordial echinoderm gut (Martynova, 1981; Markova et al., 1985).

ChEs are also present precociously during protovertebrate, Ciona ontogenesis (Durante, 1956), and they exist in the preneural chick embryo, as was shown in 1934 by Ammon and Schutte, who developed the manometric method for the measurement of these enzymes. AChE was specifically measured in most of the studies carried out during the late 1940s and 1950s (less discriminating methods were used previously). The enzyme was noticed at the gastrula stages of chick (Laasberg, 1990) and even earlier, during the 2- or 4-cell stage in maphibia (Youngstrom, 1938; see also Karczmar, 1963a.)

Gennady Buznikov and others confirmed this early work (see, for example, Buznikov et al., 1972, 2001) and demonstrated the precocious appearance of ChEs in additional forms, such as crustacea (Raineri and Falugi, 1983), the leach (Fitzpatrick-McElligott and Stent, 1981), the fish (in several instances, AChE was measured; Fluck,

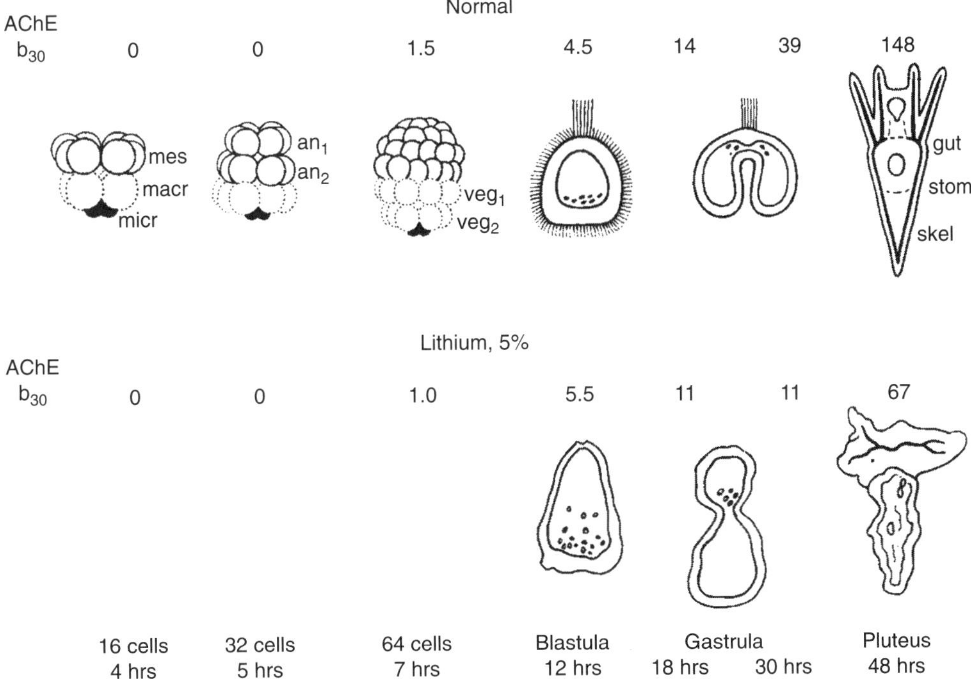

Figure 8-3. Normal and lithium-inhibited echinoderm development. AChE activities expressed in b_{30} values (Augustinsson and Gustafson, 1949). Development stages and times of their occurrence at the bottom of the figure. Sixteen-cell stage: mes, mesomoers; macr, macromeres; micr, micromeres. Thirty-two-cell stage: an_1, an_2, $animal_1$ and $animal_2$ mesomers. Sixty-four-cell stage: veg_1, veg_2, $vegetal_1$ and $vegetal_2$ macromers. Pluteus exhibits the gut, stamodaeum (stom), and skeleton (skel). (From Karczmar, 1963a, with permission.)

1978, 1982), and the rat (Metzler and Humm, 1951; for additional references, see Buznikov, 1996; Karczmar, 1963a; and Drews, 1975.) Actually, Sawyer (1944) found "no demonstrable cholinesterase" in the premotile fish *Fundulus heteroclitus* (i.e., during preneurogenesis), contrary to Fluck's (1978, 1982) subsequent findings; the reason for this discrepancy is simple: the earlier investigator used a relatively insensitive method for his measurement of ChE (he also did not discriminate between AChE and BuChE). Altogether, whenever ChEs were not found in the early, preneurogenetic embryo of any species and/or the gametes, it may be surmised that these negative findings resulted from technical failings, and that the general precocious presence of ChEs including AChE in the blastomere stage of ontogenesis or even the gametes constitutes the norm.

Studies of the early appearance of ACh, CAT, and the cholinergic receptors were initiated in the 1950s to 1980s; these studies showed that, similar to ChEs, other components of the cholinergic system appear in many forms long before the neurula stage and neurogenesis. Thus, precocious presence of CAT, ACh and/or cholinergic receptors was reported for echinoderms, ascidia, teleosts, and amphibia; in fact, in amphibia ACh existed in preblastomeric stages and in unfertilized egg (see, for example Numanoi, 1953, Reshetnikova, 1970, Fluck, 1978; Duprat et al., 1985; Ivonnet and Chambers, 1997; for further references, see Buznikov, 1973). However, the presence of cholinergic receptors could also be demonstrated directly. Binding studies (including investigations of the binding of choline; Duprat et al., 1985) demonstrated the preneural presence of catecholaminergic, cholinergic, and serotonergic receptors in many vertebrate and invertebrate species (see Buznikov, 1973). Furthermore, muscarinic receptors were found in the mesenchyme, cartilage, and myotomes of the chick limb bud and in the primordia of the mouse palate (Reich and Drews, 1983; see also Lauder and Schambra, 1999; Buznikov, 1973). ACh synthesis and choline binding were present in 5-day-old cultures of amphibian neural primordium 5 days before the onset of synaptogenesis (Duprat et al., 1985). Also, the presence of receptor-channel macromolecules in fish and mammalian oocytes was confirmed in studies using microelectrode and iontophoretic methods (Eusebi et al., 1979; Kusano et al., 1982; Dascal and Landau, 1980; Hagiwara and Jaffe, 1979); for example, Kusano et al. (1982) demonstrated the presence of ACh-induced current in the membranes of unfertilized fish oocytes. The current exhibited typical chloride-driven muscarinic response characteristics for receptor-channel macromolecules and was particularly intense in the animal hemisphere; that membrane-located receptors were involved was indicated by the finding that there were no intra-oocytic responses. Actually, the prenervous presence of nonneuronal cholinergic receptors may be surmised from the presence of ACh during the early stages of ontogenesis; also, ACh, cholinomimetics, and cholinergic antagonists affect early embryogenesis, including the cell cleavage stage of several species, and these findings are also consistent with the very early presence of cholinergic receptors (Karczmar, 1963b; Buznikov, 1990).

As already noted, cholinergic components including ChEs and the oocytic receptors or receptor-channel macromolecules may disappear following fertilization (Kusano et al.,1982; Eusebi et al., 1979; Buznikov et al., 1996). Does the presence of cholinergic components in the gametes and the fertilized oocyte denote their special role at the time? Do they disappear after they have fulfilled this role? Do cholinergic components present during ontogenesis appear de novo? Actually, there is a paucity of studies concerning the presence of cholinergic components in the crucial period between the appearance of gametes and early ontogenesis; further work may establish that actually there is a continuity, beginning with gametes and ending with ontogenetic maturation, with respect to cholinergic components. This matter will be discussed further in section BII-1 and BII-2 of this chapter.

It must be noted that this precocious, prenervous appearance of the cholinergic components not only is characteristic for the central nervous system (CNS), but also occurs during ontogenesis of the autonomic ganglia and the parasympathetic system, the neuromyal junction, and the heart (see section BIII). For example, ACh and/or ChEs appear precociously in the hearts of the rat and the chick (Sippel, 1955; Zacks, 1954, McCarty et al., 1960; see also Karczmar, 1963a, 1963b).

An interesting facet of these phenomena is that the cholinergic components present in the blastula move during subsequent ontogenesis into the animal poles of the gastrula, i.e., into the future localization of the vertebrate neurula. However, they seem occasionally to be "left behind" in nonnervous primordia; the same seems true with respect to the developmental route of cholinergic components in invertebrates (see Karczmar, 1963a; Buznikov, 1973). In fact, in some fish and additional species, the cholinergic components, particularly the cholinergic receptors, may be reduced—or even eliminated—after fertilization (Buznikov et al., 1996; Eusebi et al., 1979; Kusano et al., 1982).

2. Cholinergic Components as Morphogens, or What Else?

The early synthesis and presence of cholinergic components in the gametes as well as later, during preneuronal development, signifies clearly that these components must play a nonneuronal, nontransmittive role. What is this role? What may be the reasons for this precocious ontogenetic appearance of ChEs, ACh, CAT, and/or cholinergic receptors?

The presence of the cholinergic components in the egg, in the sperm, or in the unfertilized egg is difficult to explain, although in the case of the sperm their role may be similar to their function in the flagellar systems of certain protozoa. The cholinergic components displayed during early, preneural ontogenesis may be "descendants" of components present in the gametes and in the fertilized egg.

This explanation does not signify that the cholinergic components are "passive" during the preneural stages of the developing embryo. In fact, these components augment in the course of the preneural stages, which indicates that they play a role during this period, although the rate of this increase does not compare with that exhibited in the course of neurogenesis (cf. Karczmar, 1963a; Buznikov, 1990). This role may relate, via appropriate ontogenetic and molecular mechanisms, to their future synaptic role. In agreement with this notion, Kusano et al. (1982) suggested that the cholinergic preneural receptors "are merely a byproduct of a largely depressed genome" as the original oocytic genome has a complexity greater than the genome of its cellular progeny (in sea urchins; Galau et al., 1976). This surplus may result in a few transcripts sufficient to form the neural receptor macromolecules during subsequent development. In fact, following an injection of mRNA, the oocyte is competent to develop cholinergic responses (cf. Runft et al., 2002).

It is not inconsistent with this explanation that ChEs and the cholinergic receptors are localized in the animal pole of the unfertilized or fertilized egg and the blastomeres; this pole will subsequently form the ectoderm and mesoderm, or the neurons and the muscle; thus, the precocious presence of cholinergic components relates to their future synaptic role (Kusano et al., 1982; Fitzpatrick-McElligot and Stent, 1981; Fluck, 1978; Whittaker, 1973; for further references, see Buznikov, 1990). Particularly telling in this regard is an experiment carried out by Augustinsson and Gustafson (1949) in which the normal development of the echinoderm sea urchin was "vegetalized" by lithium. In other words, the development of the entoderm was facilitated and augmented at the expense of the ectoderm. Simultaneously, the development of AChE was more than halved, and, ultimately, neurogenesis did not occur (see Figure 8-3; Karczmar, 1963a).

The preneurogenetic components of the cholinergic system would be even less "passive" if they were involved in trophic and morphogenetic phenomena. Cholinergic agonists engender growth and proliferation of nonneuronal cells *in vitro*, and they and noncholinergic transmitters (such as serotonin) or noncholinergic agonists trigger cleavage divisions of the fertilized echinoderm, rodent, avian, and amphibian eggs; cholinergic (and serotonergic) antagonists oppose these actions. Then the cholinergic elements regulate cell movements during gastrulation and, later, they regulate organogenesis; for example, muscarinic receptors are morphogenetic and organogenetic in the case of the mouse palate and in other instances (Drews, 1975; see also Lammerding-Koppel et al., 1995; see also Nietgen et al., 1999; Wee et al., 1980). Finally, cholinergic elements exert proliferative effects on neural precursor cells (Cameron et al., 1998). However, these agents inhibit neuritic outgrowth as well as stem cell proliferation, and these negative trophic effects are regulatory in nature. Whether the growth regulation is negative or positive depends on the method and dosage of the administration of the agonists, the developmental state of the neuron, the messenger systems that are activated, and presence of glial cells (Lauder, 1993). In addition, the teratologic effects on the early embryo of cholinergic antagonists are consistent with the notion of the trophic role of the embryonic cholinergic system (see section D). These morphogenetic actions of ACh and cholinergic agonists are also illustrated by their proliferative effects on cultures of nonneuronal cells, including astrocytes and fibroblasts (Weiss et al., 1998; see also Lauder and Schambra, 1999). These activities are mediated by phospholipase C (PLC) as it activates mitogen-activated protein kinase. However, muscarinics may also promote inhibition of proliferation of certain cell types via activation of the G protein and resulting stimulation of adenylyl cyclase (Lauder and Schambra, 1999).

Accordingly, Lauder and Buznikov referred to the cholinergic components as "morphogens" (Lauder and Schambra, 1999; Lauder, 1988, 1993; Sadykova et al., 1992; see also section A). Earlier, Karczmar (1963a, 1963b) proposed, on a similar basis, the term "inductors" for ACh and cholinergic agonists as they exert morphogenetic effect. This term was used by Hans Spemann (1938) to denote morphologic systems capable of induction of morphogenetic phases or steps (see also Hamburger, 1997).

Another, older term is also relevant; this term refers to the metabotropic role of the cholinergic system. It was first used by Edith Bulbring and Joshua Burn (1949) with respect to the inotropic action of the cholinergic vagal system. Sir John Eccles (1964) expressed a similar opinion as he distinguished between transmittive or synaptotropic roles of ACh on the one hand and its metabotropic roles on the other. Finally, Joseph Blass presented a modern version of these opinions as he associated the metabotropic function of the cholinergic system with mitochondrial energy metabolism (see Blass, 2001). Jaffe and Runft (Runft et al., 2002; Jaffe, 1990) stressed the importance of activation by the cholinergic system, particularly via muscarinic receptors of the second messengers and G proteins; they related this activation to the morphogenetic cholinergic action. This activation may also involve metabotropic effects needed for the formation or maintenance of cellular filaments and cytoskeleton (see also Woolf, 1997, and Chapter 9 BVI).

The gametal presence of the components of the cholinergic system may also have a reason not related to neurogenesis and to the transmittive role of these components. They are synthesized either in the gametal somata or extracellularly, for example, in the yolk, and transported into the oocyte (see Buznikov et al., 1996). The oocytic cholinergic system may underlie the fertilization potential, which is present in several vertebrate and invertebrate forms and which serves to block polyspermy (which precludes development), as the block of the potential causes polyspermy (Hagiwara and Jaffe, 1979). The relevant point is that the ionic characteristics of the fertilization potential resemble those of the depolarizing response of the oocytic cholinergic receptor (there is some uncertainty as to the nicotinic or muscarinic nature of this response; Kusano et al., 1982; Hagiwara and Jaffe, 1979). Another early oocytic phenomenon is the activation response occurring during oocyte fertilization. Jaffe and Runft opine that a phosphoinositol (IP) cascade initiated at the IP receptors initiates free Ca^{2+} release; the muscarinic system that activates the IP cascade is likely to be involved in the fertilization activation. This release takes the form of a wave that starts at the animal pole (point of entry of the sperm) "and traverses the egg as a shallow . . . band which vanishes at the antipode some minutes later" (Gilkey et al., 1978; Jaffe, 1990); it appears that this phenomenon is common for a number of vertebrates and invertebrates (Jaffe, 1995). A related suggestion was that nicotinic gametal receptors may be needed for the intragametal fertilization process (Falugi, 1993; Jaffe, 1990).

One final comment: as appropriate for this book, the cholinergic aspects of ontogenesis are stressed. However, this emphasis should not be construed to signify that only cholinergic phenomena occur during development. In fact, dopamine and norepinephrine, serotonin, GABA and peptides, and their catabolic and anabolic enzymes exist in the unfertilized oocyte and during preneurogenetic ontogenesis (Kusano et al., 1982; Freychet, 1976; see also Buznikov, 1973; Buznikov et al., 1996). These substances elicit the responses of oocytic receptor-channel macromolecules. However, not all of the transmitters must be present during development in any given species. For example, only one catecholamine, dopamine was found during early echinodermal development (Welsh, 1968; see also Buznikov et al., 1996). It seems likely, however, that such negative findings are due to the use of insufficiently sensitive techniques.

Altogether, the oocyte and the early embryo, prior to its entry into the organogenesis period, exhibit multitransmitter phenomena. Morphogenetic differentiation of cells, that is, organogenesis, leads to a limitation of the transmitter number in a specific tissue, organ, and site. Thus, the autonomic ganglia, the neuromyal junction, or the specific CNS pathway may exhibit, preferentially, the cholinergic cells. This transmitter differentiation begins early; for example, the ratio of ACh, dopamine, and serotonin changes in the course of the mitotic phases of the second cleavage of the sea urchin egg, and similar changes were noted during early ontogenesis; the mechanism involved is unknown (Buznikov et al., 1972). This differentiation is not absolute; today's findings indicate that a neuron may release more than one neurotransmitter or modulator (see Chapter 9 BIII-2).

This section concerns the morphogen role of the cholinergic components during early, preneural ontogenesis. However, these components continue exerting this function during neurogenesis, whether early or late. This matter is discussed in section CIII, below.

BII. Development of the Cholinergic System During Neurogenesis and Organogenesis

1. Early Neurogenetic Appearance and Subsequent Rate of Development

As already pointed out (see section BI) it is not clear how the precocious, prenervous components of the cholinergic system relate to those present in the neurula of the vertebrates and early nervous systems of the invertebrate sites. There is no doubt, however, that they are present in very early neural primordia. Thus, ACh synthesis and choline binding were present in 5-days-old culture of amphibian neural primordium, that is, 5 days before the onset of actual synaptogenesis (Duprat et al., 1985). In the chick, ACh and/or CAT were found in the neuroblasts and the neural crest at 2.5 days of development (Kuo, 1939; Greenberg and Schrier, 1977), and other investigators also showed the presence of cholinergic components in the presynaptogenetic nervous system of several species (see Karczmar, 1963a; Buznikov et al., 1996; Biagioni et al., 2000). This is also true for the CNS outgrowth, the retina. In the chick, AChE appears in the inner plexiform layer between the amacrine and the ganglion cells long before synaptization, as shown by Shen (1956) in his beautiful histochemical studies of AChE. Choline acetyltransferase also appears prior to synaptization in the retinae of cat, chick, mouse, and opossum (Lindeman, 1947; De Moura Campos and Hokoc, 1999; Kliot and Shatz, 1982; Zhang et al., 2005).

As already alluded to, the activity or the concentration of several "precocious" components of the cholinergic system increases relatively slowly during early embryogenesis and augments dramatically in the course of subsequent neurogenesis. This is true with respect to ChEs, or ACh, or both for the development of insects (Minganti et al., 1977; see Karczmar, 1963a; Chadwick, 1963; Colhoun, 1963), crustacea (Raineri and Falugi, 1983), or vertebrates such as amphibia, fishes, birds, and mammals, including the rodents, the primates, and the human (Sawyer, 1944; Bonichon, 1957; Kwong et al., 2000; see Karczmar, 1963a; Drews, 1975; and Moreno et al., 1998). To cite a few examples: AChE activity increases dramatically between the late gastrula and formation of the pluteus stage of the sea urchin (Augustinsson and Gustafson, 1949), and receptor binding appeared during these stages; it must be added that, as in the case of oocytic receptors, the putative receptors disappeared in the course of subsequent development. AChE increases 15-fold in the primordial mouse brain between embryonic days 9 and 19 (Moreno et al., 1988), and CAT activity is high during late neurogenesis of the cerebellum of the monkey (Hayashi, 1987). Also, AChE increased rapidly in several vertebrate species during neurogenesis of peripheral sensory organs (i.e., retina), the vestibules, and the autonomic ganglia (Meza and Hinojosa, 1987; Giacobini, 1986). It is of interest that the human exhibits a similar pattern of development of cholinergic components: transcripts and expression for the nicotinic receptors as well as CAT appear in the human prenatal brain and the spinal cord at 4 weeks of gestation and increase very rapidly thereafter (Perry et al., 1986; Hellstrom-Lindahl and Court, 2000; Brooksbank et al., 1978).

This rapid increase in the levels or activity of cholinergic elements during neurogenesis may taper off subsequently, but, at the time of morphogenesis of cholinergic nuclei or pathways, their rapid augmentation may occur again. For example, in the human fetus, following the neurogenetic differentiation of the nucleus basalis complex, one of the important sites of cholinergic radiation in the adult, AChE activity augmented rapidly in this nucleus (Kostovic, 1986; Candy et al., 1985; Mathura et al., 1979; Armstrong et al., 1987).

The various components of the cholinergic system may not arise simultaneously in the nervous system. Actually, ChEs, including particularly AChE, appear very early in ontogenesis, that is, prior to neurogenesis (see section BI, above). In the human, the muscarinic receptors appear in the cortex or spinal cord relatively late in the differentiation of the CNS, long after the emergence of AChE and 1 to 2 months after the appearance of nicotinic receptors, CAT, AChE, and ACh (Gremo et al., 1987; Ravikumar and Sastry, 1985; Brooksbank et al., 1978; Perry et al., 1986). Active muscarinic and nicotinic receptors also appear relatively late, compared to AChE and CAT, in the rodent and human cortex, hippocampus, and brainstem; the nicotinic rodent receptors and channels are similar to their adult receptor analogs (Hohmann et al., 1985; Slotkin et al., 1987; Aracava et al., 1987; Atluri et al., 2001). Particularly, Perry et al. (1986) stressed that compared to CAT and AChE, M1 and M2 receptors emerge late in the human cortex. Similarly, muscarinic receptors emerge after the vagal innervation of the chick heart has occurred (Papanno, 1979; Galper et al., 1977; Liang et al., 1986). Also, the nicotinic and muscarinic receptors appear only on the 12th and 16th day of chick sympathetic ganglia development, as appropriate responses to microelectrode stimulation of the ganglia could be recorded at the time (Dryer and Chiapinelli, 1985). Moreover, nicotinic receptors emerge before endplate innervation during neuromyal junction development, but, again, after neuromyal appearance of AChE or ACh; at this early time these receptors do not appear in the concentrated form of the adult endplate but are spread over the myotubule surfaces (see, e.g., Kidokoro and Brass, 1985).

Ezio Giacobini (Giacobini, 1981, 1983; 1986; Hruschak et al., 1982) felt that the cholinergic elements appear during ontogenesis synchronously and that their appearance relates well to synaptogenesis and transmission. Giacobini based his general opinion on data concerning the autonomic parasympathetic and sympathetic ganglia, and this matter will be discussed below in section BII-2. However, it may be stated that this synchronicity is not characteristic during preneurogenesis or neurogenesis.

a. Comments

The neurogenetic appearance of the cholinergic elements concerns the relation of cholinergic neurogenesis to cholinergic transmission and function. During the 1930s and 1940s, it was the avowed aim of Zenon Bacq, David Nachmansohn, Charles Sawyer, and others to establish this relationship as they attempted to connect AChE with function and motility (see section A, above).

It is true that effective cholinergic transmission requires the presence of AChE and other cholinergic components. Yet, the notion of AChE being the unique marker for cholinergic transmission and function will not stand today. In fact, the ontogenetic and/or neurogenetic appearance of cholinergic elements is desynchronous, and ChEs, including AChE, generally precede other cholinergic components and the onset of synaptic transmission and function during neurogenesis. The following statement should replace the historical notion: the neurogenetic arrival of *all* cholinergic components at synaptic sites and the development of anatomical entities at these sites are sufficient and necessary for cholinergic synaptic transmission and function (see section BII-5).

Given the prenervous appearance of cholinergic components and their desynchronous neurogenesis, these components may have a role that is unrelated to synaptic transmission and function. This notion is reinforced by the gametal appearance of the cholinergic elements and their subsequent reduction. The presence of cholinergic elements in nonnervous and ephemeral tissues is also relevant to this notion (see sections BI, above, and BII-5 and BII-6, below, for a further evaluation of this concept; see also Chapter 2 B-6.)

2. Neurogenesis of the Cholinergic System

The best known cholinergic neurogenesis is that of rodents. Schambra, Lauder and their associates (Schambra et al., 1989, 1991) related the CAT stains to the germinal zones of the mouse. By staining the mitotic cells, they could describe the early migration of cholinergic cells from these zones to the forebrain. Cholinergic cells were found in all ventricular zones of the forebrain between the 11th and 14th days of gestation (mouse gestation period is 21 days). The ventral zone was the first to exhibit the CAT stain (see also Sweeney et al., 1989; Hohmann and Berger-Sweeney, 1998), as these neurons migrate into the developing neural retina. Subsequently, within 2 days of gestation, the CAT stain arrived in the germinal zones of the eye, olfactory ventricle, lateral ganglionic eminence, epiphysis, and all of the lateral ventricle walls. The cells of the eminence migrate into the caudate-putamen and globus pallidus. Several cholinergic components including AChE and CAT appeared in the septodiagonal band complex within this time period (Armstrong et al., 1987).

On the 16th day, the cholinergic cells were found in the prelimbic, pyriform, and parietal cortices, lamina terminalis, and caudatopallial angle. Then, within the next 2 gestational days, multipolar CAT-containing neurons migrated into the olfactory nucleus and tubercle, diagonal band, and medial septal nucleus, while bipolar cells appeared in the hippocampus and along fiber tracts of the corpus callosum, external capsule, fornix, and anterior commissure (see Figures 8-4 and 8-5).

According to Semba (1992), the cholinergic neurons appear somewhat earlier in the rat (gestation period 21–22 days) than in the mouse. Choline acetyltransferase stain already exists in the spinal cord, including preganglionic and dorsal horn neurons, and the brainstem (mesencephalic tegmentum) on the 11th day of gestation, while the stain appears a day or two later in the basal ganglia and striatum (Van Vulpen and Van der Kooy, 1998). Eileen Van Vulpen and Derek Vander Kooy (1998) suggested that the cholinergic neurons arriving earliest in the prestriate are localized in the patch compartment of the adult striate, while the neurons appearing later in the prestriate are found in the matrix compartment (see also Zoli et al., 1995; Rho and Storey, 2001). Day 12 marks the migration of CAT-reactive neurons to the basal forebrain; on day 17 these neurons appear in the diagonal band. On days 16 to 20, the cholinergic neurons grow rapidly and reach the septal/diagonal band, the septum, the nucleus basalis, the forebrain, and the subcortical sites. Similar data were obtained with respect to ACh (Blue and Parnaveles, 1983; Hohmann and Berger-Sweeney, 1998; Hohmann and Ebner, 1985; Yamada et al., 1986; Schlumpf et al., 1991; Aubert et al., 1996; Costa, 1993; Mathura et al., 1979; Armstrong et al., 1987; Dinopoulos et al., 1989; see also Semba, 1992 and Lauder and Schambra, 1999).[2]

Studies of the cholinergic receptors in rodent neurogenesis are not as complete as the studies of CAT; also, they are not always consistent. When compared to the arrival of CAT, nicotinic and muscarinic receptor ontogenesis may be delayed (by a day or so) in rats. Both receptor types exist in the spinal cord, brainstem and medulla, cerebellum, mesencephalon, and motoneurons by day 12, and the muscarinic and nicotinic receptors reach the mouse neocortex by days 18 to 20 (Yamada et al., 1986; Hohmann et al., 1985; see also Rho and Storey, 2000 and Tribollet et al., 2004). At that time, the receptors are present in the rat or mouse in the septodiagonal band complex, nucleus basalis, cortex and basal forebrain, inner ear, cerebellum, olfactory bulb, striatum, optic lobe, and motor nuclei; these cholinergic elements increase in number or density subsequently (Schneider et al., 1985; Heaton, 1987; Armstrong et al., 1987; Rea and Nurnberger, 1986; Yamada et al., 1986; Hohmann et al., 1985). However, some investigators claim that nicotinic receptor subtypes may be present in the neural crest and early forebrain of mice and the rat much earlier and that they appear in the mouse neocortex by day 10 of gestation (Atluri et al., 2001). In fact, by using patch-clamp experiments, Atluri et al. (2001) showed that these receptors exhibit a response similar to that of the adult receptors. It should be added that cholinergic receptor density or number may peak late prenatally and decrease during the last days of gestation and/or postnatally (see also section BII-5, this chapter).

Lehotai (1998) was one of the few investigators to study AChE as early as on the seventh day of rat gestation; AChE was present at the time in the cerebellum and spinal cord, and ChEs were

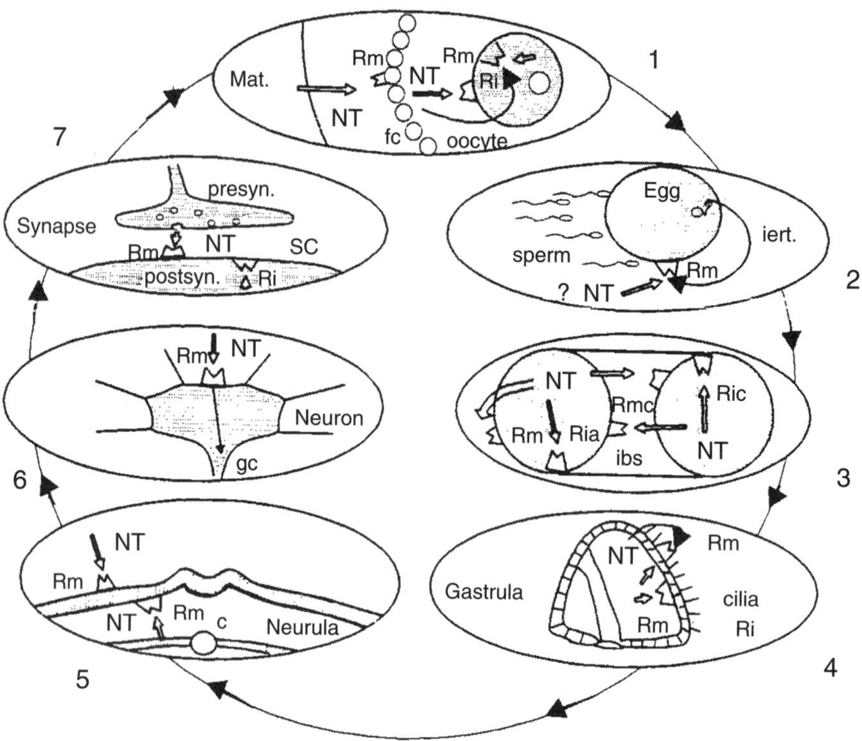

Figure 8-4. General scheme of the participation of neurotransmitters in ontogenesis: 1, gametogenesis; 2, fertilization; 3, cleavage divisions and early cell interactions; 4, local hormones of specialized physiological functions; 5, morphogenesis; 6, neurogenesis; 7, synaptic neurotransmission; NT, source of neurotransmitter; Rm, membrane receptor; Ri, intracellular receptor; Ria, intracellular receptor, participating in blastomere adhesion; Ric, intracellular receptor, participation in cleavage divisions; Mat,, maternal organism; fc, follicular cells; fert, fertilization; ibs, interblastomere space; c, notochord; gc, neuronal growth cone; postsyn, postsynaptic ending; CS, synaptic cleft. (From Buznikov et al., 1995, with permission.)

present in the dorsal root ganglion on day 12 (Koenigsberger et al., 1998). Altogether, it appears that ChEs precede CAT, ACh, and the receptors, not only in neurogenesis but during the prenervous stages of development.

In the case of avian neurogenesis, the evidence is less complete; at any rate, it follows a pattern similar to that found in rodents. Acetylcholinesterase existed during the first and second gestational days in neural folds and the neural crest of the chick (for references, see LeDouarin, 1982; LeDouarin et al., 1978, 1980; Karczmar, 1963a, 1963b; Giacobini, 1986; see also section A.) Also, CAT and ACh appear early in the neural crest of chicks and quail, though this appearance may be later than that of AChE (Kahn et al., 1980; Smith et al., 1979; Fauquet et al., 1981). Subsequently, AChE, CAT, and nicotinic and muscarinic receptors appeared in the ventral tegmental area and the mesencephalic nucleus profundus. By gestational day 12, the nicotinic and muscarinic receptors, AChE, and CAT were present in all brain parts studied, including cerebellum, thalamus, and limbic nuclei, and these cholinergic entities increased in density or activity during subsequent development (Conroy and Berg, 1998; Layer, 1991; Daubas et al., 1990).

Considerable evidence concerns cholinergic neurogenesis of humans, monkeys, and primates.

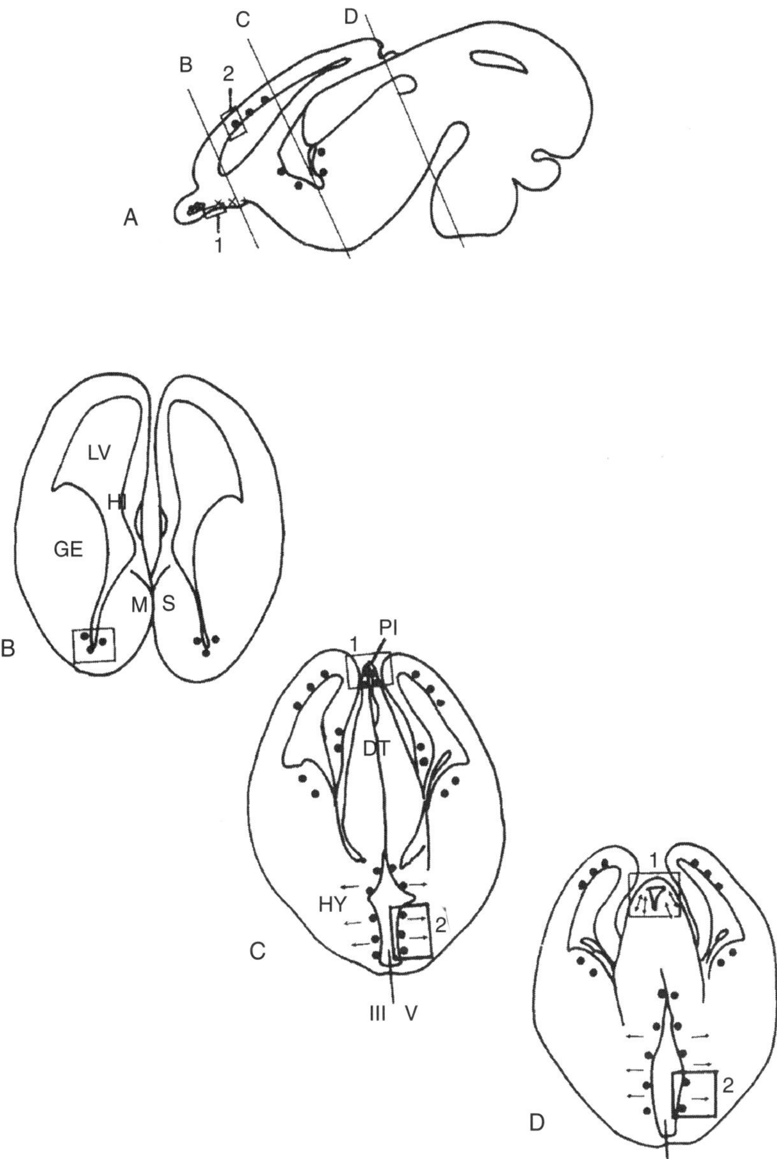

Figure 8-5. GD14. Cholinergic neurons in the mouse forebrain at gestational day 14. Line drawings illustrate: A, sagittal plane; B, C, and D, coronal planes as indicated in A. (From Schambra et al., 1989, with permission.)

Moreno et al. (1998) found both BuChE and AChE in the neural tube by the ninth day of human gestation; this is the earliest time of the reported appearance of cholinergic elements in human neurogenesis. On the ninth week of human development and on the 27th day in the rhesus monkey (gestation time 150 to 180 days), there is an outburst of AChE activity during the early stages of forebrain, limbic, and cerebellar neurogenesis (Kostovic, 1986; Kordover and Rakic, 1990; Kostovic and Goldman-Rakic, 1983; Hayashi, 1987). In the human the appearance of nucleus basalis, AChE stain precedes the emergence of AChE in the striate and the limbic system by a few days; AChE activity rapidly augmented in the nucleus basalis (Kostovic, 1986; Perry et al., 1986; Candy et al., 1985). Between weeks 10 and 11, two AChE-exhibiting bundles are formed at the nucleus basalis, one radiates to the cortex and the other to the limbic system and diagonal band.

Candy et al. (1985) and Perry et al. (1986) reported the early arrival of CAT in the human nucleus basalis and the cortex. By weeks 15 and 18, the CAT marker spreads through the cerebral hemispheres and the basal ganglia (Kwong et al., 2000). High cerebellar activity of CAT was reported for 15 to 22 weeks of human gestation by Brooksbank et al. (1978), Bull et al. (1970), Kwong et al. (2000), and Yew et al. (1999).

Schroder et al. (2001) and Agulhon et al. (1998, 1999) reported that nicotinic receptor mRNA and the receptor protein exist in the human cortex at 17 and 25 weeks, respectively, and this may be reasonable in relation to the ninth-week appearance of AChE; however, Hellstrom-Lindahl et al. (1998, 2001) found nicotinic receptors in the human spinal cord, mesencephalon, and cortex after 4 weeks of gestation, which is long before AChE arrives at these sites (see also Yew et al., 1999). The presence of muscarinic receptors was detected earlier (Egozi et al., 1986; Gremo et al., 1987; Perry et al., 1986; Kostovic, 1986; Candy et al., 1985; Ravikumar and Sastry, 1985a, 1985b; Johnston et al., 1985). High levels of muscarinic receptors exist in the human cortex and other brain parts by the 13th week of gestation (Aguilar and Lundt, 1985; Egozi et al., 1986; Gremo et al., 1987; Perry et al., 1986; Kostovic, 1986; Candy et al., 1985; Ravikumar and Sastry, 1985a, 1985b).

Cholinergic components may peak during prenatal human development and then decrease. Thus, the number and density of muscarinic receptors reach a peak in the cerebellum at 18 to 22 weeks of gestation and decrease during prenatal development. A similar pattern was noticed with respect to the nicotinic and muscarinic receptors in the hippocampus (Court et al., 1997; Brooksbank et al., 1978). This may be true also with respect to CAT: Gale (1977), Brooksbank et al. (1978), and Bull et al. (1970) found that after CAT activity reached a peak in the cerebellum, frontal cortex, and basal ganglia during the 15th to 22nd week of gestation, it declined steadily, the postnatal values being several times lower than the fetal peak (see however, Brooksbank et al., 1978; see also section BII-4).

There is a desynchrony in the development of cholinergic components during the first trimester in the human, the 11th day of rodent gestation, and the second day of gestation in the chick. As noted above, ChEs generally appear first, followed by CAT and/or ACh, with the cholinergic receptors appearing last. There are a few exceptions. In the human cerebellum, CAT may appear before ChEs. Also, in the human forebrain and cortex, muscarinic receptors appear before CAT, or, at least, their rate of increase is steeper than that of CAT (Brooksbank et al., 1978).

It must be added that particular investigators have not simultaneously studied the ontogeny and/or neurogenesis of all the cholinergic components; comparisons of the initial appearances of these components culled from experiments of single components may be not reliable. Nevertheless, it appears that during either early neurogenesis or preneurogenetic ontogenesis, cholinergic components do not appear simultaneously and/or do not develop at the same rate (see section BI). There may be differences among the prenatal patterns of cholinergic components in rodents and humans. In rodents, AChE, CAT, or cholinergic receptors continuously increase in concentration or activity during prenatal ontogenesis, though they may peak and then decrease postnatally (Rho and Storey, 2001; see section BIV), while in humans these components appear to both peak and decrease prenatally (see above; further data are necessary before this conclusion can be substantiated).

It should be noted that the subtypes of nicotinic and muscarinic receptors and AChE isoforms change during ontogenesis. Also, during mouse neurogenesis, the ratio between tetrameric and monomeric AChE changes, and so do their kinetic characteristics when compared to the characteristics of adult AChE (Aubert et al., 1996; Lehotai, 1998; Moreno et al., 1998; Large et al., 1985; Hellstrom-Lindahl et al., 1998; Inestrosa et al., 1995). There are several examples of this transition in the peripheral cholinergic nervous system. For example, as the neuromyal junction matures, there is a transition from one form of globular AChE to another (Layer et al., 1987), and from globular AChE to asymmetric (Toutant et al., 1983; see also below, this section). In other instances, synaptogenesis involves a transition from "silent" to active receptors (at the cardiovagal junction; Galper et al., 1977).

Neurogenesis, and any other type of ontogenesis, includes processes of differentiation. In other words, the appearance of an early cell type and/or components of this cell type, such as CAT, AChE, and the like, in the case of the cholinergic neurons, must then be followed by processes of maturation, that is, differentiation of the cell type in question as well as of synapses. These matters were particularly researched in the case of such sites as the heart and the neuromyal junction; thus, while not directly related to the subject of this book, these sites are referred to in detail in section BII-5, below. As pointed out in this section, several specialized proteins are involved in the processes of differentiation (see also Lopez-Coviella et al., 2000), as well as trophic factors (see section CI and CIII, below).

3. Cholinergic Neurogenesis of Sensory Brain Areas

Cholinergic neurogenesis of the sensory brain areas is of particular interest. It was proposed in the past that the cholinergic system does not participate in the sensory function (Michelson et al., 1970; Michelson, 1974); however, subsequent documentation of the cholinergicity of these areas helped to discredit this notion; see also Chapter 2 DIII). Also, the studies of sensory neurogenesis, particularly in the case of the retina, traced the arrival of cholinergic components onto the postsynaptic elements, offering an insight into cholinergic synaptogenesis.

The cholinergic components and those of other transmitter systems appear quite early—gestational days 3 to 4—in the rudiments of the retina of the chick, mouse, and other mammals, that is, in the optic vesicle originating in the first cerebral vesicle (see above; LeDouarin, 1982; Zhang et al., 2005; see however McDonald et al., 1987). AChE and BuChE stain was present in the prospective retinal ganglion cells of the chick between gestational days 3 and 4; at day 6 the stain appeared in or at the amacrine cells and soon thereafter at the sites of neuronal arborization and synaptogenesis of the synaptic contacts between the amacrine and ganglion cells at the inner plexiform layer (Shen et al., 1956; Layer et al., 1987). AChE was also found at or in the horizontal cells. Lindeman (1947) found that ACh (which he bioassayed) is present in the chick retina on the eighth gestational day; he stressed the rapidity of the subsequent increment of ACh concentration between days 18 and 20 of the development, which coincided with the appearance of the pupillary constrictor reflex.

In essential agreement with the early data, Hofman (1988) and Crisanti-Combes et al. (1978) found CAT in the retina of the chick embryos on day 6; its activity augmented rapidly subsequently. The cat (gestation time about 65 days) retina exhibited CAT immunoreactivity in the inner plexiform layer on day 56, that is, at the time of the formation of the amacrine cells (Mitrofanis et al., 1989). In the marsupial opossum, the CAT stain does not seem to appear prenatally; the stain emerges first in the amacrine cells on the 12th postnatal (in the pouch) day and augments continually till day 50; thereafter, the stain of individual amacrines does not decline, but the number of these cells decreases as they form a monolayer (De Moura Campos and Hokoc, 1999).

Nicotinic alpha 3 and alpha 8 receptors appear at the ganglion and bipolar cells at gestational days 2 and 5, respectively (Gardino et al., 1996; Hamassaki-Britto et al., 1994). This could be expected, as these cells are postsynaptic with respect to the amacrine cells, but Gardino et al. (1995) found that nicotinic receptors are also present on the second gestational day at amacrine and displaced ganglion cells (could these receptors be localized presynaptically in the case of amacrines?). Muscarinic receptors appear equally early in the inner plexiform layer of rodents including ferrets as well as in the chick (Hutchins, 1994; Hutchins and Casagrande, 1988; McKinnon and Nathanson, 1995). In the mouse, several receptor subtypes are present prenatally, and the subtype changes with ontogenesis (Nadler et al., 1999).

It should be noted that the muscarinic and nicotinic receptors may appear earlier in the retina than AChE and CAT (Drews, 1975. Thus, similar to the cerebellum, the retina constitutes a rare instance of AChE appearing in development after rather than before the other cholinergic components (see above, section BI-2); at any rate, this finding is consistent with the notion of asynchrony between the ontogenetic and/or neurogenetic appearance of cholinergic elements. It should be added that ACh applied to the chick retina induces muscarinic Ca^{2+} mobilization on the second gestational day, that is, before transmission is present (Yamashita et al., 1994).

Only sporadic information is available for other sensory brain areas. In the prenatal rat and tree shrew, AChE, alpha 3 and alpha 4 nicotinic receptors appear late in geniculocortical projections, cochlear nuclei, vestibular end organs, and primary auditory cortex (Meza and Hinojosa, 1987; Robertson et al., 1991; Hutchins and Casagrande, 1988; Morley and Happe, 2000; Happe and Morley, 2004). AChE also appears late (at the onset of the tadpole metamorphosis) in the medullary vestibular and dorsolateral nucleus during auditory neurogenesis of the prenatal frog (Kumaresan et al., 1998; also see Chapter 2 D). Finally, cholinergic innervation of the olfactory bulb occurs in rodents at birth or during early postnatal stages (see the following section.)

4. Functional Differentiation and Synaptogenesis

Final synaptic and junctional differentiation, which provide effective transmission, constitute the goal of neurogenesis; thus, understanding the association between this differentiation and cholinergic elements is of major importance. The peripheral synapses at the autonomic ganglia and at the neuromyal and cardiovagal junctions provide useful information on this matter; they will be referred to in this section jointly with the pertinent central events, even though they are not within the main scope of this book.

It should be added that the innervation of peripheral organs may not always relate to synaptogenesis and transmittive function, but rather participate in the action of ACh as a morphogen (see below, section CIII). For instance, spinal cord–originating prenatal gonadal innervation of the sea turtle concerns gonadal differentiation and its temperature dependence (Gutierrez-Ospina et al., 1999).

a. The Heart

It was reported early that urodele, chick, rabbit, guinea pig, and rat hearts exhibit ChEs long before onset of innervation, but considerably after the initiation of cardiac beat (Sippel, 1955; Kramer, 1950; see also Karczmar, 1963a). The urodele *Amblystoma* heart may contain only AChE, while both BuChE and AChE were present early in the rat and chick hearts. In the chick, AChE was present on the fourth day of development; in the chick, the cardiac beat appears on the ninth or 10th day of development, while innervation sets in around the 12th day of development (Pappano and Loffelholz, 1974; Pappano, 1977; cf. Fambrough, 1976).

Similarly to ChEs, in several species atrial and /or ventricular responses to ACh and muscarinic cholinomimetics are precocious with respect to innervation (see Karczmar, 1963b). The presence of these responses argues for the existence of receptors. Indeed, the early presence of muscarinic receptors was established by several investigators (see, for example, Galper et al., 1977). There is a controversy as to the exact time of the appearance of the receptors versus the initiation of cholinomimetic sensitivity; investigators who claimed that atrial receptors are present prior to the onset of cholinoceptivity employed the term "silent" to the receptors in question (McCarty et al., 1960; Zingoni, 1956; see Karczmar, 1963b). It is of interest that also in the case of human heart, cholinergic receptors are present long before the cardiac plexi mature and possibly before the first neuronal elements abut upon the heart (Bakaikin, 1978; obviously, there is a paucity of human data).

Choline acetyltransferase is present at low levels before the 10th day of chick embryo development. It cannot be decided whether this presence is precocious with respect to the arrival of innervation at the heart, because it is not clear whether this CAT is myogenic or neurogenic (Gifford et al., 1973). Choline acetyltransferase levels rapidly increase after day 10 (Gifford et al., 1973), which is coincidental with maturing of cardiac innervation and function. It may be added that Lissak et al. (1942) found ACh in the chick heart at 72 hours of chick development, in other words, some 7 days before the appearance according to Gifford et al. (1973) of cardiac CAT.

Clearly, AChE, cholinoceptivity, and cholinergic receptors are present in the embryonic hearts of several species long before innervation, and CAT appears in the embryonic heart before nervous regulation of the heart and cardiac function mature (or considerably earlier, if the data of Lissak et al., 1942, can be substantiated; see Crossley and Altimiras, 2000).

The cholinergic elements appear precociously with respect to the maturation of cardiac cholinergic innervation, cardiac plexus formation, and cardiac function; also, they do not appear simultaneously and they do not expand in synchrony. But, they must all be present before mature cardiac function arises, and they must undergo structural changes to make this function possible. For example, during this time period, cholinergic receptor-channel macromolecules change from "silent" or semiresponsive to fully active with respect to ACh and cholinomimetics. Similarly, the affinity of the receptor-channel macromolecules and binding characteristics with regard to ligands and their channel dynamics have to achieve maturity. Then these macromolecules must establish capacity for desensitization (Harris and Thesleff, 1971; Galper et al., 1977). Trophisms and the morphogenetic effects of innervation contribute to the cardiac synaptogenesis or cardiac junctional genesis.

b. Autonomic Ganglia

The autonomic ganglia are another system exhibiting the precocity of some emerging cholinergic components with respect to synaptogenesis and synaptic function. In the chick, AChE is present within the first 2 days of ontogenesis of the neural folds and neural crest, which are the source of autonomic ganglioblasts (for references, see LeDouarin, 1982; Karczmar, 1963a; Giacobini, 1986). In chicks and quail, CAT and ACh are also present the neural crest, possibly later than AChE (Kahn et al., 1980; Smith et al., 1979; Fauquet et al., 1981). This early emergence of cholinergic components holds true for neural crest sections, giving rise to sympathetic ganglioblasts and to parasympathetic neurons (Le Douarin, 1980, 1982, 2001).

The sympathetic ganglia obtain their cholinergic components from preganglionic spinal cord neurons and their preganglionic terminals; they derive their catecholaminergic elements from the postsynaptic neurons. Preganglionic axons enter the sympathetic ganglion chain and postganglionic axons emerge from the ganglionic cells' primordia between days 11 and 12; dendrite formation and synaptogenesis follow. The parasympathetic system similarly derives its cholinergic elements from the spinal cord.

Migrating neural crest cells start forming ganglioblasts by day 5, and this process is completed by day 7 (Yip, 1986); ganglionic cells show recognizable synapses by days 8 and 9 (Enemar et al., 1965).

Acetylcholine, AChE, and CAT can be detected in avian ganglioblasts on days 5 through 8 (Giacobini, 1986; Schneider et al., 1985); nicotinic receptors may appear even earlier (Torrao et al., 2000; Giacobini and Marchi, 1981; see, however, Dolezalova et al., 1974; Greene, 1976; Erkman et al., 2000; are these receptors pre- or postsynaptic?). However, at the end of this period only a few mature synapses may be present (Hruschak et al., 1982). Consistent with this notion, fast (nicotinic) and slow depolarizing effects of carbachol (a muscarinic response) and substance P appear only around day 12 (Dryer and Chiapinelli, 1985; Greene, 1976; Giacobini and Marchi, 1981; this was the first time that these responses were tested: Chiapinelli, personal communication).

The development of the avian parasympathetic ganglia, including the ciliary ganglion, was extensively studied by Giacobini and his collaborators (Giacobini, 1986; Landis et al., 1987; Pilar et al., 1987). Choline acetyltransferase, AChE, ACh, and high-affinity choline uptake system are present at the time of the neural crest cells' migration into the future ciliary ganglion (between the third and fourth day of development in the chick; Hammond and Yntema, 1958; Marchi et al., 1980).

Crest cell migration and ganglionic primordia formation in rodents is completed by days 11 to 13, with arrival time depending on segmental location of the crest cells. Preganglionic axons enter the sympathetic ganglion chain of the rat and postganglionic axons emerge from the ganglionic cells' primordia between days 11 and 12; dendrite formation and synaptogenesis begin on day 14 (Rubin, 1985a; 1985b; 1985c). Acetylcholinesterase, CAT, and ACh appeared earlier, although the definitive time of their initial appearance is uncertain (Coughlin et al., 1978). Their levels or activities did not change much in rodents following day 13 of gestation, though they steeply increase postnatally. That the cholinergic synaptogenesis of the sympathetic ganglia is incomplete before the 14th day of gestation is also indicated by the finding that ganglionic catecholamines and tyrosine-beta-hydroxylase appear only later (Coughlin et al., 1978); this generation of a catecholaminergic, ganglionic system seems to be the consequence of, and seems to depend on, the maturation of the presynaptic input and the presence of effective ganglionic transmission (Hruschak et al., 1982). Yet, elements of the catecholaminergic system may be present already in the neural crest (Kahn et al., 1980); this presence appears to be independent of and have a role different from that of the ganglionic catecholaminergic system.

Whether in the rodent or in the chick, preganglionic neurons clearly contribute to ganglionic CAT, ACh, and AChE (cf. Koelle, 1963). However, distinguishing between cholinergic elements derived from the neural crest and those being generated in sympathetic ganglionic neurons is difficult (Sorimachi and Kataoka, 1974; Giacobini, 1986). Additionally, the ganglia's cholinergic receptors may be presynaptic and formed on presynaptic nerve terminals, or develop on the postsynaptic ganglionic neurons.

The parasympathetic ganglia should contain higher levels and activities of the cholinergic elements, particularly CAT and AChE, than the sympathetic ganglia: the former derive cholinergic elements from the crest and its preganglionic derivatives, as well as from the cholinergic postganglionic neurons, while the latter obtain their cholinergic elements only preganglionically (see section BII-2). This holds for the adult parasympathetic and sympathetic ganglia, and it seems to be true also in the case of the embryonic ganglia (Koelle, 1963; Saelens and Simke, 1976; Giacobini, 1986).

Giacobini (cf. Hruschak et al., 1982) felt that cholinergic elements appear in synchronous manners and that they correlate with ganglionic initiation of synaptic transmission. Giacobini (1983) stressed that in the chick ciliary ganglion and its target, the iris, muscarinic or nicotinic receptors, choline uptake, CAT, and ACh appear at Hamilton-Hamburger stage $26^1/_2$ (i.e., on days 4 to 5 of chick gestation), coinciding with the innervation of these organs, the appearance of morphologically defined synaptic contacts, and the onset of transmission (Landmesser and Pilar, 1972; D'Amico-Martel, 1982). It was already pointed out that Ezio Giacobini's generalization does not hold for the central cholinergic system (see above, section BII-1). Actually, his notion may not hold for the gangia either. First of all, AChE is "out of tune" with the other cholinergic elements during gangliogenesis; for example, it appears before other cholinergic elements in the chick crest (see above); this is similar to ChE's very early appearance during neurogenesis (see section BI, above). Furthermore, several cholinergic components emerge during gangliogenesis long before the onset of synaptogenesis, transmission, and postsynaptic cholinomimetic responses. In fact, CAT and ACh are present in the crest and prior to the initiation of ganglioblast innervation by presynaptic input from spinal autonomic neurons (see above), and Giacobini (1986) recognized the presence of ACh in the neural crest. Does crest cholinergicity disappear during gangliogenesis, as it occurs in the case of certain noncholinergic neurons, or at least diminish, before reappearing (see above, section BI-1)? Do the cholinergic components of the embryonic autonomic ganglia originate de novo in the early ganglioblasts?

Furthermore, the developmental rates of the various cholinergic components differ significantly. For example, in the avian ciliary ganglion, choline uptake is near its mature, maximal value between days 4 and 5, while ACh synthesis and CAT activity are at very low levels on day 5 (Marchi et al., 1980; LeDouarin et al., 1978). Altogether, in several species the elements of the cholinergic system appear at different ontogenetic times and increase at different rates; in addition, the cholinergic elements of the ganglia and the neural crest may arise independently of each other. At any rate, cholinergic elements appear on the ontogenic scene long before the ganglionic ingrowth of the presynaptic nerves, before the maturation of synapses, and before the onset of effective transmission.

It is obvious that the presence of all cholinergic components is required for the functionability of a cholinergic synapse, whether they get to the synapse in a synchronous or desynchronous manner. However, this is by far not the only requirement. There is the need for a choline uptake system, effectiveness and homeostasis of ACh synthesis, and the existence and transport proteins concerning movement of ACh and the vesicles. The maturity of a cholinergic synapse also requires appropriate structurization of the cholinergic components, such as increased density and subneural organization of the receptors and the presence of active proteins such as CAMs (Rubin, 1985a; 1985b; 1985c; Acheson and Rutishauser, 1988; see Chapters 2 B and 3 C). Some of these requirements are referred to below, as neuromyal development and synaptogenesis are discussed.

c. The Motoneuron and the Myoneural Junction

Cholinergic elements including AChE and CAT appear early in the neuroblast primordia of the motoneurons in all species that were studied, including the human; this means of course that the cholinergic elements are present in the future motoneuron system long before the system produces axons that will grow into the mesenchyme. Only later will the mesenchyme generate myoblasts and muscle fibers; in fact, the nerve precedes the muscle with respect to the muscle differentiation of the junctional zone (Dennis, 1981; Oppenheim et al., 1986; Atsumi, 1971; Motavkin and Okhotin, 1983; Drews, 1975; Karczmar, 1963a).

That CAT appears in motoneuron primordia before axogenesis and neuromyal junction maturation should be expected, as CAT is primarily a perikaryon enzyme. But the cholinergic elements appear in the muscle long before the junction is formed. Thus, strong AChE activity as well as CAT and ACh are present in several vertebrate and human muscle rudiments (Youngstrom, 1938, 1941; for further references, see Koelle, 1963; Kupfer and Koelle, 1951; Karczmar, 1963a; Fambrough, 1976; Rotundo, 1987) and the chick myoblasts synthesize AChE before their fusion (Fluck and Strohman, 1973; Toutant et al., 1983; cf. also Rotundo, 1987). In fact, it is well known since Abdon's (1945) and Muralt's (1946) extensive research that adult skeletal muscle contains ACh that does not originate from motor nerve terminals or Schwann cells, and a stable low level of ACh is found in the muscle 4 to 6 weeks after denervation (see Hebb et al., 1964; MacIntosh and Collier, 1976; Werner and Kuperman, 1963; Miledi and Slater, 1968; Consolo et al., 1986). Also cholinergic receptors precociously appear in the developing muscle; the nicotinic receptors (alpha-bungarotoxin binding sites) appear in the rat and chick embryo during myoblast fusion into myotubes (see Salpeter, 1987, for references). In fact, when fusion was prevented in the culture by withdrawing calcium, the receptors still appeared at the myoblast surface (Prives and Patterson, 1974).

Clearly, cholinergic components appear precociously in ontogenesis of mammals and other vertebrates, both at the motoneuron (or its primordium) and the muscle end of the neuromuscular junction, long before morphological and functional maturation of the junction. This maturation requires several structural developments. Thus, myogenic AChE is initially equally spread over the myoblast and myotube in chicks, fish, or rats and gradually "concentrates," "deposits," or "accumulates" at the endplates (Dennis, 1981; Davey and Cohen, 1986). Several factors may regulate this process, which may well continue into the postnatal period (see below, this section). Apparently, the globular AChE form is solely present in chicks or rats before the onset of innervation or nerve-evoked contractions (Vigny et al., 1976), while asymmetric collagen AChE forms only start appearing with the onset of mature innervation and neuromuscular activity (Toutant et al., 1983; Rotundo, 1987; Koenig and Rieger, 1981; see Chapter 3 DII for definition of AChE forms). The maturation of the basic lamina (a meshwork of collagen and glycoprotein) and the innervation patterns involved in neuromuscular maturation seem important for differentiating AChE at postsynaptic junctions (Sanes et al., 1978; Dennis, 1981). Additionally, activating AChE may be necessary to fully integrate AChE neuromuscular junction functions, as AChE synthesized in the endoplasmic reticulum is initially inactive (Rotundo, 1987). It must be added that ontogeny of karyotic motoneuron BuChE and its transport into the terminal are largely unknown.

Similarly, distinct changes in distribution and structure occur at the neuromuscular junction with respect to cholinergic receptor ontogeny. The early wide spread of myoblast and myotube nonjunctional nicotinic receptors exhibits occasional "hot spots" or isolated clusters (Dennis, 1981). These early clusters or hot spots extend beyond the future junction. The hot spots do not induce innervation and/or formation of mature neuromyal junctions. To the contrary, the arrival of motor nerve endings induces nicotinic receptor reorganization and compartmentalized expression of their genes, concentration of clusters and the formation of endplates (Anderson et al., 1977; see also Purves and Lichtman, 1985; Purves et al., 1988; Changeux, 1993). In fact, denervation induces dispersion of the endplate in the adult (and sensitizes the denervated neuromyal junction to cholinomimetics; Bloch and Steinbach, 1981; Peng, 1987).

As if the matter were not sufficiently complex, additional factors are needed for the maturation of the neuromyal junction, as suggested by Mika Salpeter (1987). These factors are active proteins that aggregate around the immature neuromyal junction; they are studied intensely by Joshua Sanes, Jack McMahan, Jonathan Cohen, and their associates (Sanes et al., 1998; Carr et al., 1987). Among them are several cell adhesion molecules (CAMs) and substrate adhesion molecules (SAMs), discovered by Gerald Edelman (see Edelman and Chuong, 1982; Grument and Edelman, 1998), a Nobel Prize winner. Cell adhesion molecules and SAMs and related trophic factors are concerned with the movement, organization, and clustering of the cholinergic muscle receptors, whether during ontogeny or following denervation of the adult muscle (Nastuk and Gabriel, 1985; Connolly, 1985).

A member of the CAM family, N-CAM is present during ontogeny in the motor axons and their terminals, and on the surface of the myofibers; it promotes clustering of cholinergic nicotinic receptors on the surface of embryonic myotubes and organizes the endplate innervation (Rieger et al., 1985; Bixby and Reichardt, 1987). Altogether CAMs "... hold the nerve bundles together ... specify ... sites of nerve-muscle contact and ... promote ... synapse formation" (Salpeter, 1987).

Cell adhesion molecules, laminin, ascorbic acid, or other components of the junctional basement membrane do not exhaust the number of substances, which may regulate the patterning of the cholinergic receptors of the muscle and the endplate. Extracts of nervous tissues including brain may contain substances having an effect on receptor ontogenesis (see, for example, Neugebauer et al., 1985); Salpeter (1987) suggested in fact that different nerve factors may control the patterning and the membrane insertion or synthesis of the receptors; the glycoprotein sciatin, somewhat lighter than CAMs, may be one of these factors. Sciatin is chemically similar to or identical with serum transferrin—both substances seemed to increase the synthesis of the muscle cholinergic receptors and AChE and facilitate myogenesis (Oh and Markelonis, 1982; Oh et al., 1988). Sciatin is derived from the chick brain or ciliary ganglion. Both sciatin and serum transferin increase the synthesis of the endplate nicotinic receptor, and sciatin promotes the accumulation and the mature distribution of the myotubular cholinergic receptors of the chick toward the future endplate, without, it is important to note, affecting the neuromyal AChE. Similar effects were evoked by the electric organ aggregating factor; this factor also caused the formation of patches of high AChE and BuChE activity.

It is a moot question at this time as to whether or not sciatin—or tranferrin—is generated by the motoneurons (Markelonis et al., 1985). In fact, it is not clear whether or not many of the substances referred to in this section, such as CAMs or the components of the basement membrane, are present in the pertinent presynaptic neurons, in other words, the motoneurons, and, therefore, it is not certain whether they act in the anterograde fashion. Of course, rather than belonging to the categories of retrophins or anterograde acting trophins, they could be generated and act locally. It should be argued, however, that in many cases, the substances in question appear to act coincidentally with the arrival of the nerves at their developmental target and/or may antagonize the effects of denervation.

Several other active proteins, trophic factors, and other elements play a similar role; that is, they regulate the postsynaptic neuromyal differentiation, including the insertion of the receptors into the postsynaptic plasmalemma; they include several active proteins such as laminins, agrin, rapsyn and synapsin, and ascorbic acid[3] (Oh et al., 1987; Rotundo, 1987; Horowitz et al., 1989; Burden, 1998; Sanes et al., 1978, 1998, 2003; Sanes and Lichtgman, 2001; see also sections CI through CIII). Still other morphogenetic elements are trophins (see sections CI through CIII). All these factors are involved in the neuromyal ontogeny of several species, including the chick and rodents (Salpeter, 1987; Moore and Walsh, 1985; Usdin and Fischbach, 1986; 1985; Sanes et al., 1978 and 1998; Sanes and Lichtman, 2001). It should be added that concentrations of these factors generally decrease following neuromyal maturation. For example, N-CAM concentration increases during maturation of the neuromyal junction and decreases until it becomes very low postnatally, as it is converted into the adult form of CAM (Sanes et al., 1998). It increases at these sites following denervation (Rieger et al., 1985). Finally, there is the morphogenetic effect, insufficiently explored, of the nerve terminal released ACh and of the contractile activity–which precedes innervation–of the muscle.

Basal lamina/basement membrane is a collagen protein ensheathing the muscle; in the adult, it separates, with the synaptic cleft, the endplate from the nerve terminal. It is present in small amounts early in development, and it increases dramatically just before innervation of the junction and the maturation of the endplate. It participates in this maturation via the action of its active proteins such as laminin, and its concentrations coincide with the high densities of the nicotinic receptors (Sanes and Lichtman, 2001; Lichtman and Sanes, 2003; Sanes, 2003; see also Chapter 2 B-2 and B-3). It is generally opined that the membrane dictates the site where the nicotinic receptors aggregate to form a mature endplate.

Still another process or movement intervenes at the time of the patterning of the receptors. This process ensures a precise juxtaposition of postsynaptic folds and the nicotinic receptors on the one hand and appropriate nerve terminals on the other. To accomplish this exact structurization, the process also involves receptor degradation. This degradation affects both immature receptors (or "primitive" or "silent" receptors; Galper et al., 1977) and mature receptors, the net result of this phenomenon controlling nicotinic receptor density; actually, the rates of inserting nicotinic receptors onto the postsynaptic plasmalemma and the rates of their degradation differ between the immature and mature receptors (Anderson et al., 1977; Anderson, 1987). Then, the endplate "resists" additional, unneeded innervation; this has been known for some 100 years (Tello, 1907) and has been amply confirmed (Guth, 1968; Bowden and Duchen, 1976). These regulatory phenomena are a part of the neurotrophic factor hypothesis (see section CI-3e, below).

However, an ineffective synaptic transmission is present before complete maturation of the neuromuscular junction. For example, in the chick, some transmission parameters are present around 15 days before hatching excitatory endplate potentials (EPPs) (Landmesser and Morris, 1975; Salpeter, 1987, and Dennis, 1981). The presence of immature synaptic function is illustrated by the appearance of excitatory endplate potentials (EPPs), spontaneous ACh release (as confirmed by the appearance of miniature EPPs), and sensitivity to ACh and nicotinic cholinomimetics. Also, the channel kinetics and the long gating time exhibited by the primitive junctions are characteristic for immature junctions. And there is a "primitive" type of release from relatively immature terminals, the "growth" cones (Hume et al., 1983; see also Salpeter, 1987). Altogether, physiological neuromuscular junction maturity in amphibia, chicks, rodents, or humans is achieved late and is preceded developmentally by the presence of several cholinergic components (Fischbach and Schuetze, 1980; cf. also Salpeter, 1987). The structural maturation of nerve terminals and their organelles (including the endoplasmic reticulum and the synaptic vesicles), the synaptic cleft, and postsynaptic muscle surfaces (including the junctional folds) is so complex that these processes warrant long ontogenetic time; they occur late in development in humans and other species (Cull-Candy et al., 1982) and, in fact, rodents' junctions mature only postnatally (e.g., Bixby, 1981; for other references see Salpeter, 1987; Salpeter and Loring, 1985; Grinnell and Herrera, 1981; Koenig and Rieger, 1981; cf. Rotundo, 1987).[4]

As can be readily surmised, there are a number of physiological and electrophysiological differences between mature junctional and the earlier, extrajunctional receptors. First, the mobilities of the receptors differ, as the nonjunctional receptors are freely mobile in the membrane, yet become immobilized with endplate clusterization (Axelrod et al., 1976; cf. Salpeter, 1987). This event of neuromuscular development may be part of the ontogenetic decrease of fluidity at the synaptic membrane's hydrophobic core; this decrease in fluidity may underlie the increase in synaptic effectiveness and function with development (Hitzemann et al., 1984). Perhaps this immobility relates to increased resistance during development of the junctional nicotinic receptors to dispersal inducible by low external Ca^{2+} or denervation (Bloch and Steinbach, 1981; cf. Salpeter, 1987).

Second, the mature junctional receptors and/or clustered receptors differ from the immature receptors in their immunobiological structures; this is caused by the loss of 1 or 2 alpha subunits of the nicotinic receptor during receptor maturation (Hall et al., 1983; cf.; Salpeter, 1987; Anderson, 1987; see also Chapter 4). This may relate to changes in the isoelectric point and channel characteristics that also occur during maturation (cf. Salpeter, 1987).

Then, the embryonic channel-receptor macromolecules of many species exhibit a longer opening time (relaxation time) than the mature receptors, as pointed out above; chick channels seem never to achieve short opening (gating) time and remain permanently immature (Schuetze, 1980). It was suggested that the maturation of the opening time—the short gating time—relates to the formation of the junctional folds (Fischbach and Schuetze, 1980; Salpeter, 1987; see also Brenner et al., 1983; Vicini and Schuetze, 1985).

Finally, the rate of receptor turnover, that is, the rate of their insertion onto membranes and their degradation, differs between immature or extrajunctional receptors and the mature endplate receptors; the net result of the turnover controls the surface receptor density (cf. Salpeter, 1987; Anderson, 1987). In amphibia, rodents, and other species, immature receptor degradation rate compared to junctional receptors is markedly decreased during the onset of innervation and subsequent maturation of the junction (cf. Fambrough, 1979; Salpeter and Loring, 1985; and Salpeter, 1987).

5. Central Synaptogenesis

It has been emphasized (see section BII-4) with regard to the neurogenesis and synaptogenesis of the vertebrate heart, autonomic ganglia, and neuromyal junction that the presence of some cholinergic components does not necessarily coincide with synaptogenesis; rather, all such components are needed. In fact, besides the neurogenesis of such cholinergic components as AChE, CAT, and ChEs, additional cholinergic components such as factors concerning choline uptake, receptor clustering, adhesion, and vesicular cycling are required for the formation of an effective synapse. Then the presence of the trophic factors is needed for synaptogenesis. Finally, certain morphologic units, such as basal lamina in the case of the myoneural junction, may be wanted in a mature synapse or junction.

Some of these notions and processes are valid for the formation of central synapses. First, there is the notion of desynchronous arrival of cholinergic components at the synapse. In all species AChE is very precocious with respect to its appearance during neurogenesis (and, indeed, it appears prior to neurogenesis; see section BII-1; Egozi et al., 1986). Thus, AChE arrives at the rodent cortex and forebrain early, prior to the beginning of synaptogenesis, and reaches adult levels at 7 weeks postnatally, while CAT arrives at this site considerably later; the same pattern characterizes subcortical sites (Blue and Parnaveles, 1983; Hohmann and Berger-Sweeney, 1998; Hohmann and Ebner, 1985).

Another important feature of central synaptogenesis is the arrival at the future synapse of the entities concerned in the release of ACh and in the functioning of the presynaptic cell as a cholinergic neuron, such as high-affinity choline uptake (HACU) and vesicular acetylcholine transport (VAChT), as well as Nogo genes encoding myelin proteins responsible for myelin inhibition of neurite growth. These entities appear early and increase pre- and postnatally in several rodent brain parts including hippocampus (Kotas and Prince, 1981; Zahalka et al., 1993; Holler et al., 1996; Hohmann and Berger-Sweeney, 1998; Meier et al., 2003). Some of these protein factors may underlie, postnatally, brain plasticity phenomena, as proposed by Roger Nitsch and his associates (Meier et al., 2003). Also pertinent in the present context is the demonstration of the presence of HACU and VAChT early postnatally at the olfactory glomeruli; this presence correlates closely with the maturation of synapses at this site (El Far et al., 1998; LeJeune et al., 1996; Hutchins and Casagrande, 1988; see also Chapter 9 BIV).

Another system that is needed for the central (and peripheral) maturation of synapses is the system of the trophic factors (see sections CII and CIII, below). Among those factors particularly important in the present context are the so-called anterograde-acting trophics which regulate the transsynaptic growth and differentiation and neuritic outgrowth and orientation; thus, they act as morphogens (Yankner and Shooter, 1982).

Among these trophic factors and morphogens are the cholinergic components, and their role is discussed in detail in section CIII, below. At this time, let it be mentioned that the early appearance of the cholinergic components—and, in fact, components of other transmitter systems—relates to their role as morphogens and trophic and formative agents rather than to their transmittive function; consistent with this notion is the finding that the isoforms of AChE that appear early in neurogenesis are either nonfunctional or only weakly effective as catalysts (see, for example, Moreno et al., 1998; see Chapter 3 DIII). Then, the phenomenon of "peaking" and subsequent decline, whether prenatal or postnatal, of just about all cholinergic components may also relate to the timing of their trophic and tropic (see section CIII) need for synaptogenesis. Altogether, cholinergic components have to be considered not only as needed for functional synaptogenesis and transmission, but also as participants in the formation and maturation of synapses and junctions (Yew et al., 1999; Lauder and Schambra, 1999; see section CIII).

Yet, even the arrival at the future synaptic site of all the cholinergic components including HACU, VAChT, and other elements that constitute a cholinergic neuron (see Chapter 2 B), and the morphogen action of certain cholinergic components, does not suffice for creating a mature cholinergic synapse. Thus, while early synaptic organization—that is, a formation including most or all of the cholinergic components and some but not all of the synaptic anatomical entities—may be sufficient for a primitive model of transmission (Uylings et al., 1991); the synapses continue maturing and gaining in effectiveness. This involves somatic and dendritic "remodeling": in the rat neocortex, the dendritic branching (the presynaptic site) of the cholinergic cholinococeptive neuron reaches a peak at 3 weeks postnatally and declines thereafter. This "remodeling" also includes the soma of the cholinergic presynaptic neuron, the presynaptic nerve terminals, and the pattern of their juxtaposition with the dendrites. Thus, in the case of the hippocampus, the ingrowth of the septal cholinergic nerve terminals into the dendritic fields is needed for effective transmission (Rimvall et al., 1985). And, as already mentioned, Mechawar, Descarries, and their associates (Mechawar et al., 2002; Mechawar and Descarries, 2001) related synaptogenesis to the arrival at the postsynaptic dendrites of CAT- and ACh-containing varicosities; this phenomenon began in the rat at the eighth postnatal day.

There are also a number of developments at the postsynaptic site, including invagination of plasmalemma, receptor clustering, appropriate localization of a specific AChE isoform at the postsynaptic membrane, and so on (see Whittaker, 1988, 1992). Clustering of the receptors is regulated by several CAMs and SAMs; while these molecules are particularly pertinent to organization and clustering of the cholinergic muscle receptors, they also concern neuronal populations (Daniloft et al., 1989; CAMs and SAMs are discussed in detail in section CII-4); central synaptogenesis may also be regulated by a number of microtubule-associated proteins (MAPs) that are also involved in appropriate organization of the neuronal cytoskeleton (Woolf et al., 1989). Other proteins and peptides—besides the trophics—are needed for the differentiation of the central cholinergic neurons, including bone morphogenetic proteins (BMPs), which have multiple capacities beyond that of bone formation, including cholinergic neuron differentiation (see, for example, Lopez-Coviella et al., 2000; Blank et al., 2004) and cyclin-dependent kinase inhibitors (Albright et al., 2001; see also Chapter 3 CIV). All these events and the interaction between the pre- and postsynaptic elements lead to increase in the effectiveness of spontaneous end evoked release of ACh (Xie and Poo, 1986), as well to the emergence of mature characteristics of the postsynaptic potentials, miniature potentials, and so on.

The classical cholinergic site, the motoneuron junction with the endplate, is an excellent illustration of what a mature junction amounts to; the complex processes that are involved in this maturation include not only such cholinergic components as CAT, AChE, and ACh, but also a manifold array of active proteins, which are the facilitators of receptor clustering and distribution, and the basal postsynaptic lamina, which may be a source of signals that engender these proteins (Rotundo, 1987; Burden, 1998; Sanes et al., 1978, 1998; Sanes and Lichtman, 2001; see above, this section).[5] The current description of the maturation of central cholinergic synapses includes many processes and entities characterizing the maturation of the neuromyal junction.

Cell death and phenomena such as the degradation of "silent" receptors are events shared by all synapses and junctions. In fact, the diminution of a central cholinergic component following its pre- or postnatal peak may relate to cell death (see sections BII-2 and BII-4). Cell death occurs during normal peripheral and central neurogenesis (Hamburger and Keefe, 1944; Hamburger and Levi-Montalcini, 1949; Oppenheim and Haverkamp, 1988; Haverkamp, 1989; Vaca, 1988; Rabizadeh et al., 1994; see also Wyllie et al., 1980; Vaca, 1988; Rabidozeh et al., 1994). This neuronal apoptosis, elimination of some anatomic elements (such as primitive dendrites), and decrease of the concentration or activity of cholinergic components may have several causes. For example, the dendrites and/or presynaptic terminals may need to exhibit parsimony for the synapses to function properly. Competition between the pre- and postsynaptic elements for the optimal synaptic placement, as well as for the trophics that are available, may eliminate some of these elements (see section CI). In addition, metabolic postsynaptic activity contributes to the control of presynaptic growth and development and may downregulate the early development of presynaptic elements (see Vaca, 1988).

6. Comments and Conclusions

Cholinergic elements exist during development, which is long before neurogenesis. The components appear in the gametes and may reach high levels in the ovum and the sperm. Cholinergic entities do not appear simultaneously in prenervous ontogeny or during neurogenesis. Generally, ChEs and CAT appear early and cholinergic receptors late. There are exceptions to this rule, such as in the retina (see section BII-3, above) or the human cortex (see section BII-2, above). However, the data are sporadic, and further studies may be needed to eliminate any exceptions.

Synchrony during neurogenesis among the cholinergic components is nonexistent, and they do not arrive simultaneously at the synaptic sites. In fact, the divergence during development between cholinergic elements such as ChEs and the receptors may be quite marked (Egozi et al., 1986).

Cholinergic entities frequently peak pre- or postnatally during cholinergic neurogenesis and subsequently decrease in density or activity (see sections BII-2 and BII-4). This decrease may result from neuronal death and other features of synaptic maturation (see preceding section).

Section BII stressed the gametal emergence of the cholinergic components, their precocious presence during neurogenesis (i.e., long before the appearance of synapses), and their desynchronous neurogenesis. Given the results of theses findings, it can be argued that ACh, ChE, and other cholinergic components have a nontransmittive role (but perhaps not the entities characterizing ACh release processes such as HACU and VAChT; see also Chapter 2 B and C). Accordingly, they may act as morphogens or trophic factors (see section E, below), and may play other roles (see Chapter 3 DIII).

Altogether, the arrival of one or more cholinergic entities at the future synapse cannot be synonymized with the onset of cholinergic transmission. This notion is a departure from notions espoused in the 1940s by Bacq, Nachmansohn, and Sawyer (see above, section A). Cholinergic entities such as CAT and the receptors may appear before the appearance of movement and, presumably, transmission. At the same time, their appearance may not be synchronized. Egozi et al. (1986) noticed a similar divergence between the development of AChE and that of the muscarinic receptors. Finally, besides the ontogenesis of cholinergic components, the appearance of morphogenetic proteins and peptides such as CAMs, BMPs, and other factors is needed for the differentiation and maturation of cholinergic sites, including central cholinergic neurons (Black et al., 1979). Then there is the matter of the genetic control of ontogenesis and neurogenesis of the cholinergic system. This particular knowledge is barely in its infancy, although already some interesting data have begun to accumulate (Xiao et al., 2004; Liss and Roeper, 2004).

Altogether, several exciting and unexpected conclusions may be reached from the analysis of cholinergic ontogenesis and neurogenesis. Who would expect to encounter cholinergic elements in fertilized or unfertilized ovum, and why should cholinergic components develop quasi-independently of synaptogenesis?

BIII. Postnatal Neurogenesis

It was realized early by David Nachmansohn (1939) that in many animal species AChE activity of the myoneural junction and the CNS persists augmenting postnatally (see also Karczmar, 1963a). Nachmansohn described two heuristic findings. First, he noticed that early postnatally there may be "dips" in AChE activity. This finding should be related to the neurotrophic factor hypothesis (that is, reduction in cholinergic elements at certain neurogenetic periods) developed much later by Rita Levi-Montalcini, Hans Thoenen, and others (see section CI-2, below), although Nachmansohn himself ascribed these diminutions of AChE activity to the dilution of ChE-rich neurons with noncholinergic elements. Second, he observed that CNS AChE of animals that are born relatively immature, such as rats, mice, and rabbits, increases dramatically after birth, while this activity augments minimally in animals that are fairly mature at birth, and he proposed that there is a relation in animal species between the rate of postnatal change in AChE activity and the status of maturity at birth; this important generalization will be returned to subsequently in this section.

Subsequent to these early findings, other central cholinergic elements besides ChEs were studied in rodents, humans, chicks, frogs, and cats, and certain revisions with respect to Nachmansohn's postulate were made.

In all these species, by the first postnatal day all of the cholinergic components exist in the brain parts and the spinal cord (see above, section BII-2). It is of interest that the HACU and VAChT systems are among the cholinergic components that are present at birth at several brain sites, including the olfactory glomeruli (LeJeune et al., 1996); this is relevant for synaptogenesis at this site (see Chapter 2 B2 and the section below).

The postnatal pattern of change depends on species and the cholinergic component. A common pattern is the postnatal continuation of the increase in the activity and/or postnatal concentration of the cholinergic components (see also section BII-2, above). Thus, AChE and BuChE increased postnatally in the rodent forebrain, hippocampus, nucleus basalis, thalamocortex, neocortex, diencephalon and mesencephalon, and olfactory bulb, and in the chick nucleus isthmi (Hohmann and Ebner, 1985; Meisami and Firoozi, 1985; Zhang et al., 1998; Torrao et al., 2000; Hohmann and Berger-Sweeney, 1998; Lassiter et al., 1998; Brimijoin and Hammond, 1996). Then, CAT strongly increases after birth in the rat hippocampus, tegmentum, septal/diagonal band complex, forebrain, nucleus basalis, and cortex (Sofroniev et al., 1987; Large et al., 1986; Brooksbank et al., 1978; Fiedler et al., 1987; Bear et al., 1985; Hohmann and Ebner, 1985; Coyle and Yamamura, 1976; Patel et al., 1978; Sorenson et al., 1989; Ninomiya et al., 2001; Villalobos et al., 2001; Holler et al., 1996; Mechawar and Descarries, 2001). The postnatal CAT pattern of the cat geniculate is similar (Carden et al., 2000), as it is in the mouse retina (in neuroblastic and ganglion cell layer); interestingly, while the prenatal CAT development of the mouse retina does not depend on the presence of light, its postnatal development is retarded by visual deprivation (Zhang et al., 2005). This increase may not embrace all brain parts; thus, some investigators did not find any significant increase of CAT in the rodent hypothalamus and the olfactory bulb (Patel et al., 1978; Rea and Nurnbergar, 1986; Meisami, 1979). It appears that, in parallel with the increase in CAT activity, phospholipids and choline metabolism also increase postnatally (Keller et al., 1987). The same increase holds true in rodents for other cholinergic components, such as the choline uptake system and vesicular transporter (Holler et al., 1996; Aubert et al., 1996; and Mechawar and Descarries, 2001). Finally, muscarinic receptors increased in density in the rodents when measured in the whole brain or the hippocampus, and nicotinic receptors containing the alpha 7 subunit increase postnatally and reached the peak at approximately the fourth postterm week (Zhang et al., 1998; Hohmann et al., 1985; Fiedler et al., 1987, 1990).

The postnatal increase in CAT activity and its pattern relate to synaptogenesis. Mechawar and Descarries (2001) used CAT immunohistochemistry and axonal morphometry to demonstrate the rapid postnatal increase in the area, density, and length of axonal varicosities in the rat parietal somatosensory cortex. They related its peak at the first postnatal month to synaptic cytodifferentiation, as during this time the dendritic spines and branches surrounded the varicosities. This differentiation corresponds to pre- and postsynaptic specialization (see Figure 8-6; see also section BII-4). It must be added that several trophic factors increase in activity postnatally, and this phenomenon has also to do with synaptogenesis (see Mechawar and Descarries, 2001 and section BIII-5).

Another pattern of postnatal change occurs in some species and/or at some brain sites; in this case, cholinergic components peak before or at birth and decline subsequently. For example, in the cerebellar and frontal cortex and in basal ganglia, CAT activity reaches its peak at 18 to 22 weeks and declines throughout gestation as well as postnatally (Brooksbank et al., 1978; Johnston et al., 1985; Gale, 1977; Bull et al., 1970; see also section BIV). Similarly, in the human frontal cortex and the cerebellum, nicotinic receptors peak prenatally and decline thereafter. This is true also for several parts of the rodent brain (see Court et al., 1997; see above, this section): the rodent nicotinic receptors of the forebrain and the hippocampus increase early postnatally and decrease subsequently (Yamada et al., 1986; Slotkin et al., 1987). There is some inconsistency with regard to the human muscarinic receptors: while in the human cortex and forebrain, the muscarinic receptor density reaches adult levels postterm, in the striatum and the forebrain the muscarinic receptors reach their peak at birth and decrease postnatally (there is some controversy as to the pattern of change of the muscarinic receptors in the cortex; compare Ravikumar and Sastry, 1985a, 1985b; Brooksbank et al., 1978).

This postnatal decrease in cholinergic components in at least some parts of the human brain is surprising: as the human infant is born very immature compared to the newborns of any other species, a continuous, postnatal increase in the cholinergic components could have been expected instead; indeed, this increase can be predicted on the basis of Nachmansohn's and Bacq's notion (see above, section A). In fact, in the case of the marsupial opossum, which has, similarly to the human, a long postnatal development (some 70 days in the pouch), the cholinergic system of its retina increases postnatally (see section BII-2, above). It must be noted, however, that in the monkey and in humans, CAT and muscarinic receptors exhibit, at birth, high levels in the cerebellum and several other brain parts, even though their levels may decrease postnatally (see below, this section; Hayashi, 1987; Ravikumar and Sastry, 1985a, 1985b); these high natal levels of cholinergic components are consistent with the need for balance and agility in the infant monkey.

A postnatal increase in the density or activity of cholinergic components followed by a decrease constitute still another postnatal pattern of cholinergic neurogenesis. Thus, using CAT and markers for other components of the cholinergic system, Rene Quirion and his associates (Durand et al., 1999; LeJeune et al., 1996) showed that the cholinergic forebrain radiation arrived in rodents at the olfactory bulb at birth, and cholinergic components peaked during the second postnatal week and decreased subsequently; actually, this decrease continues to the fifth postnatal week. Similarly, the nicotinic receptors of the rodent forebrain increased initially postnatally and then decreased quite significantly through the 25th day of postnatal neurogenesis (Yamada et al., 1986; Slotkin et al., 1987); it should be noted that in rodents muscarinic receptors seem to follow a different postnatal pattern (see above, this section). Also, CAT levels and density of muscarinic receptors are high at birth in the cerebellum and several other brain parts of humans and the monkey, and decrease postnatally (Hayashi, 1987; Ravikumar and Sastry, 1985a, 1985b). Interestingly, AChE and its mRNA of brain regions that are noncholinergic, such as sensory thalamic relay nuclei and anterior dorsal thalamic nucleus, peaked at about 10 postnatal days and then declined (Brimijoin and Hammond, 1996).[6]

Whatever the significance of these often inconsistent findings with regard to Nachmansohn's hypothesis concerning cholinergic maturity at birth and the natal behavioral competence, it is clear that there are various patterns of pre- and postnatal neurogenesis of cholinergic components. Furthermore, the above data indicate that there is a significant desynchrony in the development of various cholinergic components. To adduce additional examples, receptor subtypes and the

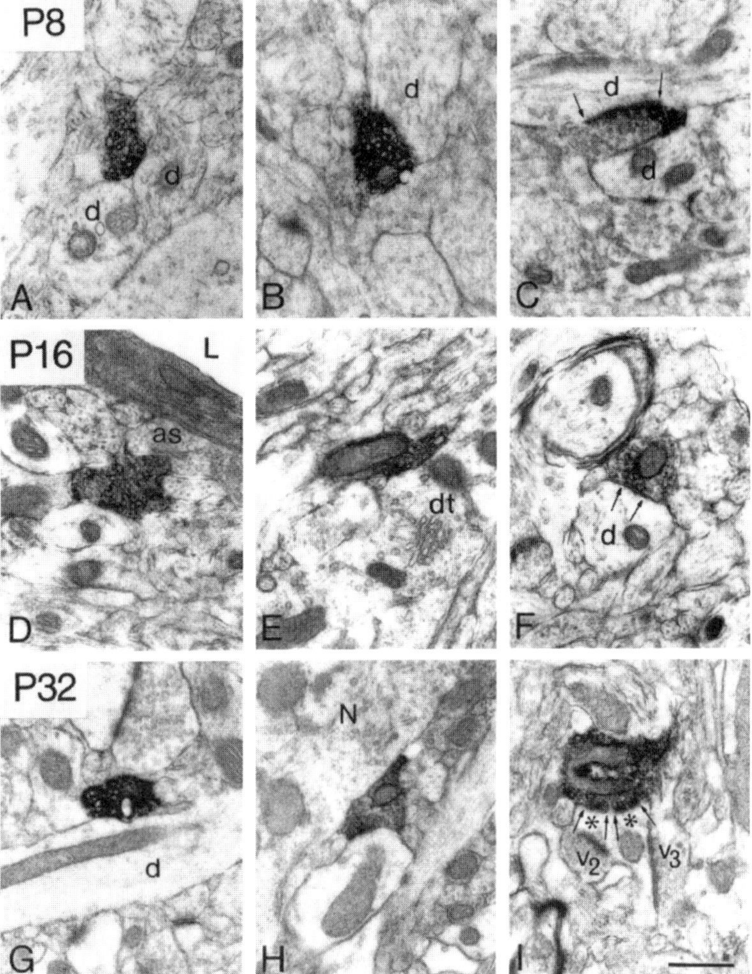

Figure 8-6. A-1. Electron photomicrographs (×30,000) of acetylcholine (ACh) (choline acetyltransferase [ChAT]—immunostained) axon varicosities from the parietal cortex at postnatal ages P8 (A to C), P16 (D to F), and P32 (G to I). The labeled profiles are identified as axon varicosities by their content in aggregated synaptic vesicles. Most also display a mitochondrion (B, E, F to I). Their average size does not significantly differ from P8 to P32. Of the 3 ACh varicosities at P8 (A, B, and C, respectively, from layers VI, II to IV, and V), only the third (C) displays a synaptic junction (between thin arrows), which is also made with a dendritic branch (d). The same is true of the 3 varicosities at P16 (D, E, and F, respectively, from layers IV, VI, and VI); the synapse in F (between thin arrows) is also made with a dendritic branch (d), and it is clearly symmetrical. In the P32 row (G, H, and I, respectively, from layers IV, VI, and VI), the synapse made by the varicosity in I (between the thin arrows) is of the perforated variety, asymmetrical, and made with a dendritic spine (asterisks), which also receives 2 other synaptic varicosities (V_2 and V_3), unlabeled. Note the diverse structural microenvironment of these ACh varicosities at all ages. In A to C, all three varicosities are partly surrounded by dendritic branches (d). The one in D comes close to a capillary (L in lumen), from which it remains separated by an astrocytes (as). In E, the ACh varicosity is juxtaposed to the base of a dendritic trunk (dt). In G and H, direct apposition to dendritic branches (d) may be observed, as well as juxtaposition to a nerve cell body (N) in H. Scale bar = .5 μmin I (applies to A to I). (From Mechawar et al., 2002, with permission.)

two types of ChEs develop desynchronously during postnatal neurogenesis of rats and chicks (see this section, above; Lassiter et al., 1998; Layer, 1983; Zhang et al., 1998). Thus, Layer et al. (1983) found that BuChE activity is higher than that of AChE in the early stages of avian neurogenesis, while AChE predominates during postnatal maturation of cholinergic pathways. Then, AChE increases postnatally in some layers of rats' visual cortex and decreases in others (Robertson et al., 1985).

Another aspect of postnatal neurogenesis is that of cerebral lateralization and sexual dimorphism; this phenomenon seems to occur postnatally in several animal species with regard to cholinergic neurons of the hippocampus (see, for example Kristofikova et al., 2004).

BIV. Comments and Conclusions as to Cholinergic Ontogeny

It is quite clear that several cholinergic markers, AChE (as well as BuChE), ACh, or CAT, and cholinergic receptors identify not only cholinergic ontogeny and synaptogenesis but certain other aspects of ontogeny, such as, for instance, the capacity of the oocytic response to sperm; this and other metabotropic roles may be fulfilled by the prenervous (Buznikov, 1984) or even oocytic presence of the cholinergic components in question. It is a moot question whether their precocious presence indicates their primordial or precursor role with respect to cholinergic synapses, or relates to the metabotropic or trophic (morphogen; see section CIII) role of these entities (see Karczmar, 1946, 1963a, 1963b; Karczmar et al., 1973). In fact, their presence, prenervous, precocious, or otherwise, in other words, at any time during developmental or adult life, may be concerned with metabotropic or trophic action. It should be added that the developmental patterns of AChE in particular may not fit at all the exigencies of neurogenesis; this generalization relates to the role of AChE in aneural organisms, at noncholinergic synapses, or in aneural tissues such as placenta, red blood cells, or plasma, and so on (see Chapter 3 DIII).

It is, of course, obvious that the major role of the components of the cholinergic system must have to do with synaptogenesis of cholinergic junctions and synapses. In fact, Ezio Giacobini and associates hypothesized with regard to the autonomic ganglia of the chick and their synaptic transmission that their maturation and function require parallel neurogenesis of the cholinergic elements (cf. Hruschak et al., 1982). Sometimes, a modified version of Giacobini's notion was presented, as it was proposed that a cholinergic entity may serve as the marker of this synaptogenesis and the ganglia's functionability. For example, Hohmann and Ebner (1985; Hohmann et al., 1985) stated that "there is a striking spatial and temporal correlation between the development of the AChE pattern and synaptogenesis in cortical and subcortical structures of the mouse." Hayashi (1987) stated that in the monkey cerebellum CAT activity is indicative of the "synaptogenesis of cholinergic mossy fibers . . . which is . . . completed during the prenatal period"; yet at this time the postnatal changes in, for example, the cholinergic receptors, are still due, and these changes are quite conspicuous in the case of such related species as man (Ravikumar and Sastry, 1985b). Actually, Hohmann and Ebner (1985), while stressing the importance of the maturation of AChE for the functionability of a synapse, also stressed a lack of parallelism between ontogenesis in AChE and CAT systems and synaptogenesis in the cortical and subcortical structures.

In fact, as discussed in this chapter, there is considerable desynchrony during development between cholinergic entities of the sympathetic and parasympathetic ganglia, the neuromyal junction and the CNS, AChE, and CAT being particularly precocious compared to the cholinergic receptors in the ganglionic development; in fact, all these components of the cholinergic system antedate in their development the appearance of the synaptic ganglionic potentials (Dryer and Chiapinelli, 1985).

The same lack of synchrony between the ontogeny of the cholinergic synapses and transmission, on the one hand, and the development of the cholinergic elements, on the other, occurs at both the peripheral and central sites. This is true in the case of the cardiac and neuromyal junction, as well as in that of the central cholinergic synapses, including the well-studied instances of neurogenesis and synaptogenesis such as in the retina (see above, section BII). In all these cases such cholinergic elements as AChE and CAT appear quite precociously with respect to cholinergic synaptogenesis and transmission. Cholinergic elements concerned with uptake of choline, cycling of vesicles, clustering of postsynaptic receptors, and so forth emerge later, and certain morphologic specializations that characterize mature, fully effective synapses and junctions may appear even later. Thus, the maturation of the neuromyal junction and its organelles is late with respect to the appearance of the cholinergic elements (cf. Salpeter, 1987; section BII-4). Furthermore, it must be emphasized that trophic, tropic, and differentiative factors contribute to this process; these phenomena were alluded to briefly in the sections BI to BIII, above, and will be discussed in detail in sections CI to CIII, below.

In fact, the presence of all the cholinergic elements as well as morphogenetic and neurogenetic specialization do not suffice for the functional maturity of a cholinergic junction or synapse. Indeed, the presence of a particular cholinergic component at a specific developmental stage may mean little without the maturity of this component's structure or nature; while these aspects of the maturation of, say, CAT or cholinergic receptors is in its infancy, yet it is already known that the form of AChE and the structure, subtypes, and distribution of the cholinergic receptors change, stepwise, during development, until they reach their mature form (see Salpeter, 1987; Rotundo, 1987; Dennis, 1981; Massoulié et al., 1993, 1998).

This does not signify that a primitive evoked response, whether to endogenous choline or cholinomimetics, cannot occur earlier, as has been known since the 1960s in the case of the chicken heart (McCarty et al., 1960; see also Karczmar, 1963a). But even this primitive response cannot be readily synchronized with cholinergic ontogenesis; in fact, it may be generalized that both cholinergic transmission and the premature responses of the postsynaptic receptors may not be readily related to specific events in the ontogenesis of the cholinergic elements.

Altogether, the development of cholinergic—or any other—transmission and synaptogenesis involves multifactorial biochemical and morphological processes that include the phenomenon of cell death and the neurotrophic factor hypothesis (see below, section CI), the achievement of the mature state of these processes being protracted well into postnatal development.

BV. Phylogenesis of the Cholinergic System

The question of the phylogenetic appearance of cholinergic components has been mentioned (see Chapter 2 B-1 and section A, above) with respect to the early proposal of Zenon Bacq and David Nachmansohn that in ontogenesis the appearance of the cholinergic system (or its components) coincides with the onset of functions such as motility. It has also been pointed out that cholinergic components such as ACh and/or ChEs are present in many species before the onset of their neurogenesis as well as in monocellular species devoid of innervation and/or motility. In fact, the bacterial presence of ACh was demonstrated by Stephenson and Rowat (1947) very soon after Nachmansohn and Bacq posited their hypothesis (see also Karczmar, 1963a).

Today, the very wide phylogenetic distribution of cholinergic components including ChEs, ACh, CAT, and/or receptors is well recognized. These components are present in all vertebrates including mammals, monotremes (Manger et al., 2002), and fishes, comprising, of course, various species of Torpedo; insects, nematodes, annelids, mollusks including cephalopods, echinoderms, sponges, and other invertebrates; and monocellular organisms including protozoa and bacteria (see Florey, 1967). The Russian investigators were particularly active in the invertebrate studies of ChEs (see Rozengart and Moralev, 2004). It should be stressed that cholinergic components appear in the neurons of innervated species, such as echinoderms and worms, in the flagellar understructures of such forms as Paramecium as known for many decades (see Beyer and Wense, 1936), in nonnervous tissues of forms endowed with nervous systems, and/or in nonnervous species.

These data vindicate the early statement that "it would be simplistic to account for the appearance of ACh system in ontogenesis or phylogenesis . . . by its being . . . the single concomitant of transmittive function" (Karczmar et al., 1973). If anything more is needed to prove this notion, ACh and CAT are present also in many plant species and in fungi (Horiuchi et al., 2003).

The Japanese authors (Horiuchi et al., 2003) state, legitimately, that their findings "suggest that ACh . . . has . . . been expressed in organism from the beginning of life, functioning as local mediator." Of course, we do not know what precisely this "local mediator function" is, although the findings presented in this chapter suggest that the cholinergic system may be active, besides as a transmitter, as a secretor, morphogen and growth regulator, and trophic factor; other functions were proposed for such cholinergic components as ChEs (see Chapter 3 DIII).

Horiuchi and his associates' (Horiuchi et al., 2003) statement that the cholinergic components play a role (but which role?) since the beginning of life is most heuristic. It relates to the findings of several investigators concerning the conserved molecular nature of such cholinergic components as AChE and nicotinic receptors and their encoding in the genes of other species, from Drosophila to human, as stressed by Jean-Pierre Toutant and his associates (Arpagaus et al., 1998; Treinin et al., 2004; see also Karczmar, 1999, 2004). Similarly, Paul Salvaterra and his associates (1993, 2004) described the presence in *Drosophila* of the cholinergic gene locus (with its transcription sites for CAT and other cholinergic markers), which is homologous with the cholinergic locus of the vertebrates, and they stressed that "the genomic organization of the cholinergic locus has been conserved in all animals where it has been studied." This ancient genetic and molecular history of cholinergic components is well documented, notwithstanding evolutionary adaptations, such as the presence of 4 AChE genes in nematodes versus only 1 in mammals.

It was demonstrated with respect to the genetics of developmental structures such as the formation of the anterior-posterior body axis or the lens that the genes in question (Hox and Pax genes, respectively) are homologous across distant phyla (arthropods and vertebrates in the case of Hox genes, and cephalopods, vertebrates, and 4 additional phyla in that of Pax genes); thus, these structures may be linked genetically with their common ancestral genes, participating in the evolution of the recent forms via adaptations and exaptations (adaptation being the primary phylogenetic response to natural selection, while exaptation is the secondary employment of this feature in a way unrelated to its original use; see Gould, 2002). It is exciting to conjecture that, similarly, the common ancestral genes for the molecular and genetic expression of cholinergic components may extend back in time to primitive monocellular organisms—pertinent data are not available at this time.

It may be even more stimulating to speculate that the original cholinergic genome acted as a morphogen that regulates growth and differentiation and which was subjected to selective adaptation as such, while its transmittive function was exaptive in nature.

C. Trophic and Growth Factors

Trophic and growth factors are proteins, peptides, gangliosides, and related agents that are involved in and necessary for neurogenesis and neuronal maintenance; while for several years following the 1940s and 1950s discoveries of Ed Bueker, Victor Hamburger, and Rita Levi-Montalcini, only few such factors were recognized, including the nerve growth factor (NGF), several dozens of these agents that belong to various biochemical categories are studied at present.

The trophics are also capable of regenerative and antidegenerative neuronal actions as well regenerative effects on prenatal axons following brain damage and experimental lesions; it is important that the regenerative neuronal actions occur in adult animals as well as adult humans. These capacities raise hope that these agents may be useful in the treatment of brain injuries and degenerative human disease (see section CI-3e and CI-4, below, and Stein, 1981).

In the context of the capacities of trophics, axonal growth and regeneration in the adult appear to be a special matter; essentially, the mammalian adult axons of the CNS do not regenerate after injury as maintained by Bandtlow (2003), although Carl Cotman, for example, is somewhat less definitive about this matter (Cotman, 1999). A number of myelin inhibitory components are concerned with the regulation of "parsimonious" axonal growth, and the activity of these components underlies inhibition of adult axonal regeneration following injury (see below, sections CIII-3e and CIV-4). It is unclear at this time whether or not the trophics can antagonize these inhibitions and facilitate adult axonal regeneration.

CI. Nerve Growth Factor

1. Discovery and Early Concepts, and Non-NGF Trophic Factors

In 1892 Ramon y Cajal (see Cajal, 1913) speculated on the existence of "neurotrophic substances" to explain his observations on the development and regeneration of nervous tissues. In the 1940s, Ed Bueker, Victor Hamburger, J. Z. Young, and Paul Weiss stressed the dependence of the development and growth or regeneration of axons on their target organs or, more generally, the "remote milieu" (Hamburger, 1952; see also Young, 1951; Hamburger and Keefe, 1944; Weiss and Hiscoe, 1948). Both Hamburger and Young came up with the heuristic idea that a metabolic "message or influence" must be transmitted to the neurites or perikarya to initiate these "trophic" effects on neurogenesis (Hamburger, 1952; Young, 1948; Levi-Montalcini and Hamburger, 1951).[7] Similarly, on the basis of their studies of nerve regeneration, J. Z. Young and Paul Weiss opined that this "message or influence," formed by the periphery (sense organs, muscle), facilitates regeneration via migrating "up the nerve fibers" (the term "retrograde trophic action" would be used today) to the neuron (Young, 1948; Weiss, 1939). Finally, Rita Levi-Montalcini, Victor Hamburger, and Stanley Cohen coined the term "nerve growth-promoting factor" (NGF) for the substance derived from the sarcoma and snake venoms (Cohen et al., 1954) and Rita Levi-Montalcini and Stanley Cohen(1958) offered direct proof and chemical identification of NGF and several other trophics.[8] Convincing evidence of the existence of such a substance or substances was provided in 1948 by Ed Bueker as he noticed that, transplanted into chick embryos, mouse sarcoma 180 caused branching and invasion into the tumor of sensory fibers and enlargement of sensory, spinal ganglia. Victor Hamburger and Rita Levi-Montalcini confirmed and expanded upon Bueker's findings, as the tumor caused growth of the sensory ganglia, autonomic ganglionic hyperplasia, and an outgrowth of the embryonic sciatic nerve into the tumor.

Venero et al. (1996) emphasized that a close relationship exists between trophic activities and cholinergic neurogenesis. This is true for the periphery, as in the case of sympathetic autonomic ganglia, ciliary ganglia, and neuromyal junctions triate, as well as for central cholinergic pathways, as first demonstrated by Thoenen and his associates (Schwab et al., 1979; see also below). In fact, the early history of growth factors relates to ganglionic neurogenesis, as Rita Levi-Montalcini and Hans Thoenen applied their neurotrophic factor hypothesis to the sympathetic autonomic ganglia and ciliary ganglia (see section BII-4). However, there are exceptions to Venero and associates' statement: trophic factors stimulate neuronal regeneration and maintain the survival of mature neurons (see, also below, section CI-2). And NGF (and other neurotrophins) affects nonnervous structures and noncholinergic neurons including those of sensory ganglia.

In their work, Rita Levi-Montalcini and Stanley Cohen used explanted ganglionic tissues to bioassay trophic actions of NGF-containing materials. These *in vitro* experiments clearly established that a chemically identifiable NGF is present in several sources and that it is responsible for the biochemical and neurogenetic effects observed *in vivo*. In addition, extracting, purification procedures, and simple chromatograms yielded large molecular protein aggregates and then complexes containing subunits (Levi-Montalcini and Angeletti, 1968; Greene and Shooter, 1980; Levi-Montalcini, 1953; Cohen and Levi-Montalcini, 1957; and Cohen, 1958). Ultimately, NGF was established as a complex protein with the sedimentation coefficient of the complex 7S; thus, the complex is frequently referred to as 7SNGF (Varon et al., 1967; Greene and Shooter, 1980). Nerve growth factor consists of 3 subunits that may be separated by ion-exchange chromatography and other methods. The beta subunit is biologically active. It is a dimer, and each dimer contains 118 amino acids. The dimer shows considerable homology with insulinlike growth factors (see Figure 8-7; Greene et al., 1971; Angeletti and Bradshaw, 1971; see also Angeletti, 1972; Hendry, 1976).

Originally, NGF was found in various sarcomas, serum, snake venoms, and the mouse iris and submaxillary gland. Snake venoms contain particularly high levels of NGF (Cohen and Levi-Montalcini, 1956; Levi-Montalcini and Angeletti, 1968; Levi-Montalcini and Hamburger, 1951; Angelettti and Bradshaw, 1971; Angeletti et al., 1971; Levi-Montalcini, 1982; see also Hendry, 1976).[9] Obviously, the notion of NGF's physiological role is not necessarily supported by its presence in snake venom or sarcomas. But Ian Hendry, Hans Thoenen, and their associates came up with a solution to the problem. They emphasized the findings that the iris and the submaxillary gland, organs that contain NGF and are innervated by the superior cervical ganglion, exert trophic, neurogenetic effect on the ganglion, and they proposed that NGF that is diffused from the gland or the iris (the target organs) may be transported in the axons, retrograde-wise, to the superior cervical ganglion neurons, where it exerts its trophic action; accordingly, they referred to the iris and the submaxillary gland as target organs.

Ian Hendry, Hans Thoenen, and their associates provided evidence that strongly supported this notion: they blocked the axonal transport, cauding the degeneration of the superior cervical ganglion. Then, using NGF radioactive tracers, they demonstrated that the NGF receptors (see below, this section) located at the nerve terminals and possibly at the perikaryon of the recipient cell "internalize" NGF into endosomal NGF vesicles via the process of endocytosis. The NGF-receptor-vesicle complex moves in retrograde fashion into the soma, where NGF is dissociated from the complex as the vesicles become acidified by a proton pump. "The actual flow of NGF occurs in membrane-bound vesicles along the micro-

Figure 8-7. Viktor Hamburger and Rita Levi-Montalcini in 1977. (From Levi-Montalcini, 1982, with permission.)

tubules . . . at a rate similar to the fast axonal flow of other molecules" (Misko et al., 1987; see also Vale and Shooter, 1984; Hendry, 1976). Accordingly, Hendry and his associates referred to NGF and related trophins as "retrophins." NGF is a retrograde-acting neurotrophin, as it originates at or in target tissues, and as its action is mediated via the growth of the nerve terminals into their target tissues and via NGF's retrograde movement into the neuron. Landreth, 1999 (see his Figure 19-1) distinguishes between target-derived trophins, paracrines and autocrines. He also refers to paracrines as trophic factors that are not derived from postsynaptic targets, but from glia and other elements. Other investigators refer to paracrines as anterograde-acting growth and morphogenetic factors exerting their effects in a peripheral fashion (see Hendry et al., 1974a, 1974b, 1983; Levi-Montalcini, 1982; Hendry, 1976; Thoenen, 1972, 1995; Davies et al., 1987; see also Landreth, 1999, 2006; Schober and Unsicker, 2001).

Demonstrating the NGF presence in the CNS constitutes an important link in providing evidence for the central role of NGF. In 1979, Greene and Sutter et al. and Benowitz demonstrated NGF's presence in cultures of gold fish brain cells, and chick embryo forebrain and tectal neuron cultures (Benowitz and Greene, 1979; Sutter et al., 1979). Also, in rodents, Hans Thoenen established the presence of NGF in the septo-hippocampal, nucleus basalis-neocortical, diagonal band-septal, and intrastriatal cholinergic pathways or systems (Thoenen et al., 1987a, 1987b; Whittemore and Seiger, 1987; Hefti and Will, 1987). Subsequent work established that as in the case of the autonomic system, NGF is absorbed, via NGF receptors, by the nerve terminals of the pertinent central neurons and then transported, retrograde-fashion, to the neuron's perikarya, where it exerts trophic actions (Schwab et al., 1979).

Sites for retrograde CNS transport of NGF are usually consistent with the central presence or expression of NGF, and the projecting axons exhibit NGF receptors (see below section 2). The final proof of the retrograde transport of NGF in the CNS via the projecting axons was provided by using radiotracer and other techniques (see Figure 8-8; Kawaja and Gage, 1991; Seiler and Schwab, 1984; Holtzman et al., 1992; see also Lucidi-Pillipi and Gage, 1993; Cuello, 1993). Hans Thoenen, Franz Hefti, and their associates showed that NGF and its receptors share, in rodents, forebrain sites with cholinergic innervation, while noncholinergic cortical projections and major dopaminergic or adrenergic sites do not exhibit NGF transport (Whittemore and Seiger, 1987; Ebbot and Hendry, 1978). Nerve growth factor-containing target tissues for the basal forebrain system, the septum, and nucleus basalis are the hippocampus, diagonal band, and neocortex. In addition, Thoenen and others also showed that NGF exists in glia and other nonneuronal cell lines (Varon et al., 1974; Thoenen and Barde, 1980).

2. The Early Story of Other Retrophins and Trophic Substances

Thoenen's team (Barde et al., 1982) identified several new, nonNGF retrophins, including the brain-derived neurotrophic factor (BDNF). Soon thereafter it became apparent that NGF and BDNF are just members of an array of neurotrophins and retrophins, as a number of extracts of various tissues including glia and astrocytes exhibited neurotrophic actions but did not contain NGF (cf. Meyer et al., 1979). Today, some 50 retrophins are known including muscle derived trophins or retrophins (see, for example, Henderson, 1988) (see Landreth, 1999, 2006). Most of them are peptides or polypeptides (McManaman et al., 1988); they act upon many central cholinergic pathways, surpassing NGF, which affects only a few central cholinergic pathways. But nonNGF, nonpeptide retrophins were also identified. These substances classically described as involved in inflammatory and immune processes are cytokines (referred to also as neurokines), including the ciliary neurotrophic factor (CNTF) and glycoprotein interleukins, including interleukin 1b.

The protein retrophins, neurotrophins, cytokines and interleukins are not the only trophic substances that exist. During the 1940s it was speculated that neurons and/or their axons may nonspecifically (not dependent on their transmitter type) exert anterograde trophic actions on nonnervous cells that regulate development or regeneration of nonnervous tissues (Schotté and Butler, 1941; Karczmar, 1946). What is particularly exciting in this book's context is that these anterograde

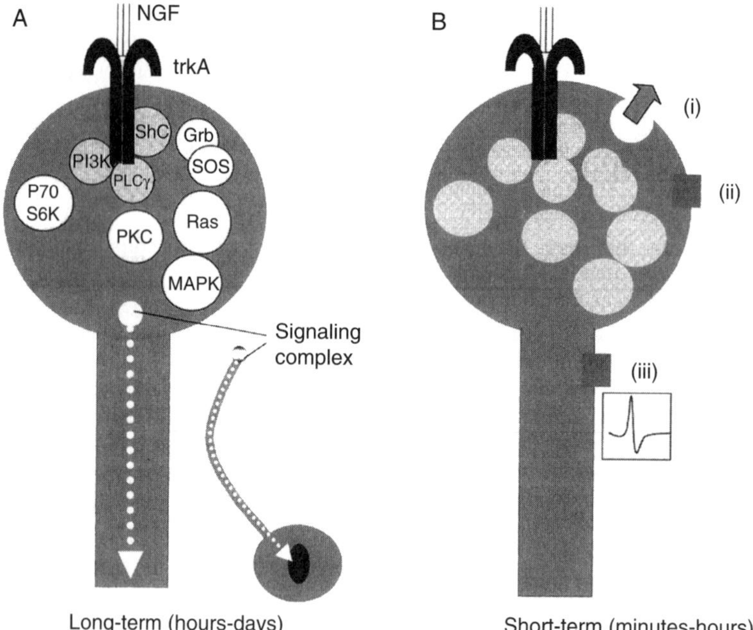

Figure 8-8. Nerve growth factor (NGF) signaling at the synapse. (A) At the nerve terminal, NGF binds to the trkA receptor. Activation of the receptor complex stimulates docking of a number of proteins; shown here are phosphatidyl inositol 3 kinase (PI3K), ShC, and phospholipase C γ (PLCγ). ShC activates the SOS/Grb complex to activate the small GTPase, Ras, and MAT kinase (MAPK). Other protein kinases, including protein kinase C (PKC) and P70-S6 kinase (P70S6K), are activated. A signaling complex is formed that is retrogradely transported along the axon to the cell body, where it can cause the phosphorylation of transcription factors and cause changes on gene expression (Friedman et al., 1999). (B) Current evidence suggests that NGF can cause short-term regulation of neurons independent of retrograde transport of the signaling complex and gene expression. Binding of NGF to Trk causes enhanced neurotransmitter release (i) and may activate various receptors (ii) and ion channels (iii) to enhance neuronal electrical activity. (From Rattray, 2001, with permission.)

nerve-derived trophic signals may include cholinergic components such as ACh or CAT and the presynaptic AChE (or postsynaptic ChEs); they are not to be confused with anterotrophic compounds (see section C-3). Trophic and morphogenetic actions of ACh were shown in the nineteen 1960s for nonvertebrates and vertebrates alike, in both the periphery and the CNS, and George Koelle (1987; personal communication) proposed a similar role for ChEs (Sellin and McArdle, 1977; Drachman, 1967; Anderson and Key, 1999; see also Buznikov, 1973 and Karczmar, 1963b). These matters are also discussed in sections A, BII, and CIII.

Then, molecules such as CAMs and SAMs may also belong to this category of trophins. While CAMs and SAMs are active at the postsynapse as differentiative and morphogenetical molecules (see section BII), they also exert neurotrophic effects. Finally, there are anterotrophins, which are referred to by Landreth (1999) as paracrines, that is, growth factors that are released by one neuron and act on another neuron or on a nonnervous tissue (such as glia; see Figure 8-9). Paracrines are discussed in section C-3a.

Finally, additional nonpeptides, nonprotein substances such as gangliosides and glycylglutamates (Koelle et al., 1987), are endowed with trophic capacities (see section C-4). In addition, there are trophic or quasi-trophic effects that depend on synaptic function. These effects provide

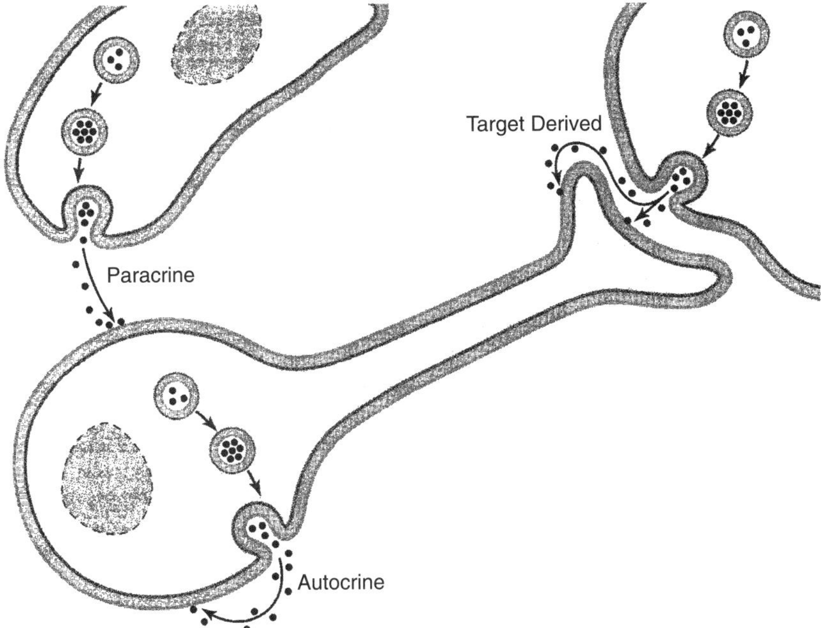

Figure 8-9. Mechanisms of growth and trophic factor support. Growth factors can be provided through autocrine mechanisms in which a cell secretes growth factors to which it in turn responds, while paracrine support is mediated by secretion of factors, which then act upon neighboring cells. Target-derived support of neurons is mediated by the growth of fibers into their target tissues. The target tissue synthesizes the growth factor and provides it to the innervating neurons. (From Landreth, 1999, with permission.)

"trans-synaptic regulation," a term coined by Julius Axelrod, Ian Hendry, and Eriminio Costa (cf. Axelrod, 1971; Costa and Guidotti, 1973; Hendry, 1973; see section CIII-2).

3. Characteristics and Functions of NGF and NGF Receptors

a. Nerve Growth Factor Actions on the Sympathetic Ganglia and Sensory Neurons

The original investigations of Hamburger, Levi-Montalcini, Thoenen, Cohen, and others concerned the ganglionic effects of NGF; their work is being pursued vigorously. While this research is out of this book's context, its salient findings are pertinent for the central trophic phenomena and are described briefly.

The diffusible—later identified as proteinaceous—NGF, which was found originally in mouse sarcomas 180 and 37 (Bueker, 1948; Levi-Montalcini, 1953) and subsequently in other materials, produces, when applied exogenously, hypertrophy of the embryonic chick sympathetic paravertebral and cranial prevertebral ganglia, as well of sensory ganglia (spinal ganglia; cf., for example, Thoenen et al., 1971; Harper and Thoenen, 1981). As expressed by Levi-Montalcini (1952; see also Levi-Montalcini and Hamburger, 1951), the sensory ganglia provide, via their hyperplasia, fibers for the invasion of the tumor tissue, while "another source of nerve supply to the tumor . . . is . . . the sympathetic ganglia; their contribution of nerves to the tumor seems even larger than that of the sensory ganglia . . . as . . . the sympathetic ganglia adjacent to the tumor underwent a hyperplasia which reached a maximum of 600 per cent." The basic effects included the embryonic hyperplasia of the ganglia, a chemotaxic influence (the neurotropic effect) on the outgrowth of the sympathetic nerve fibers, and the increase in the cell number and size. These *in vivo* observations were confirmed by several investigators and extended to newborn mice and kittens (Zaimis, 1971; Angeletti, 1972; for further references, cf. Chun and Patterson, 1977; Harper and Thoenen, 1981; Hendry, 1976; Hendry, 1977; Thoenen and Barde, 1980; Greene and Shooter, 1980; Davies et al., 1987; Black, 1978). The effectiveness of the NGF decreases rapidly postnatally as established by *in vitro* assays of NGF activity on ganglionic cultures; the question of decreased effectiveness of NGF in adulthood is more difficult to resolve (cf. Harper and Thoenen, 1981). This hyperplasia could be induced both *in vivo* and in cultures.

NGF may not be similarly effective with respect to all the ganglia, as the terminal ganglia (the "short" adrenergic neurons) appeared less sensitive to NGF than the paravertebral and prevertebral ganglia (Harper and Thoenen, 1981; see, however, Goedert et al., 1978; for the effects of the antiNGF antibodies on the terminal ganglia, see below, this section).

This morphological hyperplasia is accompanied by effects on the ganglionic metabolism, including energy metabolism (see below, this section, and section CIII-2) and metabolism concerning ganglionic transmitter systems (Bueker, 1948; Levi-Montalcini, 1953; Larrabee, 1969; Levi-Montalcini et al., 1970; Thoenen et al., 1971; Harper and Thoenen, 1981; Zaimis, 1971; Angeletti, 1972; Hendry, 1976; Patterson and Chun, 1977; Otten et al., 1978; Greene and Shooter, 1980; Davies et al., 1987; Black, 1978); for example, NGF has the capacity to promote the adrenergic phenotype (Brodski et al., 2002). This also occurs with other neurotrophins (see, for example, the dopaminergic actions of fibroblast and transforming growth factors, section CII-3, below). While these findings seem to deal on their face with an interesting pharmacological or exogenous effect, it seemed to Levi-Montalcini and Hamburger that these results closely relate to Hamburger's earlier concept of "the important role played by peripheral structures in promoting the growth and differentiation of the nerve cells that innervate these tissues" (Levi-Montalcini, 1982). These peripheral structures may be considered as target organs for the cells in question. It should be added that NGF induces certain paradoxical hyperplasias in the CNS; this point is discussed in section CI-4.

It is particularly pertinent for this book that NGF dramatically increases, *in vivo* and in culture, CAT activity and ACh levels of the ganglia. This increase corresponds to the sympathetic ganglia's progress in synaptization (Thoenen et al., 1972; Schaefer et al., 1979; Thoenen and Barde, 1980), and the effect is caused by NGF's augmentation of protein synthesis evoked by second messenger actions of the NGF receptors (see section CI-3e). However, this effect may also result from the trophic, transsynaptic action that follows am NGF-induced increase in synaptic cholinergic activity of the developing sympathetic ganglion and/or trophic action of the cholinergic components, ACh and AChE (see section CIII-2). Hendry (1976) commented that "cholinergic innervation is not required for . . . the action of NGF," and that "preganglionic nerve activity is completely independent of NGF activity." The critique of these statements must wait for the perusal of section CIII-2.

As stated before, the presence of NGF and its relatives in various types of nervous tissue, including the CNS, does not demonstrate their physiological significance. Even their effects when applied exogenously, including their hyperplastic action and their action on neuronal metabolism and the synthesis of CAT and ACh, may have a pharmacological rather than physiological meaning. Several lines of evidence resolve this problem. First, Ian Hendry, Hans Thoenen, and their associates showed the retrophic nature of NGF or NGF-like substances and demonstrated that the block of axonal transport of NGF caused a degeneration of the superior cervical ganglion (see above, section CI-1). Thus, Hans Thoenen, Rita Levi-Montalcini, and their associates showed that tissues innervated sympathetically by the sympathetic autonomic ganglia (the target organs: iris or vascular tissues) contained NGF and exhibited mRNA encoding for NGF. Then they demonstrated that in the mouse sensory trigeminal ganglion and its cutaneous target field, NGF emerged in target tissues at developmental day 11, that is, when "the earliest nerve fibers contact the target-field epithelium" (coincidentally with the appearance of NGF receptors at the axons; see below); the concentration of NGF increased by 100 percent on the 14th day of gestation (Heumann et al., 1984; Shelton and Reichardt, 1984; Davies et al., 1987). Altogether, at the periphery, the ontogenetic appearance of NGF mRNA, NGF, and NGF receptors serves as the markers for synaptogenesis of the ganglia and the neuromyal junctions (see Lichtman and Taghert, 1987); for example, the peripheral targets for developing axons contain NGF (Markus et al., 2002; Hirata et al., 2002). All this evidence supports the notion of the physiological significance of NGF.

Another line of evidence deals with the existence of specialized NGF receptors: such existence is a condition sine qua non of the physiological role of NGF. The first reports demonstrating the presence of NGF binding sites or receptors in the ganglia appeared almost simultaneously in the 1973 volume of "Proceedings of the National Academy of Sciences" (Banerjee et al., 1973; Frazier et al., 1973). Later, Rita Montalcini's and Hans Thoenen's teams and other investigators demonstrated that the trophic effects of NGF at the perikaryon require these receptors, as an injection of exogenous NGF into the cytoplasm or nucleus does not generate trophic actions (Heumann et al., 1981; Heumann, 1987; Angeletti et al., 1973; Banerjee et al., 1973; Hendry, 1976; Sutter et al., 1979; see Levi-Montalcini, 1982). Then, there seems to be a tight correlation between the developmental arrival of NGF receptors and the response of the pertinent neurons to both NGF and the antiNGF serum (see below, and see also Greene and Shooter, 1980; Herrup and Shooter, 1975; Sutter et al., 1979; Thoenen and Barde, 1980).

An important discovery was made in the 1980s by E. M. Shooter and his associates, as they showed that there are two classes of NGF receptors: high-affinity, low-capacity and low-affinity, high-capacity receptors (Vale and Shooter, 1982, 1984; Greene and Shooter, 1980; Layer and Shooter, 1983; see also Misko et al., 1987). As shown subsequently, these two types of NGF receptors differ in their distribution and in their activity (see section CI-3c and 3d, below).

Receptors activated by NGF generate processes that underlie NGF's trophic action. Early 1980s investigations concerning this matter suggested that these receptors activate cyclic nucleotides and phosphatidyl cascade (similar to the activation of muscarinic and nicotinic receptors); these and related phenomena include the release of Ca^{2+} (Heumann, 1987; Thoenen and Barde, 1980; Thoenen et al., 1987a, 1987b; Greene and Shooter, 1980; Wakelam, 1986; and Hagag et al., 1986). Also, NGF and its receptors increase the synthesis of microtubulin and other proteins (Thoenen and Barde, 1980). Further advances in this area are discussed in section CI-3b (see Figure 8-10).

Stanley Cohen must be credited with an important early finding that added to the credibility of the neurotrophic action of an endogenous NGF. He purified the NGF of the submaxillary gland and snake venoms and used it to develop a specific NGF antiserum (Cohen et al., 1954; Cohen, 1960). That this antiserum is capable of causing immunosympathectomy was rapidly established, first in neonate and adult rodents, and then during development (cf. Harper and Thoenen, 1981); in fact, when the antiserum is injected into the embryo, the destruction of the chromaffin and ganglion cells is more intense than that achieved in neonates or young adults—actually, the destructive effects of the antibody decrease rapidly postnatally, which is consistent with the rapid postnatal decrease in the effectiveness of the NGF.

Furthermore, concomitantly with its morphological effects, the antiNGF serum also causes biochemical sympathectomy, as the activity of the adrenergic enzymes and the levels of adrenergic substances such as norepinephrine are decreased by the antiserum in late embryos and in neonates (for references, cf. Harper and Thoenen, 1981; Hendry, 1976; Black, 1978). It also prevented postnatal development of the ganglionic CAT (Black et al., 1972), which is consistent with the action of NGF on the cholinergic preganglionic neurons. These effects of antiNGF serum were obtained when the serum was administered during chick and rodent late prenatal or early postnatal development; antiNGF serum was ineffective in the adult (Harper and Thoenen, 1981; Hendry, 1976; Black et al., 1977).

As in the case of NGF, the NGF antiserum appears to be less effective with respect to certain prevertebral and terminal ganglia ("short" adrenergic neurons). It was suggested by Harper and Thoenen (1981) that this may be due to the close apposition of the neurons in question with their target organs.

It is doubtful whether the effects of the antiNGF serum may be antagonized with NGF. The narrow developmental time "window" of susceptibility to the antiNGF serum and the presence of the antiNGF serum in the experimental paradigm

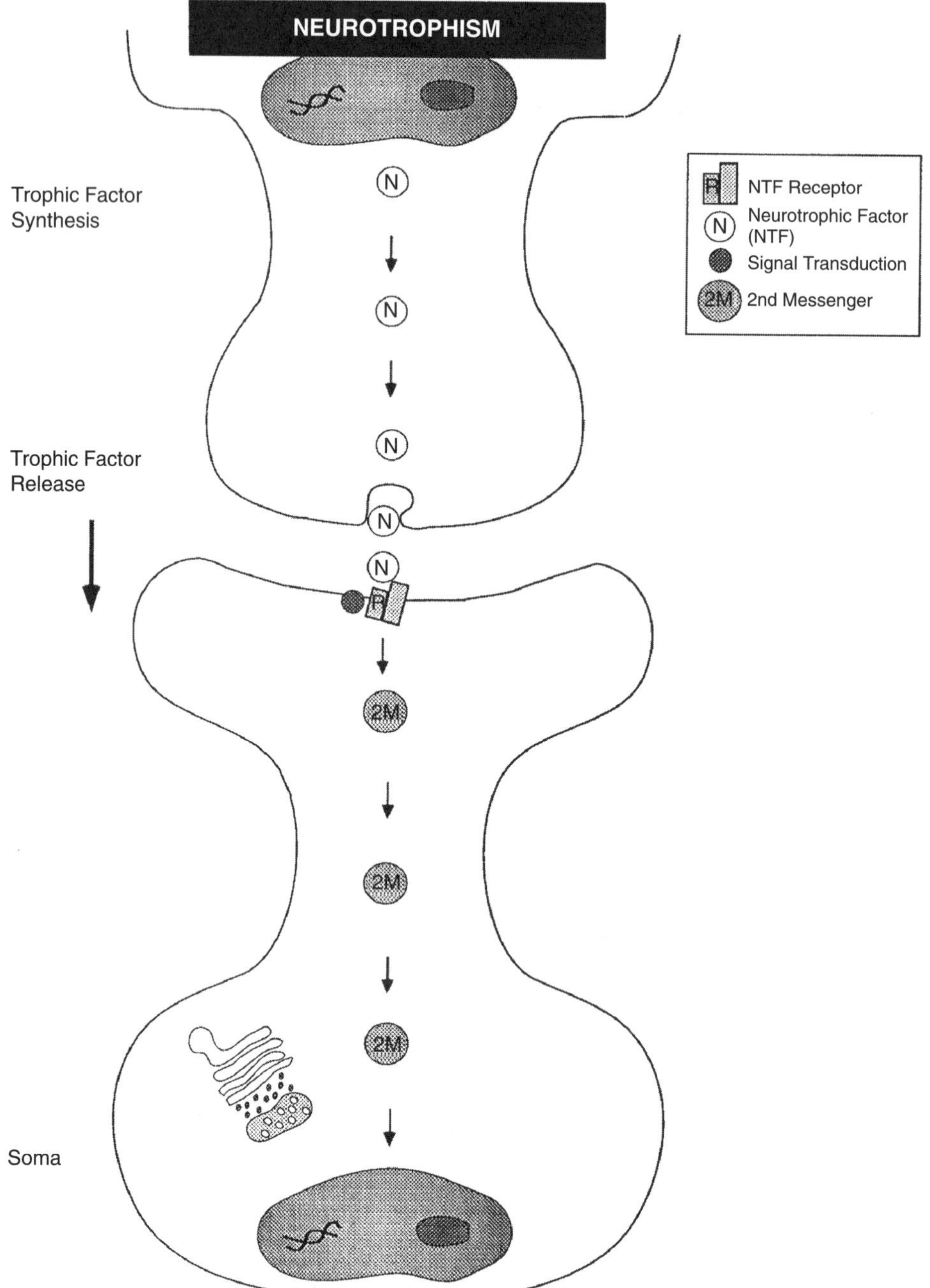

Figure 8-10. Nerve growth factor (representative of other neurotrophic factors) is produced in the target tissue. Following release, NGF is sequestered by the presynaptic neuron and retrogradely transported to the cell body, where its effects are mediated. Signal transduction is thought to be mediated by a receptor complex involving both the low- and high-affinity NGF receptor. (From Lucidi-Phillipi and Gage, 1993, with permission.)

make the exploration of the antagonism between NGF and antiNGF serum difficult.

The immunosympathectomy achieved with antiNGF serum resembles the chemical sympathectomy obtained in the neonates with 6-hydroxydopamine (6-OHDA; Angeletti and Levi-Montalcini, 1970). In fact, the effects of 6-OHDA may be antagonized by NGF (Levi-Montalcini et al., 1975), and it was suggested that 6-OHDA acts via antagonizing endogenous NGF (Black, 1978).When axotomy or ganglionic blockers caused ganglionic hypoplasia in neonate mice, NGF restored ganglionic size and function (Hendry, 1973; cf. also Hendry, 1976; Thoenen, 1975). These experiments served as models for later experiments in which NGF antagonized the degenerative effects of central lesions (see section CI-3b, below; see also Cuello, 1993).

All in all, the results obtained with the NGF antiserum strongly support the concept that NGF exists endogenously, that it acts via specific receptors, and that it is necessary for the development of sympathetic ganglion neurons. An interesting, clinching piece of evidence is as follows: NGF causes, in neonates, hypertrophy of adrenal chromaffin cells and an increase in their biochemical adrenergic markers, yet the antiserum exerts no effect on the chromaffin cells (for references, see Harper and Thoenen, 1981); thus, it is possible to differentiate between the "pharmacological" and physiological effects of NGF.

Rita Levi-Montalcini, Hans Thoenen, Ian Hendry, and others explained the developmental role of NGF in terms of the neurotrophic factor hypothesis. The original thought was that NGF trophic effects were caused by stimulation of neuronal mitosis (Levi-Montalcini and Angeletti, 1968; actually, this stimulation may also occur with respect to the glia). It was then shown that NGF-induced hyperplasia results from the augmented survival (when NGF is present) of neurons that normally degenerate during neuronal differentiation and synaptogenesis. During synaptogenesis, the neurons and nerve terminals exist in excessive numbers; they compete for neurotrophins, and the neurons that fail to obtain the neurotrophins die out (see Henderson, 1988, 1986). Altogether, NGF and other trophins regulate the processes of "programmed cell death" (see sections BII-5).

The neurotrophic factor hypothesis for the developmental role of NGF does not exclude the other actions of NGF and other neurotrophins. The older thought of the proliferative or mitogenetic actions of NGF and other trophins still holds in the case of paracrine and autocrine actions that include proliferative effects on the neuroblasts (Lindsay, 1996; Landreth, 1999; see section CI-3a, above). Then, there are their trophic effects on the mature brain ("maintenance action"); their effects on the degenerative consequences of neuronal lesions ("regenerative action"); and their capacity in promoting neuronal differentiation, such as channel formation (Mandel et al., 1988; Skaper et al., 1986). This capacity to promote neuronal differentiation includes inducting cholinergic phenotypes; it appears that these differentiative effects of NGF on cholinergic neurons are mediated by the phosphatidylinositol-3′-kinase (PIK) cascade and thus are related to muscarinic activation processes (Madziar et al., 2005). As NGF and other neurotrophins exert developmental maintenance or regenerative effects on the cholinergic neurons, they also induce the appearance of, or augment, components of the cholinergic system (i.e., CAT, AChE, or HACU; see Figure 8-11).

b. Further Findings Concerning Central NGF and NGF Receptors

Evidence for the presence of NGF and its role in the central nervous system has been accumulating since the late 1960s (see, for example, Lanser et al., 1966; see section CI-1). It appears that NGF and its receptors are ubiquitous in the CNS; the basal forebrain and its projections are the major source of the brain's NGF (see also below, this section). Shelton and Reichardt (1984) were the first to show that NGF-specific mRNA exists in the brain, and Whittemore and Seiger (1987) demonstrated that mRNA exists in specific rodent brain structures. The brain mRNA sites are consistent with the NGF brain locations, for example, in the case of the basal forebrain (Whittemore and Seiger, 1987).

However, there is a lack of information for brainstem tegmental neurons (which are cholinergic), and some brain structures do not seem to respond to NGF; then, NGF levels are low in the spinal cord (apparently the trophic status of the classic Renshaw cell circuit has not been studied; cf. Whittemore and Seiger, 1987; Landreth, 1999). Furthermore, NGF also exists in noncholinergic areas such as the cerebellum (Korsching et al.,

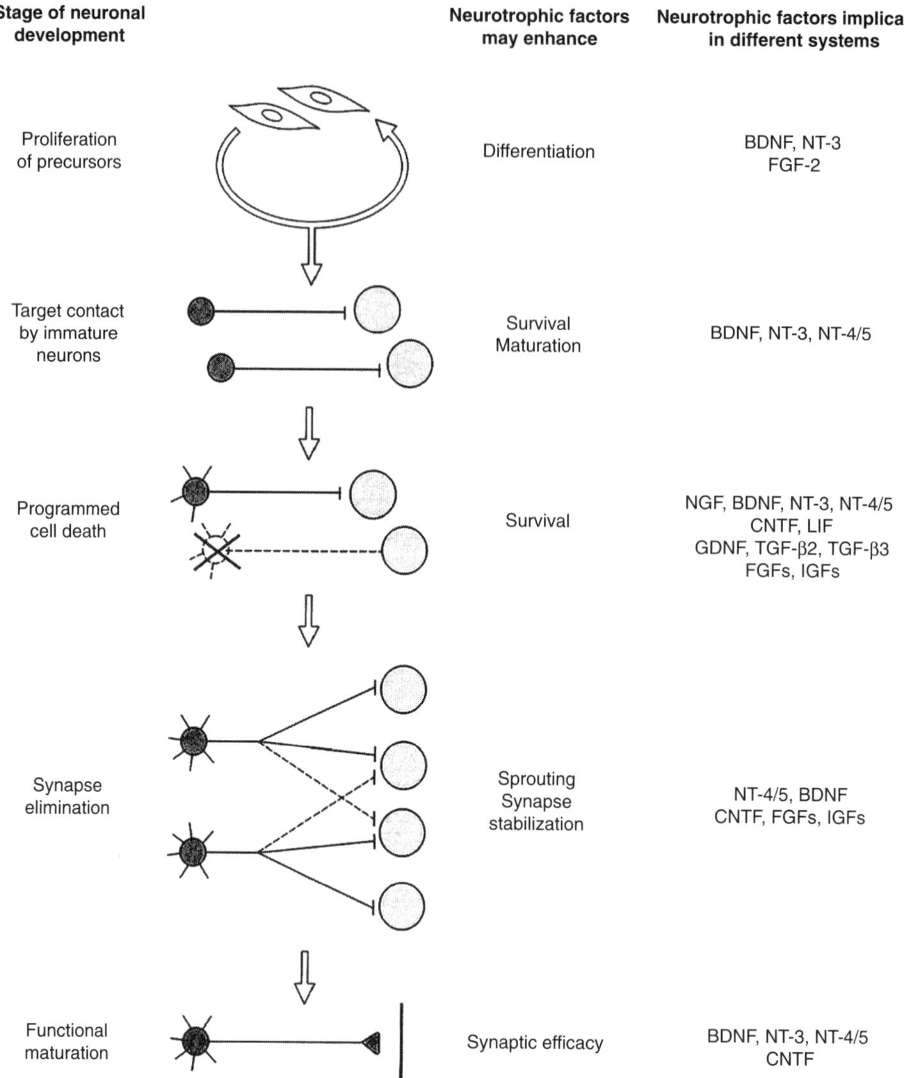

Figure 8-11. Potential roles of "neurotrophic factors" in neuronal development and maturation. Although grouped together because of their ability to promote neuronal survival, neurotrophic factors probably play other roles at different stages of development of the nervous system. The figure depicts some of these stages and indicates factors that have been reported experimentally to enhance or inhibit each step. In many instances, however, evidence for the physiological involvement of these factors is not yet available (see text). (From Henderson, 1996, with permission.)

1985) and it is synthesized by the sensory neurons or organs and the glia (see Landreth, 1999). In fact, evidence concerning the effects of the removal of the target tissues for sensory innervation on the development of sensory neurons suggests that the sensory neurons rely on NGF (and other neurotrophins) via paracrine or autocrine delivery of the latter.

Amino acid sequencing, cloning procedures, ion exchange methods , proteolytic analyses, and transcription evaluations were used with DNA and mRNA analysis to define the structure of NGF complexes present in snake venoms, mouse sarcomas and submaxillary glands, guinea pig prostate, and the like. These methods led investigators to define NGF as a complex of alpha, beta, and gamma subunits. While the stoichiometry of 2 alpha, 1 beta, and 2 gamma subunits along with 2 zinc atoms is quite common with respect to the NGF obtained from various sources (cf. Greene and Shooter, 1980), the stoichiometric composition depends on the NGF in question; also, some amino acid residues differ among the various NGFs (Beck and Perez-Polo, 1982; Werrbach-Perez and Perez-Polo, 1987; Selby et al., 1987; Meier et al., 1986; Meir et al., 1986). Accordingly, the immunological cross-reactivity among NGF complexes may be limited. The active beta subunit of 7SNGF is a dimmer, and each dimer contains 118 amino acids. The molecular weight and the sedimentation coefficient of the dimers are 26 daltons and 2.5 *S*, respectively. The "beta subunit is entirely responsible for the NGF biological effect" (Thoenen and Barde, 1980; Thoenen, 1995; see also Greene et al., 1971). It appears that this effect depends on the dimer's inner fragments, with its two-peptide sequences linked by a sulfide bridge (Mercanti et al., 1977).

The NGF complex is derived from a larger precursor chain, as first suggested by Angeletti and Bradshaw (1971). The NGF gene may be involved in the generation of 7SNGF in target organs (Selby et al., 1987). The actual mRNA-directed synthesis of NGF was mostly studied in adult tissues such as submaxillary glands or mouse sarcomas; this synthesis may be regulated by the arrival of nerve terminals at the target organ as well as by neighboring entities such as macrophages (Thoenen et al., 1988; Heumann, 1987; Greene and Shooter, 1980; Werrbach-Perez and Perez-Polo, 1987; Bradshaw, 1978).

Active NGF is split from its precursor at the precursor's N-terminal. It was proposed that 1 of the 3 subunits of NGF, the gamma subunit (which is an arginine enteropeptidase) may form NGF via its esteratic activity (see, however, Bradshaw, 1983). "The 7SNGF behaves like a typical allosteric protein in which activities of the subunits are . . . regulated . . . by their interactions" (Greene and Shooter, 1980). Once this interaction causes a release of the beta subunit's activity, this activity relies on an inner fragment of the dimer that consists of two peptide sequences linked by a sulfide bridge (Thoenen and Barde, 1980; Greene and Shooter, 1980; Mercanti et al., 1977; Landreth, 1999). The strategic structure of these NGFs and of other neurotrophins is preserved among species (Hefti et al., 1993), as "those regions of NGF [7SNGF] molecules . . . and the molecules of other neurotrophins . . . that interact with the NGF receptor have been conserved during evolution more than other parts of . . . these . . . molecules" (Thoenen and Barde, 1980). A specific amino acid sequence was proposed for NGF (Lars Olson, personal communication; see Figure 8-12).

Even though the loop domains of NGF are known (Massa et al., 2002; Pattarawarapan et al., 2002), a three-dimensional scheme for NGF binding with its two receptors has not been clarified (Bradshaw, 1978, 1983; Massa et al., 2002). Significantly, the templates that are needed for the synthesis of small molecule peptides capable of neurotrophic activity have been developed (see, for example, Pattarawarapan et al., 2002), and this discovery prognosticates that therapeutically usable NGF mimetics will soon be available.

The 7SNGF complex serves as a stable storage unit for the beta subunit and protects the unit from the arginine enteropeptidase actions of the gamma subunit (Thoenen and Barde, 1980; Server and Shooter, 1977). Though the beta subunit is reversibly bound in the complex, it does not exhibit trophic activity until the complex exposes the beta subunit at its neuronal site of action.

Following the discovery of the two types of NGF receptors in the early 1970s, subsequent investigations involved their nature and mode of action. Nerve growth factor receptors appear on the perikarya, on the nerve endings, and on or in the cell nucleus (Andres et al., 1977). Earlier, their presence was established for the embryonic and neonatal chick and rodent sympathetic, parasympathetic and sensory ganglia and central neurons (Thoenen and Barde, 1980; Frazier et al., 1974; Greene and Shooter, 1980; Max et al., 1978); later, they were found to be quite ubiquitous in the CNS. Thus, in the medial septal nucleus, nucleus basalis, and other important sites, the NGF receptors appeared on the 15th or 16th day of rat development, that is, late in prenatal neurogenesis.

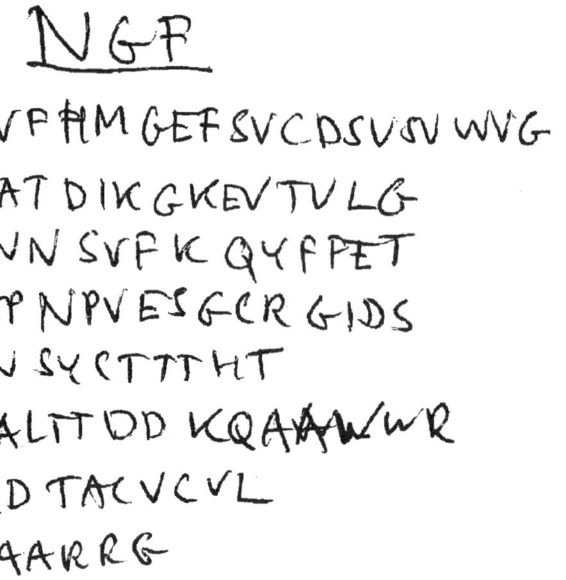

NGF

S STPVFHM GEFSVCDSVSV WVG
DKTTAT DIKGKEVTVLG
EV NINN SVFK QYFFET
KCRAP NPVESGCR GIDS
KHWN SYCTTTHT
FVKALTTDD KQAAWR
FIRID TACVCVL
SRKAARRG

Figure 8-12. In Lars Olson's handwriting, a specific amino acid sequence for NGF. (From the author's private collections.)

Similarly, Claudio Cuello and his team and other investigators obtained similar data for the forebrain, where both CAT and p75 appeared late prenatally and gained adultlike characteristics early postnatally (Cuello, 1993; Armstrong et al., 1987). Furthermore, either these receptors or their mRNA continuously augment during early postnatal stages, and then decrease (Buck et al., 1987, 1988).

Bradshaw and associates (Massague et al., 1981) achieved a major technological feat when they discovered that the two receptors differ in size. The high-affinity receptors have relative molecular masses of 145,000 and the low 85,000 (cf. Misko et al., 1987). It appears that the two receptor classes are interconvertible via initial NGF binding to low-affinity receptors (see Misko et al., 1987). It must be added that central and peripheral receptors of both types are similar or identical (cf. Frazier et al., 1974; Taniuchi et al., 1986).

c. The High-Affinity Receptors

The high-affinity receptors belong to the Trk family of proteins, which are encoded by the Trk gene family (Ultrech et al., 1999). These genes express tyrosine kinase, which is localized in a Trk receptor domain. Nerve growth factor binding to the Trk receptor induces an allosteric change in the receptor configuration that induces phosphorylation of tyrosine residues present in the intracellular regions of the Trk receptors. The phosphorylated regions dock with and activate via phosphorylations, a number of protein kinases including phospatidylinositol kinase (this confirms earlier notions on the NGF action), phospholipase C, and guanosin triphosphatase (Rattray, 2001; Wang and Wu, 2002; Haugh, 2002).

Sol Snyder's team further defined phospholipase C as phospholipase C-gamma-1 and reported that this kinase acts as a guanine nucleotide exchange factor that mediates the activation of the nuclear kinase (phosphatidylinositol-3′ kinase, P13K). According to Snyder's team, this action underlies the mitogenetic cell proliferation effect of NGF, while Jason Haugh ascribes the proliferative action to general Trk messenger pathways including Ras (Braumann et al., 1986; Ye and Snyder, 2004; Haugh, 2002).

These phenomena may also underlie the neuronal survival effect of NGF, or may represent a separate proliferative NGF action. During growth and development, there may be an NGF-like growth factor that is continually released. This could cause prolonged Trk signaling and developmental abnormalities, which are prevented by degrading receptor internalization. The pertinent findings and hypotheses were presented by Haugh, who also proposed that the endosomal Trks continue signaling, and that this signaling is different from signaling that arises from the surface receptors (Haugh, 2002; Haugh's findings also concern the epidermal growth factor, EGF, and other neurotrophins(see section CII-3).

The Trk receptors exhibit extracellular ligand-binding domains that selectively interact with individual neurotrophins; TrkA binds TrkB-BDNF and neurotrophin-4, NGF, and TrkC — neurotrophin-3. While the Trk receptors aremainly expressed in the cholinergic neurons, during neurogenesis the noncholinergic neurons of the hippocampus also exhibit these receptors (Holtzman et al., 1995).

d. The Low-Affinity Receptor

The low-affinity receptor is a 75 to 80 kDa protein and a member of the tumor necrosis receptor family (p75 protein; Bothwell, 1995; Schecterson and Bothwell, 1992; Schober and Unsicker, 2001). The earlier notion was that p75 facilitated Trk action when the two receptors were present at a site (Kaplan et al., 1991; see also Schecterson and Bothwell, 1992). However, it was later shown that p75 acts via the hydrolysis of sphingolipids and the activation of c-Jun amino terminal protein kinase (Friedman and Greene, 1999). The distinct role of p75 may be to interplay with Trks, though when Trks are absent, NGF and other neurotrophins activate p75, thus regulating the cell death program (Lewin, 1996; Carter and Lewin, 1997; Frade and Barde, 1998; see also the previous section and section BII-5). The p75 receptors characterize the cholinergic neurons, although they are present in astrocytes (Dougherty and Milner, 1999) and lesions or ischemia induce their maturation and augment p75 receptor production (Soltys et al., 2003).

Nerve growth factor and the p75 receptors are present and/or expressed in the rodent and chick ganglia, spinal preganglionic neurons, dorsal root ganglia, sensory cranial neurons and motoneurons (and the target tissues of the motor axons), brainstem and its reticular formation, nucleus basalis, diagonal band of Broca, the striate, several cortices, limbic system including the septum and hippocampus, lateral preoptic area, globus pallidum, spinal cord, and basal forebrain. The brain of the species that were infrequently studied such as the monkey also exhibited NGF and p75 receptors in all or most of the elements enumerated here (Riopelle et al., 1987). The Trk receptors are more limited in their expression, as this expression is exhibited almost exclusively in the cholinergic basal forebrain (Lindsay, 1996). Whittemore and Seiger (1987) emphasized that some of these locations for NGF and NGF receptors correspond to the target zones of the cholinergic basal forebrain system (see Figure 8-13).

e. Neurogenesis of NGF, NGF Receptors, and the Neurotrophic Trophic Factor Hypothesis

Data are not consistent with respect to the time of the ontogenetic or the neurogenetic appearance of NGF. The general consensus is that NGF appears late in neurogenesis or, indeed, only postnatally. Thus, prenatally, NGF or it mRNA was either nondetectable or present at low levels in several rodent brain systems including the septohippocampal pathway, nucleus basalis, Sensory ganglion (Amaldi et al., 1971) and cortex, while postnatally, NGF mRNA presence was distinct at these sites (Whittemore and Seiger, 1987; Auburger et al., 1987; Walker et al., 1982; Whittemore and Seiger conceded that a "distinct fetal form of NGF may also exist." Similarly, Auburger et al. (1987) reported that NGF mRNA appears late in neurogenesis, that is, before the septal cholinergic fibers reach the hippocampus; it increases at the onset of the hippocampal innervation, NGF levels becoming high at the time and increasing further postnatally when the retrograde axonal transport becomes active.

Some data indicate that NGF and/or NGF receptors appear in the brain sites and the autonomic ganglia at the critical period of synaptogenesis (15th postnatal day in the case of rodents). Altogether, Mattia-Rossi, Bonhoeffer, and their associates consider the period when NGF and its receptors emerge in the brain to coincide with the period of marked plasticity (Mattia Rossi et al.,

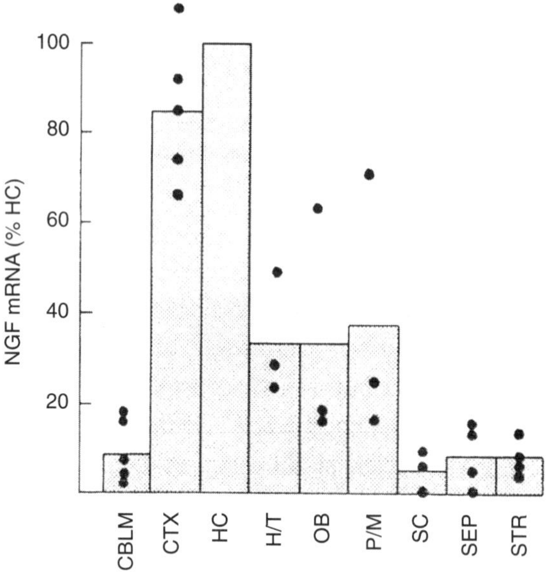

Figure 8-13. Regional distribution of NGF mRNA in the adult rat CNS. CBLM, cerebellum; CTX, cerebral cortex; HC, hippocampus; H/T, hypothalamus/thalamus; OB, olfactory bulb; P/M, pons/medulla; SC, spinal cord; SEP, septum; STR, striatum (see Whittemore and Seiger, 1987 for additional information). (From Whittemore and Seiger, 1987, with permission.)

2002; Bonhoeffer, 1996; see also Ravich and Kreutzberg, 1987). For example, Mattia Rossi et al. (2002) reported that NGF receptors are present in the visual cortex early postnatally, at the "critical period" for synaptogenesis and cortical plasticity; accordingly, at this critical time, NGF can antagonize the degenerative effects on the visual cortex of monocular deprivation.

That NGF and its receptors are present postnatally and, indeed, reach an apex quite late after birth must be stressed. NGF, NGF mRNA, and NGF receptors continue to exist postnatally after the synapses and junctions are formed. This is true for several brain parts of the adult rat and for the basal forebrain of monkeys and humans. In fact, after reaching a peak at the time of synaptogenesis, the NGF levels remain high (Eckenstein, 1988; Hefti and Mash, 1988; Schaterman et al., 1988; see also Hartikka and Hefti, 1988; Hefti et al., 1993). While Franz Hefti and associates (1993) stated that "The function of endogenous hippocampal NGF in the adult remains an enigma," these findings are consistent with the notion that NGF and other retrophins play a maintenance role in the adult CNS.

The neurogenetic arrival of NGF and its receptors is not necessarily synchronized (Johnson et al. 1971; Hendry, 1972; Richardson et al., 1986; Whittemore and Seiger, 1987; Ravich and Kreutzberg, 1987; Mattia Rossi et al., 2002; Lucidi-Phillipi and Gage, 1993; Thoenen, 1995; Bonhoeffer, 1996; Mobley et al., 1985; Segal and Greenberg, 1996; Lewin, 1996; and Lindsay, 1996; see also Reichardt and Mobley, 2004); this is similar to the lack of synchrony in the case of the development and neurogenesis of cholinergic components (see section BII). For example, NGF appears earlier in neurogenesis than the NGF receptors in the sensory dorsal root ganglia of the chick (Lindsay, 1996; Lindsay et al., 1985). Part of their desynchrony may result from the imbalance between the activity of NGF mRNA, formation of NGF, and retrograde transport of NGF (Auburger et al., 1987).

Altogether, while there is no perfect synchrony in the neurogenesis of NGF, NGF mRNA, and NGF receptors, there is a strong time relation—or relatively narrow neurogenetic time "window"—between the arrival of cholinergic input, NGF, the NGF receptors, and synaptogenesis (Davies et al.,

1987). This relationship may also include the interaction between p75 and Trks (see above), as a special time lapse between the arrivals of these two receptor types at the synaptogenetic site may contribute to the death of supernumerary neurons (Frade and Barde, 1998; see also sections BII-5, CI-1 and CI-2). And, finally, this relationship between trophic elements and neurogenesis embraces elements of the cholinergic system such as CAT (see, for example, Auburger et al., 1987). It must be added that just as the desynchrony between the neurogenesis of the cholinergic elements does not challenge the notion of their need for synaptogenesis, the desynchrony in the ontogenesis of NGF receptors, NGF, and NGF mRNA does not contradict the neuronal trophic factor hypothesis: provided these three trophic elements are present at the time of "programmed neuronal death" and synaptogenesis, an appropriate number of neurons will be preserved and synaptogenesis will be completed.

As already mentioned (section CI-3a), the early demonstration of the ganglionectomic effects of the antiNGF serum and the serum's degenerative action on the sensory and ganglionic neurons support the neuronal trophic factor hypothesis; more recent data, some of which concern the NGF receptors, on the whole substantiate the hypothesis further. Yet the data are not always consistent. For example, Lamberto Maffei and his associates showed that antibodies to NGF and to NGF receptors (both of Trk and p75 type) prevent the development of the rodent visual system (see, for example, Berardi et al., 1994; Pizzorusso et al., 1999; and Rossi et al., 2002); on the other hand, ambiguous data were obtained for the effects of antiNGF antibodies on CAT levels in the brain of developing rodents. Similarly, inconsistent data were obtained with the use of transgenic rodents lacking NGF or its receptors: some teams reported that forebrain atrophies occur in the NGF knockout rodents (where NGF activity is prevented by the expression of NGF antibodies), while other teams failed to observe this effect (Chen et al., 1997, Crowley et al., 1994; Fagan et al., 1997; Ruberti et al., 2000). However, careful studies conducted by Franco Ruberti and associates and Emanuele Pesavento, Simona Capsoni, and associates (Ruberti et al., 2000; Pesavento et al., 2002) clearly indicated that in the transgenic mouse model the cortex degenerates and the expression of the cholinergic phenotype in the basal forebrain is severely curtailed. Moreover, Hans Thoenen suggested that the negative effects occur when NGF preparations are used that that are poorly absorbed into the brain (see Thoenen et al., 1987a, 1987b). In addition, the cortical plasticity expressed by long-term potentiation (LTP) and short-term facilitation (STF; see Chapter 9 BIII) is diminished in transgenic mice. Moreover, these deficits are restored by cholinergic treatment; does this suggest that the diminution of plasticity is due to the cholinergic deficit exhibited by the transgenic mice rather than to the cortical degeneration (see Figure 8-14)? Note that Russian investigators reported the presence of autoantibodies to CNS-acting retrophins in "normal" animals and humans (Sherstnev et al., 2002).

The restoration of deficital cholinergic function by cholinergic agonists relates to trophic effects of NGF that focus on the cholinergic components; these effects are referred to as "phenotypic." Hans Thoenen, Franz Hefti, their associates, and other investigators found that in the rodent forebrain, septohippocampal pathway, neostriate, nucleus basalis, and cortex, NGF augments the levels or activities of central cholinergic neurons, AChE, CAT, HACU, VAChT, ACh release systems, and so on. These effects could be obtained *in vivo* and *in vitro*, in adult, prenatal, or neonatal animals (Altar et al., 1992; Thoenen, 1995; Thoenen and Barde, 1980; Hartikka and Hefti, 1988; see also Williams et al., 1993; Cuello, 1996; Ha et al., 1999). For example, applying NGF directly into the lateral ventricle of the adult rat causes an increase in the CAT activity of the ipsilateral cortex (Garofalo and Cuello, 1995), and Claudio Cuello commented that this effect must involve cholinergic nerve terminals radiating from the nucleus basalis magnocellularis. These effects may be readily demonstrated in perinate or neonate animals, though in adults they may be difficult to obtain *in vivo* (Whittemore and Seiger, 1987; Gnahn et al., 1983; Hefti et al., 1984).

Note that NGF does not cause these actions in all brain parts, nor does it affect every cholinergic component in each of the brain parts that was studied. Thus, NGF generally does not affect central adrenergic neurons (Schwab et al., 1979), although this is not true at the periphery (also, sympathetic invasion of the CNS can be induced by NGF; see above). Also, it may be that NGF

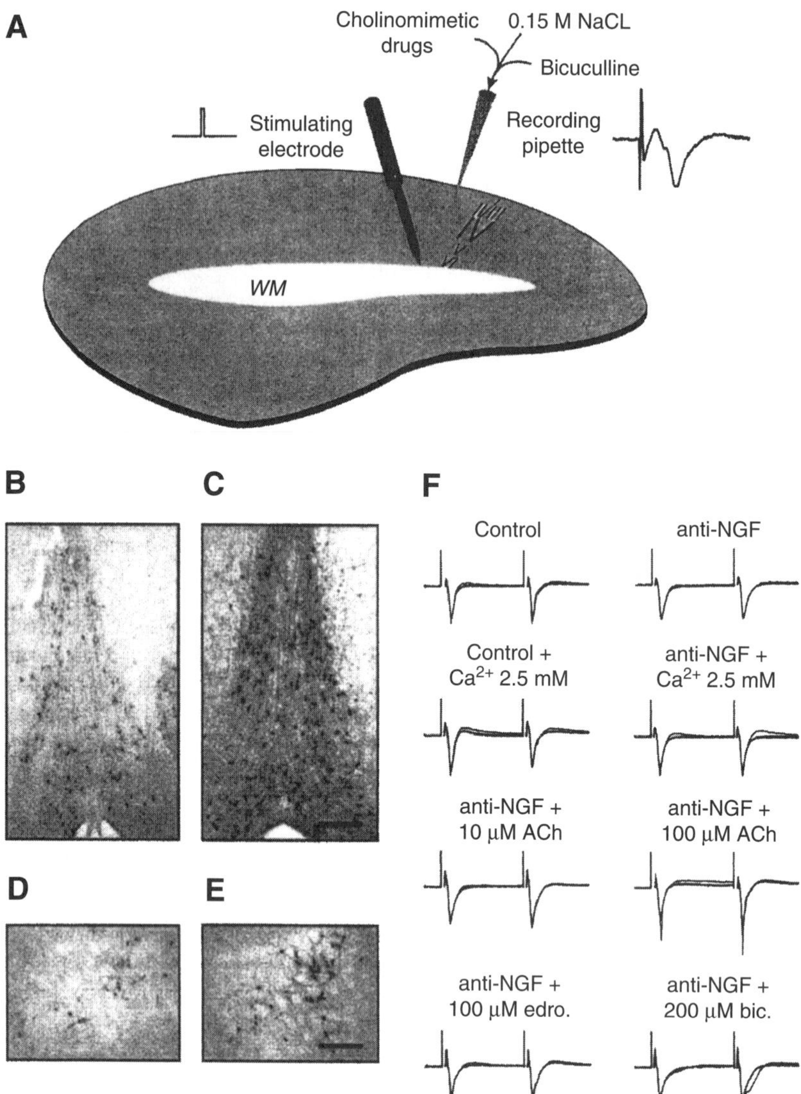

Figure 8-14. (A) Experimental design, a stimulating electrode was placed at WK/layer VI border while the recording pipette was placed in layer II/III. Drugs were locally delivered through the recording pipette. Reduction of ChAT immunostaining in anti-NGF (B) medial septum and diagonal band and (D) nucleus of Meynert, compared to corresponding areas of control mice (C and E, respectively). (F) Averaged FPs. Traces were obtained by averaging the evoked FP in each slice, at the 10th minute of baseline recording (thick line) and at the 60th minute after HFS (thin line). Scale bars, 100 μm (in C for B and C); 70 μm (in E for D and E). (From Pesavento et al., 2002, with permission.)

does not augment the cholinergic components in pedunculopontine cholinergic neurons, and NGF in the brain parts that induces a rise in CAT activity may not increase AChE action (Hartikka and Hefti, 1988; Sofroniew et al., 1989; Knusel and Hefti, 1988; Cotman, 1999). These findings relate to the notion that the developmental actions of NGF (that is, its neurotrophic neurogenetic action) and its augmentation of the levels or activity of the cholinergic components may constitute separate phenomena. However, it is possible that some of these results were obtained when the fate of the cholinergic components was already "determined" at the time of NGF application (Whittemore and Seiger, 1987). Finally, a number of trophic factors—rather than just NGF—may be needed to induce the trophic response of the brain sites that do not respond to NGF morphologically or cholinergically—this relates to the notion of the interaction between trophic factors.

As described above, the preponderance of reported NGF actions is limited to morphogenetic actions on the cholinergic neurons and to the phenotypic effects on cholinergic components, such as CAT, although NGF actions on catecholaminergic and dopaminergic neurons were also noted. Thus, transection of fimbria-fornix and of the septum induced, via action of NGF, an invasion of peripheral sympathetic innervation into the hippocampus (Levi-Montalcini, 1975; however, this action may be due to a NGF-related, nonNGF trophic factor (Loy and Moore, 1977; Whittemore and Seiger, 1987; Weskamp et al., 1986). Also, given intracerebrally to rodents near the locus coeruleus, NGF caused a penetration of large sympathetic nerve bundles into the spinal cord and their growth into the brainstem toward the NGF injection point (Levi-Montalcini, 1975, 1982, 1987).

In vivo, the neurotrophic NGF actions appear late pre- and early postnatally, which confirms the notion that NGF plays a role in regulating neuronal survival during synaptogenesis (in accordance with the neurotrophic factor hypothesis; see Hohmann and Berger-Sweeney, 1998). However, the trophic phenomena also take place in young and mature rodents, as well as in cultures. Franz Hefti, his team, and others demonstrated that NGF induces somatic, dendritic, and axonal growth of adult septal, forebrain, and cortical neurons *in vitro*. Furthermore, Claudio Cuello and others observed similar *in vivo* synaptogenetic effects of NGF on the neuronal fiber network; for example, Segal et al. (2003) demonstrated the hyperplastic effects of NGF on the somata and dendrites of cultured septal neurons obtained from adult rat brains (Hartikka and Hefti, 1988; Martinez et al., 1985; Knusel and Hefti, 1988; Hefti et al., 1984, 1993; Cuello, 1994; Qiao et al., 1998; Ha et al., 1999; Henderson, 1996). These effects suggest that NGF, besides playing an important role within the neurotrophic factor hypothesis, also maintains alive neurons in the adult animal (Hartikka and Hefti, 1988; Thoenen and Barde, 1980; Johnston et al., 1987; Altar et al., 1992; Williams et al., 1993; Oosawa et al., 1999; see also Hohmann and Berger-Sweeney, 1998). Actually, the older the animal, the less pronounced is the augmenting action of NGF on the central cholinergic components and its facilitatory effect on animal learning (Williams et al., 1993).

In the context of the NGF role in the maintenance of adult neurons, the question of the significance of NGF for aging and degenerative disease should be considered: the levels of NGF and/or NGF receptors appear to be high in aged animals that exhibit behavioral and pathological consequences of old age, yet the intracerebroventricular administration of NGF and its subtypes increases the brain levels of CAT and the CAT gene expression in old animals (for example, in the striatum; Castellanos et al., 2003); there is some controversy as to whether these levels decrease, remain the same, or increase in Alzheimer's disease (see Granholm, 2000; Lucidi-Philippi and Gage, 1993). Then, cholinoceptive receptors appear at the nerve terminals of noncholinergic neurons such as glutaminergic cells and/or the postsynaptic sites of noncholinergic neurons, including inhibitory interneurons (see section CI and Chapter 9 BI). The clarification of these matters will have a bearing on Franz Hefti's notion that NGF plays an important role in neuronal maintenance during adulthood and aging.[10]

The maintenance functions of NGF relate to its regenerative action, as amply evidenced in experiments involving cholinergic pathway lesions. Lesions to the fimbria-fornix and septo-hippocampus or decortication of adult rodents resulted in a decrease of NGF levels and degeneration of the cholinergic forebrain neurons (and increase in the septum; see below, this section); the NGF treatment prevented this degeneration and restored the cholinergic components of the forebrain. Similar evidence was obtained in primates (Koliatsos et al., 1991; Hefti, 1986; Cuello, 1996; Jonhagen, 2000; cf. Whittemore and Seiger, 1987, and Hefti and Will, 1987). In addition, Claudio Cuello and associates (Cuello, 1994) found that long-term (7 to 28 days) NGF treatment in forebrain-lesioned rodents restored HACU and CAT contents in the cortex. Nerve growth factor also increased cortical synaptogenesis of the adults, as the CAT-containing fiber network and synaptic differentiation were expanded and the number of presynaptic varicosities was increased. Furthermore, NGF induced the expression of mRNA for both types of NGF receptors. Similarly, NGF reversed degeneration of the visual cortex and prevented the shift in ocular dominance distribution of visual cortical neurons induced by monocular deprivation (see Cuello, 1996). Also, NGF or Trk antibodies prevented normal development of the visual cortex (Maffei et al., 1992; Berardi et al., 1994; and Pizzorusso et al., 1999).

Similar NGF effects were generally obtained when the lesions were induced by NGF or NGF receptor antibodies, or by neurotoxins that are selective for the cholinergic neurons (Rossner et al., 1996). But the results were not always consistent, particularly with respect to antiNGF sera or antibodies. *In vivo* injections of antiNGF serum did not always cause forebrain lesions, contrary to its effectiveness at the spinal and sympathetic ganglia. Similarly, antiNGF antibodies did not always affect CAT *in vivo* (Martinez et al., 1985). The negative results obtained *in vivo* with the antiNGF substances could be due to insufficient penetration of the antiNGF antibodies or antiserum, while in the case of the negative *in vitro* experiments, the antiNGF substances may have encountered a "predetermined" CAT activity (Whittemore and Seiger, 1987). It is of interest and consistent with the data presented here that increase in the levels of NGF and other neurotrophins is induced by lesions (for example, in the hippocampus after lesions of the entorhinal cortex; see Cotman, 1999).

The regenerative NGF actions on the degenerated neurons and/or their cholinergic constituents occur in parallel with behavioral restoration of motor or cognitive functions (Whittemore and Seiger, 1987; Freed, 1976; Cuello, 1993). Interestingly, these NGF actions on the lesioned rodent brain are similar to those of the transplants of the fetal cortex or septum (Gash et al., 1985; Daniloff et al., 1985). However, it was sometimes opined that the restorative effects on behavior of transplants in lesioned animals were not caused by neuronal and cholinergic regeneration, but by increased energy metabolism induced by the graft (Kelly et al., 1985; Gage et al., 1985; Dunnett et al., 1986; Pallage et al., 1986).

It is consistent with the notion of NGF's regenerative capacity that the destruction of several cholinergic brain neurons or pathways increases the level of NGF in the target tissues, as this effect may reflect the brain's attempt at plasticity following a lesion; for example, NGF augmentation occurs in the septum following fimbria-fornix transection. Also, the Trk NGF receptors are upregulated (Figueiredo et al., 1995a, 1995b; Crutcher and Collins, 1986; Gasser et al., 1986; Lorez et al., 1988). However, Lucidi-Philippi and Gage (1993) opined that in some cases the augmenting effect on NGF may result from lesion-induced decreases in retrograde NGF transport. Conversely, lesions of various target areas reduce the levels of NGF and/or CAT in the recipient neurons, and may cause the death of the neurons in question. For example, the removal of the target areas for the basal forebrain leads to diminution of the forebrain neuronal somata and CAT activity; NGF could reverse this effect (Rylett and Williams, 1994; Sofroniew et al., 1987; Cotman, 1999; see also Hohmann and Berger-Sweeney, 1998). Similarly, hippocampal lesions caused decreases in the septal content of NGF and CAT and either atrophy and shrinkage, or the survival of septal cholinergic cells (see also above, this section). Once septal cholinergic cell death occurred due to lesions or other maneuvers that induced deficiency in NGF or NGF receptors, the NGF treatment was useless (Cooper et al., 1996; Van der Zee and Hagg, 2002).[11] These results may have been expected, as destroying target areas and eliminating NGF content eliminate NGF access to the recipient neurons, thus causing their death.

Another aspect of regenerative NGF capacity involves the role of the NGF in adult peripheral axonal reconstruction following axonal or spinal and brain lesions. These facilitatory effects of NGF were noticed early (see Levi-Montalcini, 1982). For example, NGF used with other retrophins, alone or with fetal transplants, improves regeneration of transected dorsal column and corticospinal motor tracts; it also reduces swellings on the ends of severed axons (Sayer et al., 2002; Coumans et al., 2001; see also Mendell et al., 2001); similar regenerative axonal effects as well as expression of appropriate transcription and cell recognition factors were induced by grafting peripheral nerve onto the neostriatum (Chaisuksunt et al., 2003). Note that motor functions also returned after this treatment, and similar effects were noticed with respect to motor axons (Fine et al., 2002). It is of interest that while NGF and other neurotrophins appear to facilitate both the developmental (see above) and regenerative growth of axons, the signal transduction, that is, the induction of second messengers, differs for these two phenomena (Liu and Snider, 2001). Furthermore, a special mechanism is involved in the regenerative axonal NGF effects: while NGF induces apoptosis of most of the available Schwann cells, only a fraction of available Schwann cells participate in the NGF-induced myelination phase of axonal regeneration (Hirata et al., 2002).

Some investigators opine that NGF "promotion . . . of . . . peripheral nerve regeneration *in vivo*, particularly in adults, is controversial" (Hirata et al., 2002). These problems relate to the mulifactorial, complex character of the axonal growth phenomena. These phenomena include facilitatory mechanisms mediated by the Schwann cells (or, in the case of spinal descending axons, glia and the glia-derived neurotrophins) and a number of growth-associated proteins such as GAP43; factors that prevent uncontrolled axonal growth and regeneration such as myelin-associated inhibitors (see also section CI-4, below); the Nogos; proteins regulating the direction of the axonal regrowth, such as Netrins (see also section CI-4, below); and factors controlling myelination and remyelination. Some, if not all, of these factors interplay with NGF and other trophins. In this context, Hirata and associates (2002) found that following the transection of the adult rat sciatic nerve, intrathecal infusion of NGF significantly decreased the number of growing axons, compared to the untreated controls, and delayed the GAP43 induction. These investigators hypothesized that the axotomy causes the loss of the retrograde NGF supply from peripheral targets. This loss serves as a signal to dorsal root ganglia to evoke the production of GAP43 and other regenerative factors via retrograde transport, and this signal is eliminated by NGF administration.

4. Conclusions

The discovery of NGF is one of the most important stories concerning the cholinergic field: it defined the trophic action as a condition sine qua non of development as well as a potential for nerve and neuronal regeneration, and this is not the end of the functions of NGF which is verily a pluripotent agent, as discussed below in this section. The story is rich in drama, including the abstention of Victor Hamburger, at a crucial moment, from this area, a decision that may have cost him a Nobel Prize. And the trinity of Paul Weiss, Victor Hamburger, and Rita Levi-Montalcini, the pioneers in trophisms, is at least equal in its brilliant mode of research to any cholinergic team who were or are engaged in other areas of cholinergicity.

This exciting story of basic aspects of trophisms is currently translated into animal studies dealing with the NGF's restoration of cognition following lesions of the nucleus basalis and related regions (see, for example, Pizzo and Thal, 2004) and clinical research that involves NGF cures for injured spinal cord or brain, severed peripheral axons, and sensory afferents including the optic nerve. These studies were initiated in Scandinavia (Lars Olson and Aneta Norberg, personal communication). It helps in this use of NGF that it can be measured almost noninvasively outside of neuronal tissues, as it may be measured in human muscle biopsies and in body fluids (Kust et al., 2002).

Additional evidence suggests that NGF and other neurotrophins may be useful in pathological conditions, such as Alzheimer's disease, and psychiatric conditions, including depression and schizophrenia (Granholm, 2000; Jonhagen, 2000; Pizzo and Thal, 2004) These notions clearly illustrate that the discovery of neurotrophins and their contribution to the novel acceptance of the concept of brain plasticity have provided the groundwork for a wide range of research and potential clinical interventions (Hohmann and Berger-Sweeney, 1998).

NGF is pluripotent, as it is capable of a number of roles and functions; in fact, Rita

Levi-Montalcini (1987) stated that "the submerged areas of the Nerve Growth Factor iceberg loom very large" (see Figure 8-11).

NGF's first role is in proliferative phenomena. Levi-Montalcini's, Thoenen's, and others' early work made salient the mitotic effect of NGF on neurons, that is, a proliferative NGF action; other neurotrophins exhibit this action prominently (see next section, and Henderson, 1996).

The second and most important function NGF plays is in development and synaptogenesis, whether at the periphery or centrally; NGF regulates development of the central cholinergic pathways involved in cognitive and motor behaviors, the basal forebrain, and the septohippocampal system. As transgenic mice devoid of NGF mature, these pathways undergo degenerative neuronal changes, apoptosis, deficits in synaptogenesis, and functional and cognitive deficits. This evidence strongly supports Hans Thoenen's and Rita Levi-Montalcini's neurotrophic factor hypothesis (Thoenen et al., 1987a, 1987b). This is basically a Darwinian "competition and survival hypothesis," as it posits that at the onset of synaptogenesis, there are more neurons than are needed for the mature synapses (Thoenen et al., 1987a, 1987b). It further suggests that unnecessary neurons are eliminated as they compete for the limited amounts of NGF. In addition, an excessive number of terminals at the beginning of synaptogenesis may also be competing for survival. There may also be a cell death–inducing component that depends on the time relation between the arrivals of the two NGF receptors at the site of synaptogenesis; the p75 receptor is particularly involved in cell death at the time of synaptogenesis (see section CI-3e, above). Andrzej Szutowicz (2001) expands this notion: he hypothesizes that different forms of synaptogenic differentiation may arise depending on the interplay among NGF, its receptors, and other elements such as Nogos (see above, this section). He also posits that some of these forms may exhibit a special sensitivity for neurotoxins including beta amyloids. A decreased density of Trk receptors due to the action of neurotoxins, including beta amyloids, and the resulting hyperactivity of p75 receptors may underlie the neuronal susceptibility and the deficits in ACh levels and energy metabolism (see also Szutowicz et al., 1999; Tonnaer and Dekker, 1994). Furthermore, while antiNGF treatment (i.e., with the antiNGF serum) induces an almost complete ganglionectomy, not all central neurons disappear in the CNS following antiNGF treatment or in transgenic, nonNGF mice. Christopher Henderson (1996) suggests that, contrary to the peripheral neurons, which depend uniquely on NGF for their development, the central neurons require a number of neurotrophins for their normal growth.

Tropic effect constitutes NGF's third function. Nerve growth factor's trophic role in synaptogenesis may not be completely reflected in the neurotrophic factor hypothesis, as NGF may also contribute guidance, during development, to the axons while they extend toward the future synaptic site; this function was termed "tropic" and is distinct from the "trophic" action (Vellis and Carpenter, 1999).

NGF's regenerative capacity for peripheral motor and afferent axons constitutes the fourth NGF role (it should be stressed that contrary to young or prebirth axons, adult brain axons do not readily regenerate (see above, section CI). As already mentioned, different second messengers seem to be involved in development and regeneration, and this finding supports the notion of different natures of developmental and regenerative NGF functions (Lakshmanan, 1978; Liu and Snider, 2001; see above, section CI-3).

NGF's fifth role concerns the maintenance of neuron survival during maturity, as the presence of NGF during maturity and adulthood suggests a need for NGF long after synaptogenesis and the developmental phases of cholinergic neurons are over. Additionally, many investigators have found that this survival may be mediated by NGF's activation of the Na^+ K^+ pump (Sendtner et al,. 1988; see also above, section 3a). Altogether, NGF supports the survival *in vitro* of cholinergic cortical, nucleus basalis, and septal neurons (see, for example, Weis et al., 2001). Perhaps the long-term effect of NGF deprivation on the survival of cholinergic cells in adulthood reflects this particular role of NGF (see above, section CI-3). The maintenance role of NGF relates to its regenerative effects on the pathways and prevention of lesion-induced neuronal degeneration; this effect obtains particularly with respect to the cholinergic pathways; it is noteworthy that these effects can be obtained in adult animals. Besides NGF, other neurotrophins and gangliosides exhibit these regenerative effects (see above, section CI-3e).

The sixth role of NGF involves noncholinergic and nonneuronal effects (see above for the actions of NGF in the regeneration and maintenance of afferent neurons and sympathetic and adrenergic neurons; sections CI-1 to CI-3; see also Harper and Thoenen, 1981; Schecterson and Bothwell, 1992). The presence of high levels of NGF in sarcomas, salivary glands, and venoms of several snake species reflects the notion of noncholinergic function of NGF; studies are needed to explore the significance of NGF at these sites. It was also established that NGF exists in the skin, homeopoetic, and immunoactive cells, and exerts proliferative effects in the cultures of some of these cells (see Thoenen et al., 1987b and Bersani et al., 2000). Some evidence suggests that NGF of the skin is a target-derived trophic factor for the pain endings, and that it acts as its sensitizer; consistent with this notion, NGF levels increase during inflammation prior to hyperalgesia (see Landreth, 1999). Altogether, NGF, a factor referred to as the "cholinergic neurotrophic factor," exists in an array of noncholinergic neurons and tissues, and exerts trophic actions on some tissues and sensitizing actions on others.

In fact, for some time now (see Rabizadeh et al., 1994), nerve growth factor (and possibly other trophins) may relate to the amyloid phenoemena and the Alzheimer's disease. Some 50 years after the discovery of NGF, its presence at these sites is still enigmatic.

CII. Other Retrophic Factors

1. The Neurotrophins

As already described, Hans Thoenen's team (Barde et al., 1982) identified several nonNGF retrophins or neurotrophins (50 or so, by the modern count; Meyer et al., 1979; Landreth, 1999, 2006; Henderson, 1996) including BDNF. Retroneurotrophins are small, highly basic proteins with an approximate 13 kDA number, and they dimerize to form biologically active trophins (Ip and Yancopoulos, 1996; McDonald and Chao, 1995). Similar to NGF, the neurotrophins act on common receptor p75 and on the family of Trk receptors; depending on the trophin; they act on either TrkB (BDNF and neurotrophin 4/5 [NT 4/5]) or Trk C (neurotrophin 3 [NT3]). Like NGF, other neurotrophins are expressed in the CNS (particularly in the cholinergic basal forebrain), the peripheral autonomic system, sensory neurons, and glia; TrkB and/or TrkC receptors are expressed throughout the CNS, while TrkA's expression is limited to the basal forebrain and striatum.

Individual neurotrophins, including NGF, differ in their central trophic actions, although they all seem to affect the cholinergic basal forebrain neurons; their actions with respect to the peripheral nervous system are generally similar (see Table 8-1). The overlap of the location of several neurotrophins in the periphery and some

Table 8-1. Neurotrophin Targets

Neurotrophin	Receptor	Peripheral	Central
Nerve growth factor	TrkA	Sympathetic neurons, sensory neurons: dorsal root ganglia	Basal forebrain neurons, striatal cholinergic neurons, Purkinje cells
Brain-derived neurotrophic factor	TrkB	Sensory neurons: nodose, dorsal root neurons	Spinal motor neurons, basal forebrain cholinergic neurons, substantia nigra dopaminergic neurons, facial motor neurons, retinal ganglion cells
Neurotrophin 3	TrkC	Sympathetic neurons, sensory neurons	Basal forebrain cholinergic neurons, locus ceruleus neurons
Neurotrophin 4/5	TrkB	Sympathetic neurons, sensory neurons: nodose, dorsal root ganglia	Basal forebrain cholinergic neurons, locus cerules neurons, motor neurons, retinal ganglion cells

central areas suggests that normal development, that is, the process illustrated in the neurotrophic factor hypothesis, requires more than one trophic substance.

Brain-derived neurotrophic factor acts as a trophic factor for the development of numerous central cholinergic populations. This effect occurs at crucial developmental periods during synaptogenesis and confirms that BDNF acts in accordance with the trophic factor hypothesis, particularly since it was established that it originates in a target tissue.

Its intense effect on the noncholinergic sensory neurons should be stressed (see Table 8-1); a number of *in vitro* and *in vivo* studies showed that during rodent and chick development, BDNF promotes the survival of dorsal root ganglia and cranial sensory neurons. Thus, many investigators hypothesized that following the developmental period of trophic sensitivity, the sensory neurons rely on BDNF and other neurotrophins, including NGF, via their paracrine or autocrine delivery (see Landreth, 1999).

In knockout mice deprived of BDNF, a number of neuronal populations are eliminated and the mice die within 2 weeks of birth. However, so do the knockout mice deprived of Trk B receptors, which again illustrates the multifactorial nature of synaptic survival. Then, akin to NGF, BDNF enhances the expression of cholinergic components, facilitates the regeneration of neurons, and restores cognition following lesions of the basal forebrain in adult rodents (Pelleymounter and Cullen, 1993; Pepeu, 1993). In a similar context, BDNF helps the regeneration of the descending spinal pathways of adult rats following spinal cord injury (Novikova et al., 2002; Rylett and Williams, 1994). Also, similar to NGF, BDNF exerts paracrine and autocrine proliferative effects on the neuroblasts (Lindsay, 1996).

The actions o fNT3 and NT4/5 are similar and also resemble the actions of other neurotrophins; NT3 is particularly widely distributed in the CNS (see Table 8-1). Both these factors support the developmental survival of central cholinergic neuronal populations and of sensory neurons. The death of neuronal populations during the development of transgenic NT3 mice (NT3-null mutants) demonstrated the trophic developmental action of NT3 and the presence of NT3 during critical developmental periods of several neuronal populations (including the sensory neurons) and supports the notion that NT3 acts within the concept of the neurotrophic factor hypothesis. By extending the duration of their mitotic activity, NT3 also exhibits a proliferative action on the neural crest cells (Landreth, 1999, 2006).

Altogether, NT3 may be more active during the neuronal survival processes than NT4/5 or BDNF, as NT3 transgenic knockout mice do not survive as well as NT4/5 and BDNF transgenic mice (see Landreth, 1999). Aside from their role in sustaining neuronal survival, NT3 and NT4/5 must also partake in neuronal maintenance, as their presence and expression in various neuronal populations continues through animal adulthood, particularly in the case of sensory neurons (Lindsay, 1996). It is interesting that during rodent synaptogenesis there is a switch in the dependence of neurons from NT3 and NT4/5 to NGF (Paul and Davies, 1995; see also Henderson, 1996).

There is less evidence available regarding the potential role of NT3 and NT4/5 in preserving or regenerating the components of the cholinergic system and promoting the cholinergic phenotype. Some neurotrophins (particularly NT4/5) are expressed in the muscles that act as a target tissue for the cholinergic motoneurons; motoneurons need these neurotrophins for maintenance and/or developmental survival (Flanigan et al., 1985; see Landreth, 1999, 2006). Finally, NT3 seems to exhibit *tropic* actions, as the expression of NT3 by dentate gyrus cells serves to attract septum-derived cholinergic projections to the dentate (Robertson and Yu, 2003).

2. The Neurokines, Including Ciliary Neurotrophic Factor

The neurokines are chemically related to cytokines. They include the ciliary neurotrophic factor (CNTF), leukemia inhibitory factor (LIF), cardiotrophin 1 (CT-1), oncostatin-M (which is involved, like other kinins in hematopoietic, immunological, and inflammatory processes as well as effects on nonneural tissues, but does not seem to exert neurotrophic effects), and several interleukins. The cell surface receptors for the neurokines are complexes exhibiting a common subunit, gp130; additional subunits are specific for the various neurokines. Many cytokines, neurokines, and particularly the interleukins have been researched intensely to determine their role in inflammatory and immune processes; in fact, these studies outnumber by many times the studies concerning the neuronal trophic actions of cytokines, neurokines, and interleukins.

The early findings, which demonstrated that parasympathetic ciliary ganglion development is not affected or dependent on NGF, promoted interest in identifying a growth factor involved in parasympathetic development (see section CI-1). Ciliary neurotrophic factor was found to be such a factor. Ciliary neurotrophic factor exerts trophic actions on the ciliary ganglion, as it exists in the target tissues of this ganglion (i.e., the eye and/or the intraocular muscles). Also, CNTF exists in targets such as the cardiac and other muscles, as it regulates the development and survival of the parasympathetic ganglia and somatic motoneurons it is expressed in the CNS by glia and in the peripheral NS by Schwann cells (Ip, 1998).

Ciliary neurotrophic factor was purified from the bovine cardiac and chick intraocular muscle tissues (Ebendal et al., 1979; Barbin et al., 1984; Watters and Hendry, 1987; see also Purves and Lichtman, 1985). It is possible that two factors may be present in the ocular tissues; the smaller component (molecular size of 20,000) may be responsible for neuronal growth, and the larger component (molecular size of 50,000) may cause CAT development (Nishi and Berg, 1981; cf. also Purves and Lichtman, 1985; Watters and Hendry, 1987). Ciliary neurotrophic factor acts on membrane lipid–anchored alpha CNTF receptors, which link with two transmembrane receptors, LIFR beta and pg130 (these receptors are also involved in the action of the LIF), and its presence in the target cells identifies its role as a retrophin.

The trophic role of CNTF was established when it was demonstrated that CNTF target tissues are necessary for development, as the ganglion cells come in contact with these tissues between the eighth and14th developmental stage of the rat (Giacobini, 1986). Ian Hendry's group (Hendry et al., 1988) prepared a monoclonal antibody from the bovine heart factor, which significantly diminished CAT development of the iris during the early postnatal period; the effect of the antibody was specific, as the antibody did not affect tyrosine hydroxylase activity of the iris. Ciliary neurotrophic factor exerts trophic actions on numerous central neuronal populations including the hippocampus, retina and the basal forebrain (Ip and Yancopoulos, 1996; Fuhrmann et al., 1998). Furthermore, CNS lesions increased the level of NGF in astrocytes (Landreth, 1999). However, human and rodent mutations in the CNFT gene did not exert any marked effects on the CNS, including motoneurons. Yet, the knockout of the receptor gene for the CNFT receptor caused significant loss of the motoneurons (Landreth, 1999; Heller et al., 1996). Note that similar to NGF and a number of other neurotrophins, CNFT facilitates axonal regeneration (Frostick et al., 1998).

Neurotrophic activities of LIF were established *in vitro* and *in vivo*. *In vitro* these trophic effects were shown with respect to sensory neurons and spinal motoneurons, and LIF stimulated also—perhaps independently from its trophic action—the synthesis of ACh. In fact, *in vitro* it converts the adrenergic neurons to a cholinergic phenotype (see Landreth, 1999; Henderson, 1996). Cardiotrophin 1 is similar to LIF, as it is a survival factor for spinal motoneurons; it exerts retrophic action on the ciliary ganglion. As stressed by R. W. Oppenheim, an active researcher in spinal motoneuron survival, many kinins and other substances exert trophic developmental and survival effects on the motoneurons.

The PI 3-kinase activation of the phosphatidylinositol cascade may be the common mechanism for the spinal neuron action of kinins, though other kinases may be involved at other sites of kinin neurotrophic action (Konishi et al., 1999; Oppenheim and Haverkamp, 1988; Oppenheim et al., 1986; Dolcet et al., 2001).

Interleukins are synthesized in astrocytes, glia, and, to a lesser extent, neurons (such as hippocampal cells). They are small proteins; human and rodent interleukins vary in the length of their polypeptide chain (between 130 and 140 amino acid residues). Their receptors in mice have two subunits; one is a ligand-binding protein and the other is responsible for signaling, though humans exhibit only the signaling protein. Interleukin 3 exhibits *in vitro* and *in vivo* effects. It maintains hippocampal, septal, and basal forebrain neuron cultures obtained from embryonic, adult, or postnatal rats and mice. It also promotes the activity of CAT and VAChT *in vitro* (Kamegai et al., 1990; Kushima and Hatanaka, 1992; see also Landreth, 1999;

Konishi et al., 1999). Interleukin 3 also exhibits regenerative effects *in vivo*, as it promotes morphogenetic healing and facilitates the reformation of CAT following forebrain lesions carried out in adult rats. Similarly, interleukins support axonal regeneration; interestingly, their levels in the endoneurium of the sciatic nerve contralateral to the transected sciatic is increased following section (Ruohonen et al., 2002).

It must be emphasized that both *in vitro* and *in vivo* interleukins affect other than cholinergic cells; they promote growth of and antagonize neurotoxic and ischemic effects on GABAergic neurons of the septum and CA1 neurons of the hippocampus (Kamegai et al., 1990; see also Konishi et al., 1999). Finally, expected behavioral and neuronal deficits occurred in transgenic mice that exhibit downregulation of kinins; similar effects followed the use of antibodies to interleukins. However, behavioral and morphological abnormalities also occur in transgenic mice that overexpress interleukins (see Konishi et al., 1999). Several other cytokines seem also to affect developmental survival of cholinergic neurons; their actions were mostly studied *in vitro* (see Rothwell and Hopkins, 1995; Henderson, 1996).

3. Other Growth Factors Acting on the Nervous System

These growth factors include 4 families of trophins (see Table 8-1); the major actions or the known actions of these factors are exerted on non-nervous tissues, which is clearly indicated by their names. However, recently discovered CNS and peripheral nervous system effects of these factors are considerable, and the concentrations of 2 fibroblast growth factors, for example, are higher in certain areas of the brain than are NGF concentrations (Landreth, 1999). The trophic activities of these growth factors are exerted in anterotrophic fashion rather than in a retrophic manner, as they act as either paracrine or autocrine factors.

While fibroblast growth factors (FGFs) sustain proliferation of fibroblasts, mesoderms, and epiderms, they also exhibit regenerative and maintenance functions. Fibroblast growth factor 1 is widely distributed in the neurons, particularly motoneurons and sensory cells, cholinergic basal forebrain, striate, and subcortical neurons, while FGF2 mainly exists in the limbic system (it is also expressed in astrocytes). Both FGF1 and FGF2 are present in the mature nervous system, and, therefore, may be involved in maintenance rather than the developmental processes. Also, FGF2 exhibits a proliferative function with regard to multipotential stem cells, which subsequently give rise to cortical neurons (Temple and Quian, 1995).

In addition, FGFs facilitate peripheral axonal regeneration and the regeneration of central neurons following CNS lesions. There are at least 4 receptors of FGF1 and FGF2; they belong to a subfamily of tyrosine kinase receptors. As stated by Gary Landreth (1999), the "interaction of various FGFs with the four FGFRs and their multiple mRNA splice products is bewilderingly complex." Sometimes, both FGF1 and its receptors may exist in the same neurons, as in the case of motoneurons and substantia nigra. This suggests that this factor acts via autocrine and paracrine mechanisms, rather than as target tissue–derived retrotrophins (see Landreth, 1999). The same mechanisms are involved in the actions of transforming growth factors (TGFs; see below); thus, TGFs and FGFs may be exceptional with respect to most factors described in this section, which are retrophic in nature. Note that the action of FTF1 on the dopaminergic substantia nigra neurons exemplifies the noncholinergic actions of growth factors.

The TGFs form an extended family; only the three beta TGFs (see Table 8-1) exhibit neurotrophic and related actions. While TGF beta 1 is expressed mostly in glia and astrocytes, the two other TFG beta factors are widely present and expressed in the CNS and peripheral nervous system. The glia-derived neurotrophic factor (GDNF) is another member of the TGF family and is predominantly derived from glia, although to a lesser degree it is also expressed in astrocytes and Schwann cells (Henderson, 1996; Landreth, 1999). Neurturin, another member of the TGF superfamily, is related to GDNF and is also derived from glia. Additional TGF relatives include the bone morphogenetic proteins (BMPs); originally these factors were thought to be involved only in bone formation. The TGFs act via a unique receptor complex of two serine/threonine kinase subunits, as GDNF binds to a complex two-subunit receptor; one of the subunits is a tyrosine kinase. The BMP signaling is effected by small Smads proteins (Bau et al., 2002), although it is uncertain whether the same entities promote both neurotrophic and bone-formation BMP effects.

Bone morphogenetic proteins, beta TGFs, GDNF, and neurturin maintain survival during the development of dopaminergic neurons of the striate and the midbrain. As these growth factors are expressed in areas receiving dopaminergic radiations and in glia, they behave as target-derived factors, that is, retrophins (Landreth, 1999; Oiwa et al., 2002).

Beta TGFs also exhibit regenerative effects on injured neurons or pathways; in fact, their expression is increased following central pathway or spinal cord lesions. Their unique and paradoxical action is their antimitotic effect on neuronal proliferation during early developmental stages, and they may antagonize the proliferative actions of other growth factors during this stage of early development (Landreth, 1999). Bone morphogenetic proteins are known to exert a wide range of morphogenetic and organogenetic actions. They promote developmental survival (that is, they perform according to the neuronal trophic factor hypothesis) and synaptogenesis of several cholinergic CNS pathways and structures, including the cortex and basal forebrain. Bone morphogenetic proteins act regeneratively on lesioned brains and spinal cords; they exert phenotypic differential effects on the multipotential cells of the neural crest; and they maintain the survival of mature brain neurons (these actions were established generally for rodents; see Henderson, 1996; Landreth, 1999; and Cao et al., 2003).

The epidermal growth factor (EGF) family, platelet-derived growth factor (PDGF), insulin-like growth factor (IGF), neuregulins, protease nexin, ephrins, and other trophins are expressed in the nervous tissues, glia, Schwann cells, or platelets, as well as in such unlikely organs as the chicken gizzard (van Nieuw Amerongen, 1987; Riopelle and Riccardi, 1987; Wujek and Akeson, 1987; Hayashi et al., 1987; Landreth, 1999; Houenou et al., 1995; Zhou et al., 2001). Some of these factors, for example, EGF, may not play a major role as neuronal trophics (see Landreth, 1999). However, besides having a wide range of activities that are outside this book's scope (for example, mammary function, etc.), other growth factors such as neuregulins and particularly IGF-I, exert important neurotrophic actions.

The neuregulin family (also referred to as ERBs, Neu factors, ARIA, and HERs) is derived from a neuronal transmembrane precursor that contains an EGF-like domain. In fact, their receptors, tyrosine kinases, are homologous to those of EGFs and are generated by alternative splicing of mRNAs derived from a single gene (Ben-Baruch and Yarden, 1994). With respect to cholinergic aspects of their effects, it is interesting that a neuregulin acts as a cholinergic, nicotinic receptor adhesion factor at the myoneural junction; in fact, this neuregulin was first detected at this site (as such, it is called ARIA). Since these neuregulins are present in the motoneurons, their effect is anteroactive. Neuregulins are also highly expressed in the CNS, including cortex, during development and in maturity; they are also expressed in the peripheral sensory ganglia. While the neuregulins may exhibit maintenance effects on the cortex and other brain parts, they are mostly known for their mitogenetic, proliferative effects on all types of glia and on Schwann cells; again, these effects seem to be exerted in anterograde, paracrine fashion (Crone and Lee, 2002).

Like several other neurotrophin receptors, IGF-I's receptor possesses a tyrosine kinase domain that structurally resembles insulin receptors. IGF-I and its receptors are expressed during development and maturity in many parts of the CNS, including olfactory bulb, thalamus, hippocampus, cortex, cerebellum, and retina, as well as in the motoneurons. During development they promote *in vivo* motoneuron survival, and survival and maintenance of several brain parts. Their maintenance effect is also shown in cultures from several cell lines (De la Rosa and De Pablo, 1995; Chung et al., 2002; Torres-Aleman et al., 1994; Ye et al., 2002; see also Landreth, 1999, 2006). IGF-I also exerts regenerative effects on lesioned CNS (Chavez and LaManna, 2002).

4. Nonneurotrophic Elements Concerned with Neurogenesis and Synaptogenesis

Aside from the neurotrophins, kinins, and so forth, synaptogenesis depends on many other factors. To be effective, the cholinergic synapse requires the activity of the cholinergic gene, its expression, and additional molecular regulations. Accordingly, HACU, VAChT, factors controlling vesicular cycling, vesicular fusion with the presynaptic plasmalemma, and postsynaptic receptor clustering must be in place for synaptogenesis to be complete. These factors and the CAMs, SAMs, agrin, rapsyn, and so on, were alluded to in section BII-5 with regard to their role in the morphogenesis and maturation of the myoneural junction (see also Chapter 2 B).[12] Cell adhesion molecules and SAMs are glycoproteins that are endowed with polysialosyl residues, and their biological activity and the moment of their ontogenetic appearance depend on the polymer length of these residues (Balak et al., 1987; Franz et al., 2005). The CAMs and SAMs are present in the membranes of the nerve cells. They also appear in nonneural tissues or elements, including smooth muscle and myo-blasts, chondrocytes, Schwann cells, and glia (Gerthoffer and Singer, 2002). The processes that include these factors usually occur during the formation of the neuromyal junction (McCaig, 1986; Vaca, 1988; Sanes and Lichtman, 2001; see section BII-5).

Many of these factors, such as CAMs and SAMs, were identified during the development and maturation of the central cholinergic pathway synapses, and Dustin and Colman (2002) commented that they contribute to "a synapse . . . being . . . a stable adhesive junction between two cells." Cell adhesion molecules and related substances such as cytotactin and sciatin (Markelonis and Oh, 1981), along with SAMs, exert mechanochemical morphogenetic functions during early development and late development, and postnatally. During early ontogenesis these factors induce adhesion between neuronal plasma membranes and patterning of the neuron cells, and they promote cell division. Many of these processes can be interfered with by the use of appropriate antibodies (see, for example, Thanos et al., 1984).

Then these factors generate the differentiation of CNS germ layers, followed by the differentiation of neuronal populations (Edelman and Crossin, 1991). Moreover, either late prenatally or early postnatally, CAMs or SAMS promote neuronal growth and dendrite outgrowth of the sensory, cerebellar, and hippocampal neurons. They also cause the cessation of cell division following differentiation (Hatten, 1987; Dustin and Colman, 2002; McNamee et al., 2002; Webb et al., 2001).

Secondary CAMs (for example, Ng-CAM) and SAMs induce neuronal differentiation of the blastoderm and cause the notochordal induction of the neural plate; subsequently, they differentiate the spinal cord, the brain, and the peripheral nervous system including the ganglia. During early stages of ontogenesis in the chick ganglion cells, N-CAM appears (Thiery et al., 1982). In the cultured prenatal (E10) and early postnatal sympathetic and parasympathetic neurons, N-CAM promotes formation of neuronal clusters and the synthesis of CAT (Tuttle, 1983, 1985; Gray and Tuttle, 1987; Acheson and Rutishauser, 1988; Wong and Kessler, 1987). Since the CAMs and SAMs exist postnatally, they may participate in the maintenance processes as well as upon axonal or neuronal lesion in regeneration. Finally, the cell-to-cell adhesion generates biochemical changes (such as amino acid synthesis; Linser and Perkins, 1987). Many of these processes can be blocked by appropriate antibodies (see, for example, Thanos et al., 1984).

During development, the concentrations and ratios of CAMs (for example, between the L- and N-CAM, and NGF and CAM), their cellular localizations, and their identity undergo many changes (for instance, there are shifts between the primary CAMs and the N-CAM and L-CAM, and between primary and secondary CAMs; Sunshine et al., 1987; Edelman and Crossin, 1988).

It appears from this description that CAMs, SAMs, and related substances are particularly important for synaptogenesis; their role in this process is differentiative in nature. However, these factors also exert neurotrophic, proliferative, and regenerative actions. They are released from peripheral tissues such as Schwann cells that may act as their target sites. If so, they may act as retrophins; as they can be also released from neuronal membranes to act on the neighboring neurons, they may work in an autocrine or paracrine fashion (Landreth, 1999).

Other factors—not identified so far—may be also involved in the clustering of cholinergic receptors, whether in the course of ontogenesis of the neuromyal junction or following denervation (see below, section D-2). The question of the clustering motion of the originally dispersed receptors arises here—is this movement evoked by N-CAM or related substances? The definitive answer to this question cannot be given now, but it is quite possible that N-CAMs or related factors may affect this motion (Nastuk and Gabriel, 1985); on the other hand, substances other than CAM may be involved, such as agents involved in transsynaptic regulation (see below, section CIII; see also Connolly, 1985).

Note that CAMs and SAMs belong to multi-membered families of factors and that these factors express many trophic functions on nonneuronal cells such as smooth muscle cells (Gerthoffer and Singer, 2002). These factors and related glycoproteins and adhesion factors such as myelin-associated glycoprotein (MAG) and several Nogos are concerned with regulation of myelin and the inhibitory effects of myelin on adult axonal regeneration (Bandtlow, 2003; Bandtlow and Dechant, 2004; Bartsch, 2003; Meier et al., 2003). It should be added in this context that the effectiveness of NGF and other trophics on adult axonal regeneration and in preventing the inhibitory myelin action of regeneration is not quite clear (see above, section CI-3e).

Furthermore, CAMs and SAMs exert non-trophic, nondifferentiative functions on immune and hematopoetic systems, cancerous events, and inflammatory processes. In fact, contemporary literature is overwhelmed by references concerning these functions, and the ratio between these references on the one hand and the studies dealing with neurotrophic and synaptogenic functions of these factors on the other may be nearly 1,000 to 1. Finally, genetic factors must exert ultimate control in ontogenesis, neurogenesis, and differentiation of the cholinergic and other transmitter systems, and in the cooperative regulation of these processes by active proteins. As has been already mentioned, the knowledge of this control is still very poor.

5. Gangliosides

Gangliosides are relative newcomers to the field of trophisms, and currently they are both intensely studied and intensely argued about. Gangliosides are a family of sialic acid–containing glycosphingolipids. The hydrophilic sialic acid–containing oligosaccharide portion of their molecules protrudes from the outer membrane surface, while the hydrophobic ceramide portion is inserted into the lipid core of the membrane (cf. Sonnino et al., 1988).Their presence in the human brain was first established some 70 years ago by Ernst Klenk (1935), who coined their name. Nevertheless, this discovery was well-nigh forgotten, until Jordi Folch-Pi rediscovered gangliosides 30 years later (cf. Wiegandt, 1975). More than 40 gangliosides are chemically distinguishable (Rapport, 1981; Sonnino et al., 1988). The synthesis of new, active gangliosides is straightforward, and it is easy to vary the basic ganglioside structure by changing the number of sugars and/or the sugar molecule, the number of neuraminic acid residues, their position in the ganglioside entity, and so forth (see Figure 8-15).

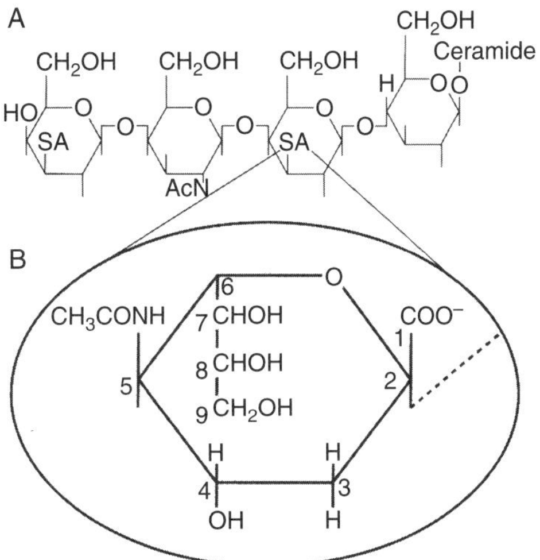

Figure 8-15. (A) The structure of a major brain ganglioside, which is termed GD1a according to the nomenclature of Svennerholm. G denotes gangliosides, D indicates disialo, 1 refers to the tetrasaccharide (Gal-GalNac-GalGlc-) backbone and distinguishes positional isomers in terms of the location of the sialic acid residues. In IUPAC-IUB nomenclature, this ganglioside is termed IV^3NeuAC, II^3NeuAc-Gg_4Cer, where the roman numerals indicate the sugar moiety (from ceramide) to which the sialic acids (NeuAC) are attached, and the Arabic numeral subscript denotes the position in the sugar moiety where Nue-AC are attached; Gg refers to the ganglio (Gal-GalNAc-Gal-Glc) series and the subscript 4 to the 4-carbohydrate backbone for the "ganglio" series. (B) The structure of sialic acid, also called N-acetyl neuraminic acid (NeuAc or NANA). Human brain gangliosides are all N-acetyl derivatives; however, some other mammalian, such as bovine, brain may contain the N-glycolyl derivatives. The metabolic biosynthetic precursor for sialylation of glycoconjugates is CMP-sialic acid, forming the phosphodiester of the 5'OH of cytidine and the 2-position of neuraminic acid. (From Agranoff et al., 1999, with permission.)

The gangliosides are components of the neuronal and axonal plasma membrane of the central and peripheral neurons; in the nervous system, they are moved anterograde from the perikarya via the axonal transport (Fishman and Brady, 1976; Ando, 1982; Ledeen et al., 1975, 1981; Ledeen and Yu, 1982). The gangliosides also exist in nonnervous tissues. They are particularly high in gray matter (3 to 4000 nM/g), although they are quite high in the white matter as well (approximately 1,000 nM/g; cf. Rapport, 1981). The gangliosidal species most abundant in animal and human brains are GM1, GM3 and GT1b (Wiegandt, 1982). They may belong to the anterotrophin family of trophins, as they are concentrated in the nerve-ending membranes and transported down the axon via the axoplasmic flow. However, they also exist in the perikaryal membrane (De Robertis et al., 1975; Ledeen et al., 1981; Yates et al., 1989).

The earliest findings regarding the biological activity of these compounds dates from the 1950s, when investigators demonstrated that gangliosides effectively antagonize axonal degeneration due to viruses (Bogoch, 1957, 1975). Van Heyningen (1958) suggested that gangliosides may serve as membrane receptors for toxins; this notion was later confirmed (cf. Svennerholm, 1975). Accordingly, many studies during that period concerned the role of gangliosides as recognition markers for the neurotoxic site of action rather than as trophics.

Several lines of evidence suggest that gangliosides act as neurotrophins, that is, factors acting within the neurotrophic factor hypothesis (see section CI-2). They may be released from nerve terminals, where they are preferentially accumulated; they can be, therefore, considered as anterograde-wise-acting neurotrophins (see above, and section CI-1; De Robertis et al., 1975). The gangliosides are present during early development and peak postnatally (Roukema et al., 1970; Willinger, 1981; Buccoliero et al., 2002). They appear at the time of neurogenesis in several brain regions, and the number of ganglioside family members increases at this time. Furthermore, the ganglioside concentrations or expressions augment during these periods (Ledeen et al., 1990; Rizzo et al., 2002).

Then, *in vitro*, exogenous gangliosides promote the growth of neurons, axons, and neurites; *in vivo*, they facilitate migration of neuroblasts during the formation of neuronal pathways such as those radiating to the cerebellum and the cortex (Miyakoshi et al., 2001). Moreover, it was shown early that antibodies to the gangliosides disrupt development and that neurogenesis of the cortex and other brain parts is blocked in mutants that do not exhibit ganglioside expression. These morphogenetic aberrations are accompanied by behavioral deficits (Willinger, 1981; Doherty and Walsh, 1987; see also Pepeu et al., 1994). However, to fulfill the desiderata of the neurotrophic factor hypothesis, direct evidence for demonstrating that gangliosides are released presynaptically and promote the elimination of superfluous neurons and nerve terminals during CNS maturation and synaptogenesis is needed; this evidence is it is incomplete for gangliosides (see sections CI-2, CII-1, and CII-2).

The regenerative trophic actions of gangliosides were first described in the 1970s by Ceccarelli and associates (cf. Ceccarelli et al, 1975), as they demonstrated that gangliosides facilitate functional and morphologic recovery following anastomosing pre- and postganglionic axons of the adult cat. This discovery was expanded by the Warsaw-Lodz-Florence-Montreal team of Claudio Cuello, Barbara Oderfeld-Nowak, and Giancarlo Pepeu. Provided exogenously, they promote in rodents morphological repair after lesions to the caudate, nucleus basalis, septum, or cortex (Wojcik, 1978; Wojcik et al., 1982; Stein et al., 1983; Oderfeld-Nowak et al., 1984, 1993; Pepeu et al., 1994; Cuello et al., 1987, 1989; Sabel et al., 1984; Stein, 1994).

Furthermore, gangliosides facilitate the recovery of hippocampal AChE and CAT following lesions to the septum (Wojcik, 1978; Oderfeld-Nowak et al., 1975, 1984, 1993; Pepeu et al., 1994), and when ACh release is blocked in the rat by lesions of the nucleus basalis, the gangliosides of the GM1 type restore that release (Florian et al., 1987). Then gangliosides repair the behavioral deficits caused by chemical and surgical lesions of cholinergic nuclei such as the nucleus basalis (see, for example, Cuello et al., 1987; Cuello, 1993). Additionally, the gangliosides prevent neuronal elimination induced in cortical and cerebellar cultures by neurotoxins (Favaron et al., 1988). The cholinergic receptors need not be involved in this phenomenon, as the gangliosides antagonized the effect of glutamate and kainate, that is, neurotoxins acting on the N-methyl-D-aspartate (NMDA) recognition site.

Several mechanisms are involved in the action of gangliosides. As signal receptor sites at the membrane, the gangliosides may be involved in "dramatic changes in the local organization of membranes" that evoke second messenger events and the resulting phosphorylations and dephosphorylations (Tettamenti et al., 1985; cf. also Sonnino et al., 1988). Indeed, gangliosides and related glycosphingolipid compounds are involved in phosphorylation phenomena; for example, they block glutamate-induced activation of protein kinase C and evoke phosphorylation activities at the neuronal membranes (Vaccarino et al., 1987; Favaron et al., 1988). Then they probably act by releasing NGF and other neurotrophins, which would activate neurotrophin receptors Trks and p75 (see above, section CI-2; Rabin et al., 2002).

Note that the gangliosides act on other than cholinergic neurons; for example, they appear to act on the NMDA receptors. In fact, they may also facilitate the release of several transmitters, including ACh and glutamate (Tanaka et al., 2004). Also, the gangliosides protect the dopaminergic striatal neurons of primates and mice from neurotoxins, and restore the serotonergic hippocampal system after it is reduced by septal lesions (Schneider et al., 1992; Gradkowska et al., 1986; see also Cuello, 1993).

Then gangliosides act on nonneural tissues. An interesting, paradoxical effect of gangliosides is exerted on Schwann cells, as they block the proliferation of these cells in culture (Sobue et al., 1988). *In vivo*, the proliferative phase of Schwann cells occurs during axonal development and axonal regeneration. This proliferation, which is needed for the initial axonal growth or regeneration, must be inhibited before further axonal growth may take place. Actually, toxins and other activators of this proliferation (but also some growth factors) block axonal growth by promoting Schwann cell production. The endogenous regulators of Schwann cell formation have not been identified. Sobue et al. (1988) hypothesize that naturally occurring gangliosides may be a part of this regulation process, but this notion has been challenged.

In view of exciting basic data concerning growth and regenerative actions of gangliosides and because the synthesis of novel gangliosides is quite easy, many pharmaceutical companies became interested in the development of clinically useful gangliosides. Currently, there is a flare-up of clinical studies—particularly in Italy and Scandinavia—concerning the clinical potential of gangliosides. Today, they are used experimentally in diabetic and other sphingolipid storage diseases, including immune neuropathies, as well as in Alzheimer's disease (Buccoliero et al., 2002; Svennerholm et al., 2002; see also Pepeu et al., 1994). There is some skepticism with regard to the clinical studies in question; this skepticism is engendered by the excessive claims accompanying some of the clinical studies.

6. Comments and Conclusions

The detection of the NGF by Rita Levi-Montalcini, Victor Hamburger, and Ed Bueker and subsequent discovery of other neurotrophins by Hans Thoenen, Ian Hendry, and others were remarkable and significant events, as they led to the detection of hundreds of neurotrophins and to an explosion of knowledge regarding the "myriad of . . . their . . . functions" (Crone and Lee, 2002). Indeed, what could be more significant than the discovery of entities without which there would be limited or no knowledge of development, maintenance, and regeneration of neurons? It should make any cholinergiker happy that so many of functions and effects of growth factors and neurotrophins specifically concern the cholinergic neurons, including those of the forebrain and septo-hippocampal pathways, cerebral and cortical cells, motoneurons, and peripheral autonomic ganglion cells.

However, let us not be chauvinistic; we must recognize that neurotrophins exert many important actions beyond their cholinergic effects. Thus, they act on noncholinergic neurons, including dopaminergic and sensory neurons. They also act on nonneuronal tissues: many neurotrophins promote mitosis of glial and Schwann cells and are involved in axonal myelinization. Furthermore, neurotrophins, the kinins, and other trophic factors exert anti-inflammatory, erythropoetic, and immunological effects. They are involved in mammary and neuroendocrine functions, and it is speculated that they may promote longevity in humans. Altogether, the neurotrophins and related factors are unique and versatile substances.

In this exceptional area, the field of synaptogenesis and junctional formation stands out for its complexity and multifactorial nature. Here, the neurotrophins and second messengers act during development within the neurotrophic factor hypothesis to reduce the wild overabundance of neurons and terminals that occurs at some stage in early neurogenesis. At these sites, the neurotrophins interact and cooperate with factors such as adhesion factors, vesicular cycling inducers, generators of ACh release, and so on (Sanes and Lichtman, 2001, Vaca, 1988; Bandtlow, 2003; see also Chapter 2 B). All of this well-synchronized complexity is needed for the development of effective transmission at the synapses and junctions. However, combined with this well-orchestrated phenomenon of synaptogenesis there is a certain randomness: for example, several ACh release processes may be available and it is not certain which of these may be primary at any given time (see Chapter 2 C). This randomness combines with the regenerative neurotrophic function in protecting brain plasticity; this plasticity is a relatively new notion that replaces the earlier concept of the rigidity of brain circuitry and nonrenewal of neurons and their contacts (Hohmann and Berger-Sweeney, 1998).

It appears that the actions of neurotrophins overlap, and that several neurotrophins cooperate whether during CNS development, neuronal maintenance, or neuronal and axonal regeneration (se Landreth, 1999). This concerted action of neurotrophins may be another example of their importance for plasticity processes, as the lack or mutagenic elimination of one neurotrophin need not signify neuronal extinction.

Clinical exploitation of these basic neurotrophic actions is barely beginning. Thus, early trials are being conducted, as could be expected in view of the regenerative and maintenance capacities of neurotrophins, in the areas of sensory pathology, motoneuron and Parkinsonian disease, spinal lesions, cognition, and Alzheimer's disease (see also Chapter 10). Less intelligible may be their trials in schizophrenia and depressions. These trials were first attempted in Sweden and Italy (see Granholm, 2000); Henderson, 1996; Landreth, 1999). When asked for advice as to how to achieve a long and alert life, Lars Olson (at a 1998 Giacobini-Becker symposium on Alzheimer's disease) drew the amino acid sequence of NGF (see Figure, 8-12).

Altogether, while there are many technical difficulties with respect to successful clinical exploitation of trophins; indeed, the experience in this area is sobering (Thoenen and Sendtner, 2002); it appears that, within the exciting cholinergic field, neurotrophins are not *ultimi servii*; their future is bright and their potential is practically infinite. New ones are continually discovered (see, for example, Lipson et al., 2003).

CIII. Anterograde-Acting Factors and Related Phenomena

1. Introduction

The notion that nerves and transmitters may exert nontransmittive functions has a long history. As mentioned earlier (see section CI-1), Ramon y Cajal speculated at the turn of the 19th century on the existence of "neurotrophic substances" that are involved in development and regeneration of nervous tissues. In the 1940s, this author (Karczmar, 1946) proposed that nerves exhibit a regenerative action on soft tissues independent of transmitters that these nerves may release. Sir John Eccles, Edith Bulbring, and Joshua Burn subsequently proposed that ACh and other transmitters exert both transmittive and "metabotropic" effects (Bulbring et al., 1954; Burn and Walker, 1954; Eccles, 1964; Bulbring and Burn, 1963; see also Csillik and Savay, 1958); the metabotropic effects include metabolic cellular effects that result from synaptic transmission. During the 1970s a related notion was raised when Julius Axelrod, Ian Hendry, Erminio Costa, Paul Greengard, Sandro Guidotti, and others opined that certain synaptic events control synaptogenesis, and then coined the term "trans-synaptic regulation" for these events (Axelrod, 1971; Costa and Guidotti, 1973; Hendry, 1973; Black and Mytilineou, 1976; Nathanson and Greengard, 1981).

Subsequently, a more expanded role was assigned to cholinergic components and cholinergic agonists and antagonists when it was posited that theses components exert trophic, morphogenetic, and developmental actions (see sections A and BI-2). The notion of the trophic actions of cholinergic components is important, as it explains the precocious, prenervous existence of CAT, ACh, and ChEs during development and their presence in the early CNS prior to synaptogenesis (see section BI-2). Accordingly, Lauder (1988) coined the term "morphogens" to refer to this early developmental action of the cholinergic components (see sections A and BI-2).

In the strict sense of the term, morphogens are supposed to be active as mitotics and morphogenetic agents during morphogenesis and in early neurogenesis, rather than in synaptogenesis (for more details regarding the actions of cholinergic agonists and antagonists, see section BI-2; see also Karczmar, 1963a). What must be considered now is whether trophic factors such as NGF and other neurotrophins, CAMs, SAMs, and so forth can account for biochemical and morphologic synaptogenesis, and whether the cholinergic components (aside from acting as morphogens) exert synaptogenic actions; in this case, the term "synaptogens" may be applied to the agents in question.

Cholinergic components including ACh and AChE are generated by neurons and released from cholinergic axons; if they act as synaptogens, they are anterotrophins (see section CI). As will be seen below, in some instances these components appear on noncholinergic neurons or noncholinergic axons and terminals, where they may act as autocrine factors (see section CI; see Figure 8-9).

2. Transsynaptic Effects and Cholinergic Components as Synaptogens

Important evidence concerning the cholinergic transsynaptic phenomena deals with the sympathetic and parasympathetic ganglia, the adrenal medulla (a modified sympathetic ganglion), and the myoneural junction. In view of the significance of these phenomena, this evidence will be briefly presented, though it is outside the direct scope of this book; of course, the evidence also relates to the CNS events, and the pertinent data are also presented in this section.

While Hendry (1973) and Thoenen (1972) used the term "trans-synaptic regulation" to refer to both growth factors and transsynaptic phenomena, Erminio Costa and Sandro Guidotti defined the term "trans-synaptic regulation" as "referring to regulation of synthesis of specific protein by persistent changes in the rate of synaptic transmission," and, even more specifically, to the "sustained release of ACh" (Costa and Guidotti, 1978). This definition excludes growth factors from the transsynaptic regulation phenomenon. Note that both mechanisms, trophic and transsynaptic, contribute to synaptogenesis. During the 1970s and 1980s, these investigators and their immediate followers employed paradigms such as denervation and pre- and postsynaptic block to evaluate the transsynaptic action; molecular methodology and direct use of cholinergic components was employed subsequently (Brimijoin and Koenigsberger, 1999).

Axelrod, Costa, and their coworkers (Axelrod, 1971; Costa and Guidotti, 1973, 1978) worked with the ganglia and/or adrenals of neonate and adult animals. The early finding that led these investigators to the transsynaptic phenomena was that cold exposure, swimming stress, and the cholinergic agonists markedly increased the activity of ganglionic and adrenal tyrosine hydroxylase; they surmised that this paradigm did not involve the growth factors but that it was due to augmented synaptic function. Studies of the transsynaptic effects on the development of the sympathetic ganglia of late prenatal and neonate rodents yielded many results that are consistent with this notion.

Some of the evidence in question concerns certain negative results. Thus, in the neonate ganglion cultures, NGF was unable to reverse the effects of denervation. Also, NGF antibodies did not affect biochemical or morphological development of the mouse embryonic superior cervical ganglion when the ganglion was cocultured with its target, the embryonic submandibular gland. This finding was consistent with the negative effects of NGF antibodies on synaptogenesis in the adult sympathetic ganglia cultures, although in this case, adding NGF enhanced the elaboration of the neurites (Johnson et al., 1972; see also Coughlin et al., 1978; Black, 1978). The consensus is that independent of the NGF and/or other growth factors, the transsynaptic regulation (which depends on the release and action of ACh) evokes tyrosine hydroxylase synthesis as well as morphogenesis of the nerve terminals, ground plexus ramification, and the augmentation of end-organ innervation density, while it reverses the postsynaptic effects of motor denervation.

However, the data are not always consistent. While the ganglionic blocker pempidine mimicked the effects of denervation in the sympathetic mouse ganglion, NGF reversed the pempidine's diminution of tyrosine hydroxylase activity (Hendry, 1973). Then NGF antibody caused immunoparasympathectomy in mice (Hendry, 1973; Hendry et al., 1988). Another ganglionic blocker, chlorisondamine, caused hypotrophy and a reduction of its CAT content in the avian ciliary ganglion and sympathetic rodent ganglia (Chiapinelli and Giacobini, 1978). Also, the block of axonal flow induced morphological changes in the developing ganglion.

These results may suggest that both the transsynaptic regulation and the NGF or other pertinent retrophins play a part in the development of target innervation (Chun and Patterson, 1977). The extent of the importance assigned to the transsynaptic regulation versus that allotted to the growth factor differs from investigator to investigator. Thus, Hendry (1973) opined that the transsynaptic function constitutes "the most important regulatory factor for the development of the mouse superior cervical ganglion during the critical first 2 weeks of life," and Black expressed a similar opinion (Black and Mytilineou, 1976; Black, 1978). However, Thoenen (1972) felt that transsynaptic regulation was considerably less important than NGF regulation for the generation of tyrosine hydroxylase and dopamine-beta-hydroxylase and for gangliogenesis.

What are the mechanisms involved in the trophic effect of the transsynaptic regulation of the ganglia and adrenal medulla? Acetylcholine and/or other components of cholinergic transmission may be engaged. Ganglionic blocking agents that interfere with biochemical and morphologic development of the ganglia block either the release or the postsynaptic effect of ACh, and also inhibit the ganglionic transmission. Costa, Guidotti, and Greengard proposed a biochemical basis for this cholinergic regulation of tyrosine hydroxylase synthesis in developing and adult ganglia. Cholinergic, nicotinic stimulation of the sympathetic ganglia or the adrenal medulla activates adenylate cyclase, leading to accumulation of cAMP; "cAMP activates protein kinase which is then translocated to the cell nucleus, where it causes gene 'depression' or disinhibition and the synthesis of new tyrosine hydroxylase" (Costa and Guidotti, 1978; Nathanson and Greengard, 1981). Reciprocally, cAMP regulates the generation of functional ACh receptors of the membrane of the ciliary ganglion (Margiotta et al., 1987). Yet, there may be other consequences of the blockade of ganglionic—and, generally, synaptic—transmission that may be pertinent for the transsynaptic regulation.

Transsynaptic regulation and the role of the released ACh in this phenomenon were studied extensively also with respect to the neuromyal junction and its development (Rotshenker and Tal, 1985; see section CII-4). How significant is the contribution of cholinergic transmission and cholinergic components to this development? Steve Thesleff was the first to describe the classical phenomena in which denervation and the neuromyal blockers, whether pre- or postsynaptic, increase the receptor number and receptor dispersion, resulting in hypersensitivity of the endplate (Thesleff, 1955, 1960; see also Duxson, 1982; Oh et al., 1987; Fernandez and Donoso, 1987a, 1987b; Brimijoin, 1987; Davis, 1987). Furthermore, late prenatal or early postnatal denervation prevented maturation and clustering of the receptors (Duxson, 1982; Barr et al., 1987). The term "differentiation" is frequently used to refer to the phenomena that occurred early postnatally in the presence of nerves, whereas denervation induces "dedifferentiation" (Schotté and Butler, 1941; Schotté and Karczmar, 1944; Karczmar, 1946).

Rodolfo Miledi (1960) carried out a crucial experiment some 40 years ago to clarify the role of transmission in the muscle denervation phenomena. He partially denervated the frog sartorius muscle; accordingly, some of the muscle fibers were denervated, but effective conduction was expressed as the full-fledged action potential and good muscle contraction were maintained. What happened was that only the denervated fibers exhibited denervation sequelae. These results suggest that a trophic influence emanating from the axon (i.e., ACh) rather than conduction and function regulates the cholinergic receptors and endplate sensitivity.

If hypersensitization results from either an absence or an incapacitation of the transsynaptic substance ACh, then any paradigm that effectively produces either of these two effects should be equipotent with denervation. However, sometimes the axonal conduction blockers (i.e., tetradotoxin) were found to be less effective in producing denervation symptomatology than the nicotinic postsynaptic blockers (i.e., alpha bungarotoxin), while blockers of axoplasmic flow (i.e., colchicine or vinblastine) seemed to be effective (Thesleff, 1960; Drachman et al., 1982; Albuquerque et al., 1974; see also Brimijoin, 1987). However, blocking conduction with either local anesthetics or tetradotoxin yielded controversial data (see Ochs, 1987; Purves and Lichtman, 1985).

These differences may result simply from differences in the effectiveness of various paradigms at blocking the release or action of ACh. For example, bungarotoxin or d-tubocurarine are long-lasting blockers of transmission and should be more effective at preventing the postsynaptic actions of released ACh than tetradotoxin or local anesthetics, which are relatively short-acting blockers. Furthermore, botulinum toxin and colchicin, the effective, long-lasting axonal flow and ACh release blockers may be even more effective than other transmission blockers for inducing hypersensitivity. However, the botulinum toxin prevents nerve stimulation–evoked ACh release, but does not prevent spontaneous release that is expressed in the presence of miniature postsynaptic potentials; thus, a "shadow" of ACh still remains in the synaptic cleft in the presence of botulinum toxin. Yet, the toxin is capable of inducing effective denervation symptomatology. Altogether, it is generally opined that spontaneous release of ACh and the concomitant miniature postsynaptic potentials cannot induce full-fledged depolarization involving the whole endplate, such as induced by evoked ACh plate, and that this "shadow" ACh does not suffice for the endplate differentiation.

However, the "morphogen" notion of postsynaptic differentiative action of ACh is not necessarily compatible with the concept that depolarization underlies the "morphogen" effect of ACh at the endplate (see Saltpeter, 1987). Then, the importance of muscle activity for postsynaptic differentiation was frequently connected with denervation phenomena (Saltpeter, 1987). For example, results obtained by Davis (1987) and Albuquerque and Guth (cf. Deshpande et al., 1976) suggest that denervation hypersensitivity may partially result from paralysis of the muscle, as electric stimulation of the denervated muscle somewhat antagonized denervation hypersensitivity. However, these results were not clear-cut, and Miska Salpeter (1987) stated that "the role of muscle activity . . . in causing and maintaining . . . the receptor . . . clusters is not clear." Of course, muscle stimulation could cause the release of muscle-localized trophins.

What is the mechanism of the denervation hypersensitivity and its morphologic equivalent, receptor unclustering? This morphogenetic effect of nerves is not restricted to the endplate or the ganglion. The phenomenon of de-differentiation occurs in other tissues and under other circumstances; thus, amputated limbs of urodele larvae that are denervated undergo mesodermal "de-differentiation" (Schotté and Butler, 1941). This phenomenon was attributed to prevention by denervation of the release of a trophic substance from the nerves (Schotté and Karczmar, 1945; Karczmar, 1946). Though these observations are old, the actual substance, the mechanism for its differentiative effect, and de-differentiation that follows its absence have not been established (see also Chapter 2 B, for the role of CAMs and SAMs in receptor clustering).

Irene Held argued against trophic action of ACh. Held, Jerry McLane, and their associates (Held, 1978; Held et al., 1987; Squinto et al., 1985; Sayer et al., 1987) found that denervation activates phosphorylation of a soluble, cytosolic muscle protein called 56K protein, which is a regulatory subunit of cAMP-dependent protein kinase type II (R-II; Zoller et al., 1979). Irene Held's work and the work of others (Sharma, 1982) led to the speculation that neuroregulation controls R-II phosphorylation, which in turn evokes the synthesis of the receptor protein. Irene Held (Held et al., 1987) related this neuroregulation to a trophic factor that she found via extractive and related procedures in the motor nerves. As hypothesized by Held, the factor regulates and prevents the synthesis of the extra receptors, and denervation symptoms arise when the factor is depleted following denervation.

Since in Irene Held's experiments, bungarotoxin and tetrodotoxin did not affect phosphorylation of the regulatory unit, she concluded that a spontaneous or evoked release of ACh was not involved in denervation phenomena and did not serve as a trophic factor. Others also sought a non-ACh factor. Sciatic nerve extracts and substances released via the axoplasmic flow from the motor nerve were among the factors studied, and it was attempted to find out whether these substances antagonize hypersensitivity and receptor dispersal induced by denervation or pre- and postsynaptic blockades. The results were inconsistent, and the substances in question were not definitively identified (Guth, 1986; Younkin et al., 1978; Davis, 1987). This 1970s and 1980s argument concerning the role of ACh and other cholinergic components versus that of noncholinergic trophic and differentiative factors and neurotrophins culminated subsequently with the description of factors such as CAMs and SAMs that regulate pre- and postsynaptic differentiation and induce maturation of the neuromyal junction (see section CII-4; Sanes and Lichtman, 1991; Sanes et al., 1998). Postsynaptic receptors may also be involved in neuromyal differentiation. For example, Joshua Sanes and his associates demonstrated that the maturation of the endplate and the receptor pattern was compromised in mutant mice incapable of replacing early receptor forms with their adult counterparts, and these important findings are consistent with the notion of the cholinergic postsynaptic receptor contribution to synaptogenesis (Missias et al., 1997).

It must be remembered that today, membrane phosphorylation processes and allosteric changes of receptor conformation are considered to be involved in depolarization hypersensitivity, receptor desensitization or inactivation, and changes in the receptor state (Changeux et al., 1992; Changeux, 1993; Greengard, 1987; see several sections, chapter 6). While Jean-Pierre Changeux considered the role of these phenomena in development (Changeux and Revah, 1987), it appears that these processes are less important for the maturation of the neuromyal junction than the action of pre- and postsynaptic differentiative factors.

More in keeping with this book's theme, ACh exerts trophic and morphogenetic actions also on the central neurons; both muscarinic and nicotinic receptors may be involved. For example, the alpha7 subunit of the nicotinic receptor regulates axonal branching and length, as well as motoneuron survival in the rat's embryonic spinal cord cultures; in fact, these two processes seem to be mutually antagonistic, as the antisense nucleotide that decreased the alpha7 protein expression caused axonal proliferation as well as apoptosis of the motoneurons (Catone and Ternaux, 2003); and cholinergic activation antagonizes cortical degenevation in anti-nerve-growth-factor mice (Pesavento et al., 2002).

The close temporal connection between the arrival of the cholinergic components (CAT, AChE, HACU, etc.; see section CII and Chapter 2 B) at the site of cholinergic synaptogenesis and the synaptic differentiation and maturation is frequently adduced as supportive of "the hypothesis of a specific role of cholinergic innervation in the maturation of its target structures" (Hohmann and Berger-Sweeney, 1998). This temporal connection exists in the case of neuronal differentiation of the neocortex, the hippocampus, and the dorsal root ganglia. Yet, as the cholinergic components arrive at the site of synaptogenesis, they do not have to arrive there synchronously or simultaneously; generally, AChE appears first and CAT or ACh second (see section BII-2). This may suggest that AChE plays a major role in synaptogenesis, or that the cholinergic components exert their synaptogenetic effects as they arrive at the synapse. Both AChE and ACh exert synaptogenetic effects, which is consistent with the second suggestion.

The presence of CAT, HACU, and so on in the dorsal root ganglia is of particular interest, as these ganglia are not cholinergic (they are GABAergic and peptidergic). Accordingly, Biagioni et al. (2000) proposed that in the dorsal root ganglia, cholinergic elements and CAT participate in synaptogenetic processes. Results obtained by Christine Hohmann may support the hypothesis of the synaptogenetic role of cholinergic components. On the first postnatal day of the rat, Hohmann electrically lesioned the diagonal band and other cholinergic neocortical afferents to induce a temporary deafferentiation of the cholinergic parts of the neocortex and a temporary depletion of AChE. While after a brief delay reinnervation occurred and CAT was restored, permanent alterations of cortical cytoarchitecture also occurred, including changes in the soma size and in the dendrite arborizations (Hohmann et al., 1991). Other forms of lesions that were induced by chemical neurotoxins (i.e., ibotenic acid and AF64A) caused similar effects (Armstrong and Pappas, 1991; see also Hohmann and Berger-

Sweeney, 1998). Expected behavioral deficits have also occurred.

Christine Hohmann commented that in view of these data and in accordance with the hypothesis of the synaptogenic role of cholinergic components, interference with cholinergic transmission via the use of antiChEs or supplementation of the cholinergic system with, for example, choline, should affect cholinergic morphogenesis and behavior. It has a bearing on the hypothesis that methyl parathion given to neonatal rats reduced operant behavior and caused some brain pathology (Gupta et al., 1985, 2000); in adult animals, such a treatment would be expected to facilitate learning. Furthermore, antiChE given late prenatally or just after weaning exerted morphogenetic actions; additional teratological effects of antiChEs were described (see Slotkin, 1999 and section D). However, results obtained by Christine Hohmann, Armstrong, Gupta, and others do not clearly support the hypothesis of the synaptogenetic action of cholinergic components. Even a temporary forebrain lesion, if applied during development, may impede development and cause pathology for reasons other than causing a temporary cholinergic deficit. Similarly, antiChEs, particularly methyl parathion, may cause neuronal pathology via their direct effect rather than via their interference with cholinergic transmission.

Many investigators attempted to influence morphogenesis using choline supplementation or deficiency. In rats, supplementing choline late in development or early postnatally increased neuronal size and branching in several parts of the hippocampus and septum (Albright et al., 1999, 2001; W. Meck, personal communication to Hohmann and Berger-Sweeney, 1998). On the other hand, choline deficiency applied during the second half of rat development caused changes in distribution and migration of the primordial neurons (Albright et al., 1999, 2001). Several investigators ameliorated visuospatial memory in rats with choline supplements given late pre- or early postnatally (see, for example, Loy et al., 1991 and Albright et al., 2001). However, choline supplementation did not increase concentration or activities of cholinergic components in other studies (see Chapter 3 CIII and CIV). It must be remembered that the effect of choline may depend on the developmental moment of choline administration; for example, Hohmann and Berger-Sweeney (1998) point out that neonates are more susceptible than adults to choline supplementation (Zeisel, 1987). At any rate, the studies of Loy et al. (1991) and others did not include measurements of brain ACh.

Further evidence demonstrates ACh action on growth cones, neurites, and nerve fibers (Weskamp et al., 1986; Lipton et al., 1988; Ruit et al., 1992). This effect is cell dependent; for example, the growth of the fibers in cultured retinal ganglion cells was inhibited by ACh and facilitated by nicotinic antagonists, while the growth of cultured amphibian dorsal root ganglion cells (which do not contain synapses) and spinal cord neurons was facilitated by ACh and directed along the ACh concentration gradient (Lipton et al., 1988; Kuffler, 1996, 2000; see also Bagioni et al., 2000). Related information concerns neuroblastoma N18TG2, which normally does not express synapsin I RNA, CAT, or muscarinic receptors, and is "morphologically immature" (Bagioni et al., 2000). Stefano Bagioni and associates demonstrated that this neuroblastoma transfected with CAT gene not only exhibited synapsin I and cholinergic components, but also showed enhanced outgrowth and branching of neuronal processes and cell differentiation. A related experiment was carried out with N1E murine neuroblastoma cells.

In addition, employing sense and antisense lines, Stephen Brimijoin, Carol Koenigsberger, and her associates (1998) established relations among the intensity of AChE expression, neurite growth, and neuronal differentiation. In view of these and other results of ACh action on neuronal soma processes and differentiation, Stefano Bagioni stated that ACh-dependent "regulation of fiber elongation is instrumental for building the cytoarchitecture of nervous system specific structures and allowing the establishment and stabilization of correct functional circuits." Depending on the neuron, this role of ACh in circuitry "stabilization" may also account for both the inhibition and the facilitation of fiber growth; the inhibition may also relate to the role of ACh in preventing overcrowding of fibers at the maturing synapse (this phenomenon would be an example of the neurotrophic factor hypothesis in action; see section CI-1). These studies also demonstrate that the trophic action of ACh is not limited to cholinergic neurons or dendrites, but extends to nonACh neurons and, possibly, nonneuronal tissues.

There is also convincing evidence that AChE and BuChE take part in morphogenesis and synaptogenesis. It must be noted that during morphogenesis and synaptogenesis, AChE may serve not only locally via its presence at the post- and presynaptic membranes of cholinergic neurons and (sometimes) noncholinergic nerve terminals, but also as a paracrine factor released from cholinergic neurons (see Chapter 3 D; Appleyard, 1994).

Evidence also supports the temporal relation between developmental appearance of AChE and synaptogenesis (see also section BII). Interestingly, sometimes AChE appears transiently during neurogenesis of noncholinergic neurons and pathways such as sensory thalamocortical relays; these pathway also exhibit transiently intense AChE activity in early postnatal rats, the monkey, and humans (Robertson and Yu, 2003; Robertson et al., 1985). This appearance coincides with synaptic formation between sensory axons and the cortex. Similar transient expression of AChE occurs during the development of the dorsal root ganglia and the spinal cord (Tennyson and Brzin, 1970; Biagioni et al., 2000; Robertson and Yu, 1993; Brimijoin and Koenigsberger, 1999; and Kostovic and Goldman-Rakic, 1983).[13] Altogether, Stephen Brimijoin opined, "transient AChE expression is no accident of development, but a means of promoting axonal outgrowth or synaptic connection" (Brimijoin and Koenigsberger, 1999).

Furthermore, AChE promotes soma and neuritic growth at both cholinergic and noncholinergic sites, including hippocampal neurons (Day and Greenfield, 2002). It also acts as an adhesion factor during synaptic or junctional formation (Bigbee et al., 1999). However, certain—but not all—AChE inhibitors, such as the bisquaternary BW2284c51, physostigmine, and iso-OMPA block neuritic growth in several model systems (Bigbee et al., 1999; Brimijoin and Koenigsberger, 1999; Layer and Willbold, 1994). It should be noted that this block is accompanied by the block of AChE release and can be induced by antiChEs that are BuChE inhibitors (see Figure 8-16 Layer and Willbold, 1994).

It must be stressed that these antiChEs did not inhibit growth via their inhibition of enzymic AChE activity, as indicated by several lines of evidence. First, other antiChEs did not induce this

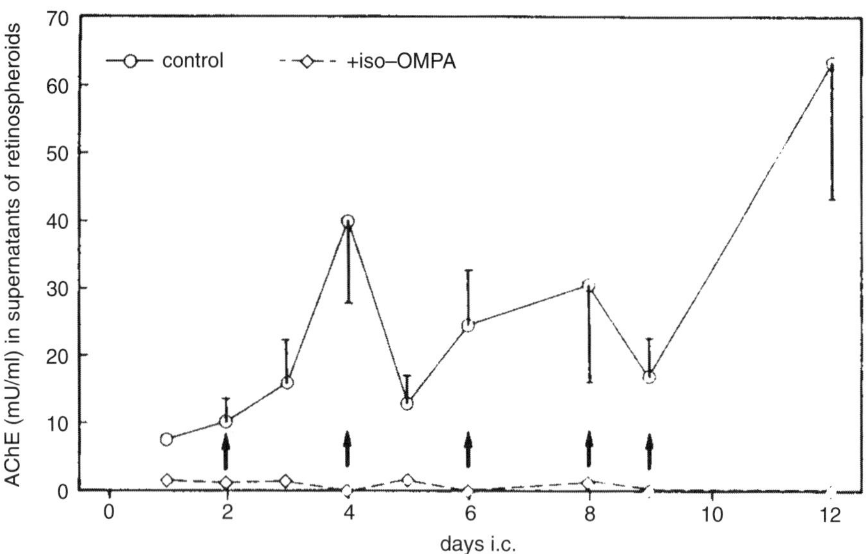

Figure 8-16. Block of release of AChE into the culture medium by the BChE inhibitor iso-OMPA indicating a regulatory relationship. Supernatants from 4 separate plates were determined in duplicate according to the Ellman method. Note that the activity before a change of media (indicated by arrows) is high due to accumulation. Standard deviation is indicated. A direct inhibition of AChE by iso-OMPA at the given concentration range is not detectable (data not shown). (From Layer and Willbold, 1994, with permission.)

block. Then, BuChE inhibitors also inhibited neuritic growth. Finally, antiChE antibodies also blocked the morphogenetic action of AChE. Layer and Bagioni (Layer and Willbold, 1994; Layer et al., 1987) presented interesting findings that were consistent with Brimijoin's hypothesis. They showed that in the embryonic neuron culture, ChE expression shifts from BuChE to AChE as the neurons switched from proliferation to differentiation and to phenotypic cholinergic commitment. More complex transitions between the two enzymes and their growth patterns occur during amphibian growth of efferent nerves, tissue proliferation during limb formation, amphibian neuromyal genesis, and amphibian dorsal root ganglia maturation. Steve Brimijoin commented that "these temporal changes imply that AChE is a postmitotic differentiation marker" . . . and "induces a factor . . . that replaces BuChE in sensory neurons when cell proliferation is complete"; similar reciprocal regulation of these two enzymes occurred in the limb and motor nerve growth (Brimijoin and Koenigsberger, 1999).

Finally, the notion of the morphogen role of AChE is supported as AChEs from various sources including humans, Torpedo, Drosophila, and mice exhibit homological domains with nonenzymic proteins, including glutactin, neurotactin, gliotactin, and neuroligin that exert trophic, morphogenetic, and adhesive action (Soreq et al., 1990; Brimijoin and Koenigsberger, 1999). These relations among the various proteins are consistent with the notion of the ancient nature and evolutionary conservation of ChE molecules. In fact, Jean Massoulié and his associates (Massoulié et al., 1993) suggested that the specialized enzymes BuChE and AChE emerged from a duplicated copy of an older BuChE gene. This conservation and the early embryonic presence of the two enzymes suggest that they exert an indispensable function that is not involved with transmission. Since this function may be developmental, Drews (1975) referred to the early AChE as "embryonic cholinesterase" and specifically related this cholinesterase to development. It should be added that, besides growth factors and cholinergic morphogens, other transmitters also may serve as trophic substances; this is true particularly for peptide transmitters such as the vasoactive intestinal peptide which appears to be present in the nervous tissues (Brennerman et al., 1985; Gozes et al., 2003).

An important issue that concerns the morphogen role of the cholinergic components is that it may relate to the ancestral role of the cholinergic system, the original role that did not involve transmission. This notion is discussed in detail in section BIV as well as in Chapter 11.

D. Teratologic and Ontogenetic Effects of Cholinergic Components and Cholinergic Agents

1. Historical Introduction

Karczmar (1963b; Karczmar et al., 1973) proposed that the "ACh system is involved in morphogenesis and differentiation" and adduced many findings in support of this notion; Drews (1975) postulated that "embryonic cholinesterase" exerts a morphogenetic role; and Lauder (1988) suggested that the components of the cholinergic system act as "morphogens" during development. As these and other components of the cholinergic system appear early in development (see sections BI and BII), they may all be referred to as "embryonic ACh," "embryonic CAT," and so forth. It could then be expected that, administered in the course of development, cholinergic agonists and antagonists, as well as antiChEs, should have a morphogenetic action. That it indeed is so was described in the preceding sections (section CIII-1 and CIII-2).

However, long before these developments and establishment of the early ontogenetic presence of cholinergic components, A. P. Mathews (1902) and Torald Sollman (1904)[14] found that in echinoderms, pilocarpine facilitates blastulation and gastrulation and atropine retards these developmental phenomena; also, these two drugs antagonized each other with respect to these effects. The motivation behind Mathews' experiments is piquant. Mathews was not interested at all in developmental actions of these drugs; he wished to falsify the contemporary (and the present) notion that these substances act at and via the autonomic junctions (this notion laid down the foundation for the theory of cholinergic transmission), and he argued that these drugs exert direct cellular actions. To prove his contention, he demonstrated that these drugs exercise effects prior to the onset of neurogenesis. His argument does not stand today with respect to autonomic (or central) cholinergic transmission; it is valid, if the term "morphogen" action is substituted for Mathews' term, the "direct" action (see section CIII-1).

Between the findings of Sollman and Mathews of the early 1900s and the current demonstration of precocious developmental appearance of cholinergic components and of their role as morphogens, a large body of investigations that became available made evident the morphogenetic role of the cholinergic system and the morphological effects of cholinergic components and cholinergic drugs. The war gas antiChEs scare added to the interest in this subject (see Chapter 7 DI and DII).

2. Teratologic and Toxic Actions of Cholinergic Components and Drugs

Cholinergic drugs induce three main types of developmental effects: they may cause abnormal development or malformations (teratologic actions), they may be toxic and produce pre- or early postnatal death or eradication of an organ, and, when cholinergic agents are applied during development, they may affect the function and behavior of the adult. Unfortunately, only a few pertinent studies were carried out with regard to this last notion; they indicated that prenatal application of antiChEs and cholinergic agonists induce changes in such adult behaviors as aggression (Karczmar et al., 1973; Karczmar, 1976). Then mice treated between days 4 and 11, postbirth, with chlorpyrifos, the insecticidal OP antiChE, showed during adulthood increased locomotion, aggression, and novelty-seeking behavior (Ricceri et al., 2003); it should be added that such treatment may evoke in rodents changes in some cholinergic components (see Richardson and Chambers, 2005). There is also some anecdotal information suggesting that OP war agents employed in the course of the Gulf War of 1991 may have caused mental and behavioral effects in the offspring of the exposed personnel; this information is unconfirmed (see Chapter 7 DII).

In several species, cholinergic drugs are toxic, teratologic, or both with respect to the early, preneurogenetic embryos. This is true for invertebrates, including echinoderms and insects (see Chadwick, 1963; Karczmar, 1963b; Karczmar et al., 1973; Gustafson and Toneby, 1971; Buznikov et al., 1972; Buznikov, 1990; see also section D-1, above, and Hayes, 1982). Thus, antiChEs cause both lethality and malformations of the gastrula in echinoderm development (Gustafson and Toneby, 1971), and the literature concerning actions of cholinergic drugs on insects is enormous, as antiChEs and cholinergic agonists are intensely used as insecticides and ovicides, and may be used as war agents (see Chapter 7 A). Let it be mentioned that, while ovicidal effects of antiChEs in insects and their toxic effects in echinoderms occur prior to neurogenesis, AChE and ACh are present in these forms at that time (see section BI).

Morphogenetic and toxic actions obtain also in early vertebrate embryos, including amphibia, aves, fishes, chick, and rodents. In all these forms, cholinergic agonists of both muscarinic and nicotinic type and antiChEs induce malformations and/or death of the early embryos; these agents can cause lethality in these species either pre- or early postnatally (Schuytema et al., 1994; Nguyen and Janssen, 2001; Meiniel, 1973, 1981; see also Hayes, 1982). Some species may be resistant. While Weyland Hayes (1982) quotes several instances of toxicity of insecticides to rat embryos and particularly pups following their administration to the pregnant dams (see also Maurissen et al., 2000), other investigators report a resistance of rodent embryos and pups to lethal actions of high doses of insecticides and other antiChEs and cholinergic agonists; or, at most, an infrequent mortality of pups was recorded (Levine and Parker, 1991; Breslin et al., 1996; Courtney et al., 1985; Lechner and Abdel-Rachman, 1984). Sporadic information suggests that bovine fetuses may also be resistant to OP insecticides (Bellows et al., 1975). On practical grounds it is important to assess the sensitivity to pesticides of the offspring of farm animals, including cows, sheep, horses, and the like; it appears that the pertinent information is scanty.

Again, anecdotal information suggests that malformations were noticed in the offspring of the personnel exposed to antiChE war gases during the Gulf War; this information is unconfirmed and probably misleading (see Chapter 7 DII).

Studies of developmental effects of combined administration of two or more antiChEs are of interest, as joint application of two or more of these compounds for their insecticidal action has become frequent (see Chapter 7 A). A study, already referred to (see above, this section), suggests that while embryo uterine implantations are "slightly" decreased by treatment of dams with Malathion or Carbaryl applied singly, they are further decreased by applying these insecticides in combination (Lechner and Abdel-Rachman, 1984).

Axial notochord, skeletal, or cartilage malformations constitute the most consistent and common effect. In the pertinent studies, antiChEs (including insecticides such as parathion) and cholinergic agonists (such as pilocarpine, ACh, and methacholine) were either given to pregnant animals, administered in utero, or injected into the yolk sac or embryo. An interesting method aimed at assessing insecticidal water pollution involves placing fish eggs in streams allegedly containing insecticides (Hiroaka and Okuda, 1984). Early, Ancel (1945) reported that physostigmine, applied late in development of the chick, caused deformation of long bones (micromelia; particularly of the leg and lower vertebrae (rumplessness) and parrot beak. In addition, syndactilism, changes in facial bone and palate, "undulating notochord," "wry neck," "short neck," malformation of the eyelids, clubbed down, fused vertebrae, neck deformities, and other changes were reported; sometimes, these malformations are referred to as axial (Meiniel, 1973; Landauer, 1977; Sullivan, 1973; Misawa et al., 1982; Rashev and Vasilev, 1982; Wittenbach and Hwang, 1984; see also Karczmar, 1963b). Cartilage changes, scoliosis, vertebral damage, and gill and fin malformations are caused by antiChEs in urodeles, amphibia, and fish (Hiraoka and Okuda, 1984; Alvarez et al., 1995; see also Karczmar, 1963b; Schotté and Karczmar, unpublished observations). It is of interest that Landauer (1977) and Meiniel (1976) managed to antagonize, by means of oximes, some but not all deformations caused by insecticides and other antiChEs in the chick.

The developmental time and length of the application of the cholinergic drugs varied a great deal in the studies listed; therefore, it is difficult to correlate the effects obtained with specific developmental phenomena. Obviously, notochord, bone, and cartilage alterations are unrelated to neurogenesis, so it is irrelevant whether or not nervous system was present at the time of the administration of the drugs in question.

The developmental transit from the notochord and the cartilage to the vertebral column is pertinent. On the basis of ultrastructural studies Meiniel (1976) concluded that cartilage is not a factor in vertebral teratology induced by antiChEs. Experimenting on the catfish, Lien et al. (1997) found that skeletal abnormalities can be induced by an insecticide applied prior to the formation of vertebrae; they suggested that these abnormalities were due to the deformation of the notochord by muscle contractions or contractures induced by the insecticide, rather than to the insecticide action on the bone. Actually, cervical somite contractions were observed in prenatal chick embryos treated with antiChE insecticides, in conjunction with twisted notochord and spinal cord (Strudel and Gateau, 1977). However, vertebral malformations were obtained when cholinergic agents were applied early postnatally or late in development, that is, after the formation of the vertebral column; also, it was suggested that different axial deformities such as neck versus torso deformities arise at different developmental times of the application of antiChEs (Misawa et al., 1982). It must be noted that even chronically administered high doses of antiChEs do not produce in the adult long-lasting contracture; rather, muscle paralysis may result under these circumstances; in fact, Landauer (1977) and Sullivan (1973) reported that neostigmine, physostigmine, and carbachol induced "muscle hypoplasia" or "paralysis" in the chick embryo (paradoxically, Landauer opined that muscle contractures may explain "the axial abnormalities"). Another possibility was raised by Wittenbach and Hwang (1984): they proposed that the insecticides cause a disruption of the notochord sheath leading to folding of the notochord and subsequent axial deformations.

Still another point that may be raised concerns the presence of cholinergic components at the time of the application of antiChEs. As already mentioned, even prior to neurogenesis the embryo exhibits cholinergic components (see section BI). Accordingly, accumulation of ACh resulting from antiChE treatment may cause malformations since negative morphogenetic effects may be induced by ACh, while inhibition of AChE may also cause malformations, as AChE exhibits trophic effects and its inhibition may cause teratology (see section CIII). However, in the case of many studies the malformations were obtained only with nearly LD50 doses of antiChEs that induced toxic symptoms to the dams (muscle fasciculations, tremors, etc.); these doses may have been higher than doses needed for complete AChE inhibition (see, for example, Nguyen and Janssen, 2001); the animals that exhibited these malformations survived as only a single treatment or several treatments separated by time intervals were employed, and the animals could recover from the treatment or treatments. Therefore, the teratology in question was probably not due to ACh accumulation or inactivation of the enzyme.

An important toxicity of OP insecticides concerns the retina. This toxicity was frequently reported for adult humans. In fact, it is so prevalent in a prefecture of Japan that it is referred to, after this region, as Saku syndrome (Dementi, 1994). However, it may also arise in adult animals following prenatal treatment with insecticide antiChEs. And many studies demonstrated that single and, particularly, chronic exposure to pesticides may induce several forms of neurotoxicity (Maurissen et al., 2000).

In view of the continuously increasing use of insecticides, particularly in combinations, more work is needed in the area of cholinergic teratology and postnatal behavioral effects of prenatal cholinergic treatment.

A related matter is that of the behavioral and environmental effects and of the influence of prenatal cholinergic treatment on cholinergic development and postnatal behavior. Behavioral effects of prenatal antiChE treatment and of environmental modifications on postnatal behavior of mice were reported early by Karczmar et al. (1973); this notion was supported for subsequent data; for example, rearing of gerbils in isolation induced changes in the density of cortical cholinergic fibers and induced behavioral alterations (Lehmann et al., 2004).

Notes

1. Originally, morphogens were defined as substances (such as retinoic acid) that created a gradient playing regulatory and morphogenetic roles in early embryos (see Weiss, 1939 and Lauder, 1988).
2. Some of the earlier investigators found CAT, AChE, and/or cholinergic receptors only postnatally in the rodent brain (see, e.g., Hohmann and Ebner, 1985), while others did not initiate their studies until late in gestation (see, e.g., Coyle and Yamamura, 1976).
3. Synapsins participate in differentiation of the neuromyal junction and in central synaptogenesis (see section BII-5 and Sanes et al., 1998).
4. Joshua Sanes and associates (1998) stated that the neuromyal junction is "the best understood of all synapses." I agree completely with the notion that this transmittive site is best understood. However, the term "synapse," which is Sherrington's nomenclature for a neuron-neuron transmittive link, is inappropriate for the neuromyal junction.
5. Naguib Mechawar, Laurent Descarries, and their associates (Mechawar and Descarries, 2001; Mechawar et al., 2002) described synapses that do not exhibit synaptic specializations, where the release of ACh occurs in a diffused manner (see Chapter 2 C; Karczmar, 2004). However, this does not contradict the notion that the final stages of synaptogenesis include the arrival of varicosities at the dendrites and intimate conformation of these various entities.
6. There may be some doubt regarding the noncholinergicity of these regions (see Chapter 2 D).
7. Victor Hamburger was a great neuroembryologist. In many of her papers, Rita Levi-Montalcini stressed and reviewed his role in establishing the concept of trophic actions and trophic factors. Rita Levi-Montalcini commented that Hamburger abandoned the field just at the time of its quantal jump, which may have cost him the Nobel Prize; instead, it was given to Rita Levi-Montalcini. Incidentally, Victor Hamburger lived to be a centenarian, his age overlapping and extending beyond the 20th century by 1 year. As for Stanley Cohen, Rita Levi-Montalcini spoke of him as a "most talented biochemist" (Levi-Montalcini, 1965; see also Oppenheim, 2001).
8. The heuristic importance of the temporal continuity and mutuality between gurus, teachers and students is well illustrated by the story of trophisms. Thus, both Hamburger and Weiss admired Ramon y Cajal's work and quoted it extensively, as they quoted each other (see, for example, Weiss, 1939); Hamburger, Young, and Weiss maintained close scientific contacts; and Elmer Bueker and Rita Levi-Montalcini were Hamburger's students at Washington University in St. Louis, while Stanley Cohen and Hans Thoenen may be considered Hamburger's spiritual grandchildren. The process went the full cycle when in 1989 Levi-Montalcini was awarded the Cajal International Diploma (subsequently, she earned the Nobel Prize). Hamburger, Young, and

Weiss maintained close scientific contacts and referred to each other's work. Finally, Hans Thoenen, Victor Hamburger, Rita Levi-Montalcini, and Stanley Cohen (who, early on, joined the St. Louis team) are coauthors of a multitude of papers.

9. Why did Rita Levi-Montalcini and Stanley Cohen look for NGF in snake venom? Indeed, NGF concentrations in the venom of snakes *Elapsidae*, *Viperidae*, and *Crotonidae* and the Gila monster are 1,000-fold higher than in mouse sarcomas. However, mouse submaxillary glands contain NGF concentrations 3 times higher than those in venoms (Hendry, 1976). This search for NGF in the venoms was serendipitous: Stan Cohen and Rita Levi-Montalcini noticed that NGF activity in venom, which they used as a part of their assay medium, is high (see Cohen and Levi-Montalcini, 1956; Cohen, 1960).
10. However, according to Hefti and associates, NGF did not affect the fibers or nerve terminals of nucleus basalis neurons (Hartikka and Hefti, 1988). These investigators also stated that the difference in the NGF sensitivity of nucleus basalis neurons on the one hand and septal neurons on the other hand shows that the septum has a greater need for NGF than the nucleus basalis. These structures receive NGF via retrograde transport from their target areas; for the septum these areas include the hippocampus, and for the nucleus basalis these areas incorporate many structures such as the neocortex and the limbic system (see Chapter 2 DII and DIII).
11. Cuello (1996) comments on the technical difficulties that make these experiments so difficult to interpret.
12. Some of the proteoglycans and glycosaminoglycans that exhibit CAM-like activities have a molecular weight of 25 to 29 kDa, which is one fifth to one tenth that of CAMs.
13. Two distinguished, pioneer investigators, the late Miro Brzin, a Yugoslavian cholinergiker, and VirginiaTennyson, a US cholinergiker, are credited with this important 1970 finding. Ivan Kostovic, an active student of human cholinergic development, was a colleague of Brzin.
14. Torald Sollman was an early US pharmacologist at Yale University who published at the end of the nineteenth century a text of pharmacology that merited several revised editions. W. B. Saunders of Philadelphia published the cumulative bibliography of these texts, which is most useful even today.

References

Abdon, N. O.: Liberation of acetylcholine from the precursor in voluntary muscles without motor-end plates. Acta Pharmacol. (Kbh) *1*: 325–335, 1945.

Acheson, A. and Rutishauser, U.: Neural cell adhesion molecule regulates cell contact-mediated changes in choline acetyltransferase activity of embryonic chick sympathetic neurons. J. Cell Biol. *106*: 479–486, 1988.

Agranoff, B. W., Benjamins, J. A. and Ifajru, A. K.: Lipids. In: "Basic Neurochemistry," G. J. Siegel, Ed.-in-chief, pp. 47–67, Lippincott-Raven, Publisher, Philadelphia, Pennsylvania, 1999.

Aguilar, J. S. and Lundtr, G. G.: Muscarinic receptor sites in human foetal brain. Neurochem. Int. *7*: 509–514, 1985.

Agulhon, C., Charnay, Y., Vallet, P., Abitbol, M., Bertrand, D. and Malafosse, A.: Distribution of mRNA for the alpha 4 subunit of the nicotinic acetylcholine receptor in the human fetal brain. Mol. Brain Res. *58*: 123–131, 1998.

Agulhon, C., Abitbol, M., Bertrand, D. and Malafosse, A.: Localization of mRNA for CHRNA7 in human fetal brain. Neuroreport *10*: 2223–2227, 1999.

Albright, C. D., Tsai, A. Y., Friedrich, C. B., Mar, M. H. and Zeisel, S. H.: Choline availability alters embryonic development of the hippocampus and septum in the rat. Dev. Brain Res. *113*: 13–20, 1999.

Albright, C. D., Mar, M.-H., Friedrich, C. B. and Brown, E. C.: Maternal choline availability alters the localization of p15Ink4B and p27Kip 1 cyclin-dependent kinase inhibitors in the developing fetal rat brain hippocampus. Dev. Neurosci. *23*: 100–106, 2001.

Albuquerque, E. X., Warnick, J. E., Sansone, F. M. and Onur, R.: The effects of vinblastine and colchicine on neural regulation of muscle. Ann. N. Y. Acad. Sci. *228*: 224, 1974.

Altar, C. A., Armanini, M., Dugich-Djordjevic, M., Bennet, G. L., Williams, R., Feinglass, S., Anicetti, V., Sinicropi, D. and Bakhi, C.: Recovery of cholinergic phenotype in the injured rat neostriatum: role of endogenous and exogenous nerve growth factor. J. Neurochem. *59*: 2167–2177, 1992.

Alturi, P., Fleck M. W., Shen, Q., Mah, S. J., Stadfelt, D., Barnes, W., Goderie, S. K., Temple, S. and Schneider, A. S.: Functional nicotinic acetylcholine receptor expression in stem and progenitor cells of the early embryonic mouse cerebral cortex. Dev. Biol. *240*: 143–156, 2001.

Alvarez, R., Honrubia, M. P. and Herraez, M. P.: Skeletal malformations induced by the insecticides ZZ-Aphox and Folidol during larval development of Rana perezi. Arch. Environ. Contam. Toxicol. *28*: 329–356, 1995.

Amaldi, P.: Polysome analysis of embryonic sensory ganglia cultured *in vitro* with the nerve growth factor. J. Neurochem. *18*: 827–832, 1971.

Ammon, R. and Schutte, E.: Über das Verhalten von Enzymen im Hühnerei wahrend der Bebrutung. Biochem. Zschr. *275*: 216–233, 1934.

Ancel, P.: L'achondroplasie. Sa realization experimentale, sa pathoghènie. Ann. Endocr. (Paris) *6*: 1–24, 1945.

Anderson, D. J.: Molecular biology of acetylcholine receptor: structure and regulation of biogenesis. In: "The Vertebrate Neuromuscular Junction," M. M. Salpeter, Ed., pp. 285–315, Alan R. Liss, New York, 1987.

Anderson, R. B. and Key, B.: Novel guidance cues during neuronal pathfinding in early scaffold of axon tracts in the rostral brain. Development *126*: 1859–1868, 1999.

Anderson, M. J. M., Cohen, W. and Zorychta, E.: Effects of innervation on the distribution of acetylcholine receptors on cultured muscle cells. J. Physiol. (Lond.) *268*: 731–756(9), 1977.

Ando, S.: Gangliosides in the nervous system. Neurochem. Int. *5*: 505–537, 1983.

Andres, R. Y., Yeng, I., and Bradshaw, R. A.: Nerve growth factor receptors: identification of distinct classes in plasma membranes and nuclei of embryonic dorsal root neurons. Proc. Natl. Acad. Sci. U.S.A. *74*: 2785–2789, 1977.

Angeletti, P. U.: Biological properties of the nerve growth factor. Adv. Exp. Biol. Med. *32*: 83–90, 1972.

Angeletti, P. U. and Levi-Montalcini, R.: Specific cytotoxic effects of 6-hydroxydopamine on sympathetic neurons. Arch. Ital. Biol. *108*: 213–221, 1970.

Angeletti, R. H. and Bradshaw, R. A.: Nerve growth factor from mouse submaxillary gland. Aminoacid sequence. Proc. Natl. Acad. Sci. U.S.A. *68*: 2417–2420, 1971.

Angeletti, R. H., Angeletti, P. U. and Levi-Montalcini, R.: Selective accumulation of [125I] labeled nerve growth factor in sympathetic ganglia. Biochemistry *46*: 421–425, 1971.

Angeletti, R. H., Hermodson, M. A. and Bradshaw, R. A.: Amino acid sequences of mouse 2.5S nerve growth factor. II. Isolation and characterization of thermolytic and peptic peptides and the complete covalent structure. Biochemistry *12*: 100–115, 1973.

Appleyard, M. E.: Non-cholinergic functions of acetylcholinesterase. Biochem. Soc. Trans. *22*: 749–755, 1994.

Aracava, Y., Deshpande, S. S., Swanson, K. L., Rapaport, H., Wonnacott, S., Lunt, G. and Albuquerque, E. X.: Nicotinic acetylcholine receptors in cultured neurons from the hippocampus and brain stem of the rat characterized by single channel recording. FEBS Lett. *222*: 63–70, 1987.

Armstrong, J. N. and Pappas, B. A.: The histopathological, behavioral and neurochemical effects of intraventricular injection of ethylcholine mustard aziridinium (AF64A) in the neonatal rat. Brain Res. Dev. Brain Res. *61*: 249–257, 1991.

Armstrong, D. M., Bruce, G., Hersh, L. B. and Gage, F. H.: Development of cholinergic neurons in the septal/diagonal band complex of the rat. Brain Res. *433*: 249–256, 1987.

Arpagaus, M., Combes, D., Culetto, E., Grauso, M., Fedon, Y., Romani, R. and Toutant, J.-P.: Four acetylcholinesterase genes in the Nematode *Caenorhabditis elegans*. In: "From *Torpedo* Electric Organ to Human Brain: Fundamental and Applied Aspects," J. Massoulié, Ed., pp. 363–367, Elsevier, Amsterdam, 1998.

Atluri, P., Fleck, M. W., Shen, Q., Mah, S. J., Stadfelt, D., Barnes, W., Goderie, S. K., Temple, S. and Schneider, A. S.: Functional nicotinic receptor expression in stem and progenitor cells of the early embryonic cerebral cortex. Development. *2450*: 143–156, 2001.

Atsumi, S.: The histogenesis of motor neurons with special reference to the correlation of their endplate formation. III. The development of motor neurons innervating the intercostal muscle in the chic embryo. Acta Anat. *80*: 504–515, 1971.

Aubert, I., Cecyre, D., Gauthier, S. and Quirion, R.: Comparative ontogenetic profile of cholinergic markers, including muscarinic and nicotinic receptors, in the rat brain. J. Comp. Neurol. *369*: 31–59, 1996.

Auburger, G., Heumann, R., Hellweg, R., Korsching, S. and Thoenen, H.: Developmental changes of nerve growth factor and its mRNA in the rat hippocampus: comparison with choline acetyltransferase. Dev. Biol. *120*: 322–328, 1987.

Augustinsson K. B. and Gustafson, T.: Cholinesterase in developing sea-urchin eggs. J. Cell Physiol. *34*: 311–321, 1949.

Axelrod, J.: Brain monoamines. Biosynthesis and fate. Neurosci. Res. Program Bull *9*: 188–196, 1971.

Axelrod, D., Ravdin, P., Koppel, D. E., Schlessinger, J., Webb, W. W., Elson, E. L. and Podleski, T. R.: Lateral motion of fluorescent-labelled acetylcholine receptors in the membrane of developing muscle fibers. Proc. Natl. Acad. Sci. U.S.A. *73*: 4594–4598, 1976.

Bacq, Z. M.: "Chemical Transmission of Nerve Impulses," Pergamon Press, Oxford, 1975.

Bakaikin, V. M.: Morphologic and histochemical organization of the pericardial nervous apparatus of vertebrates and humans. Arkh. Anat. Gistol. Embriol. *74*: 82–89, 1978.

Balak, K., Jacobson, M., Sunshine, J. and Rutishauser, U.: Neural cell adhesion molecule expression in Xenopus embryos. Dev. Biol. *119*: 540–550, 1987.

Bamburg, J. R., Derby, H. A. and Shooter, E. M.: Isolation and characterization of the 7S nerve growth factor protein from inbred strains of mice. Neurobiology *1*: 115–120, 1971.

Bandtlow, C. E.: Regeneration in the central nervous system. Exp. Gerontol. *38*: 79–86, 2003.

Bandtlow, C. E. and Dechant, G.: From cell death to neuronal regeneration, effects of the p75 neurotrophin receptor depend on interactions with partner subunits. Sci. STKE *235*: p. 24, 2004.

Banerjee, S. P., Snyder, S. H., Cuatrecasas, P. and Greene, L. A.: Binding of nerve growth factor receptor in sympathetic ganglia. Proc. Natl. Acad. Sci. U.S.A. *70*: 2519–2523, 1973.

Barbin, G., Manthorpe, M. and Varon, S.: Receptor binding: influence of ions, enzymes, and protein reagents. J. Biol. Chem. *250*: 1427–1433, 1975.

Barbin, G., Manthorpe, M. and Varon, S.: Purification of the chick eye ciliary neuronotrophic factor. J. Neurochem. *43*: 1468–1478, 1984.

Barde, Y. A., Edgar, D. and Thoenen, H.: Purification of a new neurotrophic factor from mammalian brain. EMBO J. *1*: 549–553, 1982.

Barr, G. A., Eckenrode, T. C. and Murray, M.: Normal development and effects of early deafferentiation on choline acetyltransferase, substance P and serotonin-like immunocreativity in the interpeduncular nucleus. Brain Res. *418*: 301–313, 1987.

Bartsch, U.: Neural cams and their role in the development and organization of myelin sheath. Front. Biosci. *8*: D477–490, 2003.

Bau, B., Haag, J., Schmid, E., Kaiser, M., Gebhard, P. M. and Aigner, T.: Bone morphogenetic protein-mediating receptor-associated Smads as well as common Smad are expressed in human articular chondrocytes but not up-regulated or down-regulated in osteoarthritic cartilage. J. Bone Miner Res. *17*: 2141–2150, 2002.

Bear, M. F., Carnes, K. M. and Ebner, F. F.: Postnatal changes in distribution of acetylcholinesterase in kitten striate cortex. J. Comp. Neurol. *237*: 519–532, 1985.

Beck, C. E. and Perez-Polo, J. R.: Human beta-nerve growth factor does not crossreact with antibodies to mouse beta-nerve growth factor in a two-site radioimmunoassay. J. Neurosci. Res. *8*: 137–152, 1982.

Bellows, R. A., Rumsey, T. S., Kasson, C. W., Bond, J., Warwick, E. J. and Pahnish, O. F.: Effects of organic phosphate systemic insecticides on bovine embryonic survival and development. Am. J. Vet. Res. *36*: 1133–1140, 1975.

Ben-Baruch, N. and Yarden, Y.: Neu differentiation factors: a family of alternatively spliced neuronal and mesenchymal factors. Proc. Soc. Exp. Biol. Med. *206*: 221–227, 1994.

Benowitz, L. I. and Greene, L. A.: Nerve growth factor in the goldfish brain: biological assay studies using pheochromocytoma cells. Brain Res. *162*: 164–168, 1979.

Berardi, N., Cellerino, A., Domenici, L., Fagiolini, M., Pizzorusso, T., Cattaneo, A. and Maffei, L.: Monoclonal antibodies to nerve growth factor affect the postnatal development of the cortex. Proc. Natl. Acad. Sci. U.S.A. *91*: 684–688, 1994.

Bersani, G., Iannitelli, A., Fiore, M., Angelucci, F. and Aloe, L.: Data and hypotheses on the role of nerve growth factor and other neurotrophins in psychiatric disorders. Med. Hypotheses *55*: 199–207, 2000.

Beyer, G. and Wense, T.: Über den Nachweis von Hormonen in einzelligen Tieren. I. Cholin und Acetylcholin im Paramecium. Pflügers Arch. Ges. Physiol. *237*: 417–422, 1936.

Biagioni, S., Tata, A. M., De Jaco, A. and Augusti-Tocco, G.: Acetylcholine synthesis and neuron differentiation. Int. J. Dev. Biol. *44*: 689–697, 2000.

Bigbee, J. W., Sharma, K. V., Gupta, J. J. and Dupree, J. L.: Morphogenetic role for acetylcholinesterase in axonal growth during neural development. Environ. Health Perspect. *107*: 81–87, 1999.

Bixby, J. L.: Ultrastructural observations on synapse elimination in neonatal rabbit skeletal muscle. J. Neurocytol. *10*: 81–100, 1981.

Bixby, J. L. and Reichardt, L. F.: Effects of antibodies to neural cell adhesion molecule (N-CAM) on the differentiation of neuromuscular contacts between ciliary ganglion neurons and myotubes *in vitro*. Dev. Biol. *119*, 363–372, 1987.

Black, I. B.: Regulation of autonomic development. Ann. Rev. Neurosci. *1*: 183–214, 1978.

Black, I. B. and Mytilineou, C.: Trans-synaptic regulation of the development of end organ innervation by sympathetic neurons. Brain Res. *101*: 503–521, 1976.

Black, I. B., Hendry, I. A., and Iversen, L. L.: The role of post-synaptic neurones in the biochemical maturation of presynaptic cholinergic nerve terminals in a mouse sympathetic ganglion. J. Physiol. *221*: 149–159, 1972.

Black, I. B., Coughlin, M. D. and Cochard, P.: Factors regulating neuronal differentiation. Soc. Neurosci. Symp. *4*: 184–207, 1979.

Blank, M., Triana-Baltzer, G. B., Richards, C. S. and Berg, D. K.: Alpha-protocadherins are presynaptic and axonal nicotinic pathways. Mol. Cell Neurophysiol. *26*: 530–543, 2004.

Blass, J. P.: Brain metabolism and brain disease: is metabolic deficiency the proximate cause of Alzheimer dementia? J. Neurosci. Res. *66*: 851–856, 2001.

Bloch, R. J. and Steinbach, J. H.: Reversible loss of acetylcholine receptor clusters at the developing rat neuromuscular junction. Dev. Biol. *81*: 386–391, 1981.

Blue, M. E. and Parnavales, J. G.: The formation and maturation of synapses in the visual cortex of the rat. II. Quantitative analysis. J. Neurocytol. *12*: 6976–7112, 1983.

Bogoch, S.: Inhibition of viral hemagglutination by brain gangliosides. Virology *4*: 458–466, 1957.

Bogoch, S.: History of recognition molecules in the brain. In: "Ganglioside Function: Biochemical and Pharmacological Implications," G. Porcellati, B. Ceccarelli, and G. Tettamanti, Eds., pp. 233–265, Plenum Press, New York, 1975.

Bonhoeffer, T.: Neurotrophins and activity-dependent development of the neocortex. Curr. Opin. Neurobiol. *6*: 119–126, 1996.

Bonichon, A.: Evolution de l'activité de l'acetylcholinestèrase au niveau des lobes optiques chez l'embryon de poulet. C. R. Acad. Sci. (Paris) *245*: 1345–1347, 1957.

Bothwell, M.: Functional interactions of neurotrophins and neurotrophin receptors. Annu. Rev. Neurosci. *18*: 223–253, 1995.

Bowden, R. E. M. and Duchen, L. W.: The anatomy and pathology of the neuromuscular junction. In: "Neuromuscular Junction," E. Zaimis, Ed., Springer-Verlag, Berlin, pp. 23–97, 1976.

Bradshaw, R. A.: Nerve growth factor. Ann. Rev. Biochem. *47*: 191–216, 1978.

Bradshaw, R. A.: What cloned genes can tell us about nerve growth factor. Nature *303*: 751, 1983.

Braumann, T., Jastorff, B. and Richter-Landsber, C.: Fate of cyclic nucleotides in PC12 cell cultures: uptake, metabolism, and effects of metabolites on nerve growth factor-induced neurite outgrowth. J. Neurochem. *47*: 912–919, 1986.

Brenneman, D. E., Eiden, L. E. and Siegel, R. E.: Neurotrophic action of VIP on spinal cord cultures. Peptides *6*(2): 35–39, 1985.

Brenner, H. R., Meier, Th. and Wimer, B.: Early action of nerve determines motor endplate differentiation in rat muscle. Nature *305*: 536–537, 1983.

Breslin, W. J., Liberacki, A. B. Dittenber, D. A. and Quast, J. F.: Evaluation of the developmental and reproductive toxicity of chlorpyrifos in rats. Fundam. Appl. Toxicol. *29*: 119–130, 1996.

Brimijoin, S.: Trophic roles of axoplasmic transport. In: "Nerve-Muscle Cell Trophic Communication," H. L. Fernandez and J. A. Donoso, Eds., pp. 23–39, CRC Press, Boca Raton, Fla., 1987.

Brimijoin, S. and Hammond, P.: Transient expression of acetylcholinesterase messenger RNA and enzyme activity in then developing rat thalamus studied by quantitative histochemistry and *in situ* hybridization. Neurosci. *71*: 555–565, 1996.

Brimijoin, S. and Koenigsberger, C.: Cholinesterases in neural development: new findings and toxicologic implications. Environ. Health Perspect *107*: 59–64, 1999.

Brodski, C., Schaubmar, A. and Dechant, G.: Opposing functions of GDNF and NGF in the development of cholinergic and noradrenergic sympathetic neurons. Mol. Cell Neurosic. *19*: 528–538, 2002.

Brooksbank, B. W. L., Martinez, M., Atkinson, D. J. and Balazs, R.: Biochemical development of the human brain. I. Some parameters of the cholinergic system. Dev. Neurosci. *1*: 267–284, 1978.

Buccoliero, R., Bodennec, J. and Futerman, A. H.: The role of sphingolipids in neuronal development: lessons form models of sphingolipid storage diseases. Neurochem. Res. *27*: 565–574, 2002.

Buck, C. R., Martinez, H. J., Black, I. B. and Chao, M. V.: Developmentally regulated expression of the nerve growth factor receptor gene in the periphery and brain. Proc. Natl. Acad. Sci. U.S.A. *84*: 3060–3063, 1987.

Buck, C. R., Martinez, H. J., Chao, M. V. and Black, I. B.: Differential expression of the nerve growth factor receptor gene in multiple brain areas. Brain Res. Dev. Brain Res. *44:* 259–268, 1988.

Bueker, E. D.: Implantation of tumors in the hind limb field of the embryonic chick and the developmental response of the lumbosacral nervous system. Anat. Rec. *102*: 369–389, 1948.

Bulbring, E. and Burn, J. H.: Observations on synaptic transmission by acetylcholine in the spinal cord. J. Physiol. (Lond.) *100*: 337–368, 1941.

Bulbring, E. and Burn, J. H.: Action of acetylcholine on rabbit auricles in relation to acetylcholine synthesis. J. Physiol. (Lond.) *108*: 508–524, 1949.

Bulbring, E., Kottegoda, S. R. and Shelley, A.: Cholinesterase activity in the auricles of the rabbit's heart and their sensitivity to eserine. J. Physiol. (Lond.) *123*: 204–213,1954.

Bull, G., Hebb, C. and Ratkovic, D.: Choline acetyltransferase activity in human brain tissue during development and at maturity. J. Neurochem. *17*: 1505–1516, 1970.

Burden, S. J.: The formation of neuromuscular synapses. Genes Dev. *12*: 133–148, 1998.

Burn, J. H. and Walker, J. M.: Anticholinesterases in the heart-lung preparation. J. Physiol. (Lond.) *124*: 489–501, 1954.

Buznikov, G. A.: The action of neurotransmitters and related substances on early embryogenesis. Pharm. Ther. *25*: 23–59, 1984.

Buznikov, G. A.: "Neurotransmitters in Embryogenesis," Academic Press, New York, 1990.

Buznikov, G. A.: 5-Hydroxytryptamine, catecholamines, acetylcholine, and some related substances in early embryogenesis. In "Comparative Pharmacology,"

M. J. Michelson, Ed., pp. 593–623, Intern. Encyclop. Pharmacol. Exp. Ther., vol. 2, Pergamon Press, London, 1973.

Buznikov, G. A., Sakharova, A. V., Manukhin, B. A. and Markova, L. N.: The role of neurohumours in early embryogenesis. IV. Fluorometric and histochemical study of serotonin in cleaving eggs and larvae of sea urchins. J. Embryol. Exp. Morphol. *27*: 339–351, 1972.

Buznikov, G. A., Shmukler, Y. B. and Lauder, J. M.: From oocyte to neuron: Do neurotransmitters function in the same way throughout development? Cell Mol. Neurobiol. *16*: 533–559, 1996.

Buznikov, G. A., Bezuglov, V. V., Nikitina, L. A., Slotkin, T. A. and Lauder, J. M.: Cholinergic regulation of the sea urchin embryonic and larval development. Ross. Fiziol. Zh. Im. I. M. Sechenova. *87*: 1548–1556, 2001.

Buznikov, G. A., Nikitina, L. A., Voronezhskaya, E. E., Bezuglov, V. V., Dennis Willows, A. O. and Nezlin, L. P.: Localization of serotonin and its possible role in early embryos of Tritonia diomedea (Mollusca: Nudibranchia). Cell Tissue Res. *311*: 259–266, 2003.

Buznikov, G. A., Peterson, R. E., Nikitina, L. A., Bezuglov, V. V. and Lauder, J. M.: The pre-nervous serotonergic system of developing sea urchin embryos and larvae: pharmacologic and immunocytochemical evidence. Neurochem Res. *30*: 825–837, 2005.

Cajal, S. Ramon y: "Estudios Sobre la Degeneracion y Regeneracion del Sistema Nervioso," N. Moya, Madrid, 1913.

Cameron, H. A., Hazel, T. G. and McKay, R. D.: Regulation of neurogenesis by growth factors and neurotransmitters. J. Neurobiol. *36*: 287–306, 1998.

Candy, J. M., Perry, J. M., Perry, R. H., Bloxham, C. A., Thompson, J., Johnson, M., Oakley, A. E. and Edwardson, J. A.: Evidence for the early prenatal development of cortical cholinergic afferents from the nucleus of Meynert in the human foetus. Neurosci. Lett. *61*: 91–95, 1985.

Cao, X. S., Yang, L. J., Wu, X. Y., Wu, Y. H., Zhang, L. N. and Zhang, L. F.: Changes of bone morphogenesis proteins and transforming growth factorbeta in hind-limb bones of 21 d tail-suspended rats. Space Med. Med. Eng. (Beijing) *16*: 269–271, 2003.

Carden, B. W., Datskovskaia, A., Guido, W., Godwin, D. W. and Bickford, M. E.: Development of the cholinergic, nitrergic, and GABAergic innervation of the cat dorsal lateral geniculate nucleus. J. Comp. Neurol. *418*: 65–80, 2000.

Carr, C., McCourt, D. and Cohen, J. B.: The 43-kilodalton protein of Torpedo nicotinic postsynaptic membranes: purification and determination of primary structure. Biochemistry *26*: 7090–7102, 1987.

Carter, B. D. and Lewin, G. R.: Neurotrophins live or let die: does p75NTR decide? Neuron *18*: 1187–1190, 1997.

Castellanos, M. R., Aguilar, J., Fernandez, C. L., Almaguer, W., Mejias, C. and Varela, A.: Evaluation of neurorestorative effects of the murine beta-nerve-growth factor infusions in old rat with cognitive deficit. Biochem. Biophys. Res. Commun. *312*: 867–872, 2003.

Catone, C. and Ternaux, J. P.: Involvement of the alpha 7 subunit of the nicotinic receptor in morphogenic and trophic effects of acetylcholine on embryonic spinal motoneurons in culture. J. Neurosci. Res. *72*: 46–53, 2003.

Ceccarelli, B., Aporti, F. and Finesso, M.: Effects of brain gangliosides on functional recovery in experimental regeneration and reinnervation. In: "Ganglioside Function: Biochemical and Pharmacological Implications," G. Porcellati, B. Ceccarelli, and G. Tettamanti, Eds., pp. 275–293, Plenum Press, New York, 1975.

Chadwick, L. E.: Actions on insects and other invertebrates. In "Cholinesterases and Anticholinesterase Agents," G. B. Koelle, Ed., Handbch. d. exper. Pharmakol., Ergänzungswk., vol. 15, pp. 741–798, Springer-Verlag, Berlin, 1963.

Chaisuksunt, V., Campbell, G., Zhang, Y., Schachter, M., Liberman, A. R. and Anderson, P. N.: Expression of regeneration-related molecules in injured and regenerating striatal and nigral neurons. J. Neurocytol. *32*: 161–183, 2003.

Changeux, J. P.: Allosteric proteins: from regulatory enzymes to receptors–personal recollections. Bioessays *15*: 625–634, 1993a.

Changeux, J. P.: Compartmentalized expression of acetylcholine receptor genes during motor-endplate development. Neurosci. Facts *4*: 1–3, 1993b.

Changeux, J. P., Devillers-Thiery, A., Galzi, J. L. and Revah, F.: The acetylcholine receptor: a model of an allosteric membrane protein mediating intercellular communication. Ciba. Found. Symp. *164*: 66–89, 1992.

Chavez, J. C. and LaManna, J. C.: Activation of hypoxia-inducible factor-1 in the rat cerebral cortex after transient global ischemia: potential role of insulin-like growth factor-1. J. Neurosci. *22*: 8922–8931, 2002.

Chen, K. S., Nishimura, M. C., Armanini, M. P., Crowley, C., Spencer, S. D. and Phillips, H. S.: Disruption of a single allele of the nerve growth factor gene results in atrophy of basal forebrain cholinergic neurons and memory deficits. J. Neurosci. *17*: 7288–7296, 1997.

Chiappinelli, V. A. and Giacobini, E.: Time course of appearance of á-bungarotoxin binding sites during development of chick ciliary ganglion and iris. Neurochem. Res. *3*: 465–478, 1978.

Chun, L. L. Y. and Patterson, P. H.: Role of nerve growth factor in the development of rat sympathetic neurons *in vitro*. III. Effects of acetylcholine production. J. Cell. Biol. *75*: 712–718, 1977.

Chung, Y. H., Shin, C. M., Joo, K. M., Kim, M. J. and Cha, C. I.: Age-related upregulation of insulin-like growth factor receptor type I in rat cerebellum. Neurosci. Lett. *330*: 65–68, 2002.

Cohen, S.: A nerve growth-promoting protein. In: "The Chemical Basis of Development," W. D. McElroy and B. Glass, Eds., pp. 665–676, John Hopkins Press, Baltimore, 1958.

Cohen, S.: Purification of a nerve growth protein from the mouse salivary gland and its neurocytoxic antiserum. Proc. Natl. Acad. Sci. U.S.A. *46*: 302–311, 1960.

Cohen, S. and Levi-Montalcini, R.: A nerve growth stimulating factor isolated from snake venom. Proc. Natl. Acad. Sci. U.S.A. *42*: 571–574, 1956.

Cohen, S. and Levi-Montalcini, R.: Purification and properties of a nerve growth-promoting factor isolated from mouse sarcoma 180. Cancer Res. *17*: 15–20, 1957.

Cohen, S., Levi-Montalcini, R. and Hamburger, V.: A nerve growth-stimulating factor isolated from sarcomas 37 and 180. Proc. Natl. Acad. Sci. U.S.A. *40*: 1014–1018, 1954.

Colhoun, E. H.: The physiological significance of acetylcholine in insects and observations upon other pharmacologically active substances. In: "Advances in Insect Physiology," J. W. L. Beament, J. E. Treherne and W. B.Wigglesworth, vol. 1, pp. 17–19, Academic Press, New York, 1963.

Connolly, J. A.: Microtubules, microfilaments and the transport of acetylcholine receptors in embryonic myotubes. Exp. Cell Res. *159*: 430–440, 1985.

Conroy, W. G. and Berg D. K.: Nicotinic receptor subtypes in the developing chick brain: appearance of a species containing the alpha4, beta2, and alpha5 gene products. Mol. Pharmacol. *53*: 392–401, 1998.

Consolo, S., Romano, M. Scozzesi, C., Bonetti, A. C. and Ladinsky, H.: Acetylcholine and choline contents of rat skeletal muscle determined by a radioenzymatic microassay: effects of drugs. In: "Dynamics of Cholinergic Function," I. Hanin, Ed., pp. 583–592, Plenum Press, New York, 1986.

Cooper, J. D., Skepper, J. N., Berzghi, M., Lindholm, D. and Sofroniev, M. V.: Delayed death of septal cholinergic neurons after excitotoxic ablation of hippocampal neurons during early postnatal development in the rat. Exp. Neurol. *39*: 143–155, 1996.

Costa, E. and Guidotti, A.: The role of 3′,5′-cyclic adenosine monophosphate in the regulation of adrenal medullary function. In: "New Concepts in Neurotransmitter Regulation," A. J. Mandell, Ed., pp. 135–152, Plenum Press, New York, 1973.

Costa, E. and Guidotti, A.: Molecular mechanisms mediating the transsynaptic regulation of gene expression in adrenal medulla. In: "Psychopharmacology: A Generation of Progress," M. A. Lipton, A. DiMascio, and K. F. Killiam, Eds., Raven Press, New York, 1978.

Costa, L. G.: Muscarinic receptors and the developing nervous system. In: "Receptors in the Developing Nervous System," Vol. 2, "Neurotransmitters," I. Zagon and P. J. McLaughlin, Eds., pp. 21–42, Chapman and Hall, London, 1993.

Cotman, C. W.: Axon sprouting and regeneration. In: "Basic Neurochemistry," G. J. Siegel, Ed. in Chief, pp. 589–612, Lippincott-Raven, Philadelphia, 1999.

Coughlin, M. D., Dibner, M. D., Boyer, D. M. and Black, I. B.: Factors regulating development of an embryonic mouse sympathetic ganglion. Dev. Biol. *66*: 513–528, 1978.

Coumans, J. V., Lin, T. T., Dai, H. N., MacArthur, L., McAtee, M., Nash, C. and Bregman, B. S.: Axonal regeneration and functional recovery after complete spinal cord transection in rats by delayed treatment with transplants and neurotrophins. J. Neurosci. *21*: 9334–9344, 2001.

Court, J. K. A., Lloyd, S., Johnson, M., Griffiths, M., Birdsall, N. M. J., Piggott, M. A., Oakley, A. E., Ince, P. E., Perry, E. K. and Perry, R. H.: Nicotinic and muscarinic cholinergic receptor binding in the human hippocampal formation during development and aging. Dev. Brain Res. *101*: 93–105, 1997.

Courtney, K. D., Andrew, J. E., Springer, J. Dalley, L.: Teratogenic evaluation of the pesticides baygon, carbofuran, dimethoate and EPN. J. Environ. Sci. Health B *4*: 373–406, 1985.

Coyle, J. T. and Yamamura, H. L.: Neurochemical aspects of the ontogenesis of cholinergic neurons in the rat brain. Brain Res. *118*: 329–440, 1976.

Crisanti-Combes, P. B., Dessac, B. and Calothy, G.: Choline acetyl transferase activity in chick embryo neuroretinas during development in ovo and in monolayer cultures. Dev. Biol. *65*: 228–332, 1978.

Crone, S. A. and Lee, K. F.: Gene targeting reveals multiple essential functions of the neuregulin signaling system during development of the neuroendocrine and nervous systems. Ann. NY Acad. Sci. *971*: 547–553, 2002.

Crossley, D., 2nd, and Altimiras, J.: Ontogeny of cholinergic and adrenergic cardiovascular regulation in the domestic chick (Gallus gallus). Am. J. Physiol. *279*: 1091–1098, 2000.

Crowley, C., Spencer, S. D., Nishimura, M. C., Chen, K. S., Pitts-Meek, S. and Armanini, M. P.: Mice lacking nerve growth factor display perinatal loss of sensory and sympathetic neurons yet develop basal forebrain cholinergic neurons. Cell *76*: 1001–1011, 1994.

Crutcher, K. A. and Collins, F.: Entorhinal lesions result in increased nerve growth factor-like growth-promoting activity in medium conditioned by hippocampal slices. Brain Res. *399*(2): 383–390, 1986.

Csillik, B. and Savay, G.: Die Regeneration der subneuralen Apparate der motorischen Endplatten. Acta Neuro-veg. (Wien) *19*: 41–52, 1958.

Cuello, A. C.: Trophic responses of forebrain cholinergic neurons: a discussion. In: "Cholinergic Function and Dysfunction," A. C. Cuello, Ed., pp. 265–277, Elsevier, Amsterdam, 1993.

Cuello, A. C.: Trophic factor therapy in adult CNS: remodeling of injured basalo-cortical neurons. Prosp. Brain Res. *100*: 213–220, 1994.

Cuello, A. C., Stephens, P. H., Garofalo, L., Maysinger, D. and Tagari, P. C.: Cortical damage and the effects of sialoganglioside GM1 on forebrain cholinergic neurons. In: "Cellular and Molecular Basis of Cholinergic Function," M. J. Dowdall and J. N. Hawthorne, Eds., Ellis Horwood, Ltd., Chichester, 1987.

Cuello, A. C., Garofalo, L., Koenigsberg, R. L. and Maysinger, D.: Gangliosides potentiate *in vivo* and *in vitro* effects of nerve growth factor on central cholinergic neurons. Proc. Natl. Acad. Sci. U.S.A. *86*: 2056–2060, 1989.

Cull-Candy, S. G., Miledi, R. and Uchitel, O. D.: Properties of junctional and extrajunctional acetylcholine-receptor channels in organ cultured human muscle fibres. J. Physiol. (Lond.) *333*: 251–267, 1982.

D'Amico-Martel, A.: Temporal patterns of neurogenesis in avian cranial sensory and autonomic ganglia. Am. J. Anat. *163*: 351–372, 1982.

Daniloff, J. K., Bodony, R. P, Low, W. C. and Wells, J.: Cross-species embryonic septal transplants: restoration of conditioned learning behavior. Brain Res. *346*: 176–180, 1985.

Daniloff, J. K., Crossin, K. L., Pincon-Raymond, M., Murawsky, M., Rieger, F. and Edelman, G. M.: Expression of cytotactin in the normal and regenerating neuromuscular system. J. Cell Biol. *108*: 625–635, 1989.

Dascal, N. and Landau, E. M.: Types of muscarinic response in Xenopus oocytes. Life Sci. *27*: 1423–1428, 1980.

Daubas, P., Devilliers-Thiery, A., Geoffroy, B., Martinez, S., Bessis, A. and Changeux, J.-P.: Differential expression of neuronal acetylcholine receptor alpha2 subunit gene during chick brain development. Neuron *5*: 49–60, 1990.

Davey, D. F. and Cohen, M. W.: Localization of acetylcholine receptors and cholinesterase on nerve-contacted and noncontacted muscle cells grown in the presence of the agents that block action potentials. J. Neurosci. *6*: 673–680, 1986.

Davies, A. M., Bandtlow, C., Heumann, R., Korsching, S., Rohrer, H. and Thoenen, H.: Timing and site of nerve growth factor synthesis in developing skin in relation to innervation and expression of the receptor. Nature *326*: 353–358, 1987.

Davis, H. L.: Trophic influences of neurogenic substances on adult skeletal muscles *in vivo*. In: "Nerve-Muscle Cell Trophic Communication," H. L. Fernandez, J. A. Donoso, Eds., pp. 101–145, CRC Press, Boca Raton, Fla., 1987.

Day, T. and Greenfield, S. A.: A non-cholinergic, trophic action of acetylcholinesterase on hippocamapal neurons *in vitro*: molecular mechanisms. Neurosci. *111*: 649–656, 2002.

De la Rosa, E. J. and De Pablo, F.: Cell death in early neural development: beyond the neurotrophic theory. Trends Neurosci. *23*: 454–458, 2000.

Dementi, B.: Ocular effects of organophosphates: a historical perspective of Saku disease. J. Appl. Toxicol. *14*: 119–129, 1994.

De Moura Campos, L. C. and Hokoc, J. N.: Ontogeny of cholinergic amacrine cells in the opossum (*Didelphis aurita*) retina. Int. J. Dev. Neurosci. *17*: 795–804, 1999.

Dennis, M. J.: Development of the neuromuscular junction: inductive interactions between cells. Ann. Rev. Neurosci. *4*: 43–88, 1981.

De Pablo, F. and De la Rosa, E. J.: The developing CNS: a scenario for the action of proinsulin, insulin and insulin-like growth factors. Trends Neurosci. *18*: 143–150, 1995.

De Robertis, E., Lapetina, E. G. and Fiszer de Plazas, S.: Subcellular distribution and possible role of gangliosides in the CNS. In: "Ganglioside Function: Biochemical and Pharmacological Implications," G. Porcellati, B. Ceccarelli, and G. Tettamanti, Eds., Plenum Press, New York, 105–121, 1975.

Deshpande, S. S., Albuquerque, E. X. and Guth, L.: Neurotrophic regulation of prejunctional and postjunctional membrane at the mammalian motor endplate. Exp. Neurol. *53*: 152, 1976.

Dinopoulos, A., Eadie, L. A., Dori, I. and Parnavelas, J. G.: The development of basal forebrain projections to the rat visual cortex. Exp. Brain Res. *76*: 563–571, 1989.

Doherty, P. and Walsh, F. S.: Ganglioside GM1 antibodies and B-cholera toxin bind specifically to embryonic chick dorsal root ganglion neurons but do not modulate neurite regeneration. J. Neurochem. *48*(4): 1237–1244, 1987.

Dolcet, X., Soler, R. M., Gould, T. W., Egea, J., Oppenheim, R. W. and Comella, J. X.: Cytokines promote motoneuron survival through the Janus kinase-dependent activation of the phosphatidylinositol 3-kinase pathway. Mol. Cell. Neurosci. *18*: 619–631, 2001.

Dolezalova, H., Giacobini, E., Giacobini, G., Rossi, A. and Toschi, G.: Developmental variations of choline acetyltransferase, dopamine, beta-hydroxylase and monoamine oxidase in chicken embryo and chicken sympathetic ganglia. Brain Res. *73*: 309–320, 1974.

Dougherty, K. D. and Milner, T. A.: P75NTR immunoreactivity in the rat dentate gyrus is mostly within presynaptic profiles but is also found in some astrocytic and postsynaptic profiles. J. Comp. Neurol. *407*: 77–91, 1999.

Drachman, D. B.: Is acetylcholine the trophic neuromuscular transmitter? Arch. Neurol. *17*: 206–218, 1967.

Drachman, D. B., Stanley, E. F., Pestronk, A., Griffin, J. W. and Price, D. L.: Neurotrophic regulation of two properties of skeletal muscle by impulse-dependent and spontaneous acetylcholine transmission. J. Neurosci. *2*: 232, 1982.

Drews, U.: Cholinesterase in embryonic development. Prog. Histochem. Cytochem. *7*: 1–48, 1975.

Dryer, S. E. and Chiapinelli, V. A.: An intracellular study of synaptic transmission and dendritic morphology in sympathetic neurons of the chick embryo. Brain Res. *354*: 99–111, 1985.

Dunnett, S. B., Whishaw, I. Q., Bunch, S. T. and Fine, A.: Acetylcholine-rich neuronal grafts in the forebrain of rats: effects of environment enrichment, neonatal noradrenaline depletion, host transplantation site and regional source of embryonic donor cells on graft size and acetylcholinesterase-positive fibre outgrowth. Brain Res. *378*(2): 357–373, 1986.

Duprat, A. M., Kan, P., Foulkier, F. and Weber, M.: *In vitro* differentiation of neuronal precursor cells from amphibian late gastrulae; morphological, immunocytochemical studies, biosynthesis, accumulation and uptake of neurotransmitters. J. Embryol. Exp. Morphol. *86*: 71–87, 1985.

Durante, M.: Cholinesterase in the development of the ascidian, Ciona intestinalis. Experientia *12*: 307–308, 1956.

Durand, M., Coronas, V., Jourdan, F. and Quirion, R.: Developmental and aging aspects of the cholinergic innervation of the olfactory bulb. Int. J. Dev. Neurosci. *16*: 777–785, 1999.

Dustin, M. L. and Colman, D. R.: Neural and immunological synaptic relations. Science *298*: 785–789, 2002.

Duxson, M. J.: The effect of postsynaptic block on development of the neuromuscular junction in postnatal rats. J. Neurocytol. *11*: 395–408, 1982.

Ebbott, S. and Hendry, I.: Retrograde transport of nerve growth factor in the rat central nervous system. Brain Res. *139*: 160–163, 1978.

Ebendal, T., Belew, M., Jacobson, C.-O. and Porath, J.: Neurite outgrowth elicited by embryonic chick heart: partial purification of the active factor. Neurosci. Lett. *14(7)*: 91–95, 1979.

Eccles, J. C.: "The Physiology of Synapses," Springer-Verlag, New York, 1964.

Eckenstein, F.: Transient expression of NGF-receptor-like immunoreactivity in postnatal rat brain and spinal cord. Brain Res. *446*: 149–154, 1988.

Edelman, G. M. and Chuong, C. M.: Embryonic to adult conversion of neural cell adhesion molecules in normal and staggerer mice. Proc. Natl. Acad. Sci. U.S.A. *79*: 7036–7040, 1982.

Edelman, G. M. and Crossin, K. L.: Cell adhesion molecules: implications for a molecular histology. Annu. Rev. Biochem. *60*: 155–190, 1991.

Egozi, Y., Sokolovsky, M., Schejter, E., Blatt, I., Zakut, H., Matzkel, A. and Soreq, H.: Divergent regulation of muscarinic binding sites and acetylcholinesterase in discrete regions of the developing human fetal brain. Cel. Mol. Neurobiol. *6:* 55–70, 1986.

El Far, O., O'Connor, V., Dresbach, T., Pellegrini, L., DeBello, W., Schweizer, F., Augustine, G., Heuss, C., Schaefer, T., Charlton, M. P. and Betz, H.: Protein interaction implicated in neurotransmitter release. In: "From *Torpedo* Electric Organ to Human Brain: Fundamental and Applied Aspects," J. Massoulié, Ed., pp. 129–133, Elsevier, Amsterdam, 1998.

Enemar, A. B., Faick, B. and Hakanson, R.: Observations on the appearance of norepinephrine in the sympathetic system of the chick embryo. Dev. Biol. *11*: 268–283, 1965.

Erkman, L., Matter, J., Matter-Sadzinski, L. and Ballivet, M.: Nicotinic acetylcholine receptor gene expression in developing chick autonomic ganglia. Eur. J. Pharmacol. *393*: 97–104, 2000.

Eusebi, F., Magnia, F. and Alfei, L.: Acetylcholine-elicited responses in primary and secondary mammalian oocytes disappear after fertilization. Nature *227*: 651–653, 1979.

Fagan, A. M., Garber, M., Barbacid, M., Silos-Santiago, I. and Holtzman, D. M.: A role for TrkA during maturation of striatal and basal forebrain cholinergic neurons *in vivo*. J. Neurosci. *17*: 7644–7654, 1997.

Falugi, C.: Localization and possible role of molecules associated with the cholinergic system during "non-nervous" developmental events. Eur. J. Histochem. *37*: 287–294, 1993.

Fambrough, D. M.: Development of cholinergic innervation of skeletal, cardiac, and smooth muscle. In: "Biology of Cholinergic Function," A. M. Goldberg and I. Hanin, Eds., pp. 101–160, Raven Press, New York, 1976.

Fambrough, D. M.: Control of acetylcholine receptors in skeletal muscle. Physiol. Rev. *59*: 165–227, 1979.

Fauquet, M., Smith, J., Ziller, C. and LeDouarin, N. M.: Differentiation of autonomic neuron precursors *in vitro*: cholinergic and adrenergic traits in cultured neural crest cells. J. Neurosci. *1*: 478–492, 1981.

Favaron, M., Manev, H. Alho, H., Bertolino, B., Ferret, B., Guidotti, A. and Costa, E.: Gangliosides prevent glutamate and kainate neurotoxicity in primary neuronal cultures of neonatal rat cerebellum and cortex. Proc. Natl. Acad. Sci. U.S.A. *85*: 7351–7355, 1988.

Fernandez, H. L. and Donoso, J. A.: Nerve-muscle cell trophic communication: Introductory remarks. In: "Nerve-Muscle Cell Trophic Communication," H. L. Fernandez and J. A. Donoso, Eds., pp. 1–5, CRC Press, Boca Raton, Fla., 1987a.

Fernandez, H. L. and Donoso, J. A.: Trophic mechanisms underlying nerve-muscle interactions. In: "Nerve-Muscle Cell Trophic Communication," H. L. Fernandez and J. A. Donoso, Eds., pp. 218–233, CRC Press, Boca Raton, Fla., 1987b.

Fiedler, E. P., Marks, M. J. and Collins, A. C.: Postnatal development of cholinergic enzymes and receptors in the mouse brain. J. Neurochem. *49*: 983–990, 1987.

Fiedler, E. P., Marks, M. J. and Collins, A. C.: Postnatal development of two nicotinic cholinergic receptors in seven mouse brain regions. Int. J. Dev. Neurosci. *8*: 535–540, 1990.

Figueiredo, B. C., Skup, M., Bedard, A. M., Tetzlaff, W. and Cuello, A. C.: Differential expression of p140trk, p75NGFR and growth-associated phophoprotein-43 genes in nucleus basalis magnocellularis, thalamus and adjacent cortex following neocortical infarction and nerve growth factor treatment. Neuroscience *68*: 29–45, 1995a.

Figueiredo, B. C., Pluss, K., Skup, M., Otten, U. and Cuello, A. C.: Acidic FGF induces NGF and its mRNA in the injured neocortex of adult animals. Brain Res. Mol. Brain Res. *33*: 1–6, 1995b.

Fine, E. G., Decosterd, I., Papaloizos, M., Zurn, A. D. and Aebischer, P.: GDNF and NGF released by synthetic channels support sciatic nerve regeneration across a long gap. Eur. J. Neurosci. *15*: 589–601, 2002.

Fischbach, G. D. and Schuetze, S. M.: A post-natal decrease in acetylcholine channel open time at rat end plates. J. Physiol. (Lond.) *303*: 125–137, 1980.

Fischer, H.: Vergleichende Pharmakologie von Übertragensubstanzen in tiersystematischer Darstellung. Handbch. d. exper. Pharmakol., vol. 26, Springer-Verlag, Berlin, 1971.

Fishman, P. H. and Brady, R. O.: Biosynthesis and function of gangliosides. Science *194*: 906–915, 1976.

Fitzpatrick-McElligott, S. and Stent, G. S.: Appearance and localization of acetylcholinesterase in embryos of the leech Helobdella triserialis. J. Neurosci. *1*: 901–907, 1981.

Flanigan, T. P., Dickson, J. G. and Walsh, F. S.: Cell survival characteristics and choline acetyltransferase activity in motor neurone-enriched cultures from chick embryo spinal cord. J. Neurochem. *45*(4): 1323–1326, 1985.

Florey, E.: Neurotransmitters and modulators in the animal kingdom. In: "Proc. Inter. Symp. on Comp. Pharmacol.", E. Cafruny, Ed., Fed. Proc. *26*: 1164–1178, 1967.

Florian, A., Casamenti, F. and Pepeu, G.: Recovery of cortical acetylcholine output after ganglioside treatment in rats with lesion of the nucleus basalis. Neurosci. Lett. *75*: 313–316, 1987.

Fluck, R. A.: Acetylcholine and acetylcholinesterase activity in the early embryo of medaka Oryzias latipes, a teleost. Dev. Growth Differ. *20*: 17–25, 1978.

Fluck, R. A.: Localization of acetylcholinesterase activity in young embryos of the medaka Oryzias latipes, a teleost. Comp. Biochem. Physiol. *72*: 59–64, 1982.

Fluck, R. A. and Strohman, R. C.: Acetylcholinesterase activity in developing skeletal muscle cells *in vitro*. Dev. Biol. *33*: 417–428, 1973.

Frade, J. M. and Barde, Y. A.: Nerve growth factor: two receptors, multiple functions. Bioessays *20*: 137–145, 1998.

Franz, C. K., Rutishauser, U. and Rafuse, V. F.: Polysialylated neural cell adhesion molecule is necessary for selective targeting of regenerating motor neurons. J. Neurosci. *25*: 2081–2091, 2005.

Frazier, W. A., Boyd, L. F. and Bradshaw, R. A.: Interaction of the nerve growth factor with surface membranes. Biological competence of insolubilized nerve growth factor. Proc. Natl. Acad. Sci. U.S.A.: *70*: 2931–2935, 1973.

Frazier, W. A., Boyd, L. F., Pulliam M. W., Szutowicz, A. and Bradshaw, R. A.: Properties and specificity of binding sites for ^{125}I-nerve growth factor in embryonic heart and brain. J. Biol. Chem. *249*: 5918–5923, 1974.

Freed, W. J.: The role of nerve growth factor (NGF) in the central nervous system. Brain Res. Bull. *1*: 393–412, 1976.

Freychet, P.: Interaction of polypeptide hormones with cell membrane specific receptors: studies with insulin and glucagon. Diabetologia *12*: 83–100, 1976.

Friedman, W. J. and Greene, L. A.: Neurotrophin signaling via Trks and p75. Exp. Cell Res. *253*: 1131–1142, 1999.

Friedman, H. C., Aguayo, A. J. and Bray, G. M.: Trophic factors in neuron-Schwann cell interactions. Ann. NY Acad. Sci. *883*: 427–438, 1999.

Frostick, S. P., Yin, Q. and Kemp, G. J.: Schwann cells, neurotrophic factors and peripheral nerve regeneration. Microsurgery *18*: 397–405, 1998.

Fuhrmann, S., Heller, S., Rohrer, H. and Hofmann, H. D.: A transient role for ciliary neurotrophic factor in chick photoreceptor development. J. Neurobiol. *37*: 672–683, 1998.

Gage, F. H. and Bjorklund, A.: Trophic and growth regulating mechanisms in the central nervous system monitored by intracerebral neural transplants. CIBA Found. Symp. *126*: 143–159, 1987.

Gage, F. H., Brundin, P., Isacson, O. and Bjorklund, A.: Rat fetal brain tissue grafts survive and innervate host brain following five day pregraft tissue storage. Neurosci. Lett. *60*: 133–137, 1985.

Galau, G. A., Klein, W. H., Davis, M. M., Wold, B. J., Britten, R. J. and Davidson, E. H.: Structural gene sets active in embryos and adult tissues of the sea urchin. Cell *7*: 487–505, 1976.

Gale, J. S.: Glutamic acid decarboxylase and choline acetyltransferase in human foetal brain. Brain Res. *133*: 172–176, 1977.

Galper, J. B., Klein, W. and Catterall, W. A.: Muscarinic acetylcholine receptors in developing chick heart. J. Biol. Chem. *252*: 8692–8699, 1977.

Gardino, P. F., Calaza, K. C., Hamassaki-Brito, D. E., Lindstrom, J. M., Britto, L. R. G. and Hokoc, J. N.: Neurogenesis of cholinoceptive neurons in the chick retina. Dev. Brain Res. *95*: 205–212, 1996.

Garofalo, L. and Cuello, A. C.: Pharmacological characterization of nerve growth factor and/or monosialogangloside GM1 effects on cholinergic markers in the adult lesioned brain. J. Pharmacol. Exp. Ther. *272*: 527–545, 1995.

Gash, D. M., Collier, T. J. and Sladek, J. R.: Neural transplantation: a review of recent developments and potential applications to the aged brain. Neurobiol. Aging *6*(2): 131–174, 1985.

Gasser, U. E., Weskamp, G., Otten, U. and Dravid, A. R.: Time course of the elevation of nerve growth factor (NGF) content in the hippocampus and septum following lesions of the septohippocampal pathway in rats. Brain Res. *376*: 351–356, 1986.

Gerthoffer, W. T. and Singer, C. A.: Secretory functions of smooth muscle. Mol. Interv. *2*: 447–456, 2002.

Giacobini, E.: Biochemical differentiation and development of autonomic neurons and synapses. In: "Metabolic Turnover in the Nervous System," A. Lajtha, Ed., pp. 467–488, Handbk. Neurochem., vol. 5, Plenum Press, New York.

Giacobini, E.: Formation and differentiation of synaptic contacts in the peripheral nervous system. In: "Neural Transmission, Learning and Memory," R. Caputto and C. Ajmone Marsan, Eds., pp. 113–123, Raven Press, New York, 1983.

Giacobini, E.: Development of peripheral parasympathetic neurons and synapses. In: "Developmental Neurobiology of the Autonomic Nervous System," P. M. Gootman, Ed., pp. 29–67, Humana Press, Clifton, N.J., 1986.

Giacobini, E. and Marchi, M.: Acetylcholine biosynthesis in developing cholinergic synapses. In: "Cholinergic Mechanisms," G. C. Pepeu and H. Ladinsky, Eds., pp. 1–24, Plenum Press, New York, 1981.

Gifford, P., Ouyang, G., Franke, F. R., Clarke, D. E. and Ertel, R. J.: Cholineacetyltransferase (CAT) in hearts of developing embryos. Pharmacologist *15*: 198, 1973.

Gilkey, J. C., Jaffe, L. F., Ridgway, E. B. and Reynolds, G. T.: A free calcium wave traverses the activating egg of the medaka, Oryzias latipes. J. Cell Biol. *76*: 448–466, 1978.

Gnahn, H., Hefti, F., Heumann, R., Schwab, M. E. and Thoenen, H.: NGF-mediated increase of choline acetyltransferase (ChAT) in the neonatal rat forebrain: evidence for a physiological role of NGF in the brain? Dev. Brain Res. *9*: 45–52, 1983.

Goedert, M., Otten, U. and Thoenen, H.: Biochemical effects of nerve growth factor and its antibody on the vas deferens and the adrenal medulla. Lett. *8*: 71–76, 1978.

Gould, S. J.: "The Structure of Evolutionary Theory," Belknap Press, Cambridge, Mass., 2002.

Gozes, I., Divinsky, I., Pilzer, I., Fridkin, M., Brenneman, D. E. and Spier, A. D.: From vasoactive intestinal peptide (VIP) through activity-dependent neuroprotective protein (ADNP) to NAP: a view of neuroprotection and cell division. J. Mol. Neurosci. *20*: 315–322, 2003.

Gradkowska, M., Skup, M., Kiedrowski, L., Catzolari, S. and Oderfeld-Nowak, B.: The effect of GM1 gangliosides on cholinergic and serotonergic systems in the rat hippocampus following partial denervation is dependent on the degree of fiber degeneration. Brain Res. *375*: 417–422, 1986.

Granholm, A. C.: Oestrogen and nerve growth factor—neuroprotection and repair in Alzheimer's disease. Expert. Opin. Investig. Drugs. *9*: 685–694, 2000.

Gray, D. B. and Tuttle, J. B.: [3H]acetylcholine synthesis in cultured ciliary ganglion neurons: effects of myotube membranes. Dev. Biol. *119*: 290–298, 1987.

Greenberg, J. H. and Schrier, B. K.: Development of choline acetyltransferase activity in chick cranial neural crest cells in culture. Dev. Biol. *61*: 86–93, 1977.

Greene, L. A.: Binding of alpha-bungarotoxin to chick sympathetic ganglia: properties of the receptor and its rate of appearance during development. Brain Res. *11*: 135–145, 1976.

Greene, L. A. and Shooter, E. M.: The nerve growth factor: biochemistry, synthesis and mechanism of action. Ann. Rev. Neurosci. *3*: 353–402, 1980.

Greene, L. A., Varon, S., Piltch, A. J. and Shooter, E. M.: Substructure of the beta-subunit of mouse 7S nerve growth factor. Neurobiology *1*: 37–48, 1971.

Greengard, P.: Neuronal phosphoproteins. Mediators of signal transduction. Mol. Neurobiol. *1*: 81–119, 1987.

Gremo, F., Palomba, M., Marchisio, A. M., Marcello, C., Mulas, M. L. and Torelli, S.: Heterogeneity of muscarinic cholinergic receptors in the developing human fetal brain: regional distribution and characterization. Early Hum. Dev. *15*: 165–177, 1987.

Grinnell, A. D. and Herrera, A. A.: Specificity and plasticity of neuromuscular connections: long term regulation of motoneuron function. Prog. Neurobiol. *17*: 203–282, 1981.

Grumet, M. and Edelman, G. M.: Neuron-glia cell adhesion molecule interacts with neurons and astroglia via different binding mechanisms. J. Cell Biol. *106*: 487–503, 1988.

Gupta, R. C., Rech, R. H., Lovell, K. L., Welsch, F. and Thornburg, J. E.: Brain cholinergic, behavioral, and morphological development in rats exposed in utero to methylparathion. Toxicol. Appl. Pharmacol. *77*: 405–413, 1985.

Gupta, R. C., Goad, J. T., Milatovic, D. and Dettbarn, W. D.: Cholinergic and noncholinergic brain biomarkers of insecticide exposure and effects. Hum. Exp. Toxicol. *19*: 297–308, 2000.

Gustafson, T. and Toneby, M. I.: How genes control morphogenesis: the role of serotonin and acetylcholine in morphogenesis. Am. Scientist *59*: 452–462, 1971.

Guth, L.: "Trophic" influences of nerve on muscle. Physiol. Rev. *48*: 645–687, 1968.

Gutierrez-Ospina, G., Jimenez-Trejo, F. J., Favilla, R., Moreno-Mendoza, N. A., Granados Rojas, L., Barrios, F. A., Diaz-Cintra, S. and Merchant-Larios H.: Acetylcholinesterase-positive innervation is present at undifferentiated stages of sea turtle Lepidochelis olivacea embryo gonads: implication for temperature dependent sex determination. J. Comp. Neurol. *410*: 90–98, 1999.

Ha, D. H., Robertson, R. T., Roshanaei, M. and Weiss, J. H.: Enhanced survival and morphological features of basal cholinergic neurons *in vitro*: role of neurotrophins and other potential cortically derived cholinergic trophic factors. J. Comp. Neurol. *406*: 156–170, 1999.

Hagag, N., Halegoua, S. and Viola, M.: Inhibition of growth factor-induced differentiation of PC12 cells by microinjection of antibody to *ras* p21. Nature *319*: 680–682, 1986.

Hagiwara, S. and Jaffe, S. L.: Electrical properties of egg membranes. Rev. Biophys. Bioeng. *8*: 385–416, 1979.

Hall, Z. W., Roisin, M. P., Gu, Y. and Gorin, P. D.: A developmental change in the immunobiological properties of acetylcholine receptors at the rat neuromuscular junction. Cold Spring Harbor Symp. Quant. Biol. *48*: 101–108, 1983.

Hamassaki-Britto, D. E., Gardino, P. F., Hokoc, J. N., Keyser, K. T., Karten, H. J., Lindstrom, J. M. and Britto, L. R. G.: Differential development of alpha-bungarotoxin-insensitive nicotinic acetylcholine receptors in the chick retina. J. Comp. Neurol. *347*: 161–170, 1994.

Hamburger, V.: The mitotic patterns in the spinal cord of the chick embryo and their relation to histogenetic processes. J. Comp. Neurol. *88*: 221–283, 1948.

Hamburger, V.: Development of the nervous system. Ann. N. Y. Acad. Sci. *55:* 117–132, 1952.

Hamburger, V.: A historic moment: the discovery of the chemical transmission of embryonic inductions. Dev. Neurosci. *19*: 293–296, 1997.

Hamburger, V. and Keefe, E. L.: The effects of peripheral factors on the proliferation and differentiation in the spinal cord of chick embryos. J. Exp. Zool. *96:* 223–242, 1944.

Hamburger, V. and Levi-Montalcini, R.: Proliferation, differentiation and degeneration in the Spinal Ganglia of the chick embryo under normal and experimental conditions. J. Exper. Zool, *111*: 457, 1949.

Hammond, W. and Yntema, C.: Origin of ciliary ganglia in the chick. J. Comp. Neurol. *110*: 367–389, 1958.

Happe, H. K. and Morley, B. J.: Distribution and postnatal development of alpha 7 nicotinic acetylcholine receptors in the rodent lower auditory brainstem. Brain Res. Dev. Brain Res. *153*: 29–37, 2004.

Harper, G. P. and Thoenen, H.: Target cells, biological effects, and mechanism of action of nerve growth factor and its antibodies. Annu. Rev. Pharmacol. Toxicol. *21*: 205–229, 1981.

Harris, J. B. and Thesleff, S.: Studies on tetrodotoxin resistant action potentials in derevated skeletal muscle. Acta Physiol. Scand. *83*: 382–388, 1971.

Hartikka, J. and Hefti, F.: Comparison of nerve growth factor's effects on developmental of septum, striatum, and nucleus basalis cholinergic neurons *in vitro*. J. Neurosci. Res. *21*(2–4): 365–375, 1988.

Hatten, M. E.: Neuronal inhibition of astroglial cell proliferation is membrane mediated. J. Cell Biol. *104*: 1353–1360, 1987.

Haugh, J. M.: Localization of receptor-mediated signal transduction pathways: the inside story. Mol. Interv. *2*: 292–307, 2002.

Haverkamp, L. J.: A functional framework for neurotrophic activities. Neurobiol. Aging. *10*: 537–539, 1989.

Hayashi, M.: Ontogeny of glutamic acid decarboxylase, tyrosine hydroxylase, choline acetyltransferase, somatostatin and substance P in monkey cerebellum. Brain Res. *429*: 181–186, 1987.

Hayashi, Y., Taniura, H., Miki, N.: Interaction of monoclonal antibodies with a neurite outgrowth factor from chicken gizzard extract. Brain Res. *432*(1): 11–19, 1987.

Hayes, W. J., Jr.: "Pesticides Studied in Man," Williams and Wilkins, Baltimore, 1982.

Heaton, M. B.: Influence of nerve growth factor on chick trigeminal motor nucleus explants. J. Comp. Neurol. *265*: 362–366, 1987.

Hebb, C. O., Krnjevic, K. and Silver, A.: Acetylcholine and choline acetyltransferase in the diaphragm of the rat. J. Physiol. *171*: 504–513, 1964.

Hefti, F.: Nerve growth factor promotes survival of septal cholinergic neurons after fimbrial transection. J. Neurosci. *6*: 2155–2162, 1986.

Hefti, F. and Mash, D. C.: Localization of nerve growth factor receptors in the human brain. In: "Molecular Biology of the Human Brain," E. G. Jones, Ed., pp. 119–132, Alan Liss, New York, 1988.

Hefti, F. and Will, B.: Nerve growth factor is a neurotrophic factor for forebrain cholinergic neurons: implications for Alzheimer's disease. Neural Trans. *24*, Suppl.: 309–315, 1987.

Hefti, F., Dravid, A. and Hartikka, J.: Chronic intraventricular injections of nerve growth factor elevate hippocampal choline acetyltransferase activity in adult rats with partial septohippocampal lesions. Brain Res. *293*: 305–311, 1984.

Hefti, F., Knusel, B. and Lapchak, P.: Protective effects of nerve growth factor and brain derived neurotrophic factor on basal forebrain cholinergic neurons in adult rats with partial fimbrial transections. In: "Cholinergic Function and Dysfunction," A. C. Cuello, Ed., pp. 257–263, Elsevier, Amsterdam, 1993.

Held, I. R.: Stimulation of nuclear RNA synthesis in denervated skeletal muscles. J. Neurochem. *30*: 1239–1243, 1978.

Held, I. R., Sayers, S. T., Yeoh, H. C. and McLane, J. A.: Role of cholinergic neuromuscular transmission in the neuroregulation of the autophosphorylatable regulatory subunit of cyclic AMP-dependent protein kinase type II and the acetylcholine receptor content in skeletal muscle. Brain Res. *407*: 341–350, 1987.

Heller, S., Ernsberger, U. and Rohrer, H.: Extrinsic signals in the developing nervous system: the role of neurokines during neurogenesis. Perspect. Dev. Neurobiol. *4*: 19–34, 1996.

Hellstrom-Lindahl, E. and Court, J. A: Nicotinic acetylcholine receptors during prenatal development and brain pathology in human aging. Behav. Brain. Res. *113*: 159–168, 2000.

Hellstrom-Lindahl, E., Gorbounova, O., Seiger, A., Mousavi, M. and Nordberg, A.: Regional distribution of nicotinic receptors during prenatal development of human brain and spinal cord. Dev. Brain. Res. *108*: 1147–160, 1998.

Hellstrom-Lindahl, E., Seiger, A., Kjaeldgaard, A. and Nordberg, A.: Nicotine-induced alterations in the expression of nicotinic receptors in primary cultures from human prenatal brain. Neuroscience *105*: 527–534, 2001.

Henderson, C. E.: The role of muscle in the development and differentiation of spinal motoneurons: *in vitro* studies. Ciba. Found. Symp. *138*: 172–191, 1988.

Henderson, C. E.: Role of neurotrophic factors in neuronal development. Curr. Opin. Neurobiol. *6*: 64–70, 1996.

Hendry, I. H. A.: Developmental changes in tissue and plasma concentrations of the biologically active species of nerve growth factor in the mouse, by using a two-site radioimmunoassay. Biochem. J. *128*: 1265–1272, 1972.

Hendry, I. A.: Trans-synaptic regulation of tyrosine hydroxylase activity in a developing mouse sympathetic ganglion: effects of nerve growth factor (NGF), NGF-antiserum and pempidine. Brain Res. *56*: 313–320, 1973.

Hendry, I. A.: Control in the development of the vertebrate sympathetic nervous system. In: "Reviews of Neurosciences," S. Ehrenpreis and I. J. Kopin, Eds., vol. 2, pp. 149–194, Raven Press, New York, 1976.

Hendry, I. A.: Cell division in the developing sympathetic nervous system. J. Neurocytol. *6*: 299–309, 1977.

Hendry, I. A., Stack, R. and Herrup, K.: Characteristics of the retrograde axonal transport system for nerve growth factor in the sympathetic nervous system. Brain Res. *82*: 117–128, 1974a.

Hendry, I. A., Stockel, K., Thoenen, H. and Iversen, L. L.: The retrograde axonal transport of nerve growth factor. Brain Res. *68*: 103–121, 1974b.

Hendry, I. A., Bonyhady, R. E. and Hill, C. E.: The role of target tissues in development and regeneration-retrophins. Birth Defects: Original Article Series *19*(4): 263–276, 1983.

Hendry, I. A., Hill, C. E., Belford, D. and Watters, D. J.: A monoclonal antibody to a parasympathetic neurotrophic factor causes immunoparasympathectomy in mice. Brain Res. *475*: 160–163, 1988.

Herrup, K. and Shooter, E. M.: Properties of the beta-nerve growth factor receptor in the development. J. Cell. Biol. *67*: 118–125, 1975.

Heumann, R.: Regulation of the synthesis of nerve growth factor. J. Exp. Biol. *132*: 133–150, 1987.

Heumann, R., Schwab, M. and Thoenen, H.: A second messenger required for nerve growth factor biological activity? Nature *292*: 838–840, 1981.

Heumann, R., Korsching, S., Scott, J. and Thoenen, H.: Relationship between levels of nerve growth factor (NGF) and its messenger RNA in sympathetic ganglier and peripheral target tissues. EMBO J. *3*: 3183–3189, 1984.

Hirata, A., Masaki, T., Motoyoshi, K. and Kamakura, K.: Intrathecal administration of nerve growth factor delays Gap 43 expression and early phase regeneration of adult rat peripheral nerve. Brain Res. *944*: 146–156, 2002.

Hiroaka, Y. and Okuda, H.: A tentative assessment of water pollution by medaka egg stationing method: aerial application of fenitrothion emulsion. Environ. Res. *34*: 262–267, 1984.

Hitzemann, R. J., Harris, R. A. and Loh, H. H.: Synaptic membrane fluidity and function. In "Physiology of Membrane Fluidity," M. Shinitzky, Ed., vol. 2, pp. 109–126, CRC Press, Boca Raton, Fla., 1984.

Hofmann, H. D.: Development of cholinergic retinal neurons from embryonic chicken in monolayer cultures: stimulation by glial cell-derived factors. J. Neurosci. *8*: 1361–1369, 1988.

Hohmann, C. F. and Berger-Sweeney, J.: Cholinergic regulation of cortical development and plasticity. New twists to a new story. Perspectives in Dev. Neurobiol. *5*: 401–425, 1998.

Hohmann, C. F. and Ebner, F. F.: Development of cholinergic markers in mouse forebrain. I. Choline acetyltransferase enzyme activity and acetylcholinesterase histochemistry. Dev. Brain Res. *23*: 225–241, 1985.

Hohmann, C. F., Pert, C. C. and Ebner, F. F.: Development of cholinergic markers in mouse forebrain. II. Muscarinic receptor binding in cortex. Brain Res. *355*: 243–253, 1985.

Holler, T., Berse, B., Cermak, J. M., Diebler, M.-F. and Blusztajn, J. K.: Differences in the developmental expression of the vesicular acetylcholine transporter and choline acetyltransferase in the rat brain. Neurosci. Lett. *212*: 107–110, 1996.

Holtzman, D. M., Li, Y., Parada, L. F., Kinsman, S., Chen, C., Valetta, J. S., Zhou, J., Long, J. B. and Mobley, W. C.: p 140trk mRNA marks NGF-responsive forebrain neurons: evidence that trk expression is induced by NGF. Neuron *9*: 465–478, 1992.

Holtzman, D. M., Kilbridge, J., Li, Y., Cuningham, E. T., Jr., Lenn, N. J. Clary, D. O., Reichardt, L. F. and Mobley, U. L.: TrkA expression in the CNS: evidence for the existence of several novel NGF-responsive CNS neurons. J. Neurosci. *15*: 1567–1576, 1995.

Horiuchi, Y., Kimura, R., Kato, N., Fuji, T., Seki, M., Endo, T. Kato, T. and Kawashima, K.: Evolutional study on acetylcholine expression. Life Sci. *72*: 1745–1756, 2003.

Houenou, L. J., Turner, P. L., Li, L., Oppenheim, R. W. and Festoff, B. W.: A serine protease inhibitor, protease nexin I, rescues motoneurons from naturally occurring and axotomy-induced cell death. Proc. Natl. Acad. Sci. U.S.A. *92*: 895–899, 1995.

Hruschak, K. A., Friedrich, V. I., Jr. and Giacobini, E.: Synaptogenesis in chick paravertebral sympathetic ganglia: a morphometric analysis. Dev. Brain Res. *4*: 229–240, 1982.

Hume, R. I., Role, L. W. and Fischbach, G. D.: Acetylcholine release from growth cones detected with patches of acetylcholine receptor-rich membranes. Nature *305*: 632–634, 1983.

Hutchins, J. B.: Development of muscarinic acetylcholine receptors in the ferret retina. Dev. Brain Res. *82*: 45–61, 1994.

Hutchins, J. B. and Casagrande, V. A.: Development of acetylcholinesterase activity in the lateral geniculate nucleus. J. Comp. Neurol. *275*: 241–253, 1988.

Inestrosa, N. C., Moreno, R. D. and Fuentes, M. E.: Monomeric amphiphilic forms of acetylcholinesterase appear early during brain development and may correspond to biosynthetic precursors of amphiphilic G4 forms. Neurosci. Lett. *173*: 155–158, 1994.

Ip, N. Y.: The neurotrophins and neuropoietic cytokines: two families of growth factors acting on neural and hematopoietic cells. Ann. N. Y. Acad. Sci. *840*: 97–106, 1998.

Ip, N. Y. and Yancopoulos, G.: The neurotrophins and CNTF: Two families of collaborative neurotrophic factors. Annu. Rev. Neurosci. *19*: 491–515, 1996.

Ivonnet, P. I. and Chambers, E. L.: Nicotinic acetylcholine receptors of the neuronal type occur in plasma membrane of sea urchin eggs. Zygote *5*: 277–287, 1997.

Jaffe, L. A.: First messengers at fertilization. J. Reprod. Fertil. *42*: 107–116, 1990.

Jaffe, L. A.: Calcium waves and fertilization. Ciba Found. Sym. *188*: 4–17, 1995.

Johnson, D. G., Gorden, P. and Kopi, I. J.: A sensitive radioimmunoassay for 7S nerve growth factor antigens in serum and tissues. J. Neurochem. *18*: 2355–2362, 1971.

Johnson, D. G., Silberstein, S. D., Hanbauer, I. and Kopin, I. J.: The role of nerve growth factor in the ramification of sympathetic nerve fibres into the rat iris in organ culture. J. Neurochem. *19*: 2025–2031, 1972.

Johnston, M. V., Silverstein, F. S., Reindel, F. O., Penney, J. B. and Young, A. B.: Muscarinic cholinergic receptors in human infant forebrain: [3H]quinulcidinyl benzilate binding in homogenates and quantitative autoradiography in sections. Brain Res. *35*: 195–203, 1985.

Johnston, M. V., Rutkowski, L. J., Wainer, B. H., Long, J. B. and Mobley, W. C.: NGF effects on developing cholinergic neurons are regionally specific. Neurochem. Res. *12*: 985–994, 1987.

Jonhagen, M. E.: Nerve growth factor treatment in dementia. Alzheimer Dis. Assoc. Disord. *14,* Supplt. 1: 531–538, 2000.

Kahn, C. R., Coyle, J. T. and Cohen, A. M.: Head and trunk neural crest *in vitro*: autonomic neuron differentiation. Dev. Biol. *77*: 340–348, 1980.

Kamegai, M., Niijima, K., Kunishita, T., Nishizawa, M., Ogawa, M., Neki, A., Komishi, K. and Tabira, T.: Interleukin 3 as a trophic factor for central cholinergic neurons *in vitro* and *in vivo*. Neuron *4*: 429–436, 1990.

Kaplan, D. R., Hempstead, B. L., Martin-Zanca, D., Chao, M. V. and Parcia, L. F.: The trk proto-oncogene product: a signal transducing receptor for nerve growth factor. Science *252*: 554–558, 1991.

Karczmar, A. G.: The role of amputation and nerve resection in the regressing limbs of urodele larvae. J. Exp. Zool. *103*: 401–427, 1946.

Karczmar, A. G.: Ontogenesis of cholinesterases. In: "Cholinesterases and Anticholinesterase Agents," G. B. Koelle, Ed., pp. 129–186, Handbch. d. exper. Pharmakol., Ergänzungswk., vol. 15, Springer-Verlag, Berlin, 1963a.

Karczmar, A. G.: Ontogenetic effects. In: "Cholinesterases and Anticholinesterase Agents," G. B. Koelle, Ed., pp. 799–832, Handbch. d. exper. Pharmakol., Ergänzungswk. vol.15, Springer-Verlag, Berlin, 1963b.

Karczmar, A. G.: Central actions of acetylcholine, cholinomimetics, and related drugs. In: "Biology of Cholinergic Function," A. M. Goldberg and I. Hanin, Eds., pp. 395–449, Raven Press, New York, 1976.

Karczmar, A. G.: The Tenth International Symposium on Cholinergic Mechanisms, Arcachon, France, September 1–5, 1998. CNS Drug Rev. *5*: 301–316, 1999.

Karczmar, A. G.: Conclusions and comments for the XI ISCM. In: "Cholinergic Mechanisms—Function and Dysfunction," I. Stilman, Ed. pp. 435–451, Martin Dunitz, London, 2004.

Karczmar, A. G., Srinivasan, R. and Bernsohn, J.: Cholinergic function in the developing fetus. In: "Fetal Pharmacology," L. O. Boreus, Ed., pp. 127–177, Raven Press, New York, 1973.

Kawaja, M. D. and Gage, F. H.: Nerve growth factor receptor immunoreactivity in the rat septohippocampal pathway: a light and electron microscope investigation. J. Comp. Neurol. *307*: 517–529, 1991.

Keller, F., Rimvall, K. and Waser, P. G.: Choline and acetylcholine metabolism in slice cultures of the newborn rat septum. Brain Res. *405*: 305–312, 1987.

Kelly, P. A., Gage, F. H., Ingvar, M., Lindvall, O. and Stenevi, U.: Functional reactivation of the deafferented hippocampus by embryonic septal grafts as assessed by measurements of local glucose utilization. Brain Res. *58*(3): 570–579, 1985.

Kidokoro, Y. and Brass, B.: Redistribution of acetylcholine receptors during neuromuscular formation in Xenopus cultures. J. Physiol. (Paris) *80*: 212–220, 1985.

Klenk, E.: "Biochemistry of the Developing Nervous System," H. Waelsch, Ed., pp. 397–409, Academic Press, New York, 1955.

Kliot, M. and Shatz, C. J.: Genesis of different retinal ganglion cell types in the cat. Abstr.—Soc. Neurosci. 8: 815, 1982.

Knusel, B. and Hefti, F.: Development of cholinergic pedunculopontine neurons *in vitro*: comparison with cholinergic septal cells and response to nerve growth factor, ciliary neuronotrophic factor, and retinoic acid. J. Neurosci. Res. *21*: 365–375, 1988.

Koelle, G. B.: Cytological distributions and physiological functions of cholinesterases. In: "Cholinesterases and Anticholinesterases Agents," G. B. Koelle, Ed., pp. 187–198, Handbch. d. exper. Pharmakol., Ergänzugswk., vol.15, Springer-Verlag, Berlin, 1963.

Koelle, G. B., Sanville, U. J. and Thampi, N. S.: Direct neurotrophic action of glycyl-L-glutamine in the maintenance of acetylcholinesterase and butyrylcholinesterase in the preganglionically denervated superior cervical ganglion of the cat. Proc. Natl. Acad. Sci. U.S.A. *84*: 6944–6947, 1987.

Koenig, J. and Rieger, F.: Biochemical stability of the AChE molecular forms after cytochemical staining: postnatal focalization of the 16S AChE in rat muscle. Dev. Neurosci. *4*: 249–257, 1981.

Koenigsberger, C., Hammond, P. and Brimijoin, S.: Developmental expression of acetyl-and butyrylcholinesterase in the rat: enzyme and mRNA levels in embryonic dorsal root ganglia. Brain Res. *787*: 248–258, 1998.

Koliatsos, V. E., Clatterbuck, R. E. and Nauta, H. J. W.: Human nerve growth factor prevents degeneration of basal forebrain neurons in primates. Ann. Neurol. *30*: 831–840, 1991.

Konishi, Y., Harano, T. and Tahira, T.: Neurotrophic effect of interleukin-3 (IL-3) and its mechanism of action in the nervous system. CNS Drug Rev. *5*: 265–280, 1999.

Kordover, J. H. and Rakic, P.: Neurogenesis of the magnocellular basal forebrain nuclei in the rhesus monkey. J. Comp. Neurol. *2291*: 637–653, 1990.

Korsching, S., Auburger, G., Heumann, R., Scott, J. and Thoenen. H. Levels of nerve growth factor and its mRNA in the central nervous system of the rat correlate with cholinergic innervation. EMBO J. *4*: 1389–1393, 1985.

Kostovic, I.: Prenatal development of nucleus basalis complex and related fiber systems in man: a histochemical study. Neuroscience *17*: 1047–1077, 1986.

Kostovic, I. and Goldman-Rakic, P.: Transient cholinesterase staining in the mediodorsal nucleus of the thalamus and its connections in the developing human and monkey brain. J. Comp. Neurol. *219*: 431–447, 1983.

Kotas, A. M. and Prince, A. K.: High affinity uptake of choline, a marker for cholinergic nerve terminals, is not specific in developing rat brain. Brain Res. *432*: 175–181, 1981.

Kramer, T.: Preliminary report on the innervation of the embryonic chick heart. Anat. Rec. *106*: 210, 1950.

Kristofikova, Z., Stasny, F., Bubenikova, V., Druga, R., Klaschka, J. and Spaniel, F.: Age and sex-dependent laterality of rat hippocampal cholinergic system in relation to animal models of neurodevelopment and neurodegenerative disorders. Neurochem. Res. *29*: 671–680, 2004.

Kuffler, D. P.: Chemoattraction of sensory neuron growth cones by diffusible concentration gradients of acetylcholine. Mol. Chem. Neuropathol. *28*: 199–208, 1996.

Kuffler, D.: Can regeneration be promoted within the spinal cord? PR Health Sci. J. *19*: 241–252, 2000.

Kumaresan, V., Kang, C. and Simmons, M. A.: Development and differentiation of the anuran auditory brain stem during metamorphosis: An acetylcholinesterase histochemical study. Brain Behav. Evol. *52*: 111–125, 1998.

Kuo, Z. Y.: Studies on the physiology of the embryonic nervous system IV. Development of acetylcholine in the chick embryo. J. Neurophysiol. *2*: 488–493, 1939.

Kupfer, C. and Koelle, G. B.: A histochemical study of cholinesterase during formation of motor endplate of the albino rat. J. Exp. Zool. *116*: 397–415, 1951.

Kusano, K., Miledi, R. and Stinnakre, J.: Cholinergic and catecholaminergic receptors in the Xenopus oocyte membrane. J. Physiol. (Lond.) 328: 143–170, 1982.

Kushima, Y., Nishio, C., Nonomura, T. and Hatanaka, H.: Effects of nerve growth factor and basic fibroblast growth factor on survival of cultured septal cholinergic neurons from adult rats. Brain Res. *598*: 264–270, 1992.

Kust, B. M., Copray, J. C., Browser, N., Troost, D. and Boddecke, H. W.: Elevated levels of neurotrophins in human biceps brachii tissue of amyotrophic lateral sclerosis. Exp. Neurol. *177*: 419–427, 2002.

Kwong, W. H., Chan, W. Y., Lee, K. K. H., Fan, M. and Yew, D. T.: Neurotransmitters, neuropeptides and calcium binding proteins in developing human cerebellum: a review. Histochem. J. *32*: 521–534, 2000.

Laasberg, T.: Ca++-mobilizing receptors in gastrulating chick embryo. Comp. Biochem. Physiol. *97C*: 9–12, 1990.

Lakshmanan, J.: Is there a second messenger for nerve growth factor-induced phosphatidylinositol turnover in the rat superior cervical ganglia? Brain Res. *157*: 173–177, 1978.

Lammerding-Koppel, M., Greiner-Schroeder, A. and Drews, U.: Muscarinic receptors in the prenatal mouse embryo. Comparison of M335-immunohistochemistry with [3H] quinuclidinyl benzylate autoradiography. Histochem. Biol. *103*: 301–310, 1995.

Landauer, R.: Cholinomimetic teratogens. V. The effect of oximes and related cholinesterase reactivators. Teratology *15*: 33–42, 1977.

Landis, S. C., Jackson, P. C., Fredieu, J. R. and Thibault, J.: Catecholaminergic properties of cholinergic neurons and synapses in adult ciliary ganglion. J. Neurosci. *7*: 3574–3587, 1987.

Landmesser, L. and Pilar, G.: The onset and development of transmission in the chick ciliary ganglion. J. Physiol. *222*: 691–713, 1972.

Landmesser, L. and Morris, D. G.: The development of functional innervation in the hind limb of the chick embryo. J. Physiol. *249*: 301–326, 1975.

Landreth, G. E.: Growth factors. In: "Basic Neurochemistry: Molecular, Cellular and Medical Aspects," G. J. Siegel, Ed. in Chief, pp. 383–398, Sixth Edition, Lippincott-Raven, Philadelphia, 1999.

Landreth, G. E.: Growth factors. In: "Basic Neurochemistry: Molecular, Cellular and Medical Aspects," G. J. Siegel, Ed. in Chief, pp. 471–484, Seventh Edition, Elsevier, Amsterdam, 2006.

Lanser, M. E., Carrington, J. L. and Fallon, J. F.: Survival of motoneurons in the brachial lateral motor column of limbless mutant chick embryos depends on the periphery. J. Neurosci. *6*(9): 2551–2557, 1966.

Large, T. H., Cho, N. J., DeMello, F. G. and Klein, W. L.: Molecular alteration of a muscarinic acetylcholine receptor system during synaptogenesis. J. Biol. Chem. *260*: 8873–8881, 1985.

Large, T. H., Lambert, M. P., Gremillion, M. A. and Klein, W. L.: Parallel postnatal development of choline acetyltransferase activity and muscarinic receptors in the rat olfactory band. J. Neurochem. *43*: 671–680, 1986.

Larrabee, M. G.: Metabolic effects of nerve impulses and nerve-growth factor in sympathetic ganglia. Prog. Brain Res. *31*: 95–110, 1969.

Lassiter, T. L., Barone, S., Jr. and Padilla, S.: Ontogenetic differences in the regional and cellular acetylcholinesterase and butyrylcholinesterase activity in the rat brain. Dev. Brain Res. *105*: 109–132, 1998.

Lauder, J. M.: Neurotransmitters as morphogens. Prog. Brain Res. *73*: 365–387, 1988.

Lauder, J. M.: Neurotransmitters as growth regulatory signals: role of receptors and second messengers. Trends Neurosci. *16*: 233–240, 1993.

Lauder, J. M. and Schambra, U. B.: Morphogenetic roles of acetylcholine. Environ. Health Perspect. *107*, Supplt. 1: 65–69, 1999.

Layer, P. G.: Comparative localization of acetylcholinesterase and butyrylcholinesterase during morphogenesis of the chick brain. Proc. Natl. Acad. Sci. U.S.A. *80*: 6413–6417, 1983.

Layer, P. G.: Cholinesterases preceding major tracts in vertebrate neurogenesis. Bio. Essays *12*: 412–420, 1990.

Layer, P. G.: Cholinesterases during development of avian nervous system. Cell. Mol. Neurobiol. *11*: 7–33, 1991.

Layer, P. G. and Shooter, E. M.: Binding and degradation of nerve growth factor by PC12 pheochromocytoma cells. J. Biol. Chem. *258*: 3012–3018, 1983.

Layer, P. G. and Willbold, E.: Cholinesterases in avian neurogenesis. Int. Rev. Cytol. *151*: 139–191, 1994.

Layer, P. G., Alber, R. and Sporns, O.: Quantitative development and molecular forms of acetyl- and butyrylcholinesterase during morphogenesis and synapto genesis of chick brain and retina. J. Neurochem. *49*: 175–182, 1987.

Lechner, D. M. and Abdel-Rachman, M. S.: A teratology study of carbaryl and malathion mixtures in rat. Toxicol. Environ. Health *14*: 267–278, 1984.

Ledeen, W. and Yu, R. K.: Gangliosides: structure, isolation and analysis. In: "Methods in Enzymology," Ginsburg, Ed., vol. 83, pp. 1139–1191, Academic Press, New York, 1982.

Ledeen, R. W., Skrivanek, J. A., Tirri, L. J., Margolis, R. K. and Margolis, R. U.: Gangliosides of the neuron: localization and origin. In: "Ganglioside Function: Biochemical and Pharmacological Implications," G. Porcellati, B. Ceccarelli, and G. Tettamanti, Eds., pp. 83–103, Plenum Press, New York, 1975.

Ledeen, R. W., Skrivanek, J. A., Nunez, J., Sclafani, J. R., Norton, W. T. and Farook, M.: Implications of the distribution and transport of gangliosides in the nervous system. In: "Gangliosides in Neurological and Neuromuscular Function. Development and Repair." M. M. Rapport and A. Gorio, Eds., pp. 211–223, Raven Press, New York, 1981.

Ledeen, R. W., Wu, G., Cannella, M. S., Oderfeld-Nowak, B. and Cuello, A. C.: Gangliosides as neurotrophic agents: studies on the mechanism of action. Acta Neurobiol. Exp. (Wars) *50*: 439–449, 1990.

Le Douarin, N. M.: The ontogeny of the neural crest in avian embryo chimaeras. Nature *286*: 663–669, 1980.

Le Douarin, N. M.: "The Neural Crest," Cambridge University Press, Cambridge, Mass., 1982.

Le Douarin, N. M.: Early neurogenesis in Amniote vertebrates. Int. J. Dev. Biol. *45*: 373–378, 2001.

Le Douarin, N. M., Teillet, M. A., Ziller, C. and Smith, J.: Adrenergic differentiation of the cholinergic ciliary and Remak ganglia in avian embryo after *in vivo* transplantation. Proc. Natl. Acad. Sci. U.S.A. *75*: 2030–2034, 1978.

Le Douarin, N. M., Smith, J., Teillet, M.-A., LeLievre, C. S. and Ziller, C.: The neural crest and its developmental analysis in avian embryo chimaeras. TINS, pp. 39–42, February, 1980.

Lehmann, K., Hundsdorfer, B., Hartmann, B. and Teucher-Nodt, G.: The acetylcholine fiber density of the neocortex is altered by isolated rearing and early methamphetamine intoxication in rodents. Exp. Neurol. *189*: 131–140, 2004.

Lehotai, L.: The change in different forms of AChE during the ontogenetic development of the central nervous system. Neurobiology *6*: 405–420, 1998.

LeJeune, H., Aubert, I., Jourdan, F. and Quirion, R.: Developmental profiles of various cholinergic markers in the main olfactory bulb using quantitative radioautography. J. Comp. Neurol. *373*: 433–450, 1996.

Levi-Montalcini, R.: Effects of mouse tumor transplantation on the nervous system. Ann. N. Y. Acad. Sci. *55*: 330–342, 1952.

Levi-Montalcini, R.: In-vivo and in-vitro experiments on the effects of mouse sarcoma 180 and 37 on the sensory and sympathetic system of the chick embryo. Proc. 14th Congr. Zool., Copenhagen, p. 309, 1953.

Levi-Montalcini, R.: The nerve growth: its mode of action on sensory and sympathetic nerve cells. Harvey Lect. *60*: 217–259, 1965.

Levi-Montalcini, R.: NGF: an unchartered route. In: "The Neurosciences Paths of Discovery," F. G. Worden, P. Swazey and G. Adelman, Eds., pp. 245–263, MIT Press, Cambridge, Mass., 1975.

Levi-Montalcini, R.: Developmental neurobiology and the natural history of the nerve growth factor. Ann. Rev. Neurosci. *5*: 341–362, 1982.

Levi-Montalcini, R.: The nerve growth factor 35 years later. Science *237*: 1154–1162, 1987.

Levi-Montalcini, R. and Amprino, R.: Recherches experimentales sur l'origine du ganglion ciliare dans l'embryon de poulet. Arch. Biol. *58*: 265–288, 1947.

Levi-Montalcini, R. and Angeletti, P. U.: Nerve growth factor. Physiol. Rev. *48*: 534–569, 1968.

Levi-Montalcini, R. and Hamburger, V.: Selective growth-stimulation effects of mouse sarcoma on the sensory and sympathetic nervous system of the chick embryo. J. Exp. Zool. *116*: 321–362, 1951.

Levi-Montalcini, R. and Hamburger, V.: A diffusible agent of mouse-sarcoma, producing hyperplasia of sympathetic ganglia and hyperneurotization of viscera in the chick embryo. J. Exp. Zool. *123*: 233–288, 1953.

Levi-Montalcini, R., Aloe, L., Mugnaini, E., Oesch, F. and Thoenen, H.: Nerve growth factor induces volume increase and enhances tyrosine hydroxylase synthesis in chemically axotomized sympathetic ganglia of newborn rats. Proc. Natl. Acad. Sci. U.S.A. *72*: 595–599, 1970.

Levi-Montalcini, R., Aloe, L., Mugnaini, E., Oesch, F., Thoenen, H.: Nerve growth factor induces volume increase and enhances tyrosine hydroxylase synthesis in chemically axotomized sympathetic ganglia of newborn rats. Proc. Natl. Acad. Sci. U.S.A. *72*: 595–599, 1975.

Levine, B. S. and Parker, R. M.: Reproductive and developmental toxicity studies of pyridostigmine bromide in rats. Toxicology *69*: 291–300, 1991.

Lewin, G. R.: Neurotrophins and specification of neuronal phenotype. Philos. Trans. R. Soc. Lond., B *351*: 405–411, 1996.

Liang, B. T., Hellmich, M. R., Neer, E. J. and Galper, J. B.: Development of muscarinic cholinergic inhibition of adenylate cyclase in embryonic chick heart. Its relationship to changes in the inhibitory guanine nucleotide regulatory protein. J. Biol. Chem. *261*: 9011–9021, 1986.

Lichtman, J. W. and Taghert, P. H.: Trophic factor theory matures. Nature *326*: 336–337, 1987.

Lichtman, J. W. and Sanes, J. R.: Watching the neuromuscular junction. J. Neurocytol. *32*: 767–775, 2003.

Lien, N. T., Adriaens, D. and Janssen, C. R.: Morphological abnormalities in African catfish (Clarias gariepinus) larvae exposed to Malathion. Chemosphere *35*: 1475–1486, 1997.

Lindeman, V. F.: The cholinesterase and acetylcholine content of the chick retina, with special reference to functional activity as indicated by the pupillary constrictor reflex. Am. J. Physiol. *144*: 40–44, 1947.

Lindsay, R. M.: Role of neurotrophins and Trk receptors in the development and maintenance of sensory neurons: an overview. Philos. Trans. R. Soc. Lond., B *351*: 365–373, 1996.

Lindsay, R. M., Thoenen, H. and Barde, Y.-A.: Placode and neural crest-derived sensory neurons are responsive at early developmental stages to brain-derived neurotrophic factor (BDNF). Dev. Biol. *112*: 319–328, 1985.

Linser, P. J. and Perkins, M. S.: Regulatory aspects of the *in vitro* development of retinal Muller glial cells. Cell Differ. *20*: 189–196, 1987.

Lipton, S. A., Frosch, M. P., Phillips, M. D., Tauck, D. L., Aizenman, E.: Nicotinic antagonists enhance process outgrowth by rat retinal ganglion cells in culture. Science *239*: 1293–1296, 1988.

Lipson, A. C., Widenfalk, J., Lindqvist, E., Ebendal, T. and Olson, L.: Neurotrophic properties of olfactory ensheathing glia. Exp. Neurol. *180*: 167–171, 2003.

Liss, B. and Roeper, J.: Correlating function and gene expression of individual basal ganglia neurons. Trends Neurosci. *27*: 475–481, 2004.

Lissak, M., Toro, I. and Pasztor, J.: Untersuchungen über den Zussamenhang des Acetycholingehalten and der Innervation des Herzmuskels in Gewebe-Kulturen. Pflügers Arch. Ges. Physiol. *245*: 794–801, 1942.

Liu, R. Y. and Snider, W. D.: Different signaling pathways mediate regenerative versus developmental sensory axon growth. J. Neurosci. *21*: RC 164, 2001.

Loewi, O.: Über die humorale Übertragbarkeit des Herznervenwirkung. I. Erste Mitteilung. Pflügers Arch. ges. Physiol. *189*: 239–249, 1921.

Loewi, O.: An autobiographic sketch. Perspect. Biol. Med. *4*: 3–25, 1960.

Lopez-Coviella, I., Bers, B., Krauss, R., Thies, R. S. and Blusztajn, J. K.: Induction and maintenance of the neuronal cholinergic phenotype in the central nervous system by BMP-9. Science *289*: 313–316, 2000.

Lorez, H., von Frankenber, M., Weskamp, G. and Otten, U.: Effect of bilateral, decortication on nerve growth factor content in nucleus basalis and neostriatum of adult rat brain. Brain Res. *454*: 355–360, 1988.

Loy, R. and Moore, R. Y.: Anomalous innervation of the hippocampal formation by peripheral sympathetic axons following mechanical injury. Exp. Neurol. *57*: 645–645, 1977.

Loy, R., Heyer, D., Williams, C. L. and Meck, W. H.: Choline-induced spatial memory facilitation correlates with altered distribution and morphology of septal neurons. Adv. Exp. Med. Biol. *295*: 373–382, 1991.

Lucidi-Phillipi, C. A, and Gage, F. H.: The neurotrophic hypothesis and cholinergic basal forebrain projection. In: "Cholinergic Function and Dysfunction," A. C. Cuello, Ed., pp. 241–263, Elsevier, Amsterdam, 1993.

Madziar, B., Lopez-Coviella, I., Zemelko, V. and Berse, B.: Regulation of cholinergic gene expression by nerve growth factor depends on the phosphatidylinositol-3'-kinase pathway. J. Neurochem. *92*: 767–779, 2005.

Maffei, L., Berardi, N., Domenici, L., Parisi, V. and Pizzorusso, T.: Nerve growth factor (NGF) prevents the shift in ocular dominance distribution of visual cortical neurons in monocularly deprived rats. J. Neurosci. *12*: 4651–4652, 1992.

MacIntosh, F. C. and Collier, B.: Neurochemistry of Cholinergic terminals. In: "Neuromuscular Junction," E. Zaimis, Ed., pp. 99–228, Springer-Verlag, Berlin, 1976.

Mandel, G., Cooperman, S. S., Maue, R. A., Goodman, P. H. and Brehm, P.: Selective induction of brain type II Na+ channels by nerve growth factor. Proc. Natl. Acad. Sci. U.S.A. *85*: 924–928, 1988.

Manger, P. R., Fahringer, H. M., Pettigrew, J. D. and Siegel, J. M.: The distribution and morphological characteristics of cholinergic cells in the brain of monotremes as revealed by ChAT immunohistochemistry. Brain Behav. Evol. *60*: 315–332, 2002.

Marchi, M., Hoffman, D. W., Mussini, I. and Giacobini, E.: Development and aging of cholinergic synapses. III. Choline uptake in the developing iris of the chick. Dev. Neurosci. *3*: 185–189, 1980.

Margiotta, J. F., Berg, D. K. and Dionne, V. E.: Cyclia AMP regulates the proportion of functional acetylcholine receptors on chick ciliary ganglion neurons. Proc. Natl. Acad. Sci. U.S.A. *84*: 8155–8159, 1987.

Markelonis, G. J. and Oh, T. H.: Purification of sciatin using affinity chromatography on concanavalin A-Agarose. J. Neurochem. *37*: 95–99, 1981.

Markelonis, G. J., Oh, T. H., Park, L. P., Cha, C. Y., Sofia, C. A., Kim, J. W. and Azari, P.: Synthesis of the transferrin receptor by cultures of embryonic chicken spinal neurons. J. Cell Biol. *100*: 8–17, 1985.

Markova, L. N., Buznikov, G. A., Kovacevic, N., Rakic, L., Salimova, N. B. and Volina, E. V.: Histochemical study of biogenic monoamines in early (prenervous) and late embryos of sea urchins. Int. J. Dev. Neurosci. *3*: 492–500, 1985.

Markus, A., Patel, T. D. and Snider, W. D.: Neurotrophic factors and axonal growth. Curr. Opin. Neurobiol. *12*: 523–531, 2002.

Martinez, H. J., Dreyfus, C. F., Miller Jonakait, G. and Black, I. B.: Nerve growth factor promotes cholinergic development in brain striatal cultures. Proc. Natl. Acad. Sci. U.S.A. *82*: 7777–7781, 1985.

Martynova, L. E.: Gastrulation in *Strongylocentrotus droebachinensis* sea urchin in the norm and during treatment with various substances. Sov. J. Dev. Biol. *12*: 319–315, 1981.

Massa S. M., Xie Y., Longo F. M.. Alzheimer's therapeutics: neurotrophin small molecule mimetics. J. Mol. Neurosci. *19*: 107–111, 2002.

Massague, J., Guilette, B. J., Czech, M. P., Morgan, C. J. and Bradshaw, R. A.: Identification of nerve growth factor receptor protein in sympathetic ganglia membranes by affinity labelling. J. Biol. Chem. *256*: 9419–9424, 1981.

Massoulié, J., Sussman, J., Bon, S. and Silman, I.: Structure and function of acetylcholinesterase and butyrylcholinesterase. In: "Cholinergic Function and Dysfunction," A. C. Cuello, Ed., pp. 139–146, Elsevier, Amsterdam, 1993.

Massoulié, J., Anselmet, A., Bon, S., Krejci, E., Legay, C., Morel, N. and Simon, N.: Acetylcholinesterase: C-terminal domains, molecular forms and functional localization. In: "From Torpedo Electric Organ to Human Brain: Fundamental and Applied Aspects," J. Massoulié, Ed., pp. 183–190. Elsevier, Amsterdam, 1998.

Mathews, A. P.: The action of pilocarpine and atropine on the embryos of the starfish and the sea-urchin. Am. J. Physiol. *6*: 207–215, 1902.

Mathura, C. B., Meier, G. W. and Himwich, W. A.: Experimental manipulation of acetylcholine levels in developing rat. Psychol. Rep. *45*: 811–818, 1979.

Mattia Rossi, F. M., Sala, R. and Maffei, L.: Expression of the nerve growth factor receptors TrkA and p75NTR in the visual cortex of the rat: development and regulation by the cholinergic input. J. Neurosci. *22*: 912–919, 2002.

Maurissen, J. P., Hoberman, A. M., Garman, R. H. and Hanley, T. R.: Lack of selective developmental neurotoxicity in rat pups from dams treated by gavage with chlorpyrifos. Toxicol. Sci. *57*: 225–263, 2000.

Max, S. R., Schwab, M., Dumas, M. and Thoenen, H.: Retrograde axonal transport of nerve growth factor in the ciliary ganglion of the chick and rat. Brain Res. *159*: 411–415, 1978.

McCaig, C. D.: Myoblasts and myoblast-conditioned medium attract the earliest spinal neurites from frog embryos. J. Physiol. (Lond.) *375*: 39–54, 1986.

McCarty, L. P., Lee, W. C. and Shideman, F. E.: Measurement of inotropic effects of drugs on the innervated and non-innervated embryonic chick heart. J. Pharmacol. Exp. Ther. *129*: 315–321, 1960.

McDonald, J. K., Speciale, S. G. and Parnavelas, J. G.: The laminar distribution of glutamate and choline acetyltransferase in the adult and developing visual cortex of the rat. Neuroscience *21*: 825–832, 1987.

McDonald, N. and Chao, M.: Structural determinants of neurotrophin action. J. Biol. Chem. *270*: 19669–19672, 1995.

McGeer, P. L., Eccles, J. C. and McGeer, E. G.: "Molecular Biology of the Mammalian Brain," First Edition, Plenum Press, New York, 1978.

McKinnon, L. A. and Nathanson, N. M.: Tissue specific regulation of muscarinic acetylcholine receptor expression during embryonic development. J. Biol. Chem. *270*: 20636–20640, 1995.

McManaman, J. L., Crawford, F. G., Stewart, S. S. and Appel, S. H.: Purification of a skeletal muscle polypeptide which stimulates choline acetyltransferase activity in cultured spinal cord neurons. Biol. Chem. *263*(12): 5890–5897, 1988.

McNamee, C. J., Reed, J. E., Howard, M. R., Lodge, A. P. and Moss, D. J.: Promotion of neuronal cell adhesion by members of the IgLON family occurs in the absence of either support or modification of neurite outgrowth. J. Neurochem. *80*: 941–948, 2002.

Mechawar, N. and Descarries, L.: The cholinergic innervation develops early and rapidly in the rat cerebral cortex: a quantitative immunocytochemical study. Neuroscience *108*: 555–567, 2001.

Mechawar, N., Watkins, K. C. aand Descarries, L.: Ultrastructural features of the acetylcholine innervation in the developing parietal cortex of the rat. J. Comp. Neurol. *443*: 250–258, 2002.

Meier, R., Becker-Andre, M., Gotz, R., Heumann, R., Shaw, A. and Thoenen, H.: Molecular cloning of bovine and chick nerve growth factor (NGF): delineation of conserved and unconserved domains and their relationship to the biological activity and antigenicity of NGF. EMBO J. *5*: 1489–1493, 1986.

Meier, S., Brauer, A. U., Heimrich, B., Schwab, M. E., Nitsch, R. and Savaskan, N. E.: Molecular analysis of Nogo expression in the hippocampus during development and following lesion and seizure. FASEB J. *17*: 1153–1155, 2003.

Meiniel, R.: Axial skeletal transformations in chick and quail embryos hatched from eggs treated with parathion at different stages of incubation. C. R. Seances Soc. Biol. Fil. *167*: 459–462, 1973.

Meiniel, R.: Ultrastructural studies of the evolution of the notochord and dorsal embryonic constituents derived from the mesoderm in the chicken and Japanese quail during teratogenic exposures to parathion. Arch. Anat. Microsc. Morphol. Exp. *65*: 139–163, 1976.

Meiniel, R.: Neuromuscular blocking agents and axial teratogenesis in the avian embryo. Can axial morphogenetic disorders be explained by pharmacological action upon muscle tissue? Teratology *23*: 259–271, 1981.

Meir, R., Becker-Andre, M., Gotz, R., Heumann, R., Shaw, A. and Thoenen, H.: Molecular cloning of bovine and chick nerve growth factor (NGF): delineation of conserved and unconserved domains and their relationship to the biological activity and antigenicity of NGF. EMBO J. *5*: 1489–1493, 1986.

Meisami, E.: The developing rat olfactory bulb: prospects of a new model system for the developmental neurobiology. In: "Neural Growth and Differentiation," E. Meisami and M. A. B. Brazier, Eds., pp. 183–206, Raven Press, New York, 1979.

Meisami, E. and Firoozi, M.: Acetylcholinesterase activity in the developing olfactory bulb: a biochemical study on normal maturation and influence of peripheral and central connections. Dev. Brain Res. *21*: 115–124, 1985.

Mendell, L. M., Munson, J. B. and Arvanian, V. L.: Neurotrophins and spinal plasticity in the mammalian spinal cord. J. Physiol. *533*, Pt. 1: 91–97, 2001.

Mercanti, D., Butler, R. and Revoltella, R.: A tryptic digestion fragment of nerve growth factor with nerve growth factor promoting activity. Biochim. Biophys. Acta *496*: 412–419, 1977.

Metzler, C. J. and Humm, D. G.: The determination of cholinesterase activity in whole brains of developing rats. Science *113*: 382–383, 1951.

Meyer, T., Burkart, W. and Jockusch, H.: Choline acetyltransferase induction in cultured neurons: dissociated spinal cord cells are dependent on muscle cells, organotypic explants are not. Neurosci. Lett. *11*: 59–62, 1979.

Meza, G. and Hinojosa, R.: Ontogenetic approach in cellular localization of neurotransmitters in the chick vestibule. Hear. Res. *28*: 73–85, 1987.

Michelson, M. J.: Some aspects of evolutionary pharmacology. Biochem. Pharmacol. *23*: 2211–2224, 1974.

Michelson, M. J. and Zeymal, E. B.: "Acetycholina," Hayka, Leningrad, 1970 (in Russian).

Miledi, R.: The acetylcholine sensitivity of frog muscle fibres after complete or partial denervation. J. Physiol. (Lond.) *151*(8): 1–23, 1960.

Miledi, R. and Slater, C. R.: Electrophysiology and electron microscopy of rat neuromuscular junctions after denervation. Proc. R. Soc. B *169*: 289–306, 1968.

Minganti, A., Falugi, C. and Raineri, M.: Determinationi quantitative di colineterasi in embrioni di Cirripedi (Crostacei). Rend. Acc. Naz. Lincei VIII, *63*: 598–603, 1977.

Misawa, M., Doull, J. and Uyeki, E. M.: Teratogenic effects of cholinergic insecticides in chick embryos. III. Development of cartilage and bone. J. Toxicol. Environ. Health. *10*: 551–563, 1982.

Misko, T. P., Radeke, M. J. and Shooter, E. M.: Nerve growth factor in neuronal development and maintenance. J. Exp. Biol. *132*: 177–190, 1987.

Missias, A. C., Mudd, J., Cunningham, J. M. Steinbach, J. H., Merlie, J. P. and Sanes, J. R.: Deficient development and maintenance of postsynaptic specializations in mutant mice lacking an "adult" acetylcholine receptor subunit. Development *124*: 5075–5086, 1997.

Mitrofanis, J., Maslim, J. and Stone, J.: Ontogeny of catecholaminergic and cholinergic cell distribution in the cat's retina. J. Comp. Neurol. *289*: 228–246, 1989.

Miyakoshi, L. M., Mendez-Otero, R., Hedin-Pereira, C.: The 9-0-acetyl GD3 gangliosides are expressed by migrating chains of subventricular zone neurons *in vitro*. Braz. J. Med. Biol. Res. *34*: 669–673, 2001.

Mobley, W. C., Woo, J. E., Edwards, R. H., Riopelle, R. J., Longo, F. M., Weskamp, G., Otten, U., Valletta, J. S., Johnston, M. V.: Developmental regulation of nerve growth factor and its receptor in the rat caudate-putamen. Neuron. *3*: 655–664, 1989.

Moore, S. E. and Walsh, F. S.: Specific regulation of N-CAM/D2-CAM cell adhesion molecule. EMBO J. *4*: 623–630, 1985.

Moreno, R. D, Campos, E. O., Dajas, F. and Inestrosa, N. C.: Developmental regulation of mouse brain monomeric acetylcholinesterase. J. Dev. Neurosci. *16*: 123–134, 1998.

Morley, B. J. and Happe, H. K.: Cholinergic receptors: dual roles in transduction and plasticity. Hear Res. *147*: 104–112, 2000.

Motavkin, P. A. and Okhotin, V. E.: Cholinergic neurons in the nuclear formations of the human fetal medulla oblongata. Arkh. Anat. Gistol. Embriol. *84*: 24–31, 1983.

Muralt, A. V.: "Die Signalübermittlung im Nerven," Verlag Birkhauser, Basel, 1946.

Nachmansohn, D.: Cholinestèrase dans les tissus embryonnaires. C. R. Soc. Biol. (Paris) *127*: 670–673, 1938.

Nachmansohn, D.: Cholinesterase dans le système nerveux central. Bull. Soc. Chim. Biol. (Paris) *21*: 761–796, 1939.

Nadler, L. S., Rosoff, M. L., Kalaydjian, A. E., McKinnon, L. A. and Nathanson, N. M.: Molecular analysis of the regulation of muscarinic receptor expression and function. Life Sci. *64*: 375–379, 1999.

Nastuk, M. A. and Graybiel, A. M.: Patterns of muscarinic cholinergic binding in the striatum and their relation to dopamine islands and striosomes. J. Comp. Neurol. *237*: 176–194, 1985.

Nathanson, J. A. and Greengard, P.: Cyclic nucleotides and synaptic transmission. In: "Basic Chemistry," G. J. Siegel, R. W. Albers, B. W. Agranoff, and R. Kasmon, Eds., pp. 297–310, Little, Brown, Boston, 1981.

Neugebauer, K., Salpeter, M. M. and Podleski, T. R.: Differential responses of L5 and rat primary muscle cells to factors in rat brain extract. Brain Res. *346*: 58–69, 1985.

Nguyen, L. T. and Janssen, C. R.: Comparative sensitivity of embryo-larval toxicity assays with African catfish (Clarias gariepinus) and zebra fish (Danio rerio). Environ. Toxicol. *16*: 566–571, 2001.

Nietgen, G. W., Schmidt, J., Hesse, L., Hanemann, C. W. and Durieux, M. E.: Muscarinic receptor functioning and distribution in the eye: molecular basis and implications for clinical diagnosis and therapy. Eye *13*: 285–300, 1999.

Ninomiya, Y., Koyama, Y. and Kayama, Y.: Postnatal development of choline acetyltransferase activity in the rat laterodorsal tegmental nucleus. Neurosci. Lett. *308*: 138–140, 2001.

Nishi, R. and Berg, D. K.: Two components from eye tissue that differentially stimulate the growth and development of ciliary ganglion neurons in cell culture. J. Neurosci. *1*(7): 505–513, 1981.

Novikova, L. N., Novikov, L. N. and Kellerth, J. O.: Differential effects of neurotrophins on neuronal survival and axonal regeneration after spinal cord injury in adult rats. J. Comp. Neurol. *452*: 255–263, 2002.

Numanoi, H.: Studies on the fertilization substance V. Distribution of acetylcholine esterase in egg particles of the sea urchin, Hemicentrous pulcherrhimus. Sci. Papers Coll. Gen. Educ. Univ. Tokyo *5*: 37–41, 1951.

Ochs, S.: An historical introduction to the trophic regulation of skeletal muscle. In: "Nerve-Muscle Cell Trophic Communication," H. L. Fernandez, J. A. Donoso, Eds., pp. 7–22, CRC Press, Boca Raton, Fla., 1987.

Oderfeld-Nowak, B., Wojcik, M., Ulas, J. and Potempska, A.: Effects of chronic ganglioside treatment on recovery processes in hippocampus after brain lesions in rats. In: "Ganglioside Function: Biochemical and Pharmacological Implications," G. Porcellati, B. Ceccarelli, and G. Tettamanti, Eds., pp. 197–205, Plenum Press, New York, 1975.

Oderfeld-Nowak, B., Skup, M., Ulas, J., Jezierska, M., Gradkowska, M. and Zaremba, M.: Effect of GMI ganglioside treatment on pest lesion responses of cholinergic enzymes in rat hippocampus affer various partial deafferentations. J. Neurosci. Res. *12*: 409–420, 1984.

Oderfeld-Nowak, B., Casamenti, F. and Pepeu, G.: Gangliosides in the repair of brain cholinergic neurons. Acta Biochim. Pol. *40*: 395–404, 1993.

Oh, T. H. and Markelonis, G. J.: Chicken serum transferrin duplicates the myotrophic effects of sciatin on cultured muscle cells. J. Neurosci. Res. *8*: 535–545, 1982.

Oh, T. H., Markelonis, G. J. and Shim, S. H.: Trophic influences of neurogenic substances on skeletal muscle differentiation and growth *in vitro*. In: "Nerve-Muscle Cell Trophic Communication," H. L. Fernandez and J. A. Donoso, Eds., pp. 81–99, CRC Press, Boca Raton, Fla., 1987.

Oh, T. H., Markelonis, G. J., Dion, T. L. and Hobbs, S. L.: A muscle-derived substrate-bound factor that promotes neurite outgrowth from neurons of the central and peripheral nervous systems. Dev. Biol. *127*: 88–98, 1988.

Oiwa, Y., Yoshimara, R., Nakai, K. and ItaRawa, T.: Dopaminergic neuroprotection and regeneration by neurturin assessed by using behavioral, biochemical and histochemical measurements in a model of progressive Parkinson's disease. Brain Res. *947*: 271–283, 2002.

Oosawa, H., Fujii, T. and Kawashima, K.: Nerve growth factor increases the synthesis and release of acetylcholine and the expression of vesicular acetylcholine transporter in primary cultured rat embryonic septal cells. J. Neurosci. Res. *57*: 381–387, 1999.

Oppenheim, R. W.: Neurotrophic survival molecules for motoneurons: an embarrassment of riches. Neuron *17*: 195–197.

Oppenheim, R. W.: Victor Hamburger (1900–2001): journey of a neuroembryologist to the end of the millennium and beyond. Neuron *31*: 179–190, 2001.

Oppenheim, R. W. and Haverkamp, L. J.: Neurotrophic interactions in the development of spinal cord motoneurons. Ciba Found Symp. *138*: 151–179, 1988.

Oppenheim, R. W., Houenou, L., Pincon-Raymond, M., Powell, J. A., Rieger, F. and Standish, L. J.: The development of motoneurons in the embryonic spinal cord of the mouse mutant dysgenesis (mdg/mdg): survival, morphology, and biochemical differentiation. Dev. Biol. *114*: 426–436, 1986.

Otten, U., Hatanaka, H. and Thoenen, H.: Role of cyclic nucleotides in nerve growth factor-mediated induction of tyrosine hydroxylase in rat sympathetic ganglia and adrenal medulla. Brain Res. *140*: 385–389, 1978.

Pallage, V., Toniolo, G., Will, B. and Hefti, F., Long-term effects of nerve growth factor and neural transplants on behavior of rats with medial septal lesions. Brain Res. *386*: 197–208, 1986.

Papanno, A.: Ontogenetic development of autonomic neuroeffector transmission and transmitter reactivity in embryomic and fetal hearts. Pharmacol. Rev. *29*: 3–33, 1979.

Pappano, A. J. and Loffelholz, K.: Ontogenesis of adrenergic and cholinergic neuroeffector transmission in chick embryo heart. J. Pharmacol. Exp. Ther. *191*: 468–478, 1974.

Patel, A. J., del Vecchio, M. and Atkinson, D. J.: Effect of undernutrition on the regional development of transmitter enzymes: glutamate decarboxylase and choline acetyltransferase. Dev. Neurosci. *1*: 41–53, 1978.

Pattarawarapan, M., Zaccaro, M. C., Saragovi, U. H. and Burgess, K.: New templates for syntheses of ring-fused, C(10) beta-turn peptidomimetics leading to the first reported small molecule mimic of neurotrophin-3. J. Med. Chem. *45*: 4387–4390, 2002.

Patterson, P. H. and Chun, L. L. Y.: The induction of acetylcholine synthesis in primary cultures of dissociated rat sympathetic neurons. I. Effects of conditioned medium. Dev. Biol. *56*: 263–280, 1977.

Paul, G. and Davies, A. M.: Trigeminal sensory neurons require extrinsic signals to switch neurotrophin dependence during the early stages of target field innervation. Dev. Biol. *171*: 590–605, 1995.

Pelleymounter, M. A. and Cullen, M. J.: The effects of intraseptal brain-derived neurotrophic factor or cognition in rats with MSIDB lesions. Ann. N. Y. Acad. Sci. *679*: 299–305, 1993.

Peng, H. B.: Development of acetylcholine receptor clusters induced by basic polypeptides in cultured muscle cells. In: "Neurobiology of Acetylcholine—A Symposium in Honor of Alexander G. Karczmar," N. J. Dun and R. L. Perlman, Eds., pp. 17–26, Plenum Press, New York, 1987.

Pepeu, G.: Overview and future of CNS cholinergic mechanisms. In: "Cholinergic Neurotransmission: Function and Dysfunction." A. C. Cuello, Ed., pp. 455–458. Elsevier, Amsterdam, 1993.

Pepeu, G., Oderfeld-Nowak, B. and Casamenti, F.: CNS pharmacology of gangliosides. Prog. Brain Res. *101*: 327–335, 1994.

Perry, E. K., Smith, C. J., Atack, J. R., Candy, J. M., Johnson, M. and Perry, R. H.: Neocortical cholinergic enzyme and receptor activities in the human fetal brain. J. Neurochem. *47*: 1262–1269, 1986.

Pesavento, E., Capsoni, S., Domenici, L. and Cattaneo, A.: Acute cholinergic rescue of synaptic plasticity in the neurodegenerating cortex of ant-nerve-growth-factor mice. Eur. J. Neurosci. *15*: 1030–1036, 2002.

Pilar, G., Nunez, R., McLennan, I. S. and Meriney, S. D.: Muscarinic and nicotinic synaptic activation of the developing chicken iris. J. Neurosci. *7*: 3813–3826, 1987.

Pizzo, D. P. and Thal, L. J.: Intraparenchymal nerve growth factor improves behavioral deficits while minimizing the adverse effects of intracerebroventricular delivery. Neuroscience *124*: 743–754, 2004.

Pizzorusso, T., Berardi, N., Rossi, F. M., Viegi, A., Venstrom, K., Reichardt, L. F. and Maffei, L.: TrkA activation of the rat visual cortex by antirat trkA IgG prevents monocular deprivation. Eur. J. Neurosci. *11*: 204–212, 1999.

Poulson, D. E. and Boell, E. J.: The development of cholinesterase activity in embryos of normal and genetically deficient strains of Drosophila melanogaster. Anat. Rec. *96*: 508, 1946.

Prives, J. M. and Patterson, B. M.: Differentiation of cell membranes in cultures of embryonic chick breast muscle. Proc. Natl. Acad. Sci. U.S.A. *71*: 3208–3213, 1974.

Purves, D. and Lichtman, J.W.: "Principles of Neural Development," Sinauer Associates, Boston Mass., 1985.

Purves, D., Snider, W. D. and Voyvodic, J. T.: The mammalian autonomic nervous system. Nature *336*: 123–128, 1988.

Qiao, X., Chen, L., Gao, H., Bao, S., Hefti, F., Thompson, R. F. and Knusel, B.: Cerebellar brain-derived neurotrophic factor-TrkB defect associated with impairment of eyeblink conditioning in Stargazer mutant mice. J. Neurosci. *18*: 6990–6999, 1998.

Rabin, S. J., Bachis, A. and Mochetti, I.: Gangliosides activate trk receptors by inducing the release of neurotrophins. J. Biol. Chem. *277*: 49466–49472, 2002.

Rabizadeh, S., Bittner, C. M., Butcher, L. L. and Bredesen, D. E.: Expression of the low affinity growth factor receptor enhances beta-amyloid peptide toxicity. Proc. Natl. Acad. Sci. U.S.A. *91*: 10703–10706, 1994.

Raineri, M. and Falugi, C.: Acetylcholinesterase activity in embryonic and larval development of Artemia salina (Crustacea, Phyllopoda). J. Exp. Zool. *227*: 229–246, 1983.

Raivich, G. and Kreutzberg, G. W.: The localization and distribution of high affinity ß-nerve growth factor binding sites in the central nervous system of the adult rat: a light microscopic autoradiographic study using [^{125}I] ß-nerve growth factor. Neuroscience *30*: 23–36, 1987.

Rapport, M. M.: Introduction to the biochemistry of gangliosides. In: "Neurological and Neuromuscular Function, Development and Repair," M. M. Rapport and A. Losso, Eds., pp. 15–19, Raven Press, New York, 1981.

Rashev, Z. and Vasilev, V. K.: Teratogenic effect of the pesticide preparation metathion. Vet. Med. Nauki. *19*: 79–89, 1982.

Rattray, M.: Is there nicotinic modulation of nerve growth factor? Implications for cholinergic therapies in Alzheimer's disease. Biol. Psychiatry *49*: 185–193, 2001.

Ravikumar, B. V. and Sastry, P. S.: Muscarinic cholinergic receptors in human foetal brain: characterization and ontogeny of [3H] quinuclidinyl benzilate binding sites in frontal cortex. J. Neurochem. *44*: 240–246, 1985a.

Ravikumar, B. V. and Sastry, P. S.: Cholinergic muscarinic receptors in human fetal brain: ontogeny of [H]quinuclidinyl benzilate binding sites in corpus striatum, brainstem, and cerebellum. J. Neurochem. *45*: 1948–1950, 1985b.

Rea, M. A. and Nurnberger, J. I.: Evidence for developmental synaptic regression of cholinergic afferents to the rat main olfactory bulb. Brain Res. *389*: 233–237, 1986.

Reich, A. and Drews, U.: Choline acetyltransferase in the chick limb bud. Histochemistry *78*: 383–389, 1983.

Reichardt, L. F. and Mobley, W. C.: Going the distance, or not, with neurotrophin signals. Cell *118*: 141–143, 2004.

Reshetnikova, N. A.: On the content of acetylcholine in the developing eggs of the frog, Rana temporaria (in Russian). Zh. Biochem. i Fiziol. *6*: 19–24, 1970.

Rho, J. M. and Storey, T. W.: Molecular ontogeny of major neurotransmitter receptor systems in the mammalian central nervous system: norepinephrine, dopamine, serotonin, acetylcholine, and glycine. J. Child Neurol. *16*: 271–280, 2001.

Ricceri, L., Markina, N., Valanzano, A., Fortuna, S., Cometa, M. F., Meneguz, A. and Calamandrei, G.: Developmental exposure to chlorpyrifos alters reactivity to environmental and social cues in adolescent mice. Toxicol. Appl. Pharmacol. *191*: 11189–11201, 2003.

Richardson, P. M., Issa, V. M. and Riopelle, K. J.: Distribution of neuronal receptors for nerve growth factor in the rat. J. Neurosci. *6*: 2312–2321, 1986.

Rieger, F., Grumet, M. and Edelman, G. M.: N-CAM at the vertebrate neuromuscular junction. J. Cell Biol. *101*: 285–293, 1985.

Rimvall, K., Keller, F. and Waser, P. G.: Development of cholinergic projections in organoptypic culture of rat septum, hippocampus and cerebellum. Brain Res. *351*: 267–278, 1985.

Riopelle, R. J. and Riccardi, V. M.: Neuronal growth factors from tumours of Van Recklinghausen neurofibromatosis. J. Neurol. Sci. *14*(2): 141–144, 1987.

Riopelle, R. J., Richardson, P. M. and Verge, V. M. K.: Distribution and characteristics of nerve growth factor binding on cholinergic neurons of rat and monkey forebrain. Neurochem. Res. *12*(10): 923–928, 1987.

Rizzo, A. M., Rossi, F. and Berra, B.: Glycosyltransferases in different brain regions during chick embryo development. Neurochem. Res. *27*: 815–821, 2002.

Robertson, R. T. and Yu, J.: Development of cholinergic projections to cortex: possible role of neurotrophins in target selection. In: "XI International Symposium on Cholinergic Mechanisms—Function and Dysfunction," I. Silman, H. Soreq, L. Anglister, D. Michaelson and A. Fisher, Eds., pp. 381–386, Martin Dunitz, London.

Robertson, R. T., Tijerin, A. and Gallivan, M. B.: Transient patterns of acetylcholinesterase activity in visual cortex of the rat: normal development and the effects of neonatal monocular enucleation. Dev. Brain Res. *21*: 203–214, 1985.

Robertson, R. T., Yu, B. P., Liu, H. H., Liu, N. H. and Kageyama, G. H.: Development of cholinesterase histochemical staining in cerebellar cortex: transient expression of "nonspecific" cholinesterase in Purkinje cells of the nodulus and uvula. Exp. Neurol. *114*: 330–342, 1991.

Rossi, F. M., Sala, R. and Maffei, L.: Expression of the nerve growth factor receptors TrkA and p75NTR in the visual cortex of the rat: development and regulation by the cholinergic input. J. Neurosci. *22*: 912–919, 2002.

Rossner, S., Yu, J., Pizzo, D., Warrback-Perez, K., Schliebs, R., Bigl, V. and Perez-Polo, J. R.: Effects of intraventricular transplantation of NGF-secreting cells on cholinergic basal forebrain neurons after partial immunolesion. J. Neurosci. Res. *45*: 40–56, 1996.

Rothwell, N. J. and Hopkins, S. J.: Cytokines and the nervous system II: Actions and mechanisms of action. Trends Neurosci. *18*: 130–136, 1995.

Rotshenker, S. and Tal, M.: The transneuronal induction of sprouting and synapse formation in intact mouse muscles. J. Physiol. *360*: 387–396, 1985.

Rotundo, R. L.: Biogenesis and regulation of acetylcholinesterase. In: "The Vertebrate Neuromuscular

Junction," M. M. Salpeter, pp. 247–284, Alan R. Liss, New York, 1987.

Roukema, P. A., Van Den Eijnden, D. H., Heijlman, J. and Van Der Berg, G.: Sialoglycoproteins, gangliosides and related enzymes in developing rat brain. Febs Lett. *9*: 267–270, 1970.

Rozengart, E. V. and Moralev, S. N.: Variability of substrate specificity in cholinesterases of vertebrates and invertebrates. In: "Cholinergic Mechanisms—Function and Dysfunction," I. Silman, H. Soreq, L. Anglister, D. Michaelson, and A. Fisher, pp. 685–686, Ed., Martin Dunitz, London, 2004.

Ruberti, F., Capsoni, S., Comparini, A., Di Daniel, F., Frazot, J. and Gonfloni, S.: Phenotypic knockout of nerve growth factor in adult transgenic mice reveals severe deficits in basal forebrain cholinergic neurons, cell death in the spleen, and skeletal muscle dystrophy. J. Neurosci. *20*: 2589–2601, 2000.

Rubin, E.: Development of the rat superior cervical ganglion: Ganglion cell maturation. J. Neurosci. *5*: 673–684, 1985a.

Rubin, E.: Development of the rat superior cervical ganglion: Ingrowth of preganglionic axons. J. Neurosci. *5*: 685–696, 1985b.

Rubin, E.: Development of the rat superior cervical ganglion: Initial stages of synapse formation. J. Neurosci. *5*: 697–704, 1985c.

Ruit, K. G., Elliott, J. L., Osborne, P. A., Yan, O. and Snider, W. D.: Selective dependence of mammalian dorsal root ganglion neurons on nerve growth factor during embryonic development. Neuron. *8*: 573–587, 1992.

Runft, L. L., Jaffe, L. A. and Mehlmann, L. M.: Egg at fertilization: where it all begins. Dev. Biol. *245*: 237–254, 2002.

Ruohonen, S. Jagodi, M., Khademi, M., Taskinen, H. S., Ojala, P., Olsson, T. and Roytta, M.: Contralateral non-operated nerve to transected rat sciatic nerve shows increased expression of IL-1beta, TGF-beta1, TNF-alpha, and IL-10. J. Neuroimmunol. *132*: 11–17, 2002.

Rylett, R. J. and Williams, L. R.: Role of neurotrophins in cholinergic neuron function in the adult and aged CNS. TINS *17*: 486–490, 1994.

Sabel, B. A., Dunbar, G. L. and Stein, D. G.: Gangliosides minimize behavioral deficits and enhance structural repair after brain injury. J. Neurosci. Res. *12*: 429–443, 1984.

Sadykova, K. A., Sakharova, N. I. and Makarova, L. N.: The effect of cyclic nucleotides on the sensitivity of early mouse embryos to biogenic monoamine antagonists. Ontogenes *23*: 379–384, 1992.

Saelens, J. K. and Simke, J. P.: Appendix III: acetylcholine and choline concentrations of various biological tissues. In "Biology of Cholinergic Function," A. M. Goldberg and I. Hanin, Eds., pp. 661–682, Raven Press, New York, 1976.

Salpeter, M. M.: Development and neural control of the neuromuscular junction and of the junctional acetylcholine receptor. In: "The Vertebrate Neuromuscular Junction," M. M. Salpeter, Ed., pp. 55–115, Alan R. Liss, pp. 55–115, 1987.

Salpeter, M. M. and Loring, R. H.: Nicotinic acetylcholine receptors in vertebrate muscle: properties, distribution and neural control. Prog. Neurobiol. *25*: 297–325, 1985.

Salvaterra, P., Kitamoto, T. and Ikeda, K.: Molecular specification of cholinergic neurons. In: "Cholinergic Function and Dysfunction," A. C. Cuello, Ed., pp. 167–173, Elsevier, Amsterdam, 1993.

Salvaterra, P. M., Lee, M. H. and Song, S.: Genetic regulation of cholinergic neurotransmitter phenotypes. In: "Cholinergic Mechanisms—Function and Dysfunction," I. Silman, H, Soreq, L. Anglister, D. Michaelson and A. Fisher, Ed., pp. 193–205, Martin Dunitz, London, 2004.

Sanes, J. R.: The basement membrane/basal lamina of skeletal muscle. J. Biol. Chem. *278*(15): 12601–12604, 2003.

Sanes, J. R. and Lichtman, J. W.: Induction, assembly, maturation and maintenance of a postsynaptic apparatus. Neuroscience *2*: 791–805, 2001.

Sanes, J. R., Marshall, L. M. and McMahan, U. J.: Reinnervation of muscle fiber basal lamina after removal of myofibers: differentiation of regenerating axons at original synaptic sites. J. Cell Biol. *78*: 176–198, 1978.

Sanes, J. R., Apel, E. D., Burgess, R. W., Emerson, R. B., Feng, G., Gautam, M., Glass, D., Grady, R. M., Krejci, E., Lichtman, J. W., Lu, J. T., Massoulié, J., Miner, J. H., Moscoso, L. M., Nguyen, Q., Nichol, M., Naokes, P. G., Patton, B. L., Son, Y.-J., Yancopoulos, G. D. and Zhou, H.: Development of the neuromuscular junction: genetic analysis in mice. In: "From *Torpedo* Electric Organ to Human Brain: Fundamental and Applied Aspects," Jean Massoulié, Ed., pp. 1167–1172, Elsevier, Amsterdam, 1998.

Sawyer, C. H.: Cholinesterase and behavior problem in Amblystoma. I. The relationship between the development of the enzyme and early motility. II. The effects of inhibiting cholinesterase. J. Exp. Zool. *92*: 1–27, 1943.

Sawyer, C. H.: Nature of the early somatic movements in Fundulus heteroclitus. J. Cell. Comp. Physiol. *24*: 71–84, 1944.

Sayer, S. T., Yeoh, H. C., McLane, J. A. and Held, I. R.: Decreased acetylcholine receptor content in denervated skeletal muscles infused with nerve extract. Brain Res. *407*: 341–350, 1987.

Sayer, F. T., Oudega, M. and Hagg, T.: Neurotrophins reduce degeneration of injured ascending sensory and corticospinal motor axons in adult rat spinal cord. Exp. Neurol. *175*: 282–296, 2002.

Schaefer, T. E., Schwab, E. M. and Thoenen, H.: Influence of nerve growth factor on the ontogenesis of the presynaptic cholinergic fibers in the rat superior cervical ganglion. Experientia *35*: 65, 1979.

Schambra, U. B., Sulik, K. K., Petrusz, P. and Lauder, J. M.: Ontogeny of cholinergic neurons in the mouse forebrain. J. Comp. Neurol. *288*: 101–122, 1989.

Schambra, U. B., Silver, J. and Lauder, J. M.: An atlas of the prenatal mouse brain: gestational day 14. Exp. Neurol. *114*: 145–183, 1991.

Schaterman, O. C., Gibbs, L., Lanahan, A. A., Claude, P. and Bothwell, P. M.: Expression of NGF receptor in the developing and adult primate central nervous system. J. Neurosci. *8*: 860–873, 1988.

Schecterson L. and Bothwell, M.: New roles for neurotrophins are suggested by BDNF and NT-3 mRNA expression in developing neurons. Neuron. *9*: 449–463, 1992.

Schlumpf, M., Palacios, J. M., Cortes, R. and Lichtensteiger, W.: Regional development of muscarinic cholinergic binding sites in the prenatal rat brain. Neuroscience *45*: 347–357, 1991.

Schneider, J. S., Pope, A., Simpson, K., Taggart, J., Smith, M. G. and DiStefano, L.: Recovery from experimental parkinsonism in primates with GM1 ganglioside treatment. Science *256*: 843–846, 1992.

Schneider, M., Adee, C., Betz, H. and Schmidt, J.: Biochemical characterization of two nicotinic receptors from the optic lobe of the chick. J. Biol. Chem. *260*: 14505–14512, 1985.

Schober, A. and Unsicker, K.: Growth and neurotrophic factors regulating development and maintenance of sympathetic preganglionic neurons. Intern. Rev. Cytol. *205*: 37–75, 2001.

Schotté, O. E. and Butler, E. G.: Morphological effects of denervation and amputation of limbs in urodele larvae. J. Exp. Zool. *87*: 279–322, 1941.

Schotté, O. E. and Karczmar, A. G.: Limb parameters and regeneration rates in denervated amputated limbs of urodele larvae. J. Exp. Zool. *97*: 43–70, 1944.

Schroder, H., Schutz, U., Burghaaus, L., Lindstrom, J., Kuryatov, A., Montegia, L., DeVos, R., A., Van Noort, G., Wevers, A., Nowacki, S., Happich, E., Moser, N., Arneric, S. P. and Maelicke, A.: Expression of the alpha 4 isoform of nicotinic acetylcholine receptor in the fetal human cerebral cortex. Dev. Brain. Res. *132*: 33–45, 2001.

Schuetze, S. M.: The acetylcholine channel open time in the chick muscle is not decreased following denervation. J. Physiol. (Lond.) *303*: 11–124, 1980.

Schuytema, G. S., Nebeker, A. V. and Griffis, W. L.: Toxicity of Guthion and Guthion 2S to Xenopus laevis larvae. Arch. Environ. Contam. Toxicol. *27*: 250–255, 1994.

Schwab, M., Otten, U., Agid, Y. and Thoenen, H.: Nerve growth factor (NGF) in the rat CNS: absence of specific retrograde axonal transport and tyrosine hydroxylase induction in locus coeruleus and substantia nigra. Brain Res. *168*: 473–483, 1979.

Segal, M., Landman, N. and Greenberger, V.: Central cholinergic neurons in culture: regulation of survival and function. In: "XI International Symposium on Cholinergic Mechanisms—Function and Dysfunction," I. Silman, H. Soreq, L. Anglister, D. Michaelson and A. Fisher, Eds., pp. 405–410, Marin Dunitz, London, 2003.

Segal, R. and Greenberg, M.: Intracellular signalling pathways activated by neurotrophic factors. Annu. Rev. Neurosci. *19*: 463–489, 1996.

Seiler, M. and Schwab, M. E.: Specific retrograde transport of nerve growth factor (NGF) form neocortex to nucleus basalis in the rat. Brain Res. *300*: 33–39, 1984.

Selby, M. J., Edwards, R. H. and Rutter, W. J.: Cobra nerve growth factor: structure and evolutionary comparison. J. Neurosci. Res. *18*: 293–298, 1987.

Sellin, L. C. and McArdle, J. J.: Colchicine blocks neurotrophic regulation of the resting potential in reinnervating skeletal muscle. Exp. Neurol. *55*: 483–492, 1977.

Semba, K.: Development of central cholinergic neurons. In: "Handbook of Neuroanatomy: Ontogeny of Transmitters and Peptides in the CNS," A. Bjorklund, T. Hokfelt, and M. Tohoyama, Eds., pp. 33–62, Elsevier, Amsterdam, 1992.

Sendtner, M., Gnahn, H., Wakade, A. and Thoenen, H.: Is activation of Na+K+ pump necessary for NGF-mediated neuronal survival? J. Neurosci. *8*: 458–462, 1988.

Server, A. C. and Shooter, E. M.: Nerve growth factor. Adv. Protein Chem. *31*: 339–409, 1977.

Sharma, R. K.: Cyclic nucleotide control of protein kinases. Prog. Nucleic. Acid. Res. Mol. Biol. *27*: 233–288, 1982.

Shelton, D. L. and Reichardt, L. F.: Expression of the ß-nerve growth factor gene correlates with the density of sympathetic innervation in effector organs. Proc. Natl. Acad. Sci. U.S.A. *81*: 7951–7955, 1984.

Shen, S. C., Greenfield, P. and Boell, E. J.: Localization of acetylcholinesterase in chick retina during histogenesis. J. Comp. Neurol. *106*: 433–462, 1956.

Sherstnev, V. V., Gruden, M. A., Skvortsova, V. I. and Tabolin, V. A.: Neurotrophic factors and autoantibodies to them as molecular predictors of cerebral dysfunctions. Vestn. Ross. Akad. Med. Nauk. *6*: 48–52, 2002.

Sippel, T. O.: Properties and development of cholinesterase in the hearts of certain vertebrates. J. Exp. Zool. *128*: 165–184, 1955.

Skaper, S. D., Montz, H. P. and Varon, S.: Control of Na+, K+-pump activity in dorsal root ganglionic neurons by different neuronotrophic agents. Brain Res. *386*(1–2):130–135, 1986.

Slotkin, T. A.: Development of cholinotoxicants: nicotine and chlorpyrifos. Environ. Health Perspect. *107*, Supplt. 1: 71–80, 1999.

Slotkin, T. A., Orband, K. L. and Queen, K. L.: Development of [3H] nicotine binding sites in brain regions of rats exposed to nicotine prenatally via maternal injections or infusions. J. Pharmacol. Exp. Ther. *242*: 232–237, 1987.

Smith, J., Fauquet, M., Ziller, C. and LeDouarin, N. M.: Acetylcholine synthesis by mesencephalic neural crest cells in the process of migration *in vivo*. Nature *282*: 853–855, 1979.

Smith, R. G., McManaman, J. L. and Appel, S. H.: Trophic effects of skeletal muscle extracts on ventral spinal cord neurons *in vitro*: separation of a protein with morphologic activity from proteins with cholinergic activity. J. Cell. Biol. *101*(4): 1608–1621, 1985.

Sobue, G., Taki, T., Yasuda, T. and Mitsuma, T.: Gangliosides modulate Schwann cell proliferation and morphology. Brain Res. *474*: 287–295, 1988.

Sofroniew, M. W., Pearson, R. C. and Powell, T. P.: The cholinergic nuclei of the basal forebrain of the rat: normal structure, development and experimental induced degeneration. Brain Res. *411*: 310–331, 1987.

Sollmann, T. L.: The simultaneous action of pilocarpine and atropine on the developing embryos of the sea-urchin and starfish. A contribution to the study of antagonistic action of poisons. Am. J. Physiol. *10*: 352–361, 1904.

Soltys, Z., Janeczko, K., Orzylowska-Sliwinska, O., Zarembra, M., Januszewski, S. and Oderfeld-Nowak, B.: Morphological transformations of cells immunopositive for GFAP, TrkA or p75 in the CA1 hippocampal area following transient global ischemia in the rat. A quantitative study. Brain Res. *987*: 186–193, 2003.

Sonnino. S., Ghidoni, R., Gazzotti, G., Acquotti, D. and Tettamanti, G.: New trends in gangliosides chemistry. Adv. Exp. Med. Biol. *228*: 437–464, 1988.

Sorenson, E. M., Parkinson, D., Dahl, J. L. and Chiapinelli, V. A.: Immunohistochemical localization of choline acetyltransferase in the chicken mesencephalon. J. Comp. Neurol. *28*: 641–657, 1989.

Soreq, H., Malinger, G. and Zakut, H.: Expression of cholinesterase genes in human oocytes revealed by in-situ hybridization. Hum. Reprod. *2*: 689 693, 1987.

Soreq, H., Ben Aziz, R., Prody, C. A., Seideman, S., Gnatt, A., Neville, L., Lieman-Hurwitz, J. S., Lev-Lehman, G., Ginsberg, C. and Lapidot-Lifson, Y.: Molecular cloning and construction of the coding region for human acetylcholinesterase reveals a G+ C-rich attenuating structure. Proc. Natl. Acad. Sci. U.S.A. *87*: 9688–9692, 1990.

Sorimachi, M. and Kataoka, K.: Developmental change of choline acetyltransferase and acetylcholinesterase in the ciliary and the superior cervical ganglion of the chick. Brain Res.: *70*: 123–140, 1974.

Spemann, H.: "Embryonic Development and Induction," Yale University Press, New Haven, Conn., 1938.

Squinto, S. P., McLane, J. A. and Held, I. R.: Identification of a neuroregulated phosphoprotein in skeletal muscle as the regulatory subunit of cyclic AMP-dependent protein kinase II. J. Neurosci. Res. *14*: 217–228, 1985.

Stein, D. G.: Functional recovery from brain damage following treatment with nerve growth factor. In: "Functional Recovery from Brain Damage," M. W. van Hof and G. Mohn, Eds., pp. 423–443, Elsevier/ North-Holland Biomedical Press, Amsterdam, 1981.

Stein, D. G.: Brain damage and recovery. Prog. Brain Res. *100*: 203–211, 1994.

Stein, D. G., Finger, S. and Hart, T.: Brain damage and recovery: problems and perspectives. Behav. Neur. Biol. *37*: 185–222, 1983.

Stephenson, M. and Rowatt, E.: The production of acetylcholine by a strain of Lactobacillus planatarum. J. Gener. Microbiology *1*: 279–298, 1947.

Strudel, G. and Gateau, G.: Existence of cholinergic receptors in the very young chick embryo. C. R. Hebd. Seances Acad. Sci. *284*: 4469–4472, 1977.

Sullivan, G. E.: Proceedings: paralysis and skeletal anomalies in chick embryos treated with physostigmine. J. Anat. *116*: 463–464, 1973.

Sunshine, J., Balak, K., Rutishauser, U., Jacobson, M.: Changes in neural cell adhesion molecule (NCAM) structure during vertebrate neural development. Proc. Natl. Acad. Sci. U.S.A. *84*: 5986–5989, 1987.

Sutter, A., Riopelle, R. J., Harris-Warrick, R. M. and Shooter, E. M.: Nerve growth factor receptors: characterization of two distinct classes of binding sites on chick embryo sensory ganglion cells. J. Biol. Chem. *254*: 5972–5982, 1979.

Svennerholm, L.: Interaction of cholera toxin and ganglioside G_{M1}. In: "Ganglioside Function: Biochemical and Pharmacological Implications," G. Porcellati, B. Ceccarelli, and G. Tettamanti, Eds., 191–204, Plenum Press, New York, 1975.

Svennerholm, L., Brane, G., Karlsson, I., Lekman, A., Ramstrom, I., Wikkelso, C.: Alzheimer disease—effect of continuous intracerebroventricular treatment with GM1 ganglioside and a systematic activation programme. Dement. Geriatr. Cogn. Disord. *14*: 128–136, 2002.

Sweeney, J. E., Hohmann, C. F., Oster-Granite, M. L. and Coyle, J. T.: Neurogenesis of the basal forebrain in euploid and trisomy 16 mice: an animal model for developmental disorders in Down syndrome. Neuroscience *31*: 413–425, 1989.

Szutowicz, A.: Aluminum, NO, and nerve growth factor neurotoxicity in cholinergic neurons. J. Neurosci. Res. *66*: 1009–1018, 2001.

Szutowicz, A., Jankowska, A., Blusztajn, J. K. and Tomaszewicz, M.: Acetylcholine and acetyl-CoA metabolism in differentiating SN56 septal cell line. J. Neurosci. Res. *57*: 131–136, 1999.

Tanaka, Y., Han, H., Hagishita, T., Fukui, F., Liu, G. and Ando, S.: Alpha-Sialylcholesterol enhances the depolarization-induced release of acetylcholine and glutamate in rat hippocampus: *in vivo* microdialysis study. Neurosci. Lett. *357*: 9–12, 2004.

Taniuchi, M., Schweitzer, J. B. and Johnson, E. M.: Nerve growth factor receptor molecules in rat brain. Proc. Natl. Acad. Sci. U.S.A. *83*: 1950–1954, 1986.

Tello, F.: Degenèration et regeneration des plaques motrices apres la section des nerfs. Trab. Inst. Cajal Invest. Biol. *5*: 117–149, 1907.

Temple, S., Qian, X.: bFGF, neurotrophins, and the control or cortical neurogenesis. Neuron. *15*: 249–252, 1995.

Tennyson, V. M., Brzin, M.: The appearance of acetylcholinesterase in the dorsal root neuroblast of the rabbit embryo. A study by electron microscope cytochemistry and microgasometric analysis with the magnetic diver. J. Cell Biol. *46*: 64–80, 1970.

Tettamanti, G., Sonnino, S., Ghidoni, R. Masserini, M. and Venerando, B.: Chemical and functional properties of gangliosides: their possible implications in the membrane-mediated transfer of information. In: "Physics of Amphiphiles: Micelles, Vesicles and Microemulsion," M. Corti, and V. Degiorgio, Eds., pp. 607–636, Il Nuevo Ciento, XC Corso, Soc. Italiana di Fisica, Bologna, Italy, 1985.

Thanos, S., Fujisawa, H., Bonhoeffer, F.: Elimination of ipsilateral retinotectal projections in monoophthalmic chick embryos. Neurosci. Lett. *44*: 143–148, 1984.

Thesleff, S.: The effects of decamethonium and its ethyl analogues on neuromuscular transmission. Acta Pharmacol. Toxicol. Scand. *11*: 179–186, 1955.

Thesleff, S.: Supersensitivity of skeletal muscle produced by botulinum toxin. J. Physiol. (Lond.) *151*: 598–607, 1960.

Thiery, J. P.: Mechanisms of cell migration in the vertebrate embryo. Cell Differ. *15*: 1–15, 1984.

Thiery, J. P., Duband, J. L., Rutishauser, U. and Edelman, G. M.: Cell adhesion molecules in early chicken embryogenesis. Proc. Natl. Acad. Sci. U.S.A. *79*: 6737–6741, 1982.

Thoenen, H.: Comparison between the effect of neuronal activity and nerve growth factor on the enzymes involved in the synthesis of norepinephrine. Pharmacol. Rev. *24*(2): 255, 1972.

Thoenen, H.: Neurotrophic and neuronal plasticity. Science *270*: 593–598, 1995a.

Thoenen, H.: Trends in biomedicine (author's transl.) Schweiz Rundsch. Med. Prax. *64*: 1013–1016, 1975b.

Thoenen, H. and Barde, Y.-A.: Physiology of nerve growth factor. Physiol. Rev. *60*: 1284–1335, 1980.

Thoenen, H. and Sendtner, M.: Neurotrophins: from enthusiastic expectations through sobering experiences to rational therapeutic approaches. Nat. Neurosci. *5*, Supplt. 1: 1046–1050, 2002.

Thoenen, H., Angeletti, P. U., Levi-Montalcini, R. and Kettler, R.: Selective induction by nerve growth factor of tyrosine hydroxylase and dopamine-beta-hydroxylase in the rat superior cervical ganglia. Proc. Natl. Acad. Sci. U.S.A. *68*: 1598–1602, 1971.

Thoenen, H., Saner, A., Angeletti, P. U. and Levi-Montalcini, R.: Increased activity of choline acetyltransferase in sympathetic ganglia after prolonged administration of nerve growth factor. Nature *236*: 26–28, 1972.

Thoenen, H., Auburger, R., Hellweg, R., Heumann, R. and Korsching, S.: Cholinergic innervation and levels of nerve growth factor and its mRNA in the central nervous system. In: "Cellular and Molecular Basis of Cholinergic Function," M. J. Dowdall and J. H. Hawthorne, Eds., pp. 379–388, Ellis Horwood, Chichester, 1987a.

Thoenen, H., Barde, Y.-A., Davies, A. M. and Johnson, J. E.: Neurotrophic factors and neuronal death. In: "Selective Neuronal Death," Thoenen, Ed., pp. 82–95, Wiley, Chichester, 1987b.

Thoenen, H., Bandtlow, C., Heumann, R., Lindholm, D., Meyer, M. and Rohrer, H.: Nerve growth factor: cellular localization and regulation of synthesis. Cell. Mol. Neurobiol. *8*: 35–40, 1988.

Tibbs, J.: Acetylcholinesterase in flagellated systems. Biochim. Biophys. Acta *41*: 115–122, 1960.

Tonnaer, J. A. and Dekker, A. M.: Nerve growth factor, neurotrophic agents and dementia. In: "Antidementia Agents," D. Nicholson, Ed., pp. 139–165, Academic Press, New York, 1994.

Torrao, A. S., Carmona, F. M., Lindstrom, J. and Britto, L. R.: Expression of cholinergic system molecules during development of the chick nervous system. Brain Res. Dev. Brain Res. *124*: 81–92, 2000.

Torres-Aleman, I., Pons, S. and Arevalo, M. A.: The insulin-like growth factor I system in the rat cerebellum: developmental regulation and role in neuronal survival and differentiation. J. Neurosci. Res. *39*: 117–126, 1994.

Toutant, J. P., Toutant, M., Fiszman, M. Y. and Massoulié, J. M.: Expression of the A12 form of acetylcholinesterase by developing avian leg muscle cells *in vivo* and during differentiation in primary cell cultures. Neurochem. Int. *5*: 751–762, 1983.

Treinin, M., Halevi, S. and Yassin, L.: Genetic dissection of an acetylcholine receptor involved in neuronal degeneration. In: "Cholinergic Mechanisms—Function and Dysfunction," I. Stilman, H. Soreq, L. Anglister, D. Michaelson and A. Fisher, Eds., pp. 207–312, Martin Dunitz, London, 2004.

Tribollet, E., Bertrand, D., Marguera, A. and Ragenbass, M.: Comparative distribution of nicotinic receptor subtypes during development, adulthood and aging: an autoradiographic study in the rat brain. Neuroscience *124*: 405–420, 2004.

Tuttle, J. B.: Interaction with membrane remnants of target myotubes maintains transmitter sensitivity of cultured neurons. Science *220*: 977–979, 1983.

Tuttle, J. B.: Regulation of cholinergic development by target contact. Fed. Proc. *44*: 2760–2766, 1985.

Ultrech, M. H., Wiesmann, C., Simmons, L. C., Henrich, J., Yang, M. and Reilly, D.: Crystal structures of the neurotrophin-binding domain of TrkA, TrkB and TrkC. J. Mol. Biol. *290*: 1149–1159, 1999.

Usdin, T. B. and Fischbach, G. D.: Purification and characterization of a polypeptide from chick brain that promotes the accumulation of acetylcholine receptors in chick myotubes. J. Cell Biol. *103*: 493–507, 1986.

Uylings, C. G., Van Eden, C. G., Parnaveles, J. G. and Kaalskeek, A.: The prenatal and postnatal development of the rat cerebral cortex. In: "The Cerebral Cortex of the Rat," B. Kolb and R. C. Tees, Eds., pp. 35–75, MIT Press, Cambridge, Mass., 1991.

Vaca, K.: The development of cholinergic neurons. Brain Res. *472*: 261–286, 1988.

Vaccarino, F., Guidotti, A. and Costa, E.: Gangliosicle inhibition of glutamate-mediated protein Rinase translocation in primary cultures of cerebellar neurons. Proc. Natl. Acad. Sci. U.S.A. *84*: 8707–8711, 1987.

Vale, R. D. and Shooter, E. M.: Alteration of binding properties and cytoskeletal attachment of nerve growth factor receptors in PC12 cells by wheat germ agglutinin. J. Cell Biol. *94*: 710–717, 1982.

Vale, R. D. and Shooter, E. M.: Assaying binding of nerve growth factor to cell surface receptors. Meth. Enzym. *109*: 21–39, 1984.

Van der Zee, C. E. and Hagg, T.: Delayed NGF infusion fails to reverse axotomy-induced degeneration of basal cholinergic neurons in adult p75(LNTR)-deficient mice. Neuroscience *110*: 641–651, 2002.

Van Heyningen, W. E.: Identity of the tetanus toxin receptor in nervous tissue. Nature *182*: 1809, 1958.

Van Nieuw Amerongen, A.: Nobel Prize in medicine 1986. Award for oral biology studies. Ned Tijdschr Tandheelkd, *94*(4): 168–171, 1987.

Van Vulpen, E. H. S. and Van der Kooy, D.: Striatal cholinergic interneurons: birthdates predict compartmental localization. Dev. Brain Res. *109*: 51–58, 1998.

Varon, S. S., Nomura, J. and Shooter, E. M.: Subunit structure of a high molecular weight form of the nerve growth factor from mouse submaxillary gland. Proc. Natl. Acad. Sci. U.S.A. *57*: 1782–1789, 1967.

Varon, S., Raiborn, C. and Norr, S.: Association of the antibody to nerve growth factor with ganglionic non-neurons (glia) and consequent interference with their neuron-supportive action. Exp. Cell Res. *88*: 247–256, 1974.

Vellis, de J. and Carpenter, E.: Development. In: "Basic Neurochemistry," G. J. Siegel, Ed.-in-Chief, pp. 537–563, Lippincott-Raven, Philadelphia, 1999.

Venero, J. L., Hefti, F. and Knusel, B.: Trophic effect of exogenous nerve growth factor on rat striatal cholinergic neurons: comparison between intraparenchymal and intraventricular administration. Mol. Pharmacol. *49*: 303–310, 1996.

Vicini, S. and Schuetze, S. M.: Gating properties of acetylcholine receptors at developing rat endplates. J. Neurosci. *5*: 2212–2224, 1985.

Vigny, M., Koenig, J. and Rieger, F.: The motor endplate specific form of acetylcholinesterase: appearance during embryogenesis and reinnervation of rat muscle. J. Neurochem. *27*: 1347–1353, 1976.

Villavobos, J., Rios, O. and Barbosa, M.: Postnatal development of cholinergic system in mouse basal forebrain: acetylcholinesterase histochemistry and choline-acetyltransferase immunoreactivity. Int. J. Dev. Neurosci. *19*: 495–502, 2001.

Wakelam, M. J. O., Davies, S. A., Houslay, M. D., McKay, I., Marshall, C. J. and Hall, A.: Normal p21N-*ras* couples bombesin and other growth factor receptors to inositol phosphate production. Nature *323*: 173–176, 1986.

Walker, P., Weichsel, M. E., Eveleth, D. and Fisher, D. A.: Ontogenesis of nerve growth factor and epidermal growth factor in submaxillary glands and nerve growth factor in brains of immature male mice: correlation with ontogenesis of serum levels of thyroid hormones. Pediatr. Res., *16*: 520–524, 1982.

Wang, Z. X. and Wu, J. W.: Autophosphorylation kinetics of protein kinases. Biochem. J. *368*: 947–952, 2002.

Watters, D. J. and Hendry, I. A.: Purification of a ciliary neurotrophic factor from bovine heart. J. Neurochem. *49*: 705–713, 1987.

Webb, K., Budko, E., Neuberger, T. J., Chen, S., Schachner, M. and Tresco, P. A.: Substrate-bound human recombinant L1 selectively promotes neuronal attachment and outgrowth in the presence of astrocytes and fibroblasts. Biomaterials *22*: 1017–1028, 2001.

Wee, E. L., Phillips, N., Barbiarz, B. S. and Zimmerman, E. F.: Palate morphogenesis V: effects of cholinergic agonists and antagonists on rotation in embryo culture. J. Embryol. Dev. Morph. *58*: 177–193, 1980.

Weis, C., Marksteiner, J. and Humpel, C.: Nerve growth factor and glial cell line-derived neurotrophic factor restore the cholinergic neuronal phenotype in organotypic brain slices of Meynert. Neuroscience *102*: 129–138, 2001.

Weiss, E. R., Maness, P. and Lauder, J. M.: Why neurotransmitters act like growth factor? Perspect. Dev. Neurobiol. *5*: 323–325, 1998.

Weiss, P.: "Principles of Development," University of Chicago Press, Chicago, 1939.

Weiss, P. and Hiscoe, H.: Experiments on mechanism of nerve growth. J. Exp. Zool. *107:* 315–395, 1948.

Welsh, J. H.: Distribution of serotonin in the nervous system of various animal species. Adv. Pharmacol. *6*: 171–188, 1968.

Werner, G. and Kuperman, A. S.: Actions at the neuromuscular junction. In: "Cholinesterases and Anticholinesterase Agents," G. B. Koelle, Ed., pp. 570–678, Handbch. d. exper. Pharmakol., Ergänzungswk., vol. 15, Springer-Verlag, Berlin, 1963.

Werrbach-Perez, K. and Perez-Polo, J. R.: De novo synthesis of NGF subunits in S-180 mouse sarcoma cell line. J. Neurochem. Res. *13*: 875–883, 1987.

Weskamp, G., Lorez, H. P., Keller, H. H. and Otten, U.: Cholinergic but not monoaminergic denervation increases nerve growth factor content in the adult rat hippocampus and cerebral cortex. Naunyn Schmiedebergs Arch. Pharmacol. *334*(4): 346–351, 1986.

Whittaker, J. R.: Segregation during ascidian embryogenesis of egg cytoplasmic information for tissue-specific enzyme development. Proc. Natl. Acad. Sci. U.S.A. *70*: 2096–2100, 1973.

Whittaker, V. P.: The cellular basis of synaptic transmission: an overview. In: "Cellular and Molecular Basis of Synaptic Transmission," H. Zimmermann, Ed., pp. 1–23, NATO ASI Series, vol. H21, Springer-Verlag, Heidelberg, 1988.

Whittaker, V. P.: "The Cholinergic Neuron and Its Target: The Electromotor Innervation of the Electric Ray 'Torpedo' as a Model," Birkhauser, Boston, 1992.

Whittemore, S. R. and Seiger, A.: The expression, localization and functional significance of ß-nerve growth factor in the central nervous system. Brain Res. Rev. *12*: 439–464, 1987.

Wiegandt, H.: Structure and specificity of gangliosides. In: "Ganglioside Function: Biochemical and Pharmacological Implications," G. Porcellati, B. Ceccarelli, and G. Tettamanti, Eds., pp. 3–14, Plenum Press, New York, 1975.

Wiegandt, H., Kanda, S., Inoue, K., Utsumi, K. and Nojima, S.: Studies on the cell association of exogenous glycolipids. Adv. Exp. Med. Biol. *152*: 343–352, 1982.

Williams, L. R., Rylett, R. J., Ingram, D. K., Joseph, J. A., Moises, H. C., Tang, A. H. and Mervis, R. F.: Nerve growth factor affects the cholinergic neurochemistry and behavior of aged rats. In: "Cholinergic Function and Dysfunction," A. C. Cuello, Ed., pp. 251–256, Elsevier, Amsterdam, 1993.

Willinger, M.: The expression of GM1 ganglioside during neuronal differentiation. In: "Gangliosides in Neurological and Neuromuscular Function, Development and Repair," M. M. Rapport and A. Gorio, Eds., pp. 17–27, Raven Press, New York, 1981.

Wojcik, M.: The effect of administration of gangliosides on the cholinergic system in rat brain after hippocampal lesions. In: "Abstracts, II Eur. Neurosci. Meet., Florence, Neurosci. Lett." S47, pp. 209, 1978.

Wojcik, M., Ulas, J. and Oderfeld-Nowak, B.: The stimulating effect of ganglioside injections on the recovery of choline acetyltransferase and acetylcholinesterase activities in the hippocampus of the rat after septal lesions. Neuroscience *7*: 495–499, 1982.

Wong, V. and Kessler, J. A.: Solubilization of a membrane factor that stimulates levels of substance P and choline acetyltransferase in sympathetic neurons. Proc. Natl. Acad. Sci. U.S.A. *84*: 8726–8729, 1987.

Woolf, N.: A possible role for cholinergic neurons in the basal brain and pontomesencephalon in consciousness. Consciousness and Cognition *6*: 574–596, 1997.

Woolf, N. J., Gould, E. and Butcher, L. L.: Microtubule associated protein-1 and forebrain cholinergic systems: topography and responses to thyroid hormone imbalances and axotomies. Neuroscience *28*: 611–623, 1989.

Wujek, J. R. and Akeson, R. A.: Extracellular matrix derived from astrocytes stimulates neuritic outgrowth from PC12 cells *in vitro*. Brain Res. *431*(1): 87–97, 1987.

Wittenbach, C. R. and Hwang, J. D.: Relationship between insecticide-induced short and wry neck and cervical defects visible histologically shortly after treatment of chick embryos. J. Exp. Zool. *229*: 693–697, 1984.

Xie, Z. P. and Poo, M. M.: Initial events in the formation of neuromuscular synapse: rapid induction of acetylcholine release from embryonic neuron. Proc. Natl. Acad. Sci. U.S.A. *83*: 7069–7073, 1986.

Xiao, J., Gong, S., Zhao, Y. and LeDoux, M. S.: Developmental expression of rat torsinA transcript and protein. Brain Res. Dev. Brain Res. *152*: 47–60, 2004.

Yamada, S., Kagawa, Y., Isogai, M. Tkayanagi, N. and Hayashi, E.: Ontogenesis of nicotinic acetylcholine receptors and presynaptic neurons in mammalian brain. Life Sci. *38*: 637–644, 1986.

Yamashita, M., Yoshimoto, Y. and Fukuda, Y.: Muscarinic acetylcholine responses in the early embryonic chick retina. J. Neurobiol. *25*: 1144–1153, 1994.

Yankner, B. A. and Shooter, E. M.: The biology and mechanism of action of nerve growth factor. Annu. Rev. Biochem. *51*: 845–868, 1982.

Yates, A. J., Warner, J. K., Stock, S. M. and McQuarrie, I. G.: Ganglioside synthesis and transport in regenerating sensory neurons of the rat sciatic nerve. Brain Res. 479, 277 282, 1989.

Ye, K. and Snyder, S. H.: PIKE GTPase: a novel mediator of phosphoinositide signaling. J. Cell Sci. *117*: 155–161, 2004.

Ye, P. and D'Ercole, A. J.: Insulin-like growth factor actions during development of neural stem cells and progenitors in the central nervous system. J. Neurosci. Res. 2005, (in press).

Ye, P., Li, L., Richards, R. G., DiAugustine, R. P. and D'Ercole, A. J.: Myelination is altered in insulin-like growth factor-I null mutant mice. J. Neurosci. *22*: 6041–6051, 2002.

Yew, D. T., Chan, W. Y., Luo, C. B., Zheng, D. R. and Yu, M. C.: Neurotransmitters and neuropeptides in developing human central nervous system. Biol. Signals Recept. *8*: 149–159, 1999.

Yip, J.: Migratory patterns of sympathetic ganglioblasts and other neural crest derivatives in chick embryos. J. Neurosci. *6*: 3465–3473, 1986.

Young, J. Z.: Growth and plasticity in the nervous system. Proc. R. Soc. Lond. B. Biol. Sci. *139*: 18–37, 1951.

Youngstrom, K. A.: Studies in the developing behavior of anura. J. Comp. Neurol. *68*: 357–379, 1938.

Youngstrom, K. A.: Acetylcholinesterase concentration during the development of human fetus. J. Neurophysiol. *4*: 473–477, 1941.

Younkin, S. G., Brett, R. S., Davey, B. and Younkin, L. H.: Substances moved by axonal transport and released by nerve stimulation have innervation-like effects on muscle. Science *200*: 1292, 1978.

Wyllie, A. H., Kerr, J. F. R. and Currie, A. R.: Cell death: the significance and apoptosis. Int. Rev. Cytol. *68*: 251–306, 1980.

Zacks, S. I.: Esterases in the early chick embryo. Anat. Rec. *118*: 509–537, 1954.

Zaimis, E.: The nerve growth factor (NGF) and multipotential cells. J. Physiol. *216*: 65P–66P, 1971.

Zahalka, E. A., Seidler, F. J., Lappi, S. E., Yanai, J. and Slotkin, T. A.: Differential development of cholinergic terminal markers in rat brain regions: implication for nerve terminal density, impulse activity and specific gene expression. Brain Res. *46*: 221–219, 1993.

Zeisel, S. H.: Choline availability in the neonate. In: "The Cellular Basis of Cholinergic Function," M. J. Dowdall and J. N. Hawthorne, Eds., pp. 709–719, Ellis Harwood, Chichester, 1987.

Zhang, J., Yang, Z. and Wu, S. M.: Development of cholinergic amacrine cells is visual activity-dependent in the postnatal mouse retina. J. Comp. Neurol. *484*: 33–343, 2005.

Zhang, X., Liu, C., Miao, H., Gong, Z.-H. and Nordberg, A.: Postnatal changes of nicotinic acetylcholine receptor alpha2, alpha3, alpha7 and beta2 subunits genes expression in rat brain. Int. J. Dev. Neurosci. *16*: 507–518, 1998.

Zhou, X., Suh, J., Cerretti, D. P., Zhou, R. and DiCicco-Bloom, E.: Ephrins stimulate neurite outgrowth during early cortical neurogenesis. J. Neurosci. Res. *66*: 1054–1063, 2001.

Zingoni, U.: The effect of acetylcholine and adrenin on inotropism, chronotropism and tonus of the chick embryo lacking the nervous elements. Arch. Fisiol. *56*: 226–236, 1956.

Zoli, M., Le Novere, N. and Hill, J. A., Jr.: Developmental regulation of nicotinic ACh receptor subunit mRNAs in the rat central and peripheral nervous system. J. Neurosci. *15*: 1912–1939, 1995.

Zoller, M. J., Kerlavage, A. R. and Taylor, S. S.: Structural comparisons of cAMP-dependent proteinkinases I and II from porcine skeletal muscle. J. Biol. Chem. *254*: 2408, 1979.

9

Central Cholinergic Nervous System and Its Correlates

A. Scope of This Chapter and Historical Introduction

1. Scope of the Chapter

Some of the cholinergic matters that were discussed in earlier chapters are pertinent to the understanding of the cholinergic central nervous system; these include cholinergic cells and their pathways; metabolism of acetylcholine; and central nicotinic and muscarinic receptors. This chapter paints a detailed picture of the system. The scope of this chapter is very wide, and this wide range illustrates well the diversity and the richness of the cholinergic function generally and of the central cholinergic activities specifically.

The subjects to be covered concern three organizational levels. On the neuronal level, the physiological activities of cholinergic neurons and their responses at the pre- and postsynaptic receptor sites to acetylcholine (ACh), cholinomimetics, and anticholinergic drugs are explored; additionally, cholinergic neurons' pharmacology, that is, their response to noncholinergic transmitters and drugs, is presented. Then the neurons' dependence on second messenger mechanisms and the molecular biology of the receptor sites are explored. Finally, the interactions among transmitters and modulators are described.

On the system level, this chapter describes physiological functions and responses of the cholinergic pathways or networks to cholinergic and noncholinergic transmitters and drugs. It also describes functions and responses of central areas, which involve both cholinergic and noncholinergic pathways (for example, the basal ganglia and the reticular formation) to the cholinergic and noncholinergic transmitters and drugs. Accordingly, activities and cholinergic and pharmacological responses of the electroencephalogram (EEG) and evoked potentials, respiration, hypothalamic homeostasis, and endocrine functions are depicted in detail.

Behaviorally, a higher level of organization is explored. It concerns cholinergic correlates exhibited through cognitive functions such as learning, addiction, and aggression and less frequently explored behaviors such as imprinting and vocalization. In addition the chapter addresses the pharmacology of these behaviors. Finally, as an excursion into the future, the cholinergic aspect of the mind-brain relation will also be speculated upon.

Because central cholinergic pathways and networks are ubiquitous (see Chapter 2), all functions and behaviors must exhibit cholinergic correlates. In fact, there is no known or measurable function or behavior, including mental disease, that does not exhibit cholinergic aspects, and that does not respond to cholinergic and anticholinergic drugs. However, it is a truism that these behaviors are never *purely cholinergic* in nature: they are affected by a multitude of noncholinergic transmitters, drugs, and modulators (see Karczmar, 1978a, 1978b and Glowinski and Karczmar, 1979). Indeed, there is a physiological interplay, at both pre- and postsynaptic sites, between the various transmitter and/or modulator systems, and it underlies all functions and behaviors. This interplay is reciprocal; for example, catecholamines affect the release of ACh, and vice versa. This interplay is needed for the subtle control of neuronal activities, behaviors, and functions; for example, ACh, monoamines, peptides, and amino acids interact in rapid eye movement (REM) sleep, seizures, aggression, and cognition.

2. Historical Introduction

a. Discontinuity of the Early Story

The studies of the central cholinergic system do not constitute a continuous or logical sequence of experimental events. The 19th-century findings concerning peripheral effects of physostigmine or the Calabar bean extract preceded Otto Loewi's demonstration of the presence of cholinergic transmission at the periphery; the 19th- and early-20th-century discovery of the central effects of physostigmine and atropine predate the concept and the demonstration of central cholinergic transmission. Altogether, these findings did not lead to a definition of the site and mechanism of action of these drugs. In fact, the distinguished scientists of the time period in question had as little conception of the mechanism of action of the Calabar bean, physostigmine, or atropine as the Calabar natives who used the bean in their truth ordeals (see Chapter 7 A).

Also, a considerable time gap among qualitative and quantitative site-focused studies must be noted. In the early and mid-19th century, Scottish and German pharmacologists conducted studies that were qualitative in nature; 50 years later, quantitative, site-focused studies were carried out by English, Austrian, German, and Canadian pharmacologists and neurophysiologists. It took another 30 years to conceptually relate the results obtained with physostigmine to Otto Loewi's demonstration of the peripheral cholinergic transmission. Similarly, the first quantitative studies by Gantt and Freile (1944) and Funderburk and Case (1947) of the central behavioral effects of cholinergic drugs antedate by some 10 years the demonstration of central cholinergic transmission; actually, these investigators referred to the mechanism and site of the central action of ACh as "obscure."

The 1950s demonstration of central cholinergic transmission by Henry Dale, William Feldberg, Joshua Gaddum, and John Eccles (see below, section A-2b, A-2c), the expansion of this demonstration from the spinal to supraspinal sites (cf. Krnjevic and Phillis, 1963), and the initiation of the definition of the central cholinergic pathways (cf. Chapter 2 DI) led to the multifaceted explosion of the central cholinergic lore.

b. Early Studies of Nicotine, Physostigmine, and Related Substances

The early ethnographic and postethnographic studies of the Calabar bean and its active component, physostigmine, by British government officials and missionaries to Calabar and by Edinburgh's medical men and masters of Materia Medica were described in detail in Chapter 7 A (see also Simmons, 1956). While many of the actions of these compounds that were described in Edinburgh are peripheral, it must be stressed that many of the effects evaluated in Edinburgh are central in character. For example, Hutchinson, the British consul to the "Calabar Province," reported that victims of the ordeal both "shook" and foamed at the mouth. In Edinburgh, after self-experimental ingestion of Calabar bean, Robert Christison (1855) felt "torpidity" and sleepiness; paradoxically, his "mind was so active . . . that . . . he was not conscious of sleep"—an early report of REM sleep? Another Edinburghian, Fraser (1863, 1872) experimented with animals and demonstrated that the bean extract causes respiratory depression, miosis, and paralysis, as well as hyperthermia, an early example of hypothalamic action of cholinergic drugs (see below, section BIV-2). Fraser's work on both peripheral and central interaction among the Calabar bean's active ingredients and atropine is particularly important. Fraser (1870) also described how the bean's extract prevents or abolishes strychnine convulsions.

German, British, French, and US investigators extended Fraser's findings into the second part of the 19th century. This work was helped by the extraction, purification, and crystallization of the active ingredient of the Calabar bean (see Chapter 7 A). These workers confirmed and extended Fraser's findings regarding the central actions of the bean's extract. For example, Roeber described the "complete paralysis of the nerve cells . . . concerned with . . . pain" caused by the bean's extract (1868; cf. Feldberg, 1945). However, the analgesic action of cholinergic drugs (a very interesting effect) was not picked up again until the 1940s (see below, section BV-2). In 1876, Harnack and Witkowski stressed that the active extract induces depressant actions that lead to "central paralysis" and respiratory depression. This effect was sometimes described as the second phase of action, which follows the primary, excitatory effect (Kleinwaechter, 1864; Bezold and Goetz, 1867; Harnack and Witkowski, 1876; Heubner, 1905; Rothberger, 1901; see also Feldberg, 1945; Karczmar, 1970). Investigators also confirmed Fraser's results relating to the antagonism between atropine and Calabar bean's extract.

Another early investigator who conducted studies with the bean's extract was Charles Brown-Sequard. Charles Brown-Sequard was a successor to Francois Magendie as professor at the College de France, and he is famed (or notorious) for the experiments he carried out in his old age concerning the effects in humans of monkey's genital gland extracts. But, Brown-Sequard contributed solidly to neurosciences and he described the facilitatory action of the active ingredient of the Calabar bean on convulsions evoked by spinal cord lesions (Brown-Sequard, 1860).

Brown-Sequard's finding appears contrary to Fraser's (1872) claim that both the bean's extract and strychnine act as antagonists. This claim was pursued in both the human (Kleinwaechter, 1864) and animals (Bourneville, 1867/68; see Bartholow, 1873), as these investigators were interested in using the bean as an antidote for seizures induced by atropine.

A first—but not the last—priority battle was fought at that time. An American, Roberts Bartholow, a physician and a professor at the Cincinnati Medical School, experimented with atropine-physostigmine and demonstrated the antagonism by "atropia" of the mixed paralyzant-tetanic effect of "physostigmia." He was awarded the prize of the American Medical Association for this work (Bartholow, 1873; see Karczmar, 1970). Bartholow opined that studies conducted by the French and German investigators were incomplete and unconvincing, and he claimed priority over Fraser and Bourneville's work. The fairness of Bartholow's claim was acknowledged by Fraser and Gubler (see Karczmar, 1970). To quote from Bartholow's claim: "I can quite agree . . . that Fraser's work is a model for this kind of investigations, and surpasses anything" but "in my work on Atropia . . . I distinctly announced the discovery of an antagonism between Atropia and Physostigmia," which antedated the work of Fraser and Bourneville (Figure 9-1). Another American physician, John Hudson, who is little known and not quoted today, also studied the physostigmine-atropine antagonism (Hudson, 1873) and both Hudson and Bartholow summarized, very felicitously, the characteristics of this antagonism (Karczmar, 1970).

Whether or not the bean's extract or antiChEs including physostigmine act as convulsants or anticonvulsants remains controversial. Also, there are inconsistencies concerning the effects of one drug on the LD50 value of the other (Koppanyi, 1939; Longino and Preston, 1946) and the interactions between these drugs on phenomena such as the EEG (Wescoe et al., 1948). Altogether, it appears that depending on the doses of either cholinergic drugs or convulsants, synergism or antagonism may arise (see Karczmar, 1974; section BIV-3b; Chapter 7 DI).

The mutual antagonism between atropine and physostigmine described by Fraser and others is real enough. Physostigmine was an accepted antidote of atropine coma when atropine coma was employed in the1960s as a treatment for certain depressive disorders. Today, physostigmine is the agent of choice for treating overdoses of tricyclic antidepressants that exhibit anticholinergic activity. On the other hand, Edinburgh investigators and early German and French scientists (i.e., Kleinwachterin, 1864; Bartholow, 1873; Bournevillein, 1867/68) were aware that atropine antagonized physostigmine's action or toxicity. This was a prelude for establishing atropine as an antagonist for all antiChEs, especially of the organophosphorus (OP) type (see below, this section and Holmstedt, 1963).

Many of investigators listed above studied, besides the active extract of the bean or physostigmine, alkaloids such as arecoline and pilocarpine and curari (Waser, 1953), and besides describing their peripheral actions they demonstrated their central effects. The studies concerning the central system, including those alluded to so far, were carried out with the methods available in the course of 19th century; accordingly, they concerned essentially whole animals and the overt effects of physostigmine, nicotine, or pilocarpine (cf. Holmstedt, 1988). Beginning in the 1930s, the investigators utilized more sophisticated methods such as recording limb reflex, tetanus, and clonus with appropriate levers or recording the seizure activity by means of a primitive form of EEG of the cortex or spinal cord (Schweitzer and Wright, 1937). Sometimes, these convulsions and electric

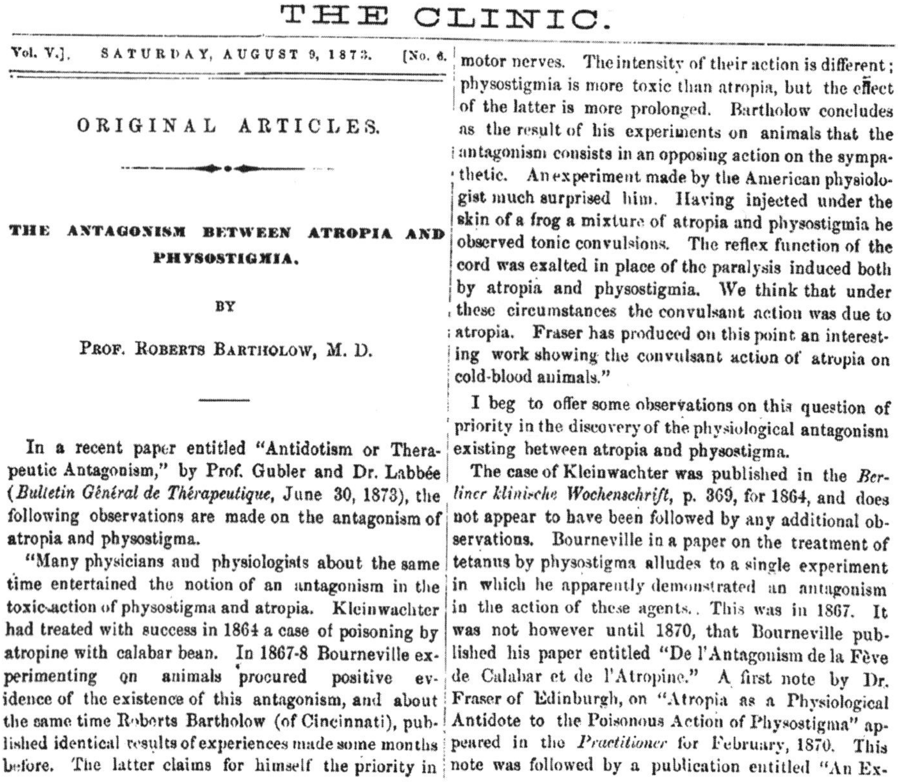

THE CLINIC.

Vol. V.]. SATURDAY, AUGUST 9, 1873. [No. 6.

ORIGINAL ARTICLES.

THE ANTAGONISM BETWEEN ATROPIA AND PHYSOSTIGMIA.

BY

PROF. ROBERTS BARTHOLOW, M. D.

In a recent paper entitled "Antidotism or Therapeutic Antagonism," by Prof. Gubler and Dr. Labbée (*Bulletin Général de Thérapeutique*, June 30, 1873), the following observations are made on the antagonism of atropia and physostigma.

"Many physicians and physiologists about the same time entertained the notion of an antagonism in the toxic-action of physostigma and atropia. Kleinwachter had treated with success in 1864 a case of poisoning by atropine with calabar bean. In 1867-8 Bourneville experimenting on animals procured positive evidence of the existence of this antagonism, and about the same time Roberts Bartholow (of Cincinnati), published identical results of experiences made some months before. The latter claims for himself the priority in motor nerves. The intensity of their action is different; physostigmia is more toxic than atropia, but the effect of the latter is more prolonged. Bartholow concludes as the result of his experiments on animals that the antagonism consists in an opposing action on the sympathetic. An experiment made by the American physiologist much surprised him. Having injected under the skin of a frog a mixture of atropia and physostigmia he observed tonic convulsions. The reflex function of the cord was exalted in place of the paralysis induced both by atropia and physostigmia. We think that under these circumstances the convulsant action was due to atropia. Fraser has produced on this point an interesting work showing the convulsant action of atropia on cold-blood animals."

I beg to offer some observations on this question of priority in the discovery of the physiological antagonism existing between atropia and physostigma.

The case of Kleinwachter was published in the *Berliner klinische Wochenschrift*, p. 369, for 1864, and does not appear to have been followed by any additional observations. Bourneville in a paper on the treatment of tetanus by physostigma alludes to a single experiment in which he apparently demonstrated an antagonism in the action of these agents. This was in 1867. It was not however until 1870, that Bourneville published his paper entitled "De l'Antagonism de la Fève de Calabar et de l'Atropine." A first note by Dr. Fraser of Edinburgh, on "Atropia as a Physiological Antidote to the Poisonous Action of Physostigma" appeared in the *Practitioner* for February, 1870. This note was followed by a publication entitled "An Ex-

Figure 9-1. Photograph of the first page of the 1873 paper of Robert Bartholow, reprinted here because of the interest in the text, and because this paper seems to be largely unknown. (From Karczmar, 1970, with permission.)

effects were demonstrated following a local application of physostigmine to the cortex or spinal cord (Sjostrand, 1937; Miller, 1937; Miller et al., 1940; cf. Feldberg, 1945; Karczmar, 1970). The work of the investigators of the 1930s was facilitated by the elucidation of physostigmine's active chemical structure and its synthesis (Polonowski and Nitzburg, 1925; Polonowski and Polonowski, 1923; Julian and Pike, 1935; see chapter 7 A).

At the same time, parallel studies were initiated with ACh itself (and choline, see Freund, 1911). Originally, the Russian investigators (see, for example, Makrosian, 1937) had described cortical actions of intravenously injected ACh. Probably, the effects were indirect and due to acetylcholine's cardiovascular effect. In 1937, Sjostrand and other investigators (see Karczmar, 1970) realized that when given systemically, both blood cholinesterases (ChEs) and the quaternary ACh's inability to pass the blood-brain barrier would prevent ACh from exercising central effects. However, several of these studies involved cortical, supraspinal, and intraventricular applications that were even being used in humans (Henderson and Wilson, 1937). EEG spikes, convulsive and twitch-like effects, and respiratory actions were noticed (see Feldberg, 1945; Karczmar, 1970; as well as the review by another early cholinergiker, Theodore Koppanyi 1948; see chapter 7, Figure 7.22). Moreover, the early observations of Christison regarding the soporific effects of physostigmine or Calabar bean extract were obtained with intraventricularly applied ACh (Henderson and Wilson, 1937). Within another few years, operant behavior was used for the first time to uncover ACh's effectiveness in facilitating a quantifiable central behavior, operant conditioning (Gantt and Freile, 1944).

As already mentioned, some of the reviewed work predates Loewi's demonstration of peripheral chemical, cholinergic transmission; this is true, of course, with respect to the work of the early Edinburgh, German, French, English, and US investigators. Many of the studies listed—such as those of Sjostrand (1937), Henderson and Wilson (1937), Macht (1924), and Gantt and Freile (1944)—antedate as well the definitive demonstration of central cholinergic transmission by Eccles, Fatt, and Koketsu (1953, 1954; see below, section A-2c), although these studies are contemporary with heuristic speculations as to the existence of central cholinergic transmission by Dale (1935), Feldberg (1945), Gaddum (Chang and Gaddum, 1933), and MacIntosh (1941). Actually, Keith and Stavraky (1935) and Schweitzer and Wright (1937) quoted Loewi, Dale, or Feldberg and suggested that their results, obtained with ACh, physostigmine, or atropine, may be explainable by the presence of chemical, cholinergic transmission in the central nervous system (CNS).

The story of research concerning ACh itself is an important component of the demonstration of chemical, cholinergic transmission. This story includes the demonstration of the endogenous presence of ACh and choline in nervous tissues of vertebrates, the description of the strict resemblance between action of ACh, at the peripheral sites and the effects of parasympathetic and neuromyal stimulation, and the remarkable pharmacologic potency of ACh as compared to that of other choline esters; Hunt and Taveau, Dale, Dixon and Reid Hunt are the main actors of this drama (see Jukes, 1984; Dale, 1933, 1952; Karczmar, 1967, 1970, 1996; see also Chapter 1 B).

c. Eccles' Conversion, the Demonstration of Central Cholinergic Transmission, and Certain Pertinent Generalizations (or, How Did Eccles Become Converted?)

Did these 1920s, 1930s, and 1940s studies set the stage for the demonstration by Eccles and his friends of the existence of cholinergic transmission at the spinal synapse between the motor nerve collateral and the Renshaw cell? It appears to this author that the studies just quoted, while known to the Canberra team, do not constitute an immediate link in the chain of reasoning that led to the demonstration in question. The actual link was the transfer of the principle involved in Loewi's demonstration of peripheral chemical, cholinergic transmission to a central synapse, that is, the transfer of the principle that synapses communicate via a chemical signal.

The principal actor in the drama, John Carew Eccles, was not free psychologically to perform the Renshaw cell experiment. Eccles was a postdoctoral student of the celebrated early 20th-century neurophysiologist Sir Charles Sherrington, who among his other scientific accomplishments stressed the importance of inhibition in the regulation of function (Eccles and Gibson, 1979; see Figure 9-2). With Sherrington, Eccles was engaged in the 1920s and 1930s in studies of the electrophysiology of neuromyal transmission. These studies laid the foundation for Eccles' belief that synaptic transmission is electric in nature (Karczmar, 2001). At the time, he was not the only investigator to embrace this notion. Indeed, in the 1940s and early 1950s, "there was . . . fairly general agreement that central synaptic transmission was likely to be electrical" (Eccles, 1964), a view forcefully expressed by the famous Russian neurophysiologist J. S. Beritoff (Beritoff and Bakuradze, 1940): "Acetylcholine does not play any essential role in the transmission of excitation or in the generation of inhibition in the central nervous system."

The following describes the logic underlying Eccles' concept of central electrical transmission. The brevity of "postulated transmitter actions" at the neuromyal and ganglionic synapses was contrasted with prolonged ACh activity at the cardiovagal sites, the demonstrated sites of its transmitter action. This contrast "gave rise to the postulate that the presynaptic action current was responsible for the initial brief excitatory action at the neuromuscular junction and sympathetic ganglion . . . while . . . the transmitter substance, acetylcholine, was responsible for the prolonged residual depolarization" that occurs at the cardiovagal sites (Eccles, 1964). According to Eccles, the current also accounted for the synaptic electrophysiological events of the CNS.

Furthermore, Eccles did not believe that the action of any chemical transmitter could be terminated rapidly enough to preserve the efficiency of

Figure 9-2. Sir John Eccles (left) and Sir Charles Sherrington, Oxford, ca. 1930. John Eccles is shown with his head bowed; a rare occasion. (From Karczmar, 2001, with permission.)

the central transmission (Eccles, 1937, 1946). This reasoning was based on his interpretation of the kinetic characteristics of the synaptic transmission, which he and Stephen Kuffler (1946, 1952) were in the position of measuring very precisely as they began at that time to use intracellular methodology in their experiments. Altogether, up to the early 1950s Eccles was a strong proponent of electric transmission, whether at the ganglia, the neuromyal junction, or the CNS.

In the early 1950s, the Austrian-born philosopher and mathematician Karl Popper visited Eccles at the University of Otago in Dunedin, New Zealand, where Eccles worked at the time. During his discussions with Popper, Eccles was impressed with Popper's notion that flexibility is a condition sine qua non for scientific creativity no less than it is for political freedom (see Popper, 1963). Many students of the events in question (see Bacq, 1975) believe that this event led Eccles to pay closer attention to the proponents of the chemical nature of transmission; they propose that the main reason for Eccles' conversion was Popper's influence.

I believe that in the discussions in question, Popper encountered in Eccles a man ready for conversion (Karczmar, 1996; 2001a, 2001b). This readiness was conditioned by a number of findings and developments, current in the mid-20th century. First, Joshua Gaddum demonstrated in 1933 the presence of ACh in the CNS (Chang and Gaddum, 1933), and William Feldberg and Martha Vogt (1948; Feldberg, 1945) showed the differential distribution and differential synthesis of ACh in the CNS—this kind of distribution would be expected of a chemical transmitter, as such a transmitter would be restricted in its distribution by the topography of its specific synapses and pathways. In fact, it also became clear in the 1940s that there is a good correlation among the distribution of ACh, choline acetyltransferase (CAT), and acetylcholinesterase (AChE) (for references, see Karczmar, 1967). Then, David Nachmansohn established in the 1940s that the enzymic action of AChE is uniquely fast (Nachmansohn, 1940, 1952; Nachmansohn and Rothenberg, 1945; see Chapter 3 DI). Still another convincing piece of evidence was the demonstration of the spontaneous release of ACh from the CNS and of the facilitation of this release by CNS stimulation or by its activity (Feldberg and Schriever, 1936; Chang et al., 1937; see also Feldberg, 1945; Karczmar, 1967).

Altogether, in the 1950s the status of central chemical transmission resembled the standing of peripheral chemical mediation prior to Otto Loewi's studies of the vagus, and Sir Henry Dale's phrase (1954) applies to both situations: "Transmission by chemical mediators was like a lady with whom the neurophysiologist was not willing to be seen in public." But Dale was ready to be seen in public with the idea of chemical central transmission, and he and Feldberg forcefully presented all the available evidence as being consistent with the hypothesis of central cholinergic transmission. It seems to this author that their arguments were more instrumental for Eccles' conversion than was Popper's influence (see also Robinson, 2001).

Indeed, these findings and concepts must have been maturing in Eccles' mind as he moved in 1953 from Dunedin, New Zealand, to Canberra. This "maturation" process can be clearly traced. In this time period, Eccles studied carefully the reprints of T. P. Feng and his associates working in the 1930s and 1940s at the Peiping (Beijing) Union Medical College, and Eccles made copious notes in pencil on the pages of their papers (see Karczmar, 1991; Figure 9-3) Now, Feng, Li, and other investigators at the Union Medical College noticed the retrograde firing, which was elicited at the amphibian neuromyal junction by physostigmine, veratrine, and guanidine. They realized that their findings may appear to support the notion that electrical transmission is effective at the motor nerve terminal, yet Feng ultimately came down in full on the side of the hypothesis that chemical, cholinergic transmission is effective at the neuromyal junction; judging from his notes, Eccles was fully cognizant of this development.

At this juncture, Eccles embarked with Paul Fatt and Kyozo Koketsu on studies of spinal motoneuron inhibition. This inhibition was considered by Lloyd (1946) as resulting from the direct impact on the motoneuron of the current generated by the motoneuron collateral. On the other hand, the inhibitory postsynaptic potential (IPSP) present at the spinal motoneuron was regarded by Renshaw (1942, 1946) as resulting from a disynaptic circuit, which contained a postulated interneuron. According to Renshaw, this cell responded to the stimulation of the collateral by a characteristic short-latency effect followed by a high-frequency, high-voltage response. (This interneuron was later called the Renshaw cell.)

Eccles and Renshaw met in 1946 at the Rockefeller Institute (cf. Eccles, 1969, 1987). Eccles greatly admired Renshaw, and Renshaw's demonstration of the disynaptic circuit ignited a spark for Eccles, as he realized that Renshaw's circuitry provided an ideal model for pharmacological analysis and for resolving the question of central chemical transmission. What helped him in his final demonstration—the crucial experiment in this story—was the advance at this time of the electrophysiological techniques (see below) for localzed stimulation of and recording from, neurons, and other methodology developed by Bo Holmstedt and his associates at the Karolinska Istitutet (Holmstedt and Skoklund, 1953). So, Eccles proceeded to demonstrate that the response of the Renshaw interneuron to the stimulation of

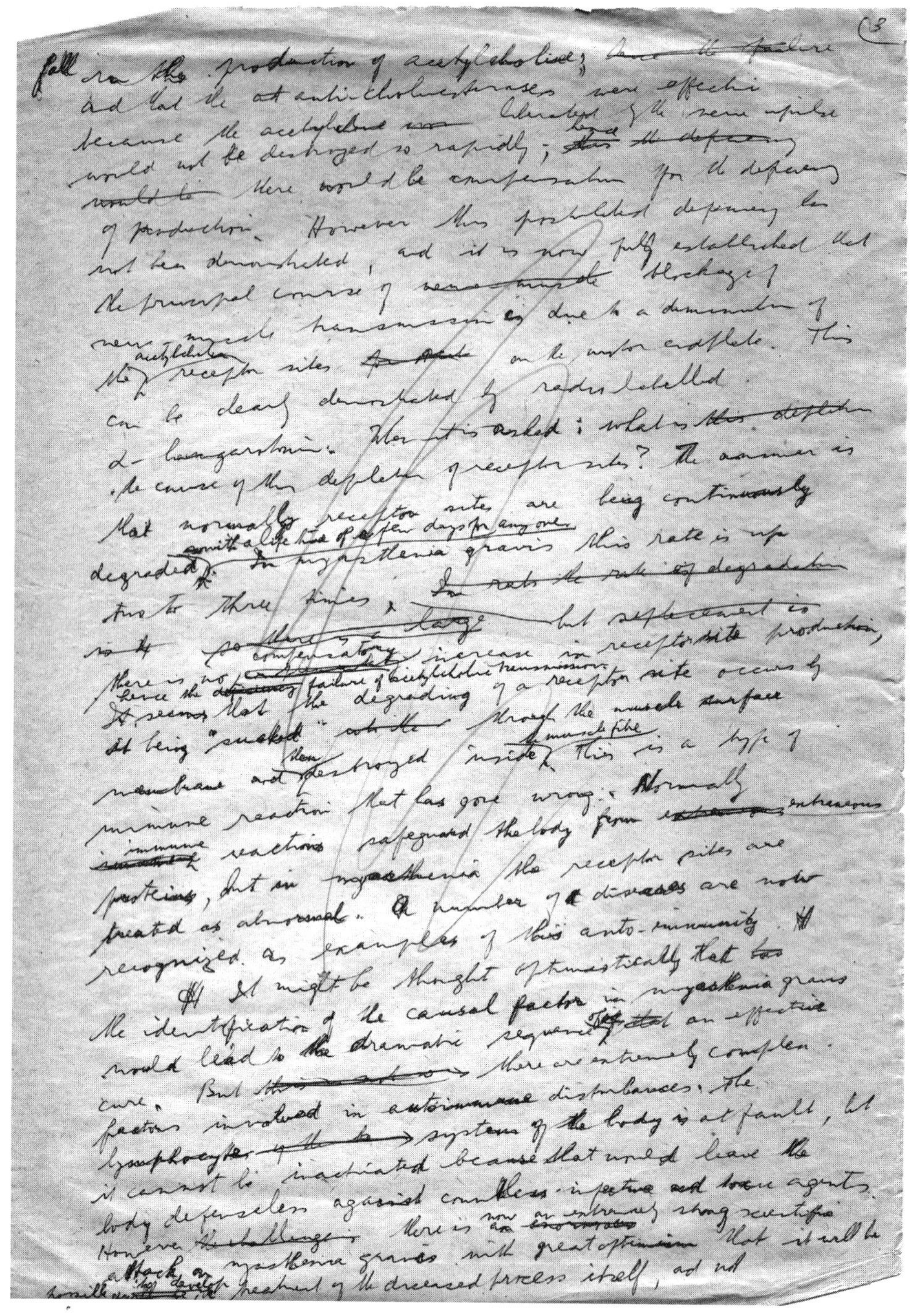

Figure 9-3. An excerpt from Sir John Eccles' handwritten pages of comments to the 1953 papers of T. P. Feng. (From the private collection of AGK.)

the collateral was, pharmacologically, a nicotinic, cholinergic response and that the Renshaw cell, in turn, evokes an IPSP in the motoneuron; finally, nicotinic antagonists blocked the Renshaw cell response but not the response of the motoneuron; the motoneuron response was blocked by strychnine. These results clearly prove the disynaptic nature of the circuit between the collateral and the motoneuron, the Renshaw cell serving as an interneuron (Eccles et al., 1953, 1954).

The Renshaw cell stimulation could be potentiated by intravenous administration of either OP antiChEs or physostigmine (physostigmine again!). It could be blocked by intravenous administration of dihydro-beta-erythroidine or its administration by the "difficult technique of intra-arterial injection into the spinal cord pioneered by Holmstedt and Skoglund" (1953; Eccles, 1987). Furthermore, the intra-arterial administration technique enabled Eccles et al. (1953, 1954) to show that both nicotine and acetylcholine mimic the Renshaw cell's responses to the stimulation of collateral. This stimulation is generally referred to as "antidromic"; actually, it imitates the routing of the physiological traffic from the motoneuron to the collateral and to the Renshaw cell (see Karczmar, 2001a, 2001b; Figures 9-4, 9-5, and 9-6).

There was subsequently a technical refinement of this story. Shortly after Eccles' Renshaw cell studies, David Curtis and John and Rosamond Eccles developed a method for the electrophoretic application of drugs and transmitters to single neurons (see Curtis and Eccles, 1958). They applied this technique to studies of the Renshaw interneuron to provide additional evidence of its cholinoceptivity and for the cholinergic nature of the collateral. Curtis and Eccles also demonstrated that the Renshaw cell responds intensely to the electrophoretic application of ACh and of nicotinic cholinomimetics and weakly to the application of muscarinic cholinomimetics; they also showed that this response as well as the response to the stimulation of the collateral is blocked by electrophoresis of d-tubocurarine (Curtis and

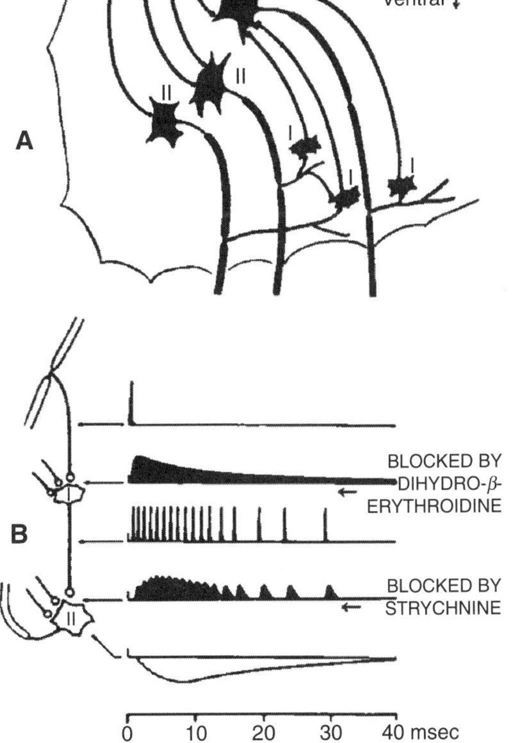

Figure 9-4. Diagram of the Renshaw cell experiment (Eccles et al., 1954). The original caption (abbreviated) reads as follows: "(A) Sketch of the neurone system in ventral horn of the spinal cord. Collaterals are given off by motor axons within the spinal cord and make synaptic contact with Renshaw interneurones (I). The axons of these interneurones make contact with the motoneurons (II), which by this system, are inhibited. Reflexly active afferents descend onto the motoneurons from the dorsal direction. (B) Diagram summarizing the postulated chain of events from the antidromic impulse in motor axons to inhibition of motoneurons. The corresponding histological structures are shown to the left (notice indicator arrows). The five events are from above downwards: (1) impulse in axon collateral; (2) time course of acetylcholine liberated at axon collateral; (3) repetitive discharge in interneurone; (4) time course of inhibitory transmitter substance liberated at interneuronal terminal; (5) hyperpolarization set up in motoneurone by inhibitory synaptic action. The summation of the synaptic action of several converging interneurons onto a motoneurone is responsible for smoothing the latter part of the motonuerone hyperpolarization." (From Eccles et al., 1954, with permission.)

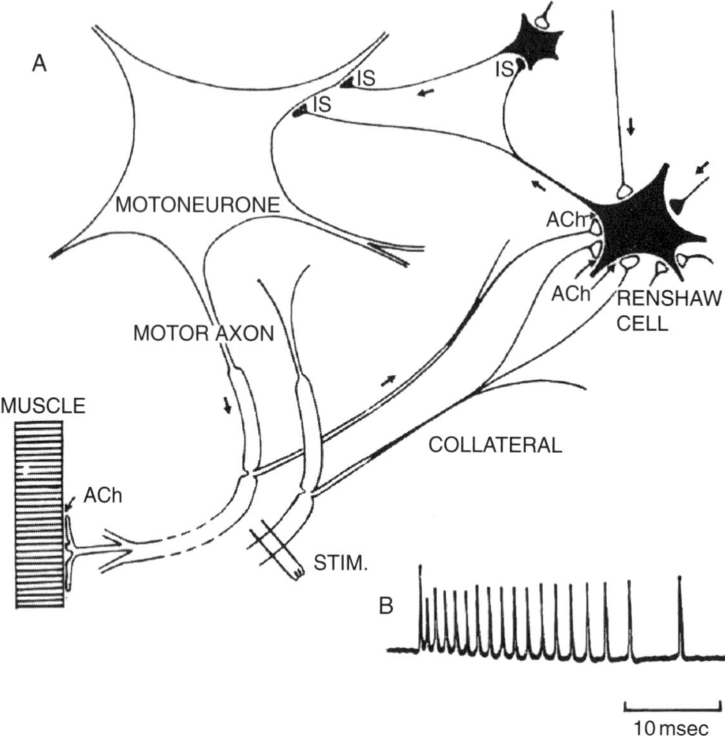

Figure 9-5. (A) Eccles' diagram of the inhibitory pathways to motoneurones by their axon collaterals and Renshaw cells. B shows extracellular recording of a Renshaw cell being excited by an antidromic volley in the motor fibers of lateral gastrocnemius muscles. (From Eccles, 1969, with permission.)

Eccles, 1958; see Curtis and Crawford, 1969). In the course of their original studies of the Renshaw cell, Eccles and his associates (1953, 1954) found systemically administered quaternary curarimimetics to be ineffective; they explained this phenomenon by the quaternary nature of these compounds, which prevented them from crossing the blood-brain barrier. This is a valid explanation, but it left a sense of incompleteness to the story, which was eliminated by the results obtained by Curtis and Eccles. It must be added that the methodological effort of the Canberra group continues unabated till today, culminating with the development of patch clamp techniques of recording from single neurons and studying the ionic characteristics of their currents.

In 1935 Dale proposed a principle that posits, in its original form, that "nerve cells of any particular type act at all their synapses by liberating the same chemical transmitter" (see Eccles, 1962). The identification of the Renshaw cell circuit vindicates Dale's principle. The motoneuron evokes an excitatory potential at the myoneural junction; if the collateral originating from the same motoneuron induced an IPSP at the motoneuron as proposed by Lloyd (1946), then Dale's principle would be invalid for the circuit in question, while the circuit proposed by Eccles and his associates is consistent with Dale's principle.

The Renshaw cell story does not end here. The Renshaw cell was identified by Renshaw and by Eccles and his collaborators via electrophysiological and pharmacological analysis. However, they did not identify it morphologically. As a result, there were doubts about the Renshaw cell's existence and the reality of Eccles' scheme. Particularly, Forrest Weight (1968) provided electrophysiological results, which he felt warranted the consideration of Lloyd's alternative hypothesis of the inhibitory response at the motoneuron. Actually, this hypothesis was the reiteration of the option presented earlier by Feng and his associates of the excitatory nerve terminal current as the generator of a postsynaptic response. Eccles referred to Weight's proposal as a "most audacious attack" and very effectively criticized it on both electrophysiological and pharmacological grounds (Eccles, 1969). Actually, at the time of Weight's proposal, Thomas and Wilson (1965) provided anatomical identification of the Renshaw cell. Additionally, Szentagothai (1958) established morphologically the presence of "motoneurones and smaller cells located ... around the emergence of motor axons from the gray matter." Finally, Jankowska and Lindstrom (1971) identified the small neurons that were generating responses characteristic for the Renshaw interneurons through intracellular injection of procion yellow.

A few caveats are in order, however. More than one transmitter or modulator may be released from a single neuron (see below, section BII); this may not invalidate Dale's principle, when it is modified to state that either one or all the transmitters released at one nerve ending of a given neuron are also released at the other nerve endings of the neuron in question. Then, while the same transmitter or transmitters are released from all the endings of a given neuron, the postsynaptic responses to the transmitter or transmitters may

Cholinergic and Inhibitory Synapses in a
Central Nervous Pathway

J.C. Eccles, P. Fatt and K. Koketsu

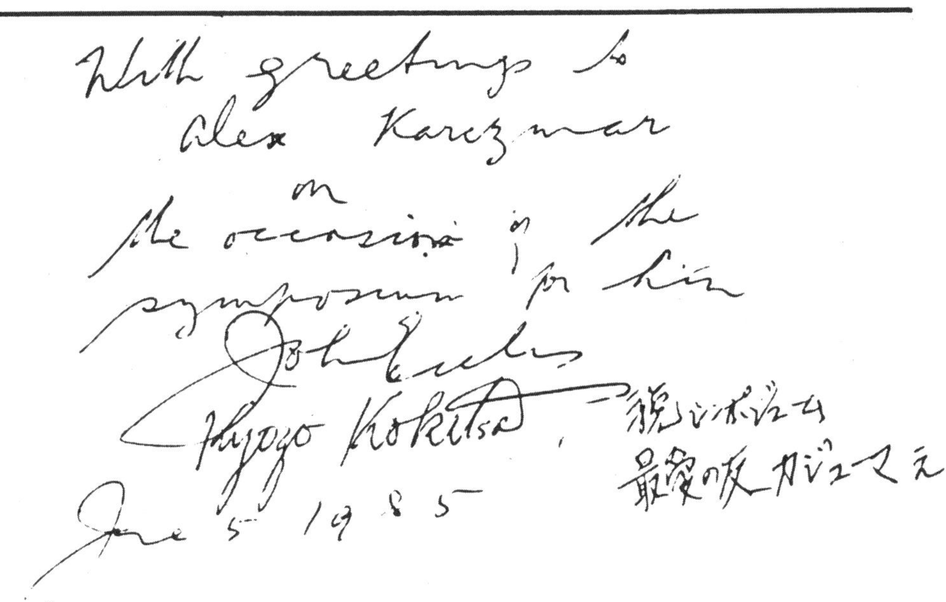

Figure 9-6. A copy of the title page from the 1953 paper of Eccles, Fatt, and Koketsu, signed in 1985 by Sir John Eccles and Kyo Koketsu. This paper, almost never quoted, was published in 1953 in the Australian Journal of Sciences. It is the early version of the 1954 paper published in the London Journal of Physiology and is the basis of Eccles' Nobel Prize. (From Karczmar, private collection.)

differ in sign, depending on the site: besides inducing the excitatory response, ACh may induce hyperpolarizations, whether at the heart, or in the ganglia, or, in fact, in the CNS (see below, section BIII-2b). However, in the case of the Ecclesian model, the change in signal from excitatory at the Renshaw cell to inhibitory at the motoneuron represents another mechanism for changing the transmission sign: this change requires the presence of an interneuron.

However, the most important set of criteria that have to be satisfied for the valid demonstration of cholinergic transmission, in the CNS or anywhere, includes the demonstration of the release of ACh from a discrete nerve ending or from a population of homogeneous nerve endings; the demonstration of the postsynaptic action of thus released ACh; and the identification, both pharmacological and electrophysiological—in term of ionic mechanisms underlying the response—of the postsynaptic reaction of the given synapse to presynaptic stimulation and to ACh (see also McLennan, 1963; McGeer et al., 1987). A definitive proof based on meeting all these criteria became available in the 1920s and 1930s with respect to autonomic and somatic transmission. Such a proof was not provided for the synapse at the Renshaw interneuron by the experiments of Eccles in the 1950s; for that matter, this kind of a definitive proof does not obtain even today, with respect to the synapse at the Renshaw cell or to any other central synapse, even among those accepted currently as cholinergic. For example, Giancarlo Pepeu's studies of the release of ACh demonstrate ACh release from a pathway rather than from a neuron; the pathway may be relatively homogeneous, but its full "purity" cannot be guaranteed (Pepeu et al., 1990). Perhaps the next-to-complete proof is available for the Renshaw cell, for synapses between the fibers generated in the medial forebrain complex and neocortical and hippocampal cells, and for retinal synapses between amacrine and ganglion cells (see section BII-2; McGeer et al., 1987; Krnjevic, 1988; Kelly and Rogawski, 1985; Kurosawa et al., 1989; Masland et al., 1984).

d. Post-Ecclesian Cholinergic Studies and Their Topics

The first topic involves further studies of ACh and CAT distribution and the propositions for the topography of central cholinergic pathways (see Chapter 2 DII). It also includes further development and improvement of push-pull cannula, pressure application of ACh and drugs, and related techniques for the measurement of ACh release, which are further discussed in this chapter. Nicholas Giarman and Giancarlo Pepeu (see Giarman and Pepeu, 1962; Pepeu, 1974) focused on using these methodologies to demonstrate the release of ACh from discrete brain sites evoked by their stimulation (Figure 9-7).

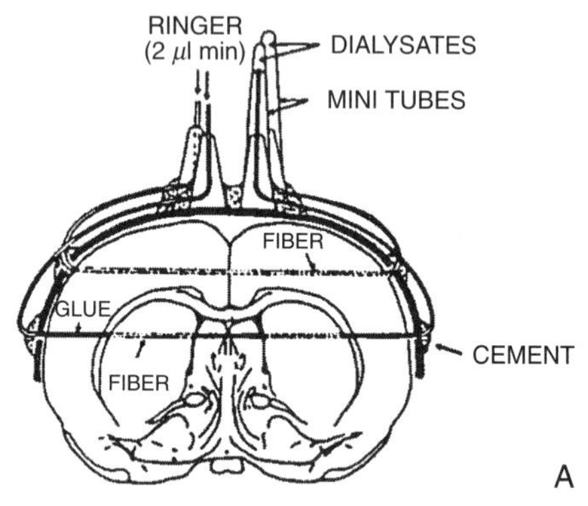

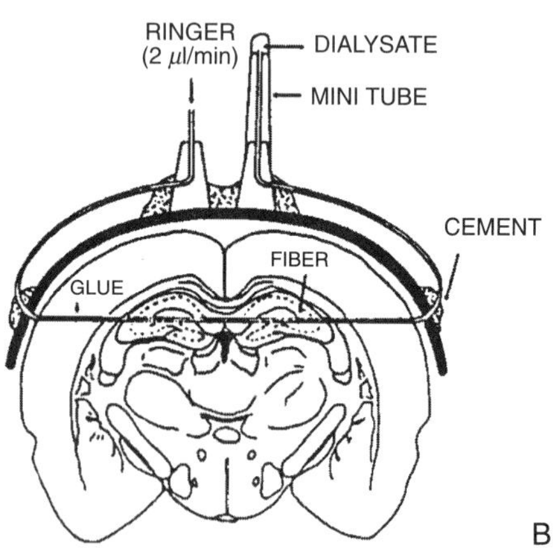

Figure 9-7. Schematic drawing of the dialysis tubings implanted at the level of the frontal cortices and striata (A), and dorsal hippocampi (B). (From Wu et al., 1988, with permission.)

Further work with the electrophoresis and pressure application of ACh and other appropriate bioactive or pharmacological agents to a single neuron also is included in this topic. Following the Canberra group's (see above) introduction of these methods, Chris Krnjevic (Krnjevic and Phillis, 1963; Krnjevic, 1969, 1988) and A. Herz (Herz and Zieglgansberger (1968) carried out pertinent work in the early 1960s. Finally, more recent findings relate to the multiplicity of nicotinic and muscarinic receptors and the demonstration of their CNS location via binding and related methods (see below, sections BI and BIII, and Chapters 5 and 6).

A long series of efforts concerned the mechanisms of pre- and postsynaptic receptor responses and the release of ACh. The early studies of J. S. Coombs, J. E. Desmedt, R. Lorente de No, John Hubbard, David Curtis, Kris Krnjevic, C. G. Phillips, and others were described by John Eccles in his monumental book of 1964; the presynaptic Ca^{2+}-dependent release of ACh and the role of Ca^{2+} were discovered by Juan del Castillo, R. Miledi, and Bernard Katz (Del Castillo and Katz, 1954; Katz and Miledi, 1965; see also sections BII and BIII).

Another topic deals with behavior. As already mentioned, behavioral cholinergic research preceded the demonstration of the existence of central cholinergic transmission (Gantt and Freile, 1944 and Funderburk and Case, 1947). Ten years later Carl Pfeiffer (see Pfeiffer and Jenney, 1957) showed the positive effects of muscarinic agonists on animal conditioning and learning. This finding led to an immediate application in humans (this was possible in the 1950s): Pfeiffer used muscarinic agonists in schizophrenic patients, apparently with some success. Pfeiffer, as some earlier investigators such as Keith and Stavraky (1935) and Gantt and Freile (1944), referred to the notion of cholinergic transmission in their interpretation of their results.

Carl Pfeiffer's findings with regard to cholinergic correlates of memory are not the first in this area. Keith and Stavraky (1935) and Gantt and Freile (1944) reached similar conclusions, and their and Pfeiffer's work was followed by similar studies of several investigators, particularly David Drachman (1978), A. Herz (1960), and Bures and his associates (Bures et al., 1964, 1983). In fact, Drachman was the first investigator to demonstrate, in the human, cholinergic correlates of memory; his other important demonstration is that loss of memory in aging exhibits cholinergic correlates (see section BV). It is true that early effects of cholinomimetics, anticholinesterases (antiChEs), and atropinics on memory and conditioning were not consistent; however, the use of better methodology brought about a clear-cut demonstration of memory facilitation by cholinergic agonists and its attenuation with atropinics. This facilitation was demonstrated for several species, including the primates (see Karczmar, 1967, 1995; Hagan and Morris, 1988). Raymond Bartus (see Bartus, 1978; Bartus et al., 1982) is an early contributor to the notions of cholinergic memory facilitation and the relation between loss of cholinergic function and geriatric memory dysfunction; it should be remembered that many investigators contributed to this notion earlier, including David Drachman, Carl Pfeiffer, Janos Bures, Giorgio Bignami (see Bignami and Rosic, 1971), and others (see Hagan and Morris, 1988).

Cholinergic effects on other behaviors and central functions were established in the 1940s, 1950s, and 1960s. Mary Pickford (1947) initiated studies of the cholinergic effects on endocrine function as she established that the cholinergic system is involved in the hypothalamically induced release of neurohypophysial hormones (see also section BIV-2a; Koelle, 1963; Sokolovsky, 1984). Later, Feldberg and Sherwood (1954) demonstrated that intraventricular administration of ACh elicited a number of motor and behavioral actions. Luigi Valzelli and L. H. Allikmets pioneered the studies of cholinergic correlates of aggression (Valzelli, 1967, 1974; Allikmets et al., 1968; Allikmets, 1974; see section BV-1).

Then there is the story of the cholinergic correlates of the EEG. Daniele Bovet, a Nobel Prize winner for his work on the synthesis and structure activity of neuromyally active compounds and sulfa drugs (see Bovet, 1988), became aware in the early 1950s of the potential of transmitter pharmacology of the CNS function for the understanding of the function of the CNS and for its clinical applications. He influenced Vincenzo Longo, at that time a budding investigator at the Istituto Superiore di Sanita in Rome, to begin systematic studies of the cholinergic and anticholinergic actions on the EEG (Longo, 1956; see section BIV-3b); one of Longo's many contributions is establishing the relation between the theta EEG rhythm and learning. Other pioneers in this area are Philip Bradley and Joel Elkes (Bradley and Elkes, 1953, 1967).

During the subsequent 30 years, evidence was continually obtained for behavioral and functional cholinergic actions (see sections BIV-3 and BIV-4). Certain studies were speculative, widely quoted at the time, and controversial. Three examples may be adduced. First, Krech, Rosenzweig, Diamond, Bennet, and their associates (cf., for example, Krech et al., 1963; for further references, see Karczmar, 1969) hypothesized that changes in brain AChE and/or butyryl cholinesterase (BuChE) reflect progress in rats' adaptive behavior, and that the levels or activities of these enzymes correspond to differences between "intelligent" and "not intelligent" rat strains. Furthermore, the investigators speculated that these changes indicated an alteration in the number of central cholinergic synapses. In their work they employed several environments varying from "poor" to "rich" in complexity and availability of training opportunities, while their criterion of adaptability and intelligence was based on the distinction between "maze bright" and "maze dull" rats.

However, there are several problems with the investigators' hypothesis (see several articles in Napier et al., 1991 and chapter 1B-3). First, they based their evidence on very small differences in BuChE and/or AChE. Generally, these differences amounted to only 2%;[1] sometimes, they involved differences between the ratios of the two enzymes or between the cortical versus subcortical enzymes (Rosenzweig, 1966). However, small changes in the levels or ratios of ChEs measured in nonlocalized brain areas cannot be taken as indices of the number of cholinergic synapses. Then, several psychologists criticized the criteria for environmental qualities and inherited adaptability employed by the investigators (Kluver, 1958); in fact, some critics referred to the California group's speculation as "reductionistic naivete" (Hirsch, 1967; see also Kluver, 1958; Feldberg, 1958; Karczmar, 1969). Yet, Krech, Rosenzweig, Diamond, and Bennet and their associates were far-seeing, as they proposed that adaptability and learning may be both inherited and nurtured, and that the cholinergic synapses are important for the phenomena in question. In the current era of the trophic factors, of the concepts of the flexibility, plasticity, and regenerative capacity of the brain, their speculation, if not the conclusions based on their own data, is validated (see, for example, Bernard et al., 1999). Indeed, their quotation of Spurzheim's 1815 statement (Rosenzweig, 1966) that "the organs . . . of the brain . . . increase by exercise"—as Lamarckian as this statement may sound—is very much a propos today (see Karczmar, 1991, 1995). As we are able today to measure CAT, ACh, and the cholinergic receptors in discrete cholinergic areas, their work should be reevaluated.

Another example of early "overexploitation" of the central cholinergic system is represented by the studies of Deutsch (Deutsch, 1971; for further references, see Hingtgen and Aprison, 1976; Karczmar, 1976). Deutsch demonstrated that the effect of di-isopropyl fluorophosphonate (DFP) on avoidance conditioning in rats depended on their stage of training: DFP facilitated learning during the early phases of learning, and caused amnesia when administered after advanced learning (Deutsch, 1971; Deutsch and Leibowitz, 1966). Deutsch and his associates speculated that there are specific cholinergic "memory" synapses in the hippocampus (in some of their experiments, DFP was applied to the hippocampus) and that during early and late stages of learning, ACh levels at the synapse were low and high, respectively. They also suggested that, "as a result of learning, the postsynaptic endings at a specific set of synapses become more sensitive to transmitters;" at that time, DFP pushes the response curve into the amnesic area. It should also be added that other investigators (Hamburg, 1967; Biederman, 1970) obtained biphasic effects on learning with physostigmine.

A number of criticisms may be raised. For example, the doses of DFP and physostigmine used by Deutsch, Hamburg, and Biederman were very high. Thus, Deutsch applied a nearly LD50 dose of DFP to the hippocampus. These doses of antiChEs may induce synaptic block either via causing an accumulation of ACh or via direct actions independent of ChE inhibition (see Chapter 7 C). Also, Deutch employed single doses of DFP rather than evaluating the dose-effect relationships. Furthermore, Deutsch and his associates did not measure levels of ACh or ChEs in the brain or elsewhere. Finally, they did not have any basis for their notion of "memory synapses" (see also Sharpless, 1964). It must be added that the knowledge of the significance of the cholinergic system for memory and learning antedates by many years Deutsch's experiments (see Karczmar, 1967).

Then Ernst Gellhorn (1944a, 1944b) proposed that the hypothalamus—his "head ganglion"—is the regulatory center of autonomic imbalance and constitutes the center for the regulation of autonomic balance. He hypothesized also that the state of the hypothalamic balance between the parasympathetic and sympathetic systems underlies emotional behavior, including "conditioned reactions." He also felt that stress or convulsion-induced activation of the hypothalamus and the resulting "augmentation of hypothalamic-cortical discharges" restores learning and conditioning when the latter is extinguished by the lack of reinforcement. Altogether, Gellhorn felt that hypothalamic imbalance is involved in the occurrence of emotions and of mental diseases, and that this notion is important clinically (see also Arnold, 1960). It must be noted that Gellhorn measured the postulated sympathetic or parasympathetic status only indirectly (in terms of changes in blood pressure and blood sugar). Gellhorn's concepts resemble those of Hess (1954) and Brodie (1957) concerning the balance between the hypothalamic "ergotrope" and "trophotrope," in other words, the balance between sympathetic and serotonergic transmission.

The evidence on which the three speculations are based may be not acceptable and the speculations themselves may not be tenable; yet, it is unfair to deem these speculations as simplistic when considering the time when the research in question was conducted and the limitations in the available methodology. Furthermore, these speculations are heuristic, as can be easily illustrated. The timeliness of the posits by Krech and Rosenzweig was already emphasized. And Brodie's and Gellhorn's notion that multitransmitter balance regulates behaviors was expanded on immediately. Stein's (1968) reward-punishment axis involves two transmitters, Dray and Straughan (1976) and Pujol et al. (1978) established the multitransmitter nature of the control of motor behavior, and Myers (1978) described multitransmitter regulation of the appetitive behavior and the homeostasis of temperature control. Perhaps most illustrative in this context are the studies of Michel Jouvet (1967) of the multitransmitter nature of REM sleep; he posited that at least three transmitters, ACh, serotonin, and norepinephrine, are involved. In fact, Jouvet referred to his investigations as the "monoamine game."

This early evidence suffices to illustrate the complexity of cholinergic functional and behavioral phenomena; later studies made this complexity even greater. There are several aspects of this complexity. First, monoamines, peptides, amino acids, and indole amines are jointly implicated in the regulation of any function or behavior. Hokfelt's most important demonstration that more than one transmitter may be released at a given cholinergic synapse contributes to this complexity (Hokfelt et al., 1978).

Then, there is the interaction between ACh and other transmitters. When Feldberg and Vogt (1948) demonstrated that the areas of presence and absence of ACh alternate throughout the brain, they speculated that other transmitters must be present in the noncholinergic areas; it could then have been surmised that cholinergic agonists or electric stimulation of cholinergic areas should affect other neurotransmitters. Regardless, it took 20 years to demonstrate this phenomenon (Varagic and Kristic, 1966; Glisson et al., 1972, 1974; see Karczmar, 1978b). Today, we know that cholinergic activation and deactivation—whether pharmacologic, electrical, or physiologic—affects the brain levels and/or turnover of norepinephrine, serotonin, GABA, and other bioactive substances (see Karczmar, 1976).

In addition, several second messenger systems including the recently discovered nitric oxide (NO) are involved in the regulation of transmitter action (see Williams et al., 1997). These systems are evoked by cholinergic activity and are necessary for cholinergic effects; furthermore, they interact. This mutual interaction among the transmitters as well as among their messenger systems adds to the complexity involved in defining the role of the cholinergic system in functions and behaviors; thus, it is difficult to decide whether a given function or behavior elicited by a cholinergic agonist results from a cholinergic effect or is due to the cholinergic action on another transmitter and/or its messenger systems.

A final addition to the complexity of cholinergic responses is the existence of multiple receptor subtypes, at both the pre- and postsynaptic sites. This concept goes much beyond the original notion of Langley, Dale, and Rosamond Eccles of the nicotinic and muscarinic receptors. It originated in the 1950s when Dimitry Kharkevich (1957) and Edward Hulme, Nigel Birdsall, Arnold Burgen, and their associates (see Hammer et al., 1980) demonstrated that several subtypes of cholinergic muscarinic receptors exist in the CNS (see Chapter 5 and sections BIII-2a, BIII-2b, BIII-2d). This was followed by a demonstration of several subtypes for central nicotinic receptors by Eric Bernard, Jean-Pierre Changeux, Ken Kennard, A. Karlin, Palmer Taylor, Jon Lindstrom, and others (see Chapter 6 and Bernard et al., 1987). Today, to define cholinergic correlates for any function or behavior one must identify receptor subtype(s) that are involved in the phenomenon in question. An even more refined understanding of the receptors requires the knowledge of their molecular biology, which is reviewed in this chapter and in Chapters 5 and 6.

The complexity and multiplicity of cholinergic mechanisms regulating central function and behavior ensures a subtle, point-to-point control of this phenomenology. At the same time, this complexity makes the analysis of the cholinergic correlates of functions and behaviors most difficult; similarly, the predictability and full understanding of the effects of cholinergic and anticholinergic drugs on any central function or behavior are not easy to achieve. These challenges only add to the significance and excitement of cholinergic studies.

BI. Cholinoceptive Responses

Cholinoceptivity is the presence of nicotinic, muscarinic, or mixed receptors and responses. Jointly with CAT, AChE, high-affinity choline uptake, and occurrence of ACh release, cholinoceptivity characterizes cholinergic synapses and pathways. It is important to emphasize the central ubiquity of cholinoceptivity; its topography is similar in the human and several animal species (Mesulam, 1990, 1995; Mesulam et al., 1989; Woolf, 1991; see Chapter 2 DI-DIII). Additionally, it must be stressed that astrocytes and other glial cells are also cholinoceptive (Hosli and Hosli, 1993).

In spite of the ubiquity of the cholinergic synapses and pathways, they are poorly represented in certain areas of the brain, including the superficial layers of the cortex and the cerebellar cortex, including cerebellar granule and Purkinje cells. These areas may contain only a few postsynaptic or presynaptic cholinoceptive sites (McCance and Phillis, 1968; Crawford et al., 1966; Krnjevic, 1969, 1974a, 1974b; Kuhar, 1978; see, however, Butcher et al., 1993 and Chapter 2 DI-DIII). On the other hand, cholinoceptive neurons are found in regions where cholinergic circuitry is not defined as yet, as in the case of the spinal interneurons (Jiang and Dun, 1986, 1987). Furthermore, cholinoceptive sites are present at noncholinergic synapses. Also, electrically conducting gap junctions show affinity to cholinergic agents, which do not, however, evoke transmission across these junctions (see, for example, Velazquez et al., 1997).

Two types of neurons are cholinoceptive, the interneurons and the principal neurons. Generally, the interneurons are members of localized networks and release inhibitory amino acids on their cholinergic stimulation. The principal neurons are usually projection neurons that originate in cholinergic pathways that release ACh or other transmitters, frequently excitatory amino acids (see Chapter 2 D; see also Van der Zee and Luiten, 1999).

Cholinoceptivity includes two families of receptors, muscarinic and nicotinic. Changeux, Lindstrom, Taylor, Wess, Birdsall, Hulme, Ladinsky, Schimerlik, and Christopoulos demonstrated different modes of action for these receptors (see Birdsall et al., 1996; Wess et al., 1996; Lena and Changeux, 1998; see also Chapters 5 and 6): while nicotinic receptors belong to a family of receptors (including GABAergic, glucinergic, and serotonergic receptors) that behave as ligand-gated ionic channels, muscarinic receptors are coupled with G proteins and second messengers (see Lambert and Nahorski, 1990 and Chapters 5 and 6).

There are many subtypes of both nicotinic and muscarinic receptors. They are located at two different neuronal locations, the pre- and postsynaptic sites; nicotinic receptors are also present at preterminal sites (Albuquerque et al., 1998a, 1998b). These two locations are described in the next section.

1. Post- and Presynaptic Cholinoceptive Responses: Where Do They Occur with Respect to Cholinergic Pathways?

a. Postsynaptic Cholinoceptivity

Teleologically, the release of acetylcholine requires postsynaptic cholinoceptive sites. Indeed, in central areas where this release was demonstrated (Pepeu, 1993), such sites abound. Postsynaptic muscarinic cholinoceptive responses are frequently shown in the infragranular neocortical layer (Price and Stern, 1983; Krnjevic, 1988), layer VI of the cat (Metherate et al., 1988a, 1988b), basal ganglia, the pyramidal cells of the hippocampus (Bird and Aghajanian, 1976; Ehlert et al., 1995), several septal areas and other limbic areas (Carette, 1997; Jones et al., 1999), and the mesopontine area (Imon et al., 1996). This is true for rodents and the cat.

Muscarinic receptors are plentiful in the CNS. All 5 muscarinic receptor subtypes and their mRNAs are represented in the central postsynaptic responses (see also Chapter 5), while M1 and M2 receptors are also present presynaptically (see Chapter 5 and below). Several muscarinic receptor subtypes may be found within any of the central cholinoceptive sites; for example, M1, M2, M3, and M4 postsynaptic sites may be found in the limbic system and in the forebrain, M1 receptors may predominate in the cortex (Quirion et al., 1993), M1 and M3 receptors in the hippocampus (Quirion et al., 1993), and M3 and M5 receptors in the cochlea (Drescher et al., 1992; see also Caulfield, 1993; Ehlert et al., 1995; Chapter 5).

While the nicotinic postsynaptic receptors are less abundant than the muscarinic receptors, they are well represented in the CNS (see Chapter 6; Lukas and Bencherif, 1992; Champtiaux and Changeux, 2004; Arneric et al., 1995). They are present in the rat hypothalamus, thalamus, brainstem, motoneurons, components of the limbic system including hippocampus, nucleus raphe magnus, and cortex (Tebecis, 1970a, 1970b; Schwartz and Kellar, 1985; Larsson and Nordberg, 1985, 1986; Nordberg et al., 1989a, 1989b; Alkondon et al., 1996; Decker et al., 1998; Ferreira et al., 2000, 2001; see also Lukas and Bencherif, 1992). They also abound at several mesencephalic sites of the bird (Chiapinelli et al., 1993). Nicotinic receptor subtypes that contain α4, β2, and α7 subunits are present in several brain areas, such as the cortex (see Lukas and Bencherif, 1992; Lena and Changeux, 1998); additional subtypes are present in the hippocampus (Alkondon and Albuquerque, 1990; Albuquerque et al., 1998a, 1998b; Lukas and Bencherif, 1992 and Bencherif et al., 1997; Mike et al., 2000; Drescher et al., 2004; see also section BIV-3c and Chapter 6).

It must be remembered that the areas that contain nicotinic and/or muscarinic receptors also exhibit noncholinergic pathways. Noncholinoceptive cells are found in the ventro-basal complex and the lateral geniculate body of the thalamus (Curtis, 1965; Curtis and Andersen, 1962; Tebecis, 1970a, 1970b). Even in sites with a dense presence of cholinoceptive postsynaptic receptors (i.e., the basal ganglia or the hippocampus), only 30% to 40% of the tested cells respond to cholinomimetics (Krnjevic, 1974a, 1974b, 1988; see also Chapter 2) and do not exhibit postsynaptic cholinoceptivity when responding to presynaptic stimulation.

Do specific cholinergic pathways differentially radiate to postsynaptic nicotinic versus muscarinic receptors? The pertinent evidence is not readily available: to answer this question, identified homogeneous radiations must be stimulated—electrically or chemically—and the specific response at appropriate postsynaptic sites characterized pharmacologically or via molecular biology techniques. Such investigation must be carried out *in situ* or in isolated preparations containing both the presynaptic input and its postsynaptic receptors. When available, the results are sometimes ambiguous; for example, in some experiments responses of presumed cholinoceptive postsynaptic receptors evoked by the stimulation of a specific pathway (for example, the reticular formation or Mesulam's Ch5 and Ch6 radiations; see Chapter 2 DIII) were not antagonized by cholinergic, whether atropinic or nicotinic, antagonists (see Kelly and Rogawski, 1985). However, in the case of the recurrent collateral pathway to the Renshaw cell, the primary nicotinic and the secondary muscarinic postsynaptic responses were clearly established (see section BIII; Curtis and Ryall, 1966); do these findings suggest that a homogeneous cholinergic input may abut on both muscarinic and nicotinic receptors?

It is easier to analyze pharmacologically functional or behavioral effects evoked by a pathway than to identify whether or not this pathway connects with a specific cholinoceptive receptor or its subtype (pertinent results are discussed in sections BIV and BV of this chapter). While many responses are nicotinic in nature, most functional and behavioral responses are muscarinic (see Karczmar, 1976). In some cases the muscarinic postsynaptic receptor subtypes could be identified. These postsynaptic sites are mainly innervated by projections of the cholinergic forebrain including radiation from the magnocellular nucleus, pedunculopontine tegmental nucleus, and septohippocampal projections (see Chapter 2 DIII). However, this does not exclude the presence of nicotinic postsynaptic receptors at these sites (Fisher et al., 1998). Nicotinic postsynaptic receptors are also present in the striatal interneuronal network. They are innervated by nigrostriatal radiations and their activation results in nicotinic activation of the mesolimbic dopaminergic system (Chapter 2 DII and DIII; Clarke, 1993a, 1993b, 1995). Nicotinic characterizations of motor function resulting from nigrostriatal radiations are discussed in section BIV-1d. Finally, the monosynaptic pathway between the retinal ganglion cells and the amacrine cells abuts on postsynaptic nicotinic receptors (see section BV-C); this finding has consequences with regard to vision (see section BV-C).

The situation may be analyzed from another viewpoint. Two kinds of circuitry can subserve the postsynaptic cholinoceptivity, the autonomic ganglia being a pertinent model (Figure 9-8). In the case of parasympathetic ganglia, the postsyn-

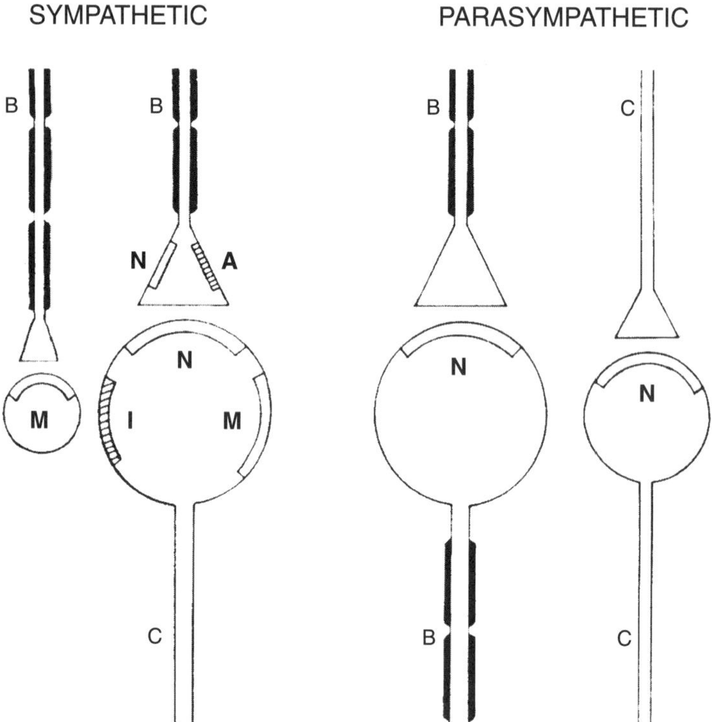

Figure 9-8. A comparison of the rabbit sympathetic and parasympathetic (ciliary) ganglions. The preganglionic terminations abut on the sympathetic ganglionic neurons (C), the interneurons or chromatic cells (with their muscarinic receptors, M), and the ganglionic parasympathetic neurons (B and C). The pre- and postsynaptic cholinergic N-receptors: I, origin of the postsynaptic inhibitory potential; A, origin of the presynaptic catecholaminergic inhibition. Note the absence of the muscarinic receptor in the case of the ciliary ganglion. (From Karczmar, 1971, with permission legend translated from the French by AGK.)

aptic cholinoceptive site is located on a cholinergic neuron, that is, a neuron that releases ACh; this circuitry constitutes a cholinergic-cholinergic disynaptic relay (C-C relay). On the other hand, in the sympathetic ganglia the cholinoceptive site is situated on a noncholinergic neuron; the sympathetic ganglia represent then a cholinergic-noncholinergic (C-non C) disynaptic relay. A non-ACh transmitter, norepinephrine is released from this ganglion relay. To distinguish between these two types of circuitry requires demonstrating, on a neuronal level or with respect to homogeneous, transmitterwise pathways, whether the stimulation of a postsynaptic cholinoceptive site causes the release of ACh or another transmitter. Again, this kind of evidence is not readily available; however, studies that employed immunoreactivity established the presence of synaptic connections between cholinergic pathways and those releasing other neurotransmitters; for example, cholinergic-dopaminergic coupling was established for the substantia nigra (Clements et al., 1991); this coupling is of a C-non C type.

Other, indirect evidence is also consistent with this model. Stimulation of predominantly but not uniquely cholinergic pathways releases, besides ACh, other transmitters: catecholamines are released following the stimulation of the neostriate (Lee et al., 1992); hippocampal stimulation yields aspartate and GABA (Raiteri et al., 1992); and nitric oxide is released via the stimulation of the thalamus (Williams et al., 1997). Furthermore, cholinomimetic agonists or antiChEs cause changes in the levels or turnovers of serotonin, catecholamines, and GABA (see Karczmar, 1978a, 1978b; see also below, sections BIII-2a, BIII-2b, BIII-2d). However, the stimulation of a CNS pathway may release non-ACh transmitters via an inadvertent stimulation of noncholinergic components of the pathway. Accordingly, the cholinomimetic or electric stimulation of the reticular formation may cause release or changes in levels of non-ACh transmitters because several neuronal systems were stimulated rather than because a two-neuron relay was excited. Also, the changes in question may be due to cholinergic stimulation of the nerve terminals of noncholinergic neurons rather than to cholinergic postsynaptic actions. Finally, multitransmitter release may occur at the neurons, which release more than one neurotransmitter (see McGeer et al., 1987, and Pepeu, 1993).

Evidence regarding the presence within the central cholinergic pathways of the relays of the parasympathetic ganglion type (C-C relays) is limited; these relays may exist in substantia innominata and elsewhere in the forebrain (Martinez-Murillo et al., 1990; Zaborszky et al., 1991; Iwamoto, 1991; Van der Zee and Luiten, 1999).

What kind of receptors and receptor subtypes may be involved in the C-C and C-non C circuitry? In the parasympathetic as well as sympathetic ganglia, the postsynaptic sites, which generate the fast response, are generally nicotinic. But the sympathetic ganglion also exhibits slow postsynaptic responses that are muscarinic and peptidergic (Karczmar et al., 1986). Today, there is no direct information regarding this matter. However, it is important to note that in several areas of the brain (including the cortex, the hypothalamus, and the hippocampus), the nicotinic and muscarinic postsynaptic sites are colocalized on noncholinergic neurons; sometimes the densities of the two receptors are equal (Van der Zee et al., 1999). It must be remembered that, similarly, muscarinic and nicotinic postsynaptic (and peptidergic sites) are present at the sympathetic ganglia, except that in this case the nicotinic receptors clearly predominate.

b. Presynaptic Cholinoceptivity and Actions of Cholinergic Autoreceptors

Cholinergic nerve terminals abut on cholinoceptive receptors of both cholinergic and noncholinergic nerve endings. Cholinoceptive receptors present at cholinergic and noncholinergic nerve terminals are referred to as auto- and heteroreceptors, respectively. Finally, both classes of receptors include nicotinic and muscarinic receptors. Their localization is possible today: radio- or stain-labeled ligands for either nicotinic or muscarinic receptors are available and their use may be combined with that of positron emission tomography (PET; see, for example, Vickroy et al., 1984a, 1984b; Ding et al., 1999); also, their presence can be related to the presence of CAT and markers for ACh transport, such as 3-H vesamicol and hemicholinium-3. It is interesting that autoreceptors may regulate not only synaptic but nonsynaptic release of ACh (see Vizi and Lendvai, 1999; see also Chapter 2 C2).

As cholinergic heteroreceptors affect release of transmitters other than ACh, they are discussed in the next section (section BIII-2b). Noncholinergic terminals are also present at cholinergic terminals. These noncholinergic terminals regulate ACh release; thus their role amounts to transmitter interaction and is discussed in the same section.

In the 1950s and 1960s, Szerb, Polak, and MacIntosh discovered that atropine facilitates the evoked or spontaneous cortical release of ACh; they claimed that this effect is due to atropine's antagonism of the muscarinic, autoreceptor-mediated inhibition of ACh release (MacIntosh and Oborin, 1953; Polak, 1965, 1970; Szerb, 1964, 1977; Pepeu et al., 1990; see also Schuetze and Role, 1987; Langer, 1977). These effects were obtained *in vitro* and *in vivo*. Subsequently, the same phenomenon was observed with respect to the striate and hippocampus, and sites innervated by the nucleus basalis magnocellularis (Sinischalchi et al., 1991; see also Pepeu et al., 1990; see section BIII-1a). It is interesting that muscarinic inhibition for the release of ACh also extends to the inhibition of the release of non-ACh transmitters from neurons that are capable of discharging two or more transmitters. This occurs in the cortex with regard to peptides such as VIP and PHI (Bartfai et al., 1988).

Do specific muscarinic receptor subtypes represent the muscarinic presynaptic autoreceptors? There is strong evidence indicating that several muscarinic receptor subtypes serve as autoreceptors and regulate ACh release. Using radio autography methods, Tonnaer et al. (1988) related the topography of the binding of several M ligands with that of hemicholinium-3, and concluded that M2 receptors are present presynaptically in the limbic and interpeduncular areas and the basal ganglia. Evidence based on the pharmacological analysis of ACh release from brain slices and synaptosomes, or in situ, suggests that M1, M3, M4, and, particularly, M2 receptor subtypes are present on the cholinergic terminals of the cortex, limbic system, and striate (Beani et al., 1984; Hass and Ellis, 1985; Lapchak et al., 1989a, 1989b; Vickroy and Cadman, 1989; McKinney et al., 1993; Quirion et al., 1993; Raiteri et al., 1989; see also Schimerlik, 1990; Ladinsky et al., 1990). The sites occupied by the various muscarinic receptor subtypes frequently overlap (Lapchak et al., 1989a, 1989b; Vickroy and Cadman, 1989). It is of interest that autoreceptors exhibit several types of muscarinic receptor subtypes; it will be seen that this multiplicity of muscarinic receptor subtypes also characterizes the presynaptic heteroreceptors and the postsynaptic receptors.

The M4 receptors may inhibit the release of acetylcholine by blocking the M and/or the Ca^{2+} current at the presynaptic autoreceptors (Nordstrom et al., 1989; Segal, 1989; Segal et al., 1989; Caulfield et al., 1993; Chapter 5). An effect on Ca^{2+} or K^+ currents may be involved in activating autoreceptors of other muscarinic subtypes (Peralta et al., 1996; Segal et al., 1989; Nordstrom et al., 1989). For further information regarding the involvement of the third messenger (nitric acid) mechanisms in presynaptic cholinoceptivity, see Chapter 5, Nathanson (1996), Kilbinger (1996), and Ma et al. (1995). It should be added that some muscarinic autoreceptors—such as the intrinsic neurons of the rat septum—may, exceptionally, facilitate upon stimulation the release of ACh (Birthelmer et al., 2003).

The nicotinic autoreceptors were described first by Koelle (1963); he assigned to them a facilitatory, "percussive" role in the release of ACh: the "priming" bolus of ACh released from the cholinergic nerve terminal retro-acts on the nicotinic autoreceptors, causing full, massive release of ACh (Figure 9-9). Koelle (1969) demonstrated that a similar effect exists centrally, for example, at the olivocochlear bundle. It appears today that the facilitatory nicotinic autoreceptor sites are present in several CNS regions, including the cortex, hippocampus, striate, and cerebellum of the cat and rodents (Lapchak et al., 1989a, 1989b; Nordberg et al., 1989a, 1989b; Araujo et al., 1988; Lukas and Bencherif, 1992; see also section BIII-2a); this is true for humans as well (Sihver et al., 1999). It is not clear which subtype of nicotinic receptor may be involved in these phenomena. The nicotine and alpha-bungarotoxin (but not tetrodotoxin; Wonnacott, 1997) binding sites were both described, and the subtype contains the $\alpha7$ subunit (Zarei et al., 1999).

Whether or not specific cholinergic pathways innervate heteroreceptors is also unknown. The activation of cholinergic muscarinic heteroreceptors generally attenuates the amplitude and/or frequency of noncholinergic excitatory and inhibitory postsynaptic potentials (EPSPs and IPSPs). This was shown for the caudate (Nabatame et al., 1988; Akaike et al., 1988), hippocampus (Segal et al., 1989), cortex (Marchi and Raiteri, 1985, 1996;

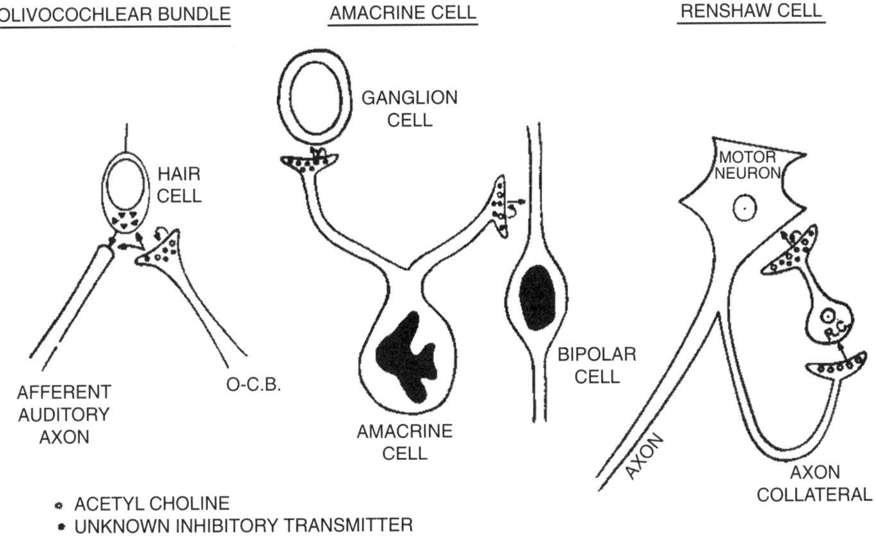

Figure 9-9. Proposed role of ACh in releasing unidentified inhibitory transmitter(s) from AChE-staining, non-cholinergic, presynaptic terminals of olivocochlear bundle, amacrine cell, and Renshaw cell. (From Koelle, 1969, with permission.)

Marchi et al., 1987), mesencephalon (Grillner et al., 1999), and spinal cord (Jiang and Dun, 1986). M2 and M3 receptors seemed to be involved. This action leads to either depression or disinhibition of transmission depending on whether the EPSP or the IPSP is affected. These effects may be a result of cholinergically mediated diminution of the release of excitatory amino acids and inhibitory catecholamines. These phenomena are an example of transmitter interaction (see section BIII-2b and BIII-2f).

There are also nicotinic heteroreceptors located at nerve terminals, as well as nicotinic preterminal, that is, axonal, nicotinic heteroreceptors. The hippocampal preterminal nicotinic receptors appear to be of the α7 subtype (Albuquerque et al., 1998a, 1998b; see section BI-1 and BI-2, below).

2. Central Postsynaptic Cholinoceptive Responses and Their Characteristics

The cholinoceptive postsynaptic responses include excitatory and inhibitory responses. They are generated by the release of ACh from the cholinergic presynaptic nerve terminal. Depending on the nature of the cholinoceptive site, they may be either excitatory or inhibitory in nature. The excitatory responses may be muscarinic, nicotinic, or mixed: the Renshaw cell exhibits a predominant nicotinic response as well as a subsidiary muscarinic response; other central cholinoceptive neurons may show, rarely, a pure muscarinic or nicotinic response and, more frequently, a predominant muscarinic and a secondary nicotinic response. However, the cholinergic presynaptic nerve terminal may release, besides ACh, other transmitters or bioactive substances, such as peptides. The sympathetic mammalian ganglion constitutes a good illustration of the mixed postsynaptic responses. Accordingly, the sympathetic ganglion serves well as a model of central cholinergic occurrences; in fact, this ganglion was referred to as "little brain" (Figures 9-10 and 9-11).

a. Central Excitatory Muscarinic and Nicotinic Responses

The development of electrophoretic methods for the application of ACh and other agents to

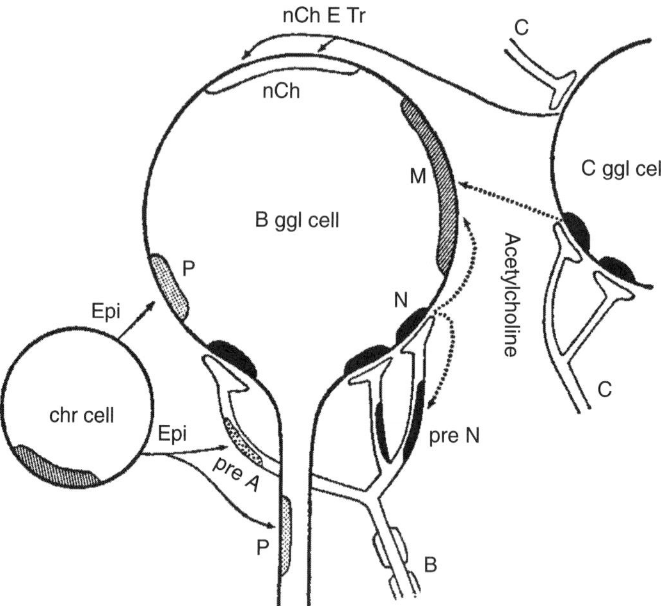

Figure 9-10. Schematic representation of the pre- and postsynaptic receptor sites of the sympathetic ganglion cells in the frog. B, C, and chr, B, C and chromaffin cells, respectively; B and C, B and C fibers, respectively; N and pre N, subsynaptic and presynaptic nicotinic sites; M and nCh, muscarinic and noncholinergic receptor sites. M site is illustrated as activated by ACh diffusing from B and C fibers, while nCh is activated by nonacetylcholine excitatory transmitters (nCH E Tr) diffusing from the C fiber to the B cell (and to the C cell). The P potential is shown as arising in the P sites of the postsynaptic membrane and of the axon. The action of epinephrine on the nerve terminal is illustrated also. It is known today that the non-ACh transmitter is a peptide. (From Karczmar and Nishi, 1971, with permission.)

localized central neurons (Curtis and Eccles, 1958; Eccles, 1964; Krnjevic, 1969) led to the identification of muscarinic and nicotinic central postsynaptic receptor sites. Eccles, Krnjevic, Curtis, and Ryall also described differences in the kinetics and other characteristics of the responses evoked at the two receptors.

Whether evoked by the application of ACh or nicotinic agonists, or by presynaptic stimulation of appropriate cholinergic pathways, nicotinic responses are fast in onset and brief in duration. On the other hand, the muscarinic responses, evoked by ACh, muscarinic agonists, or presynaptic stimulation, exhibit a delayed onset and long duration that extends beyond the duration of the stimulation or agonist application (see Krnjevic, 1969, 1974b). These differences in kinetics apply equally to the duration of the neuronal discharge rate and the duration of the depolarizing postsynaptic response (Figure 9-12). The slow, muscarinic response is blocked by systemically applied atropine; similar effect is obtained with quaternary compounds applied electrophoretically. The fast, nicotinic response is blocked by tertiary nicotinolytics or by d-tubocurarine (d-TC) if applied electrophoretically (see also below, section BIII-1). Furthermore, the delay in the onset of the slow, muscarinic response does not result from the application of muscarinic agents at the dendrites or far from the soma membrane; it is also present when the agents are applied to the soma hillock (Krnjevic, 1969; Curtis and Crawford, 1969).

Altogether, the fast nicotinic response resembles the fast EPSP obtained, whether at the autonomic ganglia or at the neuromyal junction; the slow response is similar to the slow EPSP present at sympathetic autonomic ganglia (see Figures 9-10 and 9-11). The ionic mechanisms of central fast and slow responses are similar to the ionic mechanisms of their peripheral counterparts; accord-

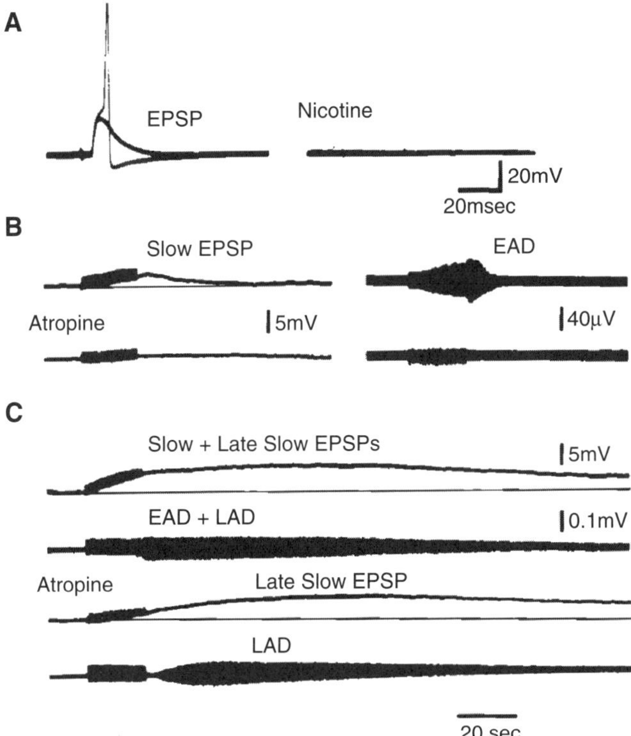

Figure 9-11. Excitatory postsynaptic potentials (EPSP, slow EPSP, and late slow EPSP) and action potential or afterdischarges (EAD and LAD), initiated by these potentials, of bullfrog sympathetic ganglion. Potentials were recorded intracellularly from two different sB cells (for A, B, and C) of a preparation, and after discharges were recorded extracellularly from another preparation. A. EPSP and action potential evoked by single preganglionic B nerve stimulation. EPSP was abolished 30 min after application of nicotine (0.12 mm). B. Slow EPSP and EAD (early after discharge) evoked by titanic stimulation (10 per second for 20 seconds) applied to preganglionic B in the presence of nicotine. Both slow EPSP and EAD were abolished by addition of atropine (0.014 mm). C. Slow plus late slow EPSPs and EAD plus LAD evoked by titanic stimulation (10 per second for 20 seconds) applied to preganglionic B and C nerves in the presence of nicotine. Slow EPSP and EAD was abolished, whereas late slow EPSP and LAD remained unaffected by addition of atropine. (From Karczmar, Nishi, and Blaber, 1972, with permission.)

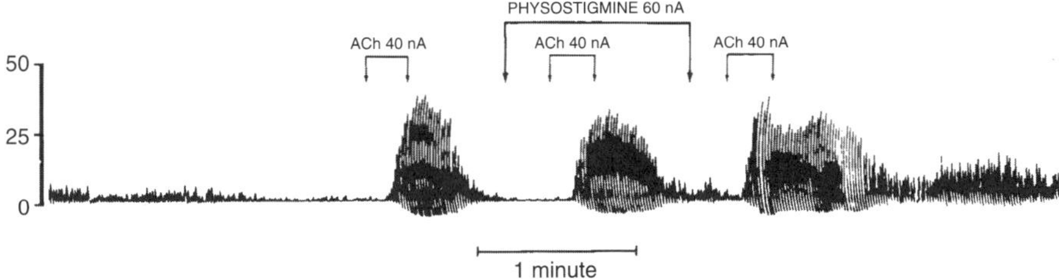

Figure 9-12. Effects of ACh and eserine applied by microntophoresis on cortical and hippocampal neurons in unanesthetized cat's "cerveau isolé." Ordinates indicate frequency of firing in counts per second. Note that the muscarinic response to acetylcholine is delayed and prolonged; it is further prolonged after administration of physostigmine. (From Krnjevic, 1969, with permission.)

ingly, the terms used for the peripheral responses, the EPSP and the slow EPSP, were adopted for the fast and slow central events.

There is still another similarity between the central and ganglionic responses. At many central sites, including the Renshaw cell (Curtis and Phillis, 1960; Curtis and Ryall, 1966), the cholinoceptive neurons show both types of responses (see Krnjevic, 1974a, 1974b, 1988, 1993). At the Renshaw cell, the muscarinic response is weak as compared to the nicotinic response, while in most instances the central muscarinic rather than the nicotinic response predominates. However, at the retinal ganglion cells (Masland and Ames, 1976; Ariel and Daw, 1982; Krnjevic, 1988) and for some thalamic, neocortical and cortical neurons (Andersen and Curtis, 1964; Stone, 1972; Van der Zee et al., 1991, 1992), the mix resembles that which obtains at the Renshaw cell. On the whole, whether present in the mix or being the only responses exhibited by a neuron, the muscarinic responses predominate in most areas of the brain (Krnjevic, 1993); this is not to disregard the solid representation of the nicotinic responses in several parts of the brain (Nordberg et al., 1989a, 1989b; Quick et al., 1999; see also chapter 6 C). Correspondingly, most functions and behaviors are muscarinic rather than nicotinic. As the central fast and slow responses are also similar to their peripheral counterparts in terms of their ionic mechanisms, the fast and slow EPSP term was adopted for the fast and slow central events.

The ionic characterization of cholinomimetic, muscarinic, and nicotinic responses optimally requires the use of video microscopy and intracellular and voltage clamp methods (Albuquerque et al., 1998a). This combined approach is not easy. However, it is clear that the fast nicotinic EPSP of the Renshaw interneuron is mediated by a decrease in resistance and an increase in cationic permeability (Zieglgansberger and Reiter, 1974). This is also true for several other nicotinic sites of the mammalian brain (see Halliwell, 1990; Lindstrom et al., 1990; Patrick et al., 1993; McCormick, 1993; Lindstrom, 1994, 1997; Lindstrom et al., 1996) and for nicotinic responses of the avian mesencephalon (Chiapinelli et al., 1993). These changes in the membrane characteristics, which occur during the response, hold for a number of nicotinic receptor subtypes (Albuquerque et al., 1996; Lindstrom, 1997; Kuryatov et al., 2000). On the other hand, Edson Albuquerque, Jon Lindstrom, and their associates established that nicotinic receptor subtypes (for example, $\alpha7$, $\alpha4\ \beta2$, and $\alpha3\ \beta4$) exhibit, in the course of their responses, different current kinetics, different antagonist and antagonist sensitivities and efficacies, and different rates of desensitization (see for example, Albuquerque et al., 1996, 1998b; Lindstrom, 1996, 1997). More in depth, Jon Lindstrom and his associates demonstrated that the transmembrane segments of the nicotinic receptors dictate the characteristics (such as efficacy) of the responses of the various receptor subtypes (Rush et al., 2002; see Chapter 6).

Several types of postsynaptic muscarinic responses (slow EPSPs; see Figures 9-10 and 9-11) occur in the CNS. At cortical rodent and human (Halliwell, 1986; McCormick and Prince, 1985) and pyramidal hippocampal neurons, slow depolarization or slow EPSP is characterized by increased resistance or decreased conductance due to the inactivation of the outward K current. This current is voltage- and Ca^{2+}-dependent but insensitive to changes in Cl concentration; it exhibits a reversal potential of about 90 mV (Krnjevic et al., 1971; Krnjevic, 1974a, 1974b, 1988; Cole and Nicoll, 1983, 1984; Karczmar, 1986); parenthetically, two ionic, voltage- and Ca^{2+}-dependent channels may exist in the CNS (Platano et al., 2005; Perez-Rosello et al., 2005). This current is analogous to the classical M current involved in the muscarinic responses of the sympathetic ganglia (Adams and Brown, 1982; Dodd et al., 1981; Halliwell, 1990; Nishi et al., 1969; McQuiston and Madison, 1999; and Okada et al., 2002; see also Chapter 5). This slow depolarization induced by muscarinics replaces after-hyperpolarization evoked by the synaptic current (McQuiston and Madison, 1999). Fukuda et al. (1989), Brown, Caulfield, and their associates (Caulfield and Brown, 1991; Robbins et al., 1991; Caulfield et al., 1992, 1993a, 1993b), and Ehlert et al. (1995) employed molecular biology methods to identify receptor subtypes that are involved in this response and concluded that the central M1 and M3 receptors are involved in the postsynaptic inactivation of the M current.

Other types of slow muscarinic responses involve blockade of voltage-independent K^+ current (McCormick and Prince, 1985; Krnjevic, 1993) and facilitation or activation of the inward-directed Na^+ current; these changes result in slow depolarization and paradoxical increase of membrane resistance ("anomalous rectification"). This phenomenon occurs in the hippocampal CA-1 neurons and elsewhere (Benardo and Prince, 1982; Halliwell and Adams, 1982; Kelly and Rogawski, 1985; Krnjevic, 1988; Halliwell, 1990). Furthermore, the block of the Ca^{2+}-evoked current is muscarinically induced in some hippocampal neurons (Cole and Nicoll, 1984a, 1984b; Caulfield et al., 1992; Caulfield, 1993; see also Chapter 8). These mechanisms may involve M2 and M4 receptor subtypes. The so-called persistent Na^+ current that may be evoked via depolarizing current following the suppression of Ca^{2+} and K^+ currents is blocked by muscarinic agonists such as carbachol (Mittman and Alzheimer, 1998). The significance of this phenomenon for the muscarinic response is not clear. Also, the stimulation of certain muscarinic receptors evokes (instead of a classical slow EPSP response) repetitive firing, which may also may appear spontaneously (see, for example, Carette, 1997; Song et al., 1998). These responses depend on the presence of specialized K^+ channels (see Chapter 5).

The second messenger mechanisms are activated during muscarinic and nicotinic excitatory responses as established first by the Hokins' Hokin and Hokin, 1959; see also Hokin and Dixon, 1993) and Paul Greengard (1987; see also Sutherland et al., 1968). The central muscarinic M2 and M4 receptors located in the cortex, striatum, and hippocampus are coupled with the G proteins of the G1/Go class, and their activation inhibits the adenylate cyclase. The M1, M3, and M5 central receptors are coupled with the G proteins of the Gq/G11 class and activate the phosphatidylinositol cascade (Caulfield et al., 1993a, 1993b; Grassi et al., 1993; Ffrench-Mullen et al., 1994; Ehlert et al., 1995; Myles and Fain, 1996; Kostenis et al., 1998; Nicole et al., 2001). However, there is controversy concerning these phenomena, particularly regarding whether M2 and M4 receptors also activate the cascade (compare, for example, Schimerlik, 1990, and McKinney, 1993; see also Richelson, 1995); also, certain muscarinic receptor subtypes may activate other second messengers, including the Ca^{2+}-calmodulin system and cGMP (cyclic nucleotide-gated channels; Kuzmiski and MacVicar, 2001). Altogether, the definition of the link between the different muscarinic receptor subtypes and the various messenger systems is still incomplete (Richelson, 1995; Ehlert et al., 1995; Kostenis et al., 1998). Interestingly, M5 activation may be induced in non-neuronal cells such as hamster ovarian cells (Singer-Lahat, 1996).

Classically, nicotinic receptors do not relate to the phosphatidylinositol cascade or the cyclic GMP, and their activity vis-à-vis cyclic AMP activity is uncertain (see Chapter 6). However, recent results indicate that nicotinic agonists (in the rat cortex and hippocampus) activate the phosphatidylinositol cascade and related Ca^{2+} fluxes (Richelson, 1995); this effect resembles that of muscarinic agonists. It is interesting that Ca^{2+} fluxes may be elicited by nicotinic receptors of the $\alpha 7$ subunit type located on central—such as hippocampal—astrocytes, which thus contribute to calcium signaling in the CNS (see, for example, Sharma and Vijayaraghavan, 2001). Furthermore, nicotinic agonists may affect the sensitivity of rat brain M1 and M3 receptors. Whether this effect relates to the phosphatidylinositol action of these agonists is unknown. Altogether, the differential role of nicotinic receptor subtypes with respect to the various second messenger systems is unclear.

The important role of couplings among cholinergic receptors and their subtypes and the second messengers and G proteins concerns the generation of postsynaptic currents (see above, this section) and the causation of overt cholinergic functions and behaviors (see section BIV). This role has not been completely clarified as yet. However, Chris Krnjevic and his associates elucidated the difference in synaptic significance and consequences of the activation of muscarinic and nicotinic receptors. Krnjevic (1969) pointed out that the synaptic, fast responses (fast EPSP) recorded classically at the neuromyal junction and at the autonomic ganglia relate, on a one-to-one basis, to effective transmission and effective fast postsynaptic potential, in contradistinction to the central slow muscarinic responses, and he demonstrated that this difference also obtains in the CNS. Accordingly, he suggested that the muscarinic responses are facilitatory and modulatory rather than obligatory in nature. Thus, they promote nicotinic responses and responses to noncholinergic transmitters, attenuate the IPSPs evoked by inhibitory transmitters, diminish the afterhyperpolarization that follows the action potential, and increase the neuronal excitability via assisting anomalous rectification and "suppressing the stabilizing effect of K outward current" (Krnjevic, 1988, 1993; Puil and Krnjevic, 1978). These phenomena promote and prolong the repetitive firing engendered by depolarizing inputs that are normally cut off by the after-hyperpolarization, and they may underlie long-term potentiation (LTP; see section BIV-3d). Krnjevic's (1993) demonstration that ACh suppresses high anoxic K^+ conductance evoked by brain ischemia and hypoxia, and prevents the resulting attenuation of neuronal excitability and repetitive firing, is consistent with the concept of the modulatory and facilitatory nature of the muscarinic activation (see section BIII-2).

It must be added that the facilitatory action of ACh may also be due to its disinhibitory effect (Kelly and Rogawski, 1985); this phenomenon arises when the activation of cholinergic pathways, including those originating in the mesencephalic formation (see Chapter 2 DI-DIII), evoke depolarization that attenuates inhibitory potentials due to other transmitters (see section BIII-2c, below).

Receptor desensitization (or inactivation) and sensitization may also be considered as modulatory phenomena, since these phenomena affect the extent of the receptor response. Nicotinic sensitization and desensitization are classical phenomena well known since the middle of the last century (Thesleff, 1955; Karczmar, 1957; Karczmar et al., 1972; see Chapter 6 and section BIII-2f, below). The desensitizing capacity of muscarinic responses was described more recently. It appears that exposing M2 receptors to ACh may cause rapid desensitization of G protein-mediated phosphorylation of GRK family receptor kinases. This in turn induces the desensitized binding of an "arrestin" protein to the receptors (Hamilton et al., 1998; see also Chapter 5).

b. Central Inhibitory Postsynaptic Cholinoceptive Responses

Randic et al. (1964) and Bradley and Wolstencroft (1965) were the first to demonstrate that ACh and cholinomimetics evoke inhibitory responses in the rodent cerebral cortex and the pons-medulla. These responses occur also in several limbic structures, basal ganglia, thalamus, cerebellar cortex, and spinal cord (Figure 9-13; see Curtis and Crawford, 1969; Vazguez and Krip, 1973; Kelly and Rogawski, 1985; Krnjevic, 1974a, 1974b, 1988; Egan and North, 1986). Though widespread, these responses are not as frequent as the excitatory nicotinic and muscarinic responses. Most of these inhibitions seem to be muscarinic in nature, as the depressant action of ACh was duplicated in the same cells by muscarinic agonists and blocked by atropine and atropinic drugs

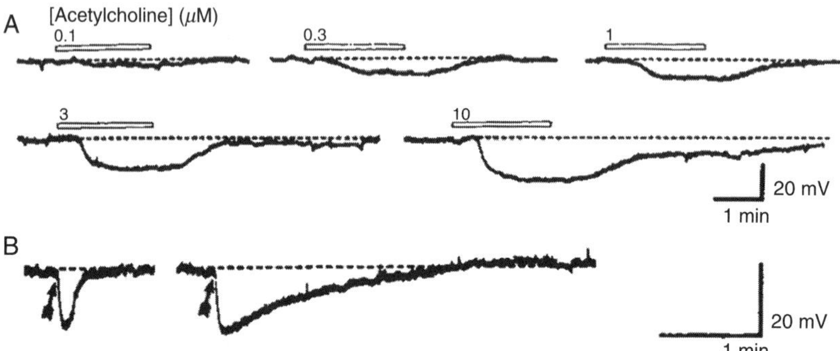

Figure 9-13. Acetylcholine (ACh) hyperpolarizes neurons of rat nucleus parabrachialis. Records show membrane potential (resting level, −60 mV). (A) The bar above each trace indicates duration of superfusion with ACh-containing solution (concentration indicated); the solution also contained neostigmine throughout this experiment. Approximately 5 min elapsed between each of the traces shown. The neurons did not fire action potentials during the hyperpolarization evoked by ACh. (B) The arrows indicate application of a pressure pulse (70 k Pa, 10 ms) to a pipette that contained ACh chloride (1 mM), the tip of which was positioned beneath the surface of the superfusion solution. Left trace, control; right trace, the same ACh application in the presence of neostigmine (1 μM). (From Egan and North, 1986, with permission.)

(Hasuo et al., 1988; McQuiston and Madison, 1999). When evoked by electrophoretically applied ACh or cholinomimetics, the inhibitory responses occur after a delay and are prolonged; these kinetics are consistent with the muscarinic nature of these responses (Curtis and Crawford, 1969; see, however, Kelly and Rogawski, 1985).

This inhibition is due to a Ca^{2+}-independent increase in K^+ conductance (Hasuo et al., 1988; McQuiston and Madison, 1999); thus, it is analogous with the muscarinic cardiovagal action (Egan and North, 1986). Nicotinically evoked inhibitions are rare; they were recorded in the rat's dorsolateral septal nucleus and cerebellar Purkinje neurons (Wong and Gallagher, 1991). Similar to muscarinic evoked inhibition, nicotinic inhibition is mediated by a Ca^{2+}-dependent increase in the K^+ conductance (Wong and Gallagher, 1991).

The nature of the inhibitory cholinoceptivity is controversial, as not all investigators agree with the notion of a direct, inhibitory cholinoceptive response. In fact, it was proposed that inhibitory response is disynaptic, as it is mediated by cholinomimetic action on an inhibitory interneuron (McQuiston and Madison, 1999; Benardo and Prince, 1982; McCormick and Prince, 1985; Gahring et al., 2004; see also Krnjevic, 1988; Karczmar, 1972a, 1972b). It is consistent with this opinion that cholinomimetic inhibition was sometimes antagonizable by strychnine and related to the release of GABA (Stone, 1972; Phillis and York, 1968; Tredway et al., 1999). It is also consistent with this notion that the inhibitory response occurs after a delay, which is characteristic of a disynaptic circuitry (A. G. Karczmar and K. Kim, 1964 unpublished data). On the other hand, sometimes ACh-evoked inhibitions occurred at sites devoid of excitatory responses to cholinomimetics; therefore, they could not be due to excitation of an interneuron followed by the inhibitory action of GABA released from the interneuron on the cell innervated by the interneuron (Dingledine and Kelly, 1977). Furthermore, microelectrode studies indicate that the inhibitory effects, whether muscarinic or nicotinic, occur directly at the cholinoceptive neuron (Wong and Gallagher, 1991). Perhaps depending on the site, inhibitory responses may be due to either direct or indirect disynaptic effects of ACh and the cholinomimetics.

There are also problems with the pharmacology of the inhibitions. When spontaneous firing recorded from the dorsolateral nucleus reticularis and ventrobasilar complex of the thalamus was inhibited by the stimulation of the mesencephalic reticular formation or via the application of ACh to the formation, the patterns and kinetics of the inhibitions obtained by these two means were similar; however, while the response to ACh could be blocked by atropine, the response to the neural stimulation could not be (Dingledine and Kelly, 1977). The significance of this evidence is obscured by the uncertainty of whether or not the pathways in question are cholinergic. In fact, in only a few instances could the inhibitory responses be related to specific cholinergic pathways. For example, Joel Gallagher, Pat Shinnick-Gallagher, and their associates demonstrated that certain IPSPs are generated by the activity of pathways originating in the medioseptal diagonal band (J. P. Gallagher, 1979 personal communication).

Finally, besides the two notions of either direct or disynaptic nature of cholinoceptive inhibitions, a third option was described: the inhibitions may result from a delayed, postexcitatory hyperpolarization (McCormick and Prince, 1987).

BII. Central Release of Acetylcholine

1. Release of ACh from and in the Brain and Isolated Tissues

Is it possible to repeat in the CNS the experiments that led to the classical demonstration of ACh release at specific peripheral cholinergic synapses? Such a demonstration must involve stimulating an identified central neuron or a homogeneous population of neurons, suspected of being cholinergic, and identifying the transmitter released from the site. This is a difficult task to accomplish *in vivo* and even in isolated preparations since the latter generally do not include a single kind of neuron. Even for the Renshaw cell, a well-identified cholinergic interneuron, this kind of demonstration is not available. However, pertinent findings begin to emerge. For example, the release of ACh was observed in cultures of identified cholinergic neurons such as those of the diagonal band of Broca or nucleus basalis of Meynert

(Takei et al., 1989). Also, ACh was released *in vivo* from specified cortical or forebrain sites following the stimulation of the nucleus basalis (Kurosawa et al., 1989; Consolo et al., 1990).

Less specifically, the release of ACh from several brain areas, not necessarily homogeneous, has been amply demonstrated since Quastel and his associates (1937) described the release of ACh-like substances from brain slices incubated in an eserinized medium. Subsequent work was carried out both *in vivo* and with isolated preparations. *In vivo*, MacIntosh with Oborin (1953), Elliott et al. (1950), Mitchell (1963) and Szerb (1964; see Karczmar, 1967 and Pepeu, 1974) applied cups or cylinders to the exposed cortex for collecting the cortical release, which contained ACh (Figures 9-14 and 9-15). Later, Gaddum's push-pull cannula, Feldberg's perfusion of intracerebral spaces, and Pepeu's microdialysis methods were employed together with precise methods for measuring ACh to evaluate ACh release from several subcortical sites (see Pepeu, 1974, 1993; Wu et al., 1988).

It is interesting that such a release could be demonstrated even for free-moving, unanesthetized animals (Wu et al., 1988; Watanabe and Shimizu, 1989; Toide, 1989). The release occurs spontaneously at cholinergically innervated central sites, and upon electric stimulation of the cholinergic sites or of appropriate afferent pathways. It can be also evoked by application of K^+ and certain neurotoxins to appropriate sites. The spontaneous release was augmented by physiological stimuli, in the case of the isolated retina stimulated by light (Masland and Livingstone, 1976; Masland and Ames, 1976) or in that of the cortical release of ACh in freely moving rats (Watanabe and Shimizu, 1989; Toide, 1989). Altogether, the cortex, basal ganglia and limbic structures (i.e., the hippocampus), cerebellum, and spinal cord release ACh *in vivo* (Pepeu, 1974; Pepeu et al., 1990; Vizi et al., 1987; Toide, 1989).

AntiChEs were frequently added in these experiments to the cup, push-pull cannula, or dialysis fluid (see Pepeu, 1974; Beani et al., 1978). However, this method may obscure the results

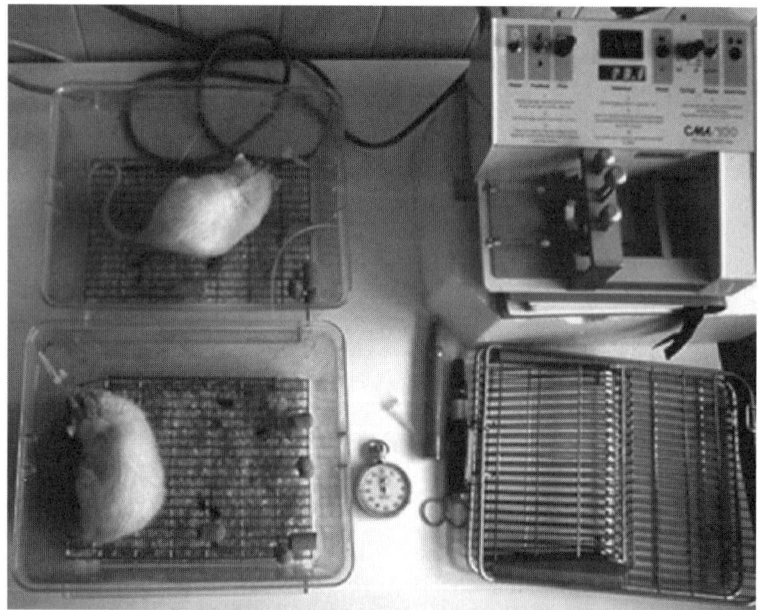

Figure 9-14. Microdialysis technique in a rat performing an operant task. Artificial cerebral spinal fluid containing a cholinesterase inhibitor was perfused at a constant flow rate of 3 μl/min. Dialysate samples were collected from the frontal cortex or hippocampus at 5-minute intervals, while the rat was at rest or pressing the lever to obtain a food reward. Acetylcholine content in the samples was quantified by a high liquid chromatography method. (By courtesy of Giancarlo Pepeu.)

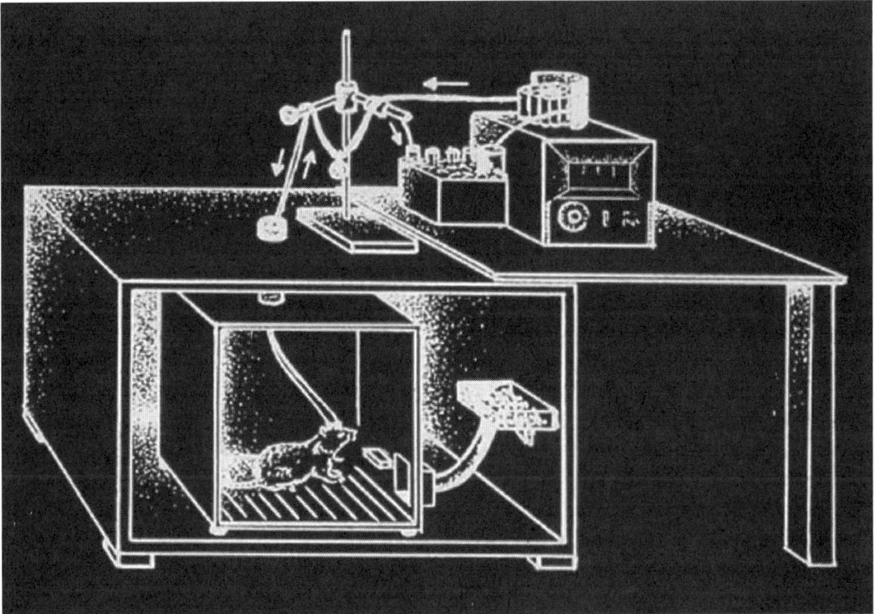

Figure 9-15. Microdialysis technique in freely moving rats. The probes, inserted in the frontal cortex or hippocampus, are connected to the perfusion pump through tubing, allowing the rats to move in the cage. Artificial cerebral spinal fluid, with or without the presence of a cholinesterase inhibitor, was perfused at a constant flow rate of 3 μl/min. Dialysate samples were collected at 5-, 15-, or 20-minute intervals, according to the need, and were directly assayed for acetylcholine by a high performance liquid chromatography method. (By courtesy of Giancarlo Pepeu.)

because antiChEs act on synthesis, turnover, and release of ACh; the latter is generally blocked by antiChEs (Hanin and Costa, 1976, and section BIII-1a, below). This inhibition and changes evoked by antiChEs in the metabolism of ACh should be considered in evaluating the pertinent results. However, in many of these experiments antiChEs were not used (see, for example Stadler et al., 1973; Gadea-Ciria et al., 1973).

As for the *in vitro* work, ACh was released from slices of rodent or rabbit cortex (Pedata et al., 1983), striatum (Jackson and Zigmond, 1988; Dutar et al., 1989), ventral tegmental area (Calabresi et al., 1989), cerebellum (see, for example, Lapchak et al., 1989a), isolated retina (Masland and Livingstone, 1976), and limbic structures, including the hippocampus (Lapchak et al., 1989a; see also Pepeu, 1993). These sites essentially agree with the sites of ACh release identified *in vivo*. The release of ACh evoked by certain toxins, potassium, or electric field stimulation also occurred in synaptosomes, cell suspensions, and/or homogenates prepared from these brain parts (Meyer et al., 1984; Lapchak et al., 1989b; Raiteri et al., 1989; cf. also Pepeu, 1993, and Giovanelli et al., 1987). As expected, the release was blocked by tetradotoxin by lowering Ca^{2+} in the medium or by the muscarinic activation of the autoreceptors (see, for example, Raiteri et al., 1989). The regulation of ACh release from the brain or isolated tissues by presynaptic auto- and heteroreceptors is discussed below (see section BIII-1 and BIII-2).

2. Release of ACh and the Notion of Cholinergicity

Among important criteria mandatory for establishing the cholinergicity of a synapse or a pathway is the demonstration of ACh's release from pertinent nerve terminals (see Chapters 1 and 2). The preceding section emphasized the difficulties encountered in meeting this criterion to a purist's satisfaction for any of the central cholinergic pathways described in Chapter 2 D, including Renshaw circuitry. Yet, the criterion was nearly met *in vivo* with respect to pathways radiating from the nucleus basalis of Meynert and the retinal networks (see Chapter 2 DII). Furthermore, this release was demonstrated with several types of isolated, relatively homogeneous preparations, as well as, in a few cases, with pure cholinergic isolates. Isolated preparations may not provide direct evidence for the physiological release of ACh from a given central pathway or from specific cholinergic synapses, yet this evidence is crucial.

It must be added that additional, important criteria of cholinergicity were satisfied with regard to the presence and function of central cholinergic transmission. Accordingly, the alleged central cholinergic neurons contain release markers such as synaptic vesicles, identified as cholinergic; synaptic vesicle-plasma membrane fusion proteins; CAT; and transporter proteins that are controlled by the "cholinergic gene locus" (see Chapter 2 B-1 and B-2; Karczmar, 1999).

Today, the caveats that were once raised with respect to the cholinergicity of specific central pathways and synapses cannot significantly disturb the quiet seas of central cholinergicity.

BIII. Pharmacology of Central Cholinergic and Cholinomimetic Responses

1. Cholinergic Pre- and Postsynaptic Pharmacology

Atropinics and nicotinolytics, cholinomimetics and antiChEs, cholinergic neurotoxins and inhibitors of ACh synthesis such as hemicholiniums, the blockers of axonal conduction such as tetrodotoxin, and ionic nerve terminal channel blockers all cause effects, both pre- and postsynaptic, that are predicated on the basis of the cholinergic, cholinolytic, or cholinomimetic nature of the phenomena in question (see Pepeu, 1974, 1993; Narahashi, 1988). In addition, ions such as Ca^{2+}, in particular, play a major part in ACh release processes.

a. Presynaptic Cholinergic Pharmacology

Physiological and pharmacological phenomena that regulate the release of ACh involve ionic channel mechanisms (Rahamimoff et al., 1980; Segal, 1989), as well as auto- and heteroreceptors (i.e., noncholinergic receptors localized on cholinergic terminals). The channel mechanisms in question involve Ca^{2+}, as known since the discoveries of Del Castillo, Miledi and Bernard Katz (Del Castillo and Katz, 1954; Katz and Miledi, 1968, 1965); today, it can be stated that "the elevation of intracellular . . . nerve terminal . . . Ca++ . . . and the membrane Ca++ voltage potential . . . are . . . necessary and sufficient to promote release" of ACh (Parnas and Parnas, 2004). In this section we are concerned mainly with the autoreceptor phenomena.

These phenomena are consistent with morphological evidence demonstrating the presence of muscarinic receptors on the terminals of cholinergic neurons (Mrzljak et al., 1993; Van der Zee and Luiten, 1999). ACh's release is autoregulated at these sites (see the preceding section and Chapters 5 and 6). Accordingly, the release is increased by atropine or scopolamine, and decreased by muscarinic cholinomimetics (i.e., oxotremorine) and antiChEs. This is true for the cortex, the basal ganglia, and the limbic structures (for example, the hippocampus) and can be demonstrated *in vivo* and *in vitro* (see the preceding section and Pepeu, 1974, 1993; Raiteri et al., 1987; Karczmar, 1986 and 1990; Albrecht et al., 1999). The control of the time course of release depends most on the M2 receptors, the Ca^{2+} influx being faster; conversely, during the termination of release, the M2 receptor-mediated processes are faster than those controlling the removal of Ca^{2+} (Parnas and Parnas, 2004).

In some brain regions (the striate and the limbic system, for example), the central presynaptic regulation depends on M2 and M4 receptors; interestingly, M1 receptors are involved in corresponding phenomena at the autonomic ganglia (Caulfield and Brown, 1991.) At these sites, specific M2 ligands are capable of antagonizing the depression of ACh release by muscarinic M2 agonists such as oxotremorine (Weiler, 1989; Lapchak et al., 1989a, 1989b). However, additional muscarinic receptors are also present at central cholinergic terminals. Muscarinic agonists inhibit ACh release in the hippocampus, several cortical areas, and the septum by blocking presynaptic Ca^{2+} channels, which results in nerve terminal depolarization (see for example, Segal, 1989; Caulfield and Brown, 1991; Dudkin et al., 1990; Quian and

Saggau, 1997). The effects of K^+ channel blockers (aminopyridines, for example) on ACh release are less certain (see Torocsik and Vizi, 1990; Marchi et al., 1999).

Nicotinic presynaptic autoreceptors are facilitatory in nature (see section BI-1 and Lindstrom, 1994, 1995, 1996). Accordingly, Lapchak et al. (1989a, 1989b, 1990; Marchi et al., 1999) found that nicotinic agonists facilitated and nicotinolytics attenuated ACh release from slices of several brain areas and synaptosomes, and *in vivo* from the hippocampus and hypothalamus. Atropinics did not affect the nicotinically induced release. Marchi and Raiteri (1996) and Marchi et al. (1999) determined that Ca^{2+} channels are involved in selecting between a nicotinic and muscarinic presynaptic effect on the release of ACh. They hypothesized that the transition to nicotinic enhancement of release occurs when cholinergic function is impaired. The finding by Eskesen and Sheardown (1993) that both nicotinolytics and atropinics attenuate miniature endplate potentials (MEPPs) and endplate potentials (EPPs) seems inconsistent with the notion that atropinics facilitate ACh release.

Inhibition of ACh synthesis by choline uptake inhibitors (i.e., hemicholiniums or appropriate antibodies), blockade of nerve terminal Ca^{2+} channels, and destruction of cholinergic neurons by cholinergic neurotoxins, for example, AF64A (Figure 9-16), lead to the inhibition of ACh release (see Marchbanks et al., 1981; Uney and Marchbanks, 1987; Pittel et al., 1992a, 1992b; Evans et al., 1993; Dudkin et al., 1990). Conversely, increasing choline uptake into the nerve terminal or decreasing its efflux facilitates ACh release (Pittel et al., 1992a, 1992b).

Altogether, cholinergic neurotoxins such as AF64A exert central actions similar to those they evoke at the cholinergic periphery. However, unexpected effects sometimes occur. For example, beta bungarotoxin, a neurotoxin, which blocks ACh release from the myoneural junction, exerts complex, multiphasic effects on the central cholinergic terminals *in vitro* (for example, in the case of rodent cortical synaptosomes) or *in vivo* (Chapell and Rosenberg, 1992).

Pertinent in this context are the second messenger phenomena that include nucleotides and their activators; phosphatidylinositol turnover (PIT); the third messengers (nitric oxide); related mechanisms (i.e., phosphorylation-dephosphorylation of synapsin, a protein involved in ACh release); and the mediatophore phenomena (see Chapter 2 C1 and C2). Accordingly, it was shown that NO, an activator of guanyl cyclase and therefore of glyceryl memephosphate (GMP), facilitates the release of ACh in the forebrain (see, for example, Prast and Philippu, 1992; Prast et al., 1995; Ohkuma et al., 1995; Leonard and Lydic, 1995). Also, PIT helps conditioning and learning via facilitating ACh release (see Van der Zee and Luiten, 1999).

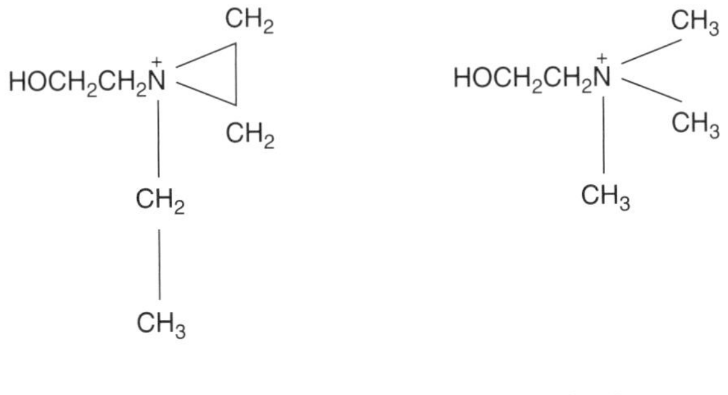

AF64A — Choline

Figure 9-16. Structures of AF64A and choline. (After Hanin, 1990, 1992, with permission.)

b. Postsynaptic Cholinergic Pharmacology

There are three types of central cholinocepticity. Some neurons including thalamic neurons show only nicotinic responses (Tebecis, 1970a, 1970b; Lindstrom, 1994, 1996); some cells (for example, hypothalamic cells) respond only muscarinically (Krnjevic, 1969); the majority of neurons show mixed sensitivity (see also section BI-2, above).

Orthodox pharmacology was established for the postsynaptic responses of these neurons. Whether superfused, applied by micropipettes, or given systemically, fast postsynaptic responses to nicotine and nicotinic agonists were blocked by nicotinolytics; additionally, the fast responses evoked by nicotinic action of microelectrophoretically applied ACh were blocked by nicotinic antagonists. On the other hand, postsynaptic responses to muscarinic cholinomimetic and muscarinic responses to ACh were blocked by antimuscarinics (for references, see Curtis and Crawford, 1969; Krnjevic, 1974a, 1974b, 1988; Karczmar, 1967; McGeer and McGeer, 1993; McGeer et al., 1987).

AntiChEs, of both carbamate and organophosphorus type, readily potentiated both ACh and cholinomimetically—provided ChE sensitive cholinomimetics were used—evoked responses of appropriate central neurons (cf., for example, Karczmar, 1967; Krnjevic, 1974). This occurred whether the responses were nicotinic or muscarinic in character.[2] Also, this potentiation could be obtained in a number of areas innervated by recognized cholinergic pathways and exhibiting significant cholinoceptivity such as the striate, cortex, thalamus, hypothalamus, and several limbic sites (see, for example, Washburn and Moises, 1992).

Receptor subtype-specific electrophysiological postsynaptic responses may be obtained in the rodent CNS with receptor subtype-specific muscarinic and nicotinic agonists including choline (Ladinsky et al., 1979; Albuquerque et al., 1996; Mike et al., 2000; Mellott et al., 2004; see also Chapters 3 CII and 8 CIII). These responses are blocked by receptor-subtype specific nicotinic and muscarinic antagonists (see, for example, McCormick, 1993; Clarke, 1993a, 1993b; Patrick et al., 1993; Ladinsky, 1993; and Lindstrom, 1994). The nicotinic receptor's subtype response was generally characterized in the oocyte expression system (see Chapter 6) rather than *in situ* or in isolated preparations. So the knowledge of the CNS distribution of nicotinic receptor subtypes is incomplete (see Chapter 6). At any rate, the central postsynaptic nicotinic and muscarinic responses share ionic characteristics with those classically established for the ganglia, as in the case of muscarinic responses obtained in slices of the amygdala (Womble and Moises, 1992).

Similar pharmacological results were obtained with responses evoked *in vitro* or *in vivo* by presynaptic electric stimulation. Accordingly, thus evoked nicotinic and muscarinic responses were readily blocked by nicotinolytics and atropinics, respectively (see, for example, Wasburn and Moises, 1992), and potentiated by antiChEs. Again, tertiary blockers were effective when given systemically, and both tertiary and quaternary blockers were active when applied electrophoretically (Curtis and Crawford, 1969; Krnjevic, 1974a, 1974b, 1988). Interestingly, functionally evoked cholinoceptive responses showed expected pharmacology. Accordingly, Zhang and Liu (1993) demonstrated *in vivo* that neurons of the thalamic nucleus parafascicularis are sensitive to microelectrophoretic application of ACh, particularly when following algetic stimulation. Both this response and the firing induced in these neurons by noxious stimulation were blocked by atropine. The nicotinic and muscarinic, neurally or functionally induced responses are blocked *in vitro* by lowering the calcium concentration of the solution or by adding tetrodotoxin or hemicholinium compounds. Taken together, these results suggest that the release of ACh from the pertinent sites and its postsynaptic action at these sites are involved in the neurally and functionally evoked responses (cf. Karczmar, 1967). These findings have a bearing on the pharmacology of functions and behaviors phenomena (see section BIV and BV).

The pharmacology of the central postsynaptic phenomena resembles the pharmacology of peripheral postsynaptic events. However, this analogy is not always perfect. Assumedly, this is because of the differences between central and peripheral cholinergic receptor subtypes. For example, alpha bungarotoxin, while capable of binding to cholinoceptive receptors of the rodent hippocampus and other brain regions (Larsson and Nordberg, 1985), cannot block central cholinergic nicotinic transmission, in contrast to what occurs at the neuromyal junction or at the ganglia. However, astrocytic nicotinic receptors seem to be blockable by alpha bungarotoxin (Sharma and Vijayaraghavan, 2001; see also section BI-2, above).

2. Noncholinergic Pharmacology and Transmitter Interactions

Noncholinergic pharmacology and transmitter interactions concern the effects of either noncholinergic drugs or noncholinergic transmitters at cholinergic nerve terminals and postsynaptic cholinoceptive sites. Conversely, ACh and cholinergic drugs affect the release of noncholinergic transmitters and their postsynaptic responses.

Effects of non-ACh transmitters on cholinoceptive sites and effects of cholinergic drugs on noncholinergic sites represent instances of transmitter interactions. In view of the ubiquity of the cholinergic pathways and their widespread impinging on the other transmitter systems, such interactions may be readily expected. Electrophysiological, pharmacological, and morphological evidence was adduced to specifically define the sites of these interactions; striate, substantia nigra, globus pallidus, and thalamus caudate are among these sites. Roth and Bunney (1976), Van der Zee and Luiten (1999), Bartholini (1980), Meldrum (1982), and others proposed the existence of complex pre- and postsynaptic interplay in the striate among the dopaminergic, cholinergic, and GABAergic neurons. Similarly complicated circuitry that may also include glutamate was proposed for the hippocampus (Costa, 1979; Costa, 1989; Van der Zee and Luiten, 1999). An interplay including several catecholaminergic, serotonergic, and aminoergic synapses was described for the dorsal tegmentum, basal forebrain, corticostriatal pathways, and pathways linking the locus coeruleus, raphe nuclei, and tegmentum (McGeer et al., 1977; Lloyd, 1976, 1978; Euvrard et al., 1977, 1979; Hobson et al., 1993; Scatton and Fage, 1986; Karczmar, 1986; Zaborszky et al., 1991). Finally, involvement of peptides in these interactions was also demonstrated (Pepeu, 1974; Harrison and Henderson, 1999). The phenomena in question may involve either pre- or postsynaptic interactions.

a. Presynaptic Noncholinergic Pharmacology

General anesthetics and depressants including ether, cyclopropane, halothane, and barbiturates as well as local anesthetics such as lidocaine (cf. Curtis and Crawford, 1969; see, however, Bazil and Minneman, 1989) decreased *in vivo* and *in vitro* hippocampal, subcortical, and cortical release of ACh (Pepeu, 1974, 1993; Inagawa et al., 2004). The extent of depression depended on the level of anesthesia and local anesthetic concentration (MacIntosh and Oborin, 1953; Mitchell, 1963). This attenuation of release was also observed with respect to brain slices. A related depressant, alcohol, exerted similar effects on ACh release (Pepeu, 1974).

In vivo, this inhibition of release was attenuated following midpontine transection; Giancarlo Pepeu (1974) suggested that the decrease was due to the "depression by the anesthetics of the cholinergic fibers originating from or controlled by the reticular formation." However, the effect may depend on the anesthetic's actions on cellular metabolism (Krnjevic, 1974a, 1974b).

Generally, the same effect is obtained with tranquilizers and opiates (Pepeu, 1974). Morphine and its congeners are particularly interesting. Crossland (1971) predicted that morphine should decrease the central release of ACh, as he found that the level of "bound" ACh was increased and that of "free" ACh was decreased by acute administration of morphine. Indeed, Crossland's prediction was true: acute (single) administration of morphine and other opiates distinctly reduced the release at several brain sites including the striate (Schoffelmeer et al., 1992). His notion is also consistent with the diminution by acute morphine administration of ACh turnover in several brain parts. This effect could also be observed in freely moving rodents (Pepeu, 1974) and in slices. The opiate-induced diminution of ACh release was antagonized by naloxone and other opiate receptor antagonists, whether *in vivo*, in slices obtained from several areas of the rodent brain including the cortex, the hippocampus, the striate, or in synaptosomal preparations (Lapchak et al., 1989a; Salto et al., 1990). Depending on the brain site, activation of Mu or delta morphine receptor subtypes was effective in attenuating ACh release and turnover (Schoffelmeer et al., 1992). However, the diminution by the opioids of ACh turnover was not obtained consistently. It should be added that kappa receptors are involved in blocking ACh release at the autonomic cholinergic nerve endings (Arenas et al., 1990; Ahmed et al., 1989).

During addiction, the effect of morphine on the release of ACh is subject to tolerance, while during withdrawal or following the administration of morphine antagonists the release is dramatically augmented (Casamenti et al., 1980; Jhamandas et al., 1971; Jhamandas and Sutak, 1976). Also, ACh pool and levels return to normal upon chronic morphinization (see section BV).

Several other drug types including tranquilizers attenuate the release of ACh. These effects frequently involve their transmitter actions (i.e., catecholaminergic antagonists like clonidine; see, for example, Zoltay and Cooper, 1993) and cofactors involved in monoamine metabolism (Koshimura et al., 1992). ACh release is also modulated by the action of various drugs on noncholinergic transmitters; thus, catecholaminergic blocking effects of some tranquilizers and catecholaminergic antagonists such as clonidine, as well as cofactors of monoamine metabolism, affect ACh release (Koshimura et al., 1992; Zoltay and Cooper, 1993).

Amphetamine and several convulsants, including both medullary and spinal stimulants, augment the release of ACh (Pepeu, 1974; Pepeu and Spignoli, 1989; Spignoli and Pepeu, 1987). This effect was observed with respect to a number of brain parts and was obtained both *in vivo* and *in vitro* (Pepeu, 1974; Bashkatova et al., 1999; Vizi, 1979). That the augmentation of ACh release by the drugs in question occurs *in vitro* suggests that this effect may not necessarily be due to the activation of central cholinergic neurons via the thalamic or mesodiencephalic pathways.

A number of other substances facilitate ACh release from the cortex and the hippocampus. Among these are derivatives of aminopyridines and related aminoacridines, pyrrolidinones, and phenylindolines (Foldes et al., 1989; Meyer and Otero, 1989; Nordberg et al., 1989a, 1989b; Buyukuysal et al., 1995; Giurgea, 1972; Cook et al., 1990; Borjesson et al., 1999). These compounds may increase ACh release via the block of K current and/or mobilization of Ca^{2+}, similar to atropinics. For example, aminopyridines act similarly to atropinics on presynaptic muscarinic receptors that are probably of M2 type (Nordberg et al., 1989a). However, Ca^{2+} overload induced by such conditions as epileptiform kindling may attenuate the release of ACh and may be reversed by Ca^{2+} channel blockers (Serra et al., 1999). Altogether, the mechanisms involved in ACh release effect for many of these compounds (i.e., pyrrolidinones and phenylindolines) are unclear.

Some compounds that promote ACh release including phenylindoline, linopirdine, and pyrrolidinones antagonize scopolamine- or atropine-induced amnesia and improve learning in animals and possibly humans (Borjesson et al., 1999; Pepeu and Spignoli, 1989). It was proposed that pyrrolidinones improve cognition because they show a special "affinity" or "tropism" for "integrative" action on cognition, memory, and learning (Giurgea, 1972; Giurgea and Salama, 1977); they are referred to as nootropic. Other factors, besides an "integrative action"—a rather vague notion—may be involved. The amnesic effect of atropinics is due to their postsynaptic action (see section BV, below); accordingly, the antagonism between atropinics and nootropics probably involves the presynaptic site where the pyrrolidinones may synergize with atropinics with regard to the presynaptic facilitatory effect on ACh release. The pertinent evidence seems to be incomplete (Pepeu et al, 1990; Pepeu and Spignoli, 1989). Finally, Li^{+}, a drug used in the treatment of depression, may also facilitate central release of ACh via its action on ACh synthesis and second messengers (Uney and Marchbanks, 1981; Jope, 1979). And, several studies (see, for example, Steingart et al., 1998) reported changes in pre- and post-synaptic sites following application of opiates and barbiturates.

b. Presynaptic Transmitter Interactions

Presynaptic transmitter interactions involve cholinergic effects on noncholinergic nerve terminals (i.e., an interaction that involves heteroreceptors; see section BI-1, above) and the converse effects, the activation of noncholinergic terminals on cholinergic presynaptic sites. The interaction is widespread because cholinergic pathways are ubiquitous and overlap or abut other transmitter pathways.

Good evidence confirms heteroreceptor facilitation of catecholamine release. For example, activation of the cholinergic neurons or application of muscarinic cholinomimetics potentiates the release of dopamine and norepinephrine from the striate, hippocampus, nucleus accumbens, and/or substantia nigra. Also, the cholinergic system may mediate facilitatory effects of dopamine receptor antagonists on dopamine release (Lehmann and Langer, 1982; Marchi and Raiteri, 1985; Marchi et al., 1987; Koda et al., 1989; Zaborszky et al., 1983; James and Massey, 1978; Vizi and Kiss, 1998; Yamada et al., 1999; Miller and Blaha, 2004). Similarly, nicotinics facilitate the release of catecholamines (including dopamine) from the hippocampus and other sites (Vizi and Kiss, 1998; Wonnacott et al., 2005). However, some information is not consistent with these results. Muscarinics decreased via their presynaptic actions, excitatory potentials elicited in the midbrain and the mesencephalon dopaminergic neurons, and the release of catecholamines from the caudate and other sites (Kemel et al., 1989; Xu et al., 1989; Vizi et al., 1989a, 1989b; Grillner et al., 1999). Additionally, some antiparkinsonian atropinics increased catecholamine release (Jackisch et al., 1993). Perhaps the effect of muscarinics on the release of catecholamines depends on the muscarinic receptor subtype: while M2 and M3 receptors are involved in facilitating the release, the activation of the M4 receptor may inhibit the release.

While activation of muscarinic M1 receptors generally blocked serotonin release from several brain parts including the caudate, nicotinics facilitated its release from the superior colliculus (Wichman et al., 1988, 1989). That indeed the cholinergic or cholinomimetic effect on the release of serotonin was localized directly at the serotonergic nerve terminals was demonstrated electrophysiologically for the spinal cord (Jiang and Dun, 1986; see also section BI-1, above). However, altogether the effects of the cholinergic system or cholinergic drugs on the central release of serotonin were inconsistent (Vizi and Kiss, 1998).

There is extensive information regarding the cholinergic control of the release of vasopressin. This control was established functionally (with respect to diuresis) in the 1930s by Mary Pickford (see section BIV of this chapter); it was subsequently demonstrated that this effect occurs at humoral terminals of supraoptic and paraventricular nuclei. This effect is nicotinically mediated at supraoptic nuclei and muscarinically at paraventricular nuclei (Ota et al., 1992). Prostaglandin may also be involved in these phenomena: Inoue at al. (1991) found that blockers of prostaglandin activation or synthesis depress the muscarinic release of vasopressin. Finally, activating the cholinergic terminals and applying cholinergic agonists facilitates the release of peptides such as thyrotropin-releasing hormone (TRH) in the retina (Mitsuma et al., 1992).

Cholinergic heteroreceptors are also involved in regulating the release of GABA, glutamate, and aspartate. Immunoreactivity studies of GABAergic and cholinergic neurons suggest that cholinergic neurons of the limbic system and the neocortex of several species abut on the dendrites, the soma, or the terminals of GABAergic cells; the responses appear to be muscarinic in nature (Beaulieu and Somogyi, 1991; Somogyi, 1988; Van der Zee and Luiten, 1999). Furthermore, electrophysiological and direct evidence indicated that in the hippocampus and amygdala the activation of cholinergic neurons might block presynaptically the release of GABA, glutamate, and aspartate (Gonzales et al., 1993; Krnjevic et al., 1980; Sugita et al., 1991; Vizi and Kiss, 1998). This effect may be mediated by M2, M3, or M4 receptors (Vizi and Kiss, 1998; Sugita et al., 1991). Also, the activation of nicotinic presynaptic heteroreceptors inhibits the release of glutamate (Rovira et al., 1983; Vizi and Kiss, 1998; see also Krnjevic, 1993).

However, increasing the release of glutamate, GABA, and other amino acids by activation of nicotinic and muscarinic presynaptic receptors was also reported (Tung et al., 1989; Russo et al., 1993; Toth et al., 1993; Radcliffe et al., 1999; Tredway et al., 1999; Hu et al., 1999). Albuquerque and his associates (Albuquerque et al., 1998a, 1998b; Alkondon et al., 1999) found that the activation of preterminal nicotinic heteroreceptors of either $\alpha7$ and $\alpha4$ $\beta2$ subtype facilitates endogenous (driven by glutaminergic neurons) GABA release, while it blocks evoked GABA release (Figures 9-17 and 9-18).

It must be remembered that, in the case of cholinergic autoreceptors, muscarinics and muscarinic actions of acetylcholine generally exert an inhibitory action on the release of ACh (see section BI-1, above). Accordingly, facilitatory actions of ACh on non-ACh transmitter release present a conceptual difficulty; especially if ionic and second messenger phenomena involved in facilitatory (in the case of heteroreceptors) and blocking

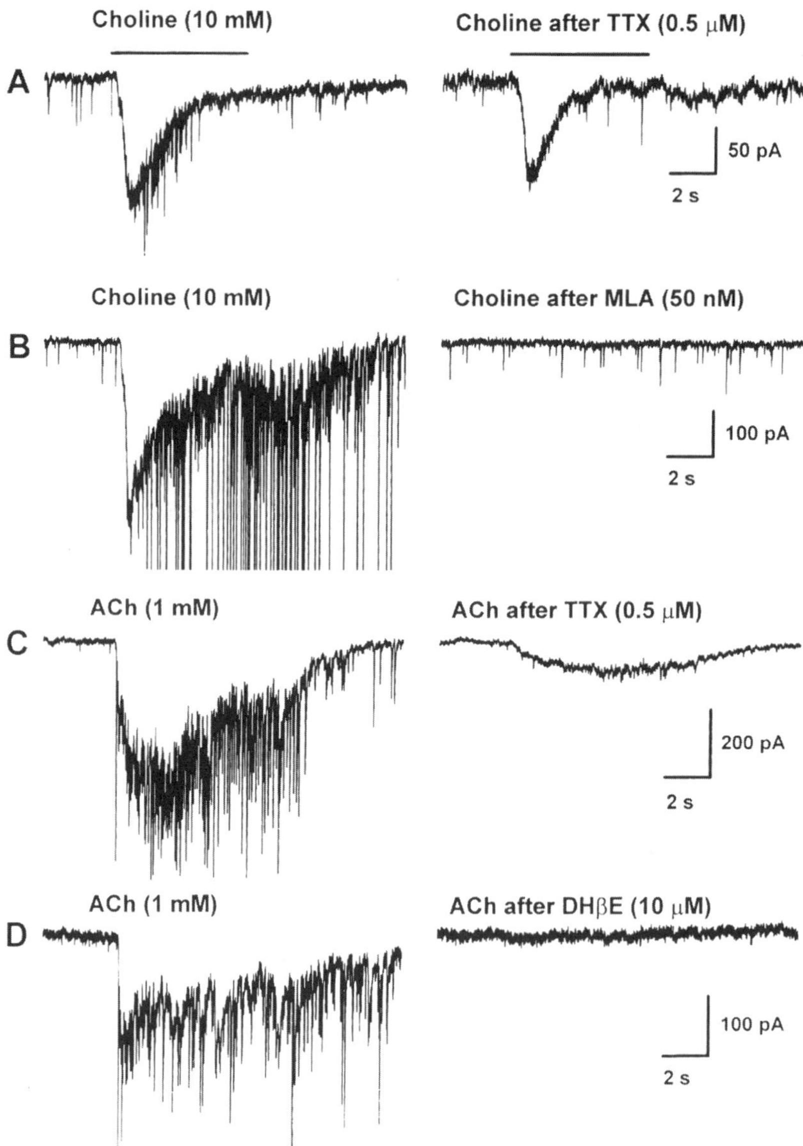

Figure 9-17. Nicotinic responses of CA1 interneurons. Responses recorded from 4 interneurons (labeled A to D) are shown. In A, choline induced a rapidly decaying whole-cell inward current accompanied by postsynaptic currents (PSCs). Superfusion of the neurons with tetrodotoxin (TTX, 0.5 μM) abolished choline-induced PSCs, but not the nicotinic current. In B, choline evoked nicotinic inward current accompanied by PSCs and fast current transients. The fast current transients are purposely clipped to unmask the other components. Superfusion with methyllycaconitine (MLA) 50 (50 nM) abolished all 3 components of the choline-evoked response. In C, ACh induced a response that consists mainly of PSCs. Superfusion of this neuron with TTX-containing solution blocked most of the response, leaving a small, slowly decaying nicotinic current, which was due to the activation of ACh receptors on the neuron from which the recordings were obtained. In D, ACh induced a similar response as in C, and this response was completely blocked by dihydro-β-erythroidine (DHβE; 10 μM). The duration of the agonist pulse is indicated by the solid line above the top traces. The interneurons studied were located at 100 μM (A), 150 μM (B and C), and 125 μM (D) from the midline of the pyramidal cell layer. Age of rats = 24 days (A, C, D) and 19 days (B). In all experiments the membrane potential was held at −60 mV (from Albuquerque, 1998a, with permission.).

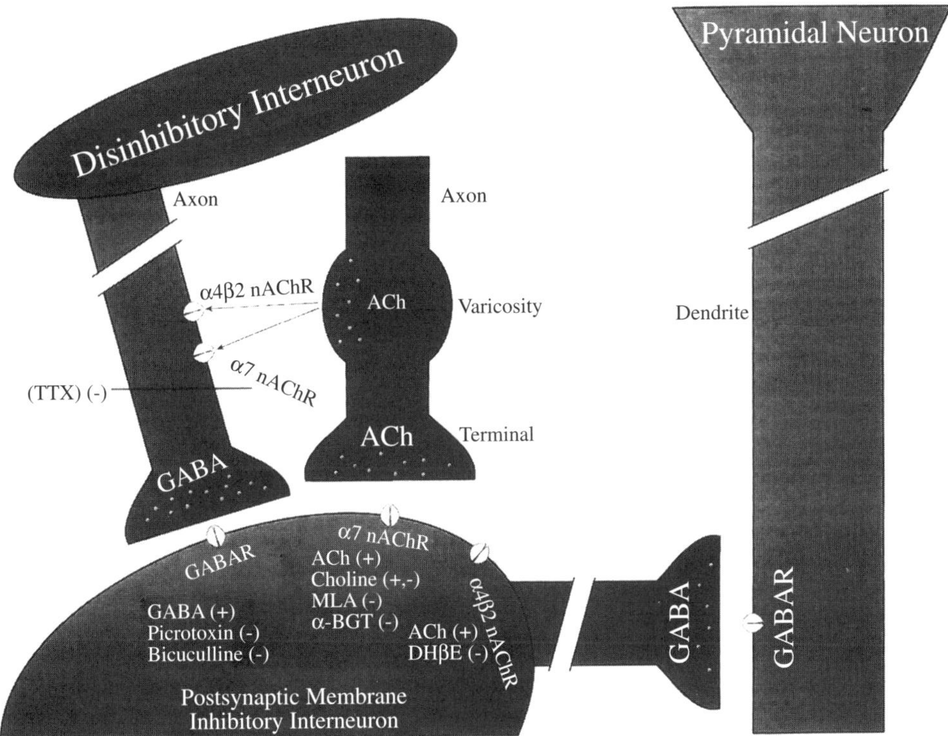

Figure 9-18. (See Figure 9-17 for abbreviations.) Scheme showing the possible sites of expression of functional α7 and α4β2 nAChRs in CA1 interneurons. The cholinergic axon is likely to come from the septum, and GABAergic axons are likely to originate from other interneurons in different layers of the CA1 region. As depicted in the far left side of the scheme, activation of α7 and α4β2 nAChRs located in presynaptic neurons can facilitate the TTX-sensitive release of GABA. Activation by GABA of $GABA_A$ receptors present in the interneuron being studied would result in elicitation of GABA-mediated PSCs, which are sensitive to blockade by $GABA_A$-receptor antagonists. The nicotinic currents recorded from the interneuron under study would have been the consequence of the activation of nAChRs present in that neuron. Currents resulting from activation of α7 nAChRs would decay within the agonist pulse and would be equally sensitive to choline and ACh as agonists and to MLA and α-BGT as antagonists. On the other hand, currents resulting from activation of α4β2 nAChRs would last longer than the agonist pulse, would be sensitive to ACh, but not to choline as an agonist, and would be blocked selectively by DHα4βE. By inference, as depicted in the far right side of the scheme, it is likely that modulation by nAChRs of the ongoing activity in the interneurons ultimately controls the excitability of the pyramidal neurons in the CA1 region. In this model, the sign (–) indicates that a given compound is capable of blocking the receptor activity, and the sign (+) indicates that the compound can activate a given receptor (from Albuquerque et al., 1998a, with permission.).

(in the case of autoreceptors) cholinergic nerve terminal actions are similar in both phenomena. However, at certain sites ACh facilitated ACh release (section BI-1). The interneuronal, indirect nature of ACh's facilitatory effect on the release of non-ACh transmitters or the nicotinic receptors' involvement in the facilitatory effect may offer a way out.

The converse matter concerns the effect of non-ACh transmitters (i.e., polypeptides, catecholamines, amino acids, transmitter precursors, and modulators) on ACh release; this effect was shown in the 1960s by Pepeu and his associates with respect to L-dopa and substance P (Pepeu, 1974). Strong evidence in this area concerns purinergic transmission: adenosine, a central neurotransmitter or modulator, inhibits *in vitro* the release of ACh from several brain structures including the cortex and the hippocampus (Dunwiddie, 1984; Scholz and Miller, 1992; Pedata et al., 1983, 1986, 1988; Nikbakht and Stone, 1949; see also Snyder, 1986). Purines, however, may affect two kinds of cholinergic nerve terminal receptors: when activated, one type of these terminals may facilitate ACh release (Ribeiro, 1996). Related purinelike compounds, methylxanthines (i.e., aminophylline and caffeine), exhibit biphasic effects of ACh release from cortical slices and antagonize the inhibitory effect of adenosine on the release (Pedata et al., 1986). The effect of adenosine may depend on both the voltage-dependent Ca^{2+} channels and sites not related to Ca^{2+} entry (Hunt and Silinsky, 1993). It is important to stress that adenosine depresses the release of not only ACh but also other transmitters, including glutamate (Schubert and Mitzdorf, 1979; Forghani and Krnjevic, 1995). Altogether, these presynaptic actions of adenosine underlie its block of LTP (Forghani and Krnjevic, 1995).

Several monoamine transmitters antagonize the release of ACh from a number of central structures, including the striate, the forebrain, the nucleus accumbens, and the hippocampus. This phenomenon is due to the topographic relationship between pertinent monoamine and cholinergic pathways in areas like the forebrain and the hippocampus (Kolasa et al., 1995; Zaborszky et al., 1991, 1999). Accordingly, serotonin and its agonists and dopamine and its agonists antagonize the release of ACh (particularly as they act on the D2 receptors); the results concerning the effects of $\alpha 1$ and $\alpha 2$ adrenoreceptor agonists on ACh release are inconsistent, and it is possible that other dopaminergic nerve terminal receptors such as D1 and sigma receptors may, to the contrary, facilitate ACh release (Stoof et al., 1987; Maura et al., 1989; Jackson et al., 1988, 1993; Carder et al., 1989; Drukarch et al., 1989; Tanganelli et al., 1989; Compton and Johnson, 1989; Vizi, 1979, 1980; Vizi et al., 1989a, 1989b; Consolo et al., 1992; Matsuno et al., 1993; Hersi et al., 1995, 2000; Obata et al., 2005). Interestingly, evidence concerning serotonin could be duplicated in the primates by noninvasive positron emission tomography (the PET scan; Dewey et al., 1993).

Amino acids also exert presynaptic effects on the release of ACh. A complex case is that of GABA and GABA agonists such as baclofen. Baclofen is clinically effective as an antispastic and as muscle relaxant, and it is used in the treatment of alcohol withdrawal; also, it depresses spinal reflexes in animals. Baclofen was primarily studied with regard to the spinal pathways and the retina; it depressed ACh release at both sites (Nistri, 1975; Massey and Neal, 1979). Additionally, it inhibits the release of other transmitters including excitatory amino acids and noradrenaline (Nistri, 1975; Dewey et al., 1993; see also Wang and Dun, 1989; Fox et al., 1978). This action may be directed at the presynaptic GABA-B receptors (Dutar and Nicoll, 1988; Wang and Dun, 1989; Lin and Dun, 1998), which, in contradistinction to the postsynaptic GABA-A receptors, are bicucullin-insensitive. It is not clear whether or not this inhibition depends on K^+ and Ca^{2+} channels (Lin and Dun, 1998). This effect of activation of GABA-B receptors on the release of several transmitters is interesting because of its, transmitterwise, nondiscriminatory nature. It appears to be homologous to the effect of GABA and GABA agonists at the neuromyal junction and in the ganglia.[3] On the other hand, excitatory amino acids such glutamate, NMDA, and related non-NMDA facilitate ACh release (for example, in the striate and hippocampus; Wood et al., 1987; Henselmans et al., 1991; Giovannini et al., 1995, 1998a, 1998b).

The early work with peptides such as substance P (Pepeu, 1974) indicates that they may increase ACh efflux in the cerveau isolé. *In vitro*, this effect was duplicated for several brain preparations including the striate; the substance P-evoked ACh release may be mediated by other peptides such as tachykinins (Arenas et al., 1991; Steinberg et al., 1995).

Several peptide hormones and related substances exert inhibitory central presynaptic actions on ACh release. These compounds include TRH, angiotensin II, endorphins and enkephalins, vasopressin and oxytocin, cholecystokinin (CCK) and its analog cerulein (or caerulein), and galanin. These substances have a wide range of distribution in the CNS. While concentrated in the hypothalamus, they are also present in the limbic system, cortex, and brainstem; they seem to overlap at most of these sites (Beinfeld and Korchak, 1985; Hoffman et al., 1982; Saito et al., 1981; Hruby et al., 1990; Zaborszky et al., 1991; McGeer and McGeer, 1993; Harrison and Henderson, 1999). The wide distribution of these substances, their presence in specialized synaptic vesicles, and their release via depolarization support the notion that aside from their hormonal actions, these substances also act as neurotransmitters and modulators (Johansson et al., 1987; Simasko and Horita, 1982; Mendez et al., 1987). Interestingly, several of the peptides coexist with ACh in the neurons. In the case of galanin, this occurs in the septohippocampal and diagonal band pathways (Lamour and Epelbaum, 1988; Lamour et al., 1988; Vogels et al., 1989).

When given systemically, via local central infusion or via perfusion of brain slices, TRH, somatostatin, CCK, and cerulein increase or evoke ACh release from the cortex, the hippocampus, and the striate (Pepeu, 1993; Giovannini et al., 1990; Shimoyama and Kito, 1989; Arneric and Reis, 1986; Magnani et al., 1987). Giancarlo Pepeu and his associates found that TRH-evoked release of ACh is weak; they emphasized that the density of TRH receptors in the striate is low (Giovannini et al., 1990; Pepeu et al., 1990; see also Ogawa and Kurota, 1983; Itoh et al., 1994). TRH-mediated release of ACh is consistent with the finding that TRH increases the turnover of ACh in the hippocampus and in the cortex (Malthe-Sorenssen et al., 1978; Brunello and Cheney, 1981) and that TRH-evoked release of ACh is blocked by tetrodotoxin (Giovannini et al., 1990; Toide et al., 1993). While the releasing effect of TRH on ACh is well founded, the effect may not always be mediated via TRH action on cholinergic nerve terminals but via its effect on GABAergic interneurons (as in the case of the septum; Brunello and Cheney, 1981). Similarly, an afferent circuitry may be involved in the cerulein-evoked facilitation of ACh release (Shimoyama and Kito, 1989). These data are of interest in the context of the clinical, stimulant, and pro-memory actions of TRH.

On the other hand, angiotensin II (but not angiotensin I) and galanin inhibited ACh release from slices of the rat entorhinal cortex and hippocampal slices, and *in vivo* from several brain sites (Consolo et al., 1991). In the case of angiotensin, this effect was directed at specific angiotensin II receptors (Barnes et al., 1989). In addition to biochemical evidence, there is also electrophysiological evidence regarding the central, presynaptic effect of galanine on excitatory cholinergic responses in the hippocampus (Dutar et al., 1989).

It is noteworthy that additional peptides (for example, bombesin and neuromedins) are present in the CNS (McGeer et al., 1987; McGeer and McGeer, 1993; Karczmar et al., 1986). Their nerve terminal localization is not certain and there is only meager information regarding their central presynaptic actions. Finally, the actual circuitry involved in the presynaptic effects of the peptides has not been definitively established (McGeer et al., 1987; McGeer and McGeer, 1993; Vogels et al., 1989).

c. Postsynaptic Noncholinergic Pharmacology

Let it be stressed that excluding the effect of antiChEs on the responses to ACh and to hydrolyzable cholinomimetics (see section BIII-1b, above), it is not easy pharmacologically to augment central cholinomimetic or cholinergic responses; this is also the situation with the ganglia and neuromyal junction. There are a few compounds (sensitizers or modulators) that have the ability to achieve augmentation at the periphery (see sections BIII-2a and BIII-2c), and some of these exhibit similar properties centrally; they include NaF, and NaF is also active centrally, as it increased and prolonged the discharge evoked by the neural stimulation of the Renshaw cell (Koketsu, 1966, 1984; Figure 9-19) and the fast potential response of hippocampal neurons to ACh (A. G. Karczmar and K. Kim, unpublished data).

It is much easier to depress postsynaptic responses. Since the late 1950s, it has been known that general and local anesthetics are capable of this action. For example, procaine reduced the sensitivity of the Renshaw cell to ACh and to synaptic stimulation (Curtis and Phillis, 1960). Lidocaine exerted similar effects on the hippocampus's responses to ACh (Segal, 1983, 1988a, 1988b). Curtis, Phillis, Crawford, and Krnjevic (see, for example, Curtis and Crawford, 1969; Krnjevic, 1972, 1974a, 1974b) opined that these depressant effects caused by local anesthetics, halogenated and nonhalogenated anesthetics, and barbiturates were nonspecific since these drugs also depressed amino acid-induced excitation of the Renshaw cell and the thalamic and cortical neurons, although only relatively high concentrations of the anesthetics suppress cortical actions of glutamate (Curtis and Phillis, 1960; Curtis and Crawford, 1969; Krnjevic, 1974a, 1974b).

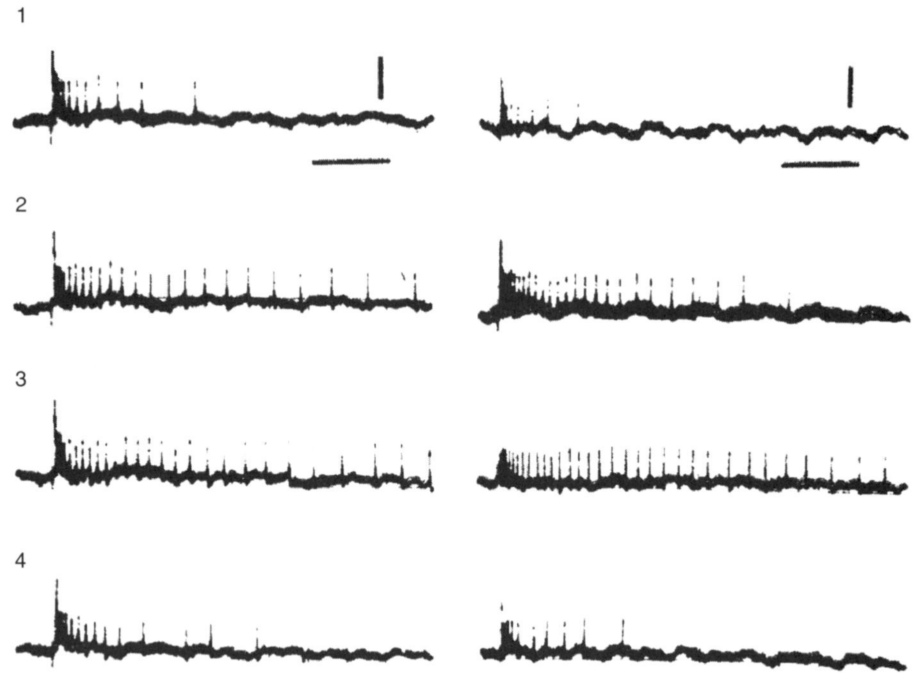

Figure 9-19. Effects of tetraethyl pyrophosphate (TEPP; 0.005–0.1 mg/kg) and NaF (20 mg/kg) on discharges of Renshaw cell of cat's spinal cord. Records 1 and 2 were obtained before and 20 minutes after TEPP was injected intravenously. NaF was given intravenously immediately after record 2 was taken; records 3 and 4 were obtained 90 seconds and 3 minutes after NaF administration. Note increased frequency after TEPP (record 2), and also a transient further increase in frequency after NaF (record 3), which is followed by a marked decrease (record 4). Records with slower sweep speed are shown in the left-hand column. Calibration: 1 mV (right- and left-hand columns); 20 milliseconds (right-hand column) and 40 milliseconds (left-hand column). (From Koketsu, 1966, with perimssion.)

Krnjevic (1974a, 1974b) proposed that the action of local and general anesthetics and barbiturates on neurally evoked responses (for example, responses generated by stimulation of the reticular formation) or reactions to cholinomimetics depend on the depressant metabolic effect of anesthetic agents rather than on specific synaptic effects and specific pre- or postsynaptic receptors. He suggested that this action is due to inhibition by anesthetics of the metabolic, ATP-dependent activity of mitochondria, the resulting attenuation of Ca^{2+} uptake, and the accumulation of intracellular free Ca^{2+} (Carafoli and Rossi, 1971; Rosenberg and Haugaard, 1973; Krnjevic and Phillis, 1963; Krnjevic, 1974a, 1974b). Furthermore, Krnjevic links the free calcium with the activation of K channels and the increase in K conductance. The muscarinic response to ACh (the slow EPSP) is partially due to the K current's inactivation (M current; see section BI-2a); hence, the increase in K conductance caused by anesthetics and barbiturates depresses the slow ACh depolarization. Then, Krnjevic and his associates found that brain anoxia or hypoxia causes a similar increase in cellular free Ca^{2+} and K+ conductance, resulting in the diminution of the slow ACh depolarization (Krnjevic, 1990, 1993; Krnjevic and Leblond, 1989).

As shown by Chris Krnjevic and his associates, the effects of dinitrophenol (DNP) relate to those of anesthetics and barbiturates. Dinitrophenol and the depressants specifically depress the slow ACh potentials without affecting the depolarizing action of glutamate (Krnjevic, 1974b; Krnjevic et al., 1978). He associated this similarity with effects of DNP and anesthetics on mitochondria-dependent K fluxes.

There are certain weaknesses in Krnjevic's notions. First, the action of anesthetics on the muscarinic response is not specific (see above, and Curtis and Crawford, 1969). Second, it may be inappropriate to compare the muscarinic response to ACh and the response to glutamate: glutamate's reaction is fast and immediate and resembles the nicotinic more than the muscarinic response. Finally, small doses of general anesthetics and barbiturates that probably could not affect mitochondrial metabolism readily depressed responses to ACh (Crawford and Curtis, 1966).

Only sporadic information is available with respect to the postsynaptic effect of several other drugs (i.e., benzodiazepines, antidepressants, and tranquilizers) on cholinergic responses (O'Regan, 1989; Bassant et al., 1988). These compounds exert inhibitory effects on ACh-evoked muscarinic excitations in septal and cortical neurons; this block may depend on their atropinic activity (Bassant et al., 1988). Finally, it appears that cholinergic, specific and semispecific neurotoxins and excitotoxins (i.e., AF64A, kainic, quinolinic, and ibotenic acids, oxadiazines and related insecticides, and NMDA) depress postsynaptic nicotinic and muscarinic responses; the analysis is incomplete (Zhao et al., 2001).

d. Postsynaptic Neurotransmitter Interaction

Similar to transmitter interaction at the nerve terminals, the postsynaptic interaction is widespread, as it is predicated on the ubiquitous contacts among neurotransmitter pathways. Indeed, in 1948 Sir William Feldberg and Martha Vogt proposed that central cholinergic transmission alternates with noncholinergic transmissions throughout the brain.

More precise evidence for postsynaptic transmitter interaction consisted of the demonstration of the presence of cholinoceptive (muscarinic and nicotinic; see Aznar et al., 2005) receptors on noncholinergic soma or dendrites. This kind of evidence became available in early research for the striatum, first with regard to catecholaminergic-cholinergic interactions and then with regard to polypeptide-cholinergic and amino acid-cholinergic interactions (see Roth and Bunney, 1976; Lee et al., 1997; and Dougherty and Milner, 1999). Similar evidence was obtained subsequently with respect to limbic and tegmental systems and locus ceruleus (Brown et al., 1990; Chen and Engberg, 1989; Crochet and Sakai, 1999; Garzon et al., 1999; Van der Zee and Luiten, 1999); for example, α7 nicotinic receptor subunits are present on serotonin hippocampal and septal neurons (on either their soma or nerve terminals), and nicotine enhances serotonin release (Aznar et al., 2005). Classically, the neurotransmitter interaction is reflected in the vectoral interplay between the postsynaptic potentials, whether inhibitory or excitatory, evoked by the pertinent transmitters (Eccles et al., 1967).

An important example of such interaction concerns facilitatory muscarinic actions of ACh on postsynaptic responses to other transmitters. The evidence for this notion was obtained in early research by Chatfield and Dempsey (1942), as they demonstrated that responses of cortical neurons to single stimulation of the thalamus are facilitated and converted into repetitive discharges by ACh infusion. Then, Chris Krnjevic (1969) postulated that the central muscarinic depolarizing action of ACh (and the evocation of the so-called plateau potentials; Kawasaki et al., 1999) is modulatory and sensitizing rather than synaptic in character. This notion was repeatedly confirmed with respect to cortical and hippocampal neurons and other central structures (see Krnjevic, 1969; 1988, and Puil and Krnjevic, 1978).

However, the facilitatory effect of ACh may not be due to its direct modulatory postsynaptic action, but to ACh-generated release of glutamate (Kawasaki et al., 1999). This phenomenon occurs at several sites including essentially noncholinergic places, such as the cerebellum, and it may happen both pre- and postsynaptically; both nicotinic and muscarinic receptors may be involved (Reno et al., 2004; De Filippi et al., 2004). Also, cholinergic stimulation of the forebrain and the concurrent release of ACh sensitize the somatosensory neurons to glutamate (Metherate et al., 1988a, 1988b) and, *in vitro*, ACh augments central excitotoxic effects of glutamate (Mattson, 1989). Anatomical evidence supports this notion; for example, postsynaptic M2 receptors abut glutaminergic neurons in the rat neocortex (Mrzljak et al., 1993).

The same facilitatory interaction occurs between ACh and polypeptides such as TRH and VIP at several brain sites. Thus, TRH and ACh act synergistically at sites such as the diagonal band and septohippocampal neurons, while actions of VIP and ACh are synergistic at the neurons of the visual cortex (Murphy et al., 1993; Lamour et al., 1988; Angelucci et al., 1990; Okada, 1991; Dutar et al., 1985); these sites abound in cholinoceptive neurons. These interactions are important for the generation of theta rhythms, EEG arousal, and visual acuity or sensitivity (see below, sections BIV-3 and BV-C).

The vectoral interaction among cholinoceptive excitatory potentials and excitatory potentials evoked by other transmitters also occurs in the case of cholinoceptive excitatory potentials and inhibitory effects of the noncholinergic transmitters. Acetylcholine reduced IPSPs evoked in the hippocampal CA1 cells by fimbrial stimulation, ACh and carbachol attenuated the current evoked by GABA agonists such as baclofen and OP antiChEs blocked certain GABA channels (Gant et al., 1987); these effects of ACh are disinhibitory in nature (Muller and Misgeld, 1989; Krnjevic, 1981; Krnjevic et al., 1980). However, Krnjevic (1981) as well as Muller and Misgeld (1989) suggested that the cholinergic disinhibition of the inhibitory responses of CA1 cells may be mediated by the presynaptic rather than postsynaptic action of ACh. Also, ACh did not seem to antagonize IPSPs evoked by iontophoretic application of GABA (Krnjevic, 1981; Muller and Misgeld, 1989; Muller et al., 1989).

Postsynaptic nicotinic effects may be also involved in disinhibitions; however, contrary to muscarinics, nicotine may synergize with the IPSPs. This was demonstrated indirectly by Simmonds (1982), who found that nicotinic antagonists blocked GABA-evoked IPSPs, and by Schwartz and Mindlin (1988) who demonstrated that nicotinic antagonists and nicotinic neurotoxins block chloride influx induced by GABA; this influx underlies the GABAergic IPSP. These findings may relate to the data indicating that amino acid sequences of certain subunits of the nicotinic and GABAergic postsynaptic receptors are homologous (Schofield et al., 1987; Grenningloh et al., 1987). This phenomenon may have a bearing on the presence of GABAergic, glycine and cholinergic receptors on the neurons of the hippocampus, septum, and neocortex and the retinal ganglion and amacrine cells; also, GABA receptors form a close topographical relationship with CAT (Chun et al., 1988; Sergeeva, 1998; Sergeeva and Haas, 2001; some GABAergic cells exhibit nicotinic receptors; Dmitrieva et al., 2001; Albuquerque et al., 1998a; 1998b; Blaker et al., 1989); furthermore, upon their stimulation, these cells corelease GABA and ACh, and cholinergic stimulation of the retina releases GABA from the retina (Albuquerque et al., 1998a; 1998b; Van der Zee and Luiten, 1999; Dmitrieva et al., 2001). The same phenomenon occurs in the septum and fascia dentata (Leranth and Frotscher, 1989; Frotscher et al., 1992).

ACh interacts as well with other inhibitory amino acids; for example, ACh modulates the nucleus tractus solitarii's responses to glycine (Talman et al., 1991).

What about the converse phenomenon, that is, the postsynaptic effect of noncholinergic transmission on cholinoceptive responses? Both the inhibitory and excitatory amino acids, and polypeptides and monoamines, affect cholinoceptivity. For example, the glutamate and/or the glutaminergic system facilitate cholinergic, muscarinic responses, while GABAergic system inhibit cortical cholinergic neurons (Wood et al., 1984) (Metherate et al., 1988a, 1988b; Fournier et al., 2004). Then, neurotensin-evoked depolarization facilitates ACh excitation at the diagonal band (Matthews, 1999). With monoamines, inhibitory effects of catechol and indole monoamines on the spontaneous firing of somatosensory neurons translate into the inhibition of these neurons' responses to ACh (see, however, Bassant et al., 1990). ACh and monoamine systems also complexly interact at hippocampal neurons, where they may jointly control the EEG theta activity (Sainsbury and Partlo, 1991; see also below, section BIV-3).

Also, bioactive substances that are not neurotransmitters are capable of inducing changes in postsynaptic responses to ACh and cholinergic drugs by affecting postsynaptic membrane potential or by changing receptor sensitivity; these substances include peptides, immunoactive substances, hormones, and sensitizers or modulators such as NaF. Thus, it appears that both gonadal steroids and corticosteroids are capable of these actions (K. Koketsu and A. G. Karczmar, 1963, unpublished data; Crayton et al., 2004; Hesen et al., 1998; see section BV-d).

Altogether, complex interaction is the mode of CNS function. Several transmitters, modulators, and/or bioactive substances including hormones and endogenous opioids are involved in any given postsynaptic outcome (see Chiodera et al., 1993).

e. Biochemical and Related Phenomena, Consistent with the Interaction Between Cholinergic and Noncholinergic Systems

Interactions like those described above have several biochemical consequences. The demonstration that antiChEs and muscarinic agonists release catecholamines from several brain parts including locus coeruleus and cortex is consistent with this notion (Varagic and Krstic, 1966; Van Gaalen et al., 1997; see Karczmar, 1976, 1993b). Also, the converse obtains, as endogenous endorphins, noncholinergic transmitters, and other bioactive substances cause changes in ACh release (see, for example, Leanza et al., 1993; section BIII-2a, above).

Another consequence of interaction of the cholinergic agents with noncholinergic transmitters is a change in levels or turnovers of these transmitters. The Loyola of Chicago team (Glisson et al., 1972, 1974; Barnes et al., 1974; see also Karczmar, 1976), Anden (1974), and Haubrich and Reid (1972) showed that OP antiChEs, physostigmine, and muscarinic agonists increased dopamine, norepinephrine, and/or serotonin brain levels; inconsistent (decreases and increases) effects of newer antiChEs were described by others (Cuadra and Giacobini, 1995; Zhu et al., 1996). Furthermore, while the turnover of serotonin was increased by arecoline or pilocarpine in the rat brain and several brain regions such as the pineal gland, pilocarpine and carbachol decreased dopamine turnover in the substantia nigra (Finocchiaro et al., 1989; Haubrich and Reid, 1972; Javoy et al., 1974). Also, cholinergic agonists including OP antiChEs affected levels and/or turnover of amino acid transmitters (Costa, 1989; Fosbraey et al., 1990), GABA and aspartate levels being increased and decreased, respectively. These findings are consistent with the presence of cholinoceptive receptors on GABAergic cells (see section BIII-2). Finally, the nicotinic agonists increased catecholamine synthesis rate (thus, very likely, their turnover also) in the nucleus accumbens, hypothalamus, and hippocampus, but not in the striate (Mitchell et al., 1989).

There is reciprocity in these interactions, as agonists and antagonists of noncholinergic systems and many bioactive substances affect the biochemical parameters of the cholinergic system; much of the pertinent work concerning the dopaminergic and peptidergic systems in particular was carried out by the Bethesda-Fidia team. For example, the agonists and antagonists of the catecholaminergic systems (noradrenergic and dopaminergic) and certain peptides and agonists of the peptidergic systems (i.e., morphine or endorphins) affect ACh turnover (Costa, 1979; Brunello and Cheney, 1981). Thus, haloperidol increased ACh turnover in the caudate and in the nucleus accumbens, and decreased it in the globus pallidus, while apomorphine, bromocriptine, dopamine, and L-dopa decreased ACh turnover in the striate and in the hippocampus (Costa, 1979; Cheney et al., 1975; Trabucchi et al., 1975; Mantovani et al., 1988; Langnickel et al., 1983; Lehman et al., 1983; Consolo et al., 1986). Noradrenergic agonists increased ACh turnover in the cortex and in the hippocampus (Robinson, 1986, 1988); this effect may be mediated by the α or $\alpha 1$ receptors (Wood et al., 1979, 1987). It is consistent with these results that the dopaminergic afferent system's relationship to the striate seems to regulate CAT levels (Kayadjanian et al., 1999).

Another type of this noncholinergic-cholinergic interaction consists of the effect of noncholinergic transmitters and/or neuroactive substances on cholinergic receptors. Thus, it was demonstrated that dopaminergic and glutamatergic systems regulate the muscarinic receptors of the striate (Kayadjanian et al., 1999). Furthermore, GABA or GABAergic agonists (i.e., muscimol) decreased ACh turnover in the hippocampus (Wood et al., 1979; Costa et al., 1983; Costa, 1989, 1991) and in the cortex (Wood, 1986; see, however, Cosi and Wood, 1988). This interaction between GABAergic and cholinergic cells is consistent with the topographical relationship between their receptors, and GABAergic receptors and CAT (see above, section BIII-2d). Interestingly, somatostatin releases both ACh and dopamine from the striate (Rakovska et al., 2003).

Endogenous bioactive substances that are not neurotransmitters also exert actions on cholinergic biochemistry. Endorphins and opiates decreased ACh turnover in the hippocampus, cortex, globus pallidus, and nucleus accumbens, areas that are innervated by cholinergic pathways (see Chapter 2 DII). However, ACh turnover was not decreased in other, also cholinergically innervated areas including the caudate and septal nuclei (Costa, 1979). Erminio Costa interpreted these results as indicating that at least some central cholinergic neurons exhibit opiate receptors. Cholecystokinin exerted a similar negative effect on ACh turnover in the cortex (Cosi et al., 1989).

Several components of the immune and related systems such as kinins and macroglobulins affect cholinergic biochemistry; for example, rodents' cortical infusion of macroglobulins significantly inhibited CAT activity (Hu et al., 1998).

f. Interactions and Modulations, and Their Significance

Reciprocal neurotransmitter interactions and interactions between endogenous bioactive substances and the cholinergic system result in electrophysiological, pharmacological, and biochemical phenomena and functional and behavioral activities (see below, section BIV). The pre- and post-synaptic neurotransmitter interactions could readily be expected on the basis of topographical overlap between cholinergic and noncholinergic pathways (see Chapter 2 DI-DIII and section BIII-2, above). These interactions include those between inhibitory and excitatory transmitters; in this context it must be remembered that ACh may induce inhibitions by acting on inhibitory interneurons (see section BI-2). Another phenomenon contributing to these interactions involves the simultaneous release of two (or more) transmitters from single a neuron. This phenomenon was discovered in the 1970s by Hokfelt and associates (Hokfelt et al., 1978, 1979; see also Chapter 2 B-5). And Victor Whittaker demonstrated that enriched cholinergic synaptosomes contain a peptide neurotransmitter, VIP, that is coreleased with ACh (Whittaker, 1992, 1998). More recently, the coexistence of cholinergic, amino acidergic, aminergic, and peptidergic transmission was documented for several of the circuits discussed in this chapter (see, for example, Lamour et al., 1988; Crawley, 1990; Van der Zee and Luiten, 1999).

The definitive description of the circuitries involved in these interactions is technically very difficult. It includes stimulation by receptor-specific agonists or by electric presynaptic stimulation of localized synaptic sites and biochemical and electrophysiological identification of the postsynaptic effects of this stimulation at localized, identified neurons; additional experimentation would concern the interactions occurring at identified presynaptic sites. Ablation or lesions of brain parts (such as thalamus) was also used to determine the locus of transmitter interaction (see, for example, Becquet et al., 1988). As difficult as this approach may be, several investigators utilized it to demonstrate that at several specific sites, including cortex and the limbic system, the interactions involve pre- and postsynaptic sites, several nicotinic and muscarinic receptor subtypes, glutamate and GABA, serotonin and processes of desensitization (Albuquerque et al., 1998a, 1998b; Alkondon et al., 1999; Lindstrom, 1994, 1996).

Neurotransmitter interactions must be differentiated from modulations (Karczmar et al., 1966, 1972; Koketsu, 1984). Modulators change the postsynaptic response of a transmitter (including ACh) without generating a potential that would vectorally interact with a response to the neurotransmitter. Postsynaptic modulation involves a change in receptor sensitivity leading to desensitization or sensitization. The term "modulation" also covers presynaptic phenomena resulting in the facilitation or inhibition of a transmitter release (see section BI-1 and Chapter 7 DI; see also Greengard, 1987; Katz et al., 1997). Modulators may act as transmitters under other conditions or at other sites.

Modulators include both endogenous and pharmacological agents as well as drugs such as nefiracetoms and neotropics (Nishizaki et al., 1998, 2000; see also below, section BV-d). Endogenous postsynaptic modulators active in the CNS include ATP and certain polypeptides, while drugs acting as postsynaptic modulators in the CNS include NaF and several anticholinesterases, such as the oxamide WIN 8078 (provided this bisquaternary drug is applied directly to the CNS) and galantamine (Karczmar, 1957; Karczmar et al., 1972; Maelicke et al., 2001; see sections BIII-2b and BIII-2c, above). Endogenous presynaptic modulators active in the CNS include many transmitters and bioactive substances such as monoamines, ACh, amino acids, and peptides. A multitude of drugs that presynaptically modulate transmitter release include anesthetics, opioids, and nootropics (see sections BIII-2a and BIII-2b, above). Currently it is not clear whether a common mechanism for all the modulatory phenomena exists, though it is likely that facilitatory (sensitizing) and desensitizing modulations result from the modulators' allosteric activities (see Lena and Changeux, 1998; Changeux et al., 1996; Maelicke, 1996; Maelicke et al., 2001). This is particularly true with respect to central nicotinic receptors, where, for example, the ligand galantamine acts as an allosterically potentiating ligand (APL), binding on the receptor site different from the agonist binding site (Schrattenholz et al., 1993; Albuquerque et al., 2004). Other cases of desensitizations depend on nonallosteric phosphorylations (Greengard, 1987). Finally, a novel type of modulation was proposed by Raad Nashimi on the basis of a novel methodology (the FRET technique); it consists of a "cross-talk" between nicotinic channels (see Karczmar, 2004).

The main objective here is to juxtapose the interaction and modulation phenomena with the notion that held from the 1920s to the 1950s chemical including cholinergic transmission; of this early notion projected the transmission as a "one-or-nothing phenomenon." For example, an effective nerve terminal release of a transmitter would lead to a full-fledged postsynaptic excitatory or inhibitory potential, the excitatory potential (the fast EPSP) by evoking a conductive action potential. The presynaptic phenomena were poorly recognized at that time (see section A, above).

The evidence that concerns presynaptic and postsynaptic transmitter interplay and modulations leads to a novel image. This image is that of a flexible neurotransmission and synaptic plasticity (Corringer et al., 2006); this new image, rather than showing a one-or-nothing response, represents a subtle, time wise, and spatially point-to-point stochastic response. The stochastic response is dependent on interaction and modulatory processes present all over the central nervous system; it constitutes the basis of brain plasticity and integrative phenomena. The ubiquitous cholinergic pathways are strategically situated with respect to these phenomena and are most important for this integration. For example, cholinergic input at the terminals of the thalamicocortical radiation, along with the inhibitory phenomena generated by GABA, integrate the receptive fields concerned with processing sensory information (Ebner and Armstrong-James, 1990).

An interesting question arises: how to quantify the afferent intersynaptic and presynaptic inputs, interactions, and modulations so as to predict the output of any particular neuron? Today, it appears that "a complex computer program . . . which does not yet begin to exist is required" for making such predictions (Karczmar, 1993a; Pepeu, 1993); however, it may prove ultimately that it will be beyond the quantitative or qualitative capacity of any future computer or database program to definitively link the neuronal output with the ultimate functional and behavioral outcome.

BIV. Functional Aspects of the Cholinergic System

Central cholinergic pathways are very widely distributed. They radiate to such important sites regulating function and behavior as the cortex and the hippocampus, the pontomedullary and the reticular systems, and the hypothalamus and the nigrostriatal axis (see, for example, Lewis and Shute and Chapter 2 DII and DIII). This distribution predicates cholinergic aspects of every function and behavior, and the research described in this chapter proves this assertion. It describes studies of the central control of several functions including respiration and cardiovascular activity as well as of behaviors such as learning and conditioning that were conducted during the pre-Loewi and pre-Eccles era (see section A-2, above; cf. also Karczmar, 1967, 1970, 1976; Schmitt, 1972; Holmstedt, 1972, 2000). The pace accelerated in the 1950s and 1960s with the work of Daniele Bovet (1953) and Jan Bures (Bures et al., 1964), and the rate of this acceleration has become exponential in the last decades.

This extensive material is subdivided into three categories (Tables 9-1 and 9-2). The first category involves organic central functions and regulations, including those generated in the hypothalamus (i.e., thermocontrol, control of thirst and hunger, etc.), and reflexogenic and motor activities. The second category concerns brain rhythms (i.e., EEG) and related functions (i.e., sleep). Finally, the third category includes behavioral, cognitive, and mental phenomena (for example, aggression, memory, and learning; see also Karczmar, 1975, 1990, 2000; Pepeu, 2000 and Giovannini, 2004).

Thus, cholinergic aspects of the overt motor behavior overlap with the mental phenomena; for example, slowdown and augmentation of the motor activity occur during different phases of

Table 9-1. Functional Effects of Cholinergic Drugs in Animals

- I. Motor behavior and other related neurological syndromes
  - A. Catalepsy
  - B. Locomotor and related actions: mobility, gnawing, self-biting, head motion, sniffing; compulsive circling; hypokinesia
  - C. Tremor
  - D. Convulsions
- II. Respiration
- III. Appetitive (hypothalamic) behaviors
  - A. Hunger and feeding: effect dependent on brain site and species
  - B. Thirst and drinking; effect dependent on brain site and species
- IV. Hypothalamic thermocontrol
  - A. Heat production
  - B. Heat loss
- V. Endocrine activities
- VI. Cardiovascular phenomena
- VII. GI activities
- VIII. EEG and brain excitability
  - A. EEG arousal and theta rhythms
  - B. Synchronization phenomena
  - C. Seizures
- IX. Sleep
  - A. REMS
  - B. SWS
- X. Chronobiology
  - A. Diurnal rhythms
  - B. Hibernation, seasonal changes
  - C. Aging
- XI. Sensorium
  - A. Nociception
  - B. Audition
  - C. Vision
  - D. Olfaction
- XII. Sexual activity

Table 9-2. Behavioral Psychological and "Mental" Functions in Animals

- I. Aggression
  - A. Emotional (affective)
  - B. Predatory
  - C. Irritable
- II. Learning and related phenomena
  - A. Conditioning
  - B. Memory (short and long term)
  - C. Habituation
  - D. Retrieval
  - E. Attention and exploration
- III. Emotional behavior and fear
- IV. Addiction, dependence, withdrawal syndrome
  - A. Opiate addiction
  - B. Alcoholism
  - C. Cocaine addiction
  - D. Nicotine addiction
- V. "Schizoid" behavior
- VI. Organism-environment interaction (OEI)
- VII. Cholinergic alert nonmobile behavior (CANMB)
- VIII. Consciousness
  - A. Self-awareness

mental arousal (see section BVI-a); however, these motor activities will be considered jointly with organic phenomena. Similarly, EEG activity and centrally evoked potentials (EEG alerting, for example) that are closely related to arousal and learning will be discussed as functional phenomena (see sections BIV-3a and BIV-3b).

"A certain bias of this author should be mentioned.... Having studied the cholinergic system for more than 40 years, this author has no desire to deprecate its role in organic and mental function ..." (Karczmar, 1988). It was remarked that "this quote ... should be ... appended to all articles in the future dealing with the neurochemical basis of CNS disorders, like the Surgeon-General's warning on cigarette packs!" (Fowler, 1988). However, this author also commented that "he ... was one of the first to stress that no behavior, normal or abnormal, is controlled by a single neurotransmitter" (Karczmar, 1978a). This statement rigorously applies to all the behaviors or functions described in these chapters. In fact, besides the "classical" noncholinergic neurotransmitters, such as amino acids, indoles, and catecholamines, new putative or "real" neurotransmitters are being described, and their contribution to functions and behaviors are listed, as in the case of orexins (see, for example, Hoang et al., 2004).

1. Central Functions and Organic Behaviors with Cholinergic Implications

a. Effects of Cholinergic Agents on Anesthesia, and Their Reflexogenic Action

Local and general anesthetics exert depressant postsynaptic effects on central cholinoceptive neurons (Roth and Bunney, 1976; see section BIII-2b). Therefore, it is teleological that in animals or humans, antiChEs that augment cholinergic transmission (at least, used in small doses) antagonize general anesthesia; in fact, short-acting antiChEs (edrophonium, for example) are clinically employed for that purpose (see, for example, Kubota et al., 1999).

The reflexogenic action of cholinergic drugs such as physostigmine has been known since the investigations conducted by the Edinburgh scientists during the 19th century (section A-2). Following their studies, Schweitzer and Wright (1937) and Wikler (1945) conducted investigations on peripherally recorded reflexes such as the knee jerk (see also Karczmar, 1967; Machne and Unna, 1963; and Holmstedt, 1959). Simultaneous studies concerned the presynaptic inhibitory pathway and its dorsal root consequence, the primary afferent depolarization (PAD) and its counterpart, the dorsal root potential (DRP; Schmidt, 1965; Barron and Mathews, 1938; Eccles, 1964). It was subsequently shown that cholinergic drugs affect PAD elicited by either the antidromic ventral root stimulation (the VR-DRP reflex) or stimulation of the adjacent dorsal root (DR-DRP reflex; Schmidt, 1965; Kiraly and Phillis, 1961; Koketsu et al., 1969; Figures 9-20 and 9-21).

The results concerning peripherally recorded reflexes are controversial. The enhancement and diminution of poly- or monosynaptic reflexes was obtained with cholinergic agonists, particularly antiChEs (see Holmstedt, 1959; Karczmar, 1967, 1976); this was true even with a simple monosynaptic reflex, the patellar reflex (compare Libet et al., 1960 and McNamara et al., 1954; Kissel and Domino, 1959). Anticholinesterases and ACh or cholinomimetics diminish the VR-DR reflex, and the DR-DR reflex is increased by antiChEs and decreased by nicotinic blockers (Kiraly and Phillis, 1961; Koketsu et al., 1969). It must be noted that ACh and cholinomimetics depolarize the dorsal root (Koketsu et al., 1969; see Figure 9-21). While this depolarization and the diminution of the DR potential is consistent with the notion of presynaptic depolarization by ACh (Schmidt, 1965), it may also result from ACh's effect on a cholinergic spinal neuron or an interneuron impinging on the appropriate afferents. Also, atropine exerted inconsistent effects on either facilitation or inhibition of the reflexes with cholinergic agonists (see Karczmar, 1976). Yoshioka et al. (1990) found that antiChEs and M receptor agonists diminish the rat monosynaptic reflex via evoking or potentiating the inhibition mediated by cutaneous afferents. However, oxotremorine consistently facilitated the monosynaptic flexor, crossed extensor, and polysynaptic extensor reflexes (Vaupel and Martin, 1973). This latter action agrees with the cholinergic nature of VR-DR, which strongly indicates a presence of cholinoceptive and/or cholinergic cells in the circuitry responsible for VR-DR. This is also indicated by a demonstration of postsynaptic cholinoceptivity of many ventral root cells, aside from the Renshaw interneuron (Jiang and Dun, 1987).

Several aspects of the matter may underlie this inconsistency. First, many of the experiments cited were carried out on intact animals, rather than chronic spinal preparations (see Vaupel and Martin, 1973; Yoshioka et al., 1990). This suggests that, besides the spinal, supraspinal effects may have been involved. In fact, the activation of the cholinoceptive or cholinergic brainstem neurons or interneurons induces, via the lateral funiculus, postsynaptic inhibition of motoneurons (Barnes and Pompeiano, 1970; Hickey and Barnes, 1975). Indeed, several types of spinal and supraspinal cholinergic actions may be involved in the effects described, and these effects may reflect the interplay between the various types of reflexogenic action. For example, an excitation of a spinal cholinoceptive cell may have interacted with inhibitory cholinergic effects, whether dependent on the activation of the Renshaw cells or on direct inhibitory cholinoceptive responses (see sections BI-1 and BI-2, above).

Also, auto- or heterocholinoceptive sites at nerve terminals may interact with postsynaptic effects. Apparently, the autoreceptor effect was involved in Yoshioka et al.'s (1990) experiments. They demonstrated that cholinergic muscarinic agonists and antiChEs might evoke or potentiate the inhibition of rat monosynaptic reflexes via their presynaptic, nerve terminal effect on the M2 receptors. And Samathanam et al. (1989) and Ohno and Warnick (1989) provided an example of a heteroreceptor presynaptic cholinergic action on the spinal reflexes: serotonin and its descending spinal pathways are involved in augmenting monosynaptic reflexes, and this serotonergic facil-

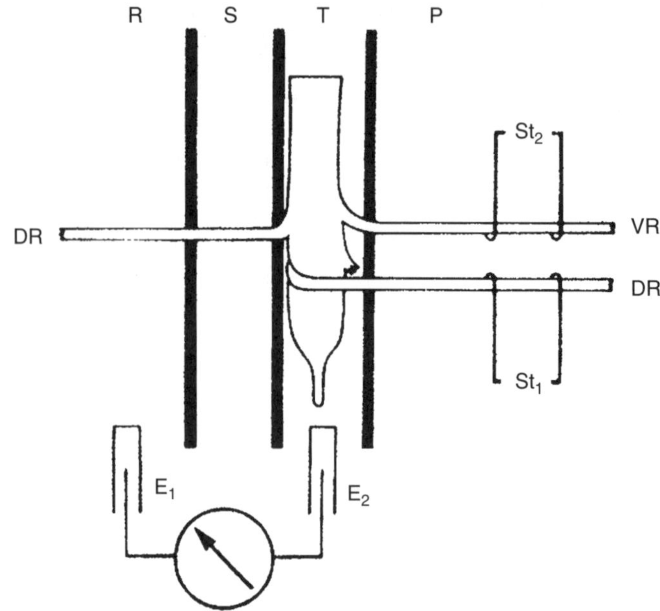

Figure 9-20. Sucrose-gap arrangement. Schematic drawing of the sucrose-gap arrangement for recording the potential changes occurring at the dorsal root nerve terminals in a hemisected toad's spinal cord. R and P represent separate pools filled with Ringer's solution and paraffin, respectively. S and T represent channels used for continuous perfusion with sucrose and various test solutions, respectively. The dorsal root (DR), the ventral root (VR), calomel electrodes (E_1 and E_2), and stimulators (S_1 and S_2) are shown. (From Koketsu et al., 1969, with permission.)

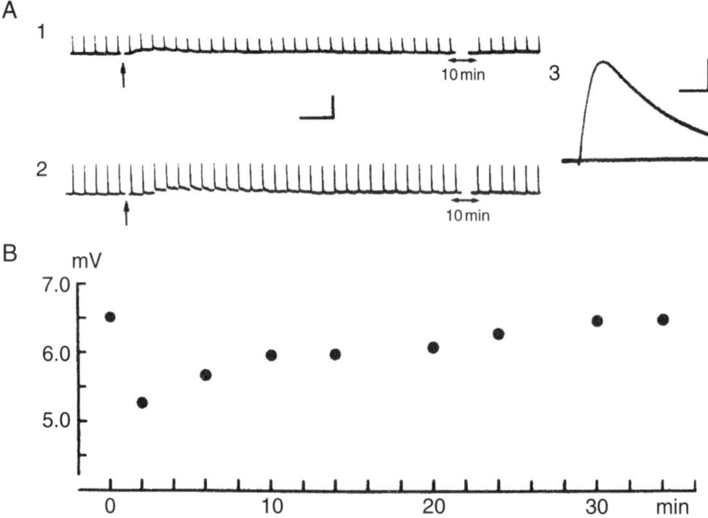

Figure 9-21. Effects in a DR-DRP preparation of long-lasting ACh perfusion (see also Figure 9-20). (A) Records 1 and 2 were obtained from two different preparations to which ACh (1×10^{-4} M) was applied at times indicated by arrows. The spike potentials seen in these records represent DR-DRPs; a single spike potential (DR-DRP) is shown with a faster time base in record 3. Recovery from the initial marked decrease in the amplitude of the DR-DRP was not complete—seen after the ACh depolarization subsided. Calibration: horizontal lines, 1 minute (records 1 and 2) and 50 milliseconds (record 3); vertical lines, 6 mV (record 1), 4 mV (record 2), and 2 mV (record 3). (B) This plot was constructed from the records obtained with a preparation to which ACh (5×10^{-5} M) was applied for 40 minutes. The amplitudes of the DR-DRPs were plotted against time (ACh was applied at 0). Note that in this case the control amplitude of the DR-DRP is almost completely restored, after 30–40 minutes of continuous application of ACh. (From Koketsu et al., 1969, with permission.)

itation may be antagonized by the inhibitory presynaptic action of the cholinergic system on the release of serotonin (see section BIII-2). Altogether, it is difficult to define the interactive nature of cholinergic action on the spinal reflexes because the understanding of pertinent circuitry is limited; however, the presence of cholinergic synapses in this circuitry is indubitable and it underlies the effects described (Chapter 2 DII and DIII).

b. Respiratory Phenomena

Since Christison's self-experimentation and the late-19th-century studies by the Edinburgh and German investigators (see above, section A-2), it has been well known that cholinergic agonists and, in particular, antiChEs affect respiration. Interestingly, the Nobel Prize winner Corneille Heymans, the man who clarified the role of the carotid sinus in the control of respiration and cardiac functions (Heymans and Neil, 1958) and who was an early investigator of antiChEs (Heymans, 1951), doubted that the effects of antiChEs on respiration were due to the inhibition of ChEs. These points of view are not tenable today (see Karczmar, 1967).

These cholinergic effects are very complex because there is an interaction among effects of the compounds in question on the bulbar respiratory center, its subcenters, and other centers including the chemoreceptor sites (Figure 9-22; see Thews, 1986). Reflex regulation of respiration, phrenic and intercostal nerves, hypoglossal and vagal neurons, and the smooth muscle of the respiratory airways is also involved (Comroe, 1943; Wills, 1963, 1970; Widdicombe, 1972; Brimblecombe, 1977; Feldman, 1986). Respiratory centers and the related sites contain cholinergic and cholinoceptive neurons (see Chapter 2 DII and DIII); in fact, ACh is released and cholinergic transmission obtains in the so-called preBotzinger Complex of the medulla and in the XII nucleus (see, for example, Shao and Feldman, 2005); these sites regulate amplitude and duration of respiratory bursts and frequency of respiration. And, Nersesian and Baklavadzhian (1989) found a few cholinoceptive neurons in the expiratory and inspiratory medullary subcenters of the cat, while higher numbers of cholinoceptive neurons were present in neighboring medullary areas including the medial pontine reticular formation and tegmental regions primarily involved in sleep and related phenomena (see section BIV-3; Kimura et al., 1990; Lydic and Baghdoyan, 1989); in fact, the neuronal networks regulating respiration interconnect with those regulating sleep and wakefulness, and certain respiratory disorders such as sleep

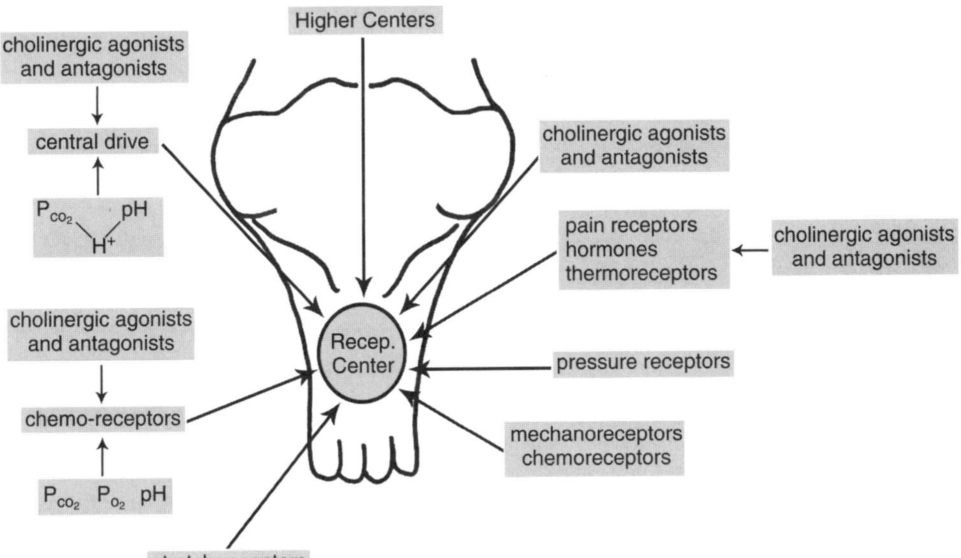

Figure 9-22. Survey of the central respiratory drives and the peripheral receptors that can modulate breathing. (From Thews, 1986, with permission.)

apnea depend on alterations of this interconnection (see section BIV-3; Haxhiu et al., 1989, 2003).

Altogether, the net respiratory effect of cholinergic drugs results from an interaction among cholinergic effects on the respiratory center, the medial pontine and tegmental areas of reticular formation, and the cholinoceptive inhibitory sites abutting the respiratory center (Nersesian and Baklavadzhian, 1989), including sites within the medial pontine and tegmental areas of reticular formation; when depression results from this interaction during cholinergically induced REM sleep, REM sleep-linked partial muscular atonia may contribute to this depression (Kok, 1993).

In addition, there is a close relation between the medullary vasomotor and respiratory centers; for example, the ventrolateral medullary respiratory centers are involved in the generation of the arterial blood pressure oscillations (Mayer waves) and the interaction between vagal activity generated by the baroreceptors and respiratory rhythm (Haxhiu et al., 1989; see also sections BIV-1d, BIV-1b, and BIV-3c, below). Also, muscarinic receptors modulate the baroreceptor reflex at the level of the solitary tract's nucleus (Benarroch, 1993, 1997a, 1997b; Jordan and Spyer, 1981; Loewy and Spyer, 1990; Jordan, 1990; see also section BIV-2b, below). Cholinergic transmission is also involved in medullary activation of the bronchomotor tone (Haselton et al., 1991) and in the direct stimulation of the bronchiolar smooth muscle. Interestingly, human cardiopulmonary transplantation may lead to bronchiolar supersensitivity to muscarinic agonists (Dinh, 1990).

When used in small, subtoxic doses, cholinergic muscarinic agonists (for example, pilocarpine and oxotremorine) and antiChEs capable of central penetration (i.e., physostigmine and OP compounds including insecticides such as Malathion and EPN and war gases or war gas prototypes such as DFP, tabun, and soman) stimulate the rate and depth of respiration despite their ability to cause bronchioles and bronchi obstruction. The same effect initially occurs with toxic doses of these drugs, which ultimately cause respiratory depression (see Chapter 7 DI). The excitatory and depressant effects on respiration are antagonized by scopolamine or atropine, while nicotinic antagonists are less effective (see Wills, 1963; Karczmar, 1967; see Chapter 7 DI). In addition, systemically administered cholinergic agonists that cannot penetrate the CNS (i.e., neostigmine) may indirectly block the respiratory centers via activating inhibitory afferents (Fleming et al., 1991; Koley et al., 1988, 1990).

The augmentatory respiratory action of the drugs is mediated by the stimulation of the chemoreceptor trigger zone of the ventral medulla (an effect related to the emetic action of these drugs; see below, section BIV-1c), the carotid sinus and respiratory centers (Heymans and Neil, 1958; Koley et al., 1988, 1990). The cholinergic drugs act on respiratory centers, since they are effective as respiratory stimulants even after the elimination of sinocarotid chemoreceptors; and since they exert a very potent biphasic effect when directly applied to pontomedullary respiratory subcenters (Karczmar, 1967, 1984; Stewart and Anderson, 1968; Feldman, 1986; see, however, Koppanyi, 1948). Similarly, respiratory depression induced by high doses of cholinergic agonists penetrating the CNS (i.e., tertiary carbamates and OP antiChEs) is caused by their central actions rather than by their ability to stimulate the bronchi and their depressant actions on intercostal (phrenic) muscles (Krivoy and Marrazzi, 1951; Krivoy et al., 1951; Rickett and Beers, 1987; see Chapter 7 DI).

Potent biphasic effects on respiration are exerted by nicotinic actions of endogenous ACh when it accumulates in the presence of antiChEs and when its muscarinic effects are blocked by atropine. Under these circumstances, ACh induces intense hyperpnea and then apnea (Koppanyi et al., 1948; Koppanyi and Karczmar, 1951; Koppanyi, 1948). However, intravenously administered nicotine may exert triphasic effects: with toxic doses of nicotine, apnea results first, followed by hyperpnea and then by apnea again. The initial apnea is reflex in nature and due to stimulation of vagal afferents, as it is eliminated by cervical vagotomy; however, the activation of pulmonary stretch receptors may also be involved in the early, nicotine-induced apnea (Koley et al., 1988, 1990). The subsequent hyperventilation may be a result of central or reflex effects mediated by the chemoreceptors (Heymans and Neil, 1958).

Obviously, cholinergic transmission plays an important role in regulating respiration, acting on known cholinergic sites within respiratory centers and circuits (Feldman, 1986; Burton et al., 1990; Shao and Feldman, 2005). It is of interest in this context that, while its is generally understood that AChE is involved in respiratory effects of antiChEs, there is some, admittedly sparse, evidence that BuChE may be implicated as well (Boudinot et al., 2005).

c. Emesis

Complex emetic processes include: reflexogenic afferent pathways originating in the labyrinthine apparatus (in the case of motion sickness) and intestine, the medullary central system or the chemoreceptor trigger zone (CTZ; Borison and Wang, 1953) located in the area postrema, the vomiting center located within the lateral medullary reticular formation, and the efferent motor pathways leading to upper gastrointestinal musculature. While reflexogenic stimulation originating in the CTZ may activate the vomiting center, dopaminergic agonists such as apomorphine may directly stimulate the central trigger zone.

That cholinergic agonists evoke emesis in hu mans has been known since the description of the use of eserine in Calabar and since the days of self-experimentation with physostigmine by the Edinburgh scientists (section A-2); emetic action of nicotine and pilocarpine and the antiemetic effect—both peripheral and central—of scopolamine and atropine were demonstrated in animal experiments quite early (Eggleston, 1916; Hatcher and Weiss, 1923; Koppanyi, 1930; Koppanyi and Karczmar, 1963). Beleslin and his associates demonstrated that the central trigger zone might be involved in emetic and antiemetic actions of cholinergic agonists and antagonists, respectively (Beleslin et al., 1981, 1989; Beleslin and Nedelkovski, 1988). Both nicotinic and muscarinic agonists (particularly those acting on M1 receptors) cause emesis by stimulating the CTZ. Beleslin and Nedelkovski (1988) postulated that muscarinic emesis is mediated by a catecholaminergic link. Interestingly, atropinics antagonize behavioral and electrophysiological actions of apomorphine (Kelland and Walters, 1992; Motles et al., 1992; see also Barnes, 1984). Accordingly, the antiemetic effect of atropinics may relate to their antiapomorphine action.

d. Motor Function

In this section, motor activities are considered as specific entities rather than components of certain behaviors, that is, independently of their willed or behavioral character, although behavioral and motor activities frequently overlap (Mori, 1995). For example, a perceptual and mental cholinergic syndrome, the cholinergic alert nonmobile behavior (CANMB) and confrontational or defensive behavior (Shapovalova, 1995) include some aspects of motor function and attention (see sections BIV-3 and BIV-4c).

The functions to be commented on now include "normal" motor activity as well as a number of neurologically abnormal motor functions, including tremors, circling, and other stereotypic activities, and catalepsy. These abnormal functions relate to human neurological disorders, such as Parkinson's disease, Huntington's chorea and dystonia. The functions in question involve the basal ganglia, the extrapyramidal system, as well as cortical, cerebellar, and limbic sites.

It should be noted that any movement involves cholinergic activity, whether at the peripheral, motoneuron, and neuromyal sites, or, more important, via the cholinergic input to layers II and III of the motor cortex (in mammals and humans; Joffe, 2004). Furthermore, the atropinic drugs antagonize motor acts when applied to motoneurons and the supraspinal areas controlling these acts, while muscarinics and other cholinergic agonists (including nicotinics) induce motor acts. For example, when applied to the striate, muscarinic agonists evoke jaw movements in several species (Mayorga et al., 1999). Moreover, cholinergic drugs induce more complex effects and affect more complex motor activities. For example, in unanesthetized animals, cholinergic muscarinic agonists (for example, arecoline and carbachol) and antiChEs arrest ongoing spontaneous locomotion and induce motor hypofunction; these effects result for augmentation of the tonic and inhibition of the phasic motor behavior (de Oliveira et al., 1997; Shapovalova, 1995). It should be added that the knockout mice deprived of M1 receptors show increased motor activity, which is in keeping with the hypoactive effects of muscarinic agonists (see Chapter 5 F). At small doses the agonists produce some degree of muscle relaxation; this relaxation is a component of CANMB (Karczmar, 1990).

Another, important normal motor activity is the tonic muscle response evoked by transient central commands or brief sensory stimuli. The cholinergic system seems to be involved in the generation of these tonic responses as in the case of eye fixation following spontaneous eye movements, which can be generated by cholinergic stimulation of reticular formation (Navarro-Lopez et al., 2004). Importance of these processes extends beyond their control of the eye movement itself, as they serve to maintain visual acuity and to provide appropriate compensatory eye movements during head motions (Delgado-Garcia, 2000, 2002).

Several muscarinic receptor subtypes (particularly of the M1 and M4 subtype) may be involved in these motor effects, and their effects may be mediated by inhibition of the striatal dopaminergic DI activity (Gomeza et al., 1999). These effects may be obtained when the appropriate agonists are given systemically or applied locally to a number of sites (i.e., the diagonal band and several sites within the basal forebrain and the parabrachial systems, the periventricular gray, the reticular formation, locus coeruleus, pedunculopontine nuclei, the caudate nucleus and the basal ganglia, and the motoneurons): such application may also produce the circling syndrome; see below, this section; see Karczmar, 1976; Reid et al., 1994a, 1994b; Elazar et al., 1995; Sterman and Fairchild, 1966; Myers, 1974; Karczmar, 2000; Stein, 1968; Hingtgen and Aprison, 1976; Luccarini et al., 1990; Brudzynski et al., 1988; Sernagor et al., 1995). It is quite dramatic to see how ongoing motor behavior of anaesthetized rats and cats is decreased or arrested by intravenous, intracarotid, or intraventricular administration of antiChEs such as DFP and physostigmine (A. G. Karczmar, 1958, unpublished observations). It is pertinent that this inhibition of motor function by antiChEs correlates with an increase in levels of ACh in the brain (Pradhan and Dutta, 1971; Sterman and Fairchild, 1966; A. G. Karczmar, unpublished data). It must be stressed that voluntary initiation of movement (e.g., reward-driven behavior) involves striatal neurons and exhibits a cholinergic correlate (Lebedev and Nelson, 1999). Applying these agents at certain neighboring brain sites may also evoke free motor behavior (Brudzynski et al., 1988).

Muscarinic (and nicotinic) hypokinesia relates to opioid-induced catalepsy (cataplexy ortonic immobility) induced in rodents by morphine and cannabinoids. Opioid-induced catalepsy is accompanied by EEG arousal, some rigidity, and increased muscle tonus. Frequently, the treated animal is capable of maintaining any position in which it was placed (Koppanyi and Karczmar, 1963). While EEG activation also obtains with cholinergic motor hypofunction, the increased muscle tonus and rigid catalepsy do not occur in cholinergic hypokinesia or catalepsy. Furthermore, cholinergically induced catalepsy is specifically linked to reduction in discharges generated at strategic sites (i.e., the striate or the locus ceruleus), although cholinergics, morphinoids, and GABAergic agents may share sites (i.e., periaqueductal sites) where they induce catalepsy (Wu et al., 1988; Monassi et al., 1999). Additionally, morphine and cannabinoids are synergistic with muscarinics in regard to the induction of catalepsy (Pertwee and Ross, 1991), whether induced by opioids or muscarinic agonists; catalepsy and cholinergic hypokinesia must be differentiated from the CANMB, which is also characterized by some atonia (see above, this section, and section BIV-3, below). It must be noted that in decorticated rats, striatal administration of muscarinic agents antagonizes opioid-induced catalepsy (Ladinsky et al., 1987).

Atropine, scopolamine, and adrenergic drugs antagonize both cholinergic hypokinesia and opioid catalepsy (Elazar and Paz, 1990; Elazar et al., 1995; Karczmar, 1976 and Karczmar and Koehn, 1980). In fact, when given in large doses, atropinic antagonists evoke hyperfunction and increased motor activity. Also, atropinics and amphetamines synergize with other agents or procedures that produce hyperkinesia such as lesions of the septum and mesencephalon, and neurotoxins such as lead (Silbergeld and Goldberg, 1976; Karczmar, 1990). On the other hand, toxic doses of atropine or atropinic drugs produce comatose states and areflexia. Once upon a time, this effect was used in the treatment of depression (Karczmar, 1979c).

Dopaminergic agonists and amphetamine antagonize hypoactivity induced by muscarinic agonists and synergize with hyperactivity induced by atropinics (Karczmar and Koehn, 1980; Silbergeld and Goldberg, 1976; Shannon and Peters, 1990). The dopaminergic effect may involve D1 receptor sites (Shannon and Peters, 1990). On the other hand, Kostenis, Wess, and their associates find that M4 receptors mediate inhibition of dopaminergically mediated hyperactivity (Gomeza et al., 1999). Thus, in catalepsy and hypokinesia and other motor activities reviewed here, the adrenergic and cholinergic systems seem to act antagonistically. A separate functional system, the punishment-reward dipole (Olds, 1958; Stein, 1968), may be concerned here; the inhibitory effect of cholinergics on the peripheral gray component of the system may contribute to the cholinergic inhibition of motor activity (see section BIV-1d and BVI, below).

Nicotine's effects on locomotor behavior were inconsistent; nicotinic hypofunction and hyperfunction were both observed (compare, for example, Fung and Lao, 1988, and Benwell and Balfour, 1992). The sites where nicotine exerts inhibitory actions may differ from the sites where it exerts facilitatory actions and where muscarinic agonists exert their inhibitory effects; facilitatory sites of nicotine's action may be present in the motor cortex (Stolerman et al., 1989). Effects of nicotine may not be primarily cholinergic in nature: they may depend on interplay between cholinergic and catecholaminergic pathways (see, for example, Sandor et al., 1991).

A classical effect of cholinergic (essentially muscarinic) agonists and antiChEs is the "aversive syndrome" or circling behavior. This interesting effect was first described in the early 1950s by Harold Himwich, Murray Aprison, and their associates (see Nathan et al., 1955; Aprison, 1965; and Hingtgen and Aprison, 1976). It follows the unilateral injection of muscarinic agonists into the basal ganglia and nigrostriatal sites. Sometimes, it may be evoked by bilateral application of these agents to the basal ganglia. This response is generally but not always contralateral or "aversive" with respect to the administration site, but it may be also ipsilateral; furthermore, the type of circling depends on the administration site of the cholinergic agonist (Kitamura et al., 1999). Aprison (1965; Hintgen and Aprison, 1976) speculated that the direction of circling is due to the endogenous asymmetry of the cholinergic system of rats as reflected by the levels of AChE in the pertinent sites: they found that circling is contralateral with respect to the level of AChE. Atropinic drugs antagonize circling. Circling is also antagonized by dopaminergic agonists, as in the case of other motor activities that are associated with release of ACh and that are described in this section; however, when the dopaminergic sites are lesioned, dopamine agonists cause circling. Dopaminergic blockers and muscarinics synergize in the production of circling (Costall et al., 1972; Karczmar, 1976; Acquas et al., 1999; Feny et al., 2001).

The circling behavior may be considered a form of stereotypic or compulsive behavior or function. Besides circling, muscarinic agonists induce additional stereotypes, including gnawing, writhing, self-biting, and purposeless, repeated motor activity, such as rearing, repeated jumps, and the like. These phenomena may be elicited by systemically administrating the drugs; they may be more readily elicited by localized application of muscarinics to the limbic and tegmental sites (see, for example, Davis et al., 1967). Interestingly, in a natural mouse habitat, the "Mouse City" (Karczmar et al., 1973), the frequency of spontaneous stereotypic acts increases after treatment with muscarinic agonists and antiChEs. Cholinergics are not always specific in the evocation of these stereotypes; for example, serotonin induces writhing (Moser, 1995). Atropinics and dopaminergic agonists antagonize cholinergically induced stereotypes.

Several cholinergic agonists and antiChEs may induce other motor function abnormalities (i.e., rigidity, tremor, and palsylike jerks such as limb and head extensions and tongue protrusions; Gunn, 1935). Both muscarinic (including, classically, oxotremorine) and nicotinic agonists are capable of inducing tremor by localized, caudate, or mesencephalic application or systemic administration (George et al., 1966; Bovet and Longo, 1951; see Karczmar, 1976); it is consistent with these data that oxotremorine loses its tremorigenic action in knockout mice deprived of M2 receptors (see Chapter 5 F). Interestingly, neurotoxins that induce dopaminergic or glutaminergic lesions (for example, 1-methyl-4-phenyl-1, 2, 3, 6-tetrahydropyridine [MPTP] and kainic acid) produce tremor and synergize in their tremorigenic action with nicotinic and muscarinic agonists (Behmand and Harik, 1992; Shinozaki et al., 1987). As in the human, they are also caused by neuroleptics and experimental or accidental damage of the nigrostriatal loop including ventromedial tegmentum, the palidothalamic relay (Pare et al., 1990), and/or the rubro-olivo-cerebello-rubral loop (see Brodal, 1980). Furthermore, whether chemically or lesion induced, the tremor is antagonized by cholinergic antagonists, dopaminergic agonists (Marco et al., 1988), and beta-adrenergic blockers (Suemaru et al., 1993).

This pharmacological profile of the chemically or lesion-induced tremor and rigidity resembles the profile of parkinsonian disease; in addition, classically, parkinsonism and chorea were considered to be the result of cholinergic-dopaminergic imbalance. Accordingly, these instances of animal tremor serve as models that are useful for the development of the therapy for parkinsonism. For example, a frequent model of this kind is the tremor induced by the muscarinic agonist tremorine and its active metabolite, oxotremorine, physostigmine, and harmine or harmaline (Bovet and Longo, 1951; Everett, 1956). Indeed, physostigmine- and muscarinically induced tremor may be more dependable than the tremor evoked by nicotine, because nicotinolytics and curarimimetics are ineffective with respect to lesion-induced tremors (Lamarre and Joffroy, 1970). It may be added that some forms of genetic dystonia exhibit lesions of cholinergic neurons of pedunculopontine and cuneiform nuclei (McNaught et al., 2004).

While the importance of the dopaminergic-cholinergic balance for regulating some of these functions and dysfunctions was stressed in this section, the nature of the pertinent sites, such as the basal forebrain, the limbic system, and the basal ganglia, indicates that other neurotransmitters (for example, GABA; Dray et al., 1975; Van Woert, 1976; and glutamate; Shinozaki et al., 1987; see also Elazar et al., 1995), modulators, second messengers, and G proteins must be involved (Jones et al., 2004). Finally, the sites pertinent for the evocation of these motor effects are also the sites important for learning, CANMB, and several EEG phenomena (see sections BIV-3 and BV).

2. Cholinergic Effects Concerning Hypothalamus and Homeostatic Function

The cholinergic phenomena to be described now include regulation of endocrine and cardiovascular function, thermoregulation, and the control of feeding and drinking; besides hypothalamus other sites such as the pineal organ may be involved (Brandstatter et al., 1995). Both thermoregulation and the control of feeding and drinking should be looked on as components of energy homeostasis, although, as will be seen, the phenomena concerning hunger and thirst on the one hand and body temperature on the other do not always correlate in a teleological manner. Similar to other central phenomena, these functions do not solely or mainly depend on their cholinergic correlates; instead, other bioamines, amino acids, polypeptides, and even ions are contributors to hypothalamic regulations. It must be noted in this respect that Kris Krnjevic (1975) proposed that neuronal metabolism and its cholinergic correlates contribute to energy regulation (see also Chapter 3 DIII).

a. Cholinergic Correlates of Endocrine Function

Two hypothalamico-hypophyseal systems are involved in the cholinergic correlates of endocrine function. The first system links the hypothalamus and neurohypophysis (posterior pituitary); it releases oxytocin and vasopressin hormones. The second system links the hypothalamus and adenohypophysis (anterior pituitary). It evokes the release of regulatory hormones, which, in turn, liberate or inhibit pituitary hormone release. Both systems are regulated by both muscarinic and nicotinic cholinergic transmission, and their second messengers, including NO and prostaglandins (Bugajski et al., 2004). It should be added that these systems are controlled not only locally, but also by other systems, such as those involved in arousal (see, for example, Bayer et al., 2005).

The first system originates with neurons located in the supraoptic and paraventricular nuclei of the hypothalamus. Their axons liberate the antidiuretic hormone (arginine/vasopressin) and oxytocin from their terminals expanding within the neurohypophysis and the Herring bodies. In 1947, Mary Pickford[4] found that physostigmine or DFP injected into the dog's supraoptic nucleus almost immediately stopped the flow of urine. Subsequently, she and Abrahams (Abrahams and Pickford, 1956) evoked uterine contractions with several antiChEs. This effect was proportional to the duration of the compounds' antiChE action, resulting in the release of oxytocin and vasopressin (antidiuretic hormone, ADH). This type of ACh action is referred to as "neurohumoral" (Koelle, 1963).

It was shown later that localized applications of ACh and cholinergic agonists to hypothalamic areas (i.e., supraoptic nuclei and paraventricular nuclei) or into the third ventricle cause similar actions. These effects were elicited very consistently by muscarinic agents when administered locally or intraventricularly or applied to isolated preparations, while systemic or local applications of atropine blocked these responses (Sakai et al., 1974; De Luca et al., 1989; Yamaguchi and Hama, 1989). In addition, the milk-ejection reflex, which is initiated by suckling and which releases oxytocin, is elicited by applying ACh locally, and the reflex is blocked by local or systemic applications of atropine (Belin and Moos, 1986; Moos and Richard, 1989; Oba et al., 1971). In fact, ACh induces discharges in the supraoptic and paraventricular nuclei cells and/or facilitates this activity when induced by appropriate physiological stimuli or reflexes (for example, the suckling reflex); these effects are atropine sensitive. Interestingly, oxytocin causes similar phenomena (Moos and Richard, 1989).

However, the hypothalamic systems are not solely muscarinic in character; both nicotinic and muscarinic hypothalamic receptors are involved in these phenomena. Hillhouse and Milton (1989a, 1989b; see also Jones et al., 1976) found that both hexamethonium and atropine antagonized vasopressin release by ACh from the isolated hypothalamus, and nicotinic rather than muscarinic agonists caused vasopressin release from the cultured rat hypothalamico-hypophyseal system, this effect being blocked by nicotinolytics (Sladek and Joynt, 1979). Specific nicotinic involvement was also indicated by electrophysiological and pharmacological analysis of the responses of supraoptic or paraventricular nuclei to nicotine (Yamaguchi and Hama, 1989). Then the nicotinic responses of the hypothalamic nuclei may mediate the well-known antidiuretic effect of the endogenous opioids (Tsushima et al., 1993).

It must be pointed out, however, that cholinergic activation of the hypothalamic systems may not be analogous with the physiological stimulation by angiotensin II of vasopressin release, as indicated by pharmacological analysis (Yamaguchi and Hama, 1989).

Ultimately it was shown that cholinergic pathways involved in the release of oxytocic and vasopressive peptides consist of several hypothalamic nuclei and the parabrachial, reticular and forebrain systems along with the magnocellular nuclei (see Koelle, 1963; Myers, 1974; Nemeroff et al., 1977; Moos and Richard, 1989; and Pow and Morris, 1989; see Chapter 2 DII and DIII). There may be a differential distribution of these pathways with respect to the two hormones in question (Pow and Morris, 1989). It must be added that arginine/vasopressin is present in other neurons (for example, those of the limbic system); ACh releases vasopressin from these neurons as well (Raber et al., 1994). Interestingly, vasopressin facilitates ACh release from rodent hippocampal slices. This may suggest that vasopressin's effect on learning is mediated by cholinergic activation (De Wied, 1969; Tanabe et al., 1999; see section BV-d).

In addition to their antidiuretic effects, cholinergic agonists produce changes in electrolytes when acting on the hypophyseal-pituitary system (De Luca et al., 1989). Cholinergics (particularly muscarinic) cause the release of the atrial natriuretic peptide (ANP) hormone from central sites within the hypothalamico-pituitary system; this release results in natriuretic and kaliuretic effects (Baldissera et al., 1989). Actually, the cholinergic system mediates the physiological release of ANP induced by blood volume expansion (Antunes-Rodrigues et al., 1993). The cholinergic pathways involved in these phenomena are the same that elicit vasopressin and oxytocin release (see above and Chapter 2 DII and DIII; Baldissera et al., 1989). In fact, there may be a topographical relation between central release of the natriuretic hormone and vasopressin.

Furthermore, the paraventricular and lateral hypothalamic nuclei are also involved in pancreatic secretion, as it was proposed that the vago-vagal reflex and cholecystokinin factors are under muscarinic control exerted by these nuclei; additional cholinergic input to the hypothalamic sites in question is provided by the lateral septum and parabrachial nucleus (see Li et al., 2003). Additional peptidic hormones (for example, hypocretin and melanin-concentrating hormone) are also excited by cholinergic agonists (see Bayer et al., 2005).

The second hypothalamico-hypophyseal system originates with the hypothalamus's hypophysiotropic zone; this system consists mainly of arcuate and periventricular nuclei situated just beneath the third ventricle. Thus, neurotransmitters and drugs introduced into the third ventricle act upon the hypophysiotropic zone. The hormones produced in the cells of these nuclei discharge into the median eminence to be conveyed, via the capillary plexus of the superior hypophysial artery and the long portal vein, to the specific endocrine target cells located in the adenohypophysis or the anterior pituitary gland; these hormones either facilitate or inhibit the release of pituitary trophic hormones. Hypothalamic release factors (HRFs) or hypophysiotropic regulatory peptides (HRPs) are the facilitators of this release, and the processes involved in the release of these peptides exhibit cholinergic correlates. The HRFs include corticotrophin release factor (CRF), prolactin inhibitory and releasing factors (PIF and PRF, respectively), gonadotropin releasing factors (GnTRFs), thyrotropin releasing factor (TLF) and growth hormone inhibitory and releasing factors (GHIF and GHRF, respectively).

Cholinergic agonists, applied directly to the hypothalamus, administered intraventricularly or systemically, release CRF, which in turn liberates the adrenocorticotropic hormone (ACTH; Hillhouse and Milton, 1989a, 1989b; Itoi et al., 2004). This augments synthesis and facilitates the secretion of the adrenal cortex's corticosteroids, including cortisol (hydrocortisone), aldosterone, and corticosterone (Davis and Davis, 1979; Makara and Stark, 1976; Calogero et al., 1989; Bugajski, 1999); weak androgenic steroids are also released from the medullary cortex. It must be remembered that cholinergic agents, particularly nicotinics, stimulate the adrenal medulla either directly or via their hypothalamic action; this is to be expected, as the medulla is a modified sympathetic ganglion. This stimulation causes, besides the release of steroids, the release of adrenal catecholamines (Koppanyi and Karczmar, 1951). Conversely, choliner-gic antagonists, GABA and hemicholiniums diminish the basal levels of these substances (Ramade and Bayle, 1989; Hillhouse and Milton, 1989a, 1989b, 1989c; Davis and Davis, 1979; and Karczmar, 1980). The nicotinic or mixed nicotinic/muscarinic agonists and nicotine inhaled during smoking induce CRF-ACTH activation; this activation is antagonized by several nicotinolytics (see Davis and Davis, 1979; and Karczmar, 1980).

It also appears that both the cholinergic system and ACh release participate with other neurotransmissions in the stress-evoked release of CRF and in CRF gene expression in the hypothalamus (Ohmori et al., 1995; Grossman and Costa, 1993). However, stress-caused release of ACh desensitizes muscarinic hypophysiotropic receptors and reduces the corticosteroid release induced by carbachol (Bugajski, 1999). Accordingly, to conclude that cholinergic agonists directly activate the hypothalamico-hypophyseal axis, there must be evidence indicating that this activation was not caused by a nonspecific stress resulting from administration of drugs and related maneuvers (Davis and Davis, 1979). In fact, Davis and Davis (1979) pointed out that as ACh is released during stress, the blocking effect of atropine on HRF release may result from atropine's antagonism of the hypothalamic action of stress-released ACh (also, large doses of atropine were used in pertinent experiments and their effect could be nonspecific or a result of atropine's anesthetic effect; Davis and Davis, 1979). Yet, while inhaling nicotine is a nonstressful procedure that increases serum levels of cortisol, hemicholinium, implanted into pigeons' third ventricles, decreased both the stress response of the hypothalamico-hypophyseal axis and the basal, resting levels of ACTH and cortisol (Ramade and Bayle, 1989; Sellini et al., 1989). Additionally, when ACh is applied to hypothalamic synaptosomes, it causes the release of a CRF-like substance capable of liberating ACTH from an isolated pituitary preparation (Edwardson and Bennet, 1974). In fact, the release of ACTH by cholinergic agonists may only occur via CRF release: Calogero et al. (1989) could not obtain this release with cholinergic agonists after the section of pituitary stalk, nor could they release ACTH by an *in vitro* application of arecoline (a mixed nicotinic-muscarinic agonist) to anterior pituicytes.

The cholinergic system and cholinergic drugs affect the pituitary secretion of prolactin. Contrary to the endocrine actions already described, this effect is inhibitory in character. Is not clear whether this inhibition results from cholinergic release of PIFs or direct cholinergic effect on the pituitary cells secreting prolactin. The type of receptor involved may depend on the animal's biological state: the surge of prolactin during the proestrous phase of the rat, its pregnancy, or lactation is under nicotinic regulation (see, for example, Kellar et al., 1989), while the elevated basal levels of prolactin that occur during stress or estrogen treatment are controlled by muscarinic receptors (Davis and Davis, 1979; see also Karczmar, 1980).

The cholinergic system affects the release of hypophysiotropic GnTRF and the luteinizing-follicle stimulating hormone releasing factor (LH-RH/FSH/RF). This was shown by Charles Sawyer, one of the pioneers of studies on the cholinergic control of the endocrine function (Sawyer et al., 1949). From adenophypophysis, the GnTRF releases the follicle stimulating hormone (FSH) and luteinizing hormone (LH, or interstitial cell stimulating hormone, ICSH), which act on the testes and ovaries to produce the sex steroids, testosterone and progesterone with estrogen. In agreement with Sawyer's notion, it appears that cholinergic agonists release LH/FSH only in the presence of the hypothalamic tissue (Fiorindo et al., 1975). Muscarinic (M1 and M2) and nicotinic agonists may be involved in GnTRF release; both inhibitory and facilitatory effects seem to obtain (Krsmanovic et al., 1998). Muscarinic receptors may be involved in these phenomena and copulation as antimuscarinic drugs or antagonists with mixed nicotinolytic-antimuscarinic effects block the copulation-ovulation reflex (Sawyer et al., 1951; Davis and Davis, 1979; see, however, Kalash et al., 1989). Interestingly, the neuronally derived septal cell line containing the cholinergic neurons (see Chapter 2 DII and DIII) exhibits jointly CAT, LH/RH, and the gonadal hormone receptors (Martinez-Morales et al., 2001) and Kasimir Blusztajn and his associates (Martinez-Morales, 2001) suggest that there may be an interaction among these entities. The septal cell line may then serve as a model for the hypothalamico-pituitary phenomena and for the interaction between release of LH/RH and cholinergic mechanisms.

The status of the natural ovulation cycle with respect to its cholinergic correlates is complex. Indeed, it is controversial whether or not muscarinics advance and atropine blocks ovulation (compare Mayerson and Palis, 1970, and Lopez et al., 1997). This inconsistency may relate to the finding that the effect of drugs on ovulation depends on the timing of their administration with the oestral cycle (Dominguez et al., 1982; Lopez et al., 1997). The Mexico City team also proposed that because of the asymmetry of CAT distribution in the hypothalamus, the effect might depend on the site of drug administration (Sanchez et al., 1994; Lopez et al., 1997). Additionally, the muscarinic activation of LH-RH may also induce the release of androgens (see section BIV-2f). Finally, nicotinic receptors appear to inhibit LH-RH/FSH/RH release (see Davis and Davis, 1979; Karczmar, 1980).

Only sporadic information is available regarding cholinergic effects on other pituitary endocrine function. Activation of the hypothalamic cholinergic system and the cholinergic agonists may release the growth hormone (or its HRF) and the TRF while muscarinic antagonists may block their liberation (Davis and Davis, 1979; Alteia et al., 1989; Thiefin et al., 1989). The release of the thyrotropic hormone or its HRF may be mediated via the vagus nerve and blocked by atropine (Thiefin et al., 1989). Similarly, only a few investigations concern the cholinergic regulation, via endocrines, of carbohydrate metabolism and insulin or glucagon secretion. This secretion may be activated by anterior hypothalamic mechanisms involving the growth hormone, and/or a vagal afferent reflex. Accordingly, Gotoh et al. (1989) and Iguchi et al. (1990) showed that when neostigmine was injected into rats' third ventricle, it elevated blood levels of insulin, glucagons, and glucose; the release of insulin was blocked by atropine or vagotomy.

Several notes of warning are due. Most of the studies listed concern the rat, and the effects may depend on the species. In fact, cholinergic stimulation or block is seemingly ineffective in humans with respect to the secretion of cortisol, prolactin, or LH (Davis and Davis, 1979). However, only a relatively small dose of physostigmine can be used in these human studies, and methylscopolamine was employed to protect the patient.

Furthermore, the interpretation of the results is difficult because of the complex and multifactorial nature of the hypothalamico-hypophyseal system. There are five main sources of complexity. First, multiple feedbacks are involved in the system's operation, including long-loop, short-loop and ultrashort-loop feedbacks. These feedbacks are generated by hormones of the peripheral glands such as corticoids, which are released by the drug action on the system, by the pituitary tropic hormones themselves, and by the HRFs (see Genuth, 1992). Cholinergic mechanisms may be involved in some of these feedbacks. For example, the cholinergic system facilitates cortisol feedback inhibition of ACTH release (Davis and Davis, 1979). Accordingly, cholinergic facilitation of ACTH release and its feedback inhibition by the corticoids may interact.

Second, a specific HRF may liberate more than one pituitary hormone (Genuth, 1992) and some HRFs and trophic hormones are present in the CNS outside the hypothalamico-hypophyseal system and may be released by cholinergic mechanisms (Chen et al., 1989; see Brownstein, 1988; McGeer et al., 1987). In fact, the neuronal networks regulating respiration interconnect with those regulating sleep and wakefulness (see section BIV-3; Haxhiu et al., 2003).

Third, the cholinergic system is not the only transmitter system involved in the phenomena. For example, the catecholaminergic, serotonergic, and GABAergic pathways radiate to a number of hypothalamic sites, and they all affect releasing factors and/or posterior and anterior pituitary hormones (see Fuxe et al., 1979; Karczmar, 1980). Accordingly, any given endocrine action results from the interplay between endocrine effects of noncholinergic and cholinergic transmitters (Hillhouse and Milton, 1989a, 1989b). Furthermore, the nonendocrine effects of transmitters interact and affect their endocrine function. For example, the cholinergic system activates dopaminergic cell bodies and causes the release of dopamine (Lichtensteiger, 1979; Ribary and Lichtensteiger, 1989); this effect may underlie the cholinergic inhibition of the prolactin secretion and other endocrine actions of cholinergic drugs (Davis and Davis, 1979; Karczmar, 1980). Similar interplay may exist between peptides, including substance P, neurotensin, enkephalins, and trophic agents (Fuxe et al., 1978; Hokfelt et al., 1978).

Fourth, these multitransmitter and multifactorial phenomena are involved in suprahypothalamic environmental, experiential, and emotional phenomena that affect the hypothalamico-hypophyseal system (see, for example, Jacobson and Sapolsky, 1991). The efferents from the hypothalamus radiate to many parts of the brain, including the thalamus and the limbic system, and generate a reciprocal effect on the hypothalamus. Even when experiments were carried out with localized electric stimulation or drug application into the hypothalamico-hypophyseal axis, the influence of these phenomena cannot be excluded; a much more complex effect occurs when the procedures employed are less localized.

Finally, it should be noted that many of the hormones released by the cholinergic system or cholinergic agonists exhibit direct CNS actions (see section BIV-2f, below) that interplay with hypothalamic cholinergic effects.

b. Centrally Controlled Cardiovascular Phenomena

Since the 19th century, scientists have discovered that upon systemic or oral administration, cholinergic agonists and physostigmine induce changes in cardiac rate and vascular pressure (see section A2). Subsequently, these effects were obtained with a number of carbamate and OP antiChEs and cholinergic agonists (particularly muscarinic but also nicotinic).

Depressor and/or bradycardic muscarinic effects involve peripheral (including ganglionic and parasympathomimetic) actions of these drugs. Furthermore, Corneille Heymans (see also section BIV-1b) suggested that muscarinic drugs and antiChEs act by sinocarotid baroreceptors (Heymans and Heymans, 1926a, 1926b; Heymans, 1951; Heymans and Neil, 1958; see also Holmstedt, 1959; Schmitt, 1972). However, it is also apparent that these cardiovascular actions of the cholinergic drugs involve the CNS (see, for example, the early studies of Delga, 1957). In fact, depressor and/or bradycardic responses were also obtained following intracisternal and hypothalamic administration of M2 type muscarinic agonists (Brezenoff and Jenden, 1969; Murugaian et al., 1989; Ni et al., 1989; Li et al., 2003); also, hypotensive responses evoked by electrically stimulating the lateral hypothalamus were blocked by atropinics (Blum et al., 1989).

ACh, muscarinic agonists, and antiChEs also evoke centrally mediated pressor actions; these actions may also be induced peripherally. In the 1930s Dikshit (1934) and Weinberg (1937) demonstrated that intraventricular administration of ACh or carbachol evokes a brief hypertension. Subsequently, Hornykiewicz and Kobinger (1956) and Schaumann (1958) produced similar but longer-lasting effects via intracisternal administration of neostigmine. Similar pressor actions were evoked by systemic administration of physostigmine and OP drugs. Careful pharmacological analysis of the data indicated that these effects were central in nature (see, for example, Paulet, 1954, and Dirnhuber and Cullumbine, 1955). Pressor and tachycardic responses were also elicited by the hypothalamic administration of nicotine (Brezenoff and Xiao, 1989). The responses were frequently biphasic: hypertension was followed by hypotension and/or bradycardia (see, for example, Brefel et al., 1995).

Ulf Svante von Euler (the Nobel Prize winner for his identification of norepinephrine as the pressor neurohumor) demonstrated in the 1930s that pressor action resulting from intracisternal administration of ACh ultimately results from the peripheral release of norepinephrine, as he could block the pressor effect by ergotoxin (see von Euler, 1956). Epinephrine may also be implicated (Brefel et al., 1995). The pressor effects of muscarinic agonists and antiChEs were related to a specific hypothalamic site by Varagic, Schmitt, Brezenoff, and Jenden and their associates (Varagic, 1955; Varagic and Krstic, 1966; Varagic et al., 1965; Lalanne et al., 1966; Brezenoff and Jenden, 1969, 1970; Brezenoff, 1972; Brezenoff and Giulano, 1982; Brezenoff and Xiao, 1989; Zhu et al., 1996). Varagic and his associates (Prostran et al., 1994) also expanded on von Euler's demonstration of cholinergically induced peripheral adrenergic outflow when they showed a concomitant glycogenolytic effect that could be obtained even in adrenalectomized animals. The cholinergically induced release of vasopressin may contribute to the cholinergic pressor action (Brefel et al., 1995).

However, additional sites are involved in central cholinergic cardiovascular regulation; actually, the hypothalamus may not be activated at all in experiments involving intracisternal, subarachnoid, or intraventricular routes of administration. Pressor effects can be obtained by microinjecting muscarinic drugs and antiChEs onto medullary forebrain and limbic cholinoceptive sites (for example, the nucleus of the solitary tract, C1 pressor, medial and ventrolateral areas, nucleus basalis maynocellularis and the hippocampus; Benarroch et al., 1986a, 1986b; Benarroch, 1993, 1997a, 1997b; Ni et al., 1989; Lacombe et al., 1989; Xia et al., 1989; Ruggiero et al., 1990; Dampney, 1994; Hori et al., 1995; Kubo et al., 1998, 1999); these vascular effects may be mediated by the medullary sites' radiations to the sympathetic preganglionic neurons, to nucleus reticularis dorsalis, or to substantia innominata and the cortex (Kubo and Misu, 1983; Lacombe et al., 1989; see also section BIV-1b, above).

Finally, vascular responses to subarachnoid or spinal administration of physostigmine or muscarinics (Takahashi and Buccafusco, 1992) and cardiac responses to the microinjection of specific M2 agonists into the interomediolateral cell column at the T1–T3 level (Sundaram et al., 1989; Sapru, 1989) indicate that spinal sites may be involved. The neurons involved were not identified; apparently, they do not include muscarinic or nicotinic receptors at the preganglionic spinal neurons (Buccafusco and Magri, 1989, 1990; Takahashi and Buccafusco, 1992; Sundaram et al., 1989).

Application of cholinergic agonists to medullary sites may evoke biphasic, pressor, or depressor effects; the specific effect depends on the site. For example, Ken Kellar and his associates demonstrated that in rats nicotine exerts depressor actions upon its microinjection into dorsal medullary (brainstem) sites (medial subnucleus of the tractus solitarius). These actions were antagonized by hexamethonium and the nicotinic receptors involved are of the $\alpha3$ subtype (Ferreira et al., 2000; Yeh et al., 2001).

There may be several reasons for cholinergic drugs' and the cholinergic system's ability to evoke both hypo- and hypertensive responses, or to have biphasic effects. Species specificity may be involved: administration of muscarinic agonists and/or antiChEs to similar hypothalamic areas in cats and dogs produces hyper- and hypotension, respectively (Mitchell et al., 1963; Shuh et al., 1935; see, however, Brefel et al., 1995).

Second, the differences may depend on the type of agonist used. For example, many OP drugs (i.e., sarin and DFP) produce hypertension, while several other OP compounds evoke hypotension (VanMeter et al., 1978; see Holmstedt, 1959; Schmitt, 1972). Furthermore, muscarinic and nicotinic drugs and agonists specific for different receptor subtypes may exert different vascular responses. Accordingly, Murugaian et al. (1989), Brezenoff and Jenden (1969), and Li et al. (2003) related the hypothalamic hypotensive and bradycardic response to muscarinic type M2 agonists and the hypothalamic hypertensive and tachycardic response to nicotinic agonists (see also Ferreira et al., 2000). Actually, nicotinic drugs induce excitatory potentials in dorsal hypothalamic neurons, and smoking and systemic absorption of nicotine produces vasopressor effect in human volunteers (Benowitz et al., 1989; Schmitt, 1972). A hypersensitive nicotinic component or upregulation of nicotinic receptors may contribute to the condition of spontaneously hypertensive rats (Brezenoff and Xia, 1989; Kubo et al., 1995; Perry et al., 1999).

Also, different hypothalamic centers may be involved in different cardiovascular responses. Brezenoff and Jenden (1969; see also Brezenoff and Xiao, 1989) obtained pressor and depressor responses with carbachol when applying it to a dorsal and posteroventral hypothalamus, respectively. Neostigmine administered into the lateral cerebral ventricle caused a muscarinic, atropine-antagonizable depressor action (Li et al., 2003), and specific M2 agonists applied to the ventrolateral hypothalamus also elicited hypotension and evoked Fos-like immunoreactivity in a number of posterior hypothalamic sites (Li et al., 1997; Murugian et al., 1989; Blum et al., 1989). In fact, Blum et al. (1989) identified the cat's depressor response as "a small locus" in the lateral hypothalamic peri-fornical region, "medial to the fields of Forel."

Another complication arises from the involvement of nonhypothalamic sites in these cardiovascular responses (see above, this section). Indeed, several medullary, midbrain (central gray) and limbic cholinoceptive sites cause either depressor, pressor, or blood flow responses depending on the specific site and on the type of the agonist (Benarroch et al., 1986a, 1986b; Benarroch, 1997a, 1997b; Ni et al., 1989; Lacombe et al., 1989; Ruggiero et al., 1990; Dampney, 1994; Colombari et al., 1994; Kubo et al., 1998, 1999; Ferreira et al., 2000).

With regard to the medulla, there is a relation between the respiratory and cardiovascular actions cholinergically evoked via the medulla (see section BIV-1b, above). Either respiratory or cardiovascular excitatory effects may be obtained by changing the site of the application of muscarinic agonists within the ventral medulla. The dissociation between these effects may also be achieved by using M1 versus M2 agonists (Nattie and Li, 1990; Takahashi and Buccafusco, 1992). Combined cardiovascular-respiratory reflexes were mediated by C afferents; they were depressed by a local administration of neostigmine. A joint respiratory-cardiovascular effect was also obtained by Chen, Koppanyi, and Karczmar; however, contrary to the results just quoted, Chen et al. (1989) blocked with atropine and Koppanyi and Karczmar (1949) augmented with antiChEs the pressor response combined with respiratory stimulation elicited by means of the stimulation of the afferent (central) vagus. Chen et al. (1989) presented evidence that this effect is mediated by the locus ceruleus and that this effect could be antagonized by atropine.

In a related context, the cholinergic system and cholinergic drugs exert effects on blood flow. The peripheral, muscarinically mediated vasodilation and concomitant increase in blood flow are well known. A cholinergic or cholinoceptive central vasodilatation also exists. For example, electric, muscarinic, and nicotinic stimulation of central cholinergic sites or pathways (i.e., the septohippocampal and cholinergic basal forebrain, substantia innominata, nucleus basalis of Meynert, and sensory afferents) induced vasodilatation in the cortex and the limbic system (Cao et al., 1989; Lacombe et al., 1989; Kurosawa et al., 1989; Scremin et al., 1988; Barbelivien et al., 1999; Tsukada et al., 1999). Atropinics and nicotinics antagonized and antiChEs augmented this vascular response (Dauphin et al., 1991; see also Van der Zee and Luiten, 1999).

Additionally, the autonomic ganglia innervate central blood vessels and their stimulation also induces central vasodilatation (Kuridze et al., 1989; Lister and Ray, 1988; Seylaz et al., 1988; see also Van der Zee and Luiten, 1999). These effects are mediated via cholinergic pathways radiating to muscarinic and nicotinic receptors present in the endothelia and smooth muscle cells of central arteries and arterioles (i.e., pial vessels and of the central capillary astrocytes). *In vitro*, these central vessels dilate when muscarinic drugs are applied (Dacey and Basset, 1993; Estrada and Krause, 1982; Kalaria et al., 1994; see also Van der Zee and Luiten, 1999), and M3 receptors are involved in endothelial-dependent vasodilation, while other muscarinic receptors, such as M5 receptors, control smooth muscle–mediated dilation (Tayebati et al., 2003). The messenger NO may also be involved in the mediation of these cholinergic actions (Shimizu et al., 1993; Van der Zee and Luiten, 1999; Chen and Lee, 1993). However, there is also intrinsic innervation of these blood vessels, and the central release of ACh, whether upon central or autonomic stimulation, may exert vasodilation at vascular sites distant from those directly innervated by these pathways (Thompson et al., 1989).

These blood-flow and cardiovascular cholinergic actions can be accompanied by an increase in glucose metabolism (Scremin et al., 1988; Vaucher et al., 1995). For example, the link between blood flow and glucose metabolism may underlie the cholinergic agonists' effects (see, for example, Blin et al., 1994, and above, this section). On the other hand, it was also reported that cholinergic agonists affect the blood flow independently of the action on glucose metabolism (Barbelivien et al., 1999). It may be added that stimulating blood pressure may accompany changes in blood flow.

Similar to other functions that exhibit cholinergic correlates, the causative link between cholinergic activity and cardiovascular phenomena is complex. Thus, cholinergic cardiovascular responses interact with endocrine phenomena, including the release of angiotensin, vasopressin, and/or aldosterone via central or peripheral cholinergic action (see sections BIV-1b and BIV-2a); non-nervous sites may be involved as well, Brown and Chen, 1990. Then, a nonspecific stress effect may be of significance as well (Davis and Davis, 1979; see section BIV-2a, above). Last but not least, the cardiovascular regulation exhibits a multitransmitter nature. That monoamines are involved in cholinergic activation of the cardiovascular system was stressed in this section; similar interaction exist between cardiovascular actions of cholinergic and peptidergic systems (Calaresu et al., 1990).

c. Centrally Controlled Gastrointestinal, Urinary, and Enteric Activity

Gastrointestinal (GI), enteric, and urinary functions are directly regulated by the peripheral parasympathetic and sympathetic systems. Supraspinal control is also needed for the normality of these functions and there is an identifiable cholinergic contribution at that level, as speculated upon in early research by Severyn Cytronberg (1927); he referred to this contribution as "vegetative."

Complex descending and ascending pathways beginning and ending in the cortex exert motor and afferent control over these functions. Gastrointestinal regulation involves the dorsal vagal complex (DVC) and the medial subnucleus of the tractus solitarius, and the pontine and tegmental areas constituting the "micturition center" are necessary for bladder control and micturition (Brodal, 1980; Benarroch, 1997a, 1997b; De Groat, 1975; De Groat and Kawatani, 1985; De Groat and Steers, 1990; Ferreira et al., 2000, 2001; see also Chapter 2 DI and DII).

These sites, including the DVC, the "micturition center," and the brainstem areas that are important for gastrointestinal and urinary functions contain cholinergic sites and pathways (Chapter 2 DII and DIII), besides exhibiting peptidergic, aminergic, and monoamine pathways (Koyama et al., 1999; De Groat and Steers, 1990; Benarroch, 1997a, 1997b; Ferreira et al., 2000, 2001; Sahibza da et al., 2002). However, few data are available concerning the role of the cholinergic system and the cholinergic drugs in supraspinal regulation of micturition or gastrointestinal function (see Camilleri, 1993). For example, the firing of the cholinergic neurons in the micturition center relates to the filling and emptying of the bladder (Koyama et al., 1999), while Ferreira et al. (2000) demonstrated that microinjection of nicotine into dorsal vagal motor nucleus and medial subnucleus of the tractus solitarius evokes either increase or decrease, respectively, of the intragastric pressure; these effects appear to be due to a4b2 nicotinic receptors (Ferreira et al., 2002). The difficulty is compounded by the paucity of information concerning direct cholinergic supraspinal efferent pathways to the spinal autonomic preganglionic neurons; interestingly, these neurons exhibit cholinoceptive sites (N. J. Dun, 2003, unpublished data). It must be added that certain GI and micturition diseases may relate to abnormalities in not only peripheral but also supraspinal cholinergic function.

d. Thermocontrol

Isenschmid and Krehl (1912, see Myers, 1974) were the first to demonstrate that thermocontrol is exerted by the hypothalamus, and Robert Myers, Peter Lomax, and Don Jenden (Lomax and Jenden, 1966) pioneered the studies of monoamine, indole, and cholinergic mediation of this phenomenon. Subsequent studies indicated that thermic responses are induced by hypothalamic administration of monoamines, peptides and amino acids, prostaglandins, and pyrogens (Myers, 1974; Clark, 1979; Lee et al., 1983; Simpson et al., 1994). Either hyper- or hypothermia obtains upon administration of these agents, depending on the site of the hypothalamic administration, on the agent, and on the species. Additionally, ionic balance is involved in thermocontrol (Freund, 1911), and related findings gave rise to Myers' notion (Myers, 1982, 1984, 1987) of a Ca^{2+}-mediated hypothalamic set point interacting with neurotransmitters and prostaglandins in the thermoregulation process.

Acetylcholine, carbachol, and antiChEs (for example, physostigmine) evoke sharp hyperthermic responses in the cat, the rat, and the monkey when applied to the anterior hypothalamus, the preoptic area, and sites rostrad and caudad from these areas (Figure 9-23; Myers and Yaksh, 1968, 1969; Myers, 1987; Bruck and Zeisberger, 1978). When applied to the posterior hypothalamus-mesencephalic interface, ACh, muscarinic cholinomimetics, and systemically administered antiChEs produce hypothermia (Clement, 1993). Atropine antagonizes the hyperthermic and hypothermic response. When applied alone or with cholinomimetics to the rat's rostral hypothalamus, atropine induces hyperthermia (Kirkpatrick and Lomax, 1970); in this experiment, atropine may have blocked the muscarinic hypothermia, liberating a nicotinic, hyperthermic response. Indeed, nicotinic agonists produce hyperthermia when applied to the posterior hypothalamus of several species (Hall and Myers, 1977).

There are species differences with regard to the thermic effects of various agents (see Myers,

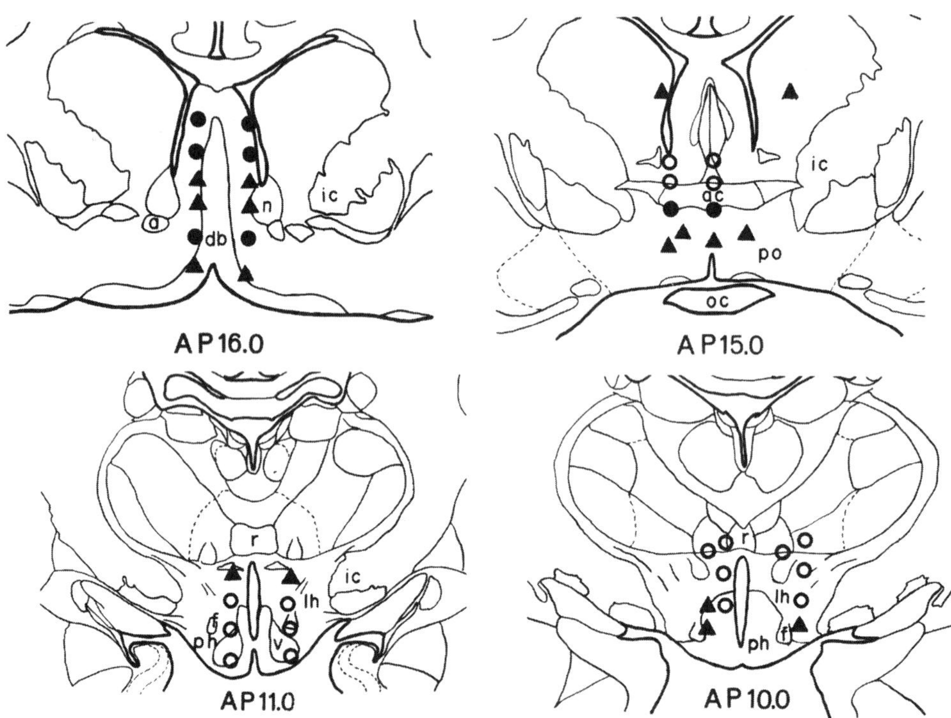

Figure 9-23. Representative frontal sections of cat brain illustrating the anatomical sites at which push-pull perfusions of RX 7260, an anticholinesterase, produce a rise in temperature. ▲, sites where push-pull perfusions of 1.0 µg/min produce hyperthermia of 0.4°C or greater; ○, sites where doses of drug 0.125–0.5 µg/min are ineffective; ●, sites where 1.0 µg/min is ineffective; a, nucleus accumbens; ac, anterior commissure; c, caudate nucleus; db, diagonal band of Broca; f, fornix; ic, internal capsule; lh, lateral hypothalamus; n, nucleus of anterior commissure; oc, optic chiams; ph, posterior hypothalamus; po, preoptic area; r, nucleus reunions; v, ventromedial hypothalamus. (From Myers, 1980, with permission.)

1974); occasionally, the results are inconsistent. For example, Lomax and Jenden (1966) obtained hypothermia when injecting muscarinic cholinomimetics into a rat's rostral or anterior hypothalamus, contrary to the results obtained by others (see above); perhaps the region where the application of cholinomimetics caused hyperthermia was caudad to the regions where the application of these drugs induced hypothermia in the hands of Lomax and Jenden (1966).

That these results reflect physiological function of the hypothalamic cholinergic sites was amply demonstrated. Thus, atropine applied to sites such as the rostral hypothalamus not only antagonizes the hyperthermic effects of cholinomimetics but also produces hypothermia in animals kept at room temperature. Furthermore, in the rhesus monkey, cooling increased and warming decreased ACh efflux from the anterior hypothalamus and preoptic area, while opposite effects were induced by cooling and warming in the mesencephalon and mid- and posterior hypothalamus (Myers, 1987). Finally, bacterial or pyrogen-induced pyrexia causes release of ACh from the anterior hypothalamus (Tangri et al., 1975). Again, the results obtained in this area are not always consistent, and there are species differences with respect to atropine's thermic actions (see Myers, 1974, 1984, and Bruck and Zeisberger, 1978).

It appears that increased heat production rather than the antagonism of heat dissipation engendered by the peripheral vasodilatation underlies the cholinergic, whether nicotinic or muscarinic, hyperthermia. Hyperthermia generated by monoamines, indoles, pyrogens, and prostaglandins is also due to augmented heat production (Rothwell, 1994).

As already stated, noncholinergic neurotransmitters, cytokines, pyrogens and bacteria, and ions participate in thermoregulation. For example, in several species (including the rhesus monkey) when applied to the hypothalamus, $\alpha 1$ and $\alpha 2$ noradrenergic and indoleaminergic agonists induce hypothermia (see Myers, 1974, 1980, 1984, 1987; Bruck and Zeisberger, 1978; and Karczmar, 1976). Pyrexia is evoked when prostaglandins (for example, PGE1) and cytokines are applied to several hypothalamic sites (Clark, 1979; Simpson et al., 1994; Rothwell, 1994). In fact, these agents may mediate bacterial pyrexia. On the other hand, when peptides are applied to the hypothalamus, they evoke either hypo- or hyperthermia (Rothwell, 1994).

These neurotransmitters and thermoactive substances interact during thermoregulation. Pyrogen- or drug-induced (for example, by alcohol; Huttunen et al., 1988) changes in body temperature affect the serotonergic, noradrenergic, and cholinergic systems (Myers, 1969, 1974, 1982), and Robert Myers felt that hypothalamic serotonin and norepinephrine "somehow work . . . in the course of thermoregulation . . . through the release of ACh" (Myers, 1974). Ions are involved as well. Many transmitters as well as peptides are released when there is a change in the "set point" due to ambient temperature variation or shifts in blood Ca^{2+} levels or in the blood Ca^{2+}/Na^{+} ratio (Lee et al., 1983). As suggested by Myers (1982, 1987; Myers and Veale, 1970) and Simpson et al. (1994), ACh may serve as a mediator between the set point of the posterior hypothalamus and the mechanisms located in the rostral hypothalamus. Indeed, there is a close similarity between the thermocontrol effects resulting from the application to the set point of Ca^{2+} ions and of ACh (Myers, 1982, 1987; Simpson et al., 1994).

e. Control of Thirst and Hunger

As known since the 1960s, both nicotinic and muscarinic agonists affect drinking and eating (Stein and Seifter, 1962; Fisher and Coury, 1962; Coury, 1967; Grossman, 1960; Stevenson, 1969). It should be emphasized that cholinergic control of thirst and hunger as well as thermal processes, heat generation, and heat dissipation point up the relationship between the appetitive behavior and energy processes and cholinergic participation in energy processes and energy homeostasis. This relationship between energy phenomena and appetitive behavior is indicated by the hypothalamic vicinity of the pertinent control sites (other central sites, such as the limbic system, are also involved in this relationship; see below) and of the relevant transmitter pathways. If this relation were strictly maintained, cholinergic thermogenesis and cholinergically induced heat production should be accompanied by an increase in food and water intake. In turn, the resulting proteinemia and an increase in blood electrolytes should lead to an additional increase in water intake. Many—but not all—findings are consistent with this notion.

Indeed, generally muscarinic agonists evoke and atropinics block food and water intake as well as pyrexia. The feeding effect of agonists may relate to their induction of a transient hypoglycemia and transient rise in insulin (Smith and Campfield, 1993; see however, Brito et al., 1993), and hunger and/or anticipation of food in rodents conditioned to expect food at a given time is accompanied by an increase in cholinergic activity in several brain sites, including the cortex and limbic system (see, for example, Ghiani et al., 1998). Also, knockout mice deficient in M3 receptors displayed a significant decrease in food intake, which reduced body weight and peripheral fat deposits; Wess and his associates (Yamada et al., 2001) demonstrated that the hypothalamus is the site involved in this M3 function (see also Chapter 5 F).

However, discordant results were also obtained in the rat (Coury, 1967), and Grossman (1964) felt that antagonistic muscarinic systems underlie thirst and hunger; furthermore, in the cat and monkey there was a divorce between the effects of muscarinics on appetitive and thermal function (see Myers, 1969, 1974; Myers et al., 1972; Karczmar, 1976, 1978a). Moreover, at certain hypothalamic sites, cholinergic agonists precipitate hypothermia rather than pyrexia (see above, section BIV-2c).

More complications exist. First, nicotine and nicotinic drugs may exert ingestive and drinking effects, which are opposite to those induced by carbachol or other muscarinic drugs. For example, Myers et al. (1972) induced a marked drinking response by injecting nicotine into the monkey's hypothalamic sites that abut the mammillary complex; on the other hand, nicotine suppressed appetite in some studies (Jessen et al., 2005). Then, noncholinergic transmitters and bioactive substances exert thermal effects and effects on thirst and hunger via their hypothalamic and limbic pathways that are also ambiguous in regard to the notion of the linkage between energy and appetitive processes; these substances include peptides (i.e., neurotensin, cholecystokinin, and substance Y), amino acids (for example, GABA), serotonin, catecholamines, and glucose (Hoebel et al., 1982; Ghiani et al., 1998; Stanley and Leibowitz, 1984; Benarroch, 1997a, 1997b; Myers et al., 1986; and Van de Kar, 1991). For example, alpha-adrenergic agonists invoke food intake, attenuate thirst, and cause hypothermia, which is not consistent with the speculation adduced above (Grossman, 1962, 1964; Yaksh and Myers, 1972; Leibowitz, 1972, 1980; Jhanvar-Uniyal et al., 1986); additionally, sometimes the alpha-adrenergic agonists blocked rather than synergized with the hunger and thirst effect of muscarinics (Callera et al., 1994, 1995). And, as in the case of thermocontrol, a Ca^{2+} regulated set point may be involved in feeling and driking (Meeker et al,. 1978). Finally, the effects of the substances in question may involve indirect rather than direct effects on feeding and drinking. For example, the peptides act peripherally on gastrointestinal secretion, which affects the feeding phenomena via vagal afferents to the hypothalamus (Smith and Gibbs, 1975; Myers and Caleb, 1981).

Inconsistencies among the data reported by various investigators may result from inconsistencies in experimental conditions. Taste and smell contribute to feeding activities and may affect hypothalamic release of substances such as ACh and catecholamines (Myers et al., 1979); these components of the feeding behavior vary from one experiment to another. Different results may be obtained with fasting or nonfasting animals; this condition may be more important in felines than in rats and monkeys. Myers (see Lee et al., 1982) attributes this difference to evolution and ethological differences, as large felines usually seek food at long intervals, while rodents and primates eat frequently. Robert Myers should know; he kept leopards and lions as pets in his household (see also Myers and Sharpe, 1968)! Another important variable is motor function; motor function may be affected by the drugs and transmitters used in the experiments, thus influencing the response to food or water. For example, peptides convert escape response to electric stimulation of the hypothalamus into feeding behavior (Zilov and Patyshakuliev, 1993).

Finally, the drugs, bioactive substances, and transmitters in question may affect thirst and hunger via their actions on the higher centers, and this effect may interplay with their hypothalamic actions (Adrianson, 1995; Oomura et al., 1988). As in the case of all hypothalamic phenomena, the thirst-hunger regulation involves other systems besides the cholinergic system, for example, peptides such as somatostatin and several gut hormones (Hajdu et al., 2003; Konturek et al., 2004).

f. Sexual Behavior

Cholinergic agents clearly affect sexual activity. These cholinergic phenomena, including erectile function, are ultimately mediated by sacral and lumbar autonomic outflows that involve cholinergic synapses; these autonomic efferents are under supraspinal control, but the central cholinergic pathways to the autonomic preganglionic sites were not clearly identified (see Chapter 2 DII and DIII). Similarly, central cholinergic pathways that control libido are not established in full detail, although the contribution of the cortex, some of the nuclei of hypothalamus, and the limbic system to the libido and sexual activity is well established. There is information that males suffering from Alzheimer's disease and treated with cholinergic drugs may exhibit increased libido (see Chapters 2 DI-DIII and 10 K; De Groat and Steers, 1990 and Bitran and Hull, 1988).

Sexual effects may be directly generated by central cholinergic receptors. However, cholinergic effects on sexual activity may be also indirect and mediated by cholinergic effects on the hypothalamic endocrine system; hence, the inclusion of the sexual phenomena in this section, which deals with the hypothalamic function (see section BIV-2a; Babichev, 1995). According to Magna Arnold, the sexual urge and love are regulated by different brain sites (Arnold, 1960).

Being biased with respect to the importance of the cholinergic system, I am glad to report that most of the findings indicate that the cholinergic system facilitates sexual activity. Abrahams and Pickford (1956) demonstrated that DFP and physostigmine implanted into preoptic nuclei induced or prolonged uterine activity and evoked signs of overt sexual activity in dogs. These effects were probably indirect and related to the cholinergically evoked release of oxytocin (see Koelle, 1963). This early work was prescient, as later it was shown that medial preoptic-anterior hypothalamic regions as well as limbic pathways are involved in psychogenic sexual functions (such as erectile response of the male; the data as to the clitoric female response are apparently unavailable; De Groat and Steers, 1990; Bitran and Hull, 1988).

Further work indicated that muscarinics (M1, M2, and M3 agonists) induce overt activity in male and female rats and monkeys (including lordosis, solicitation, "attractive" behavior, facilitation of ejaculation, etc.; these behavioral effects were blocked with atropinics (see for example, Bignami, 1966; Retana-Marquez and Velazquez-Moctezuma, 1993; Retana-Marquez et al., 1993; Lynch et al., 1999). The behavioral effects in female and male rats and monkeys were blocked with atropinics (see for example, Bignami, 1966; Retana-Marquez and Velazquez-Moctezuma, 1993; Retana-Marquez et al., 1993; Lynch et al., 1999; Dohanich et al., 1982, 1993; Soulairac and Soulairac, 1975; Rodgers and Law, 1968; Lindstrom, 1973, 1975; Kow et al., 1995). These phenomena may be related to electrophysiological effects of cholinergic agonists on several hypothalamic areas *in vitro* and *in vivo* (section BIV-2a, above, and Kow et al., 1995, 2004). However, these electrophysiological actions were M3 receptor mediated, while the actual sexual activity was evoked by M1 and M2 receptors (Retana-Marquez and Velazquez-Moctezuma, 1993). Interestingly, Heath (1972) demonstrated that orgastic behavior in humans induced by intraseptal administration of ACh was not distinguishable, subjectively or in terms of EEG, from spontaneous coitus; however, he also reported that norepinephrine induced a similar effect (Figures 9-24 and 9-25). It should be added that there is a divergence (diergism) between cholinergic hypothalamic and related pathways of mammalian males and females (Rhodes and Rubin, 1999).

The effects reported by Dorner (1979) were bizarre. He found that when the quaternary [*sic!*] antiChE pyridostigmine was neonatally administered, it "permanently" augmented sexual activity in the adult male rat; Dorner ascribed this effect to a change in brain differentiation. Another unexpected finding was that both carbachol and atropine implanted into hypothalamic and limbic sites induced lordosis in female rats (Rodgers and Law, 1968).

The facilitatory sexual effects of cholinergic, particularly muscarinic, agonists but also nicotinic agonists may be mediated strictly hormonally on the hypothalamic level and also on the brainstem level, as cholinergic agonists facilitate or induce, and cholinergic antagonists block, the release of gonadotropic hormones and/or their hypothalamic releasing factors, and coitus upgrades the $\alpha 4$ and $\alpha 7$ receptors; these bioactive substances regulate the development of gonads and their steroid hormones and the secondary sexual characteristics

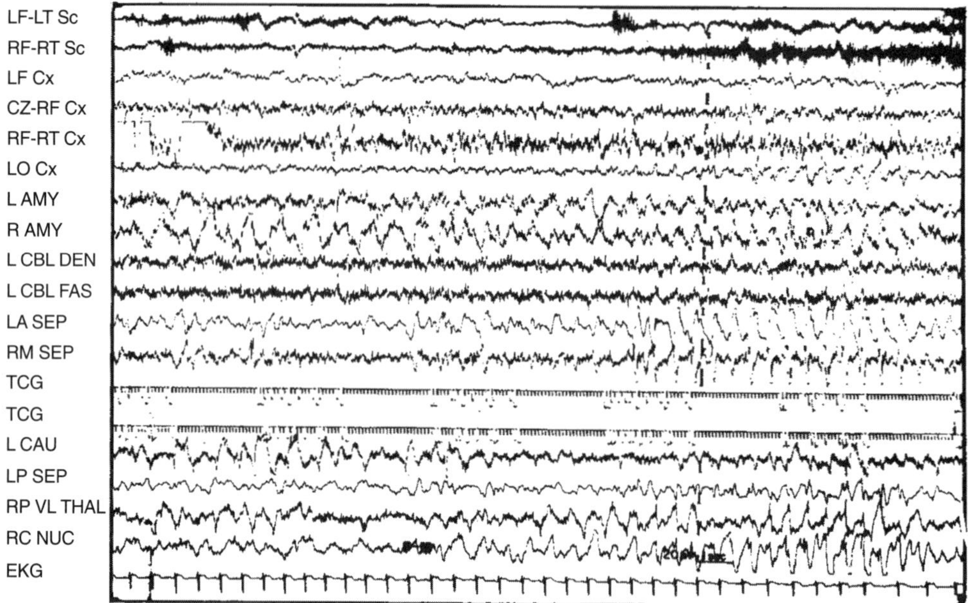

Figure 9-24. Sexual pleasure stimulus: onset of orgasm. Deep, cortical, and scalp electroencephalograms obtained from patient B-19 with onset of orgasm. LF-LT Sc, left frontal to left temporal scalp; RF-RT Sc, right frontal to right temporal scalp; LF Cx, left frontal cortex; CZ-RF Cx, central zero to right frontal cortex; RF-RT Cx, right frontal to right temporal cortex; LO Cx, left occipital cortex; L CBL FAS, left amygdala; R AMY, right amygdala; L AMY, left amygdala; L CBL DEN, left cerebellar dentate; L CBL FAS left cerebellar fastigius; LA SEP, left anterior septal region; RM SEP, right midseptal region; TCG, time code generator, machine 1, and time code generator, machine 2; L CAU, left caudate nucleus; LP SEP, left posterior septal region; RP V L THL, right posterior ventral lateral thalamus; RC NUC, right central nucleus; EKG, electrocardiogram. (Reprinted from Heath, 1972, with permission.)

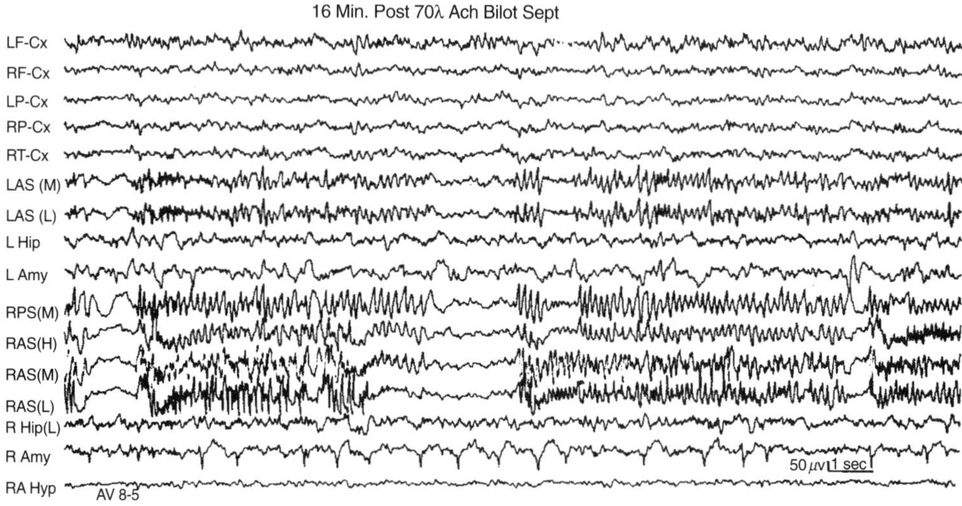

Figure 9-25. (See Figure 9-24 for abbreviations.) Deep surface electroencephalograms obtained from patient B-5 after administration of acetylcholine. Activity is typical of the high-amplitude spindling, which appeared during arousal preceding orgasm and also between orgasms. (Reprinted from Heath, 1972, with permission.)

of either sex (see section BIV-2a, above; Karczmar, 1980; Myers, 1974; Centeno et al., 2004). Finding that muscarinic facilitation of sexual activity may be dependent on the rat's gonadal state reinforces this notion. Thus, Hebert et al. (1994) found that muscarinic facilitation of overt sexual activity occurs in ovariectomized female rats primed with estrogen and/or progesterone (see Lindstrom, 1973, 1975). However, localized hypothalamic and limbic administration of cholinergic and anticholinergic drugs affect sexual behavior (see this section, above).

It is of interest that cholinergic effect on sexual function may be due not only to cholinergic release of gonadotropic hormones and resulting action of gonadal hormones but also to cholinergic-gonadal interaction on the synaptic level: these hormones facilitate cholinergically evoked neuronal potentials. In fact, the coexistence at nonhypothalamic sites such as septum of the gonadotropic hormones and their receptors, the hormones such as LH/RH, and ACh and its receptors was demonstrated (Martinez-Morales et al., 2001; S. G. Beck, 1998; N. J. Dun, 2000, unpublished data; see also section BV-4d, below).

This mixture of peripheral and central actions that are involved in sexual effects of cholinergic agents and certain drugs is a par for this topic. Actually, estrogens and progesterone of the female and androgens, including testosterone, of the male facilitate sexual activity not only via peripheral but also via central sites; this facilitation is accompanied by appropriate electrophysiological phenomena (Ruf, 1976). These phenomena are generated via the central steroid receptor sites, which are present throughout the brain, including the limbic system, the hypothalamus, and the spinal cord, and occur in a number of vertebrate species such as monkeys, primates, reptiles, and birds (Pfaff, 1973, 1988; Davidson, 1977; Lewin, 1994; McEwen et al., 1978; Holsboer, 1997; Martinez-Morales et al., 2001; see also Karczmar, 1978a, 1980; Jones, 1994; Crews et al., 1996; Balthazart et al., 1996; Myers, 1974).

The effectiveness of nicotinic agonists is controversial; although the Russian team of Kirukhin and Ryzhenkov (Kirukhin and Ryzhenkov, 1962; Ryzhenkov and Kirukhin, 1973) found that they facilitate male rats' sexual activity, Retana-Marquez et al. (1993) could not duplicate this finding.

3. Brain Rhythms, Sleep, and Related Phenomena

This section deals with electric activities of the brain represented by the EEG, evoked potentials, and related phenomena. These EEG phenomena are linked with mental activities including cognition, attention, and learning; these matters will be discussed in this section, below.

a. Endogenous Activity of Cholinergic Neurons

Induced cholinergic activities of the brain neurons such as the release of ACh and cholinoceptive responses were already described. Central cholinergic neurons also exhibit spontaneous rhythmic activities that do not depend on sensory input and may be observed in isolated preparations (Woolf and Butcher, 1986; Khateb et al., 1992; see Woolf, 1997). The forebrain, pontomesencephalic and striatal neurons and the prenatal retina's amacrine cells exhibit bursts of these spontaneous, fast activities (Lebedev and Nelson, 1999). It was proposed that these activities that arise prior to the development of sensory inputs help in the maturation of the pertinent neurons and their systems (Woolf, 1997). Some of the bursts appear in cycles regulated by the individual neurons (Woolf, 1997). In some cases, such as bursting exhibited by the pyramidal cortical neurons, the activation of cholinergic basal forebrain neurons suppresses this activity (Alroy et al., 1999). Whether or not these bursts and cycling contribute to the generation of rhythms such as the EEG is not clear at this time (see Andersen and Andersson, 1968).

b. Electroencephalogram Patterns, Evoked Potentials, and Effects of Cholinergic Agonists and Antagonists

There are two main types of EEG pattern: the slow rhythms and the synchronized pulses belong to one type of EEG activity, and the desynchronized patterns to another. The slow pulses, such as alpha and delta rhythms, and the synchronized patterns, which include recruitment, theta rhythms, spindling, augmentation, postreinforcement synchronization (PRS), and so on. depend on the reticulo-thalamico-cortical-cerebellar recruitment system, with possible contributions from the caudate (Andersen and Andersson, 1968; Buchwald et al., 1961; Steriade, 1993a, 1993b; Marczynski, 1993). Some of these patterns are discussed later in this section. The desynchronized EEG is characterized by the presence of fast-frequency rhythms, which include beta and faster activities. It complicates the matters that synchronized and desynchronized rhythms are generated by different brain parts and thus coexist; this is the case in the desynchronized pattern, which is evoked by cholinergic drugs or behavioral arousal in parallel with the theta rhythm. While there are cholinergic correlates of the slow and synchronized rhythms (Marczynski, 1993), the desynchronized patterns combined with theta rhythm exhibit particularly marked cholinergic characteristics.

Cholinergic EEG Arousal

Cortical EEG desynchronization and the theta rhythm are evoked by muscarinic and nicotinic agonists and antiChEs when given systemically or applied locally to several sites extending from the brainstem (pontine and bulbar reticular formation), to cortical areas (entorhinal cortex), the hypothalamus, and the limbic sites (Cordeau and Mancia, 1959; Marczynski, 1967; Karczmar, 1979a, 1993a, 1993b; Vertes, 1982; Caulfield et al., 1992, 1993a, 1993b; McCormick, 1992, 1993; Riekkinen et al., 1993; Van der Linden et al., 1999). The cholinergic, muscarinic, or nicotinic desynchronization consists of a shift of cortical EEG brain waves toward higher frequencies (20 to 30 Hz beta rhythms and, sometimes, gamma activity of 25 to 80 Hz), and lower-voltage wave patterns; these phenomena and the generation of the hippocampal theta, 8 to 12 Hz rhythms, constitute cholinergic EEG desynchronization or cholinergic EEG arousal. Depending on the mode and site of application, slower of faster activity, including gamma rhythms, may be induced (Van der Linden et al., 1999; Gloveli et al., 1999). Cholinergic EEG arousal is similar to EEG patterns characterizing behavioral arousal and learning; particularly, cholinergic desynchronization that obtains in the visual cortex is similar to desynchronization evoked by perceptual experience (Desmedt and Tomberg, 1995; Tomberg and Desmedt, 1998); yet, cholinergic EEG arousal differs quantitatively from behavioral EEG arousal (see below, this section, and section BIV-3c for the relationship between EEG arousal and behavioral arousal).

The desynchronizing and theta rhythm-generating effect of cholinergic agonists (cholinergic EEG arousal) was recognized in the 1930s (Jung and Kornmuller, 1938; Bonnet and Bremer, 1937; Moruzzi, 1939); their findings were greatly expanded subsequently (see Machne and Unna, 1963; Rinaldi and Himwich, 1955a, 1995b; Himwich, 1963; Karczmar, 1976, 1979a; Longo and Loizzo, 1973; Buzsaki et al., 1988). It must be stressed that cholinergic desynchronization, increase in the mean EEG frequency, and the generation of the theta rhythm were observed in all animal species including humans (Alhainen et al., 1991; Shigeta et al., 1993; Nunez, 1995a, 1995b).

The desynchronizing and theta rhythm-generating action of muscarinic agonists and antiChEs is easily antagonized by atropinics, while similar effects by nicotinics are blocked by nicotinolytics. These antagonisms may be elicited by systemic administration of these compounds' tertiary forms or by intraventricular and localized brain application of tertiary or quaternary atropinics and nicotinolytics. Actually, nicotinolytics and atropinics do not just block cholinergically elicited desynchronization but, at large doses, convert it into a special EEG state or evoke this state on their own; in either case, they elicit slow waves of high voltage that are intermingled with high-voltage spikes (Wikler, 1952; Longo, 1966; Kikuchi et al., 1999). Vincenzo Longo (1966) described this state as "hypersynchrony." This hypersynchrony differs from slow waves such as delta rhythms; furthermore, the behavioral state that accompanies atropinic hypersynchrony does not resemble in any way the states reflecting physiologically encountered slow frequencies; depending on the dose of atropine, atropinic hypersynchrony relates to hyperactivity or coma.[5]

The common denominator for the extensive distribution of cholinoceptive desynchronizing sites is constituted by cholinergic radiations that originate in the forebrain (particularly, the nucleus basalis magnocellularis; Chapter 2 DII and DIII). Accordingly, lesions of the nucleus prevent cortical arousal via electric or cholinergic stimulation of the forebrain sites and slow down the EEG, while lesions of midbrain reticular formation and the thalamus do not obliterate or block only transiently the cholinergic or electrically induced desynchronization (Buzsaki and Gage, 1989; Semba, 1991). It must be noticed, however, that desynchronization can be obtained by muscarinic stimulation of *in vitro* preparations such as cortical slices.

Muscarinic desynchronization may be due to cholinergically induced inhibition of rhythmic firing bursts and thalamic spindles, increase in the spike frequency, which relates to a decrease in the so-called spike frequency adaptation, and an interplay between the effects of ACh and those of GABA (McCormick, 1993); these phenomena are due to activation of cholinergic mesopontine nuclei, gigantocellular tegmental fields, and additional circuitry, and the resultant high-frequency response of the thalamic cells, which spreads to the cortex (Steriade, 1993a, 1993b; McCormick, 1993; Krnjevic, 1993; Steriade and Hobson, 1976). These mechanisms reflect the antagonism between desynchronization and the slow EEG patterns; this antagonism will be discussed below. The postsynaptic excitatory effect involved in muscarinic desynchronization may include a number of K^+ currents generated at the M1, M2, and M3 receptor subtypes (McCormick, 1993; Velazquez-Moctezuma et al., 1990a, 1990b; Gillin et al., 1993; Datta et al., 1991; see also section BI, above). In addition to desynchronization, these sites and mechanisms relate to theta rhythms, REM sleep (see next section), and the sleep-wakefulness dipole.

There are differences between nicotinic and muscarinic desynchronization. First, the EEG effect of nicotine, particularly when given in large doses, may be biphasic; after initial desynchronization, EEG slowing and synchronization follows; this second phase of the nicotinic action results from synaptic blocking action and/or desensitization induced by nicotine (Karczmar, 1979a, 1979b); desynchronization induced by high levels of muscarinics and antiChEs is not readily converted into flattening and EEG slowing. Different sites, as established in early research by Franco Rinaldi, Harold Himwich, Edward Domino, and their associates, generate second muscarinic and nicotinic desynchronizations. The muscarinic sites are diffuse and embrace the mesencephalicocortical ascending areas, limbic sites, and hypothalamus, while nicotine's effects are more localized and involve the midbrain reticular formation, thalamus, habenula, and tegmental areas (see the early research of Rinaldi and Himwich, 1955; Kawamura and Domino, 1969; see also Krnjevic, 1993). Nicotinic excitation results from the classical increase in K^+ conductance and glutamate release (Krnjevic, 1981, 1993), while muscarinic excitation is caused by the M receptor phenomena; additional mechanisms may be involved in the nicotinic generation of the theta rhythm (see above, this section).

The Theta Rhythm

Jung and Kornmuller (1938) were the first to describe the cholinergic generation of theta rhythms; these rhythms and their generation was then described in detail by Volia Liberson, who, as well as others, defined them as an instance of synchronization (Liberson and Cadilhac, 1954; see also Brucke and Stumpf, 1957 and Green and Arduini, 1954). Altogether, the synchronized theta rhythm combines with desynchronization in cholinergic EEG arousal. This notion is supported by ample evidence indicating that both desynchronization and the theta rhythm are induced by systemic administration of muscarinics and antiChEs as well as by their application *in vivo* to several hippocampal sites (CA1 and CA3 neurons), entorhinal cortex and stratum radiatum, the pontine tegmentum, and reticular formation (MacVicar and Tse, 1989; Dickson et al., 1994; Vertes et al., 1993; see also Karczmar, 1979a, 1979b); both cholinoceptive neurons and interneurons may be involved (Chapman and Lacaille, 1999). Also, cholinomimetics evoked theta-like bursting *in vitro*: continuously applying carbachol to rat hippocampal slice preparation produced a synchronous theta-like burst in large populations of CA3 neurons (MacVicar and Tse, 1989; Konopacki et al., 1988; Bland et al., 1988). The frequency of the theta rhythm response depends on the site generating the rhythm (Konopacki et al., 1988).

Additional results support the notion of cholinergicity of the hippocampal sites generating the theta rhythm. Fisher et al. (1998, 1999) obtained indirect evidence indicating that the theta rhythm is due to the septal release of ACh, and the septohippocampal neurons generating the theta rhythm appear to be cholinergic (Jones et al., 1998; Manns et al., 2000; Jones, 2005; see also Brazhnik and Vinogradova, 1988; Stewart and Fox, 1989). Both nicotinic and muscarinic mechanisms are involved in the generation of theta patterns. The theta rhythm was evoked by the activation of hippocampal postsynaptic M1 receptors and carbachol-induced rhythmic bursting at septal sites (Carette, 1998; Goleblewski et al., 1993). These muscarinic effects may depend on muscarinic antagonism via their depolarizing action of the slow hyperpolarization generated by repetitive stimulation; this effect may be localized at the CA1 interneurons (Egorov et al., 1999; McQuiston and Madison, 1999). Nicotinic receptors, both pre- and postsynaptic, are also involved. The presynaptic nicotinic receptors that contain the $\alpha 7$ subunit may generate the theta rhythm via mediation of Ca^{2+} influx and glutamate and GABA release (Radcliffe et al., 1999).

Vincenzo Longo's findings that atropine blocked the theta waves (Longo, 1966) has been confirmed many times (see, for example, Dhume et al., 1990). The theta rhythm sensitized to the effects of atropinics by lesions of certain hippocampal sites (Praag et al., 1997); this finding relates to the importance of noncholinoceptive hippocampal cells in the generation of the theta rhythm. As the theta rhythm has important behavioral and cognitive correlates, it is not surprising that its block also has a behavioral significance (see above, this section, and sections BIVc and BV-4, below). It should be added here that in the case of CANMB, cholinergic EEG arousal and generation of the theta rhythm are accompanied by immobility rather than motor activity (W. G. VanMeter and A. G. Karczmar, 1978; unpublished observations; Karczmar, 1979a, 1979b, and section BIV-3e).

Several investigators, notably Vanderwolf (1975), Bland (1986), Stewart and Fox (1990), and Leung (1998), differentiated in rats between 2 or 3 theta rhythms, which differed in their frequency and pattern. Vanderwolf (1975; see also Leung, 1998; Abe and Tuyosawa, 1999) associated 1 type with vocalizations and licking accompanied by "total immobility" and another with walking and rearing; Vanderwolf felt that only the theta pattern evoked by walking and rearing was blocked by the atropinics. Vanderwolf's distinction may not be valid, since his differentiation between pertinent motor activities is vague and since he refers to additional in-between theta rhythms, which may or may not be sensitive to atropine (Vanderwolf, 1975).

Actually, the alleged atropine insensitivity of components of the theta waves may reflect noncholinergic contribution to this rhythm. Stewart and Fox (1990) opined that the atropine-insensitive theta rhythm is generated via septal GABAergic rather than septal cholinergic neurons. It must be remembered that classically, Walter Hess determined that the hippocampal inputs to the amygdala and nucleus accumbens regulate locomotor activity (Hess, 1954); these inputs involve GABAergic and dopaminergic pathways; the cholinergic role in this system is not clear (see Mogenson and Yang, 1991). In addition, Cobb et al. (1999a, 1999b) found that the nicotinic system may contribute to the diversity of the theta rhythm's frequencies, as nicotinics evoked hippocampal bursting that differed in frequency from the muscarinic theta rhythm.

The notion of cholinergic generation of the theta rhythm is not unanimously accepted (see Woolf, 1997; Kichigina et al., 1999). Activation of muscarinic and nicotinic receptors may evoke nonthetalike rhythms; muscarinic and ACh-induced nontheta responses may be induced by depolarization of presynaptic muscarinic sites (Segal et al., 1989; Cobb et al., 1999). Then the theta rhythm was muscarinically induced *in vivo* in rodents with deafferented septum, provided a "critical balance between cholinergic and GABAergic circuitry" was obtained experimentally (Colom et al., 1991; Cobb et al., 1999a, 1999b). Moreover, the removal of dentate gyrus or the CA1 area did not block the effect (MacVicar and Tse, 1989). Finally, it was claimed that the theta oscillation is determined by intrinsic hippocampal mechanisms and does not require rhythmic septal input. However, in view of the strong evidence supporting the notion of significant cholinergic correlates of theta rhythms, this notion is, today, generally accepted (Karczmar, 1979a; Longo and Loizzo, 1973). The situation is not simple by any means; whether obtained via systemic administration of cholinergic agonists or via their septal, hippocampal or brainstem application, the synchronization underlying the theta rhythm is due to the cholinergic regulation of complex excitatory and inhibitory circuitry, the latter involving GABA-dependent circuitry (Krnjevic, 1993; Steriade, 1993a, 1993b; Leung, 1998; Leung and Yim, 1993; Chapman and Lacaille, 1999).

Vincenzo Longo linked theta rhythm with learning processes, and his opinion was fully supported by subsequent investigations (Longo and Loizzo, 1973; Vertes, 1982; Bland, 1986; see below, next section and section BV-d). The Moscow team of Vinogradova, Brazhnik, and others found that theta rhythm and EEG desynchronization are generated and modulated by learning processes, sensory stimulation, and attention-inducing paradigms, and that the cholinergic EEG markers reflect filtering of the sensory information (Brazhnik et al., 1993; Vinogradova et al., 1993a, 1993b, 1999; see also Dickson et al., 1994). However, the statement already made should be reiterated here: cholinergic EEG desynchronization and the accompanying theta rhythm, which constitute EEG arousal, are not synonymous with behavioral arousal; indeed, the EEG arousal patterns differ from those accompanying behavioral arousal, especially when these two EEG patterns are compared by means of power spectrum analysis (see above, this section, and section BV-4d; see also Karczmar, 1979; Fairchild et al., 1975). Yet, this "divorce" is not complete, and there is a relationship between cholinergic and behavioral EEG arousals; in fact, Mircea Steriade, a lifelong student of cholinergic alerting and arousal, opined that "although multiple, cholinergic and monoaminergic, regulatory structures are known to modulate neuronal processes in the thalamus and cerebral cortex, only the cholinergic system maintains a high excitability of the forebrain during . . . brain-active states" (Steriade, 1993b).

Synchronized, Evoked EEG Potentials

Besides theta rhythms, there are several other types of synchronized EEG patterns. Interesting but difficult-to-explain data were obtained with respect to another form of synchronization, the postreinforcement synchronization (PRS; Marczynski et al., 1968; Marczynski, 1993). PRS is a unique parieto-occipital cortex pattern recorded during reward consumption of a conditioned animal; it represents achieving an expected goal, it is obtained in humans, and it denotes the restoration of human and animal sensitivity to new clues (Marczynski, 1993; see below, this section and section BIV-3c). Postreinforcement synchronization is blocked by atropine, and this block is antagonized by antiChEs. Thus, PRS seems to exhibit cholinergic correlates, which is not consistent with cholinergic antagonism of various synchronization modes. Marczynski's (1993) complex interpretation of the PRS as resulting from a balance between two cholinergic forebrain systems and from an interaction between cholinergic and aminoergic systems may explain this inconsistency.

Another category of synchronized, evoked EEG includes evoked somatosensory cortical potentials, such as auditory potentials (the P 13 potentials of rats and P 50 potentials of humans); visual evoked potentials and reversed visual evoked potentials (VEP and PR-VEP potentials); event-related potentials or Bereitschaft (readiness) potentials and their components, the P300 and CNV waves; potentials evoked by such meaningful environmental stimuli as species-specific vocalization; as well as epicortical fields evoked by recognition or detection (detection positivity; DP), conditioned performance (skilled performance positivity; SPP), and conditioning to sensory stimulation and visceral stimulation (Pehl et al., 1998; see Marczynski, 1993). An interesting cortical potential is exhibited by the superior parietal lobule when a movement that is linked with motivation is being made—this discharge does not accompany a movement made in the absence of motivation (Steriade and McCarley, 1990).

There is strong evidence that these potentials and related phenomena such as cortical potentials evoked by limbic stimulation depend on cholinergic pathways radiating from nucleus basalis of Meynert (NBM) and the septohippocampal pathway. More evidence demonstrating the cholinergic nature of these phenomena is that they are facilitated by antiChEs, nicotinics, or muscarinics, reduced by either nicotinolytics or antimuscarinic drugs and blocked by lesions of the pathways in question (Dierks et al., 1994). It is of interest that these potentials are attenuated in Alzheimer's disease, a degenerative condition characterized by a loss of cholinergic neurons, while muscarinic, agonists antagonize this attenuation (Hollander et al., 1987; see also Chapter 10 K); the potentials in question are also diminished in trauma-induced cholinergic damage (Callaway, 1983; Wang et al., 1997; Ashkenazi et al., 1999; Arciniegas et al., 1999; Hortnagl et al., 1998; see also Smythies, 1997; Linster et al., 1999; Knott et al., 1999; Tebano et al., 1999; and Frodl-Bauch et al., 1999; Linster et al., 2005). Interestingly, these potentials may also be diminished in myasthenia gravis, a peripheral cholinergic disease (Fotiou et al., 1998).

Finally, emerging information links the ontogeny of these potentials to molecular genetic regulation of the formation of the nicotinic and muscarinic receptors (Adler et al., 1999). Thus, thalamicocortically evoked spindles (or oscillations) which seem to characterize aged rodents are blocked by cholinergic agents (Jakala and Riekkinen, 1997).

However, cholinergic agonists and antiChEs did not always facilitate the evoked potentials; in such cases, there may have been a noncholinergic or partially noncholinergic polysynaptic contribution to the link between NBM and the somatosensory cortex. Additionally, 40-Hz and faster oscillations (see above, this section) representing perceptive processes are blocked by the onset of the evoked P300 potential; yet, Nancy Woolf (1997) suggested that the 40-Hz oscillations are cholinergic in nature, thus leaving us with the dilemma of how cholinergic facilitation of both the generation of these oscillations in question and of the P300 events may occur.

Antagonism Between Cholinergic Desynchronization and Synchronized Patterns

Cholinergic desynchronization may be looked upon as antagonistic to synchronized EEG patterns; this antagonistic relation extends to the sleep-wakefulness dipole (see section BIV-3c, below). Evoked desynchronization blocks slow patterns originating in the thalamus (i.e., delta waves, thalamico-cortical augmentation and recruitment, and the slow 1- to 4-Hz and 0.3-Hz oscillations) and replaces them with faster 20- to 40-Hz rhythms (Karczmar, 1979a; Muhlethaler and Serafin, 1990; Riekkinen et al., 1990, 1993; Radek, 1993; Leung and Yim, 1993; Sasaki et al., 1975; VanMeter et al., 1978; Timofeev and Steriade, 1998; Steriade, 1993a, 1993b). Also, evoked cholinergic or behavioral desynchronization blocks synchronizing EEG activities (i.e., thalamically controlled recruitment and augmentation). Cholinergically evoked or spontaneous desynchronization also blocks EEG spindling patterns, characteristic of synchronized phases of slow sleep but which may be also generated in awakened animals by caudate and thalamic stimulation (Timofeev and Steriade, 1998).

Mircea Steriade and David McCormick stressed that dorsal geniculate nucleus (LGNd) and reticular nucleus are important for cholinergical suppression of slow waves (Steriade and Llinas, 1988; Steriade et al., 1991; Steriade, 1993a, 1993b; McCormick, 1989, 1992, 1992b, 1993). Muscarinic depolarization of LGNd neurons suppresses thalamic bursting (McCormick and Pape, 1990), while the unique hyperpolarizing action of ACh released from the basal forebrain and brainstem neurons on the cells of the reticular nucleus blocks the thalamic spindles via "decoupling of the reticular . . . nucleus . . . dendrodentritic synchronizing network" (Steriade, 1993b; see Chapter 2 and sections BI-1 and BI-2, above). The reticular nucleus is not involved in the generation of the theta rhythm, hence the lack of antagonism between desynchronization and the theta pattern. Consistently with these notions, nicotinic and muscarinic agonists block the slow rhythms and oscillations (except for the theta rhythm), and both types of antagonists reversed this action (see, for example, Radek, 1993).

The cholinergic antagonism of slow and synchronized rhythms and the cholinergic facilitation of cortical evoked potentials reflects the physiological dipole between arousal and the organism-environment interaction, on the one hand, and several synchronizing systems including thalamico-cortical-cerebellar recruitment system involved in the generation of alpha waves, spindles, and thalamico-cortical augmentation, on the other (Andersen and Andersson, 1968; Buchwald et al., 1961; Steriade, 1993a, 1993b; Steriade and McCarley, 1990). This cholinergic correlate also underlies the dipole between alerting or arousal and sleep and it illustrates the strong links among EEG desynchrony, theta rhythm, and behavioral arousal, which includes cognition (see sections BIV-3c and BVd, below). Additionally, certain desynchronizing rhythms such as the gamma activity mark cognitive phenomena such as signal processing and attention (see, for example, Van der Linden et al., 1999).

However, cholinergic desynchronization and behavioral and motor arousal may not be strictly bound. Wikler (1945, 1952) used the term "divorce" to describe the relationship between these two phenomena; in fact, Fairchild et al. (1975) used the EEG power spectrum analysis to demonstrate the nonidentity of cholinergic and behavioral EEG patterns (Fairchild et al., 1975; see also Karczmar, 1971; 1979a; see also above, this section). Also, the CANMB phase of cholinergic EEG arousal and its accompanying mental states may be characterized by immobility rather than motor activity (see section BVI). In fact, cholinergic desynchronization is preceded by a short period of slow wave patterns 8 to 12 Hz in frequency and diminution of motor activity (Karczmar, 1979a). This state occurs within seconds of intravenous or intracarotid administration of antiChEs or muscarinic agonists in freely moving, unanesthetized cats or rats, and lasts until a desynchronized state supervenes, approximately 1 to 2 minutes (W. G. Van-Meter and A. G. Karczmar, 1978, unpublished observations; Karczmar, 1979a; see also this section, above, and section BVI).

Finally, as already said with respect to other CNS phenomena and as in the case of other functions and behaviors described later in this chapter, EEG and related phenomena constitute multitransmitter phenomena that involve, besides the cholinergic system, particularly GABA and glutamate (Karczmar, 1979a, 1979b, 1981a, 1981b; Glowinski and Karczmar, 1979; Szymusiak, 1995; Marczynski, 1993; Smythies, 1997; Bevan and Bolam, 1995).

c. Awakening or Arousal, Sleep Phases, and Cholinergic Agonists and Antagonists

In the 1940s and 1950s, Moruzzi and Magoun (1949), Bremer and Chatonnet (1949), and Himwich (Rinaldi and Himwich, 1955) proposed that there is a dipole between behavioral arousal or awakening on the one hand and sleep on the other (see also Karczmar, 1967). They related this dipole to the regulatory function of the brainstem "ascending activating system," which essentially consists of the pontine reticular formation (Rinaldi and Himwich, 1955). They also described a corresponding dipole, the EEG dipole that consists of synchronization and some phases of sleep, and desynchronization, and they stressed the importance of the cholinergic system and the effectiveness of cholinergic drugs in the regulation of these phenomena. The dipole also relates to the circadian rhythms and their regulation by hypothalamic mechanisms (see section BIV-3e).

That the cholinergically evoked EEG arousal or EEG desynchronization component of the dipole does not correspond precisely to the behavioral component of the dipole of awakening was already discussed in the preceding section; see also below, sections BIV-4a and BVI. In fact, cholinergic arousal and the cholinergicity of REM sleep may reflect a special cholinergic behavioral syndrome, CANMB, as alluded to in this section, above (see also section BIV-4f).

Sleep may be divided into two main phases, rapid eye movement sleep (REM sleep) and slow wave or synchronized sleep (SW sleep). Both REM and SW sleep have been known since the days of Lucretius (see Jouvet, 1967) and Fontana (1765). Fontana (1765) was the first investigator to describe REM as deep sleep ("sonno profondo"), which generates convulsions. The EEG definition for REM and SW sleep began when Adolf Beck (1890) and Hans Berger (see Brazier, 1959) applied EEG methodology to humans. Detailed descriptions of REM and SW sleep were provided by Richard Klaue (1937) and then by Nat Kleitman and his student Eugene Nasarinsky (see Kleitman, 1963); for Narcolepsy, see Mignot, 1996. They reported that sleeping infants exhibited movement of the eyes under their closed lids, to the accompaniment of desynchronized EEG, and they described a similar phenomenon for adults. When the subjects were awakened—when their EEG showed desynchronization and their eyes moved—and quizzed, it turned out that they were dreaming in the course of the desynchronized sleep; hence, the term REM (rapid eye movement) was applied to this phase of sleep. It is of interest that similar phenomena occur during daydreaming as well as during certain presleep phases (V. Liberson, 1970, personal communication).

Ocular movement that is characteristic of REM sleep is expressed by pontogeniculate occipital (PGO) spikes. REM sleep is also accompanied by muscle atonia and theta rhythms; hence, REM sleep shows paradoxical activities (REM, desynchronization) with respect to its deep sleep nature and is called the "paradoxical" or "dream sleep." Physiological REM sleep and its activation by cholinergic agonists also mediate a number of autonomic phenomena, such as phasic changes in blood pressure and in the cardiac rate, as well as some respiratory depression accompanied by augmentation of respiratory volume; thermoregulation is almost abolished during REM sleep (Benarroch, 1997a, 1997b; Lydic and Baghdoyan, 1992).

Michel Jouvet is a pioneer student of the pharmacology and transmitters underlying REM sleep events; he defined his work as the "monoamine game" (Jouvet, 1967, 1972). His original postulate was the existence of a regulatory dipole, consisting anatomically of the Raphe nuclei and the nucleus coeruleus; biochemically, the dipole consisted of 5-HT and norepinephrine. However, Jouvet (1967) gladdened cholinergikers' souls as he stressed that atropine or scopolamine block and prevent REM sleep in several animal species. Subsequently, many investigators showed that the application of muscarinic agonists and antiChEs to several brainstem and forebrain sites of many species evokes REM sleep and all its concomitants; in fact, their effects were identical to those induced by either electrical stimulation of the appropriate sites or to "natural" REM sleep (see, for example, Pollock and Mistlberger, 2005).

Also, Jouvet's (1967) observation of the antagonistic effect of atropinics on REM sleep in animals was amply confirmed as it was established that REM sleep, whether induced muscarinically or by electric stimulation, or occurring naturally, was blocked by atropinics (see, for example, George et al., 1964; Pompeiano et al., 1967; see also Karczmar, 1979a).

Subsequent to Michel Jouvet's pioneering studies, the most detailed examination of animal REM sleep induction by muscarinic agonists was carried out by Steriade, Hobson, McCarley, Baghdoyan, Lydic, and their associates (see McCarley, 1977; Baghdoyan et al., 1987, 1993; Steriade and Hobson, 1976; Hobson and Steriade, 1986; Hobson, 1992, 2005; Steriade, 1993a, 1993b, 2003; Baghdoyan and Lydic, 1999; see also Haxhiu et al., 2003). In 1976, Mircea Steriade and Alan Hobson (1976) concluded that REM sleep involves as "the executive elements . . . the pontine . . . giant cells of the reticular formation (FTGs)" (see Figure 9-26). These cells and additional vicinal tegmental cells of the lateral dorsal tegmentum, neurons of the peribrachial tegmentum and nucleus coeruleus, and additional brainstem neurons are cholinergic. Acting via their afferents to the pontine reticular formation, these neurons induce desynchronization as they suppress the slow rhythms, such as thalamic spindles and delta waves, of SW sleep. They induce rapid eye movement as the reticular formation sends its "executive commands" to the oculomotor nuclei, and they produce atonia via activating the inhibitory pontine reticular formation and the tegmental nuclei (Kohyama et al., 1994; Yamuy et al., 1994; Shiromani et al., 1987; Hobson et al., 1993; Steriade, 1993a, 1993b).

Additional cholinergic forebrain sites, including basal forebrain, limbic system and amygdala, thalamus and hypothalamus, and brainstem including reticular formation, are also involved in cholinergic arousal, REM sleep, and SWS. In fact, these sites modulate ACh release at the cortex (Jones, 1993, 2003; Silvestri and Kapp, 1998; Demarco et al., 2004).

Interesting information was provided by Hobson and his associates (see Hobson et al., 1993), as they demonstrated in cats that carbachol may generate two types of REM sleep: its application to anterodorsal pontine tegmentum evokes "immediate but short live REM sleep," whereas its application to the peribrachial pons induces "long term," delayed REM sleep. There is also some evidence that nicotine, when applied to the sites in question, also induces REM sleep; in particular, nicotine appears to evoke PGO waves (Velazquez-Moctezuma et al., 1990a, 1990b; Datta et al., 1991; Gillin et al., 1993; Quattrochi et al., 1998). Generally, it is considered that the cholinergic, particularly muscarinic induction of REM sleep depends on ACh-induced depolarization and consequent excitation of the appropriate neurons (see, for example, Steriade, 1993b; Krnjevic, 1993). This muscarinic depolarization and REM sleep induction is mediated, both in animals and in humans, by M2 receptors, the M1 and M4 agonists and antagonists being less effective in evoking and blocking, respectively, REM sleep (Velazquez-Moctezuma et al., 1989; Gillin et al., 1991, 1993). Also, it was suggested by Angeli et al. (1993) that the M1, M2, and M4 receptors mediate different patterns of REM sleep. An opposite viewpoint was offered by Rudolfo Llinas and his associates (Leonard and Llinas, 1994), namely, that evocation of REM sleep is mediated by inhibitory, hyperpolarizing ACh action.

These findings emphasize the cholinergic aspects of REM sleep. Consistent with this notion, REM sleep deprivation causes upregulation of muscarinic receptors in pertinent brain areas of rats as well as lowering of ACh levels in the forebrain; the upregulation underlies REM sleep post-deprivation rebound (Tsai et al., 1994; Smith, 2003; see also section BVI). Furthermore, dietary maneuvering designed to decrease cholinergic activity attenuates the occurrence of REM sleep

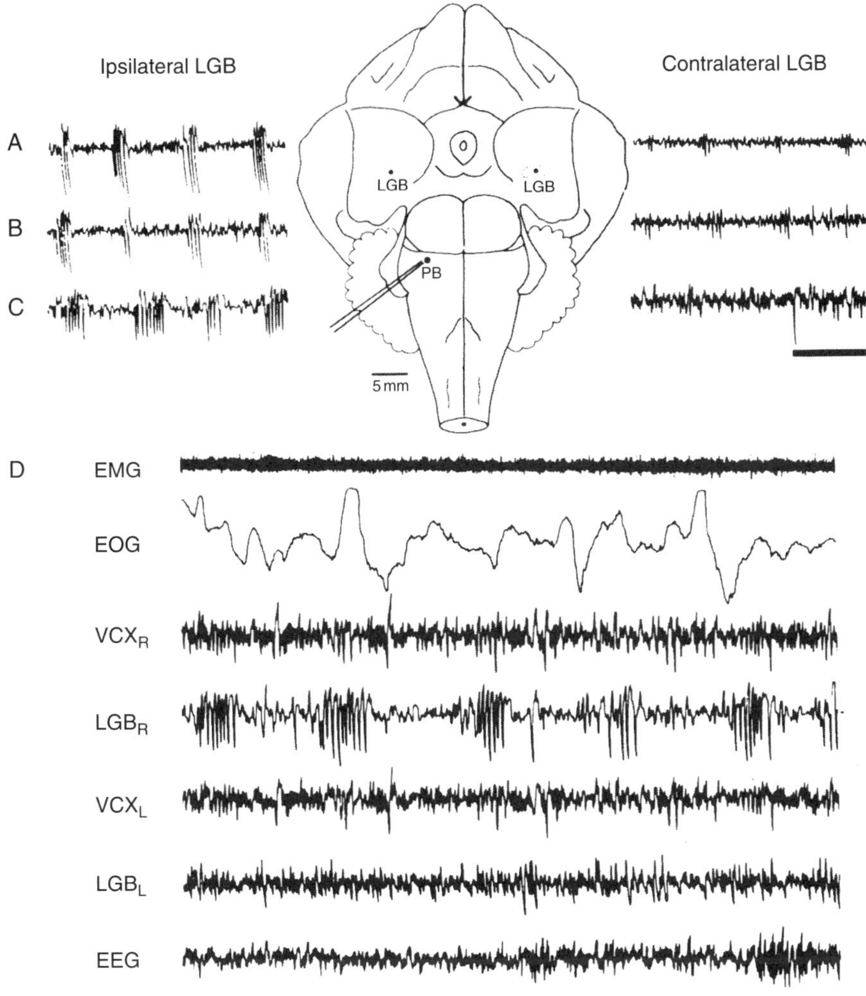

Figure 9-26. Immediate effects of carbachol. Polygraphic records (panels A to D) demonstrate PGO waves triggered 15 minutes post injection in the ipsilateral LGB (n = 4 cats). In the center of panels A to C is a schematic diagram of the ventral aspect of cat brain showing the site of carabachol injection made by a dorsal approach to the Pb and the ipsilateral LGB in which the immediate pontogeniculate occipital (PGO) effect (waves) was observed. Note in D the eye movements (EOG), stereotyped pattern of ipsilateral PGO wave clusters (LGB_R), persistent muscle tone (EMG), and an activated EEG. Also notable is the complete absence of contralateral PGO waves (LGB_L). Time scale for polygraphic recordings is 5 seconds. EMG, electromyogram; EOG, electrooculogram; EEG, electroencephalogram; $LGB_{R/L}$, lateral geniculate body right/left; $VCX_{R/L}$, visual cortex right/left. (From Hobson, 1992, with permission.)

(Szymusiak et al., 1993). Then, during REM sleep, ACh release from the cortex and the hippocampus is significantly higher than during wakefulness and three times higher in the course of REM sleep than during SW sleep. Also, CAT mRNA levels of the diagonal band increase during REM sleep (Jasper and Tessier, 1971; Kametani and Kawamura, 1990; Greco et al., 1999). Similarly, Mircea Steriade and his Cambridge team found that CAT activity in the rat limbic system is higher during wakefulness than during SW sleep, and even higher in REM sleep, and that messenger

CAT RNA is low during wakefulness and high during sleep, including REM sleep (Greco et al., 1999; Shiromani, 1998; Steriade, 1993a, 1993b, 2003). They interpreted these data as indicating that high cholinergic activity during wakefulness requires compensatory high CAT mRNA activity during sleep.

It should be added that slow sleep also features cholinergic correlates: both forms of sleep were induced by Hernandez-Peon when he applied crystals of ACh to the limbic forebrain, brainstem, and midbrain or their vicinity (see Hernandez-Peon et al., 1963). The sites in question included the Nauta's limbic circuit with the lateral preoptic area, and they subserved both electric and cholinoceptive evocation of SW sleep (Marczynski, 1967; Myers, 1974; Karczmar, 1979a). In addition, muscarinic agonists applied *in vivo* to the thalamus produce EEG spindles resembling those present in "natural" SW sleep (Muhlethaler and Serafin, 1990). Cholinergic narcolepsy may constitute a related phenomenon (Nitz et al., 1995). There is currently a relative paucity of investigations concerning cholinergic or any other aspects of SW sleep phenomena; perhaps the drama of the dream element and of the paradoxical character of REM sleep caused us to forget the issues of SW sleep.

As important as are the cholinergic correlates of REM (and SW) sleep, REM sleep, like other central phenomena, is not purely cholinergic. Michel Jouvet's dictum that REM sleep reflects the "monoamine game" is valid, although modifications of his idea may be necessary. Thus, Jouvet stressed the functional significance of norepinephrine for REM sleep; yet, it is important to note that several investigators presented evidence indicating that adrenergic action is not obligatory for induction of REM sleep. Accordingly, antiChEs readily induce REM sleep in rabbits and rats following profound depletion of catecholamines (Figure 9-27; Karczmar et al., 1970; see also Riemann et al., 1992). Perhaps the healthiest viewpoint is that a certain level of noradrenergic activity facilitates the cholinergic evocation of REM sleep (see Mastrangelo et al., 1994 and Boop et al., 1994). Furthermore, besides serotonin's contribution to sleep and arousal (Jouvet, 1967), inhibitory transmitters such as GABA and galanin block REM sleep, thus activating non-REM sleep, while glutamate is involved in arousal and REM sleep (see, for example, McGinty and Szymusiak, 2003). Also, peptides and hormones such as somatostatin are involved in the regulation of sleep (Hajdu et al., 2003).

Aside from non-ACh transmitters classically associated with REM sleep (i.e., norepinephrine, histamine, and serotonin), excitatory and inhibitory amino acids, peptides, certain trophic factors, and second messenger mechanisms (including nitric oxide) were related to REM (see, for example, Meierkord, 1994; McCormick, 1993; Szymusiak, 1995; Capece and Lydic, 1997; Williams et al., 1997; Williams and Kauer, 1997; Bevan and Bolam, 1995; Xi et al., 1999; Crochet and Sakai, 1999). The possible role of nitric oxide is suggested by the finding that thalamic release of both nitric oxide and ACh increase during REM sleep (Williams et al., 1997). Also, second messengers such as cAMP underlying particularly nicotinic—but also certain muscarinic—phenomena may be antagonistic to REM sleep generation (Capece and Lydic, 1997); however, Capece and Lydic (1997) suggested that the cAMP effect may relate to cholinergically induced rather than spontaneous REM sleep; this notion needs confirmation.

A special reference must be made to the studies of REM sleep in humans: it is always exciting when results obtained in animals are duplicated in humans; in the case of REM sleep this duplication was achieved dramatically and faithfully by Chris Gillin and his associates (Sitaram et al., 1976; Gillin et al., 1978, 1984, 1993). They employed healthy human volunteers to show that antiChEs such as physostigmine and a number of muscarinic agonists including pilocarpine and arecoline hasten the onset of REM sleep and increase the frequency of REM sleep episodes during the night (Berkowitz et al., 1990). Furthermore, muscarinic agonists, but not always antiChEs, increased the duration of REM sleep episodes, and converse effects were obtained with scopolamine and atropine (see Gillin et al., 1978, 1984; Shiromani et al., 1987; Berkowitz et al., 1990; Spiegel, 1984).

These findings are clinically important, as the changes in REM sleep in affective disorders and psychotic disease appear to have a diagnostic role; this change may also suggest that there is a cholinergic contribution to these disease states and that some consideration should be given to the use of cholinergic drugs in these disease states

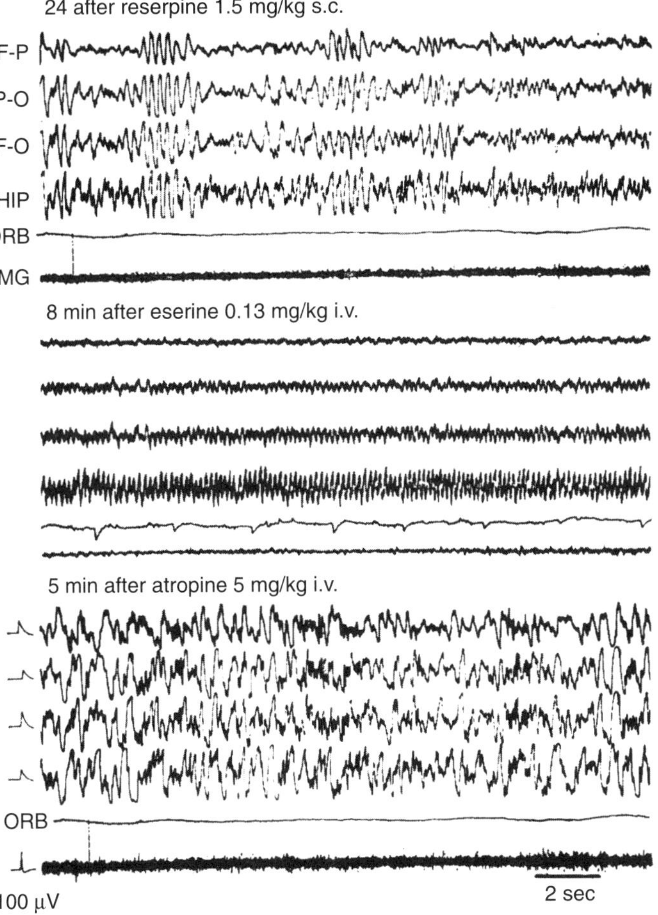

Figure 9-27. The pharmacological model of paradoxical sleep. Antagonistic effects of atropine on the EEG and EMG patterns induced by eserine in the reserpinized rabbit/cat. F-P, anterior-posterior sensory-motor cortex; P-O, posterior-sensory motor-optical cortex; F-O, frontal-sensory motor-optical cortex; HIP, dorsal hippocampus; ORB, orbital electrodes; EMG, electromyograph. (From Karczmar et al., 1970, with permission.)

(see Karczmar, 1988, 1995; Pfeiffer and Jenney, 1957).

The importance of REM sleep as a special behavioral or functional state is well defined by Mircea Steriade (1993a): "mesopontine cholinergic cells show an increased excitability in advance of the most precocious signs of brain activation during transition from resting (EEG-synchronized) sleep to [arousal] or REM sleep." This special brain activation in the course of REM sleep also relates to a behavoral/functional syndrome referred to as cholinergic non-mobile behavior (CNMD); it may be speculated that CNMB represents an "awaken phase of REM sleep" (Karczmar, 1995, and below, section BVI). In agreement with this notion, REM sleep disorder correlates with a number of diseased mental states, including memory perturbations and several forms of dementia (Ferman et al., 1999).

d. Central Nervous System Seizures and the Cholinergic System

Anticholinesterases and cholinergic muscarinic and nicotinic agonists exhibit two types of effects during overt and EEG seizures: in small doses, they act as anticonvulsants; in large, toxic doses they induce overt convulsions and EEG seizures.

Cholinergic agonists and antiChEs, of either carbamate or OP type, induce convulsions and facilitate spikes induced by strychnine when given in large doses in several mammalian and non-mammalian vertebrate species; this also occurs in humans exposed accidentally, criminally, or in war to OP war gases and insecticides (Sjostrand, 1937; Hyde et al., 1949; see also Machne and Unna, 1963; Maynert et al., 1975; Karczmar, 1974, 1979a; Lesny and Vojta, 1960; Karczmar and Koppanyi, 1953). Glenn et al. (1987), Hirvonen et al. (1993), and Cavalheiro et al. (1994) confirmed these early findings, although arecoline, even when administered in large doses, seemed to induce only EEG desynchronization (Domino et al., 1987). Furthermore, these compounds decrease the threshold for electroshock- and pentylenetetrazol- or strychnine-induced convulsions (see Maynert et al., 1975, and Karczmar, 1979a).

EEG and overt seizures occurred from systemic, intramuscular, intracarotid, or subcutaneous application of these drugs (see Maynert et al., 1975) and intraventricular administration (see, for example, Feldberg and Sherwood, 1954); they are also effective upon topical application to the cortex, or localized administration into specific brain sites, particularly limbic structures such as amygdala and hippocampus (Baker and Benedict, 1968; Cain, 1983; see also Karczmar, 1979a; Stringer and Lothman, 1991). In fact, these agents, including antiChEs, induce or facilitate seizure-like activity and discharges in isolated preparations, such as limbic slices, particularly during the blockade of inhibitory GABA-mediated potentials (Lebeda and Rutecki, 1987; Gruslin et al., 1999).

The seizures occur following complete or nearly complete inhibition of AChE and a several-fold accumulation of brain ACh, upon *in vivo* or *in vitro* administration of the antiChE (Karczmar et al., 1973; Karczmar, 1974; Tonduli et al., 1999). Furthermore, during the seizures induced in mice by electroshock or convulsants such as strychnine or pentelenetetrazole, brain ACh is depleted and ACh turnover is increased (A. G. Karczmar and G. Kindel, 1975, unpublished data; Karczmar, 1974; Serra et al., 1999). Taken together, these findings indicate that the cholinergic system is mobilized during seizures and are consistent with the epileptogenic actions of the cholinergic drugs.

Both muscarinic and nicotinic agonists induce seizures at large doses; with nicotinics the doses needed are greater than those capable of producing a tremorigenic effect. In the case of nicotinics, the effect results from direct nicotinic action on the nicotinic receptors (Domino et al., 1987), although some investigators suggest that convulsant nicotinic action results from an activation of a muscarinic link (McGown and Breese, 1990). The notion of direct involvement of nicotinic receptors in nicotinic induction of seizures is indicated by results obtained with knockout mice devoid of $\alpha 5$ receptors, as nicotinic agonists could not cause convulsions in these animals (see Chapter 6 C-3).

M1 receptors are commonly involved in muscarinically induced seizures (Cruikshank et al., 1994; Maslanski et al., 1994). Large doses of muscarinics and nicotinics, and of antiChEs, particularly of the OP type, may induce status epilepticus, which persists throughout postinjection animal life (Cavalheiro et al., 1994). Even when the status epilepticus does not obtain, cholinergically induced convulsions cause neuronal death, particularly in the hippocampus and the cortex; this may be due to a direct effect of cholinergic agonists or be a consequence of convulsions and the generation of free radicals. The brain seems to attempt to compensate for the damage via increased protein synthesis (Cavalheiro et al., 1994; Hirvonen et al., 1993; Jacobsson et al., 1999; Naffah-Mazacoratti et al., 1999). A similar loss occurs in human epilepsy and may contribute to the status epilepticus (Mello et al., 1993).

It must be emphasized that these large doses of antiChEs or cholinergic agonists produce other than convulsive actions, and these effects are relevant for the mechanism of ensuing convulsions. These effects include cardiovascular and respiratory failure, and the resulting anoxia and reduced blood flow may underlie or cause the convulsive phenomena (see above, section BIV-2b; Chapter 7 DI-1; Wills, 1963, 1970). Ample evidence, reviewed in Chapter 7 DI-1, indicates that these parameters play only a contributory role; the fact that cholinergic agonists induce seizures in isolated preparations also argues for this notion (see Lebeda and Rutecki, 1987).

This is not to say that this contribution is unimportant, and other factors may also be significant for cholinergic convulsions. For example, GABA-ergic inhibitory synapses are sensitive to anoxia, and their damage may be facilitatory with regard to the generation of convulsions. And, high noradrenergic and aminoergic NMDA-related activity induced by convulsions and/or cholinergic agonists, as well as release of cytokines, may be contributory to cholinergically induced seizures (Tonduli et al., 1999; Jankowsky and Paterson, 1999; Raveh et al., 1999).

Convulsive effects of muscarinic and nicotinic agonists are very effectively blocked by atropinics and nicotinolytics, respectively (see Karczmar, 1974, 1979a; Maynert et al., 1975; Domino et al., 1987). Furthermore, atropinics antagonize—albeit weakly—convulsions induced by strychnine and electroshock (including the electroshock applied to epileptogenic sites such as the hippocampus; see Young and Dragunow, 1993) as well as spontaneously occurring seizures (Maynert et al., 1975); similar anticonvulsant effects were obtained with nicotinolytics (Loscher et al., 2003). Also, in hippocampal slices, M1 and M2 antagonists prevented induction of discharges by cholinergic agonists and antiChEs in the presence of GABA antagonists; it is consistent with these data that muscarinic agonists (or agonists exhibiting mixed muscarinic and enicotinic agonisms, such as pilocarpine) seem to be unable to induce convulsions in knockout mice devoid of M1 receptors (see Chapter 5 F). Furthermore, atropinics synergized with diphenylhydantoin with respect to its antagonism of electrically induced seizures (Chen et al., 1968). In fact, atropine and scopolamine appeared at one time to be useful as antiepileptics in humans; their use was discontinued when the results obtained were, on the whole, inconclusive (see Wolff, 1956). Actually, atropinics elicit convulsions in animals when administered systemically at large doses or applied to the cortex; it may be added that the data concerning atropinic effectiveness as antagonist of seizures, even those induced by cholinergic agonists, are not consistent (Daniels and Spehlman, 1973; see also Maynert et al., 1975; Karczmar, 1979a).

The causation of the convulsant action of cholinergic agonists and antiChEs, and the role of the cholinergic system in seizure generation are difficult to define, particularly as the mechanisms and the sites, which are involved in seizure generation, are not localized (see McIntyre and Racine, 1986; Peterson and Albertson, 1982; Stringer and Lothman, 1990; Morin et al., 1994). It was proposed that endogenous or experimental seizures arise via sensitive loci of the forebrain, including the limbic system and certain cortical areas such as the pyriform cortex; this constitutes the epileptic neuron hypothesis. These sites are particularly easily destabilized and prone to generate seizures in response to repetitive stimuli applied to these structures, possibly via generating the long term potentiation (LTP)-like phenomenon; this facilitation occurs with muscarinic agonists and may be due to a second messenger-activated Ca^{2+} release (Okada et al., 1991). The response then spreads to other neurons; this spread is transformed into the seizure discharge via a type of recruitment referred to as kindling, that is, "progressive development of electroencephalographic and behavioral seizures" (McIntyre and Racine, 1986); the pyriform cortex may constitute the common site where the epileptiform discharge will ultimately be generated, irrespective of the original site of the kindling. These hypotheses also concern the possible role of excitatory as well as inhibitory amino acids such as glutamate, aspartate, and GABA; cholinergic agonists, including antiChEs and M1 agents, may facilitate the epileptiform action of, for example, glutaminergics (see, for example, Ryan et al., 1984; Meldrum et al., 1983; Gruslin et al., 1999; Potier and Psarropoulou, 2004). Moreover, the generation of convulsions may involve the disinhibition of GABAergic and noradrenergic synapses, as their activity may counter the induction of seizures (Kalichman, 1982; Tonduli et al., 1999). It is of interest that an aspartate antagonist attenuated pilocarpine-induced seizures (Starr and Starr, 1994).

It is important in the present context that some of these epileptogenic sites, such as the hippocampus and pyriform cortex, are rich in cholinergic synapses (see Chapter 2 DII and DIII); accordingly, at these sites the muscarinic and nicotinic agonists and antiChEs may facilitate kindling and LTP-like phenomena (Burchfiel et al., 1979; Kuryatov et al., 1997; see also above, sections BI-1 and BIV-3). Consistently with this notion, muscarinic antagonists retard the rate of the development of kindling generated in limbic structures such as amygdala or dentate (Stringer and Lothman, 1991) and act additively in this situation with amino acid antagonists (Cain et al., 1988; Raveh et al., 1999). Interestingly, a genetic mutation that induces the production of an abnormal nicotinic receptor was related recently to familial frontal lobe epilepsy; it was hypothesized that this condition is due to either receptor hypofunction resulting from increased desensitization rate or from receptor hyperfunction resulting from the augmentation of its ACh affinity (Lena and Changeux, 1998). Not clearly related to these findings is the indication that in animal models of seizures (such as kainic acid-induced recurrent epilepsy), CAT levels are increased in some and decreased in other pertinent brain areas (Baran et al., 2004).

The data are, however, inconsistent; for example, the muscarinic antagonists did not suppress the discharge in some studies and did not affect their duration in other studies (Cain et al., 1988; Stringer and Lothman, 1991); it must be noted also that the effective doses of the muscarinic antagonists were very high (Stringer and Lothman, 1991). Finally, the cholinergic agonists displayed biphasic actions on the rate of kindling (Stringer and Lothman, 1991).

This inconsistency, as well as the inconsistency noted earlier with respect to effectiveness of atropine as an anticonvulsant, relates to the earlier reference in this section as to anticonvulsant effect of small, nontoxic doses of cholinergic agonists and antiChEs. Cholinergic agonists including antiChEs reduced drug- or electroshock-induced seizures, or seizures of the genetically seizure-prone rodents (Meierkord, 1994; see also Karczmar, 1974, 1976, 1979a, 1979b, 1979c). A similar effect was demonstrated in cortical slab preparations (Vazques and Krip, 1973). In fact, physostigmine and neostigmine reduced epileptic cortical discharges in humans (this effect of the quaternary agent neostigmine is surprising; Williams and Russell, 1941). Furthermore, blockers of ACh synthesis induce or facilitate convulsions (Hoover et al, 1977; see Karczmar, 1979). The capacity of atropinics to induce seizures as mentioned in this section is consistent with this particular set of results.

What is the basis of this anticonvulsant action of small doses of cholinergic agonists? Convulsions constitute a "hypersynchrony" phenomenon (Gastaut and Fischer-Williams, 1959), and Ward and his associates (1969) stated that the hypersynchrony of the EEG seizure indicates a "massive discharge of many neurons in unison, abolishing . . . the normal integrative activity of the brain." Accordingly, Karczmar (1979a) speculated that the desynchronizing effect of sublethal doses of cholinergic agonists and antiChEs (see section BIV-3a) might antagonize this hypersynchrony. It must be mentioned that large, seizure-inducing doses of these drugs induce desynchronization prior to causing convulsions. On the other hand, high brain ACh concentration induced by antiChEs and large doses of cholinergic agonists are likely to desensitize and block postsynaptic cholinergic sites and prevent the desynchronizing, anticonvulsant role of cholinergic CNS (Lallement et al., 1992; see also Chapter 7 DI). The relation between ACh release and convulsions may also be pertinent in this context. As already mentioned, strychnine and pentylenetetrazole reduce the neuronal ACh content and picrotoxin increases ACh release (see Serra et al., 1999; Suzuki et al., 2004); do these phenomena suggest that early seizure-induced depletion of ACh attenuates the anticonvulsive function of the cholinergic system and facilitates the spread of seizures?

e. Chronobiology

It may be expected that the cholinergic system, which is so deeply involved in phases of sleep as well as in the wakefulness-alertness-sleep cycle, should also relate to circadian rhythms or chronobiological cycles, whether of diurnal or superdiurnal nature (such as seasonal and hibernation phenomena). There is much evidence as to the role of hormones, monoamines, and, particularly, indoleamines in the control of these cycles (Reghunandan et al., 1993); the participation of the cholinergic system in these cycles was pioneered by Alexander Friedman (Friedman and Walker, 1968, 1972) and Martoff (1953).

Some of the data that concern the diurnal rhythms of cholinergic parameters are inconsistent: one set of results suggests that ACh, AChE, and CAT peaks and troughs are synchronized with the diurnal rhythms of activity and nonactivity, the cycles being reversed in sign in the case of nocturnal rodents (Lin, 1973; Karczmar et al., 1973; Greco et al., 1999). Consistently with these data, Jimenez-Capdeville and Dykes (1993) and Le Bon (1992) found that ACh release from the somatosensory rat cortex increases during the night in nocturnal animals; also, REM sleep, a cholinergic phenomenon, cycles diurnally or seasonally. Saito et al. (1975), however, found an opposite rhythm; in their hands, ACh and AChE peaks occurred during the low levels of activity (incidentally, levels of ACh recorded by Saito and his associates are much lower than those reported by others; see Karczmar, 1976).

As alluded to, cholinergic parameters also change seasonally and with respect to the oestrus. Seasonal changes were reported in the case of the amphibian brain and spinal cord; for example, the cortical release of ACh appears to increase in the summer (Monnier and Herkert, 1972; see Karczmar, 1976, for further references). It was suggested that these changes relate to changes in ambient temperature—a telling factor in the case of poikilothermic amphibia—or in sexual or endocrine activity. In this latter context it is of interest that CAT levels are almost twice as high on the right than on the left side of the preoptic-anterior hypothalamic area of the rat on the first day of the oestrus, and on the left side of this area on the second day of the oestrus; this suggests that the "cholinergic system modulates, in a circadian and asymmetric way, . . . ovulation" (Sanchez et al., 1994).

Sensitivity or activity of the parasympathetic system may also cycle seasonally (Martof, 1953). Alexander Friedman (1971) suggested that the success of the epochal experiment of Otto Loewi (Loewi, 1921, 1960) was due to just this phenomenon, as this experiment was carried out in early spring (Loewi, 1921), which may be the peak of parasympathetic activity in frogs.

Is there a brain site that controls these rhythms? Is that site itself under cholinergic control? That the retinal-hypothalamic pathway that terminates at the hypothalamic suprachiasmatic and paraventricular nuclei is involved in light-sensitive, diurnal rhythm is well known; excitatory amino acids, catecholamines, and presumably other transmitters regulate the activities of this system (Benarroch, 1997a, 1997b). For example, as shown by the original student of this pathway, Julius Axelrod, the sympathetic innervation mediates the circadian function of the system with regard to melatonin secretion by the pineal gland (see Axelrod, 1974). Also, the cholinergic contribution to the function of the suprachiasmatic nuclei was emphasized (Reghunandan et al., 1993; Liu and Gillette, 1996), and the disruption of its function by nicotinic blockers was described (Rusak and Bina, 1990; Zhang et al., 1993); it is particularly telling in the present context that nicotinics block the c-fos expression in the suprachiasmatic nucleus induced by light-induced phase shifts in behavior (Zhang et al., 1993). On the other hand, there is controversy as to whether or not the muscarinic agonists affect the system (compare Kanematsu, 1994, with Colwell et al., 1993, and Liu and Gillette, 1996). Finally, in Alzheimer's disease, a condition characterized by significant loss of cholinergic neurons, there is a disturbance of circadian rhythms (Hofman and Swaab, 1992).

The situation is complex, as the generation and the persistence of the circadian rhythm may be two independent phenomena; only the latter might depend on cholinergic (nicotinic?) activity (Rusak and Bina, 1990), and the cholinergic system may affect circadian rhythms via a pathway independent of that affected by light; for example, limbic striatum shows a diurnal rhythm in ACh release regulation (Colwell et al., 1993; Jabourian et al., 2004).

BV. Mental and Behavioral Functions with Cholinergic Implications

Cholinergic aspects of various mental behaviors have been demonstrated in humans, in current studies as well as in the days when drugs were somewhat routinely given intraventricularly to human subjects (see, for example, Cushing, 1931); in the case of animals, cholinergic correlates exist even with respect to such unusual behaviors as imprinting (Kovach, 1964). Social and moral behavior of the human may constitute an exception, as the cholinergic correlates of this behavior have not be studied to this point. Yet, it may be safely hypothesized that such a correlation will be established one day, in view of the important contribution of the cholinergic system to the overall interplay between the organism and the environment (see section BVI, below).

1. Aggression

Aggression is a confrontational act of threat or injury, planned or unplanned, within a species or across species and genera; animal or human aggression also includes defensive activities. In humans, several ictal emotions accompany aggression, including fear, and it appears on the basis of animal overt behavior that the same holds for animals (see, for example, P. Gloor's discussion of Kaada's 1967 paper; see also Rochlin, 1973; Whelan, 1974; Benarroch, 1997a, 1997b, and Davidson et al., 2000).

There are many types of aggression, in humans and in animals, that fit the definition. In the case of animals, there are several types of laboratory-induced, experimental aggression. Thus, aggression may be induced by footshock or other painful stimuli, injection of chemical irritants, social isolation, limbic lesions, localized or systemic injection of certain drugs (such as cholinergic agonists, as will be discussed later, and apomorphine), extinction of reinforcement, fear, anxiety, and presence of appropriate victims (for example, presence of mice in the case of predatory rats) or foes (Figures 9-28 and 9-29). Luigi Valzelli, a pioneer of the studies of aggression used extensively as a model the violent behavior elicited in mice by social isolation; it must be added that certain subspecies of mice are genetically predisposed for aggression. The presence of a foe may evoke a defense reaction that includes either escape or tonic immobility (Hilton, 1982; de Oliveira et al., 1997; Kelly and Hake, 1970; Hutchinson et al., 1968). Several of these forms of aggression may be considered as instances of frustration-induced aggressions (Dollard et al., 1939); they are referred to as "savage syndrome" or rage. These acts of aggression occur in the absence of a teleological object of aggression and are linked to affect or emotion, particularly in the human (Davidson et al., 2000).

Furthermore, there are several forms of field or ethological aggression including territorial defense, protecting pups, predation, and establishing a mating or hierarchic position. These forms of aggression were described in the course of ethological and social investigations (Lorenz, 1963; Winslow and Camacho, 1995); these ethological aggressions may be modeled in the laboratory (Karczmar et al., 1973; Winslow and Camacho, 1995). Of course, human aggression is even more diversified, particularly with respect to planned (in the case of war) and social types of aggression (From, 1973; Rothlin et al., 1973).

Figure 9-28. Fighting episode between C57BL male mice following 2 weeks of isolation.

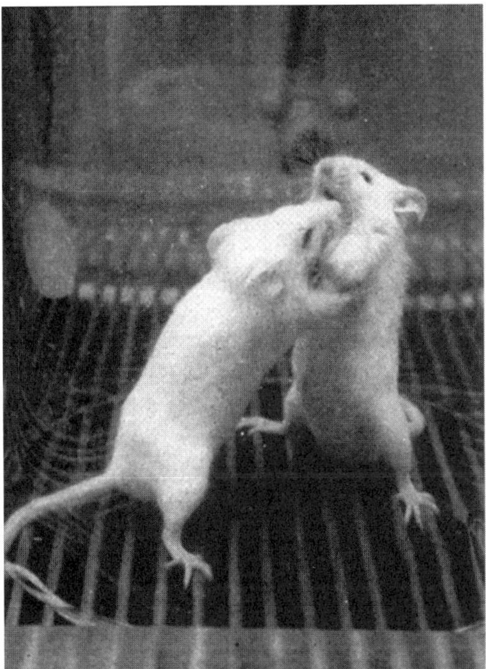

Figure 9-29. Fighting episode between C57BL male mice following electroshock applied to the feet of the mice.

Some order may be introduced into this jungle of experimental and ethological aggression types, as these types may be classified as predatory, defensive, or affective in nature (Reis, 1974). Moyer (1968) distinguished additional categories of aggression, and Davidson et al. (2000) referred to impulsive aggression, which appears to be similar to affective aggression. Predatory aggression includes aggression directed at killing rather than obtaining food, as in the case of muricidal rats, while defensive aggression may involve "idle threat" activity and immobility. Affective forms of aggression include aggression induced in animals by irritants, footshock, septal lesions, or frustration, and aggression induced in humans by fear or anger (From, 1973; Reis, 1974; Davidson et al., 2000). The animal septal lesion-induced aggression or rage relates to aggressive behavior caused in humans by accidentally incurred septal damage; technically, this response is referred to as the Kluver-Bucy syndrome (Kluver and Bucy, 1939; Luria, 1973).

It has been clearly documented in the past as well as in current studies that all aggressive behaviors, including those resulting from certain types of social interaction, as well as defensive activities, are induced or facilitated by systemic or localized administration into several hypothalamic and limbic structures of cholinergic agonists or antiChEs (and electric stimulation of these sites) and antagonized by similar application of atropinics or blockers of ACh release (it is less clear whether or not nicotinics and nicotinolytics affect aggression; Allikmets, 1974; Myers, 1974; Karczmar, 1976; Ray et al., 1989; Brudzynski et al., 1993; Srivastava et al., 1997; de Oliveira et al., 1997; Winslow and Camacho, 1995). Thus, "normal" rats are converted into muricidal rats by limbic or hypothalamic administration of muscarinic agonists (Smith et al., 1970), and aggression of affective type, whether induced by footshock or irritants, is inhibited by systemic or localized, limbic or hypothalamic injection of muscarinic antagonists (see Powell et al., 1973; Karczmar,

1976; Reis, 1974; Allikmets, 1974; Hingtgen and Aprison, 1976; Valzelli, 1974, 1978). It is of particular interest that aggression of the affective type was induced in humans following accidental or industry-related exposure to antiChEs (Devinsky et al., 1992).

A dissenting note was struck by Albert and his associates (Albert and Richmond, 1977; Albert and Chew, 1980); they reported that the limbic application of a local anesthetic induced fighting or mouse killing in rats, acts that they synonymized with rage induced by septal lesions. They concluded that rage "does not appear to be mediated by a cholinergic . . . system" (or any other neurotransmitters) and that it depends on disinhibition of pertinent limbic activities, which are not cholinergic in nature, as they found that cholinergic antagonists (and antagonists of other transmitter systems) did not induce these aggressions (it appears that they did not study the effects of the limbic application of cholinergic agonists).

It is well known that catecholaminergic and serotonergic agonists and antagonists affect aggression; particularly, dopaminergic agonists and activation of the dopaminergic system facilitate or induce aggression. This system is also involved in emotional and fear-related behavior that includes aggression (see below, section BV-f). However, the effect of these agents on aggression is not as general as that of cholinergic and anticholinergic drugs. For example, catecholaminergic antagonists block predatory but not affective aggression (see Reis, 1974), while muscarinic blockers and agonists affect all types of animal aggression. However, cholinergic drugs may be less effective with respect to certain consequences of aggressive behavior such as flight (Rodgers et al., 1990).

Additional evidence supports the notion that aggression shows marked cholinergic correlates. For example, "killer rats" exhibit higher levels of CAT and ACh in the amygdala and/or diencephalon than "nonkiller rats"; mice strains demonstrating high levels of spontaneous, ethological aggression show much higher turnover of brain ACh as compared to the nonaggressive strains; and, conversely, rats bred for cholinergic supersensitivity or for high levels of cholinergic parameters (Flinders Sensitive line of hypercholinergic rats) show markedly increased aggression (Pucilowski et al., 1990; Ebel et al., 1973; H. Yoshimura and S. Ueki, 1980, personal communication). However, experimental induction of aggression in several mice strains did not consistently increase the turnover of brain ACh (Karczmar and Kindel, 1981);

The anatomy of aggression also reflects the importance of the cholinergic system for aggression. As stressed by Kaada and others (Kaada, 1967; Bandler and Flynn, 1974; Glusman, 1974; Reis, 1974; Weiger and Bear, 1988; Davidson et al., 2000), electric stimulation of hypothalamic and limbic sites, thalamic nuclei, insular and cingulate cortex, and ventral striatum is particularly prone to elicit one or another form of aggression, including a defensive form of agonist behavior, and/or emotional behavior. All these sites are innervated by several cholinergic pathways including the NBM and the tegmental nuclei (see Chapter 2 DII). The same is true with respect to efferent pathways such as the dorsal periventricular-periaqueductal region, midbrain central gray, rostral ventrolateral medulla, and the striate and the nucleus of solitary tract; these efferent sites are involved in peripheral, whether autonomic, motor, or endocrine, expression of aggression (Bandler and Flynn, 1974; Delgado, 1969, 1974; Valzelli, 1978; Benarroch, 1997a, 1997b; Loewy and Spyer, 1990; Shapovalova, 1995; Pominova and Shapovalova, 1995). It is also pertinent that the cholinergic agonists elicit electrophysiological responses at these sites (see for example, Brudzynski et al., 1998).

Not only excitatory but also inhibitory brain sites are involved in aggression. Accordingly, a facilitatory-inhibitory balance regulates aggression and rage. The inhibitions are generated mainly by the GABAergic systems (Cannon and Britton, 1925; Kaada, 1967; Bard, 1928; Bard and Mountcastle, 1948; Glusman, 1974; Brudzynski et al., 1993, 1998; Brudzynski and Eckersdorf, 1988; see also Albert and Chew, 1980). The pathways in question course from neocortical areas and laterodorsal tegmental nuclei through amygdala to the hypothalamus and thence to brainstem sites of expression of aggression; thus, again they embrace the cholinergic radiations of the nucleus basalis magnocellularis and other basal forebrain and tegmental nuclei (see Chapter 2 DII and Brudzynski et al., 1998). Amygdala seems to be particularly involved in the balance in question, as it is important for processing emotions (Ledoux and Muller, 1997; Miot, 1998; Mucignat-Caretta et al., 2004; see also section BV-6, below). This importance of inhibitions for the regulation of aggression explains the occurrence of rage following decortication and septal ablation. Consistent with these findings, localized applications of muscarinic agonists at limbic sites that are near those where the application of these agonists induces aggression may produce suppression of ongoing aggressive behaviors (Allikmets, 1974); it is not clear whether or not, applied to the same sites, anticholinergics may elicit aggression. Similarly, cholinergic stimulation of certain amygdaloid sites elicits aggression, and their ablation blocks it (see, for example, Valzelli, 1978). What, of course, underlie these apparent inconsistencies is the heterogeneity of the amygdaloid nucleus, which Brodal (1980) refers to as amygdaloid complex, and the multiplicity of its afferent and efferent connections.

It must be stressed that aggression exhibits typical emotions and motor activities (that include advance with respect to predatory or offensive aggression and escape or tonic immobility in relation to defensive and related forms of aggression-related behavior); it also includes autonomic and endocrine effects. This multitude of effects accompanying aggression should be expected since striate, hypothalamus, limbic system, and medulla are involved in evocation of aggression and as their efferent tracts extend to the spinal cord, preganglionic neurons, and adrenal medulla (Loewy and Spyer, 1990; Benarroch, 1997a, 1997b; Pominova and Shapovalova, 1995). These phenomena include cardiovascular effects (noticed in an early study by Walter Hess; see Hess, 1954), piloerection (in animals), the release of such hormones as ACTH, and so on. The pattern of these phenomena depends on whether the aggression in question involves fear and escape or predation and offensive aggression. These hypothalamic and endocrine correlates of cholinergic aggression involve sex hormones, and this is important in view of the well-known facilitatory effects of gonadal hormones on male aggression (see, for example, Moyer, 1976; Karczmar, 1980).

In view of this well-established significance of the cholinergic system for aggression and effectiveness of cholinergic and anticholinergic drugs in facilitating and attenuating, respectively, aggression it is perplexing that several recent reviews of aggression do not refer at all to its cholinergic aspects (see, for example, Volavka, 1995; Benarroch, 1997a, 1997b; Mawson, 1999; Davidson et al., 2000).

2. Cholinergic Antinociception or Analgesia

The lore of cholinergic antinociception or analgesia—that is, either decreased pain perception or interference with nociceptive mechanisms and pathways prior to the arrival of the pain stimuli to the levels of perception—began with the demonstration in the 1930s and 1940s that carbamate antiChEs are analgesic and/or increase the analgesic effect of morphine and codeine (Pellandra, 1933; Tinel et al., 1933; Slaughter and Gross, 1940; Flodmark and Wramner, 1945; see also Karczmar, 1967; Sitaram and Gillin, 1977; Schott and Loh, 1984; and Green and Kitchen, 1986). It is notable that these effects were recorded in early studies not only in most animal species but also in humans: it was shown in the 1930s that in anesthetized subjects, systemically administered physostigmine synergized with morphine in increasing the threshold for pain induced by radiant heat applied to the forehead, and intraventricularly administered ACh relieved causalgia of the hand (Tinel, 1933; Pellandra, 1933); these findings have been confirmed many times since (see, for example, Hampf and Bowsher, 1989; Beilin et al., 2005). It must be added that the antiChE synergism with morphine is not due to antiChE-mediated increase of morphine penetration into the brain across the blood-brain barrier, antiChEs being capable of this effect (see Chapter 7 DI; see also Szerb and McCurdy, 1956; Karczmar, 1967).

Both muscarinic and nicotinic sites are involved in cholinergic analgesia. Muscarinic agonists are analgetic and/or potentiate the antinociceptive action of morphine and opioid drugs. This phenomenon occurs when muscarinics particularly, but also nicotinics, are given systemically or applied locally (for example, intrathecally) to sites concerned with conveying nociceptive signals, including spinal sites (dorsal horn and the spinothalamic tracts), periaqueductal gray, reticular sites, hypothalamus, basal ganglia, rostral ventrolateral medulla, several thalamic nuclei, and the nucleus Raphe magnus; application to the thalamus and thalamocortical pathways is particularly effective (McCormick, 1992a, 1992b; Steriade et al., 1994; Vincler and Eisenach, 2004; see also Chapter 2 DII). M1 and M2 muscarinic and α3 and α5 nicotinic brain, spinal, and trigeminal (primary sensory) receptors may be involved in analgesia (Radhakrishnan and Sluka, 2003; Vincler and Eisenach, 2004; Dussor et al., 2004). It is less clear whether these agents are effective when applied to such sites involved in pain perception as the somatosensory cortex (Zhuo et al., 1993; Haley and McCormick, 1957; Rees and Roberts, 1993; Naguib and Yaksh, 1994; Gillberg et al., 1989; Hartvig et al., 1989; Sitaram and Gillin, 1977; Green and Kitchen, 1986; Spinella et al., 1999).

Additionally, William Dewey and Norman Pedigo, leading investigators of cholinergic analgesia, found that inactive muscarinic analogs and stereoisomers of active muscarinic agonists were just as ineffective as analgesics (Dewey et al., 1975; see also Harris et al., 1969). Apparently, both M1 and M2 muscarinic receptor subtypes participate in cholinergic analgesia (Gillberg et al., 1989, 1990; Gower, 1987; Naguib and Yaksh, 1997). Furthermore, atropine and atropinic drugs antagonize these effects. In fact, sometimes atropinics also antagonized morphine's analgetic effects (see Green and Kitchen, 1986; Sitaram and Gillin, 1977), although these findings are not consistent; see below, next section; Abe et al., 2003). However, under certain circumstances atropine may induce analgesia by increasing ACh release at the appropriate targets (see section BI-1b, above). This was shown with small doses of atropine and with a propanyl atropine derivative that exhibits special affinity to the presynaptic M2 and M4 receptors (Ghelardini et al., 1990, 1997).

Ample evidence demonstrates that applied systemically or to the brain areas listed above, nicotine and synthetic or natural nicotinic agonists also induce analgesia, while the nicotinolytics antagonize this effect (Martin et al., 1990; Sahley and Berntson, 1979; Jurna et al., 1993; Rogers and Iwamoto, 1993; Sitaram and Gillin, 1977; Green and Kitchen, 1986; Iwamoto, 1989; Iwamoto and Marion, 1993; Gillberg et al., 1990; Badio and Daly, 1994). Particularly strong confirmation of the notion of the analgetic action of nicotinics was provided by Arneric and his associates as they showed that given systemically, synthetic nicotinic agonists are potent analgesics in a number of animal models with respect to several types of induced pain (Bannon et al., 1998; Decker et al., 1998). However, it is controversial whether nicotine applied intrathecally or to the spinal cord can induce analgesia (Gillberg et al., 1990). The α4 receptor subtype may be involved in nicotinic analgesia, as indicated by experiments with α4 receptor knockout mice (see Chapter 6 C).

There is a question as to the nature of the membrane response involved in cholinergically evoked analgesia. It is generally assumed that this response is depolarizing and excitatory in nature; however, an inhibitory muscarinic action (an M2 effect) was observed, in conjunction with cholinergically evoked analgesia, at the Raphe magnus nucleus (Pan and Williams, 1994). Inhibitory cholinergic effects probably arise at other central, including nicotinic, sites commonly associated with analgesia.

The majority of investigators found that antiChEs are effective as analgesics in rodents even when given at small doses, provided these doses were sufficient to inhibit central AChE (Koehn and Karczmar, 1978). However, Cox and Tha (1972) and Clement and Copeman (1984) found that the tertiary carbamate or OP antiChEs, used alone, either do not exhibit analgesic effect in rodents or are analgetic only when used in large doses.[6]

While morphine and morphinoids induce analgesia in primates, cholinergic agonists do not seem to do so; in fact, paradoxically, atropinics may be analgetic in these animals (Pert, 1975; see also Sitaram and Gillin, 1977). It is interesting that the reverse seems to obtain in amphibia; in our hands (G. L. Koehn and A. G. Karczmar, 1890, unpublished data) and in those of others (Nistri et al., 1974), in amphibia morphine and morphinoids were only marginally effective, while antiChEs were very potent (on microgram levels, when applied intraventricularly).

Different research methods that were used by various investigators may account for the diversity of results. Thus, a multitude of testing procedures was used in the human and animal studies. Then, following the analgetic treatment, animals and humans may disregard the noxious stimulus not because the treatment is effective but because the drug attenuated their motor performance. Indeed, the hot-plate test, which has as its endpoint the animal escaping from the heated plate, depends on intact motor capacity of the animal, and the immobilizing and cataleptic action of cholinergic agonists (and, in fact, also of the opioids) interferes with this particular test; actually, similar cholinoceptive sites induce both antinociception and immobility (Menescal-de-Oliveira and Hoffman, 1993; see also sections BIV-1d, above, and BVI-1, below). Other tests that have as their endpoints tail flick, flexing, or rearing require intactness of spinal reflexes; these reflexes may be attenuated by cholinergic agonists and antiChEs (see section IV-1a, above). In the case of still another test, the induction of vocalization by the pain stimulus, the animals may be made incapable of vocalization by the analgetic agent; when vocalization is induced by means other than application of chemical or physical pain stimulus, such as stimulation of certain forebrain structures, the presence of analgesia may be a matter of interpretation (Jurgens and Lu, 1993).

Evidence presented so far supports the notion of cholinergic correlates of analgesia; additional evidence strengthens the validity of this notion. Thus, in animals, analgetic stimulation of certain sites such as nucleus cuneiformis and analgetic electroacupuncture increase ACh levels, release, and/or turnover in several brain and spinal sites and in the CSF; a few concordant data were obtained in humans (He et al., 1979; Zemlan and Behbehan, 1988; see also Green and Kitchen, 1986.). Then, CAT levels in the hippocampus are reduced following evocation of pain via injection of an irritant; this finding may be construed as indicating exhaustion by nociception of the cholinergic system (Aloisi et al., 1993). Moreover, acupuncture evoked analgesia may be cholinergically mediated (Baek et al., 2005). Finally, medullary administration of hemicholinium prevented nicotine induction of analgesia, while the cholinergic neurotoxin AF64A blocked the analgetic effect of physostigmine but not of oxotremorine (Tellioglu et al., 1998; Rogers and Iwamoto, 1993). The relation between cholinergic and pain-related pathways is also pertinent in this context: cholinoceptive synapses are present in many of the sites or systems involved in pain signals and pain perception, and analgesia readily obtains with administration of muscarinic or nicotinic agonists and antiChEs to these sites (Casey, 1973; Brodal, 1980; Kandel et al., 1991; Zemlan and Behbehan, 1988; Willis, 1982, 1991; see also above, this section). However, cholinergic antinociception may also be evoked via application of pertinent drugs to sites not generally understood as directly concerned with either nociception or pain perception, such as the dorsal parabrachial region (Menescal-de-Oliveira and Hoffman, 1993).

a. Analgetic Interaction Between Cholinergic and Other Systems

Cholinergic agonists and antiChEs may evoke analgesia directly or through interaction with other transmitter systems. Indeed, the peptidergic systems including substance P pathways and the serotonergic, catecholaminergic, and aminergic systems are active at the brain sites where cholinergic analgesia obtains. Notably, noradrenergic $\alpha 2$ agonists induce analgesia when given systemically, epidurally or intraspinally; this effect may be mediated by the dorsal horn (Casey, 1973; Brodal, 1980; Kandel et al., 1991; Zemlan and Behbehan, 1988; Willis, 1982, 1991). This nociception may be mediated by cholinergic neurons when muscarinic and nicotinic antinociception is induced in the spine or pedunculopontine tegmental area (Naguib and Yaksh, 1994; Jurgens and Lu, 1993; Iwamoto and Marion, 1993; Rogers and Iwamoto, 1993; Wiklund and Hartvig, 1990).

Accordingly, noradrenergic $\alpha 2$ agonists such as clonidine release ACh into cerebrospinal fluid, and clonidine analgesia is enhanced by antiChEs and attenuated by atropine (Gordh et al., 1989; see, however, Detwiler et al., 1993). There may also be an interaction among cholinergic, peptidergic, and glutaminergic systems (involving particularly NMDA receptors) in morphine analgesia (Smith et al., 1989; Spinella et al., 1999). Furthermore, a nociceptive serotonergic descending pathway involves the periaqueductal gray and the nucleus Raphe magnus (see Willis, 1982; Brodie and Proudfit, 1986; Pan and Williams, 1994); cholinergic neurons are present in this system, muscarinic and nicotinic agonists induce analgesia when applied to these sites, and serotonergic antagonists attenuate analgesia evoked by the cholinergic agonist stimulation (Iwamoto and Marion, 1993; Ahn et al., 1999). However, discordant results were obtained with depleters of endogenous serotonin and catecholamines (see Pedigo and Dewey, 1981).

The GABAergic system is also involved in nicotinic and, particularly, muscarinic analgesia; it is pertinent that muscarinics and antiChEs inhibit the GABA receptors of the spinal dorsal horn projections (Chen and Pan, 2004).

There is also extensive research with respect to opioid-cholinergic interaction. As already pointed out in this section, cholinergic agonists and antiChEs synergize with or potentiate the analgetic effects of morphine and morphinoids; also, enkephalin synthesis and/or endorphin levels may be increased by muscarinic agonists (Capone et al., 1999; Beilin et al., 2005) and transcriptional eukephalin sites may be induced cholinergically (Weisinger et al., 1990). Also, morphine antinociception may be antagonized by nicotinolytics and atropinics, in particular by M1 and M3 antagonists (Abe et al., 2003), although this finding is not unanimous. In a related context, nicotine analgesia was reduced in mu-opioid receptor knockout mice, and this effect correlated with the reduction of the rewarding responses to nicotine in these mice (Berrendero et al., 2002).

While these findings indicate that opiate mechanisms are involved in this interaction and that, conversely, cholinergic phenomena are involved in opiate analgesia, several lines of evidence militate against this notion. Morphinoids and endogenous opioids including sigma and related opioid ligands depress the release of ACh from several brain parts and ACh turnover, and increase neuronal and extracellular levels of ACh in several brain parts; there are, however, some exceptions to this generalization (Sitaram and Gillin, 1977; Jhamandas and Sutak, 1983; Itoh et al., 1999; see also Green and Kitchen, 1986; Herken et al., 1957; Matsuno et al., 1993; Wang et al., 2004; see also section BIII-2a). These results seem to mitigate the notion of synergism between cholinergic and opioid systems, and Pedigo and Dewey (1981) opined that since "narcotics . . . suppress central cholinergic activity, it would be surprising if the antinociceptive activity of cholinomimetics were mediated through central opiate receptors or release of endogenous opiates" (see also Chan et al., 1982).

Then, cholinergic agonists used in combination with morphine potentiate the antagonistic effect of naloxone (Wong and Bentley, 1979). Furthermore, analgetic actions of locally administered beta-endorphin and morphine were not consistently antagonized by cholinergic antagonists (Spinella et al., 1999; Abe et al., 2003). Also, while both morphinoid and cholinergic analgesias are antagonized by a number of opioid antagonists including naloxone, the structure-activity relationship between opioid antagonists and their attenuation of opiate analgesia differed distinctly from the structure-activity relationship between opioid antagonists and their attenuation of cholinergic analgesia; in fact, some investigators found that atropine antagonizes cholinergic antinociception but it does not antagonize morphinergic antinociception (Wong and Bentley, 1979; Koehn et al., 1980; Naguib and Yaksh, 1994; Pedigo et al., 1975; Pedigo and Dewey, 1981; Koehn and Karczmar, 1978). Finally, at the spinal cord level, opioid antagonists may enhance rather than attenuate analgesic physostigmine effects (Fujimoto and Rady, 1989). It should be added that the finding that morphinoids but not cholinergics may be effective in some species, and vice versa, adds to the notion of the difference between cholinergic and morphinoid analgesia.

Determining whether or not there is a cross-tolerance among nociceptive opioid and cholinergic actions is crucial in this context. While the pertinent data are discordant, generally the cross-tolerance was not obtained (see Pedigo and Dewey, 1981, versus Koehn et al., 1980), and morphinoids (not cholinergics) induce analgesia in selected species (see above, this section).

What shall we deduce from these sometimes-inconsistent data? Perhaps morphinoids and endogenous opioids do not need a cholinergic link for their effect, as suggested by occasionally observed lack of atropine antagonism of opioid analgesia, while cholinergics need the presence of an opioid system for theirs. Ultimately, expanding the knowledge of endogenous opioids and their receptors may clarify this relationship (see Takemori and Portoghese, 1985).

b. Cannabinoids and Cholinergic Analgesia

Cannabinoids (including δ-9-tetrahydrocannabinol) produce analgesia when applied at spinal or supraspinal sites (Dewey, 1986; Musky et al., 1991; Smith and Martin, 1992) and affect the kinetics of CNS ACh (Dewey, 1986; Tripathi et al., 1987). It should be added that while cholinergic-cannabinoid interaction with regard to motor function was demonstrated (see section BIV-1d), such interaction was not investigated with respect to analgesia.

3. Sensorium

The past generalization was that the cholinergic system does not contribute to the afferent sensorium in vertebrates (see Koelle, 1963; Michelson, 1974; and Karczmar, 1976), yet cholinergic pathways, particularly those of the basal forebrain, abut upon, and/or cholinergic synapses are present in, the pertinent modalities, such as the retina, olivocochlear bundle, medial vestibular nucleus, olfactory bulb, pedunculopontine projections, and somatosensory, including auditory and visual cortices (see Chapter 2 DII; Rasmusson, 1993). In fact, ACh and antiChEs exert effects upon localized application to these cortices and/or central and peripheral sensory organs and may increase homeostatic cortical mechanisms (Testylier et al., 1999; Kimura et al., 1999); hence, correlates between these sensory modalities and the cholinergic system should exist, and there is some evidence that supports this notion.

Acetylcholine is the principal neurotransmitter for the olivocochlear bundle's efferents that terminate in the cochlea on the outer hair cells and in certain divisions of the vestibular nucleus. Anticholinesterases, ACh, and nicotinic and muscarinic agonists mimic the inhibition by the olivocochlear bundle of the responses of the auditory nerve terminal, although biphasic effects were also observed (Guth and Amaro, 1969; Guth et al., 1994; He and Dallos, 1999; Vetter et al., 1999; Katz et al., 2004; see also Koelle, 1963); the effects of antagonists, whether of the atropinic or nicotinolytic type, were inconsistent. The inhibitory effect of the bundle as well as of the muscarinic agonists may be mediated by the release of an inhibitory transmitter (Koelle, 1963; Guth et al., 1994). On the other hand, He and Dallos (1999) found that ACh itself induces the inhibitory effect via an outward, hyperpolarizing K^+ current, while Wikstrom et al. (1998) proposed that nicotinic receptors of a unique type (acting both ionotropically and metabolically) block the postsynaptic response of the outer hair cells via antagonizing the action of another transmitter, possibly ATP. Also, it was suggested that efferent effects upon audition may be mediated via calcium entry through a nicotinic receptor (Dawkins et al., 2005). These reports are somewhat disparate; however, they show that ACh serves to modulate the inner hair cells' mechanoelectrical transduction of the sound.

More centrally, cholinergics facilitate the evoked potentials of the primary sensory cortex induced by auditory stimulation and the consequent change in the receptive field, and cholinergic (M1) neocortical receptors are needed for refined "cortical tonotopy" (Metherate and Weinberger, 1990; Shapovalova et al., 1994; Zhang et al., 2005; see also section BIV-3b).

The topographically related equilibrium, or balance, system also shows cholinergic correlates. Acetylcholine and cholinergic agonists exert effects on the afferent neurons of the semicircular canal and the sacculae as well as on the neurons of the vestibular nuclei, and the discharge of some of the neurons of the medial vestibular nucleus is blocked by atropine (Guth et al., 1994; Takeshita et al., 1999).

The basal forebrain's efferent cholinergic pathways radiate to the visual cortex and its functional control and perception areas. They involve cholinergic structures located in associative and integrative areas of the brain, including geniculorecipient layers and superior colliculus of the monkey, humans, and other species (Rasmusson, 1993; Celesia, 1991; Tigges et al., 1997; and Dudkin et al., 1994; see also Chapter 2 DI-DIII). There are both nicotinic and muscarinic receptors at these sites. The nicotinic receptors may modulate saccadic activity (Aizawa et al., 1999; Vanni-Mercier et al., 1996) and the presynaptic muscarinic receptors may modulate visual excitation spread in the visual cortex (Kimura et al., 1999). Tigges et al. (1997) suggested that the M1 to M4 muscarinic receptor subtypes distributed in the cortices are responsible for a number of phenomena and that a transient cholinergic deprivation or lesion of the basal forebrain may interfere with visual cortical processes such as ocular dominance and orientation selectivity (Siciliano et al., 1997); some or all of these receptors may also be concerned with the cortical integration and binding of several perceptive characteristics (Colzato et al., 2005, see also section IV-1d).

Cholinergic effects on vision also include peripheral, retinal phenomena and their afferent, retinotectal, and retinogeniculate pathways. Thus, retinal ganglion cells respond to ACh and muscarinic and nicotinic agonists, in keeping with the notion of the existence of the cholinergic synapse between the amacrine and ganglion cells, and of the presence of muscarinic and nicotinic receptor subtypes on the amacrine and ganglion cells

(Keyser et al., 2000). Acetylcholine blocks the activity of the ganglionic on-units, which are activated by light and inhibited by the light-off signal, and facilitates the activity of the off-center units, which behave in a manner opposite that of the on-center ganglion cells (Straschill and Perwein, 1975; see also Chalupa and Gunhan, 2004). The ganglion cells are involved in the mechanism of dark adaptation, as their receptive fields are enlarged in dim light (see Barlow et al., 1957). Accordingly, this effect may be involved in the facilitation of the dark adaptation in humans by antiChEs (Rubin and Goldberg, 1957).

It is interesting that ChEs may be involved in this process as well as in the phenomena described by Straschill and Perwein (1975), as there are endogenous photochromic ChE inhibitors that exist, depending on photoenergy, in several forms that differ in their inhibitory potency (Kaufman et al., 1968; Friedman and Marchese, 1973). Further upstream, the cholinergic projections of the nucleus isthmi exert presynaptic, mainly nicotinic effects on the retinotectal axons, increasing the liberation of noncholinergic transmitters, and Titmus et al. (1999) opine that these nicotinic actions modulate the tectal projections of orderly, retina-generated, topographical maps. It is interesting—in an unrelated context—that muscarinic receptors may be involved in eye growth and in the regulation of the axial length of the lens (Tigges et al., 1999).

Finally, the rhinencephalon and olfactory bulb are cholinergically innervated (see Chapter 2 DII and DIII), and Castillo et al. (1999) stressed that nicotinic and muscarinic agonists exert excitatory and inhibitory effects on the bulb's mitral and interneuronal cells, respectively; this suggests that these agents may have effects on the olfactory function. In a related context, the taste stimuli coursing from the taste buds and the gustatory nerves to the thalamus (nucleus ventralis posterior) and cortical olfactory and posterior gyrus induce cortical cholinergic activity and ACh release, and these phenomena contribute to the sensorial correlates of cholinergicity.

4. Memory, Learning, and Cognition

As referred to in the introduction to this chapter (section A-2), the knowledge of the effects of cholinergic agonists and antagonists on memory, learning, and conditioning was initiated in the 1920s. The early investigators found that the cholinergic agonists, generally of the muscarinic type, exerted potent facilitatory effects, antagonizable by atropinics, with respect to the relatively simple learning paradigms they employed. However, Funderburk and Case (1947), while demonstrating facilitatory actions of pilocarpine, stressed the inhibitory actions of physostigmine on the simple Pavlovian conditioned motor response. Subsequent investigators amply confirmed these results concerning both the facilitatory and inhibitory or biphasic actions of muscarinic agonists and antiChEs; yet, the facilitatory effects of the muscarinics and ChE inhibitors predominate when these agents are employed in appropriate doses (see Hagan and Morris, 1988).

a. Methodology of the Studies of Learning, Memory, and Cognition

Modern investigators employ an extensive number of paradigms for the study of the cholinergic correlates of learning and memory. Classical conditioning involves the study of acquisition and extinction components of memory and learning, stimulus discrimination, and passive and active avoidance, the operant behavior being required as the testing modality. For the study of avoidance, simple paradigms may be used, for example, a platform consisting of the punishment and escape parts; or an automated, multiple avoidance screen (Figures 9-30 and 9-31; Karczmar and Scudder, 1969).

Auditory, visual, olfactory, taste, and other sensory signals are employed, and the unconditioned response could be a motor response or the anticipatory eye blink, this last test minimizing the need for behavioral activity; the delayed operant test, for example, delayed matching-to-sample response, requires the recognition of the stimulus or signal after a varying time period (Ferster, 1960; Karczmar, 1976; Hintgen and Aprison, 1976; Bures et al., 1983; Harvey et al., 1983; Nabeshima, 1993; Fibiger, 1991; Hudzik and Wenger, 1993; Butt and Hodge, 1995; Anglade et al., 1999; Miranda et al., 2003). In the case of the Sidman continuous trial paradigm, animals are not presented with any cue that could be associated with the shock or other punishment applied in a regular temporal sequence; they have to learn

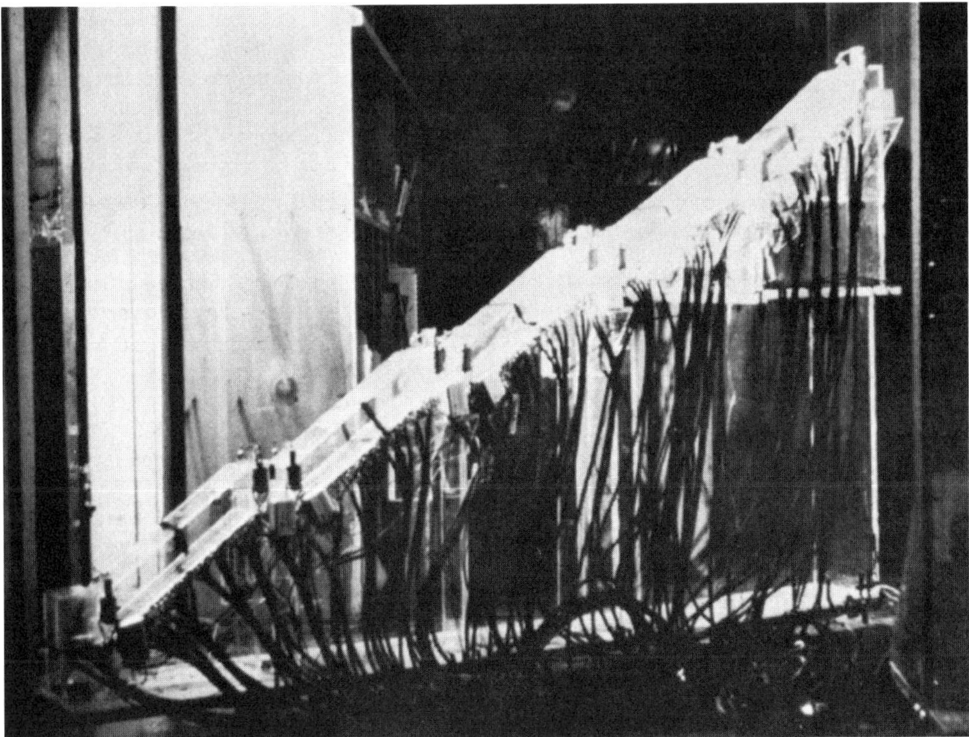

Figure 9-30. A general view of the avoidance conditioning climbing screen. The 5 climbing screens and base chambers as well as the readout panel (at left) can be seen. The wiring connects the photocells and electrifies the grids of base chambers and climbing screens. (From Karczmar and Scudder, 1969, with permission.)

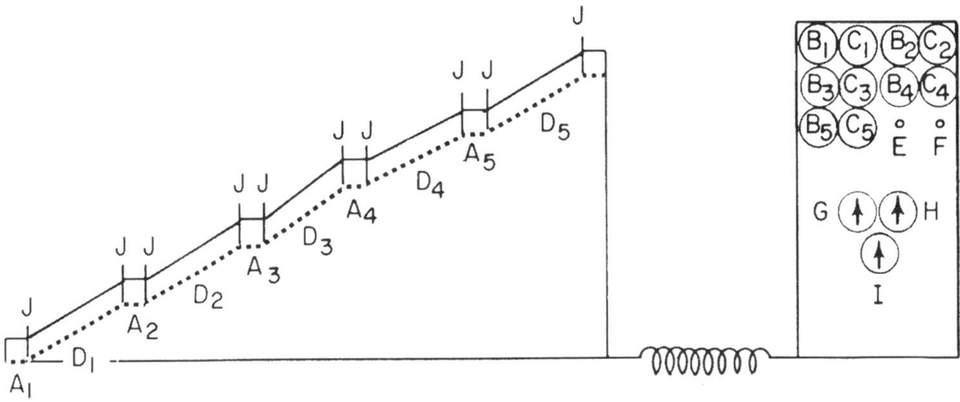

Figure 9-31. A block diagram of the avoidance conditioning climbing screen. The climbing screen is represented on the left, the operations and readout panel on the right. A, 1 to 5, consecutive base chambers; B, 1 to 5, elapsed time meters measuring the interval from the opening of the gates until the mouse begins to climb, that is, "base times"; C, 1 to 5, elapsed time meters measuring the interval during which the mouse moves from one base chamber to the next, that is, "climbing times"; D, 1 to 5, the consecutive inclined runways of the climbing screen; E, starter switch; F, reset button; G and H, timers programming the opening of the doors of the base chambers and the time period between that moment and the onset of shock, respectively; I, stimulus parameter controls; J, solenoid operated gates and photocells. (From Karczmar and Scudder, 1969, with permission.)

during the continuous trial to select an appropriate option from the several escape options, including pole jumping, a platform, and so forth. More complex are the operant approach paradigms, which involve reward- or reinforcement-dependent lever pressing or related operant behavior. Cues of varying degrees of complexity are used to show whether the animals discriminate for reinforcement or to avoid punishment; furthermore, additional dimensions of complexity may be added by manipulating the schedule of punishment and reinforcement from regular to complex and variable sequence.

Nonoperant tests include runway and maze learning types (including water maze), spatial alteration (hole board apparatus that tests, particularly, spatial orientation; Oades, 1981) and discrimination, and performing acquired, memory-dependent tasks (see, for example, Nabeshima, 1993; Fibiger, 1991; Butt and Hodge, 1995; Howard et al., 2005). Maze tests vary in complexity. A simple Y-maze paradigm includes reward-seeking or punishment-avoidance behaviors. A related procedure involves spontaneous alteration (Hughes, 1982); that is, after using a particular arm of the Y-maze, the animal has to choose a different arm during the second trial. More complex tests utilize radial and multiarm mazes with 8 or 12 arms and a number of different cues (Cressant and Granon, 2003). These and related tests involve "place learning," or spatial discrimination. After exploring available sites (8 or 12), the animals must remember which one yielded the reinforcement and, on the subsequent trial, the animal must not revisit the site where the reward was consumed. Following this test, another phase may be employed to ascertain whether the animals will procure the reward at either a novel site or the original site the second time around (Olton and Samuelson, 1976; Slangen et al., 1990). Finally, certain tests assessed cognitive states important for learning (i.e., attention; see below, this section). These tests included startle reflex measurement and the measurement of the frequency of alert (upright) body positions. In operant or nonoperant procedures, the complexity of the tasks the animal must perform can be increased to evaluate the psychological functions that may be involved in memory and learning processes (Olton et al., 1991). Finally, performance evaluations for memory-dependent tasks may be used to assess retention and memory versus learning.

Altogether, memory and conditioning tests vary from very simple to very complex.[7] It is particularly valid to study learning—and behavior generally—as a matter of ethology, that is, in field conditions, with uncaged, free-moving animals. Pertinent studies were carried out in the open field or natural habitat with primates. This kind of work is, of course, rare and limited, as drugs cannot be readily used in this case. However, some attempts were made in this direction; for example, the "Mouse City" test allows investigators to quantify learning, attention, exploration, and novelty-seeking behavior in small rodents in a pseudo-natural habitat (Figure 9-32 and Table 9-3; see Karczmar and Scudder, 1969; Karczmar et al., 1973) A new dimension of animal and human research on memory and learning was added with the use of brain lesions; this work is important in relating cognition, memory, and learning to specific transmitter pathways, including the cholinergic system, and to the brain location of the various mechanisms involved in memory and learning. In animals, the lesions are generated surgically, by inducing brain hypoxia, ischemia, or infarcts, or by employing neurotoxins, antibodies, and antagonists of neurotrophic factors. Commonly used are the surgical lesions of the nucleus basalis magnocellularis (NBM) and the septum, and ischemia-inducing carotid or bicarotid clamping (Cuello, 1993; Bartus et al., 1982; Jones et al., 1995; Butt and Hodge, 1955; Heim et al., 1994).

The neurotoxins used in animals include specific cholinotoxin, AF64A (Hanin, 1990, 1992; see Figure 9-16, section BIII), kainate (Jin, 1997), and immunotoxins (Baxter et al., 1995; Waite et al., 1999). Related toxins include excitotoxins and Al^{3+}, which are sometimes regarded as possible causative agents of Alzheimer's disease (AD; see Dunnet and Fibiger, 1993; Bilkei-Gorzo, 1994; Julka et al., 1995; Wu et al., 1998). Less specific neurotoxins (Markowska et al., 1990) include tetrahydropyridines that mimic, in animals, Parkinson's disease while causing also cognitive deficits (Schneider et al., 1999; Markowska et al., 1990) and OP antiChEs.

Finally, animals—and humans—subjected to dietary restrictions, and thiamine-deficient mice, can serve as models for the study of the cholinergic treatments of memory impairments (Nakagawasai et al., 2004).

With many toxins, components of the cholinergic system, such as CAT, ACh turnover, and the

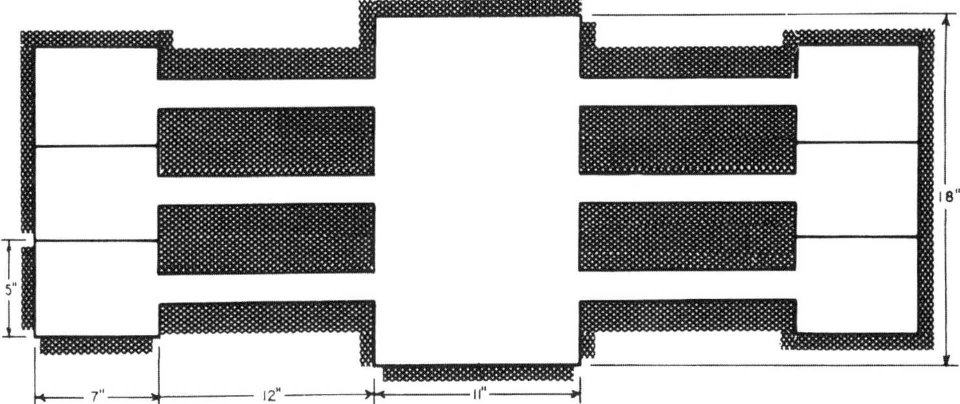

Figure 9-32. Top view of the "Mouse City." The diagram shows the large (11 in × 18 in) central compartment connected to the small (5 in × 7 in) home cages by means of tubular, 12-in-long runways. The covered compartment, home cages, and runways are illustrated in white. (From Karczmar et al., 1973, with permission.)

density of muscarinic receptors, are affected to a degree depending on the toxin in question (Hanin, 1992; Julka et al., 1995; Wu et al., 1998; Sorger et al., 1999). Some immunotoxins, such as anti-p75 nerve growth factor IgG immunotoxin and antineurotrophic factor fusion protein, seem to be as specific in lesioning cholinergic neurons as AF64A, although the data concerning their functional effects are sparse (Kwok et al., 1999). A complex effect was noted with respect to the cholinergic receptors; for example, the sensitivity of nicotinic and muscarinic receptors may increase and decrease, respectively, following AF64A treatment (Thorne and Potter, 1995).

Table 9-3. Behaviors Measured in Mouse City

- Mobility, exploration, and novelty seeking
- Digging
- Stereotypic behavior
- Sleep
- Contactual behavior
- Aggression
  - Attacking
  - Being attacked
  - Defensive activities
- Carrying objects
- Ingestion
- Sexual behavior
- Grooming
  - Self
  - Others
  - Being groomed

(For further information, see Karczmar et al., 1978.)

Sometimes, combinations of neurotoxins were used in order to affect the cholinergic as well as the noncholinergic systems; the underlying concept in this case is that aging and AD reflect multitransmitter changes (Haratounian et al., 1990). Of particular interest are animal models simulating the natural production of amyloid proteins (see, for example, Nabeshima and Nitta, 1994). Other, interesting models involve animals exhibiting genetic malfunction involving the cholinergic system such as transgenic mice that show over-expression of mRNA for AChE (Beeri et al., 1997) and knock-out mice lacking genes controlling cholinoceptive receptors (see, for example, Picciotto, 1998).

In humans, advantage is being taken of lesions due to accidents, stroke, surgery, disease including AD and Rett syndrome, and alcoholism; the memory effect of these lesions is studied during the subjects' lifetimes, and postmortem studies also provide important information on the brain sites of memory and its various forms (see Luria, 1973; Squire and Butters, 1992; Kaufman et al., 1997; and Damasio and Damasio, 1993; see also Chapter 10 A and K). Commisural damage provides an interesting example of such lesions; it is referred to below in this section, below.

Combined with the various paradigms of learning and its components, the studies of memory and learning in animals may include EEG analysis and analysis of appropriate evoked potentials (see below, this section), neuroimaging and positron emission topography (MRI and PET), as well as computational methods for assessing "synaptic strength"; these latter tests allow not only the evaluation of "gross" cholinergic transmission (that is, its presence or absence, long-term post-synaptic potentiation, etc.) but also its subtle changes and coherence at different brain sites (see, for example, Linster et al., 2003; Thiel, 2003); this kind of evaluation is capable of relating, even if speculatively, the activities and their levels at different central cholinergic sites.

Converse to lesion experiments are experiments that involve localized application of trophic factors and fetal and adult cultured cholinergic neurons (see, for example, Dickinson-Anson et al., 2003). In some countries, such as Sweden, this or related methodology is applied to humans (L. Olson, 1990, personal communication).

The tests of learning and memory in humans are of particular importance, in view of the current insurgence of interest in aging and AD. David Drachman, Klaus Unna, Arnold Ostfelt, and the Czech teams led by Hrbek and Soukopova pioneered studies in humans of the relationship between the cholinergic system on the one hand and memory and learning on the other; these investigators employed quantitative tests of digit recall and storage (Drachman and Leavitt, 1974; Drachman, 1978). Furthermore, tests based on conditioning may be readily carried out in humans, and such tests may be combined with electrophysiological monitoring of appropriate EEG parameters and brain potentials (Konorski, 1967; Woody, 1982; Crook et al., 1983; Libet, 1993). Currently employed psychometric tests involve quantitative scale assessments of cognition, intelligence, and memory (see, for example, Nyback et al., 1988; and Squire and Butters, 1992), while the clinically oriented CERAD battery of tests and geriatric assessment scale (Brink et al., 1982; Gershon et al., 1994) are particularly used in the studies of geriatrics and AD. A test of human memory that concerns visuospatial processing is of particular interest, as visuospatial orientation is faulty in aging (Meador et al., 1993). Besides conventional tests of memory and learning, novel, interesting tests of consciousness may be employed in humans for neuropsychological assessment of AD, stroke, and brain damage (see Chapter 10 K), as well as for the assessment of cognitive processes such as planning and goal-directed behavior ("Tower of London" test; see Dehaene and Changeux, 1997).

b. Kinds and Components of Memory and Learning

Results obtained in animals and humans by means of these various procedures indicated that learning and memory function are much more complex than originally envisaged; they consist of several processes, and the various paradigms referred to above may concern one or another of these processes in particular. These processes include early learning versus memory consolidation, storage, and recall or retrieval (McGaugh and Herz, 1972; Honig, 1978; Olton et al., 1979; Slangen et al., 1990; Hasselmo et al., 1995; Van der Zee and Luiten, 1999); they lead to either short- or long-term memory (or working or reference memory; Woody, 1982). In humans additional kinds of memories may be distinguished, such as "shallow output memory" (Moscovitch, 1992), that is "the memory of an event . . . that is not placed in its proper spatiotemporal context; . . . an event . . . that cannot be interpreted . . . within the context of other past or current events" (see also Cipolotti and Moscovitch, 2005), and semantic memory with its subtypes (Thompson-Schill et al., 1999).

Habituation or extinction is an important parameter of learning; animals or humans not able to habituate relatively rapidly to a cue or stimulus, which is or becomes nonvital or trivial, cannot maintain a meaningful interplay with the environment and cannot learn (see Hughes, 1982; Karczmar, 1995). Furthermore, it was demonstrated that learning includes, as an independent process, data organization, or coding, acquisition of memory and retrieval of appropriate information from encoded, stored memory (Woody, 1982).

Certain behaviors or central states that precede or accompany learning and memory processes are of importance; these include attention and its antipode, distraction, vigilance, orientation, novelty-seeking or novelty-attraction (neophilia) and exploratory behavior, habituation or extinction, and the reward-punishment balance (see Pavlov, 1940; Konorski, 1967; Olds, 1958; Stein, 1968; Karczmar, 1976; Hingtgen and Aprison, 1976; Bures et al., 1983; Decker et al., 1998; Slangen et al., 1990; Mirza and Stolerman, 1998; O'Neill et al., 2003). Also, stress may affect memory processes via changes in cholinergic transmission generated in the hippocampus (Mizoguchi, 2001).

It should be emphasized that these various processes and states appear to be specifically affected by aging and AD and that cholinergic drugs or lesions of cholinergic pathways may affect some but not others of these states (McGaughy et al., 1999). Altogether, the concatenation of the phenomena in question may be considered cognition in its broad sense (Slangen et al., 1990; see below, this section); however, the phenomena of consciousness or self-awareness are not included in this definition (see section BVI).

It is not easy to ascertain which of these specific processes may be involved in the effect on learning of a given drug, neurotoxin, or lesion, and whether or not additional, nonmemory effects are present. To cite a simplistic example, is the blocking effect of atropine or scopolamine on performance—which occurs when atropine is given after a task was learned and the animals are retested—due to atropine's action on retrieval or on memory? However, some tests may address a particular learning component, as in the case of the delayed matching-to-sample test, which is supposed to test specifically short-term memory (Ferster, 1960). Appropriate testing shows that the short-lived amnetic action induced by large doses of antiChEs is caused by the block of retrieval by these drugs (Karczmar, 1974). And, attention, neophilia, and exploration, which are behavioral components of learning processes, may be studied differentially, and these phenomena were facilitated by cholinergic, particularly muscarinic agonists, attenuated by lesions to the cholinergic basal forebrain and by atropinics, and accompanied by increased release of ACh at pertinent sites such as the hippocampus (Muir et al., 1994; Inglis et al., 1994; Jones and Higgins, 1995; Giovannini et al., 2003; see also Woolf, 1996, 1997; Smythies, 1997; Sarter et al., 2003; Voytko, 1996).

The paradigms used may involve taste and hunger and other motivational factors as reinforcing stimuli, and the drug or lesion tested may affect taste and hunger rather than memory. Similarly, drug effects that appear to involve memory and learning may actually be due to the drug's action on motor activity and sensory—visual, touch, or auditory—acuity or state. Thus, atropine may interfere with memory tests via its effects on motor activity in the case of active avoidance tests (see section BIV-1d); via its actions on hunger and thirst (see section BIV-2) in the case of tests involving food reward; via its xerostomic action in the case of tests involving liquid reward; and via its action on the nictitating membrane in the case of the blink test. Finally, depending on the extent, locus, and specificity of lesion, neurotoxins and immunotoxins may exert effects either on memory and learning processes or on such states as attention (Waite et al., 1999; Baxter et al., 1999).

c. Cholinergicity of Memory and Learning

Let us disregard the specific components of memory and learning—such as retrieval, motivation, storage, consolidation, and so on—and explore cholinergicity in the general context of these phenomena. First, it is important to demonstrate that cholinergic activity increases in the course of memory and learning phenomena. Indeed, it was demonstrated by Giancarlo Pepeu and his associates (see, for example, Giovannini et al., 1998a, 1998b) that ACh release from pertinent brain sites, including the hippocampus, increases during certain phases of learning, and this evidence was amply confirmed (see, for example, Thiel et al., 1998; Iso et al., 1999).

Nerve growth factors that improve learning may do so via preventing apoptosis of the cholinergic basal forebrain neurons (Sinson et al., 1996; Duan and Zhang, 1998; Van der Zee et al., 1999; Philips et al., 2001). In related studies, implantation of cultured cholinergic neurons into the hippocampus and forebrain of aged rats improved their learning and memory (Dickinson-Anson et al., 2003). Also, several studies demonstrated that in rats, pre- and early postnatal choline diet and presumably its effect on ACh synthesis improved the learning capacity and memory of the adult (Tees, 1999; Tees and Mohammadi, 1999; Montoya et al., 2000; Blusztajn et al., 1998, 2004; Mellott et al., 2004). Also, nootropics, drugs that without being muscarinics, nicotinics, or antiChEs exert facilitatory effects on animal learning and cognition (Giurgea and Salama, 1977; see also section BII-1), augment ACh turnover and release and CAT levels in several brain parts, including the cortex, sensitize cholinergic—possibly nicotinic—receptors, and may potentiate ACh currents generated at cholinergic receptors (Nishizaki et al., 1998; Pepeu and Spignoli, 1989; Bhattacharya et al., 1993; De Angelis and Furlan, 1995; Nakamura and Shirane, 1999; Kimura et al., 1999; Zhao et al., 2001; for a contrary opinion, see Krakauer, 1994).

Furthermore, behavioral and functional processes, including learning, affect the numbers, affinity characteristics, and/or distribution of cholinergic nicotinic and muscarinic receptors, as has been known for the last 25 years (see Edwards, 1979; Nathanson, 1987; Schuetze and Role, 1987; and Anagnostaras, 2003). These early data were expanded upon by Luiten, Van der Zee, and their associates, who employed an immunocytochemical approach to demonstrate the increase in muscarinic receptors during memory processes. It should be noted that this increase occurred in areas important for cognition such as the limbic system and its hippocampus, and the neocortex (Van der Zee et al., 1999; Van der Zee and Luiten, 1999). Wild rats, which allegedly "process . . . more numerous sensory stimuli" than the "naive," caged rats, exhibit markedly higher neocortical muscarinic immunostaining than the naive rats (Van der Zee and Luitten, 1999).

However, the story is not quite so simple. Particularly in the case of some learning tests, the increase in cholinergic activity and cholinergic receptor levels may not always relate to learning. Thus, the passive shock avoidance procedures may affect the receptors because they induce fear and immobility; in the case of active shock avoidance, the muscarinic receptor immunoactivity actually decreased in the amygdala. To complicate matters further, during associative and spatial learning the muscarinic receptors' immunoactivity increased and decreased in the hippocampal pyramidal cells and in the hippocampal interneurons, respectively (Van der Zee and Luiten, 1999). The Dutch team attributed these "counterintuitive" results to an interaction at the interneuronal hippocampal level between cholinergic and GABAergic transmission, which is important for learning (see below, this section). It must be added that Van der Zee, Luiten, and their associates emphasize that these changes in neuronal muscarinic immunoreactivity should be considered jointly with changes in the glial muscarinic receptors; the relation between the two sites might depend on the "scavenger" role of the astrocytes with respect to ACh; is this also the explanation for the marked increase in the glial muscarinic immunoreactivity that occurred in aged rats (Van der Zee and Luiten, 1999)?

It must also be noted that the forebrain pathways and sites involved in memory and learning, including radiations from the nucleus basalis magnocellularis (NBM) to the cingulated and other parts of the cortex, the tegmental nuclei, and the hippocampus, are rich in cholinergic neurons, markers for ACh synthesis and transport, and cholinoceptive receptors (see Chapter 2 DII and DIII; Mesulam, 1990; Zaborszky et al., 1999; Zaborszky, 2002). Other sites involved in memory and, particularly in habituation include dorsomedial striatum (Ragozzino, 2003; see also below, this section).

The notion of the cholinergic nature of memory and learning received strong support from studies of muscarinic knockout mice and of lesions, including those inflicted in primates (see above, this section; Harder et al., 1998; Hanin, 1990; Cuello, 1993; Voytko et al., 1994; Rossner et al., 1994; Wu et al., 1998; Tzavara et al., 2003; Winters et al., 2004). First of all, lesioning in animals of cholinergic brain systems and pathways, including NBM radiations and the limbic system—however induced: see above, this section—induces marked memory and learning deficits. In fact, the extent of OP antiChEs' induced hippocampal pathology was related to the extent of learning and memory impairment (Filliat et al., 1999). Furthermore, in many studies these deficits were paralleled by diminutions of CAT or ACh content, change in the density or number of cholinergic receptors (actually, both down- and upregulation may result from the lesions), and so forth, in the target areas of the lesioned brain parts. It is interesting that accumulated environmental damage produced, for example, by chlorinated biphenyls resulted in rats in a parallel decrease of brain CAT activity and maze performance (Provost et al., 1999). Learning deficit induced by the lesions could be antagonized, at least partially, by cholinergic muscarinic agonists and antiChEs. This deficit could also be remedied by NGF and other neurotrophins via their regenerative effect on the damaged cholinergic neurons (Cuello, 1993; Sauer et al., 1999). Also, choline supplementation of pregnant female rats protected their 1-month-old offspring from memory loss caused by pilocarpine-induced seizures (Yang et al., 2000).

Results obtained in humans suffering from accidental, surgical, or trauma-inflicted damage, as well as damage due to diseases such as AD and Rett syndrome, also support the notion of cholinergicity of memory and learning (Damasio, 1994; Damasio and Damasio, 1993). Indeed, the many postmortem studies of the brains of the elderly who exhibited senile amnesia and of patients suffering from AD confirm the notion of the relation between memory and learning and cholinergic brain systems (see also Chapter 10 A and K). Frequently there was a correlation among the site of damage, such as the hippocampus, the cholinergic character of the site, and the specific loss of cholinergic neurons (see Luria, 1973; Damasio and Damasio, 1993; Geula and Mesulam, 1994); also, the behavioral deficit induced by the lesion of these sites depended on the extent of the lesion (Wrenn et al., 1999). While the data concerning lesions are generally very supportive of the notion of cholinergicity of learning, some caveats should be posited. First, structures other than cholinergic neurons are damaged when the neurotoxins employed are only partially specific for cholinergic neurons, as is the case with, for example, quisqualate; this renders the interpretation of the nature of the effects of this drug on learning difficult. This is essentially not true for AF64A (although even in this case complete specificity may not obtain) and 192 IgG, the antibody to the low-affinity NGF receptor (Hanin, 1990, 1992; Wenk et al., 1994). Paradoxically, in the case of AD, essentially not a cholinergic disease (see Chapter 10), the employment of nonspecific neurotoxins may be particularly useful in the development of the pertinent model. In a related context, cholinergic differentiation affects noncholinergic systems, and lesions, even if specific for cholinergic radiations, may interfere with learning indirectly rather than directly, as these radiations affect functions other than learning, such as attention, vigilance, or sensory perception (see Milner et al., 1999; Dunnet and Fibiger, 1993).

Second, the effects on memory and on cholinergic parameters do not always run parallel; for example, reductions in the cortical CAT activity, following lesions of the NBM by various neurotoxins, do not always correlate with the learning deficit that they induce; the effect of the neurotoxins in question on systems other than the cholinergic system may underlie this imperfect correlation (Markowska et al., 1990; Fibiger, 1991). Rarely, lesions of the NBM affected learning only slightly.

d. Muscarinic and Atropinic Drugs and AntiChEs

It strongly supports the notion of the cholinergicity of learning and memory that cholinergic agonists facilitate and cholinergic antagonists attenuate or block learning and memory; in the case of antiChEs the memory effect coincides with increased levels of brain and/or blood AChE (Kosasa et al., 1999). It is also noteworthy that inhibitors of synthesis of ACh and/or uptake of choline also attenuate memory and learning (see, for example, Boccia et al., 2004).

In particular, muscarinic agonists and antiChEs facilitated learning in all species studied, from fish, urodele, and amphibia, to rodents, cats, dogs, and primates; also, as already mentioned, antiChEs and muscarinics attenuated memory and learning deficits induced by lesions, in both humans and animals (see Hingtgen and Aprison, 1976; Myers, 1974; Karczmar, 1976, 1990; Bartus et al., 1982; Hagan and Morris, 1988; Aigner et al., 1991; Fibiger, 1991; Ogura and Aigner, 1993; Hironaka and Ando, 1996; Inagawa, 1993; Cuello, 1993; Power, 2004a, 2004b). This was noted as well when such complex paradigms as the 8-arm radial maze was used (Inagawa, 1993).

All this evidence supports the notion of the positive effects of muscarinic agonists on memory and learning; it appears that M2 and M4 receptors are involved in these effects (see below, this section, and Tzavara et al., 2003), while the contribution of M1 receptors is less certain. In fact, the knockout mice deficient in M1 receptors may show only minimal if any memory and learning deficits; further use of muscarinic agonists specific for all muscarinic receptor subtypes are necessary to resolve this issue. It must be added that there is some inconsistency with regard to currently available cognitive effects of muscarinic agonists (see below, this section).

Effects of muscarinics on human memory and learning are of particular interest. Sitaram et al. (1979) and Berger et al. (1979) found that antiChEs and muscarinics significantly facilitate memory and learning in young subjects, while Drachman and his associates (see Drachman, 1978) thought that the effects of these drugs are minimal in the young; similarly, in aged but otherwise healthy individuals, the effect was absent or minimal (see Dean and Bartus, 1988; Borjesson et al., 1999). However, in the case of senile or AD-related memory deficit, these drugs seem to be effective, at least in the case of moderate disease status. If indeed in a young human subject the effects of muscarinics and antiChEs on memory and learning are minimal, this may relate to the demonstration that it is more difficult to cholinergically facilitate learning exhibited by "intelligent" strains of animals (such as mice) than by "dull" strains of these animals (see above, and Karczmar and Scudder, 1969; Karczmar et al., 1973). It must be noted that, contrary to the inconsistent effects of muscarinics and antiChEs in the human, the attenuation of memory and learning by atropinics was consistent in both young and elderly humans (see below).

However, sometimes the effects of muscarinics and antiChEs were negative or biphasic (see Karczmar, 1975, 1976; Hingtgen and Aprison, 1976; Hagan and Morris, 1988; Hironaka et al., 1990). The negative effects may occur when muscarinics act on brain sites and/or memory components that are not be reflected in the tests utilized (Kim and Levin, 1996). Furthermore, a cholinergic agonist's detrimental effect on learning may result from its presynaptic rather than postsynaptic effect; thus, carbachol's negative effect on learning sometimes may result from the presynaptic attenuation by carbachol of ACh release at a strategic site (Hiramatsu et al., 1998).

Biphasic and negative effects are caused particularly by antiChEs. Deutsch and his associates stressed these biphasic effects of DFP; whether facilitation of memory or amnesia was obtained depended on the state of learning (see section A-2d, above). Their conclusions were based on experiments in which high doses of DFP were used and ChEs and ACh levels or turnover not measured (see Deutsch, 1971, 1979; Deutsch and Leibowitz, 1966; Deutsch et al., 1966). Other investigators confirmed that high antiChE doses that induce brain and/or erythrocytic AChE inhibition readily block memory and learning; this block may involve the retrieval process specifically (Karczmar, 1975; Hamburg and Fulton, 1972; Geller et al., 1984, 1985). These negative effects of antiChEs and cholinergic agonists may not argue against promemory, prolearning effects of these drugs: the accumulation of ACh resulting from the use of high doses of antiChEs as well as high doses of cholinergic agonists, particularly in the case of not readily metabolizable compounds, result in a block or desensitization of cholinergic synapses, leading to the arrest of transmission and memory block. Another explanation of the antiChE-induced attenuation of memory and learning is that it may be induced by excessive cholinergic function induced by antiChEs or muscarinic agonists (see Hamm et al., 1993). It should be noted, however, that sometimes this block of learning occurred with relatively small antiChE dosing capable of not more than 20% to 30% inhibition of brain AChE (Richardson and Glow, 1967; see Hingtgen and Aprison, 1976).

Negative results may be due to pharmacological effects of the muscarinics and antiChEs that have nothing to do with their effects on learning, and this difficulty has already been referred to (see section BV-4b, above). This may be particularly true in the case of certain behavioral tests. Thus, cholinergic agonists generally facilitated passive avoidance while frequently their effect on active avoidance was minimal or even negative (G. Bignami and N. Rosic, 1969, personal communication; see Karczmar, 1975; Myers, 1974). This dichotomy may relate to the attenuation of the motor behavior by the muscarinic agonists and antiChEs (Chau et al., 1999; see section BIV-1d and BVI). Also, the antinociception action of antiChEs or muscarinics may interfere with the pain clue when the latter is used in conditioning and thus obliterate the so-called conditioned anxiety response (CAR; see above, section BIV-2; Shannon et al., 1999). Moreover, the results of the tests may depend on the site of the application of the cholinergic or anticholinergic drug (Neil and Grossman, 1970; Hingtgen and Aprison, 1976), as with certain sites the negative effect on learning that was obtained may have been due to the action of the drug on a noncholinergic system. Finally, the extent of improvement in learning, whether due to certain drugs or to training, may be inversely related to the initial level of performance (Karczmar and Scudder, 1969). Thus, the facilitatory effect may not obtain in animals exhibiting optimal levels of learning performance; for example, muscarinics or antiChEs may not improve the learning of "intelligent" mice strains that exhibit innate high learning capacity (Karczmar and Scudder, 1969).

Regarding muscarinic antagonists: negative effects of muscarinic antagonists in animals on learning are very consistent, as was their block of muscarinic facilitation of memory or learning; data obtained with knock-out mice lacking M1 receptors are also in agreement with the notion of the muscarinic correlates of memory and learning (Myers, 1974; Karczmar, 1976, 1990; Hintgen and Aprison, 1976; Hughes, 1982; Fibiger, 1991; Bartus et al., 1982; Dean and Bartus, 1988; Harder et al., 1998; Hagan and Morris, 1988; Ukai et al., 1994; Riekkinen et al., 1995; Rosat et al., 1992; Sessions et al., 1998; Curran et al., 1998; Wang et al., 1999); these effects obtain across all species, including primates; they occur upon either systemic administration of these agents or their localized application to pertinent sites (see, for example, Riekkinen et al., 1995). It is of interest that there is a synergism between atropinics and chronic alcoholism with respect to their amnetic effect (Nagahara and Handa, 1999). The effect of atropinics on learning may reflect their action on one or many components of the learning phenomena. Hughes (1982) in particular emphasized that atropinics prevent novelty-seeking behavior, which the rodents exhibit in the Y-shaped maze, novelty seeking constituting an important aspect of learning (see below, next section).

On the whole, the data support the notion of atropinic attenuation of memory and learning (in animals and humans; see below). Yet, the pertinent data are not completely consistent; for example, both facilitatory and inhibitory effects of atropinics were sometimes observed with respect to spontaneous alterations in behavior (Hagan and Morris, 1988), and a facilitatory effect on avoidance learning was observed even with small doses of atropinic drugs (Suits and Isaacson, 1968). On the other hand, it is very well documented that atropinic amnesia and learning deficits are antagonized, similar to the deficit induced by appropriate brain lesions, by muscarinics and, particularly, by antiChEs (see, for example, van der Staay and Bouger, 2005).

Learning effects of anticholinergic drugs in the human deserve special consideration. Antimuscarinics consistently produce amnesia in normal humans, as has been known since the turn of century (Gauss, 1906) and as has been confirmed amply subsequently (see, for example, Drachman and Leavitt, 1974; Drachman, 1978; Drachman and Sahakian, 1979; Dean and Bartus, 1988; Hagan and Morris, 1988; Karczmar, 1976, 1979b; Hingtgen and Aprison, 1976). This holds for a number of tests, including the paradigm concerning visuospatial memory (Meador et al., 1993). Interesting data in this context were provided by Daniel Drachman and his associates (see Drachman, 1978; Drachman and Sahakian, 1979; Drachman and Leavitt, 1974), as they used several quantitative tests of memory and learning to show that atropinics, given to normal young humans, attenuate their cognition and, indeed, render it undistinguishable from the cognition of the elderly population. Drachman's demonstration of the relation among aging, cholinergic system, and memory and learning was confirmed repeatedly (see, for example, Kurotani et al., 2003; McGeer et al., 1987), and it is of interest that during aging, cholinergic function of the hippocampus may be more impaired than that of catecholamines (Schweitzer et al., 2003). It may be added that Raymond Bartus posited in the 1980s what he referred to as a "cholinergic hypothesis" of geriatric memory impairment (Bartus et al., 1982; see also Bartus, 2000), and many investigators credited him with being the pioneer in this area. However, correlation between memory and the cholinergic system dates from the 1920s (see Karczmar, 1967, 2004), and the link between memory loss, cholinergic deficit, and aging was clearly postulated by Drachman and others some 30 years ago.

It appears from the foregoing that muscarinic postsynaptic receptors of the basal forebrain, limbic system, thalamus, and neocortex play an important role in memory and learning processes (see van der Zee and Luiten, 1999). Which of the M1 to M5 receptor subtypes are particularly involved in these processes? The M1 and M2 receptors and agonists seem to be specifically concerned with facilitation of learning, as indicated by their presence in the hippocampus, septum, thalamus, and median basal forebrain system and as suggested by their downgrading in AD, which may parallel the memory deficit of AD, and the results obtained by Jurgen and others with M2 receptor-defficient knock-out mice (Vilaro et al., 1992; Caulfield, 1993; Quirion et al., 1993; Schwartz et al., 1994; Ruske and White, 1999; Felder et al., 2001; Ladner and Lee, 1999; Bymaster et al., 2004; Mooney et al., 2004; Oki et al., 2005); The M3 receptors may be involved in behaviors other than learning (see, for example, Kow et al., 1995). Furthermore, using knockout mice lacking differentially specific muscarinic receptor subtypes, Wess, Nathanson, Hamilton, and their associates (Felder et al., 2001) proposed that the M4 receptor subtype is involved in processes of attention (as suggested by these investigators, the actual effect may concern the motor movement involved in the processes in question).

As in the case of other muscarinic (and nicotinic) effects, second messengers and phosphorylations by kinases are involved in the muscarinic, particularly M1 receptor activation of learning (and functions related to learning, such as attention). For example, such messenger and phosphorylation systems as extracellular-signal regulating kinase-mitogen-activated protein (ERK/MAPK; Adams and Sweatt, 2002; Sweatt, 2004) seem to be the components of the hippocampal contribution to memory processes including attention (Giovannini et al., 2003).

e. Nicotinics and Nicotinolytics

As shown above, the extensive research concerning memory and learning effects of muscarinics, atropinics, and antiChEs begun as early as the 1940s. Studies of nicotinics and nicotinolytics were initiated much later. In the 1960s, Daniele Bovet, Filomena Bovet-Nitti, and Alberto Oliverio[8] demonstrated that in rodents nicotine facilitates learning in several learning paradigms (see, for example, Bovet et al., 1966). As the interest in nicotine increased in view of its addictive action (see the next section), its general pharmacological profile, including its cognitive action, became of interest, and several investigators substantiated Bovet's findings (Risner et al., 1988; Rosecrans, 1989, 1995; Arneric and Williams, 1994; Levin, 1992; Warburton, 1992a, 1992b, 1994; Woodruff-Pak et al., 1994; Arendash et al., 1995; Clarke, 1995; Picciotto, 1998; McGaughy et al., 1999; Hagan and Morris, 1988). David Warburton in particular became an intense defender of the use of nicotine. Furthermore, while this effect could be demonstrated in young animals, it was particularly prominent in aged animals (Arendash et al., 1995). It is consistent with these data that nicotine did not facilitate learning of mice that were subjected to knockout mutations of nicotinic receptor subunits (Picciotto et al., 1998; Picciotto, 2003) and that knockout mice that are devoid of β2 nicotinic receptors (which are the most abundant nicotinic receptors in the brain; see Chapter 6 B and C) showed premature loss of cognition in adulthood or early aging (see Chapter 6 C). Finally, nicotine facilitated and nicotinolytics attenuated human learning and cognition (see Warburton, 1992a; Arneric and Williams, 1994). However, as is the case with muscarinics, available human (and animal) data are not extensive and are not always consistent (see Levin, 1992); sometimes, large doses of nicotine were used in the experiments in question; large doses of nicotine are capable of blocking pertinent cholinergic pathways (see Risner et al., 1988). Furthermore, the interpretation of the data was frequently distorted by the controversy surrounding nicotine addiction: the supporters of nicotine's use in humans have a tendency to emphasize and the detractors to deny its positive cognitive effect (see below, section BV-5d).

The learning or cognitive effect of nicotine may relate to a nicotine action that differs from that of muscarinic agonists; for example, Shoaib and Stolerman (1996) opined that learning facilitation by nicotine may be due to its facilitation of discrimination mediated by the dorsal hippocampus, a site rich in nicotinic receptors, while Arneric and his associates (Decker et al., 1998; Prendergast et al., 1998) demonstrated that in the monkey nicotine and nicotinic agonists facilitate learning via antagonizing the antilearning effect of distracting stimuli. Furthermore, some synthetic nicotinics—but not nicotine—may protect the cholinergic neurons, as these drugs seemed to prevent both ischemic apoptosis and the learning deficit that ischemia normally induces (Nanri et al., 1998). These data relate to the evidence that nicotinic agents may act on memory via releasing or facilitating the action of trophic factors, as well as perhaps increasing levels of several components of the cholinergic system (Hernandez and Terry, 2005).

It appears that nicotinic receptor subtypes, which are involved in the cognitive effects of nicotinic agonists, contain the α4β2 subunits; they are widely distributed in the brain, particularly in the midbrain and thalamic areas (Picciotto et al., 1995, 2000; Lena and Changeux, 1998; Arroyo-Jimenez et al., 1999; see, however, Torrao et al., 1997).

The nicotinolytics generally exerted negative effects on memory and learning; also, they prevented the antidistraction or the prevention of attention to trivial stimuli due to nicotinic agonists (see, for example, Terry et al., 1996; Clarke, 1995; Decker et al., 1998). However, the results obtained with nicotinics were not consistent; the various nicotinic antagonists used in pertinent research may deal with the different central nicotinic receptors and thus induce different effects on memory and learning (see, for example, Terry et al., 1996). Nicotinic agonists exhibit great deal of plasticity, examplified by their capacity for upgrading and mutating. This knowledge as well as the better understanding of the crystalline structure and electron images that explain receptor binding (see Chapter VI; see also Gotti and Clementi, 2004; Celie et al., 2005; and Unwin, 2005) led to synthesis by Geoffrey Dunbar, M. Bencherif, G. J. Gatto and their associates of new, cognition active nicotinic agonists with clinical potential in the treatment of AD and senile memory deficit (see Gatto et al., 2004).

f. Mechanisms of Cholinergic and Anticholinergic Actions on Memory and Learning

The mechanisms that were proposed by various investigators concern several components of the learning process, neuronal and electrophysiological phenomena, and strategic CNS sites and their interaction. In conformity with this book's context, these mechanisms are based on the notion of importance of the cholinergic corollary for memory and learning. Strong evidence for this notion was provided above. Some novel data that concern REM sleep should be added to this evidence, even though these results are controversial: intensity and duration of REM sleep could be related to learning ability and to the effectiveness of memory; conversely, REM sleep deprivation causes memory impairment and diminution of levels (and possible turnover) of forebrain ACh (Smith, 2003; see also section BIV-3). It should be added that there is a controversy as to the effect of cholinergic activity on memory and, particularly, memory consolidation during SW sleep (Power, 2004a, 2004b; Gais and Born, 2004; see also below, this section).

In an early study, Peter Carlton stressed the importance of the antihabituation component of learning; he referred the amnetic synergism between atropine and amphetamine and explained it as due to the emergence or reappearance with these drugs of irrelevant, trivial responses (dishabituation; Carlton, 1966, 1968; see also Karczmar, 1995). Subsequently, Hughes (1982) stressed the significance of the antialternation effects of atropinics: when animals placed in an Y-maze do not find the food reward at the end of one arm of the maze, they seek the reward in the other arm of the maze; this alternation is blocked by atropinics. The antialternation effect of atropinics is related to their dis-habituation effect as well as to their antagonism of the novelty seeking phenomena (neophilia). Indeed, suppression by cholinergic agonists of responses to trivia that underlies antihabituation and novelty-seeking behavior, rather than facilitation of other components of memory, may constitute the major promemory effect of the cholinergic system and the cholinergic agonists; in other words, cholinergic agonists facilitate neophilia (Hasselmo and Schnell, 1994; Hasselmo et al., 1995). Dis-habituation may also be considered as inhibition of responses to stimuli that become irrelevant, and there is evidence that this inhibition is induced by the activation of muscarinic (M1?) receptors, possibly in the dorsomedial striatum (see Ragozzino, 2003).

Deutsch (1966, 1971, 1979) made another early generalization. He speculated that learning depends on the activity of "cholinergic memory synapses," which change in sensitivity with the progress of learning and memory processes; this change would explain, according to Deutsch, the biphasic action on learning of antiChEs and accumulated ACh; Deutsch did not identify the synapses in question, nor did he and his associates measure ChEs and ACh (see above, this section).

John Eccles (1979) posited a more specific neuronal mechanism that may underlie memory formation, as he adduced evidence suggesting that dendritic sprouting underlies learning processes. His notion is supported by subsequent evidence that indicates that "enriched environment" as well as conditioning increases dendritic branching and the number of synapses in the cortex of adult rats (Escorihuela et al., 1995; see also Agranoff et al., 1999; see also Eccles, 1985). This information does not necessarily support the view of cholinergicity of learning and memory. Several cholinergic corollaries of this notion were presented, however. Rosenzweig, Bennett, Diamond, and their associates found that "intelligent" and "dumb" strains of rats as well as rats exposed to "rich" versus "poor" environment differed in levels of cortical AChE (see for example, Rosenzweig., 1966; Rosenzweig et al., 1958). The differences amounted to a few percent; the evidence presented by the California investigators was criticized on these and other grounds (see Karczmar, 1969 and Chapter 1 B-3), but it was interesting and heuristic (see a more recent publication from the team [Diamond et al., 1985] that does not concern the cholinergic system as such). A convincing cholinergic corollary of the same notion was provided by the French-Israeli team of Jean-Paul Changeux and Hermona Soreq (Beeri et al., 1997); they found that learning deficits exhibited by transgenic mice that overexpress mRNA for AChE relate to diminished formation of dendrites and spines. Then, Saar and Barkai (2003) found that modifications in intrinsic properties of cortical neurons that occur in the course of learning facilitate training in rats. While this information does not specifically implicate cholinergic neurons, Woolf (1996) emphasized that basal forebrain neurons are particularly dependent for their functioning and plasticity on neurotrophins, and she hypothesized that this plasticity underlies capacity for learning-induced increase of ACh release at cortical sites; this finding is consistent with the evidence that motor activity of rats increases their learning capacity as well as a number of cholinergic parameters, including choline uptake and the density of muscarinic receptors in the mouse hippocampus (Fordyce and Wehner, 1993; see also Agranoff et al., 1999). Furthermore, Nancy Woolf (1998), similar to Eccles, felt that dendritic branching and dendritic reorganization are components of this plasticity and showed that, specifically, these changes are mediated by ACh activation, via stimulation of the phosphatidylcholine cascade, of phosphorylation of the microtubule-associated protein MAP-2 (Johnson and Jope, 1992; Audesirk et al., 1997; Hely et al., 2001; see also Butcher and Woolf, 2003). This scheme also relates to the facilitation of memory and learning by muscarinics as their effect on M1 and other muscarinic postsynaptic brain receptors set in motion the same chain of events (see Butcher and Woolf, 2003). It should be added that Woolf, Hameroff, and others propose a similar hypothesis with respect to processes of consciousness and self-awareness (see section BVI, below). Conversely, Nancy Woolf and Larry Butcher (1989) suggested that, once neuronal and related changes needed for establishing new memories and/or new learning modes have occurred, "increasing rigidification or stabilization of the network" are required "for preservation of memory trace"; while microtubules may dominate during periods of memorization and plasticity, microfilaments may prevail during this "rigidification" (Lasek, 1981). Related views link the stabilization of synaptic changes induced by memorization and learning with molecular events such as Ca^{2+}/calmodulin-dependent protein kinase II (CaMKII) "switch" (Lisman, 2004; this mechanism does not involve specifically cholinergic synapses). An allied, early view was posited by Holger Hyden (Hyden, 1972), as he proposed that memory encoding involves formation of memory-specific proteins. Also, during processes of learning and cognition, the dynamics of brain organization seem to shift; for example, at that time, afferent influence may predominate over intracortical influences; these notions may also be concerned with brain plasticity (induced in part by the theta bursts; Huerta and Lisman, 1995; Kimura, 2000).

It should be added that these events that illustrate morphogenesis of "memory engrams" as well as brain plasticity seem to be reflected in changes in synaptic functioning and electrophysiological events, for example, changes in auditory evoked potentials that are induced by sensory preconditioning; these changes were blocked by antimuscarinics (see, for example, Maalouf et al., 1998). Similarly, Ben Libet (1979) opined that certain slow potentials that induce long-term enhancement (LTE) of synaptic transmission generate a memory trace. These notions hark back to the concept of the psychologist Donald Hebb that memory and learning result from postsynaptic sensitization (Hebb, 1949; see also Moruzzi, 1934 and Lechner and Byrne, 1998). Again, there are cholinergic correlates of this notion, as LTE observed in the sympathetic ganglia exhibits cholinergic correlates.

Long-term potentiation (LTP) is a related phenomenon, which was credited by Collier (1996), Eccles (1979), and Agranoff (Agranoff et al., 1999) with generating memory engrams; similar to LTE, LTP, which is induced by the stimulation of several limbic (including the hippocampus) and cortical sites, is enhanced by cholinergic muscarinic and nicotinic agonists (see, for example, Fuji et al., 1999; Matsuyama and Matsumoto, 2003), and it engenders a long-lasting increase of ACh release; interestingly, choline facilitates the induction of LTP (Pyapali et al., 1998). Egorov et al. (1999) ascribed this cholinergic enhancement of LTP to suppression by a muscarinic agonist of dendritic hyperpolarization evoked by repetitive stimulation; change in Ca^{2+} fluxes may be involved. Furthermore, cortical LTP is facilitated by cholinergically induced theta rhythm (Natsume and Kometani, 1997, 1999; see also Karczmar, 1996). Also, cholinergic afferents of several cortical sites or layers mediate cortical LTP (Hess and Donoghue, 1999; Patil et al., 1998). Evoked potentials related to perception are enhanced by cholinergics and attenuated by atropinics, and this facilitation, which expresses perceptual improvement, may underlie memory (see section BIV-3b).

As some of these concepts that concern LTP include the notion that LTP involves an increase in ACh release, and as some of the related phenomena are blocked by atropine, it must be remembered that atropinics can increase ACh release by inhibiting the muscarinic presynaptic block of this release (see (section BI-1). Apparently, this mechanism cannot be involved in LTP and in atropinic attenuation of memory. It must be noted, however, that some of the work concerning the memory and learning of Aplysia postulated the presynaptic induction mechanism for "associative facilitation" of memory (Lechner and Byrne, 1998).

Parenthetically, Ben Libet's and Eccles' generalization are, again, reflections of brain plasticity, as are experiments indicating that memory and information processes are regulated by a subtle interaction between multiple brain sites (see above, this section; see also Gu, 2003; Weinberger, 2003; Gold, 2003).

Another electrophysiological generalization espoused by Mircea Steriade states that EEG patterns such as arousal and desynchronization are cognomic for memory; these patterns exhibit strong cholinergic correlates (Steriade, 1993a, 1993b; Steriade and McCarley, 1990; see also sections BIV-3b and BIV-3c); it must be added that REM sleep and/or REM sleep EEG patterns also seem to be involved in memory formation (Johnson, 2005). A related hypothesis states that memory (or maintenance of information) is formed by reverberating (oscillating) neuronal activity in a recurrent network, and Angel Alonso (2005) and his associates showed that these processes are modulated or facilitated by muscarinic agonists, particularly in the case of the entorhinal cortex; this reverberation and the related phenomenon, "graded persistent activity" response to consecutive stimuli may underlie the working memory (Egorov et al., 2002). In addition, Hasselmo (2005; see also below, this Section) and his associates presented data indicating that these reverberations involve matching sensory input with decision-making and information retrieval via CA 3 and CA 1 interactions; also, they stressed the importance of these structures and of the temporal lob as a "distributed memory" model (Howard et al., 2005).

Some of the hypotheses concerning mechanisms of memory and learning also attempt to identify central sites involved in memory (see above, this section). These sites concern memory and learning generally and, specifically, their components, such as long-term versus working memory, retrieval, consolidation, and novelty-seeking behavior, as well as the prememory states, such as attention, alertness, and so on. It appears that several sites rather than a single site are involved in memory generally as well as in single components of the memory phenomena (Power, 2004a, 2004b). Thus, the NBM and other lesions affect memory and learning generally, as well as such components as attention (see Luria, 1973; Squire and Butters, 1992). Yet, it is nearly impossible to ascertain that a given memory or learning effect obtained by a localized drug application is indeed induced at the site in question, as it may be due to the effect induced by this application at other sites.

The most that can be stated at this time is that long-term or reference memory is affected by lesions of many forebrain and cortical sites, including NBM, hippocampus, amygdala, and paralimbic cortices; the use of selective immunotoxins proved particularly productive in this context (Hagan and Morris, 1988; Wilson, 1991; McGaugh et al., 1993; Galey et al., 1994; Butt and Hodge, 1995; Olton et al., 1991; Rosat et al., 1992; Everitt and Robbins, 1997; McGaughy et al., 2000; Henke et al., 2003a, 2003b; Giovannini et al., 2003; Rogers and Kesner, 2003). Short-term or working memory may be more specifically sensitive to lesions in baso-lateral amygdala, mediodorsal thalamus, and septocingulate pathway (Ohno et al., 1992; Rosat et al., 1992; Stackman and Walsh, 1995; Dougherty et al., 1998); the caudate and certain frontal cortex sites may also be involved (Liu and Su, 1993; see also Hagan and Morris, 1988). The importance of hippocampus in memory processes is evident from Kazimir Blusztajn's demonstration that learning and memory were improved in parallel with increased synaptic response to carbachol of hippocampal in slices obtained from adult offspring of female rats fed prenatally with choline compared to the memory performance and responsiveness of hippocampal slices obtained from adult offspring of choline-deficient mothers (Blusztajn et al., 1998; Montoya et al., 2000).

Several investigators claim that various sites may have to interact to generate complete memory or learning phenomena. Indeed, certain sites may act antagonistically with respect to other sites, and balance or interaction may be needed to perform a learning task (Olton et al., 1991; Rosat et al., 1992). Thus, Hasselmo (1995; Hasselmo and Barkai, 1995; Hasselmo, 2005) proposed a complex process in which the cholinergic system regulates the levels of intrinsically and extrinsically generated signal transduction, this regulation also including the switching between the hippocampus and neocortex to accomplish the transition between learning and recall. Furthermore, Hasselmo (1995) and Van der Zee and Luiten (1999) opined that the dynamics of cholinergic response to familiar versus novel input differ and that these dynamics depend on the stage of the memory and learning process. According to Van der Zee and Luiten (1999), these complex dynamics may explain the "counterintuitive" memory effects of muscarinics (see above, this section).

Certain mechanisms involving processes that are indirectly rather than directly involved in learning were referred to earlier in this section; these include attention or distraction, neophilia, coding capacity, the status of the reward-punishment dipole (see also below, section BV-e), the state of alerting or arousal, stress, and so on (see, for example, Everitt and Robbins, 1997). All these processes contain strong cholinergic correlates and are generally facilitated and attenuated by cholinergic agonists and antagonists, respectively, as already discussed in this section as well as in sections BIV-1 and BIV-3. Thus, these processes should be considered in the discussion of the mechanism or mechanisms of memory and learning. It was already pointed out (see above, this section) that it is not easy to provide evidence indicating the cholinergic role specifically in any of the functions in question, as any cholinergic agonist or antagonist that affects, for example, attention, encoding, or retrieval will also ultimately alter learning; yet studies that combine the use of lesions, special tests of learning, attentional capacities, additional, pertinent behavioral tests, and assessment and localization of ACh release may identify the brain sites of several memory components. For example, using this methodology, Giancarlo Pepeu and his associates (Giovannini et al., 2003) presented data indicating that the activation of the M1 receptor and the ERK MAPK cascade in the hippocampus (see above, this section) facilitate attention specifically (although the cortex and its cholinergic input may also constitute the strategic site for these processes; see Sarter et al., 2003; see also Voytko, 1996). Conversely, M1 agonists and antiChEs reduced distractibility in primates (O'Neill et al., 2003). However, nicotinic sites may also be involved in the generation of attention (Thiel et al., 2005).

Similar proposals were made with respect to other specific components of memory such as retrieval, encoding, and consolidation (Rogers and Kesner, 2003; Power et al., 2003; Power, 2004a, 2004b). In fact, certain lesion studies suggest that there may be a dissociation between components of memory such as attention and learning, and that learning processes may be possible in absence of certain parts of the "cholinergic learning circuitry" (Sarter et al., 2003).

There are several other parameters and components of brain function that are linked with memory and learning, including endocrines and hormones (see, for example, Tanabe et al., 2004; Crayton et al., 2004), metabolic activity and blood flow (as visualized by single photon emission computerized tomography [SPECT] and magnetic resonance imaging [MRI] methods), second messengers, and protein kinases (Agranoff et al., 1999). These parameters are not necessarily related to cholinergic activity, and transmitters other than ACh (such as serotonin) are also concerned with cognition. Much of this information is derived from the studies of invertebrates such as drosophila (Belvin and Yin, 1997) and, particularly, Aplysia, where pertinent evidence was obtained by the Nobel Prize winner Eric Kandel (see also Lechner and Byrne, 1998; Bailey et al., 1996); this information is beyond the scope of this book.

The all-encompassing theory of memory and learning that would include the processes, phenomena, and factors discussed in this section waits to be posited.

g. Conclusions: Today's Status of the Notion of Cholinergicity of Memory and Learning

The predominant cholinergic nature of memory and learning is generally accepted today (see section A-1; Karczmar, 1967; Bartus et al., 1982). Discordant opinions are rare. For example, Blokland (1995) feels "the notion that ACh plays a pivotal role in learning and memory processes seems to be overstated" and "the greater and more specific cholinergic damage [is that] fewer effects can be observed on behavioral level[s]." Blokland's evidence for these notions is not convincing.

However, the cholinergic memory and learning hypothesis does not signify that other transmitter systems do not affect memory and cognition (i.e., all other behaviors and functions reviewed in this chapter). Second messenger, nitric oxide, monoaminergic, peptidergic, and aminergic systems and several hormones are also involved in cognition (Bidmon et al., 1999, see Haroutunian et al., 1990; Karczmar, 1978a, 1978b, 1981b; Glowinski and Karczmar, 1979; Bevan and Bolam, 1995; Szymusiak, 1995; Farr et al., 1999; Kelley et al., 2003). All effects exhibited by these systems interact with respect to memory and learning and may be antagonistic or cooperative (Cassel and Jeltsch, 1995; Bidmon et al., 1999). For example, increasing serotonergic system activity may accentuate memory deficits induced by scopolamine in rats, while other monoaminergic systems may facilitate cholinergic effects on learning or memory (Haroutunian et al., 1990; Ohno and Watanabe, 1997). In addition, androgens and estrogens may facilitate learning and cholinergic agonists may augment this effect; the hormonal-cholinergic interaction may be due to neuronal phenomena or to the trophic actions of the hormones (section BIV-2f, above).

In fact, the total process of cognition, which is a broader phenomenon than learning, must be included in the pertinent considerations (see below, and section BVI). Cognition is defined as knowledge and the processes of knowing and perceiving, with learning and memory as inherent components. Since learning and memory exhibit intense cholinergic correlates, they of course facilitate cognitive processes as well, particularly since cholinergic activity appears also to facilitate the processes of attention and perception. Here, a good example is the cholinergic modulation of behaviors engendered by social contacts such as acceleration by cholinergic agonists of the rate of several modes of the "acquaintance making" behavior (Karczmar et al., 1973; Winslow and Camacho, 1995; see section BIV-3c). However, cognition is broader than learning. Effective and teleological interaction between organisms and their social and physical environment are cognition components; this interaction represents the "Welt Anschauung" and understanding of the world by animals or humans. Jointly with learning and memory, this interaction has strong cholinergic connotations. Altogether, cholinergic correlates of cognition and of several functions and behaviors are expressed in their totality by the syndrome of CANMB, which has already been referred to and which will be discussed subsequently (see section BVI).

5. Addiction

The subject of addiction deals with the role of the cholinergic system and the effect of cholinergic drugs in the addiction induced by drugs of abuse, opiates, cocaine, nicotine, and alcohol; it also concerns the mechanisms involved in addictive actions, including those of nicotine, an area that has recently begun to receive major attention.

Strong evidence implicates the cholinergic system in addiction inflicted by all these drugs. Thus, the meso-limbic reward-punishment dipole and pedunculopontine tegmental system are implicated in addiction; these sites are parts of the cholinergic pathways, and locally administered cholinomimetics activate the reward system (Picciotto, 1998, 2003; Picciotto et al., 1998, 2000; Corrigal et al., 1999; Ikemoto et al., 2003). Learning processes are also involved, and Everitt, Robbins, and their associates appropriately referred to addiction as "aberrant learning" (Everitt et al., 2001). Furthermore, the evidence clearly shows that upon either acute or chronic administration, or in the course of the self-administration paradigm, addictive drugs affect, in the brain, the release, content, and/or turnover of ACh and the components of the cholinergic system such as CAT and the cholinergic receptors.

Animal "addiction," or drug-seeking behavior, is studied by means of operant procedures in which the animal may select for ingestion a drug rather than a food or show in some other way drug preference (see below; Karczmar al., 1973; Everitt et al., 2001); with other paradigms, the animal is provided with an in-dwelling catheter and may choose to self-inject itself with a drug of abuse.

However, as is true with respect to other behaviors and functions, besides ACh, monoamine, indole, and peptide transmitters, modulators, and second messengers, whether of the nucleotide or phosphatidyl inositol type, are involved in addiction (see, for example, Deitrich and Erwin, 1975; Yi and Johnson, 1990; Karczmar, 1978a, 1978b; Grenhoff and Svenson, 1989; Ziegler et al., 1991).

a. Opiate Addiction

Various aspects of morphine administration to animals all exhibit strong cholinergic correlates. Acute administration decreased ACh release from both peripheral and central nervous system; this was true whether the release was measured in the whole brain, in brain slices obtained from several brain areas, or in synaptosomal preparations (Crossland, 1971; Pepeu, 1974; Karczmar, 1976; Salto et al., 1990; Schoffelmeer et al., 1997). Also, given acutely, opioids increase ACh brain levels and decrease its turnover in several species (see, for example, Wood, 1986; Wood and McQuade, 1987). Finally, upon their single or chronic administration, the opiates decrease the sensitivity or cause blockade of postsynaptic nicotinic receptors; this extends to opioids such as methadone and particularly involves $\alpha3\beta4$ nicotinic receptors, although the effects vary from one receptor subtype to another (Kellar et al., 1999; Xiao et al., 2001; see also Maelicke, 1996). Taken together, this evidence indicates that acute morphine administration attenuates cholinergic activity.

Chronic morphine administration to rodents and other species induces morphine-seeking behavior (Crossland, 1971); this morphine "addiction" may be demonstrated by appropriate operant paradigms. In the course of this chronic treatment, most of the cholinergic parameters, including ACh release and CAT values, tend to return to normal values (for a detailed description, see section BIII-2a, above). Interestingly, Dirken and Nijhuis (1983) found that spinal sites may be cholinergically involved in opiate addiction.

ACh release increased dramatically, in "addicted" rats or dogs, during morphine withdrawal or upon the administration of naloxone (Crossland, 1971; Jhamandas et al., 1971). This increase in ACh release is accompanied by diminution of ACh brain pool and the return of ACh turnover to control values. Also, it appears that muscarinic receptors may develop hypersensitivity during the abstinence syndrome, as the dihydropyridine-sensitive Ca^{2+} channels are markedly more active during the rat abstinence syndrome (Antkiewicz-Michaluk et al., 1990). Interesting evidence supporting the notion of the cholinergicity of withdrawal was provided by Jerry Buccafusco and his associates as they demonstrated that the expression of M2 receptors and the levels of its mRNA are increased in the course of rodent withdrawal syndrome and that prevention of this increase (by DFP, surprisingly) suppresses withdrawal symptoms (Zhang and Buccafusco, 1998). It must be noted that these changes occur during abstinence in parallel with distinct behavioral hyperactivity.

As already mentioned, noncholinergic transmitters are involved in these phenomena; for example, some of the abstinence phenomena are mediated via adrenergic alpha receptors (Beani et al., 1989), while dopamine may be involved in the cholinergically mediated opiate addiction, perhaps via the nucleus accumbens (Hikida et al., 2003).

May these effects of opiates be translated into the mechanisms of opiate addiction? The return of ACh parameters to control values in chronically treated animals is the basis of Crossland's (1971) hypothesis that this normalization underlies the development of morphine tolerance; he also posited that the explosive release of ACh during withdrawal or antagonist treatment underlies the morphine withdrawal syndrome. To support his hypothesis, Crossland (1971) presented data indicating that both the autonomic and central components of the animal withdrawal syndrome are antagonized by atropine; however, very high doses of atropine, 20 to 50 mg/kg, had to be used. These results were substantiated subsequently by Shuster, Buccafusco, and their associates (see, for example, Holland et al., 1993), who used localized, intrathecal, or intraventricular administration of muscarinic antagonists, pirenzepine being particularly effective as an antiwithdrawal drug; again, large doses of antagonists were used in these experiments. Similar data indicating the muscarinic nature of morphine withdrawal and effectiveness of atropinics in mitigating the latter were obtained in dogs by Martin and Eades (1967). It should be noted that Crossland's hypothesis pertains only to the processes of tolerance and withdrawal, and not to those of addiction and its development. Russian investigators who reported frequently on the interactions among opioids, cholinergics, catecholaminergics, and indole aminergics with respect to conditioned behavior speculated that the cholinergics facilitate inhibition of goal-directed behavior by morphine, which they related to the development of morphine addiction; that is, they claim that cholinergics facilitate the development of addiction (see, for example, Shugalev, 1990). It is interesting in this context that enkephalin- and AChE-positive sites were found to have closely related localization in several brain parts (see, for example, Graybiel and Illing, 1994).

It must be added that some investigators could not duplicate the findings concerning the antagonism of withdrawal by antimuscarinics. In fact, these drugs seemed ineffective for treating morphine withdrawal in humans (see Karczmar, 1976). This inconsistency may relate to the differences in the effectiveness of various atropinics with regard to muscarinic receptor subtypes: M2 receptor antagonists may be more potent antiwithdrawal drugs than the M1 antagonists (Buccafusco, 1991; Holland et al., 1993). Several investigators, while unable to antagonize opiate withdrawal in animals by atropine, attenuated the withdrawal with antiChEs or muscarinic and nicotinic agonists; furthermore, they managed to accentuate naloxone-induced withdrawal by using hemicholiniums, the ACh synthesis blockers (Zhang and Buccafusco, 1998; see also Way et al., 1975). In addition, the overt behavioral effects of cholinergic agonists or ACh releasers do not seem to simulate the withdrawal syndrome, as they should if Crossland's hypothesis were right.

It must be noted that morphine and cholinergics interact not only with regard to addiction but also with respect to other behaviors and functions. For example, in rodents morphine induces hypermotility, which may be antagonized by physostigmine (Stone et al., 1990); in cats, cholinergics antagonize respiratory depression induced by opioids at the brainstem level (Trouth et al., 1993). Nociceptive interactions between opioids and cholinergics were already described (see section BV-2). Also, opiates and ACh interact with catecholamines release (see section BIII-2a and BIII-2f).

b. Alcoholism

Perhaps not as dramatically as opiates, alcohol causes alcohol-seeking behavior in animals, and chronically treated animals exhibit abstinence syndrome upon withdrawal of alcohol (see, for example, Lichtigfeld and Gillman, 1994). Several operant methods are available for inducing alcohol-seeking behavior ("alcoholism") in animals. The model developed by Alfred Kahn measures mouse preference and aversion for alcohol; using this test it was demonstrated that certain genetic strains of mice "follow the bottle" even when it is placed on the cage side normally avoided by these mice (Figure 9-33; Karczmar et al., 1978).

Alcoholism induces changes in cholinergic parameters in animals. Acute or chronic administration of ethanol affects the spontaneous or stimulation-evoked release of ACh; some early data indicated that the release from cortical brain slices and slices obtained from other brain regions rich in cholinergic synapses is decreased (see Karczmar, 1976; see also section BIII-2a), an effect similar to that of opiates. But, more recent, careful studies indicated that ACh release as well as ACh synthesis (or acetyl-CoA synthetase activity) increased in several brain regions of rats treated chronically with alcohol (Kiselevski et al., 2003); this is in accordance with the early data of Rawat (1974). And, when studied with respect to specific cholinergic synapses such as the neuromyal junction or the synapse at the Renshaw cell, alcohol facilitated the release of ACh (Kalant, 1975). It must be added that alcohol withdrawal

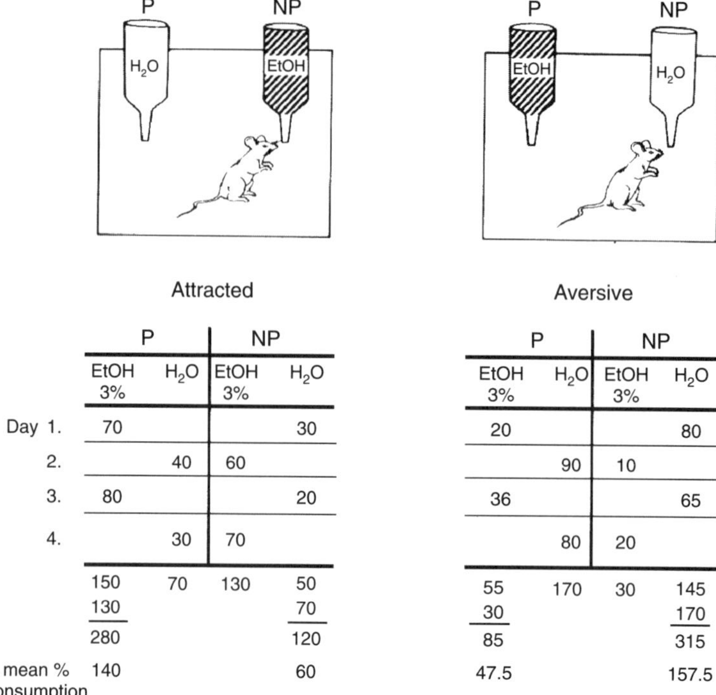

Attracted

	P: EtOH 3%	P: H_2O	NP: EtOH 3%	NP: H_2O
Day 1.	70			30
2.		40	60	
3.	80			20
4.		30	70	
	150	70	130	50
	130			70
	280			120
Σ mean % consumption on P + NP sites	140			60

Aversive

	P: EtOH 3%	P: H_2O	NP: EtOH 3%	NP: H_2O
Day 1.	20			80
2.		90	10	
3.	36			65
4.		80	20	
	55	170	30	145
	30			170
	85			315
Σ mean % consumption on P + NP sites	47.5			157.5

Figure 9-33. Diagram representation of the evaluation of alcohol behavior in mice. Hypothetical consumption of ethanol and water, expressed in percentages of daily total fluid intake, during 4 days of the second phase of the experiment (i.e., following evaluation of site preference). Left panel: alcohol-seeking behavior of an active attracted mouse. Alcohol (ETOH, 3%) was presented on the preferred (P) site on days 1 and 3 and the mouse consumed 70% and 80%, respectively, of its fluid intake from the alcohol bottle for a total of 150; on days 2 and 4, the mouse consumed 60% and 70%, respectively, of its fluid intake from the ethanol bottle presented on the nonpreferred (NP) site, for a total of 130. The two-day totals of water consumption were 70% and 50% at the preferred and nonpreferred site, respectively. Altogether the two days mean alcohol consumption of the attracted mouse at the two sites amounted to 140% (Σ) of the total fluid consumption on the basis of 200% total fluid consumption during these 2 days; the figure for water consumption for the two days was 60%. The reverse situation obtained in the case of an aversive mouse (right panel). The two figures on top of the panels illustrate the characteristic behavior of the actively attracted mouse, drinking from the alcohol bottle even when presented on the nonpreferred site (left figure), and of the aversive mouse, which followed the water bottle even when presented on the nonpreferred site (right figure). (From Karczmar et al., 1978, with permission.)

in alcohol-dependent rats induced significant increase in hippocampal release of ACh (Imperato et al., 1998).

It would be expected that the increase in ACh release should be accompanied by the increase in ACh levels, decrease in turnover, and upgrading of pre- and postsynaptic receptors. However, the pertinent data are controversial. Muscarinic receptors (or cholinergic fibers) were downgraded in the frontal and temporal cortex and in the hippocampus of alcoholic human subjects and in the striate of alcoholic rodents (Syvalathi et al., 1988; Freund and Ballinger, 1988, 1989a, 1989b, 1991; Brandao et al., 1999). On the other hand, Collins and his associates found no change in muscarinic and nicotinic receptors of alcoholic rodents (Collins et al., 1988; De Fiebre and Collins, 1993; see also Frye et al., 1995), while Madamba et al. (1995) found that the muscarinic receptors are sensitized and cholinergic transmission increased upon acute application of alcohol to the isolated hippocampus. Finally, Held et al. (1991) demonstrated in a careful study that the endplate nicotinic receptor α subunit gene is downregulated after chronic exposure of rats to alcohol. This finding is consistent with the demonstration of the regulation of alcohol consumption by the α-4 nicotinic receptor (Tritto et al., 2001).

The downgrading of the presynaptic muscarinic receptor, if it indeed occurs, and ACh release inhibition by alcohol suggests that alcoholism induces a deficit in cholinergic synthesis. Indeed, chronic alcohol treatment decreased CAT levels and AChE activity in several rodent brain parts (Miller and Rieck, 1993; Arendt et al., 1990; Arendt, 1994; Hodges et al., 1991; see, however, Moore et al., 1998). Furthermore, cholinergic transmission was attenuated in the hippocampus of the chronically treated rat (Rothberg et al., 1993; for some discordant data, see Rawat, 1974; Karczmar, 1976). Similarly, chronic alcoholism affected the cytoarchitecture, whether in the human or animals, of central cholinergic synapses (Miller and Rieck, 1993; Arendt, 1994; Kopelman, 1995; see, however, Moore et al., 1998).

Arendt (1994) stressed the importance of the degenerative effect of alcohol on the ascending activating system. This effect is of course consistent with the well-known memory decrease in animals treated chronically with alcohol, as well as in alcoholic humans and in patients suffering from the Wernicke-Korsakoff syndrome (Kopelman, 1995). Furthermore, in animals this alcohol-induced learning deficit may be alleviated by cholinergic agonists, trophic factors, or transplants rich in cholinergic synapses (Hodges et al., 1991; Arendt et al., 1990; Heaton et al., 1994). It is consistent with the notion of cholinergic deficit in alcoholism that in certain animal models cholinergic agonists decreased alcohol-seeking behavior (C. L. Scudder, A. J. Kahn, and A. G. Karczmar, 1973, unpublished results). This interesting animal effect of cholinergic agonists was apparently not replicated in alcoholic humans or humans suffering from Wernicke-Korsakoff syndrome.

It must be added that Janowsky and his associates (1989) found that "changes in anergia/inhibition, mood and pulse rate induced by intravenous physostigmine were significantly less pronounced in 26 patients with primary alcoholism than in 36 normal control subjects"; these findings led them to speculate that human alcoholics exhibit a general diminution of cholinergic function and responsiveness. Yet, there is no evidence that cholinergic agonists or antagonists affect human symptoms of alcoholism and alcohol abstinence syndrome of mice (Goldstein, 1975).

While these results are not always consistent, the general evidence strongly suggests that there is a relation between the cholinergic system and alcoholism; yet, many reviews and texts concerning alcoholism do not refer to its cholinergic correlates (see, for example, Tabakoff and Hoffman, 1995).

c. Cocaine Addiction

Similar to morphine and alcohol, cocaine induces drug-seeking behavior in humans and self-administration behavior in animals (see Goldberg and Stolerman, 1986). This behavior may result from anxiolytic and fear-reducing cocaine effects that can be demonstrated in animal paradigms and which are similar to benzodiazepines' effects (Crawley, 1981; Costall et al., 1989).

Given acutely or chronically, as well as via the self-administration paradigm, cocaine exerts several effects on cholinergic parameters, including levels of CAT and ACh, ACh turnover, and changes in muscarinic receptors and/or their binding affinity (see, for example, Mac et al., 2004); while several sites (hypothalamus, striate, ventral tegmentum, and limbic system) may participate in these phenomena, the site most involved may be nucleus accumbens, which receives dopaminergic input from the ventral tegmental area. In fact, denervation of this nucleus sensitized rats to certain behavioral effects of cocaine, presumably via denervation sensitization of postsynaptic sites of the nucleus accumbens radiation (Claye and Soliman, 1990; Wilson et al., 1994; Zeigler et al., 1991; Hikida et al., 2001; Smith et al., 2004). These and other studies suggest that ACh and dopamine act antagonistically on the nucleus accumbens circuitry; furthermore, that ablation of the cholinergic cells of the nucleus enhance cocaine addiction, while antiChEs suppress, in intact but not in lesion animals, certain addictive behaviors induced by cocaine (such as site preference; Hikida et al., 2001, 2003). However, these effects were not always consistent. For example, single administration of cocaine to rats decreased CAT levels in the hypothalamus and thalamus but increased its levels in the cortex (Claye and Soliman, 1990). Altogether, changes in cholinergic parameters seem to be related to cocaine reinforcement capacities; also, the cholinergic mediation in dopamine release may be involved in cocaine reinforcement (Shiraki et al., 1999; Mark et al., 1999).

It must be added that some of cocaine's functional effects, such as EEG arousal, the generation of theta rhythm, and analeptic action, also involve the cholinergic system, as indicated by the antagonism of these actions by scopolamine (Yabase et al., 1990). All this evidence does not lead, as yet, to a coherent image of the cholinergic correlates of cocaine addiction, particularly as certain discordant data suggest that nicotinic antagonists may augment cocaine self-administration (Corrigal et al., 1999).

d. Nicotine Addiction

Upon chronic application, nicotine induces tolerance and dependence and its withdrawal causes abstinence syndrome (see Helton et al., 1993). In fact, it has been known for centuries that smoking tobacco produces addiction; this was particularly true in the Netherlands, where smoking of tobacco reached enormous proportions in the 16th and 17th centuries (Scriverius, 1630; see also Schama, 1988).

Animals exhibit nicotine-seeking behavior, as they self-administer nicotine; also, they show discriminative preference for nicotine in this paradigm (Goldberg and Stolerman, 1986; Rosencrans, 1989). There is some controversy as to the intensity of the self-administration behavior, which may be less intense than that related to opiate or cocaine addiction (West, 1992); nevertheless the presence of nicotinic self-administration behavior is indubitable.

It was proposed (partially on the basis of lesion experiments) that the meso-limbic dopaminergic pathway, which participates in the reward-punishment system, reinforcement, and self-administration behavior and which includes peduculo-pontine and latero-dorsal and ventral tegmental nuclei and their projections in the nucleus accumbens, is involved in nicotine and other addictions; nicotinolytics block these behaviors (Glassman and Koob, 1996; Merlo Pich et al., 1998; Picciotto et al., 1998; Picciotto and Corrigal, 2002; Lanca et al., 2000; Fagen et al., 2003). A nicotinic receptor subtype that contains the β2 subunit appears to mediate the dopaminergic phenomena (see Chapter 2 DII and DIII; Picciotto, 1998; Picciotto et al., 1998, 2000; Lena and Changeux, 1998, 1999). Indeed, nicotine stimulates the cholinoceptive sites of the pertinent dopaminergic neurons or their terminals, causing dopamine release (Mifsud et al., 1989; Picciotto, 1998; Fagen et al., 2003). Fisher et al. (1998) and Shoaib (1998) presented interesting evidence indicating that the dopaminergic receptors in question differ from those that may be involved in locomotor and learning effects of dopamine. Furthermore, nicotinolytics block dopamine release (Blomqvist et al., 1992, 1993; Merlo-Pich et al. 1998). In addition, Merlo-Pich et al. (1998) found that the pleasure or reward-evoking effect of nicotine that underlies its self-administration results in c-fos expression in the mesocorticolimbic dopaminergic system (see also Ismail and el-Guebaly, 1998; Sgard et al., 1999; and Lena and Changeux, 1998). It was also suggested that the relations among nicotine, dopamine, and nicotine's addictive capacity may also be mediated by opioid receptors, but this notion is controversial (Ismail and el-Guebaly, 1998).

There are additional lines of evidence that indicate the cholinergicity of nicotine's addiction (see Laviolette and van der Kooy, 2004). The finding that chronic nicotine treatment upgrades nicotinic receptors both *in vivo* in several brain parts and in hippocampal or cortical cell cultures is consistent with the notion of nicotine's activation of reinforcement (Larsson et al., 1986; Clarke, 1933a, 1933b; Wonnacott, 1990; Barrantes et al., 1995; Dani and Zhou, 2001; Parker et al., 2004); in particular, Jon Lindstrom and his associates (Parker et al., 2004) proposed that nicotinic receptors containing α6 and β2 subunits are upgraded and their agonist affinity increased following long-term nicotine self-administration. Other investigators (see, for example, Tapper et al., 2004) found that alpha4 receptors are involved in reward processes and in nicotine addiction.

This notion that cholinergic activation of nicotinic receptors, whether linked with dopaminergic system or otherwise, underlies nicotine addiction is not consistent with the findings that chronic nicotine treatment desensitizes cholinergic nerve terminals and attenuates ACh release by nicotine; this and other nicotine desensitization phenomena may be demonstrated not only *in vivo* but also in cell culture (Bullock et al., 1994; Lapchak et al., 1989c; Yu and Wecker, 1994). However, this phenomenon relates to nicotine tolerance (or desensitization) which is an element of the nicotine's addiction capacity (Lapchak et al., 1989c; Dani and Zhou [2001] refer to the activating-desensitizing dipole of nicotinic action involved in nicotine addiction as a phenomenon of "synaptic plasticity" induced by nicotine). Finally, nicotinization synergizes with tegmental and other lesions in attenuating CAT activity (Fuxe et al., 1994).

At one time there was considerable controversy as to the addictive properties of nicotine in humans; David Warburton in particular argued against the unequivocal embracing of the Office of the US surgeon general of the position that nicotine is addictive (see Warburton, 1992b; US Department of Health and Human Services (DHHS), 1988; and more recent statements from the US DHHS); volume 108 (1992) of the journal Psychopharmacology was devoted to this argument. This controversy spilled into the field of animal data, yet it appears to this author that the results concerning the addictive properties of nicotine whether in animals or in humans are incontrovertible (Helton et al., 1993; Picciotto, 1998; Benowitz, 1999). While denying that nicotine exerts addictive properties, David Warburton stressed its positive effects, including those on memory and learning; of course, there is no denying these effects (see section BV-d, and Dursun and Kutcher, 1999).

The addictive actions of nicotine and other addictive drugs have common denominators. Thus, as mentioned above, all these drugs activate the reward side of the reward-punishment dipole and reinforce self-administration behavior. Also, nicotine may enhance alcohol consumption and preference (Collins, 1990; Collins et al., 1988; De Fiebre and Collins, 1993; Blomqvist et al., 1992; Wise, 1998; Fagen et al., 2003). Finally, these drugs, particularly nicotine and cocaine, exhibit anxiolytic properties (Costall et al., 1989). In addition, these drugs interact with respect to other functions or behaviors; for example, cocaine and nicotine exhibit cross-tolerance with respect to their hypothermic effects and their actions on open field activity.

6. Other Behaviors

While the earlier statement in this chapter that most functions and behaviors ever studied in animals exhibit cholinergic correlates is true, certain mentalizations such as moral behavior (see Churchland, 1994) and linguistic function were not studied pharmacologically and may not be readily quantifiable, whether in humans or in animals. However, our current good understanding of the complex circuitry involved in language processes (see, for example, Catani et al., 2005) will help in the future studies, and the nuclei within the basal ganglia, which are a part of the anterior forebrain song-learning circuit in birds, contain cholinergic neurons (Reiner et al., 2004; Carrillo and Doupe, 2004).

Fear, fear-related tonic immobility, and related emotional animal behaviors including anxiety, SLUD response (see below, this section), and/or aversive and stress-induced activities exhibit similar characteristics. In animals, these behaviors are mediated by limbic, particularly amygdala, hypothalamic, cortical, and several medullary sites including locus coeruleus and the nucleus of solitary tract; both nicotinic and muscarinic receptors seem to be involved (see Benarroch, 1997a, 1997b; Ledoux and Muller, 1997; de Oliveira et al., 1997; Morris et al., 1999; Pouzet et al., 1999; Sienkiewicz-Jarosz et al., 2003); they relate to such activities as aggression (see above, section BV-a). At this time it is impossible to distinguish precisely, in terms of overt and autonomic activities or sites and pathways involved, between emotions and fear on the one hand and certain types of aggression such as affective aggression on the other. At any rate, these sites are cholinoceptive, cholinergic, or significantly connected with cholinergic pathways (Monassi et al., 1999; see also Chapter 2 DII and DIII and section BV-a). Also, the behaviors in question affect cholinergic parameters in the limbic system and elsewhere; for example, chronic stress increases hippocampal release of ACh (Mizoguchi et al., 2001).

In humans, emotions, fear, anxiety, or panic and related behaviors are accompanied by autonomic cardiovascular phenomena, including changes in blood pressure, regional blood flow, baroreceptor regulation, cardiac function, as well as sweating and vocalizations; there is some evidence that these behaviors have, in humans, a cholinergic component (Battaglia and Ogliari, 2005; see also section BIV-2b; Benarroch, 1997a, 1997b; LeDoux and Muller, 1997). Furthermore, human emotions (and indeed animal emotions, particularly those of primates) have their overt counterpart in our facial expressions, as stressed already by Charles Darwin (1872); these activities are regulated cholinergic both peripherally and centrally (Delgado-Garcia, 2002).

Our anthropomorphic interpretation of certain animal behaviors as representing emotions and fear is substantiated by the fact that autonomic and other activities that accompany these behaviors are similar to those seen in the emotional human; in animals, these activities also include piloerection, the SLUD response (salivation, lachrymation, urination, and defecation), vocalizations and yawning, and tonic immobility (see, for example, Brudzynski and Eckersdorf, 1988; Baskin et al., 1994; Brudzynski, 1994; de Oliveira et al., 1997).

Several paradigms were employed in the measurement of these behaviors in animals, including the generation of "emotional" response (conditioned emotional response [CER]), evocation of site avoidance or fear symptoms by negative reinforcement or by aversive signals interposed with conditioning, stress (generated in rodents by, for example, nonescapable footshock or the "no-goal" paradigm; see below, this section; Maier, 1949; Yamanashi et al., 2001), noise (Sembulingam et al., 2005; footshock-noise combination may be considered as engendering either fear or stress; see Feiro and Gould, 2005), and induction of immobility generated by prey-predator confrontation. Interestingly, certain forms of CER may be obtained in humans (Morris et al., 1998). Other paradigms used include field behavior, startle response, site-avoiding behaviors, and stress or unavoidable punishment response (Finkelstein and Hod, 1989; Karczmar et al., 1973; Crawley, 1981; Kahn, 1969; see also section BV-d).

That the cholinergic system is involved in fear, SLUD response, aversive behavior, and related activities is indicated by the cholinergic autonomic activities that accompany these phenomena as well as the demonstration that central sites that are involved in their generation are cholinergic (see above, this section). Furthermore, antiChEs and muscarinic agonists facilitated and anticholinergic drugs blocked these phenomena (Ilyutchenov and Yeliseyeva, 1967; Brudzynski and Eckersdorf, 1988; Baskin et al., 1994; Brudzynski, 1994; de Oliveira et al., 1997); in fact, the effect of atropinics on fear resembles that of tranquilizers (Ilyutchenov and Yeliseyeva, 1967). It is important in this context that fear is one of the stimuli that induce another cholinergic phenomenon, namely, aggression (see above, this section). These results are consistent with the finding that aversive behavior induces increased release of ACh in the limbic system and in the cortex; this release may, however, relate to arousal rather than fear or aversion (Imperato et al., 1992; see also section BIV-3).

Imprinting constitutes an interesting social behavior. Atropine reduced chicken imprinting, while quaternary drugs such as neostigmine and hexamethonium had no effect (Kovach, 1964). Cholinergic characteristics of other social behaviors, such as mutual grooming, contractual activities, and maternal care, were described for mice kept in a pseudo-natural habitat ("Mouse City"; Karczmar et al., 1973). When several strains and genera of mice were compared, it appeared that high level of social activity was related to high levels of brain cholinergic activity (unfortunately, only whole-brain data were available); physostigmine facilitated, across the mice types, these activities (Scudder and A. G. Karczmar, 1973, unpublished data). Homing may be considered as still another example of social behavior, and some evidence supports the notion of cholinergic correlates of homing (Scattoni et al., 2005).

Finally, there are several central states and central diseases of the human that exhibit cholinergic correlates; these states include schizophrenia, dementias, including Alzheimer's disease (see Chapter 10), and several forms of depression, as well as organic malfunctions, including motor diseases such as tardive dyskinesia, chorea, and parkinsonian condition (see Karczmar, 1998).

In the animal model of tardive dyskinesia, cholinergic agonists ameliorate the condition, while cholinergic antagonists, particularly atropinics, augment it; also, the antagonists synergize with catecholaminergic stimulants with respect to this condition (see section BIV-1). In the case of parkinsonian disease, which is due to the striatal deficit in dopaminergic transmission and which consequently exhibits symptoms s of cholinergic hyperactivity, atropinics constituted in the past the standard treatment of this condition prior to the advent of dopaminergic therapeutics (Karczmar, 1979c). Also, atropinics are effective in animal models of parkinsonian and extrapyramidal disease, while cholinergic agonists aggravate these conditions (Van Woert, 1976).

It was already mentioned (see section A-2d; Pfeiffer and Jenney, 1957) that cholinergic agonists were employed in the treatment of hebephrenic schizophrenia. As shown some 20 years later, when schizophrenics are given atropinics to prevent tardive dyskinesia side actions of their neuroleptic treatment, the atropinics may decrease the effectiveness of neuroleptics (Singh and Kay, 1979). In the case of several animal models of schizoid behavior, cholinergic agonists improve the condition and atropinics worsen it (Table 9-4; Karczmar, 1995). An interesting animal paradigm duplicating one aspect of schizophrenia, the fixated or perseverant behavior (or behavior "without goal"), was developed by Maier (1949; Figure 9-34).

The unique feature of the "no goal" paradigm is as follows: in the case of many paradigms such as the CER, the emotional behavior is attenuated by many anti-anxiety drugs and tranquilizers, while the fixated behavior is very stubborn and only a percentage of animals abandon fixation after treatment with anxiolytics or tranquilizers; on the other hand, physostigmine and muscarinics "cure" fixation in most of the fixated animals.

Table 9-4. Animal Models of Schizoid Behavior

- Spontaneous alteration
- Cue-dependent alteration
- Flexibility versus emotional (or stereotypic) behavior
- Hypokinesia
- Emotional consequences of loss of aggressive encounters
- Amphetamine- and atropine-induced animal "psychosis"
- Nongoal behavior (fixation)
- Perseverant self-stimulation
- Emotional consequences of inescapable shock

BVI. Cholinergic Aspects of Organism-Environment Interaction—and Beyond

1. Cholinergic Alert Nonmobile Behavior, or Summary of This Section

As stressed in this chapter, the cholinergic system underlies all measurable organic as well as behavioral or mental states. In fact, cholinergic input is crucial for several mental and related states, including learning and memory, certain motor behaviors, aggression, EEG arousal, and REM sleep. Is there any coherence or teleology with respect to these functions and behaviors? Can they, or some of them, be linked together in a single entity or syndrome?

This author speculates that indeed a number of behaviors and mental states that exhibit strong cholinergic correlates do form a syndrome that is particularly pertinent for cognition and related functions. As components of this proposed syndrome, learning and memory and contingent phenomena such as attention, curiosity, neophilia, and vigilance must be listed first (see section BV-d; Mirza and Stolerman, 1998; Hagan and Morris, 1988; Karczmar, 1995, 1998, 2004; and Bartus et al., 1982). Closely linked with behavioral signs of cognition and attention are several EEG phenomena including EEG desynchronization or EEG arousal, hippocampal theta rhythms, block of alpha rhythm, rhythmic thalamic bursts and recruitment, and evoked potentials such as Bereitschaft potentials or the anticipatory waves (McCormick, 1992a, 1992b, 1993; Marczynski, 1978; see section BIV-3). Furthermore, certain synaptic phenomena such as the LTP and LTE underlie memory and learning (see Libet, 1979, 1991; see also section BV-d). Also, sensory processing and sensory modulation exhibit significant cholinergic correlates (section BIV-4e above; Rasmusson, 1993; Durieux, 1996). As described in this chapter, all these phenomena exhibit strong cholinergic correlates.

It can be readily seen that, taken together, these phenomena reflect more than cognition: they underlie organism-environment (or individual-world) interaction and cognitive processing of this interaction. These cholinergic phenomena also

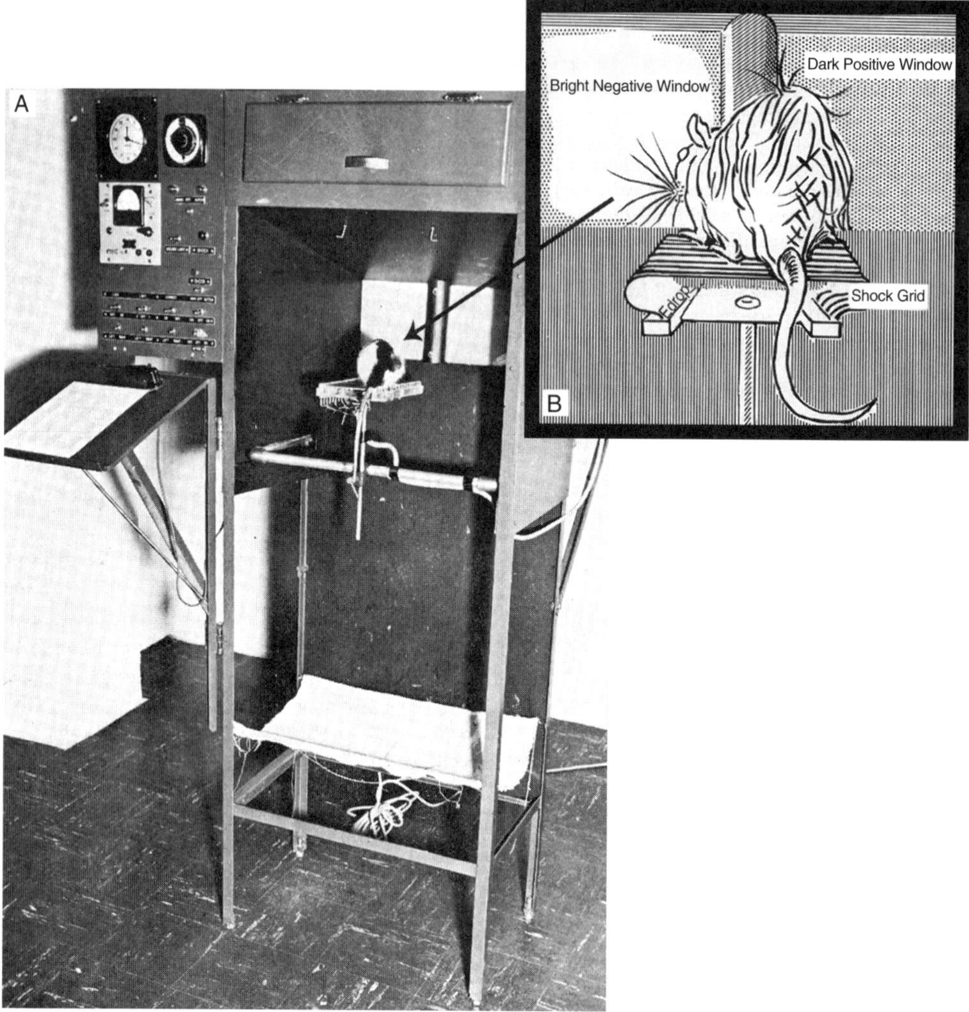

Figure 9-34. Lashley's electrified stand. The rat is placed on Lashley's electrified stand, which is fronted by two windows that are either lit or kept dark, and either locked or open. If the rat remains on the stand for more than 30 seconds, it receives a shock to its feet. During the insoluble phase of the paradigm, the windows are lit and locked at random. The rat cannot know whether the lit or the dark window will be open when jumping toward the window. If the window is locked, the rat hits its head and falls into a net (punishment mode). Under these conditions, the rat soon exhibits site preference: before receiving the shock, the rat will always jump to either the right or left window, thus avoiding the shock and punishment at least 50% of the time. (Note that there is a better percentage for reward when the rat always jumps in one direction than when the rat guesses.) During the next phase of the paradigm, the problem becomes soluble: while the dark window randomly appears on either the right or the left side, it is always open. Most of the animals remain fixated and jump toward their preferred site regardless of whether they are facing the dark or the lit window. However, the rat "knows" when the window is the reward (open) window: when the reward (dark) window is at the rat's preferred side, the rat jumps almost immediately. But it waits until nearly shock time when the punishment (lit) window is on its preferred side. Note that the rat also shows considerable "emotional behavior" at this time. To "cure" the rat from its fixation, for several rounds, it can be gently guided to the open window, or drug treatment may be attempted. (A) A full-scale view of Lashley's electrified jumping stand. (B) An expanded view of Lashley electrified (grid) stand with the rewarding (positive) and punishing (negative) windows illustrated. (From Liberson and Karczmar, 1969, with permission.)

underlie "realistic" world perception, and cholinergic deficit, whether due to mental disorders or lesions, induces "reality distortion" (Karczmar, 1988; Sarter and Bruno, 1998); accordingly, cholinergic agonists and antiChEs attenuate and atropinics augment "schizoid" behavior in animal models of schizophrenia, while antipsychotic medication that exhibits anticholinergic, particularly antimuscarinic, potency aggravates memory deficits and other symptoms of schizophrenia (see above, section BV-d; see also Karczmar, 1988, Sarter and Bruno, 1998, Bymaster et al., 2004; Minzenberg et al., 2004). It should also be noted that several studies indicate that dementive and/or schizophrenic patients exhibit a diminution of cholinergic components of their brain, although opposite findings were reported as well (Martin-Ruiz et al., 2003).

Altogether, cholinergicity of our notion of reality and of organism-environment interaction, combined with phylogenetic omnipresence of the cholinergic system (see Chapter 8 BV) indicate evolutionary significance of cholinergic system (this should have been mentioned by the late Stephen Gould in his 2002 magnum opus).

Besides the phenomena listed in this section, other cholinergic functions or behaviors are important for the effective organism-environment interaction, including hypothalamic functions and various homeostatic phenomena (see section BIV-2), and aggression, fear, nociception, and emotional responses.

As already pointed out, the cholinergically induced EEG arousal and theta rhythm differ from the EEG exhibited during "normal" behavioral arousal (see sections BIV-1d, BIV-3a, and BIV-3b). In fact, while there are similarities between behaviors and functions that accompany the cholinergically induced EEG arousal (particularly with regard to cognitive behaviors), on the one hand, and behaviors and functions that accompany the "normal" behavioral arousal, on the other, there are also differences between these two behavioral spectra. Relative immobility accompanying the cholinergic EEG arousal may be the most significant difference between this arousal and the behavioral arousal (Karczmar, 1971; Kramis et al., 1975; Vanderwolf, 1975; Shapovalova, 1995). Wikler (1945 and 1952) described this absence of motor symptoms in the course of the cholinergic EEG arousal as a "divorce" phenomenon. Indeed, the absence of motor activity in the presence of the theta rhythm, desynchronized EEG, and cognitive arousal constitutes a striking phenomenon; it is dramatic to see in unanesthetized freely moving animals the arrest of spontaneous motor behavior accompanied by obvious alertness upon the administration of cholinomimetics or antiChEs, via indwelling catheters, into one of several sites within the basal forebrain system (see section BIV-1d).

Rather than comparing the cholinergic EEG and cognitive phenomena with behavioral arousal, Karczmar (1979a) analogized these phenomena with those of another cholinergic event, REM sleep (see sections BIV-3a and BIV-3b). While the facilitation of REM sleep occurrence in unanesthetized animals and in humans was already stressed, some tests in the human suggest that dreams can be engendered in awake humans by relatively small doses of brain-penetrating antiChEs such as physostigmine (Karczmar, 1981a; this is not to recognize that, in humans, certain depressive phenomena may accompany these phenomena; see Janowsky et al., 1989). Altogether, Karczmar (1979a, 1979b, 1993a, 1993b, 1995) opined that the pertinent behavioral, functional, and EEG phenomena that are evoked by cholinergic agonists and depend on central cholinergic pathways form a unique syndrome that constitutes an awake REM sleep; he termed this syndrome cholinergic alert nonmobile behavior (CANMB). He further speculated that CANMB and its components such as cognition, attention, neophilia, sensory modulation, and vigilance (McCormick, 1992a, 1992b) underlie an effective interaction between the organism and the environment (see Table 8 in Karczmar, 1995, p. 203).[9]

2. Immense Complexity: Dualism, Reductionism, and Beyond

This section emphasized the significance of the cholinergic system for the unique behavioral and functional syndrome consisting of many components, including cognition and organism-environment interaction. Quite probably, this syndrome and its components will be definitively described within the next few years, in terms of the pertinent neuronal mechanisms, central pathways, and neurotransmitter and second messenger interactions. But the syndrome is dangerously akin to a most baffling problem: as the cholinergic system relates so closely to cognition, perception, and the interaction with the outside world, its role in matters that related to cognition such as consciousness,[10] decision making, and free will should considered. Also, the understanding of the role of the cholinergic system in matters of consciousness should be preeminent in our perennial attempts, dating from Indian and Buddhist sages, Plato and Aristotle, Fernel (see Sherrington, 1940) and Descartes, to solve the problem of the mind-body relationship.

There are two ways to consider this relationship.[11] First, there is dualism; the dualists feel, with Rene Descartes (1649a, 1649b), that the mind and body constitute parallel but distinct phenomena that may, however, have a brain site as a meeting ground (for Rene Descartes, this site was the pineal gland). Modern exponents of this notion are Karl Popper and John Eccles (Popper and Eccles, 1977); in fact, Popper and Eccles envisaged three "worlds" rather than the two of Descartes, and Eccles (1985) related one of these words to intentional processes (see below for the related work of Ben Libet). Then there are the reductionists, or monists. Traditionally, a reductionist reduces the "real" world as well as consciousness to illusions or ideas generated in the brain, as did George Berkeley. The American philosopher William James went further, as he denied the existence of consciousness; "consciousness," he said, "is the name of a nonentity." Modern reductionists identify consciousness with certain brain activities and interactions (among the limbic system, rhinencephalon, and various cortices). Thus, a leading reductionist, Dennet (1991, 1995) states that there is no "Cartesian theatre" in the brain, as the brain translates directly its parallel or reverberating processes into consciousness (see also Tipler, 1994; Smythies, 1997; Llinas, 2001, 2003; and Delgado Garcia, 2004).

It should be added that an important component of consciousness, the awareness of the flow of time (as mentioned in note 10), may be particularly difficult to define whether in reductionist or dualistic terms (see Chapter 11-2).

Whether the reductionist point of view is convincing and intelligible or not, it is important that some reductionists—full or partial—suggest that the cholinergic system informs cognition. Thus, John Smythies (1997), Francis Crick (1994), Rodolfo Llinas (Llinas and Pare, 1991), and M. E. Hasselmo (1995; see also Liljenstrom and Hasselmo, 1995) refer to cortico-thalamic and hippocampal-cortical loops and to cortical radiations of the nucleus basalis and the cholinergic forebrain as sources of cognition but not to consciousness, awareness, and decision making. Yet, Jean-Pierre Changeux (1985) suggests that peduculopontine and tegmental nuclei form a "surveillance system" that generates consciousness. Altogether, Smythies (1997) emphasized that these circuitries are cholinergic in nature and Changeux (1985) points out that the flexibility and the capacity for allosteric transitions of nicotinic receptors in particular may underlie their function in both cognition *and* consciousness.

The studies of Ben Libet (1985, 1993) and related EEG investigations (Nunez, 1995b; Deeke et al., 1976; Thatcher, 1995) are also pertinent in this context. These investigators linked EEG arousal phenomena as well as certain cortical rhythms (see Woolf, 1997, and section BIV-3a) not only to cognition but also to consciousness; these phenomena and patterns are cholinergic in nature (see this section, below). Libet (1985, 1993) used an evoked cortical EEG event, the "readiness" or "anticipatory" ("Bereitschaft") potential, to demonstrate an EEG-related awareness and free will phenomenon: he showed in the human that this potential, when evoked by peripheral or thalamic stimulation, appears some hundreds of milliseconds before conscious experience of the stimulus and of the awareness of a volitional act, while the awareness and volition in turn occur prior to the motor act. This and additional evidence (Libet, 1985; see, however, Klein, 1995, 2002) led Libet to believe that the two delays represents neuronal activities that are needed for self-awareness and for the volitional act (see also Eccles, 1979, 1994); as the cholinergic activity facilitates the potentials in question (see section BIV-3b; see, however, Stanley Klein's (2002) criticism of Libet's methodology), Libet's argument militates for the cholinergic role in the generation of consciousness. Via somewhat similar reasoning, analysis of time-related activation of the hippocampal formation and the cortex led Roger Nitsch and his associates to posit a role for the hippocampus and medial temporal lobe in conscious and nonconscious or subconscious components of semanting associative processes of human identification of objects and faces (Henke et al., 2003a, 2003b). Howard Shevrin and his associates obtained data supporting Libet's localization of the "free will" in the subconscious, and, employing evoked cortical potentials technique they found that associative learning and certain other mental phenomena occur in the subconscious (Shevrin et al., 2002; Wong et al., 2004). The cholinergic correlates of these phenomena remain to be established, but it should be stressed that hallucinatory perception exhibits a cholinergic nature (Smythies, 2005).[1]

Other reductionists embark on an increasingly complex approach. Roger Penrose[12] and S. Hameroff (Penrose, 1994; Hameroff et al., 1996; see, however, Klein, 1995, 2002) propose that certain quantal phenomena such as the collapse of coherence lead to the moment of "objective reduction"; at that moment the quantal phenomena convert into the macrocosm of our "reality," and this conversion represents the moment of consciousness. Furthermore, Hameroff and Penrose referred to certain biological cellular phenomena that may represent biological activity, which is quantal in nature; these phenomena occur, for example, at the retina and at the microtubules of certain cortical neurons.[13] They speculate that when neurons that contain microtubules in their membranes are excited, microtubular coherence ensues via modulation exerted by microtubular-associated protein-2 (MAP-2), the dephosphorylation and degradation of this linkage leading to the collapse of coherence. And this is where the notion of cholinergicity of consciousness meets the quantal events. Nancy Woolf (1997), for many years a student of cholinergic pathway maps and of their functional significance, pointed out that high levels of cortical MAP-2s occur at the cortical neurons rich in muscarinic receptors; these associations are present in certain cortical modules of the small pyramidal and nonpyramidal cells. She also emphasizes that the "unusual receptivity . . . to and dependence . . . on" the trophic factors of cholinergic forebrain neurons that radiate to these cortical sites are consistent with her notions, as these neurons have to be very active to satisfy the consciousness conditions as well as other cholinergic phenomena that are important for consciousness such as attention and cognition (Figures 9-35 and 9-36). Altogether, Woolf (1997) speculates that the muscarinic actions on MAP-2 imply cholinergic mediation of consciousness.

An even more complex approach may offer (to this author, at least) a way out. The approach in question includes postulates of physics, mathematics, computer or artificial intelligence science (AIS), and, particularly, quantal theory. Goedel's (1931) and Turing's (1937) theorems constitute important components of this approach (see, for example, Eccles, 1994; Damasio, 1994; Churchland, 1994; Penrose, 1994; Hameroff et al., 1996).[14] Their theorems relate to the obligatory inconsistency of sufficiently complex systems and to the nonalgorithmic nature of certain mathematical problems, which cannot be solved by computers, present or future, as they are not computable.[15] Penrose (1994) and Hameroff (Hameroff et al., 1996) propose that this noncomputable essence of mathematical problems extends to nonmathematical problems as well as to matters of consciousness and creativity. Furthermore, they relate this argument to their attempt to link biological quantal phenomena and consciousness, and they join forces with Nancy Woolf as they link her MAP-2 sites with the quantal phenomena that underlie, according to them, the consciousness and self-awareness phenomena (Woolf and Hameroff, 2001).

This author prefers to use the phenomenon of noncomputability and obligatory inconsistency as an escape gap. He finds that the arguments of Woolf, Penrose, and Hameroff, while brilliant, are arcane and speculative, and lack the parsimony and explanatory power that should characterize a truly heuristic hypotheses; their solution to the mind-body dilemma lacks the "Eureka!" or "Aha!" factor (Koestler, 1964). In fact, many investigators attacked the speculations in question (although frequently in a very "dense" way; see, for example, Churchland, 1994). This author (see Karczmar, 1972a, 1996) senses that the solution to the mind-body dilemma (including the matter of the cholinergic contribution to consciousness) lies in the theorem of noncompatability and inconsistency of complex problems (see note 15). That is, there is no computable missing link to the understanding of this problem, as long as one uses a "linear" approach across the mind and body. In other words, this problem may constitute, in Rudolf

[1] The concept of free will may be also criticized on epistemological and logical grounds (see Harris, 2004, Chapter 6, note 7).

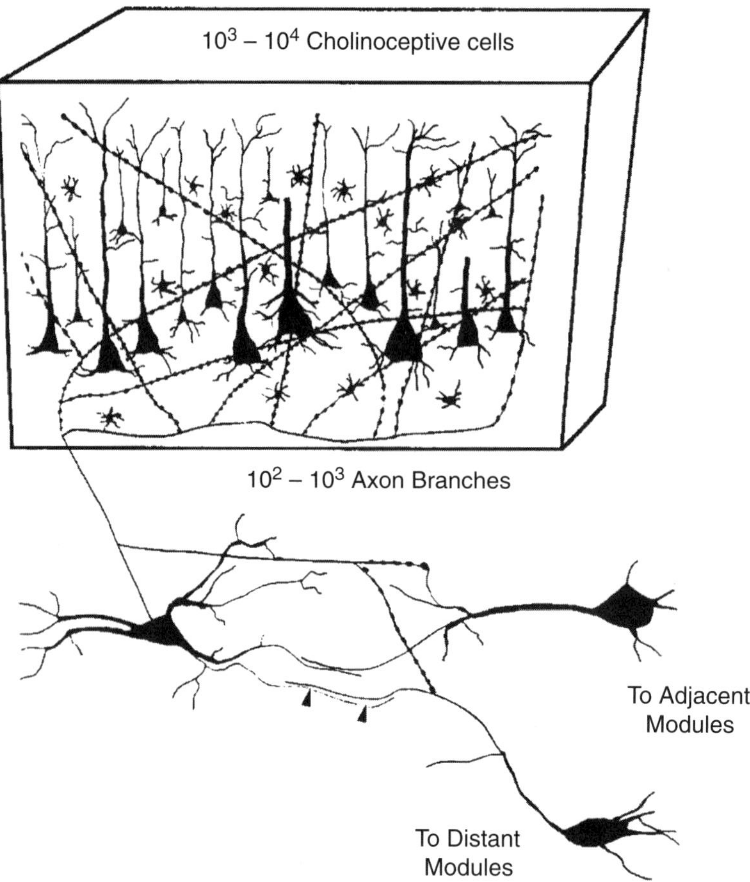

Figure 9-35. A schematic representation of the MAP-2-enriched cholinoceptive cells in a module of the cerebral cortex (top) and the cholinergic cells that innervate this and other cortical modules (bottom). Possible dendritic interactions and traditional synapses are illustrated, which link cholinergic cells with other cholinergic cells projecting to adjacent, as well as to distant, cortical modules. Arrowheads point to possible points of contact between dendrites of neighboring cholinergic cells in the bottom part of the diagram. (From Woolf, 1997, with permission.)

Carnap's sense, a nonsense or nonexistent question (see Karczmar, 1972a); a related outlook is that "there is a potential infinity of propositions in any logical or mathematical system" (Feferman, 1998, 2004). Still another way of looking at it is as follows. Scientists bring an abstract picture of a particular limited domain separating it from the unbroken wholeness. Any laws and theories formulated in relation to this domain are necessarily relative, valid only within this domain, and possible false beyond it. The activity of scientists is to extend the domain of a particular theory to its limits. When it "extends beyond those limits . . . it is falsified and an entirely new theory must be found to relate to a new domain" (Hayward, 1992). It should be added that many modern cognition philosophers are, similarly, nonreductionists (for example, Freeman Dyson), as they believe that we will "never know it all" (Feferman, 2004; see also Metzinger, 1995), particularly with regard to matters such as consciousness and self-awareness. And, the prominent US philosopher John Searle (see, for exmple, Searle, 2004[16]) considers that "most famous and influential theories . . . of mind," whether monist and materialist or dualist and idealist. "are false."

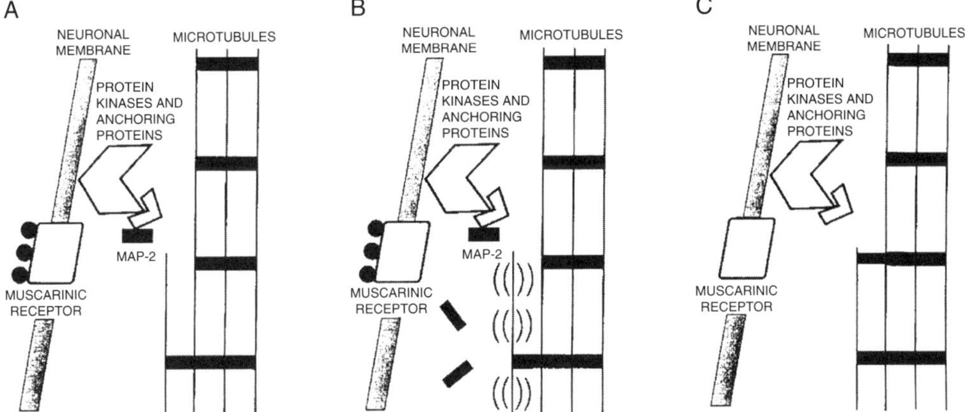

Figure 9-36. A hypothetical sequence of events linking receptor activation with MAP-2-mediated microtubular coherence and self-collapse. (A) Acetylcholine molecules (black dots) bind to the muscarinic receptor activating the anchored protein kinases that will phosphorylate MAP-2, thereby making MAP-2 unavailable for binding with local microtubules. (B) During the next several hundred milliseconds, phosphorylation of excess MAP-2 prevents it from binding with local microtubules and thereby facilitates microtubular coherence (illustrated as curved lines). (C) When the muscarinic action ends, MAP-2-phosphorylation ceases, leaving MAP-2 free to bind microtubules and possibly initiate self-collapse. (From Woolf, 1997, with permission.)

However, the problem in question is too interesting to cease being en vogue. Altogether, this "final" problem (Conan Doyle, 1893) and its cholinergic correlates reflect the cholinergic system's significance, which extends beyond the mere question of behaviors.

Notes

1. As this author criticized this particular work of Bennet, Rosenzweig, and Diamond at the 1969 FASEB Symposium on Cholinergic Mechanisms, he challenged the audience: "Is there anyone that can measure AChE so that two percent differences in the average values are significant?" One person stood up among the audience of 500. "Well, this must be doctor Bennet," said this author, and it was (see Karczmar, 1969).
2. It is easier to quantify the antiChE-induced potentiation of the muscarinic slow EPSP than that of the fast, nicotinic spike or the fast EPSP. Sometimes, this potentiation takes the form of increased firing duration, as in the case of the Renshaw cell response (see Koketsu, 1966; Karczmar and Kim, 1969, unpublished).
3. Baclofen is also potent postsynaptically, and this effect rather than its presynaptic action underlies baclofen's antispastic action (Wang and Dun, 1989; Lin and Dun, 1998; Fox et al., 1978).
4. Mary Pickford was a Scottish woman who, with John Masserman, Jose Delgado, William Feldberg, and Robert Myers, pioneered investigations of the effects of drugs and transmitters applied locally to specific brain regions or intraventricularly by means of an injection needle or "chemitrode"; see Myers, 1974; Feldberg and Myers, 1964).
5. Atropinic coma was once popular in the clinical treatment of depression (Karczmar, 1979c).
6. It must be remembered that when used in large doses, antiChEs damage the blood-brain barrier, thus increasing its permeability (Karczmar, 1967; see also Chapter 7 DI-1). This phenomenon may contribute to the analgetic effect of high doses of antiChEs as well as explain the finding of Flodmark and Wramner (1945) that neostigmine, a quaternary antiChE, as well as certain quaternary muscarinics, exerts analgetic action. This aspect of the matter is not mentioned in the pertinent research and reviews such as those of Green and Kitchen (1986) and Sitaram and Gillin, (1977).
7. These days, the pharmaceutical industry develops most intensely drugs that may be used in the human for improvement of memory, antagonism of senile memory loss and dementia, and treatment of Alzheimer's disease. The tendency, therefore, is to use fast and simple tests. Moreover, relatively

simple automated operant tests that the animals may carry out at night and yield records that may be read the next morning save a 12-hour period for the staff. When this author, who worked at the time (1956) at the pharmaceutical Sterling-Winthrop Research Institute, described this kind of test to Karl Pribram, at the time the director of the Hartford Institute of Living, Karl said, "I would not compliment these trials with the term 'learning test' "(see Pribram, 1971).

8. Daniele Bovet was a Swiss-born scientist who worked in the Pasteur Institute of Paris and subsequently in Italy; he earned the Nobel Prize in medicine in 1957 for his development of sulfa drugs, antihistaminics, and curarimimetics (see Bovet, 1953, 1988). He was also an active cholinergiker and a philosopher of the states of consciousness. His wife, Filomena Nitti, was also an active researcher and a liberal political philosopher.
9. This participation of cholinergic phenomena in organism-environment interaction links the cholinergic system with evolution since this interaction underlies the plasticity of organism vis-à-vis evolution and serves as a mechanism of adaptation (see Gould, 2002). The presence of the cholinergic system throughout the animal fila, its early evolutionary association with mobility (in protozoa; Karczmar, 1967), and the conservatism of the amino acid sequences in genes generating components of the cholinergic system such as AChE and nicotinic receptors are all consistent with this notion.
10. We all "know" what it means to be conscious, yet it is not very easy to define "consciousness." The term refers to self-awareness, consciousness of self, what it means to "feel" red, for example, and so-called intentionality (see Searle, 2004). Infrequently mentioned, yet significant for the question of consciousness, is its perception of the flow of time and of time's "arrow" (see Greene, 2004).
11. For obvious reasons, justice cannot be done to this matter in this book; some of the oeuvre quoted here (which constitutes barely a fraction of the pertinent literature) may be consulted. This author felt that, in view of the recent initiation of the studies of cholinergic implications for this problem, a brief reference to the matter is needed.
12. Roger Penrose is a great cosmologist who, jointly with Stephen Hawking, elucidated the quantal phenomena underlying the formation of black holes. Since the 1990s, he has published a number of books concerning the quantal nature of self-awareness or consciousness.
13. The brilliant University of Columbia biophysicist Selig Hecht was the early discoverer of the quantal effects at the retina as he demonstrated that the retinal rods and cones respond to 1 or 2 photones, the quantal elements of light (Hecht et al., 1941). Hecht was this author's PhD adviser.
14. While Goedel and Turing have currently become popular in neurosciences papers, this author had already quoted Goedel and Turing in the early 1970s (Karczmar, 1972a).
15. There is a more specific way to define or summarize Goedel's theorem as applying to formally specified systems of axioms (Feferman, 2004). This mode does not invalidate the inconsistency characterization of complex systems.
16. John Searle's opus of 2004 describes the philosophies of mind in a typical philosophical mode of "thought experiments" and abstract semantics, distant from the mode, pragmatism, and clarity (at least to the mathematics- or physics-minded reader) of biological and physicist presentations; yet this and other of Searle's books should excite any neuroscientist, and they serve as excellent reference sources of these many philosophies.

References

Abe, K., Taguchi, K., Kato, M., Utsunomiya, I., Chikuma, T., Hojyo, H. and Miyatake, T.: Characterization of muscarinic receptor subtypes in the rostral ventrolateral medulla and effects on morphine-induced antinociception in rats. Eur. J. Pharmacol. *465*: 237–249, 2003.

Abe, Y. and Tuyosawa, K.: Age-related changes in rat hippocampal rhythms: a difference between type 1 and type 2 theta. J. Vet. Med. Sci. *61*: 543–548, 1999.

Abrahams, V. C. and Pickford, M.: The effect of anticholinesterases injected into the supraoptic nuclei of chloralosed dogs on the release of the oxytocic factor of the posterior pituitary. J. Physiol. (Lond.) *130*: 330–335, 1956.

Acquas, E., Fenu, S., Loddo, P. and Di Chiara, G.: A within-subjects microdialysis-behavioral study of the role of striatal acetylcholine in D1-dependent turning. Behav. Brain Res. *103*: 219–228, 1999.

Adams, J. P. and Sweatt, J. D.: Molecular psychology: roles for the ERK MAP kinase cascade in memory. Annu. Rev. Pharmacol. Toxicol. *42*: 135–163, 2002.

Adams, P. R. and Brown, D. A.: Synaptic inhibition of the M-current: slow excitatory post-synaptic potential. Mechanism in bullfrog sympathetic neurones. J. Physiol. (Lond.) *332*: 263–272, 1982.

Adler, L. E., Freedman, R., Ross, R. G., Olincy, A. and Waldo, M. C.: Elementary phenotypes in the neurobiological and genetic study of schizophrenia. Biol. Psychiatry *46*: 8–18, 1999.

Adrianson, V. V.: Characteristics of the chemical sensitivity of cerebral cortical neurons at the stages of food-procuring behavior of cats as a function of quality of reinforcement. Neurosci. Behav. Physiol. *25*: 234–239, 1995.

Agranoff, B. W., Cotman, C. W. and Uhler, M. D.: Learning and memory. In: "Basic Neurochemistry," G. J. Siegel, Ed. in Chief, pp. 1027–1050, Lippincott-Raven, New York, 1999.

Ahmed, M. S., Schoof, T., Zhou, D. H. and Quarles, C.: Kappa opioid receptors of human placental villi modulate acetylcholine release. Life Sci. *45*: 2383–2393, 1989.

Ahn, D. K., Kim, Y. S. and Park, J. S.: Central-amygdaloid carbachol suppressed nociceptive jaw opening reflex in freely moving rats. Prog. Neuropsychopharmacol. Biol. Psychiatry *23*: 685–695, 1999.

Aigner, T. G., Walker, D. L. and Mishkin, M.: Comparison of the effects of scopolamine administered before and after acquisition in a test of visual recognition memory in monkeys. Behav. Neurol. Biol. *55*: 61–67, 1991.

Aizawa, H., Kobayashi, Y., Yamamoto, M. and Isa, T.: Injection of nicotine into the superior colliculus facilitates occurrence of express saccades in monkeys. J. Neurophysiol. *82*: 1642–1646, 1999.

Akaike, A., Sasa, M. and Takaori, S.: Muscarinic inhibition as a dominant role in cholinergic regulation of transmission in the caudate nucleus. J. Pharmacol. Exp. Ther. *246*: 262–272, 1988.

Albert, D. J. and Chew, G. L.: The septal forebrain and the inhibitory modulation of attack and defense in the rat: a review. Behav. Neural Biol. *30*: 357–388, 1980.

Albert, D. J. and Richmond, S. E.: Reactivity and aggression in the rat: induction by alpha-adrenergic blocking agents injected ventral to anterior septum but not in the lateral septum. J. Comp. Physiol. Psychol. *91*: 886–896, 1977.

Albrecht, C., Bloss, H. G., Jackisch, R. and Feuerstein, T. J.: Evaluation of autoreceptor-mediated control of [(3) H] acetylcholine release in rat and human neocortex. Exp. Brain Res. *128*: 383–391, 1999.

Albuquerque, E. X., Pereira, E. F. R., Bonfante-Cabarcas, R., Marchioro, M., Matsubayashi, H., Alkondon, M. and Maelicke, A.: Nicotinic acetylcholine receptors on hippocampal cells: cell compartment-specific expression and modulatory control of channel activity. In: "Cholinergic Mechanisms: From Molecular Biology to Clinical Significance," J. Klein and L. Loeffelholz, Eds., pp. 111–124, Elsevier, Amsterdam, 1996.

Albuquerque, E. X., Pereira, E. F. R., Braga, M. F. M. and Alkondon, M.: Contribution of nicotinic receptors to the function of synapses in the central nervous system: the action of choline as a selective agonist of alpha7 receptors. In: "From *Torpedo* Electric Organ to Human Brain: Fundamental and Applied Aspects," J. Massoulié, Ed., pp. 309–316, Elsevier, Amsterdam, 1998a.

Albuquerque, E. X., Alkondon, M. and Pereira, E. M. F.: Contribution of nAChRs to the function of CNS synapses: the action of choline as a selective agonist. In: "From *Torpedo* Electric Organ to Human Brain: Fundamental and Applied Aspects," J. Massoulié, Ed., pp. 401–412, Elsevier, Amsterdam, 1998b.

Albuquerque, E. A., Pereira, E. F. R., Samochocki, M., Alkondon, M., Ulmer, C. and Maelicke, A.: Nicotinic receptor modulation: implications for treatment of Alzheimer's disease and other neurological disorders. In: "Cholinergic Mechanisms: Function and Dysfunction," I. Silmand, H. Soreq, L. Anglister, D. Michaelson, and A. Fisher, Eds., pp. 297–308, Taylor and Francis, New York, 2004.

Alhainen, K. J., Riekkinen, P. J., Jr., Soininnen, H. S., Reinikainen, K. J. and Riekkinen, P. J., Sr.: Responders and nonresponders in experimental therapy of Alzheimer's disease. In: "Cholinergic Basis for Alzheimer Therapy," R. Becker and E. Giacobini, Eds., pp. 231–237, Birkhauser, Boston, 1991.

Alkondon, M. and Albuquerque, E. X.: Alpha-Cobratoxin blocks nicotinic acetylcholine receptors in rat hippocampal neurons. Eur. J. Pharmacol. *191*: 505–506, 1990.

Alkondon, M., Rocha, E. S., Maelicke, A. and Albuquerque, E. X.: Diversity of nicotinic acetylcholine receptors in rat brain. V. alpha-Bungarotoxin-sensitive nicotinic receptors in olfactory bulb neurons and presynaptic modulation of glutamate release. J. Pharmacol. Exp. Ther. *278*: 1460–1471, 1996.

Alkondon, M., Pereira, E. F., Eisenberg, H. M. and Albuquerque, E. X.: Choline and selective antagonists identify two subtypes of nicotinic acetylcholine receptors that modulate GABA release from CA1 interneurons in rat hippocampal slices. J. Neurosci. *19*: 2693–2705, 1999.

Allikmets, L. H.: Cholinergic mechanisms in aggressive behavior. Med. Biol. *52*: 19–30, 1974.

Allikmets, L. H., Vahning, V. A. and Lapin, I. P.: Effects of direct injection of mediators and chemicals influencing their metabolism into the amygdala, septum and the hippocampus of cats. Pavlov J. High. Nerv. Act. *18*: 1044–1049, 1968.

Aloisi, A. M., Albonetti, M. E., Lodi, L., Lupo, C. and Carli, G.: Decrease of hippocampal choline acetyltransferase activity induced by formalin pain. Brain Res. *629*: 167–170, 1993.

Alonso, A.: Acetylcholine and neuronal short- and long-term memory. In: "XII International Symposium on Cholinergic Mechanisms," J. Gonzalez-Ros, Ed., J. Molec. Sci. 2006 (in Press).

Alroy, G., Su, H. and Yaari, Y.: Protein kinase C mediates muscarinic block of intrinsic bursting in rat hippocampal neurons. J. Physiol. *518*, pt. 1: 71–79, 1999.

Alteia, J. A., Creagh, F., Page, M., Owens, D. R., Scanlon, M. F. and Peters, J. R.: Early-morning hyperglycemia in IDDM: acute effects of cholinergic blockade. Diabetes Care *12*: 443–448, 1989.

Anagnostaras, S. G., Murphy, G. G., Hamilton, S. E., Mitchell, S. L., Rahnama, N. P., Nathanson, N. M. and Silva, A. J.: Selective cognitive dysfunction in acetylcholine M1 muscarinic receptor mutant mice. Nat. Neurosci. *6*: 51–58, 2003.

Anden, N.-E.: Effects of oxotremorine and physostigmine on the turnover of dopamine in the corpus striatum and the limbic system. J. Pharm. Pharmacol. *26*: 738–739, 1974.

Andersen, P. and Andersson, S. A.: "Physiological Basis of Alpha Rhythms," Appleton-Century-Crofts, New York, 1968.

Andersen, P. and Curtis, D. R.: The excitation of thalamic neurones by acetylcholine. Acta Physiol. Scand. *61*: 85–99, 1964.

Angeli, P., Bianchi, S., Imeri, L. and Mancia, M.: The influence of the muscarinic receptor subtypes on the sleep-wake cycle. Farmaco *48*: 1197–1206, 1993.

Angelucci, L., Lorentz, G., Maccari, S., Patacchioli, F. R., Scaccianoce, S. and Tagliatella, G.: Profilo pharmacologico della protirelina tartrato. Ann. Ital. Med. Int. *5*: 232–244, 1990.

Anglade, F., Chapouthier, G., Dodd, R. H. and Baudouin, C.: Olfactory memory in rats, cholinergic agents and benzodiazepine receptor ligands. J. Physiol. (Paris) *93*: 225–231, 1999.

Antkiewicz-Michaluk, L., Michaluk, J., Romanska, I. and Vetulani, J.: Cortical dihydropyridine binding sites and behavioral syndrome in morphine-abstinent rats. Eur. J. Pharmacol. *180*: 129–135, 1990.

Antunes-Rodrigues, J., Marubayashi, U., Favaretto, A. L., Gutkowska, J. and McCann, S. M.: Essential role of hypothalamic muscarinic and alpha-adrenergic receptors in atrial natriuretic peptide release induced by blood volume expansion. Proc. Natl. Acad. Sci. U.S.A. *901*: 10240–10244, 1993.

Aprison, M. H.: Research approaches to problems of mental illness: brain neurohumor-enzyme systems and behavior. In: "Horizons in Neuropsychopharmacology," W. A. Himwich and J. P. Schade, Eds., pp. 48–80, Elsevier, Amsterdam, 1965.

Araujo, D. M., Lapchak, P. A., Collier, B. and Quirion, R.: Characterization of N-[3H] methylcarbamylcholine binding sites and effect of N-methylcarbamylcholine on acetylcholine release in rat brain. J. Neurochem. *51*: 292–299, 1988.

Arciniegas, D., Adler, L., Topkoff, J., Cawthra, E., Filley, C. M. and Reite, M.: Attention and memory dysfunction after traumatic brain injury: cholinergic mechanisms, sensory gating, and a hypothesis for further investigation. Brain Inj. *13*: 1–13, 1999.

Arenas, E., Alberch, J., Sanchez Arroyos, R. and Marsal, J.: Effects of opioids on acetylcholine release evoked by K+ or glutamic acid from rat neostriatal slices. Brain Res. *523*: 51–66, 1990.

Arenas, E., Alberch, J., Perez-Navarro, E., Solsona, C. and Marsal, J.: Neurokinin receptors differentially mediate endogenous acetylcholine release evoked by tachykinins in the neostriate. J. Neurosci. *11*: 2332–2338, 1991.

Arendash, G. W., Sanberg, P. R. and Sengstock, G. J.: Nicotine enhances the learning and memory of aged rats. Pharmacol. Biochem. Behav. *52*: 517–523, 1995.

Arendt, T.: Impairment in memory function and neurodegenerative changes in the cholinergic basal forebrain system induced by chronic intake of alcohol. J. Neural Transm. *44*, Supplt.:173–187, 1994.

Arendt, T., Schugens, M. M. and Bigl, V.: The cholinergic system and memory: amelioration of ethanol-induced memory deficiency by physostigmine in rat. Acta Neurobiol. Exp. *50*: 251–260, 1990.

Ariel, M. and Daw, N. W.: Retina and muscarinic effects of cholinergic drugs on receptive field properties of rabbit retinal ganglion cells. J. Physiol. (Lond.) *324*: 155–160, 1982.

Arneric, S. P. and Reis, D. J.: Somatostatin and cholecystokinin octapeptide differentially modulate the release of [3H] acetylcholine from caudate nucleus but not cerebral cortex: role of dopamine receptor activation. Brain Res. *374*: 153–161, 1986.

Arneric, S. P. and Williams, M.: Nicotinic agonists in Alzheimer's disease: does the molecular diversity of nicotine receptors offer the opportunity for developing CNS-selective cholinergic channel activators? In: "Recent Advances in the Treatment of Neurodegenerative Disorders and Cognitive Function," G. Racagni, N. Brunello, and S. Z. Langer, Eds., pp. 58–70, Int. Acad. Biomed. Res., vol. 7, Karger, Basel, 1994.

Arneric, S. P., Sullivan, J. P. and Williams, M.: Neuronal nicotinic acetylcholine receptors. In: "Psychopharmacology: The Fourth Generation of Progress," F. E. Bloom and D. J. Kupfer, Eds., pp. 95–110, Raven Press, New York, 1995.

Arnold, M. B.: "Emotion and Personality," Columbia University Press, New York, 1960.

Arroyo-Jimenez, M. M., Bourgeois, J. P., Marubio, L. M., Le Sourd, A. M., Ottersen, O. P., Rinvik, E., Fairen, A. and Changeux, J. P.: Ultrastructural localization of the alpha4-subunit of the neuronal acetylcholine nicotinic receptor in the rat substantia nigra. J. Neurosci. *19*: 6475–6487, 1999.

Ashkenazi, A., Freeman, S. and Argov, Z.: Effects of cholinergic blockers on auditory brain-stem evoked potentials in rat. J. Neurol. Sci. *164*: 124–128, 1999.

Audesirk, C., Cabell, L. and Kern, M.: Modulation of neurite branching by protein phosphorylation in cultured rat hippocampal neurons. Dev. Brain Res. *102*: 247–260, 1997.

Axelrod, J.: The pineal gland: a neurochemical transducer. Science *184*: 1341–348, 1974.

Aznar, S., Kostova, V., Christiansen, S. H. and Knudsen, G. M.: Alpha7 nicotinic receptor subunit is present on serotonin neurons projecting to hippocampus and septum. Synapse *55*: 196–200, 2005.

Babichev, V. N.: The neuroendocrine regulation of gonadotropin and prolactin secretion and the role of neuromediators. Usp. Fiziol. Nauk. *26*: 44–61, 1995 (in Russian).

Bacq, Z. M.: "Chemical Transmission of Nerve Impulses," Pergamon Press, Oxford, 1975.

Badio, B. and Daly, J. W.: Epibadine, a potent analgetic and nicotinic agonist. Mol. Pharmacol. *45*: 563–569, 1994.

Baek, Y. H., Choi do, Y., Yang, E. I. and Park, D. S.: Analgesic effect of electroacupuncture: mediation by cholinergic and serotonergic receptors. Brain Res. *1057*: 181–185, 2005.

Baghdoyan, H. A. and Lydic, R.: M2 muscarinic receptor subtype in the feline medial pontine reticular formation modulates the amount of rapid eye movement sleep. Sleep *22*: 835–847, 1999.

Baghdoyan, H. A., Rodrigo-Angulo, M. L., McCarly, R. W. and Hobson, J. A.: A neuroanatomical gradient in the pontine tegmentum for the cholinoceptive induction of desynchronized sleep signs. Brain Res. *414*: 245–261, 1987.

Baghdoyan, H. A., Spotts, J. L. and Snyder, S. G.: Simultaneous pontine and basal forebrain microinjections of carbachol suppress REM sleep. J. Neurosci. *13*: 229–242, 1993.

Bailey, C. H., Bartsch, D. and Kandel, E. R.: Toward a molecular definition of long-term memory storage. Proc. Natl. Acad. Sci. U.S.A. *93*: 13445–13452, 1996.

Baker, W. W. and Benedict, F.: Analysis of local discharges induced by intrahippocampal microinjection of carbachol or diisopropylfluorophosphate (DFP). Int. J. Neuropharmacol. *7*: 135–147, 1968.

Baldissera, S., Menani, J. W., Sotero dos Santos, L. F., Favaretto, A. L. V., Gutkowska, J., Turrin, M. Q. A., McCann, S. M. and Antunes-Rodrigues, J.: Role of the hypothalamus in the control of atrial natriuretic hormone release. Proc. Natl. Acad. Sci. U.S.A. *86*: 9621–9625, 1989.

Balthazart, J., Foidart, A., Absil, P. and Harada, N.: Effects of testosterone and its metabolites on aromatase-immunoreactive cells in the quail brain: relationship with the activation of male reproductive behavior. J. Steroid Biochem. Mol. Biol. *56*: 185–200, 1996.

Bandler, R. J. and Flynn, J. P.: Neural pathways form thalamus associated with regulation of aggressive behavior. Science *183*: 96–98, 1974.

Bannon, A. W., Decker, M. W., Holladay, M. W., Curzon, P., Donnelly-Roberts, P., Puttfarcken, P. S., Bitner, R. S., Diaz, A., Dickenson, A. H., Porsolt, R. D., Williams, M. and Arneric, S. P.: Broad-spectrum, non-opioid analgesic activity by selective modulation of neuronal nicotinic acetylcholine receptors. Science *279*: 77–81, 1998.

Baran, H., Kepplinger, H., Draxler, M. and Skotfitsch, C.: Choline acetyltransferase, glutamic acid decarboxylase and somatostatin in the kainic acid model for chronic temporal lobe epilepsy. Neurosignals *13*: 290–297, 2004.

Barbelivien, A., Bertrand, N., Besret, L., Beley, A., MacKenzie, E. T. and Dauphin, F.: Neurochemical stimulation of the rat substantia innominata increases cerebral blood flow (but not glucose use) through parallel activation of cholinergic and non-cholinergic pathways. Brain Res. *840*: 115–124, 1999.

Bard, P.: A diencephalic mechanism for the expression of rage with specific reference to the sympathetic nervous system. Am. J. Physiol. *54*: 490–515, 1928.

Bard, P. and Mountcastle, V. B.: Some forebrain mechanisms involved in expression of rage with special reference to suppression of angry behavior. Res. Publ. Assoc. Res. Nerv. Ment. Dis. *27*: 362–404, 1948.

Barlow, H. B., Fitzhugh, R. and Kuffler, S. W.: Changes in organization in the receptive fields of the cat's retina during dark adaptation. J. Physiol. (Lond.) *137*: 338–354, 1957.

Barnes, C. D. and Pompeiano, O.: A brain stem cholinergic system activated by vestibular volleys. Neuropharmacol. *9*: 391–340, 1970.

Barnes, J. H.: The physiology and pharmacology of emesis. Mol. Aspects Med. *7*: 397–508, 1984.

Barnes, J. M., Barnes, N. M., Costall, B., Horovitz, Z. P. and Naylor, R. J.: Angiotensin II inhibits the release of [3H] acetylcholine from rat entorhinal cortex *in vitro*. Brain Res. *491*: 136–143, 1989.

Barnes, L., Karczmar, A. G. and Glisson, S. N.: Effects of central accumulation of acetylcholine on levels of 3 bioamines in rabbit brain parts. Pharmacologist *16*: 241, 1974.

Barrantes, G. E., Rogers, A. T., Lindstrom, J. and Wonnacott, S.: Alpha-bungarotoxin binding sites in rat hippocampal and cortical cultures: initial characterisation, colocalisation with alpha-subunits and up-regulation by chronic nicotine treatment. Brain Res. *672*: 228–236, 1995.

Barron, D. H. and Matthews, B. H. C.: The interpretation of potential changes in the spinal cord. J. Physiol. (Lond.), *92*: 276–321, 1938.

Bartfai, T., Iverfeldt, K. and Fisone, G.: Regulation of the release of coexisting neurotransmitters. Ann. Rev. Pharmacol. Toxicol. *28*: 285–310, 1988.

Bartholini, G.: Interaction of striatal dopaminergic, cholinergic and GABAergic neurons: relation to extrapyramidal function. Trends Pharmacol. Sci. *6*: 138–140, 1980.

Bartholow, R.: The antagonism between atropia and physostigmia. Clinic *5*: 61–62, 1873.

Bartus, R. T.: Evidence for a direct cholinergic involvement in the scopolamine-induced amnesia in monkeys: effects of concurrent administration of physostigmine and methyl phenidate with scopolamine. Pharmacol. Biochem. Behav. *9*: 833–836, 1978.

Bartus, R. T.: On neurodegenerative diseases, models, and treatment strategies: lessons learned and lessons forgotten—a generation following the cholinergic hypothesis. Exp. Neurol. *163*: 495–529, 2000.

Bartus, R. T., Dean, R. L. III, Beer, B. and Lippa, A. S.: The cholinergic hypothesis of geriatric memory dysfunction. Sci. *217*: 408–417, 1982.

Bashkatova, V., Kraus, M., Prast, H., Vanin, A., Rayevsky, K. and Philippu, A.: Influence of NOS inhibitors on changes in ACH release and NO level in the brain elicited by amphetamine neurotoxicity. Neuroreport. *10*: 3155–3158, 1999.

Baskin, P. P., Gianutsos, S. and Salamone, J. D.: Repeated scopolamine injections sensitize rats to pilocarpine-induced vacuous jaw movements and enhance striatal muscarinic receptor binding. Pharmacol. Biochem. Behav. *49*: 437–442, 1994.

Bassant, M. H., Jobert, A., Dutar, P. and Lamour, Y.: Effect of psychotropic drugs on identified septohippocampal neurons. Neuroscience *27*: 911–920, 1988.

Bassant, M. H., Ennouri, K. and Lamour, Y.: Effects of iontophoretically applied monoamines on somatosensory neurons of unanaesthetized rats. Neuroscience *39*: 431–439, 1990.

Battaglia, M. and Ogliari, A.: Anxiety and panic: from human studies to animal research and back. Neurosci. Biobehav. Rev. *29*: 169–179, 2005.

Baxter, M. G., Rucci, D. J., Gorman, L. K., Wiley, R. G. and Gallagher, M.: Selective immunotoxic lesions of basal forebrain cholinergic cells: effects on learning and memory in rats. Behav. Neurosci. *109*: 714–722, 1995.

Baxter, M. G., Gallagher, M. and Holland, P. C.: Blocking can occur without losses in attention in rats with selective removal of hippocampal cholinergic input. Behav. Neurosci. *113*: 881–890, 1999.

Bayer, L., Eggermann, E., Serafin, M., Grivel, J., Machard, D., Muhlethaler, M. and Jones, B. E.: Opposite effects of noradrenaline and acetylcholine upon hypocretin/orexin versus melanin concentrating hormone neurons in rat hypothalamic slices. Neuroscience *130*: 807–811, 2005.

Bazil, C. W. and Minneman, K. P.: Clinical concentrations of volatile anesthetics reduce depolarization-evoked release of [3H] norepinephrine, but not [3H] acetylcholine, from rat cerebral cortex. J. Neurochem. *53*: 962–965, 1989.

Beani, L., Bianchi, C., Giacomelli, A. and Tamberi, F.: Noradrenaline inhibition of acetylcholine release from guinea pig brain. Eur. J. Pharmacol. *48*: 89–93, 1978.

Beani, L., Bianchi, C., Siniscalchi, A., Sivilotti, L., Tanganelli, S. and Veratti, E.: Different approaches to study endogenous acetylcholine release: endogenous ACh versus tritium efflux. Naunyn Schmiedebergs Arch. Pharmacol. *328*: 119–126, 1984.

Beani, L., Tanganelli, S., Antonelli, T., Simonato, M., Spalluto, P., Tomasini, C. and Bianchi, C.: Changes in cortical acetylcholine and gamma-butyric acid outflow during morphine withdrawal involve alpha-1 and alpha-2 receptors. J. Pharmacol. Exp. Ther. *250*: 682–687, 1989.

Beaulieu, C. and Somogyi, P.: Enrichment of cholinergic synaptic terminals on GABAergic neurons and coexistence of immunoreactive GABA and choline acetyltransferase in the same synaptic terminals in the striate cortex of the cat. J. Comp. Neurol. *304*: 666–680, 1991.

Beck, A.: O znaczeniu lokalizacji w mózgu i w rdzeniu za pomocą zjawisk elektrycznych. Thesis, Jagiellon University, Krakow, 1890.

Becquet, D., Fandon, M. and Hery, F.: Effects of thalamic lesion on the bilateral regulation of serotonergic transmission in rat basal ganglia. J. Neural Transm. *74*: 117–128, 1988.

Beeri, R., Le Novere, N., Mervis, R., Huberman, T., Grauer, E., Changeux, J.-P. and Soreq, H.: Enhanced hemicholinium binding and attenuated dendrite branching in cognitively impaired acetylcholinesterase-transgenic mice. J. Neurochem. *69*: 2441–2451, 1997.

Behmand, R. A. and Harik, S. I.: Nicotine enhances 1-methyl-4-phenyl-1, 2, 3, 6-tetrahydropyridine neurotoxicity. J. Neurochem. *58*: 776–779, 1992.

Beilin, B., Bessler, H., Papismedov, L., Weistock, M. and Shavit, Y.: Continuous physostigmine combined with morphine-based patient-controlled analgesia in the postoperative period. Acta Anaesthesiol. Scand. *49*: 78–84, 2005.

Beinfeld, M. C. and Korchak, D. M.: The regional distribution and the chemical, chromatographic, and immunological characterization of motilin brain peptides. J. Neurosci. *5*: 2502–2509, 1985.

Beleslin, D. B. and Nedelkovski, V.: Emesis induced by 4-(m-chlorophenyl-carbamoyloxy)-2-butynyltrimethylammonium chloride (McN-A-343): evidence for a predominant central muscarinic M1 mediation. Neuropharmacology *27*: 949–956, 1988.

Beleslin, D. B., Krstic, S. K., Stefanovic-Denic, K., Strbac, M. and Micic, D.: Inhibition by morphine and morphine-like drugs of nicotine-induced emesis in cats. Brain Res. Bull. *6*: 451–457, 1981.

Beleslin, D. B., Jovanovic-Micic, D., Nikolic, S. B. and Samardzic, R.: Selective effect of ethanol on norepinephrine- and nicotine-induced emesis in cats. Alcohol *8*: 499–501, 1989.

Belin, V. and Moos, F.: Paired recordings from supraoptic and paraventricular oxytocin cells in suckled rats: recruitment and synchronization. J. Physiol. (Lond.) *377*: 369–390, 1986.

Belvin, M. P. and Yin, J. C.: *Drosophila* learning and memory: recent progress and new approaches. Bioessays *19*: 1083–1089, 1997.

Benardo, L. S. and Prince, D. A.: Cholinergic excitation of hippocampal pyramidal cells. Brain Res. *249*: 315–333, 1982.

Benarroch, E. E.: The central autonomic network: Functional organization, dysfunction and perspective. Mayo Clin. Proc. *68*: 988–1001, 1993.

Benarroch, E. E.: "Central Autonomic Network: Functional Organization and Clinical Correlations," Futura, Armonk, N.Y., 1997a.

Benarroch, E. E.: The central autonomic network. In: "Clinical Autonomic Disorders," P. A. Low, Ed., pp. 17–23, Lippincott-Raven, Philadelphia, 1997b.

Benarroch, E. E., Granata, A. R., Ruggiero, D. A., Park, D. A. and Reis, D. J.: Neurons of the C1 area mediate cardiovascular responses initiated form ventral medullary surface. Am. J. Physiol. *250*: R932–R945, 1986a.

Benarroch, E. E., Granata, A. R., Giuliano, R. and Reis, D. J.: Neurons of the C1 area of rostral ventrolateral medulla mediate nucleus tractus solitarii hypertension. Hypertension *8*, Supplt. 1: I56–I60, 1986b.

Bencherif, M., Byrd, G., Caldwell, W. S., Hayes, J. R. and Lippiello, P. M.: Pharmacological characterization of RJR-2403: a nicotinic agonist with potential therapeutic benefit in the treatment of Alzheimer's disease. CNS Drug Rev. *3*: 325–345, 1997.

Benowitz, N. L.: Nicotine addiction: primary care; clinics in office practice. *26*: 611–633, 1999.

Benowitz, N. L., Jacob, P. and Yu, L.: Daily use of smokeless tobacco: systemic effects. Ann. Intern. Med. *111*: 112–116, 1989.

Benschmid, R. and Krehl, L.: Über den Einfluss des Gehims auf die Wärmeregulation. Arch. Exper. Pathol. Pharmakol. *70*: 100–134, 1912.

Benwell, M. E. and Balfour, D. J. K.: The effects of acute and repeated nicotine treatment on nucleus accumbens dopamine and locomotor activity. Br. J. Pharmacol. *105*: 849–856, 1992.

Bequet et al.: 5-HT release and ACh, 1988.

Berger, P. A., Davis, K. L. and Hollister, L. E.: Cholinomimetics in mania, schizophrenia and memory disorders. In: "Choline and Lecithin in Brain Disorders," A. Barbeau, J. H. Growdon and R. J. Wurtman, Eds., pp. 425–441, Nutrition and the Brain, vol. 5, Raven Press, New York, 1979.

Berkowitz, A., Sutton, L., Janowsky, D. S. and Gillin, J. C.: Pilocarpine, an orally active muscarinic cholinergic agonist, induces REM sleep and reduces delta sleep in normal volunteers. Psychiatr. Res. *33*: 113–119, 1990.

Bernard, E. A., Bason, D. M. W., Cokroft, V. B., Darlison, M. G., Hicks, A. A., Lai, F. A., Moss, S. J. and Squire, M. D.: Molecular biology of nicotinic acetylcholine receptors from chicken muscle and brain. In: "Cellular and Molecular Basis of Cholinergic Function," M. J. Dowdall and J. N. Hawthorne, Eds., pp. 15–32, Ellis Horwood, Chichester, 1987.

Bernard, V., Levey, A. I. and Bloch, B.: Regulation of the subcellular distribution of m4 muscarinic acetylcholine receptors in striatal neurons *in vivo* by the cholinergic environment: evidence for regulation of cell surface receptors by endogenous and exogenous stimulation. J. Neurosci. *19*: 10237–10249, 1999.

Berrendero, F., Kieffer, B. L. and Maldonado, R.: Attenuation of nicotine-induced antinociception, rewarding effects and dependence in mu-opioid-receptor knock-out mice. J. Neurosci. *212*: 10935–10940, 2002.

Bevan, M. D. and Bolam, J. P.: Cholinergic, GABAergic and glutamate-enriched inputs from the mesopontine tegmentum to the subthalamic nucleus in the rat. J. Neurosci. *15*: 7105–7120, 1995.

Bhattacharya, S. K., Upadhyay, S. N. and Jaiswal, A. K.: Effect of piracetam on electroshock induced amnesia and decrease in brain acetylcholine in rats. Indian J. Exp. Biol. *31*: 822–824, 1993.

Bidmon, H. J., Wu, J., Palomero-Gallagher, N., Oermann, E., Mayer, B., Schleicher, A. and Zilles, K.: Different nitric oxide synthase inhibitors cause rapid and differential alterations in the ligand-binding capacity of transmitter receptors in the rat cerebral cortex. Ann. Anat. *181*: 345–351, 1999.

Biederman, G. B.: Forgetting an operant response: physostigmine-produced increases in escape latency in rats as a function of time of injection. Q. J. Exp. Psychol. *22*: 384–388, 1970.

Bignami, G.: Pharmacologic influences on mating behavior in the male rat. Psychopharmacology *10*: 44–58, 1966.

Bignami, G. and Rosic, N.: The nature of disinhibitory phenomena caused by central (muscarinic) blockade. In: "Advances in Neuropharmacology," O. Vinar, Z. Votava, and P. B. Bradley, Eds., pp. 481–495, North Holland, Amsterdam, 1971.

Bilkei-Gorzo, A.: Effect of chronic treatment with aluminum compound on rat brain choline acetyltransferase activity. Pharmacol. Toxicol. *74*: 359–360, 1994.

Bird, S. J. and Aghajanian, G. K.: The cholinergic pharmacology of hippocampal pyramidal cells: a microiontophoretic study. Neuropharmacology *15*: 273–282, 1976.

Birdsall, N. J., Lazareno, S. and Matsui, H.: Allosteric regulation of muscarinic receptors. Prog. Brain Res. *109*: 147–151, 1996.

Birthelmer, A., Lazaris, A., Riegert, C., Marques Pereira, P., Koenig, J., Jeltsch, H., Jakisch, R. and Cassel, J. C.: Does the release of acetylcholine in septal slices originate from intrinsic cholinergic neurons bearing p75 (NTR) receptors? A study using 192-saporin lesions in rats. Neuroscience *122*: 1059–1071, 2003.

Bitran, D. and Hull, E. M.: Pharmacological analysis of male rat sexual behavior. Pharmacol. Rev. *11*: 365–389, 1988.

Blaker, W. D., Cheney, D. L. and Costa, E.: GABA-B vs. GABA-B modulation of septal-hippocampal interconnections. In: "Dynamics of Cholinergic Function," I. Hanin, Ed., pp. 953–971, Plenum Press, New York, 1986.

Bland, B. H.: The physiology and pharmacology of hippocampal formation theta rhythms. Prog. Neurobiol. *26*: 1–54, 1986.

Bland, B. H., Colom, L. V., Konopacki, J. and Roth, S. H.: Intracellular records of carbachol-induced theta rhythm in hippocampal slices. Brain Res. *447*: 364–368, 1988.

Blin, J., Ray, C. A., Piercey, M. F., Bartko, J. J., Mouradian, M. M. and Chase, T. N.: Comparison of cholinergic drug effects on regional brain glucose consumption in rats and humans by means of autoradiography and position emission tomography. Brain Res. *635*: 196–202, 1994.

Blokland, A.: Acetylcholine: a neurotransmitter for learning and memory? Brain Res. Brain Res. Rev. *21*: 285–300, 1995.

Blomqvist, O., Soderpalm, B. and Engel, J. A.: Ethanol-induced locomotor activity: involvement of central nicotinic acetylcholine receptors? Brain Res. Bull. *29*: 173–178, 1992.

Blomqvist, O., Engel, J. A., Nissbrandt, H. and Soderpalm, B.: The mesolimbic dopamine-activating properties of ethanol are antagonized by mecamylamine. Eur. J. Pharmacol. *249*: 207–213, 1993.

Blum, B., Israeli, J., Hart, O. and Farchi, M.: Depressor effects of muscarinic and non-muscarinic mediation induced by lateral hypothalamic stimulation in the cat. Experientia *45*: 991–996, 1989.

Blusztajn, J. K., Cermak, J. M., Holler, T. and Jackcon, D. A.: Imprinting of hippocampal metabolism of choline by its availability during gestation: implications for cholinergic neurotransmission. In: "From *Torpedo* Electric Organ to Human Brain: Fundamental and Applied Aspects," J. Massoulié, Ed., pp. 199–203, Elsevier, Amsterdam, 1998.

Boccia, M. M., Acosta, G. B., Blake, M. G. and Baratti, C. M.: Memory consolidation and reconsolidation of an inhibitory avoidance response in mice: effects of i. c. v. injections of hemicholinium-3. Neuroscience *124*: 735–741, 2004.

Bonnet, V. and Bremer, F.: Action du potassium, du calcium et d'acetylcholine sur les activités électriques, spontanées et provoquées, de l'écorce cerebrale. C. R. Soc. Biol. *136*: 1271, 1937.

Boop, F. A., Garcia-Rill, E., Dykman, R. and Skinner, R. D.: The P1: insight into attention and arousal. Pediatr. Neurosurg. *20*: 57–62, 1994.

Borison, H. L. and Wang, S. C.: Physiology and pharmacology of vomiting. Pharmacol. Rev. *5*: 193–230, 1953.

Borjesson, A., Karlsson, T., Adolfsson, R., Ronnlund, M. and Nilsson, L.: Linopirdine (DUP 996): cholinergic treatment of older adults using successive and non-successive tests. Neuropsychobiology *40*: 78–85, 1999.

Boudinot, E., Taysse, L., Daulon, S., Chatonnet, A., Champagnat, J. and Foutz, A. S.: Effects of acetylcholinesterase and butyrylcholinesterase inhibition on breathing in mice adapted or not to reduced acetylcholinesterase. Pharmacol. Biochem. Behav. *80*: 53–61, 2005.

Bovet, D.: Central action of ganglioplegic drugs. [Title translated from original Italian] Rend Ist. Sup. Sanit. *16*: 608–626, 1953.

Bovet, D.: "Une Chimie qui guérit." Editions Payot, Paris, 1988.

Bovet, D. and Longo, V. G.: The action on nicotine induced tremors of substances effective in parkinsonism. J. Pharmacol. Exp. Ther. *102*: 22–30, 1951.

Bovet, D., Bovet-Nitti, F. and Oliverio, A.: Effects of nicotine on avoidance conditioning of inbred strains of mice. Psychopharmacology *101*: 1–5, 1966.

Bradley, P. B. and Elkes, J.: The effect of atropine, hyoscyamine, physostigmine, and neostigmine on the electrical activity of the brain of conscious cat. J. Physiol. *120*: 14P, 1953.

Bradley, P. B. and Elkes, J.: The effect of some drugs on the electrical activity of the brain. Brain *88*: 77–117, 1967.

Bradley, P. B. and Wolstencroft, J. H.: Actions of drugs on single neurons of the brain stem. Br. Med. Bull. *21*: 15–18, 1965.

Brandao, F., Ribeiro-da-Silva, A. and Cadete-Leite, A.: GM1 and piracetam do not revert the alcohol-induced depletion of cholinergic fibers in the hippocampal formation of the rat. Alcohol *19*: 65–74, 1999.

Brandstatter, R., Falt, E. and Hermann, A.: Acetylcholine modulates ganglion cell activity in trout pineal organ. Neuroreport *6*: 1553–1556, 1995.

Brazhnik, E. S. and Vinogradova, O. S.: Modulation of the afferent input to the septal neurons by cholinergic drugs. Brain Res. *451*: 1–12, 1988.

Brazhnik, E. S., Vinogradova, O. S., Stafekhina, V. S. and Kitchigina, V. F.: Acetylcholine, theta rhythm and activity of hippocampal neurons in the rabbit—I. Spontaneous activity. Neuroscience *53*: 961–970, 1993.

Brazier, M. A. B.: The historical development of neurophysiology. In: "Handbook of Physiology. Section I: Neurophysiology," J. Field, Ed., H. W. Magoun, Section Ed., pp. 1–58, American Physiological Society, Washington, D.C., 1959.

Brefel, C., Lazartigues, E., Tran, M. A., Gauquelin, G., Gharib, C., Montastruc, J. L., Montastruc, P. and Rascol, O.: Central cardiovascular effects of acetylcholine in the conscious dog. Br. J. Pharmacol. *116*: 2175–2182, 1995.

Bremer, F. and Chatonnet, J.: Acetylcholine et le cortex cerebrale. Arch. Intern. Physiol. *57*: 106–109, 1949.

Brezenoff, H. E.: Cardiovascular responses to intrahypothalamic injections of carbachol and certain cholinesterase inhibitors. Neuropharmacol. *11*: 637–644, 1972.

Brezenoff, H. E. and Giulano, R.: Cardiovascular control by cholinergic mechanisms in the central nervous system. Annu. Rev. Pharmacol. Toxicol. *23*: 341–348, 1982.

Brezenoff, H. E. and Jenden, D. J.: Modification of arterial blood pressure in rats following microinjection of drugs into posterior hypothalamus. Int. J. Neuropharmacol. *8*: 593–600, 1969.

Brezenoff, H. E. and Jenden, D. J.: Changes in arterial blood pressure in rats after microinjections of carbachol into the medulla and IV ventricle of rat brain. Neuropharmacology *9*: 341–348, 1970.

Brezenoff, H. E. and Xiao, Y. F.: Acetylcholine in the posterior hypothalamus is involved in the elevated blood pressure in the spontaneously hypertensive rat. Life Sci. *45*: 1163–1170, 1989.

Brimblecombe, R. W.: Drugs acting on central cholinergic mechanisms and affecting respiration. Pharmacol. Ther. *3*: 65–74, 1977.

Brink, T. L., Yesavage, J. A. and Lum, O.: Screening tests for geriatric depression. Clin. Gerontol. *1*: 37–43, 1982.

Brito, N. A., Brito, M. N., Kettelhut, I. C. and Migliorini, R. N.: Intra-medial hypothalamic injection of cholinergic agents induces rapid hyperglycemia, hyperlactatemia and gluconeogenesis activation in fed, conscious rats. Brain Res. *626*: 339–342, 1993.

Brodal, A.: "Neurological Anatomy," Third Edition, Oxford University Press, Oxford, 1980.

Brodie, B. B.: Comments on a symposium entitled 5-hydroxytryptamine: a pharmacological analysis. In: "5-Hydroxytryptamine," pp. 64–83, Pergamon Press, Oxford, 1957.

Brodie, M. S. and Proudfit, H. K.: Antinociception induced by local injection of carbachol into the nucleus raphe magnus in rats: alteration by intrathecal injection of monoaminergic antagonists. Brain Res. *371*: 70–79, 1986.

Brown, L. A. and Chen, M.: Vasopressin signal transduction in rat type II pneumocytes. Am. J. Physiol. *158*: 301–307, 1990.

Brown, J. H., Trilivas, I., Trejo, J. and Martinson, E.: Multiple pathways for signal transduction through muscarinic cholinergic receptor. In: "Cholinergic Neurotransmission: Functional and Clinical Aspects," S.-M. Aquilonius and P.-G. Gillberg, Eds., pp. 21–29, Elsevier, Amsterdam, 1990.

Brown-Sequard, C. E.: "Lectures on the Physiology and Pathology of the Central Nervous System," J. B. Lippincott, Philadelphia, 1860.

Brownstein, M. J.: Use of molecular biological methods to study neuropeptides. NIDA Res. Monogr. *87*: 83–92, 1988.

Bruck, K. and Zeisberger, E.: Adaptive changes in thermoregulation and their neuropharmacological basis. Pharmacol. Therap. *35*: 163–215, 1987.

Brucke, F. T. and Stumpf, C.: The pharmacology of "arousal" reactions. In: "International Symposium on Psychotropic Drugs," S. Garattini and G. Ghetti, Eds., pp. 319–324, Elsevier, Amsterdam, 1957.

Brudzynski, S. M.: Ultrasonic vocalization induced by intracerebral carbachol in rats: localization and a dose-response study. Behav. Brain Res. *63*: 133–143, 1994.

Brudzynski, S. M. and Eckersdorf, B.: Vocalization accompanying emotional-aversive response induced by carbachol. Reproducibility and dose-response study. Neuropsychopharmacology *1*: 311–320, 1988.

Brudzynski, S. M., Wu, M. and Mogeson, G. J.: Modulation of locomotor activity induced by injections of carbachol into the tegmental pedunculopontine nucleus and adjacent areas in the rat. Brain Res. *451*: 119–125, 1988.

Brudzynski, S. M., Eckersdorf, B. and Golebiewski, H.: Emotional-aversive nature of the behavioral responses induced by carbachol in cats. J. Psychiatry Neurosci. *18*: 38–45, 1993.

Brudzynski, S. M., Kadishevitz, L. and Fu, X. W.: Mesolimbic component of the ascending cholinergic pathways: electrophysiological-pharmacological study. J. Neurophysiol. *79*: 1675–1686, 1998.

Brunello, N. and Cheney, D. L.: The septal-hippocampal cholinergic pathway: role in antagonism of pentobarbital anesthesia and regulation by various afferents. J. Pharmacol. Exp. Ther. *219*: 489–495, 1981.

Buccafusco, J. J.: Inhibition of the morphine withdrawal syndrome by a novel muscarinic antagonist (4-DAMP). Life Sci. *48*: 749–756, 1991.

Buccafusco, J. J. and Magri, V.: Modification of spinobulbar autonomic cholinergic systems by activation of alpha-adrenergic receptors. J. Auton. Nerv. Syst. *28*: 133–140, 1989.

Buccafusco, J. J. and Magri, V.: The pressor response to spinal cholinergic stimulation in spontaneously hypertensive rats. Brain Res. Bull. *25*: 69–74, 1990.

Buchwald, N. A., Wyers, E. J., Okuma, T. and Jeuser, G.: The "caudate spindle." I. Electrophysiological properties. Electroencephalogr. Clin. Neurophysiol. *13*: 509–518, 1961.

Bugajski, J.: Social stress adapts signaling pathways involved in stimulation of the hypothalamic-pituitary-adrenal axis. J. Physiol. Pharmacol. *50*: 367–379, 1999.

Bugajski, J., Gadek-Michalska, A. and Bugajski, A. J.: Nitric oxide and prostaglandin systems in the stimulation of hypothalamic-pituitary-adrenal axis by neurotransmitters and neurohormones. J. Physiol. Pharmacol. *55*: 679–703, 2004.

Bullock, A. E., Barke, K. E. and Schneider, A. S.: Nicotine tolerance in chromaffin cell culture: acute and chronic exposure to smoking-related nicotine doses. J. Neurochem. *62*: 1863–1869, 1994.

Burchfiel, J. L., Duchowny, M. S. and Duffy, H. F.: Neuronal supersensitivity to acetylcholine induced by kindling in the rat hippocampus. Science *204*: 1096–1098, 1979.

Bures, J., Buresova, O., Bohdanecky, Z. and Weiss, T.: The effects of physostigmine and atropine on mechanism of learning. CIBA Foundation Symposium on Animal Behavior and Drug Action, pp. 134–142, 1964.

Bures, J., Buresova, O. and Huston, J.: "Techniques and Basic Experiments for the Study of Brain Behavior," Elsevier, Amsterdam, 1983.

Burton, M. D., Johnson, D. C. and Kazemi, H.: Adrenergic and cholinergic interaction in central ventilatory control. J. Appl. Physiol. *68*: 2092–2092, 1990.

Butcher, L. L. and Woolf, N. J.: Cholinergic neurons and networks revisited. In: "The Rat Nervous System," Third Edition, Elsevier, Amsterdam, 2003.

Butcher, L. L., Oh, J. D. and Woolf, N.: Cholinergic neurons identified by in situ hybridization histochemistry. In: "Cholinergic Function and Dysfunction," A. C. Cuello, Ed., pp. 1–8, Elsevier, Amsterdam, 1993.

Butt, A. E. and Hodge, G. K.: Acquisition, retention and extinction of operant discrimination in rats with nucleus basalis magnocellularis lesions. Behav. Neurosci. *109*: 619–713, 1995.

Buyukuysal, R. L., Ulus, I. H., Aydin, S. and Kiran, B. K.: 3,4-diaminopyridine and choline increase *in vivo* acetylcholine release in rat striatum. Eur. J. Pharmacol. *281*: 179–185, 1995.

Buzsaki, G. and Gage, F. H.: The cholinergic nucleus basalis: a key structure in neocortical arousal. Experientia *57*, Supplt.: 159–171, 1989.

Buzsaki, G., Bickford, R. G., Ponomareff, G., Thal, L. G. J., Mandel, R. and Gage, G. H.: Nucleus basalis and thalamic control of neocortical activity in the freely moving rat. J. Neurosci. *8*: 4007–4026, 1988.

Bymaster, F. P., Heath, I., Hendrix, J. C. and Shannon, H. E.: Comparative behavioral and biochemical activities of cholinergic antagonists in rats. J. Pharmacol. Exp. Ther. *267*: 16–24, 1993.

Bymaster, F., McKenzie, D. L. and Felder, C. C.: New evidence for the involvement of muscarinic cholinergic receptors in psychoos. In: "Cholinergic Mechanisms. Function and Dsyfunction." I. Silman, H. Soreq, L. Anglister, D. Michaelson and A. Fisher, Ed., pp. 331–343, Martin Dunitz, London, 2004.

Cain, D. P.: Bidirectional transfer of electrical and carbachol kindling. Brain Res. *260*: 135–138, 1983.

Cain, D. P., Desborough, K. A. and McKitrick, D. J.: Retardation of amygdala kindling by antagonism of NMD-aspartate and muscarinic cholinergic receptors: Evidence for the summation of excitatory mechanisms in kindling. Exp. Neurol. *100*: 179–187, 1988.

Calabresi, P., Lacey, M. G. and North, R. A.: Nicotinic excitation of rat tegmental neurones *in vitro* studied by intracellular recording. Br. J. Pharmacol. *98*: 135–140, 1989.

Calabresi, P., Centonze, D., Pisani, A., Sancessario, G., North, R. A. and Bernardi, G.: Muscarinic IPSPs in rat striatal cholinergic interneurones. J. Physiol. *510*: 421–427, 1998.

Calaresu, F. R., McKitrick, D. J. and Weernink, E. J.: Microinjection of substance P and ACh into rat mediolateral nucleus elicits cardiovascular responses. Am. J. Physiol. *259*: 357–361, 1990.

Callaway, E.: The pharmacology of human information processing. Psychophysiology *20*: 359–370, 1983.

Callera, J. C., Saad, W. A., Camargo, L. A., Renzi, A., De Luca, L. A. and Menani, J. V.: Role of the adrenergic pathways of the lateral hypothalamus on water intake and pressor response induced by cholinergic activation of the medial septal area in rats. Neurosci. Lett. *167*: 153–155, 1994.

Callera, J. C., Saad, W. A., Camargo, L. A., Menani, J. V., Renzi, A. and Abrao-Saad, W.: Alpha-adrenergic pathways in the lateral hypothalamus are important for the dipsogenic effect of carbachol injected into the medial septal area. Braz. J. Med. Biol. Res. *28*: 240–225, 1995.

Calogero, A. E., Kamilaris, T. C., Gomez, M. T., Johnson, E. D., Tartaglia, M. E., Gold, P. W. and Chrousos, G. P.: The muscarinic cholinergic agonist arecoline stimulates the rat hypothalamic-pituitary-adrenal axis through a centrally-mediated corticotropin-releasing-hormone-dependent mechanism. Endocrinology *125*: 2445–2453, 1989.

Camilleri, M.: Autonomic regulation of gastrointestinal activity. In: "Clinical Autonomic Disorders," P. A. Low, Ed., pp. 125–132, Little, Brown, Boston, 1993.

Cannon, W. B. and Britton, S. W.: Studies on conditions of activity in endocrine glands. XV. Pseudoaffective medulliadrenal secretion. Am. J. Physiol. *72*: 283–294, 1925.

Cao, W.-H., Inanami, O., Sato, A. and Sato, Y.: Stimulation of the septal complex increases local cerebral blood flow in hippocampus in anesthetized rats. Neurosci. Lett. *107*: 135–140, 1989.

Capece, M. L. and Lydic, R.: cAMP and protein kinase A modulate cholinergic rapid eye movement sleep generation. Am. J. Physiol. *273*: R1430–1440, 1997.

Capone, F., Aloisi, A. M., Carli, G., Sacerdote, P. and Pavone, F. Oxotremorine-induced modifications of the behavioral and neuroendocrine responses to formalin pain in male rats. Brain Res. *830*: 292–300, 1999.

Carafoli, E. and Rossi, C. S.: Calcium transport in mitochondria. In: "Advances in Cytopharmacology," F. Clementi and B. Cecarelli, Eds., vol. 1, pp. 209–228, Raven Press, New York, 1971.

Carder, R. K., Jackson, D., Morris, H. J., Lund, R. D. and Zigmond, M. J.: Dopamine released from mesencephalic transplants restores modulation of striatal acetylcholine release after neonatal 6-hydroxydopamine: an *in vitro* analysis. Exp. Neurol. *105*: 251–259, 1989.

Carette, B.: Neurons in the guinea-pig lateral septum generate rhythmic bursting activities, *in vitro*, following application of carbachol. Neurosci. Lett. *239*: 93–96, 1997.

Carette, B.: Characterization of carbachol-induced rhythmic bursting discharges in newborn guinea pig lateral septum slices. J. Neurophysiol. *80*: 1042–1055, 1998.

Carlton, P. L.: Scopolamine, amphetamine and light-reinforced responding. Psychonomic Sci. *5*: 347–348, 1966.

Carlton, P. L.: Cholinergic mechanisms in the control of behavior. In: "Psychopharmacology—A Review of Progress," D. H. Efron, Ed., pp. 125–135, U.S. Government Printing Office, PHS Publication No. 836, Washington, D.C., 1968.

Carrillo, G. C. and Doupe, A. J.: Is the songbird Area X striatal, parietal or both? An anatomical study. J. Comp. Neurol. *473*: 415–437, 2004.

Casamenti, F., Pedata, F., Corradetti, R. and Pepeu, G.: Acetylcholine output from the cerebral cortex, choline uptake and muscarinic receptors in morphine-dependent, freely-moving rats. Neuropharmacology *19*: 597–605, 1980.

Casey, K. L.: Pain: a current view of neural mechanisms. Am. Sci. *61*: 194–200, 1973.

Cassel, J. C. and Jeltsch, H.: Serotonergic modulation of cholinergic function in the central nervous system: cognitive implications. Neuroscience *69*: 1–41, 1995.

Castillo, P. B., Carleton, A., Vincent, J. D. and Llado, P. M.: Multiple and opposing roles of cholinergic transmission in the main olfactory bulb. J. Neurosci. *19*: 180, 191, 1999.

Catani, M., Jones, D. K. and Fytche, D. H.: Perisylvian language networks of the human Brain. Ann. Neurol. *57*: 8–16, 2005.

Caulfield, M. P.: Muscarinic receptors—characterization, coupling and function. Pharmacol. Ther. *58*: 319–379, 1993.

Caulfield, M. P. and Brown, D. A.: Pharmacology of the putative M4 muscarinic receptor mediating Ca-current inhibition in neuroblastomaxglioma hybrid (NG 108-15) cells. Br. J. Pharmacol. *104*: 39–44, 1991.

Caulfield, M. P., Robbins, J. and Brown, D. A.: Neurotransmitters inhibit the omega-conotoxin-sensitive component of Ca current in neuroblastoma x glioma hybrid (NG 108-15) cells, not the nifedipine-sensitive component. Pflugers Arch. *420*: 486–492, 1992.

Caulfield, M. P., Robbins, J., Higashida, H. and Brown, D. A.: Postsynaptic actions of acetylcholine: the coupling of muscarinic receptor subtypes to neuronal ion channels. In: "Cholinergic Function and Dysfunction," A. C. Cuello, Ed., pp. 293–301, vol. 98, Prog. Brain Res., Elsevier, Amsterdam, 1993b.

Cavalheiro, E. A., Fernandez, M. J., Turski, L. and Naffah-Mazzacoratti, M. G.: Spontaneous recurrent seizures in rats: amino acid and monoamine determination in the hippocampus. Epilepsia *35*: 1–11, 1994.

Celesia, G. G.: Visual evoked potentials. In: "Evoked Potentials Review," C. Barber and M. Taylor, Eds., pp. 45–46, IEPS Publications, England, 1991.

Celie, P. H., Klassen, R. V., van Rossum-Fikkert, S. E., van Elk, R., van Nicrop, P., Smit, A. B. and Sixma, T. K.: Crystal structure of acetylcholine-binding protein from Butulinus truncates reveals the conserved structural scaffold and sites of variation in nicotinic acetylcholine receptors. J. Biol. Chem. *280*: 26457–26466, 2005.

Centeno, M. L., Luo, J., Lindstrom, J. M., Caba, M. and Pau, K. Y.: Expression of alpha 4 and alpha 7 nicotinic receptors in the brain stem of female rabbits after coitus. Brain Res. *1012*: 1–12, 2004.

Chalupa, L. M. and Gunthan, E.: Development of on and off retinal pathways and reticulogeniculate projections. Prog. Retin. Eye Res. *23*: 31–51, 2004.

Champtiaux, N. and Changeux, J. P.: Knockout and knockin mice to investigate the role of nicotinic receptors in the central nervous system. Prog. Brain Res. *145*: 235–251, 2004.

Chan, T. T., Izazola-Conde, C. and Dewey, W. L.: Evidence for the existence of peptide and non-peptide morphine-like materials in mouse brain: effect of an analgesic intracerebroventricular dose of acetylcholine on their levels. J. Pharmacol. Exp. Ther. *222*: 612–616, 1982.

Chang, H. C. and Gaddum, J. H.: Choline esters in tissues extracts. J. Physiol. *79*: 255–288, 1933.

Chang, H. C., Chia, K. F., Hsu, C. H. and Lim, R. K. S.: Humoral transmission of nerve impulses at central synapses. I. Sinus and vagus afferent nerves. Chin. J. Physiol. *12*: 1–36, 1937.

Changeux, J.-P.: "Neuronal Man." Oxford University Press, Oxford, 1985.

Changeux, J.-P., Bessis, A., Bourgeois, J. P., Corringer, P. J., Devilliers-Thiery, A., Eisele, J. L., Kerszberg, M., Lena, C., Le Novere, N., Picciotto, M. and Zoli, M.: Nicotinic receptors and brain plasticity. Cold Spring Harb. Symp. Quant. Biol. *61*: 343–362, 1996.

Chapell, R. and Rosenberg, P.: Specificity of action of beta-bungarotoxin on acetylcholine release from synaptosomes. Toxicon *30*: 621–633, 1992.

Chapman, C. A. and Lacaille, J. C.: Cholinergic induction of theta-frequency oscillations in hippocampal inhibitory interneurons and pacing of pyramidal cell firing. J. Neurosci. *19*: 8637–8645, 1999.

Chatfield, P. O. and Dempsey, E. W. Some effects of prostigmine and acetylcholine on cortical potentials. Am. J. Physiol. *135*: 633–640, 1942.

Chau, D., Rada, P. V., Kosloff, R. A. and Hoebel, B. G.: Cholinergic, M1 receptors in the nucleus accumbens mediate behavioral depression. A possible downstream target for fluotexine. Ann. N. Y. Acad. Sci. *877*: 769–774, 1999.

Chen, Z. and Engberg, G.: The rat nucleus paragigantocellularis as a relay station to mediate peripherally induced central effects of nicotine. Neurosci. Lett. *101*: 67–71, 1989.

Chen, F.-Y. and Lee, T. J. F.: Role of nitric oxide in neurogenic vasodilation of porcine cerebral artery. J. Pharmacol. Exp. Ther. *265*: 339–345, 1993.

Chen, G., Ensor, C. R. and Bohner, B.: Studies of drug effects on electrically-induced extensor seizures and clinical applications. Arch. Int. Pharmacodyn. *172*: 183–218, 1968.

Chen, C., Chen, M. Q., Wang, H. X. and Chen, Y.: The role of the nucleus tractus solitarius and the locus coeruleus in the abdominal vagal pressor response. Proc. Chin. Acad. Med. Sci. Peking Union Med. Coll. *4*: 142–146, 1989.

Chen, S. R. and Pan, H. L.: Activation of muscarinic receptors inhibits spinal dorsal horn projection neurons: role of GABAB receptors. Neuroscience *125*: 141–148, 2004.

Cheney, D. L., Costa, E., Hanin, I., Racagni, G. and Trabucchi, M.: Acetylcholine turnover rate in brain of mice and rats: effects of various dose regimens of morphine. In: "Cholinergic Mechanisms," P. G. Waser, Ed., pp. 217–228, Raven Press, New York, 1975.

Chiapinelli, V. A., Weaver, W. R. and McMahon, L. L.: Electrophysiology of an excitatory nicotinic synapse in the central nervous system. In: "Cholinergic Neurotransmission: Function and Dysfunction," A. C. Cuello, Ed., Abstracts, p. 56, Elsevier, Amsterdam, 1993.

Chiodera, P., Volpi, R., Capretti, L., Bocchi, R., Caffari, G., Marcato, A., Rossi, G. and Coiro, V.: Gamma-aminobutyric acid mediation of the inhibitory effect of endogenous opioids on the arginine vasopressin and oxytocin responses to nicotine from cigarette smoking. Metabolism *42*: 762–765, 1993.

Christison, R.: On the properties of the ordeal bean of Old Calabar. Mon. J. Med. *20*: 193–204, 1855.

Chun, M. H., Wassle, H. and Brecha, N.: Colocalization of [3H] muscimol uptake and choline acetyltransferase in amacrine cells of cat retina. Neurosci. Lett. *94*: 259–263, 1988.

Churchland, P. K.: "The Engine of Reason, the Seat of the Soul," MIT Press, Cambridge, Mass., 1994.

Cipolotti, L. and Moscovitch, M.: The hippocampus and remote autobiographical memory. Lancet Neurol *4*: 792–793, 2005.

Clark, W. G.: Changes in body temperature after administration of acetylcholine, histamine, morphine, prostaglandins and related agents. Neurosci. Biobehav. Rev. *3*: 179–231, 1979.

Clarke, P. B. S.: CNS nicotinic receptors: localization and relation to cholinergic transmission. In: "Cholinergic Neurotransmission: Function and Dysfunction," A. C. Cuello, Ed., pp. 77–83, Elsevier, Amsterdam, 1993a.

Clarke, P. B. S.: Nicotine dependence—mechanisms and therapeutic strategies. Biochem. Soc. Symp. *59*: 83–95, 1993b.

Clarke, P. B. S.: Nicotinic receptors and cholinergic neurotransmission in the central nervous system. Ann. N. Y. Acad. Sci., *757*: 73–83, 1995.

Claye, L. H. and Soliman, K. F.: Effects of acute cocaine administration on the cholinergic enzyme levels of specific brain regions in the rat. Pharmacol. *40*: 218–223, 1990.

Clement, J. G.: Recovery from soman-induced hypothermia is due to an increase in acetylcholine activity but not new protein synthesis. Neurotoxicology *14*: 411–416, 1993.

Clement, J. G. and Copeman, H. T.: Soman and Sarin induce a long-lasting naloxone reversible analgesia in mice. Life Sci. *34*: 1415–1422, 1984.

Clements, J. R., Toth, D. D., Highfield, D. A. and Grant, S. J.: Glutamate-like immunoreactivity is present within cholinergic neurons of the laterodorsal tegemental and pedunculopontine nuclei. In: "The Basal Forebrain," T. C. Napier, P. W. Kalivas and I. Hanin, Eds., pp. 127–142, Plenum Press, New York, 1991.

Cobb, S.: "Emotions and Clinical Medicine," Norton Publishers, New York, 1950.

Cobb, S. R., Bulters, D. O., Suchak, S., Riedel, G., Morris, R. G. and Davies, D. H.: Activation of nicotinic acetylcholine receptors patterns network activity in the rodent hippocampus. J. Physiol. *518*, pt. 1: 131–140, 1999a.

Cobb, S. R., Manuel, N. A., Marton, R. A., Gill, C. H., Collingridge, C. C. and Davies, C. H.: Regulation of depolarizing GABA (A) receptor mediated potentials by synaptic activation of GABA (B) autoreceptors in the rat hippocampus. Neuropharmacology *38*: 1723–1732, 1999b.

Cole, A. E. and Nicoll, R. A.: Acetylcholine mediates a slow synaptic potential in hippocampal pyramidal cells. Science *24*: 299–301, 1983.

Cole, A. E. and Nicoll, R. A.: Characterization of a slow cholinergic postsynaptic potential recorded *in vitro* form rat hippocampal pyramidal cells. J. Physiol. *352*: 173–188, 1984a.

Cole, A. E. and Nicoll, R. A.: The pharmacology of cholinergic excitatory responses in hippocampal pyramidal cells. Brain Res. *305*: 283–290, 1984b.

Collier, B.: Activity related changes of cholinergic transmission. In: "Ninth International Symposium on Cholinergic Mechanisms," J. Klein and K. Loeffelholz, Ed., pp. 243–249, Elsevier, Amsterdam, 1996.

Collins, A. C.: Interaction of ethanol and nicotine at the receptor level. Recent Dev. Alcohol. *8*: 221–231, 1990.

Collins, A. C., Burch, J. B., de Fiebre, C. M. and Marks, M. J.: Tolerance and cross-tolerance between ethanol and nicotine. Pharmacol. Biochem. Behav. *29*: 365–373, 1988.

Colom, L. V., Nassif-Cauderella, S., Dickson, C. T., Smythe, J. M. and Bland, B. H.: *In vivo* intrahippocampal microinfusion of carbachol and bicucullin induces theta-like oscillations in the septally deafferented hippocampus. Hippocampus *1*: 381–390, 1991.

Colombari, D. S., de Almeida Haibara, A. S., Camargo, L. A., Saad, W. A., Renzi, A., De Luca, L. A. and Menam, J. V.: Role of the medial septal area on the cardiovascular fluid and electrolytic responses to angiotensin II and cholinergic activation into the subfornical organ in rat. Brain Res. Bull. *33*: 249–254, 1994.

Colwell, C. S., Kaufman, C. M. and Menaker, M.: Phase-shifting mechanisms in the mammalian circadian system: new light on the carbachol paradox. J. Neurosci. *13*: 1454–1459, 1993.

Colzato, L. S., Fafgioli, S., Erasmus, V. and Hommel, B.: Caffeine, but not nicotine, enhances visual feature binding. Eur. J. Neurosci. *21*: 591–595, 2005.

Compton, D. R. and Johnson, K. M.: Effects of acute and chronic clozapine and haloperidol on *in vitro* release of acetylcholine from striatum and nucleus accumbens. J. Pharmacol. Exp. Ther. *248*: 521–539, 1989.

Comroe, J. H., Jr.: The effects of direct chemical and electric stimulation on the respiratory center in the cat. Am. J. Physiol. *139*: 490–498, 1943.

Conan Doyle, A.: The Adventure of the Final Problem. Strand Mag., pp. 536–557, December 1893.

Consolo, S., Wang, J. X., Fiorentini, F., Vezzani, A. and Ladinsky, H.: *In vitro* and *in vivo* studies on the regulation of cholinergic neurotransmission in striatum, hippocampus and cortex of aged rats. Brain Res. *374*: 212–218, 1986.

Consolo, S., Bertorelli, R., Forloni, G. L. and Butcher, L. L.: Cholinergic neurons of the pontomesencephalic tegmentum release acetylcholine in the basal nuclear complex of freely moving rats. Neuroscience *37*: 717–723, 1990.

Consolo, S., Bertorelli, R., Girotti, P., La Porta, C., Bartfai, T., Parenti, M. and Zambelli, M.: Pertussis toxin-sensitive G-protein mediates galanin's inhibition of scopolamine-evoked acetylcholine release *in vivo* and carbachol-stimulated phosphoinositide turnover in rat ventral hippocampus. Neurosci. Lett. *126*: 29–32, 1991.

Consolo, S., Girotti, P., Russi, G. and Di Chiara, G.: Endogenous dopamine facilitates striatal *in vivo* acetylcholine release by acting on D1 receptors localized in the striatum. J. Neurochem. *59*: 1555–1557, 1992.

Cook, L., Nickolson, V. J., Steinfels, G. F., Rohrbach, K. W. and Denoble, V. J.: Cognition enhancement by the acetylcholine releaser DuP996. Drug Dev. Res. *19*: 301–314, 1990.

Cordeau, J. D. and Mancia, M.: Evidence for the existence of an electroencephalographic synchronization mechanism originating in the lower brain stem. Electroencephalogr. Clin. Neurophysiol. *11*: 551–561, 1959.

Corrigal, W. A., Coen, K. M., Adamson, K. L. and Chow, B. L.: Manipulation of mu-opioid and nicotinic cholinergic receptors in the pontine tegmental region alter cocaine self-administration in rats. Psychopharmacology *145*: 412–417, 1999.

Corringer, P. J., Sallette, J. and Changeux, J. P.: Nicotine enhances intracellular nicotinic receptor maturation: A novel mechanism of neural plasticity? J. Physiol. Paris. *99*: 162–171, 2006.

Cosi, C. and Wood, P. L.: Lack of GABAergic modulation of acetylcholine turnover in the rat thalamus. Neurosci. Lett. *87*: 293–296, 1988.

Cosi, C., Altar, A. C. and Wood, P. L.: Effect of cholecystokinin on acetylcholine turnover and dopamine release in the rat striatum and cortex. Eur. J. Pharmacol. *165*: 209–214, 1989.

Costa, E.: Studies of the interdependence of neurotransmitter systems by microbiochemical approaches. In: "Neurotransmitters," P. Simon, Ed., Vol. 2, pp. 97–107, Advances in Pharmacol. and Ther., Pergamon Press, Oxford, 1979.

Costa, E.: Allosteric modulatory centers of transmitter amino acid receptors. Neuropsychopharmacology *2*: 167–174, 1989.

Costa, E.: The Di Mascio Lecture. The allosteric modulation of GABA receptors. Seventeen years of research. Neuropsychopharmacology *4*: 225–235, 1991.

Costa, E., Forchetti, C. M., Guidotti, A. and Wise, B. C.: Multiple signals participating in GABA receptor modulation. In: "Dale's Principle and Communication Between Neurones," N. N. Osborne, Ed., pp. 161–177, Pergamon Press, Oxford, 1983.

Costall, B., Naylor, R. G. and Oley, J. E.: Catalepsy and circling behavior after intracerebral injection of neuroleptic, cholinergic and anticholinergic agents into the caudate-putamen, globus pallidus and substantia nigra of rat brain. Neuropharmacology *11*: 645–664, 1972.

Costall, B., Kelly, M. E., Naylor, R. J. and Onaivi, E. S.: The actions of nicotine and cocaine in a mouse model of anxiety. Pharmacol. Biochem. Behav. *33*: 197–203, 1989.

Coury, J. N.: Neural correlates of food and water intake in the rat. Science *156*: 1763–1765, 1967.

Cox, B. and Tha, S. J.: The antinociception activities of oxotremorine, physostigmine and dyflos. J. Pharm. Pharmacol. *24*: 547–551, 1972.

Crawford, J. M. and Curtis, D. R.: Pharmacological studies on feline Betz cells. J. Physiol. *186*: 121–138, 1966.

Crawford, J. M., Curtis, D. R., Voorhoeve, P. E. and Wilson, V. J.: Acetylcholine sensitivity of cerebellar neurones in the cat. J. Physiol. *186*: 139–165, 1966.

Crawley, J. N.: Neuropharmacologic specificity of a simple animal model for the behavioral actions of benzodiazepines. Pharmacol. Biochem. Behav. *13*: 695–699, 1981.

Crawley, J. N.: Coexistence of neuropeptides and "classical" neurotransmitters. Functional interactions between galanin and acetylcholine. Ann. N. Y. Acad. Sci. *579*: 233–245, 1990.

Crayton, J. W., Konopka, L. M. and Karczmar, A. G.: Combination donepezil and dihydroepiandrosterone therapy for post-stroke rehabilitation. In: "Cholinergic Mechanisms: Function and Dysfunction," I. Silmand, H. Soreq, L. Anglister, D. Michaelson, and A. Fisher, Eds., pp. 543–545, Taylor and Francis, New York, 2004.

Cressant, A. and Granon, S.: Definition of a new maze paradigm for the study of spatial behavior in rats. Brain Res. Brain Res. Protoc. *12*: 116–134, 2003.

Crews, D., Godwin, J., Hartman, V., Grammar, M., Prediger, E. A. and Shepperd, R.: Intrahypothalamic implantation of progesterone in castrated male whiptail lizards (Cnemidophorus inornatus) elicits courtship and copulatory behavior and affects androgen receptor- and progesterone-receptor-mRNA expression in the brain. J. Neurosci. *16*: 7347–7352, 1996.

Crick, F.: "The Astonishing Hypothesis," Scribner, New York, 1994.

Crochet, S. and Sakai, K.: Effects of microdialysis application of monoamines on the EEG and behavioral states in the cat mesopontine tegmentum. Eur. J. Neurosci. *11*: 3738–3752, 1999.

Crook, T., Ferris, S. and Bartus, R., Eds.: "Assessment in Geriatric Psychopharmacology," Mark Pawley Associates, New Canaan, Conn., 1983.

Crossland, J.: Neurohumoral substances and drug abstinence syndromes. In: "Advances in Neuropsychopharmacology," O. Vinar, Z. Votava, and P. B. Bradley, Eds., pp. 497–507, 1971.

Cruikshank, J. W., Brudzynski, S. M. and McLachlan, R. S.: Involvement of M1 muscarinic receptors in the initiation of cholinergically induced epileptic seizures in the rat brain. Brain Res. *643*: 125–129, 1994.

Cuadra, G. and Giacobini, E.: Co-administration of cholinesterase inhibitors and idazoxan: effects of neurotransmitters in rat cortex *in vivo*. J. Pharmacol. Exp. Ther. *273*: 230–240, 1995.

Cuello, A. C.: Trophic responses of forebrain cholinergic neurons: a discussion. In: "Cholinergic Function and Dysfunction," A. C. Cuello, Ed., pp. 265–277, Elsevier, Amsterdam, 1993.

Curran, H. V., Pooviboonsuk, P., Dalton, J. A. and Lader, M. H.: Differentiating the effects of centrally acting drugs on arousal and memory: event-related potential study of scopolamine, lorazepam and diphenhydramine. Psychopharmacology (Berl.) *135*: 27–36, 1998.

Curtis, D. R.: Actions of drugs on single neurons in the spinal cord and thalamus. Br. Med. Bull. *21*: 5–9, 1965.

Curtis, D. R and Andersen, P.: Acetylcholine: a central transmitter? Nature *195*: 1105–1106, 1962.

Curtis, D. R. and Crawford, J. M.: Central synaptic transmission—microelectrophoretic studies. Annu. Rev. Pharmacol. *9*: 209–250, 1969.

Curtis, D. R. and Eccles, R. M.: The effect of diffusional barriers upon pharmacology of cells within the central nervous system. J. Physiol. *141*: 446–463, 1958.

Curtis, D. R. and Phillis, J. W.: The action of procaine and atropine on spinal neurones. J. Physiol. *153*: 17–34, 1960.

Curtis, D. R. and Ryall, R. W.: The excitation of Renshaw cells by cholinomimetics. Exp. Brain Res. *2*: 49–65, 1966.

Cushing, H.: Reaction to posterior pituitary extract (pituitrin) when introduced into cerebral ventricles. Proc. Natl. Acad. Sci. U.S.A. *17*: 163–170, 1931.

Cytronberg, S. S.: Some remarks about the influence of the vegetative system upon the gastric functions. Polska Gazeta Lekarska *9*: 12–24, 1927 (title translated from Polish).

Dacey, R. G., Jr. and Bassett, J. E.: Cholinergic vasodilation of intracerebral arterioles in rat. Am. J. Physiol. *253*: H1253–H1260, 1993.

Dale, H. H.: Nomenclature of fibres in the autonomic system and their effects. J. Physiol. *82*: 10, 1933.

Dale, H. H.: Pharmacology and nerve endings. Proc. Soc. Med. *28*: 319–322, 1935.

Dale, H. H.: "Transmission of Effects from Nerve Endings," Oxford University Press, Oxford, 1952.

Dale, H. H.: The beginnings and the prospects of neurohumoral transmission. Pharmacol. Rev. *6*: 7–13, 1954.

Damasio, A. R.: "Descartes' Error," G. P. Putnam's Sons, New York, 1994.

Damasio, A. R. and Damasio, H.: Cortical systems underlying knowledge retrieval: evidence from human lesion studies. In: "Exploring Brain Functions: Models in Neuroscience," T. A. Poggio and D. A. Glaser, Eds., pp. 233–248, Wiley and Sons, New York, 1993.

Dampney, R. A. L.: The subretrofacial vasomotor nucleus: anatomical, chemical and pharmacological properties and role in cardiovascular regulation. Prog. Neurobiol. *42*: 197–227, 1994.

Dani, J. A. and Zhou, F. M.: Synaptic plasticity and nicotine addiction. Neuron *31*: 349–352, 2001.

Daniels, J. C. and Spehlman, R.: The convulsant effects of topically applied atropine. Electroencephalogr. Clin. Neurophysiol. *34*: 83–87, 1973.

Datta, S., Calvo, J. M., Quatrocchi, J. J. and Hobson, J. A.: Long-term enhancement of REM sleep following cholinergic stimulation. Neuroreport *2*: 619–622, 1991.

Dauphin, F., Lacombe, P., Sercombe, R., Hamel, E. and Seylaz, J.: Hypercapnia and stimulation of the substantia innominata increase rat frontal cortical blood flow by different cholinergic mechanisms. Brain Res. *553*: 75–83, 1991.

Davidson, J. M.: Neurohormonal basis of male sexual behavior. Int. Rev. Physiol. Reprod. Physiol. II *13*: 225–254, 1977.

Davidson, R. J., Putnam, K. M. and Larson, C. L.: Dysfunction in the neural circuitry of emotion regulation—a possible prelude to violence. Science *289*: 591–594, 2000.

Davis, B. M. and Davis, K. L.: Acetylcholine and anterior pituitary hormone secretion. In: "Brain Acetylcholine and Neuropsychiatric Disease," K. L. Davis and P. A. Berger, Eds., pp. 445–460, Plenum Press, New York, 1979.

Davis, M., Nonneman, A. and Glickman, S. E.: Chemical stimulation of the midbrain reticular system. Psychol. Rep. *20*: 507–509, 1967.

Dawkins, R., Keller, S. L. and Sewell, W. F.: Pharmacology of acetylcholine-mediated cell signaling in the lateral line organ following efferent stimulation. J. Neurophysiol., *93*: 2541–2551, 2005.

Dean, R. L. and Bartus, R. T.: Behavioral models of aging in nonhuman primates. In: "Psychopharmacology of Aging Nervous System," L. L. Iversen, S. D. Iversen, and S. H. Snyder, Eds., pp. 325–424, Handbk. Psychopharmacol., vol. 20, Plenum Press, New York, 1988.

De Angelis, L. and Furlan, C.: The effects of ascorbic acid and oxiracetam on scopolamine-induced amnesia in a habituation test in aged mice. Neurobiol. Learn. Mem. *64*: 119–124, 1995.

Decker, M. W., Curzon, P., Holladay, M. W., Nikkel, A. L., Scott Bitner, R., Bannon, A. W., Donnelly-Roberts, D. L., Putfarcken, P. S., Kuntzweiler, C. A., Briggs, C. A., Williams, M. and Arneric, S. P.: The role of nicotinic acetylcholine receptors in antinociception: Effects of ABT-594. In: "From *Torpedo* Electric Organ to Human Brain: Fundamental and Applied Aspects," J. Massoulié, Ed., pp. 221–224, Elsevier, Amsterdam, 1998.

Deeke, L., Grotzinger, B. and Kornhuber, H. H.: Voluntary finger movements in man: cerebral potentials and theory. Biol. Cybern. *23*: 99–119, 1976.

De Fiebre, C. M. and Collins, A. C.: A comparison of the development of tolerance to ethanol and cross-tolerance to nicotine after chronic ethanol treatment in long- and short-sleep mice. J. Pharmacol. Exp. Ther. *266*: 1398–1406, 1993.

De Filippi, G., Baldwinson, T. and Sher, E.: Nicotinic receptor modulation of neurotransmitter release in the cerebellum. Prog. Brain Res. *148*: 307–320, 2004.

De Groat, W. C.: Nervous control of the urinary bladder of the cat. Brain Res. *87*: 201–211, 1975.

De Groat, W. C. and Kawatani, M.: Neural control of the urinary bladder: possible relationship between peptodergic inhibitory mechanisms and detrusor instability. Neurourol. Urodyn. *4*: 285–300, 1985.

De Groat, W. C. and Steers, W. D.: Autonomic regulation of the urinary bladder and sexual organs. In:

"Central Regulation of Autonomic Functions," A. D. Loewy and K. M. Spyer, Eds., pp. 310–333, Oxford University Press, New York, 1990.

Dehaene, S. and Changeux, J.-P.: A hierarchical neuronal network for planning behavior. Proc. Natl. Acad. Sci. U.S.A. *94*: 13293–13294, 1997.

Deitrich, R. A. and Erwin, V. G.: Involvement of biogenic amines metabolism in ethanol addiction. Fed. Proc. *34*: 1962–1968, 1975.

Del Castillo, J. and Katz, B.: The effect of magnesium on activity of motor nerve endings. J. Physiol. *124*: 553–559, 1954.

Delga, J.: Les anticholinesterasiques organophosphores. Actualites Pharmacol. *10*: 47–48, 1957.

Delgado, J. M. R.: "Physical Control of the Mind," Harper and Row, New York, 1969.

Delgado, J. M. R.: Communication with the conscious brain by means of electrical and chemical probes. In: "Factors in Depression," N. S. Kline, Ed., pp. 251–268, Raven Press, New York, 1974.

Delgado-Garcia, J. M. R.: Why move eyes if we can move the head? Brain Rse. Bull. *52*: 475–482, 2000.

Delgado-Garcia, J. M. R.: Es la cara el spejo del alma? Elementos *44*: 3–9, 2002.

Delgado Garcia, J. M.: Neurociencia para pobres. Claves de Razon Practica *102*: 43–47, 2004.

De Luca, L. A., Saad, W. A., Camargo, L. A., Renzi, A. and Menani, J. V.: Carbachol injection into the medial preoptic area induces natruresis, kaliuresis and antidiuresis in rats. Neurosci. Lett. *105*: 333–339, 1989.

Demarco, G. J., Baghdoyan, H. A. and Lydic, R.: Carbachol in the pontine reticular formation of C57BL/6J mouse decreases acetylcholine release in prefrontal cortex. Neuroscience *123*: 17–29, 2004.

Dennet, D. C.: "Consciousness Explained," Little, Brown, Boston, 1991.

Dennet, D. C.: "Darwin's Dangerous Idea," Simon and Schuster, New York, 1995.

De Oliveira, L., Hoffman, A. and Menesca de Oliveira, L.: The lateral hypothalamus in the modulation of tonic immobility in guinea pigs. Neuroreport *8*: 3489–3493, 1997.

Descartes, R.: "Les Passions de l'Ame," Henry Le Gras, Paris, 1649a.

Descartes, R.: "Discours de la Métode pour bien Conduire la Raison et Chercher la Verité dans les Sciences," 1649b (English translation: Dawsons of Pall Mall, London, 1966).

Desmedt, J. E. and Tomberg, C.: Transient phase-locking of 40 Hz electrical oscillations in prefrontal and parietal human cortex reflects the process of conscious somatic perception. Neurosci. Lett. *168*: 126–129, 1994.

Desmedt, J. E. and Tomberg, C.: Consciousness. In: "Perspectives of Event-Related Potentials Research," G. Klarames, M. Molnar, V. Csepi, I. Czigler and J. E. Desmedt, Eds., pp. 227–234, Elsevier, Amsterdam, 1995.

Detwiler, D. J., Eisenach, J. C., Tong, C. and Jackson, C.: A cholinergic interaction in alpha2 adrenoceptor-mediated antinociception in sheep. J. Pharmacol. Exp. Ther. *265*: 536–542, 1993.

Deutsch, J. A.: Substrates of learning and memory. Dis. Nerv. Syst. *27*: 20–24, 1966.

Deutsch, J. A.: The cholinergic synapse and the site of memory. Science *174*: 788–794, 1971.

Deutsch, J. A.: Physiology of acetylcholine in learning and memory. In: "Choline and Lecithin in Brain Disorders," A. Barbeau, J. H. Growdon, and R. J. Wurtman, Eds., pp. 343–350, Nutrition and the Brain, vol. 5, Raven Press, New York, 1979.

Deutsch, J. A. and Leibowitz, S. F.: Amnesia or reversal of forgetting by anticholinesterase, depending simply on time of injection. Science *153*: 1017–1018, 1966.

Deutsch, J. A., Hamburg, M. D. and Dahl, H.: Anticholinesterase-induced amnesia and its temporal aspects. Science *151*: 221–223, 1966.

Devinsky, O., Kernan, J. and Bear, D. M.: Aggressive behavior following exposure to cholinesterase inhibitors. J. Neuropsychiatry Clin. Neurosci. *4*: 189–194, 1992.

Dewey, S. L., Smith, G. S., Logan, J. and Brodie, J. D.: Modulation of central cholinergic activity by GABA and serotonin: PET studies with 11C-benztropine in primates. Neuropsychopharmacology *8*: 371–376, 1993.

Dewey, W. L.: Cannabinoid pharmacology. Pharmacol. Rev. *38*: 151–178, 1986.

Dewey, W. L., Cocolas, G., Daves, E. and Harris, L. S.: Stereospecificity of intraventricularly administered acetylmethylcholine antinociception. Life Sci. *17*: 9–10, 1975.

De Wied, D.: Effects of peptide hormones on behavior. In: "Frontiers in Neuroendocrinology 1969," W. F. Ganong and L. Martini, Eds., pp. 97–140, Oxford University Press, New York, 1969.

Dhume, R. A., Shamayev, N. N., Zhuravlev, B. V., Sudakov, K. V. and Sharma, K. N.: Effects of medial septal stimulation and its blockade with atropine on hippocampal rhythmical slow activity in free-moving rabbits. Indian J. Med. Res. *92*: 353–357, 1990.

Diamond, M. C., Johnson, R. E., Protti, A. M., Ott, C. and Kajisa, L.: Plasticity in the 909-day-old male rat cerebral cortex. Exp. Neurol. *87*: 3309–3117, 1985.

Dickinson-Anson, H., Winkler, J., Fisher, L. J., Song, H. J., Poo, M. and Gage, F. H.: Acetylcholine-secreting cells improve age-induced memory deficits. Mol. Ther. *8*: 51–61, 2003.

Dickson, C. T., Trepel, C. and Bland, B. H.: Extrinsic modulation of theta field activity in the entorhinal cortex of the anesthetized rat. Hippocampus *4*: 37–51, 1994.

Dierks, T., Frolich, L., Ihl, R. and Maurer, R.: Event related potentials and psychopharmacology. Cholinergic innervation of P300. Pharmacopsychiatry *27*: 72–74, 1994.

Dikshit, B. B.: Action of acetylcholine on the brain and its occurrence therein. J. Physiol. (Lond.) *80*: 409–421, 1934.

Ding, Y. S., Molina, P. E., Fowler, J. S., Logan, J., Volkov, D. D., Kuhar, M. J. and Carroll, F. I.: Comparative studies of epibatidine derivatives [18F] NFEP and [18F] N-methyl-NFEP: kinetics, nicotine effect. Nucl. Med. Biol. *26*: 139–148, 1999.

Dingledine, R. and Kelly, J. S.: Brain stem stimulation and acetylcholine-evoked inhibitory neurones in feline nucleus reticularis thalami. J. Physiol. *171*: 135–154, 1977.

Dinh, X. A. T.: La fonction respiratoire du transplant cardiopulmonaire. Rev. Mal. Respir. *7*: 291–299, 1990.

Dirksen, R. and Nijhuis, G. M.: The relevance of cholinergic transmission at the spinal level to opiate effectiveness. Eur. J. Pharmacol. *91*: 215–221, 1983.

Dirnhuber, P. and Cullumbine, H.: The effect of anticholinesterase agents on the rat's blood pressure. Br. J. Pharmacol. *10*: 12–15, 1955.

Dmitrieva, N. A., Lindstrom, J. M. and Keyser, K. T.: The relationship between GABA-containing cells and the cholinergic circuitry in the rabbit retina. Vis. Neurosci. *18*: 93–100, 2001.

Dodd, J., Dingledine, R. and Kelly, J. S.: The excitatory action of acetylcholine on hippocampal neurones of the guinea pig and rat maintained *in vitro*. Brain Res. *207*: 109–117, 1981.

Dohanich, G. P., Witcher, J. A., Weaver, D. R. and Clemens, L. G.: Alteration of muscarinic binding in specific brain areas following estrogen treatment. Brain Res. *241*: 347–350, 1982.

Dohanich, G. P., Ross, S. M., Francis, T. J., Fader, A. J., Wee, B. E., Brazier, M. M. and Menard, C. S.: Effects of a muscarinic antagonist on various components of female sexual behavior in the rat. Behav. Neurosci. *107*: 819–826, 1993.

Dollard, J., Miller, N. E., Doob, L. W., Mowrer, O. H. and Sears, R. R., Eds.: "Frustration and Aggression," Yale University Press, New Haven, Conn., 1939.

Dominguez, R., Riboni, L., Zipitria, D. and Revilla, R.: Is there a cholinergic circadan rhythm throughout the estrous cycle related to ovulation in the rat? J. Endocrinol. *95*: 175–180, 1982.

Domino, E. F., Simonic, A. and Krutak-Krol, H.: Muscarinic and nicotinic mechanisms of seizures induced by cholinesterase inhibitors: observations on diazepam and midazolam as antagonists. In: "Neurobiology of Acetylcholine: A Symposium in Honor of A. G. Karczmar," N. J. Dun and R. L. Perlman, Eds., pp. 473–487, Plenum Press, New York, 1987.

Dorner, G.: Hormones and sexual differentiation of the brain. In: "Sex, Hormones and Behavior," CIBA Foundation Symposia, pp. 81–101, Excerpta Medica, Amsterdam, 1979.

Dougherty, K. D. and Milner, T. A.: Cholinergic septal afferent terminals preferentially contact neuropeptide Y-containing interneurons compared to paravalbumin-containing interneurons in the rat dentate gyrus. J. Neurosci. *19*: 10140–10152, 1999.

Dougherty, K. D., Turchin, P. L. and Walsh, T. J.: Septocingulate and septohippocampal cholinergic pathways: involvement in working/episodic memory. Brain Res. *810*: 59–71, 1998.

Drachman, D. A.: Central cholinergic system and memory. In: "Psychopharmacology: A Generation of Progress," M. A. Lipton, A. DiMascio and K. F. Killam, Eds., pp. 651–662, Raven Press, New York, 1978.

Drachman, D. A. and Leavitt, J.: Human memory and the cholinergic system—a relationship to aging? Arch. Neurol. Psychiatry *30*: 113–121, 1974.

Drachman, D. A. and Sahakian, B. J.: Effects of cholinergic agents on human learning and memory. In: "Choline and Lecithin in Brain Disorders," A. Barbeau, J. H. Growdon and R. J. Wurtman, Eds., pp. 351–366, Nutrition and the Brain, vol. 20, Raven Press, New York, 1979.

Dray, A. and Straughan, D. W.: Synaptic mechanisms in the substantia nigra. J. Pharm. Pharmacol. *28*: 400–405, 1976.

Dray, A., Oakley, N. R. and Simonds, M. A.: Rotational behavior following inhibition of GABA metabolism unilaterally in the rat substantia nigra. J. Pharm. Pharmacol. *27*: 627–629, 1975.

Drescher, D. G., Upadhyay, S., Wilcox, E. and Fex, J.: Analysis of muscarinic receptor subtypes in the mouse cochlea by means of polymerase chain reaction. J. Neurochem. *59*: 765–767, 1992.

Drescher, D. G., Ramakrishman, D. A., Prescher, M. J., Chun, W., Wang, X., Meyers, S. F., Green, G. E., Sadrazodl, K., Raradaghdy, A. A., Coopat, N., Karpenko, A. N., Khan, K. M. and Hatfield, J. S.: Cloning and characterization of alpha9 Subunits of nicotinic acetylcholine receptor expressed by saccular hair cells of the rainbow trout (oncorphynehus mykiss). Neurosci. *127*: 737–752, 2004.

Drukarch, B., Schepens, E., Schoffelmeer, A. N. and Stoof, J. C.: Stimulation of D-2 dopamine receptors decreases evoked *in vitro* release of [3H] acetylcholine from rat neostriatum: role of K+ and Ca++. J. Neurochem. *52*: 1680–1685, 1989.

Duan, W. Z. and Zhang, J. T.: Stimulation of central cholinergic neurons by (-) clausenamide *in vitro*. Acta Pharmacol. Sin. *19*: 332–336, 1998.

Dudkin, K. N., Kruchinin, V. K. and Chueva, I. V.: Participation of cholinergic structures of the prefrontal and inferotemporal cortex in the processes of visual recognition in monkeys. Neurosci. Behav. Physiol. *24*: 341–350, 1994.

Dudkin, S. M., Polev, P. V. and Soldatov, N. M.: Calcium entry blockers and oxiracetam have opposite effects on the density of dihydropyridine receptors in rat cerebral cortex. Brain Res. *525*: 319–321, 1990.

Dunnet, S. B. and Fibiger, H. C.: Role of forebrain cholinergic systems in learning and memory: relevance to the cognitive deficits of aging and Alzheimer's dementia. In: "Cholinergic Function and Dysfunction," A. C. Cuello, Ed., pp. 413–420, Elsevier, Amsterdam, 1993.

Dunwiddie, T. V.: Interactions between the effects of adenosine and calcium on synaptic responses in rat hippocampus *in vitro*. J. Physiol. (Lond.) *350*: 545–559, 1984.

Durieux, M. E.: Muscarinic signalling in the central nervous system. Anesthesiology *84*: 173–189, 1996.

Dursun, S. M. and Kutcher, S.: Smoking, nicotine dependence and psychiatric disorders: evidence for therapeutic role, controversies and implications for future research. Med. Hypotheses *52*: 101–109, 1999.

Dussor, G. O., Helesic, G., Hargreaves, K. M. and Flores, C. M.: Cholinergic modulation of nociceptive responses *in vivo* and neuropeptide release *in vitro* at the level of the primary sensory neuron. Pain *107*: 22–32, 2004.

Dutar, P. and Nicoll, R. A.: Pre- and postsynaptic GABA-B receptors in the hippocampus have different pharmacological properties. Neuron *1*: 585–591, 1988.

Dutar, P., Lamour, Y. and Joubert, A.: Proprietés des cellules thalamocorticales identifiées par stimulation antidromique dans les noyaux thalamiques ventrolateral et ventrobasal chez le Rat. C. R. Acad. Sci., Ser. 3, *300*: 217–220, 1985.

Dutar, P., Lamour, Y. and Nicoll, R. A.: Galanin blocks the slow cholinergic EPSP in CA1 neurons from ventral hippocampus. Eur. J. Pharmacol. *164*: 355–360, 1989.

Ebel, A., Mack, G., Stefanovic, V. and Mandel, P.: Activity of choline acetyltransferase and acetylcholinesterase in the amygdala of spontaneous mouse-killer rats and in rats after olfactory lobe removal. Brain Res. *57*: 248–251, 1973.

Ebner, F. F. and Armstrong-James, M. A.: Intracortical processes regulating the integration of sensory information. Prog. Brain Res. *86*: 129–141, 1990.

Eccles, J. C.: Synaptic and neuro-muscular transmission. Physiol. Rev. *17*: 538–555, 1937.

Eccles, J. C.: An electrical hypothesis of synaptic and neuromuscular transmission. Ann. N. Y. Acad. Sci. *47*: 429–455, 1946.

Eccles, J. C.: Spinal neurones: Synaptic connections in relation to chemical transmitters and pharmacological responses. Proc. First Int. Pharmacol. Meeting *8*: 157–182, Pergamon Press, Oxford, 1962.

Eccles, J. C.: "The Physiology of Synapses," Springer-Verlag, New York, 1964.

Eccles, J. C.: Historical introduction. Fed. Proc. *28*: 90–94, 1969.

Eccles, J. C.: "The Human Mystery," Springer-Verlag, Berlin, 1979.

Eccles, J. C.: Mental summation: the timing of voluntary intentions by cortical activity. Behav. Brain Sci. *8*: 542–543, 1985.

Eccles, J. C.: The story of the Renshaw cell. In: "Neurobiology of Acetylcholine: A Symposium in Honor of Alexander G. Karczmar," N. J. Dun and R. L. Perlman, Eds., pp. 189–194, Plenum Press, New York, 1987.

Eccles, J. C.: "How the Self Controls Its Brain," Springer-Verlag, Berlin, 1994.

Eccles, J. C. and Gibson, W. C. "Sherrington: His Life and Thought," Spring-Verlag, Berlin, 1979.

Eccles, J. C., Fatt, P. and Koketsu, K.: Cholinergic and inhibitory synapses in a central nervous pathway. Aust. J. Sci. *16*: 50–54, 1953.

Eccles, J. C., Fatt, P. and Koketsu, K.: Cholinergic and inhibitory synapses in a pathway from motor-axon collaterals to motoneurones. J. Physiol. (Lond.) *126*: 524–562, 1954.

Eccles, J. C., Ito, M. and Szentagothai, J., Eds.: "The Cerebellum as a Neuronal Machine," Springer-Verlag, Heidelberg, 1967.

Edwards, C.: The effects of innervation on the properties of acetylcholine receptors in muscle. Neuroscience *4*: 565–584, 1979.

Edwardson, J. A. and Bennet, G. W.: Modulation of corticotrophin-releasing factor release from hypothalamic synaptosomes. Nature *250*: 425–427, 1974.

Egan, T. M. and North, R. A.: Acetylcholine hyperpolarizes cortical neurons by acting on an M2 muscarinic receptor. Nature *319*: 405–407, 1986.

Eggleston, C.: The antagonism between atropin and certain central emetics. J. Pharmacol. Exp. Ther. *9*: 11–25, 1916.

Egorov, A. V., Glovelli, T. and Muller, W.: Muscarinic control of dendritic excitability and Ca(2+) signaling in CA1 pyramidal neurons in rat hippocampal slice. J. Neurophysiol. *82*: 1909–1915, 1999.

Egorov, A. V., Hamam, B. N., Fransen, E., Hasselmo, M. E. and Alonso, A. A.: Graded persistent activity in entorhinal cortex neurons. Nature *420*: 133–134, 2002.

Ehlert, F. J., Roeske, W. R. and Yamamura, H. I.: Molecular biology, pharmacology, and brain distribution of subtypes of the muscarinic receptor. In: "Psychopharmacology: The Fourth Generation of Progress," F. E. Bloom and D. J. Kupfer, Eds., pp. 111–124, Raven Press, New York, 1995.

Elazar, P. and Paz, M.: Catalepsy induced by carbachol microinjected into the pontine reticular formation of rats. Neurosci. Lett. *115*: 226–230, 1990.

Elazar, Z., Peleg, N., Paz, M. and Ring, G.: The striatal dopaminergic catalepsy mechanism is not necessary for expression of pontine catalepsy produced by carbachol injections into the pontine reticular formation. Naunyn Schmiedebergs Arch. Pharmacol. *352*: 187–193, 1995.

Elliott, K. A. C., Swank, A. L. and Henderson, N.: Effects of anesthetics and convulsants on the acetylcholine content of the brain. Am. J. Physiol. *162*: 469–474, 1950.

Escorihuela, R. M., Tobena, A. and Fernandez, T.: Environmental enrichment reverses the detrimental action of early inconsistent stimulation and increases the beneficial effects of postnatal handling on shuttlebox learning in adult rats. Behav. Brain. Res. *61*: 169–173, 1995.

Eskesen, K. and Sheardown, M. J.: Muscarinic suppression of synaptic transmission in CA1 neurones in hippocampus slice is not mediated by M1 receptors. Abstr. Soc. Neurosci. *19*(1): 460, 1993.

Estrada, C. and Krause, D. N.: Muscarinic cholinergic receptor site in cerebral blood vessels. J. Pharmacol. Exp. Ther. *221*: 85–90, 1982.

Euvrard, C., Javoy, F., Herbert, A. and Glowinsky, J.: Effect of quipazine, a serotonin-like drug, on striatal cholinergic interneurones. Eur. J. Pharmacol. *41*: 281–289, 1977.

Euvrard, C., Premont, J., Oberlander, C., Boissier, J. and Bockaert, J.: Is dopamine-sensitive adenylate cyclase involved in regulating the activity of striatal cholinergic neurons? Naunyn Schmiedebergs Arch. Pharmacol. *309*: 241–245, 1979.

Evans, S. M., Kushner, P. D. and Meyer, E. M.: Actions of a monoclonal antibody Tor 23 on rat brain presynaptic cholinergic processes. Neurochem. Res. *18*: 339–344, 1993.

Everett, G. M.: Tremor produced by drugs. Nature *177*: 1238, 1956.

Everitt, B. J. and Robbins, T. W.: Central cholinergic system and cognition. Annu. Rev. Psychol. *48*: 649–684, 1997.

Everitt, B. J., Dickinson, A. and Robbins, T. W.: The neuropsychological basis of addictive behavior. Brain Res. Brain Res. Rev. *36*: 129–138, 2001.

Fagen, Z. M., Mansvelder, H. I., Keath, J. R. and McGehee, D. S.: Short- and long-term modulation of synaptic inputs in brain reward areas by nicotine. Ann. N. Y. Acad. Sci. *1003*: 185–195, 2003.

Fairchild, M. D., Jenden, D. J. and Mickey, M. R.: An application of long-term frequency analysis in measuring drug-specific alterations in the EEG of the cat. Electroencephalogr. Clin. Neurophysiol. *38*: 337–348, 1975.

Farr, S. A., Uezu, K., Flood, F. F. and Morley, J. E.: Septohippocampal drug interactions in post-trial memory processing. Brain Res. *847*: 221–230, 1999.

Feferman, S.: "In the Light of Logic," Oxford University Press, Oxford, 1998.

Feferman, S.: To the editors. "New York Review of Books," *51*: 61, 2004.

Feiro, O. and Gould, T. J.: The interactive effects of nicotinic and muscarinic cholinergic inhibition on fear conditioning in young and aged cC57BL76 mice. Pharmacol. Biochem. Behav. *80*: 251–262, 2005.

Feldberg, W: Present views on the mode of action of acetylcholine in the central nervous system. Physiol. Rev. *25*: 596–642, 1945.

Feldberg, W.: Discussion of the paper by M. R. Rosenzweig, D. Krech and E. L. Bennett. In: "CIBA Foundation Symposium on the Neurological Basis of Behavior," G. E. W. Wolstenholme and C. M. O'Connor, Eds., p. 358, Little, Brown, Boston, 1958.

Feldberg, W. and Myers, R. D.: Effects on temperature of amines injected into the cerebral ventricles. J. Physiol. (Lond.) *173*: 226–237, 1964.

Feldberg, W. and Schriever, H.: The acetylcholine content of the cerebro-spinal fluid of dogs. J. Physiol. *86*: 277–284, 1936.

Feldberg, W. and Sherwood, S. L.: Injections of drugs into the lateral ventricle of the cat. J. Physiol. (Lond.) *123*: 148–167, 1954.

Feldberg, W. and Vogt, M.: Acetylcholine synthesis in different regions of the central nervous system. J. Physiol. *107*: 372–381, 1948.

Felder, C. C., Porter, A. C., Skillman, T. L., Zhang, L., Bymaster, F. P., Nathanson, N. M., Hamilton, S. E., Gomeza, J., Wess, J. and McKinzie, D. L.: Elucidating the role of muscarinic receptors in psychosis. Life Sci. *68*: 22–23, 2001.

Feldman, J. L.: Neurophysiology of breathing in mammals. In: "Intrinsic Regulatory systems of the Brain," V. B. Mountcastle, F. E. Bloom, and R. Geiger, Eds., pp. 463–524, Handbk. Physiol., vol. 4, Physiological Society, Bethesda, Md., 1986.

Fenu, S., Acquas, E. and Di Chiara G.: Role of striatal acetylcholine on dopamine D1 receptor agonist-induced turning behavior in 6-hydroxydopamine lesioned rats: a microdialysis-behavioral study. Neurol. Sci. *22*: 63–64, 2001.

Ferman, T. J., Boeve, B. F., Smith, G. E., Silber, M. B., Kokmen, E., Petersen, R. C. and Ivnik, R. J.: REM sleep behavior disorder and dementia. Cognitive differences when compared with AD. Neurology *52*: 951–957, 1999.

Ferreira, M., Singh, A., Dretchen, K. L., Kellar, K. J. and Gillis, R. A.: Brainstem nicotinic receptor subtypes that influence intragastric and arterial blood pressures. J. Pharmacol. Exp. Ther. *294*: 230–238, 2000.

Ferreira, M. Jr., Sahibzada, N., Shi, M., Panico, W., Niedringhaus, M., Wasserman, A., Kellar, K. J., Verbalis, J. and Gillis, R. A.: CNS site of action and brainstem circuitry responsible for the intravenous effects of nicotine on gastric tone. J. Neurosci. *22*: 2764–2779, 2002.

Ferster, C. B.: Intermittent reinforcement of matching-to-sample in the pigeon. J. Exp. Anal. Behav. *3*: 259–272, 1960.

Ffrench-Mullen, J. M. H., Plata-Salaman, C. R., Buckley, N. J. and Danks, P.: Muscarinic modulation by a G-protein alpha-subunit of delayed rectifier K+ current in rat ventromedial hypothalamic neurones. J. Physiol. *474*: 21–26, 1994.

Fibiger, H. C.: Cholinergic mechanisms in learning, memory and dementia: a review of recent evidence. Trends Neurosci. *14*: 220–223, 1991.

Filliat, P., Baubichon, D., Burckhart, M. F., Pernot-Marino, I., Foquin, A., Masqueliez, C., Perrichon, C., Carpentier, P. and Lallelment, G.: Memory impairment after soman intoxication in rat: correlation with central neuropathology. Improvement with anticholinergic and antiglutamatergic therapeutics. Neurotoxicology *20*: 535–549, 1999.

Finkelstein, Y. and Hod, I.: Activation of the cholinergic synapse in the septo-hippocampus under stress conditions. Harefuah *117*: 154–157, 1989 (in Hebrew).

Finocchiaro, L. M., Goldstein, D. J., Finkielman, S., Nahmod, V. E. and Pirola, C. J.: Muscarinic effects on the hydroxy- and methoxyindole pathways in the rat pineal gland. J. Endocrinol. *123*: 205–211, 1989.

Fiorindo, R., Justo G., Motta, M., Simonovic, I. and Martini, L.: Acetylcholine and the secretion of pituitary gonadotrophins. In: "Hypothalamic Hormones," M. Motta, S. Grignane and L. Martini, Eds., pp. Academic Press, London, 1975.

Fisher, A. E. and Coury, J. N.: Cholinergic tracing of a central neural circuit underlying the thirst drive. Science *138*: 691–693, 1962.

Fisher, J. L., Pidoplichko, V. I. and Dani, J. A.: Nicotine modifies the activity of ventral tegmental area dopaminergic neurons and GABAergic neurons. In: "From *Torpedo* Electric Organ to Human Brain: Fundamental and Applied Aspects," J. Massoulié, Ed., pp. 209–213, Elsevier, Amsterdam, 1998.

Fisher, Y., Gahwiler, B. H. and Thompson, S. M.: Activation of intrinsic hippocampal theta oscillations by acetylcholine in rat septo-hippocampal cocultures. J. Physiol. *519*, pt. 2: 405–413, 1999.

Fleming, N. W., Henderson, T. R. and Dretchen, K. L.: Mechanisms of respiratory failure produced by neostigmine and diisopropyl fluorophosphate. Eur. J. Pharmacol. *195*: 85–91, 1991.

Flodmark, S. and Wramner, T.: The analgetic action of morphine, eserine and prostigmine studied by a modified Hardy-Wolff-Goodell method. Acta Physiol. Scand. *9*: 88–96, 1945.

Fiorindo, R., Justo, G., Motta, M., Simonovic, I. and Martini, L.: Acetylcholine and the secretion of pituitary gonadotrophins. In: "Hypothalamic Hormones," M. Motta, P. G. Crosignani and L. Martini, Eds., pp. 183–194, Academic Press, London, 1975.

Foldes, F. F., Ludvig, N., Nagashima, H. and Vizi, E. S.: The influence of aminopyridines on Ca++-dependent evoked release of acetylcholine from rat cortex slices. Neurochem. Res. *13*: 761–764, 1989.

Fonck, C., Nashmi, R., Deshpande, P., Damaj, M. J., Marks, M. J., Riedel, A., Schwarz, J., Collins, A. C. and Lester, H. A.: Increased sensitivity to agonist-induced seizures, straub tail, and hippocampal theta rhyhm in knock-in mice carrying hypersensitive alpha 4 nicotinic receptors. J. Neurosci. *23*: 2582–2590, 2003.

Fontana, F.: "Dei moti dell'iride," Lucci Publs., Lucca, 1765.

Fordyce, D. E. and Wehner, J. M.: Physical activity enhances spatial learning performance with an associated alteration in hippocampal protein kinase C activity in C57BL/6 and DBA/2 mice. Brain Res. *619*: 111–119, 1993.

Forghani, R. and Krnjevic, K.: Adenosine antagonists have differential effects on induction of long-term potentiation in hippocampal slices. Hippocampus *5*: 71–77, 1995.

Fosbraey, P., Wetheral, J. R. and French, M. C.: Neurotransmitter changes in guinea-pig brain regions following soman intoxication. J. Neurochem. *54*: 72–79, 1990.

Fotiou, F., Goulas, A., Fountoulakis, K., Koutlas, E., Hamlatzis, P. and Papakostopoulos, D.: Int. J. Psychophysiol. *29*: 303–310, 1998.

Fournier, G. N., Materi, L. M., Semba, K. and Rasmusson, D. D.: Cortical acetylcholine release and electroencephalogram activation evoked by ionotropic glutamate receptor agonists in the rat forebrain. Neuroscience *123*: 785–792, 2004.

Fowler, C. J.: Honest psychiatric bias. TIPS *9*: 422, 1988.

Fox, S., Krnjevic, K., Morris, M. E., Puil, E. and Werman, R.: Action of baclofen on mammalian synaptic transmission. Neuroscience *3*: 495–515, 1978.

Fraser, T. R.: On the characters, actions and therapeutic uses of the ordeal bean of Calabar (Physostigma venenosum Balfour). Edinburgh Med. J. *9*: 36–56, 1863.

Fraser, T. R.: On atropia as a physiological antidoter to the poisonous effects of physostigma. Practitioner *4*: 65–72, 1870.

Fraser, T. R.: An experimental research on the antagonism between the actions of physostigma and atropia. Trans. R. Soc. Edinb. *26*: 529–713, 1872.

Freund, G. and Ballinger, W. E., Jr.: Loss of cholinergic muscarinic receptors in the frontal cortex of alcohol abusers. Alcoholism *12*: 630–638, 1988.

Freund, G. and Ballinger, W. E., Jr.: Loss of muscarinic cholinergic receptors for the temporal cortex of alcohol abusers. Metab. Brain Dis. *4*: 121–141, 1989a.

Freund, G. and Ballinger, W. E., Jr.: Loss of muscarinic and benzodiazepine neuroreceptors from hippocampus of alcohol abusers. Alcohol *6*: 23–31, 1989b.

Freund, G. and Ballinger, W. E.: Loss of synaptic receptors can precede morphologic changes induced by alcoholism. Alcohol Alcohol. Supplt. *1*: 385–391, 1991.

Freund, H.: Üer das Kochsalzfeiber. Arch. Exp. Pathol. Pharmakol. *65*: 225–238, 1911.

Friedman, A.: Circumstances influencing Otto Loewi's discovery of chemical transmission in the nervous system. Pflugers Arch. Eur. J. Physiol. *25*: 85–86, 1971.

Friedman, A. H. and Marchese, A. L.: Reversal by bicuculline of the simulated state of light adaptation produced by GABA in the dark-adapted isolated eye perfused through its ophthalmic artery. Pharmacologist *15*: 162, 1973.

Friedman, A. H. and Walker, C. A.: Circadian rhythms in rat midbrain and caudate nucleus biogenic amines. J. Physiol. (Lond.) *197*: 77–85, 1968.

Friedman, A. H. and Walker, C. A.: The acute toxicity of drugs acting at cholinoceptive sites and twenty-four hour rhythms in brain acetylcholine. Arch. Toxikol. (Berl.) *29*: 39–49, 1972.

Frodl-Bauch, T., Bottlender, R. and Hegerl, U.: Neurochemical substrates and neuroanatomical generators of the event-related P300. Neuropsychobiol. *40*: 86–94, 1999.

Fromm, E.: "The Anatomy of Human Destructiveness," Holt, Rinehart and Winston, New York, 1973.

Frotscher, M., Soriano, E. and Leranth, C.: Cholinergic and GABAergic neurotransmission in the fascia dentata: electron microscopic immunocytochemical studies in rodents and primates. Epilepsy. Res. Suppl. *7*: 65–78, 1992.

Frye, G. D., Taylor, L., Grover, C. A., Fincher, A. S. and Griffith, W. H.: Acute alcohol dependence or long-term alcohol treatment and abstinence do not reduce hippocampal responses to carbachol. Alcohol *12*: 29–36, 1995.

Fuji, S., Ji, Z., Morita, N. and Sumikawa, K.: Acute and chronic nicotine exposure differentially facilitate the induction of LTP. Brain Res. *846*: 137–143, 1999.

Fujimoto, J. M. and Rady, J. J.: Intracerebroventricular physostigmine-induced analgesia: enhancement by naloxone, beta-funaltrexamine and nor-binattrophimine and antagonism by dynorphin A (1–17). J. Pharmacol. Exp. Ther. *251*: 1045–1052, 1989.

Fukuda, K., Kubo, T., Maeda, A., Akiba, I., Bujo, H., Nakai, J., Mishima, M., Higashida H., Neher, E. and Marty, A.: Selective effector coupling of muscarinic acetylcholine receptors subtypes. Trends Pharmacol. Sci. *1989*, Supplt., pp. 4–10, 1989.

Funderburk, W. H. and Case, T. J.: Effect of parasympathetic drugs on the conditioned response. J. Neurophysiol. *10*: 179–187, 1947.

Fung, Y. K. and Lao, Y. S.: Receptor mechanisms of nicotine-induced locomotor hyperactivity in chronic nicotine-treated rats. Eur. J. Pharmacol. *152*: 263–271, 1988.

Fuxe, K., Anderson, K., Hokfelt, T., Mutt, V., Ferland, L., Agnati, L. F., Ganten, D., Said, S., Eneroth, P. and Gustafson, J. A.: Localization and possible function of peptidergic neurons and their interactions with central catecholamine neurons, and central actions of gut hormones. Fed. Proc. *38*: 2333–2340, 1979.

Fuxe, K., Rosen, L., Lippoldt, A., Andbjer, B., Hasselrot, U., Finnman, U.-B. and Agnati, L. F.: Continuous infusion of nicotine and CAT. Clin. Invest. *72*: 262–268, 1994.

Gadea-Ciria, M., Stadler, H. and Lloyd, K.: Acetylcholine release within the cat strialum during the sleep-wakefulness cycle. Nature *243*: 518–519, 1973.

Gahring, L. C., Persiyanov, K., Dunn, D., Weiss, R., Meyer, E. L. and Rogers, S. W.: Mouse strain specific nicotinic acetylcholine receptor expression by inhibitory interneurons and astrocytes in the dorsal hippocampus. J. Comp. Neurol. *468*: 334–346, 2004.

Gais, S. and Born, J.: Low acetylcholine during slow-wave sleep is critical for declarative memory consolidation. Proc. Natl. Acad. Sci. U.S.A.: *101*: 1140–1145, 2004.

Galey, D., Destrade, C. and Jaffard, R.: Relationships between septo-hippocampal cholinergic activation and the improvement of long-term retention produced by medial septal electric stimulation in two inbred strains of mice. Behav. Brain Res. *60*: 183–189, 1994.

Gant, D. B., Eldefrawi, M. F. and Eldefrawi, A. T.: Action of organophosphates on GABA A receptor and voltage dependent chloride channels. Fundam. Appl. Toxicol. *9*: 698–704, 1987.

Gantt, W. H. and Freile, M.: Effect of adrenalin and acetylcholine on excitation, inhibition and neurosis. Trans. Am. Neurol. Assoc. *70*: 189–181, 1944.

Garzon, M., Vaughan, R. A., Uhl, G. R., Kuhar, M. J. and Pickel, V. M.: Cholinergic axon terminals in the ventral tegmental area target a subpopulation of neurons expressing low levels of the dopamine transporter. J. Comp. Neurol. *410*: 197–210, 1999.

Gastaut, H. and Fischer-Williams, M.: The physiopathology of epileptic seizures. In: "Handbook of Physiology," Section I: "Neurophysiology," J. Field, Ed., pp. 329–363, American Physiological Society, Washington, D.C., 1959.

Gatto, G. J., Bohme, G. A., Caldwell, W. S., Letchworth, S. R., Traina, V. M., Obinu, M. C., Laville, M., Reibaud, M., Pradier, L., Dunbar, G. and Bencherif, M.: TC-1734: an orally active neuronal nicotinic acetylcholine receptor modulator with antidepressant, neuroprotective and long-lasting cognitive effects. CNS Drug Rev. *10*: 147–166, 2004.

Gauss, C. J.: Gebürten in künstlichen Dämmerschlaf. Arch. Gynaek. *78*: 579–631, 1906.

Geller, I., Hartmann, R. J., Leal, B. Z., Haines, R. J. and Gause, E. M.: Effects of the irreversible acetylcholinesterase inhibitor Soman on match-to-sample behavior of the juvenile baboon. Proc. West. Pharmacol. Soc. *27*: 217–221, 1984.

Geller, I., Hartmann, R. J. and Gause, E. M.: Effects of subchronic administration of Soman on acquisition of avoidance-escape behavior by laboratory rats. Pharmacol. Biochem. Behav. *23*: 225–230, 1985.

Gellhorn, E.: "Autonomic Imbalance and the Hypothalamus," University of Minnesota Press, Minneapolis, 1944a.

Gellhorn, E.: The role of the autonomic nervous system in problems of rehabilitation. Fed. Proc. *3*: 266–271, 1944b.

Genuth, S. M.: The hypothalamus and pituitary gland. In: "Physiology," Third Edition, R. M. Berne and M. N. Levy, Eds., pp. 897–931, Mosby Year Book, St. Louis, Mo., 1992.

George, R., Haslett, W. L. and Jenden, D. J.: A cholinergic mechanism in the brain stem reticular formation: Induction of paradoxical sleep. Int. J. Neuropharmacol. *11*: 465–477, 1964.

George, R., Haslett, W. L. and Jenden, D. J.: The production of tremor by cholinergic drugs: Central sites of action. Int. J. Neuropharmacol. *5*: 27–37, 1966.

Gershon, S., Ferris, S. H., Kennedy, J. S., Kurtz, N. M., Overall, J. E., Pollock, B. G., Reisberg, B. and Whitehouse, P. J.: Methods for the evaluation of pharmacologic agents in the treatment of cognitive and other deficits in dementia. In: "Clinical Evaluation of Psychotropic Drugs: Principles and Guidelines," R. F. Prien and D. S. Robinson, Eds., pp. 467–499, Raven Press, New York, 1994.

Geula, C. and Mesulam, M.-M.: Cholinergic systems and related neuropathological predilection patterns in Alzheimer disease. In: "Alzheimer Disease," R. D. Terry, R. Katzman, and K. L. Blick, Eds., pp. 263–291, Raven Press, New York, 1994.

Ghelardini, C., Malmberg-Aiello, P., Giotti, A., Malcangio, M. and Bartolini, A.: Investigation into atropine-induced analgesia. Br. J. Pharmacol. *101*: 49–54, 1990.

Ghelardini, C., Galeotti, N., Gualtieri, F., Scapecchi, S. and Bartolini, A.: 3-alpha-tropanyl 2-(4Cl-phenoxy) butyrate (SM21): a review of the pharmacological profile of a novel enhancer of cholinergic transmission. CNS Drug Rev. *3*: 346–362, 1997.

Ghiani, C. A., Dazzi, L., Maciocco, E., Flore, G., Maira, G. and Biggio, G.: Antagonism by abercarnil of enhanced acetylcholine release in the rat brain during anticipation but not consumption of food. Pharmacol. Biochem. Behav. *59*: 657–662, 1998.

Giarman, N. J. and Pepeu, G.: Drug-induced changes in brain acetylcholine. Br. J. Pharmacol. *19*: 226–234, 1962.

Gillberg, P.-G., Gordh, T., Jr., Hartvig, P., Jansson, I., Petterson, J. and Post, C.: Characterization of the antinociception induced by intrathecally-administered carbachol. Pharmacol. Toxicol. *64*: 340–343, 1989.

Gillberg, P.-G., Askmark, H. and Aquilonius, S.-M.: Spinal cholinergic mechanisms. In: "Cholinergic Neurotransmission: Functional and Clinical Aspects," S.-M. Aquilonius and P.-G. Gillberg, Eds., pp. 361–370, Prog. Brain Res., vol. 98, Elsevier, Amsterdam, 1990.

Gillberg, P.-G., Sundquist, S. and Nilvebrant, L.: Comparison of the *in vitro* and *in vivo* profiles of tolterodine with those of subtype-selective muscarinic receptor antagonists. Eur. J. Pharmacol. *349*: 285–292, 1998.

Gillin, J. C., Sitaram, N., Mendelson, W. B. and Wyatt, R. J.: Physostigmine alters onset but not duration of REM sleep in man. Psychopharmacol. *58*: 111–114, 1978.

Gillin, J. C., Sitaram, N. and Wehr, T.: Sleep and affective illness. In: "Neurobiology of Mood Disorders," R. M. Post and J. Ballenger, Eds., pp. 157–189, William and Wilkins, Baltimore, 1984.

Gillin, C. J., Sutton, L., Ruiz, C., Hirsch, S., Warmann, C. and Shiromani, P. J.: Dose dependent inhibition of REM sleep in normal volunteers by bipiriden, a muscarinic antagonist. Biol. Psychiatry *30*: 151–156, 1991.

Gillin, J. C., Salin-Pascual, R., Velazquez-Moctezuma, J., Shiromani, P. and Zoltoski, R.: Cholinergic receptor subtypes and REM sleep in animals and normal controls. In: "Cholinergic Function and Dysfunction," A. C. Cuello, Ed., pp. 379–387, Elsevier, Amsterdam, 1993.

Giovanelli, L., Vannucchi, M. G. and Pepeu, G.: Effect of phosphatidylserine on acetylcholine release and calcium uptake in aged rats. In: "Cellular and Molecular Basis of Cholinergic Function," M. J. Dowdall and J. N. Hawthorne, Eds., pp. 689–695, Ellis Horwood, Chichester, 1987.

Giovannini, M. G., Casamenti, F., Nistri, A., Paoli, F. and Pepeu, G.: Effect of TRH on acetylcholine release from different brain areas investigated by microdialysis. Br. J. Pharmacol. *102*: 363–368, 1990.

Giovannini, M. G., Mutolo, D., Bianchi, L., Michelasso, A. and Pepeu, G.: NMDA receptor antagonists decrease GABA outflow from the septum and increase acetylcholine outflow from the hippocampus: a microdialysis study. J. Neurosci. *14*: 1358–1365, 1995.

Giovannini, M. G., Bartolini, L., Kopf, S. R. and Pepeu, G.: Acetylcholine release from the frontal cortex during exploratory activity. Brain Res. *784*: 218–227, 1998a.

Giovannini, M. G., Rakovska, A., Della Corte, L., Bianchi, L. and Pepeu, G.: Activation of non-NMDA receptors stimulate acetylcholine and GABA release from dorsal hippocampus: a microdialysis study in the rat. Neurosci. Lett. *243*: 152–156, 1998b.

Giovannini, M. G., Pazzagli, M., Malberg-Aiello, P., Della Corte, L., Ballini, C. and Pepeu, G.: An inhibitory avoidance test increases acetylcholine and GABA release and activates ERKs in the medial prefrontal cortex of the rat. Soc. Neurosci. Program No. 836,7, Washington, D.C., 2003.

Giurgea, C.: Vers une pharmacologie de l'activité integrative du cerveau—tentative du concept nootrope en psychopharmacologie. Actualités Pharmacol. *25*: 115–156, 1972.

Giurgea, C. and Salama, M.: Nootropic drugs. Prog. Neuropsychopharmacol. Biol. Psychiatry *1*: 235–247, 1977.

Glassman, A. H. and Koob, G. F. Neuropharmacology: psychoactive smoke. Nature *379*: 677–678, 1996.

Glenn, J. F., Hinman, D. J. and McMaster, S. B.: Electroencephalographic correlates of nerve agent poisoning. In: "Neurobiology of Acetylcholine: A Symposium in Honor of A. G. Karczmar," N. J. Dun and R. L. Perlman, Eds., pp. 503–534, Plenum Press, New York, 1987.

Glisson, S. N., Karczmar, A. G. and Barnes, L.: Cholinergic effects on adrenergic neurotransmitters in rabbit brain parts. Neuropharmacol. *11*: 465–477, 1972.

Glisson, S. N., Karczmar, A. G. and Barnes, L.: Effects of diisopropyl phosphofluoridate on acetylcholine, cholinesterase and catecholamines of several parts of rabbit brain. Neuropharmacol. *13*: 623–632, 1974.

Gloveli, T., Egorov, A. V., Schmitz, D., Heinemann, U. and Muller, W.: Carbachol-induced changes in excitability and [Ca2+]i signalling in projection cells of medial entorhinal cortex layers II and III. Eur. J. Neurosci. *11*: 3626–3636, 1999.

Gloveli, T., Egorov, A. V., Schmitz, D., Heinemann, U. and Muller, W.: Carbachol-induced changes in excitability and [Ca++] signalling in projection cells of medial entorhinal cortex layers II and III. Eur. J. Neurosci. *11*: 3626–3626, 1999.

Glowinski, J. and Karczmar, A. G.: Concluding remarks: interdependence of neurotransmitter systems in the CNS. In: "Neurotransmitters," P. Simon, Ed., pp. 145–148, vol. 2, Adv. Pharmacol. Ther., Proc. 7th Internatl. Cong. Pharmacol., Pergamon Press, Oxford, 1979.

Glusman, M.: The hypothalamic "savage" syndrome. Res. Publ. Assoc. Res. Nerv. Meut. Dis. *52*: 52–92, 1974.

Goedel, K.: Über formal unenscheidbare Säetze per Principia Mathematica und verwandter Systeme. Monatshefte f. Mathem. u. Phys. *3*: 173–198, 1931.

Gold, P. E.: Acetylcholine modulation of neuronal systems involved in learning and memory. Neurobiol. Learn. Mem. *80*: 194–210, 2003.

Goldberg, S. R. and Stolerman, I. P.: "Behavioral Analysis of Drug Dependence," Academic Press, Orlando, 1986.

Goldstein, D. B.: Physical dependence on alcohol in mice. Fed. Proc. *34*: 1953–1961, 1975.

Goleblewski, H., Eckersdorf, B. and Konopacki, J.: Muscarinic (M1) mediation of hippocampal spontaneous theta rhythm in freely moving cats. Neuroreport *4*: 1323–1326, 1993.

Gomeza, J., Zhang, L., Kostenis, E., Felder, C., Bymaster, F., Brodkin, J., Shannon, H., Xia, B., Deng, C. and Wess, J.: Enhancement of D1 dopamine receptor-mediated locomotor stimulation in M (4) muscarinic receptor knockout mice. Proc. Natl. Acad. Sci. U.S.A. *96*: 12222–12223, 1999.

Gonzales, R. A., Roper, L. C. and Westbrook, S. L.: Cholinergic modulation of N-methyl-D-aspartate-evoked [3H] norepinephrine release from rat cortical slices. J. Pharmacol. Exp. Ther. *264*: 382–288, 1993.

Gordh, T., Jr., Jansson, I., Hartvig, P., Gillberg, P.-G. and Post, C.: Interaction between noradrenergic and cholinergic mechanisms involved in spinal nociceptive processing. Acta Anesthesiol. Scand. *33*: 39–47, 1989.

Gotoh, M., Iguchi, A., Yatomi, A., Uemura, K., Miura, H., Futenma, A., Kato, K. and Sakamoto, N.: Vagally mediated insulin secretion by stimulation of brain cholinergic neurons with neostigmine in bilateral adrenolectomized rats. Brain Res. *493*: 97–102, 1989.

Gotti, C. and Clementi, F.: Neuronal nicotinic receptotrs: from structure to pathology. Prog. Neurobiol. *74*: 363–396, 2004.

Gould, S. J: "The Structure of Evolutionary Theory," Belknap Press of Harvard University Press, Cambridge, Mass., 2002.

Gower, A. J.: Effects of acetylcholine agonists and antagonists on yawning and analgesia in the rat. Eur. J. Pharmacol. *139*: 79–89, 1987.

Grassi, E., Giovanelli, A., Fucile, S., Cerasi, S., Mattei, E., Di Modugno, F. and Eusebi, F.: Acetylcholine-evoked current induces Ca++ release from InsP3-sensitive stores in mouse myotubes. In: "Cholinergic Neurotransmission: Function and Dysfunction," A. C. Cuello, Ed., Elsevier, Amsterdam, Abstracts, 1993.

Graybiel, A. M. and Illing, R. B.: Enkephalin-positive and acetylcholinesterase-positive patch systems in the superior colliculus have matching distributions but distinct developmental histories. J. Comp. Neurol. *340*: 297–310, 1994.

Greco, M. A., McCarley, R. W. and Shiromani, P. I.: Choline acetyltransferase expression during periods of behavioral activity and across natural sleep-wake states in the basal forebrain. Neuroscience *93*: 1369–1374, 1999.

Green, J. D. and Arduini, A.: Hippocampal electric activity in arousal. J. Neurophysiol. *17*: 533–557, 1954.

Green, P. G. and Kitchen, I.: Antinociception opioids and cholinergic system. Prog. Neurobiol. *26*: 119–146, 1986.

Greengard, P.: Neuronal phosphoproteins. Neurobiol. *1*: 81–119, 1987.

Greene, B.: "The Fabric of the cosmos," Alfred Knopf, New York, 2004.

Grenhoff, J. and Svensson, T. H.: Pharmacology of nicotine. Br. J. Addict. *84*: 477–492, 1989.

Grenningloh, G., Rienitz, A., Schmitt, B., Methfessel, C., Zensen, M., Begreuther, R., Gundelfinger, E. D. and Betz, H.: The strychnine-binding subunit of the glycine receptors. Nature *328*: 215–228, 1987.

Grillner, P., Bonci, A., Stevenson, T. H., Bernardi, G. and Mercuri, N. B.: Presynaptic muscarinic (M3) receptors reduce excitatory transmission in dopamine neurons of the rat mesencephalon. Neuroscience *91*: 557–565, 1999.

Grossman, A. and Costa, A.: The regulation of hypothalamic CRH: impact of *in vitro* studies on the central control of the stress response. Funct. Neurol. *8*: 325–334, 1993.

Grossman, S. P.: Eating and drinking elicited by direct adrenergic or cholinergic stimulation of the hypothalamus. Science *132*: 301–302, 1960.

Grossman, S. P.: Direct adrenergic and cholinergic stimulation of hypothalamic mechanisms. Am. J. Physiol. *202*: 872–882, 1962.

Grossman, S. P.: Behavioral effects of direct chemical stimulation of central nervous structures. Int. J. Neuropharmacol. *8*: 45–58, 1964.

Gruslin, E., Descombes, S. and Psarropoulou, C.: Epileptiform activity generated by endogenous acetylcholine during blockade of GABAergic inhibition in immature and adult rat hippocampus. Brain Res. *835*: 290–297, 1999.

Gu, Q.: Contribution of acetylcholine to visual cortex plasticity. Neurobiol. Learn. Mem. *80*: 291–301, 2003.

Gunn, J. A.: Relations between chemical constitution, pharmacological actions, and therapeutic uses, in harmine group of alkaloids. Arch. Int. Pharmacodyn. *50*: 379–396, 1935.

Guth, P. S. and Amaro, J.: A possible cholinergic link in olivo-cochlear inhibition. Int. J. Neuropharmacol. *3*: 45–58, 1969.

Guth, P. S., Dunn, A., Kronomer, K. and Norris, C. H.: The cholinergic pharmacology of the frog saccule. Hear. Res. *75*: 225–232, 1994.

Hagan, J. J. and Morris, G. M.: The cholinergic hypotheses of memory: a review of animal experiments. In: "Psychopharmacology of the Aging Nervous System," L. L. Iversen and S. H. Snyder, Eds., Handbook of Psychopharmacology, vol. 20, pp. 217–323, Plenum Press, New York, 1988.

Hajdu, J., Szentirmai, E., Obal, F., Jr. and Krueger, J. M.: Different brain structures mediate drinking and sleep suppression elicited by the somatostatin analog, octreotide, in rats. Brain Res. *994*: 115–123, 2003.

Haley, T. J. and McCormick, W. G.: Pharmacological effects produced by intracerebral injection of drugs in the conscious mouse. Br. J. Pharmacol. Chemother. *12*: 12–15, 1957.

Hall, G. H. and Myers, R. D.: Temperature changes produced by nicotine injected into the hypothalamus of conscious monkey. Brain Res. *37*: 241–251, 1977.

Halliwell, J. V.: M current in human neocortical neurons. Neurosci. Lett. *67*: 1–6, 1986.

Halliwell, J. V.: Physiological mechanisms of cholinergic action in the hippocampus. In: "Cholinergic Neurotransmission: Functional and Clinical Aspects," S.-M. Aquilonius and P.-G. Gillberg, Eds., pp. 255–272, Elsevier, Amsterdam, 1990.

Halliwell, J. V. and Adams, P. R.: Voltage-clamp analysis of muscarinic excitation in hippocampal neurones. Brain Res. *250*: 71–92, 1982.

Hamburg, M. D.: Retrograde amnesia produced by intraperitoneal injection of physostigmine. Science *156*: 973–974, 1967.

Hamburg, M. D. and Fulton, D. R.: Influence of recall on an anticholinesterase induced retrograde amnesia. Physiol. Behav. *9*: 409–418, 1972.

Hameroff, S. R., Kaszniak, A. W. and Scott, A. C. (Eds.): "Toward a Science of Consciousness," MIT Press, Cambridge, Mass., 1996.

Hamilton, S. E., Schlador, M. L., McKinnon, L. A., Chmelar, R. S. and Nathanson, N. M.: Molecular mechanisms for the regulation of the expression and function of muscarinic acetylcholine receptors. In: "From *Torpedo* Electric Organ to Human Brain: Fundamental and Applied Aspects," J. Massoulié, Ed., pp. 275–278, Elsevier, Amsterdam, 1998.

Hamm, R. J., O'Dell, D. M., Pike, B. R. and Lyeth, B. G.: Cognitive impairment following traumatic brain injury: the effect of pre- and post-injury administration of scopolamine and MK-801. Brain Res. Cogn. Brain Res. *1*: 223–226, 1993.

Hammer, R., Berrie, C. P., Birdsall, N. J. M., Burgen, A. S. V. and Hulme, E. C.: Pirenzepine distinguishes between different subclasses of muscarinic receptors. Nature *283*: 90–92, 1980.

Hampf, G. and Bowsher, D.: Distigmine and amitryptyline in the treatment of chronic pain. Anesth. Prog. *38*: 58–62, 1989.

Hanin, I.: AF64A-induced cholinergic hypofunction. In: "Cholinergic Neurotransmission: Functional and Clinical Aspects," S.-M. Aquilonius and P.-G. Gillberg, Eds., pp. 289–299, Elsevier, Amsterdam, 1990.

Hanin, I.: Cholinergic toxins and Alzheimer's disease. In: "Neurotoxins and Neurodegenerative Disease," J. W. Langston and A. B. Young, Eds., pp. 63–70, Proc. N. Y. Acad. Sci., vol. 648, New York, 1992.

Hanin, I. and Costa, E.: Approaches used to estimate brain acetylcholine turnover rate *in vivo*; effects of drugs on brain acetylcholine turnover rates. In: "Biology of Cholinergic Function," A. M. Goldberg and I. Hanin, Eds., pp. 355–377, Raven Press, New York, 1976.

Harder, J. A., Baker, H. F. and Ridley, R. M.: The role of the central cholinergic projections in cognition: implications of the effects of scopolamine on discrimination learning by monkeys. Brain Res. Bull. *45*: 319–326, 1998.

Harnack, E. and Witkowski, L.: Death by physostigmine is due to paralysis of respiratory center and spasm of bronchial muscles. Arch. Exp. Pathol. Pharmakol. *5*: 401–454, 1876.

Haroutunian, V., Santucci, A. C. and Davis, K. L.: Implications of multiple transmitter system lesions for cholinomimetic therapy in Alzheimer's disease. In: "Cholinergic Neurotransmission: Functional and Clinical Aspects," S.-M. Aquilonius and P.-G. Gillberg, Eds., pp. 333–346, Elsevier, Amsterdam, 1990.

Harris, S.: "The End of Faith", W. W. Norton and Company, New York, 2004.

Harris, L. S., Dewey, W. L., Howes, J. F., Kennedy, J. S. and Pars, H.: Narcotic-antagonist analgesics: interaction with cholinergic system. J. Pharmacol. Exp. Ther. *169*: 17–22, 1969.

Harrison, P. S. and Henderson, Z.: Quantitative evidence for increase in galanin-immunoreactive terminals in the hippocampal formation following entorhinal cortex lesions in the adult rat. Neurosci. Lett. *266*: 41–44, 1999.

Hartvig, P., Gillberg, P.-G., Gordth, T. and Post, C.: Cholinergic mechanisms in pain and analgesia. TIPS, Dec. Supplt.: 75–79, 1989.

Harvey, J. A., Gormezano, I. and Cool-Hauser, V. A.: Effects of scopolamine and methylscopolamine on classical conditioning of the rabbit nictitating membrane response. J. Pharmacol. Exp. Ther. *225*: 42–49, 1983.

Haselton, J. R., Padrid, P. A. and Kaufman, M. P.: Activation of neurons in the rostral ventrolateral medulla increases bronchomotor tone in dogs. J. Appl. Physiol. *71*: 210–216, 1991.

Hass, W. and Ellis, J.: Muscarinic receptor subtypes in the central nervous system. Int. Rev. Neurobiol. *26*: 151–199, 1985.

Hasselmo, M. E.: Neuromodulation and cortical function: modeling the physiological basis of behavior. Behav. Brain Res. *61*: 1–27, 1995.

Hasselmo, M. E.: The role of hippocampal regions CA3 and CA1 in matching entorhinal input with retrieval of associations beteen objects and context: theoretical comment on Lee et al. (2005). Behav. Neurosci. *119*: 145–1563, 2005.

Hasselmo, M. E. and Barkai, E.: Cholinergic modulation of activity-dependent synaptic plasticity in pyriform cortex and associative memory function in a network biophysical stimulation. J. Neurosci. *15*: 6592–6604, 1995.

Hasselmo, M. E. and Schnell, E.: Laminar selectivity of the cholinergic suppression of synaptic transmission in rat hippocampal CA1 region; computational modeling and brain slice physiology. J. Neurosci. *14*: 3898–3914, 1994.

Hasselmo, M. E., Schnell, E. and Barkai, E.: Dynamics of learning and recall at excitatory recurrent synapses and cholinergic modulation in rat hippocampal region CA3. J. Neurosci. *15*: 5244–5262, 1995.

Hasuo, H., Gallagher, J. P. and Shinnick-Gallagher, P.: Disinhibition in the rat septum mediated by M1 muscarinic receptors. Brain Res. *438*: 323–327, 1988.

Hatcher, R.-A. and Weiss, S.: Studies in vomiting. J. Pharmacol. Exp. Ther. *22*: 139–193, 1923.

Haubrich, R. D. and Reid, W. D.: Effects of pilocarpine or arecoline on acetylcholine levels and serotonin turnover in rat brain. J. Pharmacol. Exp. Ther. *181*: 19–27, 1972.

Haxhiu, M. A., van Lunteren, E., Deal, E. C. and Cherniak, N. S.: Role of the ventral surface of medulla in the generation of Mayer waves. Am. J. Physiol. *257*: 804–809, 1989.

Haxhiu, M. A., Mack, S. O., Wilson, C. G., Feng, P. and Strohl, K. P.: Sleep networks and the autonomic and physiologic connections with respiratory control. Front. Biosci. *8*: 946–962, 2003.

Hayward, J. H.: Scientific method and validation. In: " Gentle Bridges," J. H. Hayward and F. J. Varela, Eds., pp. 6–24, Shambala Publications, Boston, 1992.

He, D. Z. and Dallos, P.: Development of acetylcholine-induced responses in neonatal gerbil outer hair cells. J. Neurophysiol. *81*: 1162–1170, 1999.

He, L. F., Xu, S. F. and Jiang, C. C.: The role of caudate nucleus in acupuncture analgesia. Proc. Natl. Symp. Acupuncture, Maxibustion and Acupuncture Anesthesia, p. 32, Beijing, 1979.

Heath, R. G.: Pleasure and brain activity in man. J. Nerv. Ment. Dis. *15*: 3–18, 1972.

Heaton, M. B., Paiva, M., Swanson, D. J. and Walker, D. W.: Responsiveness of cultured septal and hippocampal neurons to ethanol and neurotrophic substances. J. Neurosci. Res. *39*: 305–318, 1994.

Hebb, D. O.: "The Organization of Behavior," Wiley Publishers, New York, 1949.

Hebert, T. J., Cashion, M. F. and Dohanich, G. P.: Effects of hormonal treatment and history on scopolamine inhibition of lordosis. Physiol. Behav. *56*: 835–894, 1994.

Hecht, S., Shlaer, S. and Pirenne, M. H.: Energy, quanta and vision. J. Gen. Physiol. *25*: 891–940, 1941.

Heim, C., Sieklucka, M. and Sontag, K. H.: Levemopamil injection after cerebral oligemia reduces spatial memory deficits in rats. Pharmacol. Biochem. Behav. *48*: 613–619, 1994.

Held, I. R., Sayers, S. T. and McLane, J. A.: Acetylcholine receptor gene expression in skeletal muscle of chronic ethanol-fed rats. Alcohol *8*: 173–177, 1991.

Helton, D. R., Modlin, D. L., Tizzano, J. T. and Rasmussen, K.: Nicotine withdrawal: a behavioral assessment using schedule controlled responding, locomotor activity, and sensorimotor reaction. Psychopharmacol. *113*: 205–210, 1993.

Hely, T. A., Graham, B. and Ooyen, A. V.: A computational model of dendrite elongation and branching based on MAP2 phosphorylation. J. Theor. Biol. *210*: 375–384, 2001.

Henderson, W. R. and Wilson, W. C.: Intraventricular injection of acetylcholine and eserine in man. Q. J. Exp. Physiol. *26*: 83–95, 1937.

Henke, K., Mondadori, C. R., Treyer, V., Nitsch, R. M., Buck, A. and Hock, C.: Nonconscious formation and reactivation of semantic associations by way of the medial temporal lobe. Neuropsychologia *41*: 863–876, 2003a.

Henke, K., Treyer, V., Nagy, E. T., Kneifel, S., Dursteller, M. and Nitsch, R. M.: Active hippocampus during unconscious memory. Conscious Cogn. *12*: 31–48, 2003b.

Henselmans, J. M., Hoogland, P. V. and Stoof, J. C.: Differences in the regulation of acetylcholine release upon D2 dopamine and N-methyl-D-aspartate receptor activation between striatal complex of reptiles and the neostriatum of rats. Brain Res. *566*: 8–12, 1991.

Herken, H., Maibauer, D. and Muller, S.: Acetylcholingehalt des Gehirns und Analgesia nach Einwirkung von Morphine und einigen 3-Oxymorphinanen. Arch. Exp. Pathol. Pharmakol. *230*: 313–324, 1957.

Hernandez, V. C. M. and Terry, A. V., Jr.: Repeated nicotine exposure in rats: effects on memory function, cholinergic markers and nerve growth factor. Neuroscience *130*: 997–1012, 2005.

Hernandez-Peon, R., Chavez-Ibarra, G., Morgane, P. J. and Timo-Iaria, C.: Limbic cholinergic pathways involved in sleep and emotional behavior. Exp. Neurol. *8*: 93–111, 1963.

Hersi, A. I., Richard, J. W., Gaudreau, P. and Quirion, R.: Local modulation of acetylcholine release by dopamine D1 receptors: a combined receptor autoradiography and *in vivo* dialysis study. J. Neurosci. *15*: 7150–7157, 1995.

Hersi, A. I., Kitaichi, K., Srivastava, L. K., Gaudreau, P. and Quirion R.: Dopamine D-5 receptor modulates hippocampal acetylcholine release. Brain Res. Mol. Brain Res. *76*: 336–340, 2000.

Herz, A.: Drugs and conditioned avoidance response. Int. Rev. Neurobiol. *2*: 229–277, 1960.

Herz, A. and Zieglgansberger, W.: The influence of miroelectrophoretically applied biogenic amines, cholinomimetics and procaine on synaptic excitation in the corpus striatum. Int. J. Neuropharmacol. *7*: 221–230, 1968.

Hesen, W., Karten, Y. I., van de Witte, S. V. and Joels, M.: Serotonin and carbachol induced suppression of synaptic excitability in rat CA1 hippocampal area: effects of corticosteroids receptor activation. J. Neuroendocrinol. *10*: 9–19, 1998.

Hess, G. and Donoghue, J. P.: Facilitation of long-term potentiation in layer II/III horizontal connections of rat motor cortex. Exp. Brain Res. *127*: 279–290, 1999.

Hess, W. R.: "Diencephalon," Grune and Stratton, New York, 1954.

Heubner, W.: Pharmakologisches und Chemisches über Physostigmin. Arch. Exp. Pathol. U. Pharmakol. *53*: 313–350, 1905.

Heymans, C.: Les substances anticholinesterasiques. Exposés Ann. Biochem. Med. *12*: 21–53, 1951.

Heymans, C. and Neil, E.: "Reflexogenic Areas of the Cardiovascular System," Little, Brown, Boston, 1958.

Heymans, J. F. and Heymans, C.: Sur le mecanisme de la bradycardie consecutive à l'injection de digitale, strophantine et cymarfine. J. Pharmacol. *29*: 203–222, 1926a.

Heymans, J. F. and Heymans, C.: Recherches physiologiques et pharmacologiques sur la tete isolée du chien. II. Sur l'influence repiratoire et pneumogastrique de l'adrénaline, de l'atropine, de la pituitrine, de la nicotine et des digitaliques. Arch. Int. Pharmacol. Ther. *32*: 9–41, 1926b.

Hickey, M. C. and Barnes, C. D.: Spinal cord effects of a brainstem cholinergic system. Neuropharmacology *14*: 125–138, 1975.

Hikida, T., Kaneko, S., Isobe, T., Kitabatake, Y., Watanabe, D., Pastan, I. and Nakanishi, S.: Increased sensitivity to cocaine by cholinergic ablation in nucleus accumbens. Proc. Natl. Acad. Sci. U.S.A. *98*: 13351–13354, 2001.

Hikida, T., Kitabatake, Y., Pastan, I. and Naknishi, S.: Acetylcholine enhancement in the nucleus accumbens prevents addictive behaviors of cocaine and morphine. Proc. Natl. Acad. Sci. U.S.A. *100*: 6199–6173, 2003.

Hillhouse, E. W. and Milton, N. G.: Effects of noradrenaline and gamma-butyric acid on the secretion of corticotrophin-releasing factor-41 and arginine vasopressin from the rat hypothalamus *in vitro*. J. Endocrinol. *122*: 719–723, 1989a.

Hillhouse, E. W. and Milton, N. G.: Effects of acetylcholine and 5-hydroxytryptamine on the secretion of corticotropin-releasing factor-41 and arginine vasopressin from rat hypothalamus *in vitro*. J. Endocrinol. *122*: 713–718, 1989b.

Hillhouse, E. W. and Milton, N. G.: Effect of noradrenaline and gamma-aminobutyric acid on the secretion of corticotrophin-releasing factor-41 and arginine vasopressin from the rat hypothalamus *in vitro*. J. Endocrinol. *122*: 719–723, 1989c.

Hilton, S. M.: The defence-arousal system and its relevance for circulatory and respiratory control. J. Exp. Biol. *100*: 159–174, 1982.

Himwich, H. E.: Some specific effects of psychoactive drugs. In: "Specific and Non-specific Factors in Psycho-pharmacology," M. Rinkel, Ed., pp. 3–71, Philosophical Library, New York, 1963.

Hingtgen, J. N. and Aprison, M. H.: Behavioral and environmental aspects of the cholinergic system. In: "Biology of Cholinergic Function," A. M. Godberg and I. Hanin, Eds., pp. 515–566, Raven Press, New York, 1976.

Hiramatsu, M., Murasawa, H., Mori, H. and Kameyama, T.: Reversion of muscarinic autoreceptor agonist-induced acetylcholine decrease and learning impairment by dynorphin A (1–13), an endogenous kappa-opioid receptor agonist. Br. J. Pharmacol. *123*: 920–926, 1998.

Hironaka, N. and Ando, K.: Effects of cholinergic drugs on scopolamine-induced memory impairment in rhesus monkeys. Nihon Shinkei Seishin Yakurigaku Zasshi *16*: 103–108, 1996.

Hironaka, N., Miyata, H., Takada, K. and Ando, K.: Assessment of drug effects on memory by delayed discrimination responses in rats. Nippon Yakurigaku Zasshi *95*: 309–318, 1990 (in Japanese).

Hirsch, J.: Behavior-genetic, or "experimental," analysis: the challenge of science versus the lure of technology. Am. Psychol. *22*: 118–130, 1967.

Hirvonen, M. R., Paljarvi, L. and Salvolainen, K. M.: Sustained effects of pilocarpine-induced convulsions on brain inositol and inositol monophosphate levels and brain morphology in young and old male rats. Toxicol. Appl. Pharmacol. *122*: 290–299, 1993.

Hoang, Q. V., Zhao, P., Nakajima, S. and Nakajima, Y.: Orexin (hypocretin) effects on constitutively active inward rectifier K+ channels in cultured nucleus basalis neurons. J. Neurophysiol. *92*: 3183–3191, 2004.

Hobson, J. A.: Sleep and dreaming: induction and mediation of REM sleep by cholinergic mechanisms. Curr. Opin. Neurobiol. *2*: 759–763, 1992.

Hobson, J. A.: Sleep is of the brain, by the brain and for the brain, Nature *437*: 1254–1256, 2005.

Hobson, J. A. and Steriade, M.: The neuronal basis of behavioral state control. In: "Handbook of Physiology, Section I: The Nervous System, Vol. IV. Intrinsic Regulatory Systems of the Brain," V. B. Mountcastle, F. E. Bloom and S. R. Geiger, Eds., pp. 701–823, Am. Physiol. Soc., Bethesda, Md., 1986.

Hobson, J. A., Datta, S., Calvo, J. M. and Quatrocchi, J.: Acetylcholine as a brain state modulator: triggering and long-term regulation of REM sleep. In: "Cholinergic Function and Dysfunction," A. C. Cuello, Ed., vol. 85, pp. 389–404, Elsevier, Amsterdam, 1993.

Hodges, H., Allen, Y., Sinden, J., Mitchell, S. N., Arendt, T., Lantos, P. L. and Gray, J. A.: The effects of cholinergic drugs and cholinergic-rich foetal neural transplants on alcohol-induced deficits in radial maze performance in rats. Behav. Brain Res. *43*: 7–28, 1991.

Hoebel, B. G., Hernandez, L., McLean, S., Stanley, B. G., Aulissi, P., Glimcher, P. and Margolin, D.: Catecholamines, enkephalin and neurotensin in feeding and reward. In: "The Neural Basis of Feeding and Reward," B. G. Hoebel and D. Novin, Eds., pp. 465–578, Haer Inst. Publs., Brunswick, Maine, 1982.

Hoffman, D. L., Krupp, L., Schrag, D., Nilaver, G., Valiquette, G., Kilcoyne, M. M. and Zimmerman, E. A.: Angiotensin immunoreactivity in vasopressin cells in rat hypothalamus and its relative deficiency in homozygous Brattleboro rats. Ann. N. Y. Acad. Sci. *394*: 135–141, 1982.

Hofman, M. A. and Swaab, D. F.: The human hypothalamus: comparative morphometry and photoperiodic influences. Prog. Brain Res. *93*: 133–147, 1992.

Hokfelt, T., Elde, R., Johansson, O., Ljungdahl, A., Schultzberg, M., Fuxe, K., Goldstein, M., Nilsson, G., Pernow, B., Terenius, L., Ganten, D., Jeffcoate, S. L., Rehfeld, J. and Said, S.: The distribution of peptide containing neurons in the nervous system. In: "Psychopharmacology: A Generation of Progress," M. A. Lipton, A. DiMascio and K. F. Killam, Eds., pp. 39–66, Raven Press, New York, 1978.

Hokfelt, T., Johansson, O., Ljungdahl, A., Lundberg, J., Schultzberg, M., Terenius, L., Goldstein, M., Elde, R., Steinbusch, H. and Verhofstad, A.: Histochemistry of transmitter interactions—neuronal coupling and coexistence of transmitters. In: "Neurotransmitters," P. Simon, Ed., pp. 31–143, Adv. Pharmacol. Ther., vol. 2, Pergamon Press, Oxford, 1979.

Hokin, L. E. and Dixon, J. F.: The phosphoinositide signalling system. I. Historical background. II. Effects of lithium on the accumulation of second messenger inositol 1,4,5-triphosphate in brain cortex slices. In: "Cholinergic Function and Dysfunction," A. C. Cuello, Ed., pp. 309–315, Elsevier, Amsterdam, 1993.

Hokin, L. E. and Hokin, M. R.: The mechanism of phosphate exchange in phosphatidic acid in response to acetylcholine. J. Biol. Chem. *234*: 1387–1390, 1959.

Holland, L. N., Shuster, L. C. and Buccafusco, J. J.: Role of spinal and supraspinal muscarinic receptors in the expression of morphine withdrawal symptoms in the rat. Neuropharmacology *32*: 1387–1395, 1993.

Hollander, E., Davidson, M., Mohs, R. C., Horvath, T. B., Davis, B. M., Zimishlany, Z. and Davis, K. L.: RS 86 in the treatment of Alzheimer's disease: cognitive and biological effects. Biol. Psychiatry *22*: 1067–1078, 1987.

Holmstedt, B.: Pharmacology of organophosphorus cholinesterase inhibitors. Pharmacol. Rev. *11*: 567–688, 1959.

Holmstedt, B.: Structure-activity relationships of the organophosphorus anticholinesterase agents. In: "Cholinesterases and Anticholinesterase Agents," G. B. Koelle, Ed., Handbch. d. exp. Pharmakol., Ergänzungswk., vol. 15, pp. 428–485, Springer-Verlag, Berlin, 1963.

Holmstedt, B.: The ordeal bean of Old Calabar. In: "Plants in the Development of Modern Medicine," T. Swain, Ed., pp. 303–360, 1972, Harvard University Press, Cambridge, Mass., 1972.

Holmstedt, B.: Toxicity of nicotine and related compounds. In: "The Pharmacology of Nicotine," M. J. Rand and K. Thurau, Eds., pp. 61–88, IRL Press, Washington, D.C., 1988.

Holmstedt, B.: Cholinesterase inhibitors: an introduction. In: "Cholinesterases and Cholinesterase Inhibitors," E. Giacobini, Ed., pp. 1–8, Martin Dunitz, London, 2000.

Holmstedt, B. and Skoglund, C. R.: The action on spinal reflexes of dimethylamido-ethoxy-phosphoryl cyanide, "Tabun," a cholinesterase inhibitor. Acta Physiol. Scand. *29*, Supplt. *106*: 410–427, 1953.

Holsboer, F.: Behavioral and neuroendocrine effects of neuroactive steroids. Scientific Abstracts, 36th Meeting of ACNP, p. 50, 1997.

Honig, W. K.: Studies of working memory in the pigeon. In: "Cognitive Processes in Animal Behavior," S. H. Hulse, H. Fowler and W. K. Honig, Eds., pp. 211–248, Lawrence Erlbaum Publs., Hillsdale, 1978.

Hoover, D. B., Craig, C. R. and Colasanti, B. K.: Cholinergic involvement in cobalt-induced epilepsy in the rat. Exp. Brain Res. *29*: 501–513, 1979.

Hori, H., Haruta, K., Sakamoto, N., Uemera, K., Matsubara, T., Itoh, K. and Iguda, A.: Pressor response induced by the hippocampal administration of neostigmine is suppressed by M1 muscarinic antagonist. Life Sci. *57*: 1853–1859, 1995.

Hornykiewicz, O. and Kobinger, W.: Über den Einfluss von Eserin, Tetraethylpyrophosphat (TEPP), und Neostigmin auf den Blutdruck und die pressorischen Carotissinusreflexe der Ratte. Arch. Exp. Pathol. Pharmakol. *228*: 493–500, 1956.

Hortnagl, H., Groll Knapp, E., Khanakah, G., Sperk, G. and Bubna Littitz, H.: Perception of species-specific vocalization in rats; role of the septo-hippocampal pathway and aging. Int. J. Dev. Neurosci. *16*: 715–727, 1998.

Hosli, L. and Hosli, E.: Receptors for neurotransmitters on astrocytes in the mammalian central nervous system. Prog. Neurobiol. *40*: 477–506, 1993.

Howard, M. F., Fotedar, M. S., Datey, A. V. and Hasselmo, M. E.: The temporal context model in spatial navigation and relational learning: toward a common explanation of medial temporal lobe function across domains. Psychol. Rev. *112*: 75–116, 2005.

Hruby, V. J., Chow, M.-S. and Smith, D. D.: Conformational and structural considerations in oxytocin-receptor binding and biological activity. Annu. Rev. Pharmacol. Toxicol. *30*: 501–534, 1990.

Hu, M., Walker, D. W., Vickroy, T. W. and Peris, J.: Chronic ethanol exposure increases 3H-GABA release in rat hippocampus by presynaptic muscarinic receptor modulation. Alcohol. Clin. Exp. Res. *23*: 1587–1595, 1999.

Hu, Y. Q., Liebl, D. J., Dluzen, D. E. and Koo, P. H.: Inhibition of dopamine and choline acetyltransferase concentrations in rat CNS neurons by rat alpha 1- and alpha 2-macroglobulins. J. Neurosci. Res. *51*: 541–550, 1998.

Hudson, J. Q. A.: Remarks on the physiologic action and therapeutic uses of physostigma venenosum or the ordeal bean of Calabar. Southern Med. Rec. *3*: 705–721, 1873.

Hudzik, T. and Wenger, G. R.: Effects of drugs of abuse and cholinergic agents on delayed matching-to-sample responding in the squirrel monkey. J. Pharmacol. Exp. Ther. *265*: 120–127, 1993.

Huerta, P. T. and Lisman, J. E.: Bidirectional synaptic plasticity induced by a single burst during cholinergic theta oscillation in CA1 *in vitro*. Neuron *15*: 1055–1063, 1995.

Hughes, R. N.: A review of atropinic drug effects on exploratory choice behavior in laboratory rodents. Behav. Neurol. Biol. *35*: 5–41, 1982.

Hunt, J. M. and Silinsky, E. M.: Ionomycin-induced acetylcholine release and its inhibition by adenosine at frog motor nerve endings. Br. J. Pharmacol. *110*: 828–832, 1993.

Hutchinson, R. R., Azrin, N. H. and Hunt, G.: Attack produced by intermittent reinforcement of concurrent operant response. J. Exp. Anal. Behav. *11*: 489–495, 1968.

Huttunen, P., Lapilampi, T. and Myers, R. D.: Temperature-related release of serotonin from unrestrained rats' preoptic area perfused with ethanol. Alcohol *5*: 189–193, 1988.

Hyde, J. S., Beckett, S. and Gellhorn, E.: Acetylcholine and convulsive activity. J. Neurophysiol. *12*: 17–27, 1949.

Hyden, H.: Some brain protein changes reflecting neuronal plasticity at learning. In: "Brain and Human Behavior," A. G. Karczmar and J. C. Eccles, Eds., pp. 94–110, Springer-Verlag, Berlin, 1972.

Iguchi, A., Yatomi, A., Gotoh, M., Matsunaga, H., Uemera, K., Miura, H., Satake, T., Temagawa, T. and Sakamoto, N.: Neostigmine-induced hyperglycemia is mediated by central muscarinic receptor in fed rats. Brain Res. *507*: 295–300, 1990.

Ikemoto, S., Witkin, B. M. and Morales, M.: Rewarding injections of cholinergic agonist carbachol into the ventral tegmental area induce locomotion and c-Fos expression in the retrospinal area and supramammillary nucleus. Brain Res. *969*: 78–87, 2003.

Ilyutchenov, R. Y. and Yeliseyeva, A. G.: Cholinergic mechanism of emotional fear reaction. Pavlov J. High. Nerv. Act. *17*: 330–337, 1967 (in Russian).

Imon, H., Ito, K., Dauphin, L. and McCarley, R. W.: Electrical stimulation of cholinergic laterodorsal tegmental nucleus elicits scopolamine sensitive excitatory postsynaptic potentials in medial pontine reticular formation. Neuroscience *74*: 393–301, 1996.

Imperato, A., Puglisi-Allegra, S., Grazia Scrocco, M., Casolini, P., Bacchi, S. and Angelucci, L.: Cortical and limbic dopamine and acetylcholine release as neurochemical correlates of emotional arousal in both aversive and non-aversive environmental changes. Neurochem. Int. *20*: 265S–270S, 1992.

Imperato, A., Dazzi, L., Carta, G., Colombo, G. and Biggio, G. L.: Rapid increase in basal acetylcholine release in the hippocampus of freely moving rats induced by withdrawal from long-term alcohol intoxication. Brain Res. *784*: 347–350, 1998.

Inagawa, G., Sato, K., Kikuchi, T., Nishimara, M., Shioda, M., Koyama, Y., Yamada, Y. and Andoh, T.: Opposite effects of depressant and convulsant barbiturate stereoisomers on acetylcholine release from the rat hippocampus *in vivo*. Br. J. Anaesth. *92*: 424–426, 2004.

Inagawa, K.: Cholinergic modulation of spatial working memory of mice in radial maze performance: retention curve analysis. Yakubutsu Sieshin Kodo *13*: 233–238, 1993.

Inda, M. C., Delagado-Garcia, J. M. and Carrion, A. M.: Acquisition, consolidation, reconsolidation, and extinction of eyelid conditioning responses require de novo protein synthesis. J. Neurosci. *25*: 2070–2080, 2005.

Inglis, F. M., Day, J. C. and Fibiger, H. C.: Enhanced acetylcholine release in hippocampus and cortex during anticipation and consumption of a palatable meal. Neuroscience *62*: 1049–1056, 1994.

Inoue, M., Crofton, J. T. and Share, L.: Interactions between brain acetylcholine and prostaglandins in control of vasopressin release. Am. J. Physiol. *261*: R420–R426, 1991.

Ismail, Z. and el-Guebaly, N.: Nicotine and endogenous opioids: toward specific pharmacotherapy. Can. J. Psychiatry *43*: 37–42, 1998.

Iso, H., Ueki, A., Shinjo, H., Miwa, C. and Morita, Y.: Reinforcement enhances hippocampal acetylcholine release: an *in vivo* microdialysis study. Behav. Brain Res. *101*: 207–213, 1999.

Itoh, K., Konya, H., Takai, E., Masuda, H. and Nagai, K.: Modification of acetylcholine release by nociceptin in conscious rat striatum. Brain Res. *845*: 242–245, 1999.

Itoh, Y., Ogasawara, T., Mushiroi, T., Yamazaki, A., Ukai, Y. and Kimura, K.: Effect of NS-3, a thyrotropin-releasing hormone analog, on *in vivo* acetylcholine release in rat brain: regional differences and its sites of action. J. Pharmacol. Exp. Ther. *271*: 884–890, 1994.

Itoi, K., Jiang, Y. Q., Iwasaki, Y. and Watson, S. J.: Regulatory mechanisms of corticotropin-releasing hormone and vasopressin gene expression in the hypothalamus. J. Neuroendocrin. *16*: 348–355, 2004.

Iwamoto, E. T.: Characterization of the antinociception induced by nicotine in the pedunculopontine tegmental nucleus and the nucleus raphe magnus. J. Pharmacol. Exp. Ther. *257*: 120–133, 1991.

Iwamoto, E. T.: Antinociception after nicotine administration into the mesopontine tegmentum of rats: evidence for muscarinic actions. J. Pharmacol. Exp. Ther. *251*: 412–421, 1989.

Iwamoto, E. T.: Characterization of the antinociception induced by nicotine in the pedunculopontine tegmental nucleus and the nucleus raphe magnus. J. Pharmacol. Exp. Ther. *265*: 777–789, 1993.

Iwamoto, E. T. and Marion, L.: Adrenergic, serotonergic and cholinergic components of nicotinic antinociception in rats. J. Pharmacol. Exp. Ther. *265*: 777–789, 1993.

Jabourian, M., Bourgoin, S., Perez, S., Godeheu, G., Glowinski, J. and Kemel, M. L.: Mu opioid regulation of the N-merthyl-D-aspartate-evoked release of [3H]-acetylcholine in the limbic territory of the rat striatum *in vitro*: diurnal variations and implication of a dopamine link. Neuroscience *123*: 733–742, 2004.

Jackisch, R., Huang, H. Y., Reimann, W. and Limberger, H.: Effects of the antiparkinsonian drug budipine on neurotransmitter release in central nervous system tissues *in vitro*. J. Pharmacol. Exp. Ther. *264*: 889–898, 1993.

Jackson, D. and Zigmond, M. J.: Modulation of striatal acetylcholine release by endogenous serotonin. Brain Res. *457*: 259–266, 1988.

Jackson, D., Abercrombie, E. D. and Zigmond, M. J.: Impact of L-dopa on striatal acetylcholine release: effects of 6-hydroxydopamine. J. Pharmacol. Exp. Ther. *267*: 912–918, 1993.

Jacobson, L. and Sapolsky, R.: The role of the hippocampus in feedback regulation of the hypothalamic-pituitary-adrenocortical axis. Endocrin. Rev. *12*: 118–134, 1991.

Jacobsson, S. O., Cassel, G. E. and Persson, S. A.: Increased levels of nitrogen oxides and lipid peroxidation in the rat brain after somaninduced seizures. Arch. Toxicol. *73*: 269–273, 1999.

Jakala, P. and Riekkinen, P., Jr.: Combined cholinergic and 5-HT2-receptor activation suppresses thalamocortical oscillations in aged rats. Pharmacol. Biochem. Behav. *56*: 713–718, 1997.

James, T. A. and Massey, S.: Evidence for a possible dopaminergic link in the action of acetylcholine in the rat substantia nigra. Neuropsychopharmacol. *17*: 687–690, 1978.

Jankowska, E. and Lindstrom, S.: Morphological identification of Renshaw cells. Acta Physiol. Scand. *81*: 428–430, 1971.

Jankowsky, J. L. and Paterson, P. H.: Differential regulation of cytokine expression following pilocarpine-induced seizure. Exp. Neurol. *159*: 333–346, 1999.

Janowsky, D. S., Risch, S. C., Irwin, M. and Schukit, M. A.: Behavioral hyporeactivity to physostigmine in detoxified primary alcoholics. Am. J. Psychiatry *146*: 538–439, 1989.

Jasper, H. H. and Tessier, J.: Acetylcholine liberation from the cerebral cortex during paradoxical (REM) sleep. Science *172*: 601–603, 1971.

Javoy, F., Agid, Y., Bouvet, D. and Glowinski, J.: Changes in neostriatal DA metabolism after carbachol or atropine microinjection into the substantia nigra. Brain Res. *68*: 253–260, 1974.

Jessen, A., Buemann, B., Toubro, S., Skovguard, I. M. and Astrup, A.: The appetite suppressant effects of nicotine is enhanced by caffeine. Diabetes Obes. Metab. *7*: 327–333, 2005.

Jhamandas, K. and Sutak, M.: Morphine-naloxone interaction in the central cholinergic system: the influence of subcortical lesioning and electrical stimulation. Br. J. Pharmacol. *58*: 101–107, 1976.

Jhamandas, K. and Sutak, M.: Stereospecific enhancement of evoked release of brain acetylcholine by narcotic antagonists. Br. J. Pharmacol. *78*: 433–440, 1983.

Jhamandas, K., Phillis, J. W. and Pinsky, C.: Effects of narcotic analgesics and antagonists on the *in vivo* release of acetylcholine from the cerebral cortex of the cat. Br. J. Pharmacol. *45*: 33–66, 1971.

Jiang, Z. G. and Dun, N. J.: Presynaptic suppression of excitatory postsynaptic potentials in rat ventral horn neurons by muscarinic agonists. Brain Res. *381*: 182–186, 1986.

Jiang, Z. G. and Dun, N. J.: Action of acetylcholine on spinal motoneuron. In: "Neurobiology of Acetylcholine: A Symposium in Honor of Alexander G. Karczmar," N. J. Dun and R. L. Perlman, Eds., pp. 283–293, Plenum Press, New York, 1987.

Jimenez-Capdeville, M. E. and Dykes, R. W.: Daily changes in the release of acetylcholine from rat primary somatosensory cortex. Brain Res. *625*: 152–158, 1993.

Jin, S.: AMPA- and kainate receptors differentially mediate excitatory amino-acid-induced dopamine and acetylcholine release from rat striatal slices. Neuropharmacology *36*: 1503–1510, 1997.

Joffe, M. E.: Brain mechanisms for the formation of new movements during learning: the evolution of classical concepts. Neurosci. Behav. Physiol. *34*: 5–18, 207, 2004.

Johansson, O., Hokfelt, T., Jeffcoate, S. L., White, N. and Sternberger, L. A.: Ultrastructure of TRH-like immunoreactivity. Exp. Brain Res. *38*: 1–10, 1987.

Johnson, G. V. W. and Jope, R. S.: The role of microtubule-associated protein 2 (MAP-2) in neuronal growth, plasticity and degeneration. J. Neurosci. Res. *33*: 505–512, 1992.

Johnson, J. D.: REM sleep and the development of context memory. Med. Hypotheses *64*: 499–504, 2005.

Jones, B. E.: The organization of central cholinergic systems and their functional importance in sleep-waking states. In: "Cholinergic Function and Dysfunction," A. C. Cuello, Ed., pp. 61–71, Elsevier, Amsterdam, 1993.

Jones, B. E.: Arousal systems. Front. Biosci. *8*: S438–S451, 2003.

Jones, B. E.: From waking to sleeping: neuronal and chemical substrates. Trends Pharmacol. Sci. *26*: 578–584, 2005.

Jones, D. N. and Higgins, G. A.: Effect of scopolamine on visual attention in rats. Psychopharmacology *120*: 142–149, 1995.

Jones, D. N., Barnes, J. C., Kirkby, D. L. and Higgins, G. A.: Age-associated impairment in a test of attention: evidence for involvement of cholinergic system. J. Neurosci. *15*: 7282–7292, 1995.

Jones, G. A., Norris, S. K. and Henderson, Z.: Conduction velocities and membrane properties of different classes of rat septohippocampal neurons recorded *in vitro*. J. Physiol. *517*, pt. 3: 867–877, 1999.

Jones, K. J.: Androgenic enhancement of motor neuron regeneration. Ann. N. Y. Acad. Sci. *743*: 141–164, 1994.

Jones, M. B., Siderowski, D. P. and Hooks, S. B.: The G{beta}{gamma} dimer as a novel source of subselectivity in G-protein signaling: GGL-ing at convention. Mol. Interv. *4*: 200–214, 2004.

Jones, M. T., Hillhouse, E. W. and Burden, J.: Proceedings: mechanism of action of fast and delayed corticosteroid feedback at hypothalamus. J. Endocrinol. *69*: 34P–35P, 1976.

Jope, R. S.: Effect of lithium treatment *in vitro* and *in vivo* on acetylcholine metabolism in rat brain. J. Neurochem. *33*: 487–495, 1979.

Jordan, D.: Autonomic changes in affective behavior. In: "Central Regulation of Autonomic Functions," A. D. Loewy and K. M. Spyer, Eds., pp. 349–366, Oxford University Press, New York, 1990.

Jordan, D. and Spyer, K. M.: Effects of acetylcholine on respiratory neurones in the nucleus ambiguus-retroambigualis complex of the cat. J. Physiol. (Lond.) *320*: 103–111, 1981.

Jouvet, M.: Neurophysiology of the states of sleep. In: "The Neurosciences: A Study Program," G. C. Quarton, T. Melnechuk and F. U. Schmitt, Eds., pp. 529–544, University Press, New York, 1967.

Jouvet, M.: Some monoaminergic mechanisms controlling sleep and waking. In: "Brain and Human Behavior," A. G. Karczmar and J. C. Eccles, Eds., pp. 131–160, Springer-Verlag, Berlin, 1972.

Jukes, T. H.: Nuggets on the surface. Acetylcholine. J. Appl. Biochem. *6*: 1–2, 1984.

Julian, P. L. and Pike, J.: Studies in indole series. V. Complete synthesis of physostigmine (eserine). J. Am. Chem. Soc. *51*: 755–757, 1935.

Julka, D., Sandhir, R. and Gill, K. D.: Altered cholinergic metabolism in rat following aluminum exposure: implications on learning performance. J. Neurochem. *65*: 2157–2164, 1995.

Jung, R. and Kornmuller, A. E.: Eine Methodik der Ableitung lokalisierter Potentialschwankungen aus subcorticalen Hirngebieten. Arch. Psychiatry *109*: 1–30, 1938.

Jurgens, U. and Lu, C. L.: The effects of periaqueductally injected transmitter antagonists on the forebrain-elicited vocalization in the squirrel monkey. Eur. J. Neurosci. *5*: 735–741, 1993.

Jurna, I., Krauss, P. and Baldauf, J.: Depression by nicotine of pain-related nociceptive activity in the rat thalamus and spinal cord. Clin. Investig. *72*: 65–73, 1993.

Kaada, B.: Brain mechanism related to aggressive behavior. In: "Aggression and Defense: Neural Mechanisms and Social Patterns," C. D. Clemente and D. B. Lindsley, pp. 95–216, University of California Press, Berkeley, 1967.

Kahn, A. J.: Site selection and the response to ethanol in C3H mice. Q. J. Stud. Alcohol *30*: 609–617, 1969.

Kalant, H.: Direct effects of ethanol on the nervous system. Fed. Proc. *34*: 1930–1941, 1975.

Kalaria, R. N., Homayoun, P. and Whitehouse, P. J.: Nicotinic cholinergic receptors associated with mammalian cerebral vessels. J. Auton. Nerv. Syst. *49*: S3–S7, 1994.

Kalash, J., Romita, V. and Billiar, R. B.: Third ventricular injection of alpha-bungarotoxin decreases pulsatile luteinizing hormone secretion in the ovariectomized rat. Neuroendocrinology. *49*: 462–470, 1989.

Kalichman, M. W.: Neurochemical correlates of the kindling model of epilepsy. Neurosci. Biobehav. Rev. *6*: 165–181, 1982.

Kametani, H. and Kawamura, H.: Alteration in acetylcholine release in the rat hippocampus during sleep-wakefulness detected by intracerebral dialysis. Life Sci. *47*: 421–426, 1990.

Kandel, E. R., Schwartz, J. H. and Jessel, T. M., Eds.: "Principles of Neural Science," Elsevier, New York, 1991.

Kanematsu, N.: Regulation mechanism of melatonin rhythm in the pineal gland by light: experimental studies by *in vitro* microdialysis. Hokkaido Igaku Zsshi *69*: 46–64, 1994.

Karczmar, A. G.: Antagonism between a bis-quaternary oxamide, WIN 8078, and depolarizing and competitive blocking agents. J. Pharmacol. Exp. Ther. *119*: 39–47, 1957.

Karczmar, A. G.: Pharmacologic, toxicologic, and therapeutic properties of anticholinesterase agents. In: "Physiological Pharmacology," W. S. Root and F. G. Hofman, Eds., vol. 3, pp. 163–322, Academic Press, New York, 1967.

Karczmar, A. G.: Is the central nervous system overexploited? Fed. Proc. *28*: 147–156, 1969.

Karczmar, A. G.: History of the research with anticholinesterase agents. In: "Anticholinesterase Agents," A. G. Karczmar, Ed., vol. 1, Section 13, Intern. Encyclop. Pharmacol. Ther., pp. 1–44, Pergamon Press, Oxford, 1970.

Karczmar, A. G.: Possible mechanisms underlying the so-called "divorce" phenomena of EEG desynchronizing actions of anticholinesterases. Abstracts, Regional Midwest EEG Meeting, Hines Veterans Administration Hospital, April 1971.

Karczmar, A. G.: Introduction: what we know, will know in the future, and possibly cannot ever know in neurosciences. In: "Brain and Human Behavior," A. G. Karczmar and J. C. Eccles, Eds., pp. 1–20, Springer-Verlag, Berlin, 1972a.

Karczmar, A. G.: Les synapses cholinergiques et leurs enchainements. In: "Le Système Cholinergique," G. G. Nahas, J. C. Salamagne, P. Viars, and G. Vourc'h, Eds., pp. 113–156, Librairie Arnette, Paris, 1972b.

Karczmar, A. G.: Brain acetylcholine and seizures. In: "Psychobiology of Convulsive Therapy," M. Fink, S. Kety, J. McGaugh, and T. A. Williams, Eds., pp. 251–270, Wiley and Sons, New York, 1974.

Karczmar, A. G.: Cholinergic influences on behavior. In: "Cholinergic Mechanisms," P. G. Waser, Ed., pp. 501–529, Raven Press, New York, 1975.

Karczmar, A. G.: Central actions of acetylcholine, cholinomimetics, and related drugs. In: "Biology of Cholinergic Function," A. M. Goldberg and I. Hanin, Eds., pp. 395–449, Raven Press, 1976.

Karczmar, A. G.: Multitransmitter mechanisms underlying selected function, particularly aggression, learning and sexual behavior. In: "Neuro-Psychopharmacology," P. Deniker, C. Radouco-Thomas, and A. Villeneuve, Eds., pp. 581–608, Pergamon Press, Oxford, 1978a.

Karczmar, A. G.: Summary of the session: interrelationships between various neurotransmitter systems. In: "Neuropsychopharmacology, Proceedings 10th Congress CINP," P. Deniker, C. Radouco-Thomas, and A. Villeneuve, Eds., pp. 499–501, Pergamon Press, Oxford, 1978b.

Karczmar, A. G.: Brain acetylcholine and animal electrophysiology. In: "Brain Acetylcholine and Neuropsychiatric Disease," K. L. Davis and P. A. Berger, Eds., pp. 265–310, Plenum Press, New York, 1979a.

Karczmar, A. G.: Overview: cholinergic drugs and behavior—what effects may be expected from a "Cholinergic Diet"? In: "Choline and Lecithin in Brain Disorders," A. Barbeau, J. Growdon, and R. J. Wurtman, Eds., pp. 141–175, Nutrition and the Brain, vol. 20, Raven Press, New York, 1979b.

Karczmar, A. G.: New roles for cholinergics in CNS disease. Drug Ther. *4*: 31–42, 1979c.

Karczmar, A. G.: Drugs, transmitters and hormones and mating behavior. In: "Modern Problems in Pharmacopsychiatry," T. Ban, Ed., vol. 5, pp. 1–76, Karger, Basel, 1980.

Karczmar, A. G.: Central effects of anticholinesterases with emphasis on behavior, sleep phenomena, and psyche. NAS Committee on Toxicology, Anticholinesterase Panel, Washington, D.C., December, 1981a.

Karczmar, A. G.: Multifactorial and multitransmitter basis of behaviors and their cholinergic concomitants. Psychopharmacol. Bull. *17*: 68–71, 1981b.

Karczmar, A. G.: Acute and long-lasting central actions of organophosphorus agents. Fund. Appl. Toxicol. *2*: S1–S17, 1984.

Karczmar, A. G.: Conference on dynamics of cholinergic function: overview and comments. In: "Dynamics of Cholinergic Function," I. Hanin, Ed., pp. 1215–1259, Plenum Press, New York, 1986.

Karczmar, A. G.: Schizophrenia and cholinergic system. In: "Receptors and Ligands in Psychiatry," A. K. Sen and T. Lee, Eds., pp. 29–63, Cambridge University Press, Cambridge, 1988.

Karczmar, A. G.: Physiological cholinergic functions in the CNS. In: "Cholinergic Neurotransmission: Functional and Clinical Aspects," S.-M. Aquilonius and P.-G. Gillberg, pp. 437–466, Elsevier, Amsterdam, 1990a.

Karczmar, A. G.: Conclusions and Comments, Xth International Symposium on Cholinergic Mechanisms, In: "From Torpedo Electric organ to Human Brain; Fundamental and Applied Aspects," J. Massoulié, Ed., pp. 393–400, Elsevier, Paris, 1990b.

Karczmar, A. G.: Historical aspects of cholinergic transmission. In: "The Basal Forebrain," T. C. Napier, P. W. Kalivas and I. Hanin, Eds., pp. 453–476, Plenum Press, New York, 1991.

Karczmar, A. G.: Comments to session on electrophysiological aspects of cholinergic mechanisms. In: "Cholinergic Neurotransmission: Function and Dysfunction," A. C. Cuello, Ed., Prog. Brain Res., vol. 98, pp. 279–284, Elsevier, Amsterdam, 1993a.

Karczmar, A. G.: Brief presentation of the story and present status of studies of the vertebrate cholinergic system. Neuropsychopharmacology *9*: 181–199, 1993b.

Karczmar, A. G.: Cholinergic substrates of cognition and organism-environment interaction. Prog. Neuropsychopharmacol. Biol. Psychiatry *19*: 187–211, 1995.

Karczmar, A. G.: Loewi's discovery and the XXI Century. In: "Ninth International Symposium on Cholinergic Mechanisms," J. Klein and K. Loeffelholz, Eds., pp. 1–27, Elsevier, Amsterdam, 1996.

Karczmar, A. G.: The Tenth International Symposium on Cholinergic Mechanisms. CNS Drug Rev. *5*: 301–316, 1999.

Karczmar, A. G.: Synaptic, behavioral and toxicological effects of cholinesterase inhibitors in animals and humans. In: "Cholinesterases and Cholinesterase Inhibitors," E. Giacobini, Ed., pp. 157–180, Martin Dunitz, London, 2000.

Karczmar, A. G.: Sir John Eccles, 1903–1997. Part I. Perspect. Biol. Med. *44*: 76–86, 2001a.

Karczmar, A. G.: Sir John Eccles, 1963–1997. Part II. Perspect. Biol. Med. *44*: 2001b.

Karczmar, A. G.: Conclusions: XI ISCM. In: "Cholinergic Mechanisms—Function and Dysfunction," I. Stilman, Ed., pp. 435–451, Martin Dunitz, London, 2004.

Karczmar, A. G.: Conclusions for the XII ISCM. In: "Proceedings, XII ISCM," Gonzalez-Ros, J., Ed., J. Molec Sci. 2006 (in Press).

Karczmar, A. G. and Kindel, G. H.: Acetylcholine turnover and aggression in related three strains of mice. Prog. Neuropsychopharmacol. *5*: 35–48, 1981.

Karczmar, A. G. and Koehn, G. L.: Cholinergic control of hyperkinesia. Prog. Clin. Biol. Res. *39*: 374, 1980.

Karczmar, A. G. and Koppanyi, T.: Central effects of di-isopropyl fluorophosphonate in urodele larvae. Naunyn Schmiedebergs Arch. Exp. Path. Pharmakol. *219*: 261–270, 1953.

Karczmar, A. G. and Nishi, S.: The types and sites of cholinergic receptors. In: "First International Symposium on Cell Biology and Cytopharmacology", S. Clementi and B. Cinniarelli, Eds., pp. 301–317, Raven Press, New York, 1971.

Karczmar, A. G. and Scudder, C. L.: Learning and effect of drugs on learning of related mice genera and strains. In: "Neurophysiological and Behavioral Aspects of Psychotropic Drugs," A. G. Karczmar and W. P. Koella, Eds., pp. 133–160, C. C. Thomas, Springfield, Ill., 1969.

Karczmar, A. G., Koketsu, K. and Blaber, L. C.: Multiple sites of action of oxamides, hydroxyaniliniums and NaF. Abstracts, III Internatl. Pharmacol. Congr., p. 42, 1966.

Karczmar, A. G., Longo, V. G. and Scotti de Carolis, A.: Pharmacological model of paradoxical sleep: the role of cholinergic and monoamine system. Physiol. Behav. *5*: 175–182, 1970.

Karczmar, A. G., Nishi, S. and Blaber, L. C.: Synaptic modulations. In: "Brain and Human Behavior," A. G. Karczmar and J. C. Eccles, Eds., Springer-Verlag, Berlin, 1972.

Karczmar, A. G., Scudder, C. L. and Richardson, D. L.: Interdisciplinary approach to the study of behavior in related mice types. In: "Neurosciences Research," I. Kopin, Ed., vol. 5, pp. 159–244, 1973.

Karczmar, A. G., Scudder, C. L. and Kahn, A. J.: Behavioral, genetic and neurochemical aspects of alcohol preference of mice. In: "Neuro-Psychopharmacology," P. Deniker, C. Radouco-Thomas, and A. Villeneuve, Eds., pp. 799–816, Pergamon Press, Oxford, 1978.

Karczmar, A. G., Koketsu, K. and Nishi, S., Eds.: "Autonomic and Enteric Ganglia," Plenum Press, New York, 1986.

Katz, B. and Miledi, R.: The measurement of synaptic delay and the time-course of acetylcholine release at the neuromuscular junction. Proc. R. Soc. Lond., B, Biol. Sci. *161*: 483–495, 1965.

Katz, B. and Miledi, R.: The role of calcium in neuromuscular facilitation. J. Physiol. *195*: 481–492, 1968.

Katz, E. J., Cortes, V. I., Eldefrawi, M. E. and Eldefrawi, A. T.: Chlorpypyrifos, parathion and their oxons bind to and desensitize a nicotinic receptor: relevance to their toxicities. Toxicol. Appl. Pharmacol. *146*: 227–236, 1997.

Katz, E., Elgoyhen, A. B., Gomez-Casati, M. E., Knipper, M., Vetter, D. E., Fuchs, P. A. and Glowatzki, E.: Developmental regulation of nicotinic synapses on cochlear inner hair cells. J. Neurosci. *24*: 7814–7820, 2004.

Kaufman, H., Vratsanos, S. M. and Erlanger, B. F.: Photoregulation of an enzymic process by means of a light sensitive ligand. Science *162*: 1487–1489, 1968.

Kaufmann, W. E., Taylor, C. V., Hohmann, C. F., Sanval, I. B. and Naidu, S.: Abnormalities in neuronal maturation in Rett syndrome neocortex: preliminary molecular correlates. Eur. Child Adolesc. Psychiatry *6*, Supplt.1: 15–17, 1997.

Kawamura, H. and Domino, E. F.: Differential actions of m and n cholinergic agonists on the brain stem activating system. Int. J. Neuropharmacol. *8*: 105–115, 1969.

Kawasaki, H., Palmieri, C. and Avoli, M.: Muscarinic receptor activation induces depolarizing plateau potentials in bursting neurons of the rat subiculum. J. Neurophysiol. *82*: 2590–2601, 1999.

Kayadjanian, N., Schofield, W. N., Andren, J., Sirinathsinghji, D. J. and Besson, M. J.: Cortical and nigral deafferentation and striatal cholinergic markers in the rat dorsal striatum: different effects on the expression of mRNAs encoding choline acetyltransferase and muscarinic m1 and m4 receptors. Eur. J. Neurosci. *11*: 3659–3668, 1999.

Keith, H. M. and Stavraky, G. W.: Experimental convulsions induced by administration of thujone. Arch. Neurol. Psychiatry *34*: 1022–1040, 1935.

Kelland, M. D., Sr. and Walters, J. R.: Apomorphine-induced changes in striatal and pallidal neuronal activity are modified by NMDA and muscarinic receptor blockade. Life Sci. *50*: 79–84, 1992.

Kellar, K. J., Giblin, B. A. and Lumpkin, M. D.: Regulation of brain nicotinic cholinergic recognition sites and prolactin release by nicotine. Prog. Brain Res. *79*: 209–216, 1989.

Kellar, K. J., Davila-Garcia, M. I. and Xiao, Y.: Pharmacology of neuronal nicotinic acetylcholine receptors: effects of acute and chronic nicotine. Nicotine Tob. Res. *1*, Supplt 2: 117–120, 1999.

Kelley, A. E., Andrzejewski, M. E., Baldwin, A. E., Hernandez, P. J. and Pratt, W. E.: Glutamate-mediated plasticity in corticostriatal networks: role in adaptive motor learning. Ann. N. Y. Acad. Sci. *1003*: 159–168, 2003.

Kelly, J. F. and Hake, D. F.: An extinction-induced increase in aggressive response with humans. J. Exp. Anal. Behav. *14*: 153–64, 1970.

Kelly, J. S. and Rogawski, M. A.: Acetylcholine. In: "Neurotransmitter Actions in the Vertebrate Nervous System," M. A. Rogawski and J. L. Barker, Eds., pp. 143–197, Plenum Press, New York, 1985.

Kemel, M. L., Desban, M., Glowinski, J. and Gauchy, C.: Distinct presynaptic control of dopamine release in striasomal and matrix areas of the cat caudate nucleus. Proc. Natl. Acad. Sci. U.S.A. *86*: 9006–9010, 1989.

Keyser, K. T., MacNeil, M. A., Dmitrieva, N., Wang, F., Masland, R. H. and Lindstrom, J. M.: Amacrine, ganglion, and displaced amacrine cells in the rabbit retina express nicotinic acetylcholine receptors. Vis. Neurosci. *17*: 743–752, 2000.

Kharkevich, D. A.: Differential effects of anticholinergic drugs on muscarinic receptors of different organs and localizations. Farmakol. Toksikol. *6*: 45–51, 1957.

Khateb, A., Muhlethaler, M., Alonso, A., Serafin, M., Mainville, L. and Jones, B. E.: Cholinergic nucleus basalis neurons display the capacity for rhythmic bursting activity mediated by low-threshold calcium spikes. Neuroscience *51*: 489–494, 1992.

Kichigina, V. F., Kudina, T. A., Zenchenko, K. I. and Vinogradova, O. S.: Background activity of rabbit hippocampal neurons in conditions of functional exclusion of structures, which regulate theta rhythm. Neurosci. Behav. Physiol. *29*: 377–384, 1999.

Kikuchi, M., Wada, Y., Nambu, Y., Nakajima, A., Tachibana, H., Takeda, T. and Hashimoto, T.: EEG changes following scopolamine administration in healthy subjects: quantitative analysis during rest and photic stimulation. Neuropsychobiology *39*: 219–226, 1999.

Kilbinger, H.: Modulation of acetylcholine release by nitric oxide. In: "Cholinergic Mechanisms: From Molecular Biology to Functional Significance," J. Klein and K. Loeffenholz, Eds., pp. 219–224, Elsevier, Amsterdam, 1996.

Kim, J. S. and Levin, E. D.: Nicotinic, muscarinic and dopaminergic actions in the ventral hippocampus and the nucleus accumbens: effects on spatial working memory in rats. Brain Res. *725*: 231–240, 1996.

Kimura, F.: Cholinergic modulation of cortical function: a hypothetical role in shifting the dynamics in cortical network. Neurosci. Res. *38*: 1119–1126, 2000.

Kimura, F., Fukuda, M. and Tsumoto, T.: Acetylcholine suppresses the spread of excitation in the visual cortex revealed by optical recording: possible differential effect depending on the source of input. Eur. J. Neurosci. *11*: 3597–3609, 1999.

Kimura, H., Kubin, L., Davies, R. D. and Pack, A. I.: Cholinergic stimulation of the pons depresses respiration in decerebrate cats. J. Appl. Physiol. *69*: 2280–2289, 1990.

Kimura, T., Tamura, R., Kurimoto, H. and Ono, T.: Effects of T-588, a newly synthesized cognitive enhancer, on hippocampal CA1 neurons in rat brain tissue slices. Brain Res. *831*: 175–183, 1999.

Kiraly, M. K. and Phillis, J. W.: Action of some drugs on the dorsal root potentials of the isolated toad spinal cord. Br. J. Pharmacol. *17*: 224–231, 1961.

Kirkpatrick, W. E. and Lomax, P.: Temperature change following iontophoretic injection of acetylcholine into rostral hypothalamus of the rat. Neuropharmacology *9*: 195–202, 1970.

Kirukhin, V. G. and Ryzhenkov, V. E.: Comparison of the effects of central cholinolytics, of muscarinic and nicotinic action on motor food conditioned reflexes and sexual behavior of male rats. Bull. Exp. Biol. Med. *73*: 59–62, 1962.

Kiselevski, Y., Oganesian, N., Zimatkin, S., Szutowicz, A., Angielski, S., Niezabitowski, P., Uracz, W. and Gryglewski, R. J.: Acetate metabolisms in brain mechanisms of adaptation to ethanol. Med. Sci. Monit. *9*: 78–82, 2003.

Kissel, J. W. and Domino, E. F.: The effects of some possible neurohumoral agents on spinal cord reflexes. J. Pharmacol. Exp. Ther. *125*: 168–177, 1959.

Kitamura, M., Koshikawa, N., Yoneshige, N. and Cools, A. R.: Behavioral and neurochemical effects of cholinergic and dopaminergic agonists administered into the accumbal core and shell. Neuropharmacology *38*: 1387–1407, 1999.

Klaue, R.: Die bioelektrische Tätigkeit der Grosshirnrinde im normalen Schlaf und in der Narkose durch Schlafmittel. J. Psychol. Neurol. *47*: 510–531, 1937.

Klein, S. A.: Is guantum mechanics relevant to understanding consciousness? Psychc. *2*: 1–9, 1995.

Klein, S. A.: Libet's temporal anomalies: a reassessment of the data. Consc. Cogn. *11*: 198–214, 2002.

Kleinwaechter, L.: Beobachtung über die Wirkung des Calabar-Extracts gegen Atropin-Vergiftung. Berlin Klin. Wschr. *1*: 369–371, 1864.

Kleitman, N.: "Sleep and Wakefulness," Revised Edition, University of Chicago Press, Chicago, 1963.

Kluver, H.: Discussion of the paper by M. R. Rosenzweig, D. Krech and E. L. Bennett. In: "Neurological Basis of Behavior," G. E. W. Wolstenholme and C. M. O'Connor, Eds., pp. 356–357, Little, Brown, Boston, 1958.

Kluver, H. and Bucy, P. C.: Preliminary analysis of function of the temporal lobes in monkeys. Arch. Neurol. Psychiatry *2*: 979–1000, 1939.

Knott, V., Bosman, M., Mahoney, C., Ilivitsky, V. and Quirt, K.: Transdermal nicotine: single dose effects on mood, EEG, performance, and event-related potentials. Pharmacol. Biochem. Behav. *63*: 253–261, 1999.

Koda, H., Hashimoto, T. and Kuriyama, K.: Muscarinic receptor-mediated regulation of OM-853-enhanced dopamine release in striatum of rat. Eur. J. Pharmacol. *102*: 501–508, 1989.

Koehn, G. L. and Karczmar, A. G.: Effect of diisopropyl phosphofluoridate on analgesia and motor behavior in the rat. Prog. Neuropsychopharmacol. *2*: 169–177, 1978.

Koehn, G. L., Henderson, G. and Karczmar, A. G.: Diisopropyl phosphofluoridate-induced antinociception: possible role of endogenous opiates. Eur. J. Pharmacol. *61*: 167–173, 1980.

Koelle, G. B.: Cytological distributions and phy-siological functions of cholinesterases. In: "Cholinesterases and Anticholinesterase Agents," G. B. Koelle, Ed., Handbch. d. exp. Pharmakol., Ergänzungswk., vol. 15, pp. 189–298, Springer-Verlag, Berlin, 1963.

Koelle, G. B.: Significance of acetylcholinesterase in central synaptic transmission. Fed. Proc. *28*: 95–100, 1969.

Koestler, A.: "The Act of Creation," Macmillan, New York, 1964.

Kohyama, J., Shimohira, M. and Iwakawa, Y.: Brainstem control of phasic muscle activity during REM sleep: a review and hypothesis. Brain Dev. *16*: 81–91, 1994.

Kok, A.: REM sleep pathways and anticholinesterase intoxication: a mechanism for nerve agent-induced respiratory failure. Med. Hypotheses *41*: 141–149, 1993.

Koketsu, K.: Restorative action of fluoride on synaptic transmission blocked by organophosphorous anticholinesterases. Int. J. Neuropharmacol. *5*: 247–254, 1966.

Koketsu, K.: Modulation of receptor sensitivity and action potentials by transmitters in vertebrate neurons. Jpn. J. Physiol. *34*: 945–960, 1984.

Koketsu, K., Karczmar, A. G. and Kitamura, K.: Acetylcholine depolarization of the dorsal root nerve terminals in amphibian spinal cord. Int. J. Neuropharmacol. *8*: 329–336, 1969.

Kolasa, K., Harrell, L. E. and Parsons, D. S.: The effect of hippocampal sympathetic ingrowth and cholinergic denervation on hippocampal M2 cholinergic receptors. Brain Res. *684*: 201–205, 1995.

Koley, J., Saha, J. K., Ghosh, T. K. and Koley, B. N.: Evidence behind the mechanisms of cardiac-respiratory response to nicotine. In: "The Pharmacology of Nicotine," M. Rand and K. Thurau, Eds., pp. 137–141, IRL Press, Oxford, 1988.

Koley, J., Guosh, T. K. and Koley, B.: Pharmacological and electrophysiological analysis of nicotine induced apnoea in the cat. Acta Physiol. Hung. *75*: 77–89, 1990.

Konopacki, J., Bland, B. H. and Roth, S. H.: Carbachol-induced EEG theta in hippocampal formation slices: evidence for a third generator of theta in CA3c area. Brain Res. *451*: 33–42, 1988.

Konorski, J.: "Integrative Activity of the Brain," University of Chicago Press, Chicago, Ill., 1967.

Konturek, S. J., Konturek, J. W., Pawlik, T. and Brzozowski, T.: Brain-gut axis and its role in the control of food intake. J. Physiol. Pharmacol. *55*: 137–154, 2004.

Kopelman, M. D.: The Korsakoff syndrome. Br. J. Psychiat. *166*: 154–173, 1995.

Koppanyi, T.: Studies on the pharmacology of the afferent autonomic nervous system. Arch. Intern. Pharmacodyn. Ther. *39*: 275–284, 1930.

Koppanyi, T.: The action of toxic doses of atropine on the central nervous system. Proc. Soc. Exp. Biol. Med. *40*: 244–248, 1939.

Koppanyi, T.: Acetylcholine as a pharmacological agent. Bull. Johns Hopkins Hosp. *83*: 532–561, 1948.

Koppanyi, T. and Karczmar, A. G.: The effect of cholinesterase inhibitors on the electrical excitability of the cephalic end of the vago-sympathetic trunk in dogs. Am. J. Physiol. *159*: 576–577, 1949.

Koppanyi, T. and Karczmar, A. G.: Contribution to the study of the mechanism of action of cholinesterase inhibitors. J. Pharmacol. Exp. Ther. *101*: 327–343, 1951.

Koppanyi, T. and Karczmar, A. G.: "Experimental Pharmacodynamics," Third Edition, Burgess Publ., Minneapolis, Minn., 1963.

Koppanyi, T., Karczmar, A. G. and King, T. O.: The mechanism of action of anticholinesterases. Symposium on Military Physiology, Milit. Res. and Developt. Board *4*: 271–285, 1948.

Kosasa, T., Kuriya, Y., Matsui, K. and Yamanishi, Y.: Effect of donepezil hydrochloride (E2020) on basal concentration of extracellular acetylcholine in the hippocampus of rats. Eur. J. Pharmacol. *380*: 101–107, 1999.

Koshimura, K., Ohua, T., Watanabe, Y. and Miwa, S.: Neurotransmitter releasing action of 6R-tetrahydrobiopterin. J. Nutr. Sci. Vitaminol., Special No.: 505–509, 1992.

Kostenis, E., Zeng, F.-Y. and Wess, J.: Structure-function analysis of muscarinic acetylcholine receptors.

In: "From *Torpedo* Electric Organ to Human Brain: Fundamental and Applied Aspects," J. Massoulié, Ed., pp. 265–268, Elsevier, Amsterdam, 1998.

Kovach, J. K.: Effects of autonomic drugs on imprinting. J. Comp. Physiol. Psychol. *57*: 183–187, 1964.

Kow, L. M. and Pfaff, D. W.: The membrane actions of estrogens can potentiate their lordosis behavior-facilitating genomic actions. Proc. Natl. Acad. Sci. U.S.A. *101*: 12354–12357, 2004.

Kow, L. M., Tsai, Y. F., Weiland, N. G., McEwen, B. S. and Pfaff, D. W.: *In vitro* electro-pharmacological and autoradiographic analyses of muscarinic receptor subtypes in rat hypothalamic ventromedial nucleus: implications for cholinergic regulation of lordosis. Brain Res. *694*: 29–39, 1995.

Koyama, Y., Imada, N., Kawauchi, A. and Kayama, Y.: Firing of the putative cholinergic neurons and micturition center neurons in the laterodorsal tegmentum during distention and contraction of urinary bladder. Brain Res. *840*: 45–55, 1999.

Krakauer, J.: Fitness. Playboy. *41*: 38, 1994.

Kramis, R., Vanderwolf, C. A. and Bland, B. H.: Two types of hippocampal rhythmical slow activity in both rabbit and rat: relations to behavior and effects of atropine, diethyl ether, urethane, and pentobarbital. Exp. Neurol. *49*: 58–85, 1975.

Krech, D., Rosenzweig, M. R. and Bennett, E. L.: Effects of environmental complexity and blindness on rat brain. Arch. Neurol. *8*: 403–412, 1963.

Krivoy, W. A. and Marrazzi, A. S.: Evaluation of the central action of anticholinesterases in producing respiratory paralysis. Fed. Proc. *10*: 316, 1951.

Krivoy, W. A., Hart, E. R. and Marrazzi, A. S.: Further analysis of the actions of DFP and curare on the respiratory center. J. Pharmacol. Exp. Ther. *103*: 351, 1951.

Krnjevic, K.: Central cholinergic pathways. Fed. Proc. *28*: 113–120, 1969.

Krnjevic, K.: Excitable membranes and anesthetics. In: "Cellular Biology and Toxicity of Anesthetics," R. B. Fink, Ed., pp. 3–9, William and Wilkins, Baltimore, 1972.

Krnjevic, K.: Central actions of general anaesthetics. In: "Molecular Mechanisms in General Anaesthesia," M. J. Halsey, R. A. Millar and J. A. Sutton, Eds., pp. 65–69, 1974a.

Krnjevic, K.: Chemical nature of synaptic transmission in vertebrates. Physiol. Rev. *54*: 418–540, 1974b.

Krnjevic, K.: Coupling of neuronal metabolism and electrical activity. In: "Brain Work," D. H. Ingvar and N. A. Lassen, Eds., pp. 65–78, Munksgaard Press, Kopenhagen, 1975.

Krnjevic, K.: Acetylcholine as modulator of amino-acid-mediated synaptic transmission. In: "The Role of Peptides and Aminoacids as Neurotransmitters," pp. 127–141, Alan R. Liss, New York, 1981.

Krnjevic, K.: Acetylcholine as transmitter in the cerebral cortex. In: "Neurotransmitters and Brain Function," M. Avoli, T. A. Reader, R. Dykes, and P. Gloor, Eds., pp. 227–237, Plenum Press, New York, 1988.

Krnjevic, K.: Acute hypoxia on hippocampal neurons. In: "Ionic Currents and Ischemia," J. Vereecke, F. van Bogaert, and F. Verdonck, Eds., pp. 121–143, Leuven University Press, Leuven, 1990.

Krnjevic, K.: Central cholinergic mechanisms and function. In: "Cholinergic Neurotransmission: Function and Dysfunction," A. C. Cuello, Ed., pp. 285–293, Elsevier, Amsterdam, 1993.

Krnjevic, K. and Leblond, J.: Changes in membrane currents of hippocampal neurons evoked by brief anoxia. J. Neurophysiol. *62*: 15–30, 1989.

Krnjevic, K. and Phillis, J. W.: Pharmacological properties of acetylcholine-sensitive cells in the cerebral cortex. J. Physiol. (Lond.) *166*: 328–350, 1963.

Krnjevic, K., Pumain, R. and Renaud, L.: The mechanism of excitation by acetylcholine in the cerebral cortex. J. Physiol. *215*: 247–268, 1971.

Krnjevic, K., Puil, E. and Werman, R.: Significance of 2,4-dinitrophenol action on spinal motoneurones. J. Physiol. (Lond.) *275*: 225–239, 1978.

Krnjevic, K., Reiffenstein, R. J. and Ropert, N.: Disinhibitory action of acetylcholine in the hippocampus. J. Physiol. (Lond.) *308*: 73–74, 1980.

Krsmanovic, L. Z., Mores, N., Navarro, C. E., Saeed, S. A., Arora, K. K. and Catt, K. J.: Muscarinic regulation of intracellular signaling and neurosecretion in gonadotropin-releasing hormone neurons. Endocrinology *139*: 4027–4043, 1998.

Kuba, K. and Koketsu, K.: Synaptic events in sympathetic ganglia. Prog. Neurobiol. *11*: 77–169, 1978.

Kubo, T. and Misu, Y.: The nucleus reticularis dorsalis: a region sensitive to physostigmine Neuropharmacol. *22*: 1155–1158, 1983.

Kubo, T., Ishizuka, T., Fukumori, R., Asari, T. and Hagiwara, V.: Enhanced release of acetylcholine in the rostral ventrolateral medulla of spontaneously hypertensive rats. Brain Res. *686*: 1–9, 1995.

Kubo, T., Fukumori, R., Kobayashi, M. and Yagamuchi, H.: Altered cholinergic mechanisms and blood pressure regulation in the rostral ventrolateral medulla of DOCA-salt hypertensive rats. Brain Res. Bull. *45*: 327–332, 1998.

Kubo, T., Hagiwara, Y., Sekiya, D. and Fukumori, R.: Midbrain central gray is involved in mediation of cholinergic inputs to the rostral ventrolateral medulla. Brain Res. Bull. *50*: 41–46, 1999.

Kubota, T., Hirota, K., Anzawa, N., Yoshida, H., Kushikata, T. and Matsuki, A.: Physostigmine antagonizes ketamine-induced noradrenaline release from the medial prefrontal cortex in rats. Brain Res. *840*: 175–180, 1999.

Kuffler, S. W.: Electric potential changes at an isolated nerve-muscle junction. J. Neurophysiol. *5*: 18–26, 1946.

Kuhar, M. J.: The anatomy of cholinergic neurons. In: "Biology of Cholinergic Function," A. M. Goldberg and I. Hanin, Eds., pp. 3–27, Raven Press, New York, 1976.

Kuhar, M. J.: Central cholinergic pathways: physiologic and pharmacological aspects. In: "Psychopharmacology—A Generation of Progress," A. Lipton, A. DiMascio, and K. F. Killam, Eds., pp. 199–204, Raven Press, New York, 1978.

Kuridze, N. T., Mehedlishvili, G. I., Baranidze, D. G., Gordeladze, T. and Gadamsky, R.: Osobennotsi fiziologicheskogo mekhanisma regulirovaniia mikrosyrkulatsii v kore golovnogo mozga ptits i mleko pitaiushchikkh. Fiziol. Zh. SSSR *75*: 508–514, 1989.

Kurosawa, M., Sato, A. and Sato, Y.: Well-maintained responses of acetylcholine release and blood flow in the cerebral cortex to focal stimulation of the nucleus basalis of Meynert in aged rats. Neurosci. Lett. *100*: 198–202, 1989.

Kurotani, S., Umegaki, H., Ishiwata, K., Suzuki, Y. and Iguchi, A.: The age associated changes in dopamine-acetylcholine interaction in the striatum. Exp. Gerontol. *38*: 1009–1013, 2003.

Kuryatov, A., Gerzanich, V., Nelson, M., Olale, F. and Lindstrom, J.: Mutation causing autosomal dominant nocturnal frontal lobe epilepsy alters Ca2+ permeability, conductance, and gating of human alpha-4beta2 nicotinic acetylcholine receptors. J. Neurosci. *17*: 9035–9047, 1997.

Kuryatov, A., Olale F., Cooper, I., Choi, C. and Lindstrom, J.: Human alpha6 AChR subtypes: subunit composition, assembly, and pharmacological responses. Neuropharmacology. *39*: 2570–2590, 2000.

Kuzmiski, J. B. and MacVicar, B. A.: Cyclic nucleotide-gated channels contribute to the cholinergic plateau potential in hippocampal CA1 pyramidal neurons. J. Neurosci. *21*: 8707–8714, 2001.

Kwok, K. H., Law, K. B., Wong, R. N. and Yung, K. K.: Immunolesioning of nerve growth factor p75 receptor-containing neurons in the rat brain by a novel immunotoxin: anti-p75-anti-mouse IgG-trichosanthin conjugates. Brain Res. *846*: 154–163, 1999.

Lacombe, P., Sercombe, R., Verrechia, C., Philipson, V., MacKenzie, E. T. and Seylaz, J.: Cortical blood flow increases, induced by stimulation of the substantia innominata in the unanesthetized rat. Brain Res. *491*: 1–14, 1989.

Ladinsky, H.: Acetylcholine receptors: drugs and molecular genetics. Progr. Brain Res. *98*: 103–111, 1993.

Ladinsky, H., Consolo, S. and Pugnetti, P.: A possible central muscarinic receptor agonist role for choline in increasing rat striatal acetylcholine content. In: "Nutrition and the Brain," A. Barbeau, J. H. Croydon, and R. J. Wurtman, Eds., pp. 525–538, Raven Press, New York, 1979.

Ladinsky, H., Consolo, S., Forloni, G., Fiorentini, F. and Fisone, G.: Influence of frontal decortication on drugs affecting cholinergic activity and cataleptic behavior: restoration studies. In: "Neurobiology of Acetylcholine: A Symposium in Honor of Alexander G. Karczmar," N. J. Dun and R. L. Perlman, Eds., pp. 403–416, Plenum Press, New York, 1987.

Ladinsky, H., Schiavi, G. B., Monferini, E. and Giraldo, H.: Pharmacological muscarinic receptor subtypes. In: "Cholinergic Neurotransmission: Functional and Clinical Aspects," S.-M. Aquilonius and P.-G. Gillberg, Eds., pp. 193–200, Elsevier, Amsterdam, 1990.

Ladner, C. J. and Lee, J. M.: Reduced high-affinity agonist binding at the M(1) muscarinic receptor in Alzheimer's disease brain: differential sensitivity to agonists and divalent ions. Exp. Neurol. *158*: 451–458, 1999.

Lalanne, J.-L., Schmitt, H. and Schmitt, H.: Sur le mécanisme de l'ypertension provoquée par l'eserine chez le rat. Arch. Int. Pharmacol. Therapie *163*: 444–461, 1966.

Lallement, G., Carpentier, P., Collet, A., Baubichon, D., Pernot-Marino, I. and Blanchet, G.: Extracellular acetylcholine changes in rat limbic structures during soman-induced seizures. Neurotoxicology *13*: 557–567, 1992.

Lamarre, Y. and Joffroy, A. J.: Thalamic unit activity in monkey with experimental tremor. In: "L-DOPA and Parkinsonism," A. Barbeau and F. H. McDowell, Eds., pp. 163–170, Davis Publs., Philadelphia, 1970.

Lambert, D. G. and Nahorski, S. R.: Second-messenger responses associated with stimulation of neuronal muscarinic responses expressed by a human neuroblastoma SH-SY. In: "Cholinergic Neurotransmission: Functional and Clinical Aspects," S.-M. Aquilonius and P.-G. Gillberg, Eds., pp. 31–42, Elsevier, Amsterdam, 1990.

Lamour, Y. and Epelbaum, J.: Interactions between cholinergic and peptidergic systems in the cerebral cortex and hippocampus. Prog. Neurobiol. *31*: 109–148, 1988.

Lamour, Y., Senut, M. C., Dutar, P. and Bassant, M. H.: Neuropeptides and septohippocampal neurons: electrophysiological effects and distributions of immunoreactivity. Peptides *9*: 1351–1359, 1988.

Lanca, A. J., Adamson, K. L., Coen, K. M., Chow, B. L. and Corrigal, W. A.: The pedunculopontine tegemental nucleus and the role of cholinergic neurons in nicotine self-administration in the rat: a correlative neuroanatomical and behavioral study. Neuroscience *96*: 735–742, 2000.

Langer, S. Z.: Presynaptic receptors and their role in the regulation of transmitter release. Br. J. Pharmacol. *60*: 481–477, 1977.

Langnickel, R., Bluth, R. and Oelssner, W.: Various dose-dependent influences of apomorphine on the acetylcholine turnover in striatum and mesolimbic areas of rat brain. Biomed. Biochim. Acta *42*: 937–946, 1983.

Lapchak, P. A., Araujo, D. M. and Collier, B.: Regulation of acetylcholine release from mammalian brain slices by opiate receptors: hippocampus, striatum and cerebral cortex of guinea-pig and rat. Neuroscience *31*: 313–325, 1989a.

Lapchak, P. A., Araujo, D. M., Quirion, R. and Collier, B.: Binding sites for [3H] Af-DX 116 and effect of AF-DX 116 on endogenous acetylcholine release from rat brain slices. Brain Res. *496*: 285–294, 1989b.

Lapchak, P. A., Araujo, D. M., Quirion, R. and Collier, B.: Effect of chronic nicotine treatment on nicotinic autoreceptor function and N-[3H] methylcarbamylcholine binding sites in the rat brain. J. Neurochem. *52*: 483–491, 1989c.

Lapchak, P. A., Araujo, D. M., Quirion, R. and Beaudet, A.: Chronic estradiol treatment alters central cholinergic function in the female rat: effect on choline acetyltransferase activity, acetylcholine content, and nicotinic autoreceptor function. Brain Res. *525*: 249–255, 1990.

Larsson, C. and Nordberg, A.: Comparative analysis of nicotine-like receptor-ligand interactions in rodent brain homogeneates. J. Neurochem. *45*: 24–31, 1985.

Larsson, C., Nilsson, L., Hallen, A. and Nordberg, A.: Subchronic treatment of rats with nicotine: effects on tolerance and 3H-acetylcholine and 3H-nicotine binding in the brain. Drug Alcohol Depend. *17*: 37–45, 1986.

Lasek, R. J.: Dynamic properties of cytoskeletons. Neurosci. Res. Prog. Bull. *19*: 7–32, 1981.

Laviolette, S. R. and van der Kooy, I.: The neurobiology of nicotine addiction: bridging the gap from molecules to behavior. Nat. Rev. Neurosci. *5*: 55–65, 2004.

Leanza, G., Nilsson, O. G. and Bjorklund, A.: Functional activity of intrahippocampal septal grafts is regulated by catecholaminergic host afferents as studied by dialysis of acetylcholine. Brain Res. *618*: 47–56, 1993.

Lebeda, F. J. and Rutecki, P. A.: Organophosphorus anticholinesterase-induced epileptiform activity in the hippocampus. In: "Neurobiology of Acetylcholine: A Symposium in Honor of A. G. Karczmar," N. J. Dun and R. L. Perlman, Eds., pp. 437–450, Plenum Press, New York, 1987.

Lebedev, M. A. and Nelson, R. J.: Rhythmically firing neostriatal neurons in monkey: activity patterns during reaction-time hand movements. J. Neurophysiol. *82*: 1832–1842, 1999.

Le Bon, O.: Is REM latency a dying concept? Acta Psychiatr. Belg. *92*: 131–150, 1992.

Lechner, H. A. and Byrne, J. H.: New perspectives on classical conditioning: a synthesis of Hebbian and non-Hebbian mechanisms. Neuron *20*: 355–358, 1998.

Ledoux, J. E. and Muller, J.: Emotional memory and psychopathology. Philos. Trans. R. Soc. Lond., B, Biol. Sci. *352*: 1719–1726, 1997.

Lee, H.-J., Alcorn, L. M. and Weiler, M. H.: Effects of various experimental manipulations on neostriatal acetylcholine and dopamine release. Neurochem. Res. *16*: 875–883, 1992.

Lee, T. F., Denbow, D. M., King, S. E. and Myers, R. D.: Unique contribution of catecholamine receptors in the brain of the cat underlying feeding. Physiol. Behav. *29*: 527–532, 1982.

Lee, T. F., Hepler, J. R. and Myers, R. D.: Evaluation of neurotensin's thermolytic action by icv infusion with receptor antagonists and a Ca++ chelator. Pharmacol. Biochem. Behav. *19*: 477–481, 1983.

Lee, T., Kaneko, T., Shigemoto, R., Nomura, S. and Mizuno, N.: Collateral projections from striatonigral neurons to substance P receptor-expressing intrinsic neurons in the striatum of the rat. J. Comp. Neurol. *388*: 250–264, 1997.

Lehman, J. and Langer, S. Z.: Muscarinic receptors on dopamine terminals in the rat caudate nucleus: Neuromodulation of [H3] dopamine release *in vitro* by endogenous acetylcholine. Brain Res. *248*: 61–69, 1982.

Lehman, J., Smith, R. V. and Langer, S. Z.: Stereoisomers of apomorphine differ in affinity and intrinsic activity at presynaptic dopamine receptors modulating [3H] dopamine and [3H] acetylcholine release in slices of cat caudate nucleus. Eur. J. Pharmacol. *88*: 11–26, 1983.

Leibowitz, S. F.: Central adrenergic receptors and the regulation of hunger and thirst. In: "Neurotransmitters," I. J. Kopin, Ed., Proc. Assoc. Nerv. Ment. Dis. *50*: 327–358, 1972.

Leibowitz, S. L.: Neurochemical systems of the hypothalamus: control of feeding and drinking behavior and water-electrolyte excretion. In: "Handbook of the Hypothalamus," P. J. Morgane and J. Panksepp, Eds., vol. 3, pp. 299–437, Marcel Dekker, New York, 1980.

Lena, C. and Changeux, J.-P.: Allosteric nicotinic receptors, human pathologies. In: "From *Torpedo* Electric Organ to Human Brain: Fundamental and Applied

Aspects," J. Massoulié, Ed., pp. 63–74, Elsevier, Amsterdam, 1998.

Lena, C. and Changeux, J.-P.: The role of beta 2-subunit-containing nicotinic acetylcholine receptors in the brain explored with a mutant mouse. Ann. N. Y. Acad. Sci. *868*: 611–616, 1999.

Leonard, C. S. and Llinas, R.: Serotonergic and cholinergic inhibition of mesopontine cholinergic neurons controlling REM sleep: an *in vitro* electrophysiological study. Neuroscience *59*: 309–330, 1994.

Leonard, T. O. and Lydic, R.: Nitric oxide synthetase inhibition decreases pontine acetylcholine release. Neuroreport *6*: 1325–1329, 1995.

Leranth, C. and Frotscher, M.: Organization of the septal region in the rat brain: cholinergic-GABAergic interconnections and the termination of the hippocampo-septal fibers. J. Comp. Neurol. *289*: 304–314, 1989.

Lesny, I. and Vojta, V.: Eserine activation of the electroencephalograph in children. Electroencephalogr. Clin. Neurophysiol. *12*: 742–743, 1960.

Leung, L. S.: Generation of theta and gamma rhythms in the hippocampus. Neurosci. Biobehav. Rev. *22*: 275–290, 1998.

Leung, L. S. and Yim, C. Y.: Rhythmic delta-frequency activities in the nucleus accumbens of anesthetized and freely moving rats. Can. J. Physiol. Pharmacol. *71*: 311–320, 1993.

Levin, E. D.: Nicotinic systems and cognitive function. Psychopharmacology *108*: 417–431, 1992.

Lewin, B.: "Genes," Oxford University Press, New York, 1994.

Lewis, P. R and Shute, C. C. D.: The cholinergic limbic system: projections to hippocampal formation, medial cortex, nuclei of the ascending cholinergic reticular system, and the subfornical organ and supra-optic crest. Brain *90*: 521–540, 1967.

Li, Y., Wu, X., Zhu, J., Yan, J. and Owyang, C.: Hypothalamic regulation of pancreatic secretion is mediated by central cholinergic pathways in the rat. J. Physiol. *552*, pt. 2: 571–587, 2003.

Liberson, W. T. and Cadilhac, J.: Les crises hippocampiques experimentales. Montp. Med. *45*: 515–546, 1954.

Liberson, W. T. and Karczmar, A. G.: Drug effects on behavior and evoked potentials in fixated rats. In: "Neurophysiological and Behavioral Aspects of Psychotropic Drugs," A. G. Karczmar and W. P. Koella, Eds., pp. 161–178, 1969.

Libet, B.: Slow postsynaptic actions in ganglionic functions. In: "Integrative Functions of the Autonomic Nervous System," C. McC. Brooks, K. Koizumi, and A. Sato, Eds., pp. 199–222, University of Tokyo Press, Tokyo, 1979.

Libet, B.: Unconscious cerebral initiative and the role of conscious will in the voluntary action. Behav. Brain Sci. *8*: 529–556, 1985.

Libet, B.: Introduction to slow synaptic potentials and their modulation by dopamine. Can. J. Physiol. Pharmacol. *70*: S3–S11, 1991.

Libet, B.: The neural time factor in conscious and unconscious events. In: "Experimental and Theoretical Studies of Consciousness," T. Nagel, Ed., pp. 123–146, Ciba Found. Symp. No. 174, CIBA, New York, 1993.

Libet, B., Alberts, W. W. and Perryman, J. H.: Pharmacological behavior of spinal cord potentials relative to reflex and ascending discharge. Am. J. Physiol. *198*: 414–420, 1960.

Lichtensteiger, W.: The neuroendocrinology of dopamine systems. In: "The Neurobiology of Dopamine," A. S. Horn, J. Korf and B. H. C. Westerink, Eds., pp. 491–521, Academic Press, New York, 1979.

Lichtigfeld, F. J. and Gillman, M. A.: Psychotropic analgesic nitrous oxide and neurotransmitter mechanisms involved in alcohol withdrawal state. Int. J. Neurosci. *76*: 17–33, 1994.

Liljenstrom, H. and Hasselmo, M. E.: Cholinergic modulation of cortical oscillatory dynamics. J. Neurophysiol. *74*: 288–297, 1995.

Lin, C. H.: Studies on possible correlations between cholinesterase levels in mouse brain and behavior. Ph.D. Thesis, Loyola University, Chicago, 1973.

Lin, H. H. and Dun, N. J.: Post- and presynaptic GABA-B receptor activation in neonatal rat rostral ventrolateral medulla neurons *in vitro*. Neuroscience *86*: 211–220, 1998.

Lindstrom, J.: Nicotinic acetylcholine receptors. In: "CRC Handbook of Receptors," A. North, Ed., pp. 153–175, CRC Press, New York, 1994.

Lindstrom, J.: Neuronal nicotinic acetylcholine receptors. Ion Channels. *4*: 377–450, 1996.

Lindstrom, J.: Nicotinic acetylcholine receptors in health and disease. Mol. Neurobiol. *15*: 193–222, 1997.

Lindstrom, J., Schoepfer, R. and Whiting, P.: Molecular studies of the neuronal nicotinic receptor family. Mol. Neurobiol. *1*: 281–337, 1990.

Lindstrom, J., Anand, R., Gerzanich, V., Peng, X., Wang, F. and Wells, G.: Structure and function of neuronal nicotinic receptor. In: "Cholinergic Mechanisms: From Molecular Biology to Clinical Significance," J. Klein and K. Loeffelholz, Eds., pp. 125–137, Elsevier Press, Amsterdam, 1996.

Lindstrom, L. H.: Further studies on cholinergic mechanisms and hormone-activated copulatory behavior in the female rat. J. Endocrinol. *56*: 275–283, 1973.

Lindstrom, L. H.: Cholinergic mechanisms and sexual behavior in the female rat. In: "Sexual Behavior," M. Sandler and G. L. Gessa, Eds., pp. 161–167, Raven Press, New York, 1975.

Linster, C., Wyble, B. P. and Hasselmo, M. E.: Electrical stimulation of the horizontal limb of the diagonal band of Broca modulates population EPSPs in piriform cortex. J. Neurophysiol. *81*: 2737–2742, 1999.

Linster, C., Maloney, M., Patil, M. and Hasselmo, M. E.: Enhanced cholinergic suppression of previously strengthened synapses enables the formation of self-organized representations in oolfactory cortex. Neurobiol. Learn. Mem. *80*: 302–314, 2003.

Linster, C., Sachse, S. and Galizia, C. G.: Computational modeling suggests that response properties rather than spatial position determine connectivity between olfactory glomeruli. J. Neurophysiol. *93*: 3410–3417, 2005. Epub 2005 Jan 26.

Lisman, J. E., Ed.: "The Molecular Basis for Maintenance of Synaptic Memory," Society for Neuroscience, 34th Annual Meeting, San Diego, Calif., 2004.

Lister, T. and Ray, D. E.: The role of basal forebrain in the primary vasodilation in rat neocortex produced by systemic administration of cismethrin. Brain Res. *450*: 364–368, 1988.

Liu, C.-G. and Gillette, M. U.: Cholinergic regulation of the suprachiasmatic nucleus circadian rhythm via a muscarinic mechanism at night. J. Neurosci. *16*: 744–751, 1996.

Liu, J. L. and Su, S. N.: The role of acetylcholine in the cognitive function of frontal neurons in monkeys. Sci. China, Ser. B, Chem. Life Sci. Earth Sci. *36*: 1510–1517, 1993.

Llinas, R.: "I of the Vortex: From Neurons to Self," MIT Press, Cambridge, Mass., 2001.

Llinas, R.: "El Cerebro y el Mito de Yo," Editorial Norma, Bogota, 2003.

Llinas, R. R. and Pare, D.: Of dreaming and wakefulness. Neuroscience *44*: 521–535, 1991.

Lloyd, D. P. C.: Facilitation and inhibition of spinal motoneurones. J. Neurophysiol. *9*: 421–438, 1946.

Lloyd, K. G.: The biological pharmacology of the limbic system: neuroleptic drugs. In: "Limbic Mechanisms," K. E. Livingston and O. Hornykiewicz, Eds., pp. 263–305, Plenum Press, New York, 1976.

Lloyd, K. G.: Neurotransmitter interactions related to central dopamine neurons. Essays Neurochem. Neuropharmacol. *3*: 129–207, 1978.

Loewi, O.: Über die humorale Übertragbarkeit des Herznervenwirkung. I. Erste Mitteilung. Pflugers Arch. Gesamte Physiol. Menschen Tiere *189*: 239–249, 1921.

Loewi, O.: An autobiographic sketch. Perspect. Biol. Med. *4*: 3–25, 1960.

Loewy, A. D. and Spyer, K. M., Eds.: "Central Regulation of Autonomic Function," Oxford University Press, New York, 1990.

Lomax, P. and Jenden, D. J.: Hypothermia following systemic and intracerebral injection of oxotremorine in the rat. Int. J. Neuropharmacol. *5*: 353–359, 1966.

Longino, E. H. and Preston, R. S.: The antagonism between atropine and strychnine in the mouse. J. Pharmacol. Exp. Ther. *86*: 174–176, 1946.

Longo, V. G.: Effects of scopolamine and atropine on electroencephalographic and behavioral reactions due to hypothalamic stimulation. J. Pharmacol. Exp. Ther. *116*: 198–208, 1956.

Longo, V. G.: Mechanism of the behavioral and electroencephalographic effects of atropine and related compounds. Pharmacol. Rev. *18*: 965–996, 1966.

Longo, V. G. and Loizzo, A.: Effects of drugs on hippocampal theta-rhythm. Possible relationship to learning and memory processes. In: "Brain, Nerves and Synapses," F. E. Bloom and G. H. Acheson, Eds., pp. 46–54, Proc. 5th Intern. Congr. Pharmacol., vol. 4, Karger Publs., Basel, 1973.

Lopez, E., Cruz, M. E. and Dominguez, R.: Asymmetrical effects of the unilateral implant of pilocarpine on the preoptic-anterior hypothalamic area on spontaneous ovulation of the adult rat. Arch. Med. Res. *28*: 343–348, 1997.

Lorenz, K.: "Das sogennante Böse: Zur Naturgeschichte der Aggression," Borotha-Schoeler, Vienna, 1963.

Loscher, W., Potschka, H., Wlaz, P., Danysz, W., Parsons, C. G.: Are neuronal nicotinic receptors a target for antiepileptic drug development? Studies in different seizure models in mice and rats. Eur. J. Pharmacol. *466*: 99–111, 2003.

Luccarini, P., Gahery, Y. and Pompeiano, O.: Cholinoceptive pontine reticular structures modify the postural adjustments during the limb movements induced by cortical stimulation. Arch. Ital. Biol. *128*: 19–45, 1990.

Lukas, R. J. and Bencherif, M.: Heterogeneity and regulation of nicotinic acetylcholine receptors. Int. Rev. Neurobiol. *34*: 25–131, 1992.

Luria, A. R.: "Working Brain," Basic Books, New York, 1973.

Lydic, R. and Baghdoyan, H. A.: Cholinoceptive pontine reticular mechanisms cause state-dependent respiratory changes in the cat. Neurosci. Lett. *102*: 211–216, 1989.

Lydic, R. and Baghdoyan, H. A.: Cholinergic pontine mechanisms causing state-dependent respiratory depression. News Physiol. Sci. *7*: 220–224, 1992.

Lynch, C. S., Ruppel, T. E., Domingue, R. N. and Ponthier, G. L.: Scopolamine inhibition of female sexual behavior in rhesus monkeys (Macaca mulatta). Pharmacol. Biochem. Behav. *63*: 655–661, 1999.

Ma, S., Abbound, F. M. and Felder, R. B.: Effects of L-arginine-derived nitric oxide synthesis on neuronal activity in nucleus tractus solitarius. Am. J. Physiol. *268*: R487–R491, 1995.

Maalouf, M., Miasnikov, A. A. and Dykes, R. W.: Blockade of cholinergic receptors in rat barrel cortex prevents long-term changes in the evoked potential during sensory preconditioning. J. Neurophysiol. *80*: 529–545, 1998.

Mac, D. C., Correia, E. E., Vasconcelos, S. M., Aguiar, L. M., Viana, G. S. and Sousa, F. C.: Cocaine treatment causes early and long-lasting changes in muscarinic and dopaminergic receptors. Cell. Mol. Neurobiol. *24*: 129–136, 2004.

Machne, S. and Unna, K. R. W.: Actions at the central nervous system. In: "Cholinesterases and Anticholinesterase Agents," G. B. Koelle, Ed., pp. 679–700, Handb. d. exper. Pharmakol., Ergänzungswk., vol. 15, Springer-Verlag, Berlin, 1963.

Macht, D. I.: A pharmacological analysis of the cerebral effects of atropine, homatropine, scopolamin and related drugs. J. Pharmacol. Exp. Ther. *22*: 35–48, 1924.

MacIntosh, F. C.: The distribution of acetylcholine in the peripheral and central nervous system. J. Physiol. *99*: 436–442, 1941.

MacIntosh, F. C. and Oborin, P. E.: Release of acetylcholine from intact cerebral cortex. Proc. XIX Intern. Physiol. Congr., Abstracts, Montreal, pp. 580–581, 1953.

MacVicar, B. A. and Tse, F. W.: Local neuronal circuitry underlying cholinergic rhythmical slow activity in the CA3 area of the rat hippocampal slices. J. Physiol. (Lond.) *417*: 197–212, 1989.

Madamba, S. G., Hsu, M., Schweitzer, P. and Siggins, G. R.: Ethanol enhances muscarinic cholinergic neurotransmission in rat hippocampus *in vitro*. Brain Res. *685*: 21–31, 1995.

Maelicke, A.: Nicotinic receptors of then vertebrate CNS: introductory remarks. In: "Cholinergic Mechanisms: From Molecular Biology to Clinical Significance," J. Klein and K. Loeffelholz, Eds., pp. 107–110, Elsevier, Amsterdam, 1996.

Maelicke, A., Samochocki, M., Jostock, R., Fehrenbacher, A., Ludwig, J., Albuquerque, E. X. and Zerlin, M.: Alosteric sensitization of nicotinic receptors by galantamine, a new treatment strategy for Alzheimer's disease. Biol. Psychiatry *49*: 279–288, 2001.

Magnani, M., Florian, A., Casamenti, F. and Pepeu, G.: An analysis of cholecystokinin-induced release of acetylcholine output from cerebral cortex of the rat. Neuropharmacology *26*: 1207–1210, 1987.

Maier, N. R. F.: "The Study of Behavior Without a Goal," McGraw, New York, 1949.

Makara, G. B. and Stark, E.: The effects of cholinomimetic drugs and atropine on ACTH release. Neuroendocrinology *21*: 31–41, 1976.

Makrosian, A. A.: The influence of choline and acetylcholine on chronaxy of the motor zone of the cerebral cortex. Bull. Exp. Biol. Med. *4*: 119–120, 1937 (in Russian).

Malthe-Sorenssen, D., Wood, P. H., Cheney, D. L. and Costa, E.: Modulation of the turnover of acetylcholine by intraventricular injection of thyrotropin-releasing hormone, somatostatin, neurotensin and angiotensin II. J. Neurochem. *31*: 685–691, 1978.

Manns, I. D., Alonso, A. and Jones, B. E.: Discharge properties of juxtacellularly labeled and immunohistochemically identified cholinergic basal forebrain neurons recorded in association with the electroencephalograph in anesthetized rats. J. Neurosci. *20*: 1505–1518, 2000.

Mantovani, P., Taddei, E., Gorelli, B. and Santicioli, P.: Effect of bromocriptine and ergometrin on striatal acetylcholine content in rat. Agressologie *29*: 19–21, 1988.

Marchbanks, R. M., Wonacott, S. and Rubio, M. A.: The effect of acetylcholine release on choline fluxes in isolated synaptic terminals. J. Neurochem, *36*: 379–393, 1981.

Marchi, M. and Raiteri, M.: On the presence in cerebral cortex of muscarinic receptor subtypes which differ in neuronal localization, function and pharmacological properties. J. Pharmacol. Exp. Ther. *235*: 230–233, 1985.

Marchi, M. and Raiteri, M.: Nicotinic autoreceptors mediating enhancement of acetylcholine release become operative in conditions of "impaired" cholinergic presynaptic function. J. Neurochem. *67*: 1974–1981, 1996.

Marchi, M., Pauchie, P., Gemignani, A. and Raiteri, M.: Is the muscarinic receptor that mediates potentiation of dopamine release negatively coupled to the cyclic GMP system? J. Neurosci. Res. *17*: 142–145, 1987.

Marchi, M., Lupinacci, M., Bernero, E., Bergaglia, F. and Raiteri, M.: Nicotinic receptors modulating ACh release in rat cortical synaptosomes. Neurochem. Int. *34*: 319–328, 1999.

Marco, L. A., Joshi, R. S., Brown, C., Aldes, L. D. and Chronister, R. B.: Effects of cholinergic and anticholinergic drugs on ketamine-induced lingopharyngeal motor activity. Psychopharmacology (Berl.) *96*: 484–486, 1988.

Marczynski, T. J.: Topical application of drugs to subcortical brain structures and selected aspects of electrical stimulation. Ergebn. Physiol. Biol. Chem. Exp. Pharmakol. *59*: 83–159, 1967.

Marczynski, T. J.: Neurochemical mechanisms in the genesis of slow potentials: a review and some clinical implications. In: "Multidisciplinary Perspectives in Events Related to Brain Potential Research," D. Otto, Ed., pp. 23–35, U. S. Environmental Protection Agency, Washington, D.C., 1978.

Marczynski, T. J.: Neurochemical interpretation of cortical slow potentials as they relate to cognitive processes and a parsimonious model of mammalian brain. In: "Slow Potential Changes in the Human Brain," W. C. McCallum and S. H. Curry, Eds., pp. 253–273, Plenum Press, New York, 1993.

Marczynski, T. J., Rosen, A. J. and Hackett, J. T.: Postreinforcement electrocortical synchronization and facilitation of cortical auditory potentials in appetitive instrumental conditioning. Electroenceph-alogr. Clin. Neurophysiol. *24*: 227–241, 1968.

Mark, G. P., Kinney, A. E., Grubb, M. C. and Keys, A. S.: Involvement of acetylcholine in the nucleus accumbens in cocaine reinforcement. Ann. N. Y. Acad. Sci. *877*: 792–795, 1999.

Markowska, A. L., Wenk, G. L. and Olton, D. S.: Nucleus basalis magnocellularis and memory: differential effects of two neurotoxins. Behav. Neural Biol. *54*: 13–26, 1990.

Martin, W. R. and Eades, C. G.: Pharmacologic studies of spinal cord adrenergic and cholinergic mechanisms and their relation to physical dependence on morphine. Psychopharmacologia *11*: 195–223, 1967.

Martin, T. J., Suchocki, J., May, E. L. and Martin, B. R.: Pharmacological evaluation of the nicotine's central effects by mecamylamine and pempidine. J. Pharmacol. Exp. Ther. *254*: 45–51, 1990.

Martinez-Morales, J. R., Lopez-Coviella, I., Hernandez-Jimenez, J. G., Reyes, R., Bello, A. R., Hernandez, G., Blusztajn, J. K. and Alonso, R.: Sex steroids modulate luteinizing hormone-releasing hormone secretion in a cholinergic cell line from basal forebrain. Neuroscience *103*: 1025–1031, 2001.

Martinez-Murillo, R., Villalba, R. M. and Rodrigo, J.: Immunocytochemical localization of cholinergic terminals in the region of the nucleus basalis magnocellularis of the rat: a correlated light and electron microscopic study. Neuroscience *36*: 361–370, 1990.

Martin-Ruiz, C. M., Haroutunian, V. H., Long, P., Young, A. H., Davis, K. L., Perry, E. K. and Court, J. A.: Dementia rating and nicotinic receptor expression in the prefrontal cortex in schizophrenia. Biol. Psychiatry *54*: 1222–1233, 2003.

Martof, B.: Home range and movements of the green frog, Rana clamitans. Ecology *34*: 529–543, 1953.

Masland, R. H. and Ames, A.: Responses to acetylcholine of ganglion cells in an isolated mammalian retina. J. Neurophysiol. *39*: 1220–1235, 1976.

Masland, R. H. and Livingstone, C. J.: Effect of stimulation with light on synthesis and release of acetylcholine of an isolated mammalian retina. J. Neurophysiol. *39*: 1210–1218, 1976.

Masland, H. R., Mills, J. W. and Cassidy, C.: The functions of acetylcholine in the rabbit retina. Proc. R. Soc. Lond., B, Biol. Sci. *223*: 121–391, 1984.

Maslanski, J. A., Powelt, R., Deirmengiant, C. and Patelt, J.: Assessment of the muscarinic receptor subtypes involved in pilocarpine-induced seizures in mice. Neurosci. Lett. *168*: 225–228, 1994.

Massey, S. and Neal, J. M.: The light evoked release of acetylcholine from rabbit retina *in vivo* and its inhibition by GABA. J. Neurochem. *32*: 1327–1329, 1979.

Mastrangelo, D., de Saint Hilaire-Kafi, Z. and Gaillard, J. M.: Effects of clonidine and alpha-methyl-p-tyrosine on the carbachol stimulation of paradoxical sleep. Pharmacol. Biochem. Behav. *47*: 889–900, 1994.

Matsuno, K., Matsunaga, K., Senda, T. and Mita, S.: Increase in extracellular acetylcholine level by sima ligands in rat frontal cortex. J. Pharmacol. Exp. Ther. *265*: 851–859, 1993.

Matsuyama, S. and Matsumoto, A.: Epibatidine induces long-term potentiation (LTP) via activation of alpha-4beta2 nicotinic receptors (nACHRs) *in vivo* in the intact mouse gentate gyrus: both alpha7 and alpha-4beta2 nAChRs essential to nicotinic LTP. J. Pharmacol. Sci. *93*: 180–197, 2003.

Matthews, R. T.: Neurotensin depolarizes cholinergic and a subset of non-cholinergic septal/diagonal band neurons by stimulating neurotensin-1 receptors. Neuroscience *94*: 775–783, 1999.

Mattson, M. P.: Acetylcholine potentiates glutamate-induced neurodegeneration in hippocampal cultured neurons. Brain Res. *497*: 402–406, 1989.

Maura, G., Fedele, E. and Raiteri, M.: Acetylcholine release from rat hippocampal slices is modulated by 5-hydroxytryptamine. Eur. J. Pharmacol. *165*: 173–179, 1989.

Mawson, A. R.: Stimulation-induced behavioral inhibition: a new model for understanding physical violence. Integr. Physiol. Behav. Sci. *34*: 177–197, 1999.

Mayerson, B. J. and Palis, A.: Advancement of the time for ovulation in the 5-day cyclic rat by pilocarpine. Acta Pharmacol. Toxicol. *26*, Supplt.: 68, 1970.

Maynert, E. W., Marczynski, T. J. and Browning, R. A.: The role of the neurotransmitters in the epilepsies. Adv. Neurol. *13*: 79–147, 1975.

Mayorga, A. J., Gianutsos, G. and Salamone, J. D.: Effects of striatal injections of 8-bromo-cyclic-AMP on pilocarpine-induced tremulous jaw movements in rats. Brain Res. *829*: 180–184, 1999.

McCance, I. and Phillis, J. W.: Cholinergic mechanisms in the cerebellar cortex. Int. J. Neuropharmacol. *7*: 447–462, 1968.

McCarley, R. W.: Control of sleep-waking state alteration in Felix domesticus. In: "Aspects of Behavioral Neurobiology," J. A. Ferendelli, Ed., pp. 90–128, Neuroscience Symposia, vol. 2, Bethesda, Md., 1977.

McCormick, D. A.: Cholinergic and noradrenergic modulation of thalamocortical processing. Trends Neurosci. *12*: 215–221, 1989.

McCormick, D. A.: Cellular mechanisms underlying cholinergic and noradrenergic modulation of neuronal firing in the cat and guinea pig dorsal lateral geniculate nucleus. J. Neurosci. *12*: 278–289, 1992a.

McCormick, D. A.: Neurotransmitter interactions in the thalamus and their role in neuromodulation of thalamicocortical activity. Prog. Neurobiol. *39*: 337–338, 1992b.

McCormick, D. A.: Actions of acetylcholine in the cerebral cortex and thalamus and implications for function. In: "Cholinergic Function and Dysfunction," A. C. Cuello, Ed., pp. 303–308, Elsevier, Amsterdam, 1993.

McCormick, D. A. and Prince, D. A.: Two types of muscarinic response to acetylcholine in mammalian cortical neurons. Proc. Natl. Acad. Sci. U.S.A. *82*: 6344–6348, 1985.

McCormick, D. A. and Prince, D. A.: Acetylcholine causes nicotinic excitation in the medial habenular nucleus of guinea pig, *in vitro*. J. Neurosci. *7*: 742–752, 1987.

McCormick, D. A. and Pape, H.-C.: Noradrenergic and serotonergic modulation of a hyperpolarization-activated cation current in thalamic relay neurons. J. Physiol. (Lond.) *431*: 319–342, 1990.

McCown, T. J. and Breese, G. R.: Effects of apamine and nicotinic acetylcholine receptor antagonists on inferior collicular seizures. Eur. J. Pharmacol. *187*: 49–58, 1990.

McEwen, B. S., Lewis, C. K. and Luine, V. N.: Steroid hormone action in the neuroendocrine system: When is the genome involved? In: "The Hypothalamus," S. Reichlin, R. J. Baldessarini and J. B. Martin, Eds., pp. Raven Press, New York, 1978.

McGaugh, J. L. and Herz, M. J.: "Memory Consolidation," Albion Publ., San Francisco, 1972.

McGaugh, J. L., Introini-Collison, I. B., Cahill, L. F., Castellano, C., Dalmaz, C., Parent, M. B. and Williams, C. L.: Neuromodulatory systems and memory storage: role of the amygdala. Behav. Brain Res. *58*: 81–90, 1993.

McGaughy, J., Decker, M. W. and Sarter, M.: Enhancement of sustained attention performance by the nicotinic receptor agonist ABT-418 in intact but not basal forebrain-lesioned rats. Psychopharmacology *144*: 175–182, 1999.

McGaughy, J., Everitt, B. J., Robbins, T. W. and Sarter, M. The role of cortical cholinergic afferent projections in cognition: impact of new selective immunotoxins. Behav. Brain Res. *115*: 251–263, 2000.

McGeer, P. L. and McGeer, E. G.: Neurotransmitters and their receptors in the basal ganglia. Adv. Neurol. *60*: 93–101, 1993.

McGeer, P. L., McGeer, E. G. and Hatori, T.: Dopamine-acetylcholine-GABA neuronal linkages in the extra-pyramidal and limbic systems. In: "Nonstriatal Dopaminergic Neurons," E. Costa and G. L. Gessa, Eds., Adv. Biochem. Psychopharmacol., vol. 16, pp. 397–402, Raven Press, New York, 1977.

McGeer, P. L., Eccles, J. C. and McGeer, E. G.: "Molecular Neurobiology of the Mammalian Brain," Second Edition, Plenum Press, New York, 1987.

McGinty, D. and Szymusiak, R.: Hypothalamic regulation of msleep and arousal. Front. Biosci. *8*: S1074–S1083, 2003.

McIntyre, D. C. and Racine, R. J.: Kindling mechanisms: current progress on an experimental epilepsy model. Prog. Neurobiol. *27*: 1–12, 1986.

McKinney, M., Miller, J. H. and Aagaard, P. J.: Pharmacological characterization of the rat hippocampal muscarinic autoreceptor. J. Pharmacol. Exp. Ther. *264*: 74–78, 1993.

McLennan, H.: "Synaptic Transmission," Saunders, New York, 1963.

McNamara, B. P., Murtha, E. F., Bergner, A. D., Robinson, W. M., Bender, C. W. and Wills, J. H.: Studies on the mechanism of action of DFP and TEPP. J. Pharmacol. Exp. Ther. *110*: 232–240, 1954.

McNaught, K. S., Kapustin, A., Jackson, T., Jengelley, T. A., Jnobaptiste, R., Shashidharan, P., Perl, D. P., Pasik, P. and Olanow, C. W.: Brainstem pathology in DYT1 primary dystonia. Ann. Neurol. *56*: 540–547, 2004.

McQuiston, A. R. and Madison, D. V.: Muscarinic receptor activity has multiple effects on the resting membrane potential of CA1 hippocampal interneurons. J. Neurosci. *19*: 5693–5702, 1999.

Meador, K. J., Moore, E. E., Nichols, M. E., Abney, O. L., Taylor, H. S., Zamrini, E. Y. and Loring, D. W.: The role of cholinergic systems in visuospatial processing and memory. J. Clin. Exp. Neuropsychol. *15*: 832–842, 1993.

Meeker, R., Gisolfi, C. V., Mora, F. and Myers, R. D.: Release of 45-Ca++ in the diencephalon of the monkey during feeding and drinking. Brain Res. *154*: 421–425, 1978.

Meierkord, H.: Epilepsy and sleep. Curr. Opin. Neurol. *7*: 107–112, 1994.

Meldrum, B. S.: Pharmacology of GABA. Clin. Neuropharmacol. *5*: 253–316, 1982.

Meldrum, B. S., Coucher, M. J., Badman, G. and Collins, J. F.: Antiepileptic action of excitatory amino acid antagonists in the photosensitive baboon, Papio papio. Neurosci. Lett. *39*: 101–104, 1983.

Mello, L. E., Cavalheiro, E. A., Tan, A. M., Kupfer, W. R., Pretorius, J. K., Babb, T. L. and Finch, D. M.:

Circuit mechanisms of seizures in the pilocarpine model of chronic epilepsy: cell loss and mossy fibers sprouting. Epilepsia *34*: 985–995, 1993.

Mellott, T. J., Williams, C. L., Meck, W. H. and Blusztajn, J. K.: Prenatal choline supplementation advances hippocampal development and enhances MAPK and CREB activation. FASEB J. *18*: 545–547, 2004.

Mendez, M., Joseph-Bravo, P., Cisneros, M., Vargas, M. A. and Charli, J. L.: Regional distribution of *in vitro* release of thyrotropin releasing hormone in rat brain. Peptides *8*: 291–298, 1987.

Menescal-de-Oliveira, L. and Hoffman, A.: The parabrachial region as a possible region modulating simultaneously pain and tonic immobility. Behav. Brain Res. *56*: 127–132, 1993.

Merlo Pich, E., Chiamulera, C. and Tessari, M.: Neural substrates of nicotine addiction as defined by functional maps of gene expression. In: "From *Torpedo* Electric Organ to Human Brain: Fundamental and Applied Aspects," J. Massoulié, Ed., pp. 225–228, Elsevier, Amsterdam, 1998.

Mesulam, M.-M.: Human brain cholinergic pathways. In: "Cholinergic Neurotransmission: Functional and Clinical Aspects," S.-M. Aquilonius and P.-G. Gillberg, Eds., pp. 231–241, Elsevier, Amsterdam, 1990.

Mesulam, M.-M.: Structure and function of cholinergic pathways in the cerebral cortex, limbic system, basal ganglia, and thalamus of the human brain. In: "Psychopharmacology. The Fourth Generation of Progress," F. E. Bloom and D. J. Kupfer, Eds., pp. 135–146, Raven Press, New York, 1995.

Mesulam, M.-M., Geula, C., Bothwell, M. A. and Hersh, L. B.: Human reticular formation: cholinergic neurons of the pediculopontine and laterodorsal tegmental nuclei and some cytological comparisons to forebrain cholkinergic neurons. J. Comp. Neurol. *283*: 611–633, 1989.

Metherate, R. and Weinberger, N. M.: Cholinergic modulation of responses to single tonus produces tone-specific receptive field alterations in the cat auditory cortex. Synapse *6*: 133–145, 1990.

Metherate, R., Temblay, N. and Dykes R. W.: The effects of acetylcholine on response properties of cat somatosensory cortical neurons. J. Neurophysiol. *59*: 1231–1252, 1988a.

Metherate, R., Temblay, N. and Dykes, R. W.: Transient and prolonged effects of acetylcholine on responsiveness of cat somatosensory cortical neurons. J. Neurophysiol. *59*: 1253–1275, 1988b.

Metzinger, T., Ed.: "Conscious Experience," Imprint Academic, Schoening, Padeborn, 1995.

Meyer, E. M. and Otero, D. H.: Differential effects of 4-aminopyridine on acetylcholine release triggered by K+ depolarization, veratridine, or A23187 in rat cortical synaptosomes. Neurochem. Res. *14*: 157–160, 1989.

Michelson, M. J.: Some aspects of evolutionary pharmacology. Biochem. Pharmacol. *23*: 2211–2224, 1974.

Mifsud, J.-C., Hernandez, L. and Hoebel, B. G.: Nicotine infused into the nucleus accumbens increases synaptic dopamine as measured by *in vivo* microdialysis. Brain Res. *478*: 365–367, 1989.

Mignot, E.: Perspectives in narcolepsy research and therapy. Curr. Opin. Pulm. Med. *2*: 482–487, 1996.

Mike, A., Castro, N. G. and Albuquerque, E. X.: Choline and acetylcholine have similar kinetic properties of activation and desensitization on the alpha7 nicotinic receptors in rat hippocampal neurons. Brain Res. *882*: 155–168, 2000.

Miller, A. D. and Blaha, C. D.: Nigrostriatal dopamine release modulated by mesopontine muscarinic receptors. Neuroreport *15*: 1805–1808, 2004.

Miller, F. R.: The local action of acetylcholine on the central nervous system. J. Physiol. *91*: 212–221, 1937.

Miller, F. R., Stavraky, G. W. and Woonton, G. A.: Effects of eserine, acetylcholine and atropine on the electrocorticogram. J. Neurophysiol. *3*: 131–138, 1940.

Miller, M. W. and Rieck, R. W.: Effects of chronic ethanol administration on acetylcholinesterase activity in the somatosensory cortex and basal forebrain of the rat. Brain Res. *627*: 104–112, 1993.

Milner, T. A., Hammel, J. R., Ghorbani, T. T., Wiley, R. G. and Pierce, J. P.: Septal cholinergic deafferentation of the dentate gyrus results in a loss of a subset of neuropeptide Y somata and an increase in synaptic area on remaining neuropeptide Y dendrites. Brain Res. *831*: 322–326, 1999.

Minzenberg, M. J., Poole, J. H., Benton, C. and Vinogradov, S.: Association of anticholinergic load with impairment of complex attention and memory in schizophrenia. Am. J. Psychiatry *161*: 116–124, 2004.

Miranda, M. I., Ferreira, G., Ramirez-Lugo, L. and Bermudez-Ratoni, F.: Role of cholinergic system on the construction of memories: taste memory encoding. Neurobiol. Learn. Mem. *80*: 211–222, 2003.

Mirza, N. R. and Stolerman, I. P.: Nicotine enhances sustained attention in the rat under specific task conditions. Psychopharmacology (Berl.) *138*: 266–274, 1998.

Mitchell, J. F.: The spontaneous and evoked release of acetylcholine from the cerebral cortex. J. Physiol. (Lond.) *165*: 98–116, 1963.

Mitchell, R. A., Loeschke, H. H., Massion, W. H. and Severinghaus, J. W.: Respiratory response mediated through superficial chemosensitive areas on the medulla. J. Appl. Physiol. *18*: 523–533, 1963.

Mitchell, S. N., Brazell, M. P., Joseph, M. H., Alavijeh, M. S. and Gray, J. A.: Regionally specific effects of acute and chronic nicotine on rates of catecholamine and 5-hydroxytryptamine synthesis in rat brain. Eur. J. Pharmacol. *167*: 311–322, 1989.

Mitsuma, T., Hirooka, Y., Kayama, M., Izeki, K. and Nogimori, T.: Effects of acetylcholine on the release of thyrotropin-releasing hormone for the rat retina *in vitro*. Exp. Clin. Endocrinol. *99*: 77–79, 1992.

Mittmann, T. and Alzheimer, C.: Muscarinic inhibition of persistent Na+ current in rat neocortical pyramidal neurons. J. Neurophysiol. *79*: 1579–1582, 1998.

Mizoguchi, K., Yuzurihara, M., Ishige, A., Sasaki, H. and Tabira, T.: Effect of chronic stress on cholinergic transmission in rat hippocampus. Brain Res. *915*: 108–111, 2001.

Mlot, C.: Probing the biology of emotion. Science *280*: 1005–1007, 1998.

Mogenson, G. J. and Yang, C. R.: The contribution of basal forebrain to limbic-motor integration and the mediation of motivation to action. In: "The Basal Forebrain," T. C. Napier, P. W. Kalivas, and I. Hanin, Eds., pp. 267–290, 1991.

Monassi, C. R., Leite-Panissi, C. R. and Menescal-de-Oliveira, L.: Ventrolateral periaqueductal gray matter and the control of tonic immobility. Brain Res. Bull. *50*: 201–208, 1999.

Monnier, M. and Herkert, B.: Concentration and seasonal variations of acetylcholine in sleep and waking dialysates. Neuropharmacology *11*: 479–489, 1972.

Montoya, D. A., White, A. M., Williams, C. L., Blusztajn, J. K., Meck, W. H. and Swartzwelder, H. S.: Prenatal choline exposure alters hippocampal responsiveness to cholinergic stimulation in adulthood. Brain Res. Dev. Brain Res. *123*: 25–32, 2000.

Mooney, D. M., Zhang, L., Basile, C., Senatorov, V. V., Ngsee, J., Omar, A. and Hu, B.: Distinct forms of cholinergic modulation in parallel thalamic sensory pathways. Proc. Natl. Acad. Sci. U.S.A. *101*: 320–324, 2004.

Moore, D. B., Lee, P., Paiva, M., Walker, D. W. and Heaton, M. B.: Effects of neonatal ethanol exposure on cholinergic neurons of the rat medial septum. Alcohol *15*: 219–226, 1998.

Moos, F. and Richard, P.: Paraventricular and supraoptic bursting oxytocin cells in rat are locally regulated by oxytocin and functionally related. J. Physiol. (Lond.) *408*: 1–18, 1989.

Mori, S.: Control mechanisms of locomotor movements from a viewpoint of behavioral control. Nikon Shinkei Seishin Yakurigaku Zasshi *15*: 281–287, 1995 (in Japanese).

Morin, C. L., Dolina, S., Robertson, R. T. and Ribal, C. E.: An inbred epilepsy-prone substrain of BALB/c mice shows absence of the corpus callosum, and abnormal projection to the basal forebrain, and bilateral projections to the thalamus. Cereb. Cortex *4*: 119–128, 1994.

Morris, J. S., Ohman, A. and Dolan, R. J.: Conscious and unconscious emotional learning in the human amgdala. Nature *393*: 467–460, 1998.

Morris, J. S., Ohman, A. and Dolan, R. J.: A subcortical pathway to right amygdala mediating "unseen" fear. Proc. Natl. Acad. Sci. U.S.A. *96*: 1680–1685, 1999.

Moruzzi, G.: Contribution à l'electrophysiologie du cortex moteur: facilitation après décharge et epilepsie corticales. Arch. Intern. Physiol. *49*: 33–100, 1934.

Moruzzi, G. and Magoun, H. W.: Brain stem reticular formation and activation of the EEG. Electroencephalogr. Clin. Neurophysiol. *1*: 455–473, 1949.

Moscowitch, M.: A neuropsychological model of memory and consciousness. In: "Neuropsychology of Memory," L. R. Squire and N. Butters, Eds., pp. 5–22, Guilford Press, New York, 1992.

Moser, P. C.: The effect of 5-HT3 receptor antagonists on the writhing response in mice. Gen. Pharmacol. *26*: 1301–1306, 1995.

Motles, E., Gomez, A., Tetas, M., Gonzales, M. and Acuna, C.: Cholinergic blockade with scopolamine in adult cats: effects on the behaviors evoked by apomorphine and amphetamins. Prog. Neuropsychopharmacol. Biol. Psychiatry *16*: 223–235, 1992.

Moyer, K. E.: Kinds of aggression and their physiological basis. Commun. Behav. Biol., A 2: 65–87, 1968.

Moyer, K. E., Ed.: "Physiology of Aggression," Raven Press, New York, 1976.

Mrzljak, L., Levey, A. I. and Goldman-Rakic, P. S.: Association of M1 and M2 muscarinic receptor proteins with asymmetric synapses in the primate cerebral cortex: morphological evidence for cholinergic modulation of excitatory neurotransmission. Proc. Natl. Acad. Sci. U.S.A. *90*: 5194–5198, 1993.

Mucignat-Caretta, C., Bondi, M. and Caretta, A.: Animal models of depression: olfactory lesions affect amygdala, subventricular zone, and aggression. Neurobiol. Dis. *16*: 386–395, 2004.

Muhlethaler, M. and Serafin, M.: Thalamic spindles in an isolated and perfused preparation *in vitro*. Brain Res. *524*: 17–21, 1990.

Muir, J. L., Everitt, B. J. and Robbins, T. W.: AMPA-induced excitotoxic lesions of the basal forebrain: a significant role for the cholinergic cortical system in attentional function. J. Neurosci. *14*: 2313–2326, 1994.

Muller, W. and Misgeld, U.: Carbachol reduces IK, baclofen but not IK, GABA in guinea pig hippocampal slices. Neurosci. Lett. *102*: 229–234, 1989.

Muller, W., Brunner, H. and Misgeld, U.: Lithium discriminates between muscarinic receptor subtypes on guinea pig hippocampal neurons *in vitro*. Neurosci. Lett. *100*: 135–140, 1989.

Murphy, P. C., Grieve, K. L. and Sillito, A. M.: Effects of vasoactive intestinal polypeptide on the response properties of calls in area 17 of the cat visual cortex. J. Neurophysiol. *69*: 1465–1474, 1993.

Murugaian, J., Sundaram, K. and Sapru, H.: Cholinergic mechanisms in the ventrolateral medullary depressor area. Brain Res. *502*: 287–295, 1989.

Musky, R. E., Consroe, P. and Makriyannis, A., Eds.: Pharmacological, chemical, biochemical and behavioral research on cannabis and cannabinoids. Pharmacol. Biochem. Behav. *40*: 457–708, 1991.

Myers, R. D.: Chemical mechanisms in the hypothalamus mediating eating and drinking in the monkey. Ann. N. Y. Acad. Sci. *157*: 918–933, 1969.

Myers, R. D.: "Drug and Chemical Stimulation of the Brain," Van Nostrand Reinhold, New York, 1974.

Myers, R. D.: Hypothalamic mechanisms underlying physiological set points. In: "Current Studies of Hypothalamic Function," K. Lederis and W. L. Veale, Eds., vol. 2, pp. 17–28, Karger, Basel, 1978.

Myers, R. D.: Hypothalamic control of thermoregulation—neurochemical mechanisms. In: "Handbook of the Hypothalamus," P. J. Morgane, and J. Panksepp, Eds., pp. 83–210, Marcel Dekker, New York, 1980.

Myers, R. D.: The role of ions in thermoregulation and fever. In: "Handbook of Experimental Pharmacology," A. S. Milton, Ed., pp. 151–186, Springer-Verlag, New York, 1982.

Myers, R. D.: Neurochemistry of thermoregulation. Physiologist *27*: 41–46, 1984.

Myers, R. D.: Cholinergic systems in the central control of body temperature. In: "Neurobiology of Acetylcholine: A Symposium in Honor of Alexander G. Karczmar," N. J. Dun and R. L. Perlman, Eds., pp. 391–402, Plenum Press, New York, 1987.

Myers, R. D. and McCaleb, M. L.: Peripheral and intrahypothalamic cholecystokinin act on the noradrenergic "feeding circuit" in the rat diencephalon. Neuroscience *6*: 645–655, 1981.

Myers, R. D. and Sharpe, L. G.: Chemical activation of ingestive and other hypothalamic regulatory mechanisms. Physiol. Behav. *3*: 987–995, 1968.

Myers, R. D. and Veale, W. L.: Body temperature: possible ionic mechanism in the hypothalamus controlling the set point. Science *170*: 95–97, 1970.

Myers, R. D. and Yaksh, T. L.: Feeding and temperature responses in the unrestrained rat after injection of cholinergic and aminergic substances into cerebral ventricles. Physiol. Behav. *3*: 917–928, 1968.

Myers, R. D. and Yaksh, T. L.: Control of body temperature in unanesthetized monkey by cholinergic and aminergic systems in the hypothalamus. J. Physiol. *202*: 483–500, 1969.

Myers, R. D., Hall, G. H. and Rudy, T. A.: Drinking in the monkey evoked by nicotine or angiotensin II microinjected in hypothalamic and mesencephalic sites. Pharmacol. Biochem. Behav. *1*: 15–22, 1972.

Myers, R. D., McCaleb, M. L. and Hughes, K. A.: Is the noradrenergic "feeding circuit" in hypothalamus really an olfactory system? Pharmacol. Biochem. Behav. *10*: 923–927, 1979.

Myers, R. D., Swartzwelder, P., Lee, P. T., Hepler, J. R., Denbow, D. M. and Ferrer, J. M. R.: CCK and other peptides modulate hypothalamic norepinephrine release in the rat: dependence on hunger and satiety. Brain Res. Bull. *17*: 583–597, 1986.

Myers, R. D., Beleslin, D. B. and Rezvani, A. H.: Hypothermia: role of alpha 1- and alpha 2-noradrenergic receptors in the hypothalamus of the cat. Pharmacol. Biochem. Behav. *26*: 373–379, 1987.

Myles, M. and Fain, J.: Effect of K+-induced depolarization on carbachol-stimulated inositol tetrakiphosphate accumulation in rat cerebrocortical slices. Biochim. Biophys. Acta *1310*: 19–24, 1996.

Nabatame, H., Sasa, M., Takaori, S. and Kameyama, M.: Muscarinic receptor-mediated inhibition of dopaminergic excitatory input to caudate neuron from the substantia nigra. Jpn. J. Pharmacol. *46*: 387–395, 1988.

Nabeshima, T.: Behavioral aspects of cholinergic transmission: role of basal forebrain cholinergic system in learning and memory. In: "Cholinergic Function and Dysfunction," A. C. Cuello, Ed., pp. 405–411, Elsevier, Amsterdam, 1993.

Nabeshima, T. and Nitta, A.: Memory impairment and neuronal dysfunction induced by beta-amyloid protein in rats. Tohoku J. Exp. Med. *174*: 241–249, 1994.

Nachmansohn, D.: On the physiological significance of cholinesterase. Yale J. Biol. Med. *12*: 565–589, 1940.

Nachmansohn, D.: Chemical mechanisms of nerve activity. In: "Modern Trends of Physiology and Biochemistry," E. S. G. Barron, Ed., pp. 229–276, Academic Press, New York, 1952.

Nachmansohn, D. and Rothenberg, M. A.: Studies on cholinesterase. I. On the specificity of the enzyme in nerve tissue. J. Biol. Chem. *158*: 653–666, 1945.

Naffah-Mazzacoratti, M. G., Funke, M. G., Sanabria, E. R. and Cavalheiro, E. A.: Growth-associated phosphoprotein expression is increased in supragranular regions of the dentate gyrus following pilocarpine-induced seizures in rats. Neuroscience *91*: 485–492, 1999.

Nagahara, A. H. and Handa, R. J.: Fetal alcohol-exposed rats exhibit differential response to cholinergic drugs on a delay-dependent memory task. Neurobiol. Learn. Mem. *72*: 230–243, 1999.

Naguib, M. and Yaksh, T. L.: Antinociceptive effects of spinal cholinesterase inhibition and isobolographic analysis of the interaction with miu and alpha receptor systems. Anesthesiology *80*: 1338–1348, 1994.

Naguib, M. and Yaksh, T.: Characterization of muscarinic receptor subtypes that mediate antinociception in the rat spinal cord. Anesth. Analg. *85*: 847–853, 1997.

Nakagawasai, O., Yamadera, F., Iwasaki, K., Arai, H., Taniguchi, R., Tan-No, K., Sasaki, H. and Tadano, T.: Effect of kami-untan-to on the impairment of learning and memory induced by thiamine-deficient feeding in mice. Neuroscience *125*: 233–41, 2004.

Nakamura, K. and Shirane, M.: Activation of the reticulothalamic cholinergic pathway by the major metabolites of aniracetam. Eur. J. Pharmacol. *380*: 81–89, 1999.

Nanri, M., Yamamoto, J., Miyake, H. and Watanabe, H.: Protective effect of GTS-21, a novel nicotinic receptor agonist, on delayed neuronal death induced by ischemia in gerbils. Jpn. J. Pharmacol. *76*: 23–29, 1998.

Napier, T. C., Kalivas, P. W. and Hanin, I., Eds.: "The Basal Forebrain," Plenum Press, New York, 1991.

Narahashi, T.: Mechanism of tetrodotoxin and saxitoxin action. In: "Marine Toxins and Venoms," A. T. Tu, Ed., pp. 185–210, Handbook of Natural Toxins, vol. 3, Marcel Dekker, New York and Basel, 1988.

Nathan, P., Aprison, M. H. and Himwich, H. E.: A comparison of the effects of atropine with those of several central nervous stimulants on rabbits exhibiting forced circling following the intracarotid injection od diisopropyl fluorophosphate. Confin. Neurol. *15*: 1–23, 1955.

Nathanson, N. M.: Molecular properties of the muscarinic acetylcholine receptor. Annu. Rev. Neurosci. *10*: 195–236, 1987.

Nathanson, N. M.: Regulation of expression and function of muscarinic receptors. In: "Ninth International Symposium on Cholinergic Mechanisms," J. Klein and K. Loeffelholz, Eds., pp. 165–168, Elsevier, Amsterdam, 1996.

Natsume, K. and Kometani, K.: Theta-activity-dependent and -independent muscarinic facilitation of long-term potentiation in guinea pig hippocampal slices. Neurosci. Res. *27*: 335–341, 1997.

Natsume, K. and Kometani, K.: Desynchronization of carbachol-induced theta-like activities by alpha-adrenergic agents in guinea pig hippocampal slices. Neurosci. Res. *33*: 179–186, 1999.

Nattie, E. E. and Li, A. H.: Ventral medulla sites of muscarinic receptor subtypes involved in cardiorespiratory control. J. Appl. Physiol. *69*: 33–41, 1990.

Navarro-Lopez, J. de D., Alvarado, J. C., Marquez-Ruiz, J., Escudero, M., Delgado-Garcia, J. M. and Yajeya, J.: A cholinergic synaptically triggered event participates in the generation of persistent activity necessary for eye fixation. J. Neurosci. *24*: 5109–5118, 2004.

Neil, D. G. and Grossman, S. P.: Behavioral effects of lesions or cholinergic blockade of the dorsal and ventral caudate of rats. J. Comp. Physiol. Psychol. *71*: 211–317, 1970.

Nemeroff, C. B., Konkol, R. J., Bissett, G., Youngblood, W., Martin, J. B., Brazeau, P., Rone, M. S., Prange, A. J., Jr., Breese, J. S. and Kizer, J. S.: Analysis of the disruption in hypothalamic-pituitary regulation in rats treated neonatally with monosodium-L-glutamate (MSG): evidence for the involvement of tuberoinfundibular cholinergic and dopaminergic systems in neuroendocrine regulation. Endocrinol. *101*: 613–622, 1977.

Nersesian, O. N. and Baklavadzhian, O. G.: Mikroionoforeticheskoe issledovanie vliianiia kholinerichskikh veshchestv na aktivnost' medullinykh dykhtatel'nykh neironov. Fiziol. Zh. SSSR *75*: 948–954, 1989 (in Russian).

Ni, H., Jing, L. X. and Shen, S.: The role of medial medulla in the depressive responses in pulmonary and carotid arteries to injection of acetylcholine at fourth ventricle. Shen Li Hsueh Pao *41*: 291–298, 1989 (in Chinese).

Nicole, M. M., Gallagher, M. and McKinney, M.: Visualization of muscarinic receptor-mediated phosphoinocitide turnover in the hippocampus of young and aged, learning-impaired. Long Evans rats. Hippocampus *11*: 741–746, 2001.

Nikbakht, M. R. and Stone, T. W.: Occlusive responses to adenosine A1 receptor and muscarinic M2 receptor activation on hippocampal presynaptic terminals. Brain Res. *829*: 193–196, 1999.

Nishi, S., Soeda, H. and Koketsu, K.: Unusual nature of ganglionic slow EPSP studied by a voltage clamp method. Life Sci. *8*: 33–42, 1969.

Nishizaki, T., Matsuoka, T., Nomura, T., Sumikawa, K., Shiotani, T., Watabe, S. and Yoshii, M.: Nefiracetam modulates acetylcholine receptor currents via different signal transduction pathways. Mol. Pharmacol. *53*: 1–5, 1998.

Nishizaki, T., Nomura, T., Matsuoka, T., Enikolopow, G. and Sumikawa, K.: Arachidonic acid induces a long-lasting facilitation of hippocampal transmission by modulating PKC activity and nicotinic ACh receptors. Brain Res. Mol. Brain Res. *69*: 263–272, 1999.

Nishizaki, T., Nomura, T., Matuoka, T., Kondoh, T., Enikolopo, G., Sumikawa, K., Watabe, S., Shiotani, T. and Yoshii, M.: The anti-dementia drug nefiracetam facilitates hippocampal synaptic transmission by functionally targeting presynaptic nicotinic ACh receptors. Brain Res. Mol. Brain Res. *80*: 53–62, 2000.

Nistri, A.: Further investigations into the effect of baclofen (Lioresal) on the isolated spinal cord. Experientia *31*: 1066–1067, 1975.

Nistri, A., Pepeu, G.-C., Cammelli, E., Spina, L. and De Bellis, A. M.: Effects of morphine on brain and spinal cord acetylcholine levels and nociceptive threshold in the frog. Brain Res. *80*: 199–209, 1974.

Nitz, D., Andersen, A., Fahringer, H., Ninhuis, R., Mignot, E. and Siegel, J.: Altered distribution of cholinergic cells in narcoleptic dog. Neuroreport *31*: 1521–1524, 1995.

Nordberg, A., Romanelli, L., Sundwall, A., Bianchi, C. and Beani, L.: Effect of acute and subchronic nicotine treatment on cortical acetylcholine release and on nicotinic receptors in rats and guinea pigs. Br. J. Pharmacol. *98*: 71–78, 1989a.

Nordberg, A., Hardvig, P., Lundquist, H., Antoni, G., Ulin, J. and Langstrom, B.: Uptake and distribution of (+)-(R) and (–)(S)-N-(methyl-11C)-nicotine in the brain of rhesus monkeys—an attempt to study nicotinic receptors *in vivo*. J. Neural Transm. *1*: 195–205, 1989b.

Nordstrom, O., Melander, T., Hokfelt, T., Bartfai, T. and Goldstein, M.: Evidence for an inhibitory effect of the peptide galanine on dopamine release from the rat median eminence. Neurosci. Lett. *73*: 21–26, 1989.

Nunez, P. L., Ed.: "Neocortical Dynamics and Human EEG Rhythms," Oxford University Press, Oxford, 1995a.

Nunez, P. L.: Mind, brain and encephalography. In: "Neocortical Dynamics and Human EEG Rhythms," P. L. Nunez, Ed., pp. 133–194, Oxford University Press, Oxford, 1995b.

Nyback, H., Nyman, H., Ohman, G., Nordgren, I. and Lindstrom, B.: Preliminary experiences and results with THA for the amelioration of symptoms of Alzheimer disease. In: "Current Research in Alzheimer Therapy," E. Giacobini and R. Becker, Eds., pp. 231–236, Taylor and Francis, New York, 1988.

Oades, R. D.: Types of memory or attention? Impairments after lesions of the hippocampus and limbic ventral tegmentum. Brain Res. Bull. *18*: 533–545, 1981.

Oba, T., Ota, K. and Yokoyama, A.: Inhibition of milk-ejection reflex in lactating rats by systemic administration and intracerebral implantation of atropine. Neuroendocrinology *7*: 116–126, 1971.

Obata, H., Lix, X. and Eisenach, J. C.: Alpha2-adrenoceptors activation by clonidine enhances stimulation-evoked acetylcholine release from spinal cord tissues after nerve ligation in rats. Anethesiology *102*: 657–662, 2005.

Ogawa, N. and Kurota, H.: Physiological significance of neuropeptide receptor distribution in the brain. Nippon Rinsho *41*: 1068–1070, 1983.

Ogura, H. and Aigner, T. G.: MK-801 impairs recognition memory in rhesus monkey: comparison with cholinergic drugs. J. Pharmacol. Exp. Ther. *266*: 60–64, 1993.

Ohkuma, S., Katsura, M., Guo, J. L., Hasegawa, T. and Kuriyama, K.: Involvement of peroxynitrite in P-methyl-D-aspartate and sodium nitroprusside-induced release of acetylcholine from mouse cerebral cortical neurons. Brain Res. Mol. Brain Res. *31*: 185–193, 1995.

Ohmori, N., Itoi, K., Tozawa, F., Sakai, Y., Sakai, K., Horiba, N., Demura, H. and Suda, T.: Effect of acetylcholine on corticotropin-releasing factor gene expression in the hypothalamus paraventricular nucleus of the conscious rats. Endocrinology *136*: 4858–4863, 1995.

Ohno, M. and Watanabe, S.: Differential effects of 5-HT3 receptor antagonism on working memory failure due to deficiency of hippocampal cholinergic and glutamergic transmission in rats. Brain Res. Dev. Brain Res. *101*: 273–276, 1997.

Ohno, M., Yamamoto, T. and Watanabe, S.: Involvement of cholinergic mechanisms in impairment of working memory in rats following basolateral amygdaloid lesions. *31*: 915–922, 1992.

Ohno, Y. and Warnick, J. E.; Presynaptic activation of the spinal serotonergic system in the rat by phencyclidine *in vitro*. J. Pharmacol. Exp. Ther. *250*: 177–183, 1989.

Okada, M.: Effects of a new thyrotropin releasing hormone analogue, YM-14673, on the *in vivo* release of acetylcholine as measured by intracerebral dialysis in rats. J. Neurochem. *56*: 1544–1547, 1991.

Okada, M., Wada, K., Kamata, A., Murakami, T., Zhu, G. and Kaneko, S.: Impaired M-current and neuronal excitability. *43*: 36–38, 2002.

Oki, T., Takagi, I., Inagaki, S., Taketo, M. M., Manabe, T., Matsui, M. and Yamada, S.: Quantitative analysis of binding parameters of [(3)H]N-methylscopolamine binding in central nervous system of muscarinic acetylcholine receptor knockout mice. Brain Res. Mol. Brain Res. *133*: 6–11, 2005.

Olds, J.: Self-stimulation of the brain: its use to study local effects of hunger, sex and drugs. Science *127*: 315–324, 1958.

Olton, D. S. and Samuelson, R. J.: Remembrance of places passed: spatial memory in rats. J. Exp. Psychol. Anim. Behav. Process. *2*: 97–116, 1976.

Olton, D. S., Becker, J. T. and Handelmann, G. E.: Hippocampus, space and memory. Behav. Brain Sci. *2*: 313–365, 1979.

Olton, D. S., Markowska, A., Voytko, M. L., Givens, B., Gorman, L. and Wenk, G.: Basal forebrain cholinergic system: a functional analysis. In: "The Basal Forebrain," T. C. Napier, P. W. Kalivas and I. Hanin, Eds., pp. 353–372, Plenum Press, New York, 1991.

O'Neill, J., Siembieda, D. W., Crawford, K. C., Halgren, E., Fisher, A. and Fitten, L. J.: Reduction in distractibility with AF102B and THA in macaque. Pharmacol. Biochem. Behav. *76*: 301–306, 2003.

Oomura, Y.: Chemical and neuronal control of feeding motivation. Physiol. Behav. *44*: 555–560, 1988.

O'Regan, M. H.: Xylazine-evoked depression of rat cerebral cortical neurons: a pharmacological study. Gen. Pharmacol. *20*: 469–474, 1989.

Ota, M., Crofton, J. T., Toba, K. and Shane, L.: Effect on vasopressin release of microinjection of cholinergic agonists into the rat supraoptic nucleus. Proc. Soc. Exp. Biol. Med. *201*: 208–214, 1992.

Pan, Z. Z. and Williams, J. T.: Muscarine hyperpolarizes a subpopulation of neurons by activating an M2 muscarinic receptor in rat nucleus raphe magnus *in vitro*. J. Neurosci. *14*: 1332–1338, 1994.

Pare, D., Curro Dossi, R. and Steriade, M.: Neuronal basis of the parkinsonian resting tremor: a hypothesis and its implications for treatment. Neuroscience *35*: 217–226, 1990.

Parker, S. L., Fu, Y., McAllen, K., Luo, J., MacIntosh, J. M., Lindstrom, J. M. and Sharp, B. M.: Up-regulation of brain nicotinic acetylcholine receptors in the rat during long-term self-administration of nicotine: disproportionate incidence of the alpha6 subunit. Mol. Pharmacol. *65*: 611–622, 2004.

Parnas, H. and Parnas, I.: The Ca-voltage hypothesis for neurotransmitter release—current views. In: "Cholinergic Mechanisms: Function and Dysfunction," I. Silmand, H. Soreq, L. Anglister, D. Michaelson, and A. Fisher, Eds., pp. 99–103, Taylor and Francis, New York, 2004.

Patil, M. M., Linster, C., Lubenov, E. and Hasselmo, M. E.: Cholinergic agonist carbachol enables associative long-term potentiation in pyriform cortex. J. Neurophysiol. *80*: 2467–2474, 1998.

Patrick, J., Sequela, P., Vernino, S., Amador, M., Luetje, C. and Dani, J. A.: Functional diversity of neuronal nicotinic acetylcholine receptors. In: "Cholinergic Function and Dysfunction," A. C. Cuello, Ed., pp. 113–120, Elsevier, Amsterdam, 1993.

Paulet, D.: Nouvelle contribution a l'étude de l'action pharmacologique du tetraethylpyrophosphate (TEPP). Arch. Intern. Pharmacodyn.: *97*: 157–185, 1954.

Pavlov, I. P.: "Conditioned Reflex: An Investigation of the Physiological Activity of the Cerebral Cortex," translated and edited by G. V. Anrep, Oxford University Press, Oxford, 1940.

Pedata, F., Slavikova, A., Kotas, A. and Pepeu, G.: Acetylcholine release from rat cortical slices during postnatal development and aging. Neurobiol. Aging *4*: 31–35, 1983.

Pedata, F., Giovanelli, L., De Sarno, P. and Pepeu, G.: Effect of adenosine, adenosine derivatives, and caffeine on acetylcholine release from brain synaptosomes: interaction with muscarine autoregulatory mechanisms. J. Neurochem. *46*: 1593–1598, 1986.

Pedata, F., Magnani, M. and Pepeu, G.: Muscarinic modulation of purine release from electrically stimulated rat cortical slices. J. Neurochem. *50*: 1074–1079, 1988.

Pedigo, N. W. and Dewey, W. L.: Comparison of the antinociceptive activity of intravenously administered acetylcholine to narcotic antinociception. Neurosci. Lett. *26*: 85–90, 1981.

Pedigo, N. W., Dewey, W. L. and Harris, L. S.: Determination and characterization of the antinociceptive activity of intraventricularly administered acetylcholine in mice. J. Pharmacol. Exp. Ther. *193*: 845–852, 1975.

Pehl, C., Wendl, B., Kaess, H. and Pfeiffer, A.: Effects of two anticholinergic drugs, trospium chloride and bipiriden, on motility and evoked potentials of the oesophagus. Aliment. Pharmacol. Ther. *12*: 979–984, 1998.

Pellandra, C. L.: La geneserine-morphine adjuvant de l'anesthesie génèrale. Lyon Med. *151*: 653, 1933.

Penrose, R.: "Shadows of the Mind," Oxford University Press, Oxford, 1994.

Pepeu, G.: The release of acetylcholine from the brain: an approach to the study of the central cholinergic mechanisms. In: "Progress in Neurobiology," G. A. Kerkut and J. W. Phillis, Eds., pp. 257–288, Pergamon Press, Oxford, 1974.

Pepeu, G.: Overview and future of CNS cholinergic mechanisms. In: "Cholinergic Neurotransmission: Function and Dysfunction." A. C. Cuello, Ed., pp. 455–458, Elsevier, Amsterdam, 1993.

Pepeu, G. and Giovannini, M. G.: Changes in acetylcholine extracellular levels during cognitive processes. Learn Mem. *11*: 21–27, 2004.

Pepeu, G. and Spignoli, G.: Nootropic drugs and brain cholinergic mechanisms. Prog. Neuropsychopharmacol. Biol. Psychiatry *13*: S77–S88, 1989.

Pepeu, G., Casamenti, F., Giovannini, M. G., Vannuchi, M. G. and Pedata, F.: Principal aspects of the regulation of acetylcholine release in the brain. In: "Cholinergic Neurotransmission: Functional and Clinical Aspects," S.-M. Aquilonius and P. G. Gillberg, Eds., pp. 273–278, Elsevier, Amsterdam, 1990.

Peralta, E. G., Kunkel, M. T., Cachero, T. G. and Morielli, A. D.: Modulation of K+ channels by muscarinic acetylcholine receptor subtypes. In: "The Ninth International Symposium on Cholinergic Mechanisms," J. Klein and K. Loeffelholz, Eds., Abstracts, Elsevier, Amsterdam, 1996.

Perez-Rosello, T., Figeroa, A., Salgado, H., Vilchis, C., Tecuapetla, F., Guzman, J. N., Gallaraga, E. and Bargas, J.: The cholinergic control of firing patterns and neurotransmission in rat neostratal projection neurons: role of CaV2.1 and CaV2.2+ Ca2+ channels. J. Neurophysiol., *93*: 2507–2519, 2005.

Perry, D. C., Davila-Garcia, M. I., Stockmeier, C. A. and Kellar, L. J.: Increased nicotinic receptors in brains from smokers: membrane binding and autographic studies. J. Pharmacol. Exp. Ther. *289*: 1545–1552, 1999.

Pert, A.: The cholinergic system and nociception in the primate: interactions with morphine. Psychopharmacology *44*: 131–137, 1975.

Pertwee, R. G. and Ross, T. M.: Drugs, which stimulate or facilitate central cholinergic transmission interact synergistically with delta-9-tetrahydrocannabinol to produce marked catalepsy in mice. Neuropharmacology *30*: 67–72, 1991.

Peterson, S. L. and Albertson, T. E.: Neurotransmitter and neuromodulator function in the kindled seizure and state. Prog. Neurobiol. *23*: 237–270, 1982.

Pfaff, D. W.: Luteinizing hormone-releasing factor potentiates lordosis behavior in hypophysectomized ovariectomized female rats. Science *182*: 1148–1149, 1973.

Pfaff, D. W.: "Estrogens and Brain Function," Springer-Verlag, New York, 1988.

Pfeiffer, C. C. and Jenney, E. H.: The inhibition of the conditioned response and the counteraction of schizophrenia by muscarinic stimulation of the brain. Ann. N. Y. Acad. Sci. *66*: 753–764, 1957.

Philips, M. F., Mattiasson, G., Wieloch, T., Bjorklund, A., Johansson, B. B., Tomasevic, G., Martinez-Serrano, A., Lenzlinger, P. M., Sinson, G., Grady, M. S. and McIntosh, T. K.: Neuroprotective and behavioral efficacy of nerve growth factor-transfected hippocampal progenitor cell transplants after experimental traumatic brain injury. J. Neurosurg. *94*: 765–774, 2001.

Phillis, J. W. and York, D. H.: Pharmacological studies on cholinergic inhibition of the cerebral cortex. Brain Res. *10*: 297–306, 1968.

Picciotto, M. R.: Common aspects of the action of nicotine and other drugs of abuse. Drug Alcohol Depend. *51*: 165–172, 1988.

Picciotto, M. R.: Nicotine as a modulator of behavior: beyond the inverted U. Trends Pharmacol. Sci. *24*: 493–499, 2003.

Picciotto, M. R. and Corrigal, W. A.: Neuronal systems underlying behaviors related to nicotine addiction: neural circuits and molecular genetics. J. Neurosci. *22*: 3338–3341, 2002.

Picciotto, M. R., Zoli, M., Lena, C., Bessis, A., Lallemand, Y., Le Novere, N., Vincent, P., Merlo-Pich, E., Brulet, P. and Changeux, J.-P.: Abnormal avoidance learning in mice lacking functional high-affinity nicotine receptor in the brain. Nature *374*: 65–67, 1995.

Picciotto, M. R., Zoli, M., Rimondini, R., Lena, C., Marubio, L. M., Pich, E. M. P. and Changeux, J. P.: Acetylcholine receptors containing the beta2 subunit are involved in reinforcement properties of nicotine. Nature *391*: 173–177, 1998.

Picciotto, M. R., Caldarone, B. J., King, S. L. and Zachariou, V.: Nicotinic receptors in the brain: links between molecular biology and behavior. Neuropsychopharmacology *22*: 451–465, 2000.

Pickford, M.: The action of acetylcholine in the supraoptic nucleus of chloralosed dog. J. Physiol. (Lond.) *106*: 246–270, 1947.

Pittel, Z., Cohen, S., Fisher, A. and Heldman, E.: Differential long-term effect of AF64A on [H3]ACh synthesis and release in rat hippocampal synaptosomes. Brain Res. *586*: 148–151, 1992a.

Pittel, Z., Heldman, E., Rubinstein, R. and Cohen, S.: Inhibition of choline efflux results in enhanced acetylcholine synthesis and release in the guinea-pig corticocerebral synaptosomes. Neurochem. Int. *20*: 219–227, 1992b.

Platano, D., Magli, M. C., Ferraretti, P., Gianaroli, L. and Aicardi, G.: L- and T-type voltage-gated Ca2+ channels in human granulosa cells: functional characterization and cholinergic regulation. J. Clin. Endocrin. Metab., 2005.

Polak, R. L.: An analysis of the stimulating action of atropine on release and synthesis of acetylcholine in cortical slices from rat brain. In: "Drugs and Cholinergic Mechanisms in the CNS," E. Heilbronn and A. Winter, Eds., pp. 323–338, Forsvatets Forskningsanstalt, Stockholm, 1970.

Polak, R. L.: Effect of hyoscine on the output of acetylcholine into perfused cerebral ventricles of cats. J. Physiol. *181*: 317–323, 1965.

Pollock, M. S. and Mistlberger, R. E.: Microinjection of neostigmine into the pontine reticular formation of the mouse: further evaluation of a proposed REM sleep enhancement technique. Brain Res. *1031*: 253–267, 2005.

Polonovski, M. and Nitzberg, C.: Études sur les alcaloides de la feve de Calabar. II. La geneserine, nouvelle alcaloide de la fève. Bul. Soc. Chimie *17*: 244–256, 1925.

Polonovski, M. and Polonovski, M.: Études sur les alcaloides de la fève de Calabar. XI. Quelques hypothèses sur la constitution de l'eserine. Bul. Soc. Chimie *33*: 1117–1131, 1923.

Pominova, E. V. and Shapovalova, K. B.: Participation of the cholinergic system of the dorsal and ventral striatum in the regulation of various forms of defense behavior. Neurosci. Behav. Physiol. *25*: 128–129, 1995.

Pompeiano, O.: The neurophysiological mechanism of the postural and motor events during desynchronized sleep. Res. Publ. Assoc. Res. Nerv. Ment. Dis. *45*: 351–425, 1967.

Popper, K.: "Conjectures and Refutations," Routledge and Kegan Paul, London, 1963.

Popper, K. R. and Eccles, J. C.: "The Self and Its Brain," Springer International, Berlin, 1977.

Potier, S. and Psarropoulou, C.: Modulation of muscarinic facilitation of epileptiform discharges in immature rat neocortex. Brain Res. *997*: 194–206, 2004.

Pouzet, G., Feldon, J., Veenman, C. L., Yee, B. K., Richmond, M., Nicholas, J., Rawlins, P. and Weiner, I.: The effects of hippocampal and fimbria-fornix lesions on prepulse inhibition. Behav. Neurosci. *113*: 968–981, 1999.

Pow, D. V. and Morris, J. F.: Differential distribution of acetylcholinesterase activity among vasopressin and oxytocin-containing supraoptic magnocellular neurons. Neuroscience *28*: 109–119, 1989.

Powell, D. A., Milligan, W. L. and Walters, K.: The effects of muscarinic cholinergic blockade upon shock-elicited aggression. Pharmacol. Biochem. Behav. *1*: 389–394, 1973.

Power, A. E.: Slow-wave sleep, acetylcholine, and memory consolidation. Proc. Natl. Acad. Sci. U.S.A. *101*: 1795–1796, 2004a.

Power, A. E.: Muscarinic contribution to memory consolidation: with attention to involvement of the basolateral amygdala. Curr. Med. Chem. *11*: 987–996, 2004b.

Power, A. E., Vazdarjanova, A. and McGaugh, J. L.: Muscarinic cholinergic influences in memory consolidation. Neurobiol. Learn. Mem. *89*: 178–1193, 2003.

Praag, H. van, Black, I. B. and Staubli, U. V.: Neonatal vs. adult unilateral hippocampal lesions: differential alterations in contralateral hippocampal theta rhythm. Brain Res. *768*: 233–241, 1997.

Pradhan, S. N. and Dutta, S. N.: Central cholinergic mechanisms and behavior. Int. Rev. Neurobiol. *14*: 173–231, 1971.

Prast, H. and Philippu, A.: Nitric oxide releases acetylcholine in the basal forebrain. Eur. J. Pharmacol. *216*: 139–140, 1992.

Prast, H., Fischer, H., Werner, E., Werner-Feldmayer, G. and Philippu, A.: Nitric oxide modulates the release of acetylcholine in the ventral striatum of the freely moving rat. Naunyn Schmiedebergs Arch. Pharmacol. *352*: 67–73, 1995.

Prendergast, M. A., Jackson, W. J., Terry, A. V., Jr., Decker, M. W., Arneric, S. P. and Buccafusco, J. J.: Central nicotinic receptor agonists ABT-418, ABT-089, and (–)-nicotine reduce distractibility in adult monkeys. Psychopharmacology *136*: 50–58, 1998.

Pribram, K. H.: "Languages of the Brain: Experimental Paradoxes and Principles in Neuropsychology," Brooks/Cole, Monterey, Calif., 1971.

Price, J. L. and Stern, R.: Individual cells in the nucleus basalis-diagonal band complex have restricted axonal projections to the cerebral cortex in the rat. Brain Res. *269*: 352–356, 1983.

Prostran, M., Varagic, V. M., Todorovic, Z. and Jezdimirovic, M.: The effects of physostigmine, L-arginine and NG-nitro-L-arginine methyl ester (L-NAME) on the mean arterial pressure of the rat. J. Basic Clin. Physiol. Pharmacol. *5*: 151–166, 1994.

Provost, T. L., Juarez de Ku, L. M., Zender, C. and Meserve, L. A.: Dose- and age-dependent alterations in choline acetyltransferase (ChAT) activity, learning and memory, and thyroid hormones in 15- and 30-day old rats exposed to 1.25 or 12.5 PPM polychlorinated biphenyl beginning at conception. Prog. Neuropsychopharmacol. Biol. Psychiatry *23*: 915–1928, 1999.

Pucilowski, V., Eichelman, T. S., Overstreet, D. H., Rezwani, A. H. and Janowsky, D. S.: Enhanced affective aggression in genetically bred hypercholinergic rats. Neuropsychobiology *24*: 34–41, 1990.

Puil, E. and Krnjevic, K.: Neurotransmitter actions and interactions. In: "Neuro-Psychopharmacology," P. Deniker, C. Radouco-Thomas, and A. Villeneuve, Eds., pp. 527–538, Pergamon Press, Oxford, 1978.

Pujol, J. F., Keane, P. E. and Jouvet, M.: Importance of interactions between transmitter systems in regulation of the sleep-waking cycle. In: "Neuro-Psychopharmacology," P. Deniker, C. Radouco-Thomas, and A. Villeneuve, Eds., pp. 559–568, Pergamon Press, Oxford, 1978.

Pyapali, G. K., Turner, D. A., Williams, C. L., Meck, W. H. and Swartzwelder, H. S.: Prenatal dietary choline supplementation decreases the threshold for induction of long-term potentiation in young adult rats. J. Neurophysiol. *79*: 1790–1796, 1998.

Pyck, J. C., Chang, Q., Colon-Rivera, C., Haag, R. and Gold, P. E.: Acetylcholine release in the hippocampus and striatum during place and response training. Learn Mem. *12*: 564–572, 2005.

Quastel, J. H., Tennenbaum, M. and Wheatley, A.: Choline esters formation in, and choline esterase activities of tissues "*in vitro*." Biochem. J. *30*: 1668–16781, 1937.

Quattrochi, J., Datta, S. and Hobson, J. A.: Cholinergic and non-cholinergic afferents of the caudolateral parabrachial nucleus: a role in the long-term enhancement of rapid eye movement sleep. Neuroscience *83*: 1123–1136, 1998.

Quian, J. and Saggau, P.: Presynaptic inhibition of synaptic transmission in the rat hippocampus by activation of muscarinic receptors: involvement of presynaptic calcium influx. Br. J. Pharmacol. *122*: 511–519, 1997.

Quick, M. W., Ceballos, R. M., Kasten, M., MacIntosh, J. M. and Lester, R. A.: Alpha3beta4 subunit-containing nicotinic receptors dominate function in rat medial habenula neurons. Neuropharmacology *38*: 769–783, 1999.

Quirion, R., Aubert, I., Araujo, D. M., Hersi, A. and Gaudreau, P.: Autographic distribution of putative muscarinic receptor sub-types in the mammalian brain. In: "Cholinergic Neurotransmission: Function and Dysfunction," A. C. Cuello, Ed., pp. 85–94, Elsevier, Amsterdam, 1993.

Raber, J., Pich, E. M., Koob, G. F. and Bloom, F. E.: IL-1 beta potentiates the acetylcholine-induced release of vasopressin from the hypothalamus *in vitro*, but not from the amygdala. Neuroendocrinology *59*: 208–217, 1994.

Radcliffe, K. A., Fisher, J. L., Gray, R. and Dani, J. A.: Nicotinic modulation of glutamate and GABA synaptic transmission of hippocampal neurons. Ann. N. Y. Acad. Sci. *868*: 591–610, 1999.

Radek, R. J.: Effects of nicotine on cortical high voltage spindles in rats. Brain Res. *625*: 23–28, 1993.

Radhakrishnan, R. and Sluka, K. A.: Spinal muscarinic receptors are activated during low or high frequency TENS-induced antihyperalgesia in rats. Neuropharmacology *45*: 1111–1119, 2003.

Ragozzino, M. E.: Acetylcholine actions in the dorsomedial striatum support the flexible shifting of response patterns. Neurobiol. Learn. Mem. *80*: 257–267, 2003.

Rahamimoff, R., Lev-Tov, A., Meiri, H., Rahamimoff, H. and Nussinovitch, I.: Regulation of acetylcholine release from presynatic nerve terminals. Monogr. Neural Sci. *7*: 3–18, 1980.

Raiteri, M., Marchi, M., Maura, G. and Bonanno, G.: Presynaptic regulation of acetylcholine release in the CNS. Cell Biol. Int. Rep. *13*: 1109–1118, 1989.

Raiteri, M., Marchi, M. and Paudice, P.: Presynaptic muscarinic receptors in the central nervous system. Ann. N.Y. Acad. Sci. *604*: 113–129, 1990.

Raiteri, M., Costa, R. and Marchi, M.: Effects of oxiracetam on neurotransmitter release from rat hippocampus slices and synaptosomes. Neurosci. Lett. *145*: 109–113, 1992.

Rakovska, A., Javitt, D., Raichev, P., Ang, R., Balla, A., Aspromonte, J. and Vizi, S.: Physiological release of striatal acetylcholine (*in vivo*): effect of somatostatin on dopaminergic-cholinergic interaction. Brain Res. Bull. *61*: 529–536, 2003.

Ramade, F. and Bayle, J. D.: Adaptation of the adrenocorticotropic response to chronic intermittent stress was altered by intracerebroventricular infusion of hemicholinium-3. Neuroendocrinology *50*: 165–169, 1989.

Randic, M., Sminoff, R. and Straughan, D. W.: Acetylcholine depression of cortical neurones. Exp. Neurol. *2*: 236–242, 1964.

Rasmusson, D. D.: Cholinergic modulation of sensory information. In: "Cholinergic Function and Dysfunction," A. C. Cuello, Ed., pp. 357–364, Elsevier, Amsterdam, 1993.

Raveh, L., Chapman, S. Cohen, G. Alkalay, D., Gilat, E., Rabinowitz, I. and Weissman, B. A.: The involvement of NMDA receptor complex in the protective effect of anticholinergic drugs against soman poisoning. Neurotoxicology *20*: 551–119, 1999.

Rawat, A. K.: Brain levels and turnover rates of presumptive neurotransmitters as influenced by administration and withdrawal of ethanol in mice. J. Neurochem. *22*: 915–922, 1974.

Ray, A., Sen, P. and Alkondon, M.: Biochemical and pharmacological evidence for central cholinergic regulation of shock-induced aggression in rats. Pharmacol. Biochem. Behav. *32*: 867–871, 1989.

Rees, H. and Roberts, M. H.: The anterior pretectal nucleus: a proposed role in sensory processing. Pain *53*: 121–135, 1993.

Reghunandan, V., Reghunandan, R. and Singh, P. I.: Neurotransmitters of the suprachiasmatic nucleus: role in the regulation of circadian rhythms. Prog. Neurobiol. *41*: 647–655, 1993.

Reid, M. S., Tafti, M., Gewary, J. N., Nishino, J. M., Siegel, J. M., Dement, W. C. and Mignot, E.: Cholinergic mechanisms in canine narcolepsy. I. Modulation of catalepxy via local drug administration into the pontine reticular formation. Neuroscience *59*: 511–512, 1994a.

Reid, M. S., Siegel, J. M., Dement, W. C. and Mignot, E.: Cholinergic mechanisms in canine narcolepsy. II. Acetylcholine release into pontine reticular formation is enhanced during cataplexy. Neuroscience *59*: 523–530, 1994b.

Reiner, A., Laverghetta, A. V., Meade, C. A., Cuthbertson, S. L. and Bottjer, S. W.: An immunohistochemical and pathway tracing study of the striatobasal organization of area X in the male zebra finch. J. Comp. Neurol. *469*: 239–261, 2004.

Reis, D. J.: The chemical coding of aggression in brain. In: "Neurochemical Coding of Brain Function," R. P. Drucker-Colin and R. D. Myers, Eds., pp. 125–150, Plenum Press New York, 1974.

Reno, L. A., Zago, W. and Markus, R. P.: Release of [(3)H]-L-glutamate by stimulation of nicotinic acetylcholine receptors in rat cerebellar slices. Neuroscience *124*: 647–653, 2004.

Renshaw, B.: Influence of discharge of motoneurons upon excitation of neighboring motoneurons. J. Neurophysiol. *4*: 487–498, 1942.

Renshaw, B.: Observations on interaction of nerve impulses in the gray matter and on the nature of central inhibition. Am. J. Physiol. *146*: 443–446, 1946.

Retana-Marquez, S. and Velazquez-Moctezuma, J.: Evidence that muscarinic receptor subtype mediates the effects of oxotremorine on masculine sexual behavior. Neuropsychopharmacology *9*: 267–290, 1993.

Retana-Marquez, S., Salazar, E. D. and Velazquez-Moctezuma, J.: Muscarinic and nicotinic influences on masculine sexual behavior in rats: effects of oxotremorine, scopolamine, and nicotine. Pharmacol. Biochem. Behav. *44*: 913–917, 1993.

Rhodes, E. M. and Rubin, R. T.: Functional sex differences ("sexual diergism") of central nervous system cholinergic systems, vasopressin, and hypothalamic-pituitary-adrenal axis activity in mammals: a selective review. Brain Res. Rev. *30*: 135–152, 1999.

Ribary, U. and Lichtensteiger, W.: Effects of acute and chronic prenatal nicotine treatment on central catecholamine systems of male and female rat fetuses and offspring. J. Pharmacol. Exp. Ther. *248*: 786–792, 1989.

Ribeiro, J. A., Correia-de-Sa, P., Cunha, R. A. and Sebastiao, A. M.: Purinergic regulation of acetylcholine release. In: "Cholinergic Mechanisms: From Molecular Biology to Clinical Significance," J. Klein and K. Loeffelholz, Eds., pp. 231–241, Elsevier, Amsterdam, 1996.

Richardson, A. J. and Glow, P. H.: Post-criterion discrimination behavior in rats with reduced cholinesterase activity. Psychopharmacology *11*: 435–438, 1967.

Richelson, E.: Cholinergic transduction. In: "Psychopharmacology: The Fourth Generation of Progress," F. E. Bloom and D. J. Kupfer, Eds., pp. 125–134, Elsevier, Amsterdam, 1995.

Rickett, D. L. and Beers, E. T.: Differentiation of medullary and neuromuscular effects of nerve agents. In: "Neurobiology of Acetylcholine: A Symposium in Honor of Alexander G. Karczmar," N. J. Dun and R. L. Perlman, eds., pp. 535–550, Plenum Press, New York, 1987.

Riekkinen, P., Jr., Sirvio, J., Jakala, P., Lammintausta, R. and Riekkinen, P.: Interaction between the alpha 2-noradrenergic and muscarinic systems in the regulation of neocortical high voltage spindles. Brain Res. Bull. *25*: 147–149, 1990.

Riekkinen, P., Jr., Riekkinen, M. and Sirvio, J.: Effects of nicotine on neocortical electrical activity in rats. J. Pharmacol. Exp. Ther. *267*: 776–784, 1993.

Riekkinen, P., Jr., Kuitunen, J. and Riekkinen, M.: Effects of scopolamine infusion into the anterior and posterior cingulate on passive avoidance and water maze navigation. Brain Res. *685*: 46–54, 1995.

Riemann, D., Hohagen, F., Fritsch-Montero, R., Krieger, S., Gann, H., Dressing, H., Muller, W. and Berger, M.: Cholinergic and noradrenergic neurotransmission: impact on REM sleep regulation in healthy subjects and depressed patients. Acta Psychiatr. Belg. *92*: 151–171, 1992.

Rinaldi, F. and Himwich, H. E.: Alerting responses and actions of atropine and cholinergic drugs. Arch. Neurol. Psychiatry *73*: 887–395, 1955a.

Rinaldi, F. and Himwich, H.: Cholinergic mechanisms involved in function of mesodiencephalic activating system. Arch. Neurol. Psychiatry *73*: 396–402, 1955b.

Risner, M. E., Cone, E. J., Benowitz, N. L. and Jacobs, P.: Effects of the stereoisomers on nicotine and nornicotine on schedule-controlled responding and physiological parameters of dogs. J. Pharmacol. Exp. Ther. *244*: 807–813, 1988.

Robbins, J., Caulfield, M. P., Higashida, H. and Brown, D. A.: Genotypic m3-muscarinic receptors preferentially inhibit M-currents in DNA-transfected NG108-15 neuroblastoma X hybrid cells. Eur. J. Neurosci. *3*: 820–824, 1991.

Robinson, J. D.: "A Century of Synaptic Transmission," Oxford University Press, Oxford, 2001.

Robinson, S. E.: 6-Hydroxydopamine lesion of the ventral noradrenergic bundle blocks the effect of amphetamine on hippocampal acetylcholine. Brain Res. *397*: 181–189, 1986.

Robinson, S. E.: The effect of cocaine on hippocampal cholinergic and noradrenergic metabolism. Brain Res. *457*: 383–385, 1988.

Rochlin, G.: "Man's Aggression—The Defense of Self," Gambit, Boston, 1973.

Rodgers, C. H. and Law, O. T.: Effects of chemical stimulation of the "limbic system" on lordosis in female rats. Physiol. Behav. *3*: 241–246, 1968.

Rodgers, R. J., Blanchard, D. C., Mong, L. K. and Blanchard, R. J.: Effects of scopolamine on antipredator defense reactions in wild and laboratory rats. Pharmacol. Biochem. Behav. *36*: 575–583, 1990.

Roeber, H.: Über die Wirkung des Calabarextractes auf Herz und Rückenmark. Dissertation, University of Berlin, Berlin, 1868.

Rogers, D. T. and Iwamoto, E. T.: Multiple spinal mediators in parenteral nicotine-induced antinociception. J. Pharmacol. Exp. Ther. *267*: 341–349, 1993.

Rogers, J. L. and Kesner, R. P.: Cholinergic modulation of the hippocampus during encoding and retrieval. Neurobiol. Learn. Mem. *80*: 332–342, 2003.

Rosat, R., Da Silva, R. C., Zanatta, M. S., Medina, J. H. and Izquierdo, I.: Memory consolidation of a habituation task: role of N-methyl-D-aspartate, cholinergic muscarinic and GABA-A receptors in different brain regions. Braz. J. Med. Biol. Res. *25*: 267–273, 1992.

Rosecrans, J. A.: The psychopharmacological basis of nicotine's differential effects on behavior: individual subject variability in the rat. Behav. Genet. *25*: 187–196, 1995.

Rosecrans, J. A.: Nicotine as a discriminative stimulus: a neurobehavioral approach to studying central cholinergic mechanisms. J. Subst. Abuse *1*: 287–300, 1989.

Rosenberg, H. and Haugaard, N.: The effects of halothane on metabolism and calcium uptake in mitochondria of the rat liver and brain. Anesthesiology *39*: 44–53, 1973.

Rosenzweig, M. R.: Environmental complexity, cerebral change, and behavior. Am. Psychol. *21*: 321–332, 1966.

Rosenzweig, M. R., Krech, D. and Bennett, E. L.: Brain enzymes and adaptive behavior. In: "CIBA Foundation Symposium on Neurological Basis of Behavior," G. E. W. Wostenholme and C. M. O'Connor, Eds., pp. 377–355, Little, Brown, Boston, 1958.

Rossner, S., Perez-Polo, J. R., Wiley, R. G., Schliebs, R. and Bigl, V.: Differential expression of immediate early genes in distinct layers of rat cerebral cortex after selective immunolesion of forebrain cholinergic system. J. Neurosci. Res. *38*: 282–293, 1994.

Roth, R. H. and Bunney, B. S.: Interaction of cholinergic neurons with other chemically defined neuronal systems in the CNS. In: "Biology of Cholinergic Function," A. M. Goldberg and I. Hanin, Eds., pp. 379–394, Raven Press, New York, 1976.

Rothberg, B. S., Yasuda, R. P., Satkus, S. A., Wolfe, B. B. and Hunter, B. E.: Effect of chronic ethanol on cholinergic actions in rat hippocampus: electrophysiological studies and quantification of m1–m5 muscarinic receptor subtypes. Brain Res. *631*: 227–234, 1993.

Rothberger, J. C.: Über die gegenseitigen Beziehungen zwischen Curare und Physostigmin. Arch. Ges. Physiol. *87*: 117–169, 1901.

Rothwell, N. J.: CNS regulation of thermogenesis. Crit. Rev. Neurobiol. *8*: 1–10, 1994.

Rovira, C., Ben-Ari, Y., Cherubini, E., Krnjevic, K. and Ropert, N.: Pharmacology of the dendritic action of acetylcholine and further observations on the somatic disinhibition in the rat hippocampus in situ. Neuroscience *8*: 97–106, 1983.

Rubin, L. S. and Goldberg, M. N.: Effect of sarin on dark adaptation. J. Appl. Physiol. *11*: 439, 1957.

Ruf, K. B.: Electrophysiological correlates of neuroendocrine tissues. In: "Subcellular Mechanisms in Reproductive Neuroendocrinology," F. Naftolin, K. J. Ryan and I. J. Davies, Eds., pp. 33–44, Elsevier, Amsterdam, 1976.

Ruggiero, D. A., Giuliano, R., Anwar, M., Stornetta, R. and Deis, D. J.: Anatomical substrates of cholinergic-autonomic regulation in the rat. J. Comp. Neurol. *292*: 1–53, 1990.

Rusak, B. and Bina, K. G.: Neurotransmitters in the mammalian circadian system. Annu. Rev. Neurosci. *13*: 387–401, 1990.

Rush, R., Kuryatov, A., Nelson, M. E. and Lindstrom, J.: First and second transmembrane segments of alpha3, alpha4, beta2 and beta4 nicotinic receptors subunits influence the efficacy and potency of nicotine. Mol. Pharmacol. *61*: 1416–1422, 2002.

Ruske, A. C. and White, K. G.: Facilitation of memory performance by a novel muscarinic agonist in young and old rats. Pharmacol. Biochem. Behav. *63*: 663–667, 1999.

Russo, C., Marchi, M., Andrioli, G. C., Cavazzani, P. and Raiteri, M.: Enhancement of glycine release from human brain cortex synaptosomes by acetylcholine acting at M4 muscarinic receptors. J. Pharmacol. Exp. Ther. *266*: 143–146, 1993.

Ryan, G. P., Hackman, J. C. and Davidoff, R. A.: Spinal seizures and excitatory acid-mediated synaptic transmission. Neurosci. Lett. *44*: 161–168, 1984.

Ryzhenkov, V. E., Kiriukhin, V. G.: The role of central cholinergic systems in the conditioned reflex activity and sexual behavior of male rats with different levels of androgens (Title translated from Russian). Zh. Vyssh. Nerv. Deiat. Im. I. P. Pavlova. *23*: 974–978, 1973.

Saar, D. and Barkai, E.: Long-term modifications in intrinsic neuronal properties and rule learning in rats. Eur. J. Neurosci. *17*: 2727–2734, 2003.

Sahibzada, N., Ferreira, M. Jr., Williams, B., Wasserman, A., Vicini, S., Gillis, R. A.: Nicotinic ACh receptor subtypes on gastrointestinally projecting neurones in the dorsal motor vagal nucleus of the rat. J. Physiol. *545*: 1007–1016, 2002.

Sahley, T. L. and Berntson, G. G.: Antinociceptive effects of central and systemic administration of nicotine in the rat. Psychopharmacology *65*: 279–383, 1979.

Sainsbury, R. S. and Partlo, L. A.: The effects of alpha 2 agonists and antagonists on hippocampal theta activity in freely moving rats. Brain Res. Bull. *26*: 37–42, 1991.

Saito, Y., Yamashita, I., Yamazaki, K., Okada, F., Satomi, R. and Fujieda, T.: Circadian fluctuation of brain ACh in rats. I. On the variation in the total brain and discrete brain areas. Life Sci. *16*: 281–288, 1975.

Saito, A., Goldfine, I. D. and Williams, J. A.: Characterization of receptors for cholecystokinin and related peptides in mouse cerebral cortex. J. Neurochem. *37*: 483–490, 1981.

Sakai, K. K., Marks, B. H., George, J. M. and Koestner, A.: The isolated organ-cultured supraoptic nucleus as neuropharmacological test system. J. Pharmacol. Exp. Ther. *190*: 482–491, 1974.

Salto, C., Calvet, R., Guirat, X., Solsona, C. and Marsal, J.: Opiates depress ACh and ATP release from cholinergic synaptosomes by blocking calcium uptake. Toxicol. Appl. Pharmacol. *106*: 20–27, 1990.

Samathanam, G., Duffy, P., Kalivas, P. W. and White, S. R.: A comparison of 5-hydroxytryptophan effects on rat lumbar spinal cord serotonin release and monosynaptic response amplitude. Brain Res. *501*:179–182, 1989.

Sanchez, M. A., Lopez-Garcia, J. C., Cruz, M. E., Tapi, R. and Dominguez, R.: Asymmetric changes in choline acetyltransferase activity in the preoptic-anterior hypothalamus area during the oestrous cycle of the rat. Neuroreport *5*: 433–434, 1994.

Sandor, N. T., Zalles, T., Sershen, A., Torocsik, A., Lajtha, A. and Vizi, E. S.: Effect of nicotine on dopaminergic-cholinergic interaction in the striatum. Brain Res. *567*: 313–316, 1991.

Sapru, H. N.: Cholinergic mechanisms subserving cardiovascular function in the medulla and spinal cord. Prog. Brain Res. *81*: 171–179, 1989.

Sarter, M. and Bruno, J. P.: Cortical acetylcholine, reality distortion, schizophrenia, and Lewy Body Dementia: too much or too little cortical acetylcholine? Brain Cogn. *38*: 297–316, 1998.

Sarter, M., Bruno, J. P. and Givens, B.: Attentional functions of cortical cholinergic inputs: what does it mean for learning and memory? Neurobiol. Learn. Mem. *80*: 245–256, 2003.

Sasaki, K., Matsuda, Y., Oka, H. and Mizuno, M. L.: Thalamo-cortical projections for recruiting responses and spindling-like response in parietal cortex. Exp. Brain Res. *22*: 87–96, 1975.

Sauer, H., Francis, J. M., Jang, H., Hamilton, G. S. and Steiner, J. P.: Systemic treatment with GPI 1046 improves spatial memory and reverses cholinergic atrophy in the medial septal nucleus of aged mice. Brain Res. *84*: 109–118, 1999.

Sawyer, C. H., Markee, J. E. and Towsend, B. F.: Cholinergic and adrenergic components in the neurohumoral control of the release of LH in the rabbit. Endocrinology *44*: 18–37, 1949.

Sawyer, C. H., Markee, J. E. and Everett, J. W.: Blockade of neurogenic stimulation of the rabbit adenohypophysis by banthine. Am. J. Physiol. *166*: 223–228, 1951.

Scatton, B. and Fage, D.: Excitatory amino acid influence on striatal cholinergic transmission. In: "Dynamics of Cholinergic Function," I. Hanin, Ed., pp. 981–990, Plenum Press, New York, 1986.

Scattoni, M. L., Puopolo, M., Calamandrei, G. and Ricceri, L.: Basal forebrain lesions in 7-day old rats alter ultrasound vocalizations and homing behavior. Behav. Brain Res. *161*: 169–172, 2005.

Schama, S.: "The Embarrassment of Riches. An Interpretation of Dutch Culture in the Golden Age," University of California Press, Berkeley, 1988.

Schaumann, W.: Antagonistische Wirkung von Morphin und Cholinesterase Hemmern auf das Atemzentrum. Arch. Exp. Pathol. Pharmakol. *233*: 98–111, 1958.

Schimerlik, M. I.: Structure and function of muscarinic receptors. In: "Cholinergic Neurotransmission: Functional and Clinical Aspects," S.-M. Aquilonius and P.-G. Gillberg, pp. 11–19, Elsevier, Amsterdam, 1990.

Schmidt, R. F.: The effect of drugs on the reflex paths to primary afferent fibres. In: "Studies in Physiology," D. R. Curtis and A. K. McIntyre, Eds., pp. 243–249, Springer-Verlag, Berlin, 1965.

Schmitt, H.: Actions centrales des substances parasympathomimetiques. In: "Le Système Cholinergique," G. J. Nahas, J. C. Salamagne, P. Viars and G. Vourc'h, Eds., pp. 181–228, Librairie Arnette, Paris, 1972.

Schneider, J. S., Tinker, J. P., Van Nelson, M., Menzaghi, F. and Lloyd, G. K.: Nicotinic acetylcholine receptor agonist SIB-1508Y improves cognitive functioning in chronic low-dose MPTP-treated monkeys. J. Pharmacol. Exp. Ther. *290*: 731–739, 1999.

Schoffelmeer, A. N., De Vries, T. J., Hogenboom, F., Hruby, V. J., Portoghese, P. S. and Mulder, A. H.: Opioid receptor antagonists discriminate between presynaptic mu and delta receptors and the adenylate cyclase-coupled opioid receptor complex in the brain. J. Pharmacol. Exp. Ther. *263*: 20–24, 1992.

Schoffelmeer, A. N., Hogenboom, F. and Mulder, A. H.: Kappa1- and kappa2-opioid receptors mediating presynaptic inhibition of dopamine and acetylcholine release in rat neostriatum. Br. J. Pharmacol. *122*: 520–524, 1997.

Schofield, P. R., Darlison, M. H., Fujita, N., Burt, D. R., Stephenson, F. A., Rodriquez, H., Rhee, L. M., Ramachandran, J., Reale, V. Glencorse, T. A., Seeburg, P. H. and Barnard, E. A.: Sequence and functional expression of the GABA A receptor shows a ligand-gated receptor superfamily. Nature *328*: 221–227, 1987.

Scholz, K. P. and Miller, R. J.: Inhibition of quantal transmitter release in the absence of calcium influx by a G-protein-linked adenosine receptor at hippocampal synapses. Neuron *8*: 1139–1150, 1992.

Schott, G. D. and Loh, H.: Anticholinesterase drugs in the treatment of chronic pain. Pain *20*: 201–206, 1984.

Schrattenholz, A., Godovac-Zimmermann, J., Schaeffer, H.-J. and Maelicke, A.: Photoaffinity labeling of Torpedo acetylcholine receptor by physostigmine. Eur. J. Biochem. *216*: 671–677, 1993.

Schubert, P. and Mitzdorf, U.: Analysis and quantitative evaluation of the depressive effects of adenosine on evoked potentials in hippocampal slices. Brain Res. *172*: 186–190, 1979.

Schuetze, S. M. and Role, L. W.: Developmental regulation of nicotinic acetylcholine receptors. Ann. Rev. Neurosci. *10*: 403–457, 1987.

Schwartz, R. D. and Kellar, K. J.: *In vivo* regulation of [3H] acetylcholine recognition sites in brain by nicotinic drugs. J. Neurochem. *45*: 427–433, 1985.

Schwartz, R. D. and Mindlin, M. C.: Inhibition of the GABA receptor-gated chloride ion channel in the brain by noncompetitive inhibitors of the nicotinic receptor-gated cation channel. J. Pharmacol. Exp. Ther. *244*: 963–970, 1988.

Schwartz, R. D., Callahan, M. J., Davis, R. E., Jaen, J. C., Lipinski, W., Raby, C., Sencer, C. J. and Tecle, H.: Selective muscarinic agonists for Alzheimer disease treatment. In: "Alzheimer Disease: Therapeutic Strategies," E. Giacobini and R. Becker, Eds., pp. 224–228, Birkhauser, Boston, 1994.

Schweitzer, A. and Wright, S.: Further observations on the action of acetylcholine, prostigmine and related substances on the knee jerk. J. Physiol. (Lond.) *89*: 165–197, 1937.

Schweitzer, T., Birthelmer, A., Lazaris, A., Cassel, J. C. and Jakisch, R.: 3,4-DAP-evoked transmitter release in hippocampal slices of aged rats with impaired memory. Brain Res. Bull. *62*: 129–136, 2003.

Scremin, Q. U., Allen, K., Torres, C. and Scremin, A. M.: Physostigmine enhances blood flow-metabolism ratio in neocortex. Neuropsychopharmacology *1*: 297–303, 1988.

Scriverius, P.: "Saturnalia ofte Poetisch Vasten-Avond spel vervatende het gebruyk en misbruyk vande Taback," Haarlem, Holland, 1630.

Searle, J. R.: "Mind," Oxford University Press, Oxford, 2004.

Seeger, T., Fedorova, I. K., Zheng, F., Miyakava, T., Koustova, E., Gomeza, J., Basile, A. S., Alzheimaer, C. and Wess, J.: M2 muscarinic acetylcholine receptor knock-out mice show deficits in behavioral flexibility, working memory, and hippocampal plasticity. J. Neurosci. *24*: 10117–10127, 2004.

Segal, M.: Rat hippocampus neurons in culture: response to electrical and chemical stimuli. J. Neurophysiol. *50*: 1249–1264, 1983.

Segal, M.: Synaptic activation of a cholinergic receptor in rat hippocampus. Brain Res. *452*: 79–86, 1988a.

Segal, M.: Effects of a lidocaine derivative QX-572 on CA1 neuronal responses to electrical and chemical stimuli in a hippocampal slice. Neurosci. *27*: 905–909, 1988b.

Segal, M.: Presynaptic cholinergic inhibition in hippocampal cultures. Synapse *4*: 305–312, 1989.

Segal, M., Greenberger, V. and Markram, H.: Presynaptic cholinergic action in the hippocampus. Experientia *57*, Supplt.: 88–96, 1989.

Sellini, M., Baccarini, S., Dimitriadis, E., Sartori, M. P. and Letizia, C.: Effeto del fumo sull'asse ipofisisurrene. Medicina (Firenze) *9*: 194–196, 1989.

Semba, K.: The cholinergic basal forebrain: a critical role in cortical arousal. In: "The Basal Forebrain," T. C. Napier, P. W. Kalivas and I. Hanin, Eds., pp. 197–218, Plenum Press, New York, 1991.

Sembulingam, K., Sembulingam, P. and Namasivayam, A.: Effect of Ocimum sanctum Linn on the changes in central cholinergic system induced by acute noise stress. J. Ethnopharmacol. *96*: 477–482, 2005.

Sergeeva, O. A.: Comparison of glycine- and GABA-evoked currents in two types of neurons isolated from the rat striatum. Neurosci. Lett. *243*: 9–12, 1998.

Sergeeva, O. A. and Haas, H. L.: Expression and function of glycine receptors in striatal cholinergic interneurons from rat and mouse. Neurosa. *104*: 1043–1055, 2001.

Sernagor, E., Ritter, A. and O'Donovan, M. J.: Pharmacological characterization of the rhythmic synaptic drive onto lumbosacral motoneurons in the chick embryo spinal cord. J. Neurosci. *15*: 7452–7464, 1995.

Serra, M., Dazzi, L., Caddeo, M., Floris, C. and Biggio, G.: Reversal by flunarzine of the decrease in hippocampal acetylcholine release in pentylelenetetrazole-kindled rats. Biochem. Pharmacol. *58*: 145–149, 1999.

Sessions, G. R., Pilcher, J. J. and Elsmore, T. F.: Scopolamine-induced impairment in concurrent fixed-interval responding in a radial maze task. Pharmacol. Biochem. Behav. *59*: 641–647, 1998.

Seylaz, J., Hara, H., Pinard, E., Mraovitch, S., MacKenzie, E. T. and Edvinsson, L.: Effect of stimulation of the sphenopalatine ganglion on cortical blood flow in the rat. J. Cereb. Blood Flow Metab. *8*: 875–878, 1988.

Sgard, F., Charpantier, E., Barneeoud, P and Besnard, F.: Nicotinic receptor subunit mRNA expression in dopaminergic neurons of rat brain. Ann. N. Y. Acad. Sci. *868*: 633–635, 1999.

Shannon, H. E. and Peters, S. C.: A comparison of the effects of cholinergic and dopaminergic agents on scopolamine-induced hyperactivity in mice. J. Pharmacol. Exp. Ther. *225*: 549–553, 1990.

Shannon, H. E., Hart, J. C., Bymaster, F. P., Calligaro, D. O., DeLapp, N. W., Mitch, C. H., Ward, J. S., Fink-Jensen, A., Sauerberg, P., Jeppesen, L., Sheardown, M. J. and Swedberg, M. D.: Muscarinic agonists, like dopamine receptor antagonist antipsychotics, inhibit conditioned avoidance response in rats. J. Pharmacol. Exp. Ther. *290*: 901–907, 1999.

Shao, X. M. and Feldman, J. L.: Cholinergic neurotransmission in the preBotzinger Complex modulates excitability of inspiratory neurons and regulates respiratory rhythm. Neuroscience *130*: 1069–1081, 2005.

Shapovalova, K. B.: Afferent and efferent mechanisms of the intensification of neostriatal cholinergic activity. Neurosci. Behav. Physiol. *25*: 71–80, 1995.

Shapovalova, K. B., Pominova, E. V. and Diubkacheva, T. A.: The role of activation of the intralaminar thalamic nuclei in regulating the participation of the neostriatal cholinergic system in the differentiation of sound signals by dogs. Zh. Vyssh. Nerv. Deiat. Im. I. P. Pavlova *44*: 849–852, 1994.

Sharma, G. and Vijayaraghavan, S.: Nicotinic cholinergic signaling in hippocampal astrocytes involves calcium-induced calcium release from intracellular stores. Proc. Natl. Acad. Sci. U.S.A. *98*: 3631–3638, 2001.

Sharpless, S. K.: Reorganization of function in the nervous system—use and disuse. Annu. Rev. Physiol. *26*: 357–388, 1964.

Sherrington, C. S.: "Man on his Nature," Cambridge University Press, Cambridge, 1940.

Shevrin, H., Ghannam, J. H. and Libet, B.: A neural correlate of consciousness related to repression. Conscious Cogn. *11*: 342–344, 2002.

Shigeta, M., Persson, A., Viitanen, M., Winblad, B. and Nordberg, A.: EEG regional changes during long-term treatment with tetrahydroaminoacridine (THA) in Alzheimer's disease. Acta Neurol. Scand., Supplt. *149*: 58–61, 1993.

Shimizu, T., Rosenblum, W. I. and Nelson, G. H.: M3 and M1 receptors in cerebral arterioles *in vivo*: evidence for downregulated or ineffective M1 when endothelium is intact. Am. J. Physiol. *264*: H665–H669, 1993.

Shimoyama, M. and Kito, S.: Effect of cerulein on *in vivo* release of acetylcholine from the rat striatum. Brain Res. *482*: 381–384, 1989.

Shinozaki, H., Hirate, K. and Ishida, M.: Modification of drug-induced tremor by systemic administration of kainic acid and quisqualic acid in mice. Neuropharmacology *26*: 9–17, 1987.

Shiraki, I., Sershen, H., Benuck, M., Hashim, A. and Lajtha, A.: Differences in receptor system participation between nicotine- and cocaine-induced dopamine overflow in nucleus accumbens. Ann. N. Y. Acad. Sci. *877*: 800–802, 1999.

Shiromani, P. J.: Sleep circuitry, regulation and function: lessons from c-fos, leptin and timeless. In: "Progress in Psychobiology and Physiological Psychology," A. Morrison and C. Fluharty, Eds., pp. 67–90, Academic Press, San Diego, 1998.

Shiromani, P. J., Gillin, J. C. and Henriksen, S. J.: Acetylcholine and the regulation of the REM sleep: basic mechanisms and clinical implications for affective illness and narcolepsy. Ann. Rev. Pharmacol. Toxicol. *27*: 137–156, 1987.

Shoaib, M.: Is dopamine important in nicotine dependence? J. Physiol. (Paris) *92*: 229–233, 1998.

Shoaib, M. and Stolerman, I. P.: Brain sites mediating the discriminative stimulus effects of nicotine in rats. Behav. Brain Res. *78*: 183–188, 1996.

Shugalev, N. P.: Influence of morphine on behavioral effects of pharmacological stimulation of cholinergic structures of the cat brain. Neurosci. Behav. Biol. *20*: 32–33, 1990.

Shuh, T. H., Wang, C. H. and Lim, R. K. S.: Effects of intracisternal injection of acetylcholine. Proc. Soc. Exp. Biol. *32*: 1410, 1935.

Siciliano, R., Fontanesi, G., Casamenti, F., Berardi, N., Bagnoli, P. and Domenici, I.: Postnatal development of functional properties of visual cortical cells in rats with excitotoxic lesions of basal forebrain cholinergic neurons. Vis. Neurosci. *14*: 111–123, 1997.

Sienkiewicz-Jarosz, H., Maciejak, P., Bidzinski, A., Szyndler, J., Siemiatkowski, M., Czlonkowska, A., Lehner, M. and Plaznik, A.: Exploratory activity and conditioned fear response correlation with cortical and subcortical binding of alpha4beta2 nicotinic receptor agonist [3H] epibatidine. Pol. J. Pharmacol. *55*: 17–23, 2003.

Sihver, W., Gillberg, P. G., Svensson, A. L. and Nordberg, A.: Autoradiographic comparison of [3H](–) nicotine, [3H] cytisine, and [3H] epibatidine binding in relation to vesicular acetylcholine transport sites in the temporal cortex in Alzheimer's disease. Neuroscience *94*: 685–696, 1999.

Silbergeld, E. K. and Goldberg, A. M.: Hyperactivity. In: "Biology of Cholinergic Function," A. M. Goldberg and I. Hanin, Eds., pp. 619–645, Raven Press, New York, 1976.

Silvestri, A. J. and Kapp, B. S.: Amygdaloid modulation of mesopontine peribrachial neuronal activity: implications for arousal. Behav. Neurosci. *112*: 571–588, 1998.

Simasko, S. M. and Horita, A.: Characterization and distribution of 3H (3MeHis2) thyrotropin releasing hormone receptors in rat brain. Life Sci. *30*: 1793–1799, 1982.

Simmonds, M. A.: Classification of some GABA antagonists with regard to site of action and potency in slices of cuneate nucleus. Eur. J. Pharmacol. *80*: 347–358, 1982.

Simmons, D. C.: Efik divination, ordeals and omens. S. W. J. Anthropol. *12*: 223–228, 1956.

Simpson, C. M., Ruwe, W. D. and Myers, R. D.: Prostagladins and hypothalamic neurotransmitter receptors involved in hyperthermia: a critical evaluation. Neurosci. Biobehav. Rev. *18*: 1–20, 1994.

Singer-Lahat, D., Rojas, E. and Felder, C. C.: Muscarinic receptor activated Ca++ channel in non-excitable cells. In: "Cholinergic Mechanisms: From Molecular

Biology to Clinical Significance," J. Klein and K. Loffelholz, Eds., pp. 195–199, Elsevier, Amsterdam, 1996.

Singh, M. M. and Kay, S. R.: Therapeutic antagonism between anticholinergic antiparkinsonism agents and neuroleptics in schizophrenia. Implications for a neuropharmacological model. Neuropsychobiology. *5*: 74–86, 1979.

Sinischalchi, A., Badini, I., Cintra, A., Fuxe, K., Bianchi, C. and Beani, L.: Muscarinic modulation of acetylcholine release from slices of guinea pig nucleus basalis magnocellularis. Neurosci. Lett. *140*: 235–238, 1991.

Sinson, G., Voddi, M., Flamm, E. S. and MacIntosh, T. K.: Neurotrophin infusion improves cognitive deficits and decreases neuronal cell loss after experimental brain injury. Clin. Neurosurg. *43*: 219–227, 1996.

Sitaram, N. and Gillin, J. C.: Acetylcholine: possible involvement in sleep and analgesia. In: "Brain Acetylcholine and Neuropsychiatric Disease," K. L. Davis and P. A. Berger, Eds., pp. 311–343, Plenum Press, 1977.

Sitaram, N., Wyatt, R. J., Dawson, S. and Gillin, J. C.: REM sleep induction by physostigmine infusion during sleep. Science *191*: 1281–1283, 1976.

Sitaram, N., Weingartner, H. and Gillin, J. C.: Choline chloride and arecoline: effects on memory and sleep in man. In: "Choline and Lecithin in Brain Disorders," A. Barbeau, J. H. Growdon and R. J. Wurtman, Eds., pp. 367–375, Nutrition and the Brain, vol. 5., Raven Press, New York, 1979.

Sjostrand, T.: Potential changes in the cerebral cortex of the rabbit arising from cellular activity and the transmission of impulses in the white matter. J. Physiol. (Lond.) *90*: 41–43P, 1937.

Sladek, C. D. and Joynt, R. J.: Characterization of cholinergic control of vasopressin release by the organ-cultured hypothalamico-hypophyseal system. Endocrinology *104*: 659–663, 1979.

Slangen, J. L., Earley, B., Jaffard, R., Richelle, M. and Olton, D. S.: Behavioral models of memory and amnesia. Pharmacopsychiatry *23*, Supplt. 2: 81–83, 1990.

Slaughter, D. and Gross, E. G.: Some new aspects of morphine action. Effect on intestine and blood pressure; toxicity studies. J. Pharmacol. Exp. Ther. *68*: 104–112, 1940.

Small, D. M., Gittelman, D., Simmons, K., Bloise, S. M., Parrish, T. and Mesulam, M. M.: Monetary incentives enhance prossessing in brain regions mediating top-down control of attention. Cereb. Cortex *15*: 1855–1865, 2005.

Smith, C.: The REM sleep window and memory processes. In: "Sleep and Brain Plasticity," P. Maquet, C. Smith and R. Stickgold, Eds., pp. 117–133, Oxford University Press, Oxford, 2003.

Smith, D. E., King, M. B. and Hoebel, B. G.: Lateral hypothalamic control of killing: evidence for cholinoceptive mechanism. Science *167*: 900–901, 1970.

Smith, F. J. and Campfield, L. A.: Meal initiation occurs after experimental induction of transient declines in blood glucose. Am. J. Physiol. *265*: 1423–1429, 1993.

Smith, G. B. and Gibbs, J.: Cholecystokinin: a putative satiety signal. Pharmacol. Biochem. Behav. *3*, Supplt. 1: 135–138, 1975.

Smith, J. E., Vaughn, T. C. and Co, C.: Acetylcholine turnover rates in brain rat regions during cocaine self-administration. J. Neurochem. *88*: 502–512, 2004.

Smith, M. D., Yang, X., Nha, J.-Y. and Buccafusco, J. J.: Antinociceptive effect of spinal cholinergic stimulation: interaction with substance P. Life Sci. *45*: 1255–1261, 1989.

Smith, P. B. and Martin, B. R.: Spinal mechanisms of delta-9-tetrahydrocannabinol-induced analgesia. Brain Res. *578*: 8–12, 1992.

Smythies, J.: The functional neuroanatomy of awareness: with a focus on the role of various anatomical systems in the control of intermodal attention. Conscious. Cogn. *6*: 455–481, 1997.

Smythies, J. R.: The role of acetylcholine in hallucinatory perception. Behav. Brain Sci. *28*: 773, 2005.

Snyder, S. H.: Neuronal receptors. Annu. Rev. Physiol. *48*: 461–471, 1986.

Sokolovsky, M.: Muscarinic receptors in the central nervous system. Int. Rev. Neurobiol. *25*: 139–183, 1984.

Somogyi, P.: Immunohistochemical demonstration of GABA in physiologically characterized, HRP-filled neurons and their postsynaptic targets. In: "Molecular Neuroanatomy: Techniques in the Behavioral and Neural Sciences," F. W. van Leeuven, R. M. Bujis, C. V. Pool and O. Pach, Eds., pp. 339–359, Elsevier, Amsterdam, 1988.

Song, W. J., Tkatch, T., Baranauskas, G., Ichinohe, K., Kitai, S. T. and Surmeier, D. J.: Somatodendritic depolarization-activated potassium currents in rat neostriatal cholinergic interneurons are predominantly of the A type and attributable to coexpression of Kv4.2 and Kv4.1 subunits. J. Neurosci. *18*: 3124–3137, 1998.

Sorger, D., Kampfer, I., Schliebs, R., Rossner, S., Dannenberg, C. and Knapp, W. H.: Iodo-QNB cortical binding and brain perfusion: effects of cholinergic basal forebrain lesion in rat. Nucl. Med. Biol. *26*: 9–16, 1999.

Soulairac, A. and Soulairac, M. L.: Relationship between the nervous and endocrine regulation of sexual behavior in male rats. Psychoneuroendocrinology *3*: 17–29, 1978.

Soulairac, M. L. and Soulairac, A.: Monoaminergic and cholinergic control of sexual behavior in male rat. In: "Sexual Behavior," M. Sandler and G. L. Gessa, Eds., pp. 99–116, Raven Press, New York, 1975.

Spiegel, R.: Effects of RS-86, an orally active cholinergic agonist, on sleep in man. Psychiatry Res. *11*: 1–13, 1984.

Spignoli, G. and Pepeu, G.: Interactions between oxiracetam, aniracetam and scopolamine on behavior and brain acetylcholine. Pharmacol. Biochem. Behav. *27*: 491–495, 1987.

Spinella, M., Znamensky, V., Moroz, M., Ragnauth, A. and Bodnar, R. J.: Actions of NMDA and cholinergic receptor antagonists in the rostral ventromedial medulla upon beta-endorphin analgesia elicited from the ventrolateral periaqueductal gray. Brain Res. *829*: 151–159, 1999.

Spurzheim, J. G.: "The Physiognomical System of Drs. Gall and Spurzheim," Baldwin, Cradock, and Joy, London, 1815.

Squire, L. R. and Butters, N.: "Neuropsychology of Memory," Guilford Press, New York, 1992.

Srivastava, S. K., Nath, C. and Sinha, J. N.: Evidence for antiaggressive property of some calcium channel blockers. Pharmacol. Res. *35*: 435–438, 1997.

Stackman, R. M. and Walsh, T. J.: Baclofen produces dose-related working memory impairments after intraseptal injection. Behav. Neural Biol. *61*: 181–185, 1995.

Stadler, H., Lloyd, K. G. and Gadea-Ciria, M.: Enhanced striatal acetylcholine release by chlorpromazine and its reversal by apomorphine. Brain Res. *55*: 476–480, 1973.

Stanley, B. G. and Leibowitz, S. F.: Neuropeptide Y: stimulation of feeding and drinking by injection into the paraventricular nucleus. Life Sci. *35*: 2635–2642, 1984.

Starr, M. S. and Starr, B. S.: The new competitive NMDA receptor antagonist CGP 40116 inhibits pilocarpine-induced limbic motor seizures and unconditioned motor behavior in the mouse. Pharmacol. Biochem. Behav. *47*: 137–131, 1994.

Stay, van der F. J. and Bouger, P. C.: Effects of cholinesterase inhibitors, donepezil and metrifonate on scopolamine-induced impairments in the spatial cone field orientation tasks in rats. Behav. Brain Res. *156*: 1–10, 2005.

Stedman, E. and Barger, G.: Physostigmine (eserine). Part III. J. Chem. Soc. *127*: 247–258, 1925.

Stein, L.: Chemistry of reward and punishment. In: "Psychopharmacology—A Review of Progress," D. H. Efron, Ed., pp. 105–123, U.S. Government Office, PHS Publication No. 836, Washington, D.C., 1968.

Stein, L. and Seifter, J.: Muscarinic synapses in the hypothalamus. Am. J. Physiol.*202*: 751–756, 1962.

Steinberg, R., Rodier, D., Souielhac, J., Bagault, I., Emons-Alt, X., Soubrie, P. and LeFur, G.: Pharmacological characterization of tachykinin receptors controlling acetylcholine release from rat striatum: *in vivo* microdialysis study. J. Neurochem. *65*: 2543–2548, 1995.

Steingart, R. A., Barg, J., Maslaton, J., Nesher, M. and Yanai, J.: Pre- and postsynaptic alterations in the septohippocampal cholinergic innervations after prenatal exposure to drugs. Brain Res. Bull. *46*: 203–209, 1998.

Steriade, M.: Cholinergic blockage of network- and intrinsically generated slow oscillations promotes waking and REM sleep activity patterns in thalamic and cortical neurons. In: "Cholinergic Function and Dysfunction," A. C. Cuello, Ed., pp. 345–355, Prog. Brain Res., vol. 98, Elsevier, Amsterdam, 1993a.

Steriade, M.: Modulation of information processing in thalamicocortical systems: Chairman's introductory remarks. In: "Cholinergic Function and Dysfunction," A. C. Cuello, Ed., pp. 341–342, Prog. Brain Res., vol. 98, Elsevier, Amsterdam, 1993b.

Steriade, M.: The corticothalamic system in sleep. Front. Biosci. *8*: 878–899, 2003.

Steriade, M. and Hobson, J. A.: Neuronal activity during the sleep-waking cycle. Prog. Neurobiol. *6*: 155–367, 1976.

Steriade, M. and Llinas, R. R.: The functional states of the thalamus and the associated neuronal interplay. Physiol. Rev. *68*: 649–742, 1988.

Steriade, M. and McCarley, R. W.: "Brain Stem Control of Wakefulness and Sleep," Plenum Press, New York, 1990.

Steriade, M., Curro Dossi, R. and Nunez, A.: Network modulation of a slow intrinsic oscillation of cat thalamocortical neurons implicated in sleep delta waves; cortical potentiation and brain stem cholinergic suppression. J. Neurosci. *11*: 3190–3217, 1991.

Steriade, M., Jones, E. G. and McCormick, D. A.: "Thalamus," Elsevier, Amsterdam, 1994.

Sterman, M. B. and Fairchild, M. D.: Modification of locomotor performance by reticular formation and basal forebrain stimulation in the cat; evidence for reciprocal systems. Brain Res. *2*: 205–217, 1966.

Stevenson, J. A. F.: Neural control of food and water intake. In: "The Hypothalamus," W. Haymaker, E. Anderson, and W. Nauta, Eds., pp. 524–621, Charles C. Thomas, Springfield, Ill., 1969.

Stewart, M. and Fox, S. E.: Two populations of rhythmically bursting neurons in rat medial septum are revealed by atropine. J. Neurophysiol. *61*: 982–993, 1989.

Stewart, M. and Fox, S.: Do septal neurons pace the hippocampal theta rhythm? TINS *13*: 163–168, 1990.

Stewart, W. C. and Anderson, E. A.: Effects of a cholinesterase inhibitor when injected into the medulla of the rabbit. J. Pharmacol. Exp. Ther. *162*: 309–318, 1968.

Stolerman, I. P. and Reavill, C.: Primary cholinergic and indirect dopaminergic mediation of behavioural effects of nicotine. Prog. Brain Res. *79*: 303–312, 1989.

Stone, T. W.: Cholinergic mechanisms in the rat somatosensory cerebral cortex. J. Physiol. *225*: 485–499, 1972.

Stone, W. S., Rudd, R. J. and Gold, P. E.: Glucose and physostigmine effects on morphine- and amphetamine-induced increases in locomotor activity in mice. Behav. Neural Biol. *54*: 146–155, 1990.

Stoof, J. C., Verheijden, P. F. and Leysen, J. E.: Stimulation of D2-receptors in rat nucleus accumbens slices inhibits dopamine and acetylcholine release but not cyclic AMP formation. Brain Res. *423*: 364–468, 1987.

Straschill, M. and Perwein, J.: Effects of biogenic amines and aminoacids on the cat's retinal ganglion cells. In: "Golgi Centennial Symposium: Perspectives in Neurobiology," M. Santini, Ed., p. 583, Raven Press, New York, 1975.

Stringer, J. L. and Lothman, E. W.: Maximal dentate activation: a tool to screen compounds for activity against limbic seizures. Epilepsy Res. *5*: 169–176, 1990.

Stringer, J. L. and Lothman, E. W.: Cholinergic and adrenergic agents modify the initiation and termination of epileptic discharges in the dentate gyrus. Neuropharmacology *30*: 59–65, 1991.

Suemaru, K., Gomita, Y., Furuno, K. and Araki, Y.: Effect of beta-adrenergic receptor antagonists on nicotine-induced tail-tremor in mice. Pharmacol. Biochem. Behav. *46*: 131–133, 1993.

Sugita, S., Uchimura, N., Jiang, Z. G. and North, R. A.: Distinct muscarinic receptors inhibit release of gamma-aminobutyric acid and excitatory amino acids in mammalian brain. Proc. Natl. Acad. Sci. U.S.A. *88*: 2608–2611, 1991.

Suits, E. and Isaacson, R. L.: The effects of scopolamine hydrobromide on one-way and two-way avoidance learning in rats. Int. J. Neuropharmacol. *2*: 441–446, 1968.

Sundaram, K., Murugaian, J., Krieger, A. and Sapru, H.: Microinjection of cholinergic agonists into the intermediolateral cell column of the spinal cord at T1-T3 increase heart rate and contractility. Brain Res. *503*: 22–31, 1989.

Sutherland, E. W., Robinson, G. A. and Butcher, R. W.: Some aspects of the biological role of adenosine 3', 5'-monophosphate (cyclic AMP). Circulation *37*: 279–306, 1968.

Suzuki, T., Takagi, R. and Kawashima, K.: Picrotoxin increased acetylcholine release from rat cultured embryonic septal neurons. Neurosci. Lett. *356*: 57–60, 2004.

Sweatt, J. D.: Hippocampal functions in memory. Psychopharmacology (Berl.), *174*: 99–110, 2004.

Syvalathi, E. K., Hietala, J., Roytta, M. and Gronroos, J.: Decrease in the number of rat brain dopamine and muscarinic receptors after chronic alcohol intake. Pharmacol. Toxicol. *62*: 210–212, 1988.

Szentagothai, J.: The anatomical basis of synaptic transmission of excitation and inhibition in motoneurones. Acta Morph. Acad. Sci. Hung. *8*: 287–309, 1958.

Szerb, J. C.: The effect of tertiary and quaternary atropine on cortical acetylcholine output and on the electroencephalograph in cats. Can. J. Physiol. Pharmacol. *42*: 303–314, 1964.

Szerb, J. C.: Characterization of presynaptic muscarinic receptors in central cholinergic neurons. In: "Cholinergic Mechanisms and Psychopharmacology," D. J. Jenden, Ed., pp. 49–60, Plenum Press, New York, 1977.

Szerb, J. C. and McCurdy, D. H.: Concentration of morphine in blood and brain after intravenous injection of morphine in non-tolerant, tolerant, and neostigmine-treated rats. J. Pharmacol. Exp. Ther. *118*: 446–450, 1956.

Szymusiak, R.: Magnocellular nuclei of the basal forebrain: substrates of sleep and arousal regulation. Sleep *18*: 478–580, 1995.

Szymusiak, R., McGinty, D., Fairchild, M. D. and Jenden, D. J.: Sleep-wake disturbances in an animal model of chronic cholinergic insufficiency. Brain Res. *629*: 141–145, 1993.

Tabakoff, B. and Hoffman, P. L., Eds.: "Biological Aspects of Alcoholism," Hogrefe and Huber, Seattle, Wash., 1995.

Takahashi, H. and Buccafusco, J. J.: Excitatory modulation by a spinal cholinergic system of a descending sympathoexcitatory pathway in rats. Neuropharmacology *31*: 259–269, 1992.

Takei, N., Nihonmatsu, I. and Kawamura, H.: Age-related decline of acetylcholine release evoked by depolarizing stimulation. Neurosci. Lett. *101*: 182–186, 1989.

Takemori, A. E. and Portoghese, P. S.: Affinity labels for opioid receptors. Annu. Rev. Pharmacol. Toxicol. *25*: 193–223, 1985.

Takeshita, S., Sasa, M., Ishihara, K., Matsubayashi, H., Yain, K., Okada, M., Izumi, R., Arita, K. and Kurisu, K.: Cholinergic and glutaminergic transmission in medial vestibular nucleus neurons responding to lateral roll tilt in rats. Brain Res. *840*: 99–105, 1999.

Talman, M. T., Colling, J. M. and Robertson, S. C.: Glycine microinjected into nucleus tractus solitarii of rat acts through cholinergic mechanisms. Am. J. Physiol. *260*: H1326–H1331, 1991.

Tanabe, F., Miyasaka, N., Kubota, T. and Aso, T.: Estrogen and progesterone improve scopolamine-induced impairment of spatial memory. J. Med. Dent. Sci. *51*: 89–98, 2004.

Tanabe, S., Shishido, Y., Nakayama, Y., Furushiro, M., Hashimoto, S., Terasaki, T., Tsujimoto, G. and Yokokura, T.: Effects of arginin-vasopressin fragment 4–9 on rodent cholinergic system. Pharmacol. Biochem. Behav. *63*: 459–453, 1999.

Tanganelli, S., Antonelli, T., Simonato, M, Spaluto, G., Tomasini, C., Bianchi, C. and Beani, L.: Alpha 1-adrenoreceptor-mediated increase in acetylcholine release in brain slices during morphine tolerance. J. Neurochem. *53*: 1072–1076, 1989.

Tangri, K. K., Bhargava, A. K. and Bhargava, K. P.: Significance of central cholinergic mechanisms in pyrexia induced by bacterial pyrogens in rabbits. In: "Temperature Regulation and Drug Action," J. Lomax, E. Schoenbaum, and J. Jacob, Eds., pp. 65–74, Karger, Basel, 1975.

Tapper, A. R., McKinney, S. L., Nashmi, R., Schwarz, J., Deshapnde, P., Labarca, C., Whiteaker, P., Marks, M. J., Collins, A. C. and Lester, H. A.: Nicotine activation of alpha4 receptors: sufficient for reward, tolerance, and senzitiztion. Science *306*: 1029–1032, 2004.

Tayebati, S. K., Di Tullio, M. A., Tomassoni, D. and Amenta, F.: Localization of the m5 muscarinic cholinergic receptors in rat circle of Willis and pial arteries. Neuroscience *122*: 205–211, 2003.

Tebano, M. T., Luzi, M., Palzzesi, S., Pomponi, M. and Loizzo, A.: Effects of cholinergic drugs on neocortical EEG and flash-visual evoked potentials in the mouse. Neuropsychobiology *40*: 47–56, 1999.

Tebecis, A. K.: Properties of cholinoceptive neurones in the medial geniculate nucleus. Br. J. Pharmacol. *38*: 117–138, 1970a.

Tebecis, A. K.: Studies on cholinergic transmission in the medial geniculate nucleus. Br. J. Pharmacol. *38*: 138–147, 1970b.

Tees, R. C.: The influence of rearing environment and neonatal choline dietary supplementation on spatial learning and memory in adult rats. Behav. Brain Res. *105*: 173–188, 1999.

Tees, R. C. and Mohammadi, E.: The effects of neonatal choline dietary supplementation on adult spatial and configural learning and memory in rats. Dev. Psychobiol. *35*: 226–240, 1999.

Tellioglu, T., Erin, N., Akin, S. B., Berkman, K. and Oktay, S.: Alteration of cholinergic pressor and antinociceptive response in rats pretreated with the cholinergic toxin AF64A. Gen. Pharmacol. *30*: 525–531, 1998.

Terry, A. V., Jr., Buccafusco, J. J., Jackson, W. J., Zagrodnik, S., Evans-Martin, F. F. and Decker, M. W.: Effects of stimulation or blockade of central nicotinic-cholinergic receptors on performance of a novel version of the rat stimulus discrimination task. Psychopharmacology *123*: 172–181, 1996.

Testylier, G., Maalouf, M., Butt, A. E., Miasnikov, A. A. and Dykes, R. W.: Evidence for homeostatic adjustments of rat somatosensory cortical neurons to changes in extracellular acetylcholine concentration produced by iontophoretic administration of acetylcholine and systemic diisopropylfluorophosphate treatment. Neuroscience *91*: 843–870, 1999.

Thatcher, R. W.: Tomographic electroencephalography magnetoencephalography. J. Neuroimag. *5*: 35–45, 1995.

Thesleff, S.: The effects of acetylcholine, decamethonium and succinylcholine on neuromuscular transmission in the rat. Acta. Physiol. Scand. *34*: 386–392, 1955.

Thews, G.: Pulmonary respiration. In: "Human Physiology," R. S. Schmidt and G. Thews, Eds., pp. 456–488, Springer-Verlag, Berlin, 1986.

Thiefin, G., Tach'e, Y., Leung, F. W. and Guth, P. H.: Central nervous system action of thyrotropin-releasing hormone to increase gastric mucosal blood flow in the rat. Gastroenterology *97*: 405–411, 1989.

Thiel, C. M., Huston, J. P. and Schwarting, R. K.: Hippocampal acetylcholine and habituation learning. Neuroscience *85*: 1253–1262, 1998.

Thiel, C. M., Zilles, K. and Fink, G. R.: Nicotine modulates reorienting of visuospatial attention and neural activity in human parietal cortex. Neuropsychopharmacology, *30*: 810–820, 2005.

Thomas, R. C. and Wilson, V. J.: Precise location of Renshaw cells with a new marking technique. Nature *206*: 211–213, 1965.

Thompson, L., Duckworth, J. and Bevan J.: Cryopreservation of innervation of endothelial and vascular smooth muscle function of a rabbit muscular and resistance artery. Blood Vessels *26*: 156–164, 1989.

Thompson-Schill, S. L., Aguirre, G. K., D'Esposito, M. and Farah, M. J.: A neural basis for category and modality specificity of semantic knowledge. Neuropsychologia *37*: 671–676, 1999.

Thorne, B. and Potter, P. E.: Lesion with the neurotoxin AF64A alters hippocampal cholinergic function. Brain Res. Bull. *38*: 121–127, 1995.

Tigges, M., Tigges, J., Rees, H., Rye, D. and Levey, A. I.: Distribution of muscarinic cholinergic receptor proteins m1 to m4 in area 17 of normal and monocularly deprived rhesus monkeys. J. Comp. Neurol. *388*: 130–145, 1997.

Tigges, M., Iuvone, P. M., Fernandes, A., Sugrue, M. F., Mallorga, P. J., Laties, A. M. and Stone, R. A.: Effects of muscarinic cholinergic receptor antagonists on postnatal growth of rhesus monkeys. Optom. Vis. Sci. *76*: 397–407, 1999.

Timofeev, I. and Steriade, M.: Cellular mechanisms underlying intrathalamic augmenting responses of reticular and relay neurons. J. Neurophysiol. *79*: 2716–2729, 1998.

Tinel, J., Eck, M. and Stewart, W.: Causalgie de la main guerie par l'acetylcholine. Rev. Neurol. *2*: 38–43, 1933.

Tipler, F. J.: "The Physics of Immortality," Doubleday, New York, 1994.

Titmus, M. J., Tsai, H. J., Lima, R. and Usdin, S. B.: Effects of choline and other nicotinic agonists on the tectum of juvenile and adult Xenopus frogs: a patch-clamp study. Neuroscience *91*: 753–769, 1999.

Toide, K.: Effects of scopolamine on extracellular acetylcholine and choline levels and on spontaneous motor activity in freely moving rats measured by brain dialysis. Pharmacol. Biochem. Behav. *33*: 109–113, 1989.

Toide, K., Shinoda, M., Takase, M., Iwata, K. and Yoshida, H.: Effects of a novel thyrotropin-releasing hormone analogue, JTP-2942, on extracellular acetylcholine and choline levels in the rat frontal cortex and hippocampus. Eur. J. Pharmacol. *233*: 21–28, 1993.

Tomberg, C. and Desmedt, J. E.: Human perceptual processing: inhibition of transient prefrontal-parietal 40 Hz binding at P300 onset documented in non-averaged cognitive potentials. Neurosci. Lett. *255*: 163–166, 1998.

Tonduli, L. S., Testylier, G., Marino, I. P. and Lallement, G.: Triggering of soman-induced seizures in rats: analysis with special correlation between enzymatic, neurochemical and electrophysiological data. J. Neurosci. Res. *58*: 464–473, 1999.

Tonnaer, J. A., Ernste, B. H., Wester, J. and Kelder, K.: Cholinergic innervation and topographical organization of muscarinic binding sites in rat brain: a comparative radioautographic study. J. Chem. Neuroanat. *1*: 95–110, 1988.

Torocsik, A. and Vizi, E. S.: 4-aminopyridine interrupts the modulation of acetylcholine release mediated by muscarinic and opiate receptors. J. Neurosci. Res. *271*: 228–232, 1990.

Torrao, A. S., Lindstrom, J. M. and Britto, L. R.: Distribution of the alpha 2, alpha 3, and alpha 5 nicotinic acetylcholine receptors in the chick brain. Braz. J. Med. Biol. Res. *30*: 1209–1213, 1997.

Toth, E., Vizi, E. S. and Lajtha, A.: Effect of nicotine on levels of extracellular amino acids in regions of rat brain *in vivo*. Neuropharmacology *32*: 827–832, 1993.

Trabucchi, M., Cheney, D. L., Racagni, C. and Costa, E.: *In vivo* inhibition of striatal acetylcholine turnover by L-DOPA, apomorphine and (+)-amphetamine. Brain Res. *85*: 130–134, 1975.

Tredway, T. L., Guo, J. Z. and Chiapinelli, V. A.: N-type voltage-dependent calcium channels mediate the nicotinic enhancement of GABA release in chick brain. J. Neurophysiol. *81*: 447–454, 1999.

Tripathi, H. L., Vocci, F. J., Brase, D. A. and Dewey, W. L.: Effects of cannabinoids on levels of acetylcholine and choline and on turnover rate of acetylcholine in various regions of the mouse brain. Alcohol Drug Res. *7*: 525–532, 1987.

Tritto, T., Marley, R. J., Bastidas, D., Stitzel, J. A. and Collins, A. C.: Potential regulation of nicotine and ethanol actions by alpha-4-containing nicotinic receptors. Alcohol *24*: 69–78, 2001.

Trouth, C. O., Millis, R. M., Bernard, R. G., Pan, Y., Whittaker, J. A. and Archer, P. W.: Cholinergic-opioid interactions at brainstem respiratory chemosensitive areas in cats. Neurotoxicology *24*: 459–467, 1993.

Tsai, L. L., Bergmann, B. M., Perry, B. D. and Rechtschaffen, A.: Effects of chronic sleep deprivation on central cholinergic receptors in rat brain. Brain Res. *642*: 95–103, 1994.

Tsukada, H., Yamazaki, S., Noda, A., Inoue, T., Matsuoka, N., Kakuichi, T., Nishiyama, S. and Nishimura, S.: FK960 [N-(4-acetyl-1-piperazinyl)-p-fluorobenzamide monohydrate], a novel potential antidementia drug, restores the regional blood cerebral flow response abolished by scopolamine but not by HA-996: a positron emission tomography study with anesthetized rhesus monkeys. Brain Res. *832*: 118–123, 1999.

Tsushima, H., Mori, M. and Matsuda, T.: Microinjection of dynorphin into the supraoptic and paraventricular nuclei produces antidiuretic effects through vasopressin release. Jpn. J. Pharmacol. *63*: 461–468, 1993.

Tung, C. S., Ugedo, I., Grenhoff, J., Engberg, G. and Svenson, T. H.: Peripheral induction of burst firing in locus coeruleus neurons by nicotine mediated via excitatory amino-acids. Synapse *4*: 313–318, 1989.

Turing, A. M.: On computable numbers, with an application to the Enscheidungsproblem. Proc. Lond. Math. Soc., Ser. 2, *43*: 544–546, 1937.

Tzavara, E. T., Bymaster, F. P., Felder, C. C., Wade, M., Wess, J., McKinzie, D. L. and Nomikos, G. G.: Dysregulated hippocampal neurotransmission and impaired cognition in M2 and M2/M4 muscarinic receptor knockout mice. Mol. Psychiatry *8*: 673–679, 2003.

Ukai, M., Kobayashi, T. and Kameyama, T.: Characterization of the effects of scopolamine on the habituation of exploratory activity: differential effects of oxotremorine and physostigmine. Gen. Pharmacol. *25*: 433–438, 1994.

Uney, J. B. and Marchbanks, R. M.: Inhibition of choline transport in human brain by lithium treatment. In: "Cellular and Molecular Basis of Cholinergic Function," M. J. Dowdall and J. N. Hawthorne, Eds., pp. 774–780, Ellis Horwood, Chichester, 1987.

Unwin, N.: Refined structure of the nicotinic acetylcholine receptor at 4A resolution. J. Mol. Biol. *346*: 967–989, 2005.

US Department of Health and Human Services (DHHS), "Nicotine Addiction," 1988.

Valzelli, L.: Drugs and aggressiveness. Adv. Pharmacol. *5*: 79–108, 1967.

Valzelli, L.: Neurochemical aspects of behavior. In: "Central Nervous System: Studies on Metabolic Regulation and Function," E. Genazzani and H. Herken, Eds., pp. 167–171, Springer-Verlag, New York, 1974.

Valzelli, L.: Human and animal studies on the neurophysiology of aggression. Prog. Neuropsychopharmacol. *2*: 591–610, 1978.

Van de Kar, L. D.: Neuroendocrine pharmacology of serotonergic (5-HT) neurons. Annu. Rev. Pharmacol. Toxicol. *31*: 289–320, 1991.

Van der Linden, S., Panzica, F. and Curtis, de M.: Carbachol induces fast oscillations in the medial but not in the lateral entorhinal cortex of the isolated guinea pig brain. J. Neurophysiol. *82*: 2441–2250, 1999.

Van der Staay, F. J. and Bouger, P. C.: Effects of cholinesterase inhibitors, donepezil and metrifonate on scopolamine-induced impairments in the spatial cone field orientation tasks in rats. Behav. Brain Res. *156*: 1–10, 2005.

Vanderwolf, C. H.: Neocortical and hippocampal activation in relation to behavior: effects of atropine, eserine, phenothiazines, and amphetamine. J. Comp. Physiol. Psychol. *88*: 300–323, 1975.

Van der Zee, E. A. and Luiten, P. G.: Muscarinic acetylcholine receptors in the hippocampus, neocortex and amygdala: a review of immunocytochemical localization in relation to learning and memory. Prog. Neurobiol. *58*: 409–471, 1999.

Van der Zee, E. A., Streefland, C., Strosberg, A. D., Schroeder, H. and Luiten, P. G. M.: Colocalization of muscarinic and nicotinic receptors in cholinoceptive neurons of the suprachiasmatic region in young and aged rats. Brain Res. *542*: 348–352, 1991.

Van der Zee, E. A., Streefland, C., Strosberg, A. D., Schroeder, H. and Luiten, P. G. M.: Visualization of cholinoceptive neurons in the rat neocortex: colocalization of muscarinic and nicotinic acetylcholine receptors. Mol. Brain Res. *14*: 326–336, 1992.

Van der Zee, C. E., Ross, G. M., Riopelle, R. J. and Hagg, T.: Survival of cholinergic forebrain neurons in developing p75NGFR-deficient mice. Science *274*: 1729–1732, 1999.

Van Gaalen, M., Kawahara, H., Kawahara, Y. and Westerink, B. H.: The locus coeruleus noradrenergic system in the rat brain studied by dual-probe microdialysis. Brain Res. *763*: 56–62, 1997.

VanMeter, W. G., Karczmar, A. G. and Fiscus, R. F.: CNS effects of anticholinesterases in the presence of inhibited cholinesterases. Arch. Int. Pharmacodyn. Ther. *23*: 249–260, 1978.

Vanni-Mercier, G., Debilly, G., Lin, J. S. and Pelisson, D.: The caudo ventral pontine tegmentum is involved in the generation of high velocity saccades in bursts during paradoxical sleep in the cat. Neurosci. Lett. *213*: 127–131, 1996.

Van Woert, M. H.: Parkinson's disease, tardive dyskinesia, and Huntington's chorea. In: "Biology of Cholinergic Function," A. M. Goldberg and I. Hanin, Eds., pp 583–601, Raven Press, New York, 1976.

Varagic, V. M.: The action of eserine on the blood pressure of the rat. Br. J. Pharmacol. Chemother. *10*: 349–353, 1955.

Varagic, V. M. and Krstic, M.: Adrenergic activation by anticholinesterases. Pharmacol. Rev. *18*: 796–800, 1966.

Varagic, V., Vaucher, E. and Hamel, E.: Cholinergic basal forebrain neurons project to cortical microvessels in the rat: electron microscopic study with anterogradely transported Phaseolus vulgaris leucoagglutinin and choline acetyltransferase immunocytochemistry. J. Neurosci. *15*: 7427–7441, 1955.

Varagic, V. M., Kazic, T. and Rosic, N.: Inversion of physostigmine hypertension by Paraoxon and DFP. Yugoslav Physiol. Pharmacol. Acta *4*: 113–120, 1965.

Vaupel, D. B. and Martin, W. R.: Interaction of oxotremorine with atropine, chlorpromazine, cyproheptadine, imipramine and phenoxybenzamine on the flexor reflex of the chronic spinal dog. Psychopharmacology *30*: 13–26, 1973.

Vazquez, A. J. and Krip, G.: Evidence for an inhibitory role for acetylcholine, catecholamines, and serotonin on the cerebellar cortex. In: "Chemical Modulation of Brain Function," H. C. Sabelli, Ed., pp. 137–159, Raven Press, New York, 1973.

Velazquez, J. L., Han, D. and Carlen, P. L.: Neurotransmitter modulation of gap junctional communication in the rat hippocampus. Eur. J. Neurosci. *9*: 2522–2531, 1997.

Velazquez-Moctezuma, J., Gillin, J. C. and Shiromani, P. J.: Effect of specific M1, M2 muscarinic receptor agonists on REM sleep generation. Brain Res. *503*: 128–131, 1989.

Velazquez-Moctezuma, J., Shalauta, M. D., Gillin, J. C. and Shiromani, P. J.: Microinjection of nicotine in the medial pontine reticular formation elicits REM sleep. Neurosci. Lett. *115*: 265–268, 1990a.

Velazquez-Moctezuma, J., Shalauta, M. D., Gillin, J. C. and Shiromani, P. J.: Differential effects of cholinergic antagonists on REM sleep components. Psychopharmacol. Bull. *26*: 349–353, 1990b.

Vertes, R. P.: Brain stem generation of the hippocampal EEG. Prog. Neurobiol. *19*: 159–186, 1982.
Vertes, R. P., Colom, L. V., Fortin, W. J. and Bland, B. H.: Brainstem sites for the carbachol elicitation of the hippocampal theta rhythm in the rat. Exp. Brain Res. *96*: 419–429, 1993.
Vetter, D. E., Liberman, M. C., Mann, J., Barhanin, J., Boulter, J., Brown, M. C., Saffiote-Kolman, J., Heinemann, S. F. and Elgoyhen, A. B.: Role of alpha9 nicotinic ACh receptor subunits in the development and function of cochlear efferent innervation. Neuron *23*: 93–103, 1999.
Vickroy, T. W. and Cadman, E. D.: Dissociation between muscarinic-mediated inhibition of adenylcyclase and autoreceptor inhibition of [3H] acetylcholine release in rat hippocampus. J. Pharmacol. Exp. Ther. *251*: 1039–1044, 1989.
Vickroy, T. W., Fibiger, H. C., Roeske, W. R. and Yamamura, H. I.: Reduced density of sodium-dependent [3H] hemicholinim-3 binding sites in the anterior cerebral cortex of rats following chemical destruction of the nucleus basalis magnocellularis. Eur. J. Pharmacol. *102*: 369–370, 1984a.
Vickroy, T. W., Roeske, W. R. and Yamamura, H. I.: Sodium-dependent high affinity binding of [3H] hemicholinium-3 in rat brain: a potential selective marker for presynaptic cholinergic sites. Life Sci. *35*: 2335–2343, 1984b.
Vilaro, M. T., Wiederhold, K.-H., Palacios, J. M. and Mengod, G.: Muscarinic M2 receptor mRNA expression and receptor binding in cholinergic noncholinergic cells in the rat brain: a correlative study using in situ hybridization histochemistry and receptor autoradiography. Neuroscience *47*: 367–393, 1992.
Vincler, M. and Eisenach, J. C.: Plasticity of spinal nicotinic acetylcholine receptors following spinal nerve ligation. Neurosci. Res. *48*: 139–145, 2004.
Vinogradova, O. S., Brazhnik, E. S., Stafekhina, V. S. and Kitchigina, V. F.: Acetylcholine, theta-rhythm and activity of hippocampal neurons in the rabbit—II. Septal input. Neuroscience *53*: 971–979, 1993a.
Vinogradova, O. S., Brazhnik, E. S., Kitchigina, V. F. and Stafekhina, V. F.: Acetylcholine, theta-rhythm and activity of hippocampal neurons in the rabbit—IV. Sensory stimulation. Neuroscience *53*: 993–1007, 1993b.
Vinogradova, O. S., Kitchigina, V. F., Kudina, T. A. and Zenchenko, K. I.: Spontaneous activity and sensory responses of hippocampal neurons during persistent theta-rhythm evoked by median raphe nucleus blockade in rabbit. Neuroscience *94*: 745–753, 1999.
Vizi, E. S.: Presynaptic modulation of neurochemical transmission. Prog. Neurobiol. *12*: 181–290, 1979.
Vizi, E. S.: Modulation of cortical release of acetylcholine by noradrenaline released from nerves arising from the locus coeruleus. Neuroscience *5*: 21–39, 1980.
Vizi, E. S. and Kiss, J. P.: Neurochemistry and pharmacology of the major hippocampal transmitter systems: synaptic and nonsynaptic interactions. Hippocampus *8*: 566–607, 1998.
Vizi, E. S. and Lendvai, B.: Modulatory role of presynaptic nicotinic receptors in synaptic and nonsynaptic chemical communication in the central nervous system. Brain Res. Brain Res. Rev. *30*: 219–235, 1999.
Vizi, E. S., Bernath, S., Kapocsi, J. and Serfozo, P.: The release of acetylcholine from cytoplasm is not subject to synaptic modulation. In: "Cellular and Molecular Basis of Cholinergic Function," M. J. Dowdall and J. N. Hawthorne, Eds., pp. 226–231, Ellis Horwood, Chichester, 1987.
Vizi, E. S., Harsing, L. G., Jr. and Zsilla, G.: Evidence for neuromodulatory role of serotonin in acetylcholine relase from striatal interneurons. Brain Res. *212*: 89–99, 1989a.
Vizi, E. S., Kobayashi, O., Torocsik, A., Kinjo, M., Nagashima, H., Manabe, N., Goldiner, P. L., Potter, P. E. and Foldes, F. F.: Heterogeneity of presynaptic muscarinic receptors involved in modulation of transmitter release. Neuroscience *31*: 259–267, 1989b.
Vogels, O. J., Renkawek, K., Broere, C. A., ter Laak, H. J. and van Workum, F.: Galanin-like immunoreactivity within Ch2 neurons in the vertical limb of the diagonal band of Broca in aging and Alzheimer's disease. Acta Neuropathol. *78*: 90–95, 1989.
Volavka, J.: "Neurobiology of Violence." American Psychiatric Press, Washington, D.C., 1995.
Von Euler, U. S.: "Noradrenaline," Thomas Publs., Springfield, Illinois, 1956.
Voytko, M. L.: Cognitive functions of the basal forebrain cholinergic system in monkeys: memory or attention? Behav. Brain Res. *75*: 13–25, 1996.
Voytko, M. L., Olton, D. S., Richardson, R. T., Gorman, L. K., Tobin, J. R. and Price, D. L.: Basal forebrain lesions in monkeys disrupt attention but not learning and memory. J. Neurosci. *14*: 167–168, 1994.
Waite, J. J., Wardlow, M. L. and Power, A. E.: Deficit in selective and divided attention associated with cholinergic immunotoxic lesion produced by 192-saporin; motoric-sensory deficit associated with Purkinje cell immunotoxic lesion produced by OX7saporin. Neurobiol. Learn. Mem. *71*: 325–352, 1999.
Wang, H. H., Chou, Y. C., Liao, J. F. and Chen, C. F.: Dimemorphan enhances acetylcholine release from rat hippocampal slices. Brain Res. *1008*: 113–115, 2004.
Wang, M. Y. and Dun, N. J.: Phaclofen-insensitive presynaptic inhibitory action of (+/–)-baclofen in neo-

natal rat motoneurones *in vitro*. Br. J. Pharmacol. *99*: 413–421, 1989.

Wang, Y., Nakashima, K., Shiraishi, Y., Kawai, Y., Ohama, E. and Takahashi, K.: P300-like potential disappears in rabbits with lesions in nucleus basalis. Exp. Brain Res. *114*: 288–292, 1997.

Wang, H., Carlier, P. R., Ho, W. L., Lee, N. T., Pang, Y. P. and Han, Y. F.: Attenuation of scopolamine-induced deficits in navigational memory performance in rats by bis (7)-tacrine, a novel dimeric AChE inhibitor. Zhongguo Yao Li Xue Bao *20*: 211–217, 1999.

Warburton, D. M.: Nicotine as cognitive enhancer. Prog. Neuropsychopharmacol. Biol. Psychiatry *16*: 181–191, 1992a.

Warburton, D. M.: Nicotine issues. Editorial. Psychopharmacology *108*: 393–396, 1992b.

Warburton, D. M.: The reality of beneficial effects of nicotine. Addiction *89*: 138–139, 1994.

Ward, A. A., Jr., Jasper, H. J. and Pope, A.: Clinical and experimental challenges of epilepsies. In: "Basic Mechanisms of Epilepsies," H. J. Jasper, A. A. Ward and A. Pope, Eds., pp. 1–12, Little, Brown, Boston, 1969.

Waser, P. G.: Calebassen-Curare. Helv. Physiol. Pharmacol. Acta, Supplt. *8*: 1–84, 1953.

Washburn, M. G. and Moises, H. C.: Muscarinic responses of rat basolateral amygdaloid neurones recorded *in vitro*. J. Physiol. (Lond.) *449*: 121–154, 1992.

Watanabe, H. and Shimizu, A.: Effect of anticholinergic drugs on striatal acetylcholine release and motor activity in freely moving rats studied by brain microdialysis. Jpn. J. Pharmacol. *51*: 75–82, 1989.

Way, E. L., Iwamoto, E. T., Bhargava, H. N. and Loh, H. H.: Adaptive cholinergic-dopaminergic responses in morphine dependence. In: "Neurobiological Mechanisms of Adaptation and Behavior," A. J. Mandfell, Ed., pp. 169–187, Advanc. Biochem. Psychopharmacol., vol. 13, Raven Press, New York, 1975.

Weiger, W. A., and Bear, D. M.: An approach to the neurology of aggression. J. Psychiatr. Res. *22*: 85–98, 1988.

Weight, F. F.: Cholinergic mechanisms in the recurrent inhibition of motoneurons. In: "Psychopharmacology: A Review of Progress," D. H. Efron, Ed., pp. 69–76, Public Health Publ. No. 1836, Washington, D.C., 1968.

Weiler, M. H.: Muscarinic modulation of endogenous acetylcholine release in rat neostriatal slices. J. Pharmacol. Exp. Ther. *250*: 617–623, 1989.

Weinberg, J.: Untersuchungen über die zentrale Regulation des Vasomotoren. Arch. Exp. Pathol. Pharmakol. *185*: 237–255, 1937.

Weinberger, N. M.: The nucleus basalis and memory codes: auditory cortical plasticity and the induction of specific, associative behavioral memory. Neurobiol. Learn. Mem. *80*: 268–284, 2003.

Weisinger, G., DeCristofaro, J. D. and LaGamma, E. F.: Multiple preproenkephalin transcriptional start sites are induced by stress and cholinergic pathways. J. Biol. Chem. *265*: 17389–17392, 1990.

Wenk, G. L., Stoehr, J. D., Quintana, G., Mobley, S. and Wiley, R. G.: Behavioral, biochemical, histological, and electrophysiological effects of 192 IgG-saporin injections in the basal forebrain of rats. J. Neurosci. *14*: 5986–5995, 1994.

Wescoe, W. C., Green, R. E., McNamara, B. P. and Krop, S.: The influence of atropine and scopolamine on the central effects of DFP. J. Pharmacol. Exp. Ther. *92*: 63–72, 1948.

Wess, J., Blin, N., Yun, J., Schoneberg, T. and Liu, J.: Molecular aspects of muscarinic receptor assembly and function. Prog. Brain Res. *109*: 153–162, 1996.

West, R.: Nicotine addiction: a re-analysis of the arguments. Psychopharmacology *108*: 408–410, 1992.

Whelan, R. E., Ed.: "The Neuropsychology of Aggression," Plenum Press, New York, 1974.

Whittaker, V. P.: "The Cholinergic Neuron and its Target: the Electromotor Innervation of the Electric Ray 'Torpedo' as a Model." Birkhauser, Boston, 1992.

Whittaker, V. P.: Arcachon and cholinergic transmission. In: "From *Torpedo* Electric Orrgan to Human-Brain" Fundamental and Applied Aspects," J. Massoulié, Ed., pp. 53–57, 1998.

Wichmann, T., Wictorin, K., Bjorklund, A. and Starke, K.: Release of acetylcholine and its dopaminergic control in slices from striatal grafts in the ibotenic acid-lesioned rat striatum. Naunyn Schmiedebergs Arch. Pharmakol. *388*: 622–631, 1988.

Wichmann, T., Limberger, N. and Starke, K.: Release and modulation of release of serotonin in rabbit superior colliculus. Neuroscience *32*: 141–151, 1989.

Widdicombe, J. G.: Lung reflexes. Bull. Physiopathol. Respir. *8*: 723–725, 1972.

Wikler, A.: Effects of morphine, nembutal, ether and eserine on two-neuron and multineuron reflexes in the cat. Proc. Soc. Exp. Biol. Med. *58*: 193–196, 1945.

Wikler, A.: Pharmacologic dissociation of behavior and EEG sleep patterns in dogs: morphine, n-allylmorphine and atropine. Proc. Soc. Exp. Biol. Med. *79*: 261–265, 1952.

Wiklund, L. and Hartvig, P.: Cholinergic agents in clinical anesthesiology. In: "Cholinergic Neurotransmission: Functional and Clinical Aspects," S.-M. Aquilonius and P.-G. Gillberg, Eds., pp. 399–405, Elsevier, Amsterdam, 1990.

Wikstrom, M. A., Lawoko, G. and Heilbronn, E.: Cholinergic modulation of extracellular ATP-induced calcium concentration in cochlear outer hair cells. In:

"From *Torpedo* Electric Organ to Human Brain: Fundamental and Applied Aspects," J. Massoulié, Ed., pp. 345–349, Elsevier, Amsterdam, 1998.

Williams, D. and Russell, W. R.: Action of eserine and prostigmine on epileptic cerebral discharges. Lancet *1*: 476, 1941.

Williams, J. A. and Kauer, J. A.: Properties of carbachol-induced oscillatory activity in rat hippocampus. J. Neurophysiol. *78*: 2631–2640, 1997.

Williams, J. A., Vincent, S. R. and Reiner, P. B.: Nitric oxide production in rat thalamus changes with behavioral state, local depolarization, and brainstem stimulation. J. Neurosci. *17*: 420–427, 1997.

Willis, W. D., Jr.: "Control of Nociceptive Transmission in the Spinal Cord," Springer-Verlag, New York, 1982.

Willis, W. D., Jr.: Role of the forebrain in nociception. In: "Role of Forebrain in Sensation and Behavior," G. Holstege, Ed., Progr. Brain Res., vol. 87, pp. 1–12, Elsevier, Amsterdam, 1991.

Wills, J. H.: Pharmacological antagonists of the anticholinesterase agents. In: "Cholinesterases and Anticholinesterase Agents," G. B. Koelle, Ed., pp. 883–920, Handbch. d. exper. Pharmakol., Ergänzungswk., vol. 15, Springer-Verlag, Berlin, 1963.

Wills, J. H.: Toxicity of anticholinesterases and treatment of poisoning. In: "Anticholinesterase Agents," A. G. Karczmar, Ed., pp. 355–469, Intern. Encyclop. of Pharmacol. and Ther., vol. 1, Section 13, 1970.

Wilson, F. A. W.: The relationship between learning, memory and neuronal responses in the primate basal forebrain. In: "The Basal Forebrain," T. C. Napier, P. W. Kalivas and I. Hanin, Eds., pp. 253–266, Adv. Exp. Med. Biol., vol. 295, Plenum Press, New York, 1991.

Wilson, J. M., Carroll, M. E., Lac, S. T., DiStefano, L. M. and Kish, S. J.: Choline acetyltransferase activity is reduced in rat nucleus accumbens after unlimited access to self-administration of cocaine. Neurosci. Lett. *180*: 29–32, 1994.

Winslow, J. T. and Camacho, F.: Cholinergic modulation of a decrement in social investigation following repeated contacts between mice. Psychopharmacology (Berl.) *121*: 164–172, 1995.

Winters, B. D., Robbins, T. W. and Everitt, B. J.: Selective cholinergic denervation of the cingulated cortex impairs the acquisition and performance of a conditional visual discrimination in rats. Eur. J. Neurosci. *19*: 490–496, 2004.

Wise, R. A.: Drug-activation of brain reward pathways. Drug Alcohol Depend. *51*: 13–22, 1998.

Wolff, V. H.: Die Behandlung zerebraler Anfalle mit Scopolamine. Ein Betrag zur Klinik des "Synkopalin" Syndroms. Dtsch. Med. Wochenschr. *81*: 1358–1360, 1956.

Womble, M. D. and Moises, H. C.: Muscarinic inhibition of M-current and a potassium leak conductance in neurones of rat basolateral amygdala. J. Physiol. (Lond.) *547*: 93–114, 1992.

Wong, C. L. and Bentley, G. A.: Further studies on the role of cholinergic mechanisms in the development of increased naloxone potency in mice. Eur. J. Pharmacol. *56*: 115–121, 1979.

Wong, L. A. and Gallagher, J. P.: Pharmacology of nicotinic receptor-mediated inhibition in rat dorsolateral septal neurones. J. Physiol. (Lond.) *436*: 325–346, 1991.

Wong, P. S., Bernat, E., Snodgrass, M. and Shevrin, H.: Event-related brain correlates of associative learning without awareness. J. Psychophysiol. *53*: 217–231, 2004.

Wonnacott, S.: The paradox of nicotinic acetylcholine receptor upregulation by nicotine. TIPS *11*: 216–219, 1990.

Wonnacott, S.: Presynaptic nicotinic ACh receptors. Trends Neurosci. *20*: 92–98, 1997.

Wonnacott, S., Sidhpura, N. and Balfour, D. J.: Nicotine: from molecular mechanisms to behavior. Curr. Opin. Pharmacol. *5*: 53–59, 2005.

Wood, P. L.: Pharmacological evaluation of GABAergic and glutamatergic inputs to the nucleus basalis–cortical and the septal-hippocampal cholinergic projections. Can. J. Physiol. Pharmacol. *64*: 325–328, 1986.

Wood, P. L., Moroni, F., Cheney, D. L. and Costa, E.: Cortical lesions modulate turnover rates of acetylcholine and gamma-aminobutyric acid. Neurosci. Lett. *12*: 349–354, 1979.

Wood, P. L., McQuade, P. S. and Nair, V. P.: Gabaergic and opioid regulation of the substantia innominata-cortical cholinergic pathway in the rat. Prog. Neuropsychopharmacol. Biol. Psychiatry *8*: 789–792, 1984.

Wood, P. L., Kim, H. S. and Altar, C. A.: *In vitro* assessment of dopamine and norepinephrine release in rat neocortex: gas chromatography-mass spectrometry measurement of 3-methoxytyramine and normetanephrine. J. Neurochem. *48*: 574–579, 1987.

Woodruff-Pak, D. S., Li, Y. T., Kazmi, A. and Kem, W. R.: Nicotinic cholinergic system involvement in eyeblink classical conditioning in rabbits. Behav. Neurosci. *108*: 486–493, 1994.

Woody, C. D.: "Memory, Learning, and Higher Function," Springer-Verlag, Berlin, 1982.

Woolf, N. J.: Cholinergic systems in mammalian brain and spinal cord. Prog. Neurobiol. *37*: 475–524, 1991.

Woolf, N. J.: The critical role of cholinergic basal forebrain neurons in morphological change and memory encoding: a hypothesis. Neurobiol. Learn. Mem. *66*: 258–266, 1996.

Woolf, N.: A possible role for cholinergic neurons in the basal brain and pontomesencephalon in consciousness. Conscious. Cogn. *6*: 574–596, 1997.

Woolf, N. J.: A structural basis for memory storage in mammals. Prog. Neurobiol. *55*: 59–77, 1998.

Woolf, N. J. and Butcher, L. L.: Cholinergic systems in the rat brain: III. Projections from the pontomesencephalic tegmentum to the thalamus, tectum, basal ganglia, and forebrain. Brain Res. Bull. *16*: 603–637, 1986.

Woolf, N. J. and Butcher, L. L.: Cholinergic systems in the rat brain: IV. Descending projections of the pontomesencephalic tegmentum. Brain Res. Bull. *23*: 519–540, 1989.

Woolf, N. J. and Hameroff, S. R.: A quantum approach to visual consciousness. Trends Cogn. Sci. *5*: 472–478, 2001.

Wrenn, C. C., Lappi, D. A. and Wiley, R. G.: Threshold relationship between lesion extent of the cholinergic basal forebrain and working memory impairment in the radial maze. Brain Res. *847*: 284–298, 1999.

Wu, C. F., Bertorelli, R., Sacconi, M., Pepeu, G. and Consolo, S.: Decrease of brain acetylcholine release in aging freely-moving rats detected by microdialysis. Neurobiol. Aging *9*: 357–361, 1988.

Xi, M. C., Morales, F. R. and Chase, M. H.: Evidence that wakefulness and sleep are controlled by GABA-ergic pontine mechanism. J. Neurophysiol. *82*: 2015–2019, 1999.

Xia, Q. G., Lu, L. and Li, P.: Role of rostral ventrolateral medulla in pressor response to intracerebroventricular injection of neostigmine. Shen Li Hsueh Pao *41*: 19–29, 1989 (in Chinese).

Xiao, Y., Smith, R. D., Caruso, F. S. and Kellar, K. J.: Blockade of rat alpha3beta4 nicotinic receptor function by methadone, its metabolites and structural analogs. J. Pharmacol. Exp. Ther. *299*: 366–371, 2001.

Xu, M., Mizobe, F., Yamamoto, T. and Kato, T.: Differential effects of M1- and M2-muscarinic drugs on striatal dopamine release and metabolism in freely moving cats. Brain Res. *495*: 232–242, 1989.

Yabase, M., Carino, M. A. and Horita, A.: Cocaine produces cholinergically mediated analeptic and EEG arousal effects in rabbits and rats. Pharmacol. Biochem. Behav. *37*: 375–377, 1990.

Yaksh, T. L. and Myers, R. D.: Hypothalamic "coding" in the unanesthetized monkey of noradrenergic sites mediating feeding and thermoregulation. Physiol. Behav. *8*: 251–257, 1972.

Yamada, M., Miyakawa, T., Duttaroy, A., Yamanaka, A., Moriguchi, T., Makita, R., Ogawa, M., Chou, C. J., Xia, B., Crawley, J. N., Felder, C. C., Deng, C. X. and Wess, J.: Mice lacking the M3 muscarinic acetylcholine receptor are hypophagic and lean. Nature *410*: 207–212, 2001.

Yamada, S., Harano, M., Annoh, N. and Tanaka, M.: Dopamine D (3) receptors modulate evoked dopamine release from slices of nucleus accumbens via muscarinic receptors, but not from the striatum. J. Pharmacol. Exp. Ther. *291*: 994–998, 1999.

Yamaguchi, K. and Hama, H.: Evaluation for roles of periventricular cholinoceptors in vasopressin secretion in response to angiotensin II and an osmotic stimulus. Brain Res. *496*: 345–350, 1989.

Yamanashi, K., Miyame, T., Misu, Y. and Goshima, Y.: Tonic function of nicotinic receptors in stress-induced release of l-DOPA from nucleus accumbens in freely moving rats. Eur. J. Pharmacol. *424*: 199–202, 2001.

Yamuy, J., Jimenez, I., Morales, F., Rudomin, P. and Chase, M.: Population synaptic potentials evoked in lumbar motoneurons following stimulation of the nucleus reticularis gigantocellularis during carbachol-induced atonia. Brain Res. *639*: 313–319, 1994.

Yang, Y., Liu, Z., Cermak, J. M., Tandon, P., Sarkisian, M. R., Stafstrom, C. E., Neill, J. C., Blusztajn, J. K. and Holmes, G. L.: Protective effects of prenatal choline supplementation on seizure-induced memory impairment. J. Neurosci. *20*: RC109, 2000.

Yeh, J. J., Yasuda, R. P., Davila-Garcia, M. I., Xiao, Y., Ebert, S., Gupta, T., Kellar, K. J. and Wolfe, B. B.: Neuronal nicotinic acetylcholine receptor alpha3 subunit protein in rat brain and sympathetic ganglion measured using a subunit specific antibody: regional and ontogenic expression. J. Neurochem. *77*: 336–346, 2001.

Yi, S. J. and Johnson, K. M.: Effects of acute and chronic administration of cocaine on striatal uptake, compartmentalization and release of [3H] dopamine. Neuropharmacology *29*: 475–486, 1990.

Yoshioka, K., Sakuma, M. and Otsuka, M.: Cutaneous nerve-evoked cholinergic inhibition of monosynaptic reflex in the neonatal rat spinal cord: involvement on M2 receptors and tachykininergic primary afferents. Neuroscience *38*: 195–203, 1990.

Young, D. and Dragunow, M.: Non-NMDA glutamate receptors are involved in the maintenance of status epilepticus. Neuroreport *5*: 81–83, 1993.

Yu, A. J. and Dayan, P.: Uncertainty, neuromodulation and attention. Neuron. *46*: 681–692, 2005.

Yu, Z. J. and Wecker, L.: Chronic nicotine administration differentially affects neurotransmitter release from rat striatal slices. J. Neurochem. *63*: 186–194, 1994.

Zaborszky, L.: The modular organization of brain systems. Basal forebrain: the last frontier. Prog. Brain Res. *136*: 359–372, 2002.

Zaborszky, L., Cullinan, W. E. and Luine, V. N.: Catecholaminergic-cholinergic interaction in the basal forebrain. In: "Cholinergic Function and Dysfunction," A. C. Cuello, Ed., Prog. Brain Res. *98*: 31–49, 1983.

Zaborszky, L., Cullinan, W. E. and Braun, A.: Afferents to basal forebrain cholinergic projection neurons: An update. In: "The Basal Forebrain," T. C. Napier, P. W. Kalivas and I. Hanin, Ed., pp 43–100, Plenum Press, New York, 1991.

Zaborszky, L., Pang, K., Somogyi, J., Nadasdy, Z. and Kallo, I.: The basal forebrain corticopetal system revisited. Ann. N.Y. Acad. Sci. *877*: 339–367, 1999.

Zang, Y., Dyck, R. H., Hamilton, S. E., Nathanson, N. M. and Yan, J.: Disrupted tonotopy of the auditory cortex in mice lacking M(1) cholinergic receptor. Hear Res. *201*: 144–155, 2005.

Zarei, M. M., Radcliffe, K. A., Chen, D., Patrick, J. W. and Dani, J. A.: Distribution of nicotinic acetylcholine receptor alpha7 and beta2 subunits on cultured hippocampal neurons. Neuroscience *88*: 755–764, 1999.

Zemlan, F. P. and Behbehan, M. M.: Nucleus cuneiformis and pain modulation: anatomy and behavioral pharmacology. Brain Res. *453*: 89–102, 1988.

Zhang, L. C. and Buccafusco, J. J.: Prevention of morphine-induced muscarinic (M2) receptor adaptation suppresses the expression of withdrawal symptoms. Brain Res. *803*: 114–121, 1998.

Zhang, L. X. and Liu, Z. Y.: Effects of microelectrophoretically applied ACH on the electric activities of neurons in nucleus parafascicularis of thalamus in rat. Sheng Li Hsueh Pao *45*: 55–60, 1993.

Zhang, Y., Zee, P. C., Kirby, J. D., Takahashi, J. S. and Turek, F. M.: A cholinergic antagonist, mecamylamine, blocks light-induced fos immunoreactivity in specific regions of the hamster suprachiasmatic nucleus. Brain Res. *615*: 107–112, 1993.

Zhang, Y., Dyck, R. H., Hamilton, S. E., Nathanson, N. M. and Yan, J.: Disrupted tonotopy of the auditory cortex in mice lacking M(1) cholinergic receptor. Hear Res. *201*: 144–155, 2005.

Zhao, X., Kuryatov, A., Lindstrom, J. M., Yeh, J. Z. and Narahashi, T.: Nootropic drug modulation of neuronal nicotinic acetylcholine receptors in cortical neurons. Mol. Pharmacol. *59*: 673–683, 2001.

Zhu, X. D., Cuadra, G., Brufani, M., Maggi, T., Pagella, P. G., Williams, E. and Giacobini, E.: Effects of MF-268, a new cholinesterase inhibitor, on acetylcholine and biogenic amines in rat cortex. J. Neurosci. Res. *43*: 120–126, 1996.

Zhu, D. N., Xue, L. M. and Li, P.: Effect of central muscarine receptor blockade with DKJ-21 on the blood pressure and heart rate in stress-induced hypertensive rats. Blood Press. *5*: 170–177, 1996.

Zhuo, L., Cao, X. D. and Wu, G. C.: Role of the nucleus reticularis paragigantocellularis lateralis in the pain modulatory system. Sheng Li Ko Hsueh Chin. Chan. *24*: 221–225, 1993 (in Chinese).

Ziegler, S., Lipton, J., Toga, A. and Ellison, G.: Continuous cocaine administration produces persisting changes in brain neurochemistry and behavior. Brain Res. *552*: 27–35, 1991.

Zieglgansberger, W. and Reiter, C.: A cholinergic mechanism in the spinal cord of cats. Neuropharmacology *13*: 519–527, 1974.

Zilov, V. G. and Patyshakuliev, A. P.: A pharmacological analysis of pentogastrin-modulated behavior evoked by stimulation of the ventromedial hypothalamus. Biull. Eksp. Biol. Med. *116*: 178–181, 1993 (in Russian).

Zoltay, G. and Cooper, J. R.: Dentrotoxin blocks a class of potassium channels that are opened by inhibitory presynaptic modulators in rat cortical synaptosomes and slices. Cell. Mol. Neurobiol. *13*: 59–68, 1993.

10

Links Between Amyloid and Tau Biology in Alzheimer's Disease and Their Cholinergic Aspects

George J. Siegel, Neelima Chauhan, and Alexander G. Karczmar

A. Historical Introduction

The history of Alzheimer's disease (AD) presents a difficulty in a book that is devoted to the cholinergic system. Indeed, this history emphasizes the tale of plaques and bundles; the heroes of the tale are Alois Alzheimer and Gaetano Perusini, and their discoveries concern the amyloid "cement"; this tale, told in section 1 of this introduction, does not seem to relate to cholinergicity. Similarly, as a whole, the chapter focuses on neurocellular degeneration and degenerative proteins as the basis of AD rather than on cholinergicity. Yet, there is a strong link between AD and the cholinergic system in view of the apparent preference of plaques, bundles, and degenerative proteins for the central cholinergic neurons. This matter and its consequences are discussed in section 2 of this introduction and expanded upon in section K (Wong et al., 2006).

1. History of Alzheimer's Disease and the Beta-Amyloid Cement (Amyloidosis Model)

Last scene of all,
That ends this strange eventful history,
Is second childishness and mere oblivion;
Sans teeth, sans eyes, sans taste, sans everything.

From William Shakespeare's "As You Like It,"
Act 2, Scene 7 (Jaques' monologue)

It is generally assumed that in his soliloquy Jaques refers to senility rather than to Alzheimer's disease; yet, while the "sans teeth" is not necessarily a valid description of a symptom of AD, "oblivion," sightlessness, and "sans everything" are faithful descriptions of late stages of this disease, and perhaps the Bard discovered AD before Alois Alzheimer.

If so, the only modern amendment to Shakespeare that is needed is recognizing the drastic increase in the frequency of AD cases. In 1906, Alois Alzheimer referred to a single case of "unusual disease of the cerebral cortex" that was subsequently renamed AD; some 4 million cases live currently in the United States alone, according to a survey by the Alzheimer's Association. Some statistics and fiducial analyses suggest that between 65% and 80% of those older than 80 years of age suffer from AD. If that estimate is accurate, and if the cure for AD is not forthcoming, how many AD cases will there be in a decade, or during the lifetime of the next generation, as the population of people 80 years of age and older may amount to 20% or more of the US and Western European populations? Is longevity worthwhile if it is tantamount to having an excellent chance to experience "oblivion"? In Karl Capek's 1926 play, "The Makropulos Case," a 337-year-old woman seeks an elixir that will allow her to live another 300 years; a protagonist of the play states that living so long "is an absurd idea" (see also Roush, 1996). Fortunately, in section K, hopeful information is provided that indicates that the situation may not be as bad as that.

To return to Alois: he was a Bavarian and the pupil of the great German pathologist and anatomist Friedrich Nissl. Alzheimer was 43 old when he established the hallmark of the new pathology, the presence of the cortical neurofibrils in a cortical degeneracy case of a 51-year-old woman (see Pomponi and Marta, 1992; Foerstl and Levy, 1991).[1] His pertinent statements were as follows: "The autopsy showed an . . . atrophic brain. . . . The Bielschowsky silver preparation showed very characteristic changes in neurofibrils . . . they accumulate forming dense bundles . . . and only a tangled bundle of neurofibrils indicated the site where once the neuron had been located. . . . Numerous neurons, especially in upper layers [of the cortex] disappeared. Dispersed over the entire cortex and in large numbers especially in the upper layers, miliary foci . . . could be found which depict the sites of deposition of a peculiar substance in the cerebral cortex" (Alzheimer, 1906; see also Alzheimer, 1907, 1911, 1991).

As pointed out by Pomponi and Marta (1992) Alzheimer did not emphasize the matter of the "miliary foci," nor did he use the term "senile plaques," coined by Simchowicz in 1911, to describe them. At the turn of the 19th century, these "foci" were named Redlich-Fischer plaques, after two German pathologists who described them in detail, Redlich (1898) and Fischer (1907); they© both were probably unaware of the papers by Alzheimer. Redlich's study concerned two elderly adults diagnosed with dementia, and he referred to the "miliary foci as "sclerotic"; Fischer found a "surprising" abundance of plaques, which he referred to as glandular necroses in 12 dementive cases.

In 1910 the great German psychiatrist Emil Kraepelin stated that "a particular group of cases with extremely serious cell alterations was described by Alzheimer"; Kraepelin felt that the condition in question represented a novel form of dementia, the "presenile dementia." Since that time, the condition in question has been referred to as Alzheimer's dementia or Alzheimer's disease, even though Alzheimer's study, quoted by Kraepelin, described one isolated case (Kraepelin, 1910).

And here enters Gaetano Perusini. According to Pomponi and Marta of the School of Medicine, Catholic University of Rome, Perusini was a member of "a group of Roman researchers who studied under Alois Alzheimer" (Figure 10-1). Perusini specialized in "endemic cretinism," which is common in northern Italy, Friedreich's condition, and Tabes dorsali of syphilic origin (see Pomponi and Marta, 1992).

As a member of Alzheimer's team, Perusini examined "four cases" (one of which was Alzheimer's single case) and published his results in 1910 and 1911. His 1910 work is particularly detailed and contains 79 figures, including color drawings. It was common in those days—it still is—for a professor to entrust his student with expanding on the preliminary work initiated by the team leader. In fact, Perusini states in his paper that he undertook his study "on the suggestion of Dr. Alzheimer," and this state of affairs is further illustrated by the generous references to Perusini in Alzheimer's own 1911 publication; also, Alzheimer invited Perusini to contribute to his and Nissl's review of 1910. Reciprocally, Perusini thanked his teacher and quoted him 17 times in his 1910 paper.

Why did Kraepelin not quote Perusini in his detailed reference to Alois Alzheimer's work? At any rate, the Italian investigators took care of the matter, as they frequently refer to AD as Alzheimer-Perusini disease (see Bick et al., 1987 and McMenemey, 1970).

One of Perusini's essential findings was his emphasis on "the link of neurofibrils changes with plaque formation." As can be seen, he was the first to use the term "plaque" and applied it to the accretion of plaques in neurons as well as in glia. He also related the formation of plaques to the destruction of the neurons and/or glia. Most interestingly, he hypothesized that "a kind of cement . . . glued the fibrils together," leading to the formation of plaques (Perusini, 1910)—a forecast of the beta amyloid.

It should be noted that neurofibrillatory tangles (NFTs), which contain as their major component phosphorylated isoform of the tau protein, were not recognized by Alzheimer or by Perusini (see Morris, 1996).

Subsequent work substantiated the early Alzheimer and Perusini findings and stressed the relation between the prevalence of plaques and the progress of AD (Blumenthal and Premachandra, 1990). Establishing such a relation was important; B. E. Tomlison (1982), a pioneer in dementia research, felt that "until . . . plaque's . . . role is

Figure 10-1. Gaetano Perusini (9) with Alois Alzheimer (7) in the pathological anatomy laboratory of the Deutsche Forschungsanstalt fur Psychiatrie in Munich (today the Max Planck Institute). With them are U. Cerletti (4), F. Bonfiglio (6), M. Achucarro (8), and F. H. Lewy (10). (Reprinted from Pomponi and Marta, 1992, with permission.)

precisely defined, . . . functional . . . knowledge of an important role of neuropathology and human aging will be incomplete." Actually, such a relation was proposed early by Fischer (1907) and definitive, quantitative data supporting this relation were provided by Blessed et al. (1968; see also Klatzo et al., 1965; Wisniewski, 1987).

When was the beta amyloid recognized as the "cement" for the fibrils? Actually, the amyloid deposits ("lardaceous deposits") were recognized by 18th-century anatomists, but Rokitansky and the celebrated German pathologist Rudolof Ludvig Carl Virchow (1821–1902) identified these elements as hyaline deposits a century later, and Virchow distinguished between amyloid (starch-like) and other types of hyaline (see Blumenthal and Premachandra, 1990). While Virchow referred to "senile amyloidosis" in connection with deposits exhibited in the hearts of aged subjects, amyloid deposits are present in other tissues of the aged human, including cerebrovascular sites. Bielschowski (1911) and Divry (1934) not only recognized their cerebral presence but also observed their amyloid nature (see also Glenner, 1986 and Blumenthal and Premachandra, 1990). G. Glenner (1980) recognized the presence of amyloidosis in AD, and Glenner and Wong (1984a, 1984b) recognized that neuronal amyloid plaques share antigenic determinants with beta-amyloid protein of cerebrovascular amyloid.

A significant issue is that the plaques and tangles are not the sole and specific characteristic of AD. It was recognized quite early that senile plaques and amyloid deposits are apparently present in "normal" aging (Ludwig, 1981); does their presence lead to full-blown dementia provided we live long enough (Evans, 1988; Committee on Chemical Toxicity and Aging, 1987)?

Finally, it is of interest that spontaneous senile amyloidosis is present in many animals and that it can be induced experimentally in many species, (see, for example, Higuchi et al., 1983; see also Blumenthal and Premachandra, 1990; Price, 1993). These early findings led to the distinction of the types of amyloid, and so forth, as reviewed below.

2. Cholinergic Aspects and Cholinergic Treatment of Alzheimer's Disease

The dawn of cholinergic considerations for AD arose with the realization that amyloid and related deposits, as well as AD genetics, underlie cholinergic apoptosis in several brain areas. This realization was based on the findings of the diminution in acetylcholinesterase (AChE) and choline acetyl transferase (CAT) activities. Actually, in 1941, David Glick, 1941, reviewed the early work concerning the relation of the diminution of ChE activity (he did not differentiate between AChE and BuChE; see Chapter 3 DI) and several psychiatric diseases (and "fatigue"), but he did not refer to AD specifically. Almost 40 years later, Bowen et al. (1976), Davies and Maloney (1976) and Perry et al. (1977; see also Wurtman, 1985, McGeer et al., 1984 and Mesulam, 2000) demonstrated that a significant loss of AChE and CAT activity occurs in AD. Subsequently, ample evidence was accumulated to confirm these data (see, for example, Gil-Bea et al., 2005), particularly with regard to CAT (of course, these results were obtained with respect to postmortem materials; see Mesulam, 2000; Mesulam et al., 2004). It should be stressed that, while AChE is not 100% dependable as the marker of cholinergic neurons, CAT is a very reliable identifier of these neurons and of the cholinergic pathways (Cook et al., 2004) (see Chapter 2 A). This loss is noticeable with respect to all the known cholinergic pathways of the brain: particularly relevant for the symptoms of AD are the losses that occur in the pathways that include the limbic system and the nucleus basalis magnocellularis (NBM) of Meynert (Mesulam's pathways Ch1 to Ch4; see Mesulam, 2000; Mesulam et al., 2004; see Chapter 2 DII and DIII). Interestingly, both somatostatin immunoreactivity and AChE activity were reduced in the cerebrospinal fluid of AD patients (Gomez et al., 1985); this finding may be important for noninvasive diagnosis of AD. It may be added that contrary to AChE, BuChE may increase in AD (Giacobini, 2000; 2003a). It is of interest in this context that central cholinergic neurons may be vulnerable to several diseases, including AD (see, for example, McKinney and Jacksonville, 2005)

Less clear-cut data are available with respect to cholinergic receptors. Agneta Nordberg and her associates observed early that the presynaptic muscarinic receptors are particularly affected, while the postsynaptic muscarinic receptors may be less so or they may be even spared; the presynaptic and postsynaptic central muscarinic receptors belong particularly to the M1 and M2 subtypes (Nordberg and Winblad, 1986; see also Chapter 5). Alzheimer's disease patients may also exhibit a diminution of postsynaptic nicotinic receptors of several types (Nordberg and Winblad, 1986; Whitehouse et al., 1986). There are, however, inconsistencies in the literature concerned with the state of the cholinergic receptors in AD (Adem et al., 1985). Again, certain findings concerning the cholinergic receptors may be diagnostically pertinent: nicotinic and muscarinic receptors can be distinguished in the human lymphocytes, and they may be decreased in the lymphocytes of patients suffering from AD (Adem et al., 1985).

It is important that these demonstrations of the deficits of the cholinergic components in AD relate to degeneration and death of cholinergic neurons induced by the beta amyloid cascade and its components; this relation was first described by Peter Whitehouse (Whitehouse et al., 1981, 1982, 1986) particularly for NBM and confirmed subsequently (see, for example, Kasa et al., 2004). This loss was documented for several cholinergic pathways (see Giacobini, 2000). It is significant that this loss may be minimal in the early stages of the disease. Yet, it appears that the progress of what Marsel Mesulam refers to as "tauopathy," that is, signs of amyloid pathology, may occur in human cholinergic nucleus basalis neurons earlier than diminution or elimination of their CAT activity (Mesulam et al., 2004). This concept of "tauopathy" is supported by data obtained in the human by means of micro-positron emission tomography (PET) concerning the loss of either AChE or cholinergic receptors in AD, as well as via functional magnetic resonance imaging (fMRI; see Nordberg, 2004; Thiel, 2003; see also section K for more detail). Also, results of gene association studies indicated close correlation between AD and reduction of CAT (Cook et al., 2004). In this context, it is of interest that while some degenerative brain diseases such as Loewy body disease appear to impact the cholinergic system, it seems that cholinergic deficits are absent in vascular dementia (Perry et al., 2005).

The AD senile plaques contain AChE, BuChE, and possibly other components of the cholinergic system (Moran et al., 1993). Since cholinergic neurons are attacked (indeed, preferentially) by the neurotoxic amyloids, one would expect that the amyloid plaques contain markers of the cholinergic system. However, Ezio Giacobini has another view on this matter (Giacobini, 2000 and 2003a, 2003b; see also section K). An additional link between the cholinergic system and AD was posited by Nancy Woolf and Larry Butcher as they proposed that the neurotrophin p75 receptor (see Chapter 8 CI), which is localized in the cholinergic forebrain, plays an important role in the formation of plaques and in the pathology of AD (for a more detailed discussion of this hypothesis, see below, section K, and Butcher and Woolf, 1989).

It came then to the mind of several investigators to link the loss of cholinergic neurons with the cognitive aspects of AD and with the possibility that this cognitive deficit may be remedied by the employment of cholinergic agonists and antiChEs. Indeed, the relation between the cholinergic system and cognition and memory was recognized some 60 years ago. In fact, positive effects of cholinergic agonists on animal memory and learning were described before the existence of central cholinergic transmission was demonstrated; copious confirmation of these findings became available subsequently, and pertinent information continues pouring in (see Karczmar, 1967, and Chapter 9 BV). In the 1970s, David Drachman demonstrated that cognitive deficit in human aging may be due to a loss of cholinergic activity, and he provided preliminary information indicating that physostigmine may facilitate human memory (Drachman and Leavit, 1974; Drachman, 1978; see also Chapter 9 BV). Of course, Drachman's finding concerns aging, not AD; this connection between aging and the diminution of cholinergic activity, and the "aging" effect of anticholinergics on memory and learning, are well recognized for animals and humans (see Chapter 9 BV).

All this research culminated in the first attempts to use physostigmine, first in "normal," aged subjects and then in AD patients. Actually, these first attempts were not conspicuously successful, whether in normal elderly (Davis et al., 1976) or in the AD cases (Peters and Levin, 1979). In fact, when one of the authors (AGK) met at a 1980s symposium on AD one of the first AD subjects who were treated with physostigmine, he regretfully could not be impressed with his—admittedly superficial—impression of the patient in question. Actually, even today, the effects of novel generations of antiChEs used in the treatment of AD may be apparent only when subtle tests of cognition and memory are employed.

Ezio Giacobini pointed out early that lack of definitive effects of physostigmine is due primarily to its kinetics and distribution (see Giacobini, 2000): physostigmine has a short half-life, it is poorly distributed to the CNS, and it is not sufficiently specific for AChE. Giacobini and his associates, including Robert Becker, outlined the physicochemical characteristics of antiChEs that would be clinically more successful than physostigmine in AD treatment. Following the clinical trial of the "first generation" of antiChEs employed in the treatment of AD, such as physostigmine and tetrahydro-5-aminoacridine (THA, tacrine), Giacobini's guidance led to the development of the "second generation" of antiChE agents usable in AD, such as Aricept (donepezil), rivastigmine, and galantamine; they also include an organophosphorus agent, metrifonate (see Giacobini, 2000; see also Chapter 10 K and Chapter 7 BI and BII).

This clinical progress is facilitated by the availability of animal models. While the earlier models that employed neurotoxins including the cholinotoxin AF64A and lesions may be not very pertinent for AD as it is understood today (see Karczmar, 1991, and Chapter 9 BV), several currently available animal models are very relevant; they include transgenic animals (particularly rodents) and animals in which amyloid treatment or amyloid treatment combined with hypoperfusion (Sun and Alkon, 2004) induces CNS damage; in particular, the transgenic mice model developed by M. Morgan and his associates appears to be very promising (Buxbaum, 1993, and other articles in Neuroscience Facts 4(5); see also Arendash et al., 2001).

What is the current status of antiChE therapy for AD? In the 1980s and 1990s, it became clear that antiChEs are not a cure for AD, no more than the antiChE ambenonium is a cure for myasthenia gravis; this realization was also linked with our understanding that AD is not a cholinergic disease, although it presents distinct cholinergic aspects. In fact, in that period it appeared that the initial hopes

(Giacobini, 1991) with respect to the extent of the beneficial effects of this treatment may not be fulfilled. Indeed, antiChEs exhibit a number of effects that may not be advantageous for their use in AD; for example, they may decrease ACh turnover and diminish ACh release; furthermore, a sufficient number of cholinergic neurons and/or postsynaptic cholinoceptive sites must be present to enable antiChEs to be effective. Actually, cholinergic neurons may be decreased in number or even eliminated in the late stages of AD, but cholinoceptive postsynaptic muscarinic sites present on noncholinergic neurons may remain intact even in advanced AD (Karczmar and Dun, 1988). Finally, it appeared at the time that the amelioration of cognition in antiChE-treated AD patients may be relatively limited and short lived (see, for example, Jones, 2003). On the other hand, in early AD there is a limited loss of cholinergic neurons, and this justifies cholinergic treatment during that phase of the disease (Morris and Price, 2001; see also Giacobini, 1990).

At this time, two novel aspects of antiChE therapy may lead to an upward revision of our concepts and evaluation of this therapy (see Wynn and Cummings, 2004). First, many investigators today accept Ezio Giacobini's notion that appropriate antiChEs retard the progress of AD, particularly if they are administered early (see Giacobini, 2000). Then, as first shown by Roger Nitsch, cholinergic agonists may diminish amyloid accumulation (Nitsch et al., 1992; see also Mesulam, 2004). Yet, considering this evidence as well as analyzing the relations among AD symptoms, progress in tauopathology, and progress in the disappearance of cholinergic function in AD, Marsel Mesulam (2004) concluded that "the cholinergic loss is neither a primary pathogenetic factor of AD nor the principal correlate of its clinical manifestations." These aspects will be considered in detail in section K, below.

B. Dysregulated Processing of Neurocellular Proteins

Alzheimer's disease is one of a group of age-dependent neurologic degenerative processes characterized biochemically by neuronal degeneration due to alternative processing and folding of certain proteins, leading to their aggregation and distortion of neurocellular structure. Although classically this group of diseases has been called neurodegenerations of unknown etiology, they may be better termed "neurocellular protein processing diseases" (Table 10-1). These diseases include AD, Down syndrome (DS), frontotemporal dementias (FTD), corticobasal ganglion degeneration (CBGD), Parkinson's disease (PD), multisystem atrophies (MSA), diffuse Lewy body disease (DLBD), progressive supranuclear palsy (PSP), prion diseases, motor neuron disease (MND), and the mutations producing abnormally long polyglutamine sequences. Familial forms of AD, PD, FTD, prion diseases, and MND have been identified and proved related to specific genetic mutations in about 10% or less of the affected people with each disease. In the preponderance of patients with these diseases, the incidence is sporadic and the cause unknown. The conventional wisdom holds that the causes in the sporadic forms may be a combination of genetic and environmental factors. In all the cases, the incidence increases with age, suggesting an effect of changing gene expression during aging that potentiates a particular trigger or the effect of time for accumulation of a causative factor (see the discussion on aging in Finch and Roth, 1999).

The primary proteins subjected to alternative processing and aggregation have been identified in most of these diseases (Table 10-1). The primary abnormal proteins or their products generally are aggregated into inclusion bodies. These bodies may accumulate within the cells, for example, Lewy bodies containing aggregates of α-synuclein, or as extracellular plaques, for example, amyloid plaques containing aggregates of amyloid beta peptide (Aβ) together with dystrophic cells, cell fragments, and trapped proteins that are secondarily degraded or involved in processing, immune, inflammatory, and scavenging functions. In AD, FTDs, and other tauopathies, which are associated with abnormal structure and phosphorylation of the microtubule-associated protein tau, dystrophic neurons contain intracellular paired helical filaments (PHF) of abnormal tau which are bundled into neurofibrillary tangles (NFT). Alzheimer's disease is unique in exhibiting an abundance of both extracellular amyloid plaques and intracellular NFTs. However, the particular abnormal tau isoforms and sites of phosphorylation are not identical in AD and the FTDs.

Table 10-1. Characteristic Protein Alterations and Pathological Features of Some Neurodegenerative Diseases

Disease	Protein Alterations	Mutated Gene Product	Pathological Features
Alzheimer's disease (AD); Down syndrome (DS)	Excess of Aβ; β-pleated sheet aggregates of Aβ	Amyloid precursor protein (APP); chromosome 21	Extraneuronal deposition of amyloid plaques in brain parenchyma and amyloid deposition within brain vasculature (cerebral amyloid angiopathy, CAA)
	Hyperphosphorylated tau protein (all 6 of both the 3-R-tau and 4-R-tau isoforms) at serine and threonine residues; 79 potential sites exist; about 30 phosphorylated is normal		Intraneuronal cytoplasmic deposition of paired helical filaments of tau that aggregate into neurofibrillary tangles within dystrophic neurites
Frontotemporal dementia(s) with parkinsonism linked to chromosome 17 (FTDPD-17) and variants of FTD	Hyperphosphorylation of mutated tau protein; insoluble tau (3-R or 4-R isoforms or both); abnormal ratios of 3-R and 4-R isoforms result	Tau; chromosome 17; more than 20 pathogenic mutations identified, all of which result in filaments made of hyperphosphorylated tau	Intraneuronal cytoplasmic deposition of tau-positive inclusions and neuronal loss
Other sporadic tauopathies: corticobasal ganglion degeneration (CBGD), progressive supranuclear palsy (PSP), Pick's disease (PiD) variant of FTD	Hyperphosphorylated tau; insoluble tau 4-R isoforms in PSP and CBGD; 3-R isoforms in PiD		Neuronal and glial cytoplasmic tau-positive fibrillar inclusions and neuronal loss; globose bodies, fibrillary tangles in astrocytes and oligodendrocytes in CBGD and PSP; ballooned neurons and tau-positive fibrillary tangles in neurons (Pick bodies) in PiD
Diffuse Lewy body disease (DLBD)	Alpha-synuclein; insoluble fibrillar aggregates of alpha-synuclein in β-pleated sheet structure		Lewy bodies consisting of neuronal cytoplasmic inclusions of insoluble alpha-synuclein; diffuse in cortex
Sporadic Parkinson's disease (PD)	Alpha-synuclein; insoluble fibrillar aggregates of alpha-synuclein in β-pleated sheet structure		Lewy bodies in basal ganglia
Dominant inherited PD (Park1)	Alpha-synuclein; insoluble fibrillar aggregates of alpha-synuclein in β-pleated sheet structure	Alpha-synuclein; chromosome 4q21–q23; mutated product has increased tendency to aggregate	Lewy bodies

Table 10-1. *Continued*

Disease	Protein Alterations	Mutated Gene Product	Pathological Features
Autosomal recessive juvenile PD (AR-JP, Park2)	Ubiquitin-proteasome degradation system	Parkin, an E2-dependent ubiquitin protein ligase (E3) that promotes degradation of alpha-synuclein; PARK2 gene, 6q25.2–q27	Neuronal loss in substantia nigra pars compacta and locus ceruleus without Lewy bodies
Multisystem atrophies (MSA) including striato-nigral degeneration, olivo-ponto-cerebellar degeneration, Shy-Drager syndrome	Alpha-synuclein; accrual of soluble forms; different from PD and DLBD		Oligodendroglial cytoplasmic inclusions (GCIs) and neuronal Lewy body-like inclusions of soluble forms of alpha-synuclein
Prion diseases, environmentally transmissible spongiform encephalopathies, including Creutzfeld-Jakob disease (CJD) and human variant CJD, several animal forms in cattle, sheep (scrapie), mule deer, mink	Conformational change of normal cellular prion protein (PrP^{c}) to produce insoluble aggregates of PrP^{sc} with β-pleated sheet structure; cause of the increased conformational change not known		Amyloid plaques consisting of PrP^{sc} with β-pleated sheet structure
Inherited Prion diseases, including Gerstmann-Straussler-Scheinker disease and fatal familial insomnia	Mutation in the gene for PrP potentiates the conformational change from PrP^{c} to PrP^{sc}	Homozygosity at codon 129 of the PrP gene; chromosome 20	Amyloid plaques consisting of PrP^{sc} with β-pleated sheet structure
Familial motor neuron disease (MND)	Mutant superoxide dismutase 1 (SOD1); mutant dynactin	More than 70 pathogenic mutations in gene for SOD1, all but one dominant	Neuronal and astrocytic hyaline inclusions mainly consisting of mutant SOD1; Bunina bodies
Polyglutamine diseases, including forms of spinocerebellar ataxia, Huntington's disease, Kennedy's disease, and dentatorubro-pallidoluysian atrophy	Abnormal lengths (sequences of more than 36 glutamine residues) of polyglutamine segments in certain polyglutamine-containing proteins	More than 9 different genes containing CAG trinucleotide repeat expansions in the coding region	Cytoplasmic and nuclear inclusions of insoluble, aggregated polyglutamine-containing fragments of proteins associated with other proteins

Generally, the pathogenesis of these diseases entails a complex array of toxic effects in particular neurocellular networks or regions. These toxic effects arise from the primary abnormal gene or protein and lead to apoptosis, dysregulation of multiple cellular signaling processes, derangement of synaptic, cytoskeletal, and membrane structure and function, and secondary reparative, inflammatory, and immune reactions. This chapter seeks to describe the biology of the two primary proteins, Aβ, which is derived from the amyloid precursor protein APP, and tau, apparently critical to the pathogenesis of AD, and to point out the regulatory reactions at which the two may potentially converge. The working hypothesis underlying this chapter is that excess Aβ produced by dysregulation of APP processing in particular neuronal networks causes dysregulation of many cell processes, including the processing of tau, which itself leads to another set of disrupted neuronal functions.

C. Amyloid

1. Purification and Properties of Amyloid

Amyloid (from Greek *amylon*, unmilled) refers to any of a group of chemically diverse proteins that are deposited in tissue as β-pleated sheets of linear nonbranching aggregated fibrils as seen under the electron microscope (Zhou et al., 1991). These deposits are highly insoluble, stain brown with iodine, produce a green color under polarizing light after staining with Congo red, are metachromatic with methyl violet, and fluoresce yellow after staining with thioflavine T. The chemical nature of the amyloid depends on the underlying disease. In AD, these extracellular deposits are composed of unpaired 4 to 9 nm filaments aggregated most prominently in large, highly insoluble star-shaped masses (amyloid dense cores) or as isolated bundles or diffuse oligomeric fibrils in the cortical neuropil. Similar aggregates of fibrils are found in parenchymal and meningeal vasculature. Glenner and Wong (1984a) solubilized the β-pleated sheet amyloid fibrils from AD brains in 6 M quanidine-HCl and proved the purified protein to be a 4.2 kDa polypeptide with a unique sequence. The protein was variously called amyloid beta protein because of its β-sheet structure or A4 because of its molecular mass, and is now termed Aβ. It has been found to be the same in parenchymal amyloid plaque and cerebral and meningeal vascular amyloid of AD and DS and in so-called senile plaques of normal aging brain (Glenner and Wong, 1984a, 1984b; Masters et al., 1985; Tanzi et al., 1987). However, more recent analyses indicate some structural differences between AB and AD and from age-matched cognitively normal subjects (Piccini et al., 2005). The Aβ peptide consists mainly of peptides with either 40 (Aβ40) or 42 (Aβ42) residues. The extracellular amyloid filaments were found immunologically distinct from the intraneuronal aggregates of PHF, termed NFTs, that are found in the dystrophic neurons, axons, and dendrites of neuritic plaques in AD (Selkoe et al., 1986). Later analyses proved them also to be chemically distinct and to be composed of abnormal tau filaments (see below). Accumulations of both amyloid plaques and NFTs in the brain are the hallmarks of AD.

2. Aβ Effects, Fibrillogenesis, and Amyloid Plaques

Under normal circumstances, excesses of monomeric Aβ peptides are cleared by glial-derived apolipoprotein B (ApoB) and microglial phagocytosis. However, in the presence of aging, certain genetic mutations (Table 10-2) or abnormal physiological conditions (such as high cholesterol, Niemann-Pick type C disease), clearance of excessive Aβ may not be achieved to the fullest extent, perhaps due to altered transport, saturating enzyme sites, or disturbed biochemical equilibria among the substrate, enzyme, and end product (Simons et al., 2001). Excessive monomeric Aβ further leads to oligomerization, toxicity, fibrillogenesis, and plaque formation.

Amyloid deposited in plaques is composed of fibrils of aggregated Aβ. Fibrillation of Aβ is a nucleation-dependent polymerization process, which is controlled by two kinetic parameters: the nucleation rate and the elongation or growth rate. As the kinetics of fibrillation is strongly dependent on the presence of trace amounts of fibrils, the aggregation of Aβ may be modeled as an autocatalytic reaction. Both Aβ40 and Aβ42 form fibrillar aggregates, but Aβ42 added to Aβ40 accelerates

Table 10-2. Gene Mutations and Isoforms in Alzheimer's Disease

Gene Defect	Chromosome	Mutation	Codon	Age of Onset	Aβ Phenotype
βAPP mutations (all)	21			50s	Increased production of all Aβ peptides or mainly Aβ42
Swedish		KM → NL	670/671		
Flemish		A → G	692		Marked amyloid angiopathy with hemorrhages and dementia
Dutch (HCHWA-D)		E → Q	693		Severe amyloid angiopathy; cerebral diffuse plaques; cerebral hemorrhages; no dementia
Indiana		V → I	717		
London		V → G	717		
		V → F	717		
ApoE polymorphism	19	ApoE4 allele; not causative but increases risk		60s and older	Predisposes to increased density and number of Aβ plaques and vascular deposits
Presenilin 1 mutation	14	About 45 missense mutations; 1 exon deletion		40s and 50s	Increased production of Aβ42
Presenilin 2 mutation	1	2 missense mutations		50s	Increased production of Aβ42
Down syndrome	21	Chromosome 21 trisomy		40s	Diffuse plaques; increased nonfibrillar Aβ found in childhood or teens before NFTs and dementia

the nucleation and elongation rates (Sabate et al., 2003; Yoshiike et al., 2003). The ratio of Aβ42 to Aβ40 is critical to the rate and density of aggregation. The monomeric peptides adopt either compact random coil or elongated β-strandlike structure. Assembly of oligomers of antiparallel β sheets proceeds from random coil through an alpha helix intermediate to β strand (Klimov and Thirumalai, 2003). Bound to GM1 gangliosides in cholesterol-enriched raft-like membranes, Aβ undergoes conformational transitions from α-rich to β-rich structures that form seeds promoting fibrillogenesis (Mizuno et al., 1999; Kakio et al., 2001; Yanagisawa et al., 1995; Yip et al., 2001; Chochina et al., 2001).

In summary, the fibrillogenic properties of Aβ peptides are in part a consequence of membrane composition, peptide sequence, and mode of assembly within the membrane (Waschuk et al., 2001; Van Nostrand et al., 2001; Yip and McLaurin, 2001). Extracellular amyloid fibrillar aggregates eventually accrue to themselves the fragments of dystrophic neurites, inflammatory cells, and such other proteins as serum amyloid P protein, ApoE4, α1-anti-chymotrypsin, catalase, glycoproteins, proteoglycans, cholesterol, and other lipids to become what is called the mature core plaque.

The neurotoxicity of Aβ and Aβ aggregates may be expressed through a wide array of cytologic effects that have been demonstrated *in vitro* and *in vivo*. The oligomeric form inhibits neuronal viability significantly at 10 nm and is 10 times more potent than fibrils and about 40 times more potent than unaggregated peptide (Stine et al., 2003). Toxic actions include effects on membrane transport of ions, modification of ion

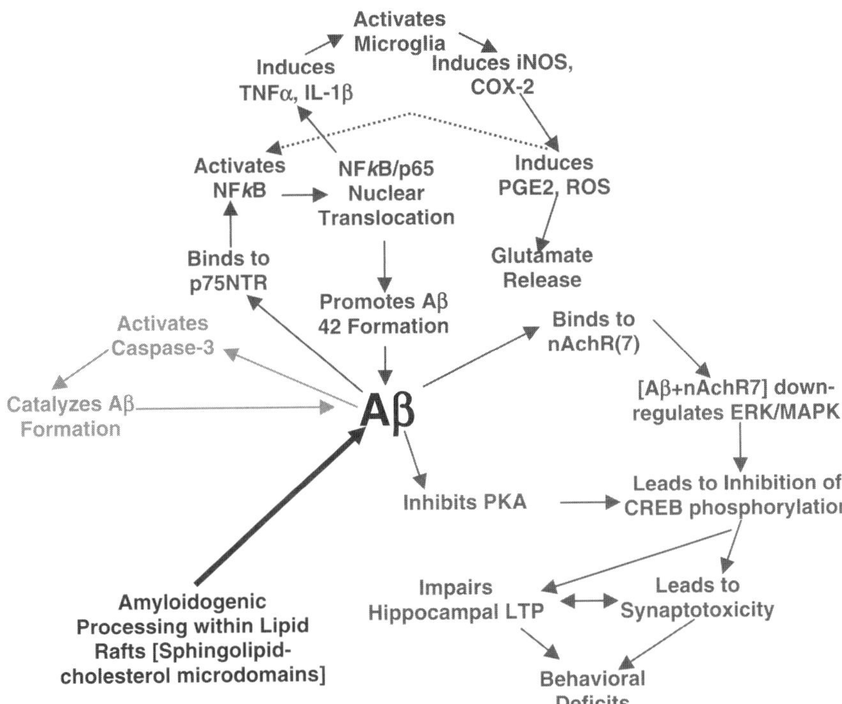

Figure 10-2. Major events of amyloidogenic vicious cycle include: (top) Aβ-induced activation of nuclear transcription factor-NFkB, which in turn activates the pro-inflammatory cytokines (TNFα, IL-1β, IFNγ), leading to subsequent induction of inflammatory molecules such as inducible nitric oxide (iNOS), cyclooxygenase 2 (COX-2), and prostaglandin E2 (PGE2). These inflammatory molecules produce reactive oxygen species (ROS), leading to oxidative damage. In addition, COX-2 and iNOS trigger further activation of NFkB. NFkB binds to the p65 subunit, translocates to the nuclear compartment, and promotes the formation of $A\beta_{42}$. Besides being a part of this vicious cycle, all these inflammatory molecules have other detrimental effects on neurons and neuronal connections, thus participating in neurodegeneration and synaptic deficits in AD; (left) activation of caspase-3, which leads to apoptosis and simultaneously potentiates further Aβ generation; (bottom right) impairment of PKA- and ERK/CREB-signaling, leading to synaptic and LTP deficits in the hippocampus that result in impairment of memory and learning; (bottom left) cholesterol, lipid, and lipoprotein receptor interactions with Aβ. (See color plate.)

channels, enhancement of calcium ion influx and glutamate toxicity, generation of reactive oxygen species, activation of inflammatory responses and apoptotic pathways, and modification of signaling pathways and of transcription and interactions with cholesterol and lipoprotein transport (Suh and Checler, 2002). Some of the better-known connections of Aβ with regulatory factors in the cell are shown in Figure 10-2, see color plate. Evidence for the potential involvement of other signaling pathways, discussed later, is still unfolding. With so many potential pathways for causing dysregulation of neurons, it is difficult to determine which are the most critical.

Evidence is accumulating that neurotoxicity of Aβ begins before accrual of amyloid plaques and is related to soluble or protofibrillar Aβ. In transgenic mouse models of AD with overexpressed mutated APP, it has been found that Ca^{2+}-dependent evoked potentials are reduced in the hippocampus at ages earlier than the formation of plaques.(Hsia et al., 1999). In the same transgenic species, SNAP-25, a presynaptic protein required for Ca^{2+}-dependent vesicle exocytosis and transmitter release was found to be severely depleted prior to plaque formation in the hilum, specifically in the inner molecular layer of the dentate gyrus and in the target CA3 to CA1 synaptic regions of the hippocampus (Chauhan and Siegel, 2002). This represents the cholinergic synaptic pathway from the mesial basal forebrain and nucleus basalis of meynert to the hippsoampus. This depletion could be reversed by injection of specific anti-Aβ antibody into the third ventricle, thus indicating that the SNAP-25 depletion was related to Aβ. SNAP-25 depletion in the same specific layers could also be demonstrated by injection of Aβ into the peri-amygdala region of wild-type mice, and this effect was potentiated by aging (Chauhan et al., 2003). Subsequently, other calcium-dependent proteins, calbindin and c-fos, were found to be reduced in granule cells of the dentate gyrus in association with learning deficits before plaque formation in transgenic mice harboring the Swedish and Indiana mutations of hAPP (see Table 10-2) (Palop et al., 2003).

In addition to toxicity of Aβ, other C-terminal fragments (CTFs) of APP (see Figure 10-2) may have toxic cellular effects and are potentially amyloidogenic. C-terminal fragments have been immunologically identified in plaques and tangles and in blood vessels from aged normal and AD brain and are produced in transgenic mice with mutated hAPP (Suh and Checler, 2002; Howland et al., 1995).

D. Amyloid Precursor Protein

1. Structure of APP

Kang et al. (1987) isolated and sequenced the cDNA clone coding for the A4 polypeptide that is derived from amyloid precursor protein (APP), and the gene for APP was localized to the long arm of chromosome 21. The 695-residue APP transcript is almost exclusively expressed in brain, mainly in neurons. Alternate APP transcripts of 751 and 770 residues that have an inserted exon encoding a Kunitz-type serine protease inhibitor (KPI) are the major isoforms expressed in all peripheral cells, peripheral nervous system, dorsal root ganglia, Schwann cells, and endothelia, and are expressed in brain as well, particularly in glia and vasculature (meningeal, parenchymal, choroid plexus), where they predominate. In rat brain, the total amount of APP proteins, including both KPI-containing and KPI-free APP isoforms, was found about 90% higher in astrocytes than in neurons (Rohan de Silva et al., 1997). APP is found in growing neurites (Masliah et al., 1992).

2. Proteolytic Cleavage of APP

Amyloid precursor protein is usually processed either by a nonamyloidogenic path involving α-secretase (after residue 687) releasing N-terminal soluble APP (sAPPα) or by an amyloidogenic path involving β-secretase (termed BACE 1) (after residue 671) and γ-secretase (after residue 710 or 712) releasing Aβ (Selkoe, 1994). Amyloid precursor protein molecules that escape α-cleavage are first subjected to β-cleavage (Kang et al., 1987), followed by γ-cleavage to release Aβ peptides (Haass et al., 1992a). Processing of APP by α-, β- and γ-secretases occurs under normal physiological conditions, indicating that all fragments of APP including Aβ peptides are part of normal physiology (Haass et al., 1992a, 1992b; Seubert et al., 1992). Only the γ-site cleavage generates the sAPPβ fragment, while α- and γ-site cleavages produce the p3 fragment. In neurons, approximately 95% of APP is cleaved by α-secretase, while the remaining 5% is subjected to β- and γ-cleavages (Simons et al., 1996, 2001). Both sAPPα and Aβ peptides at normal physiological levels have their own functions. Soluble APPα functions as a synaptotrophic molecule (Masliah et al., 1992), while normally produced Aβ potentiates tyrosine phosphorylation, increases activity of phosphoinositol-3-kinase (PI3K) and protein kinase C (PKC) (Mills and Reiner, 1999), and functions as a vascular sealant (Atwood et al., 2003). It is the excessive genesis of Aβ that is the culprit for AD pathogenesis. Therefore, reduction of excessive Aβ, not its complete removal, is critical for the therapy of AD (Smith et al., 2002).

In addition to these classic proteolytic cleavages, sometimes APP molecules are cleaved differentially, resulting in the production of amyloid fragments other than sAPPα and Aβ 40/42/43 peptides. Occasionally, after the release of the N-terminal sAPPα fragment by α-cleavage, the CTF fails to be internalized and it is released extracellularly, which can lead to toxic effects (Anderson et al., 1991; Selkoe et al., 1996).

The Aβ polypeptides consisting of either 40 or 42 amino acids (residues 671 to 711 or 671 to 713

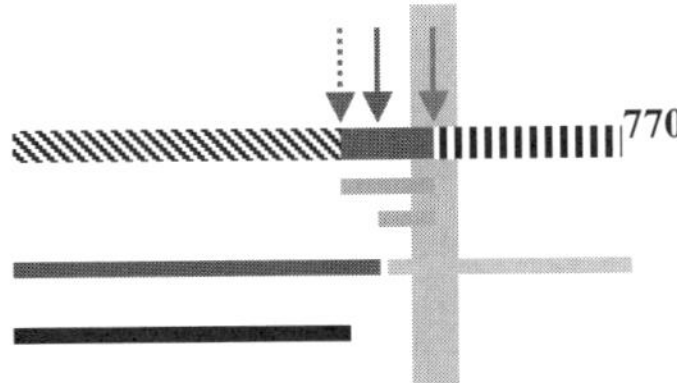

Figure 10-3. Schema showing proteolytic cleavage sites on the APP_{770} molecule and the different peptides resulting from proteolytic cleavages: bar with diagonal lines, extracellular N-terminal APP fragment; solid bar between diagonal- and vertical-line bars, Aβ-segment; bar with vertical lines, intracellular C-terminal fragment; dashed arrow, β-secretase site (after residue 671); center arrow, α-secretase site (after residue 687); right arrow, γ-secretase site (after residues 710 or 712). Second horizontal line represents full-length Aβ 40/42/43. Third horizontal line represents p3 fragment. Fourth horizontal line (left) represents sAPPα fragment (N-terminal-687). Fourth horizontal line (right) represents carboxy-terminal fragment (CTF). Bottom horizontal line represents sAPPβ fragment (N-terminal-671). Vertical bar represents the membrane-spanning segment, which comprises residues 700 to 723. (See color plate.)

of APP_{770}; see Figure 10-3, see color plate) contain sequences translated from two adjacent exons of the APP gene and thus must arise by proteolytic processing of the APP, rather than by alternative splicing of the mRNA. The identical Aβ polypeptides may derive from any of the APP isoforms synthesized in neurons, glia, peripheral cells, or vasculature.

Cleavage by α-secretase leaves an NH2-terminal ectodomain soluble sAPPα that is secreted and a membrane-bound COOH-terminal fragment CT83. Cleavage by β-secretase produces an NH2-terminal ectodomain soluble sAPPβ that is secreted and a membrane-bound COOH-terminal CT99. Either CT83 or CT99 is a substrate for the γ-secretase, which hydrolyzes a peptide bond within the membrane-spanning portion at either residue 710 or residue 712. The γ-cleavage of CT99 (after β-cleavage) produces Aβ and CT57–59 (amyloid intracellular domain, AICD) while γ-cleavage of CT83 (after α-cleavage) leaves the CT57–59 fragment as well as the 10 kDa peptide called p3. Formation of Aβ40 or Aβ42 depends on the β- and γ-secretases acting in sequence. Cleavage within the Aβ sequence by α-secretase prevents formation of Aβ. Most of the Aβ is of 40 residues and a small proportion is of 42, although Aβ42 is more prone than Aβ40 to fibrillation and Aβ42 predominates in amyloid core plaques (Kinoshita et al., 2002).

3. Trafficking and Axonal Transport of APP

Amyloid precursor protein, as it is synthesized, is transported through the endoplasmic reticulum (ER) to the Golgi and trans-Golgi network (termed "secretory pathway") to be inserted in the plasmalemma. Amyloid precursor protein is also reinternalized and recycled from the plasmalemma through the endosomal-lysosomal (endocytic or recycling) pathway. In addition, some clathrin-coated vesicles bud off the trans-Golgi to enter the endosomal-lysosomal path directly. Identifying the precise locations of the secretase activities and the compartment(s) in which Aβ is formed are the subjects of ongoing research.

The activity of β-secretase is demonstrated in the secretory path, and the bulk of β-secretase cleavage, as studied in a transgenic mouse model (see later), takes place in the secretory path with very little β-cleavage through the endocytotic path (Steinhilb et al., 2002), while acid-dependent γ-secretase activity together with the mature presenilin1 (PS1)-nicastrin complex required for y-secre tase activity colocalizes with APP in the lysosomal outer membranes (Pasternak et al., 2003). The ER is not a major site for γ-secretase cleavage of the C99 terminal of APP, even though the PS1 protein may be localized to the ER (Maltese et al., 2001). Cleavage activity of γ-secretase depends on maturation of the enzyme as it acquires bound nicastrin during its passage through the ER. Thus, the β-cleavage may be in the ER and the γ-cleavage in the endosomal-lysosomal path. It is thought that excess quantities of β-cleaved APP in the trans-Golgi network are exposed to γ-cleavage in the endosomal-lysosomal system.

There is evidence for Aβ accumulation both in the trans-Golgi and in the endosomal-lysosomal compartments in cultured cells (Koo and Squazzo, 1994; Wild-Bode et al., 1997; Xu et al., 2000). It is possible that Aβ found in the secretory path may be due to the activity of a neutral pH-dependent γ-secretase activity demonstrated in the ER and not related to presenilins (Wilson et al.,

2002). The early endosomal compartment with its associated proteases was found to be greatly enlarged in pyramidal neurons in prefrontal cortices of postmortem brain from sporadic AD subjects as compared to brain from age-matched subjects who had no brain disease or who had other neurodegenerative diseases. This implies greatly increased endosomal activity in the AD brain, which may be related to Aβ accumulation in sporadic AD (Cataldo et al., 1997).

Certain critical issues in the biology of APP and pathogenesis of AD concern the transport of APP, the secretases and Aβ and the compartments they traverse. There is evidence for transport of PS1 and BACE1 from entorhinal cortex to hippocampus in transgenic mice based on removal of the perforant pathway afferents which results in about 50% reductions in these enzymes in the dentate gyrus (Sheng et al., 2003). In mutant PS1 transgenic mice, densities of synaptophysin-containing vesicles and mitochondria were reduced in neuritic processes of hippocampal neurons while APP, PS1 and synaptophysin levels were found reduced in sciatic nerves, consistent with a reduction in fast transport, in the transgenic mice. Alterations in axoplasmic transport obviously may compromise neuronal structure and function in AD in many ways (Pigino et al., 2003).

Functions of APP

a. Physiologic Effects of Secreted APP

The role of APP in the normal physiology of the brain or other tissues is not fully understood. Current proteomic technology has identified 21 proteins co-immunoprecipitated with APP from AD brain lysates All the identified proteins could be classified as those related to axonal transport, vesicle formation and transport, cytoskeletal structure, chaperone and adaptor functions, implying that APP could also be related to any of these structural or transport functions (Cottrell et al., 2005). The structure of APP and its recycling through endosomes have features of a cell surface receptor (Kang, 1987) and its association with caveolae-like vesicles enriched also with glypiated protein suggest a function in transmembrane signaling for GPI-linked glycoproteins (Bouillot et al., 1996). Amyloid precursor protein may function as a platelet-released inhibitor of coagulation factor XI and as an anticoagulant (Smith et al., 1990; Schmaier et al., 1993). Long isoforms of APP contain a 56-residue sequence inserted at residue 289 in the extracellular segment that is homologous to the Kunitz family of serine protease inhibitors, for example, aprotonin, bovine pancreatic trypsin inhibitor, and the human plasma protein inter-alpha-trypsin inhibitor (Ponte et al., 1988; Tanzi et al., 1988; Kitaguchi et al., 1988). Protease nexin-II (PN-2) (Seguchi et al., 1999) which inhibits proteolysis and coagulation factor XIa, was found identical to KPI-containing APP. This protein is contained in platelet granules and is secreted upon platelet activation (Van Nostrand et al., 1989; Oltersdorf et al., 1989). Thus, the long isoforms of secreted APP may function both as extracellular protease inhibitors and as platelet regulators of coagulation. Secreted forms of APP may derive from any of the isoforms with or without the inserted KPI motif. In humans, both the soluble KPI-containing and the KPI-free derivatives are readily detected in CSF from both AD cases and controls (Palmert et al., 1989). Binding of fibrillar Aβ to PN-2/APP enhances fivefold the inhibition of coagulation factor XIa, and PN-2/APP along with Aβ are elevated in cerebral vessels in hereditary cerebral hemorrhage with amyloidosis-Dutch type (HCHWA-D) (Table 10-2); these conditions might favor hemorrhage (Wagner et al., 2000). There are at least two amyloid precursorlike proteins, identified as APLP1 and APLP2, that have properties of proteinase inhibition and coagulation factor inhibition similar to but less potent than those of PN-2/APP. These proteins probably form a family of proteinase inhibitors and may compensate for each other to some extent (Van Nostrand et al., 1990, 1994).

An intracellular C-terminal portion of 57 or 59 residues, termed the amyloid intracellular domain (AICD), is produced by γ-cleavage of the APP molecule (Figure 10-3). The amyloid intracellular domain has been shown to translocate to the nucleus after interaction with an adapter protein, Fe65, in the cytoplasm. This complex is transcriptionally active. The amyloid intracellular domain may function in regulating phosphoinositide-mediated Ca^{2+} signaling through modulation of ER Ca^{2+} stores (O'Neill et al., 2001). However, some reports suggest possible toxic effects of C-terminal fragments (Suh and Checler, 2002).

Amyloid precursor protein knockout mice have shown reduced local microglial activation and improved neuronal survival after transection injury to the medial forebrain bundle (DeGiorgio et al., 2002; Harper et al., 1998).

However, APP null mice also exhibit subtle locomotor dysfunction and forelimb weakness (Heber, 2000). Blocking APP synthesis *in vitro* decreases adhesiveness of neurons on collagen and laminin suggesting APP is involved in cell membrane interactions with extracellular matrix (Coulson, 1997). APP exhibits both neuroprotective effects at lower levels and toxic effects at higher levels, depending also on the isoforms expressed (Mucke, 1996). APLP2-null mice appear healthy until age 22 months. However, of mice made null for both APLP2 and APP, 80% die before the first week, indicating that APP or APLP2 is required for early development and that these proteins can substitute for each other functionally to some extent (von Koch et al., 1997).

b. Apolipoprotein E Polymorphism and APP Processing

Apolipoproteins (ApoE) are lipid carrier molecules involved in the transport and distribution of cholesterol and lipids from liver to all extra-hepatic tissues via low-density lipoprotein (LDL)-receptor-mediated mechanisms (Brown and Goldstein, 1986). In brain, ApoE is not involved in the transport of cholesterol to and from the brain but instead is involved in the redistribution of cholesterol to different subcellular compartments and membranes of neurons. This redistribution proceeds together with membrane remodeling during regeneration of neurites, axons, and synapses (Mahley, 1988; Handelmann et al., 1992; Poirier et al., 1993; Nathan et al., 1994; Masliah, 1995). Apolipoprotein E also is known to be an Aβ-scavenging molecule that regulates extracellular Aβ concentration through ApoE-receptor internalization and the endosomal-lysosomal system (Poirier, 2000).

Apolipoprotein E is a 34 kDa protein (Mahley, 1988) encoded by a polymorphic gene located on chromosome 19 (Lin-Lee et al., 1985). The three different isoforms of ApoE (ApoE2, ApoE3, ApoE4) are respectively coded by three separate alleles (ε2, ε3, ε4) that are inherited in a codominant fashion at a single genetic locus (Zannis and Breslow, 1981). These isoforms differ in amino acids at positions 112 and 158 (Mahley, 1988). The most common isoform, ApoE3, has cysteine at position 112 and arginine at position 158; ApoE2 has cysteine at both positions, whereas ApoE4 lacks cysteine at both positions and contains arginine instead (Han et al., 2003). Therefore, ApoE4 cannot undergo intramolecular or intermolecular disulfide cross-linking (Selkoe and Lansbury, 1999).

The role of ApoE4 in AD was first discovered by Strittmatter and coworkers, who showed a correlation between a genetic linkage site on chromosome 19 of the individuals affected with AD (Strittmatter et al., 1993). The ApoE4 allele was found present in approximately 65% of the cases with late-onset familial AD and in 50% or more of the sporadic late-onset AD cases (Roses et al., 1995; Saunders, 2000). However, the precise mechanisms by which the increased proportion of the ε4 to other alleles of ApoE might lead to AD are not fully understood.

Physiological effects of ApoE4 are isoform specific and have been attributed to the lack of cysteine at positions 112 and 158. Since ApoE4 cannot undergo intramolecular or intermolecular disulfide cross-linking, it is thought that this particular isoform may fail to internalize extracellular Aβ to endosomes or lysosomes and hence may not clear extracellular Aβ efficiently (Selkoe and Lansbury, 1999). Isoform-specific effects of ApoE in increasing Aβ levels and neuritic plaque formation in transgenic PDAPP mice (see Table 10-3) with different human ApoE isoforms are shown to be in accord with those relationships found in AD, ε4 being more pronounced than ε3, and ε3 more pronounced than ε2 (Hartman et al., 2001; Fagan et al., 2002). In contrast, neurotrophic and synaptotrophic effects of ε2 and ε3 are more pronounced than those of ε4 (Nathan et al., 1994).

c. Cholesterol and APP Processing

The brain is an organ rich in cholesterol (Dietschy and Turley, 2001), and most of it is derived by neosynthesis in the brain itself. Since the brain is protected by the blood-brain-barrier (BBB), it does not compete with circulating lipoproteins for cholesterol to the extent that peripheral tissues do (Simons et al., 2001). The brain apolipoproteins are not involved in the transport of cholesterol to and from the brain but in the redistribution of cholesterol within the brain during axonal-synaptic remodeling (Mahley,

Table 10-3. Widely Used Transgenic Models of Alzheimer's Disease and Tauopathy

Transgenic Models of AD	Transgene	Neurodegenerative Changes	Behavioral Deficits
APP Transgenic Mice			
APP Indiana PDAPP	Human APP695 cDNA with V717F	Aβ plaques; synaptic loss; astrogliosis; microgliosis	Some behavioral impairment (Games et al., 1995)
APP Swedish Tg2576	Human APP695 cDNA with K670M/N671L	Numerous Aβ plaques; synaptic loss; astrogliosis; microgliosis; lipid and oxidative damage	Spatial reference memory impairment (Hsiao et al., 1996)
APP Swedish APP23	Human APP751 cDNA with K670M/N671L	Numerous Aβ plaques; Aβ in CSF; cerebral amyloid angiopathy; synaptic loss; astrogliosis; microgliosis	Impairment in water maze performance and passive avoidance test (Sturchler-Pierrat, 1997; Boncristiano et al., 2002)
APP London	Human APP751 cDNA with V717I	Cerebral plaques; cerebral amyloid angiopathy; cholinergic fiber degeneration	Decreased exploration; increased neophobia; increased male aggressive behavior (Moechars et al., 1999)
APP Swedish+Indiana TgCRND8	Human APP751 cDNA with both K670M/N671L and V717F	Early onset of neuritic pathology and dense cored plaques at 3 to 5 months	Impairment in acquisition and learning and water maze performance (Chishti et al., 2001)
Presenilin Transgenic Mice			
PS1 M146L	Human PS1 cDNA with M146L	No neuropathology evident up to 2.5 years of age	No behavioral abnormalities (Duff et al., 1996)
PS1 M146V	Human PS1 cDNA with M146V	No neuropathology evident up to 2.5 years of age	No behavioral abnormalities (Duff et al., 1996)
PS1 A246E	Human PS1 cDNA with A246E	No neuropathology evident up to 2 years of age; followed by neuronal apoptosis and seizures	No behavioral abnormalities (Schneider et al., 2001)
PS2 N141I	Human PS2 cDNA with N141I	Age-dependent increase in Aβ42; higher levels of insoluble Aβ	No behavioral abnormalities (Oyama et al., 1998)
ApoE Transgenic Mice			
ApoE4	Human ApoE gene (ε4 allele)	CNS axonopathy; hyperphosphorylation of tau	Muscular degeneration; motor deficits (Umans et al., 1999)
Tau Transgenic Mice			
Tau P301L-JNPL3	Longest human tau cDNA with 4 repeats P301L	Fibrillary gliosis; axonal degeneration; neuronal lesions similar to FTDP-17	Severe motor and behavioral disturbances (Lewis et al., 2000)
Tau R406W	Longest human tau cDNA with R406W	Accumulation of insoluble tau and congophilic hyperphosphorylated tau inclusions in aged brain	Impaired associative memory; impaired contextual fear conditioning (Tatebayashi et al., 2002)
Tau V337M	Longest human tau cDNA with V337M	Hippocampal neuronal degeneration and deposition of PHF tau filaments	Impaired spontaneous locomotion; no changes in water maze performance (Tanemura et al., 2002)

Table 10-3. *Continued*

Transgenic Models of AD	Transgene	Neurodegenerative Changes	Behavioral Deficits
Multiple Cross Transgenic Mice			
APP+PS1	Human APP695 K670M/N671L and PS1 A246E	Numerous cerebral Aβ deposits at an early age; astrogliosis; microgliosis	Spatial reference memory impairment early on (Gordon et al., 2002; Borchelt et al., 1997)
APP+3Repeat tau	Human APP695 K670M/N671L and PS1 M146L	Somatodendritic accumulation of transgenic tau; increased tau phosphorylation but no evidence of tangles	No behavioral deficits observed (Boutajangout et al., 2002)
APP+4Repeat tau TAPP	Human APP695 K670M/N671L and tau P301L	Cerebral Aβ deposits; neurofibrillary tangle pathology in limbic system and olfactory cortex	Motor disturbances including hind-limb weakness; decreased grooming (Lewis et al., 2001)
APP+PS1+tau	Human APP695 K670M/N671L, PS1 M146V, P301L	Age-dependent progressive development of plaques and tangles; synaptic degeneration	Early LTP deficits (Oddo et al., 2003)

1988). Removal of excess brain cholesterol occurs mainly via conversion of cholesterol to 24-hydroxycholesterol, which can pass the BBB (Bjorkhem et al., 1998). Cholesterol levels in specific compartments of cells are tightly regulated since membrane cholesterol affects membrane functions by regulating physico-chemical properties of the cell membrane (Simons et al., 2001). All membrane-associated proteins including APP are likely to be affected by the lipid composition of the membrane to which they are anchored. Cholesterol may alter the activity of enzymes or receptor proteins embedded within the membrane by modulating membrane fluidity (Racchi et al., 1997), or by a mechanism independent of its membrane-ordering effect (Schroeder et al., 1991).

The significance of cholesterol transport to amyloidogenesis is proved clearly in the inherited Neimann-Pick type C diseases (NPC1 and NPC2). These diseases are due to mutations in one of two proteins required for cholesterol and/or fatty acid homeostasis and are associated with cholesterol accumulation in the late endosomal compartment (Davies et al., 2000; Naureckiene et al., 2000). The brain contains depositions of amyloid plaques and NFTs of abnormally phosphorylated tau as in AD (Sawamura et al., 2001). Insoluble Aβ accumulates together with cholesterol in the late endosomes in NPC mutant cells (Yamazaki et al., 2000). Pharmacologic inhibition of cholesterol transport from the endosomal-lysosomal compartment to the ER leads to retention of cholesterol along with accumulation of PS1 and PS2 colocalized with Aβ42 in endosomal vesicular compartments as in cells deficient in the NPC1 cholesterol transport protein (Runz et al., 2002). It is of interest that the NC1 protein shares homology with hydroxymethylglutaryl CoA reductase, the enzyme inhibited by the statin drugs used to lower blood cholesterol.

Membrane colocalization of α-secretase and substrate APP are critical for the activity α-secretase. An α-secretase cleavage requires that APP be inserted into a membrane, and the site of APP cleavage is a fixed distance from the membrane rather than at a specific amino acid sequence (Maruyama et al., 1991; Sisodia, 1992). A change in the lipid ordering of the membrane may perturb the steric relationship of the enzyme to APP and may thereby inhibit α-cleavage. Impedance of membrane fluidity due to cholesterol loading also appears to inhibit the required contact between the enzyme and the target protein (Bodovitz and Klein, 1996). In addition to the effect of cholesterol on membrane lipid ordering, elevated levels of cholesteryl esters are associated with increased Aβ production. Inhibitors of acyl-coenzyme A: cholesterol acyltransferase, the ER enzyme that catalyzes synthesis of cholesteryl esters and is activated by elevated free cholesterol, reduce Aβ generation in neuronal and other cell lines(Puglielli et al., 2001).

Cholesterol-enriched lipid rafts are a major site for amyloidogenic processing of APP. Significant portions of Aβ along with presenilins are associated with detergent-insoluble-glycolipid (DIG)-enriched fractions from rat brain cortical gray, hippocampal neurons and cultured neurons (Lee et al., 1998; Morishima-Kawashima and Ihara, 1998). Copatching of APP and the β-site cleavage enzyme to plasma membrane and segregation away from nonraft markers dramatically increase Aβ production in a cholesterol- and endocytosis-dependent manner (Ehehalt et al., 2003). Reductions in total cellular cholesterol in various neuronal types inhibits Aβ production and in some cases increases sAPPα production (Simons et al., 1998, 2001; Kojro et al., 2001). Current data suggest two pools of APP and that APP within the raft clusters undergoes amyloidogenic cleavage while APP outside the rafts undergoes α-cleavage (Simons et al., 1998, 2001; Ehehalt et al., 2003).

However, either excess cholesterol or acute depletion of cholesterol may be harmful to the cell. Acute cholesterol depletion disrupts the clusters of SNAREs required for exocytosis (Lang et al., 2001); blocks the internalization of clathrin-coated endocytic vesicles (Rodal et al., 1999; Subtil et al., 1999); and disperses PIP(4,5)P2 from low-density membrane receptor domains, which inhibits hormone-stimulated PtdIns signaling (Pike and Miller, 1998). Therefore, acute depletion of cholesterol in brain by lipophilic statins may have adverse effects on brain function (for discussion of membrane dynamics and cholesterol, see Chauhan, 2003).

5. Mutations of APP

The known mutated genes or chromosomes causing or definitely predisposing to AD are listed in Table 10-2. Transgenic animal models useful in investigating mutations causing AD are listed in Table 10-3 (see also Wong et al., 2001). The mutations in APP all alter the proteolysis of APP in such a way as to augment production of Aβ peptides, particularly that of the more highly amyloidogenic Aβ42. Alzheimer's disease-causing mutations in APP near the β-site potentiate β-cleavage, leading to elevation of both Aβ40 and Aβ42 (Citron et al., 1992; Cai et al., 1993), while mutations near the γ-site specifically increase levels of Aβ42 (Suzuki et al., 1994; Suh and Checler, 2002).

Down syndrome, trisomy 21, is a chromosomal dysgenesis syndrome consisting of a variable constellation of abnormalities caused by triplication or translocation of chromosome 21. There result 3 copies of the APP gene and excessive or dysregulated expression of APP and production of Aβ peptides. Alzheimer's disease with characteristic deposition of abundant amyloid plaques and NFTs develops invariably in these individuals by the age of 40 years. Immunoreactivity for APP is found in postmortem cortex of fetuses, neonates, and infants among control and DS brains but disappears in childhood in both groups and reappears in adulthood only in the DS group. These data suggest that APP is a growth factor in early development and that there is an excessive expression of APP in adulthood in DS (Arai et al., 1997). Increased soluble Aβ42 is found in postmortem brains of DS as early as gestational age 21 weeks and intraneuronal immunostaining for Aβ is found in hippocampus and cortex of young (3 years) DS patients earlier than the formation of amyloid plaques or NFTs (Teller et al., 1996; Gyure et al., 2001; Mori et al., 2002). This is further evidence that toxicity leading to AD ensues first from protofibrillar Aβ, which then results in NFTs and amyloid plaques. DS and AD markers map to mouse chromosomes 16 and 17 (Cheng et al., 1988).

E. α-Secretase

This enzymatic activity entails properties of certain cell surface membrane-bound metalloproteases of the ADAM (a disintegrin and metalloprotease) family. Two candidates are tumor necrosis factor-α (TNF-α)-converting enzyme (TACE or ADAM-17) and ADAM-10. These enzymes are also implicated in the Notch signaling pathway. Cleavage of APP on the C-terminal side of residue 16 of the Aβ peptide (between Lys16 and Leu17 within the Aβ domain) generates an 83-residue C-terminal fragment (CT83) and the large secreted ectodomain sAPPα (Figure 10-2). Thus, α-secretase cleaves APP within the Aβ sequence and precludes amyloidogenesis or excessive formation of Aβ peptides, plaque formation, and the array of effects subsequent to excessive Aβ formation.

α-Cleavage activity is found at the cell surface and in the trans-Golgi network. α-Secretase appears to have an unregulated (basal or constitutive) portion and a portion regulated by protein kinase C. The fraction of cellular α-secretase activity regulated by protein kinase C (PKC) has also been attributed to the mentioned ADAM family members (Buxbaum et al., 1998). There is evidence for α-cleavage of APP at the cell surface (Ikezu et al., 1998); this activity has also been identified as ADAM-10 (Kojro et al., 2001). While sAPPα is produced at the plasmalemma, there is evidence that the portion of cellular α-secretase APP cleavage catalyzed by TACE/ADAM-10 *and* regulated by PKC occurs in the trans-Golgi membranes, a site at which it is thought the bulk of β-secretase activity (see below) also occurs. It was observed that PKC activation increases sAPPα and decreases sAPPβ by altering the relative rates of α- and β-cleavages within the same organelle. Since it is the case that once the α-cleavage has occurred the formation of Aβ is precluded, PKC-regulated α-secretase activity in the trans-Golgi network may be a strategic site for pharmacologic intervention (Skovronsky et al., 2000; Levites et al., 2003).

A small portion, about 1% to 5%, of total neuronal APP is associated with caveolae, which are cholesterol-enriched (not lipid rafts or clathrin-coated pits) plasma membrane invaginations that concentrate signaling elements, including multiple tyrosine kinase receptors. The caveolae fraction of neuronal plasma membranes comprises 5% to 10% of the plasma membrane protein and is thought to be organized for signal transduction (Bouillot et al., 1996; Wu et al., 1997). APP in the caveolae is bound by its cytoplasmic domain to caveolin-1, a cholesterol-binding protein, and α-secretase processing is promoted by recombinant overexpression of caveolin-1 in intact cells, resulting in increased secretion of the soluble extracellular domain of sAPPα (Ikezu et al., 1998). The potential significance of this pool to total APP processing is not clear (Hailstones et al., 1998; Parkin et al., 1999).

F. β-Secretase

Vassar et al. (1999) cloned a human transmembrane aspartyl protease that had all the known characteristics of the β-secretase, and this enzyme was designated BACE, for "beta-site APP-cleaving enzyme." The enzyme shares significant sequence similarity with members of the pepsin subfamily of aspartyl proteases. The structure of BACE indicates that it is a type 1 transmembrane protein with the active site on the lumenal side of the membrane (with respect to the ER), where β-secretase cleaves APP.

By in situ hybridization, expression of BACE mRNA in rat brain is observed at higher levels in neurons than in glia, supporting the idea that neurons are the primary source of the extracellular Aβ deposited in amyloid plaques. BACE in neurons is localized by immunocytochemistry to the Golgi and endosomes. It traverses a secretory pathway to the plasmalemma where it is internalized to endosomes. BACE is targeted by its C-terminus dileucine signal to the late endosomal/lysosomal compartments to be degraded by the lysosomal pathway (Young HK et al., 2005). BACE overexpression in cultured cells induces cleavage only at the known β-secretase positions, Asp1 and Glu11, and decreases α-cleavage. A second aspartyl protease capable of cleaving APP at the β-site was designated BACE2 or ASP2 (Yan et al., 1999; Sinha et al., 1999; Hussain et al., 1999). BACE2 is localized to glia. While both enzymes might be involved in producing Aβ, Cai et al. (2001) established that BACE1 is the principal neuronal protease required to cleave APP at the +1 and +11 sites that generate N termini of Aβ.

Other specific pathophysiologic manifestations of BACE1 activity or deficiency are not known. Both BACE enzymes may have substrates other than APP. In experimental knockout mice deficient in BACE1, brains lacked Aβ and β-secretase-cleaved APP C-terminal fragments. While earlier studies did not discern phenotypic abnormalities (Luo et al., 2001; Roberds et al.,

2001), more recent work did reveal evidence of morbidity in BACE1 null mice, including postnatal mortality and retarded growth, hyperactivity and abnormal sodium-gated channels in surviving mice; BACE2 null animals on the other hand appeared normal (Dominquez et al., 2005). No evidence for either genetic linkage or allelic association between BACE and AD nor for coding sequence mutations in the open reading frame of the BACE gene was found (Nicolaou et al., 2001). This is in contrast to the situation with γ-secretase, mutations of which are associated with familial AD (see below).

G. γ-Secretase

1. Localization of γ-Secretase Activity

Immunochemical analyses have localized PS1 and PS2 to similar intracellular compartments (ER and Golgi complex) and perinuclear membranes. It has also been observed that PS1 immunoreactivity in the plasma membrane is concentrated in regions of cell-cell contact, suggesting that PS1 may be a cell adhesion molecule (Takashima et al., 1996). However, localization in the ER and Golgi membranes may be related to sites of synthesis of these proteins, particularly in overexpressing cell models, rather than to sites of their enzymatic activity.

Wilson et al. (2002) analyzed the production of several forms of secreted and intracellular Aβ in mouse cells lacking PS1, PS2, or both proteins. Although most Aβ species were abolished in PS1/PS2 –/– cells, the production of intracellular Aβ42 generated in the ER/intermediate compartment was unaffected by the absence of these proteins, either singly or in combination. It was concluded that another, as yet unknown, γ-secretase activity must be responsible for cleavage of APP within the early secretory compartments.

2. γ-Secretase Membrane Protein Complex

γ-Secretase is considered the founding member of a class of intramembrane-cleaving aspartyl proteases that hydrolyze a peptide bond within the hydrophobic environment of a lipid bilayer (Kopan and Goate, 2000; Wolfe and Selkoe, 2002). Besides its role in the production of Aβs, γ-secretase catalyzes the proteolytic release of the intracellular domains of Notch, βAPP, and other transmembrane receptors, such as sterol regulatory element-binding proteins (SREBPs) and an epidermal growth factor tyrosine kinase receptor for neuregulins (ErbB-4) (Lee et al., 2002).

PS1 and PS2 share 67% amino acid homology, but it is not clear whether they have different substrates since they do reside in different complexes (Kopan and Goate, 2000). While presenilin is the active protease site, its activity is tightly regulated by its structure within the membrane complex. In studies of PS1, Kimberly and colleagues (2000) have found that this complex is composed of 4 identified proteins that copurify, coprecipitate, and together are sufficient to produce the active enzyme. The complex consists of PS1, nicastrin, Aph-1, and Pen-2, presenting a total molecular mass of 200 to 250 kDa. Thus, the γ-secretase may be more accurately termed the γ-secretase protein complex (GSPC).

The mechanism of the γ-secretase activity is being intensively investigated. The γ-secretase is one of a group of intramembrane-cleaving proteases (I-CLiPs) which "must create a microenvironment for water and the hydrophilic residues for catalysis within the membrane, then bend or unwind their substrates to make the amide bonds susceptible to hydrolysis" (Wolfe and Kopan, 2004). The biological activity of PS1 involves its autoproteolytic cleavage into its N-terminal and C-terminal fragments that remain associated in the complex as a heterodimer. The active site is thought to be at the interface of the two segments. PS1 contains two conserved aspartate residues (Asp257, Asp385) that are situated in its transmembrane (TM) loops 6 and 7, respectively (Kimberly et al., 2003). Site-directed mutation of either of these residues to alanine in PS1 substantially reduces Aβ production and increases the amounts of the carboxy-terminal fragments of APP that are the substrates of γ-secretase, as measured in cell culture and cell-free microsomes (De Jonghe et al., 2001). Either of the Asp-to-Ala mutations also prevents the normal autoproteolysis of PS1 in the TM6 to TM7 cytoplasmic loop to produce the N- and C-terminal segments (Wolfe et al., 1999). The current data indicate that Pen-2 preferentially binds to PS1 holoprotein and is required for autoproteolyis of PS1, while this function is regulated by Aph-1 (Luo et al., 2003).

The level of heterodimerization of the two fragments of PS1 in the membrane is tightly regulated by other limiting factors, which are not entirely understood. The combination of the 4 proteins in the GSPC has been found sufficient to modify the regulation of the PS heterodimer levels and to thus augment γ-secretase activity in mammalian cells. Coexpression of all 4 proteins "unleashes γ-secretase" and potentiates product formation. The model is one in which nicastrin, Aph-1, and Pen-2 interact with PS to permit its autoproteolysis, with the active site of the mature protease located at the cytoplasmic interface between the two PS TM segments (Kimberly et al., 2003).

As mentioned above, γ-secretase activity in the early secretory pathway does not depend on PS. In addition, while PS itself may be demonstrated in the early ER and Golgi membranes, the γ-secretase substrates are localized in late secretory compartments and plasmalemma. This apparent paradox is resolved by finding that maturation and/or trafficking of nicastrin through the Golgi and formation of the nicastrin-PS complex are associated with PS-dependent γ-secretase activity in the late secretory compartment (Siman and Velji, 2003; Herreman et al., 2003; Arawaka et al., 2002). Fluorescence microscopy methods show that the strongest interactions between APP and PS1 are at the cell surface (Berezovska et al., 2003).

3. Presenilin Mutations

As outlined earlier, accumulation of Aβ protein in the cerebral cortex is an early and invariant event in the pathogenesis of AD. The final step in the generation of Aβ from APP is an apparently intramembranous proteolysis of APP by γ-secretase(s). The most common cause of autosomal dominant familial AD is mutation of the genes encoding presenilins 1 and 2 on chromosomes 14 and 1, respectively (Table 10-2). These mutations all alter γ-secretase activity to increase the production of the highly amyloidogenic Aβ42 isoform. About 100 missense mutations and 1 exon deletion of PS1 have been found, some of them causing familial AD before the age of 30 years. About 6 mutations on PS2 gene have been identified. Most mutations on PS1 occur at amino acids conserved in the two proteins and are clustered within their transmembrane domains (Citron et al., 1997; Russo et al., 2000).

The PS1 gene product facilitates function of the LIN12/Notch proteins involved in cell-cell recognition, signal transduction between plasmalemma and nucleus, and cell development during embryogenesis. Absence of the PS1 gene is lethal, and PS1-null embryos display abnormal somite development, neuronal deficits, and growth retardation. The particular mutations in PS1 that cause familial AD apparently do not involve functions of PS1 that are manifested during embryogenesis, since PS1-null mice can be rescued from embryonic lethality by either wild-type or AD-linked PS1 mutated genes (see Suh and Checler, 2002; Tandon and Fraser, 2002).

Almost all the pathogenic mutations of presenilins associated with familial AD affect the processing of APP such that γ-cleavage is augmented, leading to increased amounts of Aβ40 and more so of Aβ42 at the expense of physiologically secreted soluble APP products. The augmentation in catalytic activity caused by the mutation is called a "toxic gain in function."

However, deficiency and mutations in the presenilins may lead to manifestations of dysregulated metabolism, in addition to excessive production of Aβ, that may be operative in AD pathogenesis. PS1 and PS2 are known to positively regulate activation of the phosphatidylinositol 3-kinase (PI3)/Akt and the MEK/ERK pathways while these pathways are suppressed by AD-linked PS1 mutations and PS1 or PS2 deficiencies *in vitro*. The PI3/Akt pathway phosphorylates glycogen synthase kinase-3 (GSK3β), thereby inactivating the latter, and reduces tau phosphorylation (see later) whereas PS1 or PS2 deficiency increases tau phosphorylation. The *in vitro* data implicate platelet-derived growth factor receptors which are at the membrane surface, thus upstream of the signaling pathways (Kang et al., 2005).

Large increases in levels of the Aβ peptides have been measured in brain, plasma, or secreted medium of fibroblasts from patients with PS-related familial AD (Scheuner et al., 1996). The increased γ-cleavage activity and overproduction of Aβ peptides has been demonstrated in all biological experimental models in which the mutated PS gene is overexpressed, including transfected cells and transgenic mice (Borchelt et al., 1996; Duff et al., 1996; Harrison and Beher, 2003). Transfection with wild-type PS genes does not alter levels of APP, Aβ peptides, or α- and β-secretase activities.

H. Learning and Memory: Effects of Amyloid and Mutations of APP and PS1

Many studies have shown that transgenic mice overexpressing mutated human genes for APP or presenilins linked to inherited AD gradually develop deficits in learning and memory functions as they grow older and the degree of deficit is correlated with accumulation of amyloid. The beginning of memory deficits can, however, be discerned at young ages before the accumulation of plaques. Decreased Ca-dependent evoked potentials in hippocampal slices from such animals can also be seen at early ages as can depletion of presynaptic SNAP25, a protein required for vesicle exocytsis (Chauhan and Siegel, 2002). The early changes are thought to be due to pre-fibrillar or diffusible Aβ.

Using 3 groups of transgenic mice (see Table 10-3) carrying the presenilin A246E mutation, the amyloid precursor protein K670N/M671L mutation, or both mutations, Dineley et al. (2002) showed that coexpression of both mutant transgenes resulted in accelerated Aβ accumulation, first detected at 7 months in the cortex and hippocampus, compared to the APP or PS1 transgene alone. On the other hand, contextual fear learning, a hippocampus-dependent associative learning task, but not cued fear learning, was impaired in mice carrying both mutations or the APP mutation, but not the PS1 mutation alone. The impairment was manifested at 5 months of age, preceding detectable plaque deposition, and increased with age. While Aβ and/or the mutated proteins exert many neurotoxic effects that may disrupt cellular regulatory pathways, it is likely that the ultimate effect related to learning is on gene regulation. A number of immediate early genes (Arc, Nur77, Zif268) and plasticity-reelated genes (GluR1, CaMKIIα, Na,K-ATPase α3 isoform) critical for learning and memory were found normal in young mice harboring either or both of the mutant APP and PS1 genes. But the expression of these genes decreased as the mice grew older and accumulated amyloid plaques. The gene repression was correlated with amyloid burden (Dickey et al., 2004).

I. Tau Biology

1. Pathology

One of the two principal hallmarks for the pathologic diagnosis of AD is the presence of intraneuronal neurofibrillary tangles. The NFTs consist of bundles of silver-staining intraneuronal cytoplasmic fibers that are paired, helically wound filaments (PHF) of about 10 nm often mixed with 15-nm straight filaments. The NFTs are frequently present in the perikarya, whose axons project to sites of amyloid/neuritic plaques. The neuronal networks earliest and most involved are the entorhinal-hippocampal perforant and the basal forebrain (nucleus basalis of Meynert)-hippocampal-neocortical pathways (Terry et al., 1999; Selkoe and Lansbury, 1999; Brion et al., 2001).

The NFTs are also found in different distributions in the brain in the absence of amyloid plaques in the group of diseases called tauopathies (Table 10-1). It is now known that the NFTs in all cases are composed of abnormally phosphorylated twisted filaments of tau, a low-molecular-weight microtubule-associated protein (MAP; Vogelsberg-Ragaglia et al., 1999). However, the particular tau isoforms and their sites of phosphorylation as found in the NFTs are characteristic for the particular class of diseases (Table 10-1).

Small numbers of amyloid plaques and NFTs are found in postmortem brains and olfactory bulbs of nondemented aged individuals, and there is controversy as to which comes first and whether the density of NFTs or plaques can be correlated with cognitive scores or onset and progression of dementia (Kovacs et al., 1999). A problem with such attempted correlations is that the triggering or toxic effects of Aβ on neuronal biology in general or on tau in particular may well be evident prior to the formation of anatomically distinct aggregates or plaques. The presence of NFTs and plaques in postmortem brains of nondemented aged persons leads to the idea that these formations to a small extent are by-products of physiologic processes, while the pathogenesis of AD is the result of an exaggeration and dysregulation of these processes. Postmortem AD brain almost always contains abundant amyloid plaques and neurofibrillary tangles (NFTs), while brains from tauopathy subjects contain only NFTs (Table 10-1). In transgenic mice harboring mutated human tau genes, the injection of exogenous Aβ fibrils into the brain (Gotz et al., 2001) or the cotransgenic expression of mutated human APP (Lewis et al., 2001; Perez et al., 2005) exaggerates the development of NFTs, whereas the overexpression of the mutated tau gene does not increase the Aβ formation. These data taken together indicate that in AD, Aβ is the initial trigger. Yet it ought to be kept in mind that these mutations also have effects on signaling pathways other than production of Aβ.

2. Tau Isoforms and Phosphorylation in AD

Tau proteins in adult brain comprise 6 isoforms translated from a single gene on chromosome 17. The isoforms have either 3 or 4 repeats of microtubule-binding domains that bind to tubulin monomers, promoting their assembly into microtubules that form the neuronal microtubule network. Three of the isoforms have 3 repeats (3-R), and 3 have 4 repeats (4-R). The assembly and disassembly of microtubules are involved in the dynamics of neuronal function and structure, and their regulation must also be sensitive to rapid and finely tuned changes.

Microtubules maintain the cell shape and are the tracks for axonal transport. Tau proteins also link microtubules to other cytoskeletal proteins. Thus, interference with tubulin will disturb the cytoskeletal structure and axonal transport. Tau isoforms are mainly expressed in neurons but may function in glia as well (Ferrer et al., 2003a, 2003b). The function of the tau isoforms is regulated by their phosphorylation at multiple sites by kinases and dephosphorylation by phosphatases. When the isoforms are abnormally excessively phosphorylated, they cannot regulate the assembly of microtubules for axonal transport nor maintain linkages to other cytoskeletal proteins, thus leading to neuronal dystrophy. The abnormally phosphorylated tau isoforms themselves undergo conformational changes, aggregation, and formation of PHFs forming the intracellular NFTs. The tau in normal adult brain biopsied tissue is phosphorylated at 12 sites, while the tau found in PHFs is phosphorylated at an additional 9 sites.

The isoforms that are hyperphosphorylated in neurons and oligodendroglia are disease specific. In AD, all 6 isoforms are hyperphosphorylated and aggregated into the NFTs, whereas in corticobasal degeneration and progressive supranuclear palsy only the 4-R, and in Pick's disease only the 3-R, isoforms are hyperphosphorylated and aggregated. The tau isoforms in astrocytes accumulated similarly in all (Arai et al., 2001).

a. Glycogen Synthase Kinase 3β (GSK3β)

Glycogen synthase kinase-3β (GSK3β) is crtical to regulating many cellular functions (Hardt and Sadoshima, 2002). It was originally studied in conjunction with its regulation of glycogen synthesis, but it is now known to phosphorylate tau as a major substrate, and it has been termed GSK/tau kinase. In this chapter, the designation GSK3β will be retained. The major substrates for GSK3β phosphorylation identified in brain are tau, β-catenin, and kinesin light chain (KLC), related to axonal transport. In squid axoplasm, GSK3β phosphorylates KLC and disrupts fast axoplasmic transport (see later) (Morfini et al., 2002). There is also evidence for GSK phosphorylation of the cytoplasmic portion of APP Thr743 *in vitro* (Aplin et al., 1996). Of potential significance to the axonal pathology in AD is the recent demonstration that active GSK3β also phosphorylates, thereby inhibiting, collapsing response mediator protein (CRMP-2) which is a microtubule assembly regulatory protein promoting axon formation and important in neuronal polarization (Yoshimura et al., 2005). This provides another potential mechanism for dysregulated GSK kinase activity to produce neuronal pathology in AD (Doble and Woodgett, 2003).

In its active dephosphorylated form, GSK3β phosphorylates tau, which inhibits the ability of tau to promote microtubule self-assembly and to organize microtubules into ordered arrays required for normal dynamic structural functions of neuronal processes (Utton et al., 1997; Wagner et al., 1996; Ishii et al., 2003). It should be kept in mind that tau may be phosphorylated under various conditions by several other kinases, including Cdk5, ERK-1,2, MAPK, p38K, and c-Jun-N-terminal kinase (JNK), but that phosphorylation of tau by GSK is most extensive (Anderton et al., 2001; Zhu et al., 2002).

GSK3β is an unusual protein-serine kinase that is active in resting cells and is inhibited by being phosphorylated in response to activation of several distinct pathways, including those acting by elevation of 3′ phosphorylated phosphatidylinositol lipids and adenosine 3′-5′-monophosphate (cAMP), PI3K/Akt (PKB) and MEK/ERK (Kang et al., 2005; Aoukaty and Tan, 2005). Dephosphorylation by protein phosphatase, of which there are several classes, restores its kinase activity. Protein phosphatases are themselves subject to inhibitory regulation by several inhibitor proteins which are in turn positively or negatively regulated by several kinases including GSK itself (Bibb and Nestler, 2006). GSK net kinase activity is therefore the resultant of an intricate interplay of other kinases, phosphatases and inhibitor proteins (Iqbal et al., 2002). GSK3β is a substrate for PI3-kinase and protein kinase B (Akt) which have roles in survival of diverse cell types. PS1 or PS2 deficiency or mutation suppresses activation of both PI3K/Akt and MEK/ERK signaling pathways either of which suppression results in tau hyperphosphorylation (see above). Insulin and growth factors also activate Akt/PKB which inhibits GSK3β by phosphorylation. In addition, GSK3β is distinctly regulated by, and is a core component of, the Wnt pathway (see later). GSK can be regulated also by the non-kinase effectors, axin, GSK-binding protein and disheveled-1.

One of the first sets of biochemical links between GSK3α or GSK3β and abnormal tau from AD brain was discovered by Hanger and colleagues (1992). They found that phosphorylation of tau by GSK3 altered the mobility of and prevented binding of a specific monoclonal antibody to human tau and that recombinant tau phosphorylated by GSK3 aligned on SDS-PAGE with the abnormally phosphorylated tau (PHF-tau) associated with the paired helical filaments in AD brain. They localized GSK3 within granular structures in pyramidal cells and suggested that GSK3α and GSK3β may have a role in the production of PHF-tau in AD.

In postmortem AD brain, active GSK3β was found distributed in neurons with tanglelike inclusions in the entorhinal cortex, and the distribution extended to other brain regions coincident with that of neurofibrillary changes. This finding is consistent with the possible role of abnormally active GSK3β in the formation of abnormally phosphorylated tau and the consequent NFTs and dystrophic neurons in AD brain (Pei et al., 1999). However, PKB/Akt, was also found increased in AD brain in association with increases in phosphorylated tau (Pei et al., 2003). GSK3β immunoreactivity was associated with cells showing granulovacuolar degeneration and with tau, although it was not enriched in tau fractions separated from AD brain (Leroy et al., 2002). Phosphorylated GSK was also found colocalized in neurons, NFTs, and glia with abnormal tau, suggesting partial inactivation of GSK in the regulation of the cascade and possibly to protect tau-containing cells from apoptosis (Ferrer et al., 2002).

In transgenic mice (2576) harboring the Swedish APP mutation linked to familial AD, it was found that feeding propentofylline (PPF) decreased levels both of tau phosphorylation and of active GSK3β and increased levels of inactive phosphorylated GSK3β (Chauhan and Siegel, 2005) while in another experiment under similar conditions PPF feeding resulted in the induction of NGF, reduction in amyloid burden and a shift from β- to α-processing of APP (Chauhan and Siegel, 2003). NGF is a major trophic factor for the cholinergic mesial basal forebrain-hippocampus circuitry principally involved in AD. These observations link the phosphorylation of tau with an increase in the ratio of active to inactive GSK3β and with increased formation of amyloid due to the APP mutation and further suggest that these effects of the mutated transgene may be reversed by inducing NGF in the cholinergic basal forebrain neuronal circuitry (Chauhan and Siegel, 2005).

In another transgenic mouse model (Tet/GSK3β) harboring an overexpressing GSK3β gene but without the mutated tau gene, the mice developed learning deficits, while the overexpression of GSK was found in hippocampal and cortical neurons in association with tau hyperphosphorylation, neuronal death, and reactive gliosis, but there was no tau filament formation (Hernandez et al., 2002). Thus, the structure of the tau may be critical to the outcome of the modified phosphorylation.

Studies with knockout mice deficient in Reelin or ApoE have shown increases in both GSK3β activity and tau phosphorylation (Ohkubo et al., 2003). This result may be significant to the role of certain ApoE isoforms in the protection versus pathogenesis of AD.

Lithium ion, which induces phosphorylation and hence inhibition of GSK3β, also inhibits GSK3β phosphorylation of tau (Lee et al., 2003). In cultures of various neuronal cell types, lysophosphatidic acid produces neurite retraction correlated with increased phosphorylation of tau mediated by activation of GSK3β (Sayas et al., 2002). In familial tauopathies, increased expression has been found of various kinases in neuronal populations with NFTs and hyperphosphorylated tau (Ferrer et al., 2003). The abnormally phosphorylated tau isoforms in the tauopathies are not the same as in AD (Table 10-1).

b. Cyclin-Dependent Kinase 5

Tau may be aberrantly glycosylated and abnormally phosphorylated by cyclin-dependent kinase 5 (CDk5) (Liu et al., 2002). In a study of transgenic mice overexpressing both the CDk5 activator p25 and the mutant (P301L) human tau, tau was hyperphosphorylated at several sites in the double transgenics, and there was a highly significant accumulation of aggregated tau in brainstem and cortex. This was accompanied by increased numbers of silver-stained NFTs. Insoluble tau was also associated with active GSK. Thus, CDk5 can initiate a major impact on tau pathology progression that probably involves several kinases (Bian et al., 2002; Noble et al., 2003).

Definitive data concerning potential effects of Cdk5 activity in relation to Aβ-induced toxicity were obtained in a study of fetal rat cortical neurons. It was found that a selective Cdk5 inhibitory peptide derived from p25 by calpain proteolysis specifically inhibits p25/Cdk5 kinase activity, while suppressing Aβ1-42-induced aberrant tau phosphorylation and Aβ1-42-induced neuronal apoptosis (Zheng et al., 2005). However, possible effects of Cdk5 on GSK or on phosphatases have not been excluded in vertebrate neurons, leaving it possible that Cdk acts indirectly via GSK or perhaps both kinases act together.

A model for concerted Cdk5 and GSK actions in tau hyperphosphorylation is suggested by findings with regard to fast anterograde axoplasmic transport (FAT) in squid axoplasm. In this case, activation of GSK3β or inhibition of Cdk5 reduces the rate of FAT in association with increased phosphorylation of kinesin light chain (KLC) which results in reduced kinesin-based motility. It was found that inhibition of Cdk5 leads to activation (or deinhibition) of protein phosphatase 1 (PP1) and activation of GSK3β with subsequent phosphorylation of KLC by the GSK3β. These authors suggest that the activated PP1 dephosphorylates thereby activating GSK and that a similar mechanism may be operative in AD both to impair axoplasmic transport and to hyperphosphorylate tau (Morfini et al., 2002).

One of the major questions pertaining to the role of Cdk5 or GSK in AD pathogenesis is which of multiple signaling systems predominates in their dysregulation in this disease. Cdk5 (Song et al., 2005; Sarker and Lee, 2004) and GSK (Kang et al., 2005; Aukati and Tan, 2005) both can be regulated by MEK/ERK and by PI3/Akt pathways. A difference which may be critical is that inhibition of ERK in human neuroblastoma (Song et al., 2005) or of PI3K/Akt in myoblasts (Sarker and Lee, 2004) blocks enhancement of Cdk5 activity. These signaling effects on Cdk5 activity if true for differentiated neurons would be expected to lead to reduced phosphorylation in the presence of presenilin mutation or deficiency which has been demonstrated to suppress both pathways while increasing tau phosphorylation *in vitro* (Kang et al., 2005). In contrast, GSK activity is increased by suppression of both PI3K/Akt and MEK/ERK pathways which would lead to increased tau phosphorylation, as indeed found.

To resolve this apparent paradox, one might hypothesize that suppression of PI3K/Akt and MEK/ERK pathways in AD brain leads to inhibition of Cdk5 and activation of a phosphatase with subsequent de-inhibition of GSK, as in the squid axoplasm described above (Morfini et al., 2002). This conclusion would contradict the results with fetal cortical neurons in which inhibition of Cdk5 suppressed Aβ-induced aberrant phosphorylation (see Zheng et al. above). The latter contradiction could be resolved on the presumption that Cdk5 actually is stimulating the tau phosphorylation by GSK3b, for which there is evidence (see later). Alternatively, it could be resolved if there were identified a putative Cdk5-activated P-GSK phosphatase activity in neurons. The actual effects of AD on all the possibly relevant signaling pathways in brain are not yet known. The pertinent experiments with inhibition or knock-out of Cdk5 and GSK3β in transgenic mice harboring the various APP, presenilin and tau mutations are arduous and remain to be performed. Presently available data do not exclude either theory regarding Cdk5 or GSK3β.

Dysregulation of Cdk5 may affect other systems besides tau. As described above, the Cdk5/p35 complex also phosphorylates β-catenin which effect in turn enhances binding to presenilin 1. Inhibition of Cdk/p35 reduces the amount of PS1 bound to β-catenin (Kesavapany et al., 2001). Thus, theoretically, Cdk/p35 may modify both regulation of tau phosphorylation and PS regulation of β-catenin transcriptional activity (see later).

Both Cdk5 and GSK3β are examples of proline-dependent protein kinases (PDPK). Tau may be phosphorylated at various sites by PDPKs and non-PDPKs. Sengupta et al. (1997) found that, similar to the activity of GSK3β, the tau phosphorylating activity of Cdk5 is stimulated if tau were first prephosphorylated by any of several non-PDPKs (PKA, PKC, CK-1, CaM-kinase II). In addition, prephosphorylation of tau by Cdk5 stimulated both the rate and extent of a subsequent phosphorylation catalyzed by GSK-3β, whereas no significant stimulation of phosphorylation by Cdk5 was observed when GSK-3β was followed by Cdk5. By contrast, prephosphorylation of tau by Cdk5 served to inhibit subsequent phosphorylation catalyzed by PKC and CK-1, but not by PKA or CaM-kinase II. These results indicate complex regulation of tau phosphorylation at different sites by kinases. Most significantly with respect to linking tau hyperphosphorylation to amyloidogenesis in AD, Cdk5 potentially has the combined effects of phosphorylating tau and of upregulating GSK3β phosphorylation of tau by priming tau. Cdk5 may have the additional effects of downregulating the activities of PKC and CK-1.

In summary, GSK3β activity is inhibited by Wnt, PI3K/Akt/(PKB), MEK/ERK, ILK, PKA, PKC, LiCl, and certain ribosomal-S6 kinases and is activated by tyrosine kinase. There is evidence that Cdk5 activity is enhanced by PI3K/Akt and MEK/ERK. Suppression of these pathways therefore would be expected to increase GSK3β and decrease Cdk5 activities. Both PI3K and MEK pathways are suppressed in presenilin deficient neurons *in vitro* in association with hyperphosphorylation. Despite opposing effects of the signaling pathways on their activities, both GSK3β and Cdk5 are associated with tau hyperphosphorylation in various different *in vitro* experimental paradigms pertinent to AD pathogenesis. Information on the effects of these multiple signaling pathways on the regulation of these kinases and on the relevant secretases and APP mutations in *in vivo* models is lacking. Based on current available information, either or both kinases may be involved in aberrant tau phosphorylation and PHF formation.

J. Potential Regulatory Interactions in the Pathogenesis of Alzheimer's Disease

The following sections seek to identify patterns within current information about the regulatory processes that may connect the abnormal processing of amyloid and tau in AD. Since most of the information is dynamic and derived from experimental model systems, rather than from AD brain, the conclusions point to mechanisms that are possibly but not necessarily causative in the actual disease.

1. WNT/GSK3β/β-Catenin/E-Cadherin Pathways and Presenilin

A central regulatory pathway that may unite key features of AD is the Wnt/GSK3β/β-catenin path. We first will define the players. Wnt proteins are secreted glycoproteins that play important and diverse roles in cell polarity, cell proliferation, and development. Their roles in development have been probed in fruit flies, nematodes, zebrafish, frogs, and mice. Wnts bind to the Frizzleds (Fz) family of transmembrane receptors and regulate multiple pathways, the best known being the Wnt/β-catenin pathway. The Frizzleds are members of the superfamily of heterotrimeric guanine nucleotide-binding protein (G protein)-coupled receptors (GPCRs) and regulate Ca^{2+} signaling and cGMP through G protein-coupled effectors such as phospholipase Cβ and phosphodiesterase (Woodgett, 2001).

Some Wnts in association with Fz promote the accumulation of the Armadillo homologue β-catenin, which, when translocated to the nucleus, regulates transcription of specific genes (Figure 10-4, see color plate). GSK3β is one known critical regulator of this signaling mechanism. GSK3β (described above) forms a complex with axin and the tumor suppressor APC. This complex, called the destruction complex, controls the stability of β-catenin. Axin is a scaffold molecule that sequesters GSK3β. When bound to axin and activated, GSK3β phosphorylates APC and β-catenin and stimulates interaction between β-catenin and β-Trcp, a regulator of E3 ubiquitin ligase, resulting in sub-sequent degradation of β-catenin in proteasomes. The result of activating GSK3β is to degrade β-catenin. β-catenin is found associated with PS in membranes but, as noted earlier, it is dissociable and not required for γ-secretase activity.

In the canonical Wnt pathway, binding of Wnt to Fz activates disheveled-1 (dvl-1), which is thought to recruit a GSK-binding protein that binds GSK and disrupts the interaction between

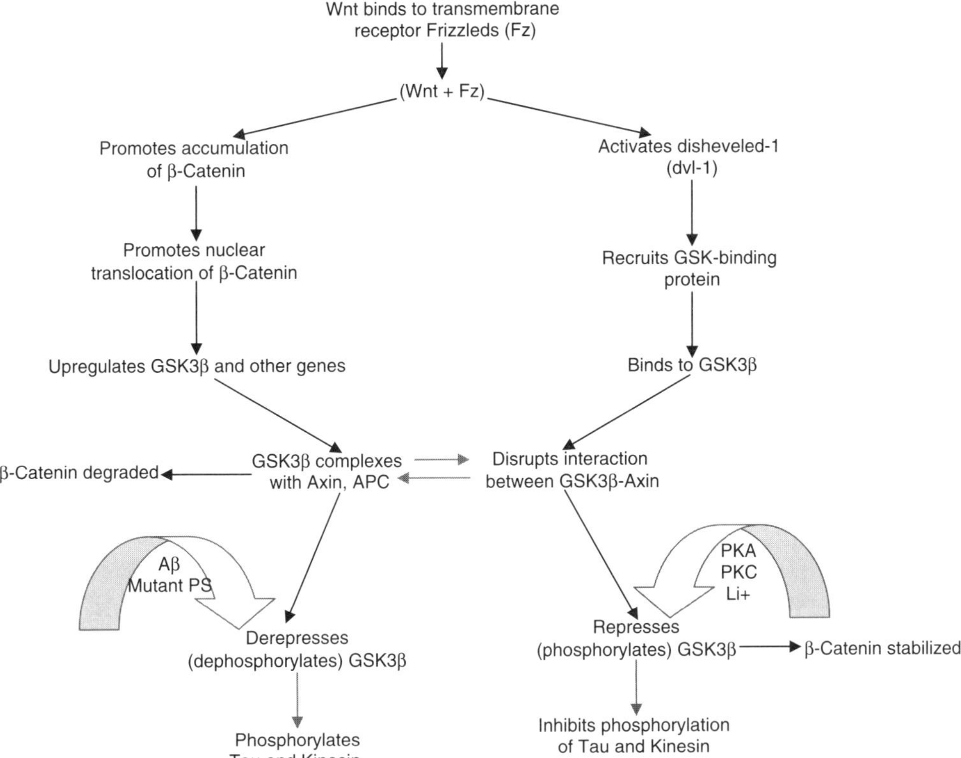

Figure 10-4. Schema showing GSK3β signaling possibly involved in linking amyloid, tau, and kinesin biology. Signaling pathway at the top indicates Wnt-GSK3β cascade under normal circumstances. Signaling events in the middle indicate dynamic balance between phosphorylation and dephosphorylation of tau under normal circumstances. Signaling events at bottom left indicate overactivation of GSK3β leading to overphosphorylation of tau and kinesin under abnormal conditions. Signaling events at bottom right indicate factors that attenuate overphosphorylation of tau and kinesin. (See color plate.)

GSK3β and axin (Figure 10-4). This disruption inhibits the activity of GSK3β, thereby stabilizing β-catenin. Stabilized β-catenin is translocated to the nucleus, where it associates with members of the lymphoid enhancer factor (LEF)/T-cell factor (TCF) family and activates specific genes. However, there are gaps in the understanding of this signaling system. Inhibition of GSK3β is not sufficient to activate LEF/TCF, and other factors are required to facilitate dissociation of GSK3β from axin (Hardt and Sadoshima, 2002).

Two known functions of β-catenin linked to presenilin activity, not necessarily to γ-cleavage, are as an element in the Wnt signaling pathway that leads to β-catenin activation of transcription and as an adhesion molecule linking epithelial-cadherin molecules, E-cadherin, at adherens junctions to the cytoskeleton. β-catenin translocation and its function in regulating transcription are in turn subject to modification by Wnt, presenilin/E-cadherin, and another kinase system, cyclin-dependent kinase 5 (cdk5). Each of these pathways is implicated by data from various experimental models in Aβ toxicity, tau phosphorylation, and AD pathogenesis.

a. Presenilin and the Wnt/GSK Pathway

In CHO or HEK cells, transfection with Wnt increases cytoplasmic and nuclear β-catenin levels and β-catenin transcriptional activity. Double transfection with Wnt and wild-type (wt) PS1 attenuates the effects of Wnt on both β-catenin levels and transcriptional activity. Similar results are obtained when cells are transfected with the downstream effector dvl with or without PS1. When cells overexpress FRAT, the effector downstream to dvl, there are again increased levels of β-catenin in cytoplasm and nucleus with increased transcriptional activity. However, cotransfection with PS1 has no effect on the FRAT-induced increase in β-catenin but inhibits FRAT-induced transcription even though the levels of β-catenin are not changed. Li^+ inhibits GSK phosphorylation of β-catenin and increases transcriptional activity. PS1 has no effect on the levels of β-catenin but inhibits the increased transcriptional activity resulting from Li^+ inhibition of GSK.

Therefore, from these experiments, PS1 attenuates the β-catenin stabilization induced by Wnt/dvl signaling but not the stabilization induced by the downstream FRAT or Li^+-inhibition of GSK phosphorylation of β-catenin, although PS1 inhibits the transcriptional activity induced through both the Wnt/dvl and FRAT/GSK segments. This suggests that PS1 acts at two sites, one on the final common pathway of β-catenin transcriptional activation and another on the Wnt/dvl segment that reduces β-catenin stability and nuclear translocation independently of the FRAT/GSK segment and GSK-phosphorylation (Killick et al., 2001).

Killick and colleagues also studied the effects of several PS1 mutations (mt) on the β-catenin signaling path. All the mtPS1 were less effective than wtPS1 in attenuating Wnt- or dvl-induced β-catenin stabilization or transcriptional activity. In contrast, when cells were transfected with FRAT or exposed to Li^+, the cotransfection with wtPS1 or any of the mtPS1 had no effect on the FRAT- or Li^+-induced β-catenin stabilization, but all the mtPS1 as well as the wtPS1 inhibited the FRAT- or Li^+-induced transcriptional activity. These authors conclude that wtPS1 may antagonize transcriptional activity of β-catenin by acting at all levels of the Wnt pathway. Differences between the wt and mtPS1s associated with AD become apparent when wtPS1 attenuates the Wnt/dvl path but not the FRAT/GSK path. One exception was the AD-related delta 9 exon deletion, which had somewhat less inhibition of transcriptional activity compared to that of wtPS1 at all levels (Killick et al., 2001).

Kang and coworkers have adduced evidence for β-catenin phosphorylation by two kinases independent of the Wnt path and that PS1 deficiency results in loss of this kinase activity together with increased β-catenin stability in the presence or absence of Wnt. In this view, PS1 acts as a scaffold bringing the β-catenin and kinases together (Kang et al., 2002).

Experimental evidence indicates that deficiency of PS1 in cells results in altered response to Wnt signaling, impaired ubiquitination of β-catenin, and increased β-catenin stability. The accumulated cytosolic β-catenin leads to a β-catenin/LEF-dependent increase in cyclin D1 transcription and accelerated entry into the S phase of the cell cycle. Conversely, PS1 specifically represses LEF-dependent transcription in a dose-dependent manner. The effect of PS1 deficiency on β-catenin response can be reversed by reintroducing PS1 expression or overexpressing axin, but not by expressing a PS1 mutant that does not bind β-catenin (PS1Deltacat) or by two different familial AD mutants. In contrast, PS1Deltacat restores Notch-1 proteolytic cleavage and Aβ generation in PS1-deficient cells, indicating that the PS1 role in modulating β-catenin levels can be separated from its role in facilitating γ-secretase cleavage of Aβ precursor protein and in Notch-1 signaling. From these data, it is clear that the presenilin mutation may have parallel and independent effects on γ-cleavage and on Wnt signaling (Soriano et al., 2001).

b. Presenilin and the E-Cadherin/ β-Catenin Pathway

PS1 binding to E-cadherin stimulates E-cadherin binding of β-catenin and promotes association of this junctional complex to the cytoskeleton. E-cadherin inhibits β-catenin/TCF signaling and transcriptional activation without changing the total levels of β-catenin. It is thought that a small fraction of the β-catenin exists as a "signaling-competent" pool that can bind to both the cytoplasmic domain of E-cadherin and to TCF. In this view, the small signal pool is regulated by E-cadherin independently of the total cytoplasmic and nuclear pools. PS1 affects this signaling-competent pool independently of its effects upstream on the Wnt pathway and, by promoting the binding of β-catenin to E-cadherin, depletes the signaling-competent component (Gottardi et al., 2001; Killick et al., 2001). Thus, the E-cadherin/PS1 reactions attenuate the transcriptional activity of β-catenin. There is growing evidence for another kinase system that regulates β-catenin and which is also involved in AD (see section on cdk5 below).

While there exist multiple putative mechanisms or pathways, most evidence indicates some connections among Wnt, PS, β-catenin, kinases, and E-cadherins, in which normal PS downregulates β-catenin levels, nuclear translocation, or signaling competency at different sites in the signaling network to finally downregulate β-catenin effects on transcriptional activity (Marambaud et al., 2002; Killick et al., 2001; Kawamura et al., 2001; Soriano et al., 2001; Fraser et al., 2001; De Ferrari and Inestrosa, 2000; Van Gassen et al., 2000; Nishimura et al., 1999).

AD-related mutations or deficiencies in PS exert differential effects on β-catenin stability and transcriptional activity (Killick et al., 2001). Dysregulation involving β-catenin has been demonstrated to potentiate toxicity due to Aβ. Activation of PKC protects rat hippocampal neurons from Aβ-induced toxicity in association with inhibition of GSK3β activity and accumulation of cytosol β-catenin. The Wnt activators Wnt 3a and lithium ion mimic the PKC activation, thus implicating the Wnt/β-catenin pathway in the protection mechanism (Garrido et al., 2002). In this view, mutant PS1 that fails to regulate the degradation of β-catenin in this Wnt-activated pathway would also potentiate the toxic effects of Aβ as well as the production of excessive Aβ through its gain of γ-cleavage activity.

β-catenin may be the target of multiple kinases, and the effect of phosphorylation of β-catenin undoubtedly depends on the residue phosphorylated. For example, phosphorylation of β-catenin at the Thr393 site by another kinase, protein kinase CK2, leads to proteasome resistance and increased transcriptional activity of β-catenin (Song et al., 2003).

c. Cyclin-Dependent Kinase 5 and β-Catenin

Cyclin-dependent kinase 5 (cdk5)/P35 is a system that may play a positive role in regulating β-catenin transcriptional activity. This kinase phosphorylates β-catenin at Tyr654 and decreases β-catenin binding to E-cadherin, increases binding to TATA-binding protein and increases β-catenein/TCF transcriptional activity. Also, inhibition of cdk5 phosphorylation of β-catenin reduces its β-catenin binding to PS1 (Killick et al., 2001).

Regulation of β-catenin by cdk/P35 assumes great importance because of accumulating evidence that implicates cdk5 in AD pathogenesis in several ways (Maccioni et al., 2001). Cdk5 and the regulatory proteins p35, p25, and p39 are involved in tau phosphorylation, which regulates microtubule assembly, and are concentrated at the leading edge of axonal growth cones. Cdk5 and its neuron-specific activator p35 are essential molecules for neuronal migration, axonal extension, and laminar configuration of the cerebral cortex. Observations from cultured neurons show that Aβ induces deregulation of cdk5 and that cdk5 is involved in tau hyperphosphorylation promoted by fibrillary Aβ. Cdk5 inhibitors protect hippocampal neurons against both tau anomalous phosphorylations and neuronal death.

P25, a truncated form of the required activator p35, accumulates in neurons in the brains of patients with AD (Patrick et al., 1999). This accumulation correlates with an increase in cdk5 kinase activity. Unlike p35, p25 is not readily degraded, and binding of p25 to cdk5 constitutively activates cdk5, changes its cellular location, and alters its substrate specificity. *In vitro*, the activity of p25/Cdk5 can be selectirely inhibited by a specific peptide cleavage product of p25, as mentioned above (Zheng et al., 2005). *In vivo* the p25/cdk5 complex hyperphosphorylates tau, which reduces tau's ability to associate with microtubules. Moreover, expression of the p25/cdk5 complex in cultured primary neurons induces cytoskeletal disruption, morphological degeneration, and apoptosis. These authors conclude that cleavage of p35, followed by accumulation of p25, may be involved in the pathogenesis of cytoskeletal abnormalities and neuronal death in neurodegenerative diseases.

In summary, dysregulation of β-catenin function induced by mutant presenilin or factors acting upon presenilin may lead to amyloid formation and abnormal tau phosphorylation with attendant neuronal dystrophy through signaling pathways that connect presenilin to Wnt, β-catenin, and Cdk5. These nodal points that may serve to connect abnormal processing of amyloid and tau are in addition to that of GSK3β, the activation of which bymutant presenilin also leads to abnormal phosphorylation of both tau and kinesin (see below).

2. Apoptosis, Caspase, Calsenilins, and Presenilins

There is evidence for programmed cell death, apoptosis, in brains from AD patients. AD brains also exhibit increased levels of activated caspase-3, a critical element in the cascade leading to apoptosis, particularly around amyloid plaques and NFTs. Aβ can induce apoptosis in mouse cortical neurons (Yuan and Yankner, 1999). In cultured neurons, Aβ stimulates apoptosis, apparently via c-Jun N-terminal kinase (JNK) (Troy et al., 2001). One mechanism, that may lead to apoptosis in AD involves calsenilin, which is associated with presenilin (Leissring et al., 2000, 2002).

Calsenilin is a substrate for caspase-3 and is a member of the Ca^{2+} sensor family involved in Ca^{2+} signaling pathways (Lilliehook et al., 2002; Choi et al., 2001). Calsenilin is strongly associated with cytoplasmic surfaces of intracellular membranes of ER and Golgi, is localized within neuronal perikarya and neuronal processes in hippocampus and cerebellum, and is developmentally regulated. This protein interacts with the carboxyl terminals of both PS1 and PS2 and colocalizes and coprecipitates with presenilins from mouse brain (Zaidi et al., 2002). It increases susceptibility of neuroglioma cells to apoptosis associated with release of calcium ion stores from ER (Lillliehook et al., 2002). Increased release of calcium ion can induce apoptosis. However, in Xenopus oocytes, calsenilin reversed the mutant presenilin-mediated enhancement of calcium ion release (Leissring et al., 2000). While PS, calsenilin, calcium ion, and caspase may be involved in apoptosis, the relationship to pathogenesis of AD is not well understood.

Calsenilin, as well as certain other proteins that are found together with PS, β-catenin, and presenilin-associated protein, is not required for the γ-secretase function of PS (Esler et al., 2002). It may be noted that PS may bind to or interact with various proteins of ER and membrane channels or receptors (Morohashi et al., 2002). Not all of these associations are necessarily part of the γ-secretase function or of the pathogenesis of AD, however.

3. Presenilin, GSK3β and Axoplasmic Transport

Takashima et al. (1998) found that tau and GSK3β bind to the same region of PS1 and that the PS1 binding domain on tau is the repeat region that binds also to microtubules. Mutations in PS1 that cause Alzheimer's disease increase the ability of PS1 to bind GSK3β and, correspondingly, increase its tau-directed kinase activity. (see under Tau) On this basis, it was proposed that the increased association of GSK3β with mutant PS1 leads to increased phosphorylation of tau (Takashima et al., 1998). As discussed above, Pigino et al. (2003) found that GSK3β activity is increased and associated with increased phosphorylation of kinesin and disruption of fast axoplasmic transport in mutant PS1 transgenic mice. Thus normal PS1 may facilitate fast axonal transport. These authors concluded that mutations in PS1 may compromise neuronal function by affecting GSK3β activity and kinesin-1-based motility (Pigino et al., 2003).

GSK prefers substrates that have first been phosphorylated at some site (primed) by other kinases and it may be inhibited by small molecule inhibitors, e.g., aloisines, paullones, and maleimides (Doble et al., 2003). This feature makes it a plausible target for pharmacologic regulation as potential therapy in AD (Woodgett, 2001). GSK3β has been purified and crystallized (Aoki et al., 2000) and its gene has been cloned (Lau et al., 1999).

4. Aβ, Tau, Apoptosis, and GSK3β

Takashima and coworkers (1996) observed that treatment of cultured rat hippocampal neurons with Aβ (25–35) inactivated PI3K, and activated glycogen synthase kinase-3β while enhancing tau phosphorylation and neuronal death. GSK3α or mitogen-activated protein kinase (MAP kinase) were not affected by the Aβ peptide. Treatment with GSK3β antisense oligonucleotide inhibited the enhancement of tau phosphorylation induced by Aβ (25–35) exposure. Another observation, little discussed at that time, was that Aβ treatment also induced accumulation of APP in the cytosol and decreased secreted APP released to the medium along with the increased GSK3β activity and neuronal death. These effects also were blocked by GSK-antisense oligonucleotide, and it was proposed that Aβ impaired axonal transport of APP (Takashima et al., 1995). Now it is recognized that GSK3β also phosphorylates kinesin and interferes with fast axonal transport that involves APP (see section D.3).

In human umbilical vein endothelial cells, Aβ42 is proapoptotic, producing activation of caspase-3, while inhibiting Akt/PKB phosphorylation associated with reduced phosphorylation of the downstream substrates GSK3β and eNOS and reduced NO production. Activation of Akt reversed Aβ toxicity, showing that Aβ toxicity is mediated at least in part through disruption of the Akt/GSK3β pathway (Suhara et al., 2003). These observations indicate that Aβ toxicity with respect to tau phosphorylation and apoptosis can be mediated through its effects to activate or derepress GSK3β. In addition, tau is proteolyzed rapidly by caspases in Aβ-treated neurons, and the resulting product is assembled into tau filaments more extensively than is the wild-type tau; a similar truncated tau is recognized immunologically in AD brain (Gamblin et al., 2003).

Reduced PI3-kinase/PKB activity can induce apoptosis in various central nervous system (CNS) and peripheral nervous system (PNS) neurons in culture in association with activated GSK3β activity, and these effects can be blocked by selective inhibition of GSK3β by small molecule inhibitors being explored as potential pharmacologic tools (Cross et al., 2001).

It is clear that Aβ produces neurotoxicity and apoptosis in association with inhibition of phosphorylation of GSK, thus enabling GSK3β activity, in various models. For example, activation of PKC by phorbol 12-myristate 13-acetate protects rat hippocampal neurons from Aβ toxicity via inhibition of GSK3β activity. This effect leads to the accumulation of cytoplasmic β-catenin and transcriptional activation through β-catenin/T-cell factor/lymphoid enhancer factor-1 (TCF/LEF-1), of Wnt target genes, engrailed-1 (en-1), and cyclin D1 (cycD1). In contrast, inhibition of Ca^{2+}-dependent PKC isoforms activated GSK3β and offered no protection from Aβ neurotoxicity. Wnt-3α and lithium salts, classical activators of the Wnt pathway, mimic PKC activation. Thus, regulation of members of the Wnt signaling pathway by Ca^{2+}-dependent PKC isoforms may be important in controlling the neurotoxic process induced by Aβ (Garrido et al., 2002; see section 5.1.b).

In rat hippocampal neurons, Aβ derepresses GSK3β, which leads to increased phosphorylation of tau (Takashima et al., 1998a). Aβ injected intracisternally in rabbit brain produces activation of GSK3β and caspase-3, leading to mitochondrial injury, tau phosphorylation, and nuclear translocation of NF-κB (Ghribi et al., 2003). Moreover, Li+ inhibition of GSK3β blocks the accumulation of Aβ in mouse brain overexpressing APP (Phiel et al., 2003).

In summary, there is the setup for a vicious cycle in which mutant γ-secretase or increased production of Aβ from mutant APP or inefficient clearance of Aβ produced physiologically leads to derepression of GSK3β, which in turn leads to further increases in Aβ accumulation and abnormal tau and kinesin phosphorylation (see Figure 10-4).

5. GSK3β, Tau, and Amyloidogenic Processing

If GSK is in fact the node that links all the components of AD, then links must also exist to α- and β-secretase activities and to Aβ toxicity that are not dependent on mutated or absent PS. Protein kinase C (PKC) is known to increase nonamyloidogenic α-secretase cleavage of APP, producing secreted APP (sAPPα) while therefore reducing β-cleavage and Aβ formation. Both PKC and GSK3β are components of the Wnt signaling cascade. Hence, Wnt signaling may provide the connection.

Amyloidogenic processing may also be linked to abnormal tau phosphorylation. Overexpression of a member of the Wnt signaling pathway, disheveled (dvl-1), increases sAPPα production. Both human dvl-1 and Wnt-1 also reduce the phosphorylation of tau by GSK3β. The disheveled action on APP is mediated via both c-Jun N-terminal kinase (JNK) and protein kinase C (PKC)/mitogen-activated protein (MAP) kinase but not via p38 MAP kinase. These data position dvl-1 upstream of both PKC and JNK. Murine dvl-1 was also found to inhibit GSK3β-mediated phosphorylation of tau in cotransfected CHO cells (Wagner et al., 1997). Therefore, both the shift in balance between α-secretase- and β-secretase-processing of APP and the abnormal tau and probably kinesin phosphorylation are potentially linked through Wnt signaling and GSK3β regulation (Mudher et al., 2001).

6. Nerve Growth Factor, Muscarinic Cholinergic Receptors, and GSK3β

Anatomically and biochemically, the pathology in AD is distributed in specific brain regions and neuronal pathways. This regional specificity also must be accounted for by a unifying hypothesis. The neuronal pathways earliest involved in AD are the cholinergic neurons of the nucleus basalis of Meynert that project to the hippocampus and neocortical regions, the entorhinal nuclei that project to the hippocampus and, later, hippocampal projections to the neocortex in the frontal and parietal regions, sparing the primary motor and sensory cortex, occipital cortex, basal ganglia, and cerebellum (see Chapter 2 DI–DIII). The specific regional distribution of biochemical pathology related to amyloidogenesis and tau phosphorylation (PHF/NFT) in AD arises predominantly in cholinergic systems, although eventually other systems and neurotransmitters are involved (Selkoe and Lansbury, 1999; Chauhan and Siegel, 2002). A critical issue is to identify the biochemical mechanisms that render these particular neuronal networks most sensitive to the dysregulation in the amyloid, tau, and kinesin molecular systems.

Nerve growth factor (NGF), a member of the NGF superfamily of neurotrophic factors (see Chapter 8 CI–CII), is known to be crucial to cholinergic neuronal development, synapse formation, and regulation in neurons of the nucleus basalis of Meynert (NBM) that terminate in the hilar region of the hippocampus and in cholinergic regulation of APP processing (Siegel and Chauhan, 2000; Rossner et al., 1998). These appear to be the first pathways to degenerate in AD.

NGF-mediated kinase pathways inhibit GSK activity, and it has been found that Li^+, a potent inhibitor of GSK3β, diminishes Aβ40 and Aβ42 secretion from transfected COS7 cells in a dose-dependent manner associated with inhibition of GSK3β activity (Sun et al., 2002). Phiel et al. (2003) demonstrated in studies of mouse brain that therapeutic concentrations of Li^+ blocked production of Aβ by inhibiting γ-cleavage of APP while not inhibiting Notch protein processing. Li^+ reduced accumulation of Aβ peptides in brains of mice with overexpression of APP. Hence, it is presumed that γ-cleavage of APP is potentiated by active GSK3β. If so, the protective effect of NGF and facilitation of α-processing opposed to amyloidogenesis may result from NGF inhibition of GSK3β.

Inhibition of GSK3β also has the effect of activating some Wnt target genes through the accumulation of cytoplasmic β-catenin and transcriptional activation via β-catenin /TCF/LEF-1. Li^+ and PKC activation by phorbol esters, both of which conditions inhibit GSK3β activity, have similar effects on the β-catenin path and both protect hippocampal neurons from apoptosis (Garrido et al., 2002). Thus, the synergistic effect of M1 Cholinergic receptor activation on NGF regulation of amyloidogenesis, accumulation of Aβ peptides, and apoptotic effects of Aβ may all be transduced through inhibitory regulation of GSK3β via the latter's effects on γ-cleavage, apoptotic activation, and transcriptional activation of Wnt proteins.

7. NGF Regulation of Aβ

In young Tg2576 mice, which overexpress an APP mutation related to familial AD, whole brain NGF mRNA is reduced in association with greatly increased Aβ and decreased sAPPα. Feeding these animals propentofylline, which induces NGF mRNA, resulted in increasing NGF mRNA and sAPPα while reducing the Aβ levels (Chauhan and Siegel, 2003). Moreover, as discussed earlier, in a separate experiment it was demonstrated that feeding PPF under the same conditions resulted in decreased tau phosphorylation and an increase in the ratio of inactive to active GSK3β (Chauhan and Siegel, 2004). These data taken together suggest that normally NGF promotes α-secretase at the expense of amyloidogenic processing, that the excessive Aβ produced by the overexpressed mutant APP gene downregulates NGF and that NGF downregulates GSK3β leading to decreased tau phopsphorylation. In the presence of the over-expressing mutated APP gene, however, a vicious cycle is set up in which excess Aβ depresses NGF and the reduced NGF level permits Aβ to increase further. Inducing NGF reverses the mutant APP effects at least partly.

Activation of muscarinic M1/M3 acetylcholine receptor (AChR)-associated postsynaptic pathways shifts the balance in APP proteolysis to enhance α-secretase activity and sAPPα secretion and to decrease the β-secretase amyloidogenic processing. NGF is synergistic with M1 receptors in promoting atropine-sensitive neurite outgrowth. NGF is also synergistic with M1-selective agonists in increasing sAPP secretion from rat cerebral cortex and in activating sAPPα secretion from PC12 cells via both MAPK and PKC-dependent pathways (Haring et al., 1998).

In addition, selective M1 agonists result in reducing phosphorylation of tau while promoting nonamyloidogenic processing of APP (Fisher et al., 2000; Forlenza et al., 2000). Since it is known that PKC/MAPK inhibits GSK3β (see above), the activation of PKC/MAPK by the synergistic action of M1 receptor stimulation by ACh in the presence of NGF can at once regulate α-secretase and tau phosphorylation inversely in cholinergic neurons. The ACh-M1/NGF/PKC activation that in turn inhibits GSK3β would confer neuronal or network specificity onto this inverse relationship between α-secretase processing and tau phosphorylation. The inverse relationship is imposed by the fact that NGF-mediated phosphorylation of PKC/MAPK activates the path toward α-secretase while the phosphorylation of GSK3β is inhibitory toward its phosphorylation of tau and kinesin.

NGF may also regulate GSK3β through NGF-stimulated integrin-linked kinase (ILK), the pharmacologic inhibition or genetic depletion of which reduces NGF-induced neurite outgrowth and reduces NGF-mediated phosphorylation of both Akt/PKB and GSK3β in rat pheochromocytoma cells and dorsal root ganglion neurons, thus leading to hyperphosphorylation of tau. Stimulation of ILK has the opposite effects. ILK, as with PKC/MAPK, permits NGF to exert inverse effects on tau phosphorylation and, in this case, on neuritic outgrowth (Mills et al., 2003).

NGF trophic effects are not confined to cholinergic neurons alone, nor are its actions mediated exclusively through GSK3β. bFGF also may induce APP (Gray and Patel, 1993).

A similar inverse regulatory relationship can be found between NGF protection from apoptosis and inhibition of GSK3β. Growth factor-dependent survival of a variety of mammalian cells is dependent on the activation of PI3K and its downstream effector, the protein kinase Akt/PKB, which in turn phosphorylates and inhibits GSK3β. Overexpression of catalytically active GSK3β induces apoptosis in PC12 cells, whereas dominant-negative GSK3β prevents apoptosis that otherwise follows inhibition of PI3K. GSK3β thus plays a critical role in regulation of apoptosis and represents a key downstream target of the PI3K/Akt/PBK survival-signaling pathway by which its inhibition can be regulated by NGF (Pap and Cooper, 1998).

In summary, NGF may regulate GSK3β via several kinase pathways: PKC/MAPK, ILK/Akt/PKB, and PI3K/Akt/PKB. Phosphorylation of all these kinases is activating, except for the unique instance of GSK3β, which is inhibited by its phosphorylation. In all these cases, NGF exerts inverse effects, positive regarding α-secretase activity and decreased amyloidogenesis, increased neurite outgrowth, or protection from apoptosis, on the one hand, and GSK3β inhibition, on the other. Although upregulated GSK3β activity is required for the apoptosis that is induced by inhibition of the PI3K/Akt/PKB path, it can also be shown that apoptosis induced by withdrawal of NGF does not depend on GSK3β activity, thus indicating that NGF also acts via other pathways (Rabizadeh et al., 1994; Crowder and Freeman, 2000).

8. Aβ Regulation of NGF

Aβ produces a vicious cycle of events in brain, and these may be through effects of Aβ on regulatory reactions. It is known that Aβ in sublethal concentrations suppresses cAMP-response element binding (CREB) protein phosphorylation and downstream activation of the brain-derived neurotrophic factor (BDNF) promoter in cortical neurons. In addition, CREB phosphorylation is found early in granule cell differentiation prior to expression of BDNF and neurotrophin-3 (NT-3), which are in the NGF family, and is therefore upstream of neurotrophin regulation. Thus, CREB is a candidate for a site to transduce Aβ effects to NGF expression (Chauhan and Siegel, 2003).

There is emerging information that links CREB and GSK3β or Wnt. CREB is activated to induce gene transcription through phosphorylation at certain residues by PKA and other kinases. *In vitro*, PKA phosphorylation of CREB increases the latter's binding affinity to gene elements while, in contrast, phosphorylation of CREB by GSK3β, which proceeds after that by PKA, decreases CREB binding affinity and inhibits the PKA-stimulated CREB binding to DNA (Bullock and Habener, 1998; Tyson et al., 2002). GSK3β is required for phosphorylation of CREB at a serine residue (S129), which is involved in the cAMP control of CREB and gene expression (Fiol, 1994; Swarthout, 2002). GSK3β inhibitory regulation of CREB binding to DNA has been demonstrated in differentiated human neuroblastoma cells.

Activation of Akt/PKB (which is downstream to PI3K) or exposure to Li^+, both of which inhibit GSK3β, serve to increase CREB DNA binding affinity, while overexpression of GSK3β blocks the increases in CREB DNA binding otherwise induced by certain growth factors (e.g., EGF, ILGF-1, forskolin) and cAMP. Of interest is that another antidepressant medication that is an anticonvulsant and is believed to stabilize mood, sodium valproate, also inhibits GSK3β and reverses GSK3β inhibitory effects on CREB activation (Grimes and Jope, 2001).

Wnt activation increases both ILK activation and ILK induction of cyclin D1, which involves CREB signaling, and this induction of cyclin D1 is blocked by overexpression of GSK3β. Inhibition of PI3K and its downstream activator Akt/PKB also reduces ILK induction of cyclin D1 (D'Amico et al., 2000). As mentioned above, Wnt activation promotes β-catenin stability and transcriptional activity, while mutated PS1 or PS1 deficiency leads to reduced β-catenin and/or transcriptional activity. Observations based on a yeast system show that β-catenin interacts with CREB, acting synergistically with CREB-binding protein to stimulate transcription (Takemaru and Moon, 2000). Thus, a decrease in β-catenin stability through decreased Wnt activation or increased GSK3β activity is expected to result in decreased CREB transcriptional activity as well.

9. Amyloid and Tau Interactions

Intracisternal injections of Aβ in rabbits produced evidence for apoptosis, including activation of caspase and tau phosphorylation in the hippocampus. Li^+ treatment blocked the caspase activation but did not prevent DNA damage or tau hyperphosphorylation, indicating that the amyloid-induced tau hyperphosphorylation was independent of GSK3β in this experimental model (Ghribi et al., 2003). Contrasting results have been obtained in cultured hippocampal neurons. Takashima and colleagues (1996) reported that Aβ25–35 exposure induced an inactivation of PI3K and an activation of GSK3β in rat hippocampal culture. Wortmannin, an inhibitor of PI3K, also activated GSK3β, leading to an enhancement of tau phosphorylation and neuronal death in hippocampal culture. Further experiments with hippocampal neurons in culture showed that this Aβ peptide fragment activates GSK3β activity and tau phosphorylation and that this effect of Aβ is blocked by

GSK3β-antisense oligonucleotide, indicating an effect of Aβ on transcriptional activation (Takashima et al., 1998a, 1998b).

There may be more than one mechanism for Aβ effects on tau phosphorylation, but all the data taken together indicate that excessive Aβ itself induces dysregulation of signaling pathways, particularly of Wnt and GSK3β, tau hyperphosphorylation, certain transcriptional effects, and neuronal death. Activation of GSK3β or of Cdk5 by Aβ may lead to phosphorylation of tau and of kinesin light chain (see above), both of which disrupt axonal transport and cytoskeletal structure.

There is controversy as to whether the production of Aβ or of abnormal tau processing is the initial trigger in AD and as to how the constituents of amyloid plaques and NFTs are related pathogenically to each other and to the synaptic pathology and cognitive deficits in AD. It is clear that NFTs and abnormal tau phosphorylation in the absence of excess Aβ or amyloid plaques may produce neuronal degeneration and cognitive losses, as observed in the tauopathies. Also, excess Aβ or amyloid plaques in the absence of abnormal tau or NFTs may cause neuronal degeneration and cognitive losses, as seen in some transgenic AD mouse models overexpressing mutated APP or presenilin (Games et al., 1995; Duff et al., 1996; Hsiao et al., 1996; Borchelt et al., 1997) and to some variable degrees in normal aged humans. However, these transgenic mouse models are not complete representations of AD in humans. In familial AD, mutations in APP or in enzymes processing APP produce both amyloid plaques and NFTs.

Studies with transgenic mice harboring mutated human tau bear on this issue. Mice expressing human mutated tau develop NFTs without amyloid plaques, while the double mutant animals with mutant tau and APP develop amyloid deposits as expected and marked enhancement of NFT pathology as compared to NFTs in the single tau mutation (Lewis et al., 2001; Perez, 2005). Injections of Aβ fibrils into brain of homozygous transgenic mice expressing human wild-type tau fail to produce NFTs, while injections of Aβ fibrils into brain of mutant tau transgenic mice increase by fivefold the number of NFTs in the amygdala neurons projecting to the injection site (Gotz et al., 2001). These results indicate that Aβ can potentiate NFT formation in the presence of the abnormal, but not normal, tau, while tau does not appear to potentiate amyloid plaque formation in the presence of mutated APP. In addition, Aβ accumulates in brain of early childhood Down syndrome prior to NFTs or plaques, and in familial AD, the abnormal APP processing is associated with both amyloid plaques and NFTs.

Taken all together, these data support the hypothesis that in AD, Aβ is the initial trigger that leads to NFT formation and that in human AD, there exists an as-yet-undiscovered unique factor related to tau gene or tau processing that permits the Aβ effect to potentiate neurofibrillary pathology in humans. In this view, the abnormality probably precedes the hyperphosphorylation. While there is controversy with respect to the kinase responsible for tau hyperphosphorylation, the preponderance of available information on regulatory sequences suggests that GSK3β is the nodal point between amyloidogenesis and tau hyperphosphorylation that evntually leads to the classic hallmarks of AD pathology. There appear to be several set-ups for vicious cycles in which Aβ dysregulates GSK3β and other factors, the dysregulation of which leads to further production of Aβ and other wide-spread consequences to neuronal biology, including abnormalities in tau and transport functions (Figures 10.2 and 10.4).

K. Therapy of Alzheimer's Disease and Future Directions

1. Acetylcholinergic Pharmacotherapy

Understanding of the genetics and biology of amyloid and tau and their intimate effects on regulatory processes in the cell has advanced considerably in the past few years. However, there are no available means at present for preventing or reversing the inexorable progression of AD. The only pharmacotherapy approved currently by the Food and Drug Administration (FDA) is the use of acetylcholinesterase (AChE) inhibitors. In this section the anticholinesterase (antiChE) and cholinergic therapy will be described, followed by discussion of the future possibilities and directions of AD therapy.

a. Cholinergic Therapy

It should be stressed that, today, several diagnostic, cholinergically related methods are available to assess Marsel Mesulam's "tauopathy" concept, that is, the concept of early onset of cholinergic loss in AD (as well as in animal models of AD) via the use of micro-PET scan methods (see also section A-2). These methods are generally based on PET estimation of AChE and both muscarinic and nicotinic cholinergic receptors (the muscarinic receptors are, however, not necessarily lost in AD; see section A-2, above). So far, the data on the whole support Mesulam's concept (Thiel, 2003; Kemp et al., 2003; Xie et al., 2004; Mason and Mathis, 2003; Fujita et al., 2003; Nordberg, 2003, 2004; Tsukada et al., 2004). Functional magnetic resonance imaging (fMRI) begins to contribute to this diagnostic value of cholinergically correlated neuroimaging methods, and, in fact, to the demonstration of the effectiveness in AD of antiChE treatment (Thiel, 2003; Tanaka et al., 2003; Kessler, 2003; Saykin et al., 2004).

Today, several antiChEs and several muscarinic agonists are used to ameliorate the dementia component of AD or are being tested in clinical trials for their potential use in AD. The main reason for the clinical use of either antiChEs or cholinergic agonists is the well-documented loss in AD of cholinergic neurons, particularly in the nucleus basalis of Meynert (NBM) and its neocortical projections, and hippocampus, which are structures critical to memory, learning, and cognition. The pathology and physiology supporting this rationale as well as the limitations of the use of antiChEs in AD are adduced in section A.

The antiChEs available for clinical use or clinical trials are either synthetics or naturally occurring alkaloids. They include donepezil, galantamine, rivastigmine, heptastigmine, huperzine A, and tacrine. These agents are reversible or pseudoreversible inhibitors of AChE (and, to a varying extent, butyryl ChE); they are congeners of the carbamate antiChE physostigmine. Irreversible antiChEs of the organophosphorus type such as metrifonate are also employed (see, for example, Cummings et al., 2001; see also Chapter 7 A, BI, and BII and also Giacobini, 2000). After mediocre success with the use of physostigmine in AD, these drugs were developed following criteria posited by Ezio Giacobini and Robert Becker; these criteria include specificity for brain forms of AChEs, good penetration into the brain, relatively long half-life, and limited or, ideally, absent side actions and toxicities (Becker and Giacobini, 1991; Giacobini and Becker, 1988; Giacobini, 2000).

It should be added that, rather than looking for antiChEs that are specific for AChE, a different strategy with respect to antiChE treatment of AD could be embraced: as AChE activity is diminished in AD, while BuChE activity may be spared, Giacobini (2000, 2003a) suggested that an antiChE that inhibits both enzymes or, indeed, exhibits significant antiBuChE activity may be useful in AD, as it should accumulate ACh in AD cases (see section A and Chapter 3 DIII).

Giacobini stressed that antiChE treatment-induced accumulation of ACh may impose a negative feedback on the release of endogenous ACh (see also Karczmar and Dun, 1988). Accordingly, he suggested that "a slow dose titration" of the antiChE and/or the use of slow-release agents (drugs that become active upon metabolically induced release of the active agent) may reduce this limiting feedback effect. He also posited that decreased clinical response of AD patients to an antiChE, which is noticed generally 30 to 36 weeks after the initiation of therapy, may not depend on the progress of the disease but on the development of tolerance to the drug. He describes several treatment strategies that may limit the development of tolerance and he opines that antiChEs may differ with respect to their capacity to induce tolerance (Giacobini, 2000).

Tetrahydroaminoacridine (tacrine), the first antiChE to be approved by the FDA for AD treatment, was withdrawn from the market in the 1990s, as it frequently induced hepatic toxicity. The subsequent intense synthesis of new antiChEs was prompted and facilitated by several developments.

First, multidimensional quantitative assessments became available to measure, in AD, various components of cognition, behavior, special adaptation, and emotional responsivity, as well as to assess the caregiver milieu and the quality of life. Leon Thal has been a pioneer in this quantitative testing of the progress of the disease (Thal et al., 1983; see also Tractenberg et al., 2002). Several rating scales are available, including those developed by the National Institute of Neurological and Communicative Disorders and Stroke and the Alzheimer's Disease and Related Disorders Association (NINCDS-ADRDA); the scales establish the criteria for diagnosis of AD and for monitoring its progress (see, for example, Morris, 1996; Battistin and Gerstenbrand, 1990).

New tests may be added in the future. The Disorders of Consciousness Scale, which relies on observation protocols and appropriate tests of neurobehavioral efficiency, and the earlier related scales, such as the Western Neuro Sensory Stimulation Profile, are novel diagnostic procedures ordinarily used for recovery from stroke and coma, but which seem to be usable for estimation not only of cognition but also of consciousness and self-awareness in AD (see Bender-Pape et al., 2005; see also Chapter 9 BV-4). Also, special EEG methods such as coherence analysis may serve as effective diagnostic and testing tools (Adler et al., 2003).

At the outset of the treatment trials, the evaluation of the effectiveness of AD treatment was relatively subjective and the results or conclusions were often inconsistent. Thus, the early use of physostigmine in AD led to conclusions varying from very positive to quite negative. The development of testable, quantitative methods of evaluation that could be subjected to statistical analysis provided a boost for the development of new anti-AD agents.

Second, as already pointed out, the clinical appraisal of the results of the early attempts at antiChE treatment of AD was inconsistent and frequently pessimistic. In the 1990s, many investigators felt that antiChEs employed in AD treatment have approximately similar and very modest effects in reducing the rate of progressive memory deficits and that they do not alter the eventual outcome. A somewhat more positive attitude developed subsequently as statistically significant results became available, and a certain degree of consensus seemed to have been reached as to the "real" effect of antiChEs in AD. The general sense is that when used in early AD antiChEs such as donepezil, galantamine, and rivastigmine reduce, initially, the neuropsychiatric symptoms, "stabilize" the condition, and delay by perhaps a year the appearance of symptoms that would be present in untreated or placebo-treated cases at the time in question (Anand et al., 1998; Wynn and Cummings, 2004; see also Giacobini, 2000); indeed, abrupt withdrawal of the antiChE therapy in certain forms of dementia seems to cause a decline in cognition (Minett et al., 2003). Other possible benefits of antiChE treatment are sporadically reported, such as increased regional cerebral blood flow (Lojkowska et al., 2003; see also Chapter 9 BIV-2).The major benefits are seen in relatively improved behavior measurements and delays in nursing home placement compared to placebo controls or untreated groups. Certain aspects of AD besides cognition, such as apathy and agitation, also seem to be improved by the use of antiChEs (Boyle and Malloy, 2004; Profenno and Tariot, 2005).

Yet, it is also opined that the results at best are limited, and that major financial and research efforts should be addressed at therapy other than antiChE therapy (New York Times, 2004; Jones, 2003; see also section A-2). It must be stressed that the majority if not all of the evaluations of the effectiveness of antiChEs in AD were sponsored by pharmaceutical companies, and it was sometimes claimed that the reports in question favored the sponsor (Hogan et al., 2004).

It must be added that quantitative evaluation of the matter is not easy. For example, quantitative evaluation frequently compares the data obtained in the treated individuals with the projection of the progress of the disease in the placebo-treated patients; in view of the well-known variability of the progress of the disease, this comparison may not be reliable.[2] It must be remembered that, in early AD, many or even the majority of cholinergic neurons remain intact and can serve as a basis for the cholinergic effect of antiChEs (see section A).

Third, certain special effects of antiChEs must be considered in this context. Thus, antiChE-induced accumulation of ACh may activate, via the action of ACh on M1 muscarinic receptors, a cholinergic gene promoter (this effect was established for the AChE gene promoter by Nitsch et al., 1998). AntiChE activation of M1 receptors (whether directly or via accumulating ACh) may be favorable in the treatment of AD via transcriptional effects; it must be noted that muscarinic receptors are probably spared in AD (see section A) and therefore are potentially available for regulatory activation. Finally, as has been known since the 1970s (see, e.g., Glisson et al., 1972, 1974), antiChEs cause effects on noncholinergic neurons, such as catecholaminergic and serotonergic cells, and these effects may be beneficial in AD (see Chapter 9 BIII and Svensson and Giacobini, 2000).

But the most important effect of antiChEs with respect to the biology and treatment of AD concerns their action on the amyloid metabolism; Roger Nitsch's initial observations on this question were referred to in section A of this chapter (Nitsch et al., 1992). The abnormal or dysregulated protein processing that constitutes the pathogenesis of AD relates to abnormal processing of amyloid precursor protein (APP) and of tau protein, as described in this chapter. A number of reports, including those of Roger Nitsch and Paul Kasa (with Magdolina Pakaski), indicate that antiChEs acting possibly via M1 and M3 receptors modulate the APP processing pathway, promoting a nonamyloidogenic secretory pathway (Svensson and Giacobini, 2000; Pakaski et al., 2000). Other, amyloid metabolism-related actions of antiChEs were also reported; these include an effect on alpha-secretase and protection against amyloid-induced apoptosis (in the case of human neuroblastoma), possibly via induction of anti-apoptotic proteins (Zimmerman et al., 2004; Orozco et al., 2004; Arias et al., 2004, 2005). This type of neuroprotection may vary from one antiChE to another, and it may be mediated by nicotinic receptors (Arias et al., 2005; Laviolette and van der Kooy, 2004).

Support for the notion of the antiChE or cholinergic regulation of the Aβ peptide and APP secretion and metabolism is provided by the evidence for the negative relationship between the cholinergic system and amyloid accretion. Thus, the Montreal team of Rene Quirion and Claudio Cuello demonstrated that lesions of the forebrain cholinergic system increase APP synthesis and induce its accumulation in the cortex and cerebrospinal fluid (see, for example, Auld et al., 1998). Conversely, application of Aβ peptide directly to the rat NBM or its use in rat neuronal cultures decreases ACh synthesis and choline uptake and causes cholinergic hypofunction (Auld et al., 1998; Harkany et al., 1998; see also Giacobini, 2000); in addition, nicotinics and antiChEs seem to be capable of antagonizing Aβ peptide neurotoxicity in several in vitro preparations. These effects are the basis for the animal model of AD (see section A-2). In this context, transgenic mice expressing human APP also show increased levels of AChE (Sberna et al., 1998). AChE and BuChE are colocalized with Aβ peptide deposits in brains of patients suffering from AD (see section A), and Ezio Giacobini suggested that the incorporation of AChE into amyloid fibrils promotes their neurotoxic properties (Giacobini, 2000).

Certain caveats are in order. The work quoted above was carried out in rodents, sometimes *in vivo*, but mostly *in vitro*; it is not certain that the results in question apply to human AD. Then, the data concerning the inhibitory effects of antiChEs on amyloid formation are not always consistent, and, occasionally, antiChEs such as tacrine seem to induce rather than reduce amyloidosis (Chong and Suh, 1996; Mazzucchelli et al., 2003). Then, the data concerning the inhibitory effects of antiChEs on amyloid formation are not always reliable, and occasionally, antiChEs, such as tacrine, seem to induce rather than reduce amyloidosis (Chong and Suh, 1996; Mazzucchelli et al., 2003). Commerting on the effects of tacrine, Ezio Giacobini suggested that "tacrine may reduce levels of neurotoxic beta-amyloid peptide as well as neuroprotective soluble ATP, and thereby influence the formation of beta-amyloid peptide aggregates" (Giacobini, 2000).[3]

Finally, it must be remembered that even when effective in early or moderate AD, antiChEs lose their "stabilizing" effect within 1 year or so from the initiation of the treatment; at best, their effect is limited, even in early AD. These findings may suggest that the anti-amyloid action that the antiChEs exhibit in rodents does not obtain in the human; the findings also suggest that the ameliorative antiChE effect in AD is not curative but rather "symptomatic" (see Jones, 2003). AntiChEs are not effective in late AD; by that time neurotoxic phenomena of AD may be irreversible and the substrates for antiChE action such as cholinergic synapses and receptors are probably eliminated. It must be added that AD cannot be considered a primarily cholinergic disease, as was already maintained some time ago by this author (Karczmar and Dun, 1988; Karczmar, 1991), and as opined recently by Marsel Mesulam (2004; see section A 2, above).

To turn now to cholinergic agonists: Nicotinic receptors seem to be gradually eliminated in AD (see section A-2), and thus it may be not expected that nicotinic agonists would be useful in the treatment of AD; yet, there is some development in this area (Nordberg, 2000; see also several chapters in Giacobini and Becker, 1988, and Becker and Giacobini, 1991). On the other hand, both pre- and postsynaptic central muscarinic receptors (of the M1 and M3 subtypes) are spared in AD, and a concerted effort is being made to develop appropriate muscarinic agonists. It was proposed that besides facilitating or initiating cholinergic transmission, these compounds may have, similar to antiChEs, a beneficial effect on amyloidosis (Fisher, 1997; Fisher et al., 1998).

One more consideration is appropriate in the present context. Routinely, drugs other than cholinergic agonists are employed in therapy of AD. Thus, the use of antidepressants is common; the antidepressants employed include tricyclics such as amitriptyline and oxepin and other antidepressants such as bupropion. Also antispasmodics, antiemetics, and antihistaminics are utilized. The problem is that, besides the anticholinergic antispasmodics, antihistaminics, antiemetics, and, particularly, tricyclics such as amitryptyline exhibit quite potent anticholinergic effects; anticholinergics exert anticognitive and antimemory actions, as reviewed in detail in Chapter 9 BV. Accordingly, it could be expected that the use of these drugs in AD cases treated with cholinergic therapy may diminish the cholinergic efficacy in this condition. That this indeed is so is indicated in several studies (see, for example, Lu and Tune, 2003); comprehensive evaluation of this problem is under way in several hospitals (personal communication, J. Bieri, C. Palla, and P. Drahos, 2004).

It should be added that antiChEs were employed and found relatively useful in dementias of non-Alzheimer nature, such as Parkinson's disease and Lewy bodies dementias (see, for example, Burn and McKeith, 2003). Finally, cholinergic agonists and antiChEs were used in certain human conditions that are not relevant to their use in AD; thus, they are employed as antidotes in the case of toxicity arising from the use of tricyclic antidepressants and to accelerate recovery from anesthesia (see Karczmar, 1979, 1995, see also Chapter 11-1). Furthermore, both nicotinic and muscarinic agonists may facilitate, in animals and in elderly humans who do not suffer from AD, memory, learning, and cognition, although the pertinent data are inconsistent (see for example, Blacker, 2005) (see also Chapter 9 BV). Occasionally, more exotic uses of antiChEs and cholinergic agonists were reported sporadically; for example, according to Fujiki et al. (2005), donepezil attenuated experimental cerebral infarction induced in rats.

b. Other Therapies and Future Directions

Observational evidence has accumulated suggesting the possible benefits of some nonsteroidal anti-inflammatory agents (McGeer et al., 1996), statins (Wolozin et al., 2000; Jick et al., 2000; Crisby et al., 2002), and antioxidants (Martin, 2003) in reducing the progression of AD. The potential effects of anti-inflammatory agents in brain of AD patients may be independent of their action on cyclooxygenase activity or inflammation and may stem instead from their inhibition of γ-secretase (Eriksen et al., 2003). One randomized, double-blind, prospective trial completed recently showed that rofecoxib (25 mg/d) or low-dose naproxen (440 mg/d) did not slow cognitive decline in patients with mild to moderate AD over 1 year (Aisen et al., 2003). Other similar studies are in progress.

Statins, which inhibit hydroxymethylglutaryl CoA reductase and are used clinically to lower blood cholesterol, have been found effective in reducing β-amyloid peptides in transgenic animals and *in vitro* (Frears et al., 1999; Fassbender et al., 2001). In the AD double APP mutation transgenic model TgCRND8, statins both increased the proportion of α- to β-cleavage and reduced total expression of the human mutated APP, thus reducing the total of Aβ-peptides while increasing sAPPα quantitatively (Chauhan et al., 2004). It is known that cholesterol decreases α-secretase activity (Bodovitz and Klein; 1996). These effects may ensue from reducing cholesterol content of neurocellular membranes and altering cholesterol trafficking and the balance between free cholesterol and cholesterol esters, and/or by decreasing mevalonate levels with the consequent reduced prenylation of certain small G proteins that are involved in plasmalemma-signaling pathways affecting α-secretase function. But the optimum dosages, duration of treatment, and ratios of risk to benefit for long-term and early treatment or prevention of dementia, possibly needed at young adult stages, have not been determined for any of these various agents, including the statins and anti-inflammatory agents (Etminan et al., 2003).

The *N*-methyl-D-aspartate (NMDA) receptor antagonist, memantine, has been tested in clinical trials on the theory that blockade of "weak" NMDA receptors at some low level of inhibitor may reduce calcium influx and associated excitotoxicity in cholinergic neurons in the hippocampus before interfering with memory and learning (Rogawski and Wenk, 2003). A meta-analysis revealed modest benefits in cognitive tests in moderate to severe AD but not in the clinical impression of change (Areosa and Sherriff, 2003).

An exciting prospect is the potential benefit of immunization against Aβ based on the pronounced efficacy of anti-Aβ in reducing amyloid burdens in transgenic mice (Schenk et al., 1999; Janus et al., 1999; Weiner et al., 2000; Bard et al., 2000; DeMateos et al., 2001) and reducing or preventing memory and learning deficits in various paradigms in transgenic mice (Janus et al., 1999; Morgan et al., 2000, 2003). In addition, passive immunization by injection of antibodies into 3rd ventricle of transgenic mice has proved effective in reducing amyloid, reversing presynaptic changes and gliosis (Chauhan and Siegel, 2002, 2003, 2005). It is now needed to prove this effective in primate brain. And, a monkey model was developed by Gandy et al. (2004). A clinical trial of systemic active immunization against Aβ in humans had to be discontinued because 6% of 300 subjects developed meningoencephalitis (Schenk, 2002; Lee et al., 2005). Although dosing was interrupted, many patients developed significant anti- Aβ antibody titers. Epitope mapping of serum samples from these patients showed that the predominant response was "exquisitely specific" for the free amino terminus of Aβ regardless of the presence or absence of meningoencephalitis. Autopsied cases revealed regions of brain with uncharacteristically low plaque burden suggesting that immunization is effective as shown by the transgenic animal studies (Lee et al., 2005). It should be added that encephalitis was not induced in Gandy's model. A clinical trial with passive systemic immunization using anti-Aβ is in progress.

Increased reactivity of T cells to Aβ has been found in normal aging humans and subjects with AD (Monsonego et al., 2003). Patients with late-onset AD were found to have lower serum levels of antibody to Aβ42 than had sera from elderly controls, suggesting a possible impairment of B cell production of antibodies to Aβ42 in the AD group (Weksler et al., 2002). Studies with transgenic animals suggest that prolonged active immunization is required to ameliorate behavioral deficits (Austin et al., 2003). Yet, sustained high levels of plasma antibodies to Aβ from systemic immunization may adversely affect cerebral vessels containing amyloid and thereby cause hemorrhage and perivascular inflammatory responses (Pfeifer et al., 2002a, 2002b). Passive immunization through intracerebroventricular (ICV) infusion of anti-Aβ, which may avoid systemic T cell and macrophage responses and exposure of vasculature luminal surface to antibodies, results in clearance of amyloid and reversal of inflammatory responses with no microhemorrhage in transgenic mice this approach constitutes another potential means of antibody administration (Chauhan et al., 2004).

As discussed above, NGF is a critical cholinergic trophic substance that promotes growth of cholinergic neurons and fibers, most notably within the basal forebrain nuclei of Meynert in the CNS; these neurons degenerate in AD (see section A-2 this chapter). This information suggested the employment of trophic factors in the treatment of AD. Preclinical studies with primates showed safety and efficacy of NGF gene delivery to brain with ex vivo genetically modified cells. A clinical trial was mounted in which autologous fibroblasts genetically modified ex vivo to produce and secrete human NGF were stereotactically implanted into the basal forebrain in 8 patients. After 22 months of follow-up in 6 individuals, no adverse effects had occurred and a 50% decrease in the rate of cognitive decline was measured at 18 months while PET scans showed significant increases in cortical metabolic rate (Tuszynski et al., 2005). Presently, a clinical trial is in progress with implanted NGF gene linked with adeno-associated viral vector for *in vivo* genetic modification. Other trials of the use of trophins in AD were undertaken in Scandinavia and Italy (Lars Olson, personal communcation; Hefti and Weiner, 1986; Russel, 1988; Massa et al., 2002; see also Butcher and Woolf, 1989; see also Chapter VIIICI, CII and CIII). The final verdict is far from being reached, for the success of this treatment is as yet unavailable. And, Larry Butcher and Nancy Woolf launched an interesting counterattack on the suggestions in question, as well as posited new treatment possibilities (Butcher and Woolf, 1999 and 2003; Woolf et al., 1989). They pointed out that "abnormal structural profiles," represented for example by "filopodia-like extensions form the . . . neuronal . . . soma" appear early in several brain parts of AD patients. They associated these aberrant morphologies with a change in the proteins of the neuronal cytoskeletal framework (see, for example, Wolozin et al., 2000; Wolozin and Davis, 1987; and Section B, above), including an increase in the microtubule-associated proteins (MAPs; see also Chapter VIII BII). Finally, they related these morphopathologies and protein changes to the action of trophins, including NGF, astrocytes and the p75 neurotrophin receptor (see Chapter CI). Earlier (Butcher and Woolf, 1989) they hypothesized that rather than acting as an antagonist of apoptosis (see Chapter VIIICI), NGF and other trophins "accelerate the pathological cascade" of AD. Subsequently, they stressed (Butcher and Woolf, 2003) the apoptotic function of p75 neurotrophin receptor, as it may, under certain circumstances, increase apoptosis; they point out that NGF and the neurotrophn receptor may be, indeed, increased in AD (see Atterwill and Bowen, 1986 and Rabidazeh, 1993). Accordingly, they suggest that rather than using trophins, future therapies of AD should aim at diminishing NGF synthesis and blocking NGF receptors! Indeed, therapies that are not directed to eliminating the primary causative agent may produce limited symptomatic improvement but cannot cure the disease.

At this writing, many prospective, randomized, double-blinded clinical studies as well as animal investigations are under way, including: non-steroidal anti-inflammatory drugs (rofecoxib, naproxen, celecoxib, ibuprofen, dapsone), antioxidants and oxygen radical scavengers (alpha-tocopherol, selegeline, idebenone, Ginko biloba extract), estrogen replacements, estrogen plus progesterone, AMPA glutamate receptor modulators and glutamate receptor (NMDA) antagonists or neuroprotective agents (it is interesting that some derivatives or stereoisomers of antiChEs may serve as such (Li et al., 2005)); serotonin and nicotinic receptor agonists, GABA receptor antagonist, calcium flux regulators (memantine and its second genearation derivatives (Lipton, 2004)); statins (simvastatin, atorvastatin), agents such as folate, B6 and E12 that lower plasma homocysteine levels, immunomodulatory agent (cyclophosphamide), and antigonadotrophins (such as neuprolide). One study seeks to improve CSF drainage through flow-regulated ventriculoperitoneal shunting in order to remove putative toxins. Many of these measures are palliative, and frequently the results so far obtained are not encouraging (see, for example, Thal et al., 2003). In addition, there are studies in progress to image and quantify amyloid plaques and muscarinic and nicotinic receptors in AD brain with PET and SPECT to aid in early diagnosis, prior to onset of dementia, and to follow the course of the illness and of treatment effects.

Based on what is known of the molecular pathology, research in therapeutic strategies is addressed to finding biological agents or synthesizing molecules: that may control certain of the regulatory factors such as GSK3β and Cdk5 involved in APP and tau processing and their functions; that inhibit the β- and γ-secretases which produce Aβ; that regulate the balance between α- and β-processing of APP; that prevent Aβ from accumulating by binding to and blocking Aβ from aggregating; or that dissolve amyloid plaques with copper-zinc chelating (Cherny et al., 2001) and peptide trapping agents, that regulate cholesterol compartmentation and interaction with Aβ; and that regulate secondary immune and inflammatory reactions in AD. These subjects have been frequently reviewed (Suh and Checler, 2002; Janus, 2003; Marks and Berg, 2003; Gandy, 2003; Bush, 2003; Martin, 2003; Harrison, 2003; Etminan, 2003; DeKosky, 2003).

In addition, understanding genetic polymorphism may help determine which individuals may respond to particular drugs (Frank and Hargreaves, 2003). The research into these areas is very exciting and promises to yield novel methods for treatment based on the underlying pathologic processes and genetic regulation of responses but there is much yet to be learned about the physiology and pathology of all the cellular processes involved in AD.

Notes

1. Nowadays, it is not generally recognized that Alois Alzheimer contributed significantly to another degenerative disease, Huntington's chorea (Alzheimer, 1911).
2. It must be noted that the effectiveness in question involves statistically evaluated changes in scales; here Steve (Bernard B.) Brodie's dictum comes to mind: "If one needs statistics to demonstrate a difference, then perhaps there is no difference."
3. Ezio Giacobini refers to increased levels of BuChE in AD, as he suggests that this enzyme should be considered in AD therapy (Giacobini, 2003a, 2003b); his reasoning is not entirely clear (see this chapter and chapter 3 DIII).

References

Adem, A. Nordberg, A., Bucht, G. and Winblad, B.: Nicotinic and muscarinic binding sites on lymphocytes of Alzheimer patients. In: "Alzheimer's and Parkinson's Diseases," A. Fisher, I. Hanin, and C. Lachman, Eds., pp. 337–344, Plenum Press, New York, 1985.

Adler, G., Brassen, S. and Jajcevic, A.: EEG coherence in Alzheimer's dementia. J. Neural. Transm. *110*: 1051–1058, 2003.

Aisen, P. S., Schafer, K. A., Grundman, M., Pfeiffer, E., Sano, M., Davis, K. L., Farlow, M. R., Jin, S., Thomas, R. G. and Thal, L. J.: Alzheimer's Disease Cooperative Study. Effects of rofecoxib or naproxen vs placebo on Alzheimer disease progression: a randomized controlled trial. JAMA *289*(21): 2819–2826, 2003.

Alzheimer, A.: Über eine einartige Erkränkung der Hirnrinde. XXXVII Versammlung südwestdeutscher Irrenärzte in Tübingen, 4 Nov., 1906. Neurologishes Centralblatt, p. 1134, 1906.

Alzheimer, A.: Über eine einartige Krankheitsfälle des späteren Alters. Zeitshrft. Ges. Neurologie u. Psychiatrie *4*: Heft 3, 1907.

Alzheimer, A.: Über die anatomische Grundlage der Hutingtonschen Chorea und der choreatischen Bewegungen überhaupt. Ztschrft. Ges. Neurol. Psychiat. *3*: 566, 1911.

Alzheimer, A.: On certain peculiar diseases of old age. In: "History of Psychiatry," Translated, with an Introduction by H. Foerstl and R. Levy, Alpha Academic Press, London, vol. 2, pp. 71–101, 1991.

Anand, R., Hartman, R. and Messina, J.: Long-term treatment with rivastigmine continues to provide benefits for up to one year. Fifth International Geneva-Springfield Symposium on Advances in Alzheimer Therapy, Geneva, Abstract 18, 1998.

Anderson, J. P., Esch, F. S., Keim, P. S., Sambamurti, K., Lieberburg, I. and Robakis, N. K.: Exact cleavage site of Alzheimer amyloid precursor in neuronal PC-12 cells. Neurosci. Lett. *128*(1): 126–128, 1991.

Anderton, B. H., Betts, J., Blackstock, W. P., Brion, J. P., Chapman, S., Connell, J., Dayanandan, R., Gallo, J. M., Gibb, G., Hanger, D. P., Hutton, M., Kardalinou, E., Leroy, K., Lovestone, S., Mack, T., Reynolds, C. H. and Van Slegtenhorst, M.: Sites of phosphorylation in tau and factors affecting their regulation. Biochem. Soc. Symp. *2001*(67): 73–80, 2001.

Aoki, M., Iwamoto-Sugai, M., Sugiura, I., Sasaki, C., Hasegawa, T., Okumura, C., Sugio, S., Kohno, T. and Matsuzaki, T.: Expression, purification and crystallization of human tau-protein kinase I/glycogen synthase kinase-3beta. Acta Crystallogr. D Biol. Crystallogr. *56*, Pt. 11: 1464–1465, 2000.

Aoukati, A. and Tan, R.: Role for glycogen synthase kinase-3 in NK cell cytotoxicity and X-linked lymphoproliferative disease, J. Immunol. *174*: 4551–4558, 2005.

Aplin, A. E., Gibb, G. M., Jacobsen, J. S., Gallo, J. M. and Anderton, B. H.: *In vitro* phosphorylation of the cytoplasmic domain of the amyloid precursor protein by glycogen synthase kinase-3beta. J. Neurochem. *67*(2): 699–707, 1996.

Arai, Y., Suzuki, A., Mizuguchi, M. and Takashima, S.: Developmental and aging changes in the expression of amyloid precursor protein in Down syndrome brains. Brain Dev. *19*(4): 290–294, 1997.

Arai, T., Ikeda, K., Akiyama, H., Shikamoto, Y., Tsuchiya, K., Yagishita, S., Beach, T., Rogers, J., Schwab, C. and McGeer, P. L.: Distinct isoforms of tau aggregated in neurons and glial cells in brains of patients with Pick's disease, corticobasal degeneration and progressive supranuclear palsy. Acta Neuropathol. (Berl.) *101*(2): 167–173, 2001.

Arawaka, S., Hasegawa, H., Tandon, A., Janus, C., Chen, F., Yu, G., Kikuchi, K., Koyama, S., Kato, T., Fraser, P. E. and St. George-Hyslop, P.: The levels of mature glycosylated nicastrin are regulated and correlate with gamma-secretase processing of amyloid beta-precursor protein. J. Neurochem. *83*(5): 1065–1071, 2002.

Arendash, G. W., King, D. L., Gordon, M. N., Morgan, D., Hatcher, J. M., Hope, C. L. and Diamond, D. M.: Progressive, age-related behavioral impairments in transgenic mice carrying both mutant amyloid precursor protein and presenilin-1 transgenes. Brain Res. *891*: 42–53, 2001.

Areosa, S. A. and Sherriff, F.: Memantine for dementia. Cochrane Database Syst. Rev. (3): CD003154, 2003.

Arias, E., Ales, E., Gabilan, N. H., Cano-Abad, M. F., Villarroya, M. Garcia, A. G. and Lopez, M. G.: Galantamine prevents apoptosis induced by beta-amyloid and thapsigargin: involvement of nicotine acetylcholine receptors. J. Pharmacol. Exper. Therap. *310*: 387–994, 2004.

Arias, E., Gallego-Sandin, S., Villarroya, M., Garcia, A. G. and Lopez, M. G.: Unequal protection afforded by the acetylcholinesterase inhibitors galantamine, donepezil and rivastigmine: role of nicotinic receptors of SH-SY5H neuroblastoma cells. J. Phamacol. Exper. Therap. 311, 2005.

Atterwill, C. K. and Bowen, D. M.: Neurotrophic factor for central cholinergic neurons is present in both normal and Alzheimer brain tissue. Acta Neuropthol. (Berlin) *69*: 341–342, 1986.

Atwood, C. S., Perry, G., Smith, M. A.: Cerebral hemorrhage and amyloid-beta. Science *299*(5609): 1014, 2003.

Auld, D. S., Kar, S. and Quirion, R.: Beta-amyloid peptides as direct cholinergic neuromodulators: a missing link? TINS *21*: 43–49, 1998.

Austin, L., Arendash, G. W., Gordon, M. N., Diamond, D. M., DiCarlo, G., Dickey, C., Ugen, K. and Morgan, D.: Short-term beta-amyloid vaccinations do not improve cognitive performance in cognitively impaired APP + PS1 mice. Behav. Neurosci. *117*(3): 478–484, 2003.

Battistin, L. and Gerstenbrand, F., Eds.: "Aging Brain and Dementia," Wiley-Liss, New York, 1990.

Becker, R. and Giacobini, E., Eds.: "Cholinergic Basis for Alzheimer Therapy," Birkhauser, Boston, 1991.

Bender-Pape, T. L., Heinemann, A. W., Kelly, J. P., Giobbie Hurder, A. and Lundgren, S.: A measure of neurobehavioral functioning after coma. I. Theory, reliability and validity of the disorders of consciousness scale. J. Rehabil. Res. Dev., *42*: 1–17, 2005.

Berezovska, O., Ramdya, P., Skoch, J., Wolfe, M. S., Bacskai, B. J. and Hyman, B. T.: Amyloid precursor protein associates with a nicastrin-dependent docking site on the presenilin 1-gamma-secretase complex in cells demonstrated by fluorescence lifetime imaging. J. Neurosci. *23*(11): 4560–4566, 2003.

Bian, F., Nath, R., Sobocinski, G., Booher, R. N., Lipinski, W. J., Callahan, M. J., Pack, A., Wang, K. K. and Walker, L. C. Axonopathy, tau abnormalities, and dyskinesia, but no neurofibrillary tangles in p25-transgenic mice. J. Comp. Neurol. *446*(3): 257–266, 2002.

Bick, K., Amaducci L. and Pepeu, G.: "The Early Story of Alzheimer Disease," Liviana Press, Padova, 1987.

Bielschowski, M.: Zur Kentniss der Alzheimerschen Krankheit. J. Psychol. Neurol. *18*: 273–292, 1911.

Bjorkhem, I., Lutjohann, D., Diczfalusy, U., Stahle, L., Ahlborg, G. and Wahren, J.: Cholesterol homeostasis in human brain: turnover of 24S-hydroxycholesterol and evidence for a cerebral origin of most of this oxysterol in the circulation. J. Lipid Res. *39*(8): 1594–1600, 1998.

Blacker, D.: Mild cognitive impairment—no benefit from vitamin E, little from donepezil. N. Engl. J. Med. *352*: 2439–2441, 2005.

Blessed, G., Tomlison, B. E. and Roth, M.: The association between quantitative measures of dementia and senile changes in the cerebral grey matter of elderly subjects. Brit. J. Psychiatry *114*: 797–811, 1968.

Blumenthal, H. T. and Premachandra, B. N.: Bridging the aging-disease dichotomy. I. The amyloidosis model. Perspect. Biol. Med. *33*: 402–420, 1990.

Bodovitz, S. and Klein, W. L.: Cholesterol modulates alpha-secretase cleavage of amyloid precursor protein. J. Biol. Chem.: *271*(8): 4436–4440, 1996.

Boncristiano, S., Calhoun, M. E., Kelly, P. H., Pfeifer, M., Bondolfi, L., Stalder, M., Phinney, A. L., Abramowski, D., Sturchler-Pierrat, C., Enz, A., Sommer, B., Staufenbiel, M. and Jucker, M.: Cholinergic changes in the APP23 transgenic mouse model of cerebral amyloidosis. J. Neurosci. *22*(8): 3234–3243, 2002.

Borchelt, D. R., Thinakaran, G., Eckman, C. B., Lee, M. K., Davenport, F., Ratovitsky, T., Prada, C. M., Kim, G., Seekins, S., Yager, D., Slunt, H. H., Wang, R., Seeger, M., Levey, A. I., Gandy, S. E., Copeland, N. G., Jenkins, N. A., Price, D. L., Younkin, S. G. and Sisodia, S. S.: Familial Alzheimer's disease-linked presenilin 1 variants elevate Abeta1–42/1–40 ratio *in vitro* and *in vivo*. Neuron *17*(5): 1005–1013, 1996.

Borchelt, D. R., Ratovitski, T., van Lare, J., Lee, M. K., Gonzales, V., Jenkins, N. A., Copeland, N. G., Price, D. L. and Sisodia, S. S.: Accelerated amyloid deposition in the brains of transgenic mice coexpressing mutant presenilin 1 and amyloid precursor proteins. Neuron *19*(4): 939–945, 1997.

Bouillot, C., Prochiantz, A., Rougon, G. and Allinquant, B.: Axonal amyloid precursor protein expressed by neurons *in vitro* is present in a membrane fraction with caveolae-like properties. J. Biol. Chem. *271*(13): 7640–7644, 1996.

Bowen, D. M., Smith, C. B., White, P. and Davison, A. N.: Neurotransmitter-related enzymes and indices of hypoxia in senile dementia and other abiotrophies. Brain *49*: 459–495, 1976.

Boyle, P. A. and Malloy, P. F.: Treating apathy in Alzheimer's disease. Dement. Geriatr. Cogn. Disord. *17*: 91–99, 2004.

Brion, J. P., Anderton, B. H., Authelet, M., Dayanandan, R., Leroy, K., Lovestone, S., Octave, J. N., Pradier, L., Touchet, N. and Tremp, G.: Neurofibrillary tangles and tau phosphorylation. Biochem. Soc. Symp. *2001*(67): 81–88, 2001.

Bullock, B. P. and Habener, J. F.: Phosphorylation of the cAMP response element binding protein CREB by cAMP-dependent protein kinase A and glycogen synthase kinase-3 alters DNA-binding affinity, conformation, and increases net charge. Biochemistry *37*(11): 3795–3809, 1998.

Burn, D. J. and McKeith, I. G.: Current treatment of dementia with Lewy bodies and dementia associated with Parkinson's disease. Mov. Disord. *18*, Supplt. 6: 872–879, 2003.

Bush, A. I.: The metallobiology of Alzheimer's disease. Trends Neurosci. *26*: 207–214, 2003.

Butcher, L. L. and Woolf, N. J.: Neurotrophic agents may exacerbate the pathologic cascade of Alzheimer's disease. Neurobiol. Aging *10*: 557–570, 1989.

Buxbaum, J. D.: Animal models of Alzheimer's disease. Neurosci. Facts *4*: 1, 1993.

Buxbaum, J. D., Liu, K. N., Luo, Y., Slack, J. L., Stocking, K. L., Peschon, J. J., Johnson, R. S., Castner, B. J., Cerretti, D. P. and Black, R. A.: Evidence that tumor necrosis factor alpha converting enzyme is involved in regulated alpha-secretase cleavage of the Alzheimer amyloid protein precursor. J. Biol. Chem. *273*(43): 27765–27767, 1998.

Cai, H., Wang, Y., McCarthy, D., Wen, H., Borchelt, D. R., Price, D. L. and Wong, P. C.: BACE1 is the major beta-secretase for generation of Abeta peptides by neurons. Nat. Neurosci. *4*(3): 233–234, 2001.

Cai, X. D., Golde, T. E., Younkin, S. G.: Release of excess amyloid beta protein from a mutant amyloid beta protein precursor. Science *259*(5094): 514–516, 1993.

Cataldo, A. M., Barnett, J. L., Pieroni, C. and Nixon, R. A.: Increased neuronal endocytosis and protease delivery to early endosomes in sporadic Alzheimer's disease: neuropathologic evidence for a mechanism of increased beta-amyloidogenesis. J. Neurosci. *17*(16): 6142–6151, 1997.

Chauhan, N. B.: Membrane dynamics, cholesterol metabolism and Alzheimer's disease, J. Lipid Res. *44*: 2019–2029, 2003.

Chauhan, N. B. and Siegel, G. J.: Reversal of amyloid beta toxicity in Alzheimer's disease model Tg2576 by intraventricular antiamyloid beta antibody. J. Neurosci. Res. *69*(1): 10–23, 2002.

Chauhan, N. B. and Siegel, G. J.: Effect of PPF and ALCAR on the induction of NGF- and p75-mRNA and on APP processing in Tg2576 brain. Neurochem. Int. *43*(3): 225–233, 2003.

Chauhan, N. B. and Siegel, G.J.: Efficacy of anti-Aβ antibody isotypes used for intracerebroventricular immunization in TgCRND8. Neurosci. Lett. *375*: 143–147, 2005.

Chauhan, N. B., Siegel, G. J. and Feinstein, D. L.: Propentofylline attenuates tau hyperphosphorylation in Alzheimer's Swedish mutant model Tg2576. Neuropharmacology. Jan; *48*(1): 93–104, 2005.

Chauhan, N. B., Lichtor, T. and Siegel, G. J.: Aging potentiates Abeta-induced depletion of SNAP-25 in mouse hippocampus. Brain Res. *982*(2): 219–227, 2003.

Chauhan, N. B., Siegel, G. J. and Feinstein, D. L.: Effects of lovastatin and pravastatin on amyloid processing and inflammatory response in TgCRND8 brain. Neurochem. Res. *29*: 1897–1911, 2004.

Cheng, S. V., Nadeau, J. H., Tanzi, R. E., Watkins, P. C., Jagadesh, J., Taylor, B. A., Haines, J. L., Sacchi, N. and Gusella, J. F.: Comparative mapping of DNA markers from the familial Alzheimer disease and Down syndrome regions of human chromosome 21 to mouse chromosomes 16 and 17. Proc. Natl. Acad. Sci. U.S.A. *85*(16): 6032–6036, 1988.

Cherny, R. A., Atwood, C. S., Xilinas, M. E., Gray, D. N., Jones, W. D., McLean, C. A., Barnham, K. J., Volitakis, I., Fraser, F. W., Kim, Y., Huang, X., Goldstein, L. E., Moir, R. D., Lim, J. T., Beyreuther, K., Zheng, H., Tanzi, R. E., Masters, C. L. and Bush, A. I.: Treatment with a copper-zinc chelator markedly and rapidly inhibits beta-amyloid accumulation in Alzheimer's disease transgenic mice. Neuron. *30*: 665–676, 2001.

Chochina, S. V., Avdulov, N. A., Igbavboa, U., Cleary, J. P., O'Hare, E. O. and Wood, W. G.: Amyloid beta-peptide 1–40 increases neuronal membrane fluidity: role of cholesterol and brain region. J. Lipid Res. *42*(8): 1292–1297, 2001.

Choi, E. K., Zaidi, N. F., Miller, J. S., Crowley, A. C., Merriam, D. E., Lilliehook, C., Buxbaum, J. D. and Wasco, W.: Calsenilin is a substrate for caspase-3 that preferentially interacts with the familial Alzheimer's disease-associated C-terminal fragment of presenilin 2. J. Biol. Chem. *276*(22): 19197–19204, 2001.

Chong, Y. H. and Suh, Y. H.: Amyloidogenic processing of Alzheimer's amyloid precursor protein *in vitro* and its modulation by metal ions and tacrine. Life Sci. *59*: 545–557, 1996.

Citron, M., Oltersdorf, T., Haass, C., McConlogue, L., Hung, A. Y., Seubert, P., Vigo-Pelfrey, C., Lieberburg, I. and Selkoe, D. J.: Mutation of the beta-amyloid precursor protein in familial Alzheimer's disease increases beta-protein production. Nature *360*(6405): 672–674, 1992.

Citron, M., Westaway, D., Xia, W., Carlson, G., Diehl, T., Levesque, G., Johnson-Wood, K., Lee, M., Seubert, P., Davis, A., Kholodenko, D., Motter, R., Sherrington, R., Perry, B., Yao, H., Strome, R., Lieberburg, I., Rommens, J., Kim, S., Schenk, D., Fraser, P., St. George Hyslop, P. and Selkoe, D. J.: Mutant presenilins of Alzheimer's disease increase production of 42-residue amyloid beta-protein in both transfected cells and transgenic mice. Nat. Med. *3*(1):67–72, 1997.

Committee on Chemical Toxicity and Aging: "Aging in Today's Environment," National Academy Press, Washington, D.C., 1987.

Cook, L. J., Ho, L. W., Wang, L., Terrenoir, E., Brayne, C., Evans, J. G., Xuereb, J., Cairns, N. J., Turic, D., Hollingwrth, P., Moore, P. J., Jehu, L., Archer, N., Walter, S., Foy, C., Edmondson, A., Powell, J., Lovestone, S., Williams, J. and Rubinsztein, D. C.: Candidate gene association studies of genes involved in neuronal cholinergic transmission in Alzheimer's disease suggests choline acetyltransferase as a candidate deserving further study. Am. J. Med. Genet. B Neuropsychiatr. Genet. *132B*: 5–8, 2004.

Cottrell, B. A., Galvan, V., Banwait, S., Gorostiza, O., et al.: A pilot proteomic study of amyloid precursor interactors in Alzheimer's disease. Ann Neurol. *58*: 277–289, 2005.

Crisby, M., Carlson, L. A. and Winblad, B.: Statins in the prevention and treatment of Alzheimer disease. Alzheimer Dis. Assoc. Disord. *16*(3):131–136, 2002.

Cross, D. A., Culbert, A. A., Chalmers, K. A., Facci, L., Skaper, S. D. and Reith, A. D.: Selective small-molecule inhibitors of glycogen synthase kinase-3 activity protect primary neurones from death. J. Neurochem. *77*(1): 94–102, 2001.

Crowder, R. J. and Freeman, R. S.: Glycogen synthase kinase-3 beta activity is critical for neuronal death caused by inhibiting phosphatidylinositol 3-kinase or Akt but not for death caused by nerve growth factor withdrawal. J. Biol. Chem. *275*(44): 34266–34271, 2000.

Cummings, J. L., Nadel, A., Masterman, D. and Cyrus, P. A.: Efficacy of metrifonate in improving the psychiatric and behavioral disturbances of patients with Alzheimer's disease, J. Geriatr. Psychiatry Neurol. *14*: 101–108, 2001.

D'Amico, M., Hulit, J., Amanatullah, D. F., Zafonte, B. T., Albanese, C., Bouzahzah, B., Fu, M., Augenlicht, L. H., Donehower, L. A., Takemaru, K., Moon, R. T., Davis, R., Lisanti, M. P., Shtutman, M., Zhurinsky, J., Ben-Ze'ev, A., Troussard, A. A., Dedhar, S. and Pestell, R. G.: The integrin-linked kinase regulates the cyclin D1 gene through glycogen synthase kinase 3beta and cAMP-responsive element-binding protein-dependent pathways. J. Biol. Chem. *275*(42): 32649–32657, 2000.

Davies, P. and Maloney, A. J. F.: Selective loss of central cholinergic neurons in Alzheimer's disease. Lancet 2: 1402, 1976.

Davies, J. P., Chen, F. W. and Ioannou, Y. A.: Transmembrane molecular pump activity of Niemann-Pick C1 protein. Science *290*: 2295–2298, 2000.

Davis, K. L., Hollister, L. E. and Tinkleberg, J. R.: Physostigmine: effects on cognition and affect in normal subjects. Psychopharmacology *51*: 23–27, 1976.

De Ferrari, G. V. and Inestrosa, N. C.: Wnt signaling function in Alzheimer's disease. Brain Res. Brain Res. Rev. *33*(1): 1–12, 2000.

DeGiorgio, L. A., Shimizu, Y., Chun, H. S., Cho, B. P., Sugama, S., Joh, T. H. and Volpe, B. T.: APP knockout attenuates microglial activation and enhances neuron survival in substantia nigra compacta after axotomy. Glia. *38*(2): 174–178, 2002.

De Jonghe, C., Esselens, C., Kumar-Singh, S., Craessaerts, K., Serneels, S., Checler, F., Annaert, W., Van Broeckhoven, C. and De Strooper, B.: Pathogenic APP mutations near the gamma-secretase cleavage site differentially affect Abeta secretion and APP C-terminal fragment stability. Hum. Mol. Genet. *10*(16): 1665–1671, 2001.

DeKosky, S. T.: Pathology and pathways of Alzheimer's disease with an update on new developments in treatment. J. Am. Geriatr. Soc. *51*(5), Supplt. Dementia: S314–320, 2003.

Dietschy, J. M. and Turley, S. D.: Cholesterol metabolism in the brain. Curr. Opin. Lipidol. *12*(2): 105–112, 2001.

Dineley, K. T., Xia, X., Bui, D., Sweatt, J. D. and Zheng, H.: Accelerated plaque accumulation, associative learning deficits, and up-regulation of alpha 7 nicotinic receptor protein in transgenic mice co-expressing mutant human presenilin 1 and amyloid precursor proteins. J. Biol. Chem. *277*(25): 22768–22780, 2002.

Divry, P.: De la nature de l'alteration fibrillaire d'Alzheimer. J. Belge Neurol. Psychiat. *34*: 197–201, 1934.

Doble, B. W. and Woodgett, J. R.: GSK-3: tricks of the trade for a multi-tasking kinase. J. Cell Sci. *116*:1175–1186, 2003.

Dominguez, D., Tournoy, J., Hartmann, D., Huth, T., Cryus, K., Deforce, S., Serneels, L., Camacho, I. E., Marjauy, E., Craessaerts, K., Roebroek, A. J., Schwake, M., D'Hooge, R., Back, P., Kalinke, U., Moechars, D., Alzheimer, C., Reiss, K., Saftig, P. and De Strooper, B.: Phenotypic and biochemical analyses of BACE1- and BACE2-deficient mice. J. Biol. Chem. *280*: 30797–30806, 2005.

Drachman, D. A.: Central cholinergic system and memory. In: "Psychopharmacology: A Generation of Progress," M. A. Lipton, A. DiMascio, and K. F. Killam, Eds., pp. 651–662, Raven Press, New York, 1978.

Drachman, D. A. and Leavitt, J.: Human memory and the cholinergic system—a relationship to aging? Arch. Neurol. Psychiat. *30*: 113–121, 1974.

Duff, K., Eckman, C., Zehr, C., Yu, X., Prada, C. M., Perez-tur, J., Hutton, M., Buee, L., Harigaya, Y., Yager, D., Morgan, D., Gordon, M. N., Holcomb, L., Refolo, L., Zenk, B., Hardy, J. and Younkin, S.: Increased amyloid-beta42(43) in brains of mice expressing mutant presenilin 1. Nature *383*(6602): 710–713, 1996.

Ehehalt, R., Keller, P., Haass, C., Thiele, C. and Simons, K.: Amyloidogenic processing of the Alzheimer beta-amyloid precursor protein depends on lipid rafts. J. Cell Biol. *160*(1): 113–123, 2003.

Eriksen, J. L., Sagi, S. A., Smith, T. E., Weggen, S., Das, P., McLendon, D. C., Ozols, V. V., Jessing, K. W., Zavitz, K. H., Koo, E. H. and Golde, T. E.: NSAIDs and enantiomers of flurbiprofen target gamma-secretase and lower Abeta 42 *in vivo*. J. Clin. Invest. *112*(3): 440–449, 2003.

Enz, A., Amstutz, R., Bodekke, H., Gmelin, G. and Malanowski, J.: Brain selective inhibition of acetylcholinesterase: a novel approach to therapy for Alzheimer's disease. In: "Cholinergic Function and Dysfunction," A. C. Cuello, Ed., pp. 431–438, Elsevier, Amsterdam, 1993.

Esler, W. P., Kimberly, W. T., Ostaszewski, B. L., Ye, W., Diehl, T. S., Selkoe, D. J., Wolfe, M. S.: Activity-dependent isolation of the presenilin- gamma-secretase complex reveals nicastrin and a gamma substrate. Proc. Natl. Acad. Sci. U.S.A. *99*(5): 2720–2725, 2002.

Etminan, M., Gill, S. and Samii, A.: Effect of non-steroidal anti-inflammatory drugs on risk of Alzheimer's disease: systematic review and meta-analysis of observational studies. BMJ *327*(7407): 128, 2003.

Evans, J. G.: Ageing and disease. In: "Research and the Ageing Population," D. Evered and J. Whalen, Eds., Ciba Found. Symp. No. 124, Wiley, Chichester, 1988.

Fagan, A. M., Watson, M., Parsadanian, M., Bales, K. R., Paul, S. M. and Holtzman, D. M.: Human and murine ApoE markedly alters A beta metabolism before and after plaque formation in a mouse model of Alzheimer's disease. Neurobiol. Dis. *9*(3): 305–318, 2002.

Farmery, M. R., Tjernberg, L. O., Pursglove, S. E., Bergman, A., et al.: Partial purification and characterization of γ-secretase from post-mortem human brain. J. Biol. Chem. *278*: 24277–24284, 2003.

Fassbender, K., Simons, M., Bergmann, C., Stroick, M., Lutjohann, D., Keller, P., Runz, H., Kuhl, S., Bertsch, T., von Bergmann, K., Hennerici, M., Beyreuther, K. and Hartmann, T.: Simvastatin strongly reduces levels of Alzheimer's disease beta-amyloid peptides Abeta 42 and Abeta 40 *in vitro* and *in vivo*. Proc. Natl. Acad. Sci. U.S.A. *98*(10): 5856–5861, 2001.

Ferrer, I., Barrachina, M. and Puig, B.: Glycogen synthase kinase-3 is associated with neuronal and glial hyperphosphorylated tau deposits in Alzheimer's disease, Pick's disease, progressive supranuclear palsy and corticobasal degeneration. Acta Neuropathol (Berl.) *104*(6): 583–591, 2002.

Ferrer, I., Barrachina, M., Tolnay, M., Rey, M. J., Vidal, N., Carmona, M., Blanco, R. and Puig, B.: Phosphorylated protein kinases associated with neuronal and glial tau deposits in argyrophilic brain disease. Brain Pathol. *13*(1): 62–78, 2003a.

Ferrer, I., Pastor, P., Rey, M. J., Munoz, E., Puig, B., Pastor, E., Oliva, R. and Tolosa, E.: Tau phosphorylation and kinase activation in familial tauopathy linked to deln296 mutation. Neuropathol. Appl. Neurobiol. *29*(1): 23–34, 2003b.

Finch, C. E. and Roth, G. S.: Biochemistry of Aging. In: "Basic Neurochemistry", Sixth Edition, G. J. Siegel, B. W. Agranoff, R. W. Albers, S. K. Fisher, and M. D. Uhler, Eds., pp. 614–633, Philadelphia: Lippincott-Raven, 1999.

Fiol, C. J., Williams, J. S., Chou, C. H., Wang, Q. M., Roach, P. J. and Andrisani, O. M.: A secondary phosphorylation of CREB341 at Ser129 is required for the cAMP-mediated control of gene expression. A role for glycogen synthase kinase-3 in the control of gene expression. J. Biol. Chem. *269*(51): 32187–32193, 1994.

Fischer, O.: Miliary Nekrosen mit drueszinger Wucherungen der Neurofibrillen, eine regelmaessige Veraenderung der Hirnrinde bei seniler Demenz. Monatsschrft. Psych. U. Neurol. *22*: 361–372, 1907.

Fisher, A.: Muscarinic agonists for the treatment of Alzheimer's disease: progress and perspectives. Eur. Opin. Invest. Drugs *6*: 1395–1411, 1997.

Fisher, A., Brandeis, R., Haring, R., Eshar, N., Heldman, E., Karton, Y., Eisenberg, O., Meshulam, H., Marciano, D., Bar-Ner, N. and Pittel, Z.: Novel m1 muscarinic agonists in treatment and delaying the progression of Alzheimer's disease: an unifying hypothesis. In: "From *Torpedo* Electric Organ to Human Brain: Fundamental and Applied Aspects," J. Massoulié, Ed., pp. 227–340, Elsevier, Amsterdam, 1998.

Fisher, A., Michaelson, D. M., Brandeis, R., Haring, R., Chapman, S. and Pittel, Z.: M1 muscarinic agonists as potential disease-modifying agents in Alzheimer's disease. Rationale and perspectives. Ann. N. Y. Acad. Sci. *920*: 315–320, 2000.

Foerstl, H. and Levy, R., Eds.: Introduction to Alzheimer's paper "On certain diseases of Old Age," Alpha Academic Press, London, *2*: 71–73, 1991.

Forlenza, O. V., Spink, J. M., Dayanandan, R., Anderton, B. H., Olesen, O. F. and Lovestone, S.: Muscarinic agonists reduce tau phosphorylation in non-neuronal cells via GSK-3beta inhibition and in neurons. J. Neural. Transm. *107*(10): 1201–1212, 2000.

Frank, R. and Hargreaves, R.: Clinical biomarkers in drug discovery and development. Nat. Rev. Drug Discov. *2*(7): 566–580, 2003.

Fraser, P. E., Yu, G., Levesque, L., Nishimura, M., Yang, D. S., Mount, H. T., Westaway, D. and St. George-

Hyslop, P. H.: Presenilin function: connections to Alzheimer's disease and signal transduction. Biochem. Soc. Symp. *2001*(67): 89–100, 2001.

Frears, E. R., Stephens, D. J., Walters, C. E., Davies, H. and Austen, B. M.: The role of cholesterol in the biosynthesis of beta-amyloid. Neuroreport *10*(8): 1699–1705, 1999.

Fujita, M., Ichise, M., van Dyck, C. H., Zoghbi, S. S., Tamagnan, G., Mukhin, A. G., Bozkurt, A., Seneca, N., Tipre, D., DeNucci, C. C., Iida, H., Vaupel, D. B., Horti, A. G., Koren, A. O., Kimes, A. S., London, E. D., Seibyl, j. P., Baldwin, R. M. and Innis, R. B.: Quantification of nicotinic acetylcholine receptors in human brain using [1231]5-I-A-85380 SPET. Eur. J. Nucl. Med. Mol. Imaging *30*: 1620–1619, 2003.

Fujiki, M., Kobayashi, H., Uchida, S., Inoue, R. and Ishii, K.: Neuroprotective effect of donepezil, a nicotinic acetylcholine receptor activator, on cerebral infarction in rats. Brain Res. *1043*: 2236–2241, 2005.

Gamblin, T. C., Chen, F., Zambrano, A., Abraha, A., Lagalwar, S., Guillozet, A. L., Lu, M., Fu, Y., Garcia-Sierra, F., LaPointe, N., Miller, R., Berry, R. W., Binder, L. I. and Cryns, V. L.: Caspase cleavage of tau: linking amyloid and neurofibrillary tangles in Alzheimer's disease. Proc. Natl. Acad. Sci. U.S.A. *100*(17): 10032–10037, 2003.

Games, D., Adams, D., Alessandrini, R., Barbour, R., Berthelette, P., Blackwell, C., Carr, T., Clemens, J., Donaldson, T., Gillespie, F., et al.: Alzheimer-type neuropathology in transgenic mice overexpressing V717F beta-amyloid precursor protein. Nature *373*(6514): 523–527, 1995.

Gandy, S.: Estrogen and neurodegeneration. Neurochem. Res. *28*(7): 1003–1008, 2003.

Gandy, S., DeMattos R. B., Lemere, C. A., Heppner, F. L., Leerone, J., Aguzzi, A., Ershler, W. B., Dai, J., Fraser, P., St. George-Hyslop, P., Holtzman, D. M., Walker, L. C. and Keller, E. T.: Alzheimer Aa vaccination of Rhesus monkeys (Macaca Mulatta). Alzheimer Dis. Assoc. Disord. *303*: 44–46, 2004.

Garrido, J. L., Godoy, J. A., Alvarez, A., Bronfman, M. and Inestrosa, N. C.: Protein kinase C inhibits amyloid beta peptide neurotoxicity by acting on members of the Wnt pathway. FASEB J. *16*(14): 1982–1984, 2002.

Ghribi, O., Herman, M. M. and Savory, J.: Lithium inhibits Abeta-induced stress in endoplasmic reticulum of rabbit hippocampus but does not prevent oxidative damage and tau phosphorylation. J. Neurosci. Res. *71*(6): 853–862, 2003.

Giacobini, E.: Cholinergic receptors in human brain: effects of aging and Alzheimer's disease. J. Neurosci. Res. *27*: 548–560, 1990.

Giacobini, E.: The second generation of cholinesterase inhibitors: pharmacological aspects. In: "Cholinergic Basis for Alzheimer Therapy," R. Becker and E. Giacobini, pp. 247–296, Birkhaeuser, Boston, 1991.

Giacobini, E.: Cholinesterase inhibitors: from the Calabar bean to Alzheimer therapy. In: "Cholinesterases and Cholinesterase Inhibitors," E. Giacobini, Ed. pp. 181–226, Martin Dunitz, London, 2000.

Giacobini, E., Ed.: "Butyrylcholinesterase. Its Function and Inhibitors," Martin Dunitz, London, 2003a.

Giacobini, E.: Cholinergic function and Alzheimer's disease. Int. J. Geriatr. Psychiatry *18*, Supplt. 1: S1–S5, 2003b.

Giacobini, E. and Becker, R., Eds.: "Current Research in Alzheimer Therapy," Taylor and Francis, New York, 1988.

Gil-Bea, F. J., Garcia-Alloza, M., Dominquez, J., Marcos, B. and Ramirez, M. J.: Evaluation of cholinergic markers in Alzheimer's disease and in model of cholinergic deficits. Neurosci. Lett. *375*: 37–41, 2005.

Glenner, G.: Amyloid deposits and amyloidosis. N. Engl. J. Med. *302*: 1283–1292, 1980.

Glenner, G.: Editorial: reactive systemic amyloidosis in cystic fibrosis and other disorders associated with chronic inflammation. Arch. Pathol Lab. Med. *110*: 873–877, 1986.

Glenner, G. G. and Wong, C. W.: Alzheimer's disease: initial report of the purification and characterization of a novel cerebrovascular amyloid protein. Biochem. Biophys. Res. Commun. *120*: 885–890, 1984a.

Glenner, G. G. and Wong, C. W.: Alzheimer's disease and Down's syndrome: sharing of a unique cerebrovascular amyloid fibril protein. Biochem. Biophys. Res. Commun. *122*: 1131–1135, 1984b.

Glick, D.: The nature and significance of cholinesterase. In: "Comparative Biochemistry: Intermediate Metabolism of Fats, Carbohydrate Metabolism, Biochemistry of Choline," H. B. Lewis, Ed., pp. 213–233, Jacques Cattell Press, Lancaster, Penn., 1941.

Glisson, S. N., Karczmar, A. G. and Barnes, L.: Cholinergic effects on adrenergic neurotransmitters in rabbit brain parts. Neuropharmacology *11*: 465–477, 1972.

Glisson, S. N., Karczmar, A. G. and Barnes, L.: Effects of diisopropyl phosphorofluoridate on acetylcholine, cholinesterase and catecholamines of several parts of rabbit brain. Neuropharmacology *13*: 623–632, 1974.

Gomez, S., Davies, P., Faivre-Bauman, A., Valade, D., Jeannin, C., Rondot, P. and Puymirat, J.: Acetylcholinesterase activity and somatostatin-like immunoreactivity in lumbar cerebrospinal fluid of demented patients. In: "Alzheimer's and Parkinson's Diseases," A. Fisher, I. Hanin, and C. Lachman,

Eds., pp. 317–322, Plenum Press, New York, 1985.

Gottardi, C. J., Wong, E. and Gumbiner, B. M.: E-cadherin suppresses cellular transformation by inhibiting beta-catenin signaling in an adhesion-independent manner. J. Cell Biol. *153*(5): 1049–1060, 2001.

Gotz, J., Chen, F., van Dorpe, J. and Nitsch, R. M.: Formation of neurofibrillary tangles in P301l tau transgenic mice induced by Abeta 42 fibrils. Science *293*(5534): 1491–1495, 2001.

Gray, C. W. and Patel, A. J.: Induction of beta amyloid precursor protein isoform mRNAs by bFGF in astrocytes. Neuroreport *4*(6): 811–814, 1993.

Grimes, C. A. and Jope, R. S.: CREB DNA binding activity is inhibited by glycogen synthase kinase-3 beta and facilitated by lithium. J. Neurochem. *78*(6): 1219–1232, 2001.

Gyure, K. A., Durham, R., Stewart, W. F., Smialek, J. E. and Troncoso, J. C.: Intraneuronal abeta-amyloid precedes development of amyloid plaques in Down syndrome. Arch. Pathol. Lab. Med. *125*(4): 489–492, 2001.

Haass, C., Schlossmacher, M. G., Hung, A. Y., Vigo-Pelfrey, C., Mellon, A., Ostaszewski, B. L., Lieberburg, I., Koo, E. H., Schenk, D., Teplow, D. B., et al.: Amyloid beta-peptide is produced by cultured cells during normal metabolism. Nature *359*(6393): 322–325, 1992a.

Haass, C., Koo, E. H., Mellon, A., Hung, A. Y. and Selkoe, D. J.: Targeting of cell-surface beta-amyloid precursor protein to lysosomes: alternative processing into amyloid-bearing fragments. Nature *357*(6378): 500–503, 1992b.

Hailstones, D., Sleer, L. S., Parton, R. G. and Stanley, K. K.: Regulation of caveolin and caveolae by cholesterol in MDCK cells. J. Lipid Res. *39*(2): 369–379, 1998.

Han, X., Cheng, H., Fryer, J. D., Fagan, A. M. and Holtzman, D. M.: Novel role for apolipoprotein E in the central nervous system: modulation of sulfatide content. J. Biol. Chem. *278*(10): 8043–8051, 2003.

Handelmann, G. E., Boyles, J. K., Weisgraber, K. H., Mahley, R. W. and Pitas, R. E.: Effects of apolipoprotein E, beta-very low density lipoproteins, and cholesterol on the extension of neurites by rabbit dorsal root ganglion neurons *in vitro*. J. Lipid Res. *33*(11): 1677–1688, 1992.

Hanger, D. P., Hughes, K., Woodgett, J. R., Brion, J. P. and Anderton, B. H.: Glycogen synthase kinase-3 induces Alzheimer's disease-like phosphorylation of tau: generation of paired helical filament epitopes and neuronal localisation of the kinase. Neurosci Lett. *147*(1): 58–62, 1992.

Hardt, S. E. and Sadoshima, J.: Glycogen synthase kinase-3beta: a novel regulator of cardiac hypertrophy and development. Circ. Res. *90*(10): 1055–1063, 2002.

Haring, R., Fisher, A., Marciano, D., Pittel, Z., Kloog, Y., Zuckerman, A., Eshhar, N. and Heldman, E.: Mitogen-activated protein kinase-dependent and protein kinase C-dependent pathways link the m1 muscarinic receptor to beta-amyloid precursor protein secretion. J. Neurochem. *71*(5): 2094–2103, 1998.

Harkany, T., O'Mahoney, S. and Kelly, J. P.: Beta-amyloid {Phe [(So3H) 24] 25–35} in rat nucleus basalis induces behavioral dysfunctions, impairs learning and memory and disrupts cholinergic innervation. Behav. Brain Res. *90*: 133–145, 1998.

Harper, S. J., Bilsland, J. G., Shearman, M. S., Zheng, H., Van der Ploeg, L. and Sirinathsinghji, D. J.: Mouse cortical neurones lacking APP show normal neurite outgrowth and survival responses *in vitro*. Neuroreport *9*(13): 3053–3058, 1998.

Harrison, T. and Beher, D.: Gamma-secretase inhibitors—from molecular probes to new therapeutics? Prog. Med. Chem. *41*: 99–127, 2003.

Hartman, R. E., Wozniak, D. F., Nardi, A., Olney, J. W., Sartorius, L. and Holtzman, D. M.: Behavioral phenotyping of GFAP-apoE3 and -apoE4 transgenic mice: apoE4 mice show profound working memory impairments in the absence of Alzheimer's-like neuropathology. Exp. Neurol. *170*(2): 326–344, 2001.

Hefti, F. and Weiner, W. J.: Nerve growth factor and Alzheimer's disease. Ann. Neurol. *20*: 275–281, 1986.

Hernandez, F., Borrell, J., Guaza, C., Avila, J. and Lucas, J. J.: Spatial learning deficit in transgenic mice that conditionally over-express GSK-3beta in the brain but do not form tau filaments. J. Neurochem. *83*(6): 1529–1533, 2002.

Herreman, A., Van Gassen, G., Bentahir, M., Nyabi, O., Craessaerts, K., Mueller, U., Annaert, W. and De Strooper, B.: Gamma-secretase activity requires the presenilin-dependent trafficking of nicastrin through the Golgi apparatus but not its complex glycosylation. J. Cell Sci. *116*, Pt. 6: 1127–1136, 2003.

Higuchi, K., Matsumara, A. and Honma, A.: Senile amyloidosis in senescence accelerated mice. Lab. Invest. *48*: 231–240, 1983.

Hogan, D. B., Goldlist, B., Naglie, G. and Patterson, C.: Comparison studies of cholinesterase inhibitors for Alzheimer's disease. Lancet Neurol. *3*: 622–626, 2004.

Howland, D. S., Savage, M. J., Huntress, F. A., Wallace, R. E., Schwartz, D. A., Loh, T., Melloni, R. H., Jr., DeGennaro, L. J., Greenberg, B. D., Siman, R., Swanson, M. E. and Scott, R. W.: Mutant and native human beta-amyloid precursor proteins in transgenic mouse brain. Neurobiol. Aging *16*(4): 685–699, 1995.

Hsia, A. Y., Masliah, E., McConlogue, L., Yu, G. Q., Tatsuno, G., Hu, K., Kholodenko, D., Malenka, R. C., Nicoll, R. A. and Mucke, L.: Plaque-independent disruption of neural circuits in Alzheimer's disease mouse models. Proc. Natl. Acad. Sci. U.S.A. *96*(6): 3228–3233, 1999.

Hsiao, K., Chapman, P., Nilsen, S., Eckman, C., Harigaya, Y., Younkin, S., Yang, F. and Cole, G.: Correlative memory deficits, Abeta elevation, and amyloid plaques in transgenic mice. Science *274*(5284): 99–102, 1996.

Hussain, I., Powell, D., Howlett, D. R., Tew, D. G., Meek, T. D., Chapman, C., Gloger, I. S., Murphy, K. E., Southan, C. D., Ryan, D. M., Smith, T. S., Simmons, D. L., Walsh, F. S., Dingwall, C. and Christie, G.: Identification of a novel aspartic protease (Asp 2) as beta-secretase. Mol. Cell. Neurosci. *14*(6): 419–427, 1999.

Ikezu, T., Trapp, B. D., Song, K. S., Schlegel, A., Lisanti, M. P. and Okamoto, T.: Caveolae, plasma membrane microdomains for alpha-secretase-mediated processing of the amyloid precursor protein. J. Biol. Chem. *273*(17): 10485–10495, 1998.

Ishii, T., Furuoka, H., Muroi, Y. and Nishimura, M.: Inactivation of integrin-linked kinase induces aberrant Tau phosphorylation via sustained activation of glycogen synthase kinase 3beta in N1E-115 neuroblastoma cells. J. Biol. Chem. *278*(29): 26970–26975, 2003.

Iqbal, K., Alonso Adel, C., El-Akkad, E., Gong, C. X., Haque, N., Khatoon, S., Pei, J. J., Tsujio, I., Wang, J. Z. and Grundke-Iqbal, I. Significance and mechanism of Alzheimer neurofibrillary degeneration and therapeutic targets to inhibit this lesion. J. Mol. Neurosci. *19*(1–2): 95–99, 2002.

Janus, C.: Vaccines for Alzheimer's disease: how close are we? CNS Drugs. *17*(7): 457–474, 2003.

Jick, H., Zornberg, G. L., Jick, S. S., Seshadri, S. and Drachman, D. A.: Statins and the risk of dementia. Lancet *356*(9242): 1627–1631, 2000.

Jones, R. W.: Have cholinergic therapies reached their clinical boundary in Alzheimer's disease? Int. J. Geriatr. Psychiatry *18*, Supplt. 18: S7–S13, 2003.

Kakio, A., Nishimoto, S. I., Yanagisawa, K., Kozutsumi, Y. and Matsuzaki, K.: Cholesterol-dependent formation of GM1 ganglioside-bound amyloid beta-protein, an endogenous seed for Alzheimer amyloid. J. Biol. Chem. *276*(27): 24985–24990, 2001.

Kamal, A., Almenar-Queralt, A., LeBlanc, J. F., et al.: Kinesin-mediated axonal transport of a membrane compartment containing beta-secretase and presenilin-1 requires APP. Nature *414*: 643–648, 2001.

Kang, D. E., Soriano, S., Xia, X., Eberhart, C. G., De Strooper, B., Zheng, H., Koo, E. H.: Presenilin couples the paired phosphorylation of beta-catenin independent of axin: implications for beta-catenin activation in tumorigenesis. Cell *110*(6): 751–762, 2002.

Kang, J., Lemaire, H. G., Unterbeck, A., Salbaum, J. M., Masters, C. L., Grzeschik, K. H., Multhaup, G., Beyreuther, K. and Muller-Hill, B.: The precursor of Alzheimer's disease amyloid A4 protein resembles a cell-surface receptor. Nature *325*(6106): 733–736, 1987.

Karczmar, A. G.: Pharmacologic, toxicologic, and therapeutic properties of anticholinesterase agents. In: "Physiological Pharmacology," W. S. Root and F. G. Hofman, Eds., vol. 3, pp. 163–322, Academic Press, New York, 1967.

Karczmar, A. G.: New roles for cholinergics in CNS disease. Drug Ther. *4*: 31–42, 1979.

Karczmar, A. G.: SDAT models and their dynamics. In: "Cholinergic Basis for Alzheimer Therapy," R. Becker and E. Giacobini, Eds., pp. 141–152, Birkhaeuser, Boston, 1991.

Karczmar, A. G.: Cholinergic substrates of cognition and organism-environment interaction. Prog. Neuropsychopharmacol. Biol. Psychiatry *19*: 187–211, 1995.

Karczmar, A. G. and Dun, N. J.: Effects of anticholinesterases pertinent for SDAT treatment but not necessarily underlying their clinical effectiveness. In: "Current Research in Alzheimer Therapy," E. Giacobini and R. Becker, Eds., pp. 15–29, Taylor and Francis, New York, 1988.

Kasa, P., Sr., Papp, H., Kasa, P., Jr., Pakaski, M. and Balaspiri, L.: Effects of amyloid-beta on cholinergic and acetylcholinesterase-positive cells in cultured basal forebrain neurons of embryonic rat brain. Brain Res. *998*: 73–82, 2004.

Kawamura, Y., Kikuchi, A., Takada, R., Takada, S., Sudoh, S., Shibamoto, S., Yanagisawa, K. and Komano, H. Inhibitory effect of a presenilin 1 mutation on the Wnt signalling pathway by enhancement of beta-catenin phosphorylation. Eur. J. Biochem. *268*(10): 3036–3041, 2001.

Kemp, P. M., Holmes, C., Hoffman, S., Wilkinson, S., Zivanovic, M., Thom, J., Bolt, L., Fleming, J. and Wilkinson, D. G.: A randomized placebo controlled study to assess the effects of cholinergic treatment on muscarinic receptors in Alzheimer's disease. J. Neurol. Neurosurg. Psychiatry *74*: 1567–1570, 2003.

Kesavapany, S., Lau, K. F., McLoughlin, D. M., Brownlees, J., Ackerley, S., Leigh, P. N., Shaw, C. E. and Miller, C. C.: p35/cdk5 binds and phosphorylates beta-catenin and regulates beta-catenin/presenilin-1 interaction. Eur. J. Neurosci. *13*(2): 241–247, 2001.

Kessler, R. M.: Imaging methods for evaluating brain function in man. Neurobiol. Aging *24*, Supplt. 1: S21–S35, 2003.

Killick, R., Pollard, C. C., Asuni, A. A., Mudher, A. K., Richardson, J. C., Rupniak, H. T., Sheppard, P. W., Varndell, I. M., Brion, J. P., Levey, A. I., Levy, O. A., Vestling, M., Cowburn, R., Lovestone, S. and Anderton, B. H.: Presenilin 1 independently regulates beta-catenin stability and transcriptional activity. J. Biol. Chem. *276*(51): 48554–48561, 2001.

Kimberly, W. T., Xia, W., Rahmati, T., Wolfe, M. S. and Selkoe, D. J.: The transmembrane aspartates in presentilin 1 and 2 are obligatory for gamma-secretase activity and amyloid beta-protein generation. J. Biol. Chem. *275*: 3173–3178, 2000.

Kimberly, W. T., LaVoie, M. J., Ostaszewski, B. L., Ye, W., Wolfe, M. S. and Selkoe, D. J.: Gamma-secretase is a membrane protein complex comprised of presenilin, nicastrin, Aph-1, and Pen-2. Proc. Natl. Acad. Sci. U.S.A. *100*(11): 6382–6387, 2003.

Kinoshita, A., Whelan, C. M., Smith, C. J., Berezovska, O. and Hyman, B. T.: Direct visualization of the gamma secretase-generated carboxyl-terminal domain of the amyloid precursor protein: association with Fe65 and translocation to the nucleus. J. Neurochem. *82*(4): 839–847, 2002.

Kitaguchi, N., Takahashi, Y., Tokushima, Y., Shiojiri, S. and Ito, H.: Novel precursor of Alzheimer's disease amyloid protein shows protease inhibitory activity. Nature *331*(6156): 530–532, 1988.

Klatzo, I., Wisniewski, H. M. and Streiker, E.: Experimental production of neurofibrillary degeneration. J. Neuropathol. Exp. Neurol. *24*: 1187–1199, 1965.

Klimov, D. K. and Thirumalai, D.: Dissecting the assembly of abeta(16–22) amyloid peptides into antiparallel Beta sheets. Structure (Camb.) *11*(3): 295–307, 2003.

Kojro, E., Gimpl, G., Lammich, S., Marz, W. and Fahrenholz, F.: Low cholesterol stimulates the non-amyloidogenic pathway by its effect on the alpha - secretase ADAM 10. Proc. Natl. Acad. Sci. U.S.A. *98*(10): 5815–5820, 2001.

Koo, E. H. and Squazzo, S. L.: Evidence that production and release of amyloid beta-protein involves the endocytic pathway. J. Biol. Chem. *269*(26): 17386–17389, 1994.

Kopan, R. and Goate, A.: A common enzyme connects notch signaling and Alzheimer's disease. Genes Dev. *14*(22): 2799–2806, 2000.

Kovacs, T., Cairns, N. J. and Lantos, P. L.: Beta-amyloid deposition and neurofibrillary tangle formation in the olfactory bulb in ageing and Alzheimer's disease. Neuropathol. Appl. Neurobiol. *25*(6): 481–491, 1999.

Kraepelin, E.: "Lehrbuch der Psychiatry," II Band. Johann Ambrosius Barth Verlag, Leipzig, 1910.

Lang, T., Bruns, D., Wenzel, D., Riedel, D., Holroyd, P., Thiele, C. and Jahn, R.: SNAREs are concentrated in cholesterol-dependent clusters that define docking and fusion sites for exocytosis. EMBO J *20*(9): 2202–2213, 2001.

Lau, K. F., Miller, C. C., Anderton, B. H. and Shaw, P. C.: Molecular cloning and characterization of the human glycogen synthase kinase-3beta promoter. Genomics *60*(2): 121–128, 1999.

Laviolette, S. R. and van der Kooy, I.: The neurobiology of nicotine addiction: bridging the gap from molecules to behavior. Nat. Rev. Neurosci. *5*: 55–65, 2004.

Lazarov, O., Morfini, G. A., Lee, E. B., Farah, M. H., Szodora, A., DeBeer, S. K., Koliatsos, V. E., King, S., Lee, V. M., Wong, P. C., Price, D. L., Brady, S. T. and Sisodia, S. S.: Axonal transport, amyloid precursor protein, kinesin-1, and the processing apparatus:revisited. J. Neurosci. *25*: 2386–2395, 2005.

Lee, C. W., Lau, K. F., Miller, C. C. and Shaw, P. C.: Glycogen synthase kinase-3 beta-mediated tau phosphorylation in cultured cell lines. Neuroreport *14*(2): 257–260, 2003.

Lee, H. J., Jung, K. M., Huang, Y. Z., Bennett, L. B., Lee, J. S., Mei, L. and Kim, T. W.: Presenilin-dependent gamma-secretase-like intramembrane cleavage of ErbB4. J. Biol. Chem. *277*(8): 6318–6323, 2002.

Lee, S. J., Liyanage, U., Bickel, P. E., Xia, W., Lansbury, P. T., Jr. and Kosik, K. S.: A detergent-insoluble membrane compartment contains A beta *in vivo*. Nat. Med. *4*(6): 730–734, 1998.

Leissring, M. A., Yamasaki, T. R., Wasco, W., Buxbaum, J. D., Parker, I. and LaFerla, F. M.: Calsenilin reverses presenilin-mediated enhancement of calcium signaling. Proc. Natl. Acad. Sci. U.S.A. *97*(15): 8590–8593, 2000.

Leissring, M. A., Murphy, M. P., Mead, T. R., Akbari, Y., Sugarman, M. C., Jannatipour, M., Anliker, B., Muller, U., Saftig, P., De Strooper, B., Wolfe, M. S., Golde, T. E. and LaFerla, F. M.: A physiologic signaling role for the gamma -secretase-derived intracellular fragment of APP. Proc. Natl. Acad. Sci. U.S.A. *99*(7): 4697–4702, 2002.

Leroy, K., Boutajangout, A., Authelet, M., Woodgett, J. R., Anderton, B. H., Brion, J. P.: The active form of glycogen synthase kinase-3beta is associated with granulovacuolar degeneration in neurons in Alzheimer's disease. Acta Neuropathol. (Berl.) *103*(2): 91–99, 2002.

Levites, Y., Amit, T., Mandel, S. and Youdim, M. B.: Neuroprotection and neurorescue against Abeta toxicity and PKC-dependent release of nonamyloidogenic soluble precursor protein by green tea polyphenol (-)-epigallocatechin-3-gallate. FASEB J. *17*(8): 952–954, 2003.

Lewis, J., McGowan, E., Rockwood, J., Melrose, H., Nacharaju, P., Van Slegtenhorst, M., Gwinn-Hardy, K., Paul Murphy, M., Baker, M., Yu, X., Duff, K., Hardy, J., Corral, A., Lin, W. L., Yen, S. H., Dickson,

D. W., Davies, P., Hutton, M.: Neurofibrillary tangles, amyotrophy and progressive motor disturbance in mice expressing mutant (P301L) tau protein. Nat. Genet. *25*(4): 402–405, 2000. Erratum in: Nat. Genet. *26*(1): 127, 2000.

Lewis, J., Dickson, D. W., Lin, W. L., Chisholm, L., Corral, A., Jones, G., Yen, S. H., Sahara, N., Skipper, L., Yager, D., Eckman, C., Hardy, J., Hutton, M. and McGowan, E.: Enhanced neurofibrillary degeneration in transgenic mice expressing mutant tau and APP. Science *293*(5534): 1487–1491, 2001.

Li, W., Pi, R., Chan, H. H., Lee, N. T., Tsang, H. W., Pu, Y., Chang, D. C., Li, C., Luo, J., Li, Z., Xue, H., Carlier, P. R., Pang, Y., Tsim, K. W., Li, M. and Han, Y.: Novel dimeric AChE inhibitor Bis(7)-Tacrine but not donepezil, prevents glutamate-induced neuronal apoptosis by blocking NMDA receptors. J. Biol. Chem., *280*: 18179–18188, 2005.

Lilliehook, C., Chan, S., Choi, E. K., Zaidi, N. F., Wasco, W., Mattson, M. P. and Buxbaum, J. D.: Calsenilin enhances apoptosis by altering endoplasmic reticulum calcium signaling. Mol. Cell. Neurosci. *19*(4): 552–559, 2002.

Lin-Lee, Y. C., Kao, F. T., Cheung, P. and Chan, L.: Apolipoprotein E gene mapping and expression: localization of the structural gene to human chromosome 19 and expression of ApoE mRNA in lipoprotein- and non-lipoprotein-producing tissues. Biochemistry *24*: 3751–3756, 1985.

Lipton, S. A.: Paradigm shift in NMDA receptor antagonist drug development. Molecular mechanism of uncompetitive inhibition by memantine in the treatment of Alzheimer's disease and other neurological disorders. J. Alzheimers Dis. *6*, Supplt. 6: S61–S74, 2004.

Liu, F., Iqbal, K., Grundke-Iqbal, I. and Gong, C. X.: Involvement of aberrant glycosylation in phosphorylation of tau by cdk5 and GSK-3beta. FEBS Lett. *530*(1–3): 209–214, 2002.

Lojkowska, W., Ryglewicz, D., Jedrzejczak, T., Mine, S., Jakubowska, T., Jarosz, H. and Bochynska, A. J.: The effect of cholinesterase inhibitors on the regional blood flow in patients with Alzheimer's disease and vascular dementia. Neurol. Sci. *216*: 119–126, 2003.

Lu, S. and Tune, L. E.: Chronic exposure to anticholinergic medication adversely affects the course of Alzheimer's disease. Am. J. Geriatr. Psychiatry *11*: 458–461, 2003.

Ludwig, F. C.: Senescence: pathology facing medicine's ultimate issue. Arch. Pathol. Lab. Med. *105*: 445–451, 1981.

Luo, W. J., Wang, H., Li, H., Kim, B. S., Shah, S., Lee, H. J., Thinakaran, G., Kim, T. W., Yu, G. and Xu, H.: PEN-2 and APH-1 coordinately regulate proteolytic processing of presenilin 1. J. Biol. Chem. *278*(10): 7850–7854, 2003.

Luo, Y., Bolon, B., Kahn, S., Bennett, B. D., Babu-Khan, S., Denis, P., Fan, W., Kha, H., Zhang, J., Gong, Y., Martin, L., Louis, J. C., Yan, Q., Richards, W. G., Citron, M. and Vassar, R.: Mice deficient in BACE1, the Alzheimer's beta-secretase, have normal phenotype and abolished beta-amyloid generation. Nat. Neurosci. *4*(3): 231–232, 2001.

Maccioni, R. B., Otth, C., Concha, I. I. and Munoz, J. P.: The protein kinase Cdk5: structural aspects, roles in neurogenesis and involvement in Alzheimers pathology. Eur. J. Biochem. *268*(6): 1518–1527, 2001.

Mahley, R. W.: Apolipoprotein E: cholesterol transport protein with expanding role in cell biology. Science *240*(4852): 622–630, 1988.

Maltese, W. A., Wilson, S., Tan, Y., Suomensaari, S., Sinha, S., Barbour, R. and McConlogue, L.: Retention of the Alzheimer's amyloid precursor fragment C99 in the endoplasmic reticulum prevents formation of amyloid beta-peptide. J. Biol. Chem. *276*(23): 20267–20279, 2001.

Marambaud, P., Shioi, J., Serban, G., Georgakopoulos, A., Sarner, S., Nagy, V., Baki, L., Wen, P., Efthimiopoulos, S., Shao, Z., Wisniewski, T. and Robakis, N. K.: A presenilin-1/gamma-secretase cleavage releases the E-cadherin intracellular domain and regulates disassembly of adherens junctions. EMBO J. *21*(8): 1948–1956, 2002.

Marks, N. and Berg, M. J.: APP processing enzymes (secretases) as therapeutic targets: insights from the use of transgenics (Tgs) and transfected cells. Neurochem. Res. *28*(7): 1049–1062, 2003.

Martin, A.: Antioxidant vitamins E and C and risk of Alzheimer's disease. Nutr. Rev. *61*(2): 69–73, 2003.

Maruyama, K., Kametani, F., Usami, M., Yamao-Harigaya, W. and Tanaka, K.: "Secretase," Alzheimer amyloid protein precursor secreting enzyme is not sequence-specific. Biochem. Biophys. Res. Commun. *179*(3): 1670–1676, 1991.

Masliah, E.: Mechanisms of synaptic dysfunction in Alzheimer's disease. Histol. Histopathol. *10*(2): 509–519, 1995.

Masliah, E., Mallory, M., Ge, N. and Saitoh, T.: Amyloid precursor protein is localized in growing neurites of neonatal rat brain. Brain Res. *593*: 323–337, 1992.

Mason, N. S. and Mathis, C. A.: Positron emission tomography radiochemistry. Neuroimaging Clin. N. Am. *13*: 671–687, 2003.

Massa, S. M., Xie, Y. and Longo, F. M.: Alzheimer's therapeutics: neurotrophin small molecule mimetics. J. Mol. Neurosci. *19*: 107–111, 2002.

Masters, C. L., Simms, G., Weinman, N. A., Multhaup, G., McDonald, B. L. and Beyreuther, K.: Amyloid plaque core protein in Alzheimer disease and Down syndrome. Proc. Natl. Acad. Sci. U.S.A. *82*(12): 4245–4249, 1985.

Mazzucchelli, M., Porello, E., Viletti-Pietra, C., Govoni, S. and Racchi, M.: Characterization of the effect of ganstigmine (CHF2819) on amyloid precursor protein metabolism in SH-SY5Y neuroblastoma cells. J. Neural Transm. *110*: 935–947, 2003.

McGeer, P. L., McGeer, E. G., Suzuki, J., Dolman, C. E. and Nagai, T.: Aging, Alzheimer's disease and the cholinergic system of the basal forebrain. Neurology *34*: 741–745, 1984.

McGeer, P. L., Schulzer, M. and McGeer, E. G.: Arthritis and anti-inflammatory agents as possible protective factors for Alzheimer's disease: a review of 17 epidemiologic studies. Neurology *47*(2): 425–432, 1996.

McKinney, M. and Jacksonville, M. C.: Brain cholinergic vulnerability: relevance to behavior and disease. Biochem. Pharmacol. *70*: 1115–1124, 2005.

McMenemey, W. H.: Alois Alzheimer and his disease. In: "Alzheimer's Disease," G. E. Wolstenholme and M. O'Connor, Eds. Ciba Found., Churchill, London, 1970.

Mesulam, M. M.: Neuroanatomy of cholinesterases in the normal human brain and in Alzheimer's disease. In: "Cholinesterases and Cholinesterase Inhibitors," E. Giacobini, Ed., pp. 121–137, Martin Dunitz, London, 2000.

Mesulam, M. M.: The cholinergic lesion of Alzheimer's disease: Pivotal factor or side show? Learn. Mem. *11*: 43–49, 2004.

Mesulam, M. M., Shaw, P., Mash, D. and Weintraub, S.: Cholinergic nucleus basalis tauopathy emerges early in the aging-MCI-AD continuum. Ann. Neurol. *55*: 815–828, 2004.

Mills, J. and Reiner, P. B.: Regulation of amyloid precursor protein cleavage. J. Neurochem. *72*(2): 443–460, 1999.

Mills, J., Digicaylioglu, M., Legg, A. T., Young, C. E., Young, S. S., Barr, A. M., Fletcher, L., O'Connor, T. P. and Dedhar, S.: Role of integrin-linked kinase in nerve growth factor-stimulated neurite outgrowth. J. Neurosci. *23*(5): 1638–1648, 2003.

Minett, T. S., Thomas, A., Wilkinson, L. M., Daniel. S. L., Sanders, J., Richardson, J., Littlewood, E., Myint, P., Newby, J. and McKeith, I. G.: What happens when donepezil is suddenly withdrawn? An open label trial in dementia with Lewy bodies and Parkinson's disease with dementia. Int. J. *Geriatr.* Psychiatry *18*: 988–993, 2003.

Mizuno, T., Nakata, M., Naiki, H., Michikawa, M., Wang, R., Haass, C. and Yanagisawa, K.: Cholesterol-dependent generation of a seeding amyloid beta-protein in cell culture. J. Biol. Chem. *274*(21): 15110–15114, 1999.

Moechars, D., Dewachter, I., Lorent, K., Reverse, D., Baekelandt, V., Naidu, A., Tesseur, I., Spittaels, K., Haute, C. V., Checler, F., Godaux, E., Cordell, B. and Van Leuven, F.: Early phenotypic changes in transgenic mice that overexpress different mutants of amyloid precursor protein in brain. J. Biol. Chem. *274*(10): 6483–6492, 1999.

Monsonego, A., Zota, V., Karni, A., Krieger, J. I., Bar-Or, A., Bitan, G., Budson, A. E., Sperling, R., Selkoe, D. J. and Weiner, H. L.: Increased T cell reactivity to amyloid beta protein in older humans and patients with Alzheimer disease. J. Clin. Invest. *112*(3): 415–422, 2003.

Morfini, G., Pigino, G., Beffert, U., Busciglio, J. and Brady, S. T.: Fast axonal transport misregulation and Alzheimer's disease. Neuromolecular Med. *2*(2): 89–99, 2002.

Mori, C., Spooner, E. T., Wisniewsk, K. E., Wisniewski, T. M., Yamaguch, H., Saido, T. C., Tolan, D. R., Selkoe, D. J., Lemere, C. A.: Intraneuronal Abeta42 accumulation in Down syndrome brain. Amyloid. *9*(2): 88–102, 2002.

Moran, M. A., Mufson, E. J. and Gomez-Ramos, P.: Colocalization of cholinesterases with beta amyloid protein in aged and Alzheimer's brains. Acta Neuropathol. *85*: 362–369, 1993.

Morgan, D.: Learning and memory deficits in APP transgenic mouse models of amyloid deposition. Neurochem. Res. *28*(7): 1029–1034, 2003.

Morgan, D., Diamond, D. M., Gottschall, P. E., Ugen, K. E., Dickey, C., Hardy, J., Duff, K., Jantzen, P., DiCarlo, G., Wilcox, D., Connor, K., Hatcher, J., Hope, C., Gordon, M. and Arendash, G. W.: A beta peptide vaccination prevents memory loss in an animal model of Alzheimer's disease. Nature *408*(6815): 982–985, 2000.

Morishima-Kawashima, M. and Ihara, Y.: The presence of amyloid beta-protein in the detergent-insoluble membrane compartment of human neuroblastoma cells. Biochemistry *37*(44): 15247–15253, 1998.

Morohashi, Y., Hatano, N., Ohya, S., Takikawa, R., Watabiki, T., Takasugi, N., Imaizumi, Y., Tomita, T. and Iwatsubo, T.: Molecular cloning and characterization of CALP/KChIP4, a novel EF-hand protein interacting with presenilin 2 and voltage-gated potassium channel subunit Kv4. J. Biol. Chem. *277*(17): 14965–14975, 2002.

Morris, J. C. and Price, A L.: Pathologic correlates of nondemented aging, mild cognitive impairment, and early-stage Alzheimer's disease. J. Mol. Neurosci. *17*(2): 101–118, 2001.

Morris, R. G.: "A Cognitive Neuropsychology of Alzheimer Dementia," Oxford U. Press, Now York, 1996

Mudher, A., Chapman, S., Richardson, J., Asuni, A., Gibb, G., Pollard, C., Killick, R., Iqbal, T., Raymond, L., Varndell, I., Sheppard, P., Makoff, A., Gower, E., Soden, P. E., Lewis, P., Murphy, M., Golde, T. E., Rupniak, H. T., Anderton, B. H. and Lovestone,

S.: Dishevelled regulates the metabolism of amyloid precursor protein via protein kinase C/mitogen-activated protein kinase and c-Jun terminal kinase. J. Neurosci. *21*(14): 4987–4995, 2001.

Nathan, B. P., Bellosta, S., Sanan, D. A., Weisgraber, K. H., Mahley, R. W. and Pitas, R. E.: Differential effects of apolipoproteins E3 and E4 on neuronal growth *in vitro*. Science *264*(5160): 850–852, 1994.

Naureckiene, S., Sleat, D. E., Lackland, H., Fensom, A., Vanier, M. T., Wattiaux, R., Jadot, M. and Lobel, P.: Identification of HE1 as the second gene of Niemann-Pick C disease. Science *290*: 2298–2301, 2000.

New York Times, Editorial: Do Alzheimer's drugs work? Editorial Desk, Section 4, p. 10, April 11, 2004.

Nicolaou, M., Song, Y. Q., Sato, C. A., Orlacchio, A., Kawarai, T., Medeiros, H., Liang, Y., Sorbi, S., Richard, E., Rogaev, E. I., Moliaka, Y., Bruni, A. C., Jorge, R., Percy, M., Duara, R., Farrer, L. A., St. George-Hyslop, P. and Rogaeva, E. A.: Mutations in the open reading frame of the beta-site APP cleaving enzyme (BACE) locus are not a common cause of Alzheimer's disease. Neurogenetics *3*(4): 203–206, 2001.

Nishimura, M., Yu, G., Levesque, G., Zhang, D. M., Ruel, L., Chen, F., Milman, P., Holmes, E., Liang, Y., Kawarai, T., Jo, E., Supala, A., Rogaeva, E., Xu, D. M., Janus, C., Levesque, L., Bi, Q., Duthie, M., Rozmahel, R., Mattila, K., Lannfelt, L., Westaway, D., Mount, H. T., Woodgett, J., St. George-Hyslop, P., et al. Presenilin mutations associated with Alzheimer disease cause defective intracellular trafficking of beta-catenin, a component of the presenilin protein complex. Nat. Med. *5*(2): 164–169, 1999.

Nitsch, R. M., Slack, B. E., Wurtan, R. J. and Growdon, J. H.: Release of amyloid precursor derivatives stimulated by activation of muscarinic acetylcholine receptors. Science *258*: 304–307, 1992.

Nitsch, R. M., Rossner, S. and Albrecht, C.: Muscarinic acetylcholine receptors activate the acetylcholinesterase gene promoter. J. Physiol. *25*: 257–264, 1998.

Noble, W., Olm, V., Takata, K., Casey, E., Mary, O., Meyerson, J., Gaynor, K., LaFrancois, J., Wang, L., Kondo, T., Davies, P., Burns, M., Veeranna, Nixon, R., Dickson, D., Matsuoka, Y., Ahlijanian, M., Lau, L. F. and Duff, K.: Cdk5 is a key factor in tau aggregation and tangle formation *in vivo*. Neuron *38*(4): 555–565, 2003.

Nordberg, A.: The effect of cholinesterase inhibitors studied with brain imaging. In: "Cholinesterases and Cholinesterase Inhibitors," E. Giacobini, Ed., pp. 237–247, Martin Dunitz, London, 2000.

Nordberg, A.: Toward an early diagnosis and treatment of Alzheimer's disease. Int. Psychogeriatr. *15*: 223–227, 2003.

Nordberg, A.: Functional studies of cholinergic activity in normal and Alzheimer's disease states by imaging technique. Prog. Brain Res. *145*: 301–310, 2004.

Nordberg, A. and Winblad, B.: Brain nicotinic and muscarinic receptors in normal aging and dementia. In: "Alzheimer's and Parkinson's Diseases," A. Fisher, I. Hanin, and C. Lachman, Eds., pp. 95–108, Plenum Press, New York, 1986.

Oddo, S., Caccamo, A., Shepherd, J. D., Murphy, M. P., Golde, T. E., Kayed, R., Metherate, R., Mattson, M. P., Akbari, Y. and LaFerla, F. M.: Triple-transgenic model of Alzheimer's disease with plaques and tangles: intracellular Abeta and synaptic dysfunction. Neuron *39*(3): 409–421, 2003.

Ohkubo, N., Lee, Y. D., Morishima, A., Terashima, T., Kikkawa, S., Tohyama, M., Sakanaka, M., Tanaka, J., Maeda, N., Vitek, M. P. and Mitsuda, N.: Apolipoprotein E and Reelin ligands modulate tau phosphorylation through an apolipoprotein E receptor/disabled-1/glycogen synthase kinase-3beta cascade. FASEB J. *17*(2): 295–297, 2003.

Oltersdorf, T., Fritz, L. C., Schenk, D. B., Lieberburg, I., Johnson-Wood, K. L., Beattie, E. C., Ward, P. J., Blacher, R. W., Dovey H. F. and Sinha, S.: The secreted form of the Alzheimer's amyloid precursor protein with the Kunitz domain is protease nexin-II. Nature *341*(6238): 144–147, 1989.

O'Neill, C., Cowburn, R. F., Bonkale, W. L., Ohm, T. G., Fastbom, J., Carmody, M. and Kelliher, M.: Dysfunctional intracellular calcium homoeostasis: a central cause of neurodegeneration in Alzheimer's disease. Biochem. Soc. Symp. *2001*(67): 177–194, 2001.

Orozco, C., de Los Rios, C., Arias, E., Leon R. Garcia, A. G., Marco, J. L., Villarroya, M. and Lopez, M. G.: ITH4012 (ethyl 5-amino-6,7,8,9-tetrahydro-2-methyl-4-phenylbenzol(1,8) naphtythyridine-3-carboxylate. a novel acetylcholinesterase inhibitor with "calcium-promotor" and neuroprotective properties. J. Pharmacol. Exper. Therap. *310*: 987–994, 2004.

Oyama, F., Sawamura, N., Kobayashi, K., Morishima-Kawashima, M., Kuramochi, T., Ito, M., Tomita, T., Maruyama, K., Saido, T. C., Iwatsubo, T., Capell, A., Walter, J., Grunberg, J., Ueyama, Y., Haass, C. and Ihara, Y.: Mutant presenilin 2 transgenic mouse: effect on an age-dependent increase of amyloid beta-protein 42 in the brain. J. Neurochem. *71*(1): 313–322, 1998.

Pakaski, M., Rakonczay, Z., Fakla, I., Papp, H. and Kasa, P.: *In vitro* effects of metrifonate on neuronal precursor protein processing and protein kinase C level. Brain Res. *863*: 266–270, 2000.

Palmert, M. R., Podlisny, M. B., Golde, T. E., Cohen, M. L., Kovacs, D. M., Tanzi, R. E., Gusella, J. F., Whitehouse, P. J., Witker, D. S., Oltersdorf, T., et al.: The beta amyloid protein precursor: mRNAs, membrane-associated forms, and soluble derivatives. Prog. Clin. Biol. Res. *317*: 971–984, 1989.

Palop, J. J., Jones, B., Kekonius, L., Chin, J., Yu, G. Q., Raber, J., Masliah, E. and Mucke, L.: Neuronal depletion of calcium-dependent proteins in the dentate gyrus is tightly linked to Alzheimer's disease-related cognitive deficits. Proc. Natl. Acad. Sci. U.S.A. *100*(16): 9572–9577, 2003.

Pap, M. and Cooper, G. M.: Role of glycogen synthase kinase-3 in the phosphatidylinositol 3-Kinase/Akt cell survival pathway. J. Biol. Chem. *273*(32): 19929–19932, 1998.

Papp, H., Pakaski, M., Kasa, P.: Presenilin-1 and the amyloid precursor protein are transported bidirectionally in the sciatic nerve of adult rat. Neurochem. Int. *41*(6): 429–435, 2002.

Parkin, E. T., Turner, A. J. and Hooper, N. M.: Amyloid precursor protein, although partially detergent-insoluble in mouse cerebral cortex, behaves as an atypical lipid raft protein. Biochem. J. *344*, Pt. 1: 23–30, 1999.

Pasternak, S. H., Bagshaw, R. D., Guiral, M., Zhang, S., Ackerley, C. A., Pak, B. J., Callahan, J. W. and Mahuran, D. J.: Presenilin-1, nicastrin, amyloid precursor protein, and gamma-secretase activity are co-localized in the lysosomal membrane. J. Biol. Chem. *278*(29): 26687–26694, 2003.

Patrick, G. N., Zukerberg, L., Nikolic, M., de la Monte, S., Dikkes, P. and Tsai, L. H.: Conversion of p35 to p25 deregulates Cdk5 activity and promotes neurodegeneration. Nature *402*(6762): 615–622, 1999.

Pei, J. J., Braak, E., Braak, H., Grundke-Iqbal, I., Iqbal, K., Winblad, B. and Cowburn, R. F.: Distribution of active glycogen synthase kinase 3beta (GSK-3beta) in brains staged for Alzheimer disease neurofibrillary changes. J. Neuropathol. Exp. Neurol. *58*(9): 1010–1019, 1999.

Pei, J. J., Khatoon, S., An, W. L., Nordlinder, M., Tanaka, T., Braak, H., Tsujio, I., Takeda, M., Alafuzoff, I., Winblad, B., Cowburn, R. F., Grundke-Iqbal, I. and Iqbal, K.: Role of protein kinase B in Alzheimer's neurofibrillary pathology. Acta Neuropathol. (Berl.) *105*(4): 381–392, 2003.

Perez, M., Ribe, E., Rubio, A., Lim, F., et al.: Characterization of a double (amyloid precursor protein-tau) transgenic: tau phosphorylation and aggregation. Neuroscience *130*: 339–347, 2005.

Perry, E. K., Gibson, P. H., Blessed, G., Perry, R. H. and Tomlison, B. E.: Neurotransmitter enzyme abnormalities in senile dementia. J. Neurol. Sci. *34*: 247–265, 1977.

Perry, E. K., Ziabreva, I., Perry, R. H., Aarsland, D. and Ballard, C.: Absence of cholinergic deficits in "pure" vascular dementia. Neurology *64*: 132–133, 2005.

Perusini, G.: Über klinisch und histologisch eigenmartige psychische Erkränkung der späteren Lebensalters. In: "Histologische und Histopataologische Arbeiten," F. Nissl and A. Alzheimer, Eds., Bd. 3, pp. 197–351, 1910.

Perusini, G.: Sul valore nosografico di alcuni reperti istopatologici caractteristici per la senilita. Revista Ital. Neuropatol. Psichiat. Elettroter. *4*: 145–151, 1911.

Peters, B. and Levin, H. S.: Effects of physostigmine and lecithin on memory in Alzheimer's disease. Ann. Neurol. *6*: 219–221, 1979.

Pfeifer, L. A., White, L. R., Ross, G. W., Petrovitch, H. and Launer, L. J.: Cerebral amyloid angiopathy and cognitive function: the HAAS autopsy study. Neurology *58*(11): 1629–1634, 2002a.

Pfeifer, M., Boncristiano, S., Bondolfi, L., Stalder, A., Deller, T., Staufenbiel, M., Mathews, P. M. and Jucker, M.: Cerebral hemorrhage after passive anti-Abeta immunotherapy. Science *298*(5597): 1379, 2002b.

Phiel, C. J., Wilson, C. A., Lee, V. M. and Klein, P. S.: GSK-3alpha regulates production of Alzheimer's disease amyloid-beta peptides. Nature *423*(6938): 435–439, 2003.

Piccini, A., Russo, C., Gliozzi, A., Relini, A., et al.: β-Amyloid is different in normal aging and in Alzheimer disease. J. Biol. Chem. *280*: 34186–34192, 2005.

Pigino, G., Morfini, G., Pelsman, A., Mattson, M. P., Brady, S. T. and Busciglio, J.: Alzheimer's presenilin 1 mutations impair kinesin-based axonal transport. J. Neurosci. *23*(11): 4499–4508, 2003.

Pike, L. J. and Miller, J. M.: Cholesterol depletion delocalizes phosphatidylinositol bisphosphate and inhibits hormone-stimulated phosphatidylinositol turnover. J. Biol. Chem. *273*(35): 22298–22304, 1998.

Poirier, J.: Apolipoprotein E and Alzheimer's disease: a role in amyloid catabolism. Ann N. Y. Acad. Sci. *924*: 81–90, 2000.

Poirier, J., Baccichet, A., Dea, D. and Gauthier, S.: Cholesterol synthesis and lipoprotein reuptake during synaptic remodelling in hippocampus in adult rats. Neuroscience *55*(1): 81–90, 1993.

Pomponi, M. and Marta, M.: On the suggestion of Dr. Alzheimer I examined the following four cases. Dedicated to Gaetano Perusi. Aging Clin. Exp. Res. *5*: 135–139, 1992.

Ponte, P., Gonzalez-DeWhitt, P., Schilling, J., Miller, J., Hsu, D., Greenberg, B., Davis, K., Wallace, W., Lieberburg, I. and Fuller, F.: A new A4 amyloid mRNA contains a domain homologous to serine proteinase inhibitors. Nature *331*(6156): 525–527, 1988.

Price, D. L.: The Macaca mulatta model of Alzheimer's disease. Neurosci. Facts *4*: 17–18, 1993.

Profenno, L. A. and Tariot, P. N.: Pharmacologic management of agitation in Alzheimer's disease. Dement. Geriatr. Cogn. Disord. *17*: 65–77, 2005.

Puglielli, L., Konopka, G., Pack-Chung, E., Ingano, L. A., Berezovska, O., Hyman, B. T., Chang, T. Y., Tanzi, R. E. and Kovacs, D. M.: Acyl-coenzyme A: cholesterol acyltransferase modulates the generation of the amyloid beta-peptide. Nat. Cell Biol. *3*(10): 905–912, 2001.

Rabidazeh, S., Oh, S., Ord, J., Zhong, L.-T., Yang, J., Bittner, C. M., Butcher, L. L. and Dresdesen, D. E.: Induction of apoptosis by low-affinity NGF receptor. Science *261*: 345–348, 1993.

Rabizadeh, S., Bittner, C. M., Butcher, L. L. and Bredesen, D. E.: Expression of the low affinity growth factor receptor enhances beta-amyloid peptide toxicity. Proc. Natl. Acad. Sci. U.S.A. *91*: 10703–10706, 1994.

Racchi, M., Baetta, R., Salvietti, N., Ianna, P., Franceschini, G., Paoletti, R., Fumagalli, R., Govoni, S., Trabucchi, M. and Soma, M.: Secretory processing of amyloid precursor protein is inhibited by increase in cellular cholesterol content. Biochem. J. *322*, Pt. *3*: 893–898, 1997.

Redlich, E.: Über miliare Sklerose der Hirnrinde bei seniler Atrophie. Jahrbuche Psych. u. Neurol. *17*: 361–372, 1898.

Roberds, S. L., Anderson, J., Basi, G., Bienkowski, M. J., Branstetter, D. G., Chen, K. S., Freedman, S. B., Frigon, N. L., Games, D., Hu, K., Johnson-Wood, K., Kappenman, K. E., Kawabe, T. T., Kola, I., Kuehn, R., Lee, M., Liu, W., Motter, R., Nichols, N. F., Power, M., Robertson, D. W., Schenk, D., Schoor, M., Shopp, G. M., Shuck, M. E., Sinha, S., Svensson, K. A., Tatsuno, G., Tintrup, H., Wijsman, J., Wright, S. and McConlogue, L.: BACE knockout mice are healthy despite lacking the primary beta-secretase activity in brain: implications for Alzheimer's disease therapeutics. Hum. Mol. Genet. *10*(12): 1317–1324, 2001.

Rodal, S. K., Skretting, G., Garred, O., Vilhardt, F., van Deurs, B. and Sandvig, K.: Extraction of cholesterol with methyl-beta-cyclodextrin perturbs formation of clathrin-coated endocytic vesicles. Mol. Biol. Cell. *10*(4): 961–974, 1999.

Rogawski, M. A. and Wenk, G. L.: The neuropharmacological basis for the use of memantine in treatment of Alzheimer's disease. CNS Drug Rev. *9*(3): 275–308, 2003.

Rohan de Silva, H. A., Jen, A., Wickenden, C., Jen, L. S., Wilkinson, S. L. and Patel, A. J.: Cell-specific expression of beta-amyloid precursor protein isoform mRNAs and proteins in neurons and astrocytes. Brain Res. Mol. Brain Res. *47*(1–2): 147–156, 1997.

Roses, A. D., Saunders, A. M., Corder, E. H., Pericak-Vance, M. A., Han, S. H., Einstein, G., Hulette, C., Schmechel, D. E., Holsti, M., Huang, D., et al.: Influence of the susceptibility genes apolipoprotein E-epsilon 4 and apolipoprotein E-epsilon 2 on the rate of disease expressivity of late-onset Alzheimer's disease. Arzneimittelforschung. *45*(3A): 413–417, 1995.

Rossner, S., Ueberham, U., Schliebs, R., Perez-Polo, J. R. and Bigl, V.: The regulation of amyloid precursor protein metabolism by cholinergic mechanisms and neurotrophin receptor signaling. Prog. Neurobiol. *56*(5): 541–569, 1998.

Roush, W.: Live long and prosper? Science *273*: 42–46, 1996.

Runz, H., Rietdorf, J., Tomic, I., de Bernard, M., Beyreuther, K., Pepperkok, R. and Hartmann, T.: Inhibition of intracellular cholesterol transport alters presenilin localization and amyloid precursor protein processing in neuronal cells. J. Neurosci. *22*(5): 1679–1689, 2002.

Russell, R. W.: Brain "transplants," neurotrophic factors and behavior. Alzheimer Dis. Assoc. Disord. *2*: 77–95, 1988.

Russo, C., Schettini, G., Saido, T. C., Hulette, C., Lippa, C., Lannfelt, L., Ghetti, B., Gambetti, P., Tabaton, M., Teller, J. K.: Presenilin-1 mutations in Alzheimer's disease. Nature *405*(6786): 531–532, 2000.

Sabate, R., Gallardo, M. and Estelrich, J.: An autocatalytic reaction as a model for the kinetics of the aggregation of beta-amyloid. Biopolymers *71*(2): 190–195, 2003.

Sarker, K. P. and Lee, K. Y.: L6 myoblast differentiation is modulated by Cdk5 via the PI3K-Akt-p70s6K signaling pathway, Oncogene. *23*: 6064, 2004.

Saunders, A. M.: Apolipoprotein E and Alzheimer disease: an update on genetic and functional analyses. J. Neuropathol. Exp. Neurol. *59*(9): 751–758, 2000.

Sawamura, N., Gong, J. S., Garver, W. S., Heidenreich, R. A., Ninomiya, H., Ohno, K., Yanagisawa, K. and Michikawa, M.: Site-specific phosphorylation of tau accompanied by activation of mitogen-activated protein kinase (MAPK) in brains of Niemann-Pick type C mice. J. Biol. Chem. *276*: 10314–10319, 2001.

Sayas, C. L., Avila, J. and Wandosell, F.: Regulation of neuronal cytoskeleton by lysophosphatidic acid: role of GSK-3. Biochim. Biophys. Acta *1582*(1–3): 144–153, 2002.

Saykin, A. J., Wishart, H. A., Rabin, L. A., Flashmen, L. A., McHugh, T. L., Mamot, A. C. and Santulli, R. B.: Cholinergic enhancement of frontal lobe activity in mild cognitive impairment. Brain *127*: 1574–1583, 2004.

Sberna, G., Saez-Valero, J. and Li, Q.-X.: Acetylcholinesterase is increased in the brains of transgenic mice expressing the C-terminal segment (CT100) of the beta-amyloid protein precursor of Alzheimer's disease. J. Neurochem. *71*: 723–731, 1998.

Schenk, D.: Amyloid-beta immunotherapy for Alzheimer's disease: the end of the beginning. Nat. Rev. Neurosci. *3*(10): 824–828, 2002.

Scheuner, D., Eckman, C., Jensen, M., Song, X., Citron, M., Suzuki, N., Bird, T. D., Hardy, J., Hutton, M., Kukull, W., Larson, E., Levy-Lahad, E., Viitanen, M., Peskind, E., Poorkaj, P., Schellenberg, G., Tanzi, R., Wasco, W., Lannfelt, L., Selkoe, D. and Younkin, S.: Secreted amyloid beta-protein similar to that in the senile plaques of Alzheimer's disease is increased *in vivo* by the presenilin 1 and 2 and APP mutations linked to familial Alzheimer's disease. Nat. Med. *2*(8): 864–870, 1996.

Schmaier, A. H., Dahl, L. D., Rozemuller, A. J., Roos, R. A., Wagner, S. L., Chung, R. and Van Nostrand, W. E.: Protease nexin-2/amyloid beta protein precursor. A tight-binding inhibitor of coagulation factor IXa. J. Clin. Invest. *92*(5): 2540–2545, 1993.

Schneider, I., Reverse, D., Dewachter, I., Ris, L., Caluwaerts, N., Kuiperi, C., Gilis, M., Geerts, H., Kretzschmar, H., Godaux, E., Moechars, D., Van Leuven, F. and Herms, J.: Mutant presenilins disturb neuronal calcium homeostasis in the brain of transgenic mice, decreasing the threshold for excitotoxicity and facilitating long-term potentiation. J. Biol. Chem. *276*(15): 11539–11544, 2001.

Schroeder, F., Nemecz, G., Wood, W. G., Joiner, C., Morrot, G., Ayraut-Jarrier, M. and Devaux, P. F.: Transmembrane distribution of sterol in the human erythrocyte. Biochim. Biophys. Acta *1066*(2): 183–192, 1991.

Seguchi, K., Kataoka, H., Uchino, H., Nabeshima, K. and Koono, M.: Secretion of protease nexin-II/ amyloid beta protein precursor by human colorectal carcinoma cells and its modulation by cytokines/ growth factors and proteinase inhibitors. Biol. Chem. *380*(4): 473–483, 1999.

Selkoe, D. J.: Cell biology of the amyloid beta-protein precursor and the mechanism of Alzheimer's disease. Annu. Rev. Cell Biol. *10*: 373–403, 1994.

Selkoe, D. J., Abraham, C. R., Podlisny, M. B. and Duffy, L. K.: Isolation of low-molecular-weight proteins from amyloid plaque fibers in Alzheimer's disease. J. Neurochem. *46*(6): 1820–1834, 1986.

Selkoe, D. J., Yamazaki, T., Citron, M., Podlisny, M. B., Koo, E. H., Teplow, D. B. and Haass, C.: The role of APP processing and trafficking pathways in the formation of amyloid beta-protein. Ann. N. Y. Acad. Sci. *777*: 57–64, 1996.

Selkoe, D. J. and Lansbury, P. J., Jr.: Biochemistry of Alzheimer's disease and prion diseases. In: "Basic Neurochemistry," Sixth Edition, G. J. Siegel, B. W. Agranoff, R. W. Albers, S. K. Fisher, and M. D. Uhler, Eds., pp. 950–968, Lippincott-Raven, Philadelphia, 1999.

Sengupta, A., Wu, Q., Grundke-Iqbal, I., Iqbal, K. and Singh, T. J.: Potentiation of GSK-3-catalyzed Alzheimer-like phosphorylation of human tau by cdk5. Mol. Cell Biochem. *167*(1–2): 99–105, 1997.

Seubert, P., Vigo-Pelfrey, C., Esch, F., Lee, M., Dovey, H., Davis, D., Sinha, S., Schlossmacher, M., Whaley, J., Swindlehurst, C., et al.: Isolation and quantification of soluble Alzheimer's beta-peptide from biological fluids. Nature *359*(6393): 325–327, 1992.

Shen, Z.-X.: The significance of the activity of CSF cholinesterases in dementias. Med. Hypotheses *47*: 363–376, 1996.

Sheng, J. G., Price, D. L. and Koliatsos, V. E.: The beta-amyloid-related proeteins presenilins 1 and BACE1 are axonally transported to nerve terminals in the brain. Exp. Neurol. *184*: 1053–1057, 2003.

Siegel, G. J. and Chauhan, N. B.: Neurotrophic factors in Alzheimer's and Parkinson's disease brain. Brain Res. Brain Res. Rev. *33*(2–3): 199–227, 2000.

Siman, R. and Velji, J.: Localization of presenilin-nicastrin complexes and gamma-secretase activity to the trans-Golgi network. J. Neurochem. *84*(5): 1143–1153, 2003.

Simchowicz, T.: Histologische Sytudien über die senile Demenz. Nissl-Alzheimer Arbeiten *4*: 267–444, 1911.

Simons, M., Keller, P., De Strooper, B., Beyreuther, K., Dotti, C. G. and Simons, K.: Cholesterol depletion inhibits the generation of beta-amyloid in hippocampal neurons. Proc. Natl. Acad. Sci. U.S.A. *95*(11): 6460–6464, 1998.

Simons, M., Keller, P., Dichgans, J. and Schulz, J. B.: Cholesterol and Alzheimer's disease: is there a link? Neurology *57*(6): 1089–1093, 2001.

Sisodia, S. S.: Beta-amyloid precursor protein cleavage by a membrane-bound protease. Proc. Natl. Acad. Sci. U.S.A. *89*(13): 6075–6079, 1992.

Sinha, S., Anderson, J. P., Barbour, R., Basi, G. S., Caccavello, R., Davis, D., Doan, M., Dovey, H. F., Frigon, N., Hong, J., Jacobson-Croak, K., Jewett, N., Keim, P., Knops, J., Lieberburg, I., Power, M., Tan, H., Tatsuno, G., Tung, J., Schenk, D., Seubert, P., Suomensaari, S. M., Wang, S., Walker, D., Zhao, J., McConlogue, L. and John, V.: Purification and cloning of amyloid precursor protein beat-secretase from human brain. Nature *402*: 537–540, 1999.

Sisodia, S. S., Koo, E. H., Hoffman, P. N., Perry, G. and Price, D. L.: Identification and transport of full-length amyloid precursor proteins in rat peripheral nervous system. J. Neurosci. *13*(7): 3136–3142, 1993.

Skovronsky, D. M., Moore, D. B., Milla, M. E., Doms, R. W. and Lee, V. M.: Protein kinase C-dependent alpha-secretase competes with beta-secretase for cleavage of amyloid-beta precursor protein in the trans-Golgi network. J. Biol. Chem. *275*(4): 2568–2575, 2000.

Smith, M. A., Drew, K. L., Nunomura, A., Takeda, A., Hirai, K., Zhu, X., Atwood, C. S., Raina, A. K., Rottkamp, C. A., Sayre, L. M., Friedland, R. P. and Perry, G.: Amyloid-beta, tau alterations and mitochondrial dysfunction in Alzheimer disease: the chickens or the eggs? Neurochem. Int. *40*(6): 527–531, 2002.

Smith, R. P., Higuchi, D. A. and Broze, G. J., Jr.: Platelet coagulation factor XIa-inhibitor, a form of Alzheimer amyloid precursor protein. Science *248*(4959): 1126–1128, 1990.

Song, D. H., Dominguez, I., Mizuno, J., Kaut, M., Mohr, S. C. and Seldin, D. C.: CK2 phosphorylation of the armadillo repeat region of beta-catenin potentiates Wnt signaling. J. Biol. Chem. *278*(26): 24018–24025, 2003.

Song, J. H., Wang, C. X., Song, D. K., Wang, P., et al.: Interferon gamma induces neurite outgrowth by up-regulation of p35 neuron-specific cyclin-dependent kinase 5 activator via activation of ERK1/2 pathway. J Biol. Chem. *280*: 12896–12901, 2005.

Soriano, S., Kang, D. E., Fu, M., Pestell, R., Chevallier, N., Zheng, H. and Koo, E. H.: Presenilin 1 negatively regulates beta-catenin/T cell factor/lymphoid enhancer factor-1 signaling independently of beta-amyloid precursor protein and notch processing. J. Cell Biol. *152*(4): 785–794, 2001.

Steinhilb, M. L., Turner, R. S. and Gaut, J. R.: ELISA analysis of beta-secretase cleavage of the Swedish amyloid precursor protein in the secretory and endocytic pathways. J. Neurochem. *80*(6): 1019–1028, 2002.

Stine, W. B., Jr., Dahlgren, K. N., Krafft, G. A. and LaDu, M. J.: *In vitro* characterization of conditions for amyloid-beta peptide oligomerization and fibrillogenesis. J. Biol. Chem. *278*(13): 11612–11622, 2003.

Strittmatter, W. J., Saunders, A. M., Schmechel, D., Pericak-Vance, M., Enghild, J., Salvesen, G. S. and Roses, A. D.: Apolipoprotein E: high-avidity binding to beta-amyloid and increased frequency of type 4 allele in late-onset familial Alzheimer disease. Proc. Natl. Acad. Sci. U.S.A. *90*(5): 1977–1981, 1993.

Sturchler-Pierrat, C., Abramowski, D., Duke, M., Wiederhold, K. H., Mistl, C., Rothacher, S., Ledermann, B., Burki, K., Frey, P., Paganetti, P. A., Waridel, C., Calhoun, M. E., Jucker, M., Probst, A., Staufenbiel, M. and Sommer, B.: Two amyloid precursor protein transgenic mouse models with Alzheimer disease-like pathology. Proc. Natl. Acad. Sci. U.S.A. *94*(24): 13287–13292, 1997.

Subtil, A., Gaidarov, I., Kobylarz, K., Lampson, M. A., Keen, J. H. and McGraw, T. E.: Acute cholesterol depletion inhibits clathrin-coated pit budding. Proc. Natl. Acad. Sci. U.S.A. *96*(12): 6775–6780, 1999.

Suh, Y. H. and Checler, F.: Amyloid precursor protein, presenilins, and alpha-synuclein: molecular pathogenesis and pharmacological applications in Alzheimer's disease. Pharmacol. Rev. *54*(3): 469–525, 2002.

Suhara, T., Magrane, J., Rosen, K., Christensen, R., Kim, H. S., Zheng, B., McPhie, D. L., Walsh, K. and Querfurth, H.: Abeta42 generation is toxic to endothelial cells and inhibits eNOS function through an Akt/GSK-3beta signaling-dependent mechanism. Neurobiol. Aging *24*(3): 437–451, 2003.

Sun, M. K. and Alkon, D. L.: Cerebral hypoperfusion and amyloid-induced synergistic impairment of hippocampal CA1 synaptic efficacy and spatial memory in young adult rats. J. Alzheimers Dis. *6*: 355–366, 2004.

Sun, X., Sato, S., Murayama, O., Murayama, M., Park, J. M., Yamaguchi, H. and Takashima, A.: Lithium inhibits amyloid secretion in COS7 cells transfected with amyloid precursor protein C100. Neurosci. Lett. *321*(1–2): 61–64, 2002.

Suzuki, N., Cheung, T. T., Cai, X. D., Odaka, A., Otvos, L., Jr., Eckman, C., Golde, T. E. and Younkin, S. G.: An increased percentage of long amyloid beta protein secreted by familial amyloid beta protein precursor (beta APP717) mutants. Science *264*(5163): 1336–1340, 1994.

Svensson, A.-L. and Giacobini, E.: Cholinesterase inhibitors do more than inhibit cholinesterase. In: "Cholinesterases and Cholinesterase Inhibitors," E. Giacobini, Ed., pp. 227–235, Martin Dunitz, London, 2000.

Swarthout, J. T., Tyson, D. R., Jefcoat, S. C. Jr., Partridge, N. C. and Efcoat, S. C., Jr.: Induction of transcriptional activity of the cyclic adenosine monophosphate response element binding protein by parathyroid hormone and epidermal growth factor in osteoblastic cells. J. Bone Miner. Res. *17*(8): 1401–1407, 2002. Erratum in: J. Bone Miner. Res. *17*(10): 1917, 2002.

Takashima, A., Yamaguchi, H., Noguchi, K., Michel, G., Ishiguro, K., Sato, K., Hoshino, T., Hoshi, M. and Imahori, K.: Amyloid beta peptide induces cytoplasmic accumulation of amyloid protein precursor via tau protein kinase I/glycogen synthase kinase-3 beta in rat hippocampal neurons. Neurosci. Lett. *198*(2): 83–86, 1995.

Takashima, A., Noguchi, K., Michel, G., Mercken, M., Hoshi, M., Ishiguro, K. and Imahori, K.: Exposure of rat hippocampal neurons to amyloid beta peptide (25–35) induces the inactivation of phosphatidyl inositol-3 kinase and the activation of tau protein kinase I/glycogen synthase kinase-3 beta. Neurosci. Lett. *203*(1): 33–36, 1996.

Takashima, A., Honda, T., Yasutake, K., Michel, G., Murayama, O., Murayama, M., Ishiguro, K. and

Yamaguchi, H.: Activation of tau protein kinase I/ glycogen synthase kinase-3beta by amyloid beta peptide (25–35) enhances phosphorylation of tau in hippocampal neurons. Neurosci Res. *31*(4): 317–323, 1998a.

Takashima, A., Murayama, M., Murayama, O., Kohno, T., Honda, T., Yasutake, K., Nihonmatsu, N., Mercken, M., Yamaguchi, H., Sugihara, S. and Wolozin, B.: Presenilin 1 associates with glycogen synthase kinase-3beta and its substrate tau. Proc. Natl. Acad. Sci. U.S.A. *95*(16): 9637–9641, 1998b.

Takemaru, K. I. and Moon, R.T.: The transcriptional coactivator CBP interacts with beta-catenin to activate gene expression. J. Cell Biol. *149*(2): 249–254, 2000.

Tanaka, Y., Hanya, H., Sakurai, H., Takasaki, M. and Abe, K.: Atrophy of the substantia innominata on magnetic resonance imaging predicts response to donepezil treatment in Alzheimer's disease patients. Dement. Geriatr. Cogn. Disord. *16*: 119–1125, 2003.

Tandon, A. and Fraser, P.: The presenilins. Genome Biol. *3*(11): reviews 3014, 2002.

Tanemura, K., Murayama, M., Akagi, T., Hashikawa, T., Tominaga, T., Ichikawa, M., Yamaguchi, H. and Takashima, A.: Neurodegeneration with tau accumulation in a transgenic mouse expressing V337M human tau. J. Neurosci. *22*(1): 133–141, 2002.

Tanzi, R. E., Bird, E. D., Latt, S. A. and Neve, R. L.: The amyloid beta protein gene is not duplicated in brains from patients with Alzheimer's disease. Science *238*(4827): 666–669, 1987.

Tanzi, R. E., McClatchey, A. I., Lamperti, E. D., Villa-Komaroff, L., Gusella, J. F. and Neve, R. L.: Protease inhibitor domain encoded by an amyloid protein precursor mRNA associated with Alzheimer's disease. Nature *331*(6156): 528–530, 1988.

Tatebayashi, Y., Miyasaka, T., Chui, D. H., Akagi, T., Mishima, K., Iwasaki, K., Fujiwara, M., Tanemura, K., Murayama, M., Ishiguro, K., Planel, E., Sato, S., Hashikawa, T. and Takashima, A.: Tau filament formation and associative memory deficit in aged mice expressing mutant (R406W) human tau. Proc. Natl. Acad. Sci. U.S.A. *99*(21): 13896–13901, 2002.

Teller, J. K., Russo, C., DeBusk, L. M., Angelini, G., Zaccheo, D., Dagna-Bricarelli, F., Scartezzini, P., Bertolini, S., Mann, D. M., Tabaton, M. and Gambetti, P.: Presence of soluble amyloid beta-peptide precedes amyloid plaque formation in Down's syndrome. Nat. Med. *2*(1): 93–95, 1996.

Terry, R. D., Masliah, E. and Hansen, L. A.: The neuropathology of Alzheimer disease and the structural basis of its cognitive alterations. In: "Alzheimer Disease," Second Edition, R. D. Terry, R. Katzman, K. L. Bick, and S. S. Sisodia, Eds., pp. 187–206, Lippincott, Williams and Wilkins, Philadelphia, 1999.

Thal, L. J., Fuld, P. A., Masur, D. M. and Sharpless, N. S.: Oral physostigmine and lecithin improve memory in Alzheimer's disease. Ann. Neurol. *13*: 491–496, 1983.

Thal, L. J., Grundman, B., Berg, J., Ernstrom, K., Margolin, E., Pfeiffer, E., Weiner, M. F., Zamrini, E. and Thomas, R. G.: Idebenone treatment fails to slow cognitive decline in Alzheimer's disease. Neurology *61*: 1498–1502, 2003.

Thiel, C. M.: Cholinergic modulation of learning and memory in the human brain as detected with functional neuroimaging. Neurobiol. Learn. Mem. *80*: 233–244, 2003.

Tomlison, B. E.: Plaques, tangles and Alzheimer's disease. Psychol. Med. *12*: 449–459, 1982.

Tractenberg, R. E., Gamst, A., Thomas, R. G., Patterson, M., Schneider, L. S. and Thal, L.J.: Investigating emergent symptomatology as an outcome measure in a behavioral study of Alzheimer's disease. J. Neuropsychiatry Clin. Neurosci. *14*: 303–310, 2002.

Troy, C. M., Rabacchi, S. A., Hohl, J. B., Angelastro, J. M., Greene, L. A. and Shelanski, M. L.: Death in the balance: alternative participation of the caspase-2 and -9 pathways in neuronal death induced by nerve growth factor deprivation. J. Neurosci. *21*(14): 5007–5016, 2001. Erratum in: J. Neurosci. *21*(16): 1b, 2001.

Tsukada, H., Nishiyama, S., Fukumoto, D., Ohba, H., Sato, K. and Kakiuchi, T.: Effects of acute acetylcholinesterase inhibition on the cerebral cholinergic neuronal system and cognitive function: functional imaging of the conscious monkey brain using animal PET in combination with microdialysis. Synapse *52*: 1–10, 2004.

Tuszynski, M. H., Thal, L., Pay, M., Salmon, D. P., U, H. S., Bakay, R., Patel, P., Blesch, A., Vahlsing, H. L., Ho, G., Tong, G., Potkin, S. G., Fallon, J., Hansen, L., Mufson, E. J., Kordower, J. H., Gall, C. and Conner, J.: A phase 1 clinical trial of nerve growth factor gene therapy for Alzheimer disease. Nat. Med. *11*: 551–555, 2005.

Tyson, D. R., Swarthout, J. T., Jefcoat, S. C. and Partridge, N. C.: PTH induction of transcriptional activity of the cAMP response element-binding protein requires the serine 129 site and glycogen synthase kinase-3 activity, but not casein kinase II sites. Endocrinology *143*(2): 674–682, 2002.

Umans, L., Serneels, L., Lorent, K., Dewachter, I., Tesseur, I., Moechars, D. and Van Leuven, F.: Lipoprotein receptor-related protein in brain and in cultured neurons of mice deficient in receptor-associated protein and transgenic for apolipoprotein E4 or amyloid precursor protein. Neuroscience *94*(1): 315–321, 1999.

Utton, M. A., Vandecandelaere, A., Wagner, U., Reynolds, C. H., Gibb, G. M., Miller, C. C., Bayley,

P. M., Anderton, B. H.: Phosphorylation of tau by glycogen synthase kinase 3beta affects the ability of tau to promote microtubule self-assembly. Biochem. J. *323*, Pt. *3*: 741–747, 1997.

Van Gassen, G., De Jonghe, C., Nishimura, M., Yu, G., Kuhn, S., St. George-Hyslop, P. and Van Broeckhoven, C.: Evidence that the beta-catenin nuclear translocation assay allows for measuring presenilin 1 dysfunction. Mol. Med. *6*(7): 570–580, 2000.

Van Nostrand, W. E., Wagner, S. L., Suzuki, M., Choi, B. H., Farrow, J. S., Geddes, J. W., Cotman, C. W. and Cunningham, D. D.: Protease nexin-II, a potent antichymotrypsin, shows identity to amyloid beta-protein precursor. Nature *341*(6242): 546–549, 1989.

Van Nostrand, W. E., Wagner, S. L., Farrow, J. S. and Cunningham, D. D.: Immunopurification and protease inhibitory properties of protease nexin-2/amyloid beta-protein precursor. J. Biol. Chem. *265*(17): 9591–9594, 1990.

Van Nostrand, W. E., Schmaier, A. H., Neiditch, B. R., Siegel, R. S., Raschke, W. C., Sisodia, S. S. and Wagner, S. L.: Expression, purification, and characterization of the Kunitz-type proteinase inhibitor domain of the amyloid beta-protein precursor-like protein-2. Biochim. Biophys. Acta *1209*(2): 165–170, 1994.

Van Nostrand, W. E., Melchor, J. P., Cho, H. S., Greenberg, S. M. and Rebeck, G. W.: Pathogenic effects of D23N Iowa mutant amyloid beta-protein. J. Biol. Chem. *276*(35): 32860–32866, 2001.

Vassar, R., Bennett, B. D., Babu-Khan, S., Kahn, S., Mendiaz, E. A., Denis, P., Teplow, D. B., Ross, S., Amarante, P., Loeloff, R., Luo, Y., Fisher, S., Fuller, J., Edenson, S., Lile, J., Jarosinski, M. A., Biere, A. L., Curran, E., Burgess, T., Louis, J. C., Collins, F., Treanor, J., Rogers, G. and Citron, M.: Beta-secretase cleavage of Alzheimer's amyloid precursor protein by the transmembrane aspartic protease BACE. Science *286*: 735–741, 1999.

Vogelsberg-Ragaglia, V., Trojanowski, J. Q., Lee, V. M.-Y. In: "Alzheimer Disease," Second Edition, R. D. Terry, R. Katzman, K. L. Bick, and S. S. Sisodia, Eds., pp. 359–372, Lippincott, Williams and Wilkins, Philadelphia, 1999.

Von Koch, C. S., Zheng, H., Chen, H., Trumbauer, M., Thinakaran, G., van der Ploeg, L. H., Price, D. L. and Sisodia, S. S.: Generation of APLP2 KO mice and early postnatal lethality in APLP2/APP double KO mice. Neurobiol. Aging *18*(6): 661–669, 1997.

Wagner, M. R., Keane, D. M., Melchor, J. P., Auspaker, K. R., Van Nostrand, W. E.: Fibrillar amyloid beta-protein binds protease nexin-2/amyloid beta-protein precursor: stimulation of its inhibition of coagulation factor XIa. Biochemistry *39*(25): 7420–7427, 2000.

Wagner, U., Utton, M., Gallo, J. M. and Miller, C. C.: Cellular phosphorylation of tau by GSK-3 beta influences tau binding to microtubules and microtubule organisation. J. Cell Sci. *109*, Pt. *6*: 1537–1543, 1996.

Wagner, U., Brownlees, J., Irving, N. G., Lucas, F. R., Salinas, P. C. and Miller, C. C.: Overexpression of the mouse dishevelled-1 protein inhibits GSK-3beta-mediated phosphorylation of tau in transfected mammalian cells. FEBS Lett. *411*(2–3): 369–372, 1997.

Waschuk, S. A., Elton, E. A., Darabie, A. A., Fraser, P. E. and McLaurin, J. A.: Cellular membrane composition defines A beta-lipid interactions. J. Biol. Chem. *276*(36): 33561–33568, 2001.

Weksler, M. E., Relkin, N., Turkenich, R., LaRusse, S., Zhou, L. and Szabo, P.: Patients with Alzheimer disease have lower levels of serum anti-amyloid peptide antibodies than do healthy elderly individuals. Exp. Gerontol. *37*: 943–948, 2002.

Whitehouse, P. J., Price, D. L., Clark, A. W., Jr., Coyle, J. T. and DeLong, M. R.: Alzheimer's disease: evidence for selective loss of cholinergic neurons in the nucleus basalis. Ann. Neurol. *10*: 122–126, 1981.

Whitehouse, P. J., Price, D. L., Struble, R. G., Clark, A. W., Coyle, J. T. and DeLong, M. R.: Alzheimer's disease and senile dementia: loss of neurons in the basal forebrain. Science *215*: 1237–1239, 1982.

Whitehouse, P. J., Martino, A. M., Antuono, P. G., Lowenstein, P. R., Coyle, J. T., Price, D. L. and Kellar, K. J.: Nicotinic acetylcholine binding sites in Alzheimer's disease. Brain Res. *371*: 146–151, 1986.

Wild-Bode, C., Yamazaki, T., Capell, A., Leimer, U., Steiner, H., Ihara, Y. and Haass, C.: Intracellular generation and accumulation of amyloid beta-peptide terminating at amino acid 42. J. Biol. Chem. *272*(26): 16085–16088, 1997.

Wilson, C. A., Doms, R. W., Zheng, H. and Lee, V. M.: Presenilins are not required for A beta 42 production in the early secretory pathway. Nat. Neurosci. *5*(9): 849–855, 2002.

Wisniewski, H. M.: Clinical, pathological and biochemical aspects of Alzheimer's disease. In: "Alzheimer's and Parkinson's Diseases," A. Fisher, I. Hanin and C. Lachman, Eds., pp. 25–36, Plenum Press, New York, 1987.

Wolfe, M. S. and Kopan, R.: Intramembrane proteolysis: theme and variations. Science *305*: 1119–1123, 2004.

Wolfe, M. S. and Selkoe, D. J.: Biochemistry. Intramembrane proteases—mixing oil and water. Science *296*(5576): 2156–2157, 2002.

Wolfe, M. S., Xia, W., Ostaszewski, B. L., Diehl, T. S., Kimberly, W. T., Selkoe, D. J.: Two transmembrane aspartates in presenilin-1 required for presenilin endoproteolysis and gamma-secretase activity. Nature *398*(6727): 513–517, 1999.

Wolozin, B. and Davis, P.: Alzheimer-related neuronal protein A68: specificity and distribution. Ann. Neurol. *22*: 521–526, 1987.

Wolozin, B., Kellman, W., Rousseau, P., Celesia, G. G. and Siegel, G.: Decreased prevalence of Alzheimer disease associated with 3-hydroxy-3-methyglutaryl coenzyme A reductase inhibitors. Arch. Neurol. *57*: 1439–1443, 2000.

Wong, P. C., Cai, H., Borchelt, D. R. and Price, D. L.: Genetically engineered models relevant to neurodegenerative disorders: their value for understanding disease mechanisms and designing/testing experimental therapeutics. J. Mol. Neurosci. *17*(2): 233–257, 2001.

Wong, P. L., Li, T. and Price, D.L.: Neurobiology of Alzheimer's Disease. In: "Basic Neurochemistry: Molecular, Cellular and Medical Aspects Aspects," G. J. Siegel, R. N. Albers, S. T. Brady, D. L. Price, Eds., 7th edition, Chap. 47, London: Elsevier/ Academic Press, 2006.

Woodgett, J. R.: Judging a protein by more than its name: GSK-3. Sci STKE. 2001(100): RE12, 2001.

Woolf, N. J., Gould, E. and Butcher, L. L.: Nerve growth factor receptor is associated with cholinergic neurons of the basal forebrain but not the pontomesencephalon. Neuroscience *30*: 30–45, 1989.

Wu, C., Butz, S., Ying, Y. and Anderson, R. G.: Tyrosine kinase receptors concentrated in caveolae-like domains from neuronal plasma membrane. J. Biol. Chem. *272*(6): 3554–3559, 1997.

Wurtman, R. J.: Alzheimer's disease. Sci. Amer. *252*: 62–74, 1985.

Wynn, Z. J. and Cummings, J. L.: Cholinesterase inhibitor therapies and neuropsychiatric manifestations of Alzheimer's disease. Dement. Geriatr. Cogn. Disord. *17*: 100–108, 2004.

Xie, G., Gunn, R. N., Dagher, A., Daloze, T., Plourde, G., Backman, S. B., Diksic, I. and Fiset, P.: PET quantification of muscarinic cholinergic receptors with [N-metyl]-benztropine and application to studies of propofol-induced unconsciousness in healthy human volunteers. Synapse *51*: 91–101, 2004.

Xu, D., Joglekar, A. P., Williams, A. L. and Hay, J. C.: Subunit structure of a mammalian ER/Golgi SNARE complex. J. Biol. Chem. *275*(50): 39631–39639, 2000.

Yamazaki, T., Chang, T. Y., Haass, C. and Ihara, Y.: Accumulation and aggregation of amyloid beta-protein in late endosomes of Niemann-pick type C cells. J. Biol. Chem. *276*: 4454–4460, 2001.

Yan, R., Bienkowski, M. J., Shuck, M. E., Miao, H., Tory, M. C., Pauley, A. M., Brashier, J. R., Stratman, N. C., Mathews, W. R., Buhl, A. E., Carter, D. B., Tomasselli, A. G., Parodi, L. A., Heinrikson, R. L. and Gurney, M. E.: Membrane-anchored aspartyl protease with Alzheimer's disease beta-secretase activity. Nature *402*(6761): 533–537, 1999.

Yanagisawa, K., Odaka, A., Suzuki, N. and Ihara, Y.: GM1 ganglioside-bound amyloid beta-protein (A beta): a possible form of preamyloid in Alzheimer's disease. Nat. Med. *1*(10): 1062–1066, 1995.

Yip, C. M. and McLaurin, J.: Amyloid-beta peptide assembly: a critical step in fibrillogenesis and membrane disruption. Biophys. J. *80*(3): 1359–1371, 2001.

Yip, C. M., Elton, E. A., Darabie, A. A., Morrison, M. R. and McLaurin, J.: Cholesterol, a modulator of membrane-associated Abeta-fibrillogenesis and neurotoxicity. J. Mol. Biol. *311*(4): 723–734, 2001.

Yoshiike, Y., Chui, D. H., Akagi, T., Tanaka, N. and Takashima, A.: Specific compositions of amyloid-beta peptides as the determinant of toxic beta-aggregation. J. Biol. Chem. *278*(26): 23648–23655, 2003.

Yoshimura, T., Kawano, Y., Arimura, N., Kawabata, N., et al.: GSK-3beta regulates phosphorylation of CRMP-2 and neuronal polarity. Cell *120*: 137–149, 2005.

Yuan, J. and Yankner, B. A.: Caspase activity sows the seeds of neuronal death. Nat. Cell Biol. *1*(2): E44–45, 1999.

Zaidi, N. F., Berezovska, O., Choi, E. K., Miller, J. S., Chan, H., Lilliehook, C., Hyman, B. T., Buxbaum, J. D. and Wasco, W.: Biochemical and immunocytochemical characterization of calsenilin in mouse brain. Neuroscience *114*(1): 247–263, 2002.

Zannis, V. I. and Breslow, J. L.: Human very low density lipoprotein apolipoprotein E isoprotein polymorphism is explained by genetic variation and post-translational modification. Biochemistry *20*(4): 1033–1041, 1981.

Zheng, Y.-L., Kesavapany, S., Gravell, M., Hamilton, R. S., et al.: A Cdk5 inhibitory peptide reduces tau hyperphosphorylation and apoptosis in neurons. The EMBO Journal *24*: 209–220, 2005.

Zhou, H., Pfeifer, U. and Linke, R.: Generalized amyloidosis from beta 2-microglobulin, with caecal perforation after long-term haemodialysis. Virchows Arch. A Pathol. Anat. Histopathol. *419*(4): 349–353, 1991.

Zhu, X., Lee, H. G., Raina, A. K., Perry, G. and Smith, M. A.: The role of mitogen-activated protein kinase pathways in Alzheimer's disease. Neurosignals *11*(5): 270–281, 2002.

Zimmermann, M., Gardoni, F., Marcello, E., Colciaghi, F., Borroni, B., Pardovani, A., Cattabeni, F. and Di Luca, M.: Acetylcholinesterase inhibitors increase ADAM10 activity by promoting its trafficking in neuroblastoma cell lines. J. Neurochem. *90*: 1489–1499, 2004.

11

Envoi

This book is a journey along the trails of cholinergic research on the vertebrate central nervous system. It also presents the highlights of the present status of this research, whether directly concerned with central nervous system cholinergicity and its pathways or dealing with its components, such as cholinesterases (ChEs), cholinergic receptors, acetylcholine (ACh) metabolism, and neuronal elements of ACh release.

Two kinds of final comments concerning central cholinergicity are presented here. First, remarks will be made as to the tasks that remain to be accomplished within the various fields of the cholinergic system and which are rather mundane; that is, the state-of-art methodology for the field in question is such that the accomplishment of these tasks is only a matter of time. Second, questions are raised that deal with matters that are far beyond the present status of cholinergicity and its methodology. Some of these questions are raised in the concluding remarks for each chapter of this book, but these matters are appropriate for this final chapter, so at the risk of redundancy these questions are presented here.

A. Immediate, Mundane Concerns

1. Clinical Aspects

It may be said—regretfully on the part of a devoted cholinergiker—that the clinical use of cholinergics is not on par with the importance of cholinergic system for neurotransmission, brain function, and behavior. Yet, cholinergic and anticholinergic drugs are used as muscle relaxants and in the treatment of glaucoma, myasthenia gravis, bladder dysfunction and incontinence, bronchiolar and intestinal dysfunctions, and chronic obstructive pulmonary disease, areas that are outside the scope of this book (see, for example, Stothers, 2004; Barnes, 2004; Karczmar, 1979). More in keeping with the tone of this book, it should be stressed that cholinergic agonists are employed in the treatment of Alzheimer's disease, Parkinsonism, and other degenerative diseases (see Chapter 10; see also Barbeau, 1962, 1978), shortening the recovery from anesthesia (including the recovery from curarimimetics; Oh et al., 2004), and antidoting overdose with tricyclics; and as adjuvants to anesthesia. It is interesting that anticholinesterases (antiChEs) are used in the treatment of venomous snake poisoning (see, for example, Bawaskar and Bawaskar, 2004). In the past, atropine coma was employed, quite successfully, in the management of depression; since the atropinic overdose is readily treated today, perhaps this therapy should be revisited (see Karczmar, 1979).

Certain potentials of cholinergic drugs should be explored. More attention should be devoted, in the case of Alzheimer's disease, to the ameliorative actions of antiChEs and cholinergic agonists on the amyloid cascade (see Chapter 10 K). In a related context, these drugs are currently employed in non–Alzheimer's disease memory dysfunctions; since in these conditions cholinergic neurons remain mostly intact, this therapy should be theoretically effective and this use should be systematized. Then, cholinergically related trophic factors such as nerve growth factor (NGF) are at present used in degenerative diseases as well as in certain psychoses, and this use should be further explored and encouraged (Angelucci et al., 2004).

Finally, Karl Pfeiffer proposed in nineteen forties that muscarinics and antiChEs may be helpful in the treatment of schizophrenia, and, in fact he used some of these agents, with some success, in the therapy of hebephrenics. Some twenty yeas ago this author adduced evidence supporting Pfeiffer's notion (Karczmar, 1988), as did, more recently, Dean and his associates (2003).

2. Cholinergic Pathways

Though the identification of the central cholinergic pathways is nearly complete, a few outstanding problems remain. These include the definition of ascending and descending spinal cholinergic pathways. Also, Marsel Mesulam's description of Ch7 and Ch8 cholinergic sectors may be incomplete, and several sectors seem to work in series toward the same aim, the activating effects of the ascending cholinergic system (see Pepeu, 1993; Chapter 2 DIII). Is this a case of redundancy, or of different delineations of the sectors in question?

Furthermore, the current definition of the association of the different pathways with their behavioral outcome, while advanced, is incomplete, as in the case of Ch5 and Ch6 sectors innervating various thalamic neurons.

A longer-range problem is represented by the colocalization and corelease of ACh with other transmitters (particularly peptides) and concerns identifying neurotransmitter interaction at post- and presynaptic sites. This problem has been a part of the cholinergic lore since Thomas Hokefelt and others' research during the 1970s (see Chapter 9 BII-2 and BIII-2f), and while there are tools competent for providing pertinent evidence, the remaining task is long and laborious.

Choline acetyltransferase (CAT) immunohistochemistry and other methods for localizing cholinergic cells, as well as similar methods used for identification of other transmitters, are capable of resolving these problems. Yet, it is interesting to approach the problem via visualizing the cholinergic transmitter itself. The methods for localizing ACh were developed (see Chapter 2 B-2), but they proved unsatisfactory; perhaps there will be a happy ending to this matter soon.

3. Cholinesterases

Remarkable success has been achieved with structurization of both butyryl cholinesterase (BuChE) and AChE, the three-dimensional image of their active sites, the polymorphism of ChEs, and the mechanisms of their catabolic function. The available three-dimensional images of ChEs, their active sites and catalytic gorges, and their anchoring subunits are well-nigh unbelievable! Jean Massoulié, Mona Soreq, Joel Sussman, Israel Silaman, Richard Rotundo, Nibaldo Inestrosa, Avigdor Shafferman, Palmer Taylor, and their associates made great strides in this area as well as with identifying isoforms and variants of ChEs. Mona Soreq presented a novel concept when she illustrated the ChE adaptability that leads to forming variants such as the stress-related variant (see Chapter 3 DIII). The remaining tasks include identification of additional variants and complete definition of the localization and functional role of these variants.

A more complicated problem may be an explanation for the role of AChE in nonneuronal tissues (such as red blood cells of several vertebrate species), particularly ephemeral tissues such as placenta, and in the embryo prior to neurogenesis (see the next section for discussion of this last point). A related and perhaps even more difficult problem concerns the role of BuChE. What does BuChE do in the neuronal or nonneuronal tissues? Several speculations have been offered, particularly by Ezzio Giacobini (see Chapter 3 DIII), but they do not hit the crux of the matter.

4. Development and Phylogenesis

Development and phylogenesis comprise an area that concerns more than the delineation of cholinergic neurogenesis. Zenon Bacq, David Nachmansohn, and others (see Chapter 8 A) introduced this area when they endeavored to establish the role of AChE in functions such as motility, and the need for AChE in synaptic transmission. Studies of cholinergic development continue to raise questions regarding the functional roles of ChEs and other cholinergic components. Since they appear in ontogenesis long before neurogenesis is initiated (in the 2- or 4-cell stage in some species), what is the role of these components in ontogenesis? What light does this precocity throw on the roles of cholinergic components elsewhere than in ontogenesis (see Chapter 8 BIV and D-2)? An additional conundrum is added to this question, since cholinergic components appear early in phylogenesis, including in monocellular organisms such as bacteria and protozoa; in fact, their presence is ubiquitous not only across the animal but also plant phyla (see Chapter 8 BV). Indeed, in 1963 Victor Whittaker referred to ACh as "the most ancient mediator." In flagellates, the presence of cholinergic components at the base of the flagella of the cholinergic system relates to the movement of flagellae—George Koelle suggested that BuChE plays a similar role in the smooth muscle (AChE is present in the skeletal muscle as well).

While this perplexing question regarding the precocious ontogeny and early phylogeny of cholinergic components has not been answered, certain plausible hypotheses have been presented. For example, cholinergic components, including ACh and ChEs, may act as morphogens (see Karczmar, 1963; Chapter 8 BI and BIV). This trophic action relates cholinergic components to trophic factors all the more since the trophic factors act on cholinergic targets. The discovery of trophic factors by Victor Hamburger and Rita Levi-Montalcini is one of the watersheds of cholinergically related research. The role of the NGF in neuronal and axonal growth precipitated intense pharmaceutical and academic research concerning the discovery of additional trophic factors and identifying their role. Perhaps 50 trophic factors, mostly peptides, have been identified, and additional factors are discovered just about each month. New mechanisms of trophisms will be described, and trophics endowed with affinity to specific targets within the brain, acting on either the somata or axons, will be discovered.

In view of their morphogenetic, regenerative, and neuronal maintenance potential, it is somewhat disappointing that the trials of trophics in brain illnesses carried out mainly in Sweden and Italy did not show their clear-cut effectiveness in these conditions. It may be surmised that this may change in the near future.

Finally, it is electrifying to speculate that the cholinergic genome acted initially as a morphogen, in ancestral living forms and, later, across the phyla, regulating cellular growth and differentiation. The expression of the genome as a morphogen was subjected subsequently to selective adaptation, while its transmittive function is exaptive in nature. Altogether, present-day ChEs may represent adaptations of an ancestral enzyme[1] (see Chapter 8 BV).

5. Metabolism of ACh

Since the research of the 1940s and 1950s (see Chapters 2 B and 3 A), investigators have known that CAT, choline and its uptake, acetate, and coenzyme A constitute the system that synthesizes ACh. Following the realization of the significance of choline as possibly the limiting factor in this synthesis, Konrad Loeffelholz, Jochen Klein, and Stanislas Tucek became concerned with the other side of the coin: under certain conditions choline's synthesis and nerve terminal uptake may become excessive and choline may become an agonist. Choline's special action as an agonist at cholinergic preterminals has recently been stressed (Chapter 9 BI-1 and BI-2); thus, there is a need for a system that would serve to maintain choline homeostasis; future research is necessary to confirm this notion and to define the components of the system.

Many attempts have been made to provide choline or ACh precursors that could facilitate cholinergic function and to maintain cholinergic function in aged animals or humans. Recently, Roger Wurtman, Kazimierz Blusztajn, and Lynn Wecker have been active in this area. If the notion of endogenous choline homeostasis is correct, and if it holds also for other endogenous precursors of ACh, this may prove to be a difficult task.

6. The Cholinergic Neuron and ACh Release

The cholinergic neuron owns an exceedingly complex active protein system that is involved in transporting cholinergic components such as ACh and the vesicles (for more about the consequences of this complexity, see the next section.) This system combines with the system subserving ACh synthesis to provide for ACh release, which can adapt to many environmental conditions. Some of the flexibility required for responding to environmental needs is provided by cholinergic and noncholinergic nerve terminal regulation of the release.

Enormous success, partially resulting from the use of molecular biology methods, brought about a detailed description of the release system's components (Chapter 2 B and C). Even so, more detail should become available in the future; this additional information is particularly needed with regard to the interactive harmony that underlies the synthesis and release of ACh and the flexibility of the systems in question vis-à-vis environmental demands.

Still another problem to be solved concerns the classical versus unorthodox images of the final steps in the release process (see Chapter 2 C). Are the synaptic vesicles the only instruments of the quantal release, as proposed classically by Victor Whittaker? Ample evidence supports this classical tenet; yet, several unorthodox hypotheses including the postulate of cytoplasmic rather than vesicular ACh release have their adherents (see Chapter 2 C-1 and C–2 and Victor Whittaker's Foreword). The controversy among proponents of the various modes of release is sometimes almost violent. Further research is needed to resolve this matter. Is it possible that various release processes may coexist? May the mode of release depend on conditions, such as high demand for release?

7. Receptology

The receptor research during the last 50 or 60 years has led us a long way from Langley's and Dale's simple view of monotypic muscarinic and nicotinic receptors, as the brain exhibits many subtypes of nicotinic and muscarinic receptors at the presynaptic and postsynaptic sites. While much information is available with respect to brain distribution of nicotinic and muscarinic receptor subtypes (see Chapters 5 C, 6 C; 7 DII and DIII; and 9 BI-2, BIV, and BV), further work is needed to completely define their brain location.

In context, our knowledge of discrete functions and behaviors associated with each transmitter subtype is certainly incomplete. Sometimes, several transmitter subtypes have been linked with a single function or behavior (see Chapter 9 BI-2, BIV, and BV). Is this because single functions and behaviors are not dealt with in these cases? Is it because the "final shaping" of the activities needs the activation of several receptor subtypes?

Modern knowledge of receptor subtypes led to the development of agonists and antagonists that may be specific for human diseases. This development will be aided with a more comprehensive understanding of the three-dimensional receptor subtype structures and the employment of quantitative structure-activity relationship (QSAR) and Hansch analysis for a perfect fit between the drug and the active sites of the receptors (see Chapter 4 D and E). As for the nicotinic receptor, the crystallization of the molluscan nicotinic receptor, its structure, and the structure of its extracellular domain are available, but the crystal structure of mammalian extracellular domain has not been accomplished (see Chapter 6 B). This achievement and especially obtaining the structure of the human nicotinic receptor will have an enormous applied significance.

As mentioned earlier, the activation of several receptor subtypes is needed for the "final shaping" of a given function or behavior. The clarification of this matter should be the final step in the search for the therapeutic use of cholinergic agonists and antagonists.

8. Channels

Ion channels provide the response of the pre- or postsynaptic membrane to ACh (and other transmitters). Research concerning channel proteins and the genes encoding these proteins has been progressing quickly, as there are a number of peripheral and central disease states caused by channelopathies (Hiremath et al., 2003). Recently, the complete structure of several voltage-gated channels, such as those conveying the fast ACh response, was determined (Chapter 6 B). As with the receptors, these discoveries will provide templates for drug design and for providing specific drugs to evaluate cholinergic transmission and for the therapies of channelopathies.

9. Central Cholinergicity

Problems remain to be solved on both the synaptic and functional or behavioral levels. As for the synaptic level, it was frequently demonstrated that a cholinergic radiation abuts directly on a noncholinergic postsynaptic membrane. However, sometimes a cholinergic-cholinergic link is present, and it is not easy to technically demonstrate such a presence. In a related context, technical obstacles are great with regard to determining *all* the transmitter (including ACh) inputs into a cholinergic neuron, whether at its somatic or nerve terminal site; similarly, it is not easy to determine how many and which transmitters share with ACh access to a given noncholinergic neuron (see Chapter 9 BIII-1). In other words, much remains to be done with respect to the definition of transmitter interactions at presynaptic or postsynaptic sites.

This question of transmitter interaction relates to cholinergic response modulations, whether post- or presynaptic (in the case of presynaptic modulation, the modulation regulates the release of transmitters, including ACh). Transmitter interaction results in vectoral modulation between the excitatory and inhibitory responses, as proposed by Sir John Eccles (see Eccles, 1964). But, there are also modulations that are not caused by vectoral neurotransmitter interactions, as in the case of receptor inactivation and sensitization (see Chapter 9 BIII-2c). The status of modulation at a given synapse and at a given time is unclear. In other words, the post- or presynaptic outcome as illustrated by the generation of a productive spike and actual release of ACh cannot be readily predicted.

Finally, noncholinergic and cholinergic drugs maximally effective at any post- or presynaptic site may have not been developed (see above). Perhaps the most obvious lacuna concerns the presynaptic facilitators of ACh release other than atropine and atropinics (the ACh-releasing capacity of atropinics is counterbalanced by their postsynaptic effects and their actions outside of the CNS; see Chapter 9 BIII-2a). There may be a potential for these drugs in cases of cholinergic hypofunction, as the antidepressive action of nootropics may result from their facilitation of ACh release. However, this mechanism is poorly documented for nootropics, and other drugs of this type (including novel pyrimidines) will undoubtedly be developed in the future.

There are other applied aspects of central cholinergicity that should be further exploited. Accordingly, better analgetic muscarinic agonists should be developed; while they may not be appropriate for use as everyday painkillers, such as acetaminophen, there are indications that these analgesics may be useful in surgery and in attenuating effective morphine doses. Finally, while unsuccessful attempts were made to control withdrawal symptoms with atropinics (see Chapter 9 BV-5), additional trials should be carried out in view of the demonstrated cholinergicity of abstinence and the availability of muscarinic subtype-specific antagonists.

Several future tasks should be emphasized with respect to functions and behaviors that are endowed with cholinergic correlates. At one time, this author opined that cholinergic correlates were established for all measurable behaviors, particularly in animals (see, for example, Karczmar, 1990). While indeed cholinergicity of many functions and behaviors has been established (see Tables 9-1 to 9-4), this author's early opinion that *all* behaviors have cholinergic correlates should be revised and read as follows: "Though cholinergic correlates have been established for *most* measurable behaviors in animals, the cholinergicity of many human behaviors have not been investigated." Very little is known with regard to animals and to the cholinergic correlates of their social behaviors. For example, territorial protection and defense and social behaviors, such as contactual behavior and grooming, may be studied in pseudonatural laboratory habitats, such as Mouse City (Chapter 9 BV-4), and in zoos. However, little information is currently available describing the cholinergicity of these behaviors.

Olfaction is also a part of social behavior. Pathway-wise, it is a part of the rhinencephalon and of the limbic system. Accordingly, olfaction exhibits cholinergic correlates (see Chapter 2 DII-DIII and Chapter 9 V-c). Yet, there is little information as to the cholinergic and anticholinergic effects on olfactory function; further work is needed in this area.

It is important to fill these lacunae: in toto, cholinergic behaviors constitute the cholinergic alert nonmobile behavior (CANMB; see Chapter 9 BVIa), which, as postulated by this author, underlies animal and human adaptability to the environment and its changes. Social behaviors and olfaction are important components of CANMB, and their cholinergic correlates should be defined.

Moreover, emotional attitudes, particularly the state of being shy, are also pertinent with respect to social behaviors; they are not quantified as yet in animals, although they are measured in humans and observed in primates (and in the human associated with the limbic system). These emotional states should be measured and assessed as to their cholinergicity in animals; it is not impossible to do likewise in humans.

Development of language is a behavior that is a part of learning; presumably, this behavior is exhibited by the human and is not present or present to a limited extent only in animals. As postulated by Noam Chomsky (1968), acquisition of speech and grammar forms are brain based and thus amenable to pharmacological evaluation; this may be impossible in humans but it would be very important if it could be done.

Finally, the acquisition of CANMB may be an evolutionary step for vertebrates and the human (see Chapter 9 BVIa). Further pursuits of cholinergic phylogeny (Chapter 2 A) are worth undertaking to clarify the phylogenetic descent of the CANMB. As for self-awareness, this matter belongs to the "far beyond" subjects of central cholinergicity and is referred to in the next section.

B. The "Far Beyond" Matters

1. "Accountability" of Cholinergic Functions and Behaviors

First, there is the question of "accountability" of cholinergic functions and behaviors. Leaving aside the morphogenetic and trophic as well as tropic functions of the cholinergic system and focusing only on transmittive processes that culminate in a given behavior, these processes are most complex. We may take their outcome for granted, so it is worthwhile to illustrate this complexity by summarizing the pertinent events. At the very onset, this process consists of the synthesis and catabolism of ACh; there are multicomponent phenomena. Then, there is the release of ACh and its constituents, such as choline uptake, cycling and transport of vesicles and ACh, and the modes of its exocytosis. Further, there is the nerve terminal regulation—inhibition and facilitation—of the release; this regulation is multitransmitter in nature. The channel phenomena are involved on the postsynaptic level; these phenomena control whether the released ACh is inhibitory or produces an excitatory potential. Vectoral multitransmitter events follow, and they decide whether the release of ACh events leads to a transmittive spike. The transmittive spike, if it occurs, activates both cholinergic and noncholinergic chain of events, including a multitude of receptors and their subtypes, second messenger mechanisms and modulations, and receptor sensitization and receptor inactivation. Several semiparallel chains of this nature that include many cholinergic and noncholinergic pathways ultimately "shape" a given behavior or function.

All this complexity works harmoniously and is flexible. The outcome changes from moment to moment, to assure the optimal organism-environment interaction and adaptability and also to be responsive to inner drives and emotions.

In Chapter 2 C, I referred specifically to one of the steps of this complexity, the step concerning the release of ACh, and I opined that I am not sure that there can be a computer program that will define this step and that will allow us to predict whether this step will be made under specific circumstances (see Pepeu, 1993). What, then, about the whole process that ultimately results in a given function or behavior? Yet, a science likes to be predictive as to the outcome of phenomena and it considers that it has a handle on a phenomenon when it becomes possible to construct a mathematical formula describing the phenomenon in question. Is then cholinergicity (or the phenomenology of any other transmitter) still in its prescientific mode?

2. Consciousness in Its Self-Awareness Mode, or: How Does It Feel to See Red?

The second question is that of consciousness in its self-awareness mode. Or: how does it feel to see red?[2] The related question is that of "free will,"[3] and it must be stressed that this notion is beyond the phenomena of memory and cognition, even though memory is a substantive part of self-awareness. Yet, even today, some neuroscientists relate the problem of consciousness and self-awareness simply to the identification of the pertinent brain regions (Miller, 2005). The problem may be defined as that of body-consciousness relationship. As described in Chapter 9 BVI, there is an intense controversy between the reductionists, who feel that self-awareness is as material and as brain dependent as any motor function (strict behaviorists such as B. F. Skinner and J. B. Watson), and the Cartesian dualists, who consider self-awareness as an expression of the brain but distinct from other brain functions (this is a two- or even three-worlds notion; see Popper and Eccles, 1977). John Searle (2004) may be unique among the discussants of mind and consciousness, as he opines that both the reductionists ("monists") and dualists essentially talk nonsense, and that the mind (and/or consciousness) is simply a property of matter (see Chapter 9 BVI). Yet, the modern reductionists are sophisticated indeed. For example, Ben Libet based his notions of free will on the methodology of cortical evoked potentials and the capacity to measure within milliseconds cortical and volitional events, as he proposed that the time periods separating cortical potential evoked by sensory stimulation, subjective awareness, and motor acts represent awareness and free will (see Chapter 9 BVI).

An even greater refinement and interdisciplinary approach characterizes certain modern investigations of consciousness and self-awareness. Thus, Roger Penrose "reduces" consciousness to quantal transition modes accompanying the transition from quantal to macroscopic events (Penrose, 1994; Metzinger, 1995).[4] Very interestingly, Nancy Woolf adapted Roger Penrose's quantal phenomenon to the cholinergic cytoarchitecture of the cortex and to the notion of the cholinergicity of consciousness and self-awareness (see Figure 9-36). It is also interesting that Nancy Woolf's notion is in part based on the demonstrated plasticity of the neurons in question (Woolf, 1997).

Nancy Woolf's notion must be dear to every cholinergiker; however, I feel that at this time the notions presented by her, Roger Penrose, Ben Libet, and other modern reductionists do not bring about complete understanding of consciousness. Particularly, their notions of self-awareness do not appear to me to be intuitive and parsimonious, parsimoniousness being an accepted criterion of an acceptable scientific hypothesis; many investigators are also not at ease with this problem, even when they avow to be "materialists" (see, for example, Dennet, 2001, 2003). What strikes me also is the lack of commonality of languages used in the description of "neuroscientific" on the one hand and "consciousness phenomena" on the other. Or, to put it more scientifically, the questions asked within one category of phenomena are not couched in the language of the other category; nevertheless, when the questions are transplanted from one language to another, they become nonquestions and there are no answers to them. Furthermore, it appears that the notions of self-awareness and free will, as presented by Libet, Penrose, and Woolf, cannot be verified; yet, verifiability is the criterion sine qua non of a scientific hypothesis (Popper, 1995; see also Karczmar, 1972 and Chapter 9 BVI).

Altogether, it is my belief that cholinergicity or any other transmitter phenomenon cannot be, today, related to consciousness, although the latter may be considered an observable fact; as explained in detail in Chapter 9 BVI, matters of self-awareness belong to the category of occurrences for which explanatory options are infinite and/or illogical.

It should be stressed—to add another issue to this already immensely complicated subject—that, besides several aspects of self-awareness listed here (see also note 10 in Chapter 9), consciousness exhibits still another, special facet, which is the conscious perception of the flow of time and of the time's "arrow" (see also note 10 to Chapter 9); again, this facet involves a dualism between the "real world" and consciousness. In this case it is clear that there is no time's arrow and no concept of the flow and direction of time in classical (Newtonian) physics, as time in classical physics is entirely symmetrical and there is no difference between "past" and "future" (see, for example, Greene, 2004). One could suspect that the phenomenon of entropy and the seemingly steady increase of disorder do exhibit the time's arrow, but the matter is not so simple, as is also the case with the question of time's arrow in quantal physics (see, for example, Greene, 2004). If there is no concept of time in physics (perhaps in any kind of physics), then time's arrow is a sole prerogative of consciousness. And, even if entropy and/or quantal physics are asymmetrical with regard to time and exhibit the past and the future, how does the time's arrow of the quantal or entropic phenomena translate into consciousness of the time's arrow?

All this does not mean that the question of the mind-consciousness relationship raised first by the Greek philosophers and pursued ever since should not be asked: Aristotle referred to consciousness as "awesomely strange" (see Hecht, 2003), and since his days the knowledge of consciousness constitutes a "final problem" in the domain of humanness. Accordingly, the work of the investigators quoted in this chapter and Chapter 9 BVI is brilliant and is not in vain. But an entirely new methodology and epistemology are needed to answer this question that is most perennial and most important to humanity.

C. The Last Word

This book illustrates the peaks, gradually higher and higher, that have been reached in the course of the history of cholinergic research, but more and greater peaks appear in our sights and are still to be scaled (Figure 11-1, see color plate.).

Notes

1. Jean Massoulié (personal communication) stresses that the term "ancestral cholinesterase" may not be warranted, at least not today. He points out that ChEs of many animal species and plants were not cloned; thus, he feels that it is not possible to know whether these enzymes derive from an ancestral cholinesterase, common to all living groups. Yet, it appears to this author that the speculation regarding an "ancestral cholinesterase," if proven, resolves many difficulties in explaining the role of ChEs in cases described in this section and in Chapter 3 D.
2. For a definition of consciousness, see note 10 in Chapter 9. Consciousness or self-awareness must be distinguished from feelings of social self-identity (see Taylor, 1989). The mode of social self-identity and the feeling of what it is to be a human agent or a person depends on many sensory, social, and mental functions, as well as consciousness; these functions,

Figure 11-1. Distant peaks. (From Goodwin Ahlberg, 1998, with permission.) (See color plate.)

including the latter, exhibit cholinergic correlates; yet social self-identity is not consciousness.

3. Sir John Eccles tried to define (in a commonsense way) the existence of free will, that is, of a phenomenon that is not a reflex, simple or complex as the case may be. Let us imagine and perform a completely trivial action, that is, an action of no survival or other functional or behavioral consequence to the perpetrator, such as dropping a pencil at one spot rather than at another: we will drop the pencil one way or another way this supports the contention of free will. As in other instances of free will or self-consciousness, there are many arguments against the alleged significance of Eccles' example.
4. The quantum referred to here is a physical entity and has nothing to do with the quantal release of ACh.

References

Angelucci, F., Mathe, A. A. and Aloe, L.: Neurotrophic factors and CNS disorders: findings in rodent models of depression and schizophrenia. Prog. Brain Res. *146*: 151–165, 2004.

Barnes, P. J.: The role of anticholinergics in chronic obstructive pulmonary disease. Am. J. Med. *117*, Supplt. 12A: 24S–32S, 2004.

Barbeau, A.: Pathogenesis of Parkinson's disease: a new hypothesis. Can. Med. Assoc. J. *87*: 802–807, 1962.

Barbeau, A.: The last ten years of progress in clinical pharmac-ology of extrapyramidal system. In: "Psychopharmacology: A Generation of Progress," M. A. Lipton, A. DiMascio and K. F. Killam, Eds., pp. 771–776, Raven Press, New York, 1978.

Bawaskar, H. S. and Bawaskar, P. H.: Envenoming by the common karait (Bungarus caeruleus) and Asian cobra (Naja naja): clinical manifestations and their management in a rural setting. Wilderness Environ. Med. *15*: 257–266, 2004.

Chomsky, N.: "Language and the Mind," Harcourt Brace and World, New York, 1968.

Dean, B., Bymaster, F. P. and Scarr, E.: Muscarinic receptors in schizophrenia, Curr. Mol. Med. *3*: 419–426, 2003.

Dennet, D. C.: "Consciousness Explained," Little, Brown, Boston, 2001.

Dennet, D. C.: "Freedom Evolves," Viking, New York, 2003.

Eccles, J. C.: "The Physiology of Synapses," Springer-Verlag, New York, 1964.

Greene, B.: "The Fabric of the Cosmos," Alfred Knopf, New York, 2004.

Hecht, J. M.: "Doubt," Harper, San Francisco, 2003.

Hiremath, M., Stewart, R. R. and Liu, Y.: NINDS supports neural channelopathy research. Neurosci. Q.: 12–13, 2003.

Karczmar, A. G.: Ontogenetic effects. In: "Cholinesterases and Anticholinesterase Agents," G. B. Koelle, Ed., pp. 799–832, Handbch. d. exper. Pharmakol., Erganzungswk. vol. 15, Springer-Verlag, Berlin, 1963.

Karczmar, A. G.: Introduction: what we know, will know in the future, and possibly cannot ever know in neurosciences. In: "Brain and Human Behavior," A. G. Karczmar and J. C. Eccles, Eds., pp. 1–20, Springer-Verlag, Berlin, 1972.

Karczmar, A. G.: New roles for cholinergics in CNS diseases. Drug Ther. *4*: 31–42, 1979.

Karczmar, A. G.: Schizophrenia and cholinergic system. In: "Receptors and Ligands in Psychiatry," A. G. Sen and T. Lee, Eds., pp. 29–63, Cambridge University Press, Cambridge, 1988.

Karczmar, A. G.: Physiological cholinergic functions in the CNS. In: "Cholinergic Neurotransmission: Functional and Clinical Aspects," S.-M. Aquilonius and P.-G. Gillberg, Eds., pp. 437–466, Elsevier, Amsterdam, 1990.

Metzinger, T., Ed.: "Conscious Experience," Ferdinand Schoeningh, Paderborn, 1995.

Miller, G.: What is the biological basis of consciousness? Science *309*: 79, 2005.

Oh, A. Y., Kim, S. D. and Kim, C. S.: Early and late reversal of rocuronium with pyridostigmine during sevoflurane anaesthesia in children. Anaesth. Intensive Care *32*: 649–652, 2004.

Penrose, R.: "Shadows of the Mind," Oxford University Press, Oxford, 1994.

Pepeu, G.: Overview and future of CNS cholinergic mechanisms. In: "Cholinergic Neurotransmission: Function and Dysfunction," A. C. Cuello, Ed., pp. 455–458, Elsevier, Amsterdam, 1993.

Popper, K. R.: "The Logic of Scientific Discovery," Basic Books, New York, 1995.

Popper, K. R. and Eccles, J. C.: "The Self and Its Brain," Springer International, Berlin, 1977.

Searle, J. R.: "Mind," Oxford University Press, Oxford, 2004.

Stothers, L.: Reliability, validity, and gender difference in the quality of life index of the SEAPI-QMM classification system. Neurourol. Urodyn. *23*: 223–228, 2004.

Taylor, C.: "Sources of Self: The Making of Modern Identity," Harvard University Press, Cambridge, Mass., 1989.

Whittaker, V. P.: Cholinergic neurohormones. Comp. Endocrinol. *2*: 182–208, 1963.

Woolf, N.: A possible role for cholinergic neurons in the basal brain and pontomesencephalon in consciousness. Conscious Cogn *6*: 574–596, 1997.

Index

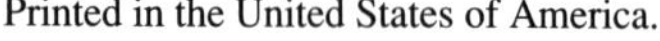
Printed in the United States of America.